Study Guide/Solutions Manual for Jones's Organic Chemistry

Study Guide/Solutions Manual for Jones's Organic Chemistry

MAITLAND JONES, Jr.
and HENRY L. GINGRICH

Princeton University

W · W · NORTON & COMPANY  NEW YORK · LONDON

PRINTED IN THE UNITED STATES OF AMERICA

ISBN 0-393-97155-4

W.W. Norton & Company, Inc., 500 Fifth Avenue, New York, N.Y. 10110
http://web.wwnorton.com

W.W. Norton & Company, Ltd., 10 Coptic Street, London, WC1A 1PU

1 2 3 4 5 6 7 8 9 0

This book is dedicated to the generations of undergraduates upon whom we have inflicted this collection of problems, solutions, and advice. For the most part, they have borne this burden with remarkable good cheer.

Contents

Introduction

In this Study Guide we try to go beyond straightforward "bare bones" answers to problems. Once one gets past simple questions, problem solving in organic chemistry becomes very hard to teach. The answers to many problems are intuition-intensive, and therefore it becomes difficult to explain in explicit terms how to proceed. Nonetheless there are things that can be said that may help; there are techniques of problem solving that can be learned. In this collection of comments on solutions to all of the problems that are not solved in the text itself, we try to help you to do that. As in the text, we seek in these pages to show you how to approach problems in organic chemistry in general, not just how we got from here to there in specific cases.

The exercises in these chapters become much easier when we have an idea of where we are going. What exactly are we trying to do in this problem? What tasks must be accomplished? What bonds are we trying to make or break? What rings must be closed or opened? Such questions seem simple, but it is amazing how few people really start problems with the simple question: "What happens in this reaction?" Analyze! Once a goal is in mind, the path to that goal becomes much, much easier. In a sense, a good problem solver has learned, first and foremost, to avoid "thrashing." We know that is a flip remark, but it is, nonetheless, true. A person solving a tough problem is like a bacterium swimming up a food gradient—he or she (or it) is following a pathway that "feels good." We will try to show you how to do that in this Study Guide, but there can be no denying that experience is important, and experience can be gained only by practice. Practice and more practice will teach you what feels good in terms of problem solving—of how to swim up that food gradient—but we will try to give you some hints along the way.

The problems solved in the Study Guide will recapitulate each chapter, and thus will generally start off with the easier examples and then go on to tougher stuff. Don't worry if the hard problems at the end of the sections do not come so easily; they are meant to tax you, to demand some hard work and careful thought. Some difficult problems will be dealt with best over time. If a problem resists solution, and some will, come back to it after a while; let your subconscious work on it for a while. Most research chemists carry unsolved problems around in their heads for a long time, sometimes for years, returning to them now and then. There is nothing wrong with emulating that process. People think at vastly different rates, and it is a rare situation that requires a *rapid* solution of a problem. (Hour exams may be an exception, unfortunately.)

Many of us who actually do organic chemistry for a living (believe it or not there are such people) typically get great pleasure from problem solving. We hope that you will be similarly stimulated. In a fundamental way that is what we humans are about. We have evolved to be curious and to turn over rocks to see what is underneath. Perhaps, thinks our ancestral hunter-gatherer, I will find something good to eat! From such imperatives we humans have become problem solvers, and it gives us pleasure to work out what's happening in unknown situations.

The two of us have solved a good many problems together, and it is great fun. Here is one favorite example, which makes an important point about problem solving. This lesson is so simple as to be trivially obvious and yet at the same time so profound as to be most difficult in practice. We had been working for some time on the synthesis of the pterodactyl-shaped compound shown below.

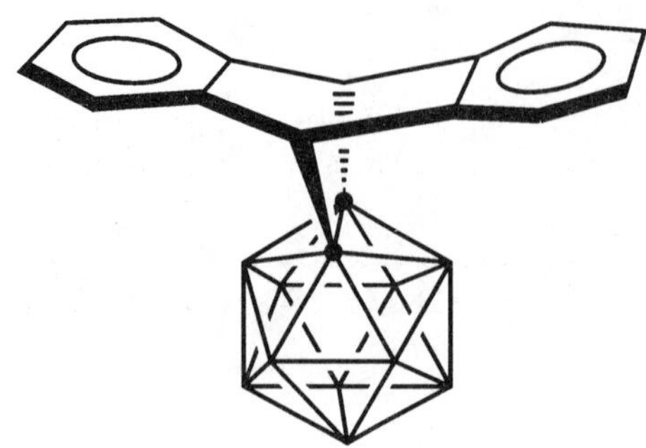

One of us (MJ) had become entrapped in devising increasingly clever "solutions" that a series of graduate students and undergraduates had not been able to make work—and for good reasons. Those clever solutions were complicated, and extraordinarily hard to carry out in a practical sense. While MJ was away one July, HG had the wit to avoid all the foolish "cleverness" and to do what we beseech our students to do—to "think simple." HG went back to basics, did the work himself, and solved the problem. MJ arrived back in Princeton and was presented with a vial containing exquisitely beautiful crystals of the long-sought compound. What's the lesson? Don't be too clever. If you ground yourself in the basics, analyze what you want to do, and then apply those basics, you will prosper.

Remember, think simple.

Maitland Jones, Jr.
Henry L. Gingrich

Study Guide/Solutions Manual for Jones's Organic Chemistry

Chapter 1 Outline

1.1 Atoms and Atomic Orbitals
1.2 Covalent Bonds and Lewis Structures
1.3 Introduction to Resonance Forms
1.4 More on Atomic Orbitals

Atoms, Atomic Orbitals, and Bonding

In this chapter, you are learning a bit about atomic structure and acquiring skills that you will need throughout your study of organic chemistry. In a sense, you are learning vocabulary and grammar that will enable you to write sentences a little later, and eventually to compose whole paragraphs and short stories. The problems of this chapter concentrate on tool building and require less thought and imagination than those of later chapters. That does not mean that they are unimportant. Even though much of this chapter may review what you already know, please do not skip past it until you are certain that you can write Lewis structures easily, determine the position of charges without error, and use the arrow formalism to write resonance forms with ease. These skills will be as necessary in Chapter 26 as they are now.

Problem 1.1 A bond dipole will result when two atoms of different electronegativities are attached to each other. This problem requires you only to look up the electronegativities of the two atoms in the bond. The electronegativity (Table 1.2, p 15) of each atom in the bond is shown in parentheses. As you see below, the atom of greater electronegativity will be at the negative end of the dipole. The answer uses an arrow to represent the direction of the dipole, δ^+ ⟶ δ^-.

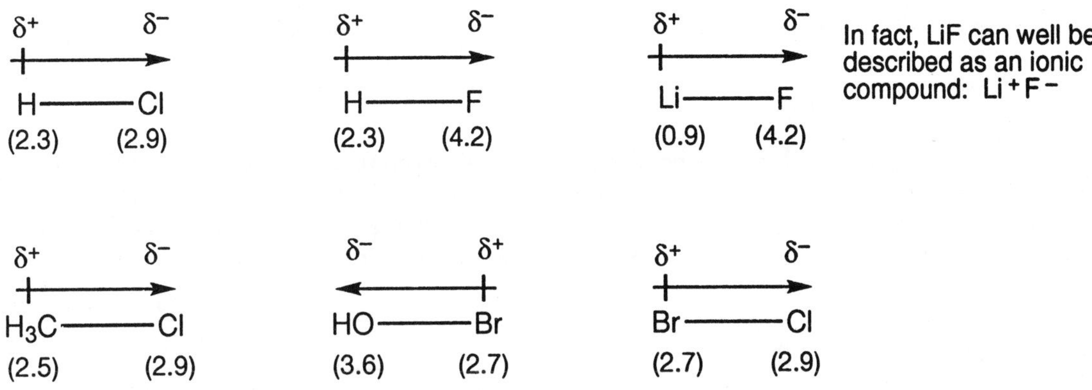

Problem 1.2 This classic problem is asked in many textbooks. It requires you to estimate the net dipole in a molecule by summing the bond dipoles. In carbon tetrachloride (CCl_4), the four dipoles cancel, and there is no net dipole. In chloroform ($CHCl_3$), the dipoles do not cancel, and a net dipole exists.

Here, the four dipoles cancel..............................but in this molecule they do not

Problem 1.3 In each part of this problem, we will first determine the number of electrons available for bonding on each atom. For atoms in the second row of the periodic table, this will be the atomic number of the atom less the two 1*s* electrons. Then we will see how many bonds between atoms are possible using these electrons. Finally, be careful to write the left over nonbonding (lone pair) electrons as dots.

(b) H_2Be
Beryllium ($_4Be$) has two electrons available for bonding (4 – 2 1*s* electrons), and each hydrogen contributes one electron. Two beryllium-hydrogen single bonds are formed. There are no nonbonding electrons.

H• •Be• •H ⟶ H:Be:H = H—Be—H

(c) BH_3
As in BF_3 (a), boron ($_5B$) has three electrons available for bonding, and each hydrogen contributes one electron. Three boron-hydrogen single bonds are formed.

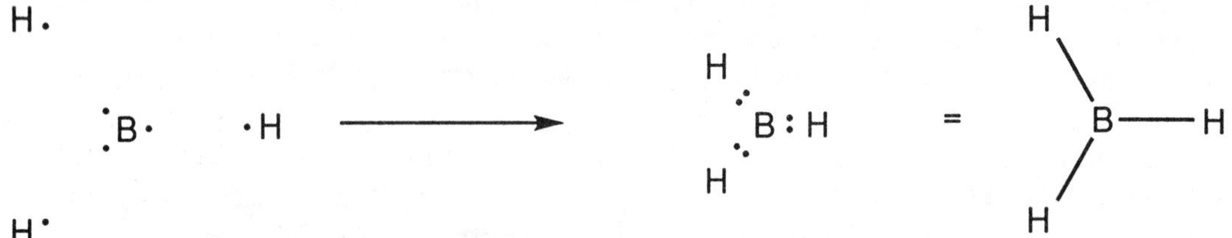

(e) $HOCH_3$
Oxygen ($_8O$) has six electrons available for bonding (8 – 2 1*s* electrons). Carbon forms three covalent single bonds with the three hydrogens, which leaves one carbon electron available for covalent bond formation between carbon and oxygen. Oxygen forms the carbon-oxygen bond as well as one covalent bond with hydrogen, leaving oxygen with two pairs of nonbonding electrons.

(f) $H_2N—NH_2$
Nitrogen has five electrons available for bonding. In this molecule, each nitrogen forms covalent bonds with two hydrogens and a third with the other nitrogen. This uses three of the available electrons, leaving a nonbonding pair of electrons on each nitrogen.

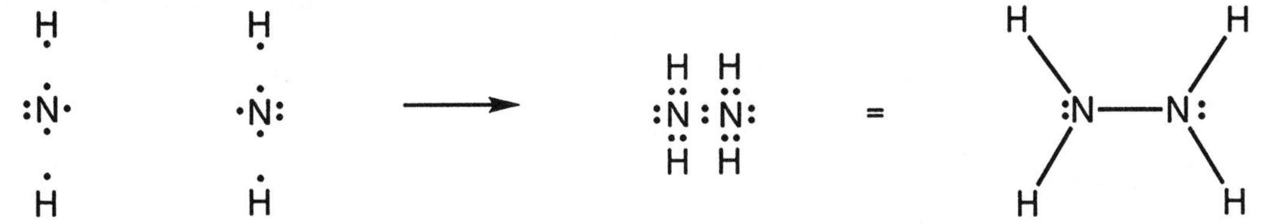

Problem 1.4
(c) Br
($_{35}$Br) is in the fourth row of the periodic table, so we can ignore the 28 1*s*, 2*s*, 2*p*, 3*s*, 3*p* and 3*d* electrons. This leaves seven electrons. Like the other halogens, (F, Cl, and I), Br has three nonbonding pairs and a single odd electron.

(d) OH
Oxygen ($_8$O) has six electrons available for bonding. One electron is used in forming a covalent bond with hydrogen, leaving two pairs of nonbonding electrons and a single odd electron on oxygen.

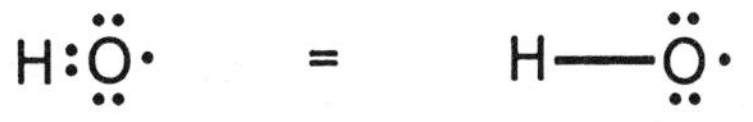

H:Ö· = H—Ö·

(e) NH_2
Nitrogen has five electrons available for bonding. Two covalent bonds are formed to hydrogens, leaving a nonbonding pair and a single odd electron remaining on nitrogen.

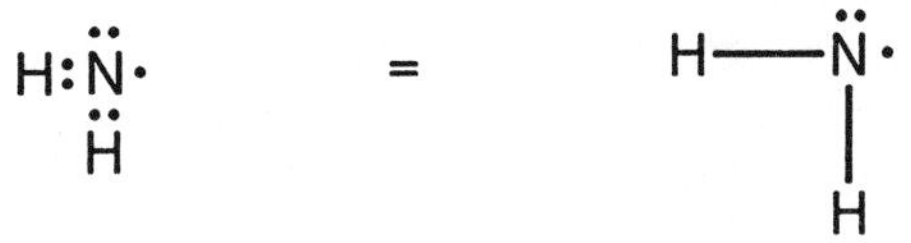

(f) $H_3C—N$
There are two possible answers to this one, and both kinds of molecule are known. Carbon has four electrons available for bonding. Three are used in forming single bonds to the three hydrogens, and the fourth is used in the single bond to nitrogen. Nitrogen has five electrons available for bonding. One covalent bond is formed to carbon, leaving four electrons. These can be used either as two pairs of nonbonding electrons or as one nonbonding pair and two odd electrons

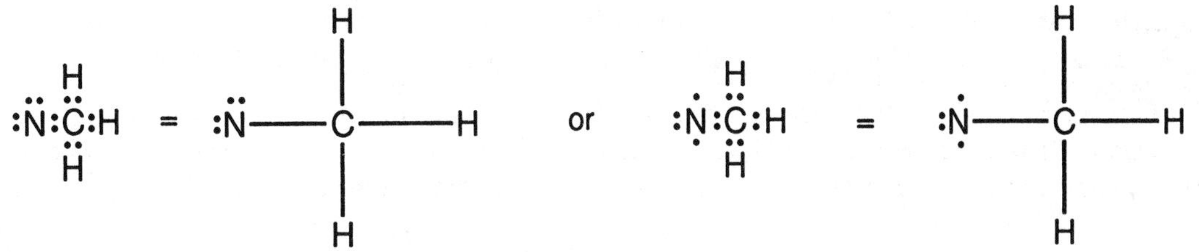

or

Problem 1.5 The question itself helps with the hard part—working out the connectivity of the atoms. Once again, this is an exercise in electron counting. First, determine the number of electrons available for bonding (atomic number less two 1*s* electrons for most atoms, or a single 1*s* electron for hydrogen), then make single bonds. Finally, we look to see where multiple bonds can be formed.

(a) F_2CCF_2
Carbon has four electrons available for bonding and fluorine seven, including a single odd electron. As in ethylene (Fig. 1.13), each carbon forms three covalent bonds, two with fluorines and one with the other carbon atom. Thus, there are three nonbonding pairs remaining on each fluorine, and a single electron left over on each carbon. These are shared in a second covalent carbon-carbon bond.

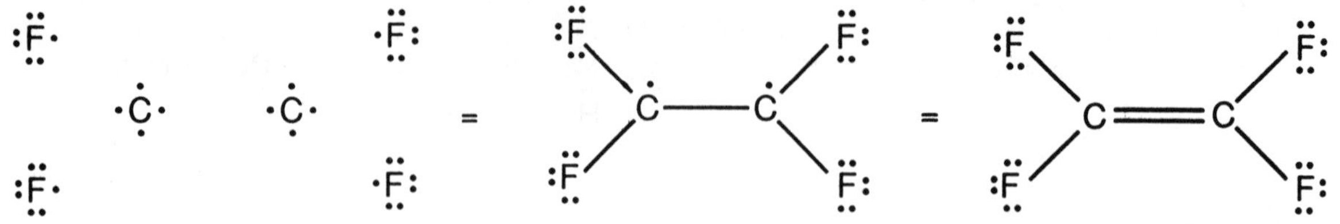

(c) H_2CO
Once again, carbon has four electrons available for bonding, and oxygen six. Carbon uses two electrons to form bonds to hydrogen, and one to bind to oxygen. Oxygen uses one electron in the bond to carbon, leaving one unshared electron on carbon, and five on oxygen. Formation of a second carbon-oxygen bond leaves two nonbonding electron pairs on oxygen.

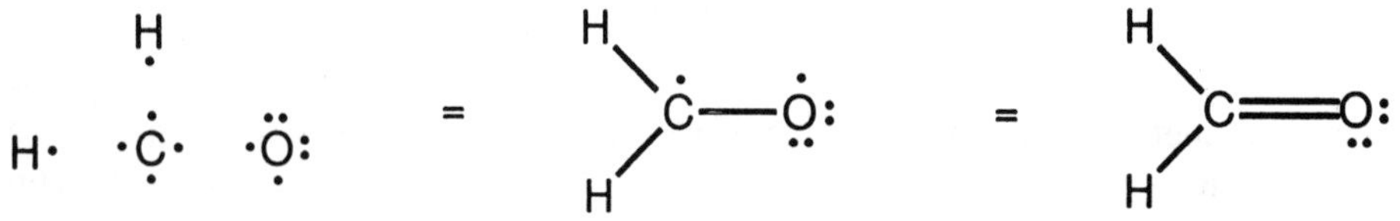

(d) H_2CCO
Each carbon has four available electrons, and the single oxygen six. The left-hand carbon uses three electrons to form three covalent bonds to the two hydrogens and the adjacent carbon. The right-hand carbon uses two electrons in forming covalent bonds with the left-hand carbon and the oxygen, leaving the left-hand carbon with one electron, the right-hand carbon with two, and the oxygen with five. Formation of a second carbon-carbon bond and a second carbon-oxygen bond completes the picture, leaving the oxygen with two nonbonding pairs of electrons.

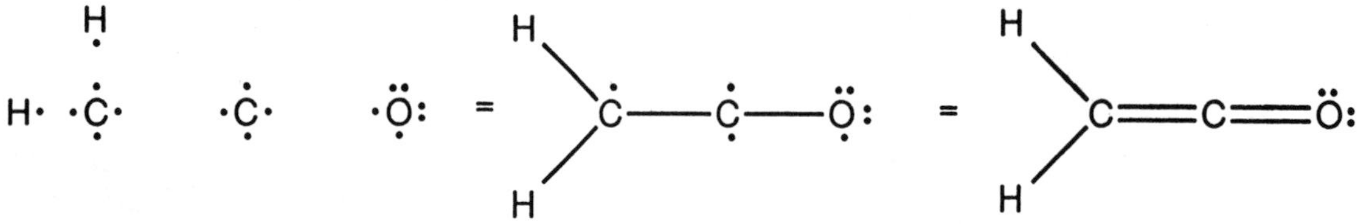

(e) $H_2CCHCHCH_2$
Each of the four carbons has four electrons available for bonding. The two terminal carbons form two covalent bonds with the two hydrogens and a third covalent bond with the adjacent carbon. Each internal carbon forms two covalent bonds with the adjacent carbons and a third to a hydrogen. This leaves one electron on each carbon, allowing the formation of two additional carbon-carbon bonds.

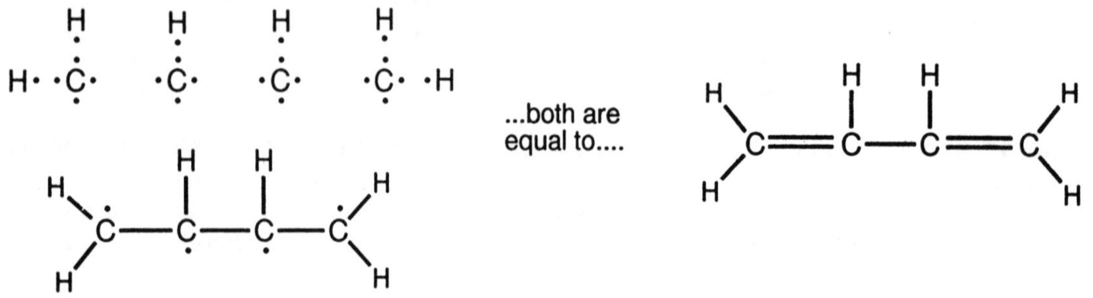

...both are equal to....

(g) $H_3COCOOH$
Each carbon has four available electrons for bonding, and each oxygen has six. The left-hand carbon (a) forms three bonds to hydrogen and a fourth to one oxygen (b). Oxygen (b) forms two bonds to a pair of carbons. The remaining carbon (c) forms covalent bonds with oxygen (b) and the two other oxygens (d). Each oxygen (d) forms a covalent bond with carbon (c), and one oxygen (d) forms a bond to hydrogen. This process leaves one odd electron on carbon (c) four electrons on the oxygen (d) bound to hydrogen, and five electrons on the non-hydrogen bound oxygen (c). Oxygen (b) has four electrons remaining. Formation of a carbon (c) oxygen (d) bond completes the picture.

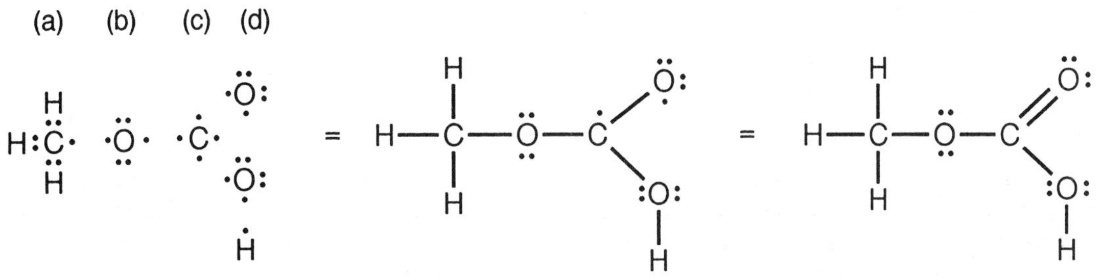

(a) (b) (c) (d)

Problem 1.6 This problem is just like Problem 1.4 except that you need to make an adjustment for the charge. First, calculate the number of available electrons on the neutral atom, then add one electron for a negative charge or subtract one electron for a positive charge.

(a) HO^-
Neutral oxygen ($_8O$) has six electrons available for bonding ($8 - 2\ 1s = 6$). Therefore, negatively charged oxygen must have seven electrons.

neutral negative

One covalent bond can be made to the lone hydrogen, which supplies a single electron.

(b) $^-BH_4$
Neutral boron ($_5B$) has three electrons available for bonding ($5 - 2\ 1s$). Therefore, negatively charged boron must have four electrons, allowing four covalent bonds to be made to the four hydrogens, each of which supplies a single electron. Notice that there is no pair of electrons on the negatively charged boron atom. In most negatively charged species, there is a nonbonding pair of electrons. This molecule is an exception.

neutral negative

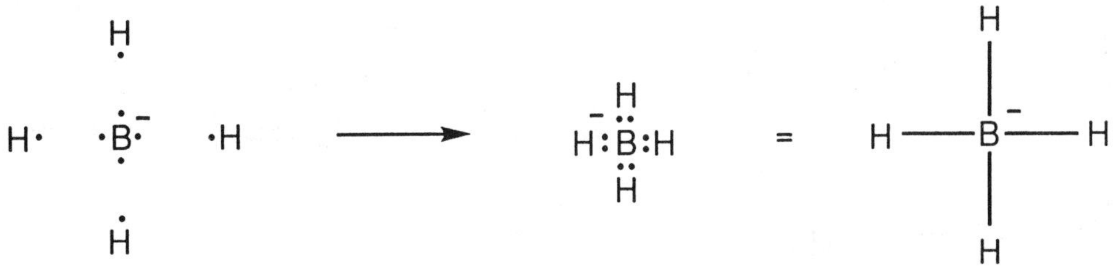

(c) $^{+}NH_4$

Neutral nitrogen ($_7N$) has five electrons available for bonding (7 – 2 1*s*). Positively charged nitrogen must have four electrons for bonding, allowing four single bonds to the four hydrogens, each of which supplies its single electron.

:N· ·N·+

Neutral Positive

H· ·N·+ ·H (with H above and below) ⟶ H:N:H (H above and below, +) = H—N—H (H above and below, +)

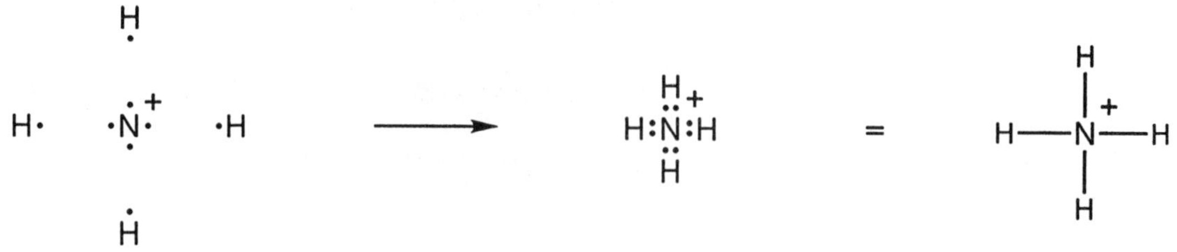

(d) ^{-}Cl

Neutral chlorine ($_{17}Cl$) has seven electrons available for bonding (ignore the 10 1*s*, 2*s*, and 2*p* electrons). Therefore, negatively charged chlorine must have 8 electrons.

:Cl· :Cl:⁻

Neutral Negative

(e) $^{+}CH_3$

Neutral carbon ($_6C$) has four electrons available for bonding (6 – 2 1*s*). Positively charged carbon must have only three electrons for bonding, allowing three single bonds to the hydrogens, each of which supplies its single electron.

·C· ⟶ +C· ⟶ +C· ·H (H above and below) = +C:H (H above and below) = +C—H (H above and below)

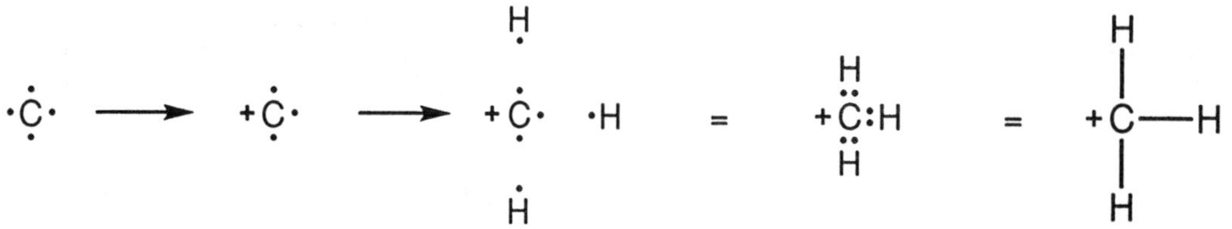

Problem 1.7

In these examples we will first show a full Lewis structure in which each bonding electron appears as a dot, then a more schematic Lewis structure in which bonds are shown as lines and nonbonding electrons as dots. These structures will be followed by the charge calculation.

(a) :C:H H = :C(H)(H)

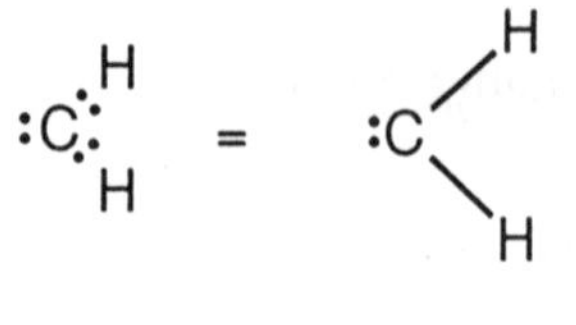

$_6C$ **6 protons = 6 positive charges**

2 1*s* electrons
2 nonbonding electrons
2 shared electrons

6 negative charges

neutral

(b) ·C:H (H above and below) = ·C—H (H above and below)

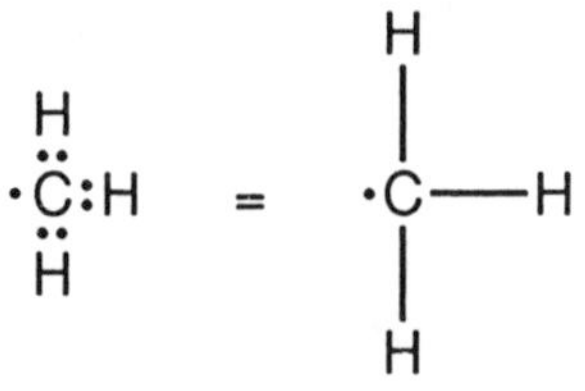

$_6C$ **6 protons = 6 positive charges**

2 1*s* electrons
1 nonbonding electrons
3 shared electrons

6 negative charges

neutral

(c)

$\cdot\ddot{C}:H$ = $\cdot\ddot{C}—H$

$_6C$

6 protons = 6 positive charges

2 1*s* electrons
3 nonbonding electrons
1 shared electron

6 negative charges

neutral

(d)

$^-:\ddot{\underset{..}{O}}:H$ = $^-:\ddot{\underset{..}{O}}—H$

$_8O$

8 protons = 8 positive charges

2 1*s* electrons
6 nonbonding electrons
1 shared electron

9 negative charges

net 1^-

(e)

H_3O^+ (Lewis structure: :O:H with H above and below, + charge) = :O⁺—H with H above and below

$_8O$

8 protons = 8 positive charges

2 1*s* electrons
2 nonbonding electrons
3 shared electrons

7 negative charges

net 1^+

(f)

$H_2\dot{C}—\dot{\ddot{O}}:$ = $H_2C{=}\ddot{O}:$

$_8O$

8 protons = 8 positive charges

2 1*s* electrons
4 nonbonding electrons
2 shared electrons

8 negative charges

neutral

$_6C$

6 protons = 6 positive charges

2 1*s* electrons
0 nonbonding electrons
4 shared electrons

6 negative charges

neutral

(g)

$^-:\ddot{N}:H_2$ = $^-:\ddot{N}(H)—H$

$_7N$

7 protons = 7 positive charges

2 1*s* electrons
4 nonbonding electrons
2 shared electrons

8 negative charges

net 1^-

(h)

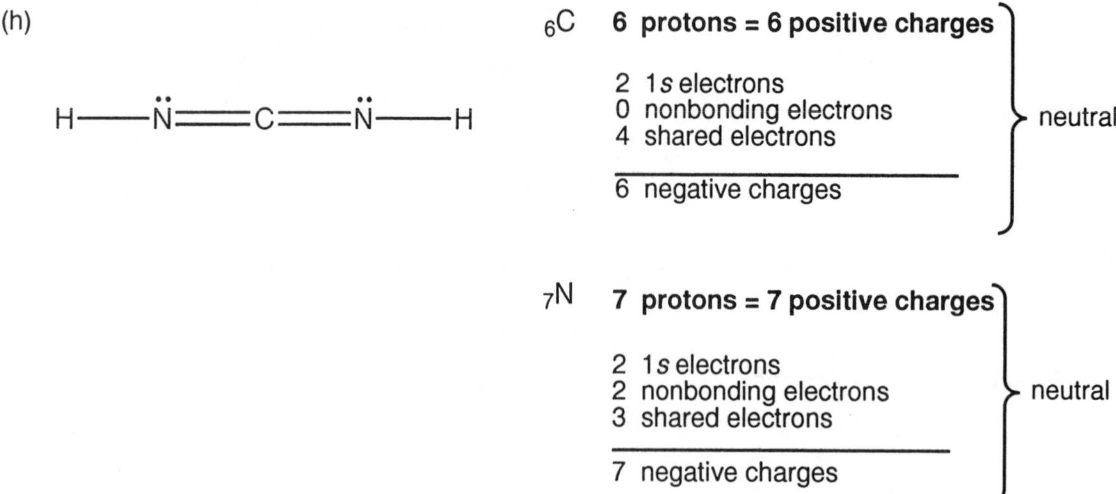

Problem 1.8 The task here is to work out the number of nonbonding electrons (if any) on the charged atom. Each answer first shows the neutral atom, then the atom with an electron added or removed to get the proper charge. Finally, electrons are used to make the bonds to the available hydrogen atoms or other groups. In (a) for example, we first see carbon with four bonding electrons ($_6$C; 6 electrons – 2 1s electrons = 4 bonding electrons), then with one electron removed to get $^+$C; finally two of the remaining three electrons form single bonds to the two available hydrogens. This leaves $^+CH_2$ with a single nonbonding electron.

(a) $^+CH_2$

·C· → +·C· —add 2 H·→ +·C· ·H → $^+$C(H)—H

neutral positive

(b) $CH_3\bar{C}H_2$

·C· → ·C:$^-$ —add 2 H· 1 H_3C·→ H_3C—C(H)(H):$^-$

neutral negative

(c) $H\bar{C}{=}CH_2$

·C· → ·C:$^-$ —add 1 H· and :CH_2→ $H\bar{\ddot{C}}{=}CH_2$

neutral negative

(d) H_3O^+

·Ö: ·Ö:$^+$ —add 3 H·→ H—$^+$O(H)(H):

neutral positive

(e) HO^-

·Ö: $^-$·Ö: —add 1 H·→ H—Ö:$^-$

neutral negative

(f) $\overset{+}{N}H_2$

·N̈· neutral ·N̈· positive (+) add 2 H· → H—N̈—H (+)

(g) $\overset{-}{N}H_2$

·N̈· neutral ⁻·N̈· negative add 2 H· → H—N̈—H (−)

(h) $H_3C—C{\equiv}\overset{+}{N}—H$

·N̈· neutral ·N̈· positive (+) add 1 H· and $H_3C—\dot{C}·$ → $H_3C—C{\equiv}\overset{+}{N}—H$

Problem 1.9 Figure 1.20 shows the structure of nitromethane, $H_3C—NO_2$. By analogy, we can write a structure for nitric acid, $HO—NO_2$.

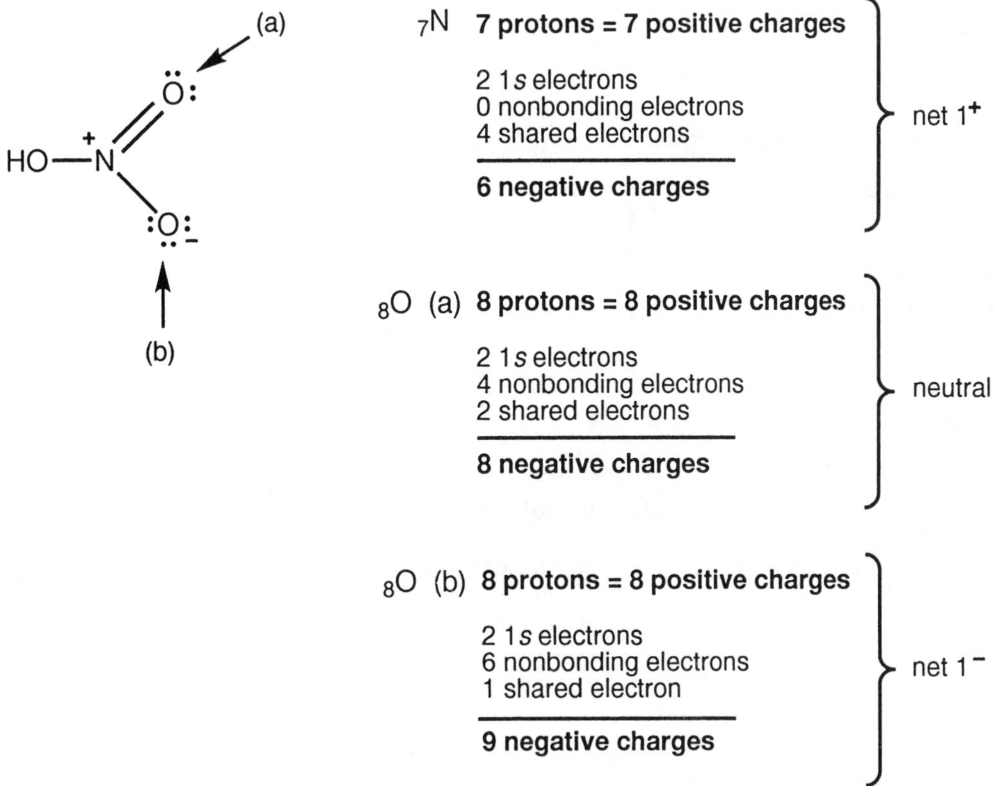

Problem 1.10 Notice that the lower arrow pushes a nonbonding pair of electrons on one oxygen to displace a bonding pair (shown only as a line in the drawing). The displaced pair winds up as a nonbonding pair on the other oxygen, and the displacing pair as a new bond between the lower oxygen atom and nitrogen.

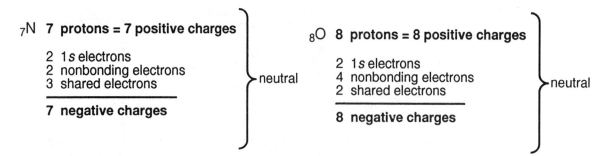

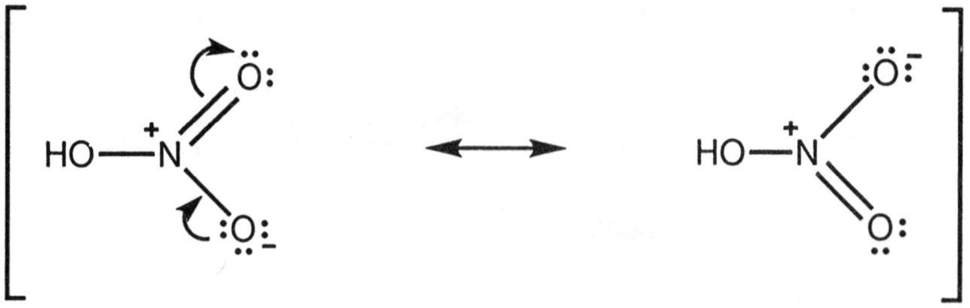

Problem 1.12

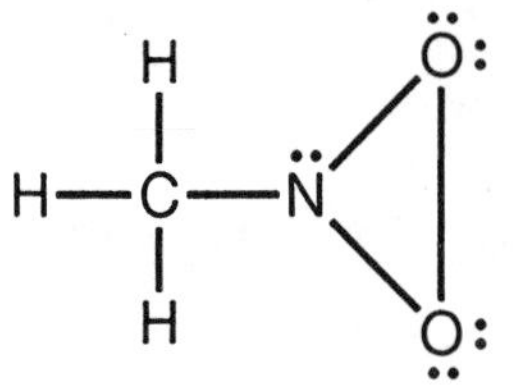

$_6$C **6 protons = 6 positive charges**

2 1*s* electrons
0 nonbonding electrons
4 shared electrons

6 negative charges — neutral

$_7$N **7 protons = 7 positive charges**

2 1*s* electrons
2 nonbonding electrons
3 shared electrons

7 negative charges — neutral

$_8$O **8 protons = 8 positive charges**

2 1*s* electrons
4 nonbonding electrons
2 shared electrons

8 negative charges — neutral

Problem 1.13

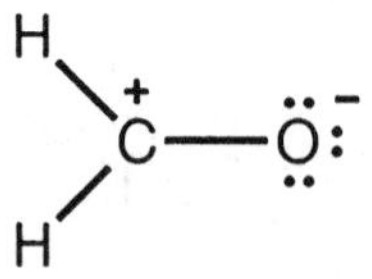

$_6$C **6 protons = 6 positive charges**

2 1*s* electrons
0 nonbonding electrons
3 shared electrons

5 negative charges — net 1 $^+$

$_8$O **8 protons = 8 positive charges**

2 1*s* electrons
6 nonbonding electrons
1 shared electron

9 negative charges — net 1 $^-$

Problem 1.15 *Remember:* the double-barbed, curved arrows move pairs of electrons, and we must be careful neither to violate the rules of valence nor to move atoms. A new convention appears in the last example. When "pushing" single electrons, a single-barbed, curved arrow is used.

(a)

(b)

(c)

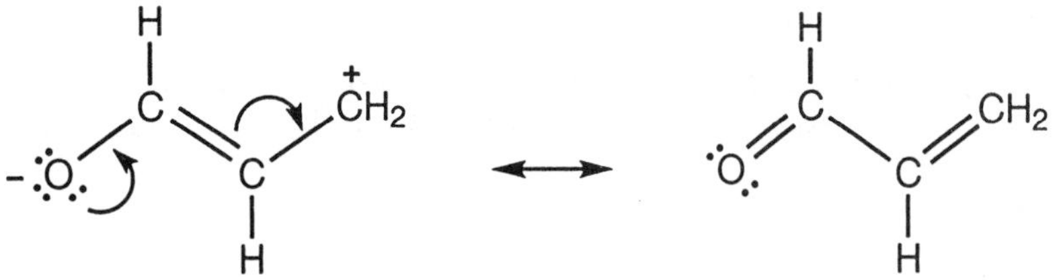

(d) This is the tricky part to this problem. It is sooooo tempting to push the arrow as shown:

But this makes no sense: There is no low-lying empty orbital on nitrogen to accept an electron pair and no pair of electrons on boron to push! The form on the right violates the rules of valence twice.

(e)

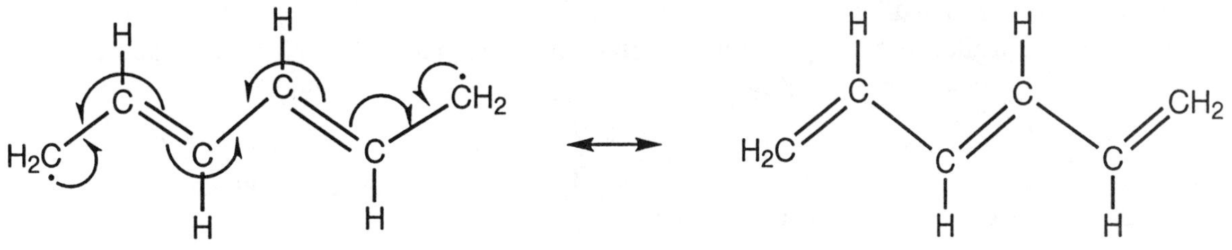

In this example, notice the use of *single-barbed*, or "fishhook" arrows to show the motion of single electrons!

Problem 1.16

(a)

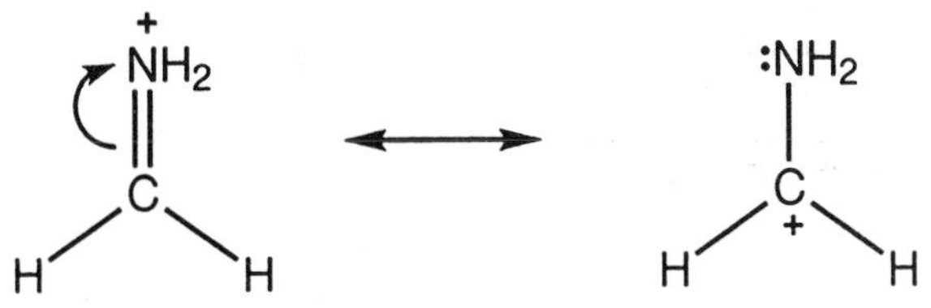

(b)

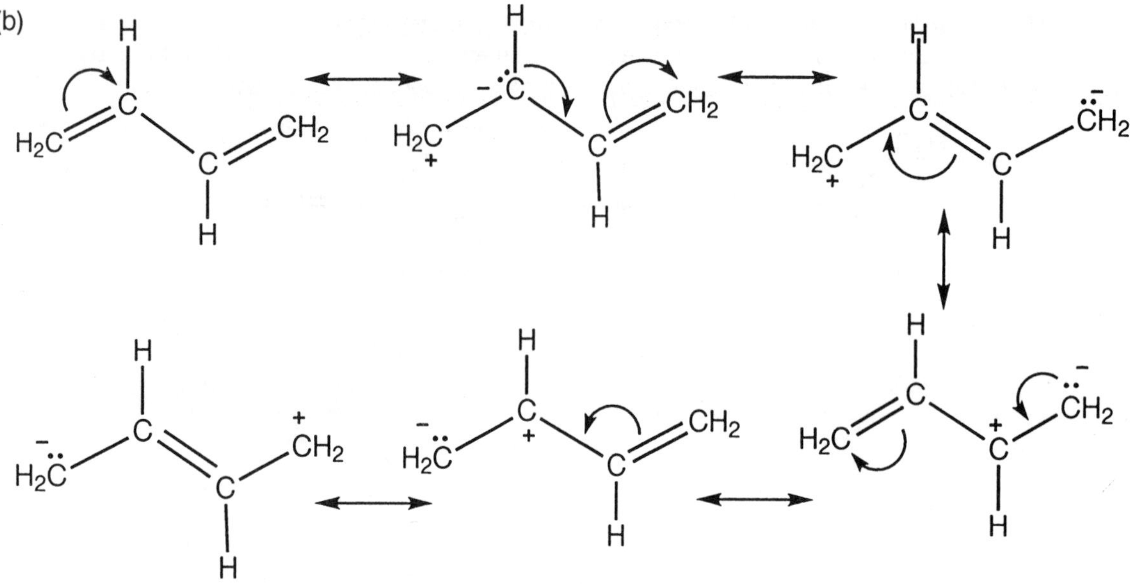

None of the resonance forms in part b, except the first, uncharged one, is very good. Each contains one fewer bond than the first, and each requires substantial charge separation.

(d)

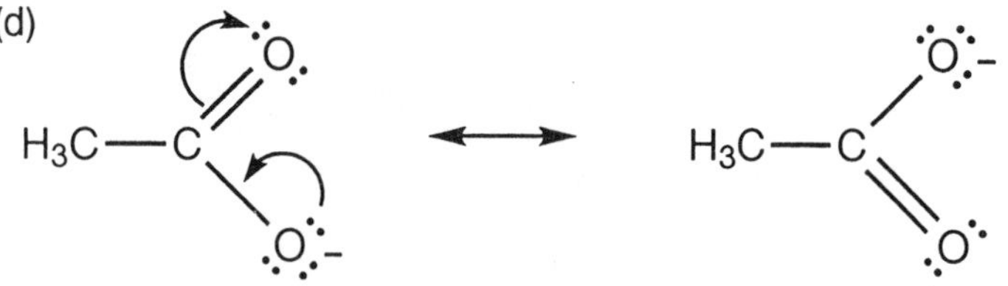

(e)

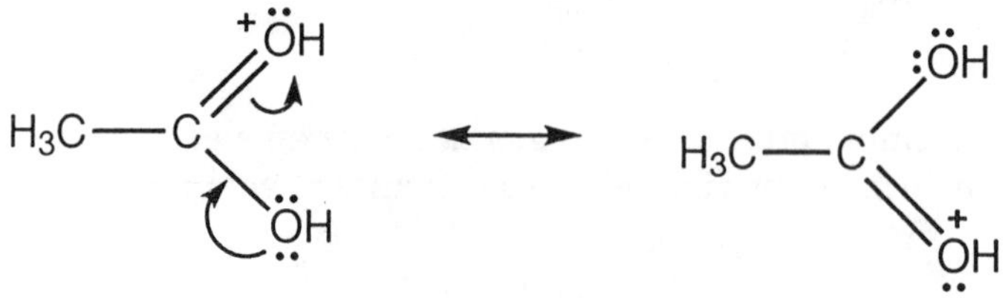

Problem 1.18 See Problem 1.17. If the fifth and sixth electrons in a carbon atom occupied the same orbital with parallel spins they would have the same values for all four quantum numbers, a violation of the Pauli principle. If they occupy the same orbital, their spin quantum numbers (s) *must* be opposite (paired).

Problem 1.19 In this row we fill the $n = 3$ energy levels, first filling the 3s orbital, then moving on to the higher energy 3p and 3d orbitals.

$_{11}$Na $1s^2\ 2s^2\ 2p_x^{\ 2}\ 2p_y^{\ 2}\ 2p_z^{\ 2}\ 3s$
$_{12}$Mg $1s^2\ 2s^2\ 2p_x^{\ 2}\ 2p_y^{\ 2}\ 2p_z^{\ 2}\ 3s^2$
$_{13}$Al $1s^2\ 2s^2\ 2p_x^{\ 2}\ 2p_y^{\ 2}\ 2p_z^{\ 2}\ 3s^2\ 3p_x$
$_{14}$Si $1s^2\ 2s^2\ 2p_x^{\ 2}\ 2p_y^{\ 2}\ 2p_z^{\ 2}\ 3s^2\ 3p_x\ 3p_y$
$_{15}$P $1s^2\ 2s^2\ 2p_x^{\ 2}\ 2p_y^{\ 2}\ 2p_z^{\ 2}\ 3s^2\ 3p_x\ 3p_y\ 3p_z$
$_{16}$S $1s^2\ 2s^2\ 2p_x^{\ 2}\ 2p_y^{\ 2}\ 2p_z^{\ 2}\ 3s^2\ 3p_x^{\ 2}\ 3p_y\ 3p_z$
$_{17}$Cl $1s^2\ 2s^2\ 2p_x^{\ 2}\ 2p_y^{\ 2}\ 2p_z^{\ 2}\ 3s^2\ 3p_x^{\ 2}\ 3p_y^{\ 2}\ 3p_z$
$_{18}$Ar $1s^2\ 2s^2\ 2p_x^{\ 2}\ 2p_y^{\ 2}\ 2p_z^{\ 2}\ 3s^2\ 3p_x^{\ 2}\ 3p_y^{\ 2}\ 3p_z^{\ 2}$

Additional Problem Answers

Problem 1.20 In these answers, the arrows shown will always produce the resonance form immediately to the right or below. That is, these answers are to be read left to right or top to bottom. Notice the extensive use of the double headed resonance arrow.

(a) carbonate

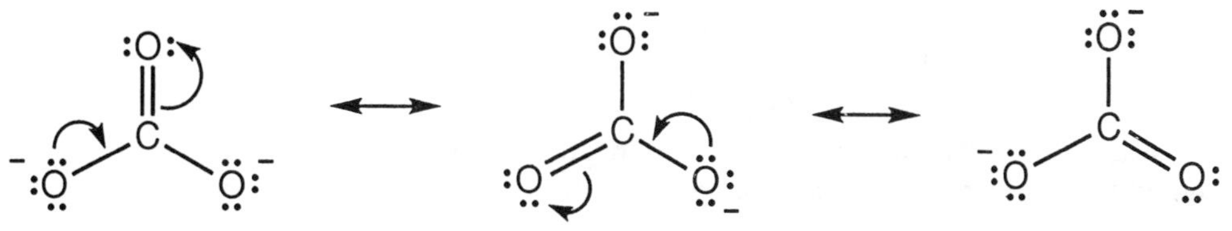

(b) sulfate

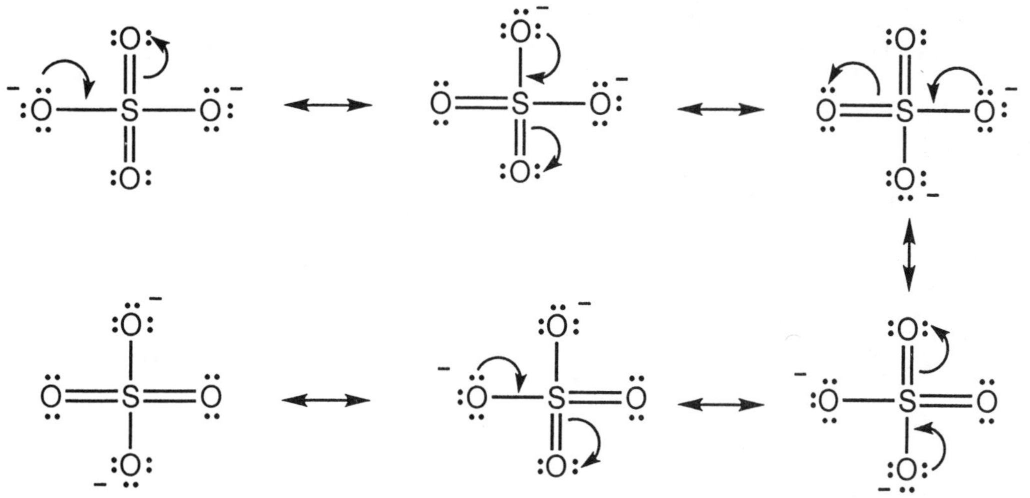

There are also resonance forms in which sulfur bears a positive charge and three oxygen atoms are negative. Can you draw these?

(c) nitrate

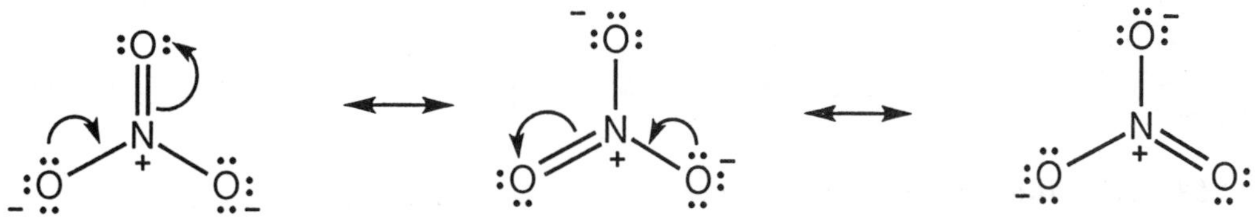

(d) guanidinium

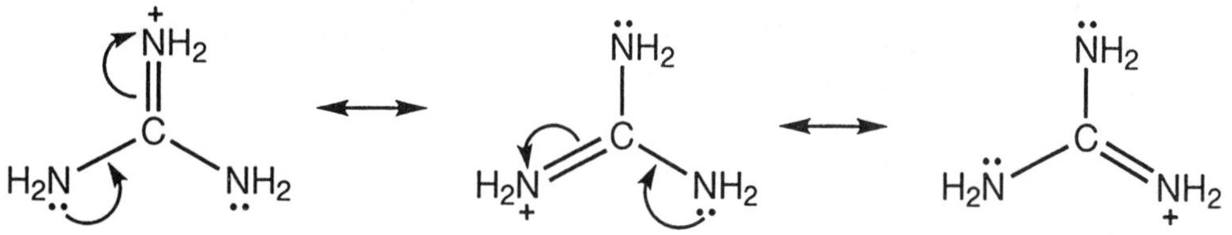

(e) vinyl ammonium
There are no important resonance forms for this species. There is no pair of electrons on nitrogen, and five bonds cannot be formed to nitrogen. Once again, it is tempting to "push the arrow" but in

this case it is best to resist that temptation. There is no empty orbital on nitrogen to receive the electrons we are trying to push.

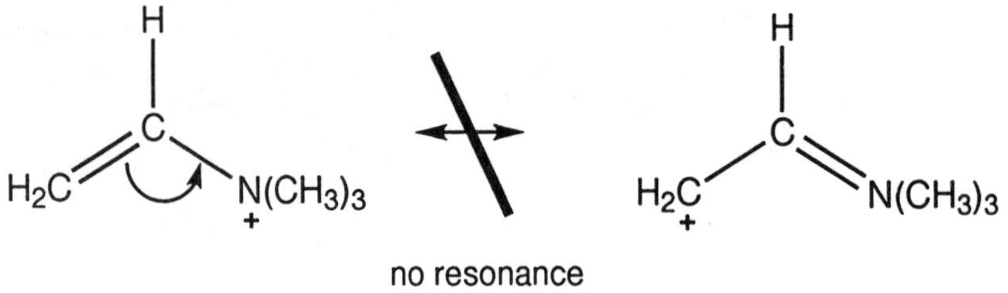

no resonance

Problem 1.21 In drawing resonance forms for these molecules it is sometimes hard to know when to stop, as less and less stable stuctures are produced. We have perhaps gone too far on occasion. Can you see which forms are likely to be especially unstable, and therefore minor contributors?

(a)

(b)

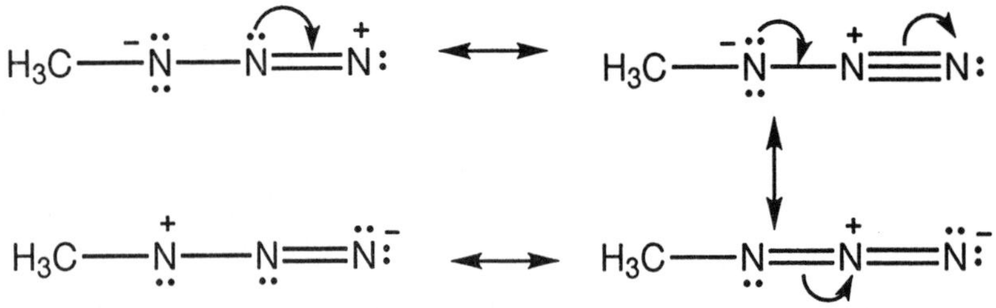

(c)

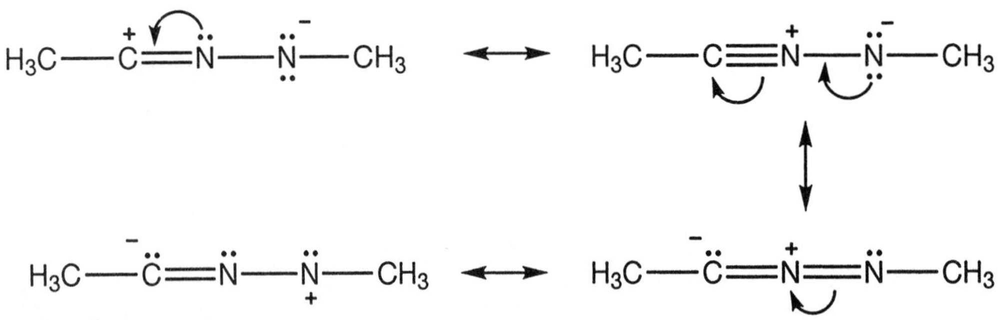

(d) each type of charge needs to be the main atoms

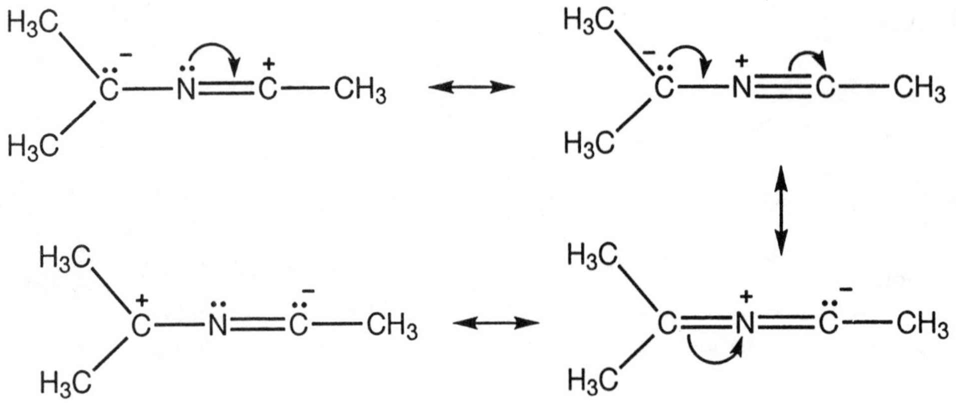

(e)

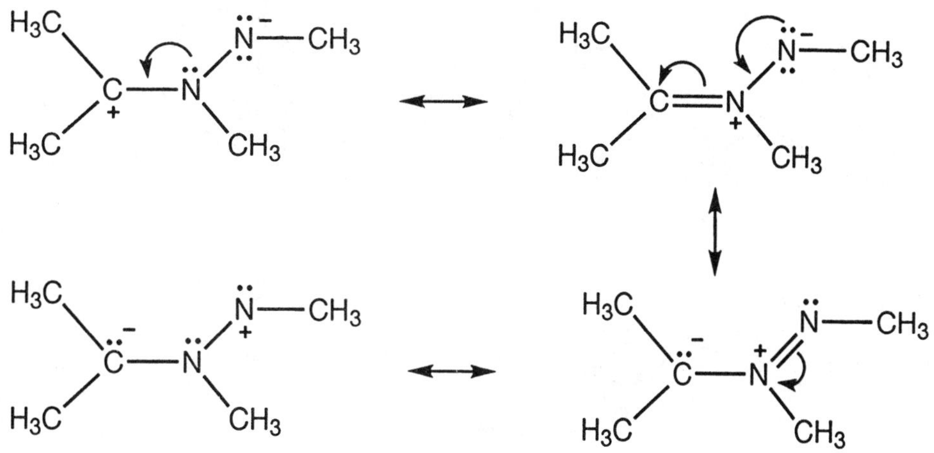

(f)

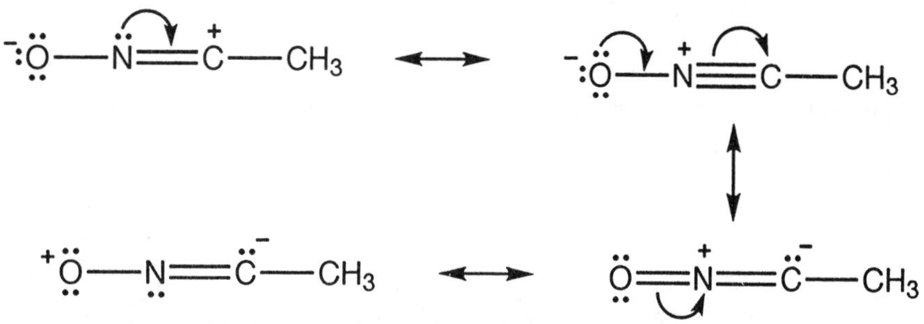

Problem 1.22

(a)

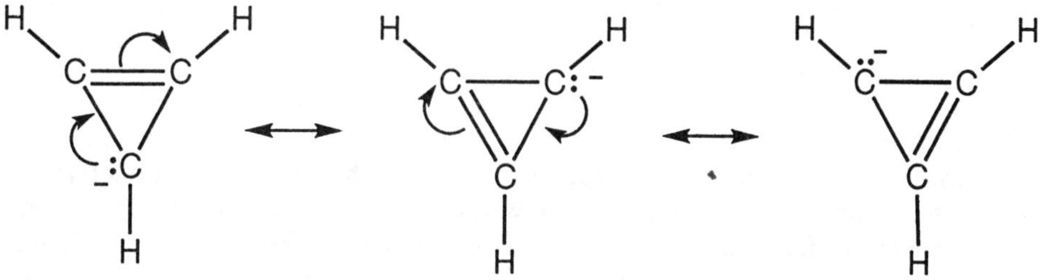

(b)

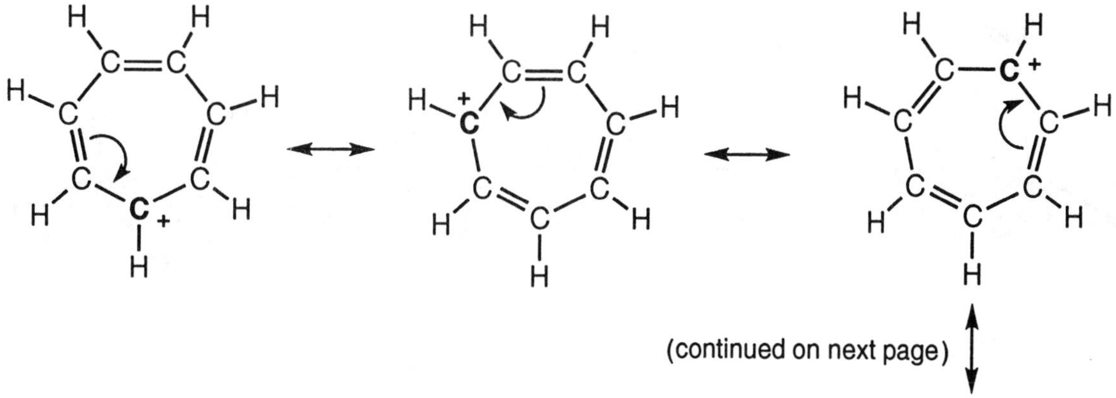

(continued on next page)

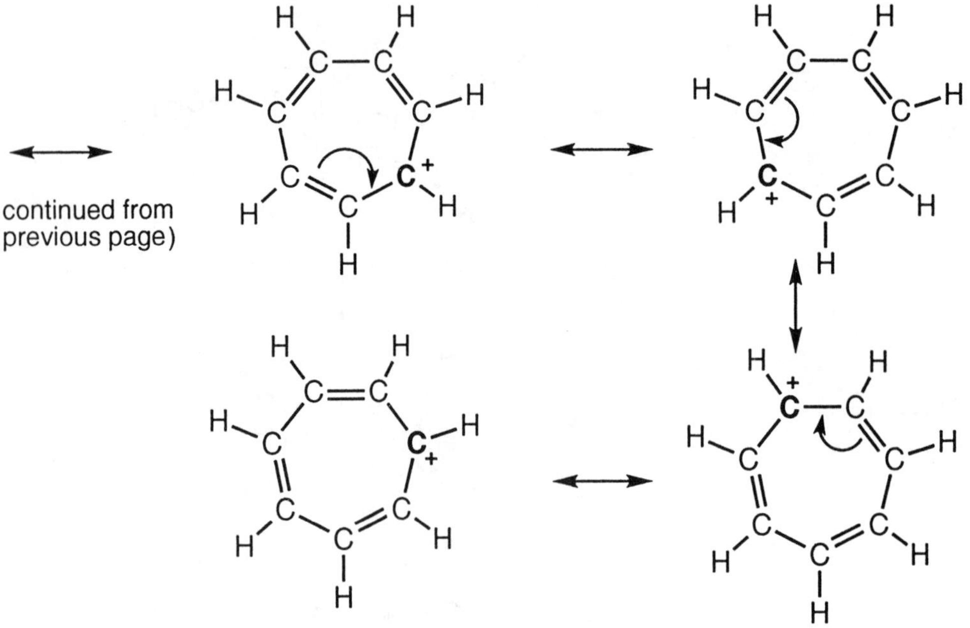

(continued from previous page)

In parts (a) and (b) all ring atoms share the charge. This sets you up for part (c) in which they do not:

(c)

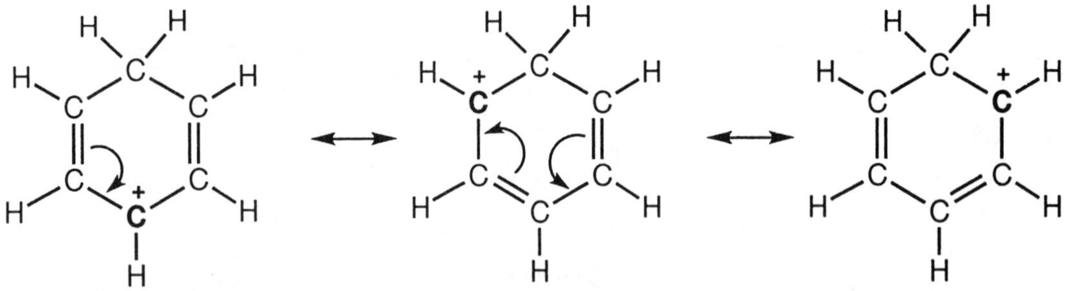

Notice the difference between (a) and (b) of this question and (c). In the first two molecules, every carbon helps share the charge. In (c), in which a CH_2 group interrupts the connectivity of 2*p* orbitals, only three carbons share the charge. We will talk more about such molecules in the future, but note now the difference between these two kinds of system. In cyclic molecules in which every ring atom has a *p* orbital, each atom shares the charge. In systems in which there is an insulator (an atom or atoms without a *p* orbital), not every ring atom will share the charge.

Problem 1.23

(a)

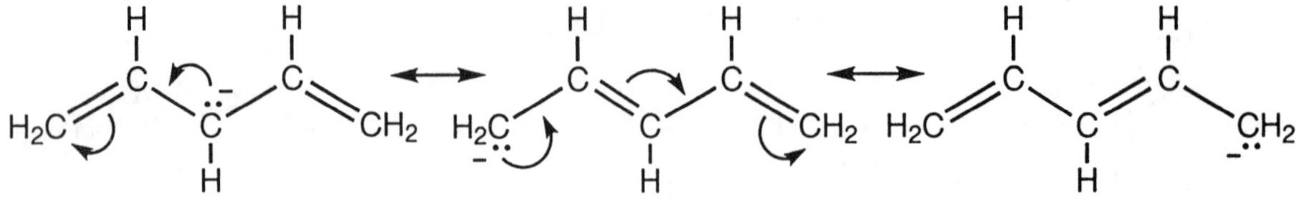

(b)

Now look carefully at these "two" species. They are exactly the same! Each has exactly the same set of three resonance contributors. Tricky, tricky, tricky.

Problem 1.24

The neutral resonance form can be produced this way:

The cyclic, "no charge" resonance form *is* a resonance form, but it is not a very good one. *Remember*: No atoms may be moved in drawing resonance forms—only electrons. Resonance forms are different *electronic* structures. The long bond in the cyclic form is not a strong bond because the two oxygen atoms are very far apart.

Problem 1.25 These molecules all look alike, but a bit of electron counting shows that the charges on the noncarbon atom are different.

(a) For oxygen, the atomic number is 8, which means that there are eight positive charges in the nucleus. The oxygen atom in the three-membered ring has a total of eight electrons: two 1*s* electrons (never shown), four nonbonding electrons, and a share in two bonds for an additional two electrons. This oxygen is neutral. For $_8$O, 2 (1*s*) + 4 (nonbonding) + 2 (shared) = 8 electrons The other electron counts are:

(b) For $_7$N, 2 (1*s*) + 4 (nonbonding) + 2 (shared) = 8 electrons, so N is net 1^-.

(c) For $_{35}$Br, the calculation is tougher. Bromine is in the same column as fluorine, $_9$F, and will have the same number of valence electrons. There are 28 "core" electrons (n = 1, 2, 3 levels, 1*s*, 2*s*, 2*p*, 3*s*, 3*p*, and 3*d*), leaving 7 remaining. In addition, this bromine atom has 4 nonbonding and 2 shared electrons for a total of 6, and a grand total of 28 + 6 = 34. The Br is net 1^+.

(d) For $_{16}$S there are 2 (1*s*) + 2 (2*s*) + 6 (2*p*) + 4 (nonbonding) + 2 (shared) = 16 electrons. This S is neutral. Notice that atoms in the same column of the periodic table are similar. For example, both oxygen and sulfur are neutral.

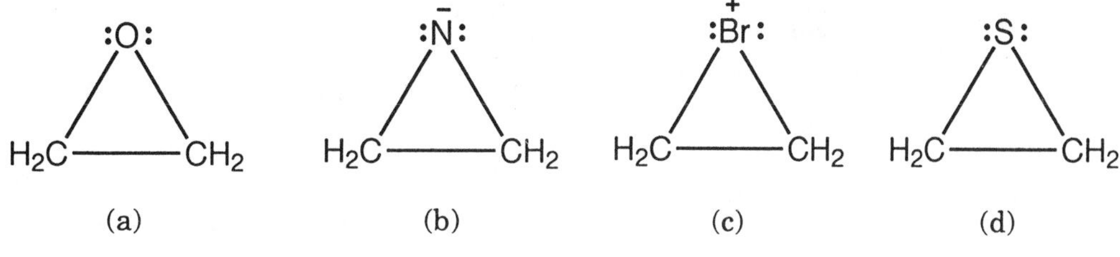

Problem 1.26 This problem reinforces the notion that atoms in the same column of the periodic table with the same number of electron dots in their Lewis structure will have the same charge. If you count for oxygen across the first three structures, you get

(a)

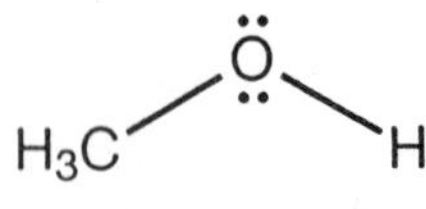

$_8O$ 2 (1*s*)
4 (nonbonding)
2 (shared)

8 total

O neutral

(b)

$_8O$ 2 (1*s*)
6 (nonbonding)
1 (shared)

9 total

O negative

(c)

$_8O$ 2 (1*s*)
2 (nonbonding)
3 (shared)

7 total

O positive

The charge determination for sulfur comes out the same:

(d)

$_{16}S$ 2 (1*s*)
2 (2*s*)
6 (2*p*)
4 (nonbonding)
2 (shared)

16 total

S neutral

(e)

$_{16}S$ 2 (1*s*)
2 (2*s*)
6 (2*p*)
6 (nonbonding)
1 (shared)

17 total

S negative

(f)

$_{16}S$ 2 (1*s*)
2 (2*s*)
6 (2*p*)
2 (nonbonding)
3 (shared)

15 total

S positive

Similarly, nitrogen and phosphorus atoms in the same column of the periodic table, are also identically charged in identical bonding situations.

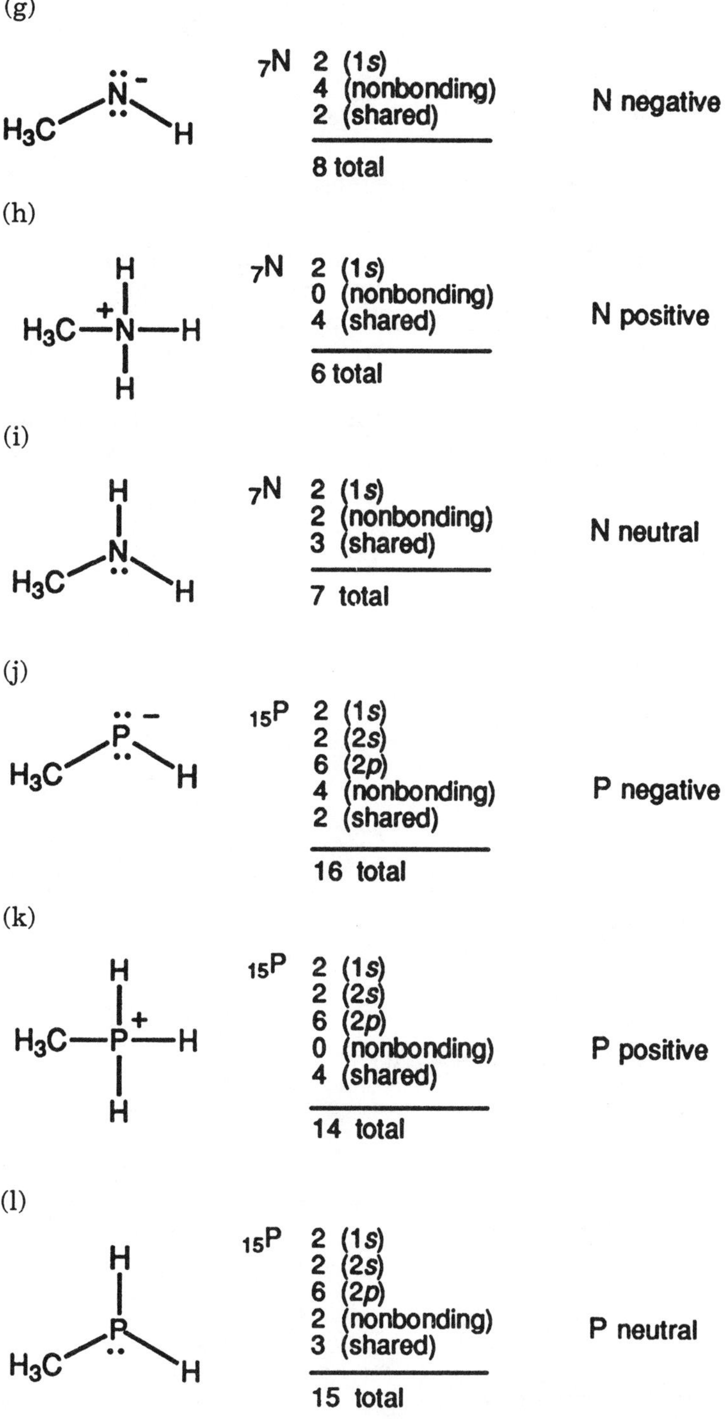

Problem 1.27 Start by drawing Lewis dot structures for the atoms. For oxygen ($_8O$), there will be six electrons available for bonding (8 less the two 1*s* electrons), for fluorine ($_9F$), there will be seven electrons available for bonding (9 less the two 1*s* electrons), and for nitrogen ($_7N$), there will be five (7 less the two 1*s* electrons). For O_2 and F_2, in which there is a single bond between the two atoms,

there will be five and six electrons left over, respectively. For N_2, in which there is a triple bond between the atoms, there will only be a single pair of electrons left on each nitrogen.

$_8$O = ·Ö· ·Ö· ·Ö· = ·Ö—Ö·

$_9$F = :F̈· :F̈· ·F̈: = :F̈—F̈:

$_7$N = N̈: N̈: :N̈ = :N≡N:

Problem 1.28 As shown in the chapter, the ground state of carbon is $1s^2\ 2s^2\ 2p_x\ 2p_y$. There are many possible excited states including $1s^2\ 2s^2\ 2p_x^{\ 2}$ and $1s^2\ 2s\ 2p_x\ 2p_y\ 2p_z$. In these species, two electrons have been brought closer together than is optimal, or one electron has been promoted to a higher energy orbital.

Problem 1.29 For these ions we first figure out the configuration of the neutral atom. Then we add or remove electrons as necessary to accommodate the charge.

$_{11}$Na = $1s^2\ 2s^2\ 2p_x^{\ 2}\ 2p_y^{\ 2}\ 2p_z^{\ 2}\ 3s$
..... so $_{11}$Na$^+$ with one fewer electron will be:

$1s^2\ 2s^2\ 2p_x^{\ 2}\ 2p_y^{\ 2}\ 2p_z^{\ 2}$

$_9$F = $1s^2,\ 2s^2,\ 2p_x^{\ 2},\ 2p_y^{\ 2},\ 2p_z$
..... so $_9$F$^-$, with one more electron will be:

$1s^2,\ 2s^2,\ 2p_x^{\ 2},\ 2p_y^{\ 2},\ 2p_z^{\ 2}$

$_{20}$Ca = $1s^2\ 2s^2\ 2p_x^{\ 2}\ 2p_y^{\ 2}\ 2p_z^{\ 2}\ 3s^2\ 3p_x^{\ 2}\ 3p_y^{\ 2}\ 3p_z^{\ 2}\ 4s^2$
.....so $_{20}$Ca^{++}, with two fewer electrons must be:

$1s^2,\ 2s^2,\ 2p_x^{\ 2},\ 2p_y^{\ 2},\ 2p_z^{\ 2},\ 3s^2,\ 3p_x^{\ 2},\ 3p_y^{\ 2},\ 3p_z^{\ 2}$.

Problem 1.30

$_{19}$K = $1s^2\ 2s^2\ 2p_x^{\ 2}\ 2p_y^{\ 2}\ 2p_z^{\ 2}\ 3s^2\ 3p_x^{\ 2}\ 3p_y^{\ 2}\ 3p_z^{\ 2}\ 4s$ = [Ar] $4s$
$_{20}$Ca = [Ar]$4s^2$
$_{21}$Sc = [Ar]$4s^2 3d$
$_{22}$Ti = [Ar]$4s^2\ 3d^2$
$_{23}$V = [Ar]$4s^2\ 3d^3$
$_{24}$Cr = [Ar]$4s^2\ 3d^4$ (In fact, $_{24}$Cr is [Ar]$4s\ 3d^5$)
$_{25}$Mn = [Ar]$4s^2\ 3d^5$
$_{26}$Fe = [Ar]$4s^2\ 3d^6$
$_{27}$Co = [Ar]$4s^2\ 3d^7$
$_{28}$Ni = [Ar]$4s^2\ 3d^8$
$_{29}$Cu = [Ar]$4s^2\ 3d^9$ (In fact, $_{29}$Cu is [Ar]$4s\ 3d^{10}$)
$_{30}$Zn = [Ar]$4s^2\ 3d^{10}$
$_{31}$Ga = [Ar]$4s^2\ 3d^{10}\ 4p_x$
$_{32}$Ge = [Ar]$4s^2\ 3d^{10}\ 4p_x\ 4p_y$
$_{33}$As = [Ar]$4s^2\ 3d^{10}\ 4p_x\ 4p_y\ 4p_z$
$_{34}$Se = [Ar]$4s^2\ 3d^{10}\ 4p_x^{\ 2}\ 4p_y\ 4p_z$
$_{35}$Br = [Ar]$4s^2\ 3d^{10}\ 4p_x^{\ 2}\ 4p_y^{\ 2}\ 4p_z$
$_{36}$Kr = [Ar]$4s^2\ 3d^{10}\ 4p_x^{\ 2}\ 4p_y^{\ 2}\ 4p_z^{\ 2}$

Problem 1.31

$_{14}Si = 1s^2\ 2s^2\ 2p_x^{\ 2}\ 2p_y^{\ 2}\ 2p_z^{\ 2}\ 3s^2\ 3p_x^{\ \uparrow}\ 3p_y^{\ \uparrow}$

$_{15}P = 1s^2\ 2s^2\ 2p_x^{\ 2}\ 2p_y^{\ 2}\ 2p_z^{\ 2}\ 3s^2\ 3p_x^{\ \uparrow}\ 3p_y^{\ \uparrow}\ 3p_z^{\ \uparrow}$

$_{16}S = 1s^2\ 2s^2\ 2p_x^{\ 2}\ 2p_y^{\ 2}\ 2p_z^{\ 2}\ 3s^2\ 3p_x^{\ \uparrow\downarrow}\ 3p_y^{\ \uparrow}\ 3p_z^{\ \uparrow}$

Hund's rule states that for orbitals of equal energy, such as the three 3*p* orbitals, the electronic configuration with the greatest number of parallel spins will be the lowest in energy. Electrons with parallel spins (same spin quantum number *s*) cannot occupy the same orbital, and are therefore kept apart, minimizing electron–electron repulsion. We faced the same problem in determining the spin state of a carbon atom. Here $_6C = 1s^2 2s^2 2p_x^{\ \uparrow} 2p_y^{\ \uparrow}$ is preferred to $1s^2 2s^2 2p_x^{\ 2}$ or $1s^2 2s^2 2p_x^{\ \downarrow} 2p_y^{\ \uparrow}$.

Problem 1.32

In oxygen, the last two electrons fill the $2p_y$ and $2p_z$ orbitals and have unpaired spins (Hund's rule). In this case the ESR machine will find two unpaired spins.

$_8O = 1s^2\ 2s^2\ 2p_x^{\ 2}\ 2p_y^{\ \uparrow}\ 2p_z^{\ \uparrow}$

In O^+ there will be one fewer electron, and the ESR machine will still find the three unpaired electrons in the $2p_x$, $2p_y$, and $2p_z$ orbitals.

$_8O^+ = 1s^2\ 2s^2\ 2p_x^{\ \uparrow}\ 2p_y^{\ \uparrow}\ 2p_z^{\ \uparrow}$

In O^{2-} there will be two more electrons than in neutral O. The electronic configuration will be $_8O^{2-} = 1s^2, 2s^2, 2p_x^{\ \uparrow\downarrow}, 2p_y^{\ \uparrow\downarrow}, 2p_z^{\ \uparrow\downarrow}$ and the ESR machine will see no unpaired electrons.

In neutral neon ($_{10}Ne$), there are 10 electrons, so in $_{10}Ne^+$ there will be only 9. The electronic configuration will be $_{10}Ne^+ = 1s^2, 2s^2, 2p_x^{\ \uparrow\downarrow}, 2p_y^{\ \uparrow\downarrow}\ 2p_z^{\ \uparrow}$. Once again, the ESR machine will find a single unpaired electron.

Fluoride, F^-, has the electronic configuration of Ne. All electrons are paired and the ESR machine will seek in vain for an unpaired spin.

$_9F^- = 1s^2, 2s^2, 2p_x^{\ \uparrow\downarrow}, 2p_y^{\ \uparrow\downarrow}\ 2p_z^{\ \uparrow\downarrow}$

Problem 1.33 Both carbon and oxygen are neutral

$_6C$	$_8O$
2 (1*s*)	2 (1*s*)
2 (shared)	4 (shared)
2 (nonbonding)	4 (nonbonding)
6 total	8 total
C is neutral	O is neutral

As oxygen is more electronegative than carbon, the dipole will be

$$\overset{+\!\!\longrightarrow}{C{=}O}$$
$$\delta^+ \quad \delta^-$$

The second Lewis structure is

$$:C{=}\ddot{\underset{..}{O}} \longleftrightarrow \overset{-}{:C}{\equiv}\overset{+}{O}:$$

In the new Lewis structure the carbon is negative and the oxygen positive:

$_6$C		$_8$O	
	2 (1s)		2 (1s)
	3 (shared)		3 (shared)
	2 (nonbonding)		2 (nonbonding)
	7 total		7 total
	C is negative		O is positive

So, to the extent that this second resonance form is important, the dipole will be in the opposite direction:

δ− δ+
:C≡O:

The dipoles in the two Lewis structures (two resonance forms) will tend to cancel each other out. The result is a very small observed dipole.

Problem 1.34 This molecule also has two important resonance forms, but the dipole is in the same direction in each and will reinforce. Formaldehyde will have a larger dipole moment than carbon monoxide.

δ+ δ−
H_2C=O: ⟷ $H_2\overset{+}{C}$—$\ddot{O}$:$^-$

Problem 1.35 Structure **B** will have no dipole moment, as the dipoles cancel. Therefore, the observation of a dipole moment for CH_2F_2 eliminates **B** as a possibility. However, in planar structure **C** the dipoles reinforce. This molecule will have a dipole moment, and cannot be distinguished from **A** on this basis.

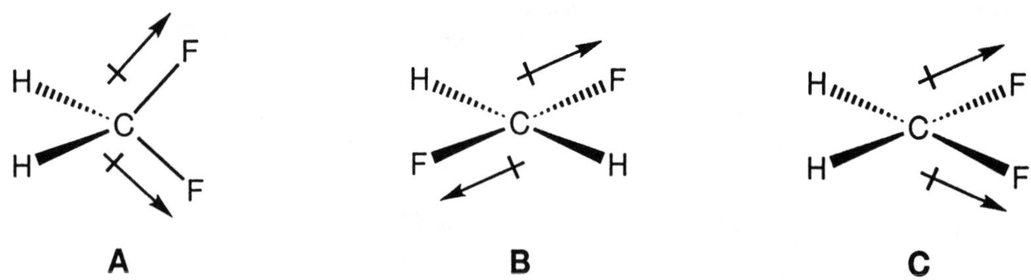

Problem 1.36 It all starts normally enough, so n must increase monotonically. The 1s shell fills with two electrons so $s = \pm 1/2$. Lithium (Li) and beryllium (Be) are normal as the 2s shell fills. However, there seem to be no 2p orbitals. So l must not be the same as in our universe. In the new universe, $_5$B is similar to $_1$H and $_3$Li, and $_6$C similar to $_2$He and $_4$Be. The 3s shell must be filling with these two atoms.

$_1$H = 1s
$_2$He = 1s^2
$_3$Li = 1s^2, 2s
$_4$Be = 1s^2, 2s^2

(as no 2p orbitals are available in this universe, start to fill the 3s shell in boron)

$_5$B = 1s^2, 2s^2, 3s
$_6$C = 1s^2, 2s^2, 3s^2

Now we find six atoms filling, $_7$N through $_{12}$Mg. These must be the three 3*p* orbitals.

$_7N = 1s^2, 2s^2, 3s^2, 3p$
$_8O = 1s^2, 2s^2, 3s^2, 3p^2$
$_9F = 1s^2, 2s^2, 3s^2, 3p^3$
$_{10}Ne = 1s^2, 2s^2, 3s^2, 3p^4$
$_{11}Na = 1s^2, 2s^2, 3s^2, 3p^5$
$_{12}Mg = 1s^2, 2s^2, 3s^2, 3p^6$

Apparently, l in this universe has the value $l = 0, 1, 2, 3....(n - 2)$.
So, when $n = 1, l = 0, n = 2, l = 0$ (no 2*f* orbitals), $n = 3, l = 1, m_l = -1...0...+1, s = \pm 1/2$

Now the 4*s* shell must fill.

$_{13}Al = 1s^2, 2s^2, 3s^2, 3p^6, 4s$
$_{14}Si = 1s^2, 2s^2, 3s^2, 3p^6, 4s^2$

Now the 4*p* orbitals fill to complete the periodic table as shown

$_{15}P = 1s^2\ 2s^2\ 3s^2\ 3p^6\ 4s^2\ 4p$
$_{16}S = 1s^2\ 2s^2\ 3s^2\ 3p^6\ 4s^2\ 4p^2$
$_{17}Cl = 1s^2\ 2s^2\ 3s^2\ 3p^6\ 4s^2\ 4p^3$
$_{18}Ar = 1s^2\ 2s^2\ 3s^2\ 3p^6\ 4s^2\ 4p^4$
$_{19}K = 1s^2\ 2s^2\ 3s^2\ 3p^6\ 4s^2\ 4p^5$
$_{20}Ca = 1s^2\ 2s^2\ 3s^2\ 3p^6\ 4s^2\ 4p^6$

Problem 1.37

(a)

1*s*	2*s*	2*p*
$n = 1$	$n = 2$	$n = 2$
$l = 0$	$l = 0$	$l = 1$
$m_l = 1, 0, -1$	$m_l = 1, 0, -1$	$m_l = 2, 1, 0, -1, -2$
$s = \pm 1/2$	$s = \pm 1/2$	$s = \pm 1/2$

3*s*	3*p*	3*d*
$n = 3$	$n = 3$	$n = 3$
$l = 0$	$l = 1$	$l = 2$
$m_l = 1,0,-1$	$m_l = 2, 1, 0, -1, -2$	$m_l = 3, 2, 1, 0, -1, -2, -3$
$s = \pm 1/2$	$s = \pm 1/2$	$s = \pm 1/2$

(b)

1*s* can hold 6 electrons
2*s* can hold 6 electrons
2*p* can hold 10 electrons
3*s* can hold 6 electrons
3*p* can hold 10 electrons
3*d* can hold 14 electrons

(c)

1 $1s_x$
2 $1s_x\ 1s_y$
3 $1s_x\ 1s_y\ 1s_z$
4 $1s_x^2\ 1s_y\ 1s_z$
5 $1s_x^2\ 1s_y^2\ 1s_z$
6 $1s_x^2\ 1s_y^2\ 1s_z^2$
7 $1s_x^2\ 1s_y^2\ 1s_z^2\ 2s_x$
8 $1s_x^2\ 1s_y^2\ 1s_z^2\ 2s_x\ 2s_y$
9 $1s_x^2\ 1s_y^2\ 1s_z^2\ 2s_x\ 2s_y\ 2s_z$
10 $1s_x^2\ 1s_y^2\ 1s_z^2\ 2s_x^2\ 2s_y\ 2s_z$
11 $1s_x^2\ 1s_y^2\ 1s_z^2\ 2s_x^2\ 2s_y^2\ 2s_z$
12 $1s_x^2\ 1s_y^2\ 1s_z^2\ 2s_x^2\ 2s_y^2\ 2s_z$
13 $1s_x^2\ 1s_y^2\ 1s_z^2\ 2s_x^2\ 2s_y^2 2s_z^2\ 2p_a$
14 $1s_x^2\ 1s_y^2\ 1s_z^2\ 2s_x^2\ 2s_y^2\ 2s_z^2\ 2p_a\ 2p_b$

(d) Element **22** completes the filling of the $2p$ level in this universe. Remember that in this weird place the $2p$ level can hold 10 electrons. Element **23** will begin to fill the $3s$ shell and will fall beneath elements **1** and **7**.

22 $1s_x^2\ 1s_y^2\ 1s_z^2\ 2s_x^2\ 2s_y^2\ 2s_z^2\ 2p_a^2\ 2p_b^2\ 2p_c^2\ 2p_d^2\ 2p_e^2$

23 $1s_x^2\ 1s_y^2\ 1s_z^2\ 2s_x^2\ 2s_y^2\ 2s_z^2\ 2p_a^2\ 2p_b^2\ 2p_c^2\ 2p_d^2\ 2p_e^2\ 3s_x$

1	2	3	4	5	6										
7	8	9	10	11	12	13	14	15	16	17	18	19	20	21	22
23															

Chapter 2 Outline

2.1 The Covalent Bond: Hydrogen (H_2)
2.2 Bond Strength
2.3 Linear and Square H_4
2.4 Something More: The Molecular Orbitals of Tetrahedral H_4

2

Molecules, Molecular Orbitals, and Bonding

Many of the early problems in this chapter require simple combinations of orbitals to produce new orbitals, and then the placement of electrons in the orbitals. The key things to remember are that orbitals can interact in both constructive and destructive ways; $H_{1s} + H_{1s} = \Phi_B$ and $H_{1s} - H_{1s} = \Phi_A$ are prototypal examples discussed in the chapter. The interaction of orbitals is often shown in a graphical way:

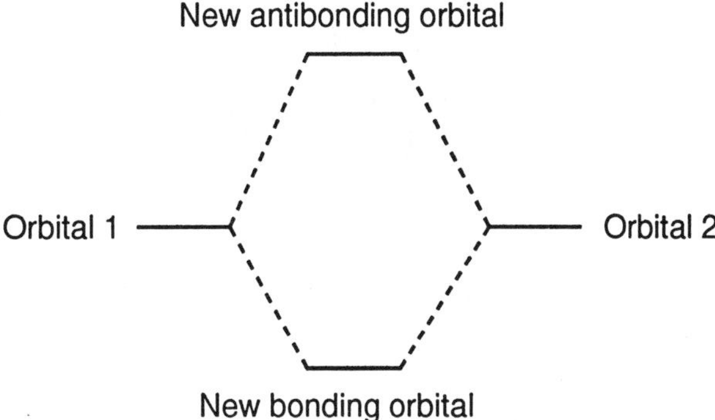

Each orbital can hold a total of two electrons. Remember that placing single electrons with parallel spins in different orbitals of equal energy gives a lower energy species than the one made by placing two electrons with paired (opposite) spins in a single orbital. We move from fairly simple examples to quite sophisticated systems in these problems, but the principles remain the same. Notice the complete absence of mathematics.

Problem 2.1 The simplest molecule is "H_2 minus something." The H_2 molecule contains only two protons and two electrons. As loss of a proton doesn't leave a molecule behind, that "something" can only be an electron. The simplest molecule must be H_2^+. Another electron cannot be lost to give something even simpler because H_2^{2+} is not a molecule. In H_2^{2+} there would be no electrons to bind the two nuclei.

Problem 2.2 Two carbon 2*s* orbitals can interact in a bonding (2*s* + 2*s*) or antibonding (2*s* – 2*s*) way. The node in the antibonding orbital is shown as a bar:

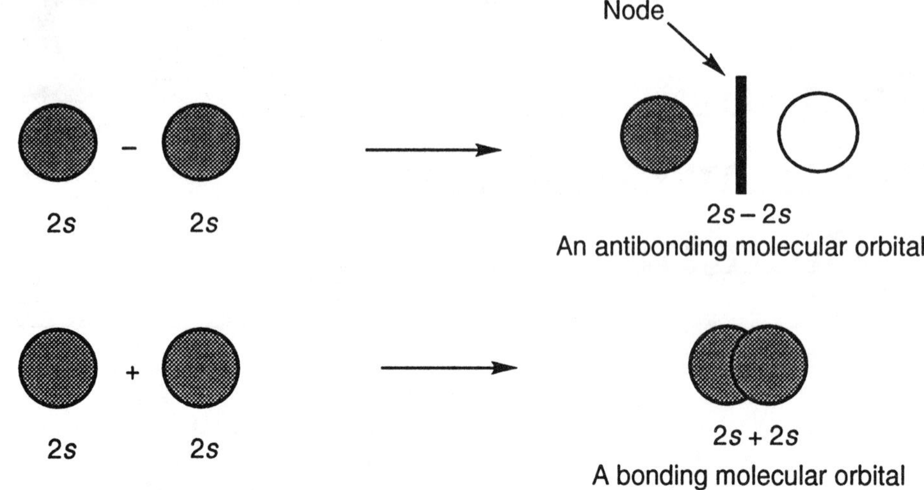

Problem 2.7 The diagram for He_2^+ can be easily derived from the diagram in Figure 2.13 by removal of one electron. *Remember*: Construction of molecular orbitals from atomic orbitals does not depend on the number of electrons. The electrons are placed in the appropriate orbitals later. In this case, we first build the molecular orbitals of He_2 from two He 1*s* orbitals. In Figure 2.13, we put in four electrons to construct He_2. In this problem, you need only put in three electrons to make He_2^+.

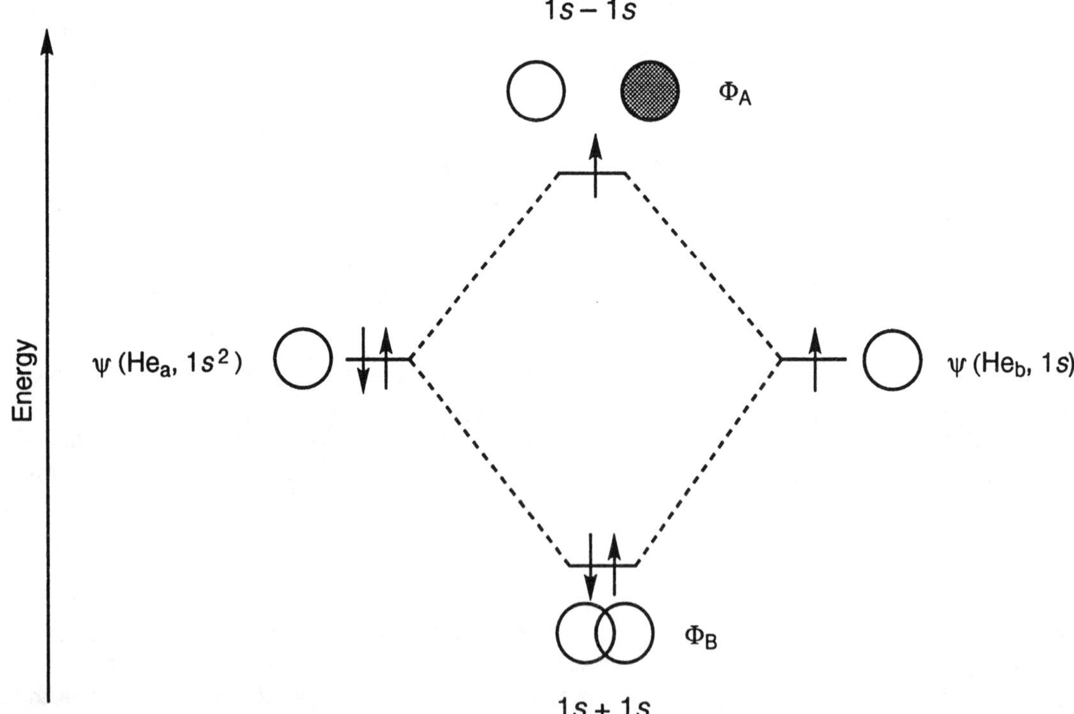

Problem 2.9 The energy of an electron in an orbital depends on both stabilizing and destabilizing forces. Whenever two electrons occupy an orbital, electron–electron repulsion is a destabilizing factor and raises the energy of the electrons in that orbital. An electron in a filled bonding orbital is moved higher in energy, closer to the energy of its constituent atomic orbitals. An electron in a

filled antibonding orbital is also increased in energy, but moves away from the energy of its constituent atomic orbitals. The net result is that an electron in a filled bonding molecular orbital is stabilized less than an electron in a filled antibonding molecular orbital is destabilized.

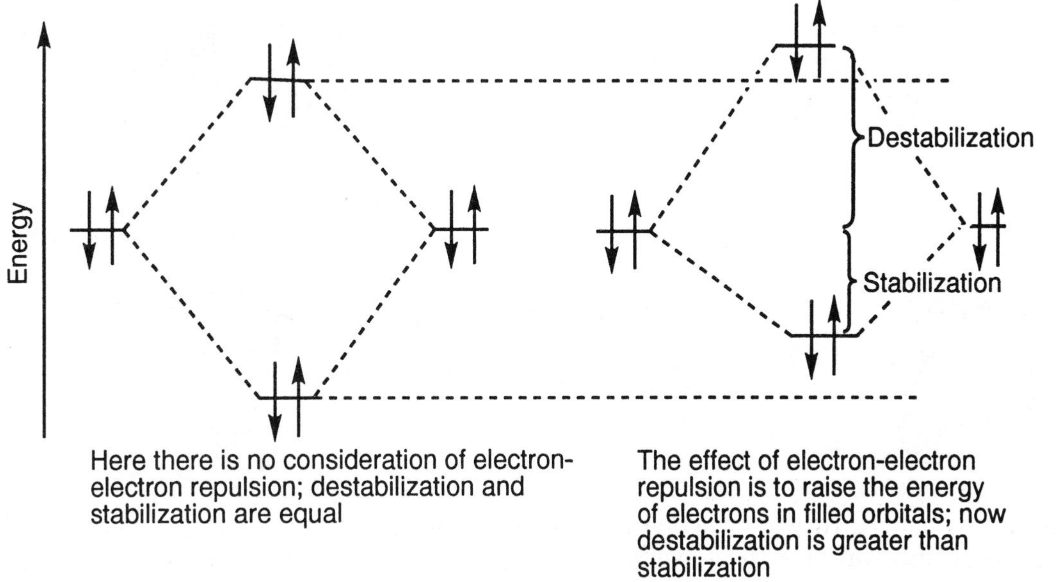
Energy
Destabilization
Stabilization
Here there is no consideration of electron-electron repulsion; destabilization and stabilization are equal
The effect of electron-electron repulsion is to raise the energy of electrons in filled orbitals; now destabilization is greater than stabilization

Problem 2.11 As the lobe destined to become the apex of the tetrahedron is moved up, a new interaction, either bonding or antibonding, appears with the lobe originally across the ring.

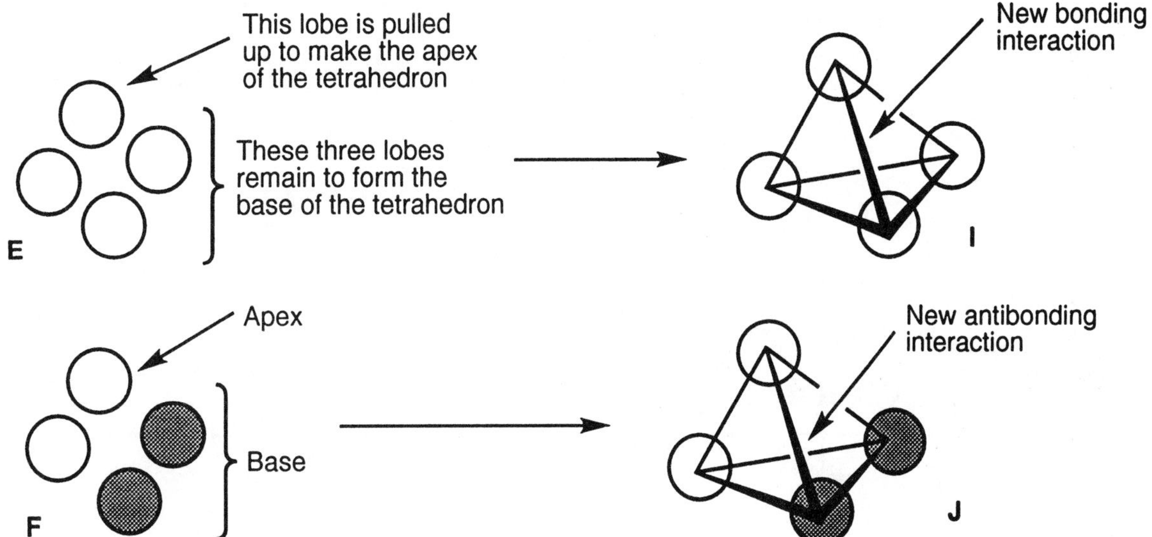
This lobe is pulled up to make the apex of the tetrahedron
These three lobes remain to form the base of the tetrahedron
E
New bonding interaction
I
Apex
Base
F
New antibonding interaction
J

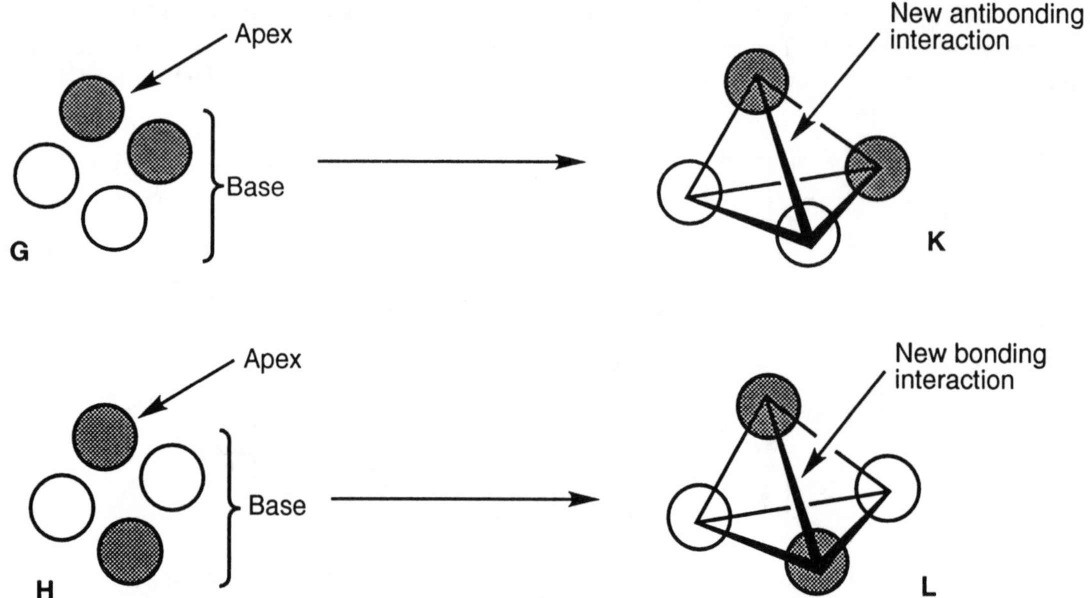

Problem 2.12 Neutral H_4 will have four electrons, one from each hydrogen. Therefore H_4^+ must have only three electrons. Notice the operation of Hund's rule in the diagrams for H_4. Two electrons of the same spin are placed in equi-energetic orbitals (**F** and **G**, or **J** and **K**).

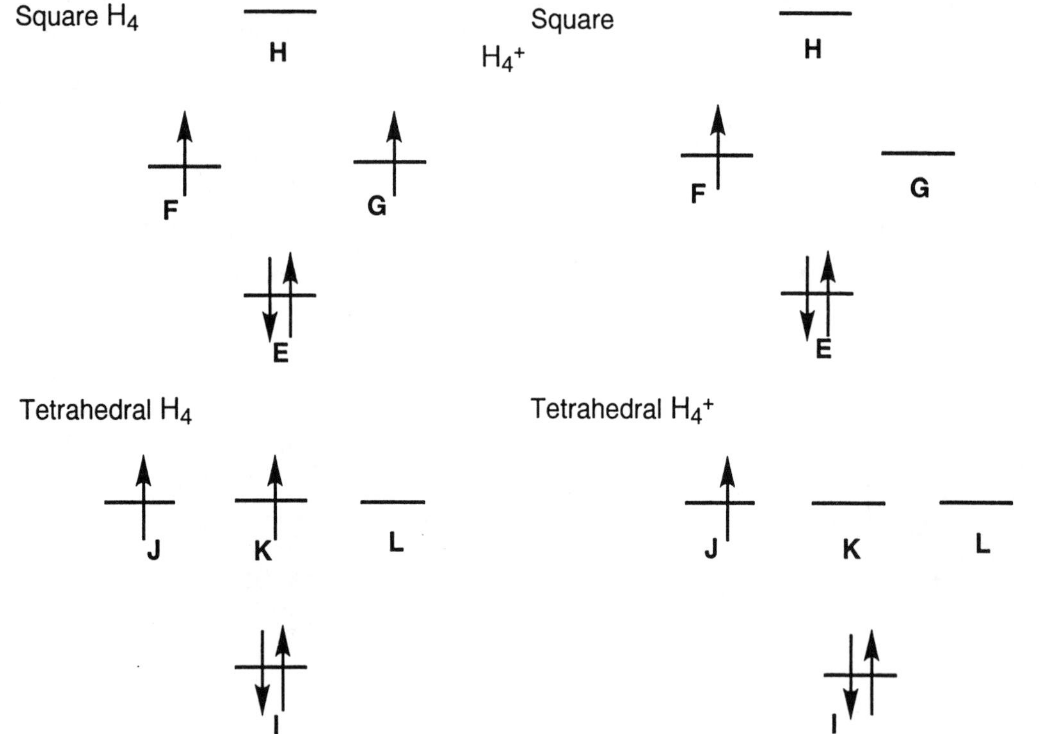

Problem 2.13 In **F** and **G,** a new antibonding interaction appears as the tetrahedron is built from the square (Problem 2.11). Accordingly, the energy of these orbitals is increased as the transformation to the tetrahedron takes place. In **H** and **E** a new bonding interaction appears, so

the energy drops. The new orbitals **J, K,** and **L** all have a single new node and, thus, are of the same energy (Fig. 2.22).

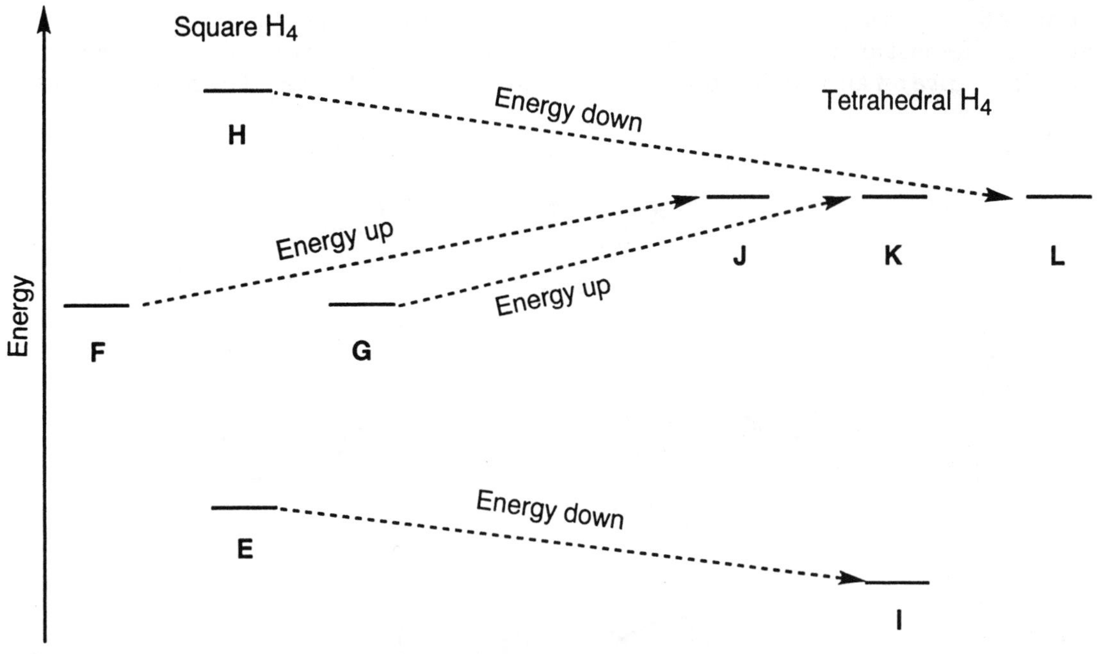

Problem 2.14 Linear, square, and tetrahedral H_4^{2+} are all two-electron species, so only the lowest molecular orbital is occupied. As tetrahedral H_4^{2+} has the lowest energy molecular orbital of the three species, it is the most stable form.

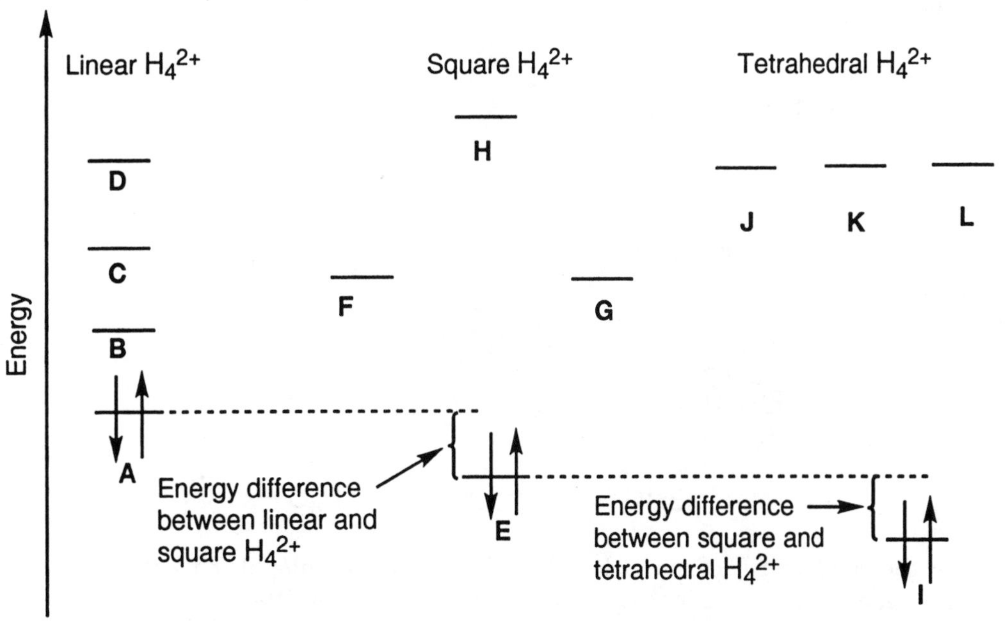

Additional Problem Answers

Problem 2.15 This straightforward problem is designed to remind you that a pair of 2*p* orbitals can interact in a constructive way, 2*p* + 2*p*, to give a bonding molecular orbital and in a destructive way, 2*p* – 2*p*, to give an antibonding orbital. Note the new node in 2*p* – 2*p*, shown as a bar in the figure.

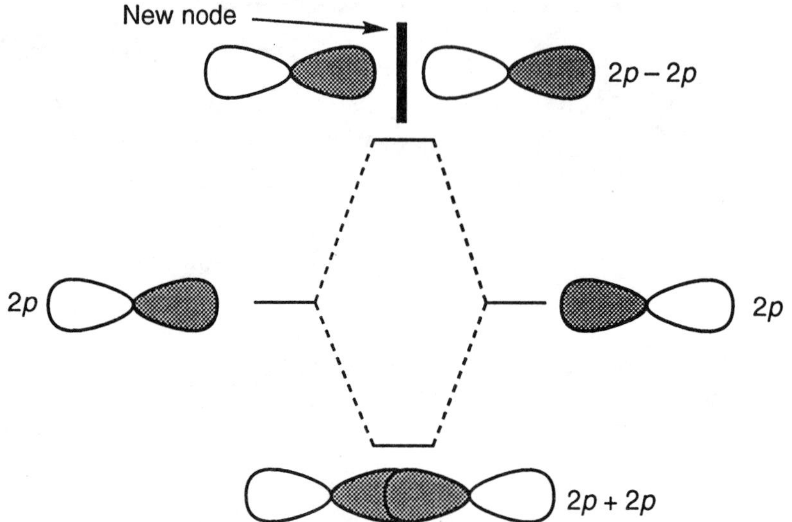

Problem 2.16 Yes, this is a net-zero interaction. If you look at the two orbitals from this perspective, you can easily see the offsetting bonding and antibonding interactions. These orbitals are orthogonal.

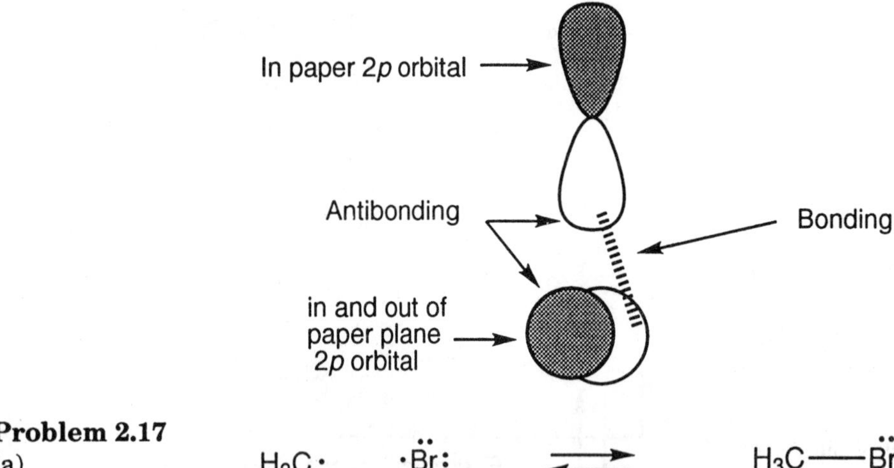

Problem 2.17

(a) $$H_3C\cdot \quad \cdot\ddot{\underset{\cdot\cdot}{Br}}: \rightleftharpoons H_3C\text{—}\ddot{\underset{\cdot\cdot}{Br}}:$$

Table 2.1 gives the bond energy of the carbon–bromine bond, 66 kcal/mol. This reaction involves pure bond making, an exothermic process. The reaction would be written:

$$H_3C\cdot \quad \cdot\ddot{\underset{\cdot\cdot}{Br}}: \rightleftharpoons H_3C\text{—}\ddot{\underset{\cdot\cdot}{Br}}: \qquad \Delta H^\circ = -66 \text{ kcal/mol}$$

(b) $H_3C—\ddot{\underset{..}{Cl}}: \rightleftharpoons H_3C\cdot \ \cdot\ddot{\underset{..}{Cl}}:$

By contrast, this reaction involves only the breaking of the carbon–chlorine bond. This reaction is endothermic by about 79 kcal/mol. The reaction would be written

$$H_3C—\ddot{\underset{..}{Cl}}: \rightleftharpoons H_3C\cdot \quad \cdot\ddot{\underset{..}{Cl}}: \qquad \Delta H^\circ = +79 \text{ kcal/mol}$$

(c) $H_2C{=}CH_2 + H—H \rightleftharpoons H_2C(H)—CH_2(H)$

This reaction is more complicated. Bonds are both made and broken. Assume that the four carbon–hydrogen bonds in the starting material are exactly balanced by the four carbon–hydrogen bonds in the product.

Bonds broken in starting material (kcal/mol)		Bonds made in products (kcal/mol)	
C=C	148	C—C	90
H—H	104	two C—H	192
	252 total		282 total

This reaction is exothermic by about 30 kcal/mol (282 – 252 = 30). The reaction would be written

$$H_2C{=}CH_2 + H—H \longrightarrow H_2C(H)—CH_2(H) \qquad \Delta H^\circ = -30 \text{ kcal/mol}$$

(d)

$$H_2C{=}CH_2 \quad H—\ddot{\underset{..}{Cl}}: \rightleftharpoons H_2C(H)—CH_2(:\ddot{\underset{..}{Cl}}:)$$

This reaction is also complicated. Bonds are both made and broken.

Bonds broken in starting material (kcal/mol)		Bonds made in products (kcal/mol)	
		C—C	90
$H_2C{=}CH_2$	148	C—H	96
H—Cl	103	C—Cl	79
	251 total		265 total

This reaction is exothermic by about 14 kcal/mol (265 – 251 = 14). The reaction should be written

$$H_2C{=}CH_2 \quad H—\ddot{\underset{..}{Cl}}: \longrightarrow H_2C(H)—CH_2(:\ddot{\underset{..}{Cl}}:) \qquad \Delta H^\circ = -14 \text{ kcal/mol}$$

Problem 2.18

(a) As there are three orbitals going into our calculation, there must be three coming out. The H_3 molecule will have three molecular orbitals.

(b) First of all, remember that the problem tells us how to construct H_3. Place the new H in between the two hydrogens of H—H. The interaction of Φ_B with 1*s* will yield two new molecular orbitals, **1** and **3**.

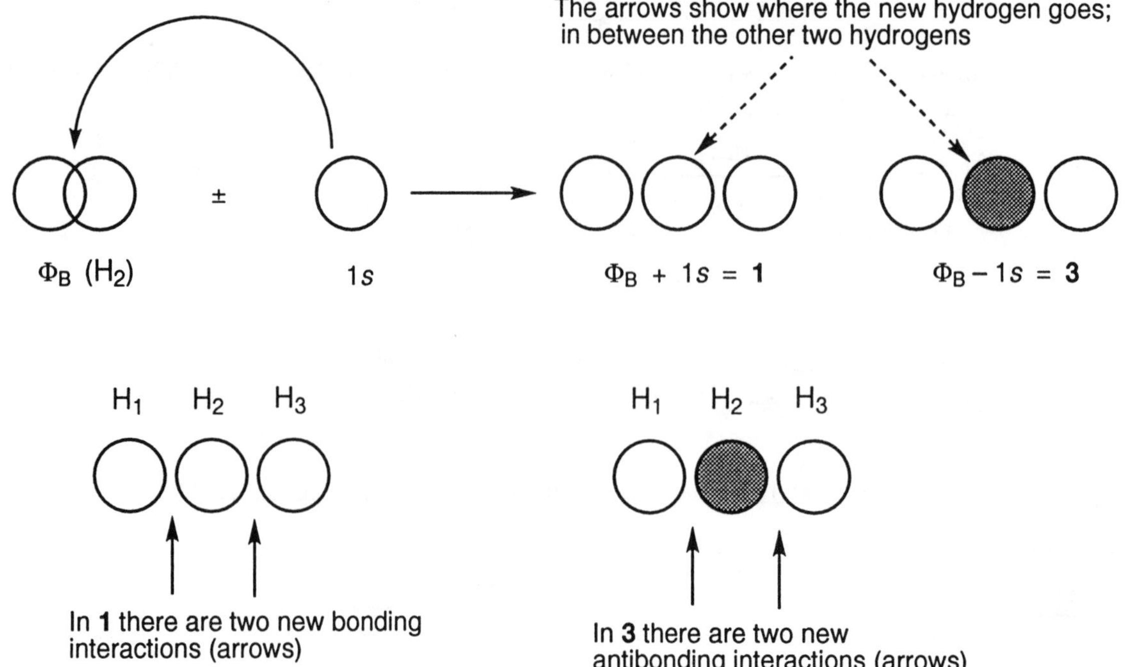
The arrows show where the new hydrogen goes; in between the other two hydrogens
Φ_B (H_2)
±
1s
Φ_B + 1s = **1**
Φ_B − 1s = **3**
H_1 H_2 H_3
H_1 H_2 H_3
In **1** there are two new bonding interactions (arrows)
In **3** there are two new antibonding interactions (arrows)

However, Φ_A will not interact with a hydrogen 1*s* orbital placed between the two hydrogens; this is a net-zero interaction, as the new bonding interaction is exactly canceled by the new antibond.

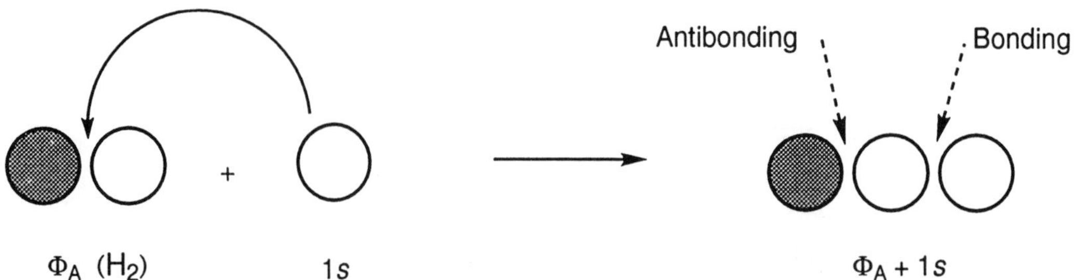
Antibonding
Bonding
+
Φ_A (H_2)
1s
Φ_A + 1s
New antibonding exactly cancels new bonding-
there is no net interaction between the orbitals

So, the three orbitals for HHH are **1** and **3**, and the old Φ_A of H_2, modified only by the moving apart of the two hydrogens. Let's call this one **2**. The center dot shows the position of the middle hydrogen. The sign of the wave function at this point is zero.

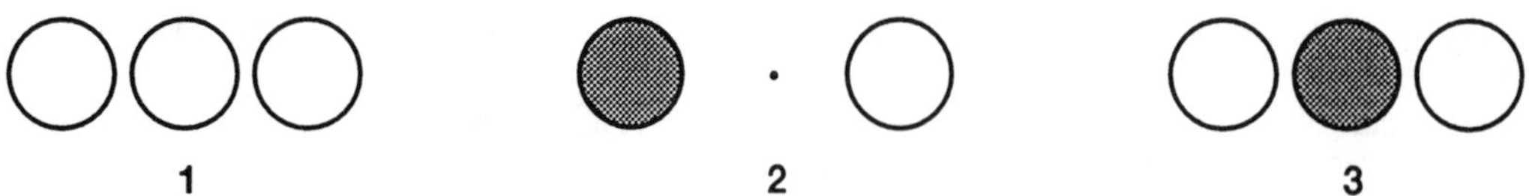
1
2
3

(c) You can easily order these new orbitals by simply counting the nodes.

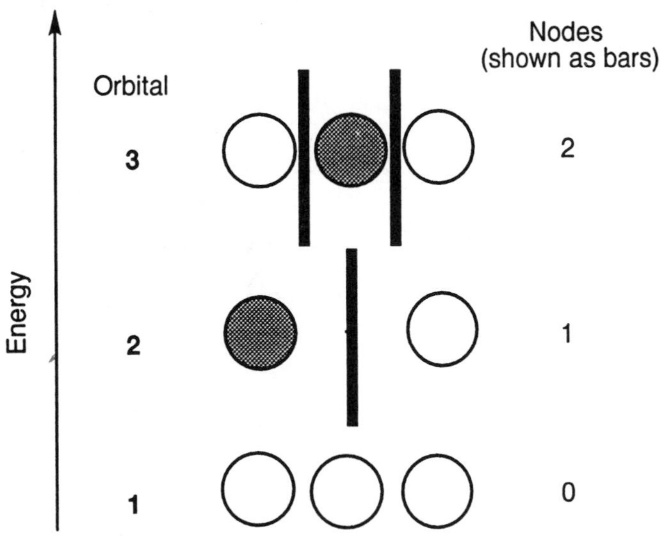

Problem 2.19
(a) All you need to do is to bend the orbitals for HHH into a triangle to generate the molecular orbitals for triangular H_3. The direction of energy change is determined by noting whether a new bonding or antibonding interaction is created.

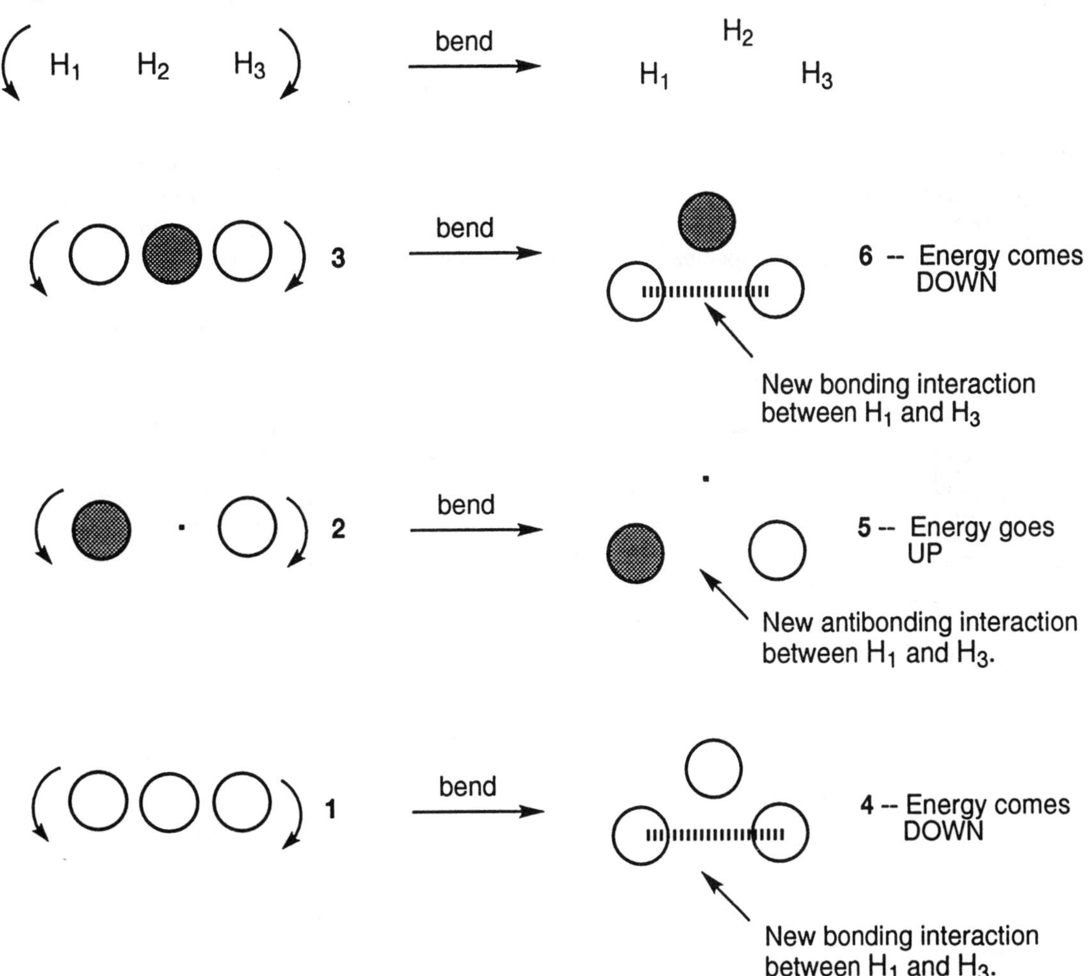

(b) Apply what the answer to (a) tells us. This answer shows how the energies of **1**, **2**, and **3** will change as bending occurs to make **4**, **5**, and **6**. Notice that orbitals **5** and **6**, each with one node (shown as a bar in the figure), are placed at the same energy.

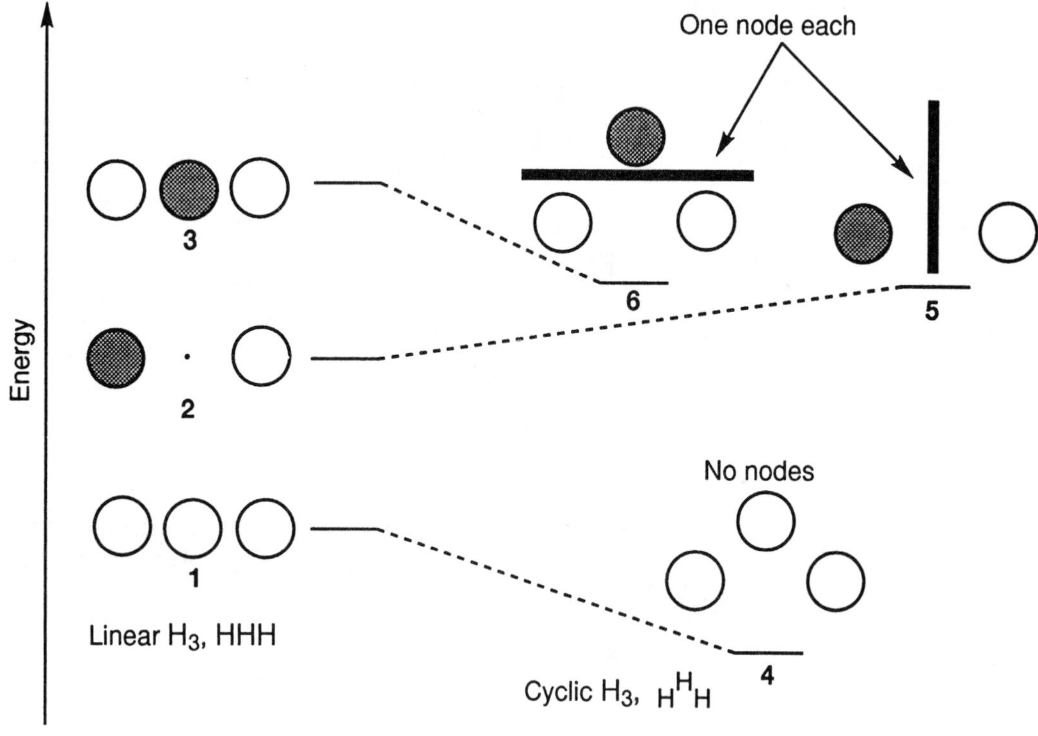

Problem 2.20 In any H_3^+ molecule, there will be two electrons. In neutral H_3 there are three electrons, one from each hydrogen, and H_3^+ will have one fewer. So, only the lowest molecular orbital will be occupied. As the diagram for Problem 2.19 shows, this orbital is lower for triangular H_3 than for linear H_3. The bent species will be more stable than the linear molecule.

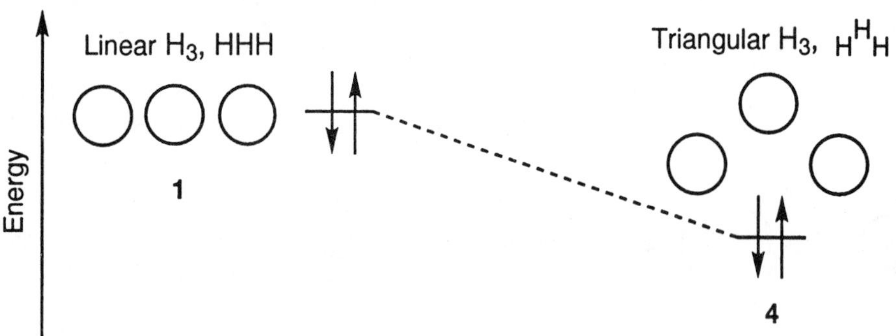

Problem 2.21 The situation is directly parallel to the formation of HHH, made in Problem 2.18. Once again, there will be three molecular orbitals. Two new orbitals, **A** and **B**, are formed by the interaction of a single 2*p* orbital with (2*p* + 2*p*).

$2p + 2p$ ± $2p$ → $(2p + 2p) + 2p$ $(2p + 2p) - 2p$

A **B**

The third orbital, **C**, comes from (2*p* – 2*p*), which does not interact with a 2*p* orbital placed in the middle. Once again, this is a net-zero situation as the bonding and antibonding interactions exactly cancel.

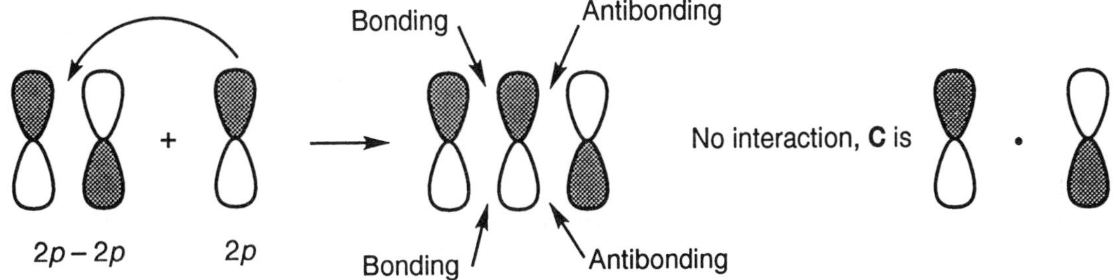

The new orbitals **A**, **B**, and **C** can be ordered by counting nodes:

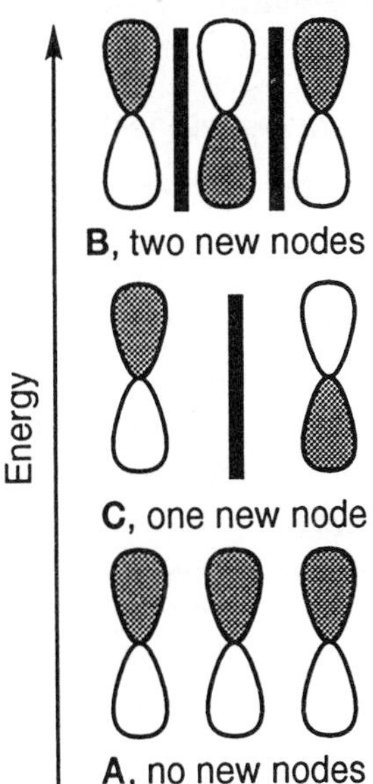

Problem 2.22 Mimic the treatment of HHHH in the text by placing the two pairs of $2p$ orbitals end-to-end. Let $(2p + 2p)$ interact with the equi-energetic $(2p + 2p)$ and $(2p - 2p)$ with $(2p - 2p)$, which yields four new molecular orbitals **A**, **B**, **C**, and **D**.

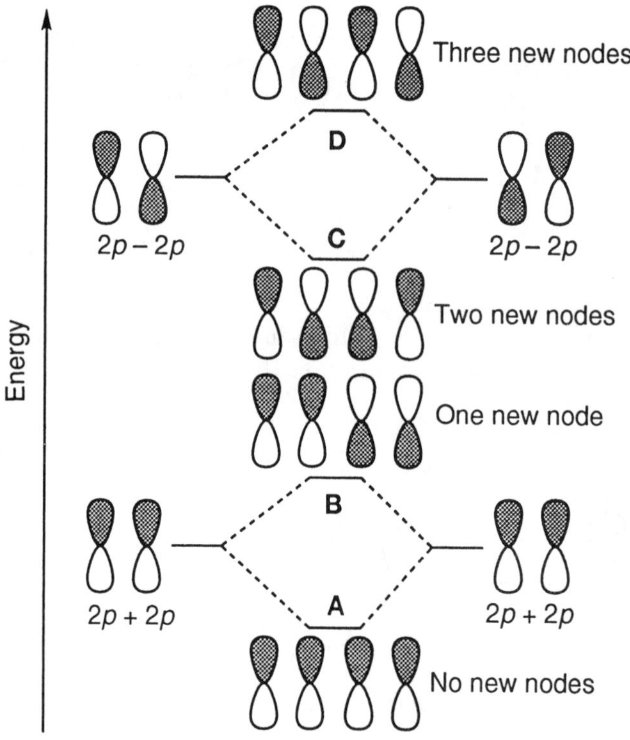

The new orbitals appear in order of their energies. Once again you order the orbitals by counting the new nodes formed.

Problem 2.23 The combination of Φ_B with Φ_B will lead to **E** and **G**.

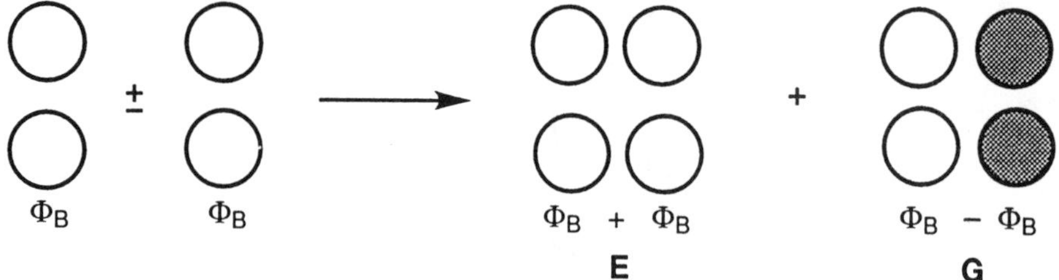

Similarly, the combination of Φ_A with Φ_A gives **F** and **H**.

Φ_A ± Φ_A → $\Phi_A + \Phi_A$ + $\Phi_A - \Phi_A$

F **H**

Nothing to it.

Problem 2.24

(a) Atomic orbitals of carbon — Molecular orbitals of hydrogen

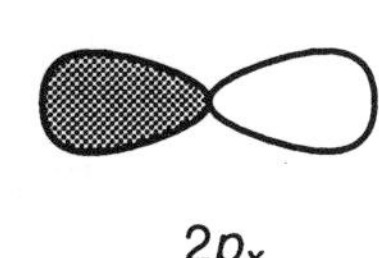

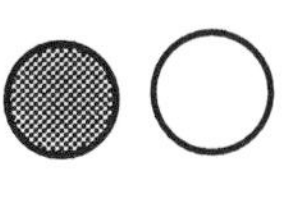

$2s$ $2p_y$ $2p_x$ $2p_z$ Φ_B Φ_A

(b) As four carbon atomic orbitals and two hydrogen molecular orbitals go into the calculation, there must be six CH_2 molecular orbitals.

(c)

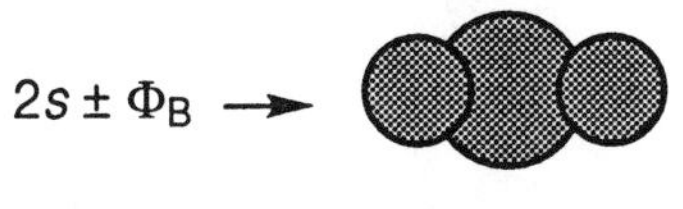

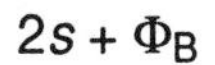

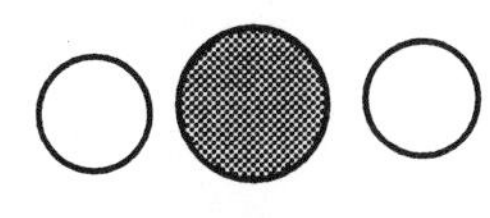

$2s - \Phi_B$

$2p_y$ $2p_z$

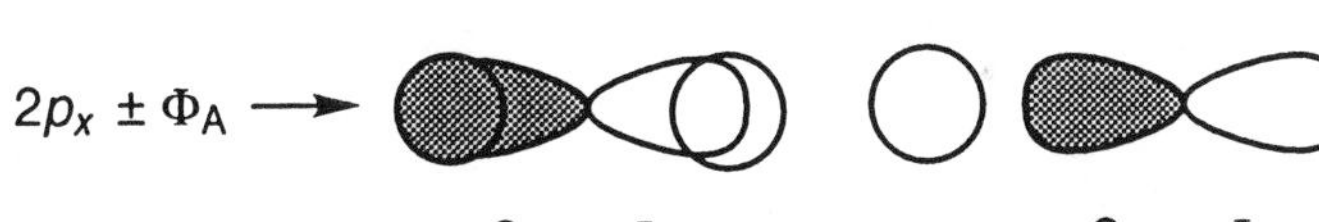

$2p_x + \Phi_A$ $2p_x - \Phi_A$

Unchanged in the calculation – these have net-zero (orthogonal) interactions with both H_2 molecular orbitals

(d and e)

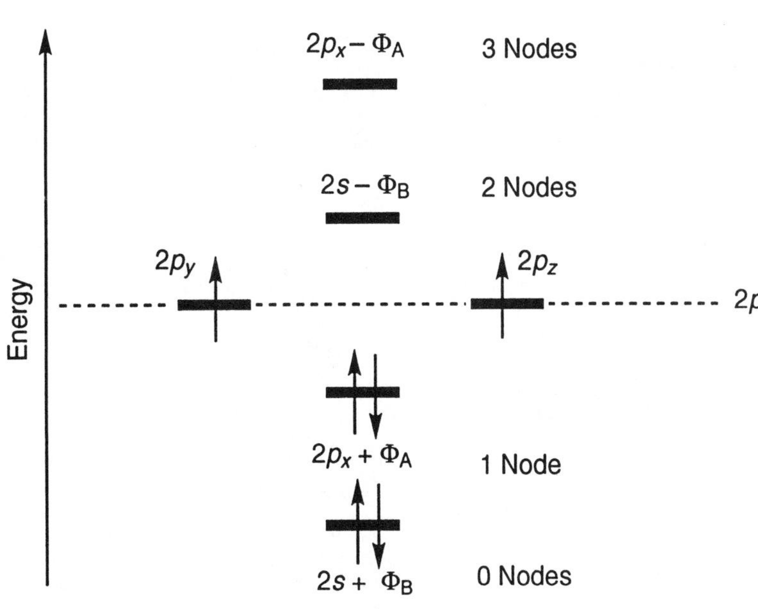

Problem 2.25

(a) The orbitals suggested are the molecular orbitals of triangular H_3 (**4**, **5**, and **6** from Problem 2.19) and the atomic orbitals of nitrogen, $2s$, $2p_x$, $2p_y$, and $2p_z$.

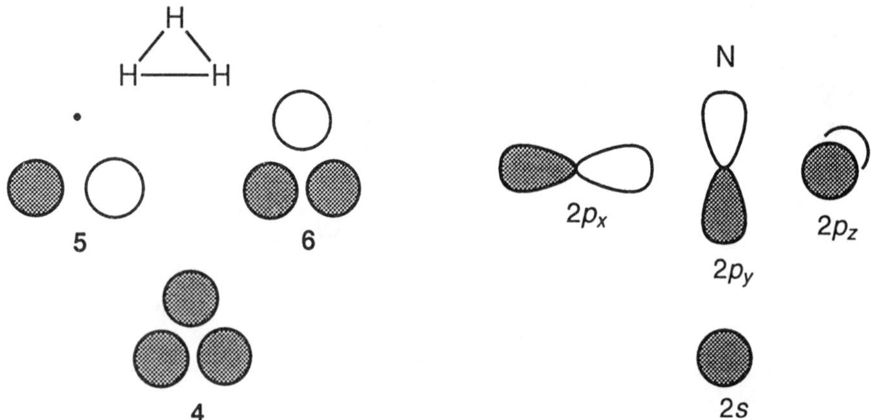

(b) The combination of three molecular orbitals for H_3 and four atomic orbitals for N must yield seven orbitals for NH_3.

(c)

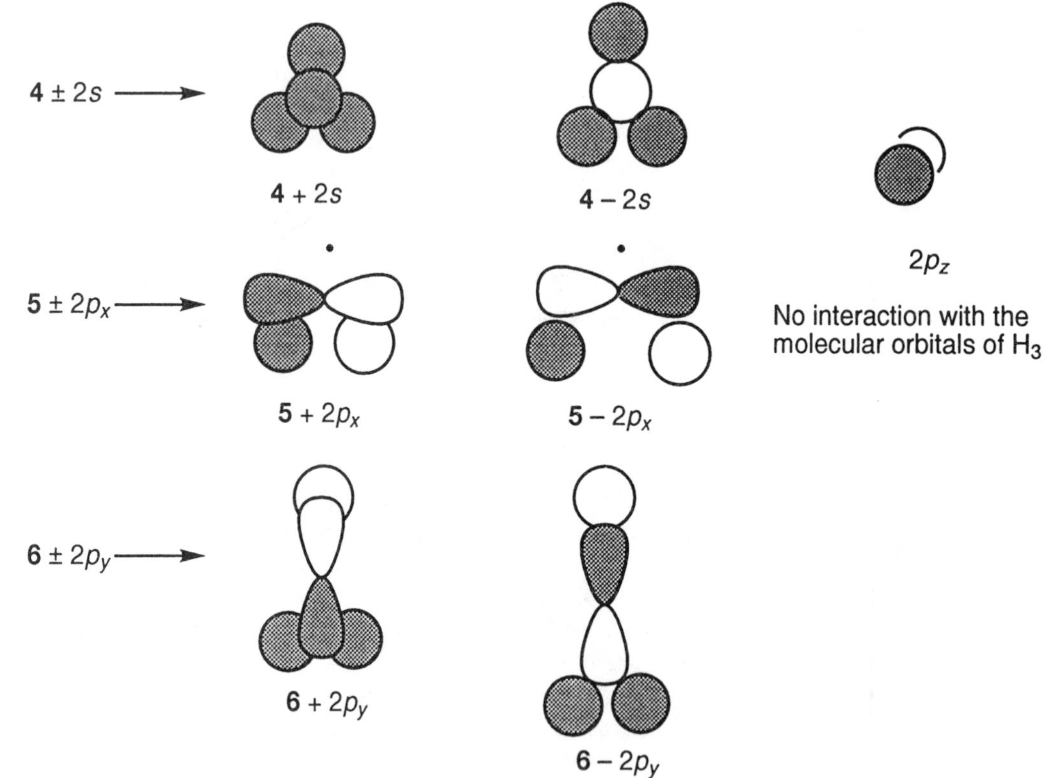

(d)

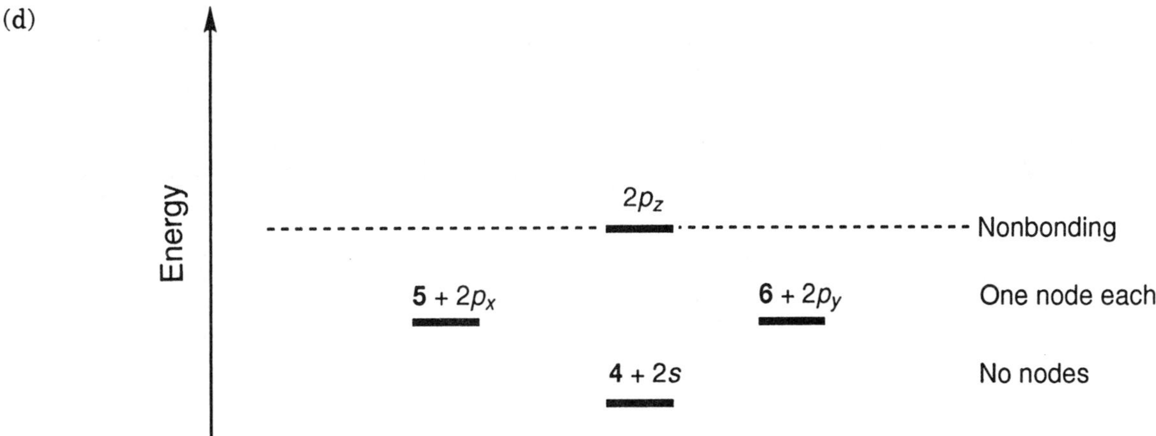

(e) There will be a total of eight electrons, three from the three hydrogens and five from nitrogen.

2pz

5 + 2px

6 + 2py

4 + 2s

Problem 2.26

(a) These molecular orbitals are H_2 bonding and antibonding (Φ_B, Φ_A) and the three molecular orbitals of triangular H_3 (**4**, **5**, and **6**).

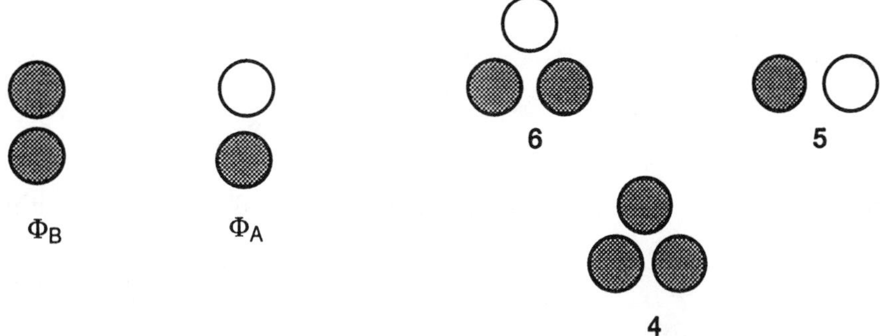

(b) The H_5 molecule must have five molecular orbitals, as it is constructed from five H 1*s* orbitals.
(c) Only the lowest of the cyclic H_3 orbitals (**4**) can interact with the Φ_B of H_2. This produces two new molecular orbitals (**A** and **B**). The remaining three orbitals; two from cyclic H_3 (**5** and **6**) and the antibonding orbital of H_2 (Φ_A) make up the other three molecular orbitals of H_5 (**C, D,** and **E**).

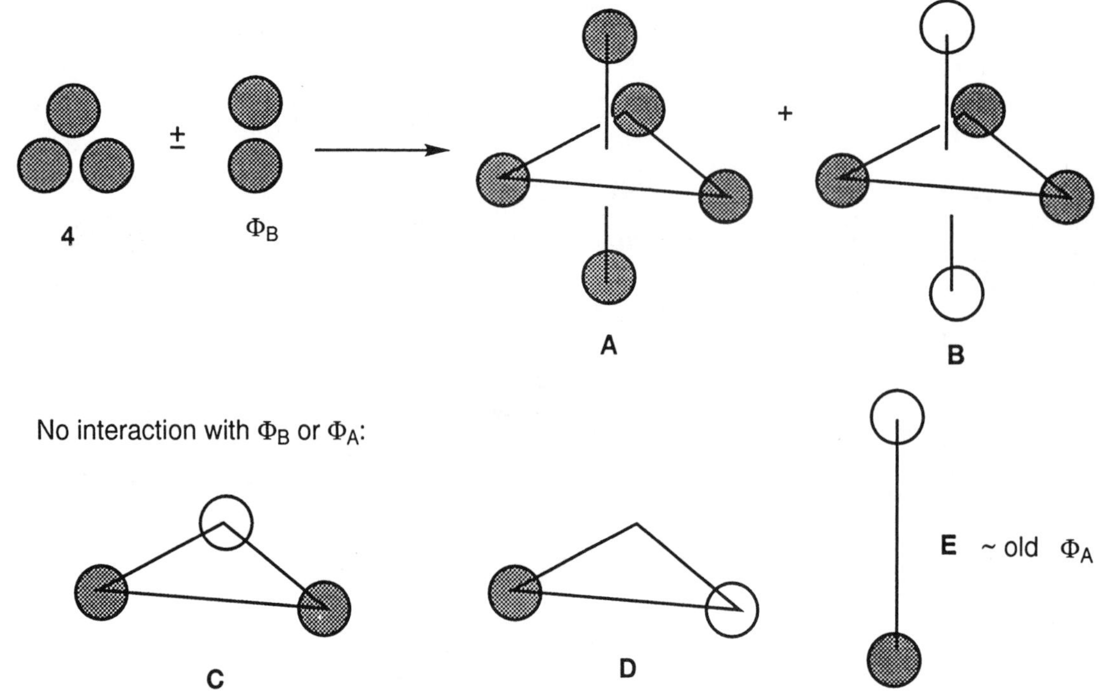

(d and e) Assume molecular orbitals with equal numbers of nodes have equal energy. Five electrons go in as shown, obeying Hund's rule.

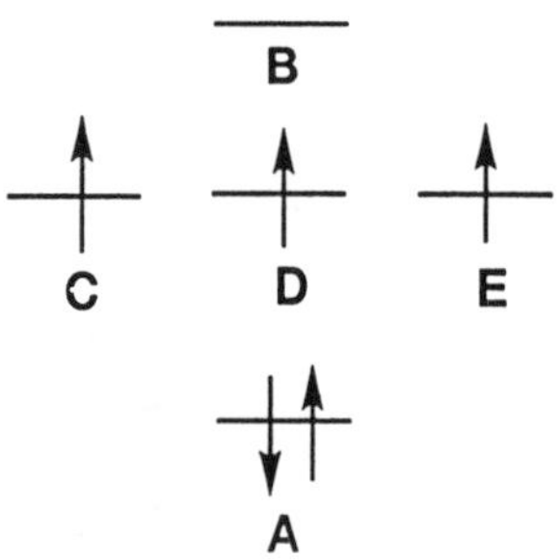

Problem 2.27
(a) We use the molecular orbitals of square H_4 (**E**, **F**, **G**, and **H**) and the atomic orbitals of carbon.

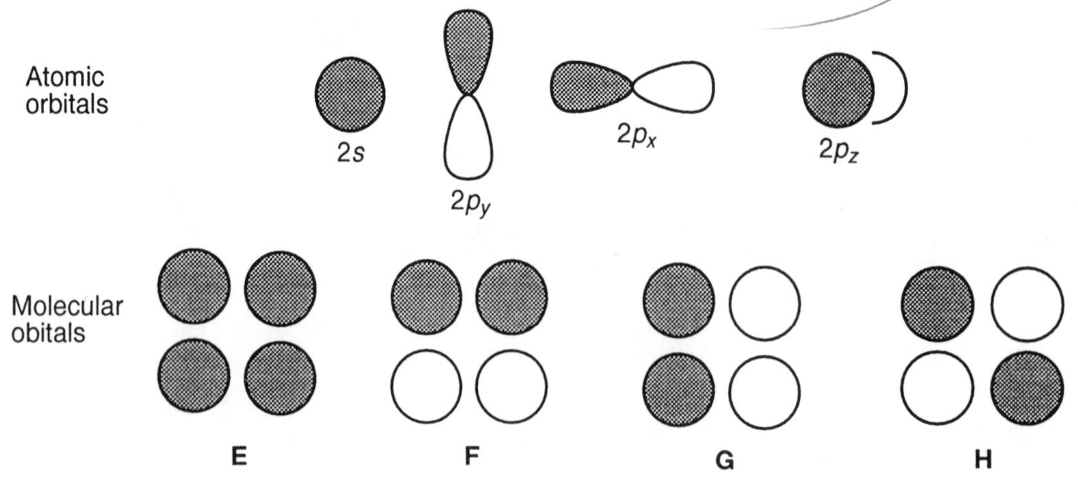

(b) Eight orbitals go into the calculation, so eight must come out. Planar methane will have eight molecular orbitals.

(c)

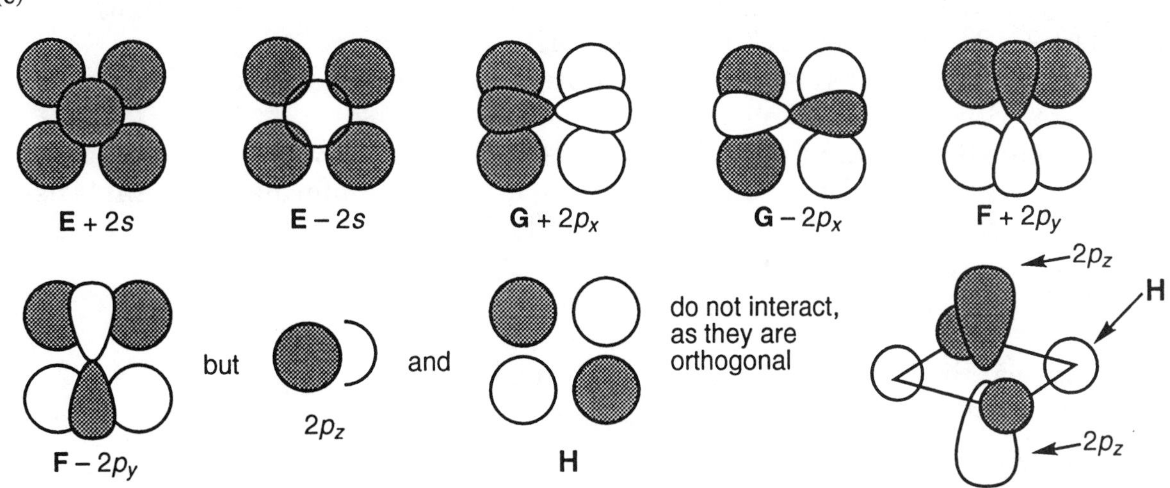

(d and e) The ordering is done by counting nodes. Here **E** + 2s has no nodes and will be the lowest energy molecular orbital. Both **F** + 2p_y and **G** + 2p_x have a single node and will be at the same energy. The noninteracting 2p_z orbital is nonbonding. There will be a total of eight electrons, four from carbon and four from the four hydrogens.

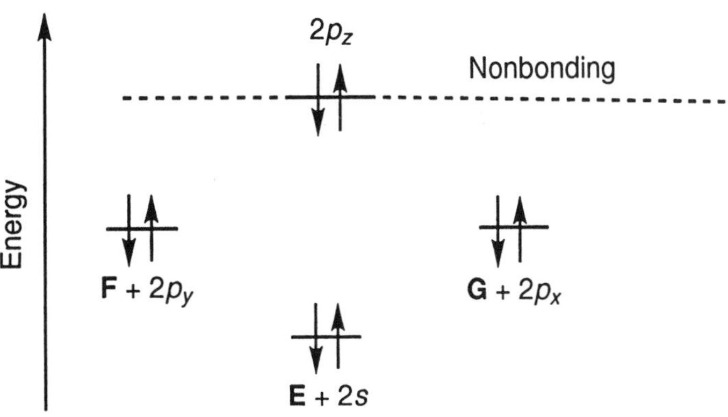

Chapter 3 Outline

3

Alkanes

This set of problems provides practice in using the various hybridization schemes developed in this chapter. The major theme of the chapter is structure, and the specific focus is on the structure of alkanes. Special attention is paid to writing and naming isomeric alkanes and to using Newman projections to help us to visualize molecules in three dimensions. Practice in using the various coded two-dimensional representations of three-dimensional molecules is provided, and ring compounds are introduced.

Problem 3.1 Here are four representations for butane and pentane. In (a), all atoms are shown and all bonding electrons are put in as dots. In (b), only the atoms survive as the bonding electrons are transformed into two-electron lines. In (c), a schematic representation is shown, and in (d), the ultimate schematic representation, not even the atoms are shown. Here your imagination must supply carbons at every vertex and terminus, and the appropriate number of hydrogens must be added.

Butane

(a)

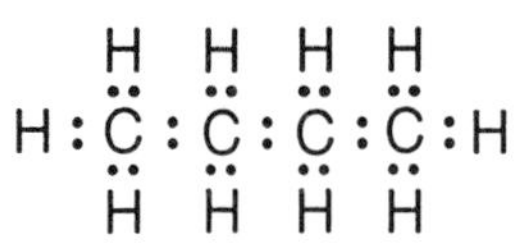

(b)

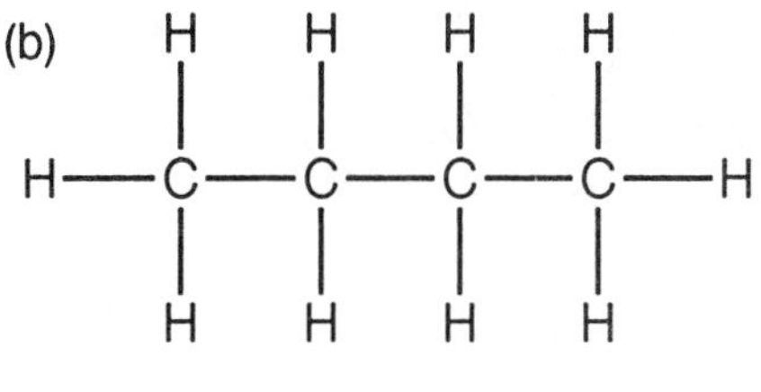

(c) $CH_3CH_2CH_2CH_3$

(d)

Pentane

(a)

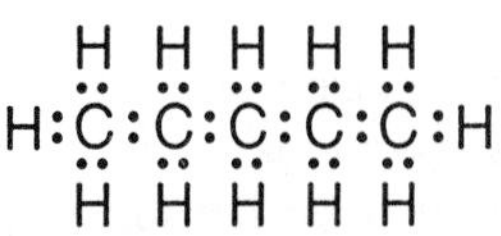

(b)

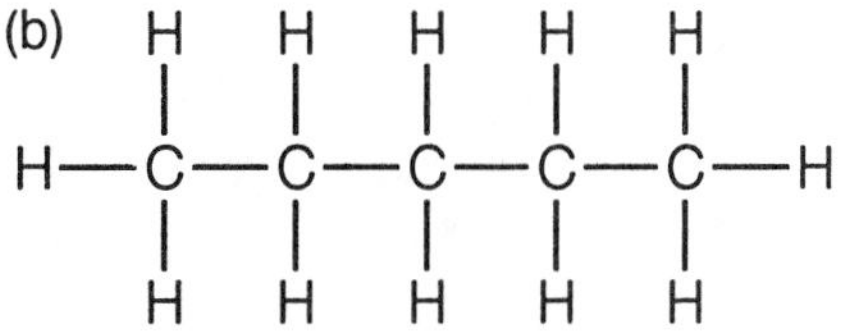

(c) $CH_3CH_2CH_2CH_2CH_3$

(d)

Problem 3.3 As discussed in Chapter 2 (p. 55), two orbitals interact most strongly when they overlap well. The amount of overlap depends (among other things) on proximity; the closer two orbitals are, the better they overlap. The overlap of the orbitals making up the carbon–hydrogen bonds, two *filled* orbitals, is net destabilizing, as the filled antibonding molecular orbital is destabilized more than the filled bonding molecular orbital is stabilized (Problems 2.6 and 2.9,

pp. 52 and 55). The further apart two overlapping filled orbitals are, the more poorly they overlap, and the lower the destabilization.

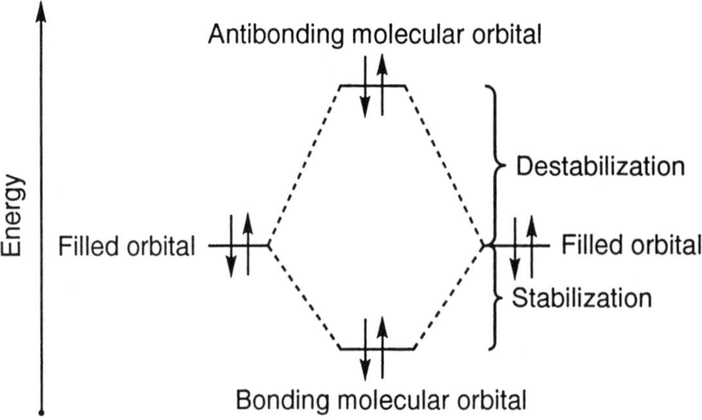

Destabilization > Stabilization

Problem 3.4 As the bonding and antibonding interactions exactly cancel, there is *no* net interaction between the orbitals. In this orientation, a 2*p* and an *s* orbital do not interact, they are orthogonal.

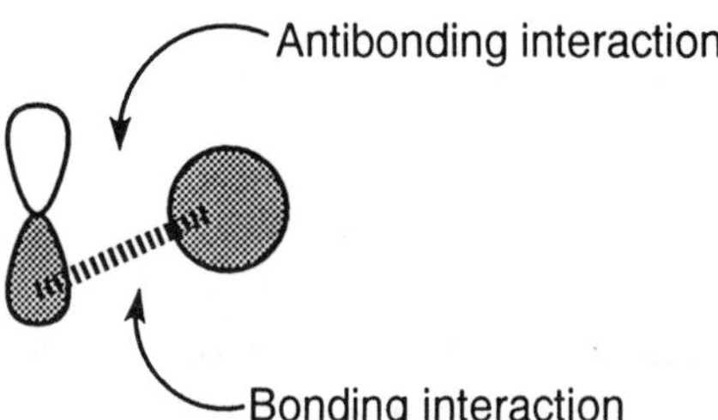

Problem 3.5 In BeH_2, the empty $2p_y$ and $2p_z$ orbitals are oriented at 90° to each other and to the pair of *sp* hybrid orbitals.

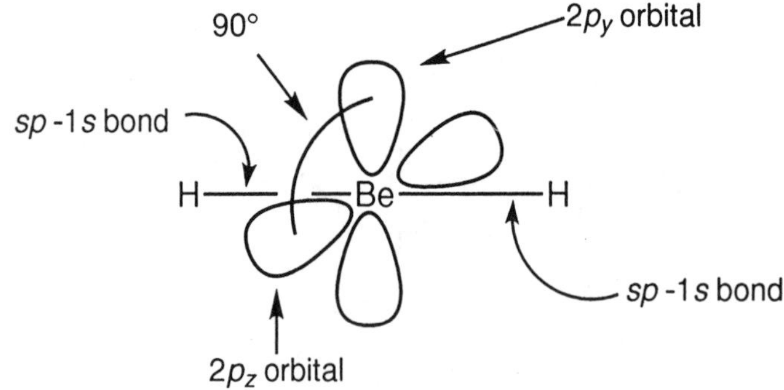

Problem 3.7 The electronic configuration of boron is $_5B = 1s^2\ 2s^2\ 2p_x$, and therefore there are three electrons available for bonding (5 – 2 1*s*) with the three fluorines. Fluorine's electronic configuration is $_9F = 1s^2\ 2s^2\ 2p_x^{\ 2}\ 2p_y^{\ 2}\ 2p_z$, and there are seven electrons available for bonding (9 – 2 1*s*), including one odd electron.

We combine the boron $2s$, $2p_x$, and $2p_y$ orbitals to form three sp^2 hybrid orbitals, each of which can contain one of boron's three bonding electrons. Each sp^2 hybrid forms a single covalent bond with the orbital containing the odd electron on fluorine. The boron $2p_z$ orbital is not used in this bonding scheme.

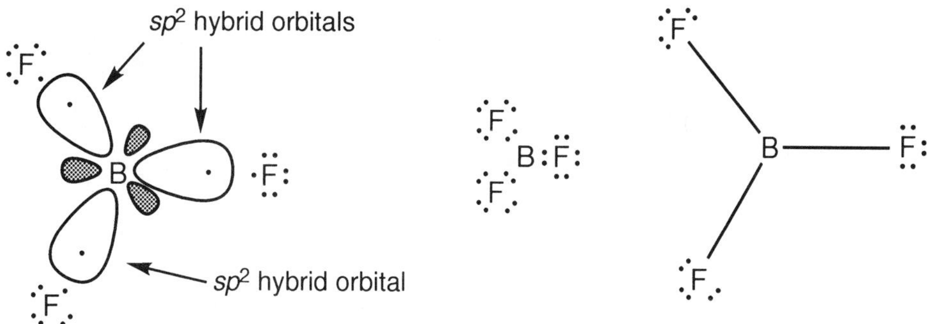

Problem 3.8 The unused empty $2p_z$ orbital on boron extends above and below the xy plane of the three boron–hydrogen bonds.

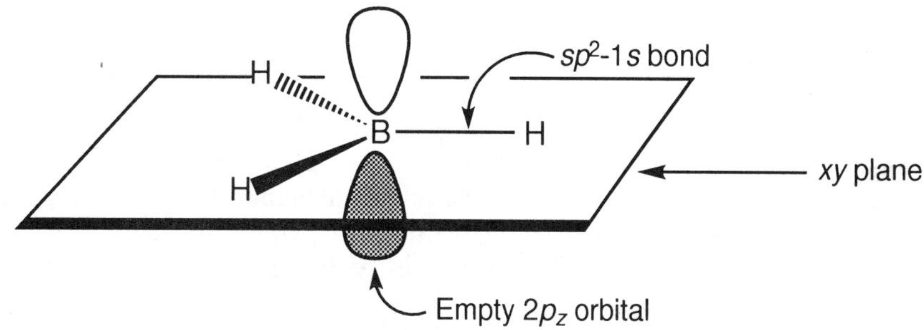

Problem 3.9 There many things wrong with the 90° structure. For the methyl anion $^-$:CH_3, the carbon–hydrogen bonds are made from $2p - 1s$ overlap, which you have seen is an inefficient form of bonding. The rear lobes of the $2p$ orbitals are unused and thus "wasted," for example. Second, in this structure, the electrons in the carbon–hydrogen bonds are rather close together and electron–electron repulsion is not minimized. For an exactly parallel discussion, see the treatment of "unhybridized methane" (p. 68).

Problem 3.10 Exactly half-way between the two pyramids the molecule must be flat, and therefore the carbon atom is sp^2 hybridized. One pyramidal molecule cannot pass to the other without going through a planar form.

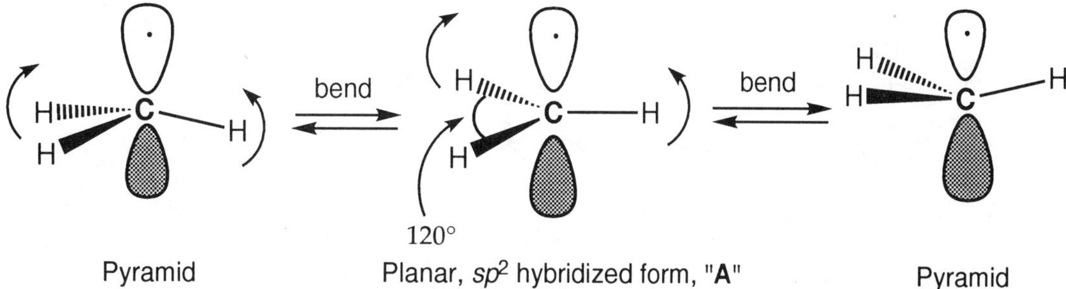

Problem 3.11 Look down the carbon–carbon bond of ethyl chloride (CH_3—CH_2—Cl) with the methyl group in front. You see three carbon–hydrogen bonds attached to the front carbon. In the

rear you see the carbon represented as a circle attached to two hydrogens and a chlorine. There are three staggered forms of equal energy.

Ethyl chloride

Newman projection

The three equivalent forms are interconverted by 120° rotations about the carbon-carbon bond.

The second compound, 1,2-dichloroethane (Cl—CH_2—CH_2—Cl), is more complicated. As you look down the central carbon–carbon bond, you see in front a carbon attached to two hydrogens and a chlorine. In back, the carbon represented as a circle is also attached to two hydrogens and a chlorine. There are three forms interconverted by rotation, **A**, **B**, and **B′**. Forms **B** and **B′** are of equal energy, but **A** is different. As **A** keeps the two relatively large chlorine groups as far apart as possible, it is lower in energy than **B** or **B′**.

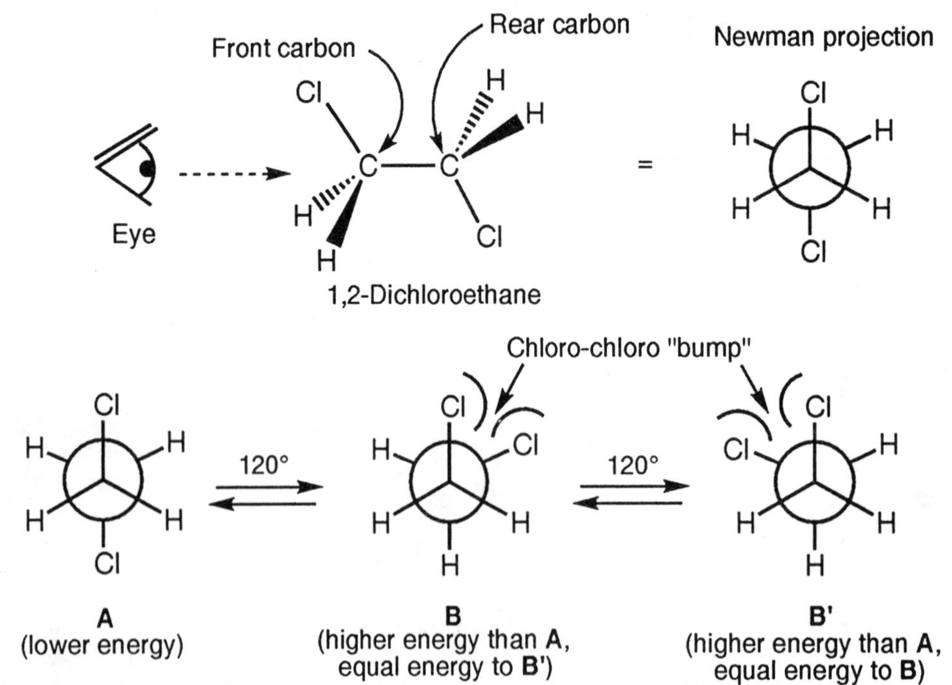

Problem 3.12 A transition state was first encountered in Problem 3.10 (p. 83). The planar form separating the two pyramidal forms of the methyl radical is a transition state. Transition states are often shown in brackets, [TS].

Problem 3.14

$(Me)_2CH_2$ = propane = $EtCH_3$
= $CH_3CH_2CH_3$ = EtMe = MeEt

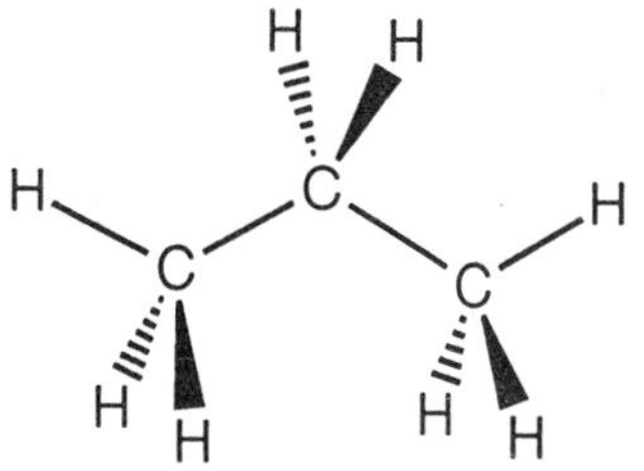

$(CH_3)_4C$ =

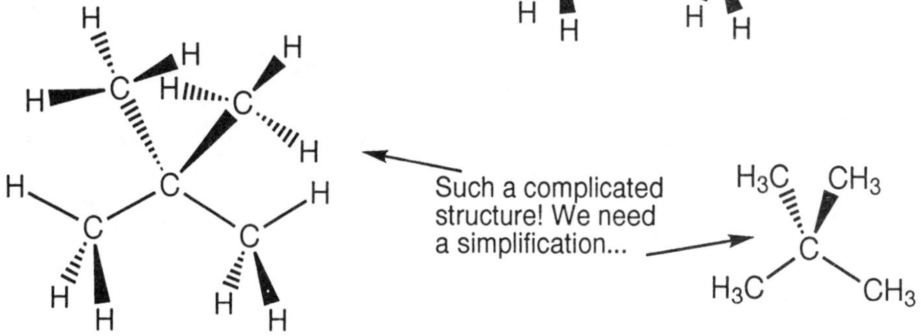

Such a complicated structure! We need a simplification...

$(CH_3)_3CH$ =

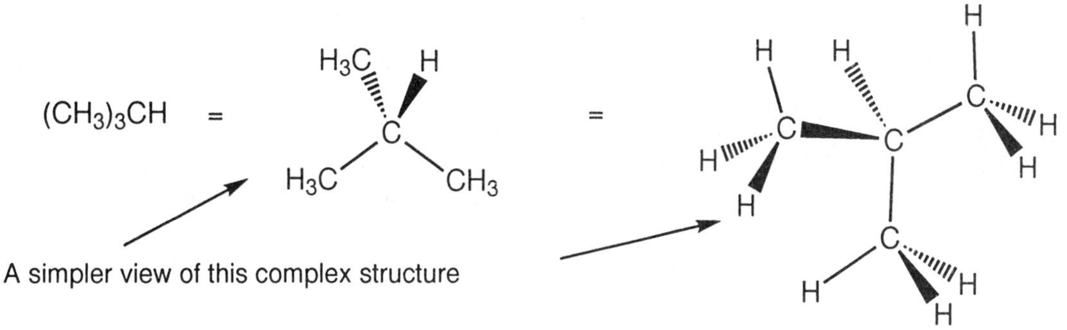

A simpler view of this complex structure

$(Et)_2$ = butane =

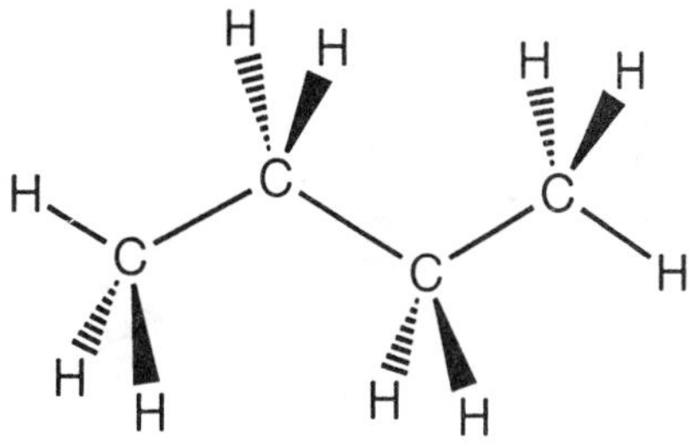

Problem 3.15 Your models will serve you better than words here. Replacement of the "end" hydrogen to give **A** and the "corner" hydrogen to give **B** leads to the same thing. The C—C—C right angle apparent in structure **B** is not real. The take home lesson here is, *Always Remember*: Organic molecules are three-dimensional. Never trust the two-dimensional surface.

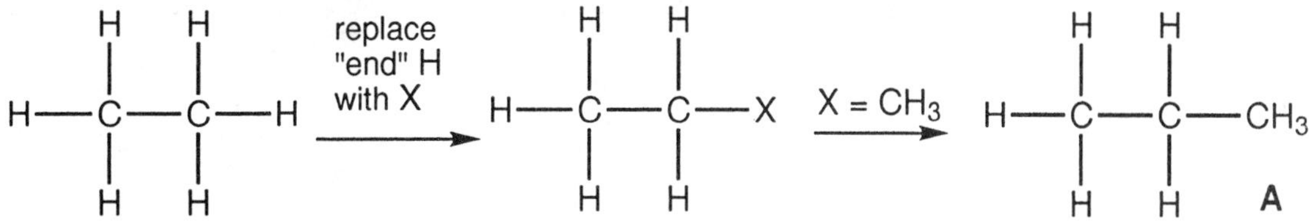

replace "end" H with X

X = CH_3

A

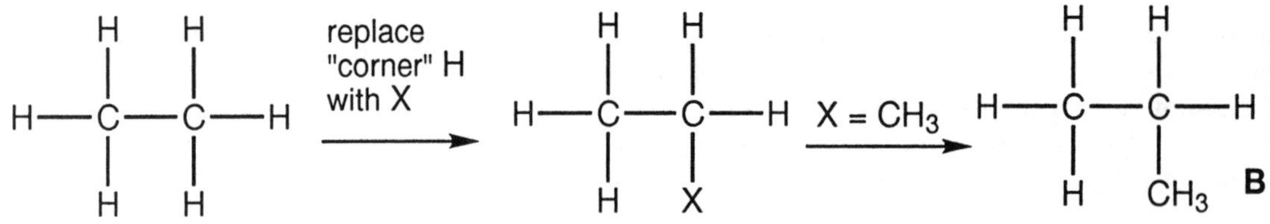

Both **A** and **B** are in fact the same, $CH_3CH_2CH_3$, propane =

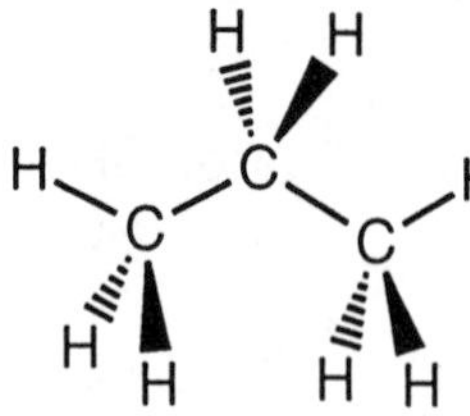

Problem 3.16 The only trick here is to remember to keep all carbon–hydrogen and carbon–hydroxyl bonds staggered.

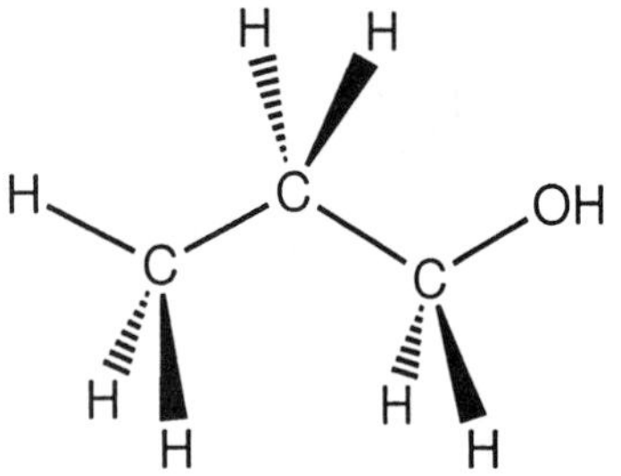

$CH_3–CH_2–CH_2—OH$ 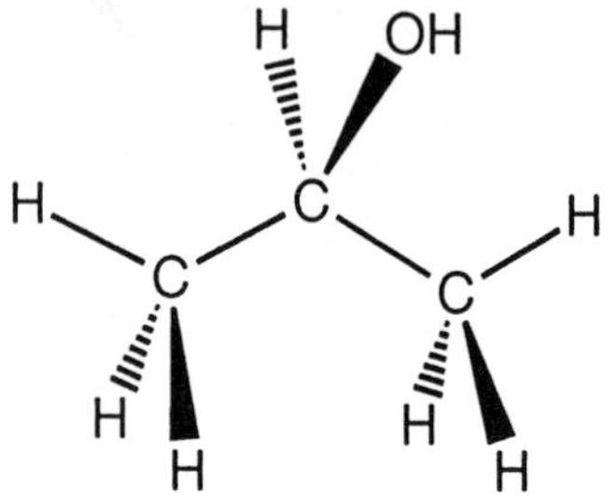 $CH_3–CH(OH)–CH_3$

Problem 3.18 For the first molecule, the Newman projection is constructed as usual. The front carbon bears two methyl groups and a hydrogen, and the rear carbon, shown as a circle, two hydrogens and one methyl group. Start at 0° with an eclipsed form and then proceed by 60° rotations of the rear carbon to generate the other Newman projections.

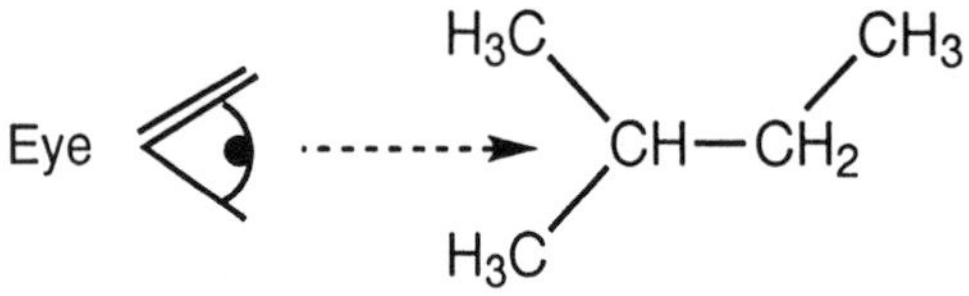

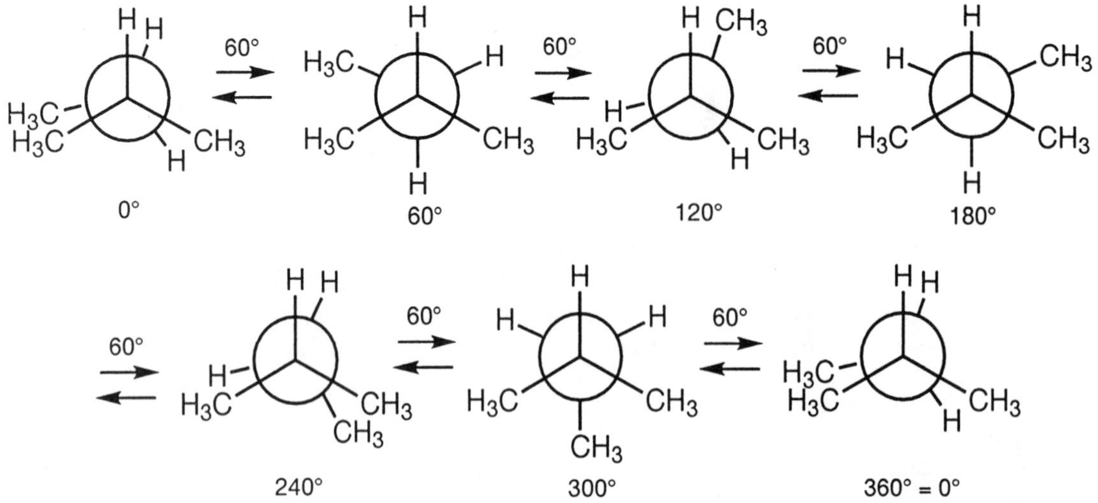

0° 60° 120° 180°

240° 300° 360° = 0°

The second molecule is more symmetrical, as both carbons bear two methyl groups and a single hydrogen. The 0° and 360° forms are identical, and contain two methyl–methyl eclipsed interactions. These will be the highest energy conformations. A 60° rotation leads to a staggered molecule with three methyl–methyl *gauche* interactions. The 120° and 240° transition states have one methyl–methyl eclipsed interaction and two methyl–hydrogen eclipsed interactions and will be lower in energy than the 0° and 360° conformations. Lowest energy of all is the 180° form, which contains only two methyl–methyl *gauche* interactions. The graph shows the relative energies.

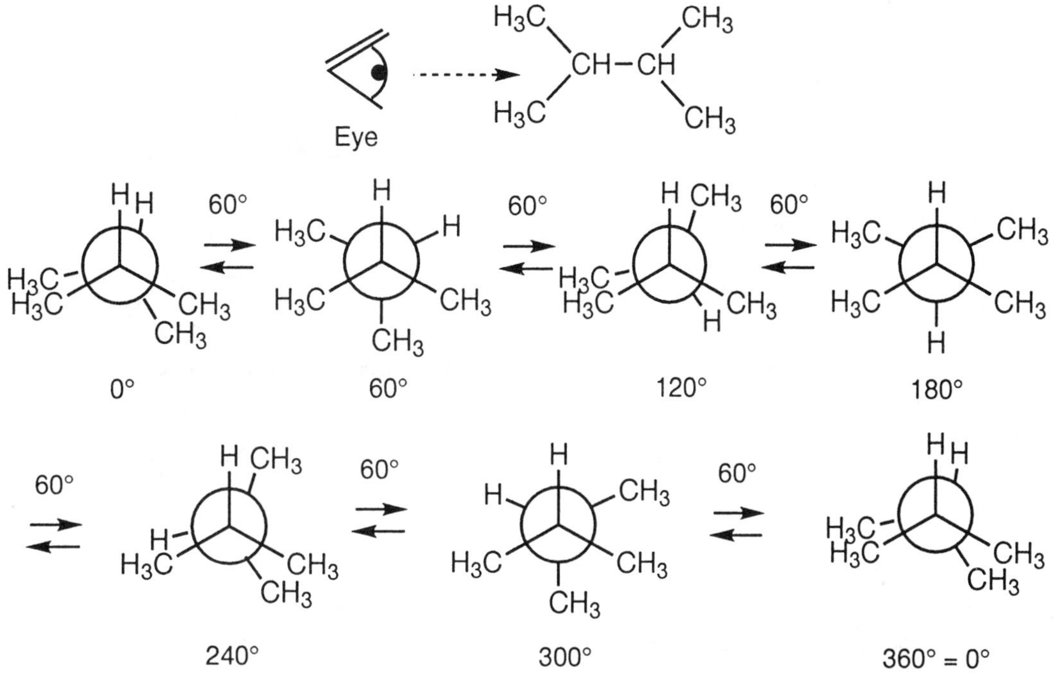

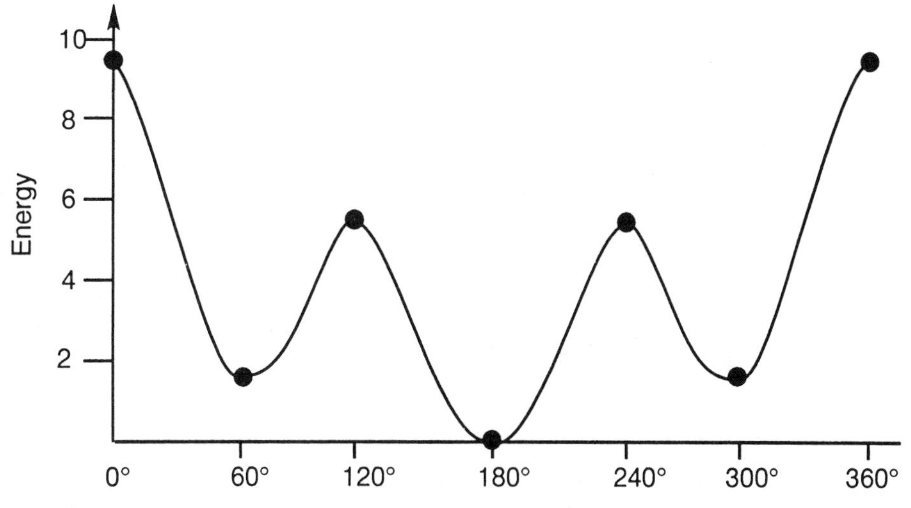

$(H_3C)_2CH-CH(CH_3)_2$

Eye

0° ⇌(60°) 60° ⇌(60°) 120° ⇌(60°) 180° ⇌(60°) 240° ⇌(60°) 300° ⇌(60°) 360° = 0°

Energy

10, 8, 6, 4, 2

0° 60° 120° 180° 240° 300° 360°

Problem 3.19 Any compound containing a carbon attached to four other carbons contains a "quaternary" carbon. The quaternary carbon is shown in boldface type in the figure.

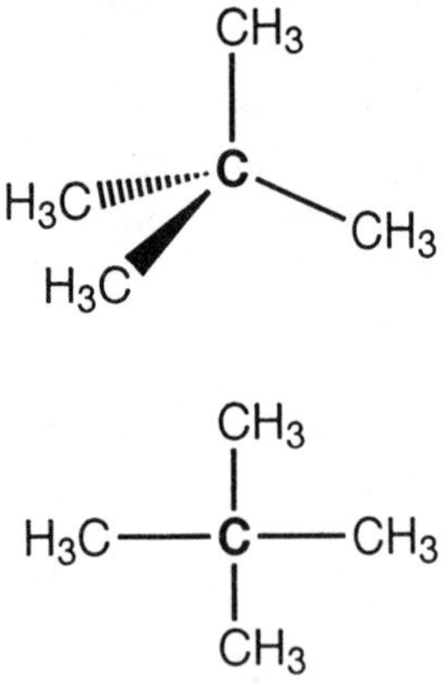

2,2-Dimethylpropane
(neopentane)

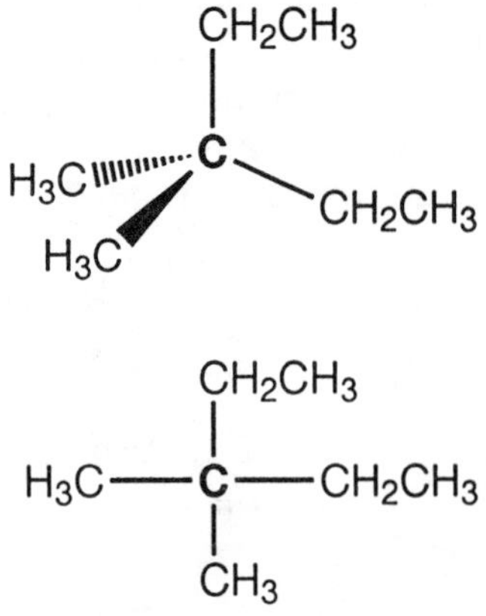

3,3-Dimethylpentane

Problem 3.20 The answers are drawn in two schematic ways. Notice that no attempt is being made to show three dimensionality here.

Pentanes

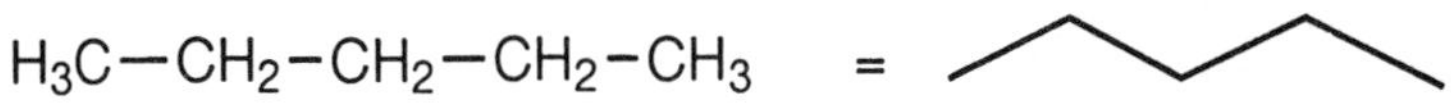

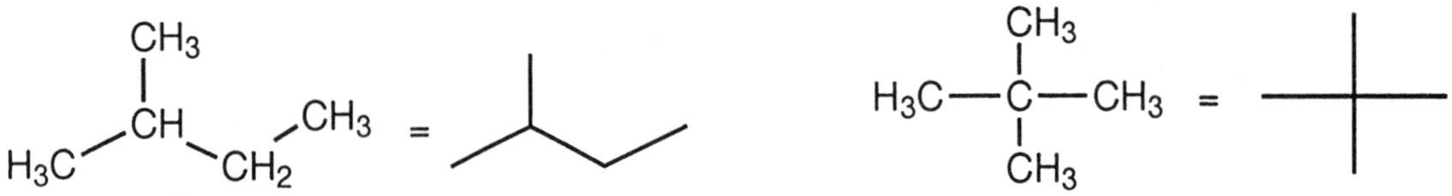

Hexanes

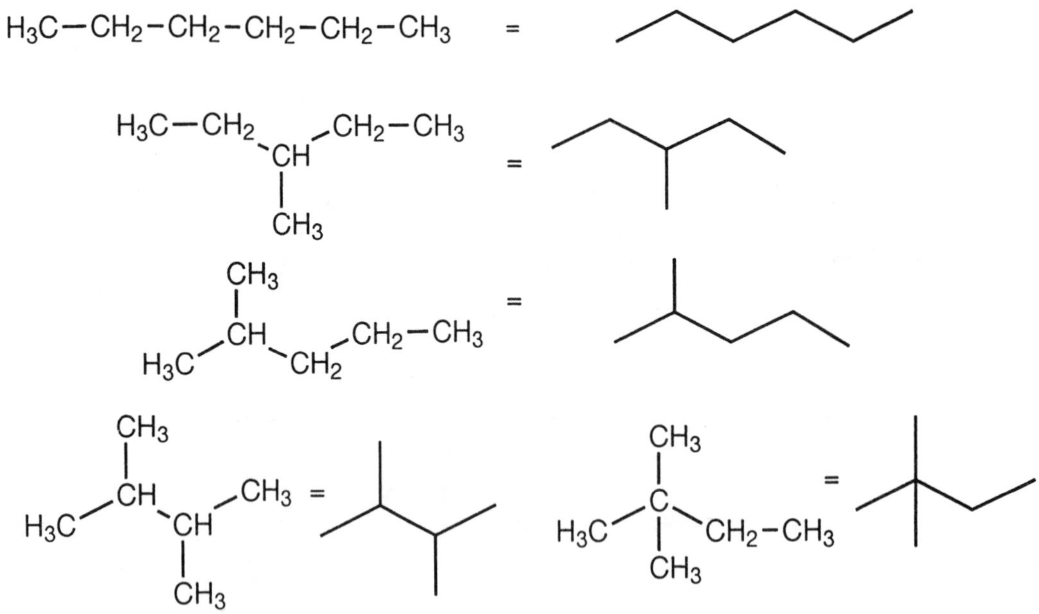

Problem 3.21

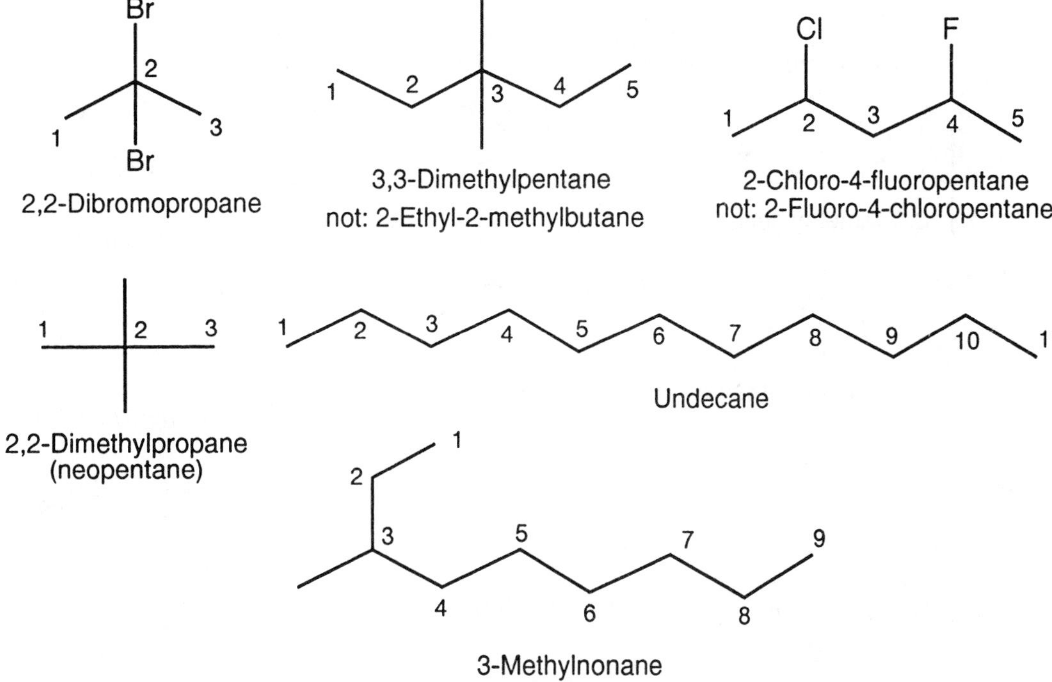

Problem 3.22

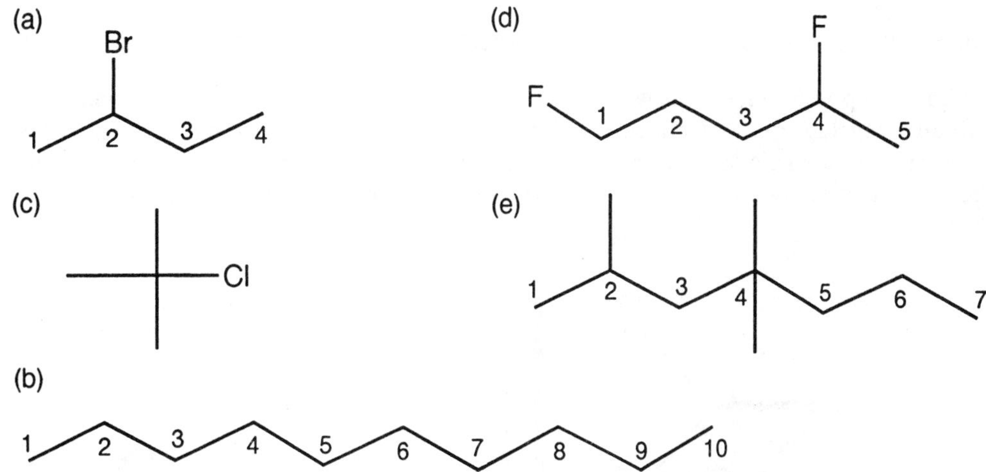

Problem 3.23 Here we use only the most schematic representations for the molecules.

Octane

2-Methylheptane

3-Methylheptane

4-Methylheptane

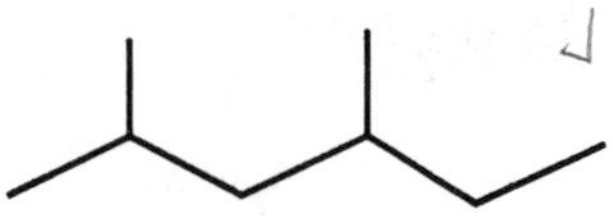

2,4-Dimethylhexane

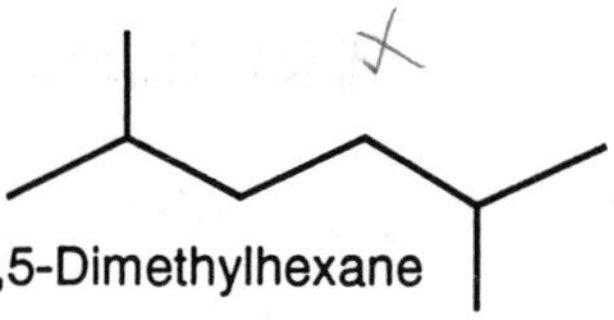

2,5-Dimethylhexane

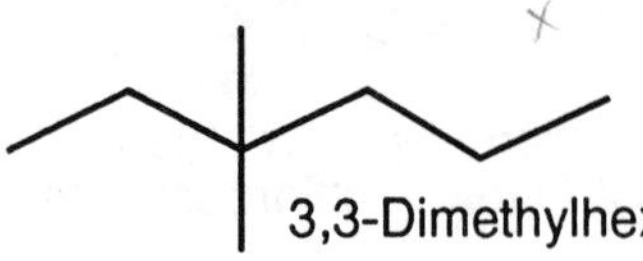

3,3-Dimethylhexane

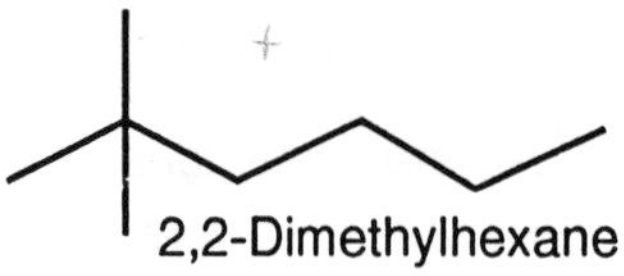

2,2-Dimethylhexane

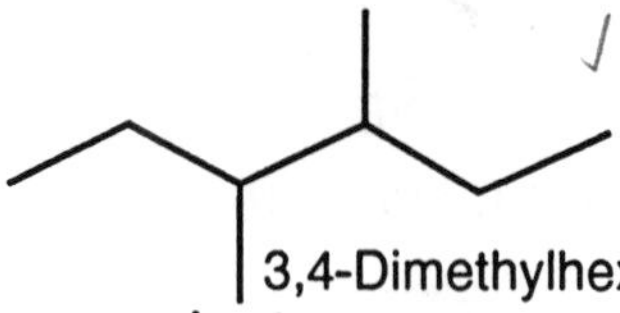

3,4-Dimethylhexane

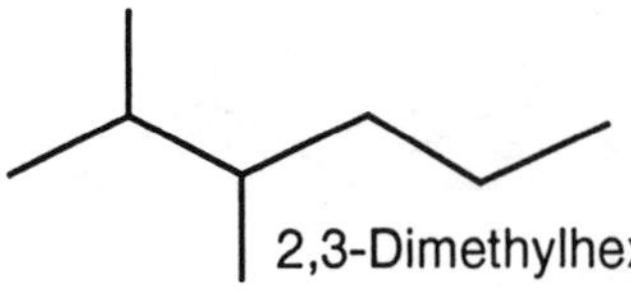

2,3-Dimethylhexane

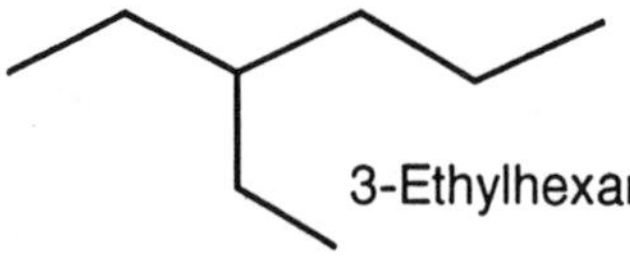

3-Ethylhexane

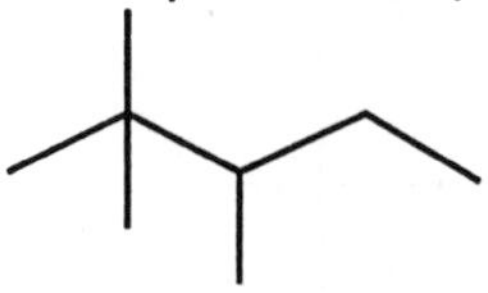

2,2,3-Trimethylpentane

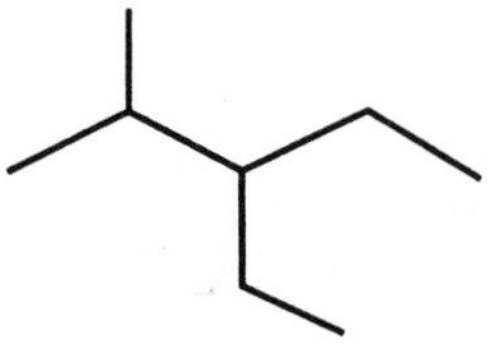

3-Ethyl-2-methylpentane

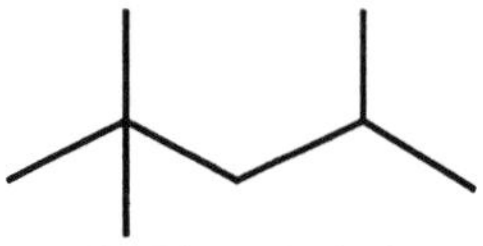

2,2,4-Trimethylpentane

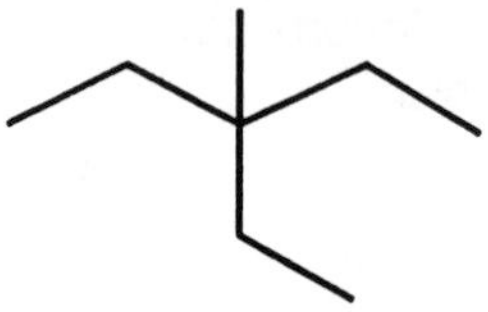

3-Ethyl-3-methylpentane

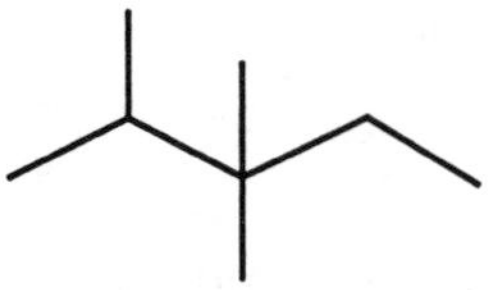

2,3,3-Trimethylpentane

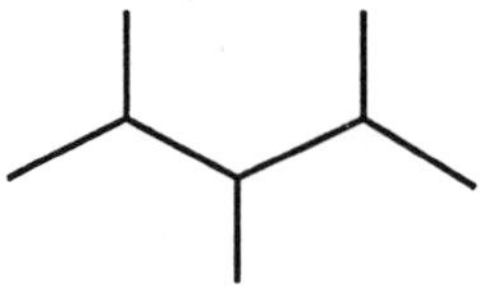

2,3,4-Trimethylpentane

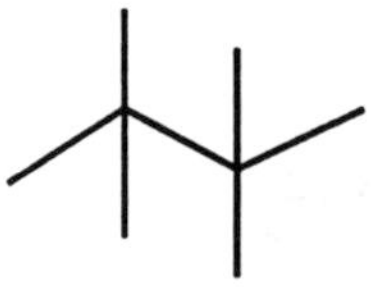

2,2,3,3-Tetramethylbutane

Problem 3.24 This problem looks forward to Chapters 4 and 5. There really are two isomers of *trans*-1,2-dimethylcyclopropane. No number of translational or rotational operations will suffice to change one mirror image into the other. These two isomers are related in the same way that your right and left hands are. By all means use your models to be certain of this answer.

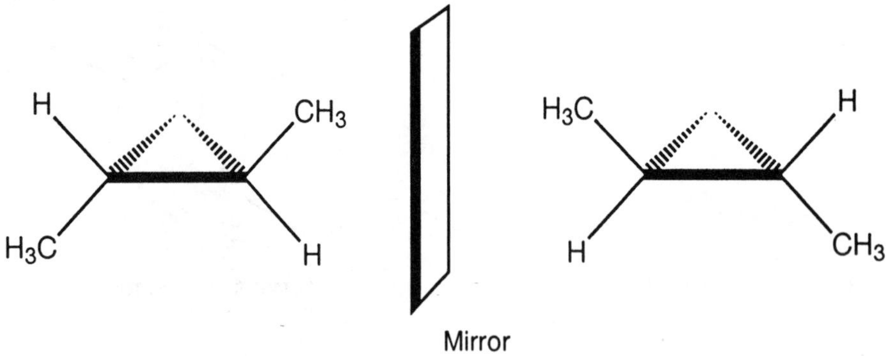

H CH_3 H_3C H

H_3C H H CH_3

Mirror

Additional Problem Answers

Problem 3.29 This problem provides vital practice in drawing Lewis structures.

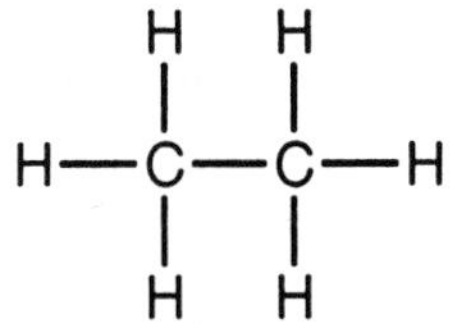

Ethane

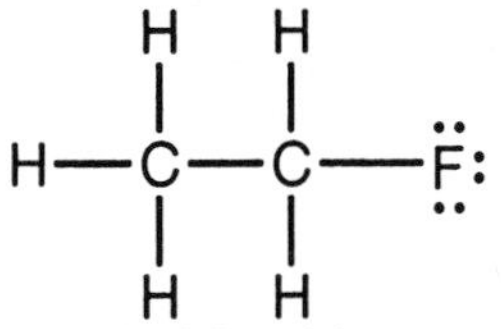

Ethyl fluoride

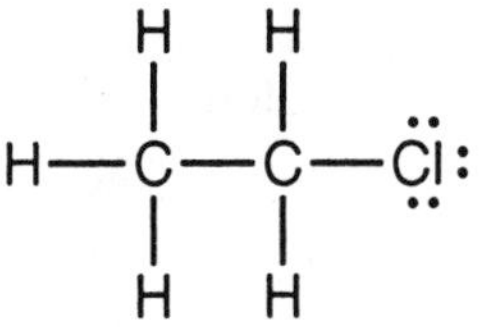

Ethyl chloride

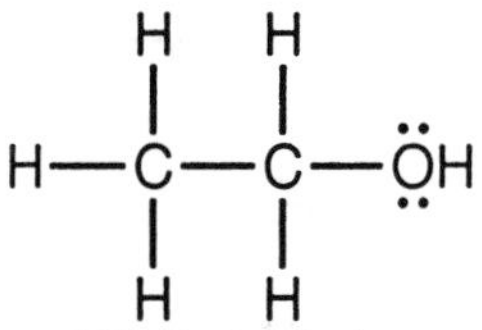

Ethyl alcohol

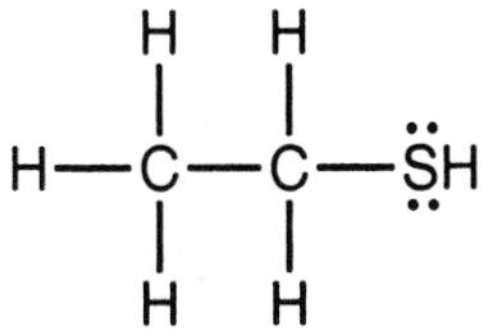

Ethyl mercaptan

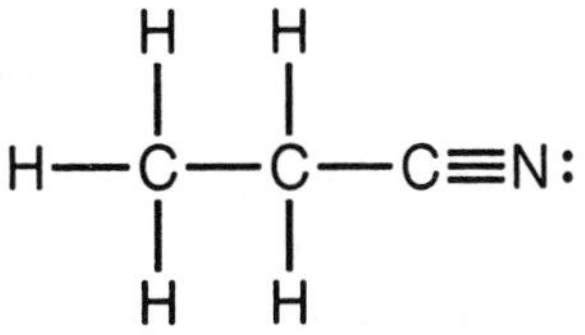

Ethyl cyanide

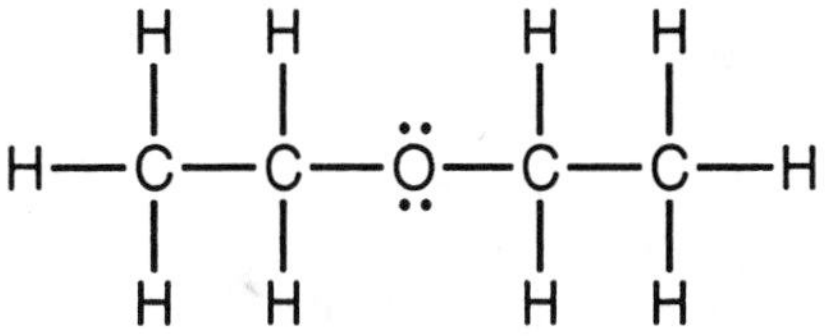

Diethyl ether

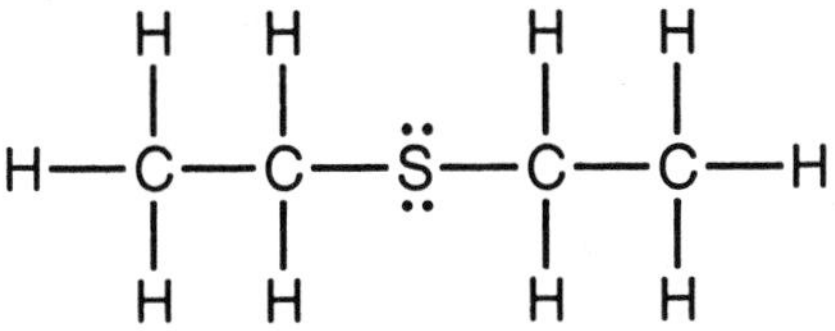

Diethyl sulfide

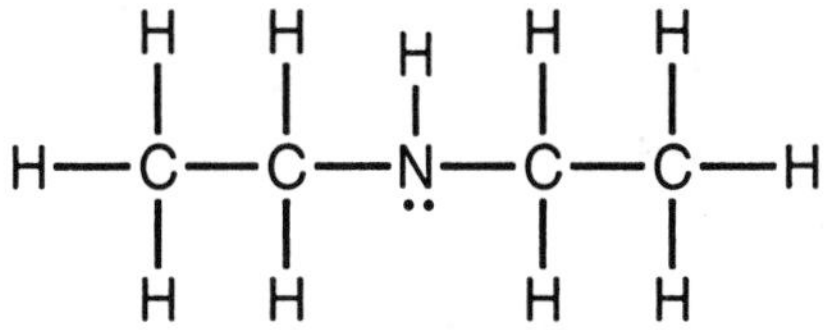

Diethylamine

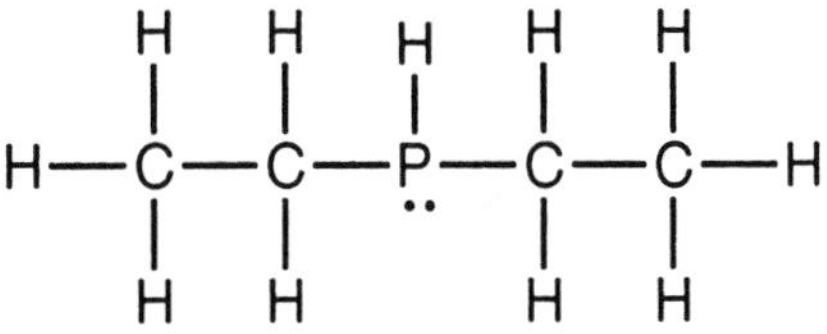

Diethylphosphine

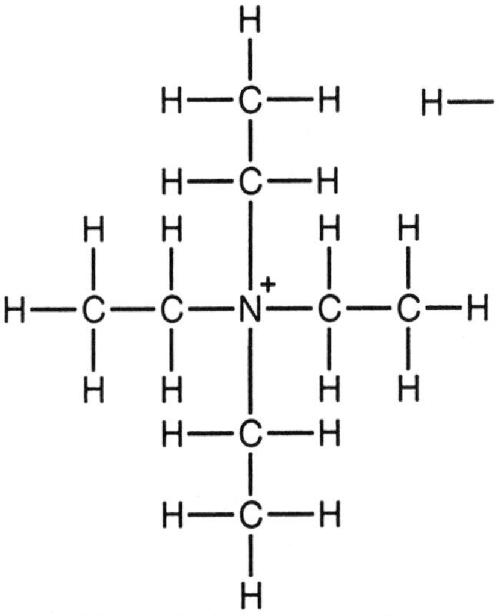

Tetraethylammonium ion

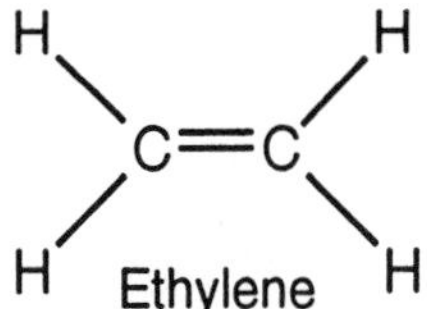

Ethylene (ethene)

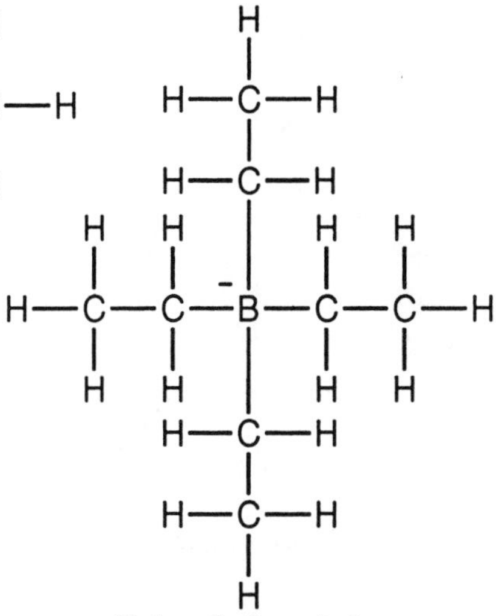

Tetraethylborate ion

H—C≡C—H

Acetylene (ethyne)

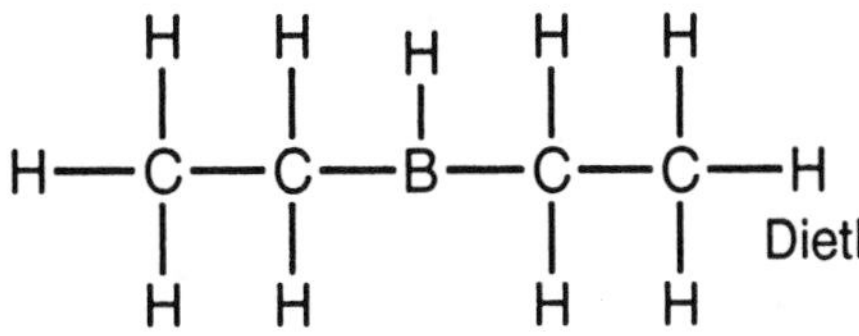

Diethylborane

In the remaining answers we lapse into a slightly more abstract code in which the "hydrocarbon" parts of the structure, the methyl and ethyl groups, are only schematically drawn.

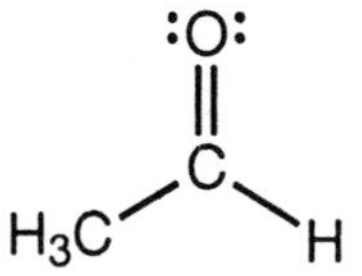

Acetaldehyde

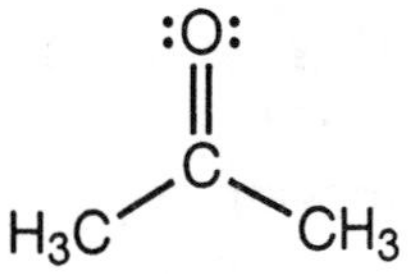

Dimethyl ketone (acetone)

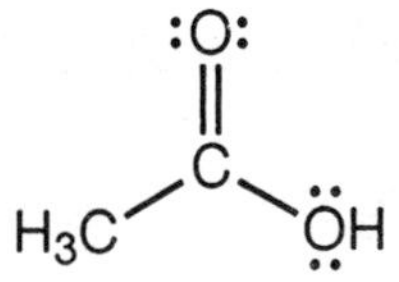

Acetic acid

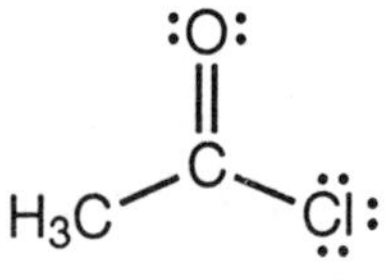

Acetyl chloride

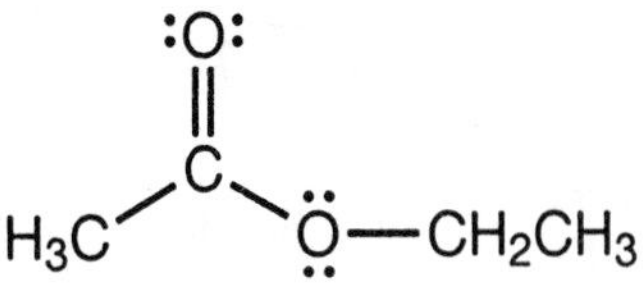

Ethyl acetate

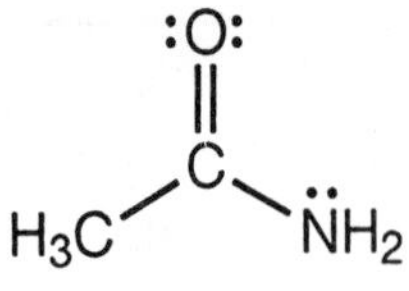

Acetamide

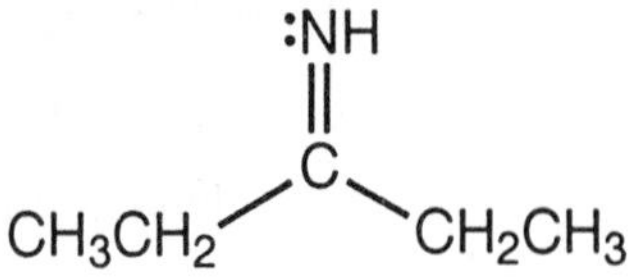

Imine of diethyl ketone

Problem 3.30 There are eight isomers of $C_5H_{11}Br$ (see Fig. 3.46). To find them all, first draw the three possible pentane isomers; pentane, isopentane, and neopentane. Then, see how many different monobromo isomers are available from each pentane isomer. Be careful to avoid duplication.

From pentane

1-Bromopentane

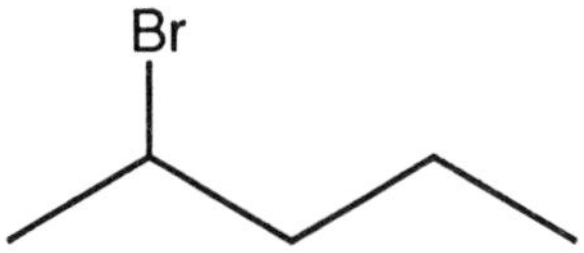

2-Bromopentane

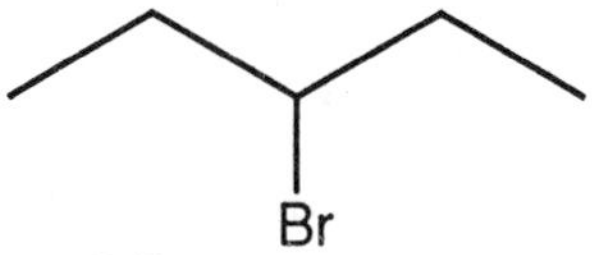

3-Bromopentane

From isopentane

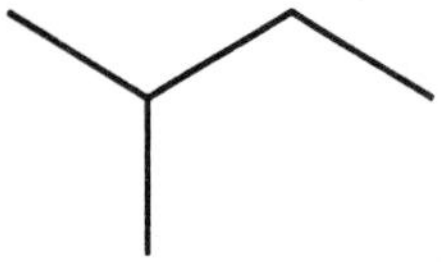

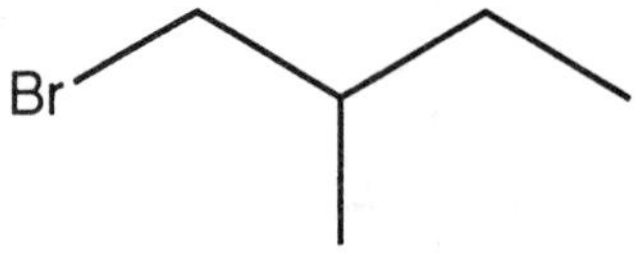

1-Bromo-2-methylbutane

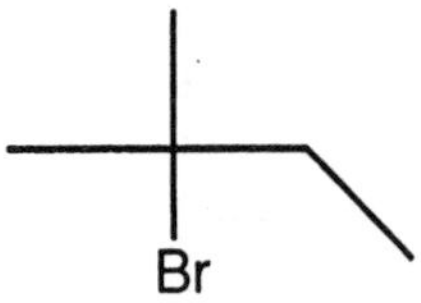

2-Bromo-2-methylbutane

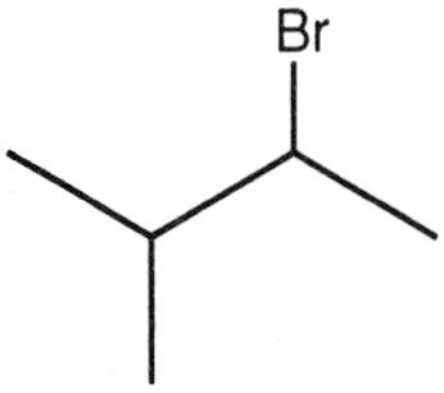

2-Bromo-3-methylbutane

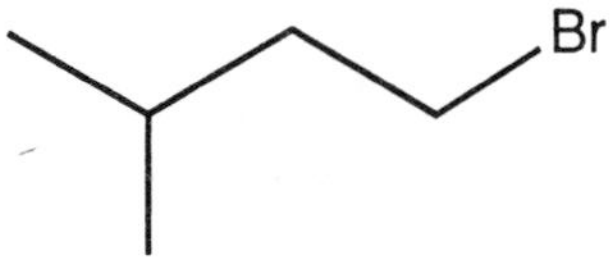

1-Bromo-3-methylbutane

From neopentane

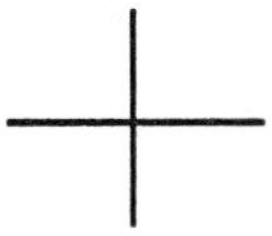

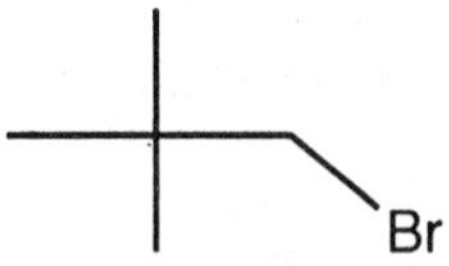

1-Bromo-2,2-dimethylpropane

Problem 3.31 Once again, first draw the five possible hexanes: hexane, 2-methylpentane, 3-methylpentane, 2,2-dimethylbutane, and 2,3-dimethylbutane. Then we see how many different monochloro isomers are available from each hexane isomer.

From hexane

Cl

1-Chlorohexane

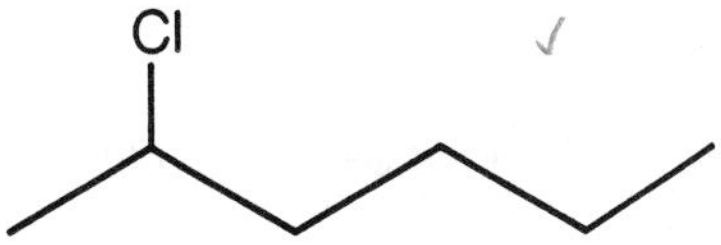

Cl

2-Chlorohexane

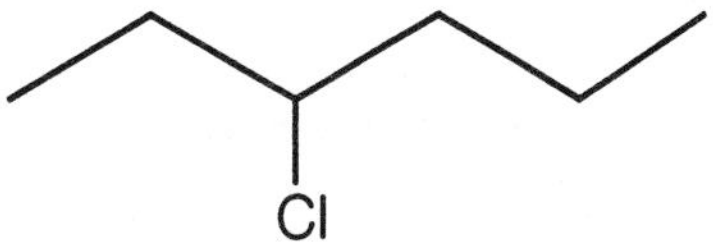

Cl

3-Chlorohexane

From 2-Methylpentane

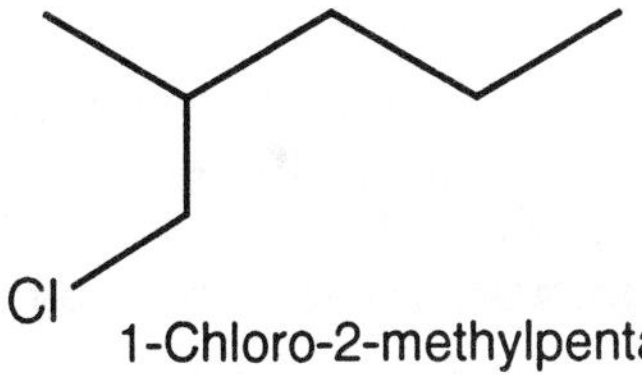

Cl

1-Chloro-2-methylpentane

Cl

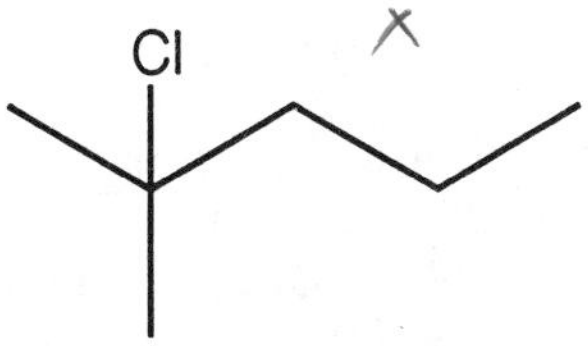

2-Chloro-2-methylpentane

Cl

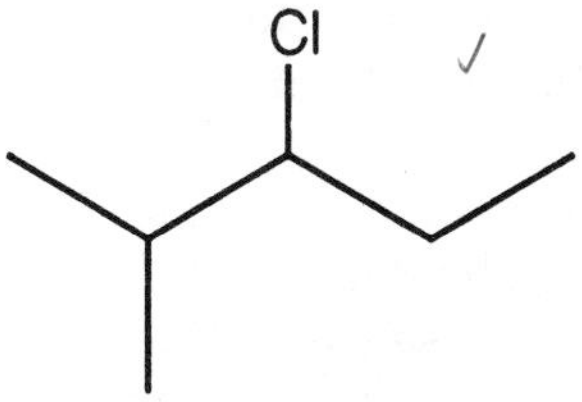

3-Chloro-2-methylpentane

Cl

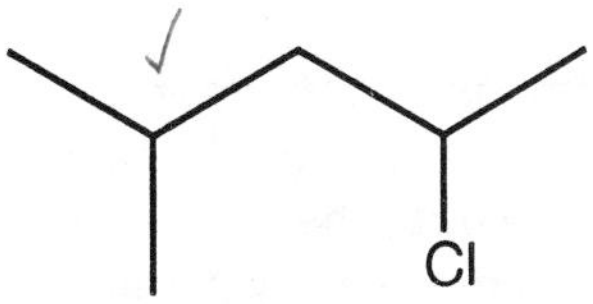

2-Chloro-4-methylpentane

Cl

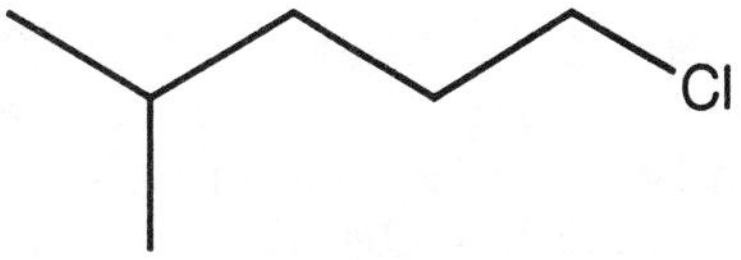

1-Chloro-4-methylpentane

From 3-Methylpentane

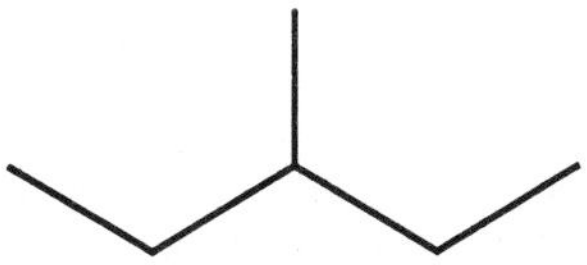

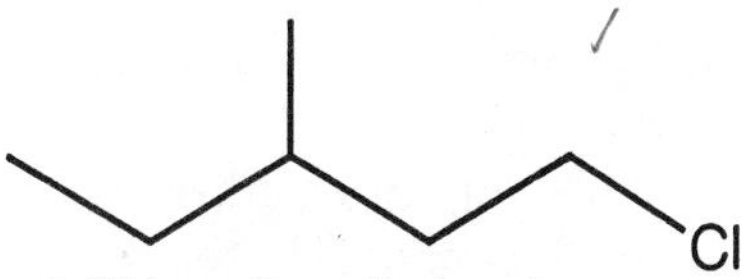

Cl

1-Chloro-3-methylpentane

Cl

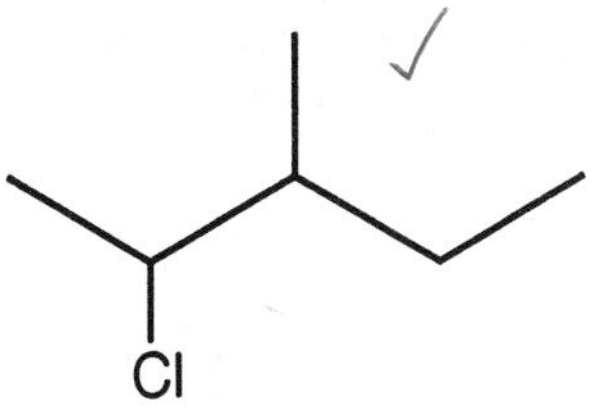

2-Chloro-3-methylpentane

Cl

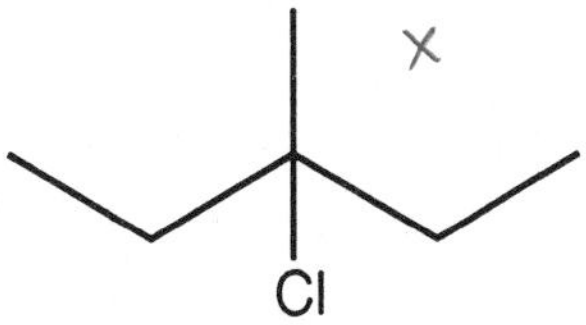

3-Chloro-3-methylpentane

Cl

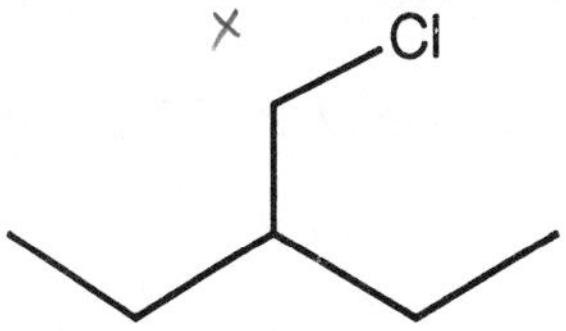

3-Chloromethylpentane

From 2,2-Dimethylbutane

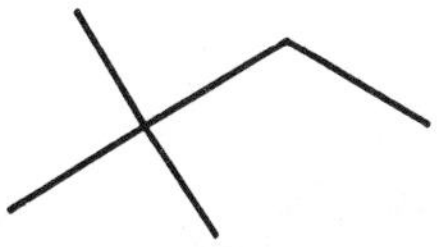

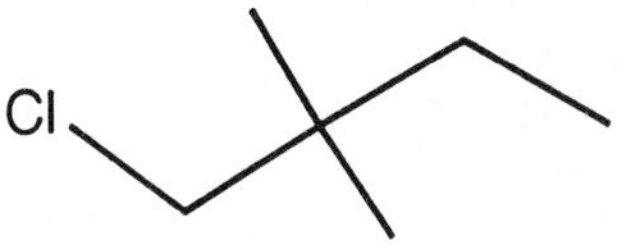

Cl

1-Chloro-2,2-dimethylbutane

Cl

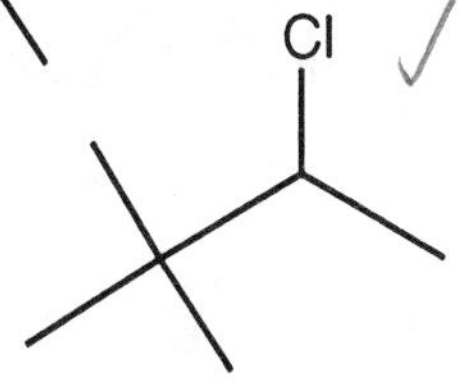

3-Chloro-2,2-dimethylbutane

Cl

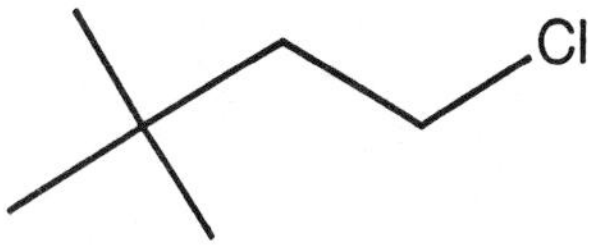

1-Chloro-3,3-dimethylbutane

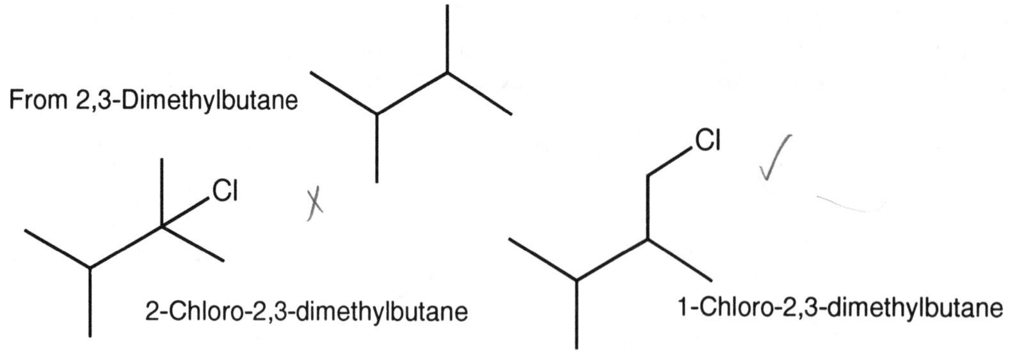

From 2,3-Dimethylbutane

2-Chloro-2,3-dimethylbutane

1-Chloro-2,3-dimethylbutane

Problem 3.32
(a and b) In the first two compounds carbon is attached to three other groups, H, H, and O in (a) and C, C, and H in (b). Three hybrid orbitals are needed, and so sp^2 hybridization is appropriate.
(c) Here carbon is attached to only two other groups (H and the other C) and so two hybrid orbitals are needed. The hybridization is *sp*.
(d) In this molecule, carbon is attached to four identical chlorine atoms. Four hybrids are needed and the hybridization is exactly sp^3.

Problem 3.33 Extrapolate from the boiling points of dodecane ($C_{12}H_{26}$, bp 216.3 °C) and eicosane ($C_{20}H_{42}$, bp 343 °C). So, 343 – 216.3 = 126.7. The compound C_{15} is three-eighths of the way from C_{12} to C_{20}, so we take three-eighths of 126.7 = 47.5. The boiling point of C_{15} should be the boiling point of C_{12} plus this number, 216.3 + 47.5 = 263.8 °C. This procedure works reasonably well, as the real value is found to be 270.6 °C.

Problem 3.34 As noted in Section 3.13 (p. 107), symmetry is important in determining melting points. Highly symmetrical molecules pack well into crystal lattices, and more energy is required to break up the lattice than for molecules that do not pack so well. Thus, the highly symmetrical neopentane melts 113 °C higher than pentane. However, branched-chain hydrocarbons without high symmetry tend to have lower melting points than straight-chain hydrocarbons because the branching interferes with regular packing in the crystal. Accordingly, isopentane melts 30 °C lower than pentane.

Problem 3.35
(a) Ethane yields only two compounds. You can either replace two hydrogens on one carbon or one hydrogen on each carbon. Those are the only possibilites.

$$H_3C—CH_3 \xrightarrow{\text{replace two H with X}} H_3C—CHX_2 \quad \text{and} \quad XH_2C—CH_2X$$

(b) Propane yields four compounds:

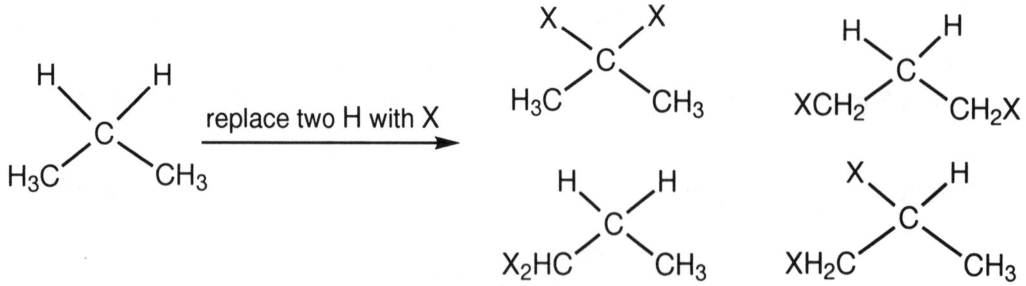

$(H)(H)C(H_3C)(CH_3)$ —replace two H with X→ $(X)(X)C(H_3C)(CH_3)$; $(H)(H)C(XCH_2)(CH_2X)$; $(H)(H)C(X_2HC)(CH_3)$; $(X)(H)C(XH_2C)(CH_3)$

(c) Butane yields six new compounds.

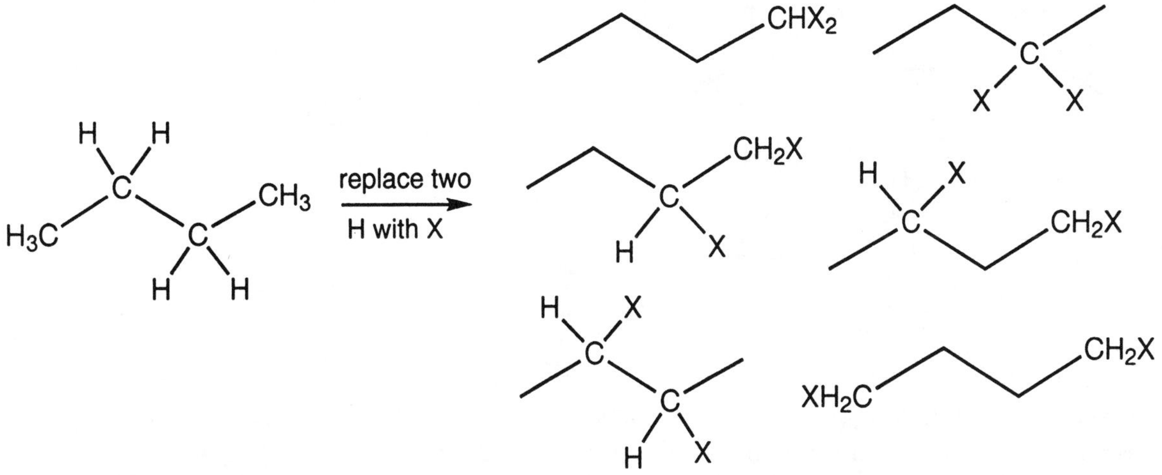

[Structure: butane, H_3C–CH_2–CH_2–CH_3, converted by "replace two H with X" into six dihalobutanes: $CH_3CH_2CH_2CHX_2$; $CH_3CH_2CX_2CH_3$; $CH_3CH_2CHXCH_2X$; $CH_3CHXCH_2CH_2X$; $CH_3CHXCHXCH_3$; $XH_2CCH_2CH_2CH_2X$]

Notice the changing code level in the representations of these compounds.

Problem 3.36 Start with the 0° form, an eclipsed energy maximum (a transition state). Rotate the back carbon clockwise in 60° increments to generate the stable, staggered conformations (60°, 180°, and 300°) and the other transition states (120° and 240°).

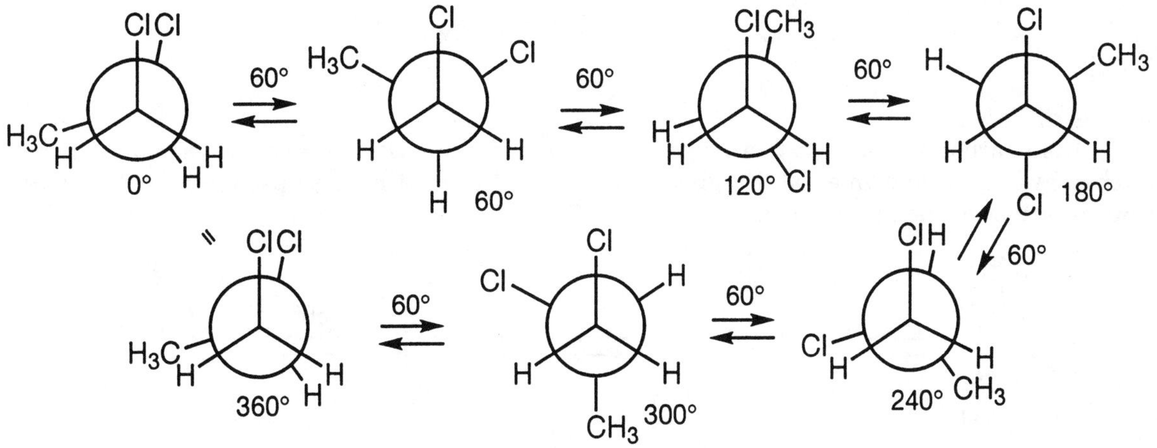

[Newman projections at 0°, 60°, 120°, 180°, 240°, 300°, and 360° (= 0°), interconverted by 60° rotations]

Problem 3.37 Look down the C(1)—C(2) bond of 2-methylpentane with the methyl group in front.

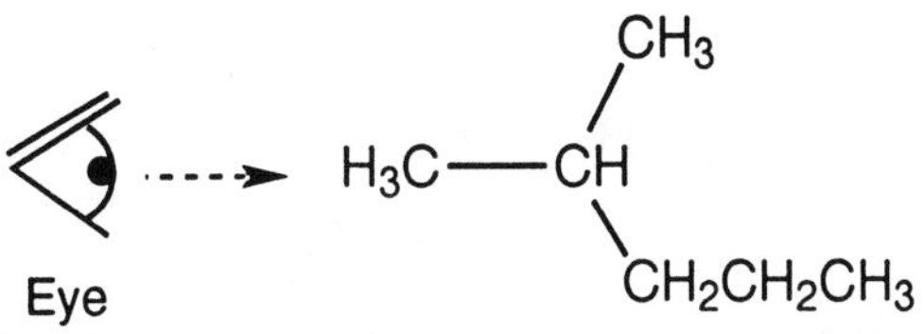

[Structure: Eye - - - → H_3C—$CH(CH_3)$—$CH_2CH_2CH_3$]

On the front carbon you see three hydrogens. On the rear carbon you see one hydrogen, a methyl group, and a propyl (Pr) group. Start at 0°, an eclipsed form, and then proceed by 60° rotations of the rear carbon.

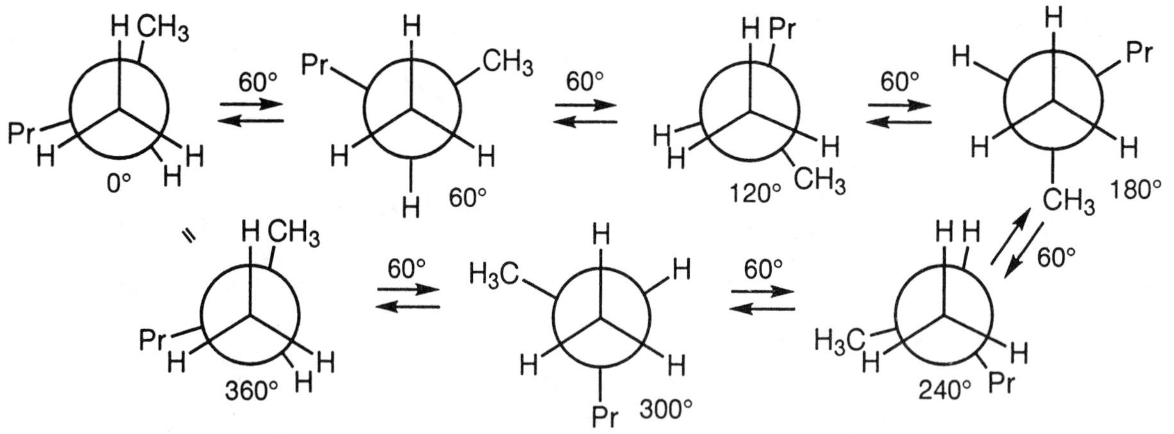

The three eclipsed conformations (0°, 120°, and 240°) are equi-energetic. These will be the highest energy conformations as they contain eclipsed methyl–hydrogen, propyl–hydrogen, and hydrogen–hydrogen interactions. The three staggered conformations (60°, 180°, and 300°) are also of equal energy. They will be lower in energy than the eclipsed conformation, as they contain only *gauche*, not eclipsed interactions. Now look down the C(2)—C(3) bond of 2-methylpentane with the methine (CH) carbon in front.

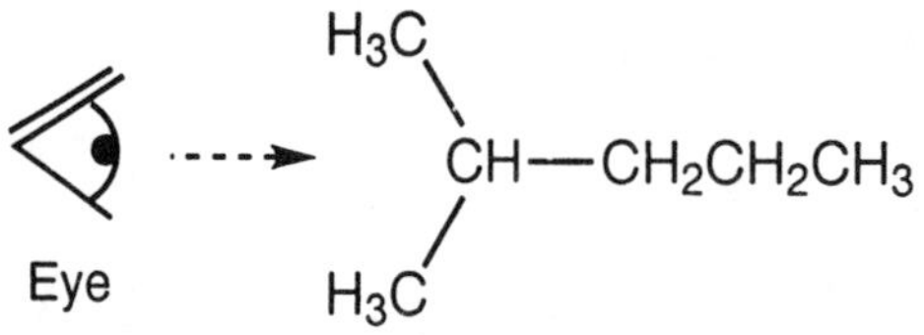

On the front carbon you see two methyl groups and a hydrogen. On the rear carbon you see two attached hydrogens and an ethyl (Et) group. Again, start at 0° with an eclipsed form and proceed by 60° rotations of the rear carbon.

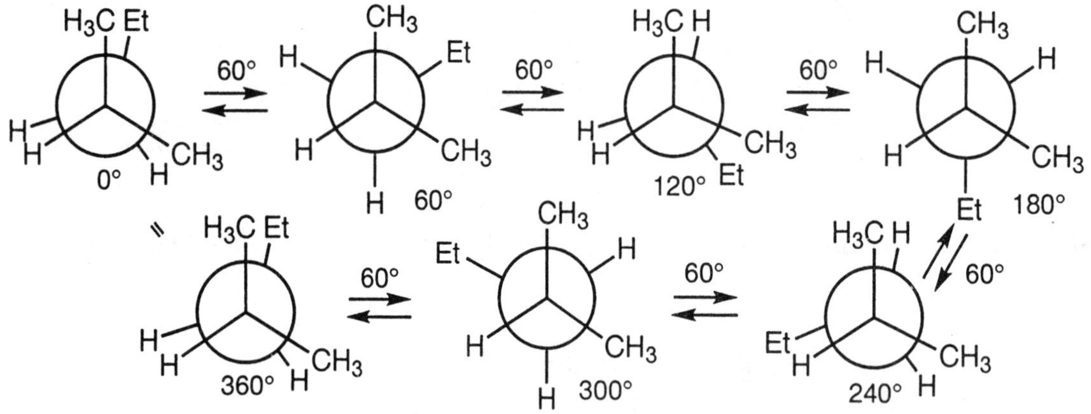

The situation is more complex from this view. The eclipsed conformations (0°, 120°, and 240°) are still energy maxima. The 0° and 120° conformations each contain a methyl–ethyl, a methyl–hydrogen, and hydrogen–hydrogen eclipsed interaction and are equi-energetic. These are probably higher in energy than the eclipsed 240° form, which contains two methyl–hydrogen and one ethyl–hydrogen interactions. The staggered conformations (60°, 180°, and 300°) represent energy minima. The 180° and 300° conformations, which contain one methyl–ethyl *gauche* interaction, are equi-energetic and are lower in energy than the 60° conformation in which there are two methyl–ethyl *gauche* interactions.

Problem 3.38 Start by drawing the two conformations. If size were all that mattered, the two would surely be very close in energy.

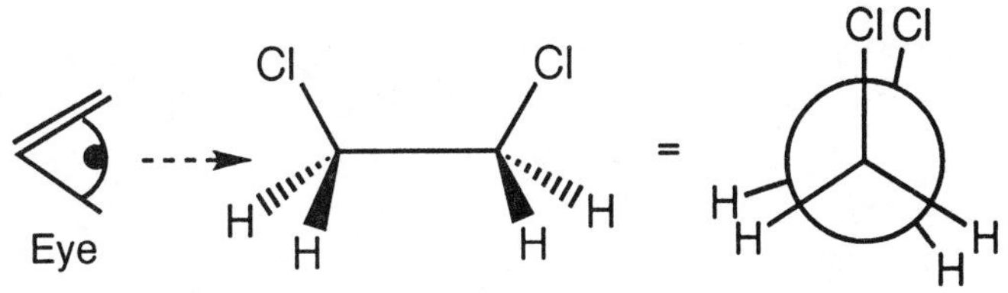

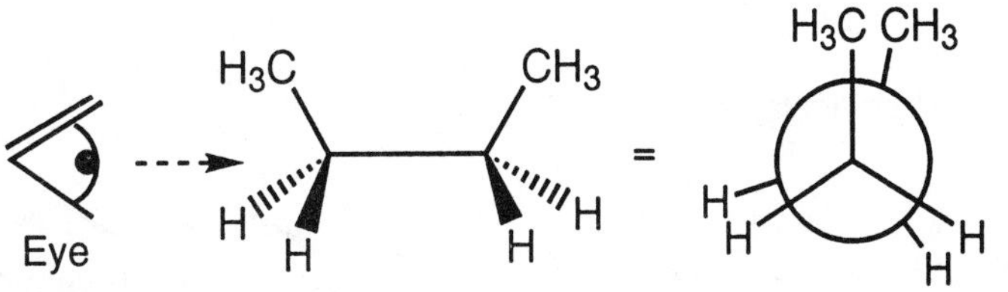

Cl Cl
Eye
H H H H
=
Cl Cl
H H H H

H_3C CH_3
Eye
H H H H
=
H_3C CH_3
H H H H

However, the carbon–chlorine bond is much more polar than a carbon–methyl bond. Note that in the eclipsed form of 1,2-dichloroethane shown the two C—Cl dipoles are lined up. This molecule will be strongly destabilized through charge–charge opposition.

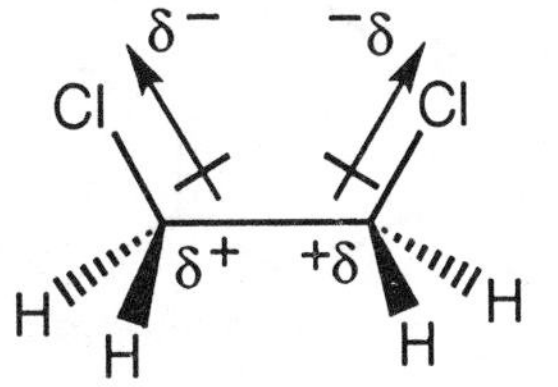

δ− −δ
Cl Cl
δ+ +δ
H H H H

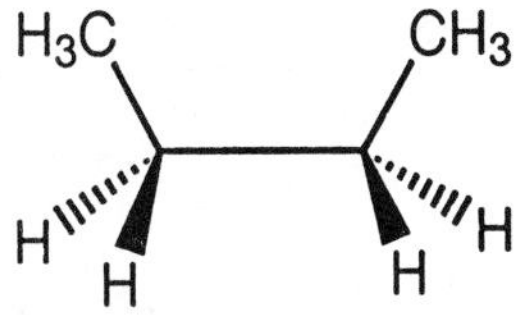

H_3C CH_3
H H H H

Problem 3.39

Pentane

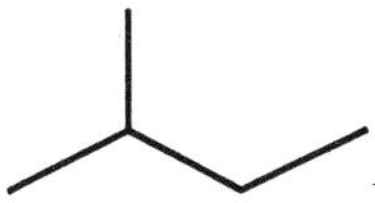

2-Methylbutane

2,2-Dimethylpropane

Hexane

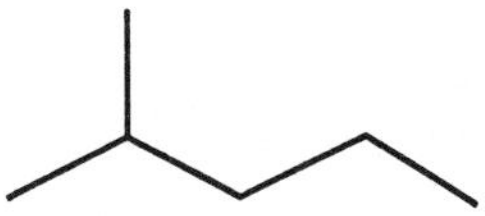

2-Methylpentane

3-Methylpentane

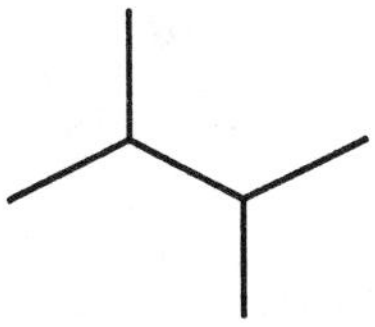

2,3-Dimethylbutane

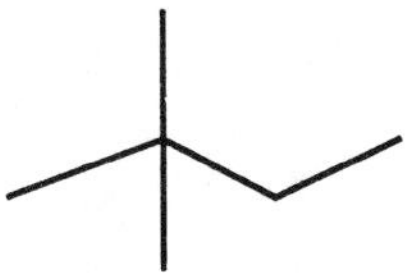

2,2-Dimethylbutane

Problem 3.40

Nonane

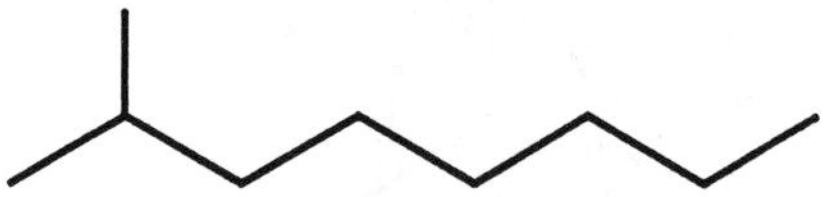

3,4-Dimethylheptane

2-Methyloctane

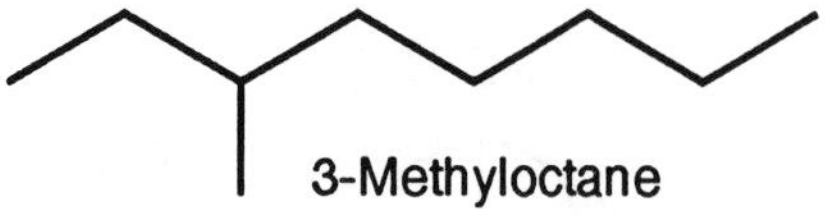

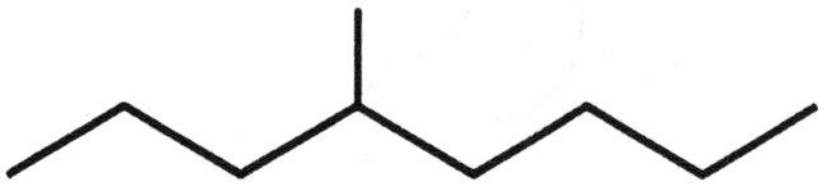

3,5Dimethylheptane

3-Methyloctane

4,4-Dimethylheptane

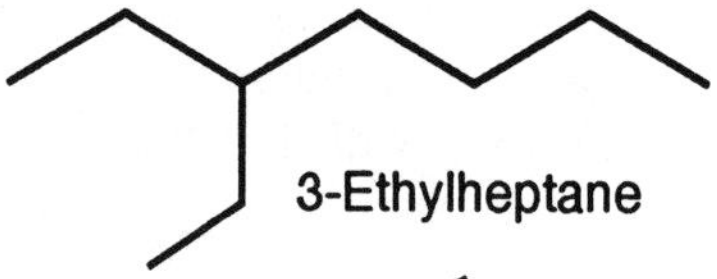

4-Methyloctane

3-Ethylheptane

2,2-Dimethylheptane

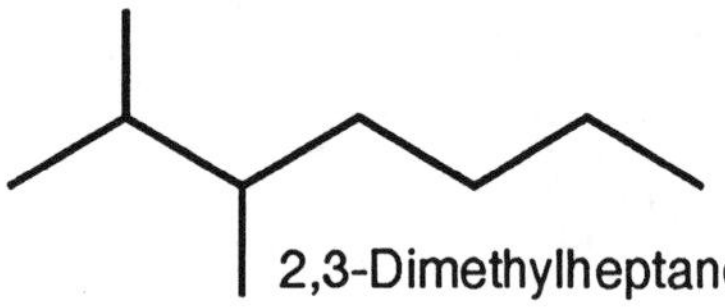

4-Ethylheptane

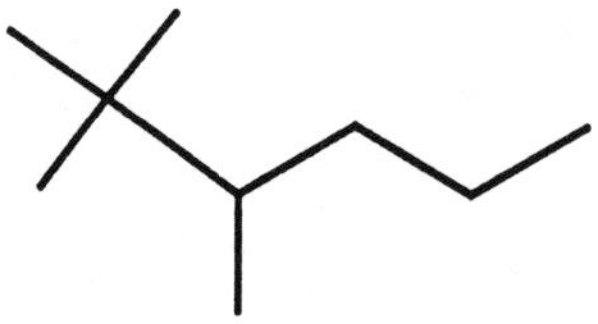

2,3-Dimethylheptane

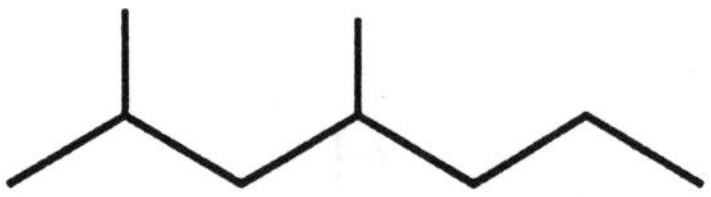

2,4-Dimethylheptane

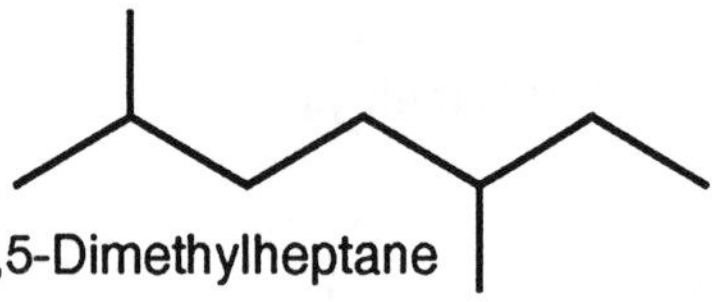

2,2,3-Trimethylhexane

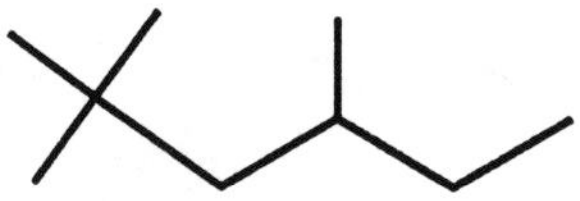

2,5-Dimethylheptane

2,2,4-Trimethylhexane

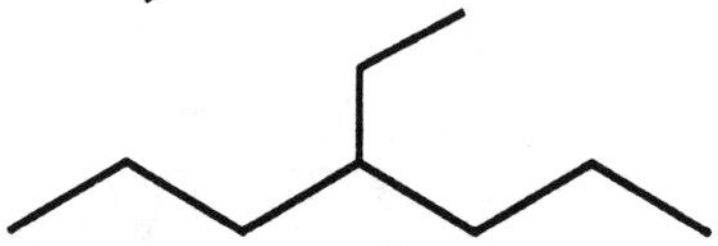

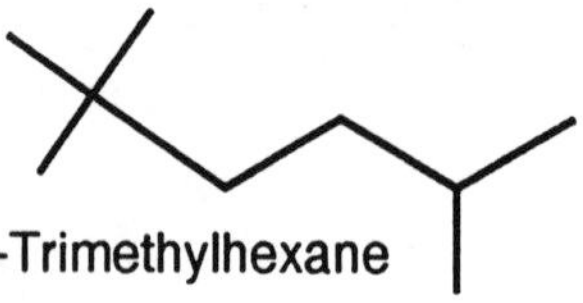

2,6-Dimethylheptane

2,2,5-Trimethylhexane

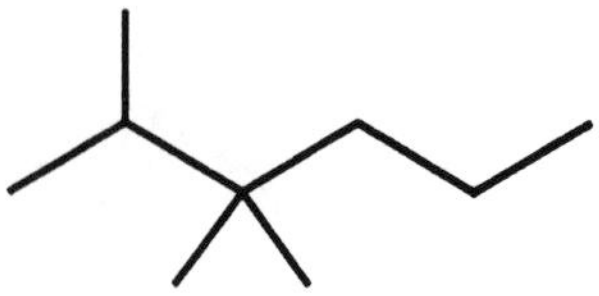

3,3-Dimethylheptane

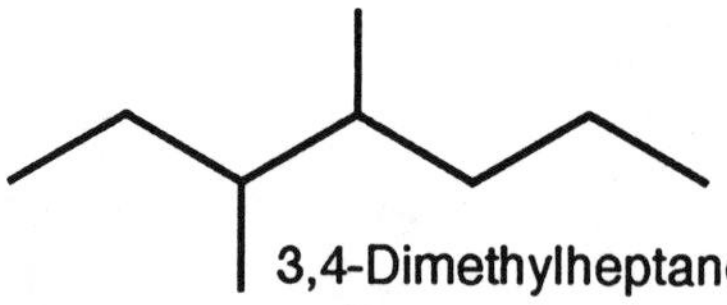

2,3,3-Trimethylhexane

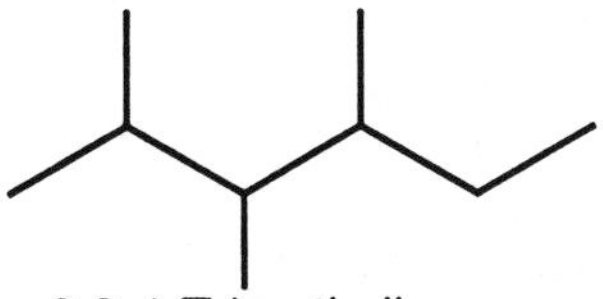

2,3,4-Trimethylhexane

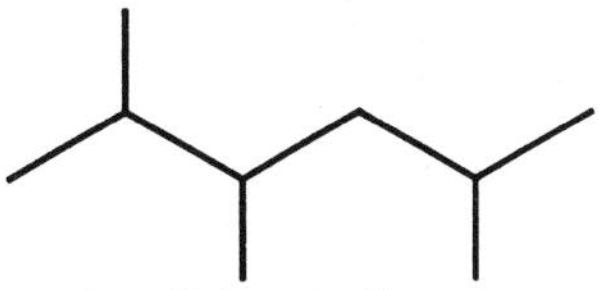

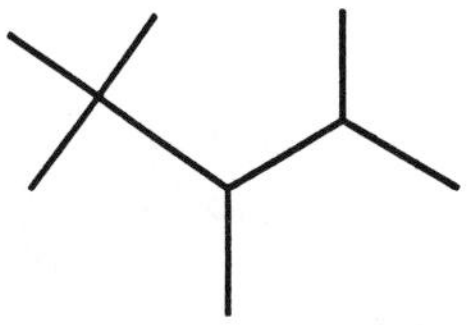

2,2,3,4-Tetramethylpentane

2,3,5-Trimethylhexane

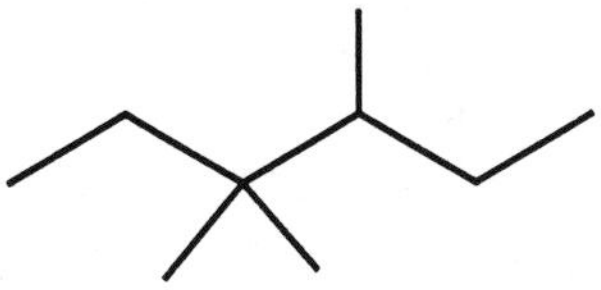

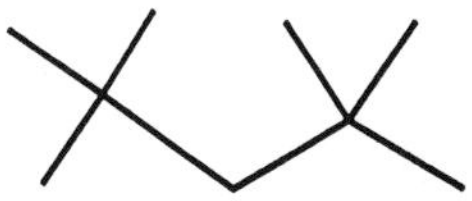

2,2,4,4-Tetramethylpentane

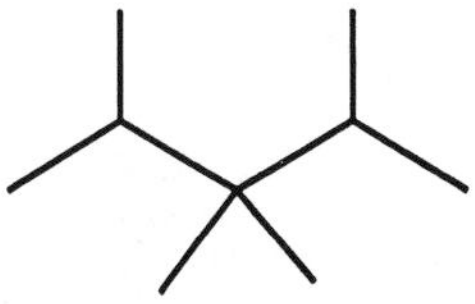

3,3,4-Trimethylhexane

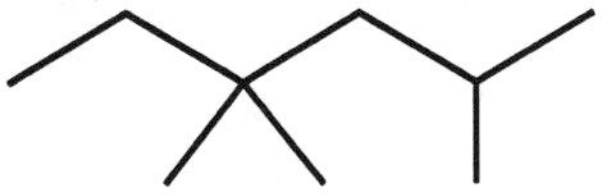

2,3,3,4-Tetramethylpentane

2,4,4-Trimethylhexane

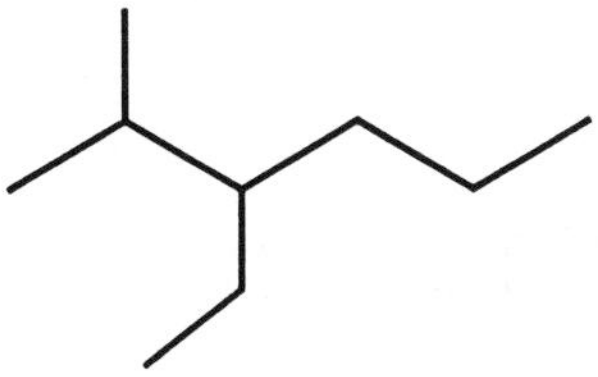

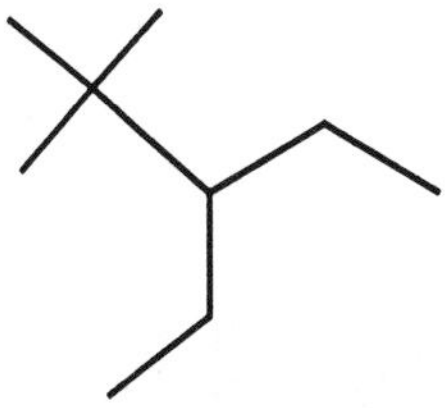

3-Ethyl-2,2-dimethylpentane

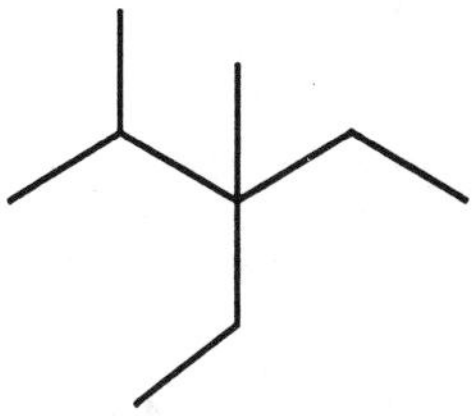

3-Ethyl-2-methylhexane

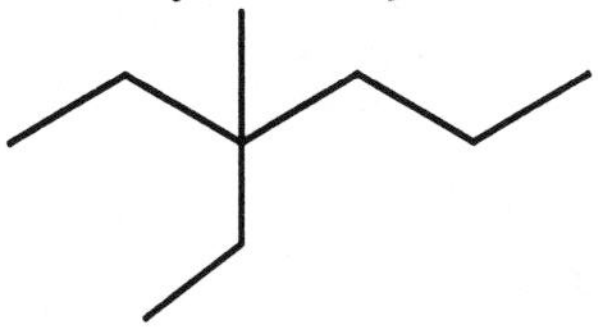

3-Ethyl-3-methylhexane

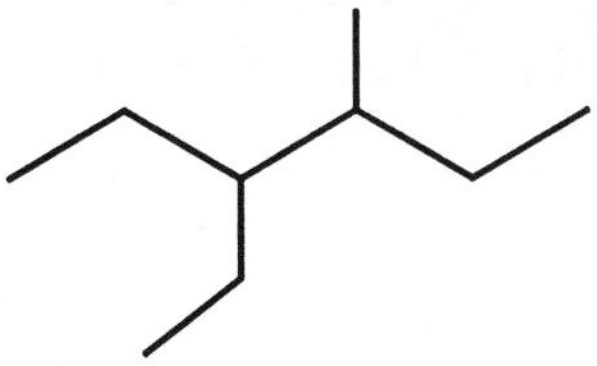

3-Ethyl-2,3-dimethylpentane

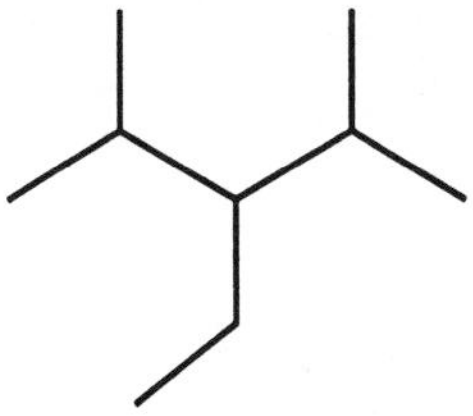

3-Ethyl-4-methylhexane

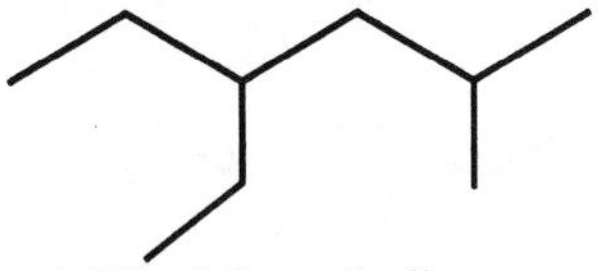

3-Ethyl-2,4-dimethylpentane

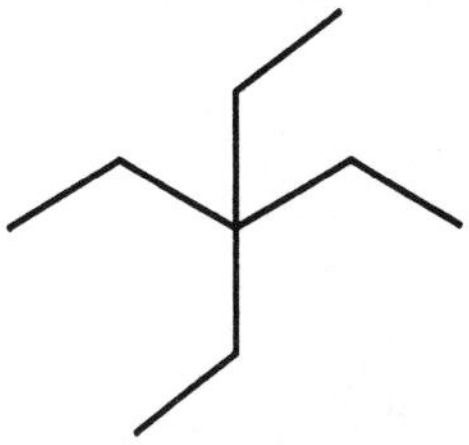

4-Ethyl-2-methylhexane

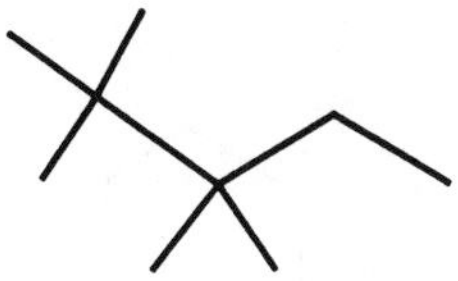

2,2,3,3-Tetramethylpentane

3,3-Diethylpentane

Problem 3.41

(a)

3-Isopropyl-2,5-dimethylhexane
not 3-Isobutyl-2,4-dimethylpentane
not 4-Isopropyl-2,5-dimethylhexane

(b)

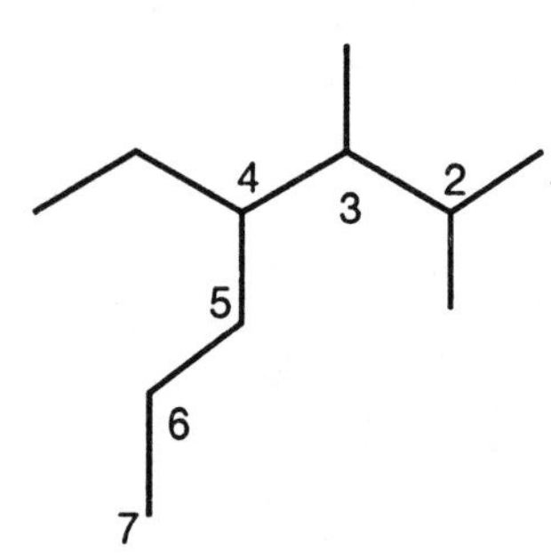

4-Ethyl-2,3-dimethylheptane
not 2,3-Dimethyl-4-propylhexane
not 4-Ethyl-5,6-dimethylheptane

(c)

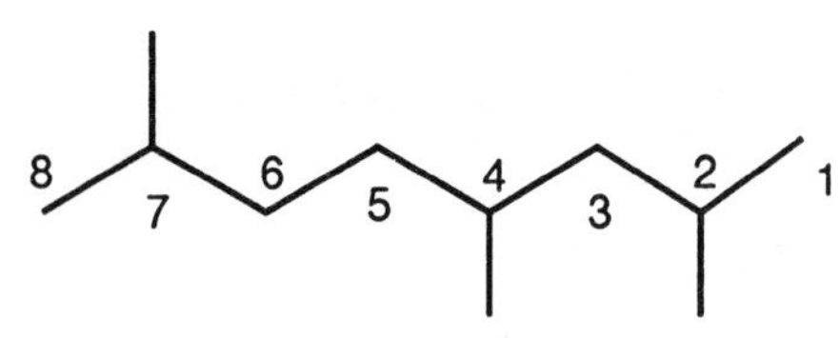

2,4,7-Trimethyloctane
not 2,5,7-Trimethyloctane

(d)

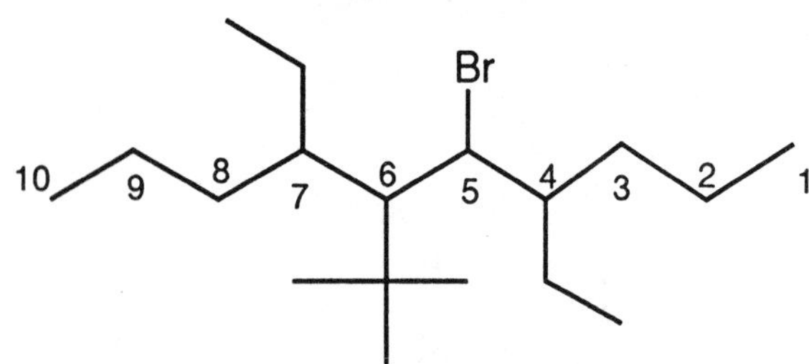

5-Bromo-6-*tert*-butyl-4,7-diethyldecane
not 6-Bromo-5-*tert*-butyl-4,7-diethyldecane
not 4,7-Diethyl-6-bromo-5-*tert*-butyldecane

Problem 3.42

(a) F

(b) Br Br

(c)

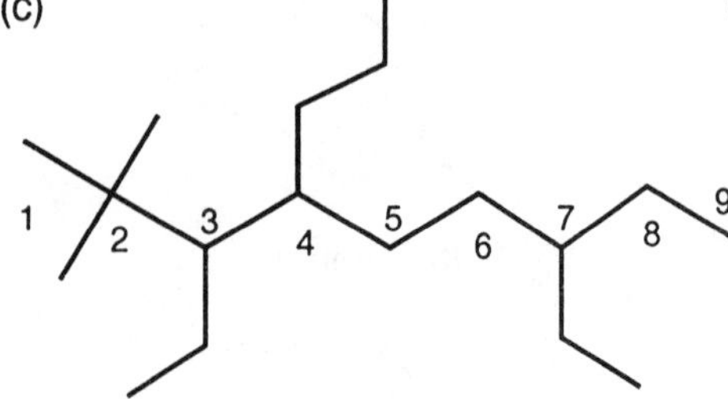

(d) Cl I

Problem 3.43

(a)

Really: **2,2-Dimethylbutane**
The wrong name fails to find the longest "straight chain."

(b)

Really: **3,7-Dimethylnonane**
Once again, the wrong name uses a too short "longest straight chain"

(c)

Really: **1-Bromo-3-ethylhexane**
Guess what the wrong name does?

(d)

F

Really: **3-Ethyl-5-fluoro-2,2,8-trimethylnonane**

Problem 3.44 There are two equal possibilites for the "longest straight chain."

3-Ethyl-2-methylpentane or 3-Isopropylpentane

However, there is another rule to resolve this problem. (There are always other rules!) If chains of equal length compete for selection as the main chain in a saturated branched alkane, then the choice goes to the chain that has the greatest number of side chains.

Problem 3.45 This problem gives you a chance to apply the new rule of Problem 3.44. The correct name has two substituents.

2-Methyl-4-propylheptane not 4-Isobutylheptane

Chapter 4 Outline

4.1 Alkenes: Structure and Bonding
4.2 Derivatives, Isomers, and Names of Alkenes
4.3 Nomenclature
4.4 The Cahn–Ingold–Prelog Priority System
4.5 Relative Stability of Alkenes: Heats of Formation
4.6 Double Bonds in Rings
4.7 Physical Properties of Alkenes
4.8 Alkynes: Structure and Bonding
4.9 Relative Stability of Alkynes: Heats of Formation
4.10 Derivatives and Isomers of Alkynes
4.11 Triple Bonds in Rings
4.12 Physical Properties of Alkynes
4.13 Molecular Formulas and Degrees of Unsaturation
4.14 Something More: Acidity of Alkynes

Alkenes and Alkynes

4

This chapter is devoted almost entirely to structure, and the following problems reflect that emphasis. Here we explore the structural consequences of sp^2 and sp hybridization in alkenes and alkynes. There is practice in finding isomers in both cyclic and acyclic molecules. Stereochemistry becomes especially important in the alkenes and ring compounds, and there are several opportunities in the following problems for you to work on stereochemical aspects of these kinds of molecules.

Questions of energy and stability also arise. The π bonds contributing to the double and triple bonds encountered here are weaker than the σ bonds emphasized in the earlier chapters. There will be a number of chances to make assessments of relative energies in the problems that follow.

Problem 4.3 In the 90° form the two *p* orbitals are of course also at 90°. In this arrangement, there is no overlap between the two orbitals because the bonding and antibonding interactions exactly cancel.

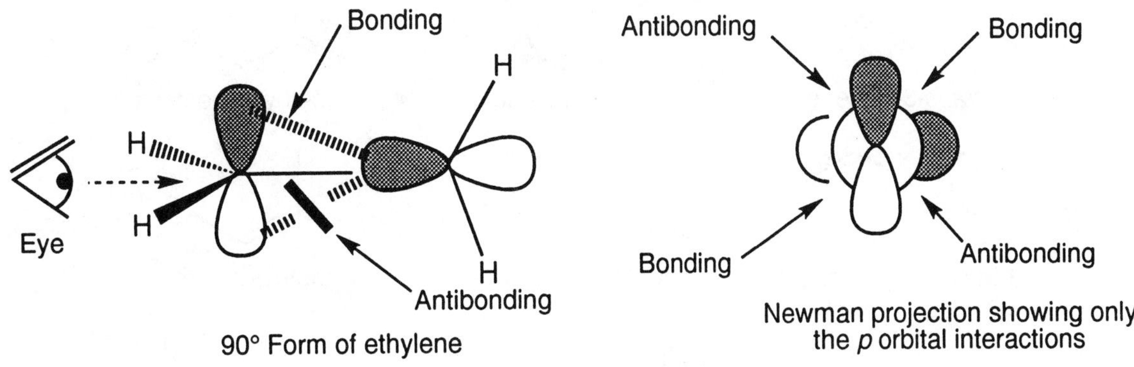

90° Form of ethylene

Newman projection showing only the *p* orbital interactions

Problem 4.4 Only two of these molecules can exist in cis/trans (*Z*/*E*) forms.

H H H

(E) Form *(Z)* Form *(Z)* Form *(E)* Form

The others have only one possible form. These alkenes are flat and the "different" isomers shown below are really identical, as one can simply be turned over to make the other. If this isn't clear, and if the drawing doesn't seem obvious to you, by all means make models of the "two" forms and show that they are identical by superimposing one on the other.

Problems 4.5 and 4.6

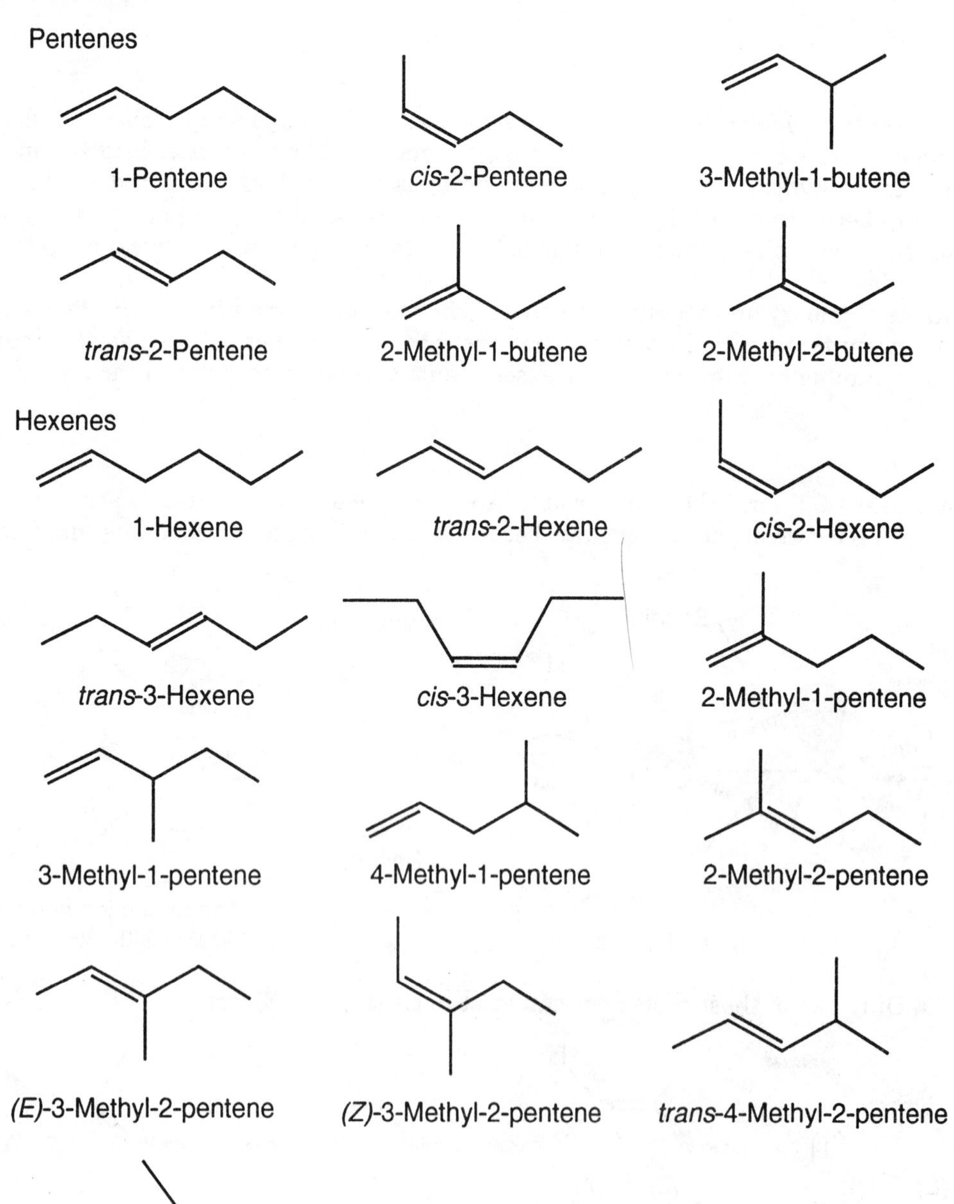

2,3-Dimethyl-1-butene 2-Ethyl-1-butene 3,3-Dimethyl-1-butene 2,3-Dimethyl-2-butene

Problem 4.7

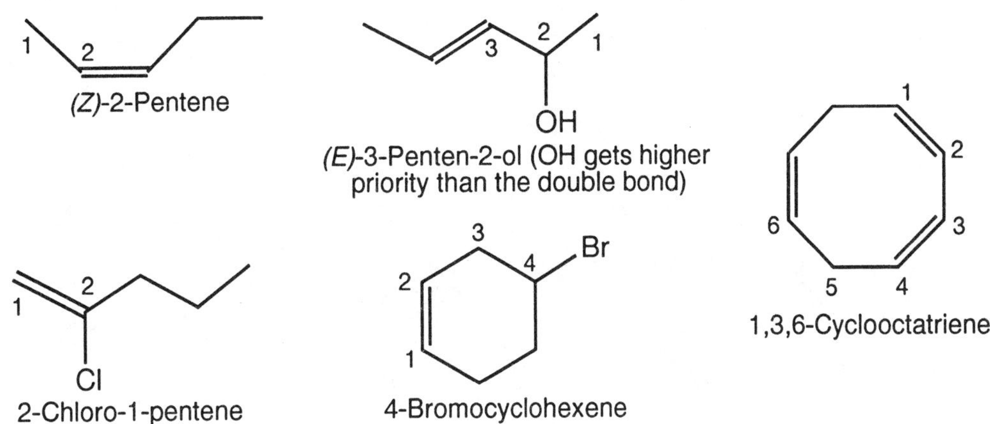

Problem 4.9 On the right-hand carbon Cl (atomic number = 17) is higher priority than H (atomic number = 1). On the left-hand carbon the branched chain group is higher priority than the straight chain. The tie is broken at the second carbon. In the straight chain the second carbon is attached to C,H,H, whereas the second carbon of the branched chain is attached to C,C,H. So, the molecule on the left with the higher priority groups on opposite sides is *(E)* and the molecule on the right, with the higher priority groups on the same side of the double bond is *(Z)*.

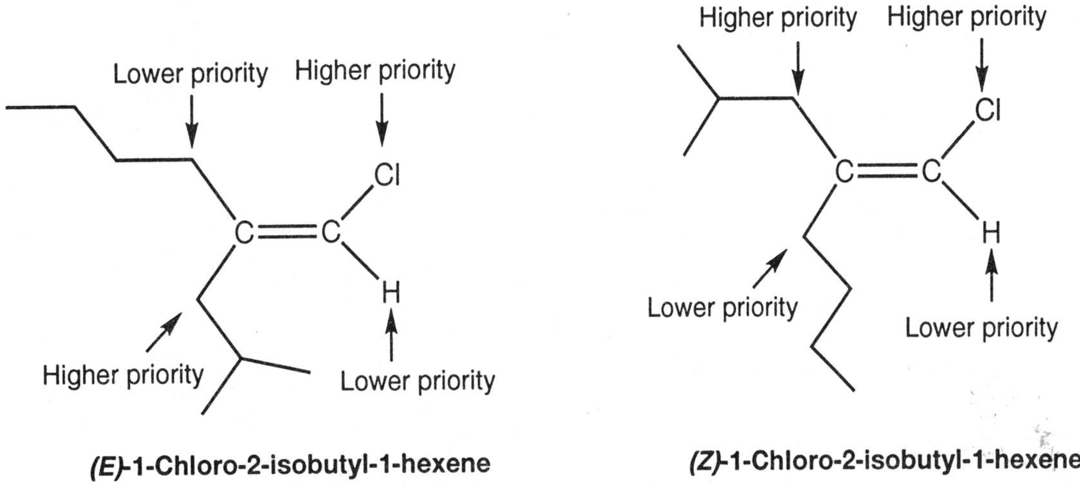

***(E)*-1-Chloro-2-isobutyl-1-hexene** ***(Z)*-1-Chloro-2-isobutyl-1-hexene**

Problem 4.10 In (a) the higher priority groups are fluorine and ethyl, so the *(E)* isomer has those groups on opposite sides of the double bond. In (b) the higher priority groups are propyl and ethyl, and the *(E)* isomer must have them on opposite sides of the double bond. In (c) the higher priority groups are I and Cl. The *(Z)* isomer will have them on the same side of the double bond.

(a) F

(b)

(c) Cl I F Br

Problem 4.11 In (a), the higher priority groups are CH_3 and F. The *(Z)* isomer has them on the same side; the *(E)* isomer has them on opposite sides. In (b), the higher priority groups are ethyl (CH_3CH_2) and amino (NH_2). In (c), the higher priority groups are the methylene (CH_2) groups starting the ring. In (d), the higher priority groups are deuterium (D) and the ring carbon bearing the methyl (CH_3) group.

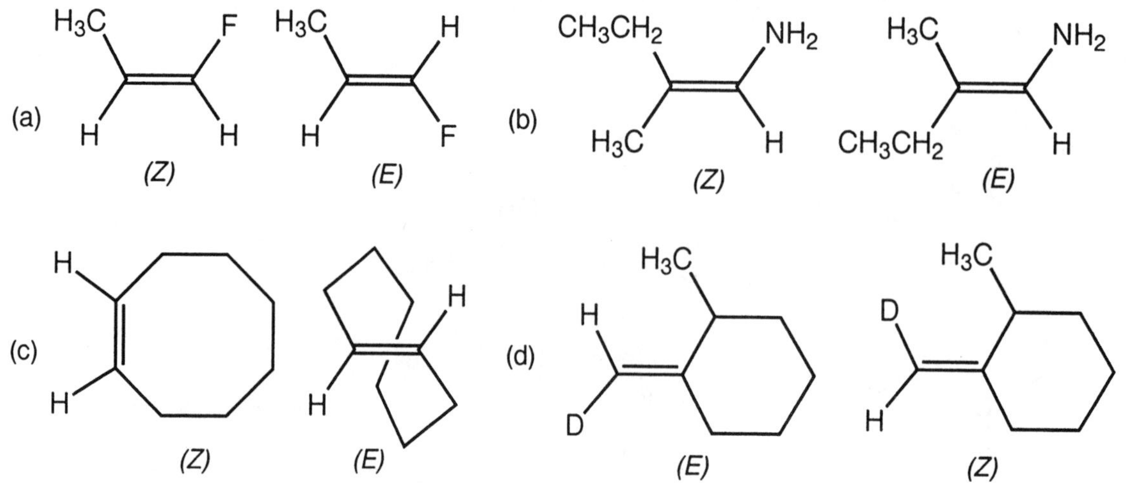

Problem 4.12 The procedure is to classify each carbon-carbon σ bond in the molecules as sp^3-sp^3, sp^3-sp^2, or sp^2-sp^2. The more low energy bonds in the molecule, the better.

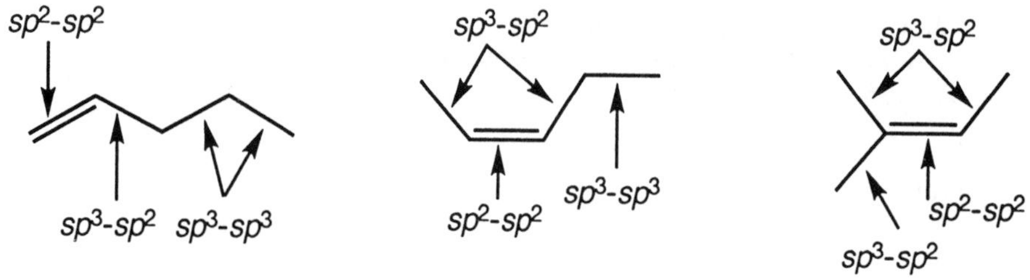

	sp^2-sp^2	sp^3-sp^2	sp^3-sp^3	
	1	1	2	Least stable
	1	2	1	
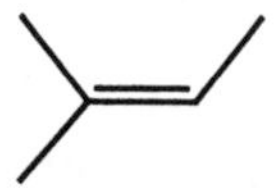	1	3	0	Most stable

Problem 4.13

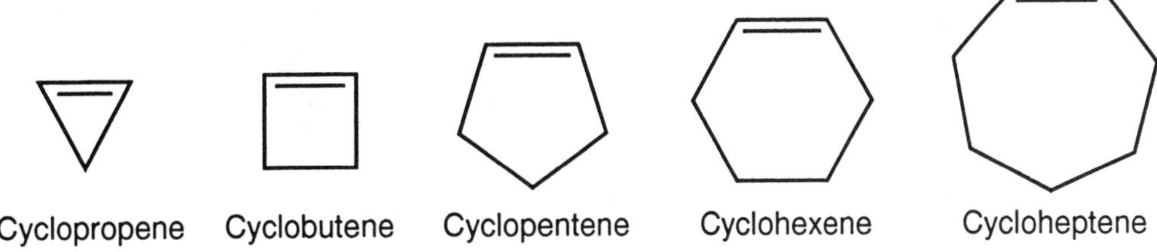

Problem 4.14 *Remember:* in the last three compounds the ene gets priority over the *yne.*

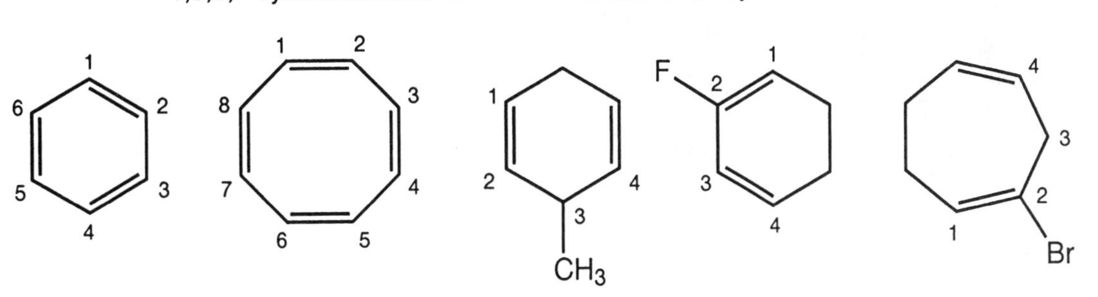

Problem 4.16

$\Delta G = -RT \ln K$. In this case

$\Delta G = 11.4$ kcal/mol
$R = 1.986$ cal/deg.mol
$T = 298$ K

$RT = 592$ cal/mol $= 0.592$ kcal/mol

$$11.4 = -(0.592)\ln K$$
$$\ln K = -19.26$$
$$K = 2.3 \times 10^{-8}$$

Problem 4.17 The form shown in Figure 4.50 is *(Z)*. On the left-hand carbon of the double bond the higher priority group is CH_2 and the lower priority group is H. On the right-hand side of the double bond the higher priority group is CH_2-CHC_2 and the lower priority group is CH_2-CH_2C. As the higher priority groups are on the same side of the double bond, the compound is *(Z)*.

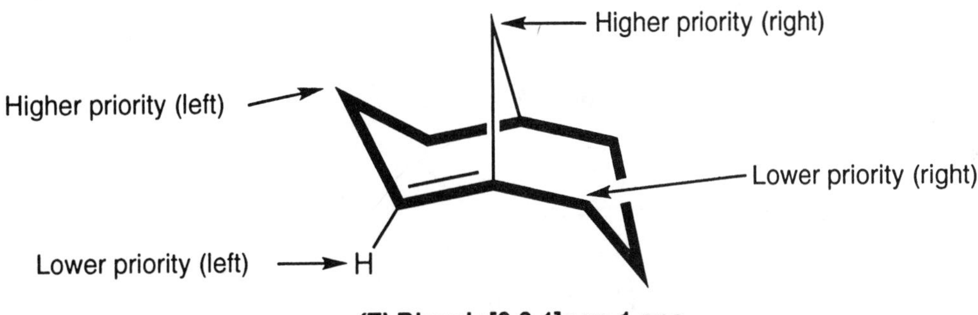

***(Z)*-Bicyclo[3.3.1]non-1-ene**

Problem 4.19 The picture is exactly the same as for alkenes (Fig. 4.13, p 122), except that there are two p bonds at 90° to each other. The diagram shows only one orbital interaction diagram, and indicates both π bonds schematically.

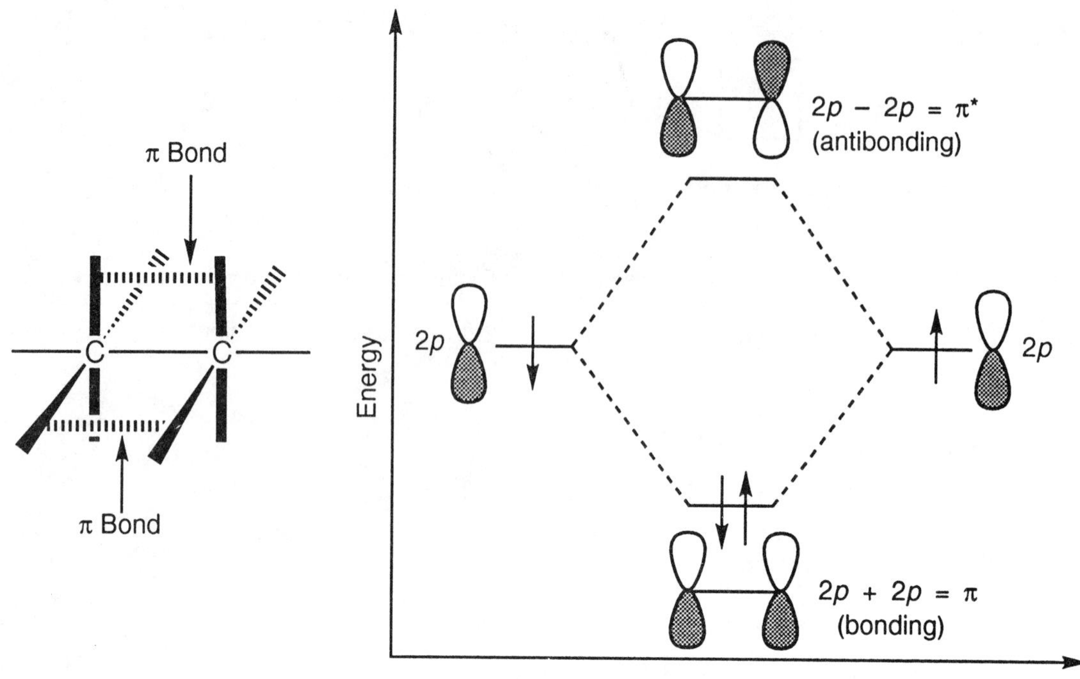

Problems 4.20 and 4.21

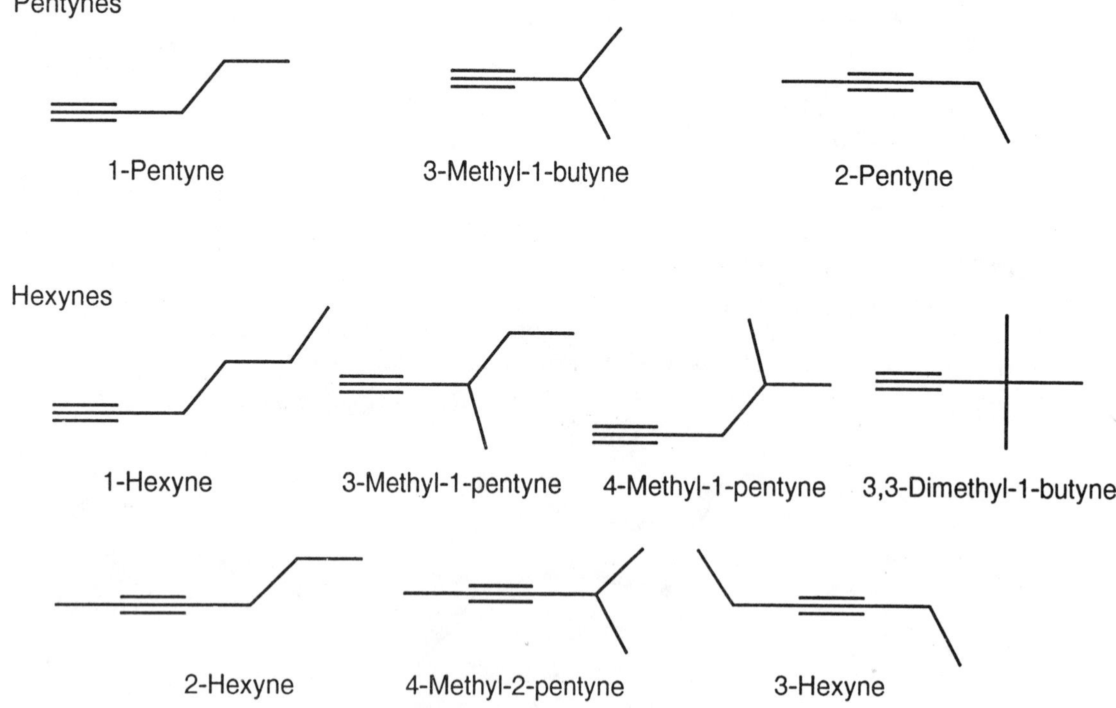

Heptynes

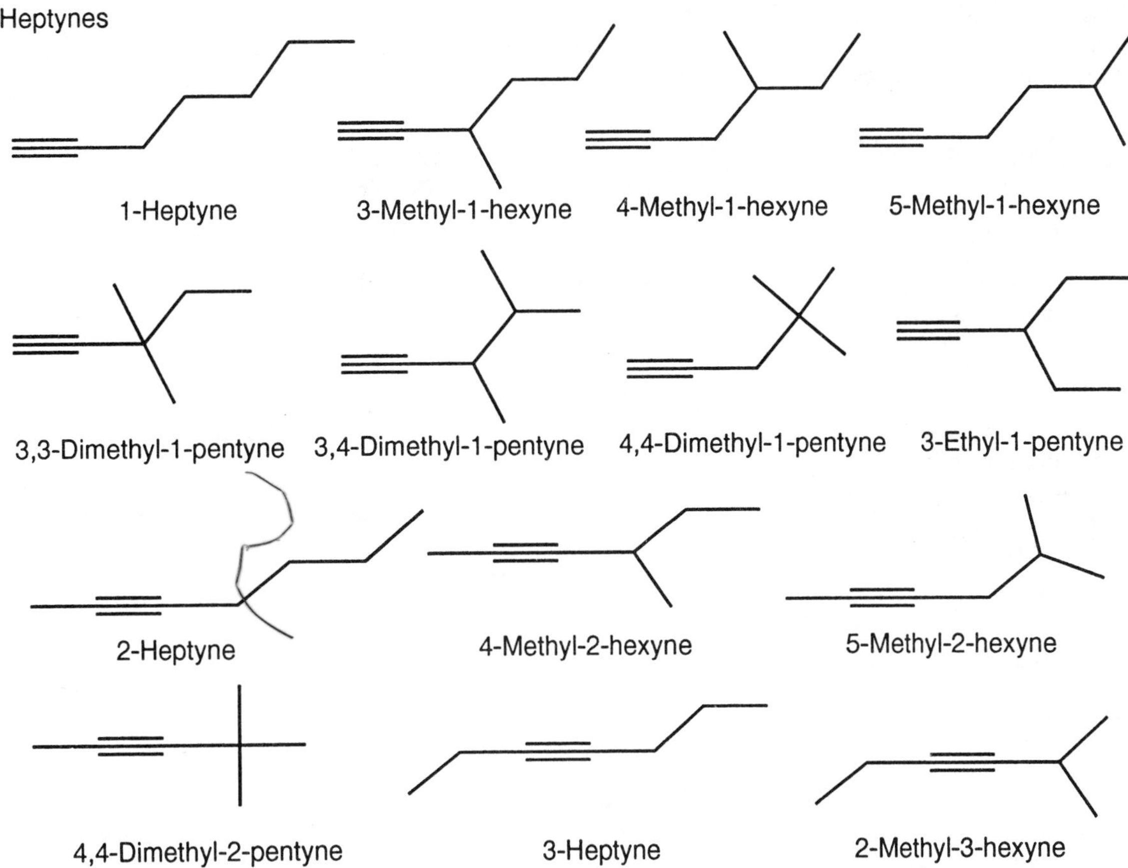

Problem 4.22

(a) C_5H_6 (C_nH_{2n-4}) For C_5H_{2n+2}, $2n + 2 = 12$

Calculation: (12 − 6)/2 = 6/2 = 3 degrees of unsaturation, which means there must be a total of three π bonds and/or rings. Some possibilities are

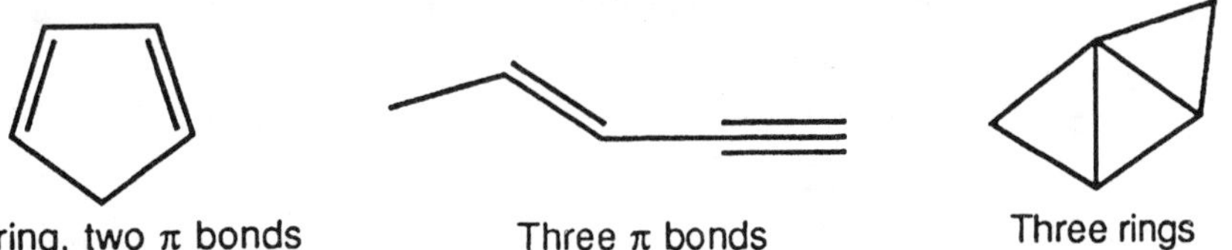

One ring, two π bonds Three π bonds Three rings

(b) C_7H_8 (C_nH_{2n-6}) For C_7H_{2n+2}, $2n + 2 = 16$

Calculation: (16 − 8)/2 = 8/2 = 4 degrees of unsaturation, which means there must be a total of four π bonds and/or rings. Some possibilities are

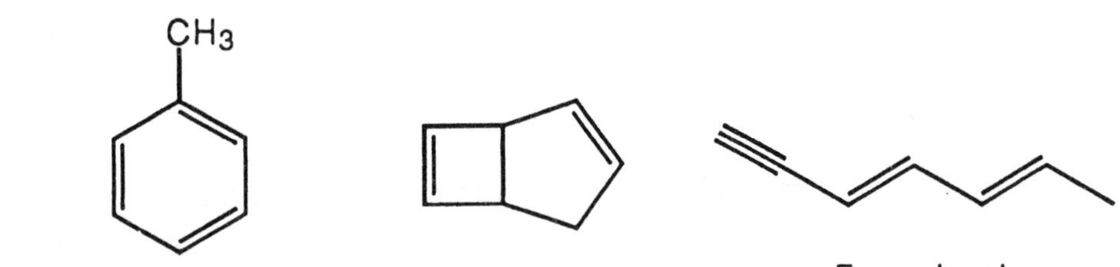

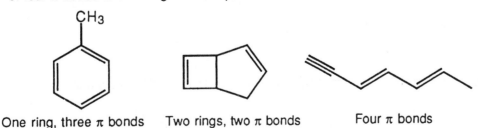

One ring, three π bonds Two rings, two π bonds Four π bonds

(c) $C_{10}H_{10}$ (C_nH_{2n-10}) For $C_{10}H_{2n+2}$, $2n + 2 = 22$
Calculation: $(22 - 10)/2 = 12/2 = 6$ degrees of unsaturation. This means there must be a total of six π bonds and/or rings. Some possibilities are

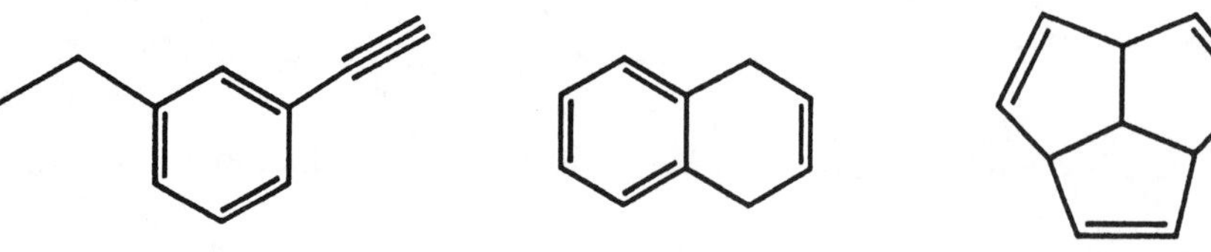

One ring, five π bonds — Two rings, four π bonds — Three rings, three π bonds

(d) $C_5H_8Br_2$ For C_5H_{2n+2}, $2n + 2 = 12$
Calculation: $[12 - (8 + 2)]/2 = 2/2 = 1$ degree of unsaturation. This means there must be a total of one π bond and/or ring. Some possibilities are

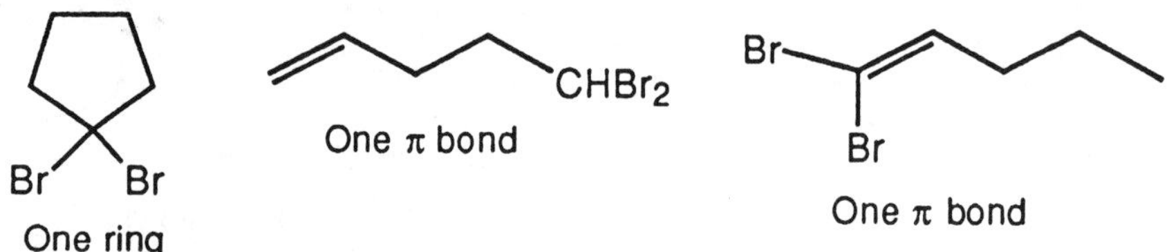

One ring — One π bond — One π bond

Additional Problem Answers

Problem 4.23 First, find the weight of carbon present in the carbon dioxide. This must also be the weight of carbon present in the sample compound (MW = molecular weight).

$$\text{Wt (C)} = \frac{\text{MW (C)}}{\text{MW (CO}_2\text{)}} \times \text{Wt (CO}_2\text{)} = \frac{12.011 \text{ g/mol}}{44.009 \text{ g/mol}} \times 16.90 \text{ mg} = 4.61 \text{ mg}$$

Similarly, the weight of hydrogen can be calculated from the weight of water:

$$\text{Wt (H)} = \frac{2 \times \text{MW (H)}}{\text{MW (H}_2\text{O)}} \times \text{Wt (H}_2\text{O)} = \frac{2 \times 1.008 \text{ g/mol}}{18.015 \text{ g/mol}} \times 3.46 \text{ mg} = 0.39 \text{ mg}$$

Note that the sum of the weights of carbon and hydrogen equals the weight of the sample. The sample must have contained only carbon and hydrogen.

The weight percents of carbon and hydrogen can now be easily calculated.

$$\%\text{C} = \frac{4.61 \text{ mg}}{5.00 \text{ mg}} \times 100 = 92.2\% \qquad \%\text{H} = \frac{0.39 \text{ mg}}{5.00 \text{ mg}} \times 100 = 7.8\%$$

Problem 4.24

(a) C_5H_{10} MW = 70.135 g/mol
C 5 x 12.011 g/mol = 60.055 g/mol, H 10 x 1.008 g/mol = 10.080 g/mol

$$\%\text{C} = \frac{60.055 \text{ g/mol}}{70.135 \text{ g/mol}} \times 100 = 85.63\% \qquad \%\text{H} = \frac{10.080 \text{ g/mol}}{70.135 \text{ g/mol}} \times 100 = 14.37\%$$

(b) C_9H_6ClNO MW = 179.606 g/mol

C 9 x 12.011 g/mol = 108.099 g/mol
H 6 x 1.008 g/mol = 6.048 g/mol
Cl 1 x 35.453 g/mol = 35.453 g/mol
N 1 x 14.007 g/mol = 14.007 g/mol
O 1 x 15.999 g/mol = 15.999 g/mol

$$\%C = \frac{108.099 \text{ g/mol}}{179.606 \text{ g/mol}} \times 100 = 60.19\%$$

$$\%H = \frac{6.048 \text{ g/mol}}{179.606 \text{ g/mol}} \times 100 = 3.37\%$$

$$\%Cl = \frac{35.453 \text{ g/mol}}{179.606 \text{ g/mol}} \times 100 = 19.74\%$$

$$\%N = \frac{14.007 \text{ g/mol}}{179.606 \text{ g/mol}} \times 100 = 7.80\%$$

$$\%O = \frac{15.999 \text{ g/mol}}{179.606 \text{ g/mol}} \times 100 = 8.91\%$$

Problem 4.25 The "missing" weight percent is oxygen, in this case 23.50%. Now, assume a 100 g sample of the compound in question and compute the number of moles of each element present in a sample of this size. If the compound contains 70.58% carbon, 100 g of sample will contain 70.58 g, or 5.88 mol.

$$C = \frac{70.58 \text{ g}}{12.011 \text{ g/mol}} = 5.88 \text{ mol}$$

Similarly, we can determine the number of moles of H and O:

$$H = \frac{5.92 \text{ g}}{1.008 \text{ g/mol}} = 5.87 \text{ mol} \qquad O = \frac{23.50 \text{ g}}{15.999 \text{ g/mol}} = 1.47 \text{ mol}$$

Therefore, a formula that expresses the relative molar proportions of carbon, hydrogen, and oxygen is $C_{5.88}H_{5.87}O_{1.47}$.

Now you need to convert this formula into one in which the elements are present in whole number ratios. Divide through by the element present in the smallest amount, in this case oxygen.

$$C = \frac{5.88}{1.47} = 4.00 \qquad H = \frac{5.87}{1.47} = 3.99 \qquad O = \frac{1.47}{1.47} = 1.00$$

This calculation yields an empirical formula of C_4H_4O. As C_4H_4O has a molecular weight of 68 g/mol, it can't be the molecular formula in this case because you know that the molecular weight is about 135 g/mol. However, simply multiplying by 2 gives the correct molecular formula of $C_8H_8O_2$, MW = 136 g/mol.

Problem 4.26 A C_5 saturated alkane (C_nH_{2n+2}) would have 12 hydrogens. So, in C_5H_8 there are (12 – 8) = 4/2 = two degrees of unsaturation. These compounds must have a total of two rings and/or π bonds. Here are just a few examples of possible compounds.

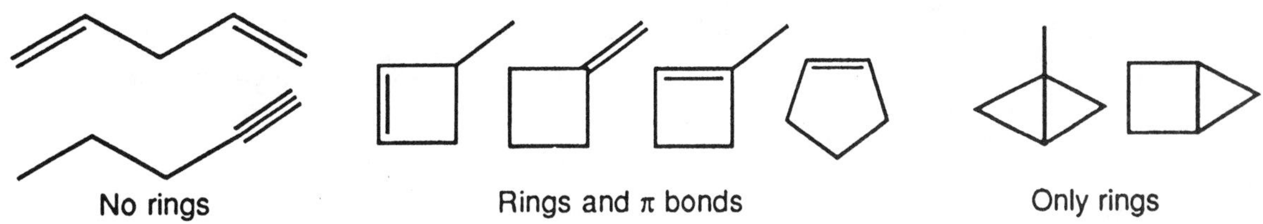

Problem 4.27

1-Heptene

trans-2-Heptene

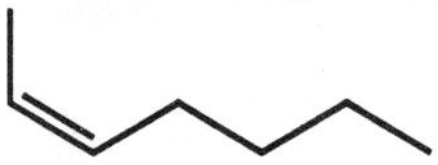

cis-2-Heptene

trans-3-Heptene

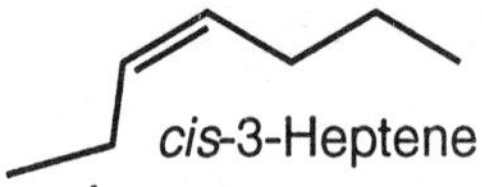

cis-3-Heptene

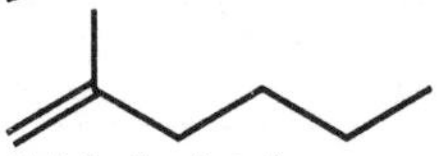

2-Methyl-1-hexene

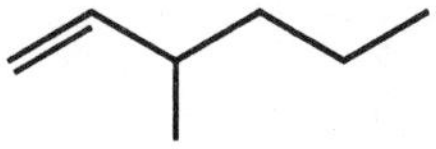

3-Methyl-1-hexene

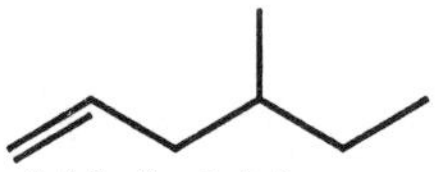

4-Methyl-1-hexene

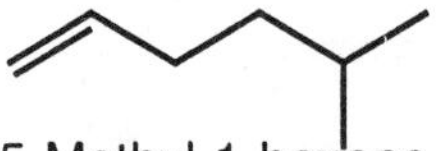

5-Methyl-1-hexene

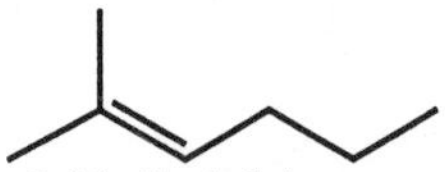

2-Methyl-2-hexene

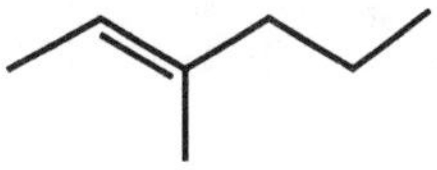

(*E*)-3-Methyl-2-hexene

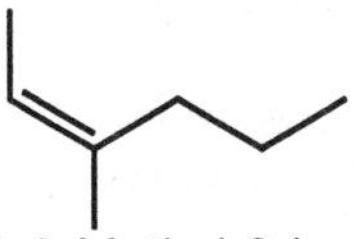

(*Z*)-3-Methyl-2-hexene

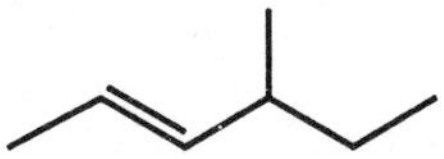

trans -4-Methyl-2-hexene

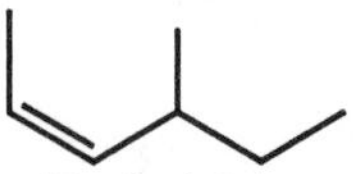

cis-4-Methyl-2-hexene

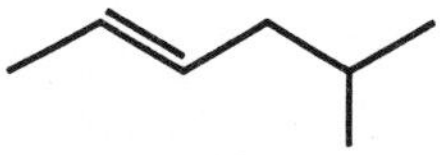

trans-5-Methyl-2-hexene

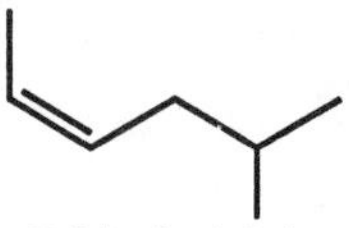

cis-5-Methyl-2-hexene

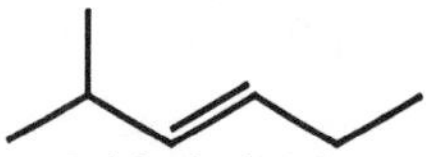

trans-2-Methyl-3-hexene

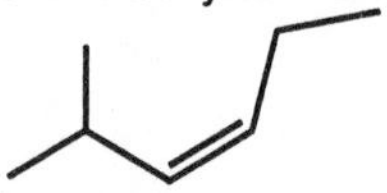

cis-2-Methyl-3-hexene

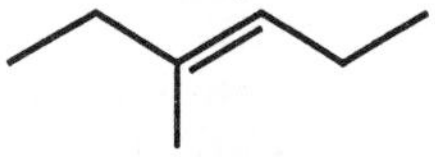

(*E*)-3-Methyl-3-hexene

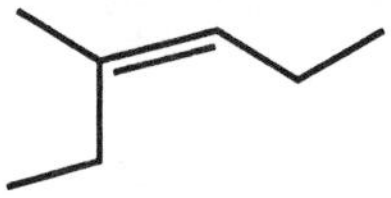

(*Z*)-3-Methyl-3-hexene

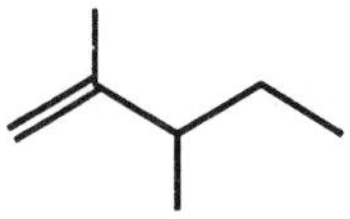

2,3-Dimethyl-1-pentene

2,4-Dimethyl-1-pentene

3,3-Dimethyl-1-pentene

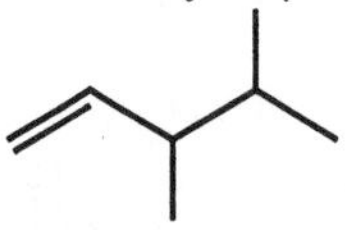

3,4-Dimethyl-1-pentene

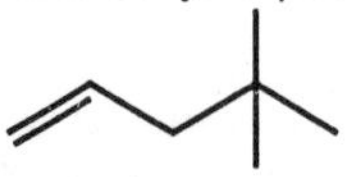

4,4-Dimethyl-1-pentene

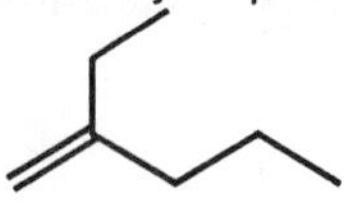

2-Ethyl-1-pentene

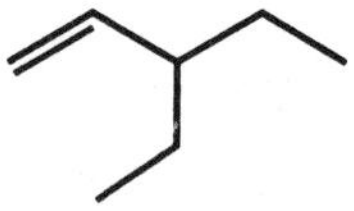

3-Ethyl-1-pentene

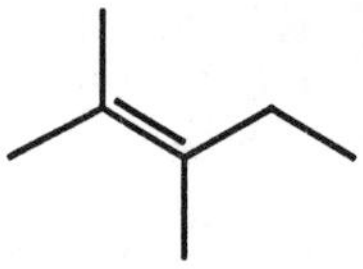

2,3-Dimethyl-2-pentene

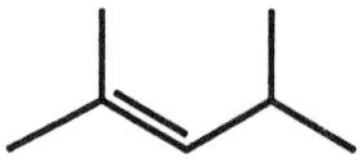

2,4-Dimethyl-2-pentene

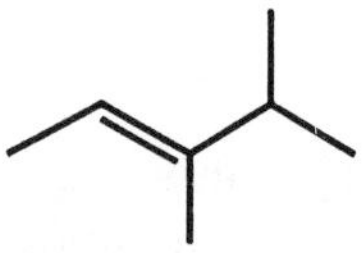

(*E*)-3,4-Dimethyl-2-pentene

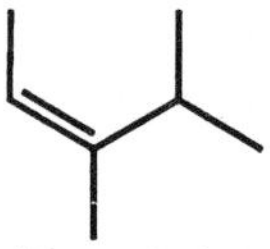

(*Z*)-3,4-Dimethyl-2-pentene

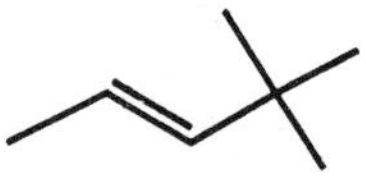

trans-4,4-Dimethyl-2-pentene

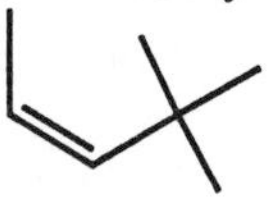

cis-4,4-Dimethyl-2-pentene

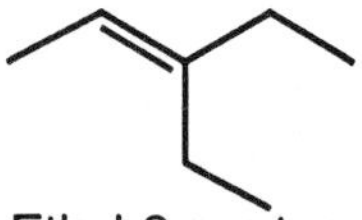

3-Ethyl-2-pentene

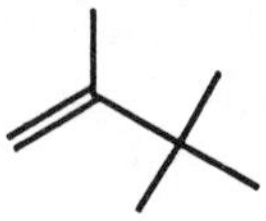

2,3,3-Trimethyl-1-butene

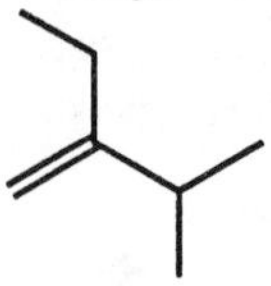

2-Ethyl-3-methyl-1-butene

Problem 4.28

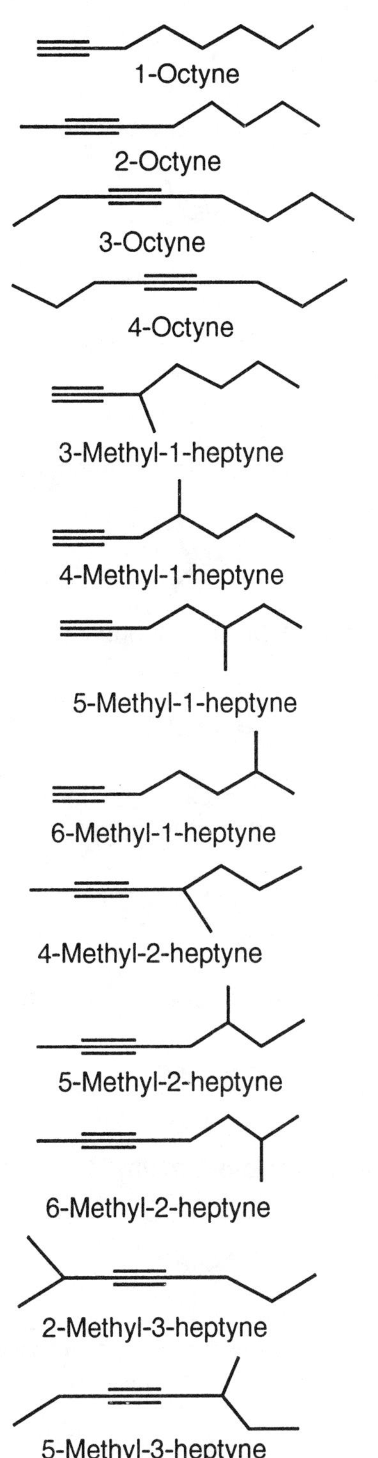
1-Octyne
2-Octyne
3-Octyne
4-Octyne
3-Methyl-1-heptyne
4-Methyl-1-heptyne
5-Methyl-1-heptyne
6-Methyl-1-heptyne
4-Methyl-2-heptyne
5-Methyl-2-heptyne
6-Methyl-2-heptyne
2-Methyl-3-heptyne
5-Methyl-3-heptyne

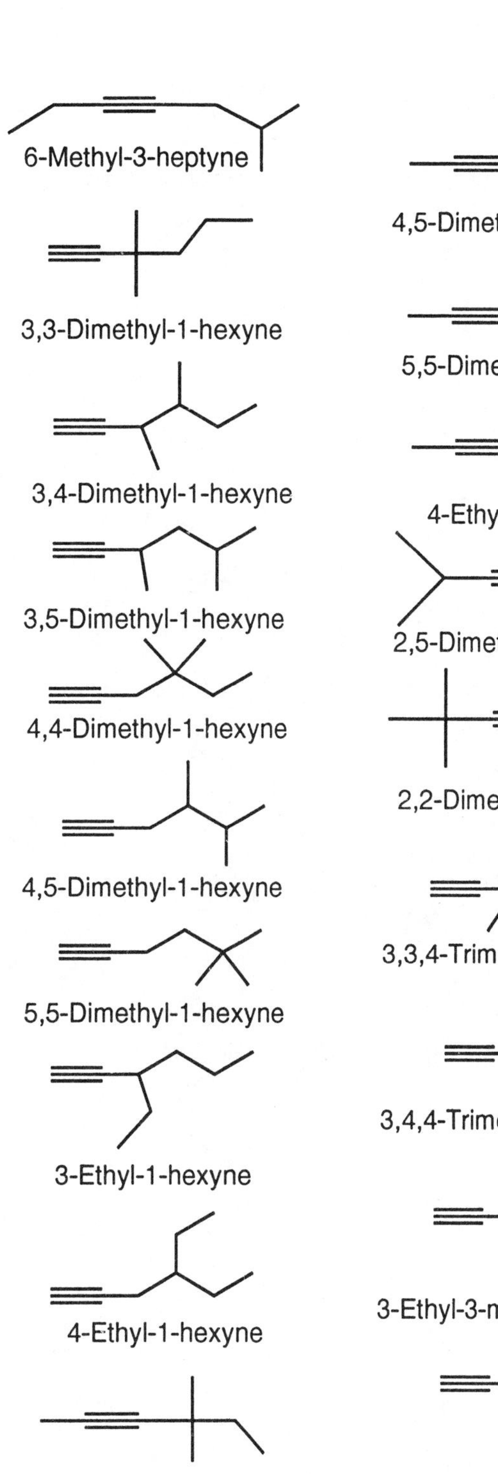
6-Methyl-3-heptyne
3,3-Dimethyl-1-hexyne
3,4-Dimethyl-1-hexyne
3,5-Dimethyl-1-hexyne
4,4-Dimethyl-1-hexyne
4,5-Dimethyl-1-hexyne
5,5-Dimethyl-1-hexyne
3-Ethyl-1-hexyne
4-Ethyl-1-hexyne
4,4-Dimethyl-2-hexyne

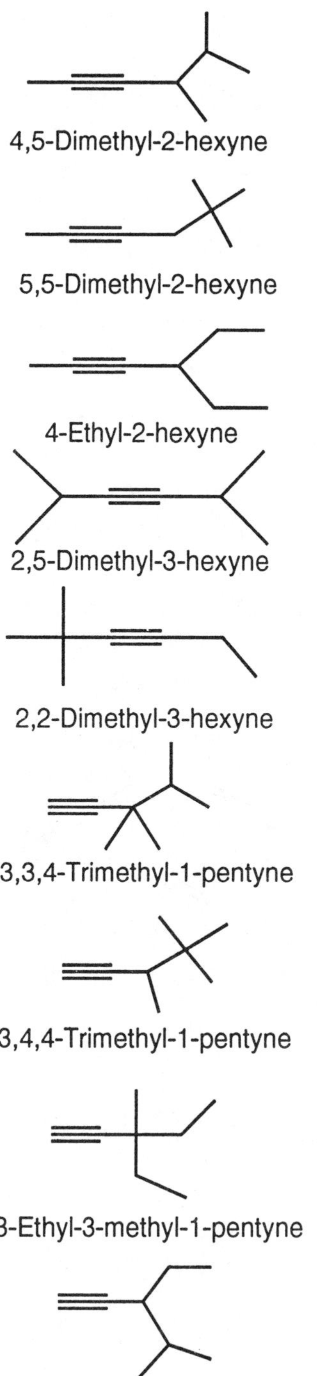
4,5-Dimethyl-2-hexyne
5,5-Dimethyl-2-hexyne
4-Ethyl-2-hexyne
2,5-Dimethyl-3-hexyne
2,2-Dimethyl-3-hexyne
3,3,4-Trimethyl-1-pentyne
3,4,4-Trimethyl-1-pentyne
3-Ethyl-3-methyl-1-pentyne
3-Ethyl-4-methyl-1-pentyne

Problem 4.29

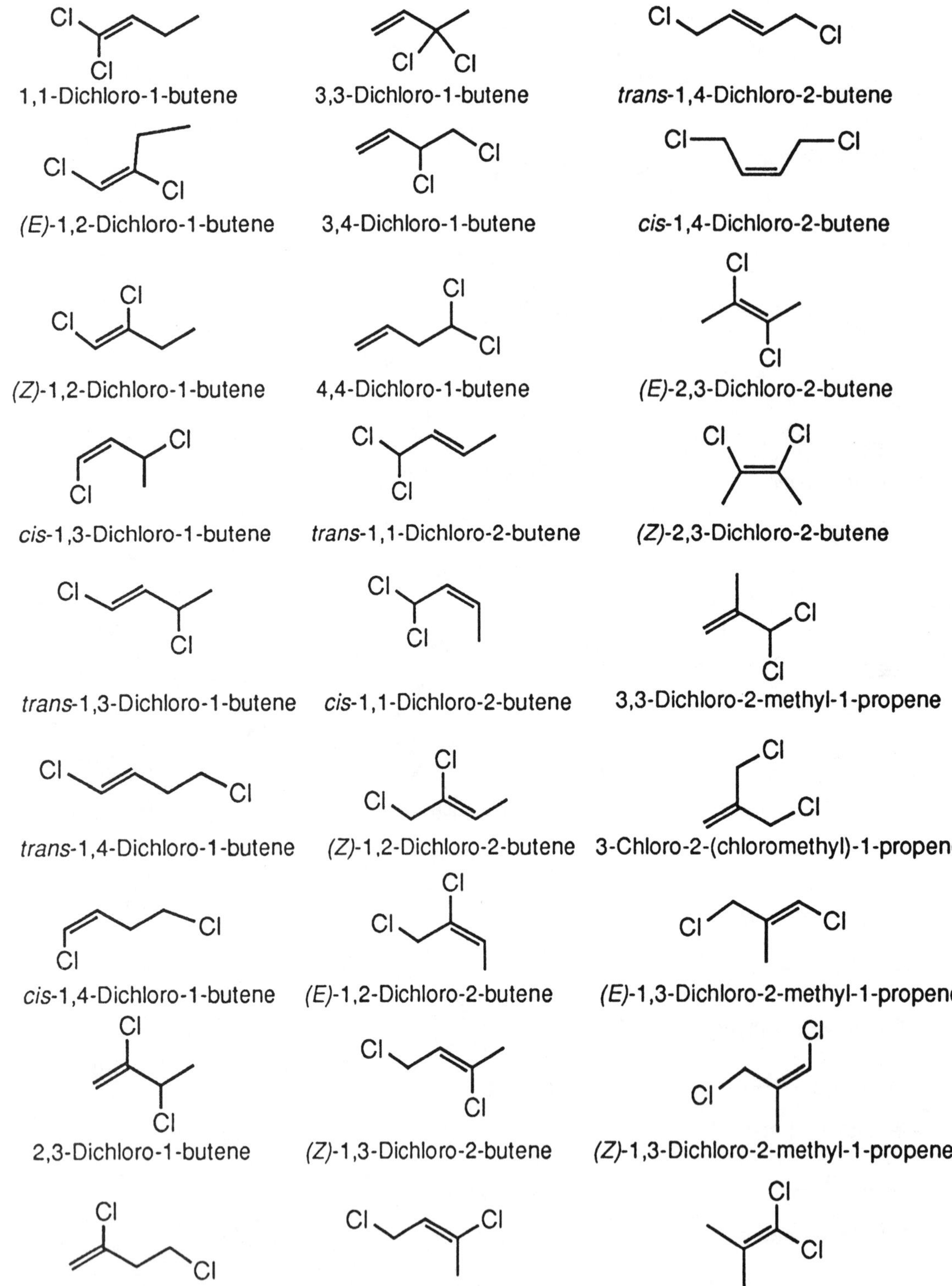

Problem 4.30 (a) *trans*-5-Iodo-2,7-dimethyl-3-nonene (the "di" is ignored, so iodo comes before dimethyl), (b) (*Z*)-2-chloro-3-ethyl-3-hexene, (c) 4-bromo-5-isopropyl-2-octyne, (d) *cis*-4,4-dimethyl-2-hepten-5-yne (-ene gets priority over -yne).

Problem 4.31

(a)

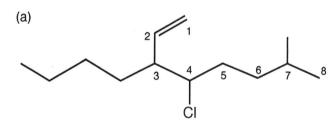

(b)

(c)

(d)

Problem 4.32

(a)

3-Butyl-4-chloro-7-methyl-1-octene

The name is based on the longest chain containing the double bond even though it is not the longest chain in the molecule

(b)

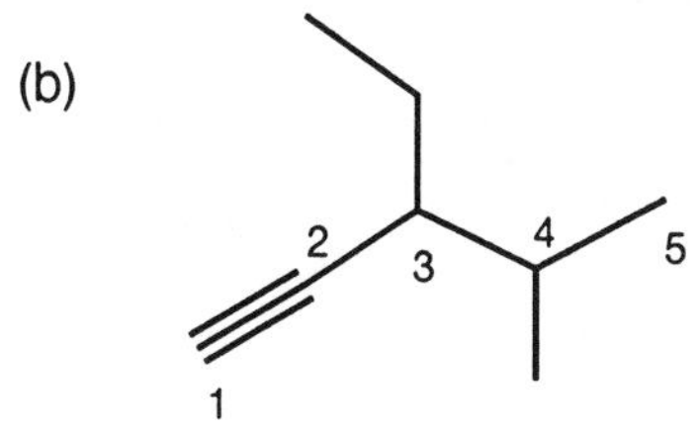

3-Ethyl-4-methyl-1-pentyne

When chains of equal length compete for selection as the main chain, the choice goes to the chain that has the greatest number of side chains

(c)

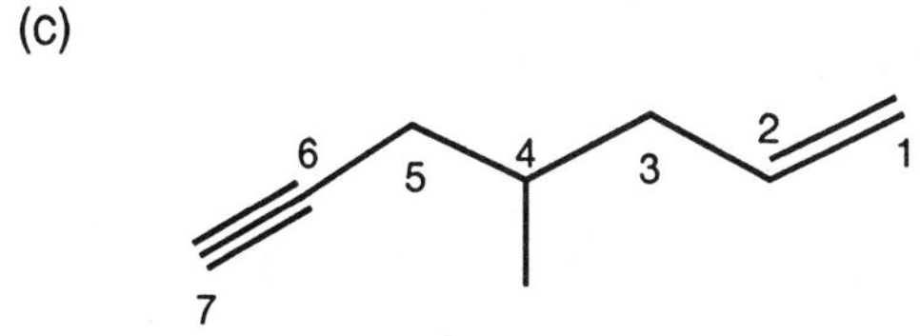

4-Methyl-1-hepten-6-yne

When the numbering scheme produces two names in which the lower number could go to either the -ene or the -yne, the -ene has priority

(d)

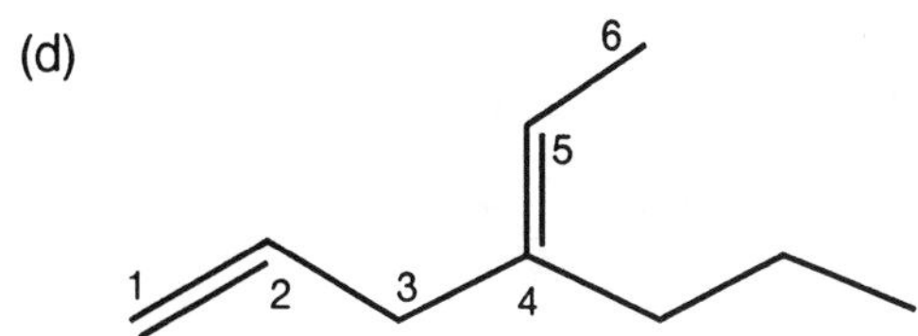

(E)-4-Propyl-1,4-hexadiene

In numbering the diene chain "1,4" is lower than "2,5"; note also that the allyl group has a higher priority than the propyl group. Hence the *(E)* designation

Problem 4.33 The first thing to do is to determine the number of degrees of unsaturation in $C_4H_6Br_2$. The related saturated alkane is C_4H_{10}. The bromines in the compound are univalent and can be treated as hydrogens for the purpose of counting. So, 10 – 8 = 2/2 = one degree of unsaturation. The compounds in question must contain only one ring or one π bond. Let's take this question step by step. There are three noncyclic chains to consider: propene, 1-butene, and 2-butene. Two bromines can be arranged in the following ways:

Propene-based molecules

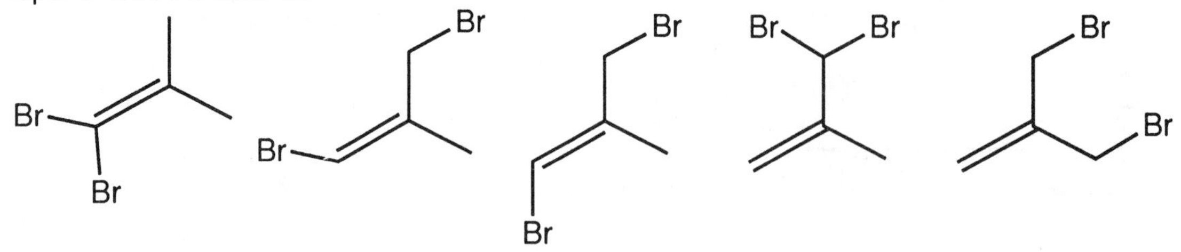

1-Butene-based molecules

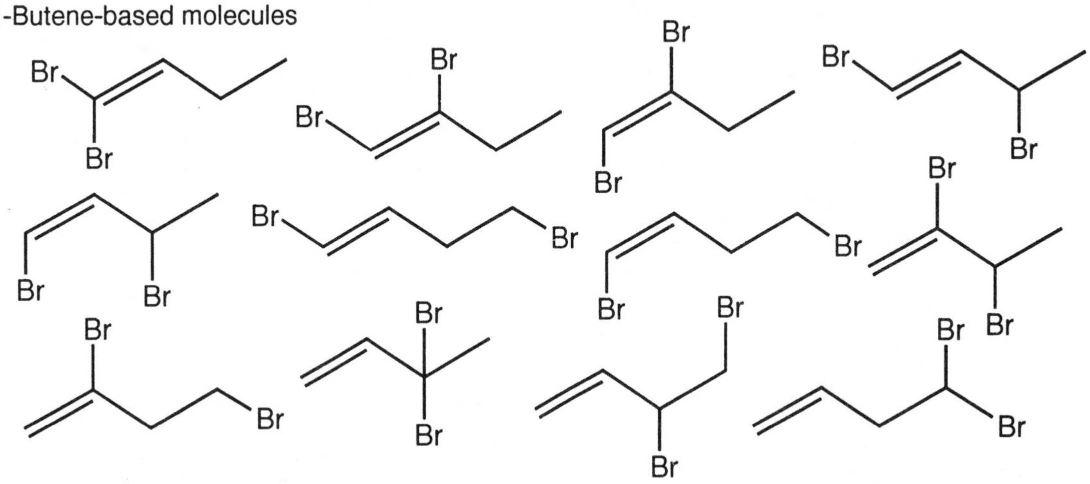

2-Butene-based molecules

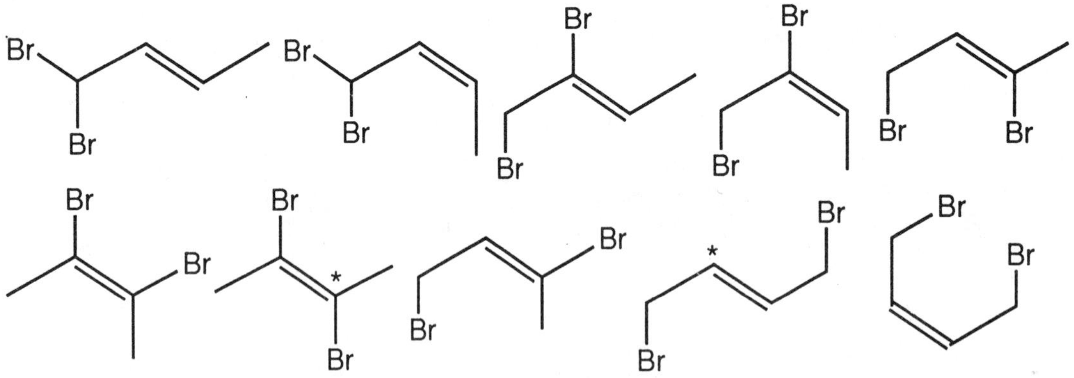

Only the molecules with the asterisk have zero dipole moments.

There are five possible molecules containing a four-membered ring, but only the indicated isomer has a zero dipole moment.

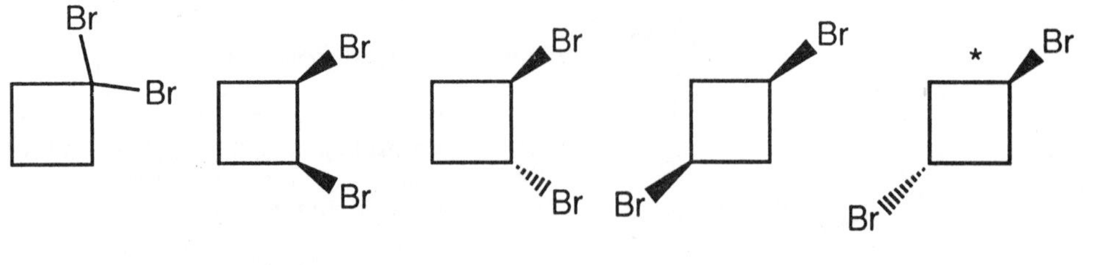

There are nine molecules containing a three-membered ring, but none has a zero dipole moment.

$CHBr_2$ CH_2Br Br CH_2Br Br

Br CH_2Br Br CH_3 Br CH_3 Br Br

Br Br CH_3 Br Br Br CH_3 Br Br CH_3

So we find 14 cyclic isomers. It has been claimed (*J. Chem. Ed.* **1992,** *69,* 452 - an article on finding isomers that is well worth a look) that there are 15. Who's wrong?

Problem 4.34 The saturated hydrocarbon is C_4H_{10}. The formula is (10 – 6) = 4/2 = 2. There are two degrees of unsaturation in this molecule, and so the possible combinations are, two rings, one ring and one π bond, and two π bonds. The possible molecules are given below, with an asterisk showing the ones with four different carbon atoms.

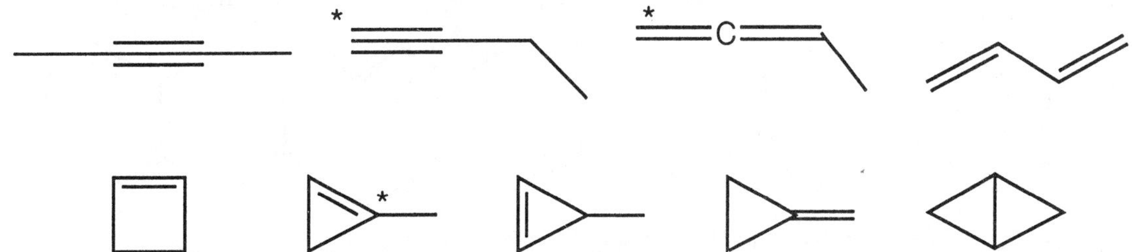

Problem 4.35 First, determine the degrees of unsaturation. The chlorine is univalent and can be treated as a hydrogen, so the related saturated alkane is C_4H_{10}, where (10 – 6) = 4/2 = two degrees of unsaturation. You must look for compounds containing two π bonds, one π bond and one ring, or two rings.

There are 10 isomers containing two π bonds:

Cl Cl Cl Cl C

Cl C Cl C Cl

Cl Cl Cl

There are also 10 compounds with one ring and one π bond:

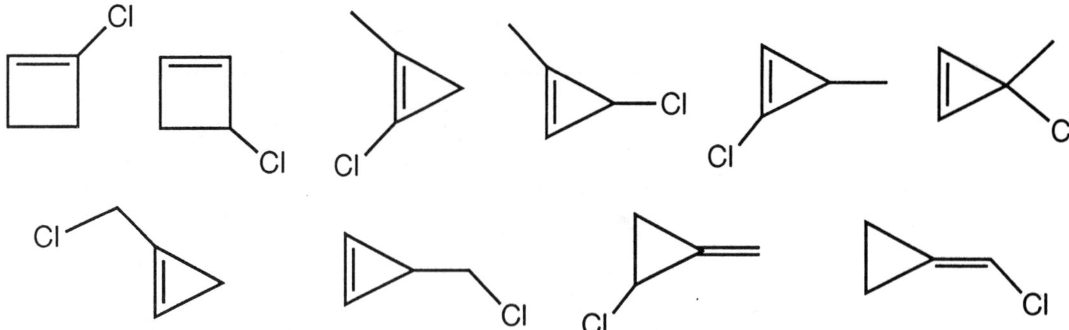

Finally, there are three structures containing two rings and no π bonds:

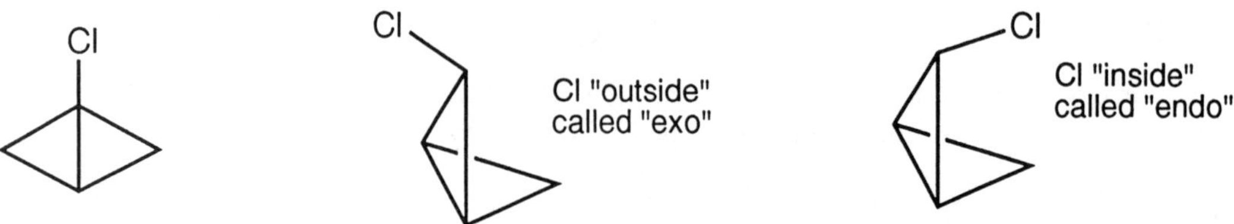

Problem 4.36 The saturated alkane would have 22 hydrogens ($C_{10}H_{22}$). The formula shows that there are three degrees of unsaturation in this compound (22 – 16) = 6/2 = 3. If there must be only two rings, and if the rings must be six-membered, there must be one π bond. The possibilities are:

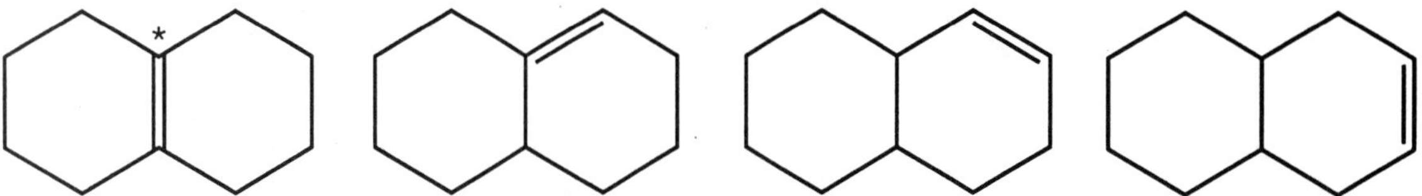

The asterisk shows the one with only three kinds of carbon.

Problem 4.37. An sp^2 hybridized carbon atom is shown below. In the figure two sp^2 hybrids are shown schematically, one coming towards you (solid wedge), the other retreating (dotted wedge). In the first figure the four valence electrons of carbon are shown as dots, one in each of the three sp^2 hybrids, and one in the unhybridized $2p_z$ orbital. The C-H bonds are shown as overlapping 1*s* and sp^2 orbitals and then as schematic "line" bonds.

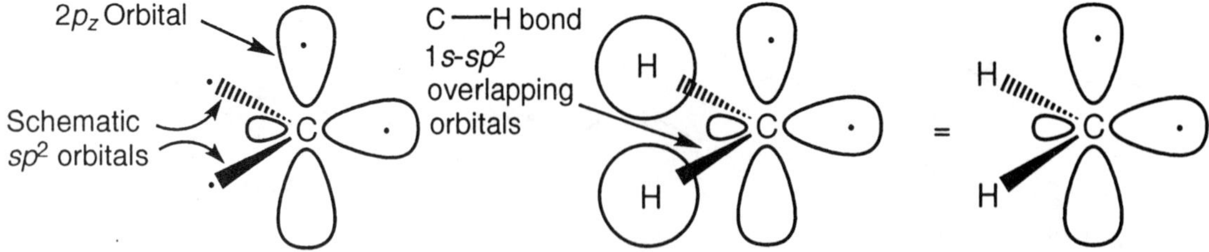

An oxygen atom has six valence electrons, and so two of the sp^2 hybrids must be doubly occupied.

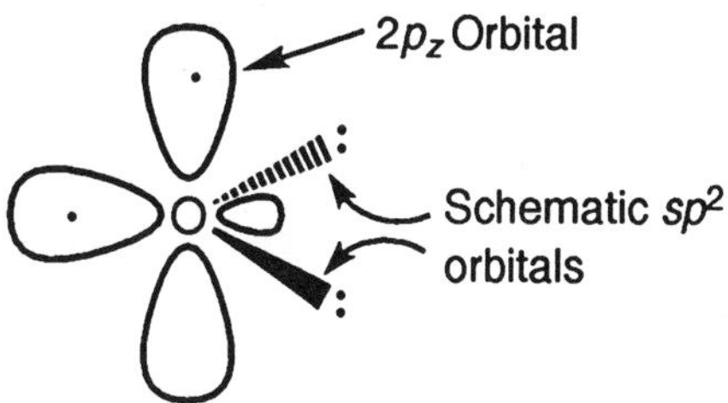

In our scheme we will form one σ bond through sp^2-sp^2 overlap and a π bond through $2p_z$-$2p_z$ overlap. Don't forget that these overlapping orbitals produce both a bonding combination (σ and π) and an antibonding combination, (σ^* and π^*). The result is a double bond, σ and π, shown as a pair of identical lines between C and O.

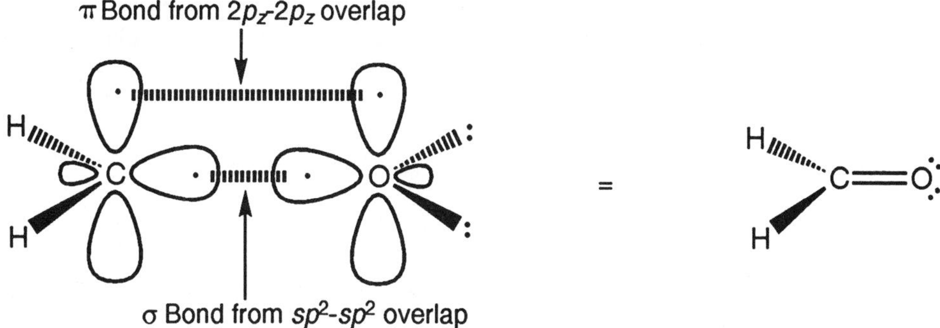

Problem 4.38

(a) We can use the sp^2 hybrid orbitals to bond to the hydrogen and the two neighboring carbons. The figure first shows the planar six-membered ring with the carbons and hydrogens and then focusses on one sp^2 hybridized carbon and the bonds to it.

Each carbon in the ring is identical, so let's now look at the six singly occupied $2p_z$ orbitals. In the left-hand figure the lower lobes of the $2p_z$ orbitals on the back two carbons are not shown, and the C-H bonds are not drawn in. In the right-hand figure we draw in three π bonds made from the usual $2p_z$-$2p_z$ overlap.

(b) Clearly, if there is only one C-C bond distance in this molecule there cannot be simple, alternating single and double bonds, which must be different lengths. However, we made an arbitrary decision in the figure above to let certain $2p_z$ orbitals overlap and ignore others. Why not do it another way?

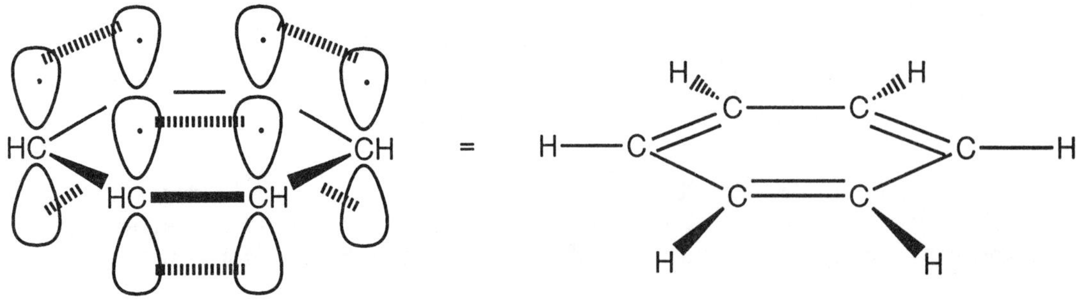

In fact, all the $2p_z$ orbitals are the same and must overlap equally with their right-hand and left-hand neighboring $2p_z$ orbitals. At this point we can't do too much better than this - just note that there is excellent overlap among these $2p_z$ orbitals above and below the plane of the ring. We shall examine this molecule in detail in Chapter 13.

Problem 4.39. The two carbocations in question would be:

(a) 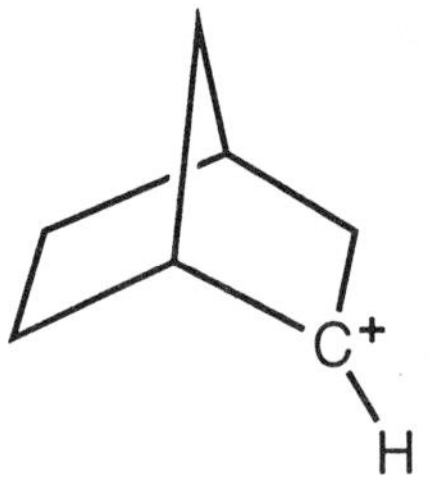 (b)

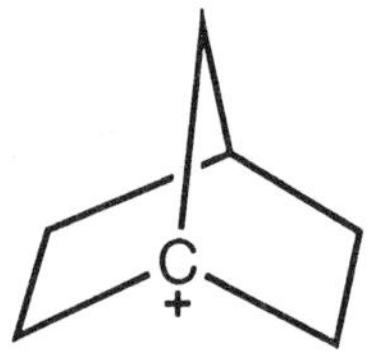

There is no problem accommodating a planar carbocation in (a):

(a) 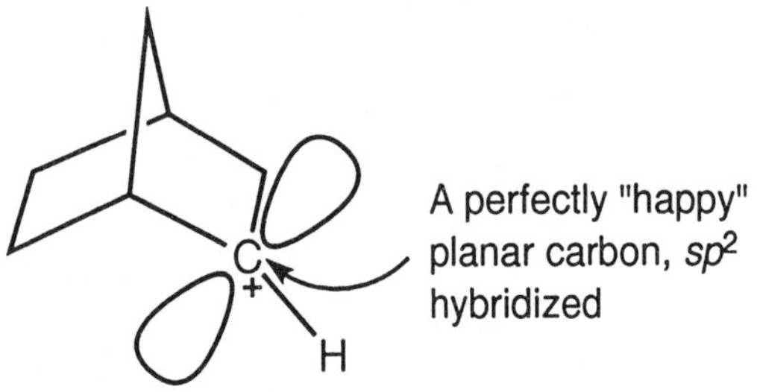 A perfectly "happy" planar carbon, sp^2 hybridized

However, the drawing of (b) is harder to see (a model should make everything crystal clear). There is no way to flatten this molecule out at this "bridgehead" position. As carbocations are most stable when planar, this introduces a necessary instability into this cation, and makes it difficult to form.

(b) 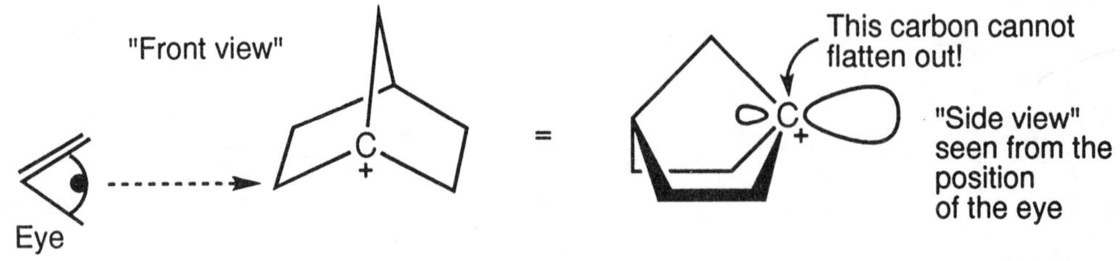 "Front view" Eye

This carbon cannot flatten out!

"Side view" seen from the position of the eye

Chapter 5 Outline

5.1 Chirality
5.2 Properties of Enantiomers: Physical Differences
5.3 The (*R*/*S*) Convention
5.4 The Physical Basis of Optical Activity
5.5 Properties of Enantiomers: Chemical Differences
5.6 Interconversion of Enantiomers by Mobile Equilibria: Gauche Butane
5.7 Diastereomers: Molecules Containing More Than One Stereogenic Atom
5.8 Physical Properties of Diastereomers: Optical Resolution
5.9 Determination of Absolute Configuration (*R* or *S*)
5.10 Stereochemical Analysis of Ring Compounds (A Beginning)
5.11 Review of Isomerism
5.12 Something More: Chirality without "Four Different Groups Attached to One Carbon"

5 Stereochemistry

In this chapter, we deal with the details of molecular structure. There can be no hiding the fact that stereochemistry is difficult for many people, but it will yield to careful work and practice. Moreover, at this point we must reemphasize the idea that you cannot simply read this material and hope to get it. That approach is hopeless - repeat, ***hopeless***. In this chapter we see the consequences of the three-dimensional nature of molecules, and from now on you will have to be able to visualize molecules in three dimensions. Do not trust the two dimensions available to this book, or the blackboard, or a piece of paper.

The notion that organic chemistry is to be "read with a pencil," has been emphasized many times and nowhere is that admonition more appropriate than here. Indeed, you should work through these problems with your models at the ready; there will be many points when they will be helpful or even essential.

Why is stereochemistry so important? Soon we are going to look at reactions in detail, and try to work out how and why they proceed in the ways they do. As you will see, stereochemistry plays a huge role in this analysis. Working out the mechanistic details requires an ability to keep the stereochemical nuances of structure in mind. In turn, when we come to organic synthesis, the construction of molecules from simpler starting materials, we will always have to keep stereochemistry in mind. It is not enough to devise a synthesis that builds a molecule with its constituent parts more or less in the right place, they must be exactly right; perfectly positioned. Stereochemical control is vital, and to achieve that you must understand the subject itself well.

Practice, practice, practice! The following problems, none of them especially difficult, give you a chance to start. Later, more complicated stereochemical analyses will accompany our discussions of many of the mechanisms of organic reactions.

Problem 5.1 There is nothing to be done in problems such as these but to draw out the mirror images and see if they are superimposable. If they are, the molecule in question is achiral; if not, it is chiral.

Both of these molecules are chiral. The mirror images are not superimposable on the originals.

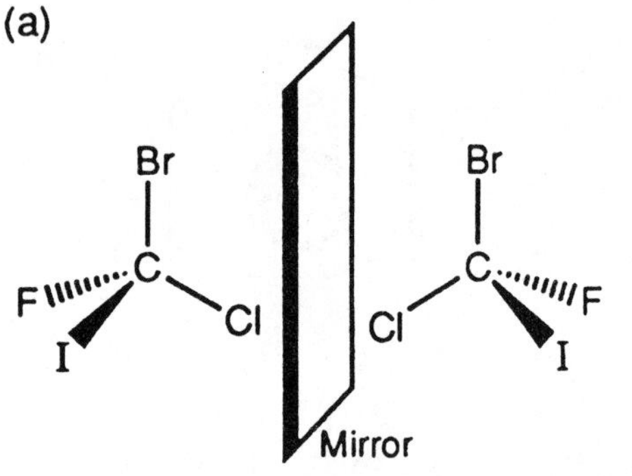

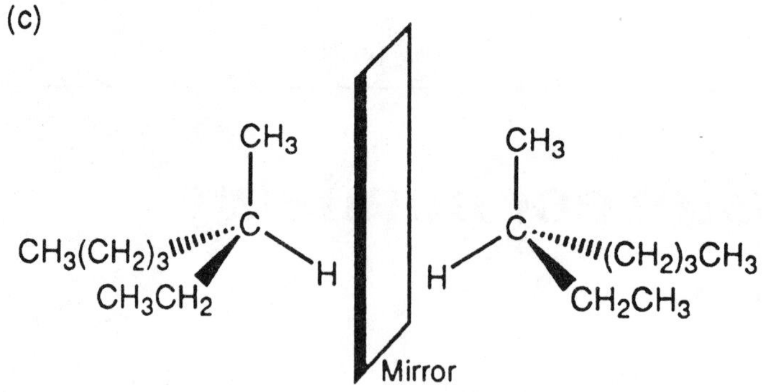

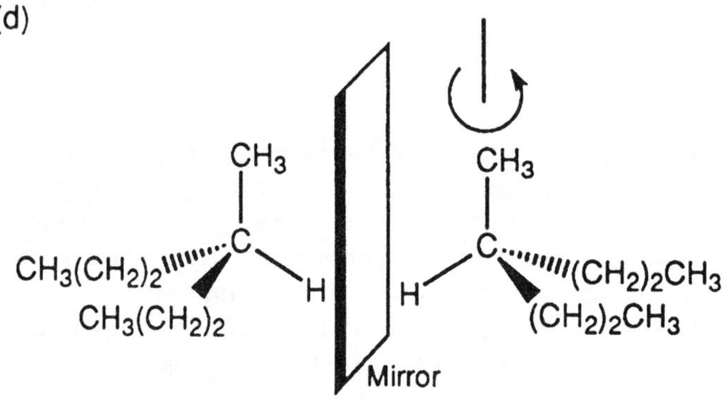

Both of these molecules are achiral. To superimpose the mirror images of the right-hand molecule (d) simply rotate 180° around the carbon-methyl bond. For the left-hand molecule (b) rotate in the plane of the paper as shown.

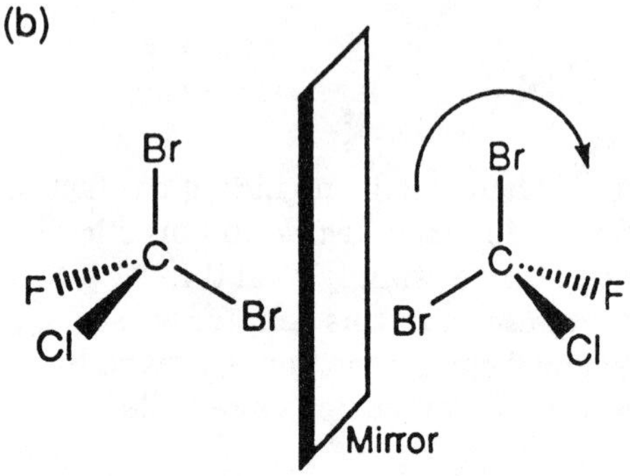

Problem 5.2

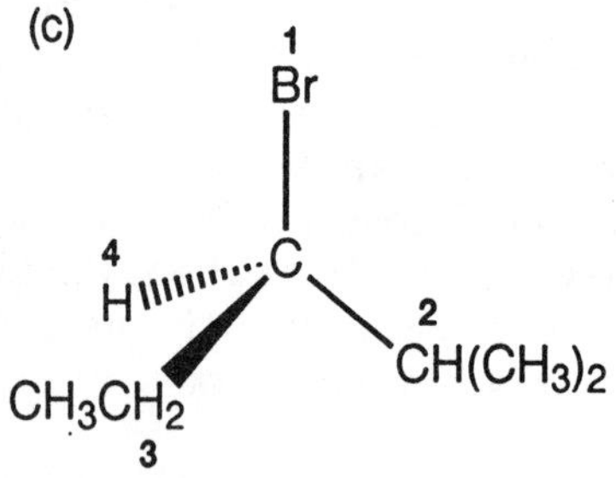

(a) H is **4** and O is **1** on the basis of atomic number. The tie between CH_3 and CD_3 is broken by the greater atomic weight of D over H.

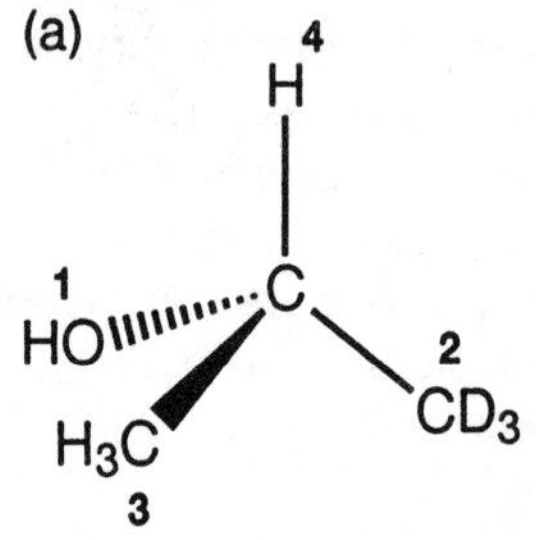

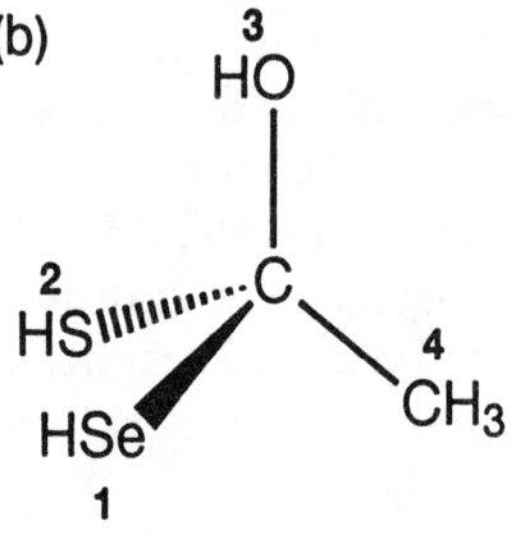

(b) Priorities can be assigned strictly by atomic number.

(c) Atomic number makes H priority **4** and Br priority **1**. The tie between the two carbons is broken by working out along the chain. The ethyl carbon is attached to C,H,H (**3**) and the isopropyl carbon to C,C,H (**2**).

(d)

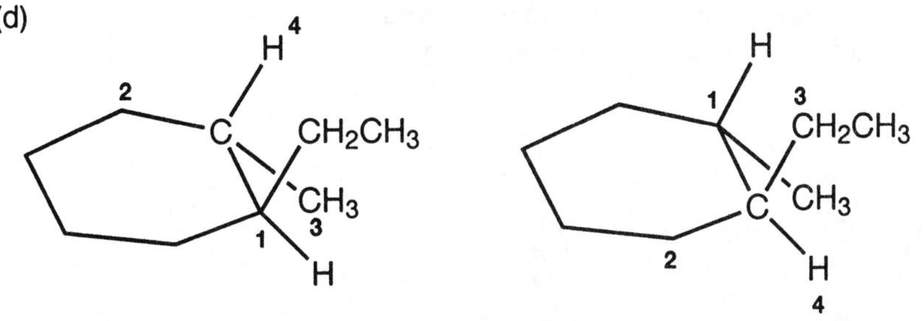

H 4 2 C CH2CH3 CH3 3 1 H

H 1 3 CH2CH3 C CH3 2 H 4

(d) There are two stereogenic carbons in this molecule. For the rear carbon H is priority **4**, but the other three substituents are carbon. The methyl carbon (attached to H,H,H) is priority **3**, the ring CH_2 group priority **2** (attached to C,H,H) and the other ring carbon **1** (attached to C,C,H). A similar tie-breaking procedure serves to assign priorities to the front carbon.

Problem 5.3 This problem may seem easy, even mindless. But watch out! First of all, drawing mirror images requires practice, and ring compounds introduce strange difficulties for many people. It is well worth your time to practice drawing mirror images a bit. Invent some more problems for yourself.

(a)

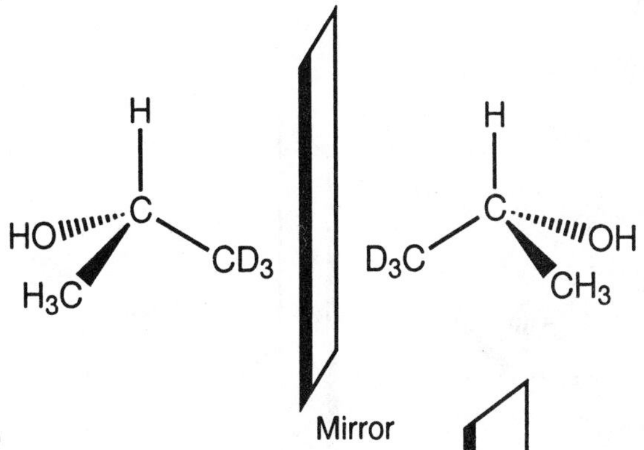

H HO C CD3 H3C | Mirror | H D3C C OH CH3

(b)

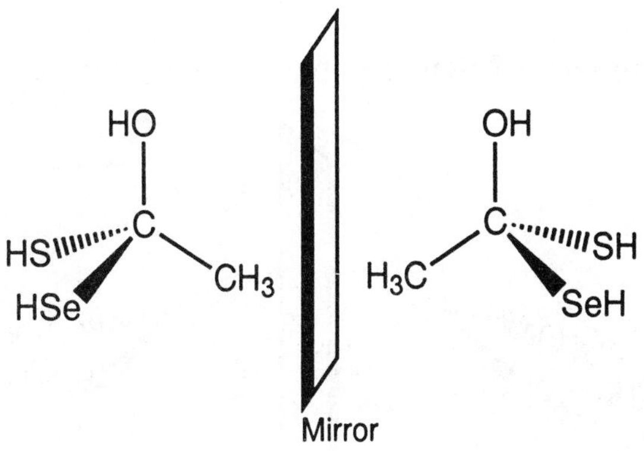

HO HS C CH3 HSe | Mirror | OH H3C C SH SeH

(c)

Br H C CH(CH3)2 CH3CH2 | Br (CH3)2CH C H CH2CH3

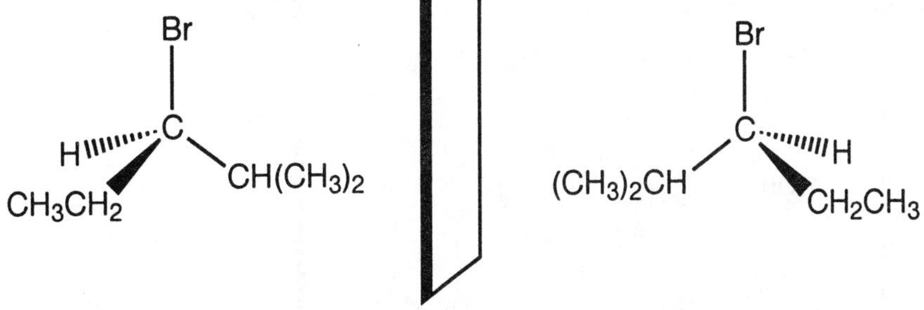

(d)

H CH2CH3 CH3 H | H CH3CH2 H3C H

Mirrors

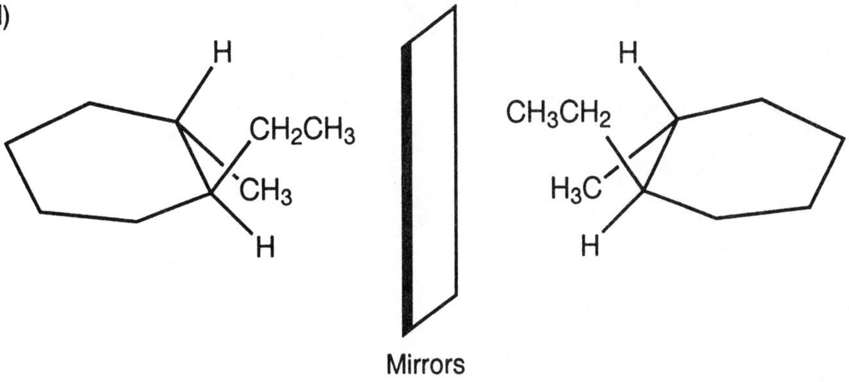

Problem 5.4 Remember, look down the C—priority **4** bond (arrow), and connect the groups **1**→**2**→**3**. The priorities were assigned in the usual way, through the Cahn-Ingold-Prelog protocol.

(a) 4 H; 1 HO; (S) C; 2 CD_3; H_3C 3

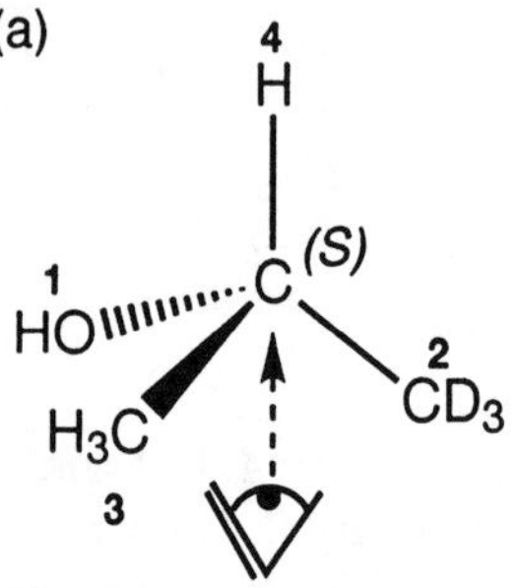

(b) 3 HO; 2 HS; (R) C; HSe 1; 4 CH_3

(c) 1 Br; 4 H; C (R); 2 $CH(CH_3)_2$; CH_3CH_2 3

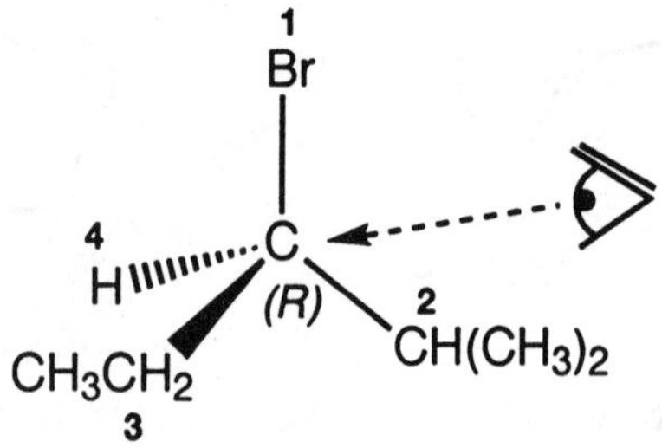

(d) H; 1; 3 CH_2CH_3; C; CH_3; (S); 2; H 4

4 H; 2; (S) C; CH_2CH_3; CH_3 3; 1; H

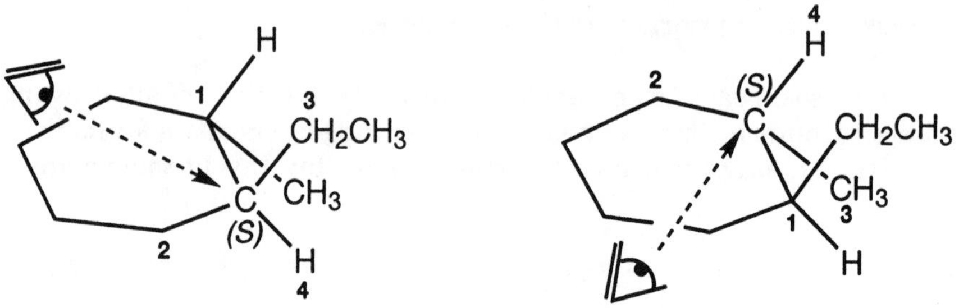

Problem 5.5 Here is more practice in drawing mirror images.

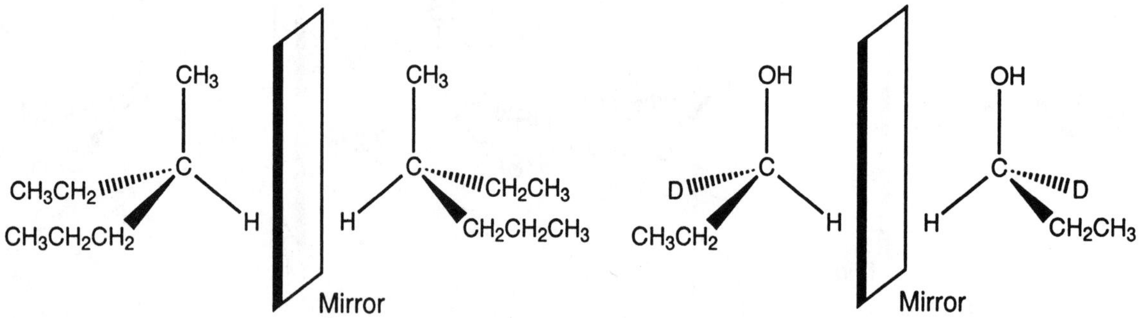

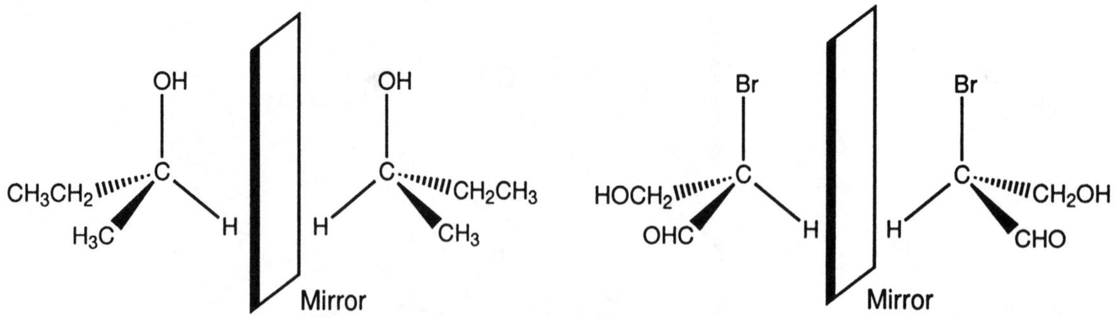

CH_3; CH_3CH_2; C; $CH_3CH_2CH_2$; H | Mirror | CH_3; C; CH_2CH_3; H; $CH_2CH_2CH_3$

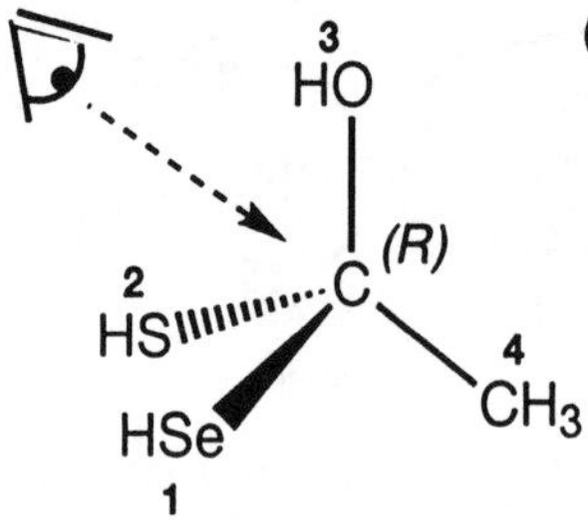

OH; D; C; CH_3CH_2; H | Mirror | OH; C; D; H; CH_2CH_3

OH; CH_3CH_2; C; H_3C; H | Mirror | OH; C; CH_2CH_3; H; CH_3

Br; $HOCH_2$; C; OHC; H | Mirror | Br; C; CH_2OH; H; CHO

Problem 5.6 The net rotation will be zero, as the two enantiomers will rotate the plane of plane-polarized light by equal amounts in opposite directions.

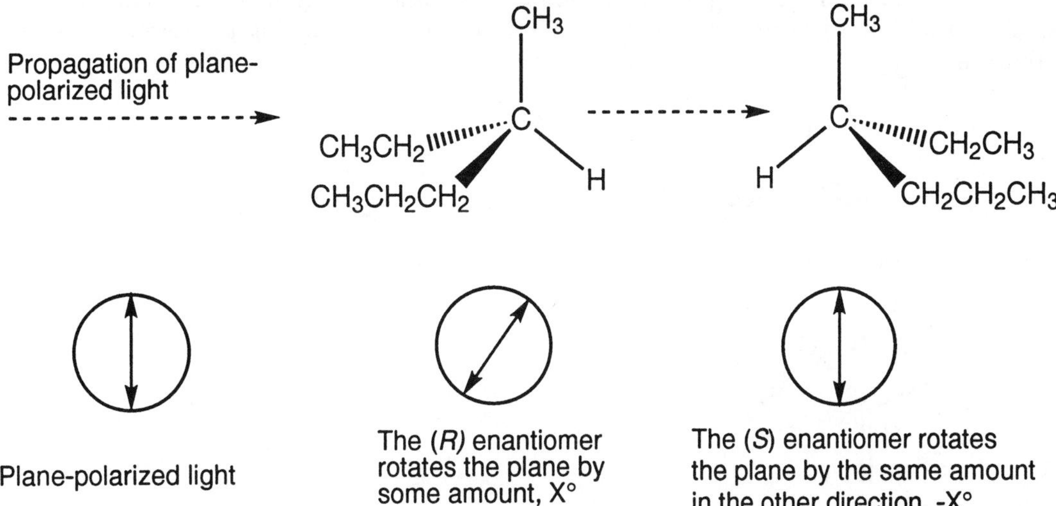
Propagation of plane-polarized light
CH_3
CH_3CH_2
C
$CH_3CH_2CH_2$
H
CH_3
C
CH_2CH_3
H
$CH_2CH_2CH_3$
Plane-polarized light
The (*R*) enantiomer rotates the plane by some amount, X°
The (*S*) enantiomer rotates the plane by the same amount in the other direction, -X°

Problem 5.8 A real shell will work best for both Problem 5.8 and 5.9, but here is a picture of a shell and its mirror image. They are not superimposable.

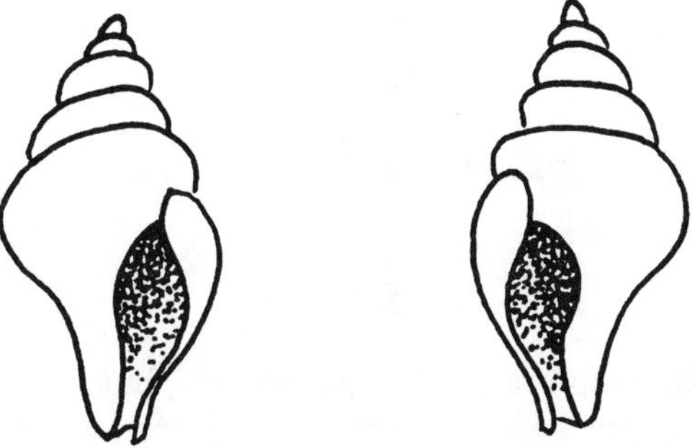

Problem 5.9 Here is the interaction of an inverted shell and a pair of hands. The interactions with the entiomeric pair of hands are still different. Turning the shell makes no difference whatsoever.

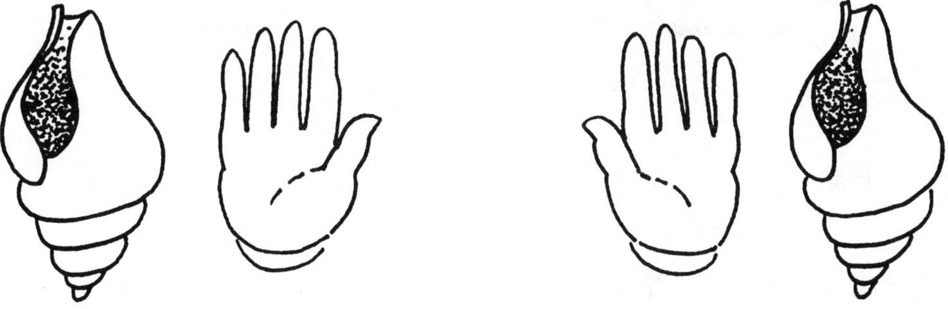

Problem 5.11 This is a nuts and bolts question, but, as we have said many times, cyclic systems are sometimes inexplicably difficult. So here is a chance to practice dealing with rings and priority assignment. The priorities are assigned as shown. H is (**4**), and the doubly bonded carbon is (**1**). The C-C=O gets a higher priority (**2**) than the C-C=C (**3**) because of the higher atomic number of oxygen than carbon.

(R)-(–)-Carvone *(S)*-(+)-Carvone

Problem 5.12 Ha! The mirror image of "*R*—***S***" is "***R***—*S*." Most people say, too quickly, "*S*—*R*." The boldface "*S*" is reflected as "***R***".

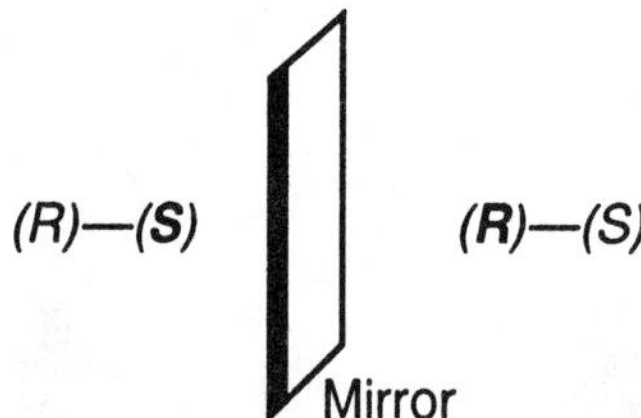

Problem 5.14 Remember, diastereomers are "stereoisomers, but not mirror images." cis/trans Isomers in ring compounds or alkenes are diastereomers. Here are two typical pairs of diastereomers:

Problem 5.15 Compounds **A** and **B** are not mirror images, so they are not enantiomers. The mirror image of **A** is **A′** not **B**. **A** and **B** (and **A′** and **B**) are stereoisomers, however, so they must be diastereomers.

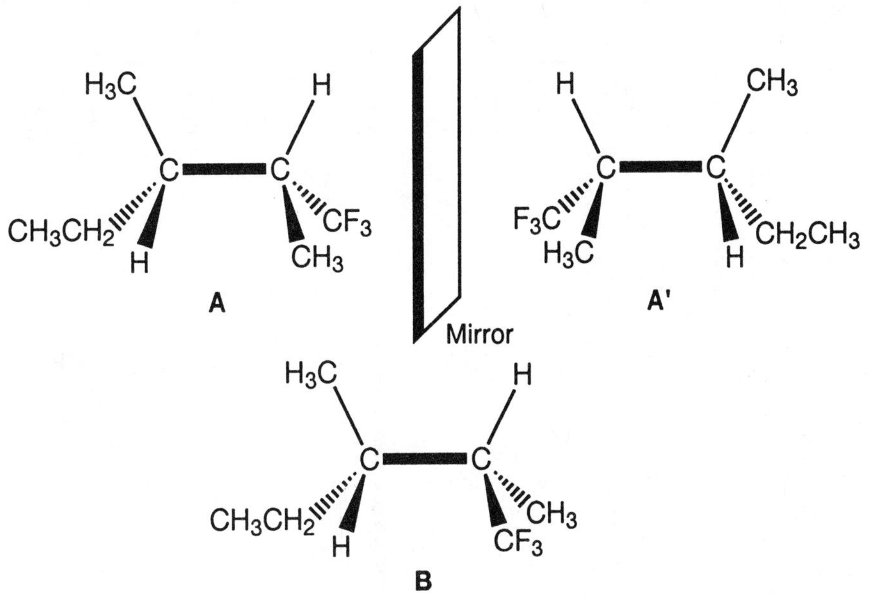

H_3C H H CH_3
C—C C—C
CH_3CH_2 H CF_3 CH_3 F_3C H_3C H CH_2CH_3
A Mirror A'

H_3C H
C—C
CH_3CH_2 H CH_3 CF_3
B

Problem 5.16 There is no problem here. The designations "*S*" and "*R*" come from the *individual* priority assignments for the two stereogenic carbons in **A**. In the hypothetical bond-forming process we have developed, the right-hand carbon comes from (*S*)-2-chloro-1,1,1-trifluoropropane in which it is the Cl that is priority **1**. In the new compound, **A**, this chlorine is no longer there. Now the priority **1** is the trifluoromethyl group, and a proper priority count shows that the right-hand carbon is "*R*."

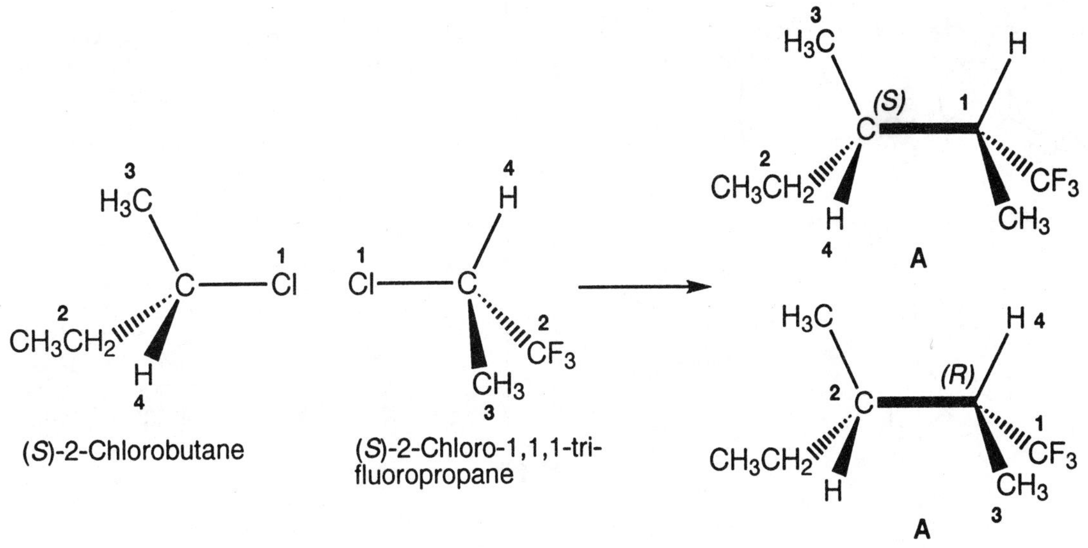

(*S*)-2-Chlorobutane

(*S*)-2-Chloro-1,1,1-trifluoropropane

A (*S*)

A (*R*)

Problem 5.17, 5.19 As the hint tells us, it is best to draw these molecules as eclipsed forms. Of course, they really exist predominately in their energy minimum staggered conformations, but the analysis is easier if they are drawn eclipsed. Two of these compounds, 2,2,3,3-tetrabromobutane (a) and 2,2-dibromo-3,3-dichlorobutane (b), are achiral.

(a)

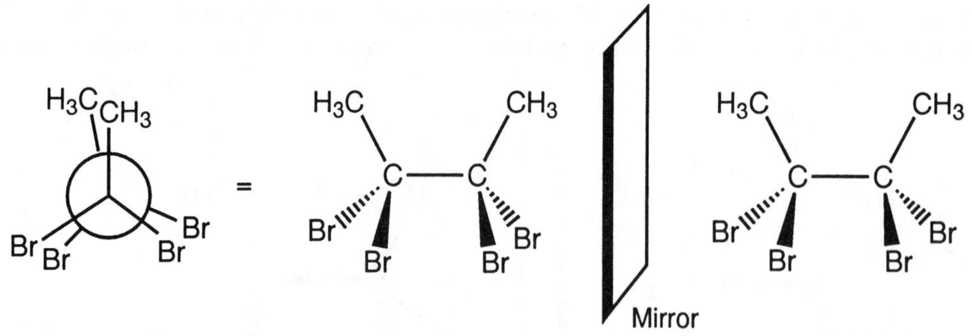

H3C CH3 / Br Br Br Br = H3C CH3 C—C Br Br Br Br | Mirror | H3C CH3 C—C Br Br Br Br

To superimpose the two isomers, simply slide one figure over.

(b)

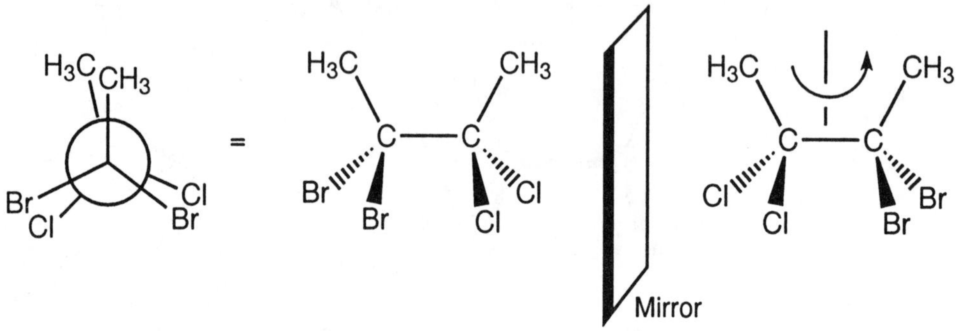

H3C CH3 / Br Cl Cl Br = H3C CH3 C—C Br Br Cl Cl | Mirror | H3C CH3 C—C Cl Cl Br Br

To superimpose the two isomers rotate as shown, and slide one figure over.

One of these compounds, 2,3-dibromo-2,3-dichlorobutane (c), has three stereoisomers - a pair of enantiomers and a meso compound. Priorities are assigned in the usual way.

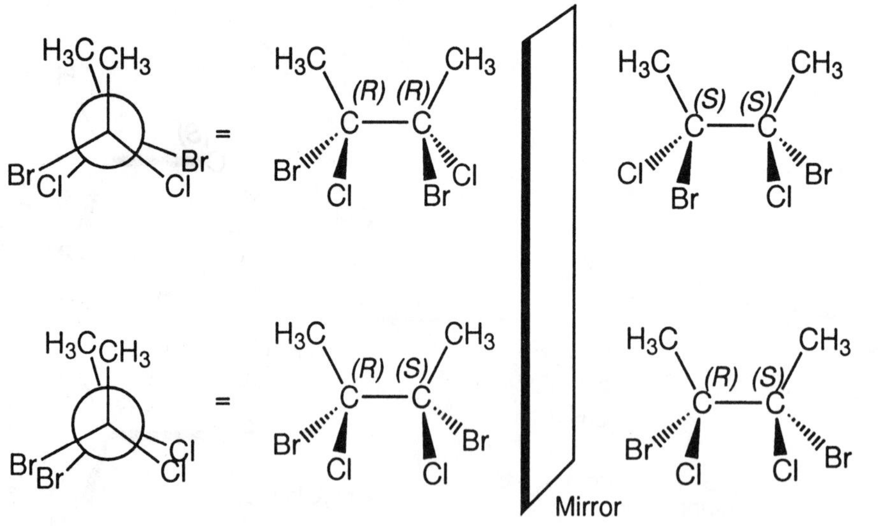

H3C CH3 / Br Cl Br Cl = H3C CH3 (R) (R) C—C Br Cl Cl Br | H3C CH3 (S) (S) C—C Cl Br Br Cl

These isomers are enantiomers; the mirror image is not superimposable on the original

H3C CH3 / Br Br Cl Cl = H3C CH3 (R) (S) C—C Br Cl Br Cl | H3C CH3 (R) (S) C—C Br Cl Cl Br

This is a meso compound as the mirror image is superimposable on the original

Mirror

Problem 5.21 The (*R*) and (*S*) designations are assigned from the following priorities:

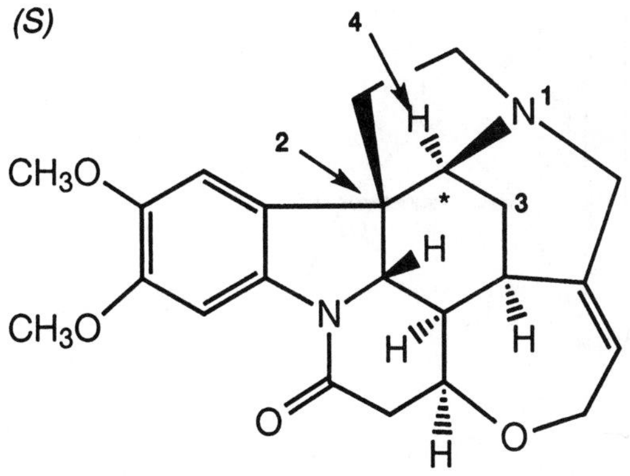

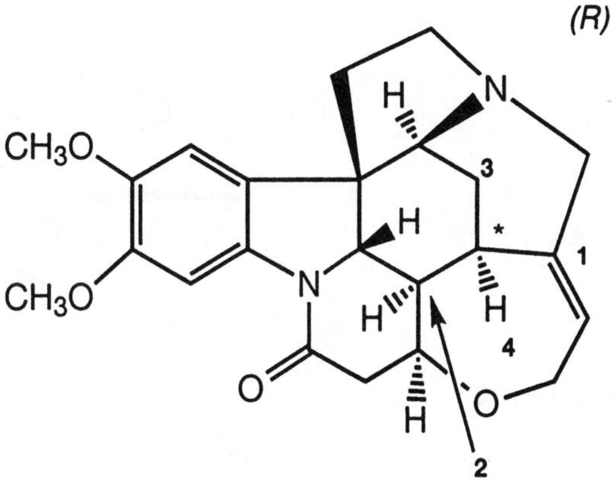

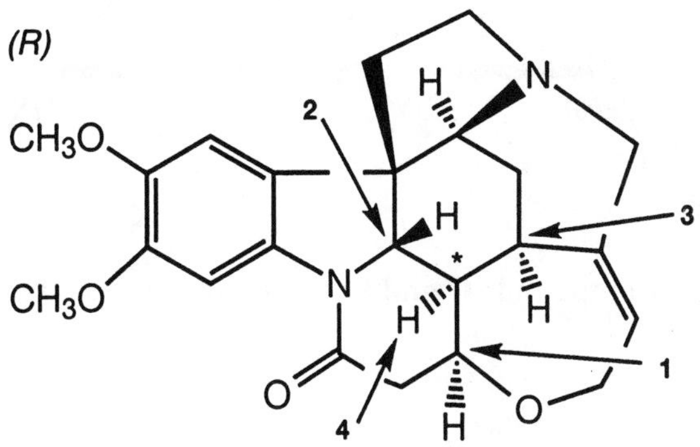

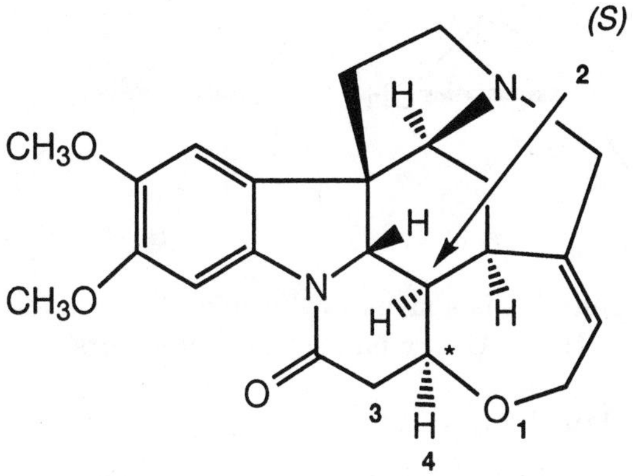

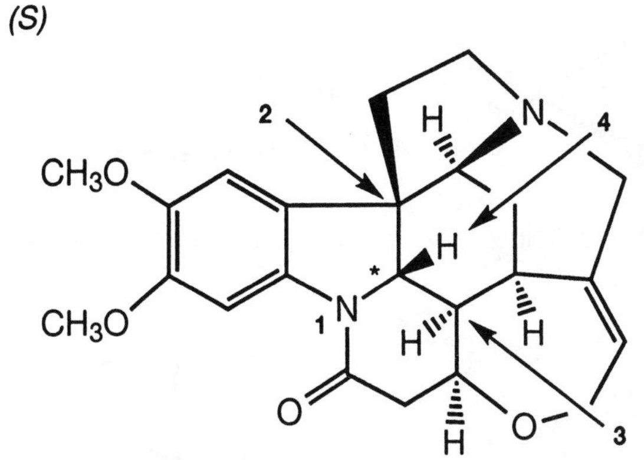

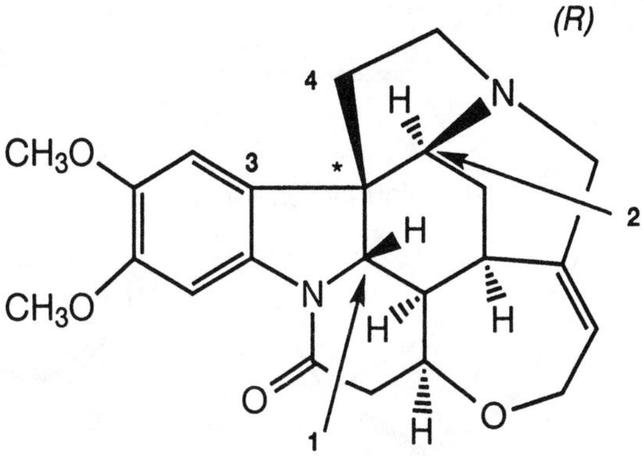

(S) (R)

(R) (S)

(S) (R)

Problem 5.22 Priorities are assigned to the left- and right-hand carbons in the following way:

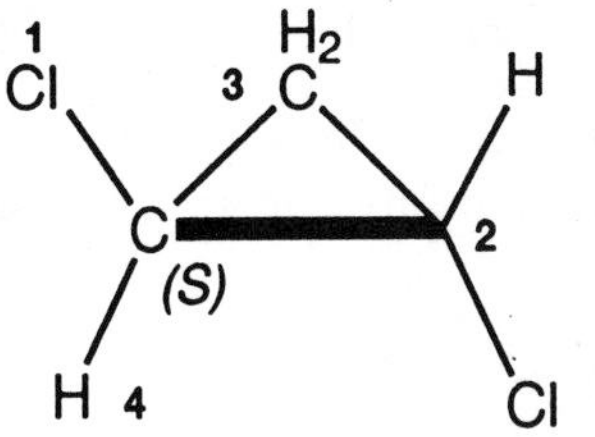

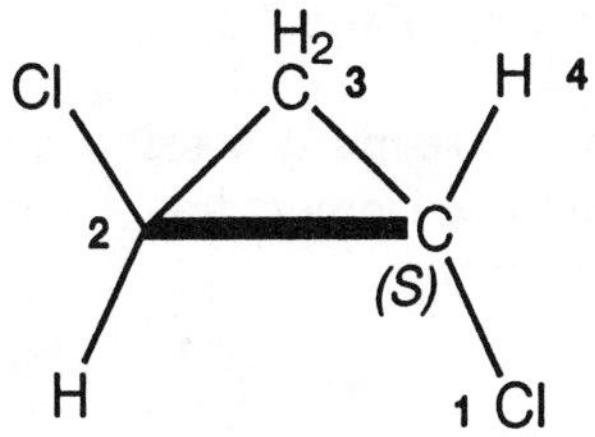

Of course, the reflection of the "*S S*" trans compound must be "*R R*."

Problem 5.23

A B C C'

Compound **A** is a structural isomer of **B**, **C**, and **C′**. Compounds **C** and **C′** are enantiomers; **B** and **C**, and **B** and **C′** are pairs of diastereomers.

Problem 5.24

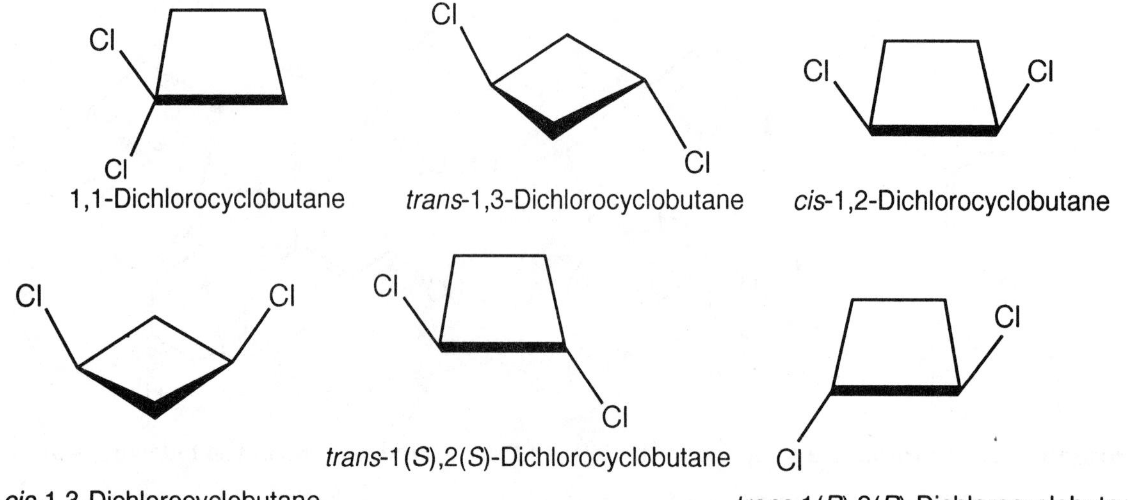

1,1-Dichlorocyclobutane; *trans*-1,3-Dichlorocyclobutane; *cis*-1,2-Dichlorocyclobutane; *cis*-1,3-Dichlorocyclobutane; *trans*-1(*S*),2(*S*)-Dichlorocyclobutane; *trans*-1(*R*),2(*R*)-Dichlorocyclobutane

Problem 5.25 There are many possible answers to this question. Here is just one. Each of the following two structural isomers has five methyl groups, one methylene group, one methine group and one quaternary carbon.

Problem 5.30

HOOC
COOH
NO_2
O_2N
Mirror
COOH
HOOC
O_2N
NO_2
Rotate left-hand ring 180° around the junction bond
these are the same
HOOC
NO_2
COOH
O_2N
turn whole molecule 180°
COOH
HOOC
O_2N
NO_2

Problem 5.31 Were hexahelicene to be planar, the atoms of the "end" rings would have to occupy the same space. Accordingly, one ring (dark lines) slips over the other (dotted lines). A coil, or helix is formed.

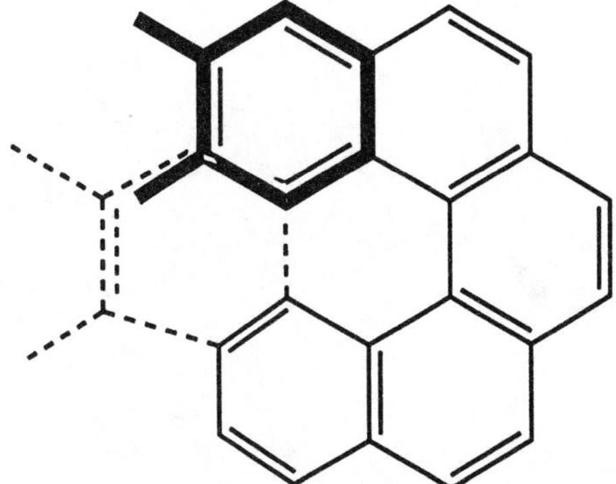

Helices are chiral, and can spiral in a right-handed or left-handed way. The right-handed helix and the left-handed helix are mirror images:

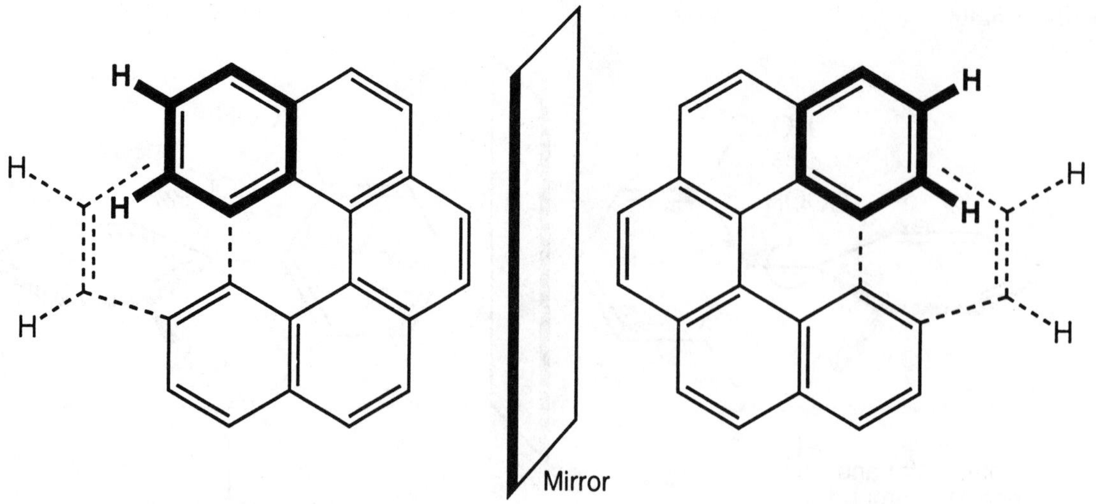

Mirror

Additional Problem Answers

Problem 5.32 This one is not bad at all; there are only two.

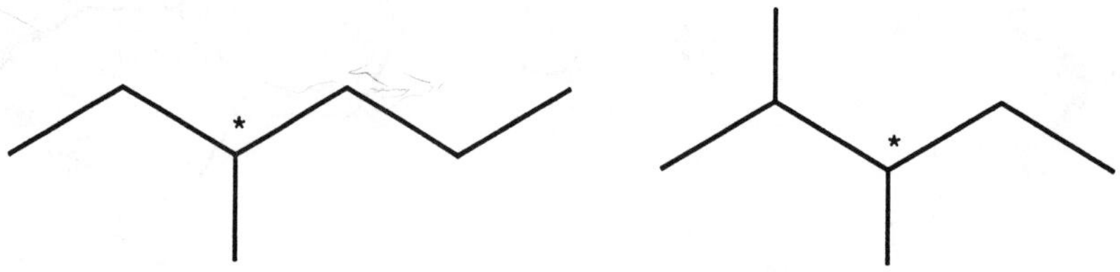

3-Methylhexane

2,3-Dimethylpentane

Problem 5.33 The lowest priority group will be H. Methyl is next lowest, ethyl next lowest, and either propyl or isopropyl will be the highest.

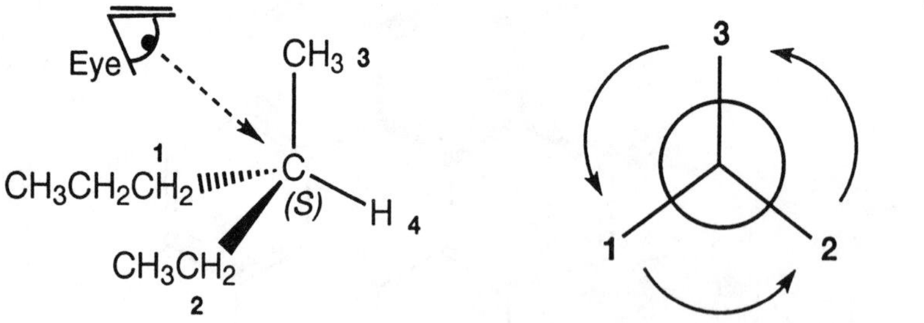

Eye

CH_3 3

1

$CH_3CH_2CH_2$ C (*S*) H 4

CH_3CH_2

2

(*S*)- Arrows counterclockwise

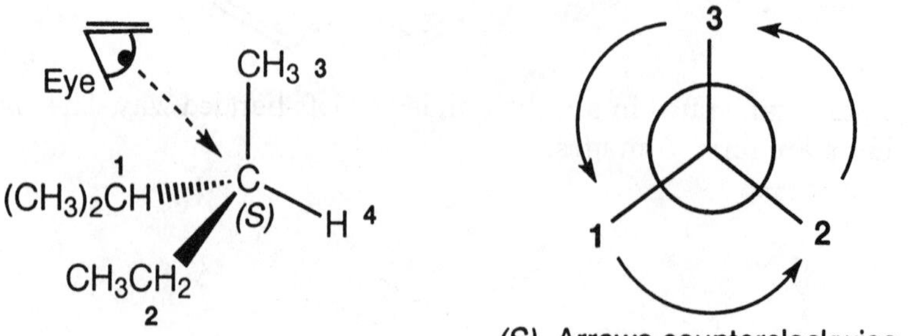

Eye

CH_3 3

1

$(CH_3)_2CH$ C (*S*) H 4

CH_3CH_2

2

(*S*)- Arrows counterclockwise

Problems 5.34 and **5.35** There are five possibilities, and only one is properly named as a "pentane."

3-Methylheptane

3,4-Dimethylhexane

2,3-Dimethylhexane

2,4-Dimethylhexane

2,2,3-Trimethylpentane

=

3 CH_3

2 CH_3CH_2 C H 4

$(H_3C)_3C$ 1

=

3 2 1

(R)

Problem 5.36 The total number of isomers is 15; three of which have two stereogenic carbons (shown in boxes).

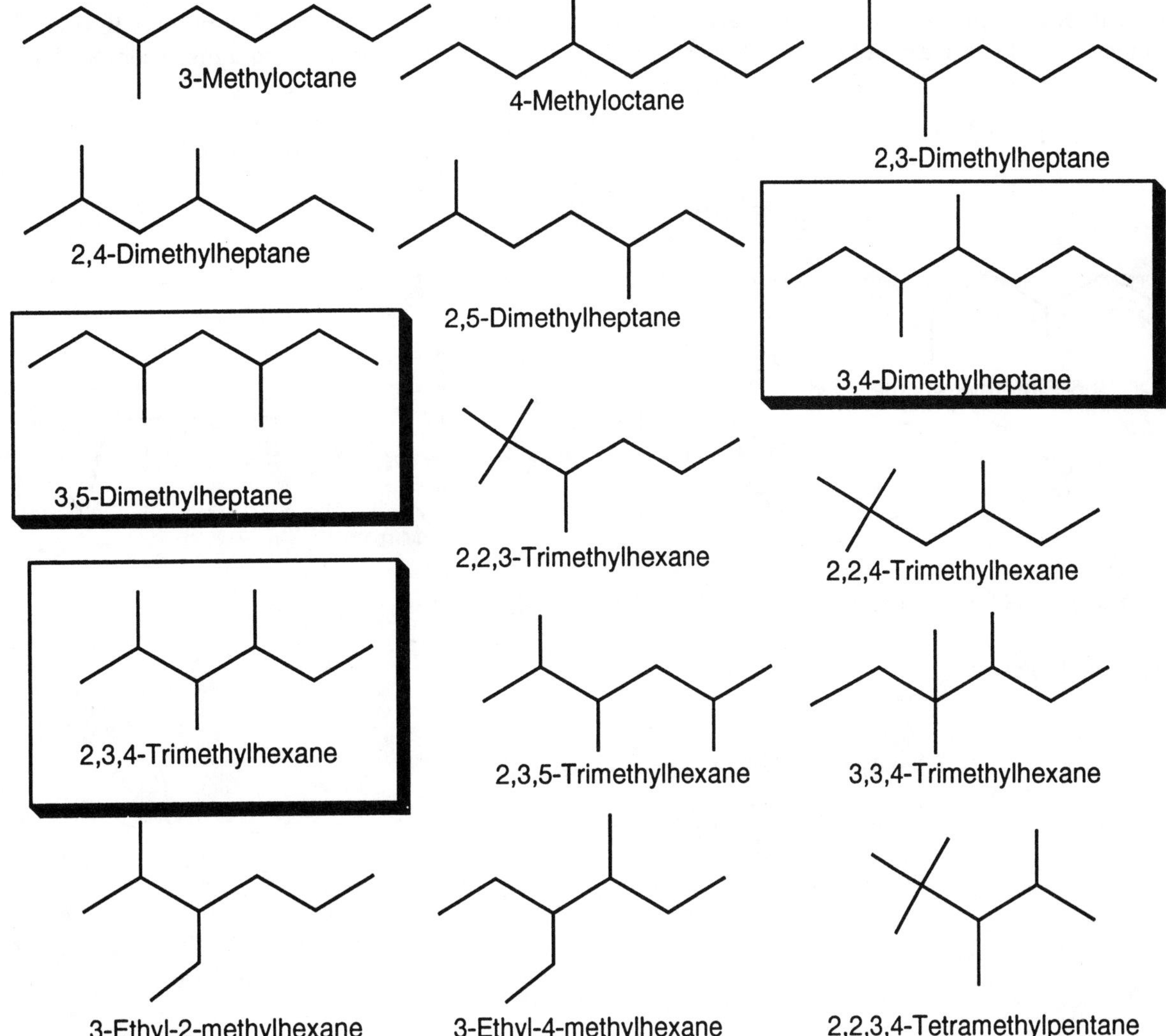

Problem 5.37 3,4-Dimethylheptane has two stereogenic carbons, so the maximum number of stereoisomers is $2^2 = 4$. These appear as two pairs of enantiomers. Me = methyl, Et = ethyl, and Pr = propyl.

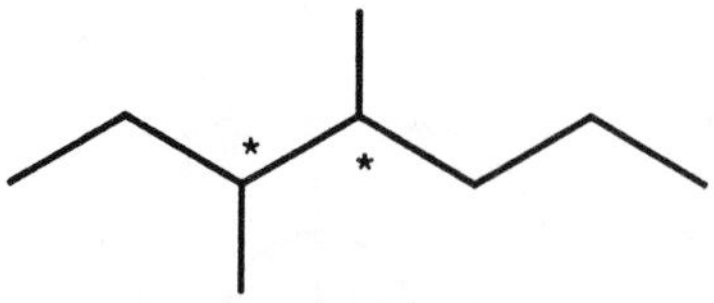

3,4-Dimethylheptane

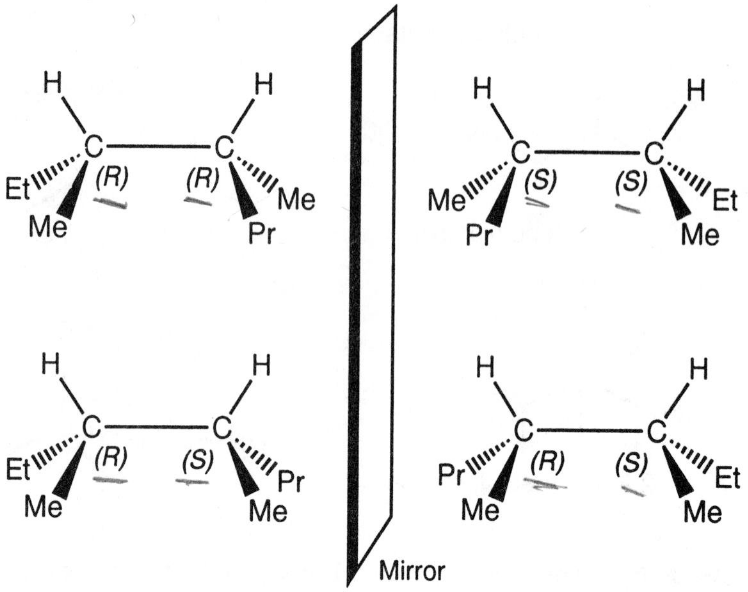

Mirror

3,5-Dimethylheptane also has two stereogenic carbons, but there is a plane of symmetry in this molecule, and there are only three stereoisomers: one pair of enantiomers and a meso compound.

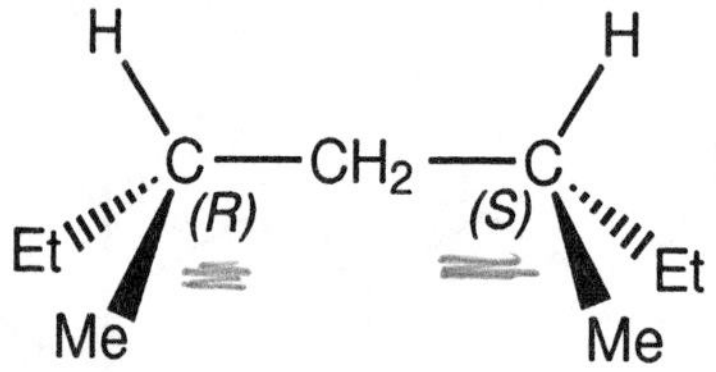

A meso compound - superimposable on its mirror image

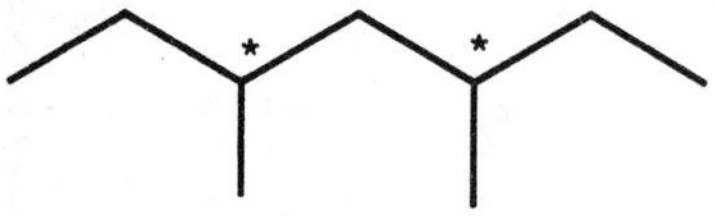

3,5-Dimethylheptane

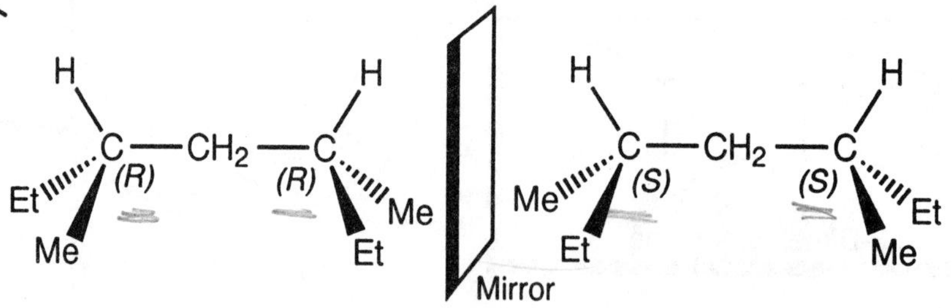

Mirror

Problem 5.38 There is only one hexene but there are six heptenes.

From the hexenes

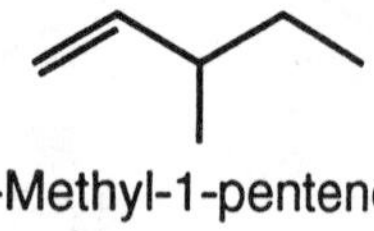

3-Methyl-1-pentene

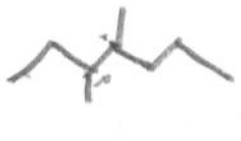

From the heptenes

3-Methyl-1-hexene

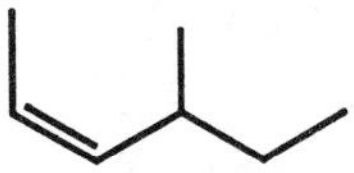

cis-4-Methyl-2-hexene

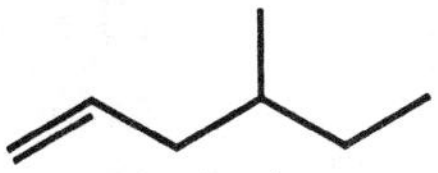

4-Methyl-1-hexene

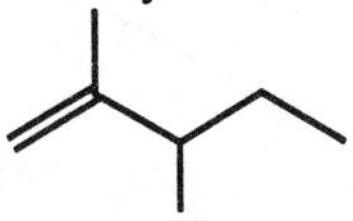

2,3-Dimethyl-1-pentene

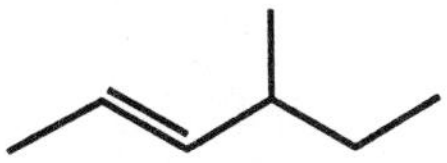

trans-4-Methyl-2-hexene

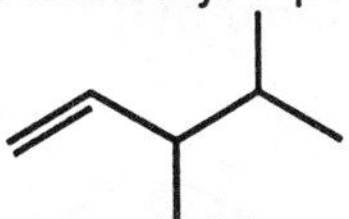

3,4-Dimethyl-1-pentene

Problem 5.39 There is one chiral hexyne; there are four heptynes, and no fewer than 13 octynes.

Hexynes

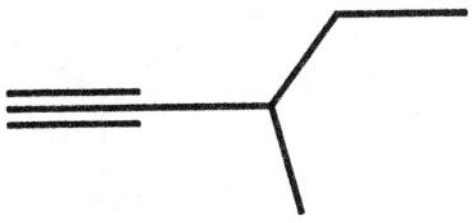

3-Methyl-1-pentyne

Heptynes

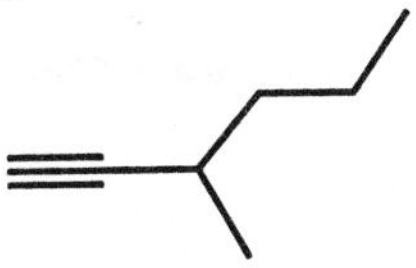

3-Methyl-1-hexyne 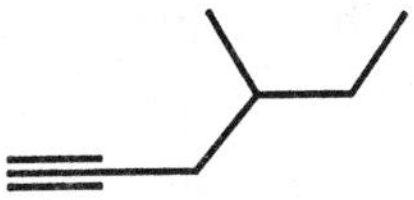 4-Methyl-1-hexyne 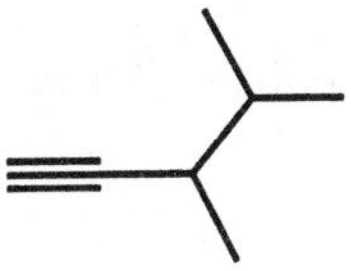 3,4-Dimethyl-1-pentyne 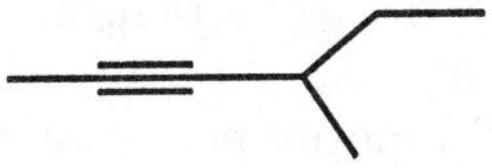 4-Methyl-2-hexyne

Octynes

3-Methyl-1-heptyne  4-Methyl-1-heptyne  5-Methyl-1-heptyne 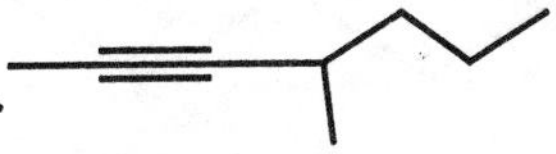 4-Methyl-2-heptyne

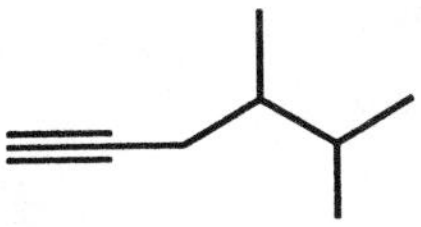

4,5-Dimethyl-1-hexyne 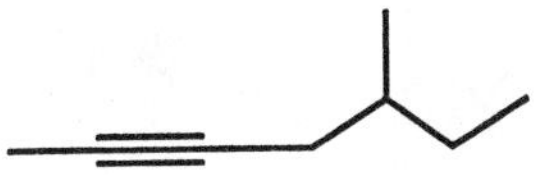 5-Methyl-2-heptyne 5-Methyl-3-heptyne 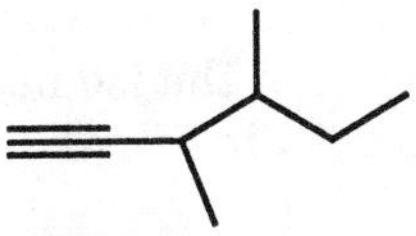 3,4-Dimethyl-1-hexyne

3,5-Dimethyl-1-hexyne 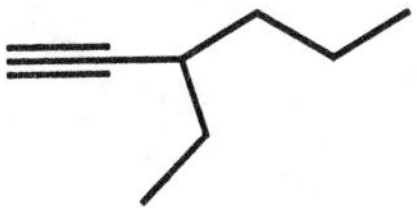 3-Ethyl-1-hexyne 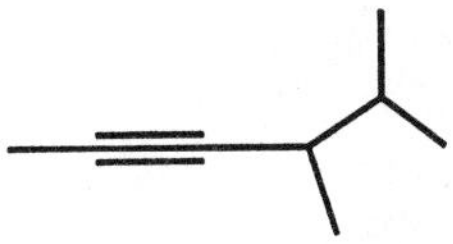 4,5-Dimethyl-2-hexyne 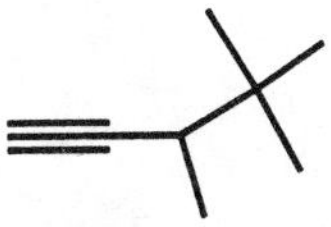 3,4,4-Trimethyl-1-pentyne

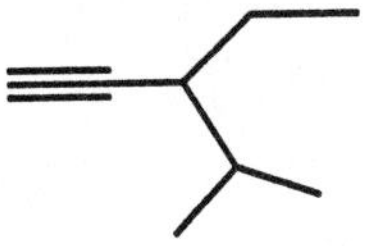

3-Ethyl-4-methyl-1-pentyne

Problem 5.40 Two of the isomers formed from hexane are chiral.
From hexane:

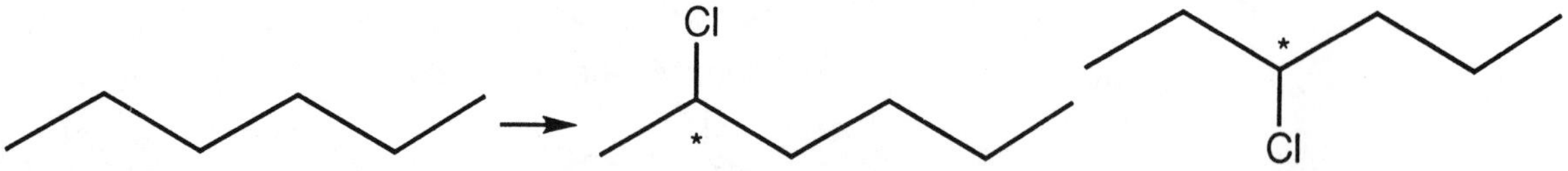

2-Chlorohexane 3-Chlorohexane

Three of the molecules formed from 2-methylpentane are chiral:
From 2-methylpentane:

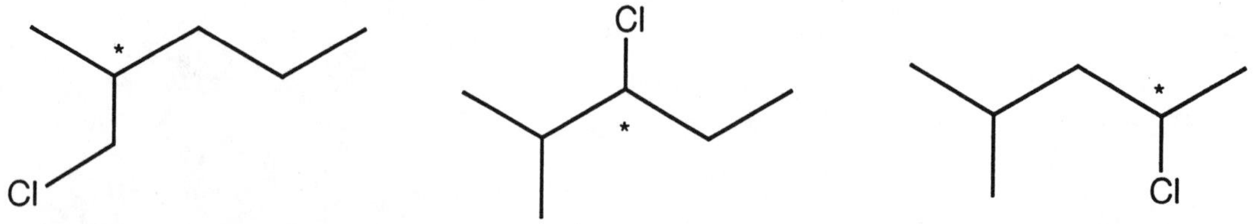

1-Chloro-2-methylpentane 3-Chloro-2-methylpentane 2-Chloro-4-methylpentane

Two of the isomers formed from 3-methylpentane are chiral, and one isomer has two stereogenic carbons.
From 3-methylpentane:

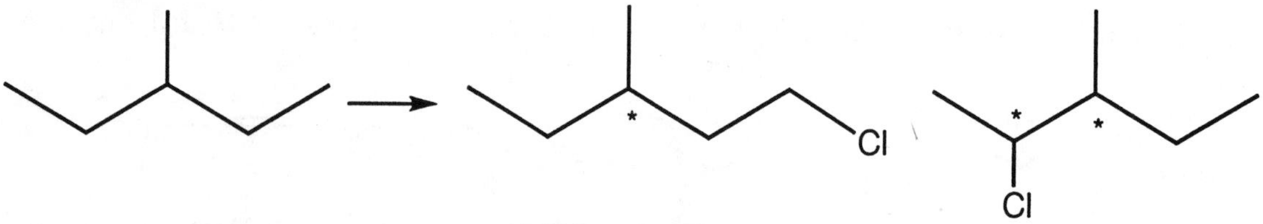

1-Chloro-3-methylpentane 2-Chloro-3-methylpentane

Only one of the isomers from 2,2-dimethylbutane is chiral:
From 2,2-dimethylbutane:

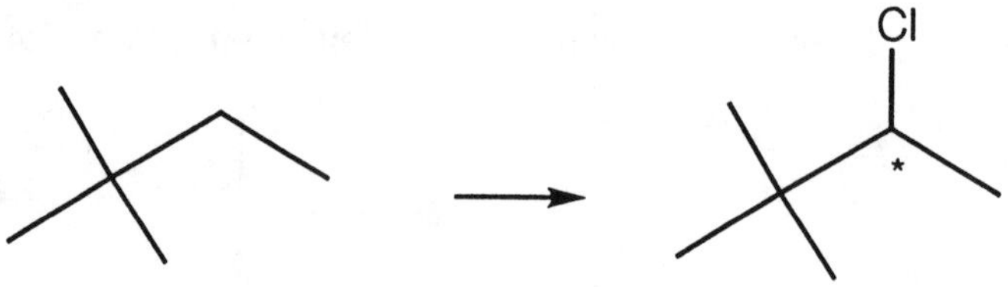

2-Chloro-3,3-dimethylbutane

Similarly, only one of the isomers from 2,3-dimethylbutane is chiral:
From 2,3-dimethylbutane:

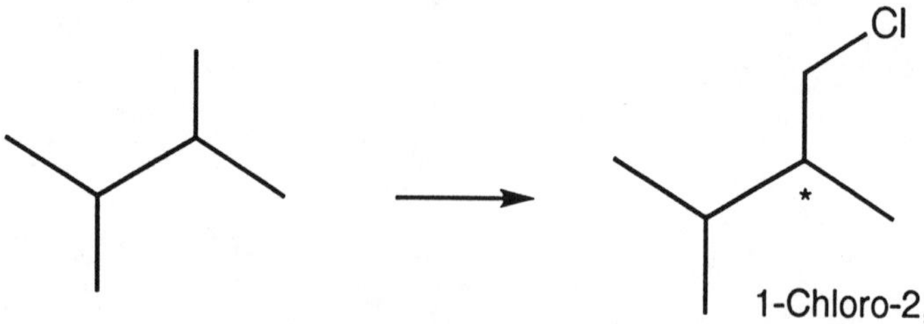

1-Chloro-2,3-dimethylbutane

Problem 5.41 There is only one isomer to consider, 2-chloro-3-methylpentane. There will be four stereoisomers; two pairs of enantiomers.

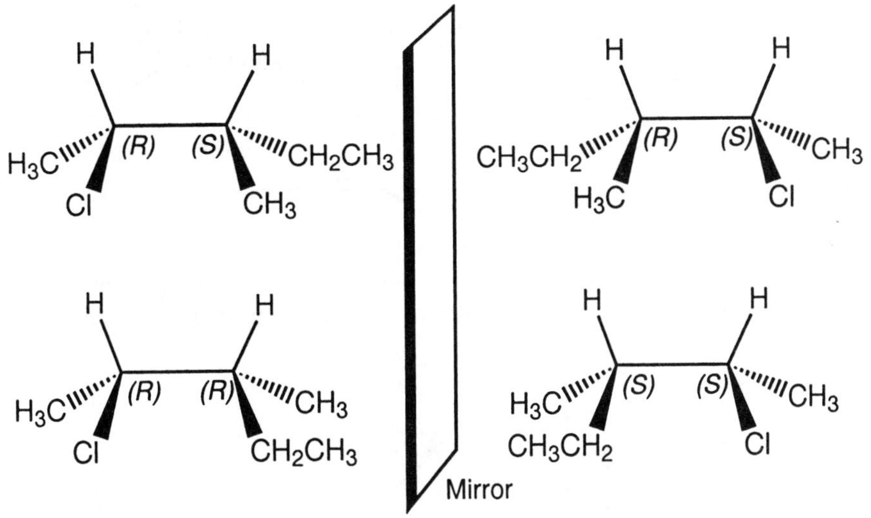

H H H H
H_3C (R) (S) CH_2CH_3 CH_3CH_2 (R) (S) CH_3
Cl CH_3 H_3C Cl

H H H H
H_3C (R) (R) CH_3 H_3C (S) (S) CH_3
Cl CH_2CH_3 CH_3CH_2 Cl
Mirror

Problem 5.42 Only compound (b) can exist in (*E*) and (*Z*) forms. Compounds (a) and (d) have stereogenic carbon atoms:

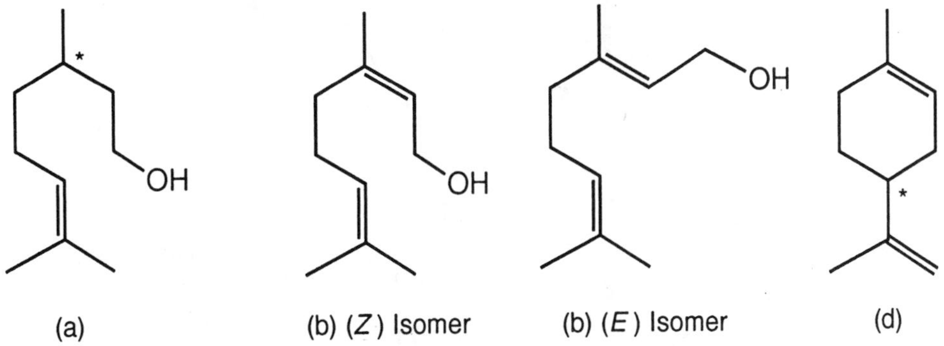

OH OH OH

(a) (b) (*Z*) Isomer (b) (*E*) Isomer (d)

Technically, (d) could exist in (*E*) and (*Z*) forms. However, a "trans" double bond in a six-membered ring produces too much strain, and the isomer shown is the only one practically possible.

Problem 5.43 Base the tetrahedra on the stereogenic carbons. Determining priorities is certainly no problem in the acyclic molecules, but may be more difficult in the cyclic species. *Remember*: The ring makes no special difference in this problem. For example, the priorities would be exactly the same in molecule **A**. Why should anything be altered by closing the ring? It isn't.

(a)

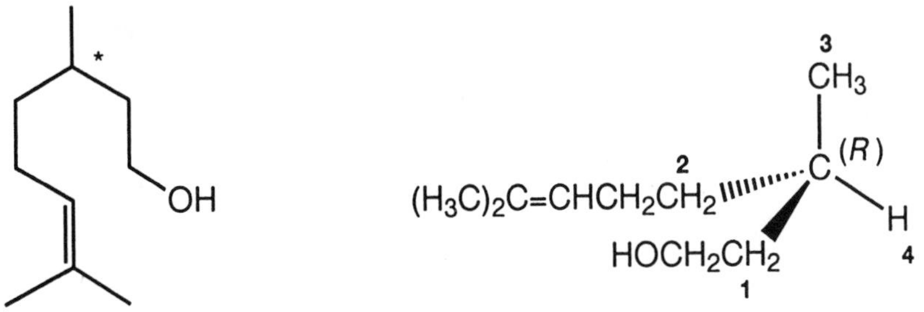

OH

3 CH_3
2 (R)
$(H_3C)_2C{=}CHCH_2CH_2$ C H
$HOCH_2CH_2$ 4
1

(d)

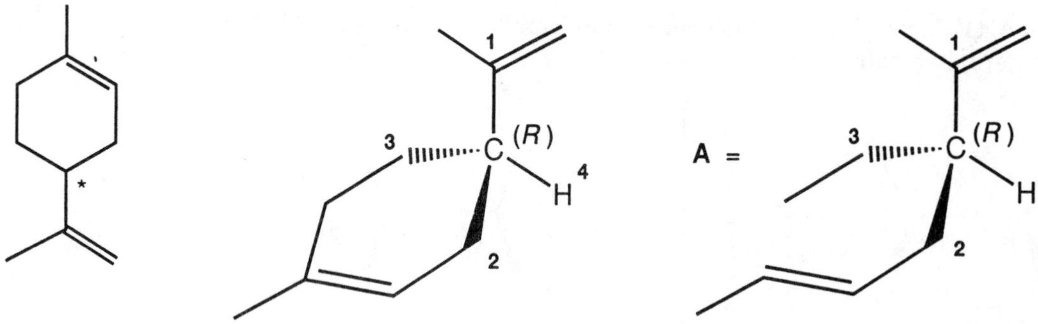

A =

Problem 5.44 This molecule comes in cis and trans versions. In each case, there is a pair of enantiomers. There is a total of four stereoisomers, and the figure shows the enantiomeric and diastereomeric relationships.

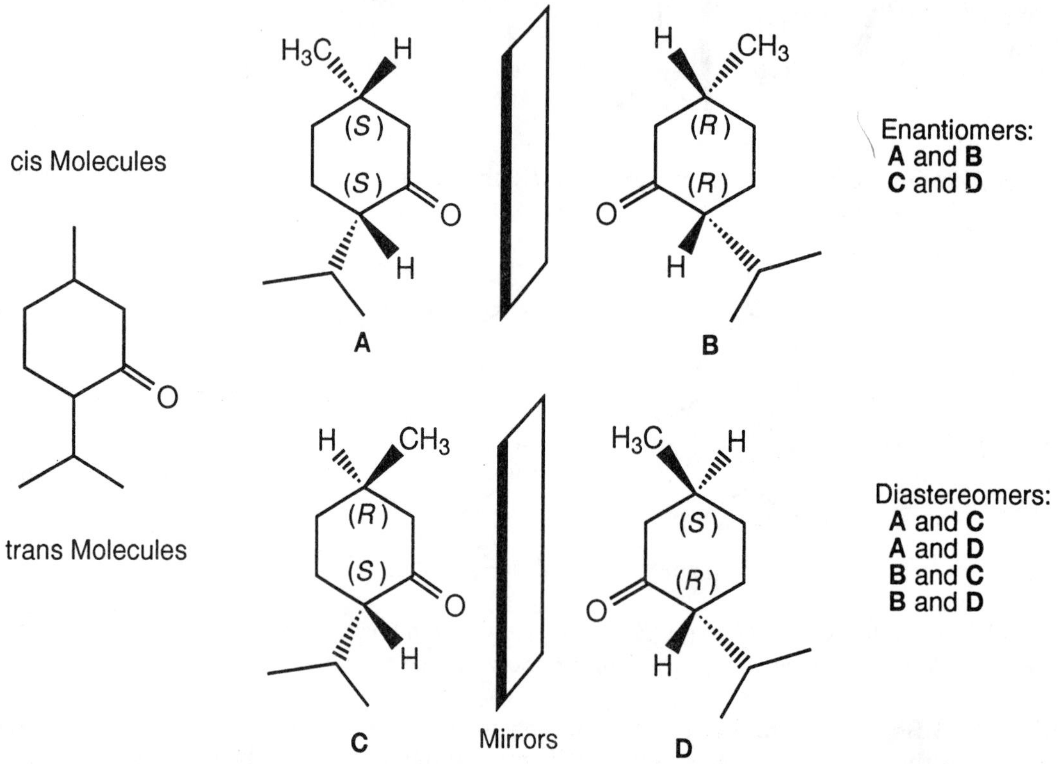

cis Molecules

trans Molecules

Enantiomers:
A and **B**
C and **D**

Diastereomers:
A and **C**
A and **D**
B and **C**
B and **D**

Mirrors

Problem 5.45 There are only four:

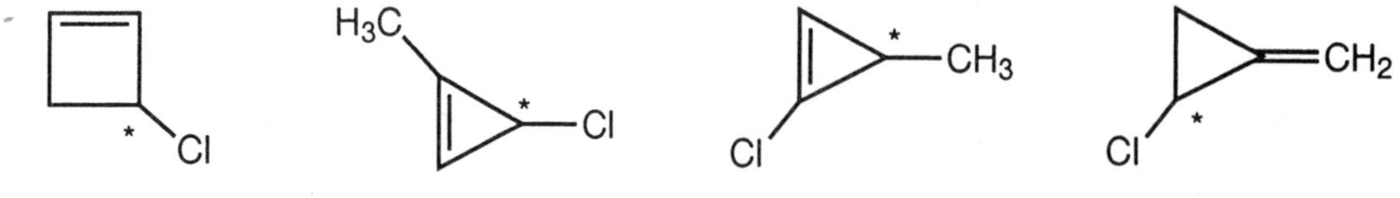

Problem 5.46 Hydrogen is clearly the lowest priority (**4**), and equally clearly, Cl will have the highest priority (**1**) whenever it is on the stereogenic carbon. The doubly bonded carbon will have a higher priority (**2**) than the CH_2 or CH_3 group (**3**).

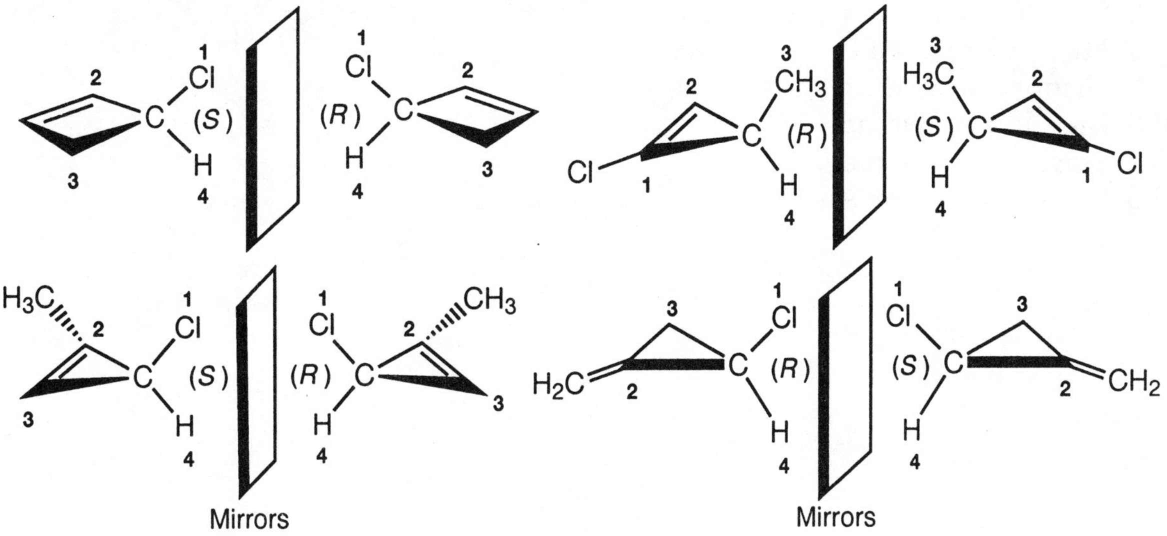

Mirrors Mirrors

Problem 5.47 This time there are a lot. Six of the nine are chiral.

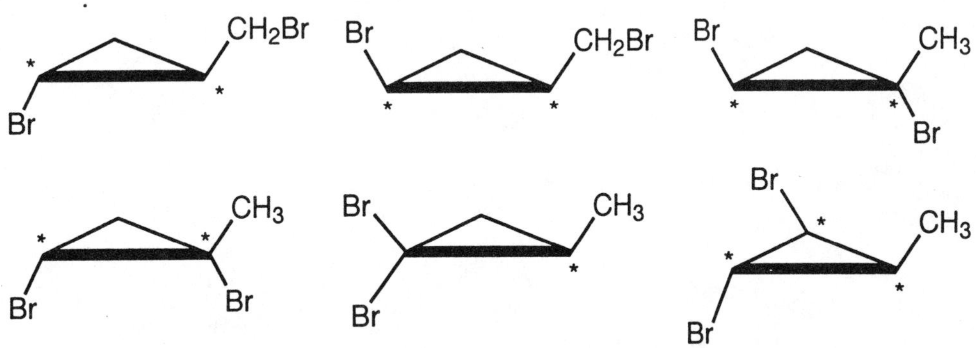

Problem 5.48 Hydrogen is the lowest priority (**4**). The methyl group, (CH_3) will be the next lowest (**3**), and CH_2 will be lower (**2**) than CBr_2 (**1**). The two enantiomers are shown below:

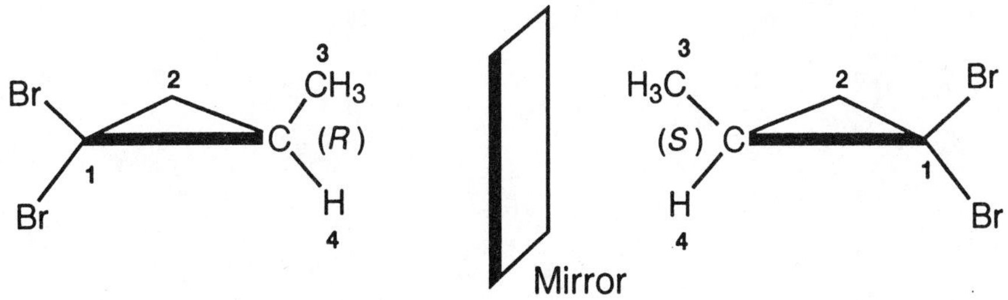

Mirror

Chapter 6 Outline

6.1 Rings and Strain
6.2 Quantitative Evaluation of Strain Energy
6.3 Stereochemistry of Cyclohexane: Conformational Analysis
6.4 Monosubstituted Cyclohexanes
6.5 Disubstituted Ring Compounds
6.6 Bicyclic Compounds
6.7 Polycyclic Systems
6.8 Something More: Adamantane and Diamond

6

Rings

Problem 6.2 First look at planar cyclopentane in which the five top carbon-hydrogen bonds (as well as the bottom carbon-hydrogen bonds) are eclipsed. Once again, offset your eye just a bit so you can see the eclipsed bonds and atoms in the back. As in cyclobutane (Fig. 6.9), puckering of the five-membered ring relieves some of the eclipsing of the carbon-hydrogen bonds.

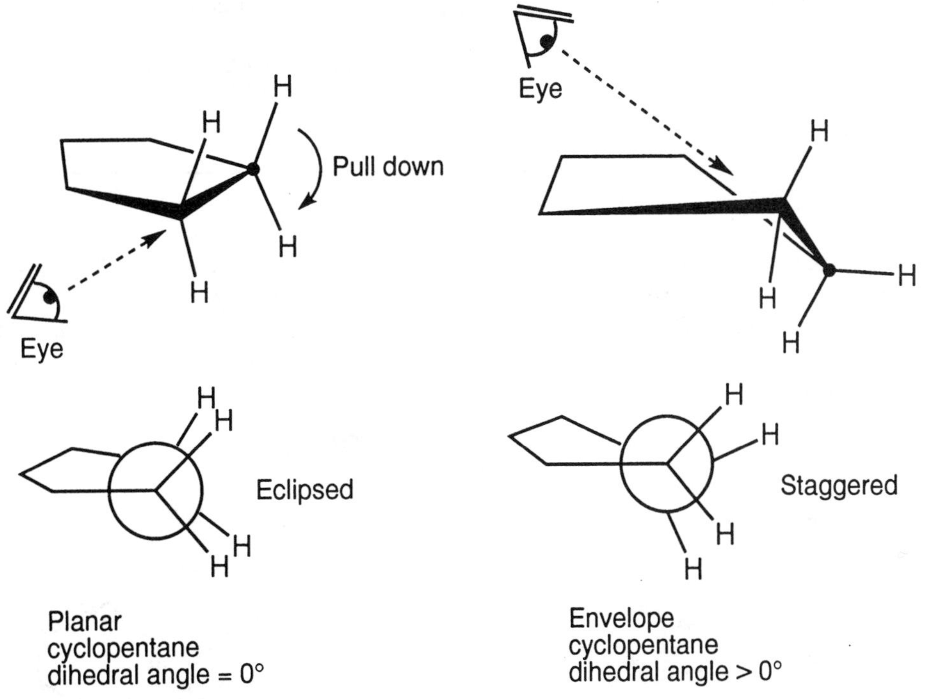

Note how puckering opens up the dihedral angle between the carbon-hydrogen bonds.

Problem 6.4 Presumably, taking the difference in heat of formation between two compounds that differ only by one methylene group will give the heat of formation of a single methylene. In practice, one takes the average of many such determinations. In this example: heptane (ΔH_f° = – 44.8) – hexane (ΔH_f° = –39.9) gives –4.9; octane (ΔH_f° = –49.8) – heptane (ΔH_f° = –44.8) gives –5.0; nonane (ΔH_f° = –54.5) – octane (ΔH_f° = –49.8) gives –4.7; decane (ΔH_f° = –59.6) –nonane (ΔH_f° = –54.5) gives –5.1. The average of these determinations is –4.9 kcal/mol.

Problem 6.5 This question is admittedly vague, and your answer may vary depending on what molecule(s) you choose. Use your models, and angle, torsional, and "across-the-ring" (or "transannular") van der Waals strains will appear. The energy-minimum structure will be a compromise in which the sum of all these strains is minimized.

Problem 6.6 This figure resembles Figure 6.23. The three isomeric alkanes all react with oxygen to give $7CO_2 + 8H_2O$. The product of combustion of all three hydrocarbons is exactly the same. Therefore, the differences in their heats of combustion give the differences in their energies.

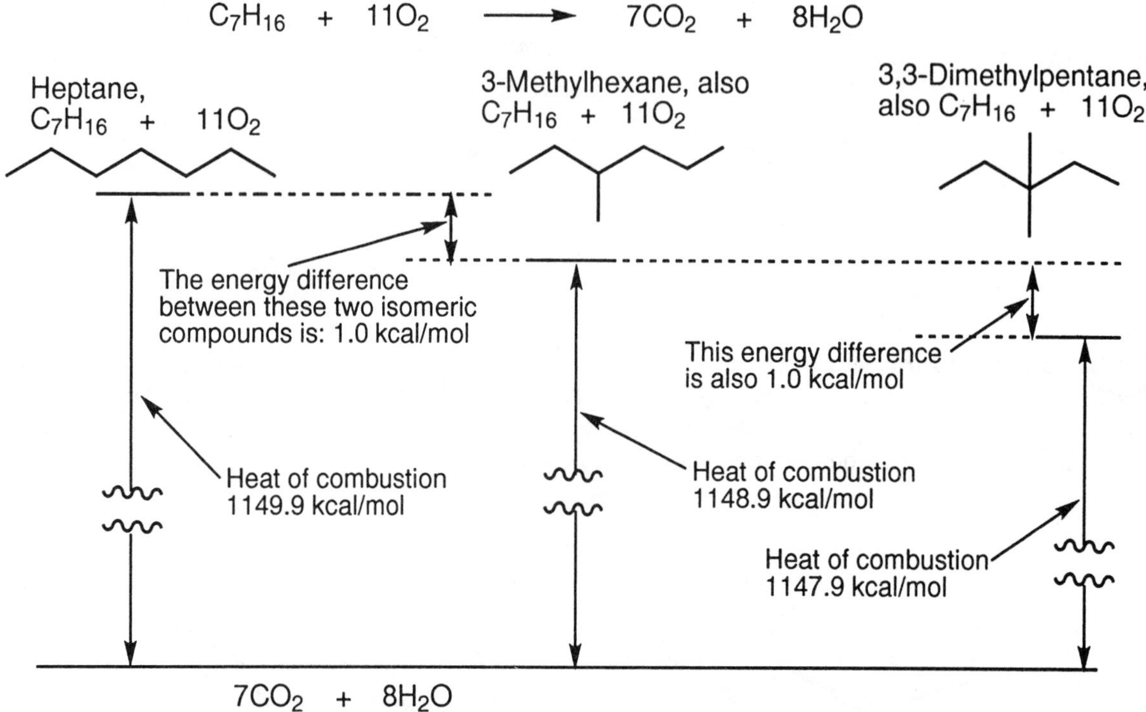

Problem 6.9 A *gauche* butane interaction involves destabilizing methyl-methyl "bumpings." That's not exactly what is happening in the *gauche* interactions in axial methylcyclohexane. In the case of methylcyclohexane, we have $CH_3 — CH_2C$ interactions instead. We should expect only similarity to the *gauche* butane value, not identity, and this is what we see (1.74 kcal/mol is close to 1.2 kcal/mol, but not identical to it).

Axial hydrogen

One of two *gauche* methyl-ring interactions, which involve "bumping" between CH_3 and the ring $CH_2–C$

Axial hydrogen

Eye

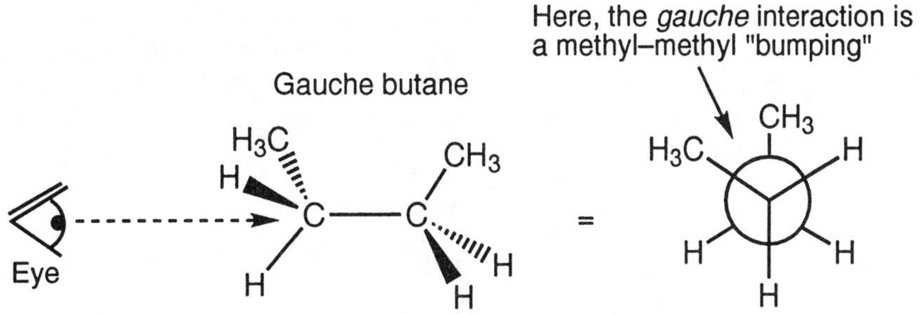

Here, the *gauche* interaction is a methyl–methyl "bumping"

Gauche butane

Eye

Problem 6.10 $\Delta G = -RT\ln K$, where R is about 2 cal/deg.mol, which is equal to about 0.002 kcal/deg.mol; T = 25 °C = 298 K; and ΔG = 2.8 kcal/mol.
$2.8 = -(0.002)(298)\ln K$
$-4.70 = \ln K$
$9.1 \times 10^{-3} = K$
So, 2.8 kcal/mol translates into a mixture of products in which the ratio is 99.1: 0.9.

Problem 6.12 In 1-isopropyl-1-methylcyclohexane either the methyl or isopropyl group can be equatorial, but not both. Your intuition might tell you that the conformation with the larger isopropyl group equatorial will be preferred, but you will need to consult Table 6.4 to do the quantitative calculation. Table 6.4 tells us that a cyclohexane with an equatorial isopropyl group is more stable than a cyclohexane with an axial isopropyl group by 2.61 kcal/mol. Similarly, a cyclohexane with an equatorial methyl is 1.74 kcal/mol more stable than a cyclohexane with an axial methyl group. The ring-flip of 1-isopropyl-1-methylcyclohexane involves the interconversion of equatorial and axial isopropyl and methyl groups.

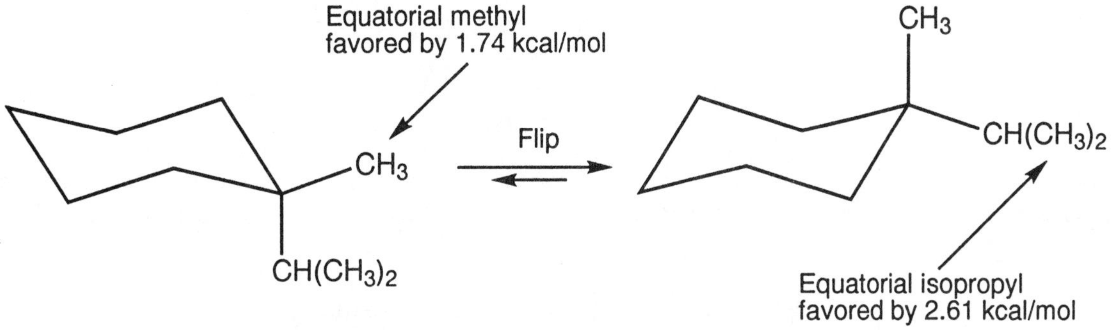

Equatorial methyl favored by 1.74 kcal/mol

Flip

Equatorial isopropyl favored by 2.61 kcal/mol

We calculate: 2.61 – 1.74 = 0.87 kcal/mol in favor of the isomer with the equatorial isopropyl group.

Problem 6.13 As the Newman projection shows, this dihedral angle is 180°.

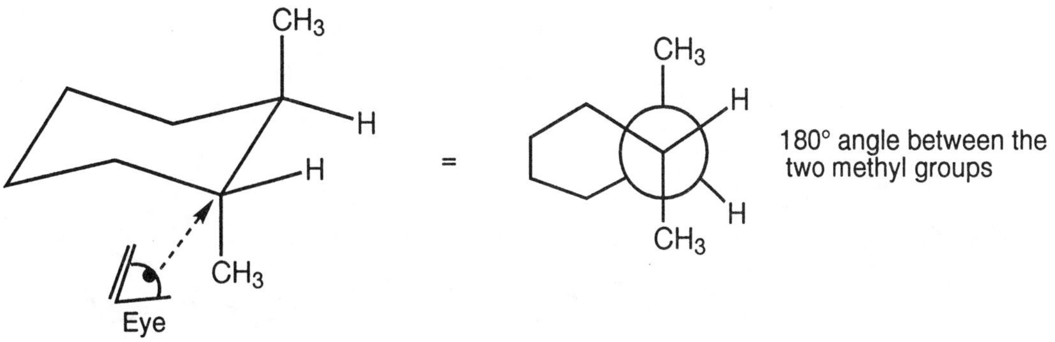

180° angle between the two methyl groups

Eye

Problem 6.14 In *trans*-1,2-dimethylcyclohexane, the ring-flip converts the diequatorial stereoisomer into the diaxial form. These two molecules are conformational diastereomers, not enantiomers. There can be no racemization in this ring flip. Be certain you see the difference between this process and that shown in Figure 6.42 for similar ring-flipping of *cis*-1,2-dimethylcyclohexane.

Enantiomers

H CH3 CH3 H | H3C H3C H H

Diastereomers flip | flip Diastereomers

CH3 H H CH3 | H CH3 H CH3

Mirror

Enantiomers

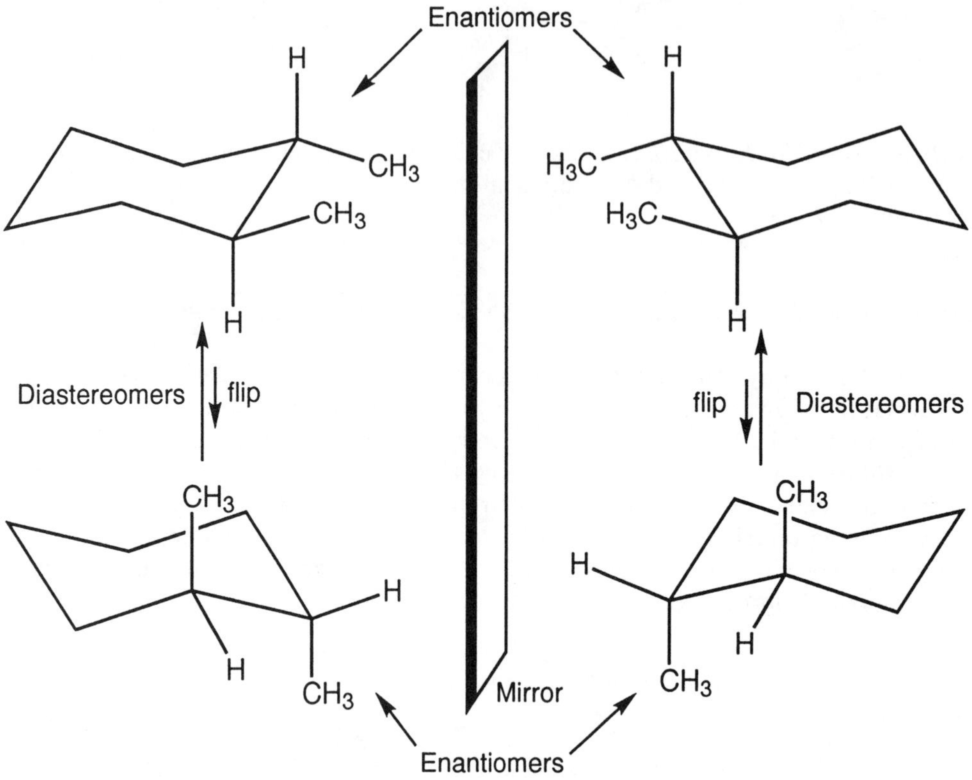

Problem 6.15 The two stereogenic carbons are shown in bold and lightface. The priorities used to determine (*R*) or (*S*) are also shown in bold and lightface, as is appropriate for each carbon.

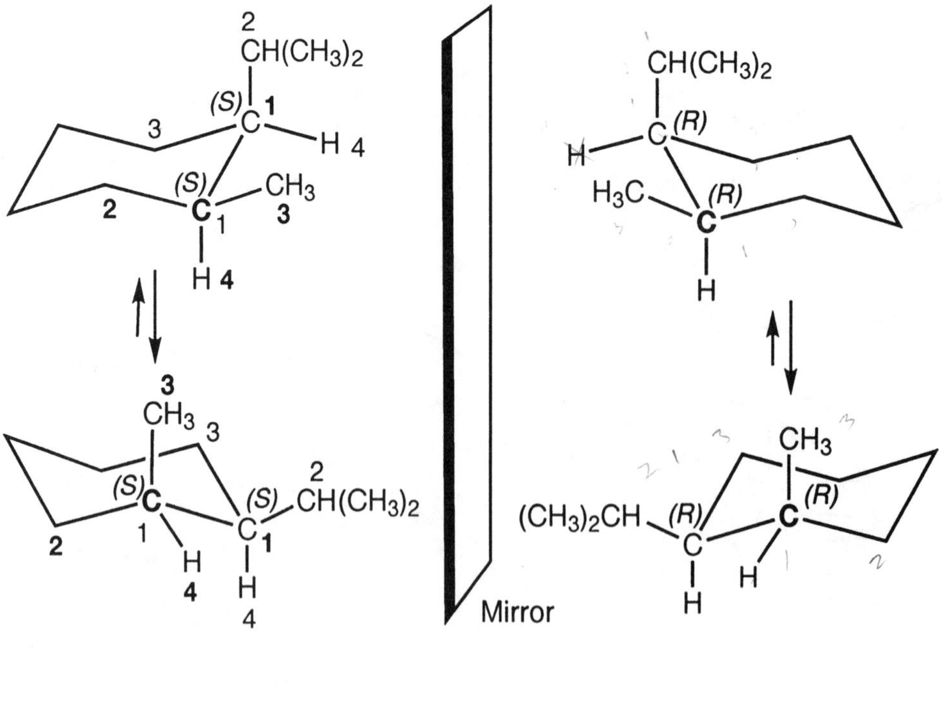

2 CH(CH3)2 (S) C 1 H 4 3 (S) C 1 CH3 3 2 H 4 | CH(CH3)2 C (R) H H3C (R) C H

3 CH3 3 (S) C 1 2 H 4 (S) CH(CH3)2 2 C 1 H 4 | CH3 (CH3)2CH (R) C H (R) C H

Mirror

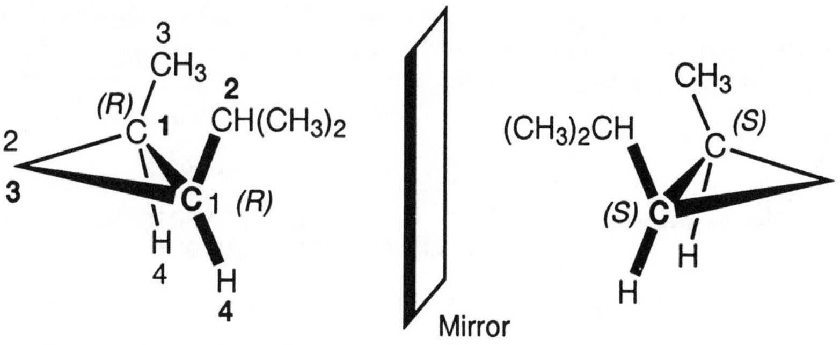

Problem 6.16

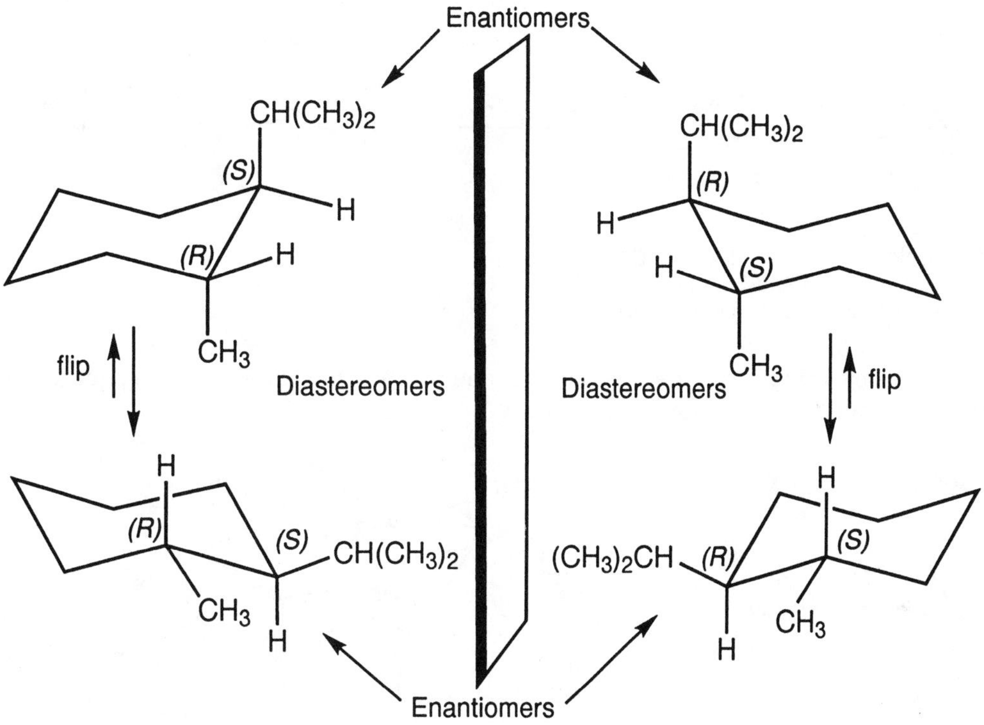

Problem 6.18 Table 6.4 shows that a cyclohexane with an equatorial methyl group is favored over a cyclohexane with an axial methyl group by 1.74 kcal/mol. In this ring-flip, two equatorial methyl groups interconvert with two axial methyl groups. Accordingly, we expect the diequatorial isomer to be favored by at least $2 \times 1.74 = 3.48$ kcal/mol. There will be an additional serious destabilizing interaction between the two 1,3-diaxial methyl groups. The total energy difference between the two is about 5.5 kcal/mol.

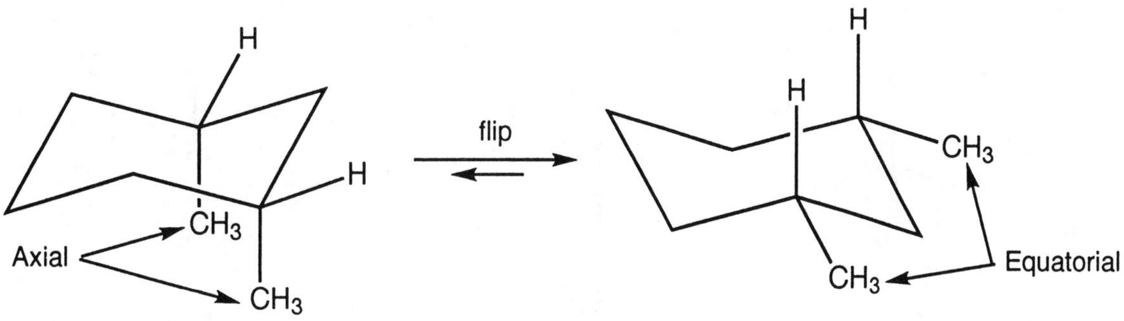

Problem 6.20 The three molecules at the left of Figure 6.53 all contain small three- or four-membered rings. Models will show this most clearly, but here is one example of how it is not possible to span a trans fusion in these molecules:

H

There is no easy way to span these positions with just one or two carbons!

H

Problem 6.21 Get out your models! In the trans compound, ring-flip requires that the second ring be connected through a pair of axial bonds. There are not enough atoms in the chain to do this.

H

Too long a distance to span

H

For this molecule to "flip," the two darkened bonds must become axial; it is not possible to connect the second ring in this way

By contrast, in the cis compound, the junction is made through one axial and one equatorial bond. Ring flip interchanges these two, and is quite easy.

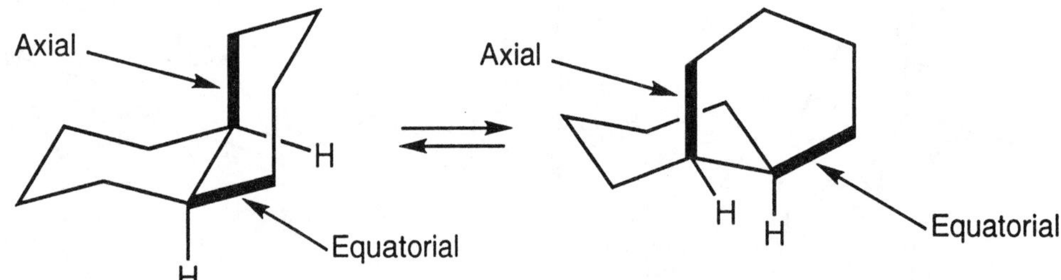

Problem 6.23 The directions for constructing bicyclic molecules are in the chapter. In (a) we connect the bridgeheads (always start with the bridgehead atoms) with two three-carbon bridges and one no-carbon bridge. In (b) all the bridges are the same, and in (c) there is a six-carbon bridge, a one-carbon bridge, and a no-carbon bridge. Be careful in this example to note that the bridge junction is trans. In (b) we place the fluorine at the 1-position, the bridgehead. In (c) we count around the longest bridge first. This makes the free cyclopropane position the 9-position.

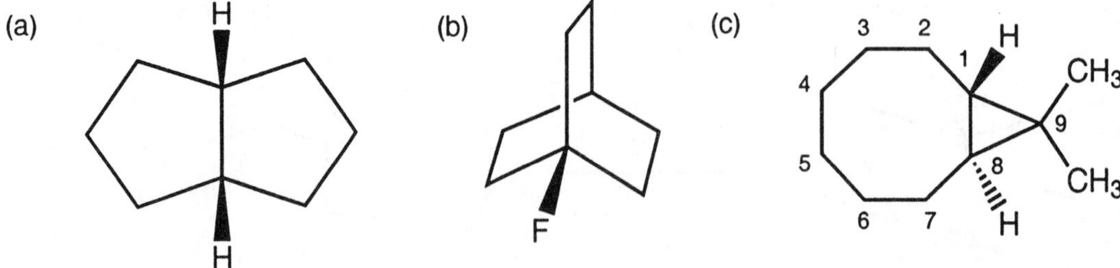

Problem 6.24

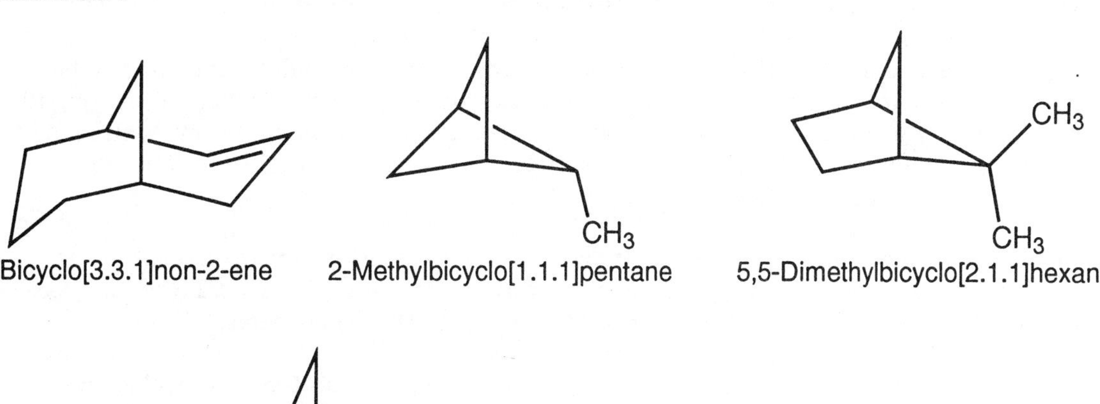

Bicyclo[3.3.1]non-2-ene   2-Methylbicyclo[1.1.1]pentane   5,5-Dimethylbicyclo[2.1.1]hexane

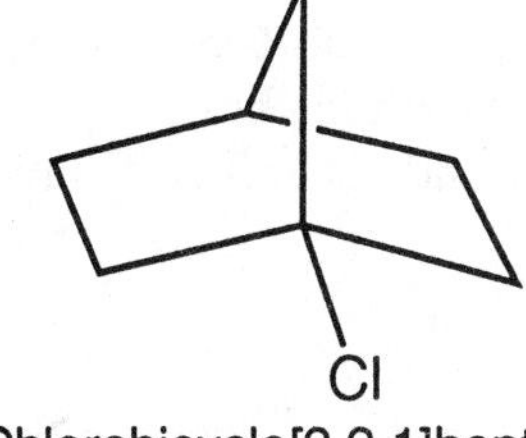

1-Chlorobicyclo[2.2.1]heptane

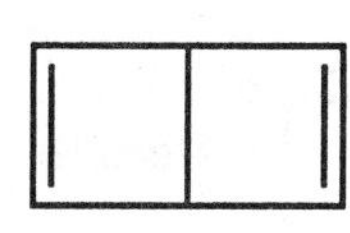

Bicyclo[2.2.0]hexa-2,5-diene

Problem 6.25 The empirical formula of diamond is C.

Problem 6.26 There are three possibilities for tetramantane. This problem is nearly impossible to do without models at this point in your career.

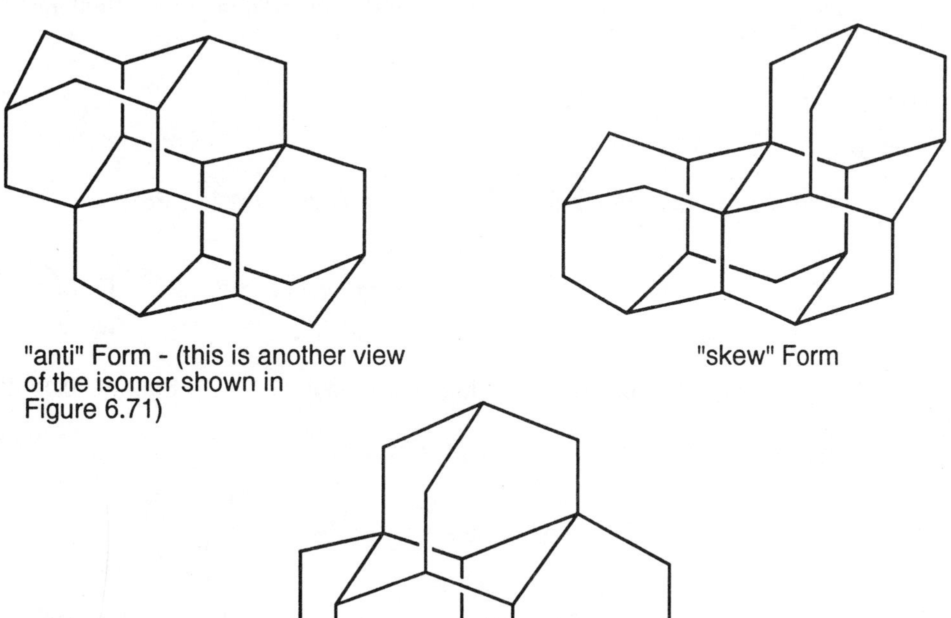

"anti" Form - (this is another view of the isomer shown in Figure 6.71)

"skew" Form

"iso" Form

Additional Problem Answers

This chapter deals almost exclusively with the structures of ring compounds, and six-membered rings (cyclohexanes) are examined especially closely. The structural consequences of the ring flip of cyclohexanes (really just a series of rotations around carbon-carbon bonds) are emphasized, and the following problems should give you lots of practice in drawing, flipping, and evaluating substituted cyclohexanes.

Problem 6.27 First determine the number of degrees of unsaturation. For C_5, the saturated alkane would have 12 hydrogens. Application of the formula leads to 12 – 8 = 4/2 = two degrees of unsaturation. There must be a total of two π bonds or rings in these molecules.

Start with acyclic molecules. These must contain either two double bonds or one triple bond. Here the problem is likely to be to find the molecules in which one carbon is shared by two double bonds. It will be almost certainly tough to see that one of these compounds is chiral (Chapter 5). Chirality is indicated with an asterisk (*). Here they are:

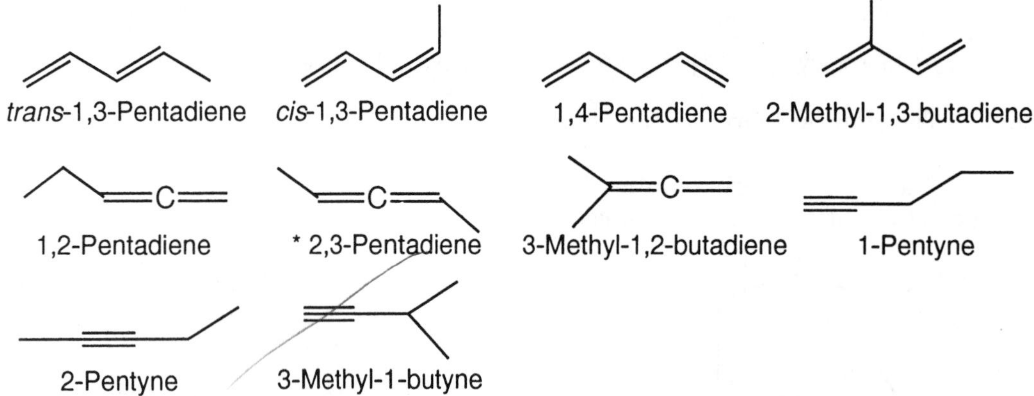

Next find the compounds containing five- and four-membered rings. One is chiral.

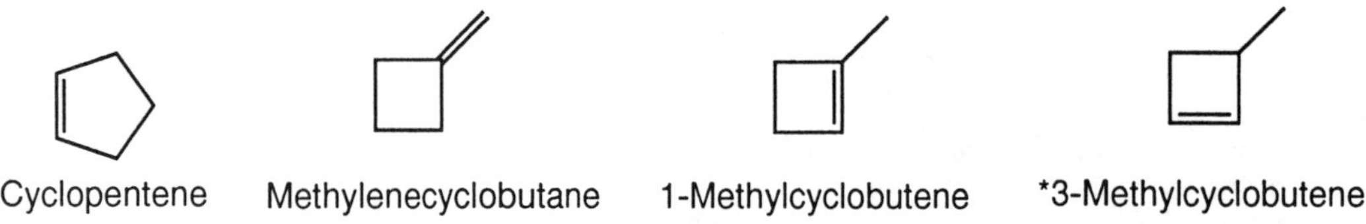

Now find the cyclopropenes. There are five, and one is chiral.

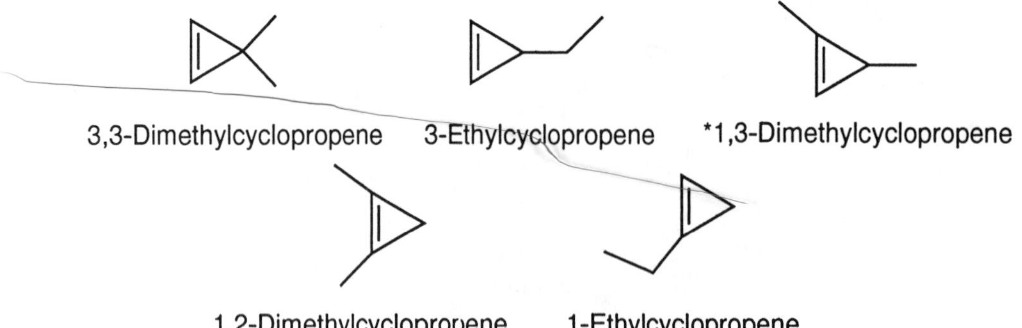

Now find the cyclopropanes. There are only three and one is chiral.

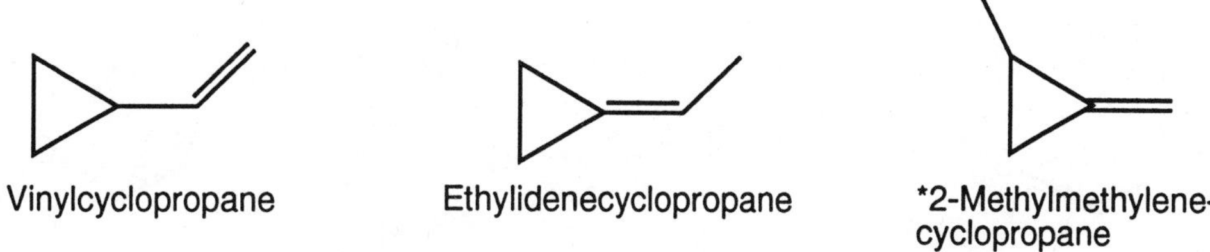

Finally, and most difficult, find the spiro and bicyclic isomers: There are five, one of which can exist in two forms called exo and endo. If you got these, you are really sharp.

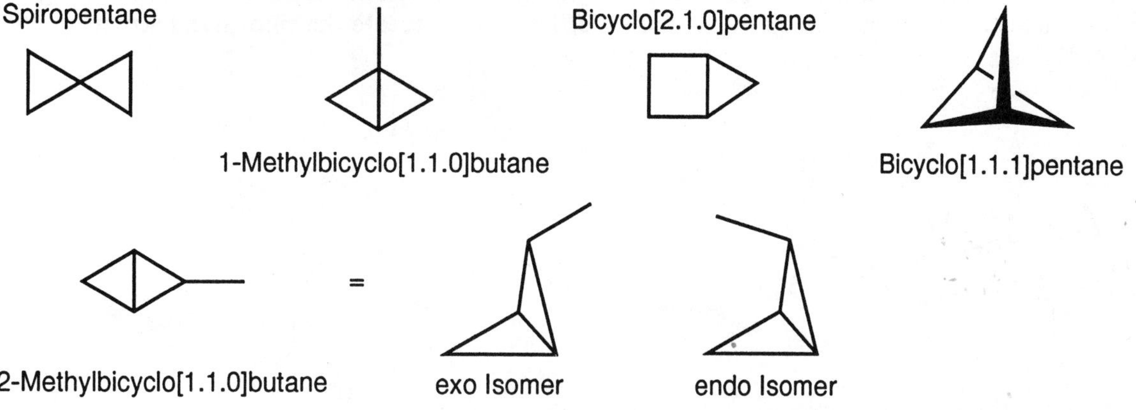

Problem 6.28 The cyclic molecules should be easy, but the 1,2-diene may cause problems unless you remember, look up, or work out the results of Problems 5.28 and 5.29 (pp. 185-186). The groups at the ends of the diene are in perpendicular planes, and this molecule is chiral.

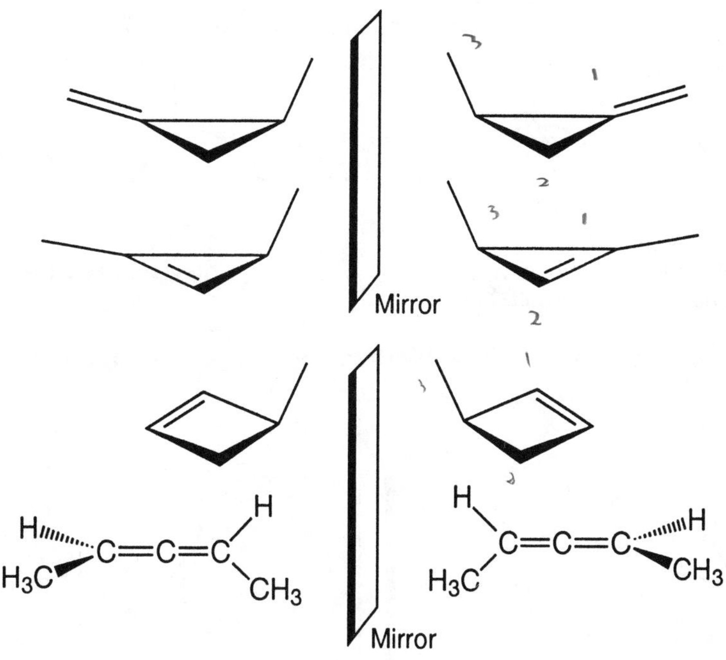

Problem 6.29 This problem should be a simple matter of assigning priorities. The molecules in question are all (S). The stereogenic carbons are shown as "C".

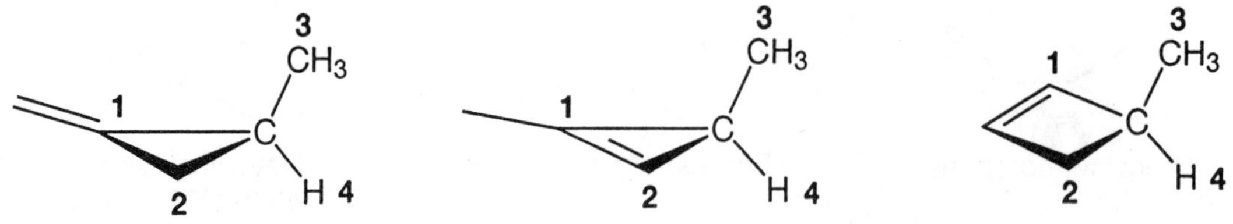

3 CH3 1 C 2 H 4 (S)

3 CH3 1 C 2 H 4 (S)

3 CH3 1 C 2 H 4 (S)

Problem 6.30 In the cis isomer, one methyl group must be equatorial and the other axial. Ring "flip" of one isomer produces an identical molecule. To see this, rotate the right-hand structure 180° as shown.

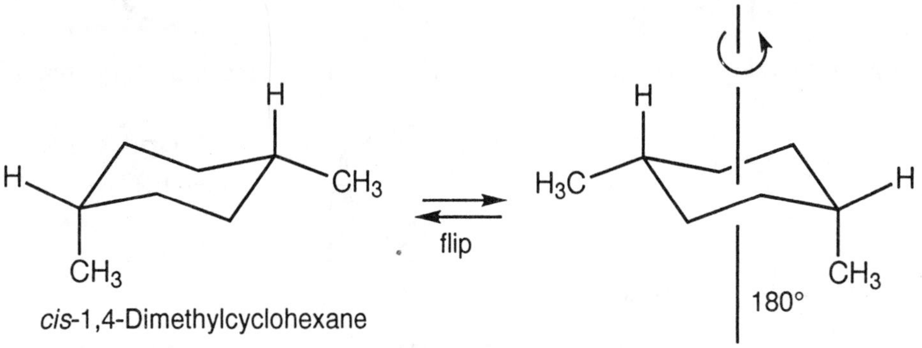

H CH3 H CH3 flip H3C H H CH3 180° These two forms are identical

cis-1,4-Dimethylcyclohexane

In the trans isomer both methyl groups are equatorial or both are axial. The molecule with the two groups equatorial will be far more stable.

These two forms are diastereomers

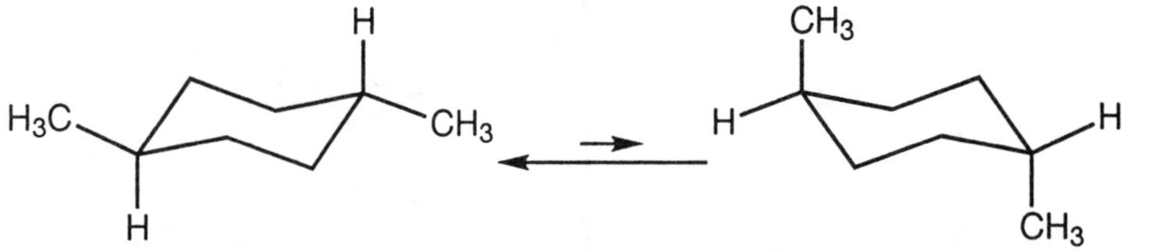

H H3C CH3 H CH3 H H CH3

trans-1,4-Dimethylcyclohexane diequatorial form = more stable

Diaxial form = less stable

Neither the cis nor trans compound is chiral. Both mirror images are superimposable on the original.

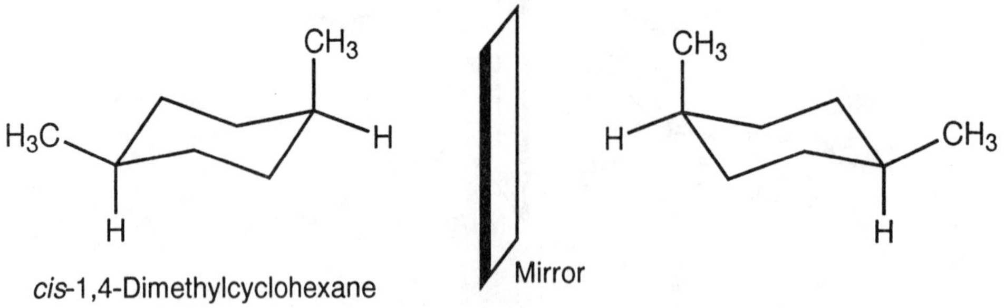

CH3 H3C H H Mirror CH3 H CH3 H

cis-1,4-Dimethylcyclohexane

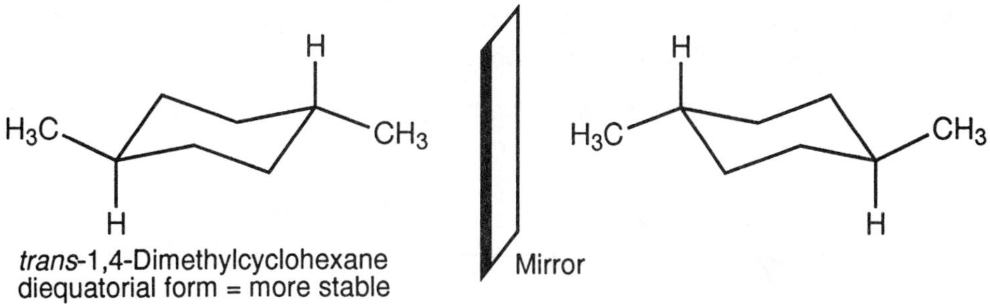

H, H_3C, CH_3, H

trans-1,4-Dimethylcyclohexane
diequatorial form = more stable

Mirror

H, H_3C, CH_3, H

Problem 6.31 In the cis form, it will be the diastereomer with the large isopropyl group equatorial that is favored.

These two forms are diastereomers

H, H, $CH(CH_3)_2$, CH_3

cis-1-Isopropyl-4-methylcyclohexane
(more stable diastereomer)

flip

H, H_3C, H, $CH(CH_3)_2$

(less stable diastereomer)

Similarly, in the trans form it is the diastereomer with both groups equatorial that is more stable.

These two forms are diastereomers

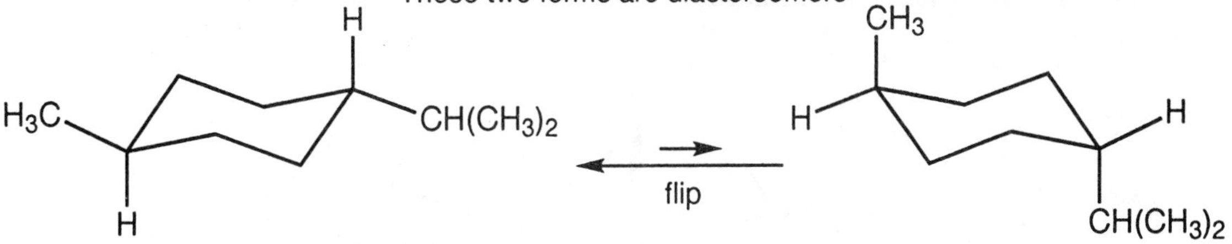

H, H_3C, $CH(CH_3)_2$, H

trans-1-Isopropyl-4-methylcyclohexane
(more stable diastereomer)

flip

CH_3, H, H, $CH(CH_3)_2$

This diaxial form is less stable

Once again, neither molecule is chiral:

In each case the mirror image is superimposable on the original

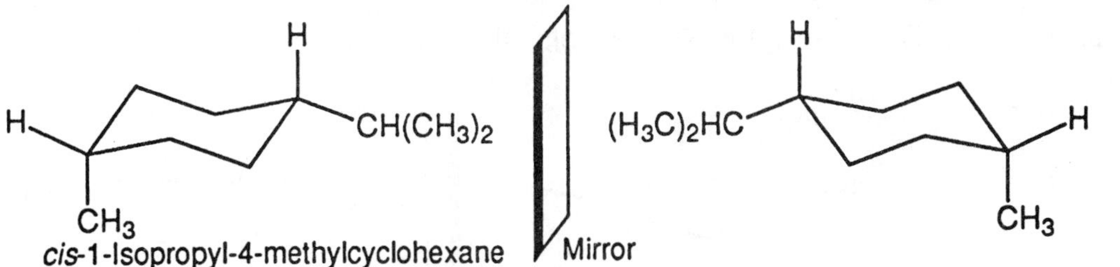

H, H, $CH(CH_3)_2$, CH_3

cis-1-Isopropyl-4-methylcyclohexane
(the more stable diastereomer)

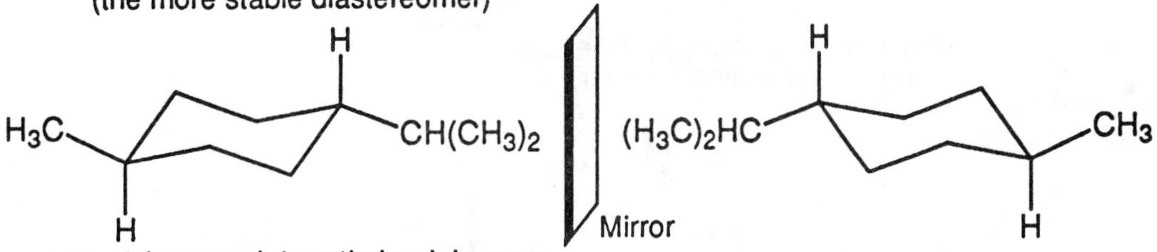

Mirror

H, $(H_3C)_2HC$, H, CH_3

H, H_3C, $CH(CH_3)_2$, H

trans-1-Isopropyl-4-methylcyclohexane
(the more stable diastereomer)

Mirror

H, $(H_3C)_2HC$, CH_3, H

Problem 6.32 The cis compound is more stable with the two methyl groups equatorial. Both the diequatorial and diaxial compounds are meso compounds.

These two forms are diastereomers - each is a meso compound

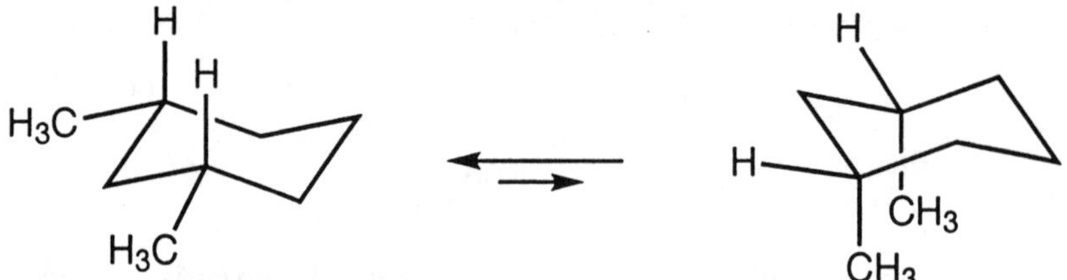

cis-1,3-Dimethylcyclohexane (more stable diastereomer)

(Less stable diastereomer)

The trans molecule "flips" into itself.

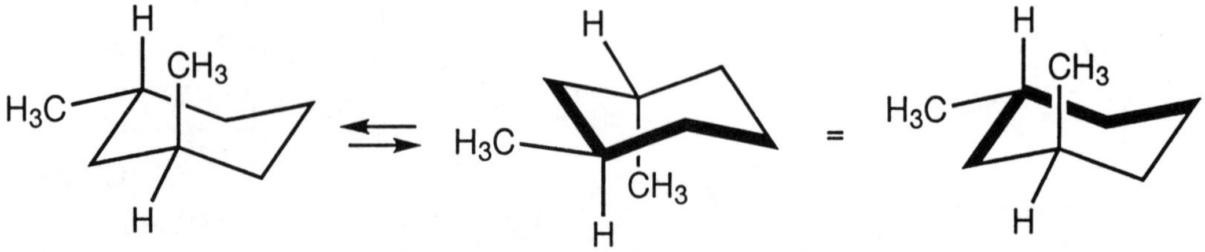

trans-1,3-Dimethylcyclohexane

These two "forms" are identical -just turn this one over, "back-to-front"; the bold bonds show the front-to-back transformation

This compound is chiral, as its mirror image is not superimposable upon the original:

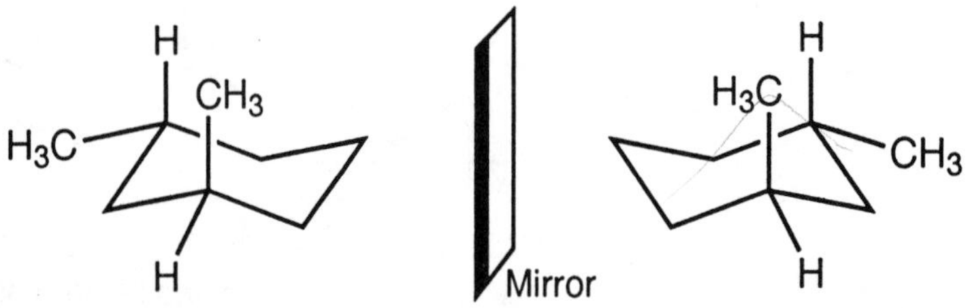

Mirror

Problem 6.33 The more stable conformation of *cis*-1-isopropyl-3-methylcyclohexane is the one with the two alkyl groups equatorial. Ring 'flip" generates the less stable conformation with the two alkyl groups axial. Each of these two diastereomers is chiral.

Neither mirror image is superimposable on the original—both diastereomers are chiral:

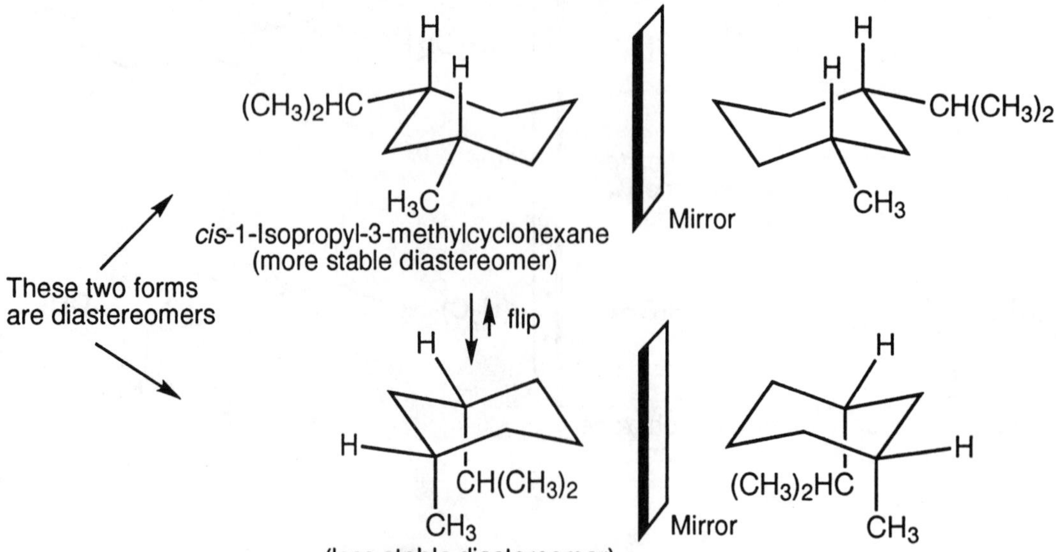

cis-1-Isopropyl-3-methylcyclohexane (more stable diastereomer)

These two forms are diastereomers

flip

Mirror

(less stable diastereomer)

The situation is the same in the trans molecule. The more stable conformation has the large ispropyl group equatorial. Ring "flip" generates the less stable conformation with the large isopropyl group axial. Each of these two diastereomers is chiral.

Neither mirror image is superimposable on the original—both diastereomers are chiral:

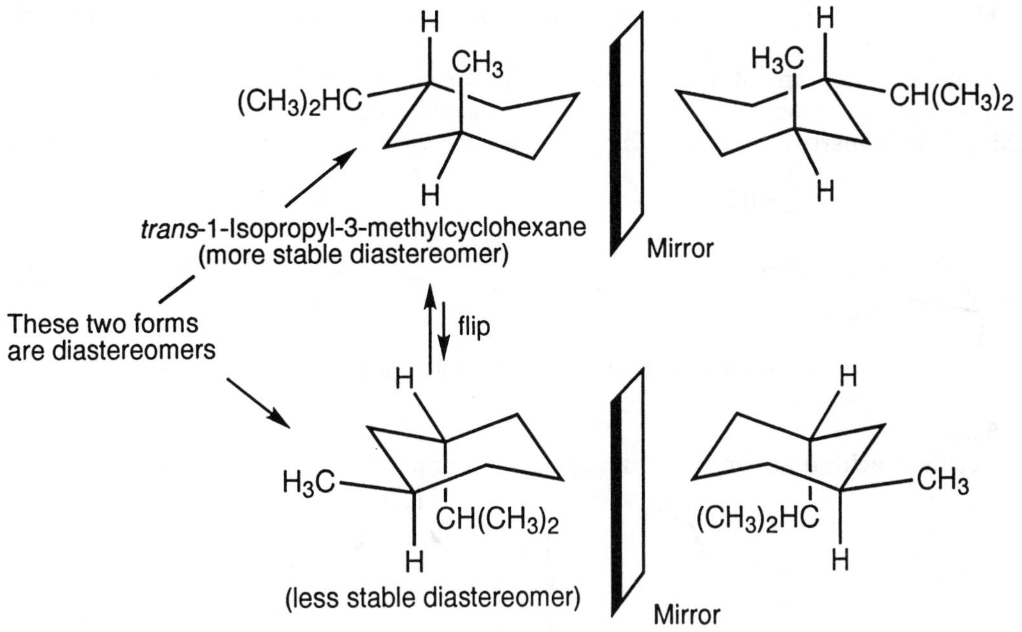

trans-1-Isopropyl-3-methylcyclohexane (more stable diastereomer)

These two forms are diastereomers

flip

(less stable diastereomer)

Mirror

Mirror

Problem 6.34 *cis*-1,4-Dimethylcyclohexane "flips" into itself. Thus there can be no energy difference between the two isomers.

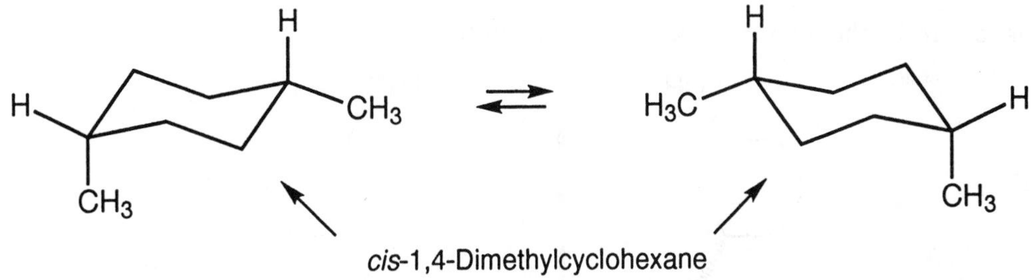

cis-1,4-Dimethylcyclohexane

In *trans*-1,4-dimethylcyclohexane, the molecule with two equatorial methyl groups will be more stable by twice the energy difference between equatorial and axial methylcyclohexane, or 2 x 1.74 = 3.48 kcal/mol (Table 6.4, p. 210).

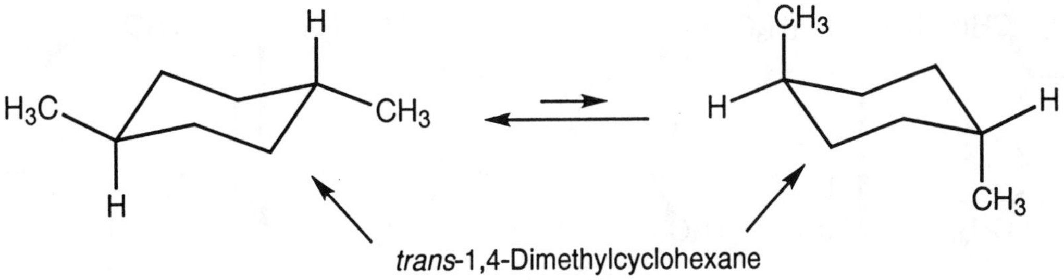

trans-1,4-Dimethylcyclohexane

Problem 6.35 Let's do the trans compound first. In this molecule both substituents are equatorial or both are axial. Table 6.4 tells us that an axial methyl group is destabilizing by 1.74 kcal/mol, and that an axial isopropyl group is destabilizing by 2.61 kcal/mol. Accordingly, the chair with both groups equatorial will be more stable than the chair with both groups axial by the sum of these numbers, 4.35 kcal/mol.

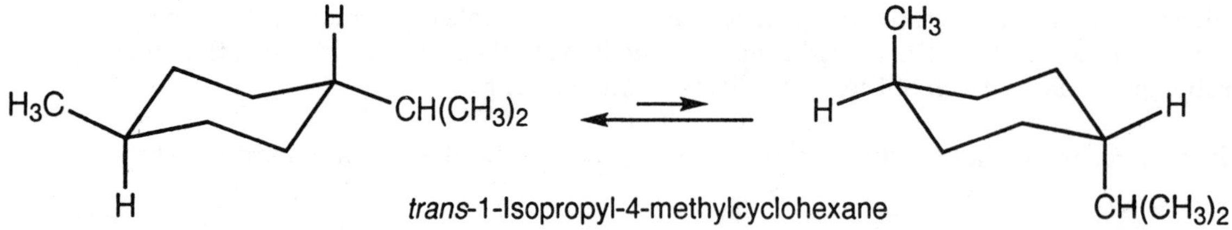

H3C CH(CH3)2 H H CH3 H H CH(CH3)2

trans-1-Isopropyl-4-methylcyclohexane

In the cis isomer of 1-isopropyl-4-methylcyclohexane, one chair has the methyl group axial, and the other has the isopropyl group axial. The right-hand diastereomer will be more stable by 2.61 – 1.74 = 0.87 kcal/mol. It is energetically more favorable to have the larger isopropyl group equatorial.

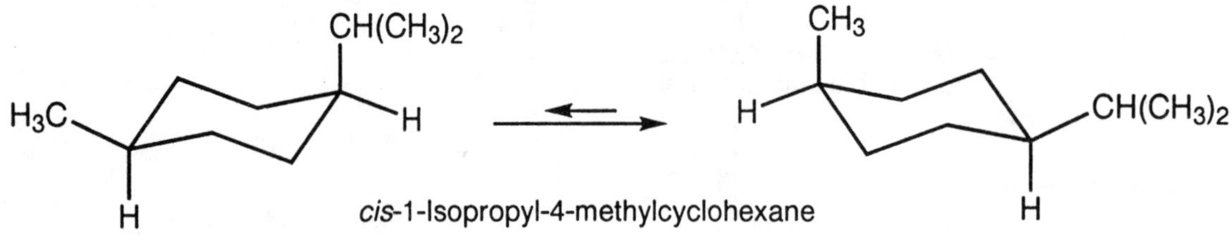

cis-1-Isopropyl-4-methylcyclohexane

Problem 6.36

(a) This molecule is achiral as the mirror image is superimposable on the original.

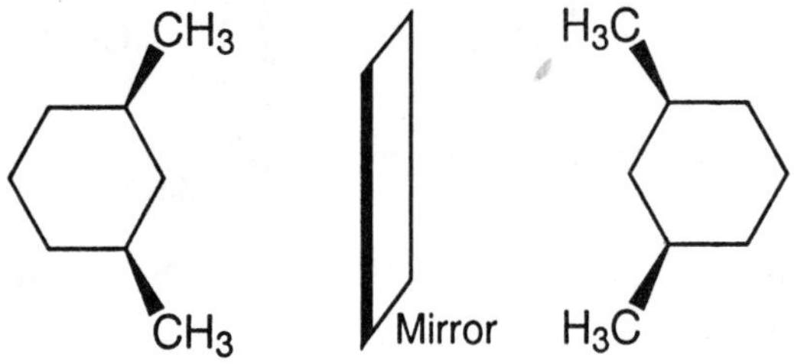

(b) This one is chiral—the mirror image is nonsuperimposable.

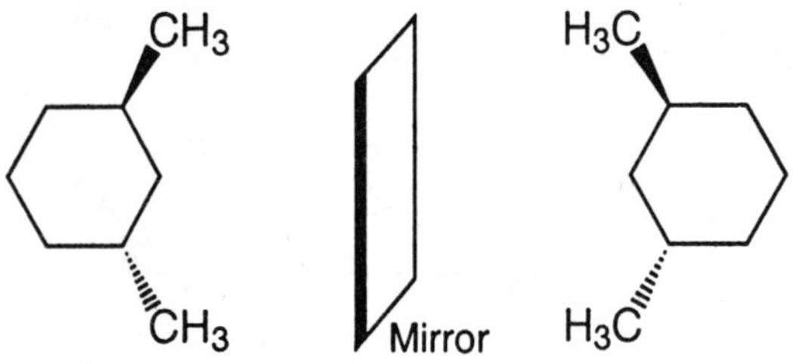

(c and d) Both are chiral.

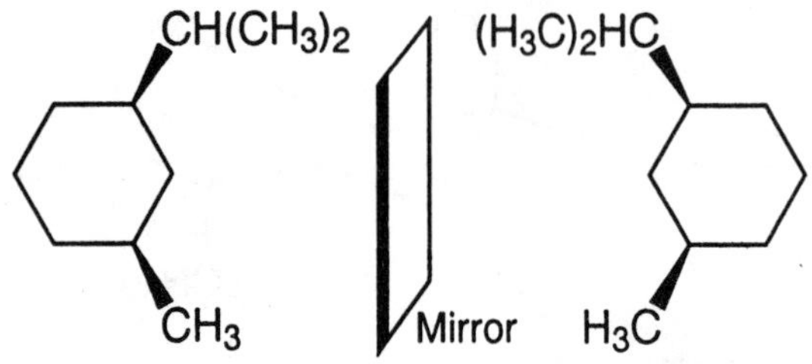

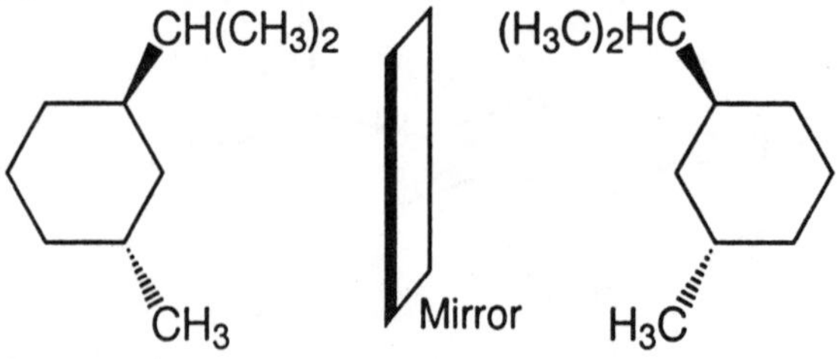

(e—h) All are achiral.

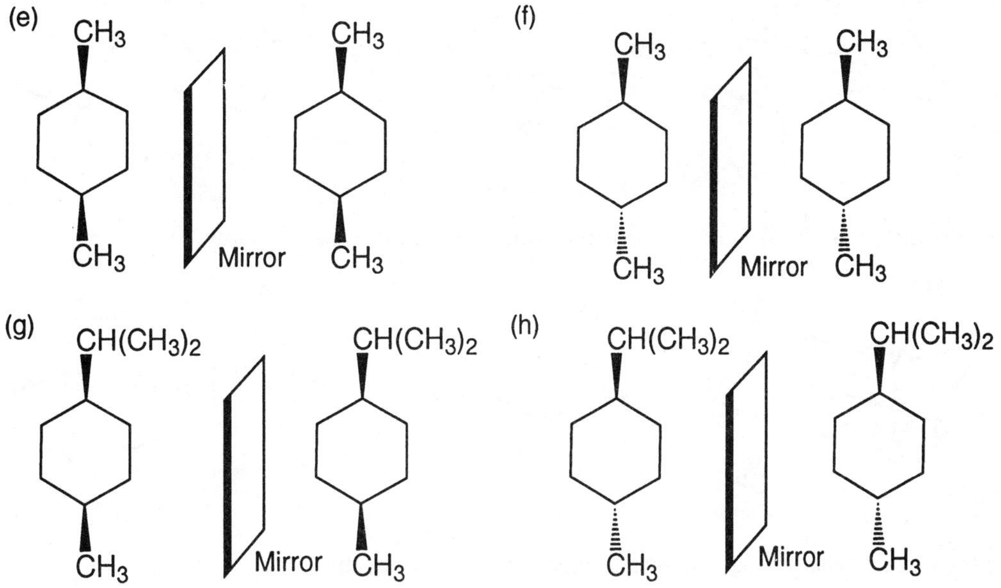

Problem 6.37
(a) We start with a tough call. The conformation with two groups equatorial (Cl and CH_3) is the more stable one, even though the largest group, isopropyl, must be axial.

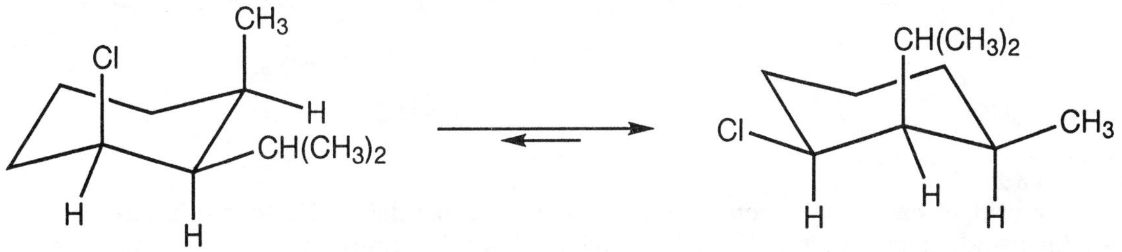

(b) This one is easy. The conformation with all three groups equatorial is the more stable one.

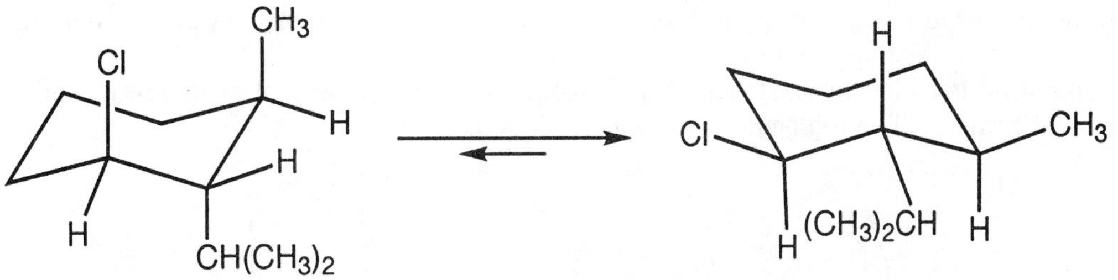

(c and d) Once again, the conformations with two groups equatorial are more stable than the conformations with two groups axial.
(c)

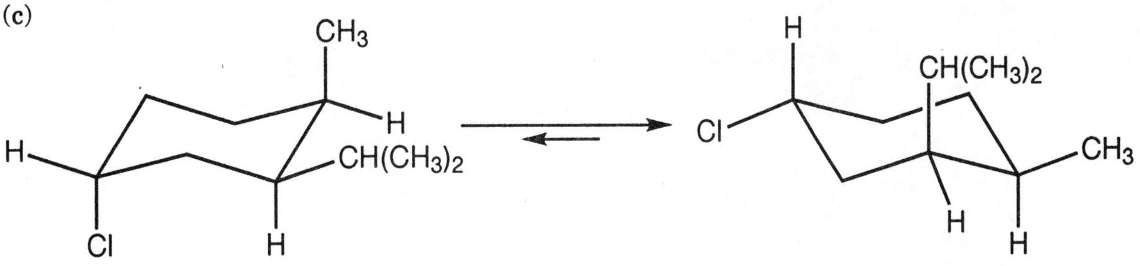

(d)

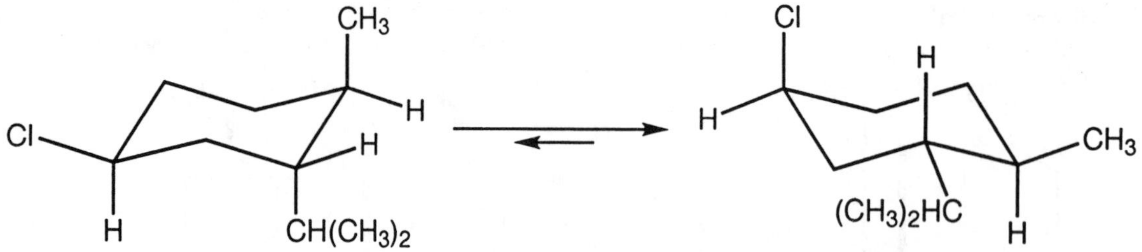

(e) The conformation with all three groups equatorial is the more stable one.

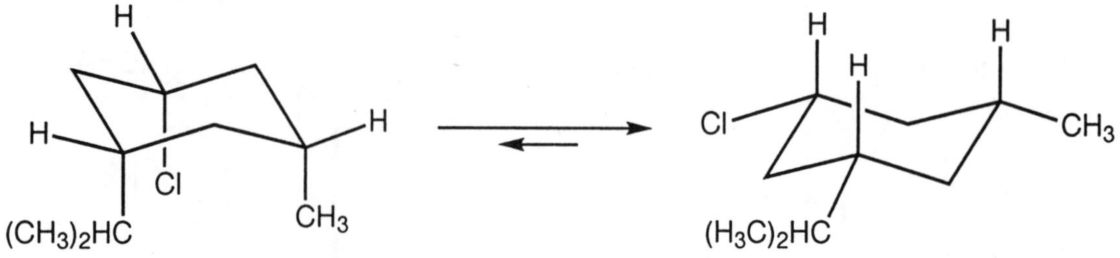

(f) The conformation with two groups equatorial is the more stable one.

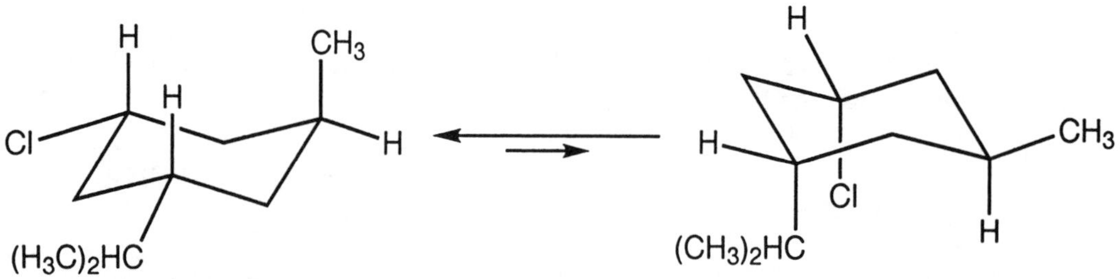

Problem 6.38

(a) All three bridges have two carbons, so the base name is bicyclo[2.2.2]octene. We start counting at the bridgehead, and give the double bond the lowest possible number, so the final name is bicyclo[2.2.2]oct-2-ene.

(b) We count first around the larger ring, so this compound is 2,2-difluorobicyclo[2.1.0]pentane.

(c) This compound is a bicyclo[3.1.1]heptene. We count around the largest bridge first, and give the OH the lowest number. The molecule is bicyclo[3.1.1]hept-3-en-2-ol.

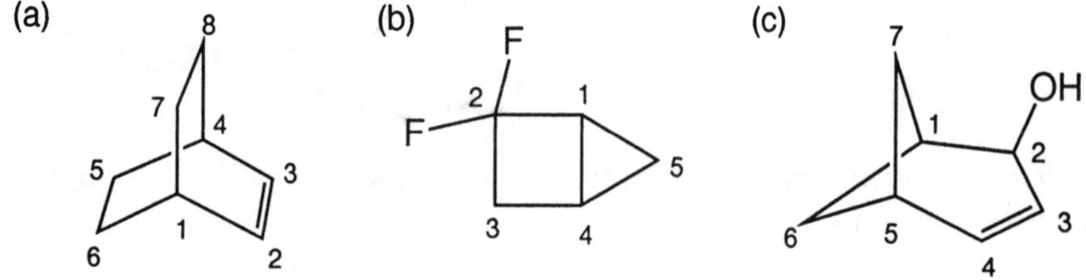

Problem 6.39 In (a), (b), and (c) remember to start counting at the bridgehead. In (c), no numbers are necessary for the methyl groups because this is the only possible hexamethyl compound.

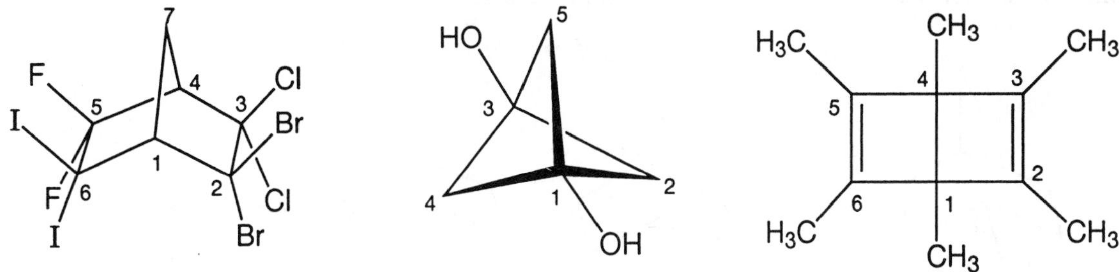

Problem 6.40 First, draw a chair with the oxygen taking the place of one of the ring carbons of cyclohexane. Then add the substituents, putting them all in the more favorable equatorial positions.

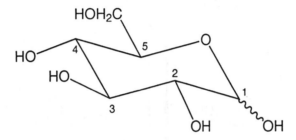

The OH at C(1) can be either "up", in the equatorial position, or "down" in the axial position.

Problem 6.41 There are three different carbons in bicyclo[2.2.1]heptane, and a methyl group could be attached to each of them. At the 1 or 7 position there is only one possible way of attaching a group, but at the 2-position the methyl (or any group) could point toward the short bridge or the longer bridge. These positions are called "exo" and "endo," respectively.

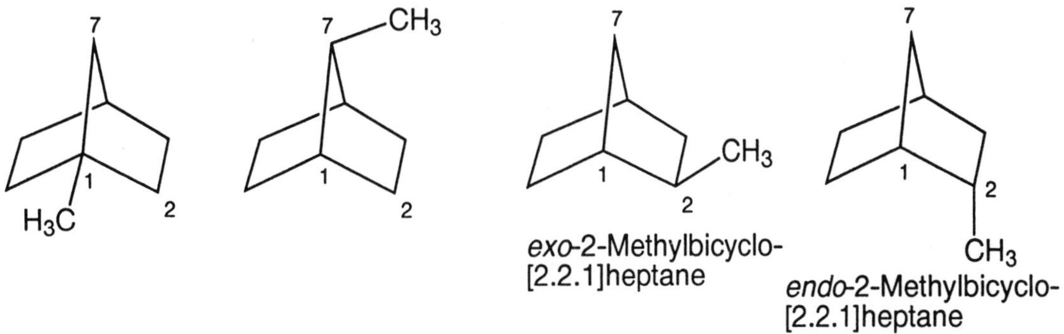

exo-2-Methylbicyclo-[2.2.1]heptane

endo-2-Methylbicyclo-[2.2.1]heptane

Problem 6.42

(a) No, there are no exo and endo forms here, as the two sides of the molecule are identical. There are only two structural isomers of methylbicyclo[2.2.2]octane.

1-Methylbicyclo[2.2.2]octane

All three structures are 2-methylbicyclo[2.2.2]octane

(b) Yes, 2-methylbicyclo[2.2.2]octane is chiral. Its mirror image is not superimposable on the original.

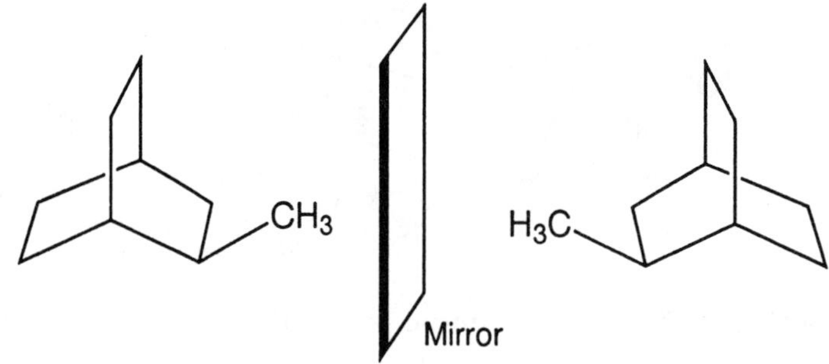

Chapter 7 Outline

7.1 Alkyl Halides: Nomenclature and Structure
7.2 Reaction Mechanism: Brønsted Acids and Bases
7.3 The Arrow Formalism
7.4 Lewis Acids and Lewis Bases
7.5 HOMO–LUMO Interactions
7.6 Reactions of Alkyl Halides: The Substitution Reaction
7.7 Substitution, Nucleophilic, Bimolecular: The S_N2 Reaction
7.8 Substitution, Nucleophilic, Unimolecular: The S_N1 Reaction
7.9 Overview of the S_N2 and S_N1 Reactions
7.10 The Unimolecular Elimination Reaction: E1
7.11 The Bimolecular Elimination Reaction: E2
7.12 What Can We Do with These Reactions? Synthetic Utility
7.13 Something More: The E1cB Reaction

7

Substitution and Elimination Reactions: The S_N2, S_N1, E2, and E1 Reactions

Chapter 7 emphasizes chemical change for the first time. Four different "building block" reactions are studied in detail: the S_N2 and S_N1 substitutions and the E2 and E1 eliminations. The problems in this chapter focus on relatively simple aspects of these four reactions. There will be many elaborations and complicated examples in complex settings, but those will come later. These reactions are so fundamental that we will be referring to them throughout the remainder of this book. It is crucial that these reactions become so familiar that when the elaborations appear, when complicated reactions of molecules are the issue, the mechanisms of these four reactions are second nature. So, here the emphasis will be on the design of substitution and elimination reactions, the prediction and rationalization of rates and products, and the synthetic consequences of these four reactions. The stereochemical consequences of the reactions are of prime importance throughout.

In the earlier chapters, we have been learning grammar and vocabulary. In Chapter 7, we will write some sentences with the words we have learned, and put them together with the use of the rules of grammar we have worked out. The problems in this chapter involve more sentences, and a few simple paragraphs. However, the future will see us writing some complicated essays. We will still need the grammar and vocabulary, and we shall still need to be able to write straightforward sentences when more complicated questions arise.

Problem 7.1 Conjugate acids and conjugate bases are related through the gain or loss of a single proton. The conjugate acids of the molecules shown are always "the molecule plus one proton."

(a) $H_3\overset{+}{O}$ (b) HOH (c) $\overset{+}{N}H_4$ (d) $CH_3\overset{+}{O}H_2$ (e) $H_2C{=}\overset{+}{O}H$ (f) CH_4

Sometimes there are choices to be made as to where to put the proton in making the conjugate acid. In this set, only $H_2C{=}O$ really presents such a choice. Another conjugate acid is $H_3C{—}O^+$, although this is much less stable than the one given above. (Why?)

Problem 7.2 The conjugate bases of these molecules will always be "the molecule less one proton."

(a) HO^- (b) O^{2-} (c) $^-NH_2$ (d) CH_3O^- (e) $^-CH_3$ (f) $^-OSO_2OH$ (g) $^-OSO_2O^-$

Problem 7.4 The HOMO is the filled 1*s* orbital on hydrogen and the LUMO is the empty $2p_z$ orbital of the methyl cation.

Problem 7.6 In these substitution reactions, the HOMO is always a filled orbital on the substituting group and the LUMO an empty, σ^* orbital of the molecule being substituted.

(a) HOMO: Filled nonbonding orbital of HO^-
LUMO: Empty σ^* orbital of CH_3—I.

(b) HOMO: Filled nonbonding orbital on sulfur.
LUMO: Empty σ^* orbital of a C—N bond.

(c) HOMO: Filled nonbonding orbital on C of cyanide.
LUMO: Empty σ^* orbital of the C—Cl bond.

(d) HOMO: Filled nonbonding orbital on F^-
LUMO: Empty σ^* orbital of the C—N_2^+ bond.

(e) HOMO: Filled nonbonding orbital on N.
LUMO: Empty σ^* orbital of the C—I bond.

Problem 7.8 An argument has just been made (p. 251) that steric factors are important in determining the rate of the S_N2 reaction. The more substituted the substrate, the slower the reaction. In tertiary substrates, the three R groups guard the rear of the C-L bond so efficiently that the S_N2 reaction completely fails. There should be no surprise that the size of the entering nucleophile is important as well. The larger the nucleophile [$(CH_3)_3C$—O^- > CH_3—O^-], the greater the steric interactions, and the slower the reaction.

Problem 7.10 Strain in the starting material will raise its energy and any change of this kind will lower the activation energy for a reaction, if there is no change in the energy of the transition state, a most unlikely prospect.

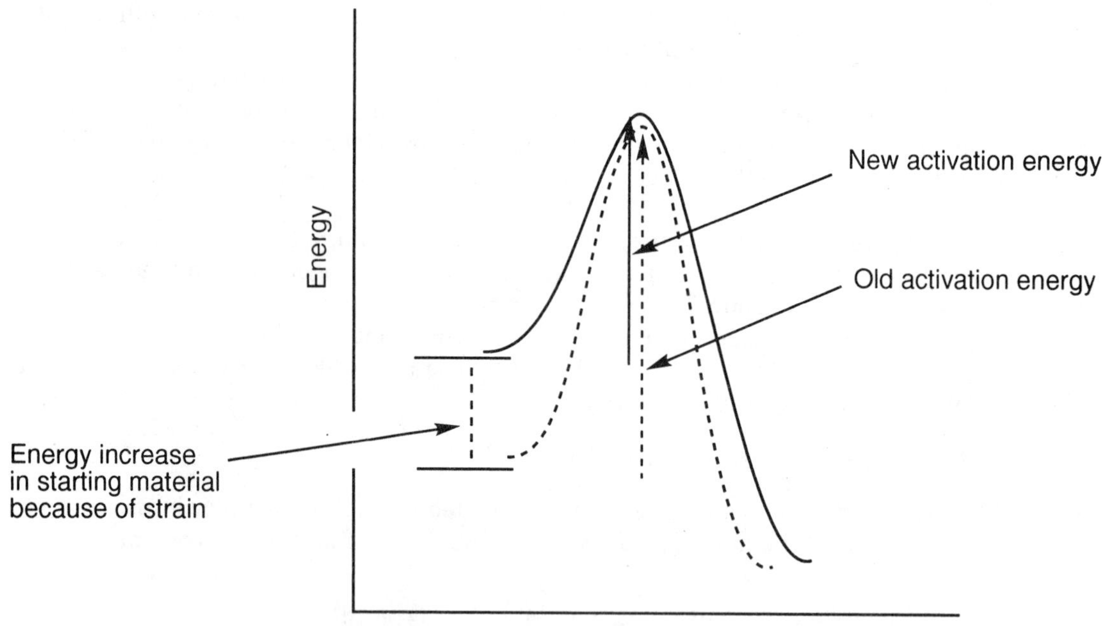

In fact, in this case the energy of the transition state is raised *more* than is the energy of the starting material. The hint tells us to consider the structures of both starting material and transition state. Let's focus on deviations from ideal angles. In the starting material, the cyclopropane "wants" something like 109°, but is restricted to 60°. There are some 49° of angle strain. In the transition state, the sp^2 carbon "wants" 120°, but is restricted to 60° (see Fig. 7.36). There are 60° of angle strain, and the transition state is more destabilized than the starting material by angle strain effects. The result is an increase in activation energy even though the starting material is destabilized by angle strain.

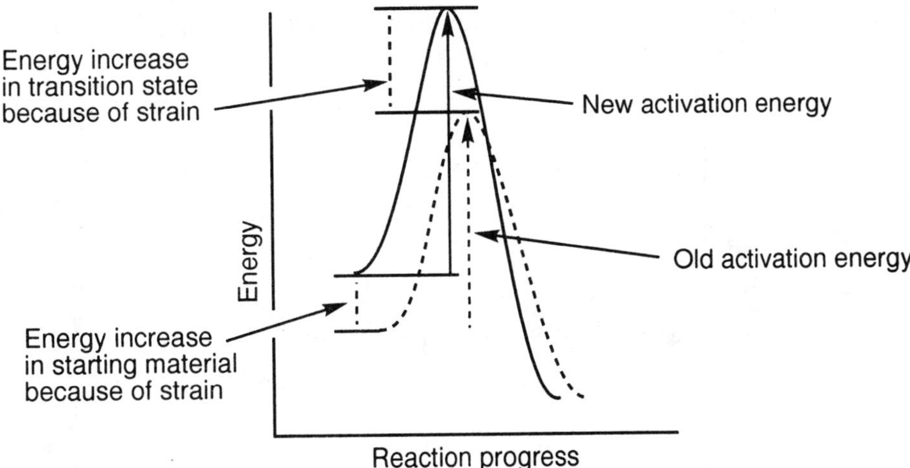

Problem 7.12 It is amazing how hard this question is. Perhaps our focus on displacement reactions makes it difficult to see the good old fashioned Brønsted acid-base chemistry. The strong base removes the proton from the alcohol to give the alkoxide ion, RO^-. You need to be able to consider more than one reaction at the same time.

R—Ö—H :B̄ ⇌ R–Ö:⁻ + H–B

Problem 7.13 Localized charge is bad. Delocalized charge is better. In the sulfonates, three oxygen atoms help bear the negative charge. This delocalization makes the sulfonate a relatively stable anion, and a good leaving group. The resonance forms show it well.

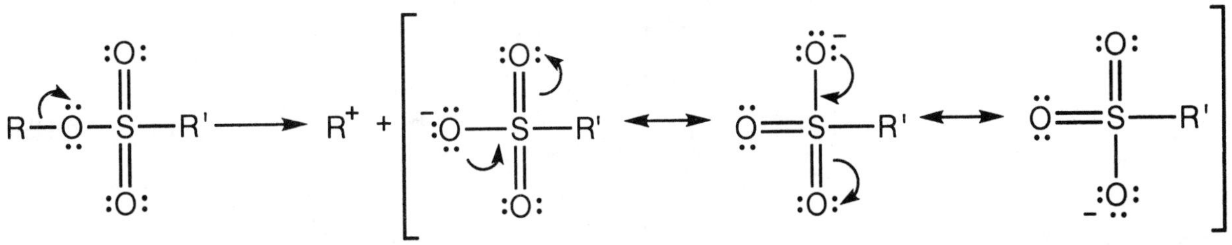

Problem 7.14 In this problem, you are asked to outline an arrow formalism for a reaction you have not yet seen, although you will certainly see this reaction later in this chapter. You need to reason backward from the product, visualize what must have happened, and then supply the arrows. The elements of HBr are eliminated to give the alkene isobutylene. The formula C_4H_8 suggests this much. Here is the product along with an arrow formalism for the process.

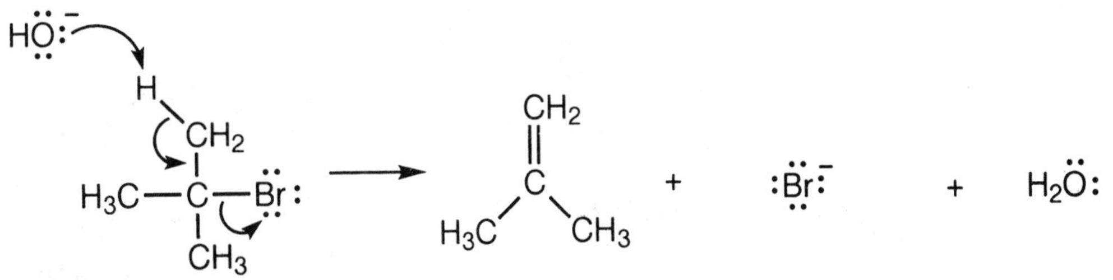

Problem 7.15 The S_N1 reaction is usually more favorable for tertiary halides such as the one shown at the top of Figure 7.60. However, this particular secondary halide leads to an especially stable carbocation. Localized charge is bad, and delocalized charge is better. The "cyclohexatriene"

ring provides a way to delocalize the charge generated on ionization. This delocalization makes this secondary carbocation relatively stable, and therefore relatively easy to form.

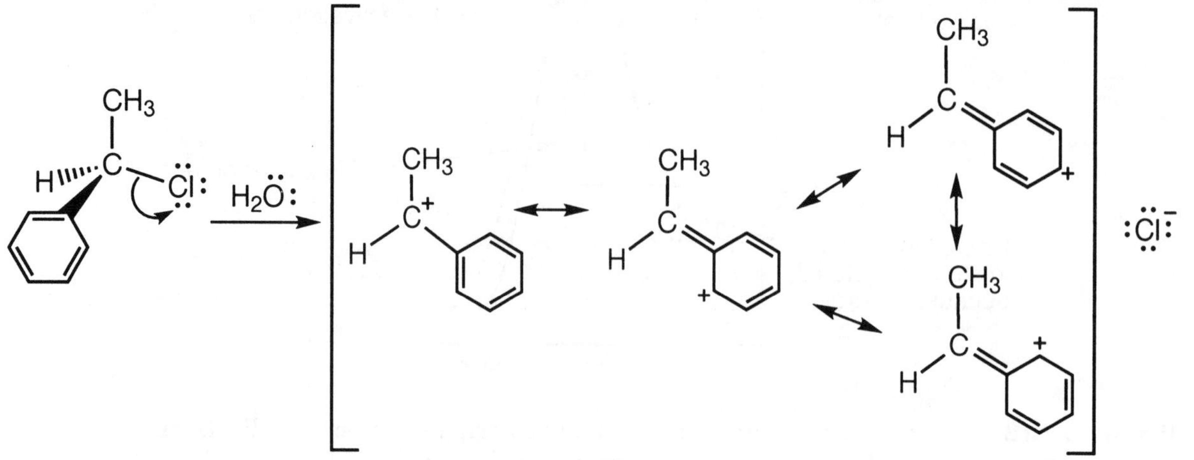

Several resonance forms make for a stable species

Problem 7.18

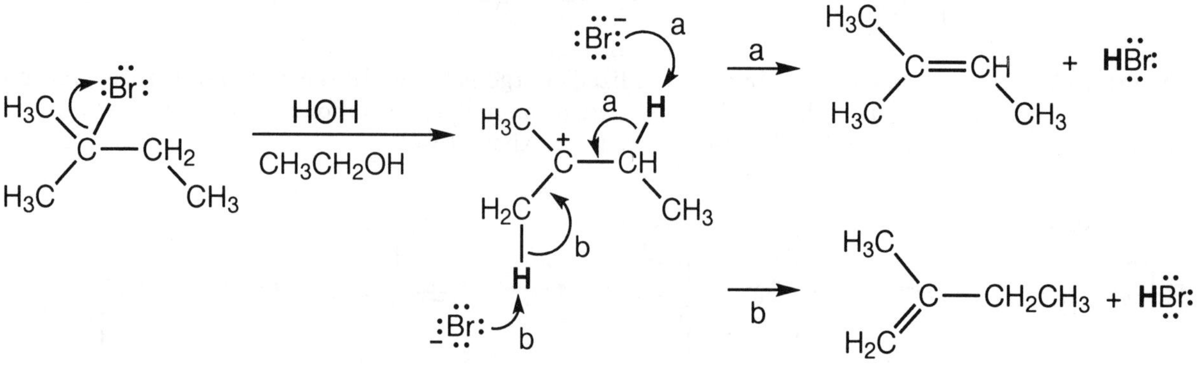

A key point here is the assistance in loss of H^+ by a base, here shown as bromide ion.

Problem 7.19 The S_N1 reaction involves ionization followed by product formation through addition of a nucleophile. The carbocation can be captured by any of the nucleophiles present. If bromide ion captures the carbocation, starting material is regenerated. The other nucleophiles present are ethyl alcohol (CH_3CH_2OH) and water. Reaction of these molecules with the carbocation leads ultimately to an alcohol and an ethyl ether:

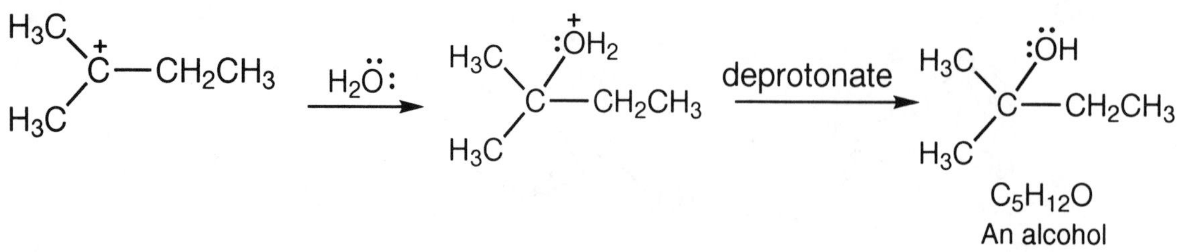

$C_5H_{12}O$
An alcohol

$$(H_3C)_2\overset{+}{C}{-}CH_2CH_3 \xrightarrow{CH_3CH_2\ddot{O}H} (H_3C)_2C(CH_2CH_3){-}\overset{+}{\ddot{O}}(H)CH_2CH_3 \xrightarrow{\text{deprotonate}} (H_3C)_2C(CH_2CH_3){-}\ddot{O}CH_2CH_3$$

$C_7H_{16}O$
An ethyl ether

Problem 7.20 The interesting case here is the last, bicyclic molecule. Ionization to the bridgehead cation cannot take place, even though the ion is tertiary. The bicyclic system ensures that the intermediate cation cannot become planar, and that destabilization is enough to stop the reaction. The lesson here is that you can never suspend thought. You have to look critically at each reaction intermediate in every reaction.

(a) $(CH_3CH_2)_3C{-}\ddot{B}r:$ —$H_2\ddot{O}:$, S_N1→ cation; S_N1, $H_2\ddot{O}:$, deprotonate → $(CH_3CH_2)_3C{-}\ddot{O}H$; E1, deprotonate → alkene

(b) $H_2\ddot{O}:$, S_N1 → cation; S_N1, $H_2\ddot{O}:$, deprotonate → :OH product; E1, deprotonate → alkenes

Minor + Major

(c) $H_2\ddot{O}:$, S_N1 (blocked) → No reaction

Problem 7.21 This problem is hard. First of all, the order of cation stability is tertiary > secondary >> primary > methyl. Ordinarily, a primary halide such as isobutyl chloride would not react in S_N1 fashion. In this case, the presence of silver ion allows the formation of a partial positive charge (δ^+) on the primary carbon as the silver ion "pulls" on the chloride.

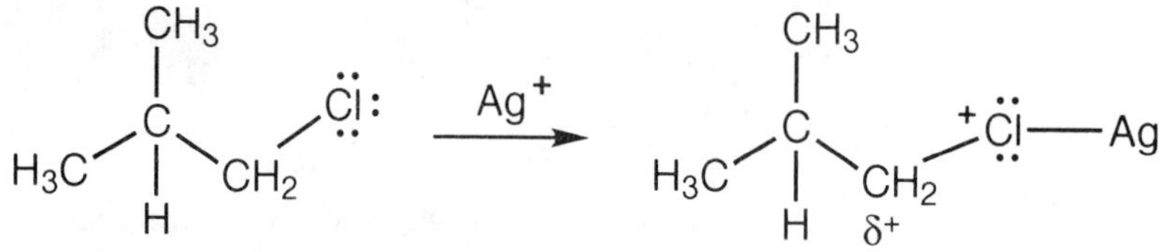

Now, if the hydrogen with its pair of electrons (hydride) migrates as the chloride leaves, a relatively stable tertiary carbocation is formed. Capture by acetate gives *tert*-butyl acetate (S_N1), and deprotonation gives isobutene (E1).

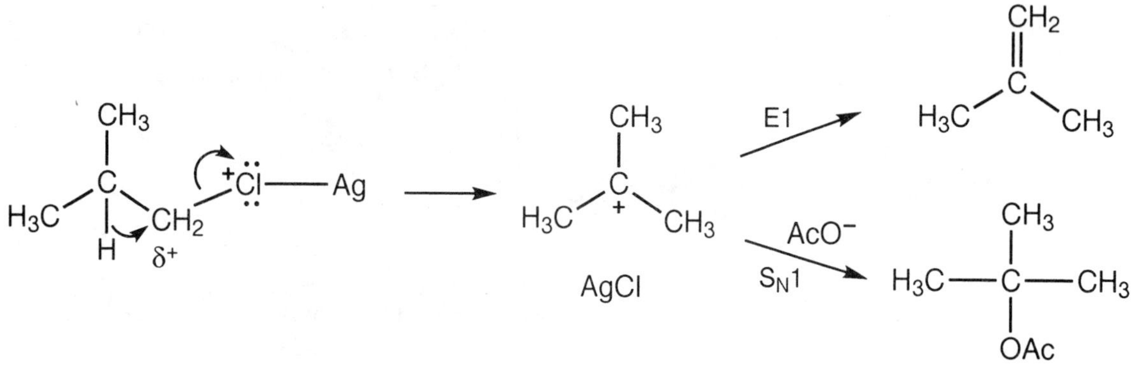

How is one to do such a strange problem? You will see many similar "rearrangements" in later chapters, but there can be no denying that right now, with little or no experience in this sort of thing, the migration is hard to spot. One suggestion is to work backward, asking the question, How can *tert*-butyl acetate be formed? The answer to this is simple, From the *tert*-butyl cation and acetate ion, AcO^-. The next question is, How can the *tert*-butyl cation be formed from something that might come from isobutyl chloride? Now you might be pardoned for imagining that the isobutyl cation is formed and that the hydride migrates. This answer involves a tiny error, as the hydrogen actually probably moves as the chloride begins to leave (this avoids formation of the most unstable primary carbocation), but it is a long way toward getting the question right. In many hard questions (and lots of easy ones), the road to success is to work backward.

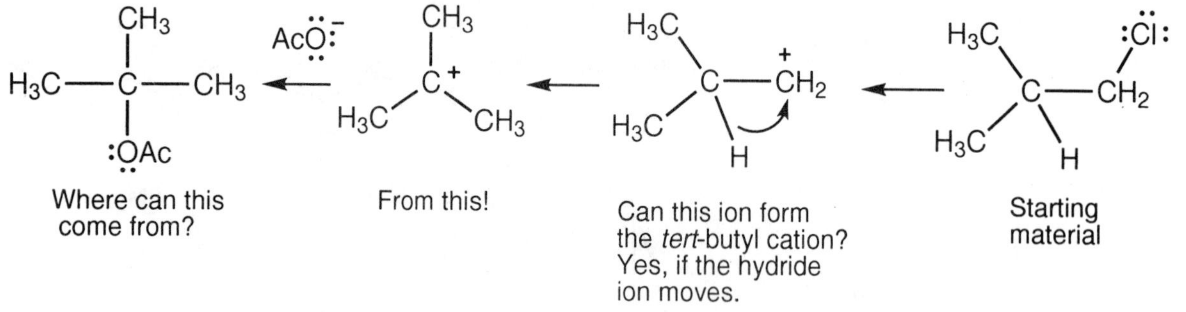

Where can this come from?

From this!

Can this ion form the *tert*-butyl cation? Yes, if the hydride ion moves.

Starting material

Problem 7.22. There is a second E2 elimination possible in which one of the six equivalent methyl hydrogens is lost.

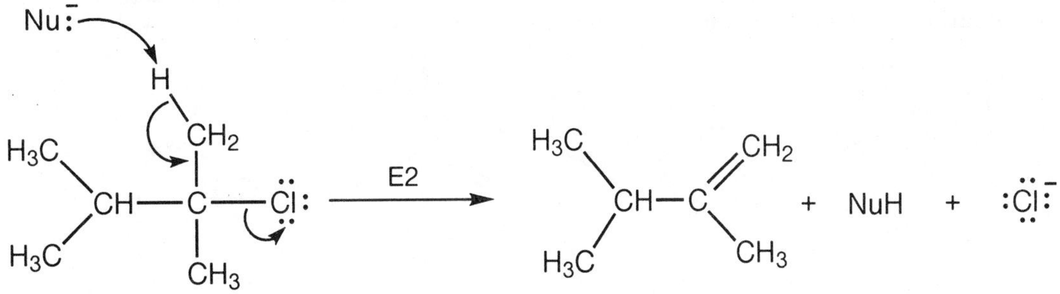

Problem 7.24 The possible isomers are presented in both schematic and three-dimensional form. In the schematics, the heavy dot means "chlorine up."

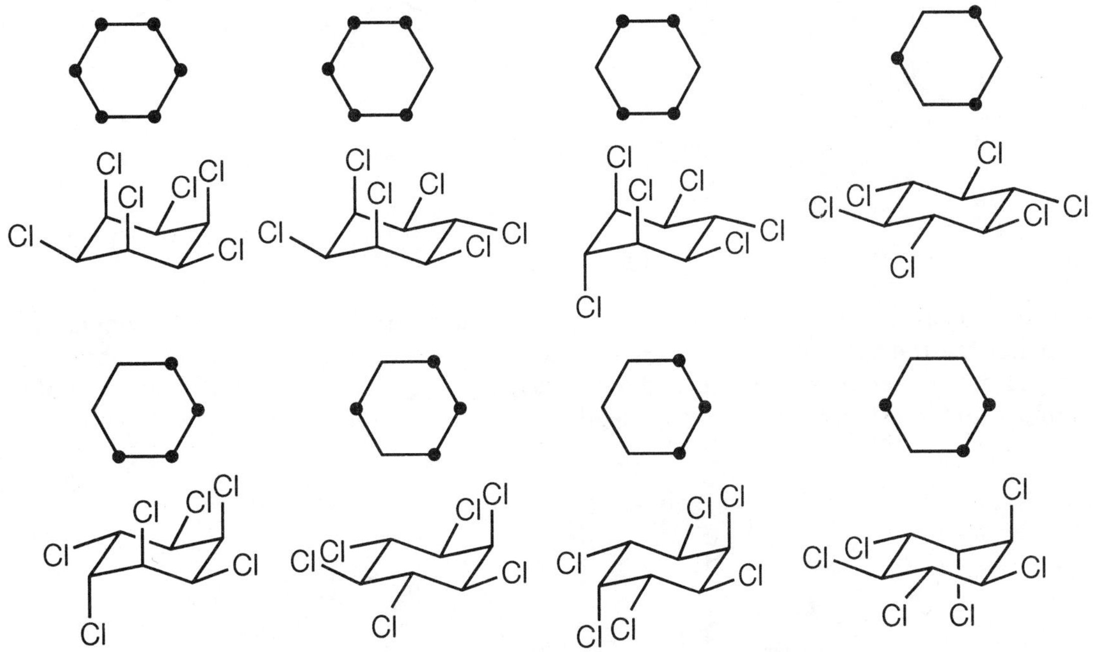

Problem 7.25 The hint should aim you in the right direction. You are looking for the ideal E2 reaction, and that requires a 180° alignment of the breaking carbon-hydrogen and carbon-chlorine bonds. In one of the isomers, there is no possible 180° anti elimination. A 180° anti elimination requires axial carbon-hydrogen and carbon-chlorine bonds. The rate of elimination of HCl from this isomer is about 3000 times slower than from the others.

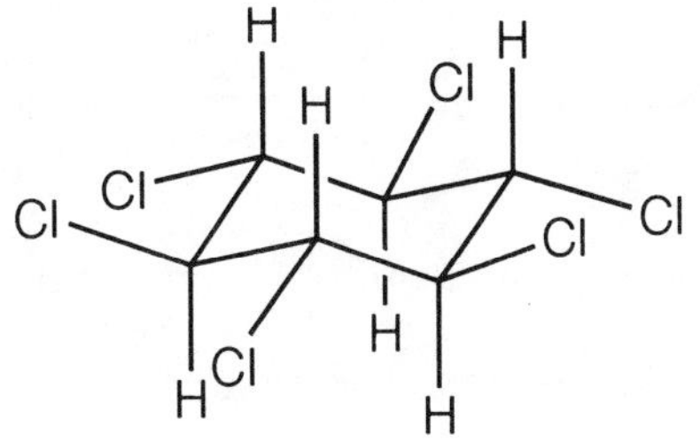

Problem 7.27 In the first molecule, no S_N2 reaction is possible at the tertiary bromide, so the products will be formed entirely through the E2 reaction. The more substituted material will be the major product. In the second molecule, the S_N2 reaction is possible and the products will include both substitution (alcohol) and elimination products (alkenes).

(a) Br — E2 (S_N2 not possible) → [alkene] > [alkene]

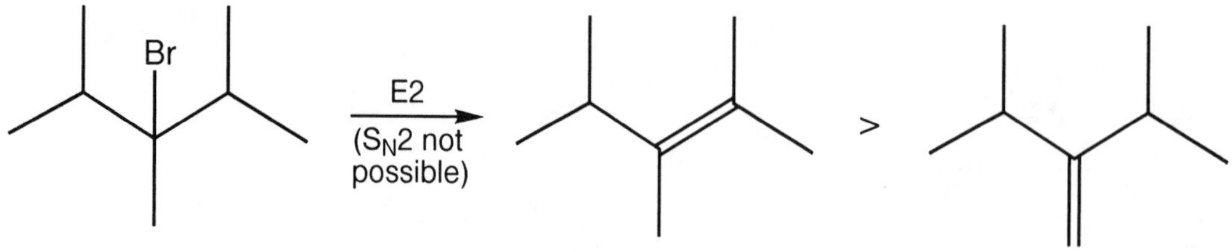

(b) Br — E2 → [alkene] > [alkene] > [alkene]; S_N2 → OH

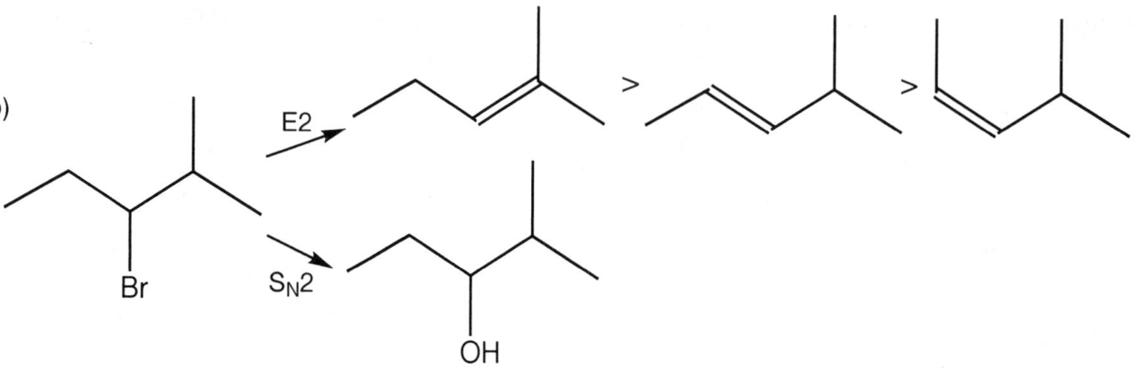

The third molecule poses a classic question in organic chemistry. This bridgehead bromide can undergo neither the S_N2 nor the E2 reaction. There is no way for an incoming nucleophile to attack the rear of the carbon-bromine bond, so the S_N2 reaction cannot occur. The E2 reaction would lead to an impossibly twisted alkene, a "bridgehead alkene" (Chapter 4, p. 139), and will not occur.

(c) $\ddot{:}Nu^-$ Br

S_N2 reaction impossible

The orbitals making up this "double bond" do not, in fact, overlap at all well, and the E2 reaction does not occur

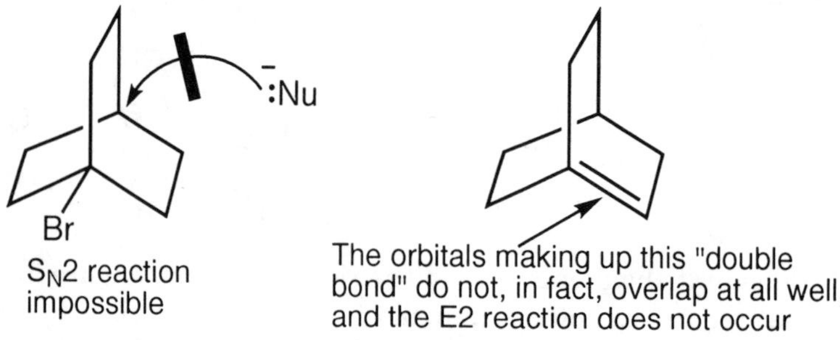

Problem 7.28 Ammonia (pK_a = 38) is a much stronger acid than is an alkene (pK_a ~ 45). Therefore, the amide ion $^-NH_2$ will not be strong enough to make significant amounts of carbanion through reaction with ethylene.

$$:\ddot{N}H_2^- + H_2C{=}CH_2 \rightleftharpoons :NH_3 + H\ddot{C}^-{=}CH_2$$

pK_a = 45, weaker acid pK_a = 38, stronger acid

Problem 7.29 This problem is an exercise in chosing among competing substitution and elimination reactions.

(a) $HO^- + CH_3Br \longrightarrow HO{-}CH_3$ (Simple S_N2)

(b) $HO^- + (CH_3)_3CBr \longrightarrow (CH_3)_2C{=}CH_2$

E2 only. The S_N2 is impossible at the tertiary carbon

(c) $H_2O + (CH_3)_3CBr \longrightarrow (CH_3)_3C{-}OH + (CH_3)_2C{=}CH_2$

A combination of E1 and S_N1 reactions is likely for this tertiary bromide

Problem 7.30. This is your first synthesis question. ***Always, always do synthesis questions backward.*** Ask yourself, "What reaction can make the target molecule?" Do not start with the overall problem of taking a starting material all the way through to the target. This early in your studies synthesis questions may well be easy enough to do forward. Later, however, these problems will be hard enough so that doing them backward will probably be necessary. The first four examples involve S_N2 reactions, and (e) involves an S_N1 reaction.

(a) N₃-substituted chain ⟵ ($Na^+ \ {}^-N_3$) ⟵ I-substituted chain

(b) CN-substituted structure ⟵ ($K^+ \ {}^-CN$) ⟵ I-substituted structure

(c) alkyne product with $C{\equiv}C{-}CH_3$ ⟵ ($Na^+ \ {}^-C{\equiv}C{-}CH_3$) ⟵ I-substituted chain

$Na^+ \ {}^-C{\equiv}C{-}CH_3 \longleftarrow HC{\equiv}C{-}CH_3$ (NH_3, $Na^+ \ {}^-NH_2$)

(d) $(CH_3)_3C{-}O{-}CH_3 \longleftarrow (CH_3)_3C{-}O^- + I{-}CH_3$

$(CH_3)_3C{-}O^- \ Na^+ \longleftarrow (CH_3)_3C{-}OH + Na$

$(CH_3)_3C{-}OH \longleftarrow (CH_3)_3C{-}I$ (H_2O)

(e) $(CH_3)_3C{-}N_3 \longleftarrow Na^+ \ {}^-N_3 + (CH_3)_3C{-}I$ (polar nonnucleophilic solvent)

Problem 7.31

(a) 2-Bromobutane will be faster in the S_N1 reaction (secondary carbocation more stable than a primary carbocation) and 1-bromobutane faster in the S_N2 reaction (primary bromide less sterically hindered than the secondary bromide).

(b) Cyclopentyl chloride will be faster in the S_N1 reaction and 1-chloropentane faster in the S_N2 reaction for the same reasons as in (a).

(c) 1-Iodopropane will be faster in the S_N2 reaction because iodide is a better leaving group than is chloride. Neither molecule will undergo the S_N1 reaction because a primary carbocation must be formed.

(d) *tert*-Butyl iodide will be faster in the S_N1 reaction because the tertiary carbocation is more stable than a secondary carbocation and therefore formed more easily. Isopropyl iodide will be faster in the S_N2 reaction because it is less sterically hindered. Approach to the rear of the carbon-iodine bond is less hindered in the secondary system than in the tertiary system.

(e) Neither *tert*-butyl iodide nor *tert*-butyl chloride will react at all in the S_N2 reaction. *tert*-Butyl iodide will be more reactive in the S_N1 reaction because iodide is a better leaving group than is chloride.

Problem 7.32 For a good E1cB reaction, a poor leaving group and a highly acidic hydrogen are needed.

Want an acidic H so that it can be easily removed by the nucleophile.

Want a poor leaving group so that the E1 and E2 reactions will be slow.

The following molecule fills the bill. Methoxide is a poor leaving group and the two $COOCH_3$ (methyl ester) groups will stabilize the negative charge on an adjacent carbon by resonance, thus making the loss of a proton easier. There are many other possible answers, of course.

Acidic hydrogen

Poor leaving group

Additional Problem Answers

Problem 7.33 The best procedure here is to draw out full Lewis structures for the starting materials and possible conjugate bases and then look for differences in energy between the products.

(a) $CH_3CH_2—C≡N: \longrightarrow CH_3\bar{C}H—C≡N:$ or $^-:CH_2CH_2—C≡N:$

(b) $HO—CR_2—H \longrightarrow HO—\bar{C}R_2$ or $^-:O—CR_2—H$

(c) $CH_3CH_2—C(=O)—OR \longrightarrow CH_3\bar{C}H—C(=O)—OR$ or $^-CH_2CH_2—C(=O)—OR$

In part (a), one of the two possible anions is stabilized by resonance, the other is not. In one case, the negative charge is shared by two atoms, a carbon and a nitrogen. This structure will be much more stable than the other possibility in which the negative charge is localized on a single carbon. Part (c) is similar; one anion is delocalized, the other is not.

(a) $[CH_3\ddot{C}H^{-}-C\equiv N: \longleftrightarrow CH_3CH=C=\ddot{N}:^{-}]$ more stable than $^{-}\ddot{C}H_2CH_2-C\equiv N:$

(c) $[CH_3\ddot{C}H^{-}-C(=O)OR \longleftrightarrow CH_3CH=C(O^{-})OR]$ more stable than $^{-}\ddot{C}H_2CH_2-C(=O)OR$

In (b), the anion on the more electronegative oxygen is more stable than the anion on the less electronegative carbon.

(b) $^{-}:\ddot{O}-CHR_2$ more stable than $H\ddot{O}-\ddot{C}R_2^{-}$

Problem 7.34 The choices are

(a) $(CH_3)_3C-C(=O)OH \xrightarrow{HA} (CH_3)_3C-C(=\overset{+}{O}H)OH$ or $(CH_3)_3C-C(=O)\overset{+}{O}H_2$

(b) $(H_3C)_2C=\ddot{O} \xrightarrow{HA} (H_3C)_2C=\overset{+}{O}H$ or $(H_3C)_2CH-\ddot{O}^{+}$

(c) $(H_3C)_2C=CHCH_3 \xrightarrow{HA} (H_3C)_2\overset{+}{C}-CH_2CH_3$ or $H_3C-CH(CH_3)-\overset{+}{C}HCH_3$

In parts (a) and (b) one of the two forms is resonance stabilized and thus more stable than the other, in which there is a localized positive charge.

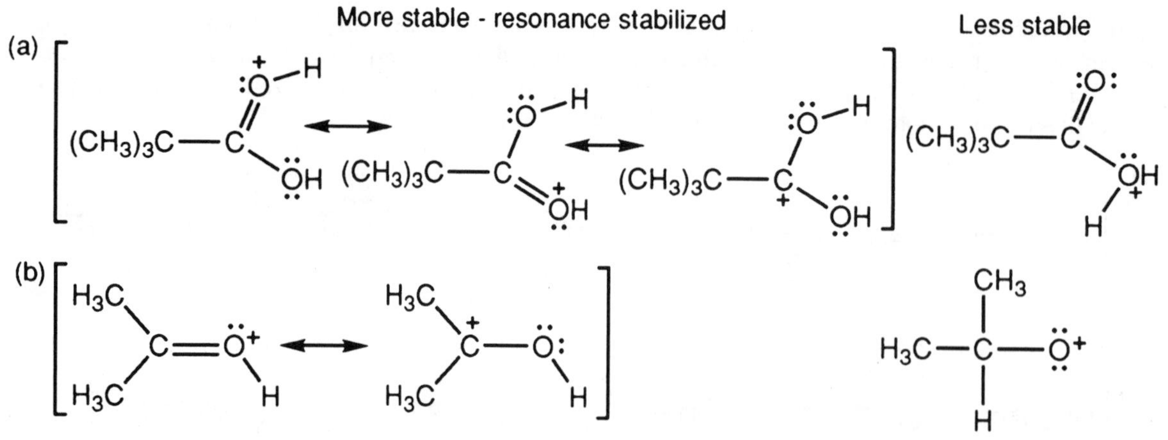

(a) More stable - resonance stabilized / Less stable

(b)

In part (c) the choice is between the more stable tertiary carbocation and the less stable secondary carbocation.

(c) more stable than

Problem 7.35 The purpose of this problem is to provide practice in using the arrow formalism to map out where new bonds are made and old ones are broken. These are shown as bold lines in the following figures. In (a), the base hydroxide plucks a proton from one methyl group. Water and a resonance-stabilized carbanion are formed, as the new oxygen-hydrogen bond is made and the carbon-hydrogen bond is broken.

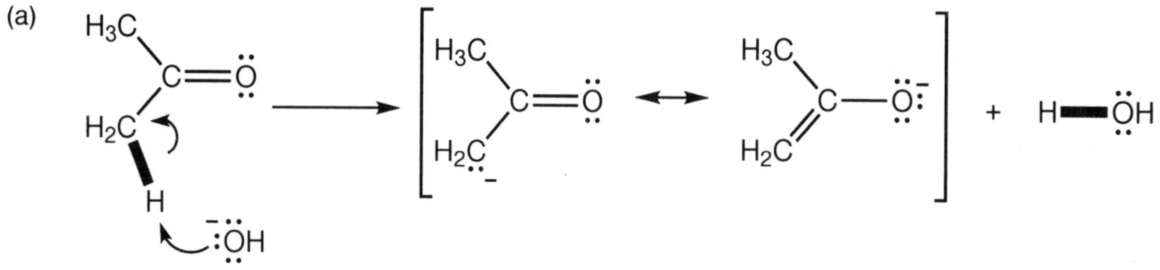

(a)

In (b) a new oxygen-carbon bond is formed as the hydroxide adds to the carbon-oxygen double bond.

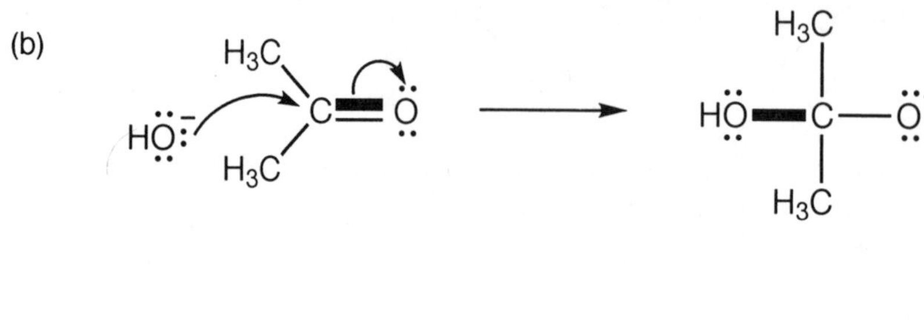

(b)

In (c), the oxygen of the carbon-oxygen double bond acts as base toward the acid H_3O^+. An oxygen-hydrogen bond is both made and broken in this reaction.

(c)

$(CH_3)_3C-C(=O)OH + H-OH_2^+ \longrightarrow (CH_3)_3C-C(=O^+H)OH + :OH_2$

Problem 7.36 This problem should be fairly easy, except for (d) and (e), which ask you to devise intramolecular S_N2 reactions. Nothing is fundamentally changed from the early parts of this problem, but it always seems difficult in the beginning to think "intramolecularly." In parts (a) and (b), there is nothing more than a straightforward displacement of a good leaving group (here chosen as iodide, but many others would do as well) by the appropriate nucleophile.

(a) $\xrightarrow{^-:N_3}$ N_3 + $:I:^-$

(b) $\xrightarrow{^-:CN:}$ $CN:$ + $:I:^-$

Part (c) is similar, except that a charged species, an ammonium ion, is formed.

(c) $\xrightarrow{HN(CH_3)_2}$ $\overset{+}{N}(CH_3)_2$ H $:I:^-$

Part (d) can be answered reasonably in two ways. In one, we use a cyclic amine to do an intermolecular S_N2 reaction. In the other, we design an intramolecular version of the S_N2 reaction closely resembling (c).

(d) N–H $\xrightarrow[S_N2]{CH_3\text{-}I:}$ $\overset{+}{N}$ H CH_3 $:I:^-$ $\xleftarrow[S_N2]{heat}$ NH CH_3 I:

In part (e) we must design another intramolecular S_N2 reaction. The nucleophile required must be formed first from an alcohol by treatment with a base such as sodium hydride.

(e) OH I: $\xrightarrow{NaH}$ H_2 + $O:^-$ I: $\xrightarrow{S_N2}$ O + $:I:^-$

Problem 7.37 Parts (a)-(c) are straightforward displacements of a leaving group by a nucleophile. Part (a) leads to a quaternary immonium ion.

(a)

CH_3CH_2—I: → CH_2CH_3 :I:$^-$

(b)

CH_3CH_2-O:$^-$ I: → OCH_2CH_3 + :I:$^-$

(c)

:I:$^-$ Br: → :I: + :Br:$^-$

Part (d) requires you to be careful about stereochemistry. The S_N2 reaction always occurs with inversion, so the cis starting material becomes a trans product as iodide is displaced by cyanide.

(d)

:I $^-$:CN: CH_3 → :NC CH_3 + :I:$^-$

In (e), there can be no reaction, as the substrate is tertiary and can undergo no S_N2 reaction.

(e)

:I:

A tertiary iodide - no reaction in the S_N2 process!
The back side of the C–I bond is completely blocked.

Problem 7.38 Recall that in the transition state for the S_N2 reaction, the carbon at which displacement occurs is approximately sp^2 hybridized. The transition state for S_N2 displacement in allyl systems benefits from delocalization through overlap of the alkene π orbitals with the $2p$ orbital at the "central" carbon.

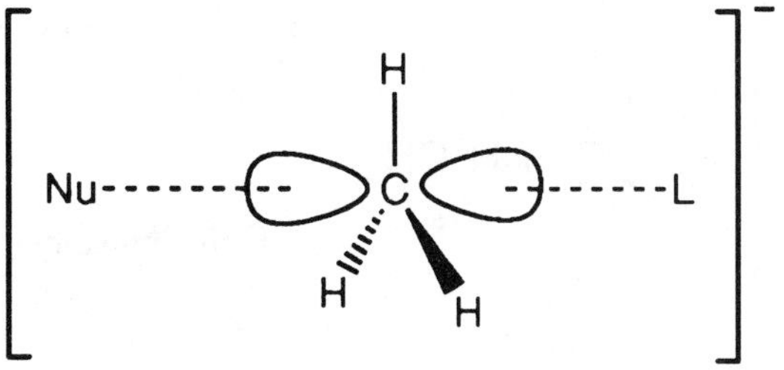

Transition state for typical S_N2 displacement reaction of CH_3-L

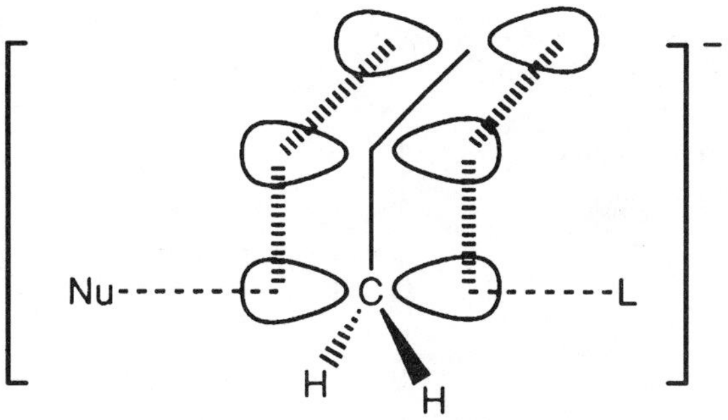

Transition state for S_N2 displacement reaction in an allyl system; note the overlap with the π orbitals of the double bond

Problem 7.39 In (a), a pair of diastereomeric alcohols will be produced as ionization produces a carbocation to which water can add at either the "top" or "bottom" lobe of the 2*p* orbital. The intermediate oxonium ions are deprotonated by water to give the alcohol products.

(a)

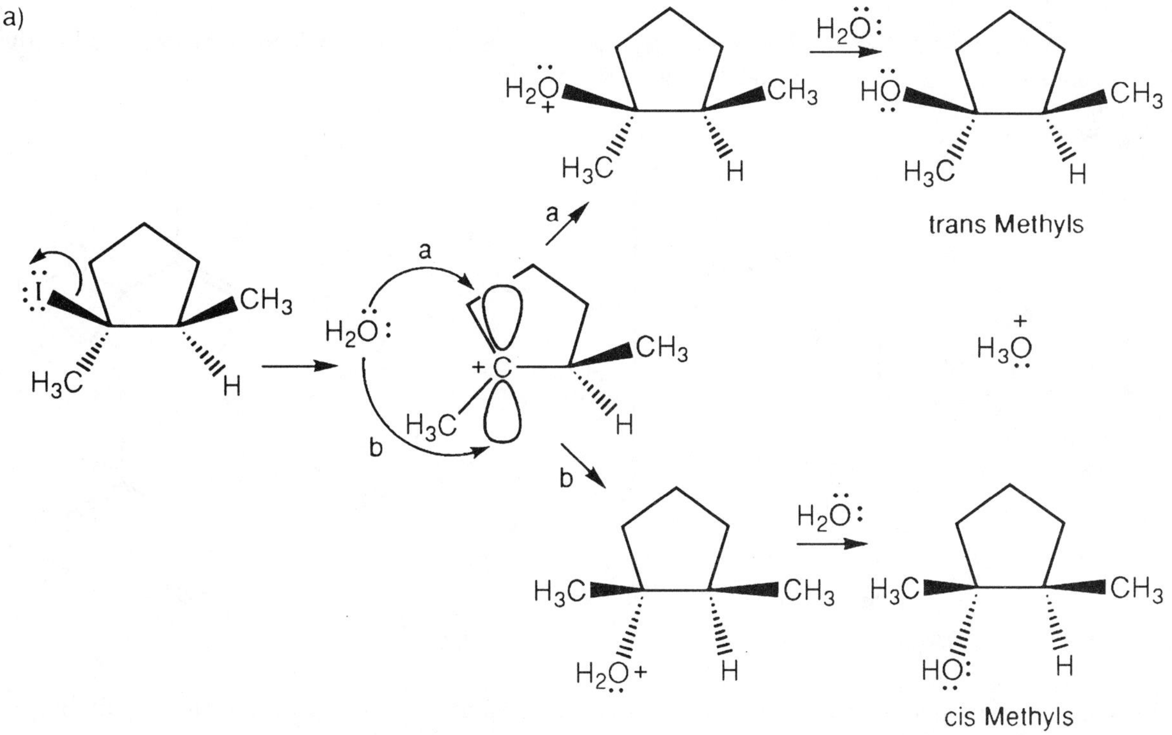

trans Methyls

cis Methyls

In (b), ionization to a secondary carbocation is likely to be accompanied by hydride shift to give the more stable tertiary carbocation. Both these species can be captured by ethyl alcohol to give, after deprotonation, the ethyl ethers.

(b)

CH_3CH_2-OH

H

CH_3CH_2-OH

O-CH_2CH_3

OCH_2CH_3

+ CH_3CH_2-OH_2

H

CH_3CH_2-OH

CH_3CH_2

H

O:

CH_3CH_2-OH

CH_3CH_2

+ CH_3CH_2-OH_2

In (c), a resonance-stabilized carbocation is formed on ionization. This cation can be captured at the two carbons sharing the positive charge to give two products.

(c)

H_2O:

H

Br:

H_2O:

O

H

O

H

+ H_3O^+

O

H

H

H_2O:

H

O:

(d) Ionization will give the *tert*-butyl cation. This intermediate can be captured by any available nucleophile. As there are three nucleophiles present, there will be three products.

-C-OH_2

deprotonation steps

-C-OH

addition steps

ionization step

-C-Br:

C+

:Br:−

H_2O:

CH_3-OH

CH_3CH_2-OH

H

-C-OCH_3

-C-OCH_3

H

-C-OCH_2CH_3

-C-OCH_2CH_3

Problem 7.40
(a) The primary bromide will react faster than the secondary bromide in an S_N2 displacement because the secondary position is more sterically hindered than the primary position. The entering nucleophile has more difficulty in approaching the rear of the secondary carbon-bromine bond than that of the primary carbon-bromine bond.

(b) Iodide is a better leaving group than bromide and so will be more easily displaced.

(c) The closer the *tert*-butyl group is to the leaving group, the more it will shield the rear of the departing bond. Accordingly, $(CH_3)_3CCH_2CH_2$-I will be more reactive than $(CH_3)_3CCH_2$-I.

(d) Primary allylic bromides will be more reactive than simple primary bromides, as the transition state for displacement will be stabilized (see Table 7.4, p. 251, and Problem 7.38). So, the compound with the double bond will be more reactive in the S_N2 displacement than will the simple primary bromide.

Problem 7.41
(a) The tertiary iodide will ionize to the relatively stable tertiary carbocation, whereas the secondary iodide must give the relatively unstable secondary carbocation. The tertiary compound will be faster.

(b) Iodide is a better leaving group than bromide, and *tert*-butyl iodide will be the faster compound.

(c) Ionization of the allylic iodide leads to a resonance-stabilized, relatively stable carbocation, whereas ionization of the simple primary iodide must give a most unstable primary carbocation. The allylic compound will react much faster.

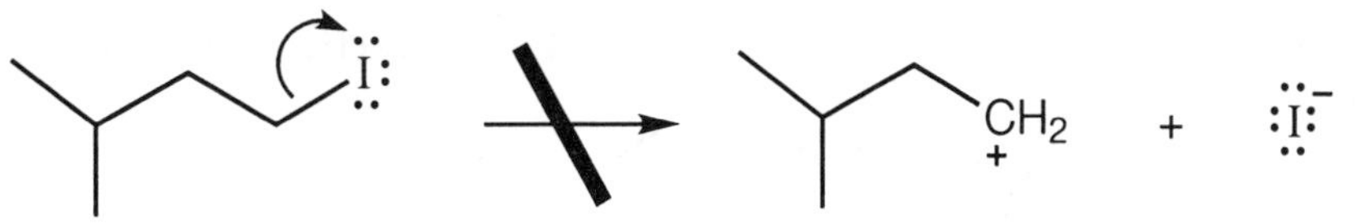

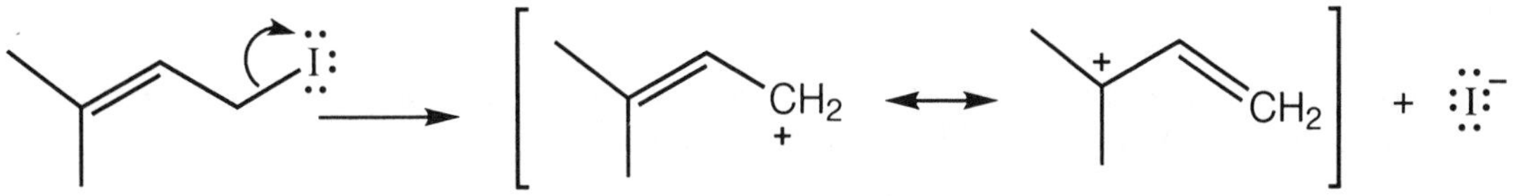

Note that the rate-determining ionization is not the step that determines the structure of the product.

Problem 7.42 In principle, three ethers are possible, *tert*-butyl ethyl ether, di-*tert*-butyl ether, and diethyl ether.

$CH_3CH_2—\ddot{O}—C(CH_3)_3$ $(CH_3)_3C—\ddot{O}—C(CH_3)_3$ $CH_3CH_2—\ddot{O}—CH_2CH_3$

In practice, only two of these can be made from the starting materials given. *tert*-Butyl iodide cannot be used as a substrate in the S_N2 reaction because it is a tertiary halide and the S_N2 reaction will not proceed with tertiary halides (an E2 elimination will occur instead). We are restricted to using ethyl iodide as the halide, and this leads to only two of the three possible ethers.

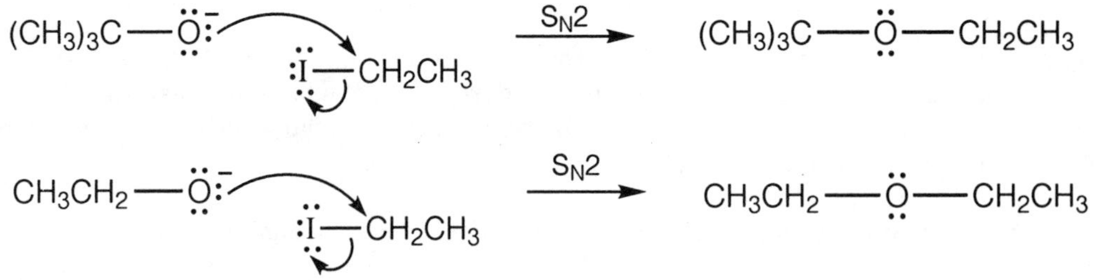

$(CH_3)_3C—\ddot{O}:^- + :\ddot{I}—CH_2CH_3 \xrightarrow{S_N2} (CH_3)_3C—\ddot{O}—CH_2CH_3$

$CH_3CH_2—\ddot{O}:^- + :\ddot{I}—CH_2CH_3 \xrightarrow{S_N2} CH_3CH_2—\ddot{O}—CH_2CH_3$

Problem 7.43

(a) The primary bromide will surely be more reactive than the tertiary bromide in the S_N2 reaction.

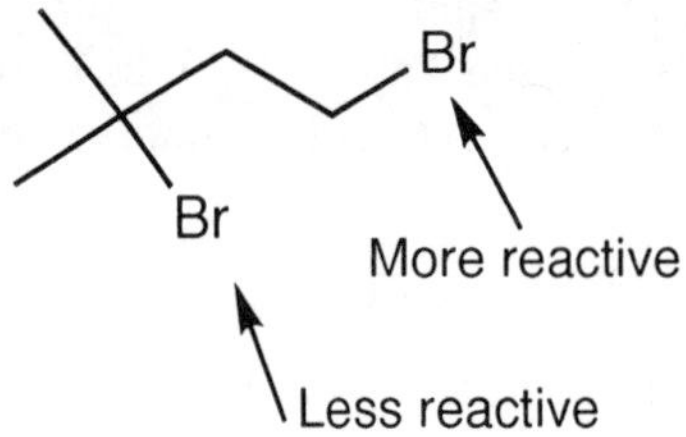

Br
Br
More reactive
Less reactive

(b) It will be the allylic iodide that is the more reactive, as the transition state for S_N2 displacement will benefit from delocalization (Table 7.4, p. 296, Problem 7.38, p. 251).

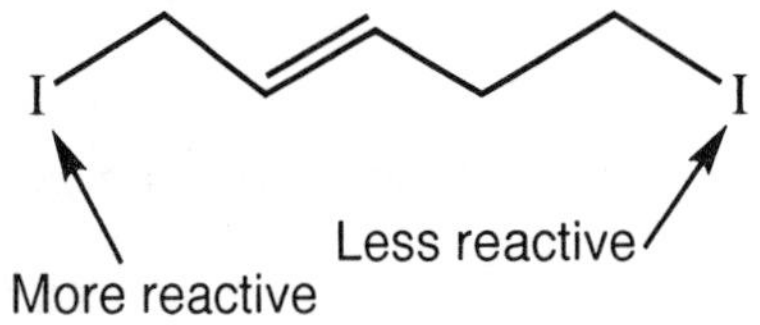

I
I
More reactive
Less reactive

(c) Bromide is a better leaving group than chloride, and will be more active in the S_N2 reaction.

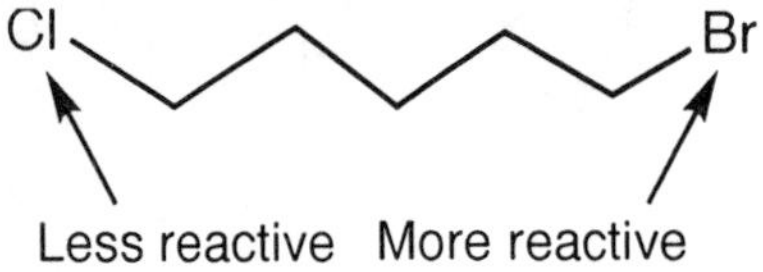

Cl
Br
Less reactive
More reactive

Problem 7.44

(a) The tertiary bromide will ionize to a relatively stable tertiary carbocation. It will be much more reactive in the S_N1 reaction than will the primary bromide.

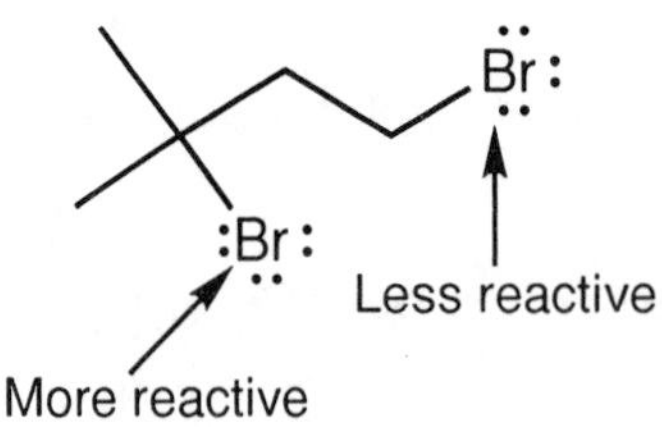

:Br:
:Br:
Less reactive
More reactive

(b) Ionization of the allylic iodide would give a resonance-stabilized carbocation. This will be preferred to ionization of the primary iodide.

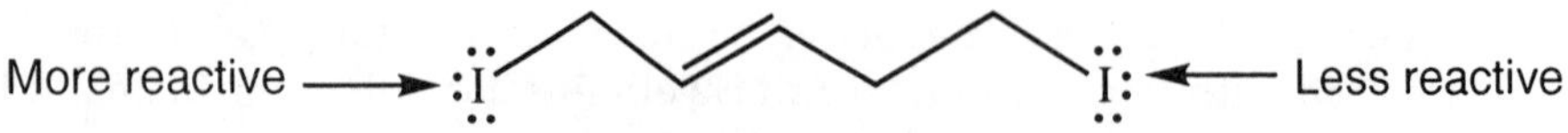

More reactive → :I: :I: ← Less reactive

(c) Ionization of either bromide would lead to a tertiary carbocation. However, only one of the carbocations can become planar. As carbocations are most stable when planar, it will be that bromide that is lost more easily.

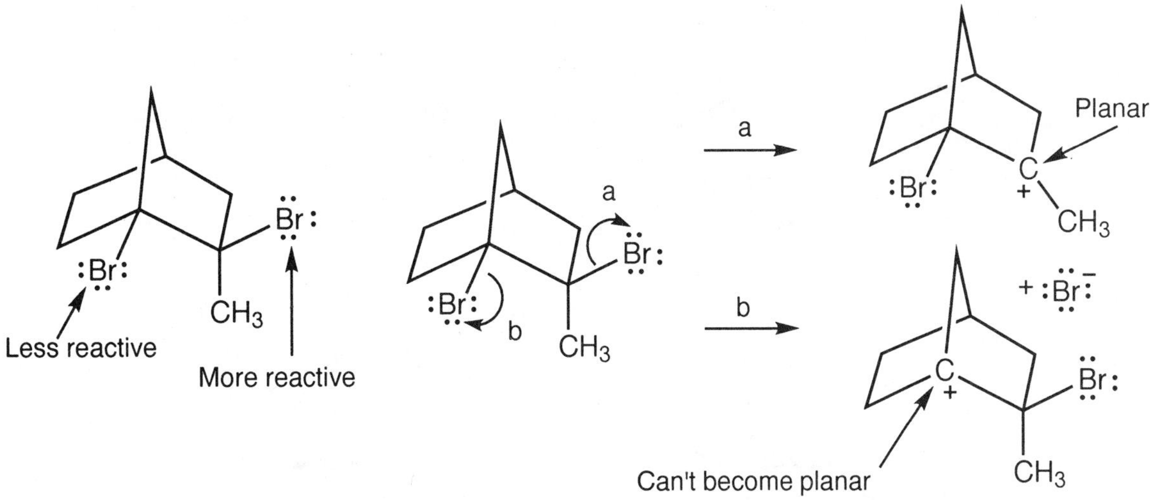

Less reactive

More reactive

Planar

Can't become planar

Problem 7.45 The first thing to do is to draw the two possibile iodides. What does that squiggly bond really mean? The two possibilities are

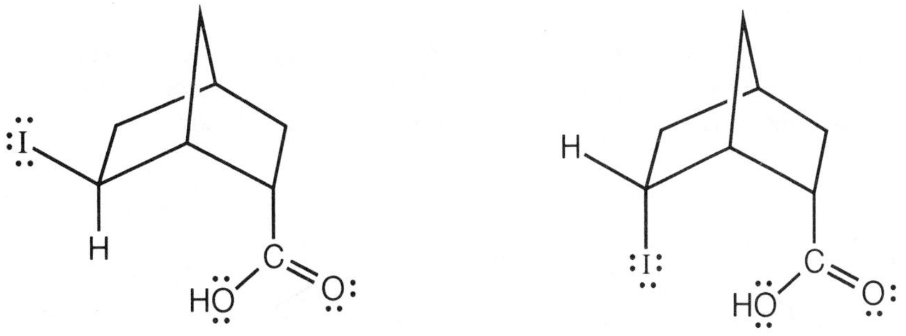

Notice that only one of the salts has a carboxylate in position to do a back-side S_N2 displacement of iodide. This displacement leads to the compound $C_8H_{10}O_2$. The other alkoxide would have to do a front-side S_N2 reaction, and, as we know, this never happens. So, the molecule that undergoes the intramolecular S_N2 reaction has the iodine in the "exo" position and the molecule that cannot do the S_N2 reaction must have the iodine "endo."

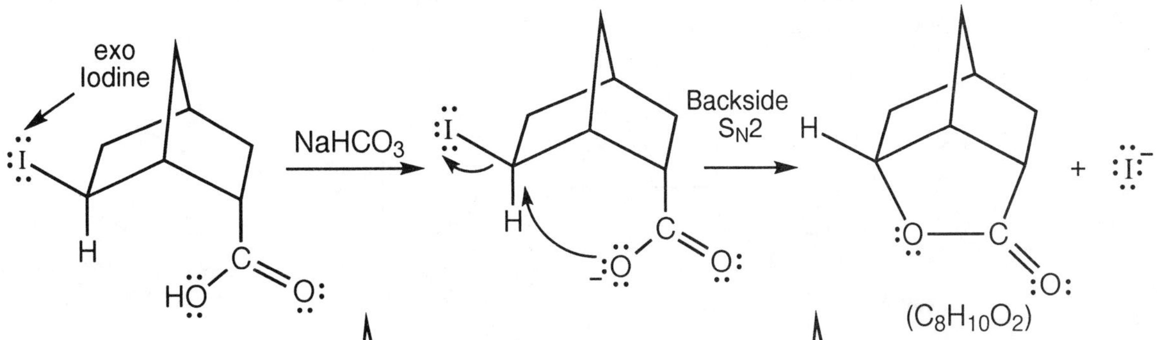

exo Iodine

$NaHCO_3$

Backside S_N2

($C_8H_{10}O_2$)

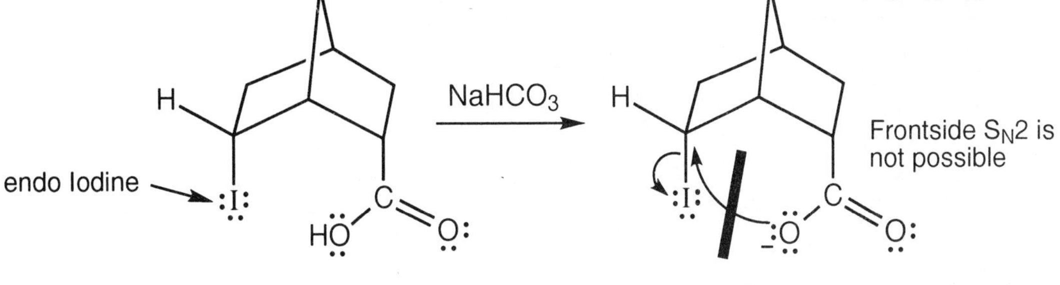

endo Iodine

$NaHCO_3$

Frontside S_N2 is not possible

Problem 7.46 The same point is made in this problem as in the last one. The impossibility of a frontside S_N2 reaction renders one stereoisomer unreactive. The isomer that can undergo a backside S_N2 reaction does so to produce the new compound.

H_2 + backside S_N2 C_5H_8O + :I:⁻

Na⁺ H:⁻

H_2 +

Frontside S_N2 impossible

Na⁺ H:⁻

Problem 7.47

(a) There are two possibilities for the E2 reaction. The more substituted Saytzeff product will be preferred, as the transition state for its formation will benefit from the partial formation of the trisubstituted double bond (p. 133, 282). The arrows for formation of the major product are the only ones shown.

CH_3CH_2-OH

H_a H_b CH_3CH_2-O:⁻

Loss of H_a leads to the major product + Loss of H_b leads to the minor product

(b) Once again it is the more substituted, Saytzeff product that is the major one. There are two different hydrogens that could be lost, and it is removal of H_a, leading to the more stable, trisubstituted double bond that is favored (arrows).

H_b Br H_a ⁻:O-CH_2CH_3 CH_3CH_2-OH

Loss of H_a leads to the major product + Loss of H_b leads to the minor product

(c) Here, there is only one kind of hydrogen that can be removed. However, the product can be either *cis*-2-pentene or *trans*-2-pentene. The more stable (less steric "bumping") trans compound will be preferred.

CH_3CH_2-Ö:⁻ H :I: CH_3CH_2-ÖH

More stable, major product + Less stable, minor product + :I:⁻

(d) In this molecule there can be no E2 elimination. Although the paper permits you to draw both an arrow formalism and the alkene product, in reality the *p* orbitals of the "alkene" do not overlap, and no double bond can be formed! Remember "Bredt's rule".

CH_3CH_2-Ö:⁻ H :I: CH_3CH_2-ÖH =

No real overlap here- and no double bond!

(e) This reaction is a standard Hofmann elimination. For this leaving group, it will be the less substituted alkene that is the major product (loss of H_a).

CH_3CH_2-Ö:⁻ H_a $\overset{+}{S}(CH_3)_2$ → Major product + CH_3CH_2ÖH + $S(CH_3)_2$

$\overset{+}{S}(CH_3)_2$ H_b :Ö-CH_2CH_3 → Minor product + CH_3CH_2ÖH + $S(CH_3)_2$

Problem 7.48

(a) This reaction should lead to three products. Two are straightforward. Ionization will give initially the secondary carbocation, and there are two possible losses of hydrogen to give an alkene. It will be the more stable alkene (the trisubstituted one in this case) that dominates (Saytzeff elimination).

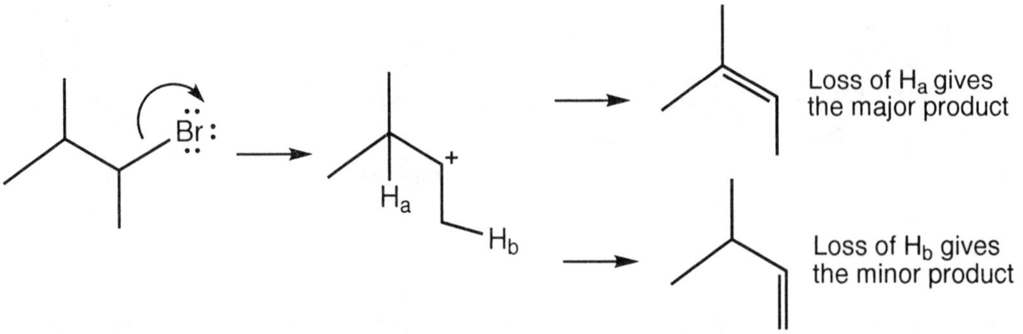

Loss of H_a gives the major product

Loss of H_b gives the minor product

However, we can also expect rearrangement through hydride shift to give a more stable tertiary carbocation, and there are also two different hydrogens (H_c and H_d) that can be lost to give an alkene. One of the product alkenes is the same as the major one formed from the secondary carbocation, but the other one is new. All three will be produced.

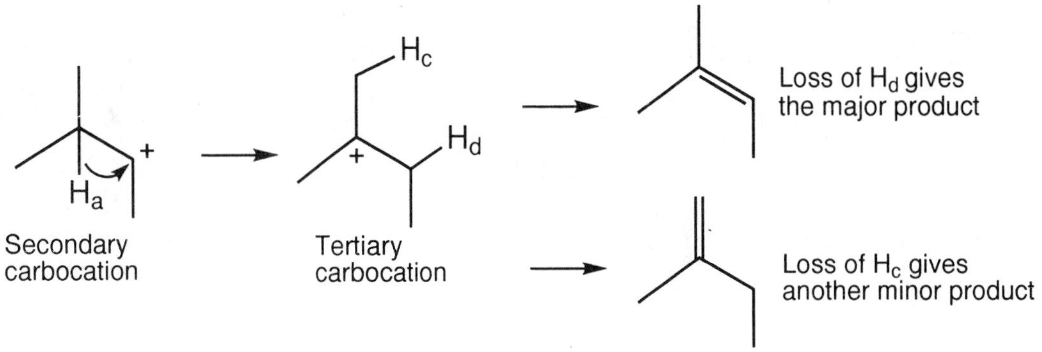

Secondary carbocation

Tertiary carbocation

Loss of H_d gives the major product

Loss of H_c gives another minor product

(b) This reaction is a straightforward Saytzeff elimination. The tertiary carbocation will be formed on ionization, and two protons can be removed to give different alkenes. The more substituted, more stable, alkene will be the major product.

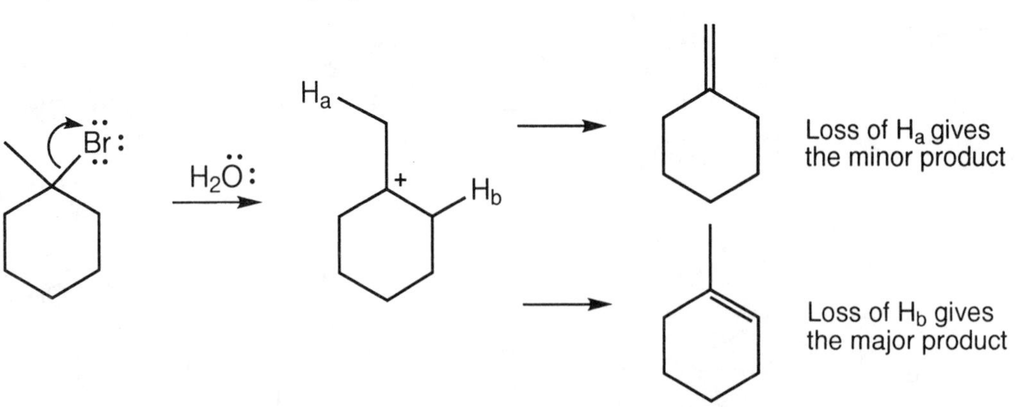

Loss of H_a gives the minor product

Loss of H_b gives the major product

(c) There can be no E1 reaction, as a primary carbocation must be formed.

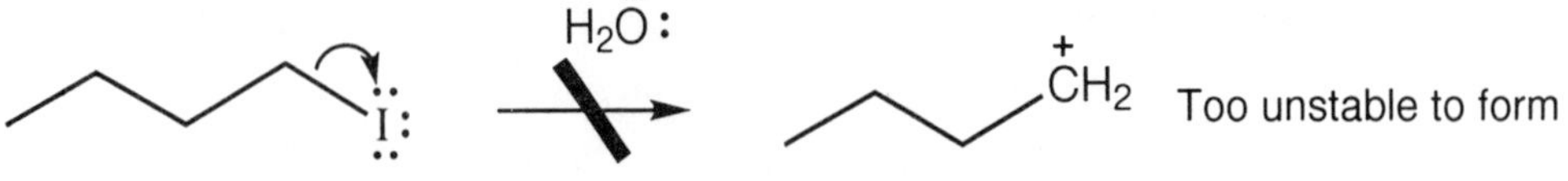

$\overset{+}{C}H_2$ Too unstable to form

(d) Even though the leaving group is a "Hofmann" leaving group, this E1 reaction will give the more substituted alkene (Saytzeff elimination) as the major product. The intermediate is the same as in (b), and there can be no substantial change in product distribution.

$S(CH_3)_2$ $H_2\ddot{O}:$ H_a H_b

Loss of H_a gives the minor product

Loss of H_b gives the major product

Problem 7.49 (a) The "obvious" mechanism is an intramolecular S_N2 reaction to transfer the methyl group from oxygen to carbon.

SO_2 :O: CH CH_3 OTs S_N2 SO_2 :O: HC—CH_3 OTs

(b) If the reaction is bimolecular, two molecules must be involved in the transition state for the reaction. The mechanism cannot be the simple unimolecular *intramolecular* displacement we wrote in (a). Perhaps the methyl group is transferred in a normal *intermolecular* S_N2 reaction.

SO_2 CH_3 :CH OTs S_N2 SO_2 CH_3 CH OTs SO_2 CH_3 :CH OTs SO_2 CH_3 HC—CH_3 OTs S_N2 SO_2 O: HC—CH_3 OTs

(c) Why is this intermolecular process more favorable energetically than the intramolecular reaction? There is a good nucleophile and an excellent leaving group. However, the transition state for the S_N2 reaction is ideally linear, and a linear arrangement of nucleophile, substrate carbon, and leaving group cannot be attained in the cyclic process. Apparently, this is enough to make the bimolecular reaction more favorable.

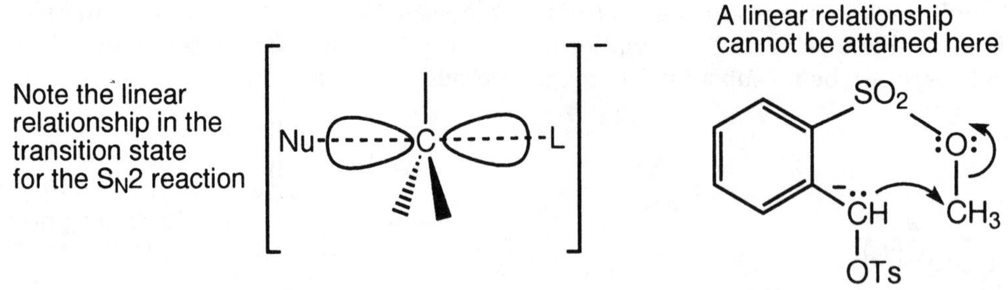

Note the linear relationship in the transition state for the S_N2 reaction

Nu- - - - C - - - - -L

A linear relationship cannot be attained here

SO_2 :O: CH_3 CH OTs

Problem 7.50

(a) Start with a good three-dimensional drawing of menthyl chloride. In the more stable chair form there is no hydrogen oriented at the optimal 180° angle for E2 elimination. In the less stable chair there is only one hydrogen in the proper, 180°, orientation to the chloride for an E2 elimination. Thus, there is only one product.

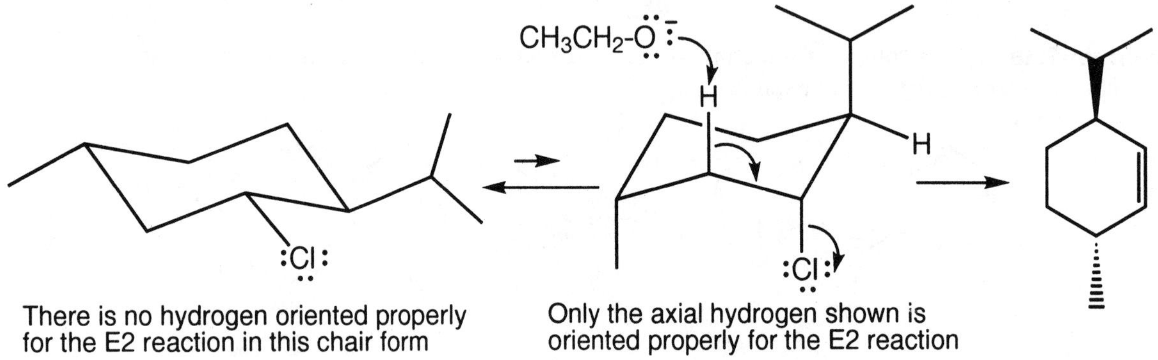

CH_3CH_2-O:

:Cl:

There is no hydrogen oriented properly for the E2 reaction in this chair form

Only the axial hydrogen shown is oriented properly for the E2 reaction

The more stable chair conformation for neomenthyl chloride has the chlorine axial. In this molecule there are *two* axial hydrogens that can be lost in a "perfect" 180° E2 reaction. As usual, the more stable product is the major one formed (Saytzeff elimination).

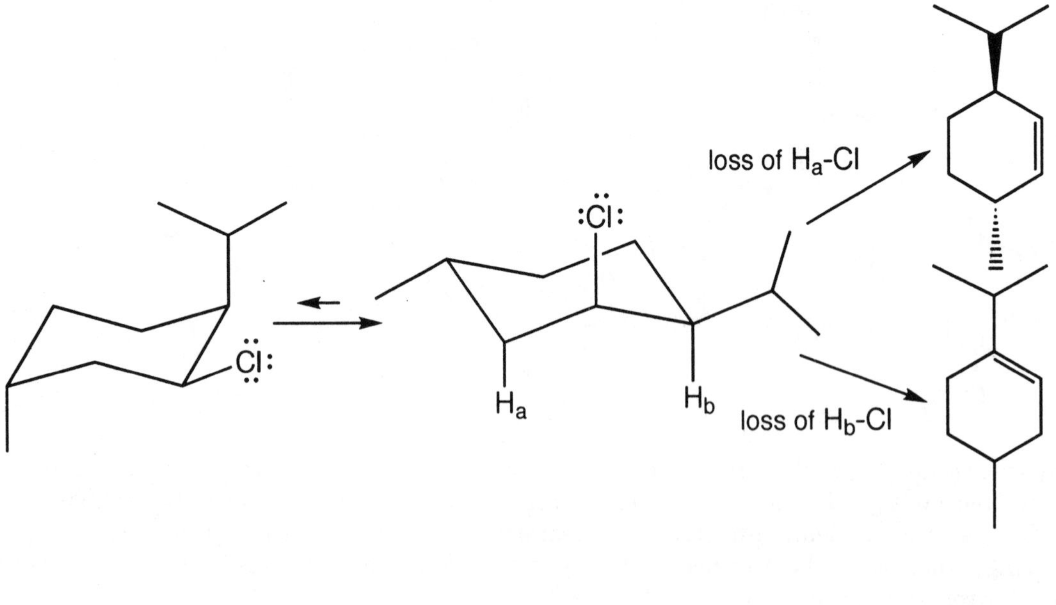

loss of H_a-Cl

:Cl:

H_a H_b

loss of H_b-Cl

(b) This reaction is S_N1, and ionization will generate a planar carbocation that can lose either of two hydrogens to give the two products shown in the problem. The more stable, trisubstituted alkene will be favored (Saytzeff elimination).

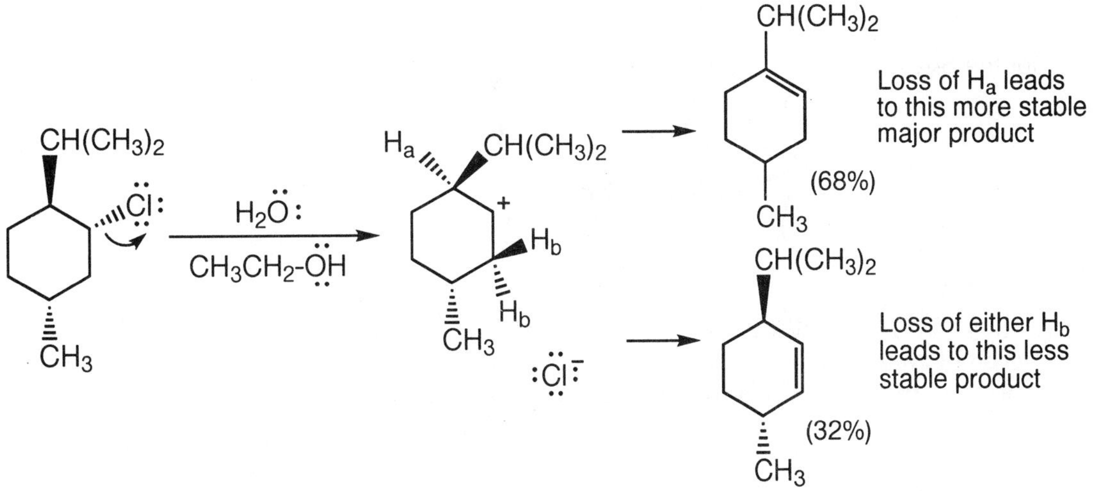
CH(CH3)2
:Cl:
H2O:
CH3CH2-OH
CH3
Ha
CH(CH3)2
+
Hb
Hb
CH3
:Cl:−
Loss of Ha leads
to this more stable
major product
(68%)
Loss of either Hb
leads to this less
stable product
(32%)

Problem 7.51 In the ionization the angles around the central carbon expand from approximately 109° (sp^3) to approximately 120° (sp^2). The large *tert*-butyl groups are further apart in the planar, sp^2 carbocation than they are in the starting material. The larger the R group the more strain relief there will be on ionization, and the faster the reaction.

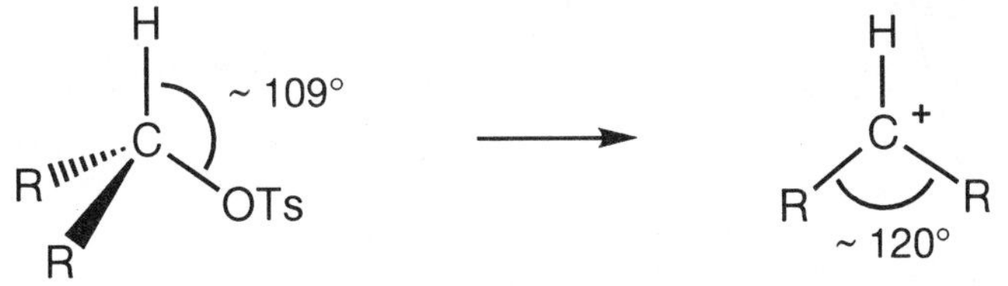
~ 109°
OTs
~ 120°

Here is an Energy versus Reaction progress diagram for the two reactions.

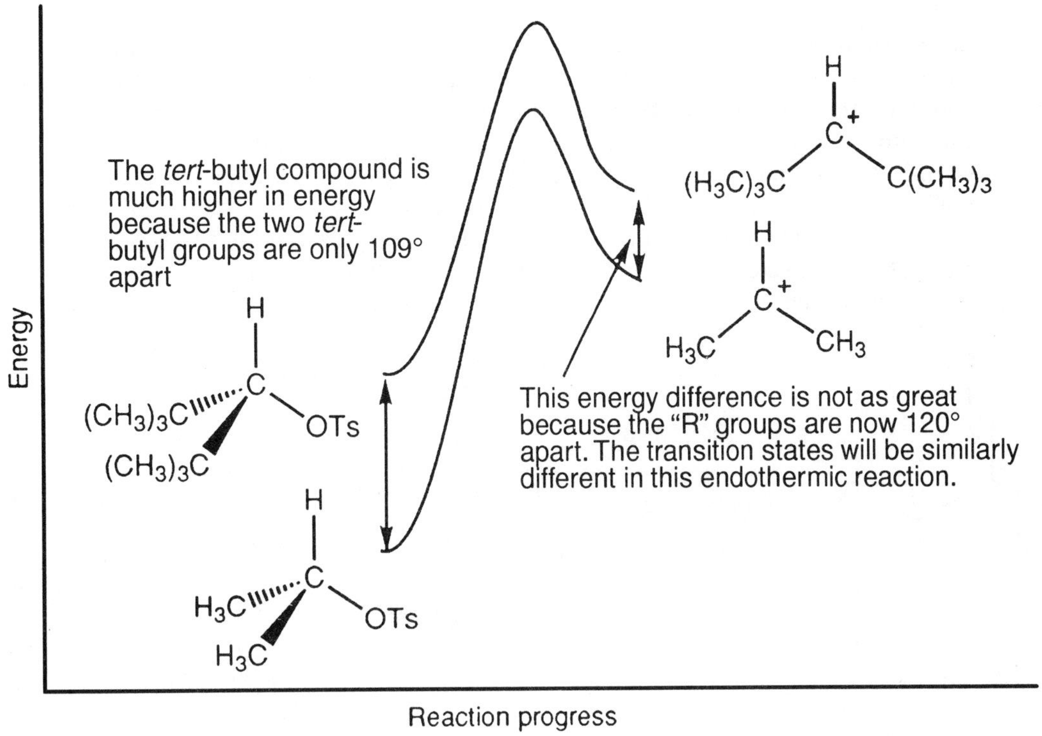
The *tert*-butyl compound is
much higher in energy
because the two *tert*-
butyl groups are only 109°
apart
(CH3)3C
OTs
H3C
(H3C)3C
C(CH3)3
CH3
This energy difference is not as great
because the "R" groups are now 120°
apart. The transition states will be similarly
different in this endothermic reaction.
Energy
Reaction progress

Problem 7.52 The stabilization by resonance of the cation formed from ionization of the starting halides depends on overlap between an exo-ring $2p$ orbital and the π orbitals of the ring.

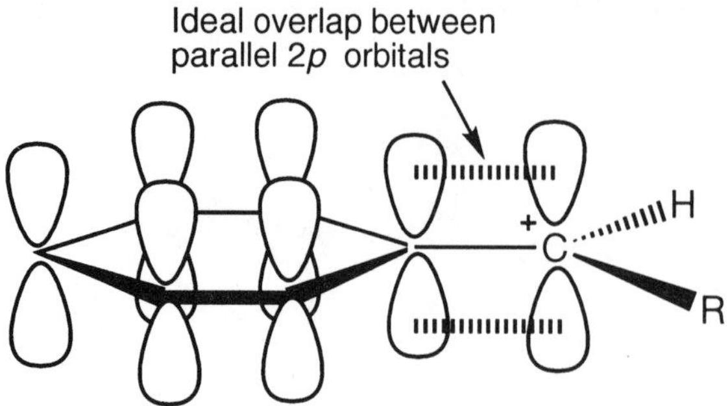

However, when "R" is large, it "bumps" with the nearby hydrogens. This destabilizing interaction can be relieved by rotation about a carbon-carbon bond, but this decreases resonance stabilization in the carbocation and makes the intermediate more difficult to form.

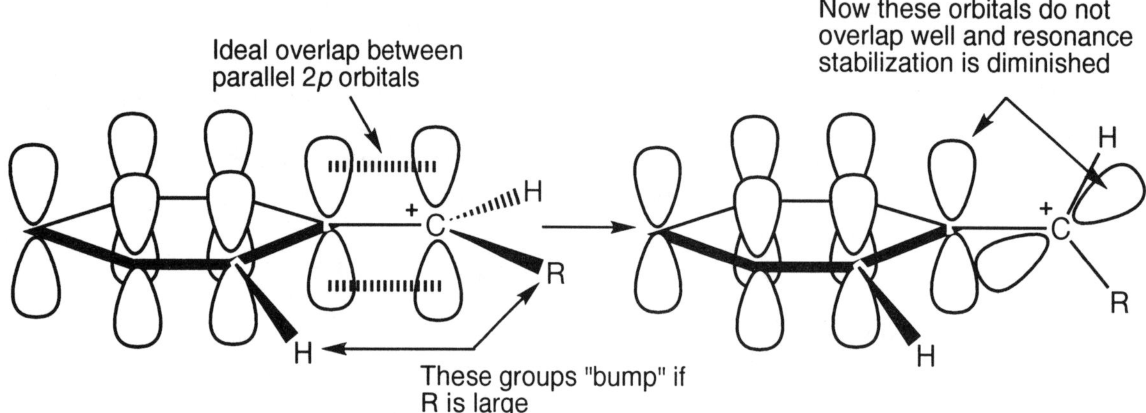

Accordingly, the bigger the "R" group the more difficult it is to form the carbocation.

Problem 7.53 Task one is to see how the unwanted compound **B** is formed. It is pretty clearly the result of an S_N2 displacement with inversion. The S_N2 and E2 reactions are both sensitive to steric effects. However, the steric requirements for deprotonation (E2) are far less severe than those for the S_N2 reaction, which requires the approach of a nucleophile to the rear of the departing carbon-leaving group bond. In this case, our hopes of making alkene are being thwarted by the S_N2 displacement of bromide to give the dialkoxy compound. If we increase the size of the reactants, we should be able to prejudice the reaction in favor of the sterically less demanding E2 reaction. We can't do anything about the size of the substrate, the bromide, but we could increase the size of the base. *tert*-Butoxide, a branched alkoxide base, would be a good choice.

H_3CO … Br $\xrightarrow[HOC(CH_3)_3]{Na^+\ ^-OC(CH_3)_3}$ H_3CO **A** + H_3CO … $OC(CH_3)_3$

Major from E2 — Minor from S_N2

Problem 7.54. This procedure takes advantage of iodide being both an excellent nucleophile and an excellent leaving group (Table 7.6, figure 7.46). Initially the small amount of iodide displaces chloride (S_N2) to give an intermediate iodide that is more reactive in the S_N2 reaction than the original chloride. The nucleophile now displaces iodide (S_N2 again) to generate a molecule of product and *regenerate* iodide ion. The iodide now can recycle, displacing another chloride, etc.

Na⁺ :I:⁻ ; R–CH₂CH₂CH₂–Cl —(S_N2)→ R–CH₂CH₂CH₂–I + Nu:⁻ Na⁺ —(S_N2)→ R–CH₂CH₂CH₂–Nu + :I:⁻

Iodide as nucleophile

Iodide as leaving group

Regenerated!

Chapter 8 Outline

8 Equilibria

Problem 8.1 Here is a simple, plug-in-the-numbers problem. This problem is designed to help you to develop a feeling for the relationship between ΔG and K so that you can make quick estimates in either direction.

$$\Delta G° = -2.3RT \log K; \text{ at } 25\ °\text{C}, 2.3RT = 1.364 \text{ kcal/mol}$$
$$\text{if } K = 7, \Delta G° = -(1.364) \log 7 = -1.15 \text{ kcal/mol}$$
$$\text{if } K = 14, \Delta G° = -(1.364) \log 14 = -1.56 \text{ kcal/mol}$$

Problem 8.2 This problem is more drill. Here you only need to know the bond dissociation energies of the various types of bonds involved in these reactions. Table 8.2 will help you. These reactions are all exothermic ($\Delta H°$ is negative).

(a)

C—I	= 57 kcal/mol (bond broken)
C—O	= 92 kcal/mol (bond made)
$\Delta H°$	= –35 kcal/mol

(b)

H—H	= 104.2 kcal/mol (bond broken)
C=C (π)	= 66 kcal/mol (bond broken)
2 × C—H (σ)	= 200.6 kcal/mol (bonds made)
$\Delta H°$	= –30.4 kcal/mol

(c)

Br—Br	= 46.1 kcal/mol (bond broken)
C=C (π)	= 66 kcal/mol (bond broken)
2 × C—Br	= 140 kcal/mol (bonds made)
$\Delta H°$	= –27.9 kcal/mol

(d)

H—Cl	= 103.2 kcal/mol (bond broken)
C=C (π)	= 66 kcal/mol (bond broken)
C—H	= 100.3 kcal/mol (bond made)
C—Cl	= 85 kcal/mol (bond made)
$\Delta H°$	= –16.1 kcal/mol

Problem 8.3 In every case the reaction is R—H ⟶ R• + •H. There can be no difference in the energy of the hydrogen atom formed in every reaction, and the differences must lie in the R• partner, the radical. As it takes less energy to form the more substituted radical (R•), the more substituted radicals must be more stable than the less substituted radicals. One can even make a rough estimate as to the amount of the increased stability.

$$H{-}CH_3 \longrightarrow H\bullet + \bullet CH_3 \qquad \Delta H° = 104.8 \text{ kcal/mol}$$

$$H{-}CH_2CH_3 \longrightarrow H\bullet + \bullet CH_2CH_3 \qquad \Delta H° = 100.3 \text{ kcal/mol}$$

[primary radical more stable than methyl radical by 4.5 kcal/mol]

$$H{-}CH(CH_3)_2 \longrightarrow H\bullet + \bullet CH(CH_3)_2 \qquad \Delta H° = 96.0 \text{ kcal/mol}$$

[secondary radical more stable than primary radical by 4.3 kcal/mol]

$$H{-}C(CH_3)_3 \longrightarrow H\bullet + \bullet C(CH_3)_3 \qquad \Delta H° = 93.3 \text{ kcal/mol}$$

[tertiary radical more stable than secondary radical by 2.7 kcal/mol]

Problem 8.5

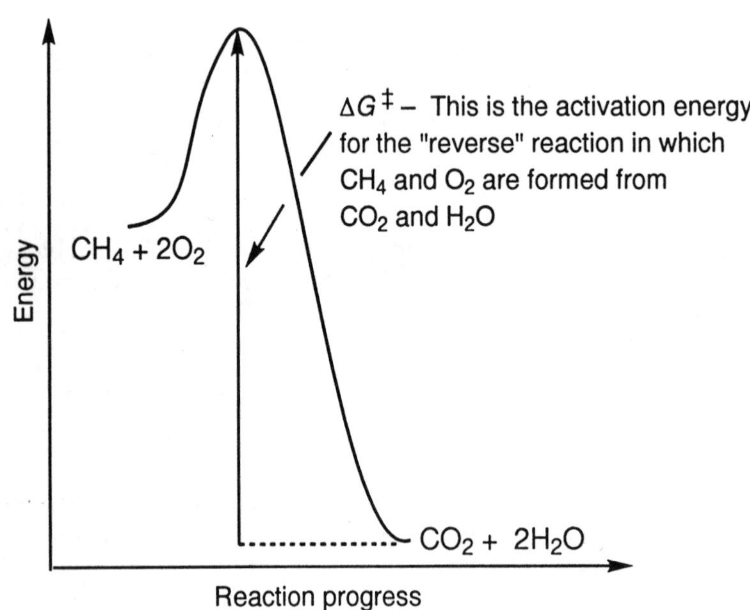
ΔG‡ – This is the activation energy
for the "reverse" reaction in which
CH4 and O2 are formed from
CO2 and H2O
CH4 + 2O2
CO2 + 2H2O
Energy
Reaction progress

Problem 8.8

For the reaction A + B $\rightleftharpoons$ C + D

The rate of the forward reaction, v, is $v = k[A][B]$
The rate of the reverse reaction, v', is $v' = k'[C][D]$
The question tells us that at equilibrium $v = v'$, so...

$k[A][B] = k'[C][D]$, or $k/k' = [C][D]/[A][B] = K$

Problem 8.9 In this S_N1 reaction, the slow step is ionization to give the cation. Capture of the ion by the two nucleophiles present will be faster.

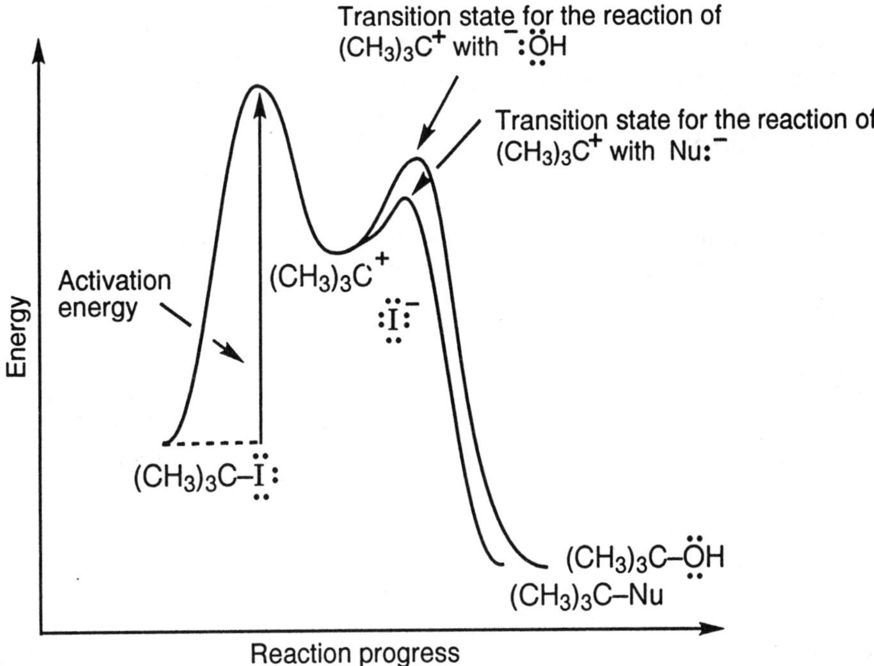

Problem 8.10 We hope this is a no-brainer by now. If not, be sure to go back into the chapter and read about transition states and activation energies.

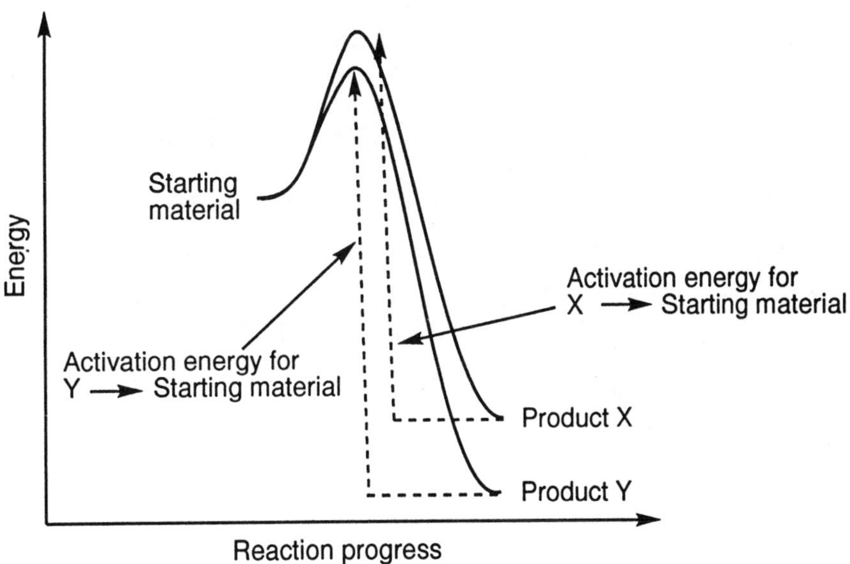

Additional Problem Answers

Problem 8.12
(a) Le Chatelier's principle says that a system at equilibrium responds to stress in such a way as to relieve that stress. To drive the equilibrium toward the alkyl iodide, we need only increase the concentration of one of the "starting materials", the alkyl chloride, alkyl bromide, or sodium iodide. Alternatively, one of the "products," the alkyl iodide or sodium bromide or chloride, could be removed. As the product is removed, equilibrium will be reestablished, and more alkyl iodide formed.
(b) As sodium chloride and sodium bromide are insoluble in acetone, they will precipitate from solution as they are formed. This will shift the equilibrium to the right, increasing the concentration of the desired alkyl iodide.

Problem 8.13
(a) If we can remove either cyclohexene or water as it is formed, the equilibrium will be shifted to the right, toward cyclohexene. In practice, the lower-boiling cyclohexene is usually distilled from the reaction mixture as it is formed.
(b) In order to drive the reaction to the left, towards cyclohexanol, it is convenient to increase the concentration of water, the cheapest reagent. Increasing the concentration of cyclohexene would also work, but would cost you more.

Problem 8.14(a) The formula to use is $\Delta G° = -2.3RT \log K$. At 25 °C, $2.3RT = 1.364$ kcal/mol. So, when $K = 1.4 \times 10^{-3}$, $\Delta G° = -(1.364)\log 1.4 \times 10^{-3}$ and $\Delta G° = +3.9$ kcal/mol. The carbonyl (C=O) form is favored by this amount.
(b) When $K = 2.8 \times 10^4$, $\Delta G° = -(1.364)\log 2.8 \times 10^4$ and $\Delta G° = -6.1$ kcal/mol. The hydrate form is favored by this amount.

Problem 8.15 As the system is at equilibrium, the percentages allow the calculation of the equilibrium constants for the two reactions. That the three compounds are all equilibrating does not mean that we cannot treat the two "halves" of the reactions separately. So, for the equilibrium between bicyclo[2.2.2]octane and bicyclo[3.3.0]octane we first normalize to 100%.

32.95 + 3.66 = 36.61. There is 32.95/36.61 = 90% bicyclo[3.3.0]octane and 10% bicyclo[2.2.2]–octane.

For the equilibrium between bicyclo[3.3.0]octane and bicyclo[3.2.1]octane: 63.35 + 32.95 = 96.3. There is 63.35/96.3 = 66% bicyclo[3.2.1]octane, and 34% bicyclo[3.3.0]octane.

At equilibrium there is 90% bicyclo[3.3.0]octane and 10% bicyclo[2.2.2]octane, so the equilibrium constant K is 9.

As $\Delta G° = -2.3RT\log K$, and at 25 °C, $2.3RT = 1.364$ kcal/mol, for $K = 9$, $\Delta G° = -(1.364)\log 9, = -1.3$ kcal/mol.

At equilibrium, there is 66% bicyclo[3.2.1]octane and 34% bicyclo[3.3.0]octane, so the equilibrium constant, K, is 1.94. For $K = 1.94$, $\Delta G° = -(1.364)\log 1.94, = -0.39$ kcal/mol.

Problem 8.16
Once again, the formula to use is: $\Delta G° = -2.3RT \log K$. At 25 °C, $2.3RT = 1.364$ kcal/mol.
So, $-1.51 = -(1.364)\log K$, $\log K = 1.11$; $K = 12.8$.

Problem 8.17

(a) F—F + $H_2C{=}CH_2$ ⇌ $H_2C(F)$—$CH_2(F)$

Bonds broken: 38 kcal/mol 66 kcal/mol Bonds made: 2 x 108 = 216 kcal/mol
The reaction is exothermic by 216 – 104 = 112 kcal/mol

(b) I—I + $H_2C{=}CH_2$ ⇌ $H_2C(I)$—$CH_2(I)$

Bonds broken: 36 kcal/mol 66 kcal/mol (π bond) Bonds made: 2 x 57 = 114 kcal/mol
The reaction is exothermic by 114 – 102 = 12 kcal/mol

(c) I· + $(CH_3)_2C{=}CH_2$ ⇌ $(CH_3)_2\dot{C}$—$CH_2(I)$

Bond broken: 66 kcal/mol Bond made: 57 kcal/mol
The reaction is endothermic by 66 – 57 = 9 kcal/mol

(d) $(CH_3)_2\dot{C}$—$CH_2(Cl)$ + H—Cl ⇌ $(CH_3)_2C(H)$—$CH_2(Cl)$ + Cl·

Bond broken: 103.2 kcal/mol Bond made: 93.3 kcal/mol
The reaction is endothermic by 103.2 – 93.3 = 9.9 kcal/mol

Problem 8.18 Of course, there are many possible answers to this question. One merely must design a reaction in which the "products" are less stable than the starting materials. Here is one possibility.

I—I + H_3C—CH_3 ⇌ 2 H_3C—I

Bonds broken: 36 kcal/mol 90 kcal/mol Bonds made: 2 x 57 = 114 kcal/mol
The reaction is endothermic by 126 – 114 = 12 kcal/mol

Problem 8.19 One only has to use the formula $\Delta G = \Delta H - T\,\Delta S$ to do this problem. The take-home lesson is the increased contribution of entropy at higher temperature. In example (b), for example, the reaction is exergonic at 25 °C, but endergonic at 200 °C.

(a) at 25 °C

$\Delta G° = \Delta H - T\,\Delta S$
$\Delta G° = (-14.0 \text{ kcal/mol}) - (298 \text{ deg})(-15.8 \text{ cal/deg mol})$
$\Delta G° = (-14.0 \text{ kcal/mol}) + 4.71 \times 10^3 \text{ cal/mol}$
$\Delta G° = (-14.0 + 4.71) \text{ kcal/mol} = -9.3 \text{ kcal/mol}$

at 200°C

$\Delta G° = (-14.0 \text{ kcal/mol}) - (473 \text{ deg})(-15.8 \text{ cal/deg mol})$
$\Delta G° = (-14.0 \text{ kcal/mol}) + 7.47 \times 10^3 \text{ cal/mol}$
$\Delta G° = (-14.0 + 7.47) \text{ kcal/mol} = -6.5 \text{ kcal/mol}$

(b) at 25 °C

$$\Delta G^\circ = \Delta H - T\,\Delta S$$
$$\Delta G^\circ = (-6.0 \text{ kcal/mol}) - (298 \text{ deg})(-15.8 \text{ cal/deg mol})$$
$$\Delta G^\circ = (-6.0 \text{ kcal/mol}) + 4.71 \times 10^3 \text{ cal/mol}$$
$$\Delta G^\circ = (-6.0 + 4.71) \text{ kcal/mol} = -1.3 \text{ kcal/mol}$$

at 200 °C

$$\Delta G^\circ = (-6.0 \text{ kcal/mol}) - (473 \text{ deg})(-15.8 \text{ cal/deg mol})$$
$$\Delta G^\circ = (-6.0 \text{ kcal/mol}) + 7.47 \times 10^3 \text{ cal/mol}$$
$$\Delta G^\circ = (-6.0 + 7.47) \text{ kcal/mol} = +1.5 \text{ kcal/mol}$$

Problem 8.20 The diagrams for the E1 and S_N1 reactions will closely resemble each other. There is a carbocation intermediate, and the initial ionization will be endothermic. The exothermicity or endothermicity depends only on the relative energies of the starting compound and the alkene product.

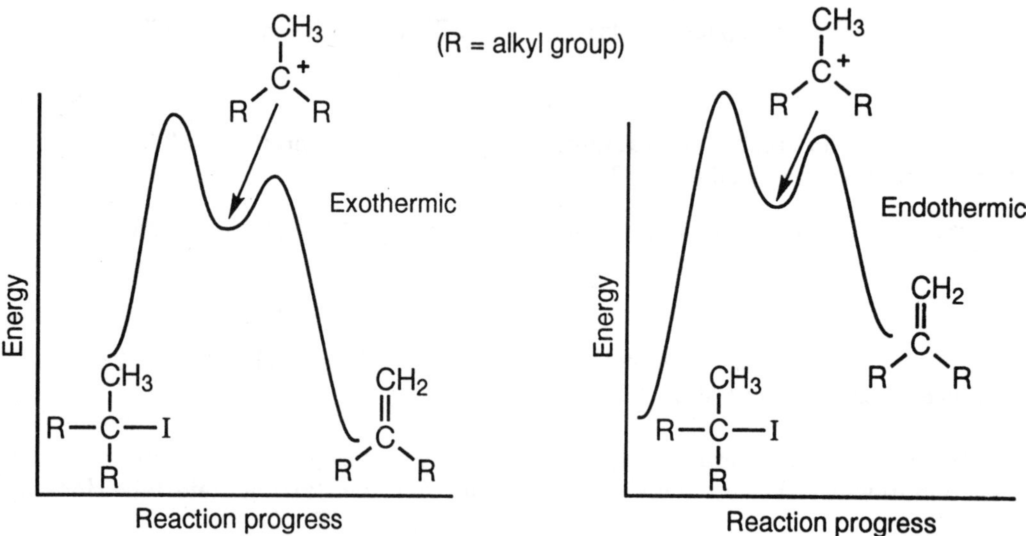

The picture for the E2 reaction is more simple, and resembles that for the S_N2 reaction.

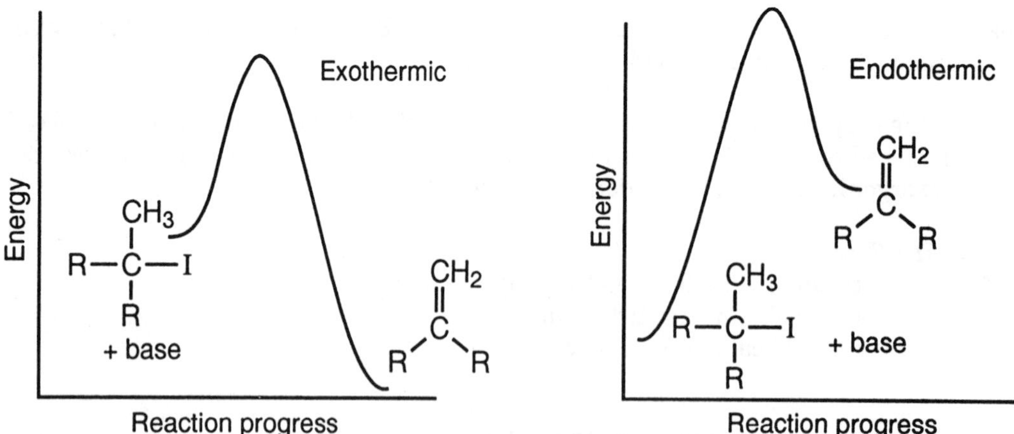

Problem 8.21

(a) In each case, there will be three compounds present at the end of the reaction, **A**, **B**, and **C**. There can be no **D** or **E** present because these are transition states, and have no lifetime. Only energy minima can be isolated. It is a common mistake to ignore intermediate **B**. Compound **B** is real and can be isolated.

(b) In case (1), the major compound will be **C**. In case (2), it will be **A**. In each case **B** will be present in the smallest amount. The relative amounts are determined by the relative energies of the three compounds.
(c) The rate–determining step in each case is the endothermic formation of **B** from **A**.
(d)

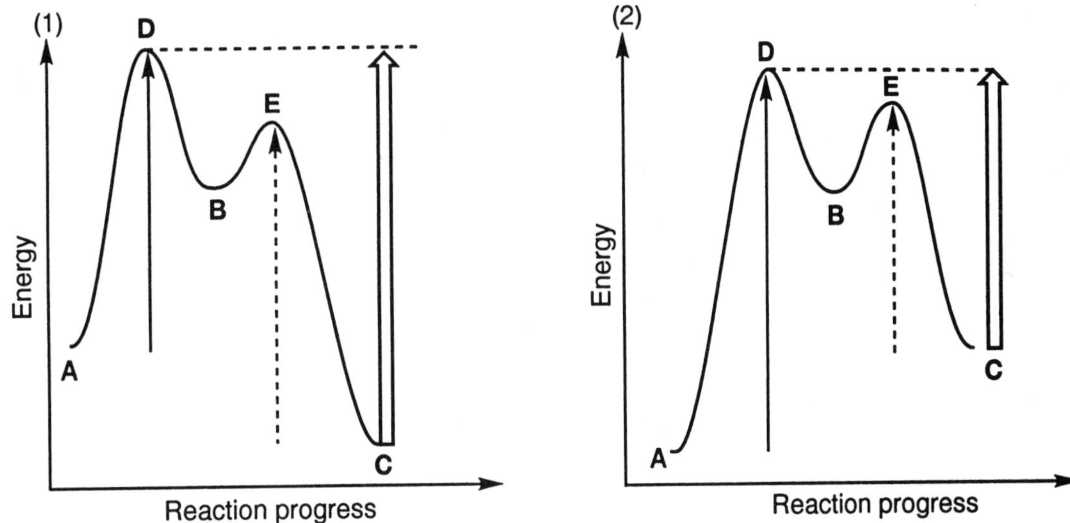

The solid arrows ——▶ show the activation energies for **A** ——▶ **B** and **A** ——▶ **C.**
The dashed arrows - - - - -▶ show the activation energies for **C** ——▶ **B.**
The fat arrows ⇨ show the activation energies for **C** ——▶ **A.**
(e) The rate-determining step is always the step with the highest transition state, so for **A** ——▶ **C** this is the step **A** ——▶ **B**, with the activation energy shown by the solid arrow.

Problem 8.22 In principle, this statement might be wrong. Just because we know that a tertiary carbocation is more stable than a secondary carbocation does not mean that the transition states for formation of the cations *must* have the same relative energies as in figure (a). In principle, the diagram *might* look like figure (b).

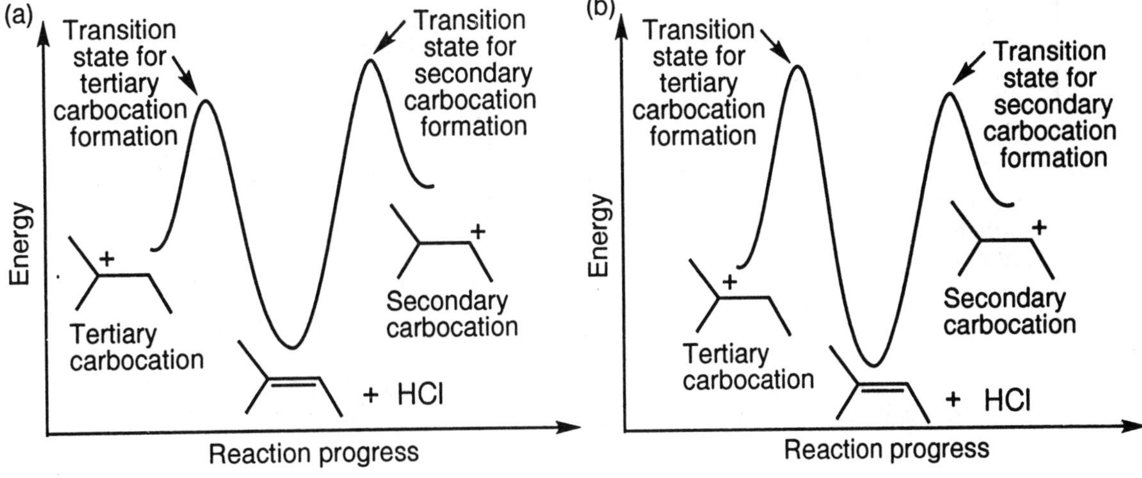

Now, why is it likely that the situation in (a) will be the correct one? That is, why should the transition state for formation of the more stable, tertiary carbocation be lower in energy than the transition state for formation of the less stable secondary carbocation? Draw out the transition states for cation formation.

Transition state for tertiary carbocation formation

Transition state for secondary carbocation formation

In the transition state for tertiary carbocation formation there is a partially formed tertiary carbocation, symbolized by δ^+. In the transition state for secondary carbocation formation there is a partially formed secondary carbocation, also shown as δ^+. The factors that make a tertiary carbocation more stable than a secondary carbocation will be present in the transition states for cation formation in which positive charges are partially developed. Accordingly, it is not surprising to find that the transition state with a tertiary δ^+ is more stable than the transition state with the secondary δ^+.

Problem 8.23 In both of these diagrams, the Hammond postulate is violated. For an exothermic reaction (a), the transition state should resemble starting material. For a thermoneutral reaction (b), the transition state should be exactly intermediate between starting material and products.

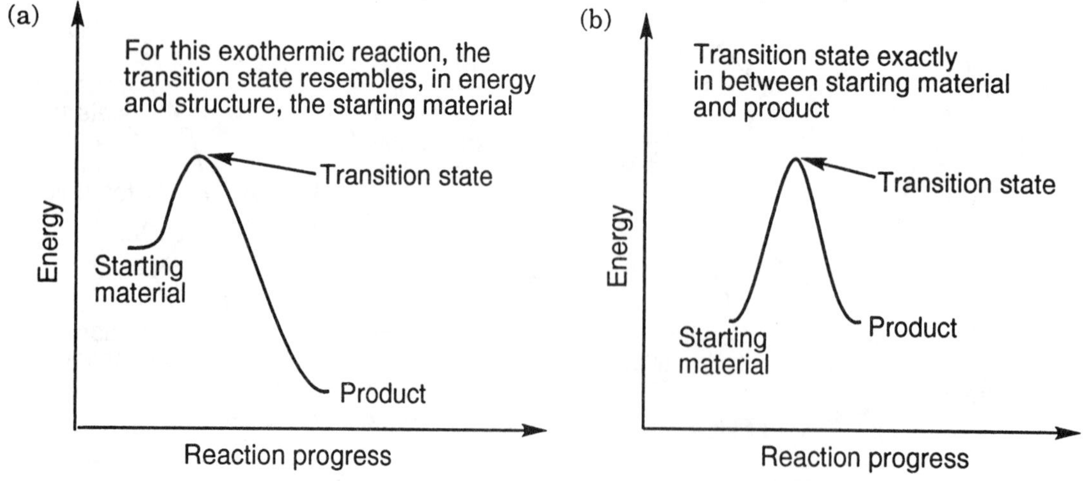
(a)
For this exothermic reaction, the transition state resembles, in energy and structure, the starting material
Transition state
Energy
Starting material
Product
Reaction progress
(b)
Transition state exactly in between starting material and product
Transition state
Energy
Starting material
Product
Reaction progress

Chapter 9 Outline

9.1 Electronegativity
9.2 Mechanism of the Addition of Hydrogen Halides to Alkenes
9.3 Regiochemistry
9.4 Resonance Effects
9.5 Review of Resonance: How to Write Resonance Forms
9.6 Resonance and the Stability of Carbocations
9.7 Inductive Effects on Addition Reactions
9.8 H–X Addition Reactions: Hydration
9.9 Dimerization and Polymerization of Alkenes
9.10 Hydroboration
9.11 Utility of Hydroboration: Alcohol Formation
9.12 Rearrangements During H–X Addition to Alkenes
9.13 Something More: Protonation of Ethylene

9

Additions to Alkenes 1

Problem 9.1 As always, our estimate of the exo- or endothermicity of this reaction is made by comparing the bond energies of the bonds broken and made in the reaction. In this case, the bonds broken are the π bond of the alkene, and the σ bond of hydrogen chloride. The bonds made are the carbon-hydrogen and carbon-chlorine bonds in the product chloride.

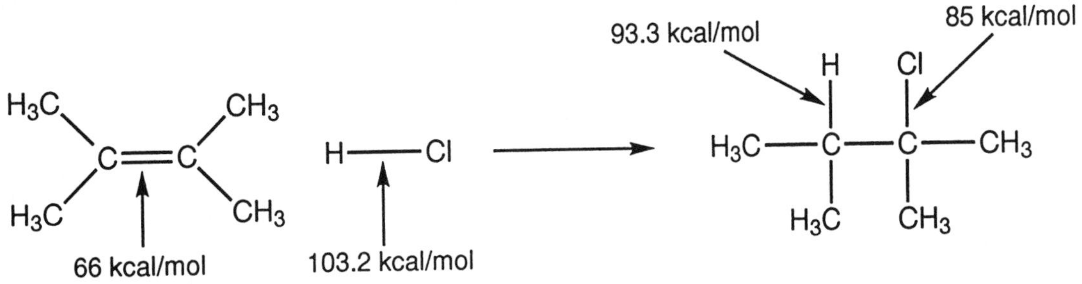

93.3 kcal/mol

85 kcal/mol

66 kcal/mol

103.2 kcal/mol

The bonds to be broken are worth 169.2 kcal/mol, but the bonds made are worth 178.3 kcal/mol. Accordingly, the reaction is exothermic by about 9 kcal/mol (178.3 – 169.2).

Problem 9.3 The π molecular orbitals of allyl are constructed from three parallel carbon 2p orbitals. The procedure is essentially the same as for linear H_3, done in Chapter 2 (p. 64). The only difference is that the orbitals of H_3 were made from hydrogen 1s orbitals, whereas allyl is built up from carbon 2p orbitals. An easy way to do it is to interact the π and π^* orbitals of ethylene with a third 2p orbital placed in between the first two:

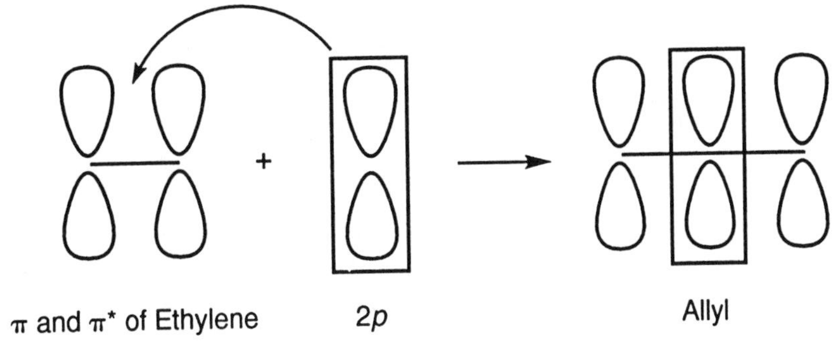

π and π^* of Ethylene

2p

Allyl

The bonding molecular orbital of ethylene, π, interacts with the new $2p$ orbital producing two new allyl molecular orbitals, Φ_1 and Φ_3.

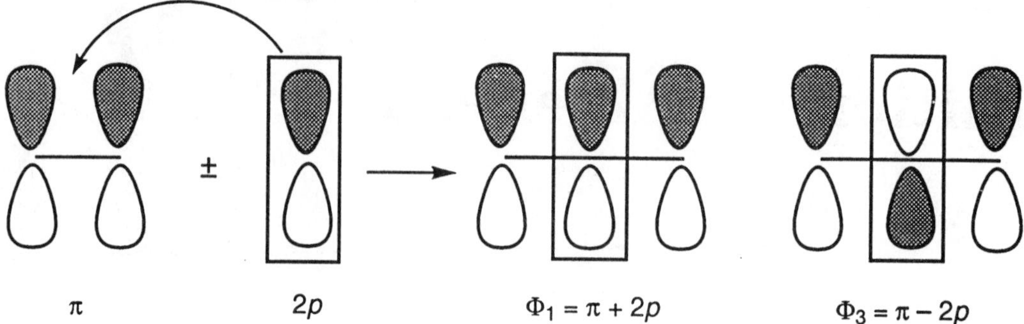

The antibonding molecular orbital of ethylene does not interact with the new $2p$ orbital at all as the new bonding interactions are exactly cancelled by the new antibonding interactions; Φ_2 is the result.

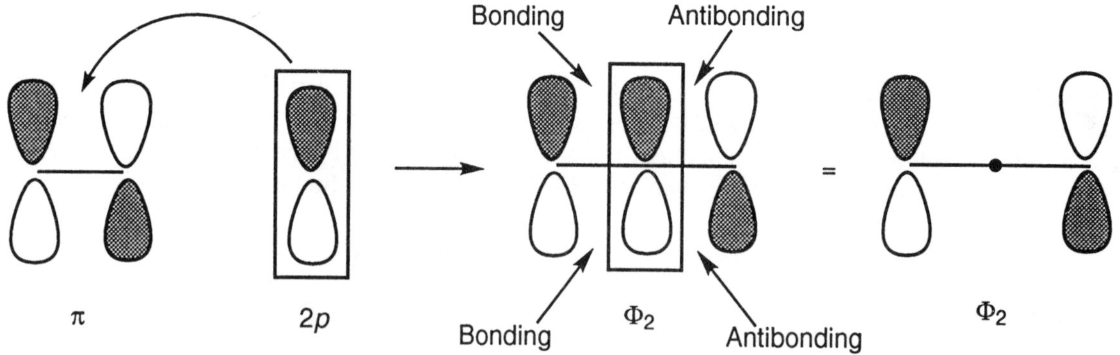

So, the three new allyl molecular orbitals are Φ_1, Φ_2, and Φ_3. They are ordered in energy by counting the new nodes. Of course, the node present in all $2p$ orbitals remains as well.

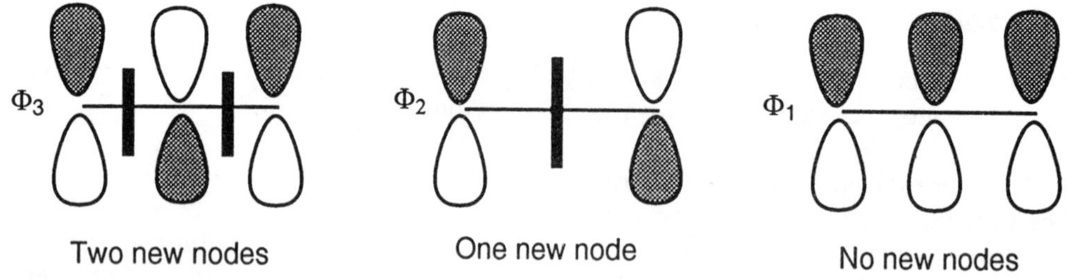

Problem 9.4 As there are two π electrons in the allyl cation, there must be three in the allyl radical, and four in the allyl anion.

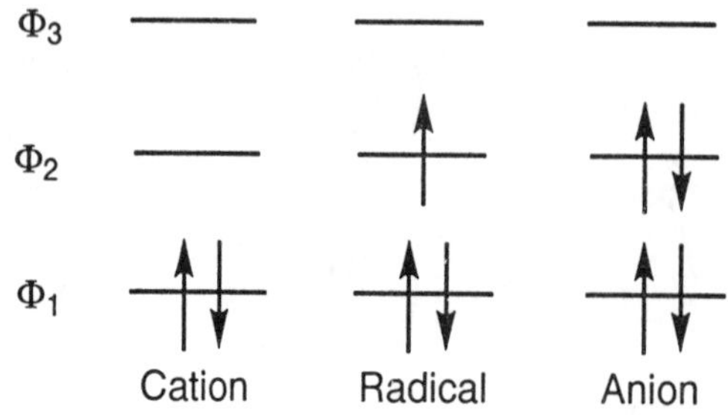

Problem 9.5 The easiest way to make these molecular orbitals is to repeat the technique used in Chapter 2 for cyclic H_3 and to transform the π molecular orbitals of allyl into those of the cyclic molecule by bending.

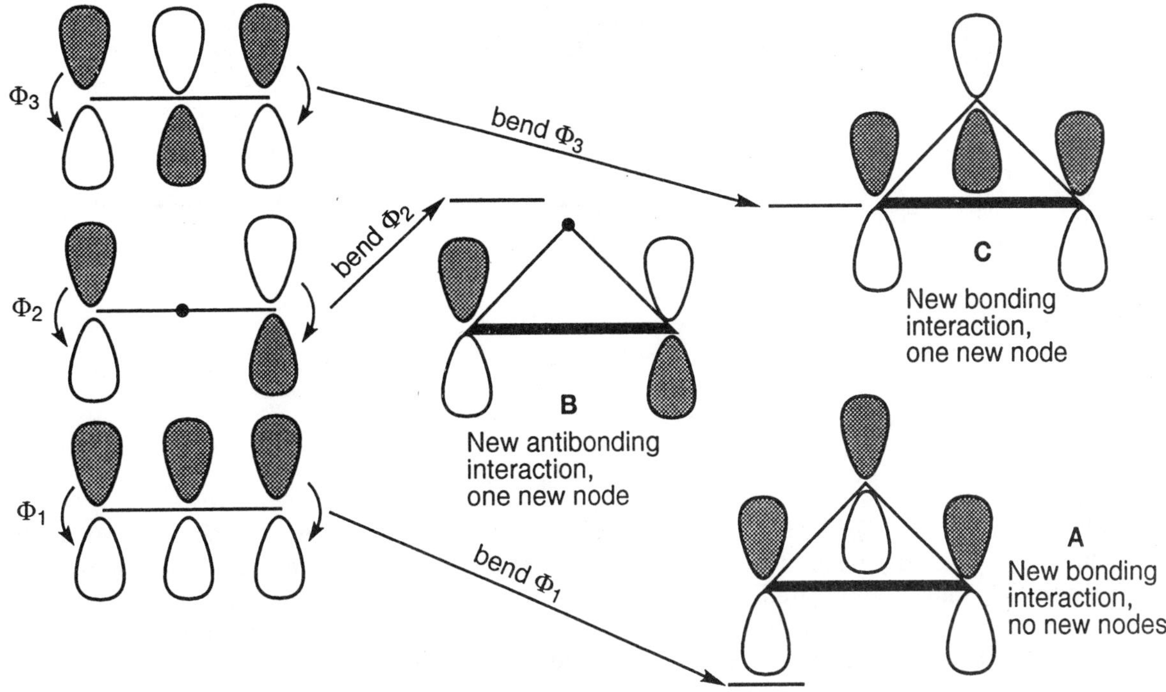

As orbitals Φ_1 and Φ_3 transform into the cyclic orbitals **A** and **C,** new bonding interactions appear between the end lobes and the energy decreases. As orbital Φ_2 bends, however, a new antibonding interaction appears, and the energy increases. As in cyclic H_3, the two highest energy molecular orbitals each have a single node and the two are of equal energy.

Problem 9.6

(a)

(b)

(d)

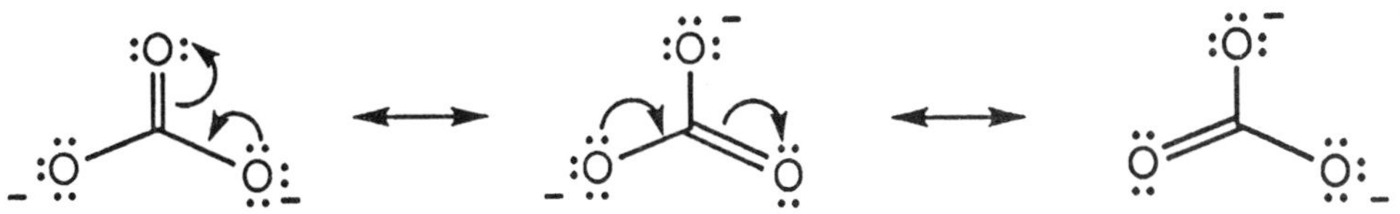

(e)

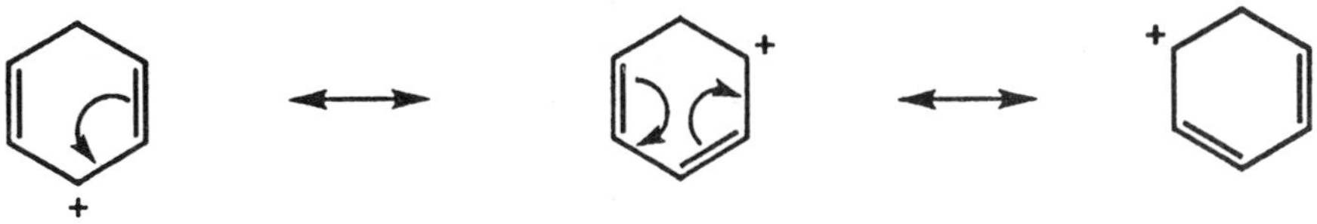

(g)

Problem 9.7

(a)

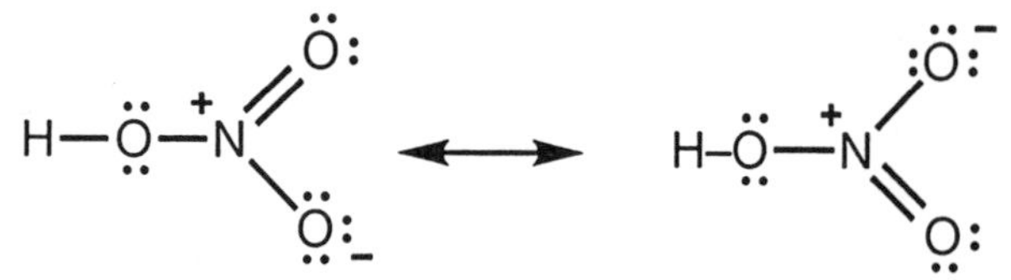

(b)

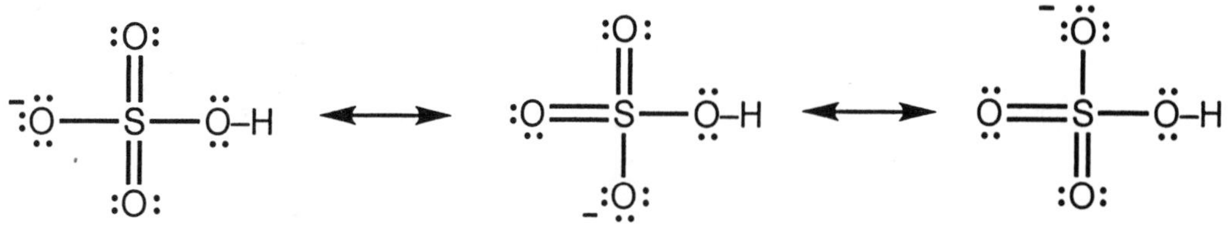

(c)

Problem 9.8 Items (b), (d), (f) are not resonance forms. Item (b) is dealt with in the chapter itself. In (d), atoms have clearly been moved; *cis*- and *trans*-2-butene are diastereomers, not resonance forms. Item (f) is more complicated. Here, *on paper*, the resonance double arrow is as correct as, for example, the resonance relationship shown in (e). However, geometry raises its ugly head, and there is really little or no overlap between the *p* orbitals making up this deceptive, "imitation" allyl system.

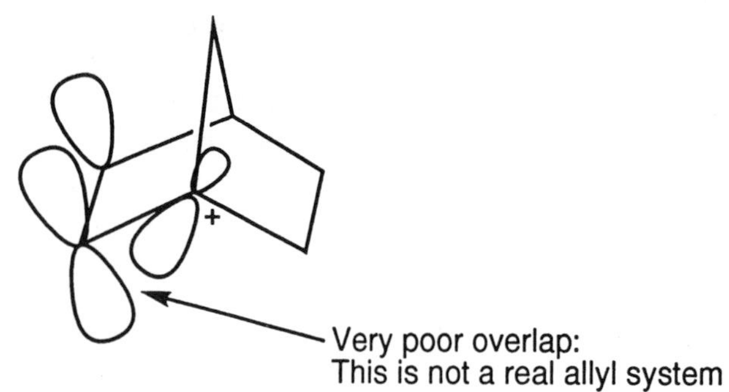
+
Very poor overlap:
This is not a real allyl system

Problem 9.9 In (a) the negative charge is better off on the more electronegative oxygen than on carbon.

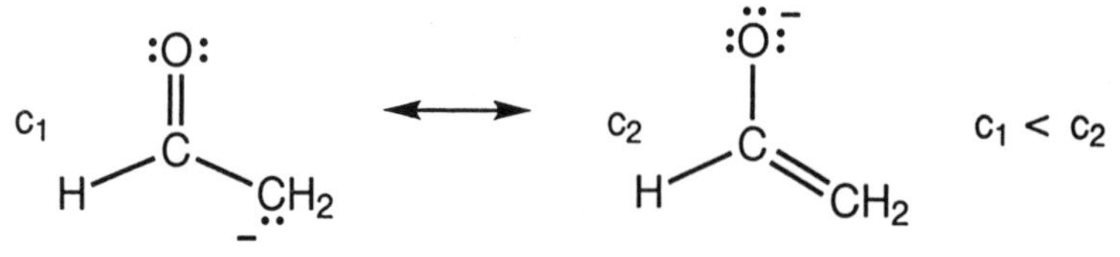

c_1 $\longleftrightarrow$ c_2 $c_1 < c_2$

In (b), the two forms are identical, and must be equally weighted.
In (c), it is the tertiary carbocation that is more stable than the primary carbocation, and this form will be the more heavily weighted one.

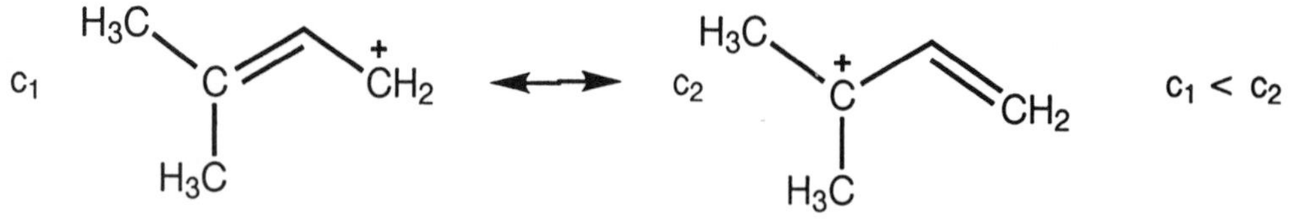

c_1 $\longleftrightarrow$ c_2 $c_1 < c_2$

In (d), it is the traditional "carbonyl" carbon-oxygen double bond that is the better form. It contains more bonds and has no charge separation.

c_1 ⟷ c_2 $c_1 > c_2$

Problem 9.10 "Yes and no" might be the best answer. Certainly one can draw a resonance form for each carbon-hydrogen bond of the three methyl groups; this makes a total of nine. However, orbitals must overlap for there to be real delocalization. All nine carbon-hydrogen bonds cannot be properly lined up at the same time. A look at one methyl group makes the point.

Good overlap

Eye

Poor overlap

Newman projection

There is good overlap of this C–H bond with the empty 2*p* orbital

There is poor overlap of these C–H bonds with the empty *p* orbital

Problem 9.12 It is weaker. Overlap is less good, as the orbitals involved are not optimally lined up.

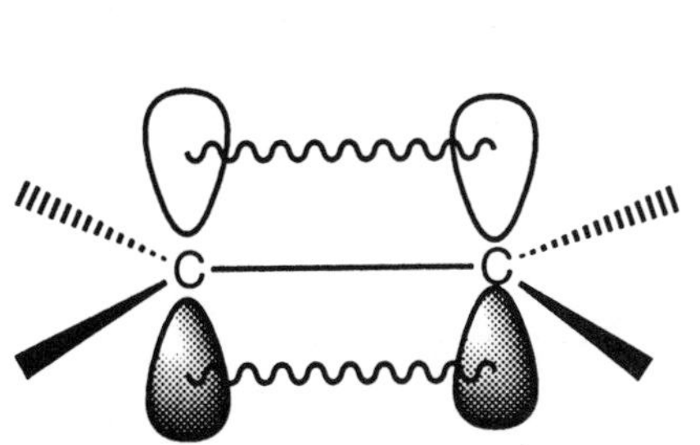

The π bond in ethylene is made up of 2*p*/2*p* overlap. The two 2*p* orbitals are parallel and overlap well

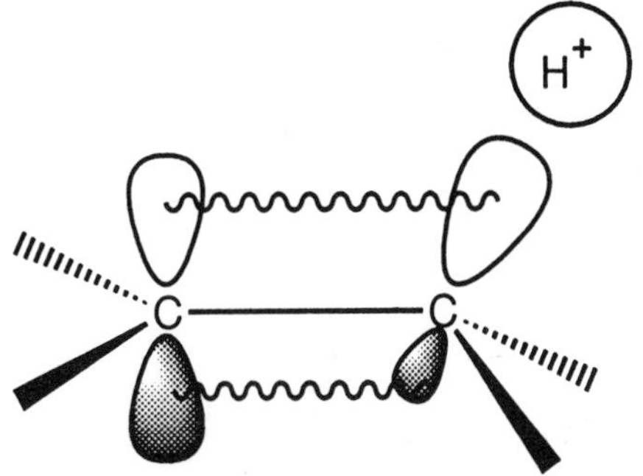

In the "hyperconjugative" resonance form for the ethyl cation, the "double" part of the double bond is not made up of 2*p*/2*p* overlap, but by 2*p*/sp^3 overlap. These orbitals do not overlap as well as the two 2*p* orbitals, and the bond will be weaker

Problem 9.13 Protonation of ethylene must lead to a primary carbocation, a most unstable species. Accordingly, one can expect this reaction to be difficult. Recall the Hammond postulate (Chapter 8, p. 320): In this highly endothermic reaction, the transition state will strongly resemble the product, the primary carbocation, and will, therefore, be very high in energy. For another view of the protonation of ethylene, see Section 9.13.

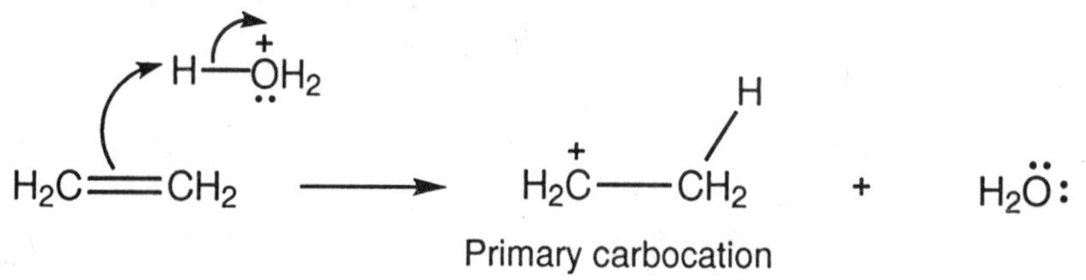

$H_2C{=}CH_2 + H{-}\overset{+}{O}H_2 \longrightarrow H_2\overset{+}{C}{-}CH_2H + H_2\ddot{O}:$

Primary carbocation

Problem 9.14 Protonation of 2-ethyl-1-butene takes place to give the more stable, tertiary carbocation rather then the less stable primary carbocation.

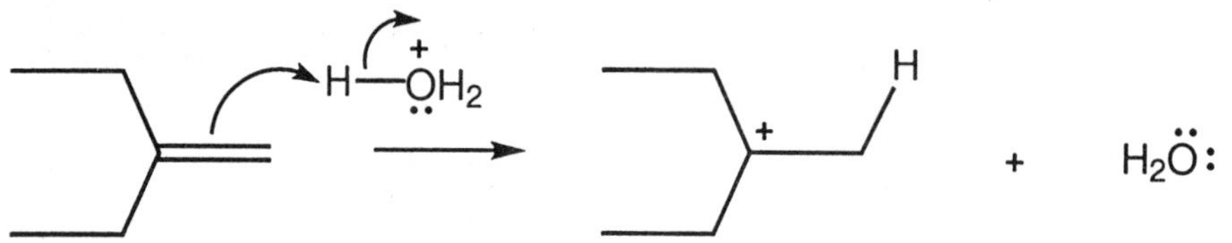

The tertiary carbocation can add to another alkene, again producing the more stable, tertiary carbocation.

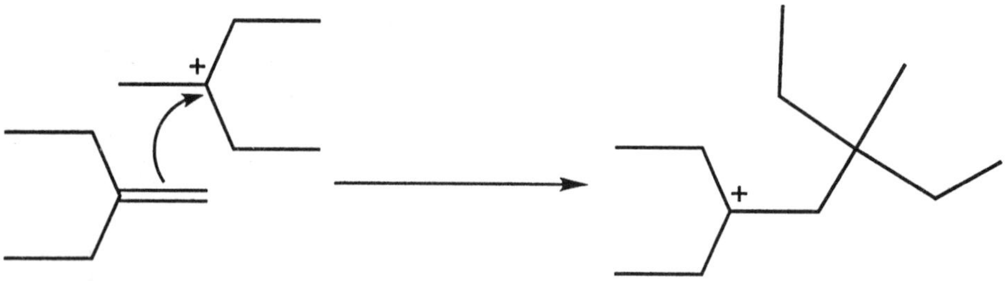

There are three different products that can be formed by loss of a proton from this new cation:

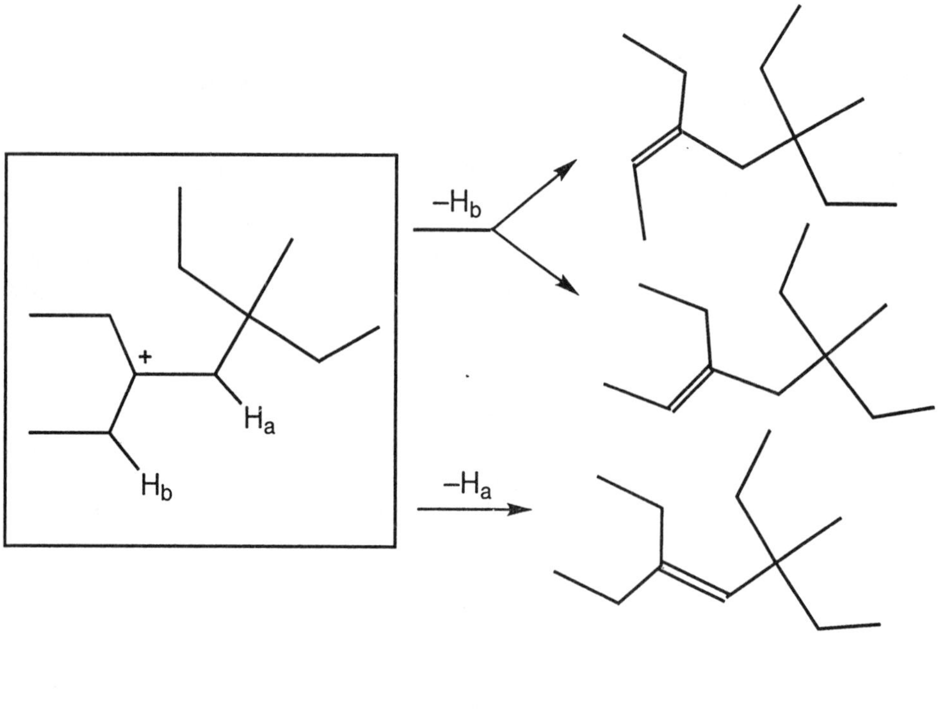

$-H_b$

H_a

H_b

$-H_a$

Problem 9.15 In $_5B$, there are five positive charges in the nucleus. In BH_3 or BF_3, boron has a pair of 1*s* electrons and a share in the electrons in the three two-electron σ bonds for a total of five negative charges. Boron is neutral as the positive and negative charges exactly balance. The sp^2 hybridization uses the boron $2s$, $2p_x$, and $2p_y$ atomic orbitals leaving the $2p_z$ unused, and in this case, empty.

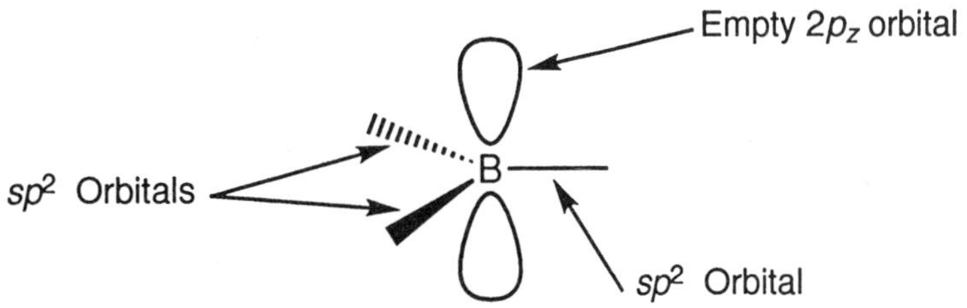

Empty $2p_z$ orbital

*sp*² Orbitals

*sp*² Orbital

Problem 9.18 This is not a mechanistic question; it only asks from which direction (top or bottom) the two bromines attach. In this case, the structures of the products make it clear that one bromine comes from the "top" and the other from the "bottom," in what is called "anti" addition.

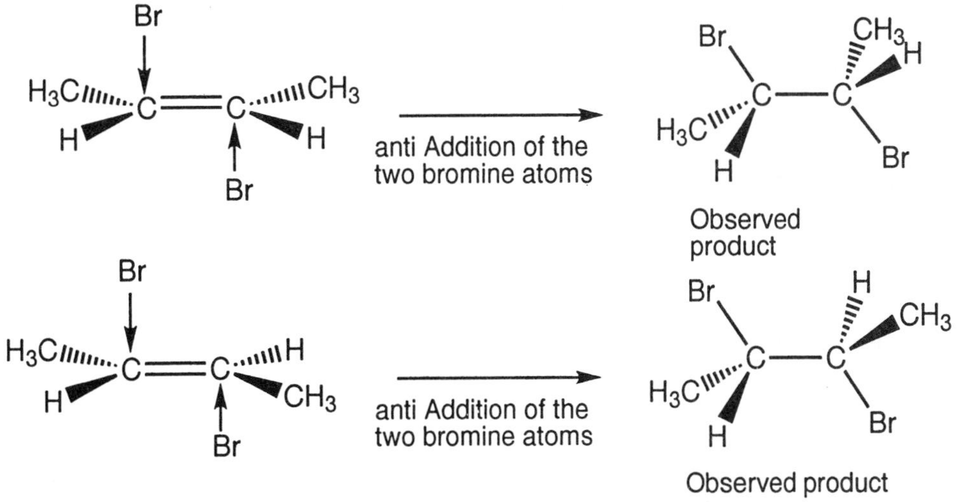

anti Addition of the two bromine atoms

Observed product

anti Addition of the two bromine atoms

Observed product

Had addition of the two bromines been from the same side, in syn fashion, just the opposite results would have been seen:

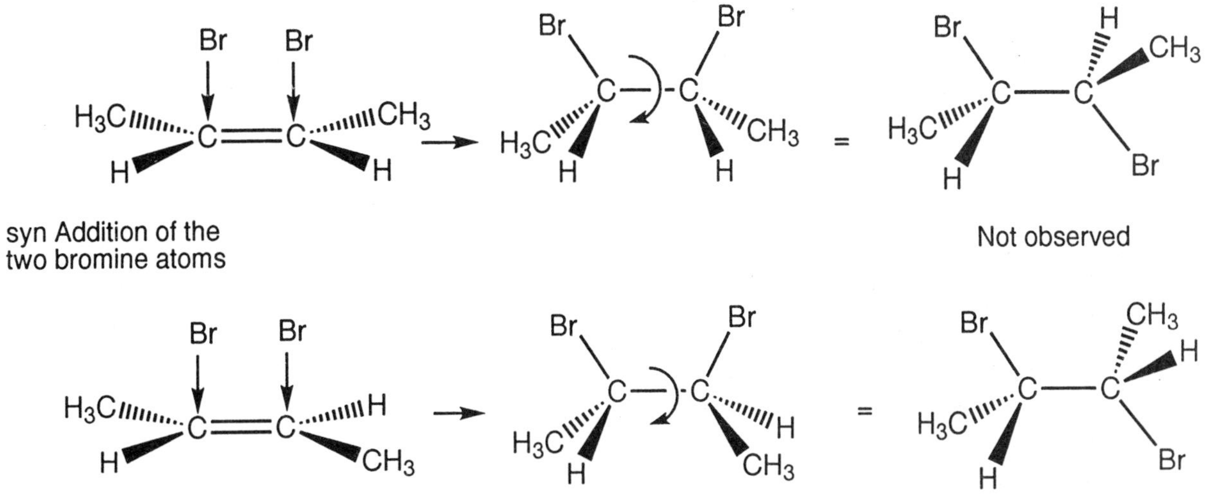

syn Addition of the two bromine atoms

Not observed

Problem 9.19 There are many possibilities, although most will require some sort of isotopic label. The question states that any appropriate starting material is available, so that's not a problem. Here's one in which a deuterated cycloalkene is used.

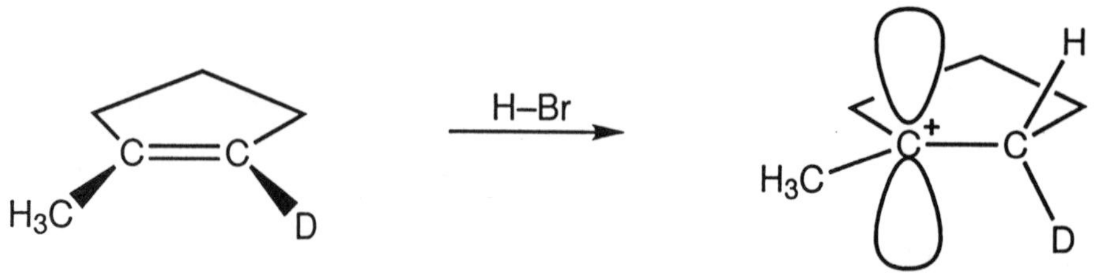

H–Br

Addition can occur to either lobe of the 2*p* orbital to give either syn or anti addition. A careful determination of the structure of the product will allow a determination of the direction of addition.

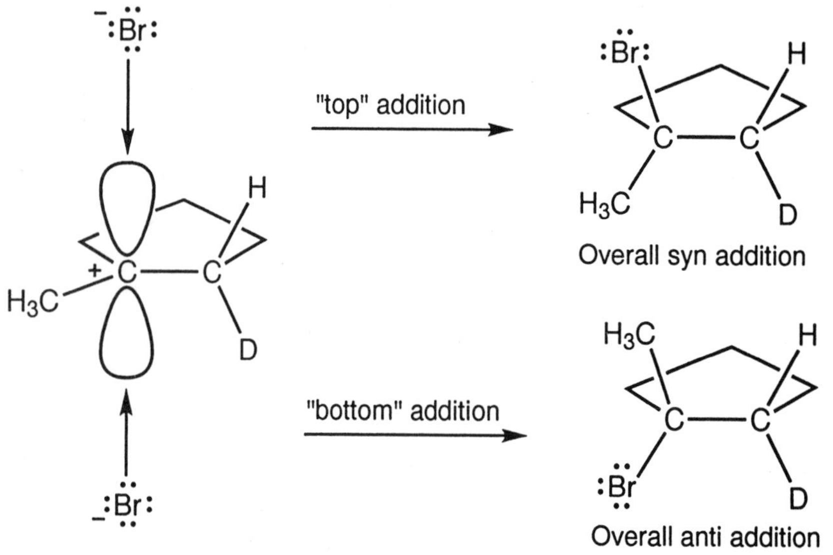

"top" addition

Overall syn addition

"bottom" addition

Overall anti addition

Problem 9.21 The boron-oxygen bonds have partial double bond character, and are therefore especially strong. One filled orbital on oxygen overlaps with the empty 2*p* orbital on boron. In resonance terms, there are two important forms contributing to the structure.

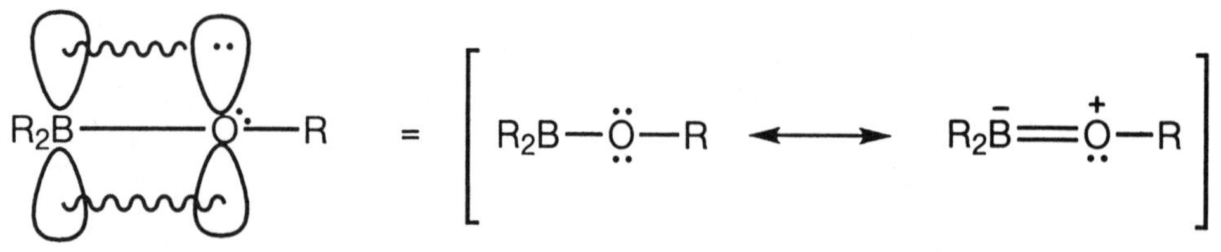

$R_2B—\ddot{O}—R$ = [$R_2B—\ddot{O}—R$ ⟷ $R_2\bar{B}=\overset{+}{O}—R$]

Problem 9.23 There are several straightforward parts to this question, which only require that you remember one synthetic procedure.

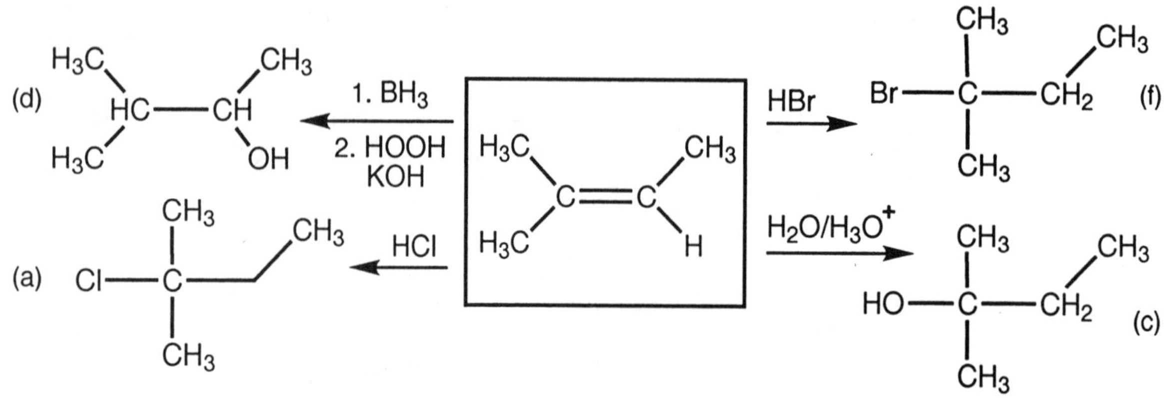

(d) 1. BH_3 2. HOOH, KOH

(a) HCl

(f) HBr

(c) H_2O/H_3O^+

The remaining three molecules demand one more level of sophistication as they cannot be made directly, but require a second transformation involving a molecule already constructed.

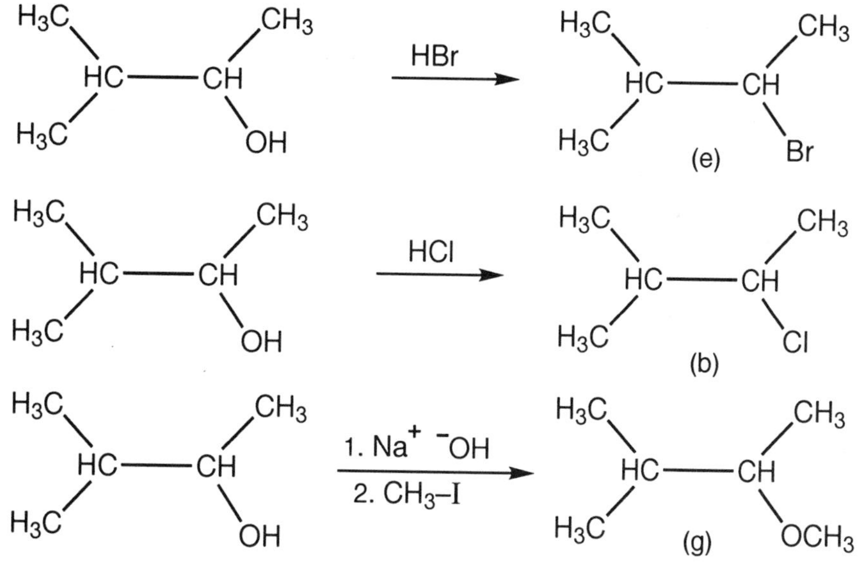

Problem 9.24 The Lewis acid is of course the empty $2p$ orbital of the carbocation. The Lewis base is the filled σ orbital of the carbon-hydrogen bond.

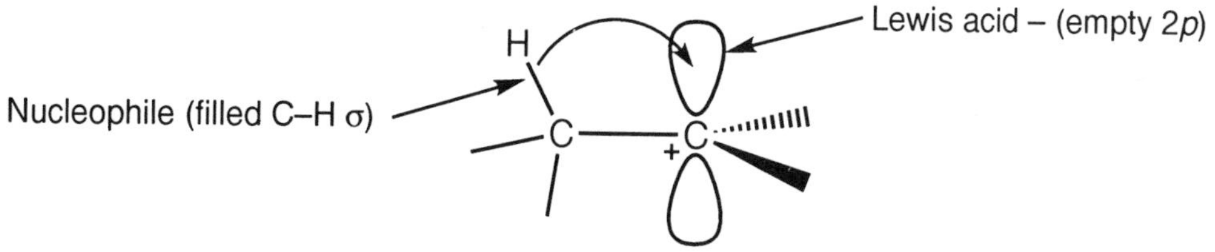

Problem 9.25 In the transition state, the methyl group must be half-migrated from one carbon to the other.

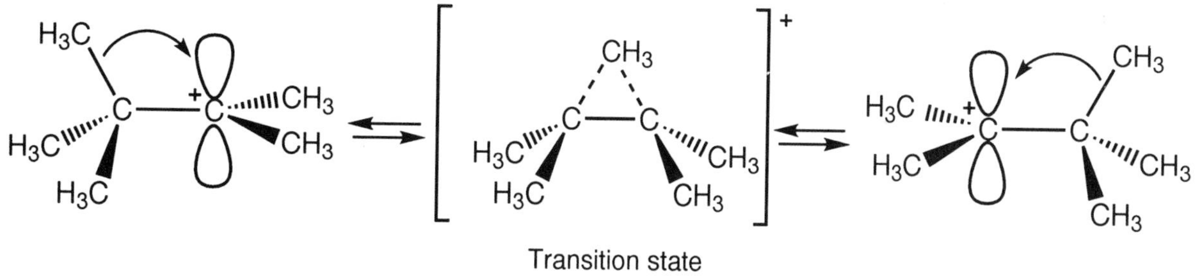

The orbitals involved are shown below. There is a triangle of overlapping hybrid orbitals, very similar to the case for cyclic H_3 in which a triangular arrangement of three $1s$ orbitals is used (Chapter 2, p. 64).

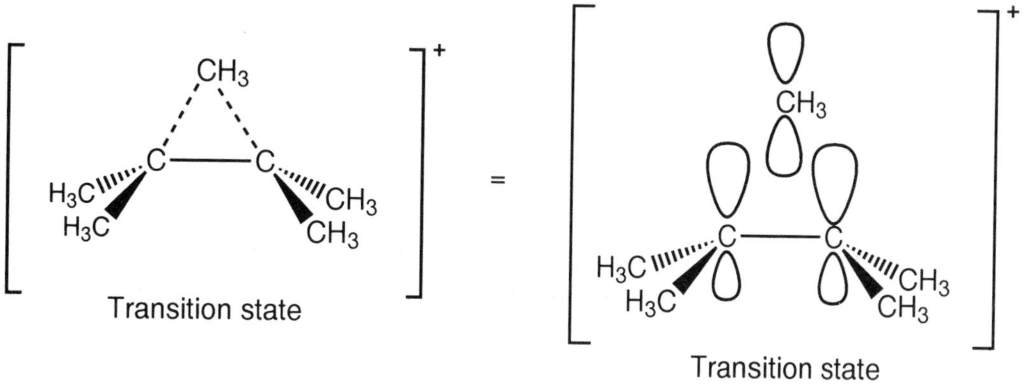

As for cyclic H_3, there will be three molecular orbitals arranged as shown in the diagram below. In the cation, there are two electrons to be placed in the orbitals, and they will quite nicely occupy the single low-energy bonding molecular orbital. As the transition state looks to be favorable energetically, this reaction should be an easy one.

Energy

Nonbonding

Transition state molecular orbitals

Additional Problem Answers

Problem 9.27 The π molecular orbital system for allyl consists of three orbitals. These were constructed in the chapter (p. 341-42), but by now it should be possible to write them out without the construction process. It is especially useful to recall that the number of new nodes will increase 0....1....2. In the π system of the cation there are only two electrons. In the neutral radical there must be one more, and in the negatively charged anion, two more.

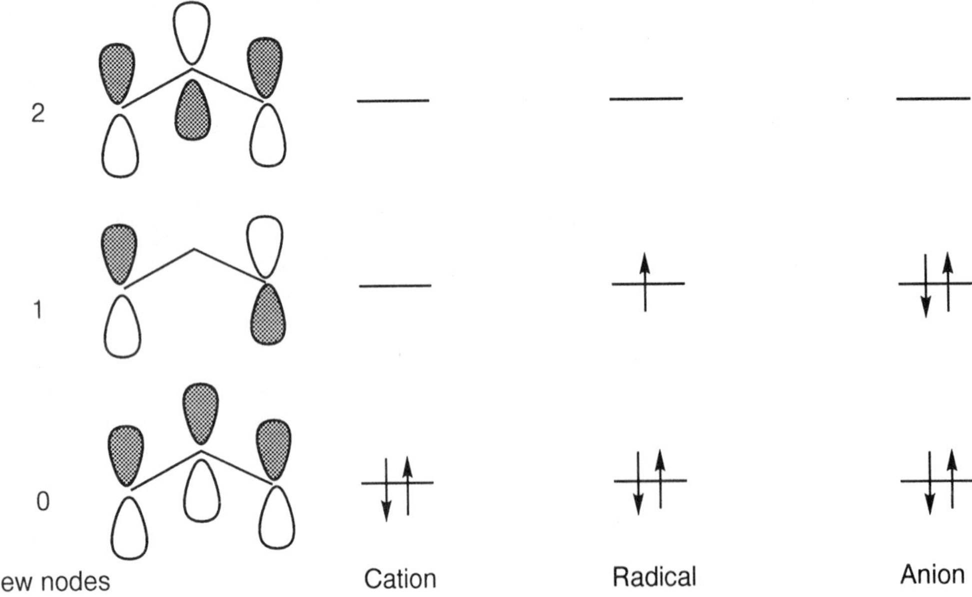

Problem 9.28 Here are the top views of allyl and pentadienyl.

Allyl	Pentadienyl
	+ − + − +
+ − +	+ − 0 + −
+ 0 −	+ 0 − 0 +
	+ + 0 − −
+ + +	+ + + + +

What generalizations can you make? Some things you know already. For example, you know that the number of molecular orbitals must match the number of atomic orbitals being combined. You also know that the number of nodes must increase monotonically (one more node each time) starting from the lowest molecular orbital with no nodes other than the node present in any orbital made from 2*p* orbitals.

Here are some generalizations:

1. As long as you start at the left of our orbital with (+), the right-hand ends of the orbitals alternate sign, (+), (–), (+), (–), etc. Presumably, if Nature is a kind, symmetrical creature, this will always be true.

2. The highest molecular orbital is always alternating (+) and (–). You already know the lowest energy molecular orbital is "all +." So, you can specify the lowest and highest molecular orbital with no trouble.

3. The middle molecular orbital is always (+), (0), (–), (0), (+), and so on.

4. The pattern of zeros (atoms at which the sign of the wave function is zero) is symmetrical, and the number of zeros increases by one to the middle molecular orbital, then decreases by one.

Perhaps you can find other "rules." These observations, together with some faith that Nature is symmetrical at this level, allow us to write molecular orbitals for acyclic, "linear" fully conjugated systems quite easily. But don't extend this too far! These "rules" won't work for systems containing even numbers of carbons, or cyclic molecules. For such systems you will need other observations.

Problem 9.29 Let's apply the observations from Problem 9.28. First, there will be seven molecular orbitals and the nodes will increase monotonically. Next, the bottom (lowest energy) and top (highest energy) molecular orbitals must be as in drawing (a). We can next add the middle molecular orbital as in (b). Next, add the ends of all the molecular orbitals (c), and fill in the orbital second lowest in energy. It has only a single node and is always easy to draw (d). Now the work begins. The next-to-highest molecular orbital will have a zero in the same place as the next-to-lowest, and the two other molecular orbitals will each have two "zeros". Given that information, it is not so hard to fill the rest in (e). At the end, check to see that the proper nodal pattern (0..1..2..3..) appears.

(a)	(b)	(c)	(d)	(e)	Nodes
+ − + − + − +	+ − + − + − +	+ − + − + − +	+ − + − + − +	+ − + − + − +	6
		+ −	+ −	+ − + 0 − + −	5
		+ +	+ +	+ − 0 + 0 − +	4
	+ 0 − 0 + 0 −	+ 0 − 0 + 0 −	+ 0 − 0 + 0 −	+ 0 − 0 + 0 −	3
		+ +	+ +	+ + 0 − 0 + +	2
		+ −	+ + + 0 − − −	+ + + 0 − − −	1
+ + + + + + +	+ + + + + + +	+ + + + + + +	+ + + + + + +	+ + + + + + +	0

Problem 9.30

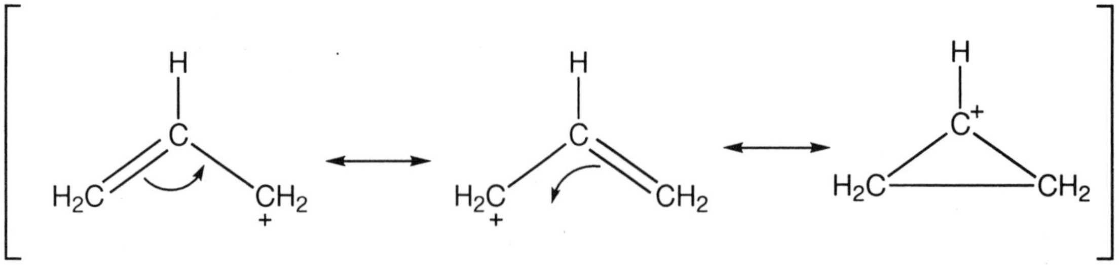

In each case, the curved arrow leads to the resonance form on the right.

Be careful not to move atoms in your answer. If atoms change position we are dealing with chemical equilibrium, not resonance. The third resonance form is a cyclic structure, a three-membered ring. But it is no ordinary three-membered ring! Not only is there a long, high energy bond, but this bond is a π bond, made from the long-range overlap of two $2p$ orbitals. This bond is certainly weak, and will contribute little to the structure of the allyl cation. The practical result is that the charge is shared between the two end carbons, but not the third, central carbon.

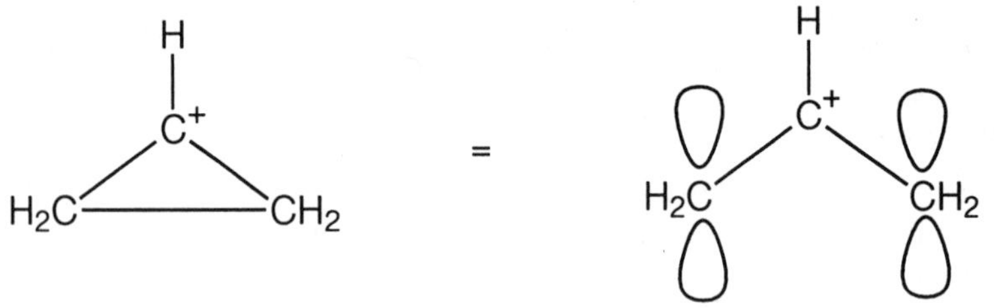

Problem 9.31

(a)

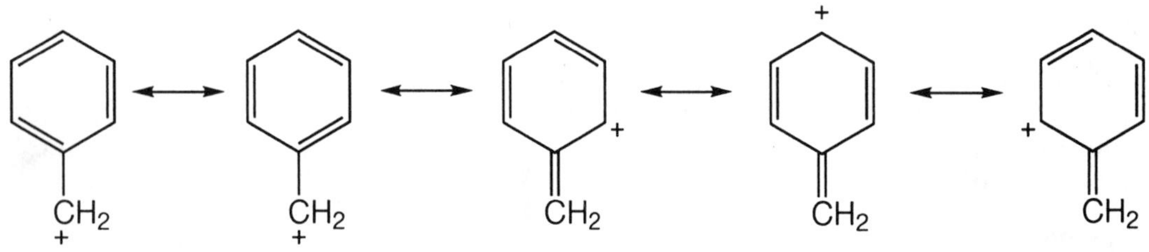

(b)

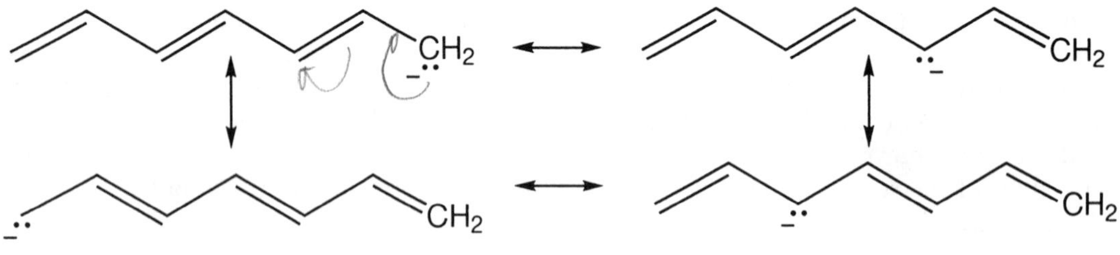

(c)

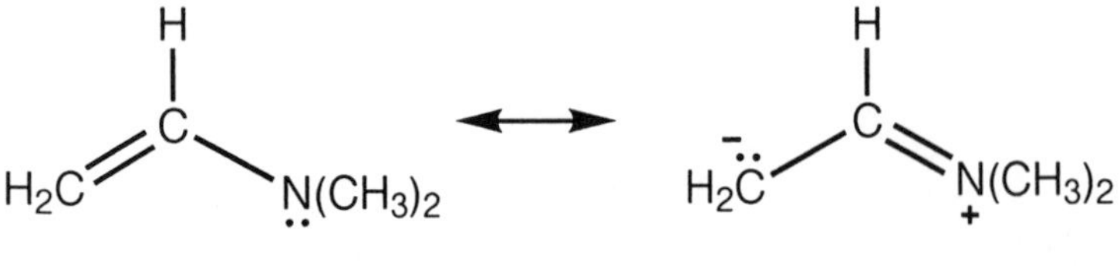

(d)

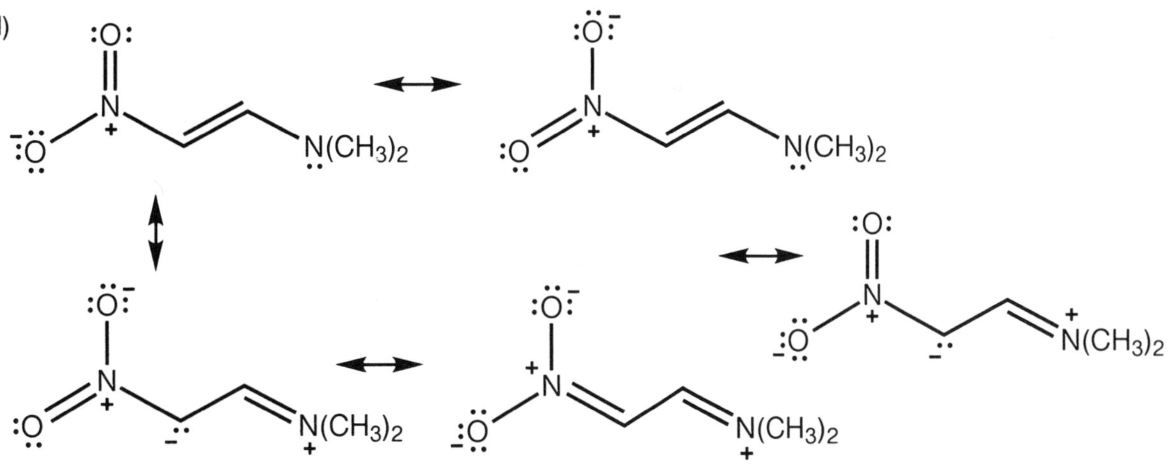

Problem 9.32

(a) These structures are surely not a pair of resonance forms. Write in the hydrogens and you will see that one must be moved in order to distribute the charge as shown. These structures might be in chemical equilibrium, but they are not resonance forms.

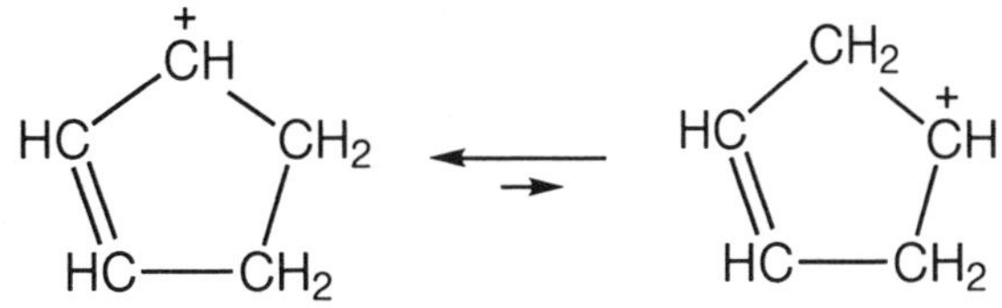

(b) No problem. This is resonance. Only electrons have been moved.

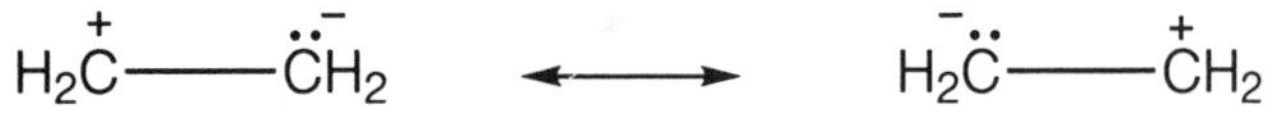

(c) As in part (a), an atom has been moved and this cannot be resonance. It might be an equilibrium.

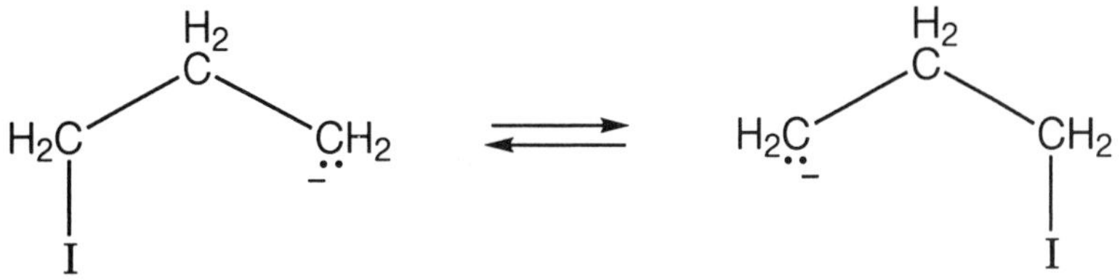

(d) This is resonance. Only electrons have been moved.

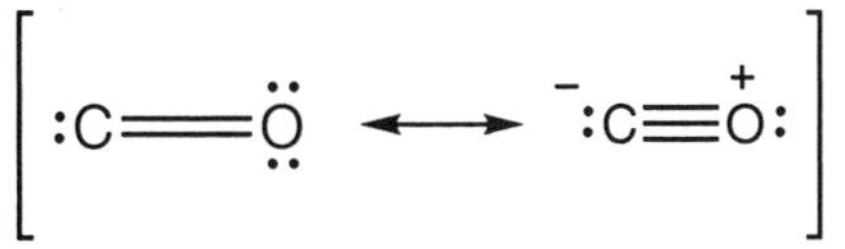

Problem 9.33 First, draw good Lewis structures for this pair.

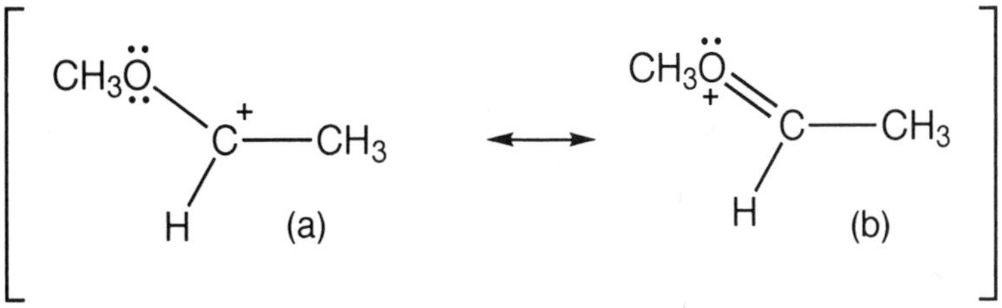

(a) (b)

This problem is tougher than than it first appears to be. At first glance, one might assume that resonance form (a) would be the more important one because (b) has a positive charge on oxygen, a more electronegative atom than carbon. However, form (b) has one more bond than (a) and in (a) the carbon atom is electron-deficient. In (b) both the oxygen and carbon have complete octets of electrons, making this an important resonance form.

Problem 9.34(a) This part is simple. The secondary carbocation will be formed far more easily than the primary carbocation, and this will always lead to the isopropyl-X compound.

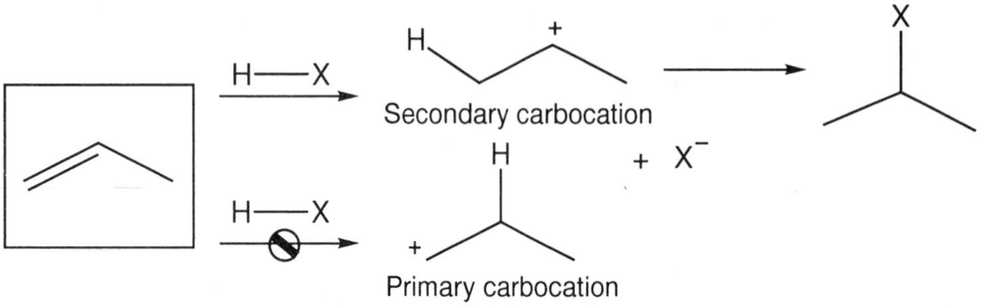

H—X
Secondary carbocation
+ X^-
H—X
Primary carbocation

(b) Protonation will give the more stable, resonance-stabilized carbocation rather than the simple secondary carbocation in which charge is localized. Addition of the anion leads to the product shown.

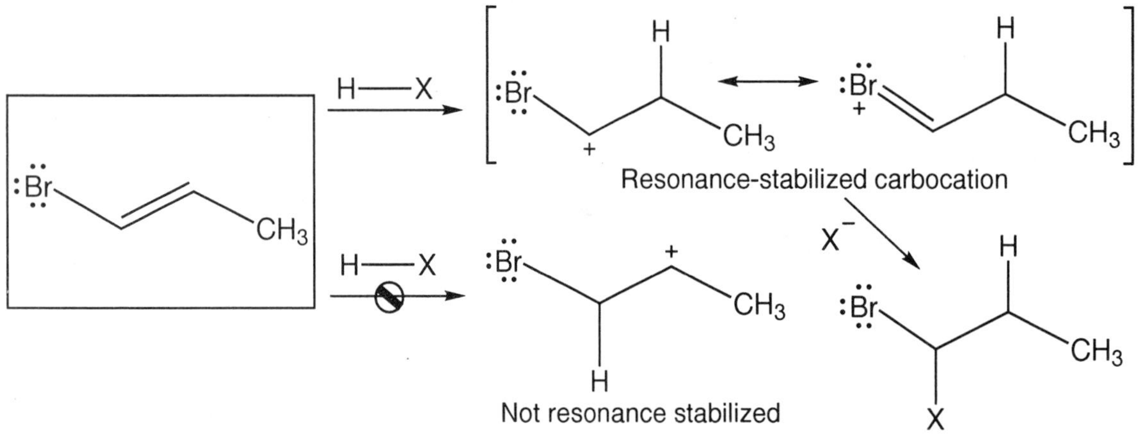

H—X
:Br: CH3
Resonance-stabilized carbocation
X^-
H—X
Not resonance stabilized

(c) Here, too, one possible cation is nicely stabilized by resonance, and the other is not. The resonance-stabilized cation is the more stable one of the pair.

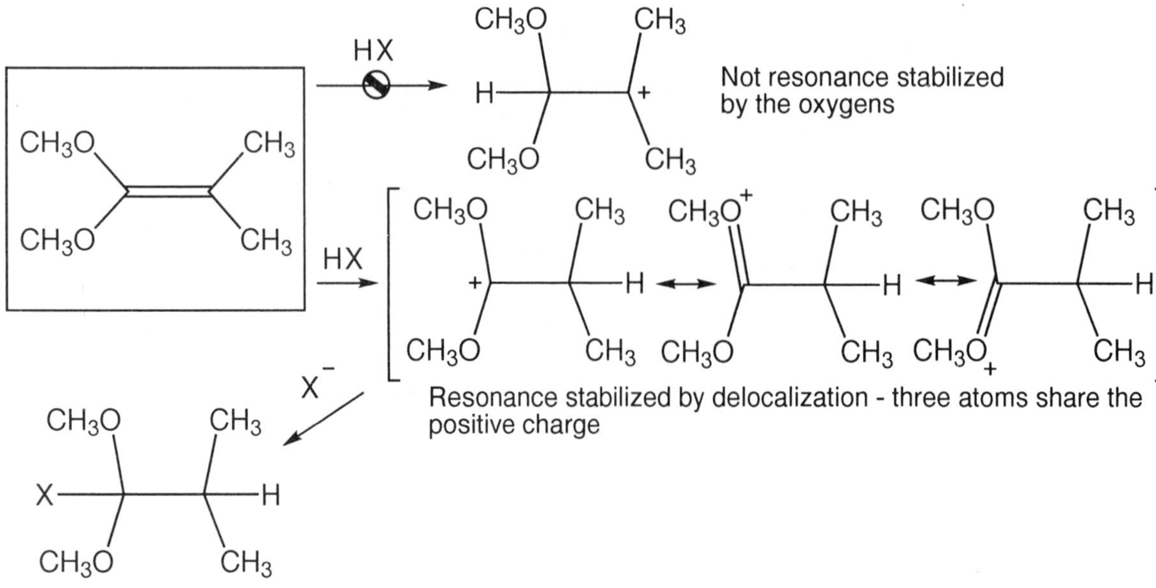

HX
Not resonance stabilized by the oxygens
HX
Resonance stabilized by delocalization - three atoms share the positive charge
X^-

(d) The answer is similar in this case. One possible cation is resonance stabilized and favored, the other is localized and therefore substantially higher in energy. It will not be formed. If you're getting the idea that one answer to *lots* of questions in organic chemistry is "It's resonance stabilized," you're on the right track.

This carbocation is stabilized by resonance; the charges in parenthesis show the positions sharing the charge; if this isn't obvious, by all means draw out the individual forms

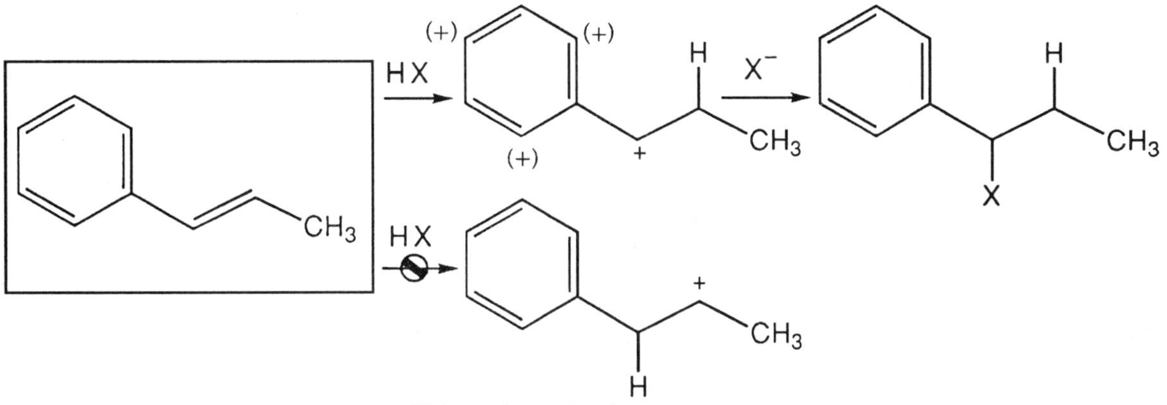

(+) (+) H HX X⁻ H CH3 CH3 (+) + X CH3 HX + CH3 H

This carbocation is not resonance stabilized

(e) This problem is the only tricky one of the group. Fluorine is a very electronegative atom, and therefore exceptionally electron withdrawing. The carbons to which the fluorines are attached bear a substantial partial positive charge. Protonation will take place so as to keep the positive charge as far from this partially positive carbon as possible.

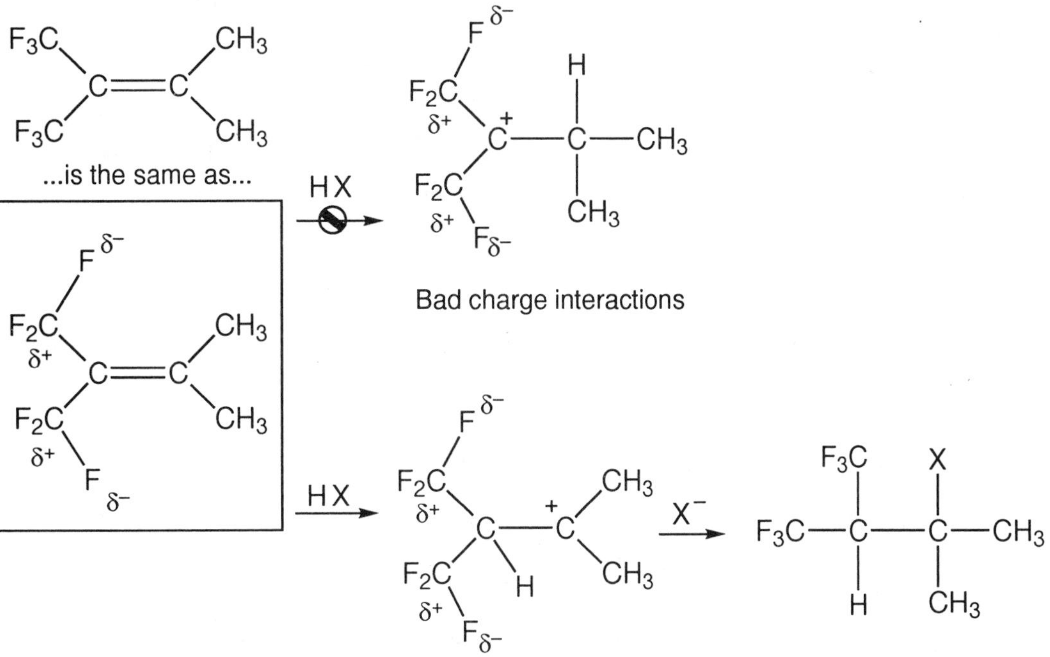

...is the same as...

Bad charge interactions

Problem 9.35 For cyclohexene, protonation generates a secondary carbocation that acts as Lewis acid toward another cyclohexene, and on, and on, and on....

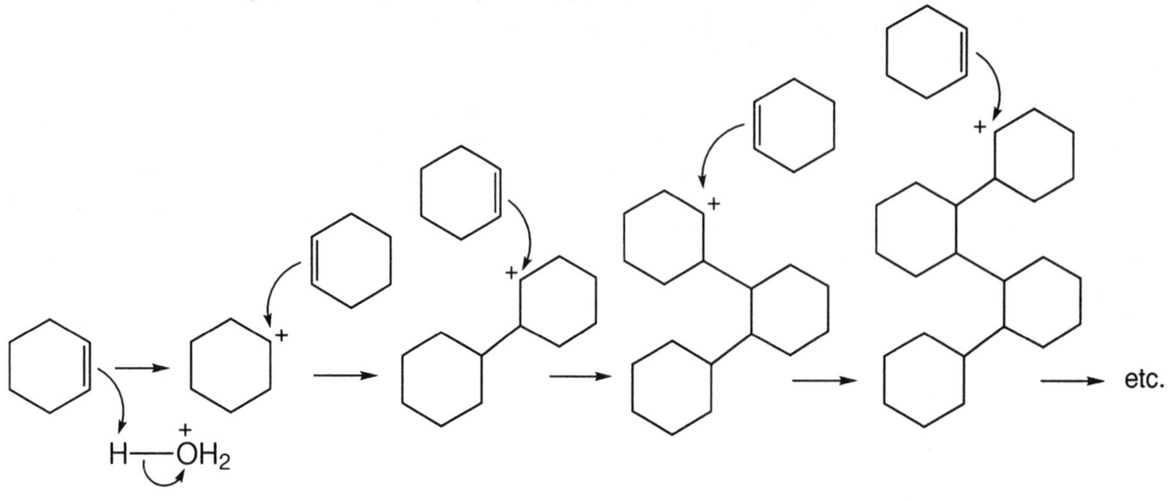

$H—\overset{+}{O}H_2$ → etc.

Similarly, for styrene:

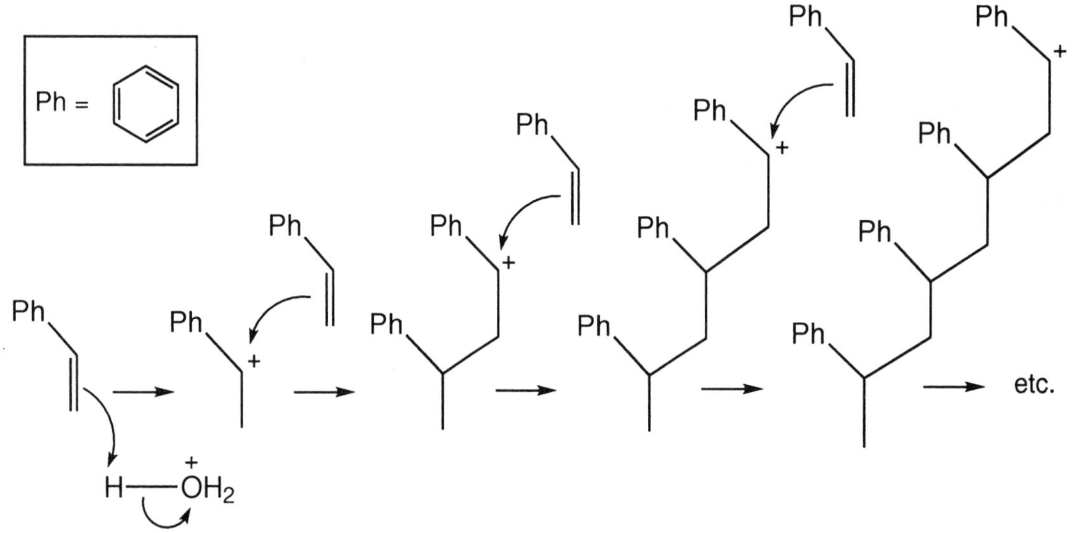

Ph =

$H—\overset{+}{O}H_2$ → etc.

Problem 9.36 The first problem is to decide in which direction butadiene will protonate. There really is no choice; one protonation leads to a resonance-stabilized carbocation, the other to a simple localized primary carbocation.

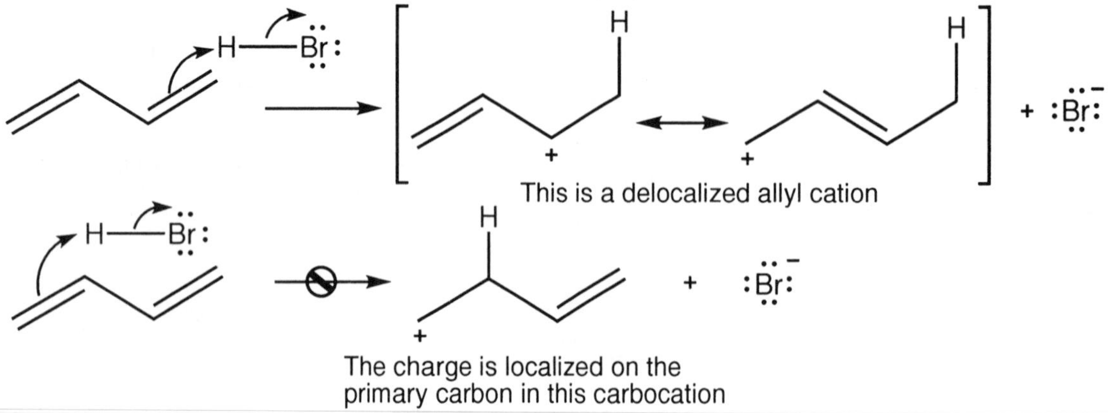

H—Br: → + :Br:⁻

This is a delocalized allyl cation

H—Br: + :Br:⁻

The charge is localized on the primary carbon in this carbocation

Two carbons share the positive charge in the resonance-stabilized, allyl cation. Be careful in the drawing below not to fall into the trap of thinking of the two resonance forms as separate entities. Two carbons share the charge in a *single* structure, summarized at the right of the figure with the "dashed bond" structure.

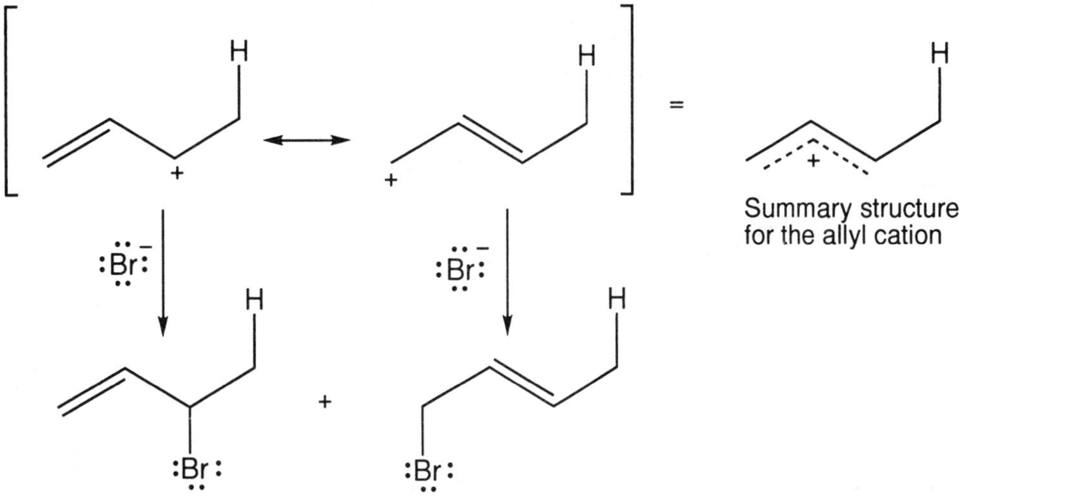

Summary structure for the allyl cation

Problem 9.37 Protonation gives a planar, secondary carbocation. Addition of bromide ion from the top of the cation gives one enantiomer; addition from the bottom gives the other. As these two pathways are equivalent, the two products must be formed in equal amounts. This reaction gives a racemic mixture.

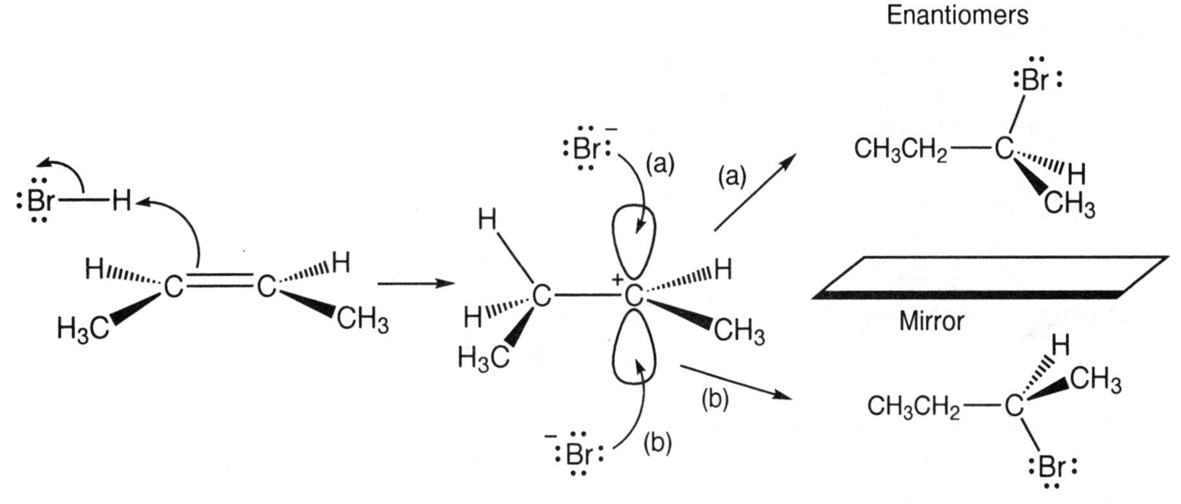

Enantiomers

Mirror

There is something else that the carbocation can do besides react with the bromide ion. Like all cations, it can lose a proton to give an alkene, if there is a proton attached to an adjacent carbon atom. In this case there are two such protons, H_a and H_b. H_a can be lost from two rotational isomers of the carbocation to give the starting alkene, *cis*-2-butene and its diastereomer, *trans*-2-butene. Alternatively, H_b can be lost to give 1-butene. Any base in the reaction can assist in proton removal. The figure shows bromide acting as the base.

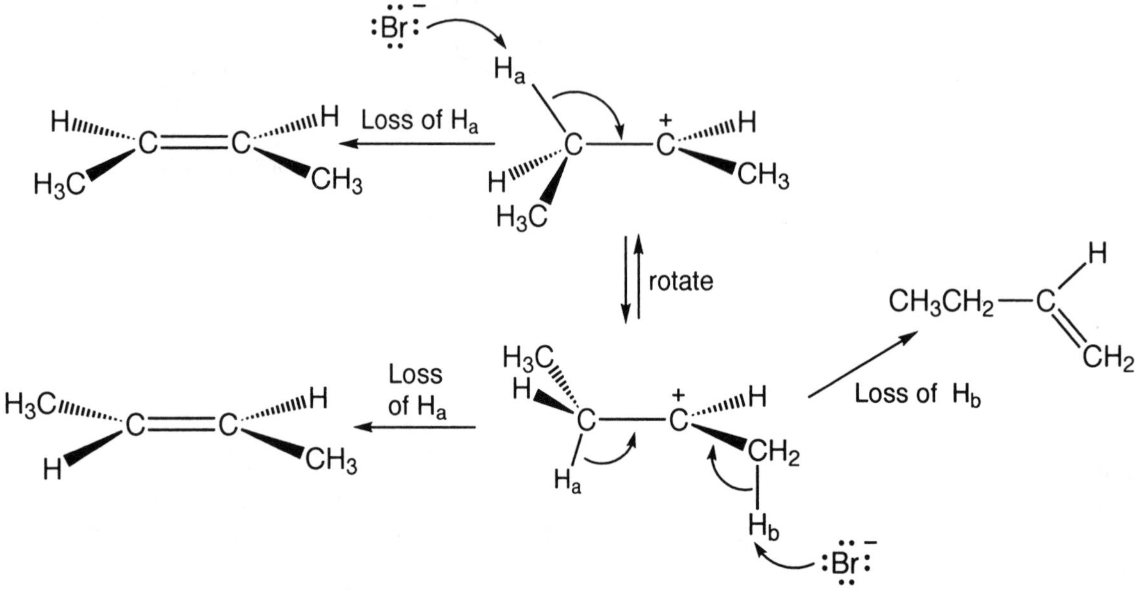

Loss of H_a

rotate

Loss of H_a

Loss of H_b

Problem 9.38

(a) This reaction is closely related to hydration. Alcohol replaces water, but the mechanistic process is the same. Protonation leads to a tertiary carbocation (not the less stable primary carbocation) that is captured by alcohol. A final deprotonation gives the product.

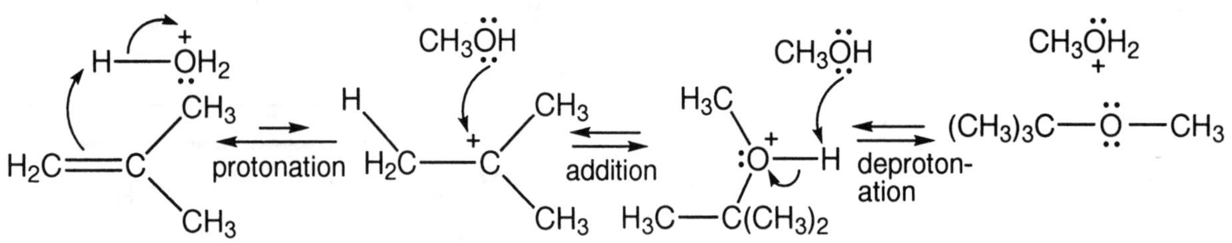

protonation addition deprotonation

(b) This problem is just a common example of (a), the acid-catalyzed addition of an alcohol to an alkene. In this case, protonation occurs to give the resonance-stabilized carbocation, not the localized secondary carbocation.

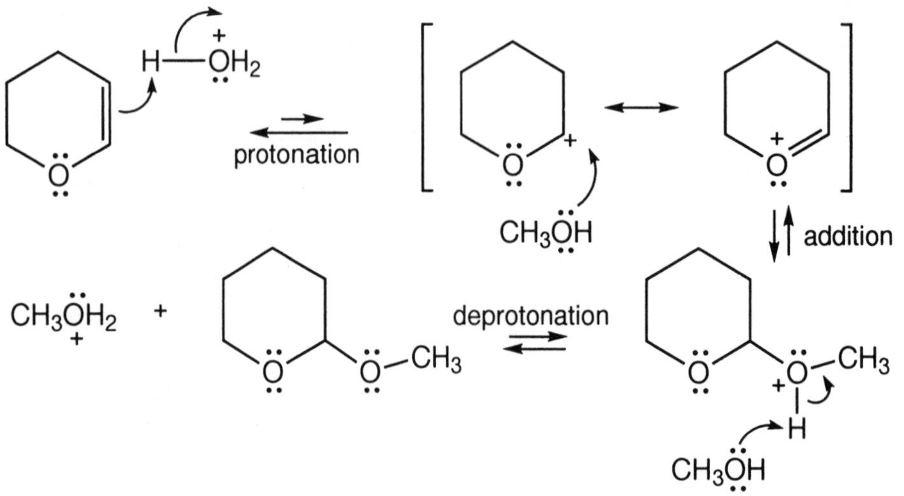

protonation addition deprotonation

Problem 9.39 The first part of this reaction is exactly like hydration except that hydrogen sulfide (H_2S), plays the part of water. The product is a thiol, (RSH). Notice the pattern of three steps: protonation, addition, and deprotonation that is present in many addition reactions.

$(CH_3)_2C{=}C(CH_3)_2$ + $H{-}OH_2^+$ ⇌ (protonation) $(CH_3)_2CH{-}C^+(CH_3)_2$ + $:SH_2$ ⇌ (addition) $(CH_3)_2CH{-}C(CH_3)_2{-}S^+H_2$ ⇌ (deprotonation, $:SH_2$)

RSH = $(CH_3)_2CH{-}C(SH)(CH_3){-}CH_3$ + $:SH_3^+$

A thiol

As this product, RSH, builds up, it can begin to compete with H_2S for the carbocation. The product of this reaction is the thioether (RSR), the other product shown in the problem.

$(CH_3)_2C{=}C(CH_3)_2$ + $H{-}SH_2^+$ ⇌ (protonation) $(CH_3)_2CH{-}C^+(CH_3)_2$ + RSH ⇌ (addition) $(CH_3)_2CH{-}C(CH_3)_2{-}S^+(R)H$ ⇌ (deprotonation, RSH)

RSR = $(CH_3)_2CH{-}C(SR)(CH_3){-}CH_3$ + RSH_2^+

Problem 9.40 There are two possibilities for protonation of the carbon-carbon double bond in this molecule, called a "ketene." The more favorable one leads to a resonance-stabilized carbocation. Capture of this ion by water leads to the observed product. Capture of the less stable carbocation by water would give the compound that is not observed in the reaction.

$H_2C{=}C{=}O$ (Ketene) + $H{-}OH_2^+$ (a) or $H_2O^+{-}H$ (b); path (a) ⊘→ $H(H_2C^+)C{=}O$ → → $H(H_2C{-}OH)C{=}O$

(continued from previous page)

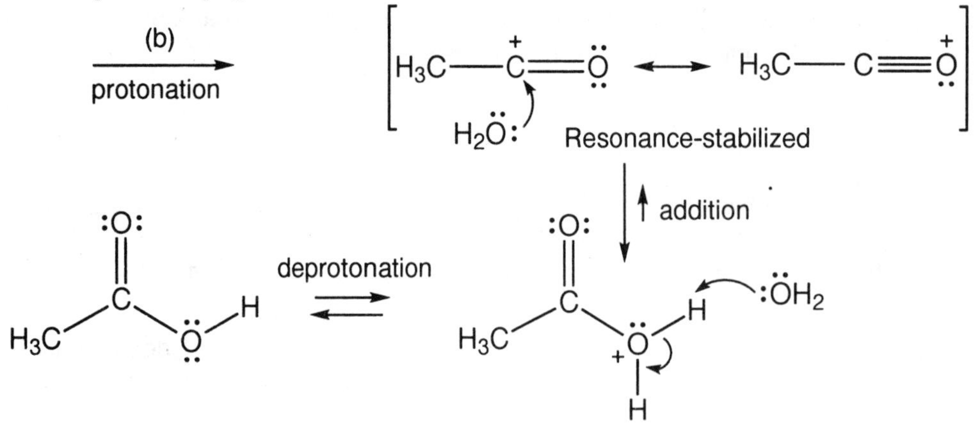

(b) protonation

$H_3C—\overset{+}{C}=\ddot{O}$ ⟷ $H_3C—C\equiv\overset{+}{O}$

$H_2\ddot{O}$: Resonance-stabilized

addition

deprotonation

$:\ddot{O}H_2$

Problem 9.41 Both of these reactions involve multiple hydroborations, one or more of which are intramolecular.

(a) Let R = $(CH_3)_2CHC(CH_3)_2$, and draw the molecule in a suggestive way. In doing this we are only taking advantage of the free rotations about carbon-carbon single bonds.

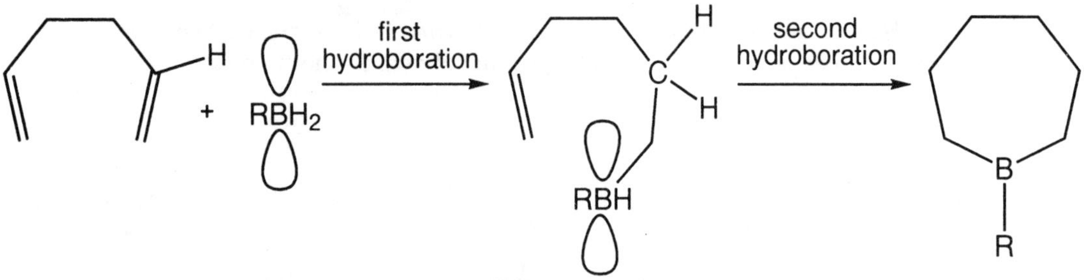

+ RBH_2 —first hydroboration→ RBH —second hydroboration→

(b)

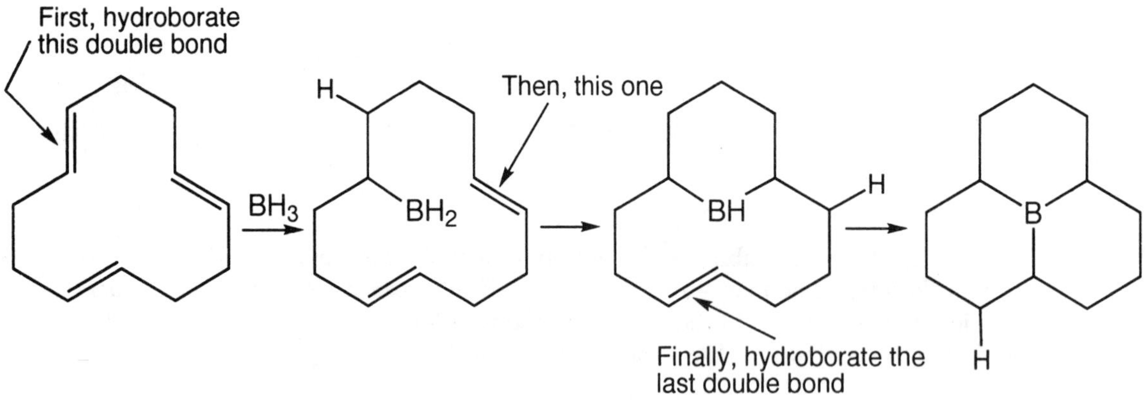

First, hydroborate this double bond

BH_3 → BH_2 Then, this one → BH → B

Finally, hydroborate the last double bond

Problem 9.42 We know that the original hydroboration occurs in a syn fashion.

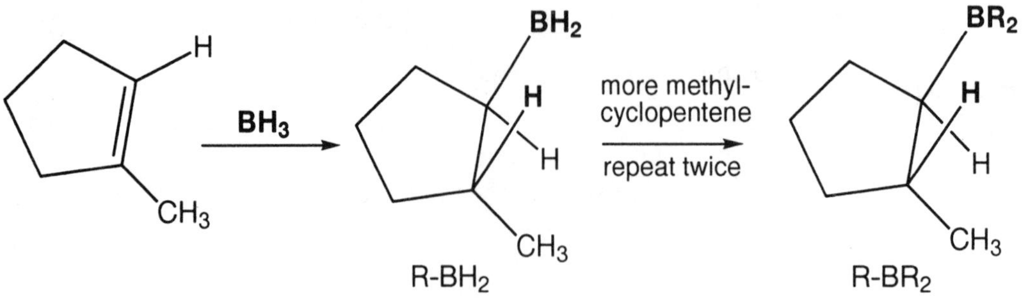

BH_3 → $R\text{-}BH_2$ —more methylcyclopentene, repeat twice→ $R\text{-}BR_2$

As the product alcohol has the same stereochemistry as the original borane, the overall replacement of BH_2 with OH must occur with retention, whatever the mechanism.

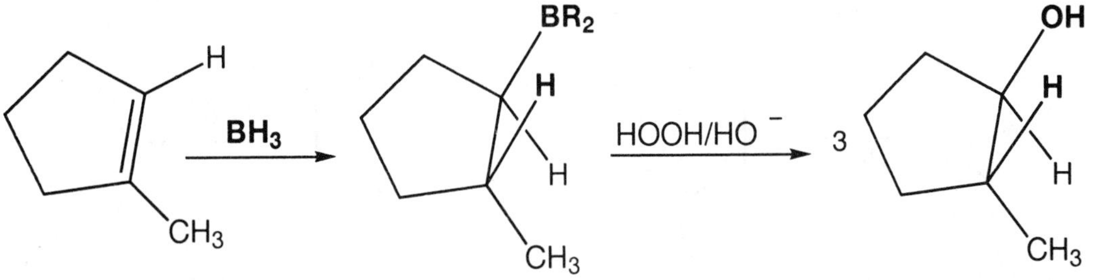

Problem 9.43 3-Methyl-2-butanol is a secondary alcohol and it could react with hydrogen bromide and hydrogen chloride by either an S_N1 or S_N2 process (or both). In any case, the first step is protonation of the alcohol.

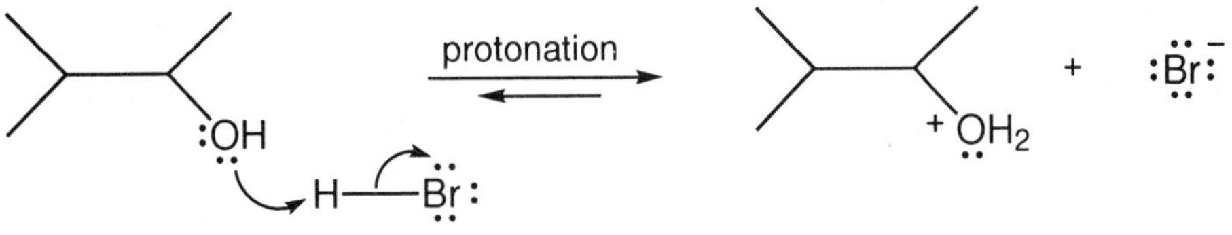

If the leaving group water is displaced by halide ion (shown here as bromide) in an S_N2 process, there should be no problem. (We might well expect this to be a relatively slow S_N2 reaction - Why?)

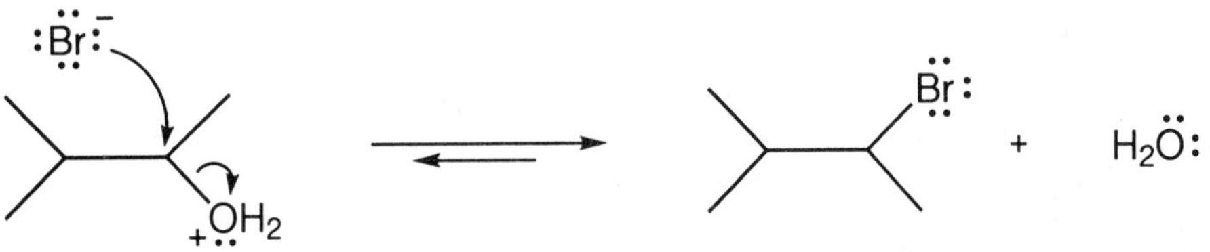

It is also possible that water could leave the protonated alcohol in an S_N1 reaction to give a secondary carbocation. If this cation is captured by halide ion, there is no problem. However, the secondary carbocation could also rearrange to the more stable tertiary carbocation through a hydride shift. If this happens, capture by halide will give 2-halo-2-methylbutanes.

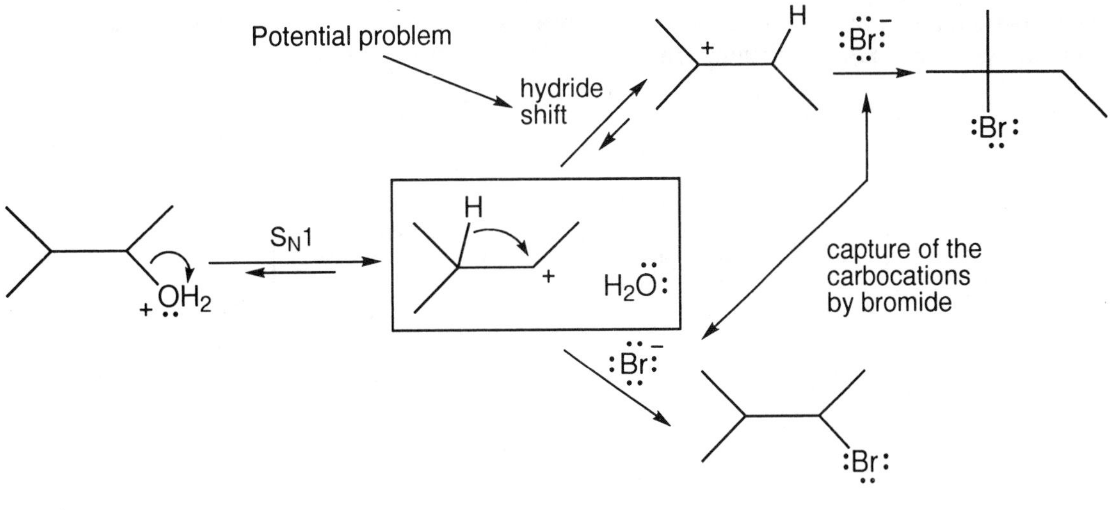

So, depending on the extent of the S_N1 component of these reactions, the desired 2-halo-3-methylbutanes could be contaminated with the isomeric 2-halo-2-methylbutanes.

Problem 9.44 Two parts of this problem are extremely easy, and require only that you remember a pair of simple, Markovnikov addition reactions.

H_3O^+ / H_2O → (a); HBr → (b)

Another part, (c), involves hydroboration to give anti-Markovnikov addition. As yet, you have no way of adding hydrogen bromide directly in an anti-Markovnikov sense, so the synthesis of (d) requires a further reaction of (c):

1. BH_3 2. H_2O_2, HO^- → (c); HBr → (d)

Similarly, (d) can be used to make (e) and (f) through S_N2 reactions.

(d) — (e) $Na^+ \ ^-N_3$ (f) $Na^+ \ ^-OCH_3$ → (e) N_3, (f) OCH_3

Sodium methoxide can be made from methyl alcohol by treatment with sodium hydride (NaH) or other strong base. There could be much E2 elimination in this process, however, as methoxide is a strong base. An alternative synthesis of (f) involves use of the alkoxide related to (c) as the displacing agent in an S_N2 reaction. This synthesis is a better route to (f).

(c) — NaH → H_2 + O^- Na^+ — CH_3I → (f) OCH_3

Problem 9.45 The best first step is protonation of the right-hand double bond so as to give the tertiary carbocation, the most stable carbocation possible in this system.

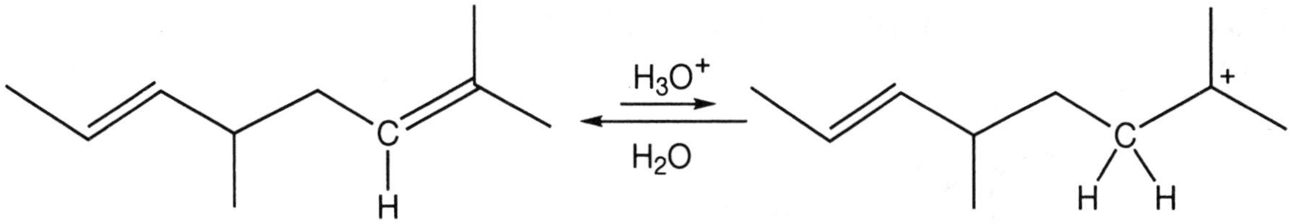

The product is a cyclic molecule, so clearly a ring must be closed. In this case, the carbocation (Lewis acid) adds to the internal double bond (Lewis base) to form a new carbon-carbon bond and close a six-membered ring. Don't be offended at the simplicity of this analysis; "The starting material is acyclic, but the product is cyclic, therefore a ring must be closed." It is extraordinarily important to think this way when doing problems.

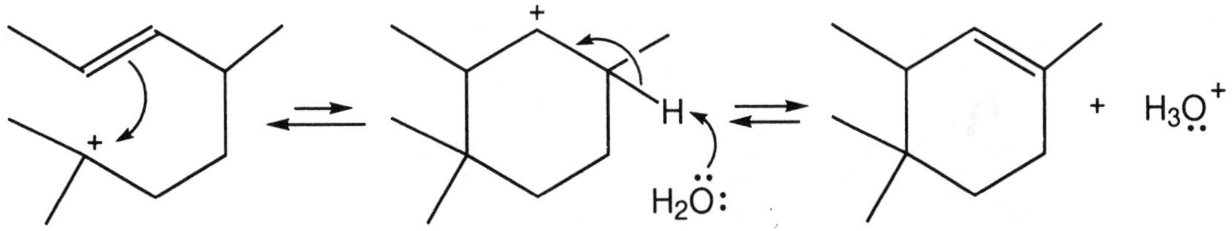

Here an important, and vexing point arises. In the last step, there is another possible proton loss to give **A**, a molecule not formed in the reaction.

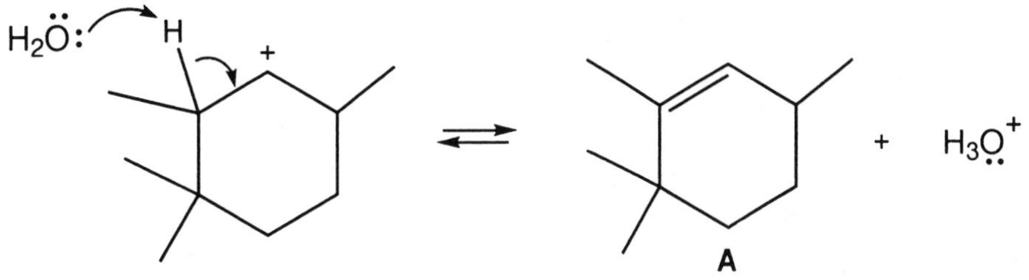

How is one to know which one to chose? The answer to this kind of question is frustrating, because sometimes the only answer is, Because of the structure of the product. There is no obvious (to us, anyway) reason that the final deprotonation goes the way it does. Can one just say, Because the product formed is more stable than the one that isn't formed? Well, you can say that, but it doesn't add anything, at least unless you can explain *why* the product shown is more stable. We are left to reason backward. Another way of putting this is to point out that it is not a fair question to ask which alkene would be formed from the cyclized carbocation. It is fair to show the product and ask how it is formed.

Problem 9.46 Once again, start by protonating to give the most stable possible carbocation.

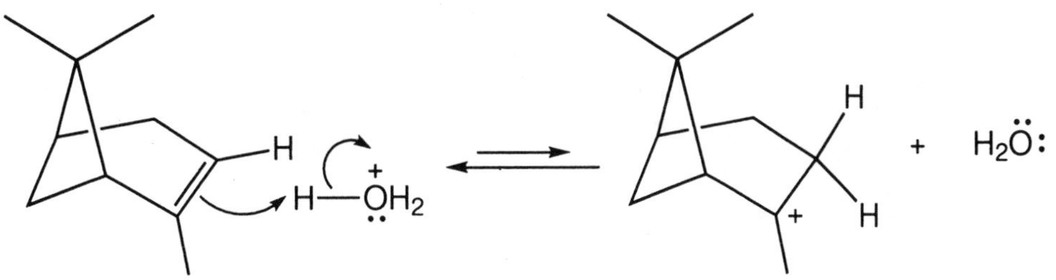

Now look at the products. Clearly a ring has been opened. And it is not difficult to see why. The starting material contains a four-membered ring, and the strain of that ring is eliminated (no pun) if the ring opens. Just as a carbon-hydrogen bond adjacent to a carbocation can break, so can a carbon-carbon bond. That is what happens here. Notice that the carbocation produced is still tertiary.

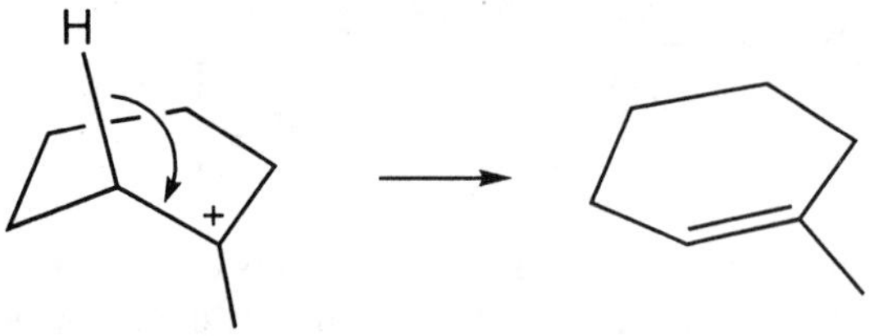

A familiar deprotonation

A less familiar, but closely related, carbon-carbon bond breaking

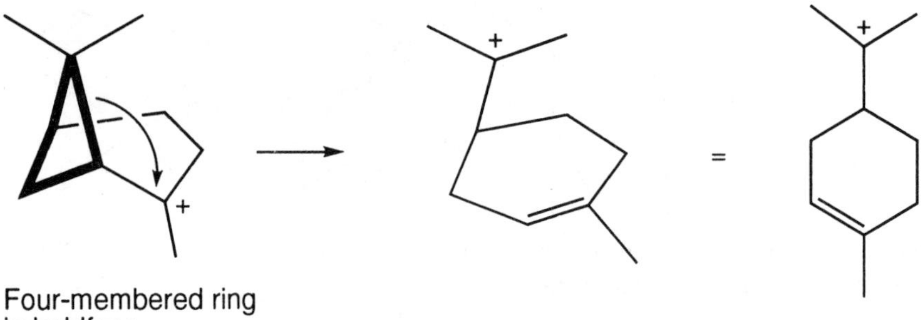

Four-membered ring in boldface

Now what? There are two possible proton losses (H_a and H_b), and they lead to the two observed products.

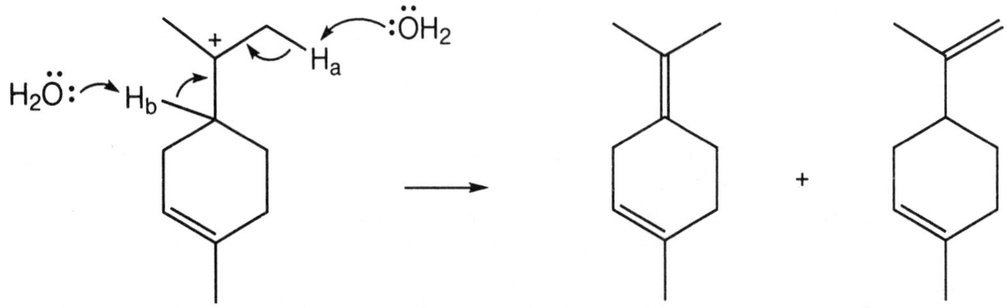

Loss of H_b Loss of H_a

Problem 9.47 For the third time, the first step is protonation to give the tertiary carbocation. The "easy" product is formed by capture of this ion by chloride ion.

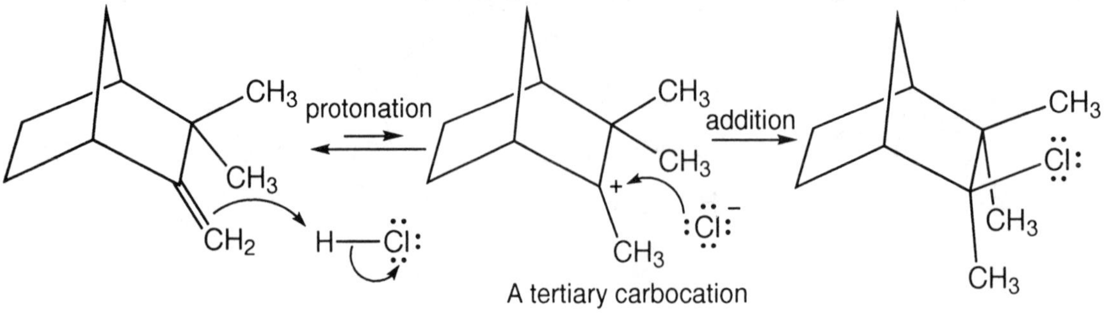

protonation addition

A tertiary carbocation

Finding the mechanism for the "hard" product is really difficult at this point. Here's how. First of all, recognize that something really strange has gone on. Apparently methyl groups have been wandering all over the place. Let's try to find another mechanism because these wholesale migrations are surely unlikely. Remember that carbons can migrate in carbocationic reactions (Wagner-Meerwein rearrangement). In the first-formed carbocation, there is a carbon atom beautifully poised for migration. The figure shows the line up between the empty orbital of the carbocation and the carbon-carbon bond.

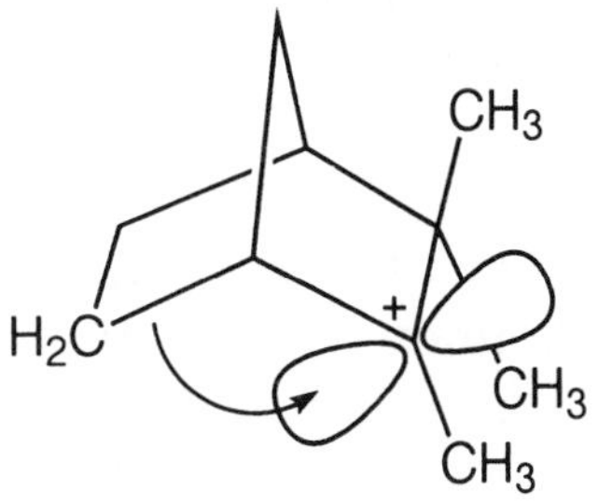

Migration of carbon does the trick, but it takes some spacial reorganizing to see it. Some labels are left in the drawing to show the relationship between the two pictures of the new cation.

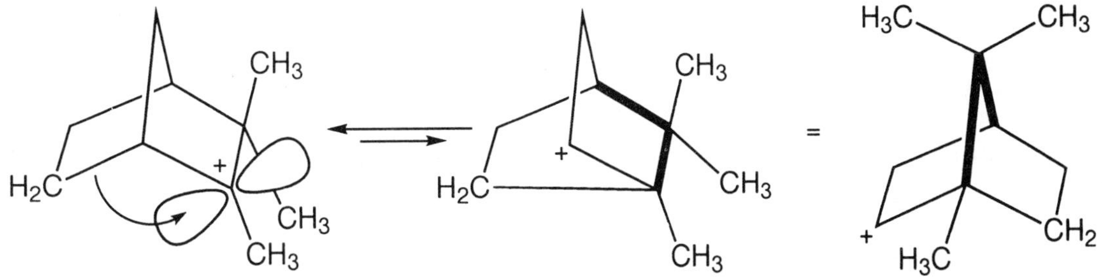

Capture by chloride gives the product.

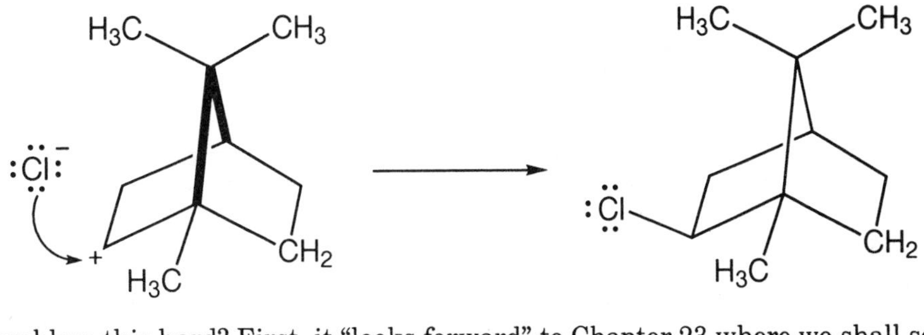

Why give a problem this hard? First, it "looks forward" to Chapter 23 where we shall see that the situation is even more complicated than it seems here. Second, it is not <u>that</u> hard, and it provides useful practice in spatial manipulation. Finally, and probably most important, problems like this are fun, and mimic quite well what "real" chemists have to think about. When we find a strange compound in a reaction we must seek a reasonable route for its formation, and sometimes that involves new chemistry that's hard for anyone to see at first. There is no reason to deny you this pleasure.

Chapter 10 Outline

10.1 Addition of X_2 Reagents
10.2 Hydration Through Mercury Compounds: Oxymercuration
10.3 Other Addition Reactions Involving Three-Membered Rings: Oxiranes and Cyclopropanes
10.4 Dipolar Addition Reactions
10.5 Addition Reactions of Alkynes: H–X Addition
10.6 Addition of X_2 Reagents to Alkynes
10.7 Hydration of Alkynes
10.8 Hydroboration of Alkynes
10.9 Hydrogenation of Alkynes: syn Hydrogenation
10.10 Reduction by Sodium in Ammonia: Anti Hydrogenation
10.11 Something More: Triplet Carbene Reactions

Additions to Alkenes 2; Additions to Alkynes

10

Problem 10.1 Why do we press this point so relentlessly? You should get into the habit of asking this question about any new reaction. The analysis is always the same: compare the bond energies of the bonds broken and made in the reaction. The bonds broken are the π bond of ethylene and the σ bond of hydrogen. Two new carbon-hydrogen σ bonds are made. The reaction is exothermic by about 30 kcal/mol, and ΔH is negative for this reaction.

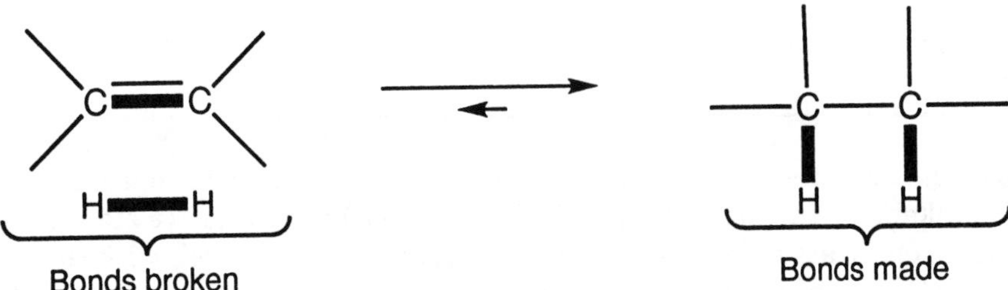

π bond = 66 kcal/mol   σ bond = 104 kcal/mol   Two C–H σ bonds = about 200 kcal/mol

200 – 170 = 30 kcal/mol exothermic, ΔH = – 30 kcal/mol

Problem 10.2

It is relatively easy for β-pinene to be adsorbed on the metal surface if the bridge bearing the two methyl groups is "up", away from the surface

CH_3

H_3C

H_2

CH_3

H_3C

Adsorbed hydrogen

H

H

CH_2

Metal surface

H

CH_3

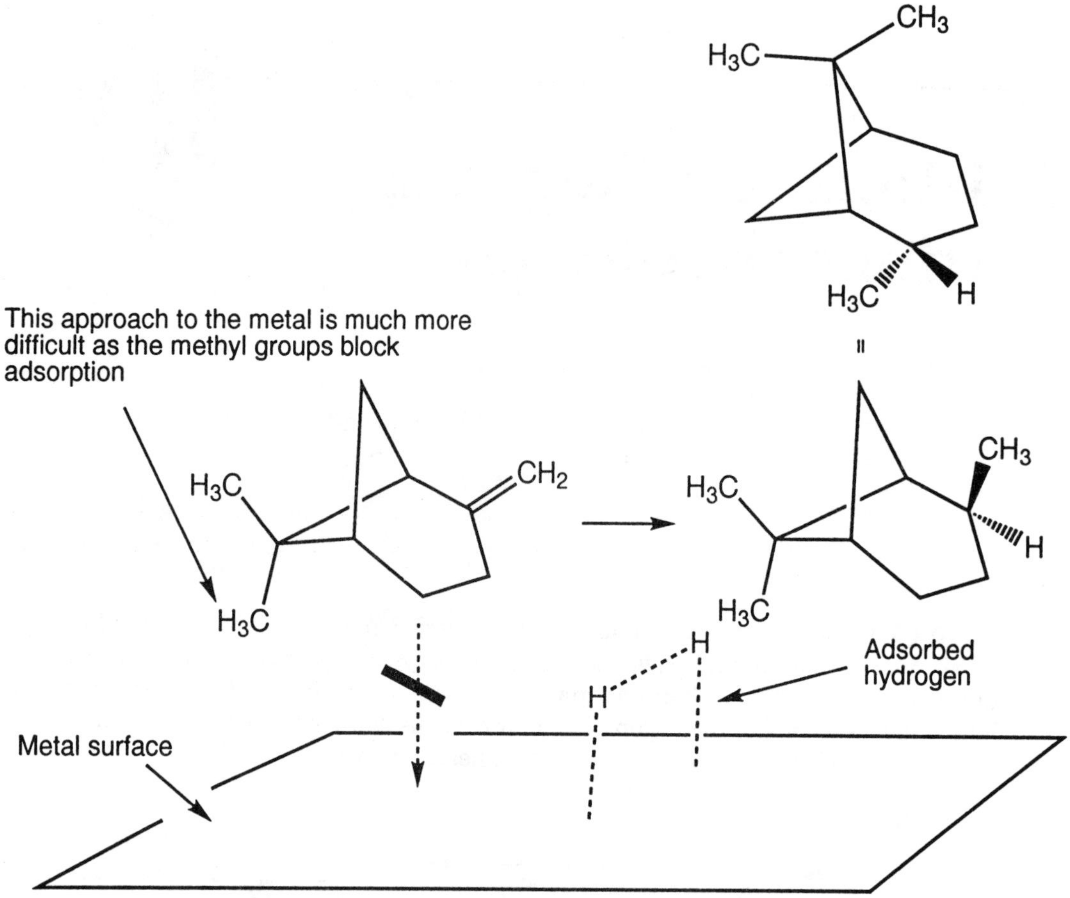

This approach to the metal is much more difficult as the methyl groups block adsorption

H_3C CH_2 H_3C CH_3 H_3C H_3C CH_3 H_3C H H

Metal surface

Adsorbed hydrogen

Problem 10.3 The arithmetic is easy, and, as in Chapter 4 (p. 133), it appears that the more substituted an alkene is, the more stable it is. The values for the relative stabilities of the hexene isomers derived from heats of formation (ΔH_f°) compare reasonably well with those from heats of hydrogenation ($\Delta H^\circ_{H_2}$), although one might worry a bit about the difference for 2,3-dimethyl-2-butene.

Isomer	$\Delta H^\circ_{H_2}$ (kcal/mol)	Relative Energy (kcal/mol)	Relative energy for ΔH_f° (kcal/mol)[a]
1-hexene	–29.6	0	0
cis-3-hexene	–27.3	–2.3	–1.2
trans-3-hexene	–26.4	–3.2	–2.1
2-methyl-1-pentene	–28.0	–1.6	
2,3-dimethyl-2-butene	–26.6	–3.0	–6.6

[a] From Chapter 4

Problem 10.4 The reaction does not occur spontaneously because there is an activation energy barrier (Chapter 8, p. 311). It is not enough that the starting material, here alkene plus hydrogen, be higher in energy than the product, here alkane. There must be energy enough to surmount the barrier. The role of the catalyst is to provide a lower energy mechanistic pathway leading to product; a trail on a lower slope of the activation mountain (Fig. 8.20, p. 315).

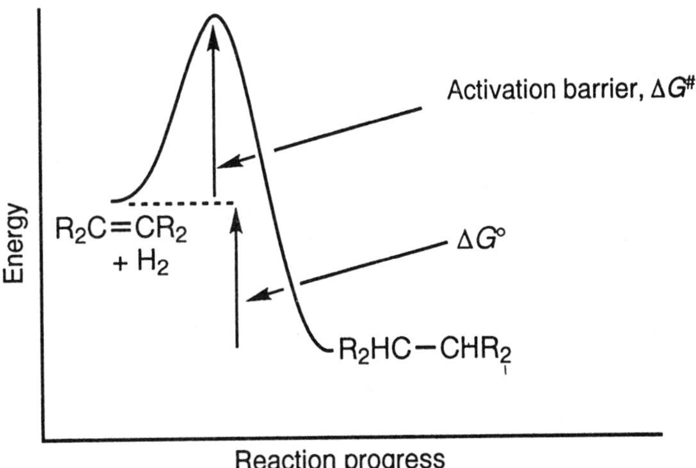

Problem 10.6 Another easy one, possibly even too easy for you to trust your answer. The bulky adamantyl groups simply shield the rear of the carbon-bromine bonds in the bromonium ion, making S_N2 attack difficult.

Problem 10.8 First, the carbon-oxygen double bond of any carboxylic acid is polarized in such a way (C, positive; O, negative) as to stabilize a developing negative charge on the hydroxyl oxygen. This effect is absent in ethyl alcohol.

Second, acetic acid ionizes to give the resonance-stabilized acetate ion, in which two oxygen atoms help bear the negative charge. By contrast, ethyl alcohol must give ethoxide ion in which the negative charge is carried by one oxygen only. The resonance-stabilized, delocalized anion is much more stable and therefore formed more quickly. Watch out! Remember the Hammond postulate and why it works (Chapter 8, p. 320).

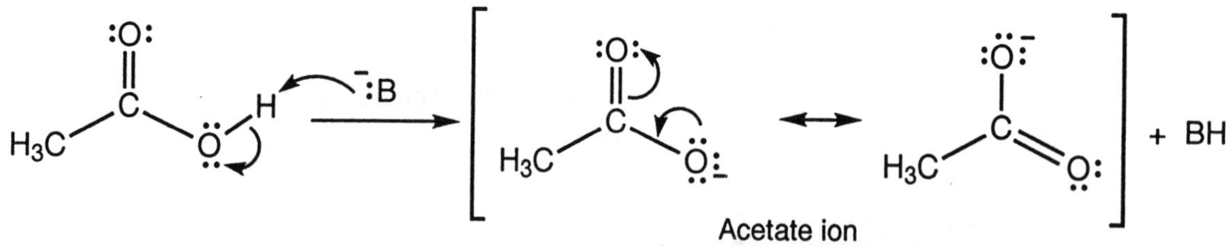

Acetate ion
(resonance stabilized)

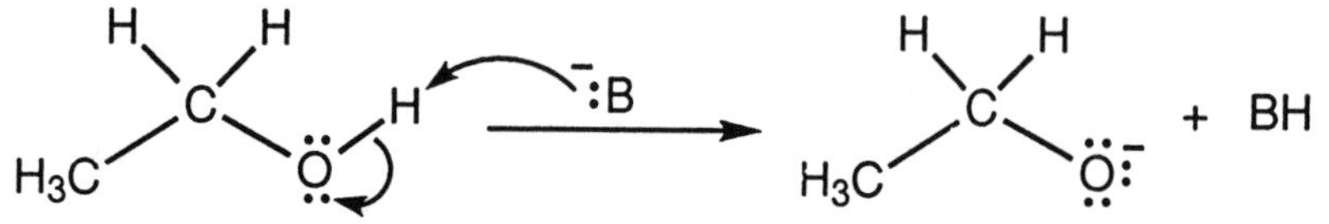

Ethoxide ion
(here the charge is localized)

Problem 10.10

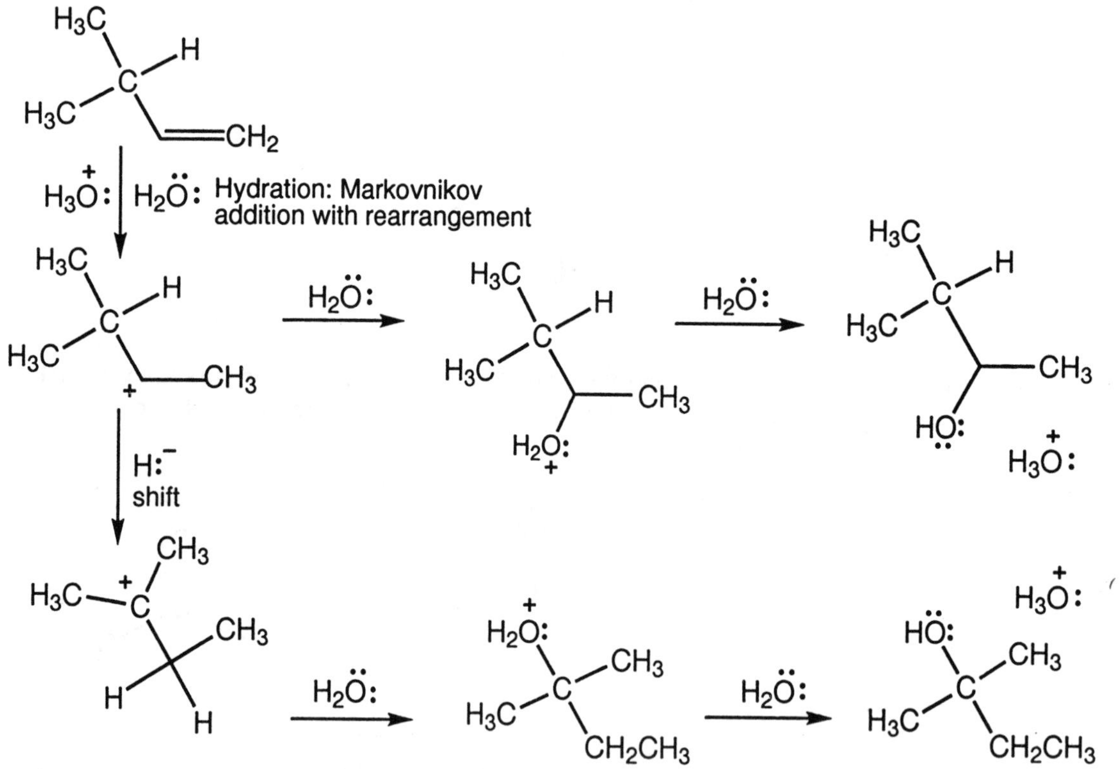

Hydration: Markovnikov addition with rearrangement

H:⁻ shift

Oxymercuration: Markovnikov addition without rearrangements.

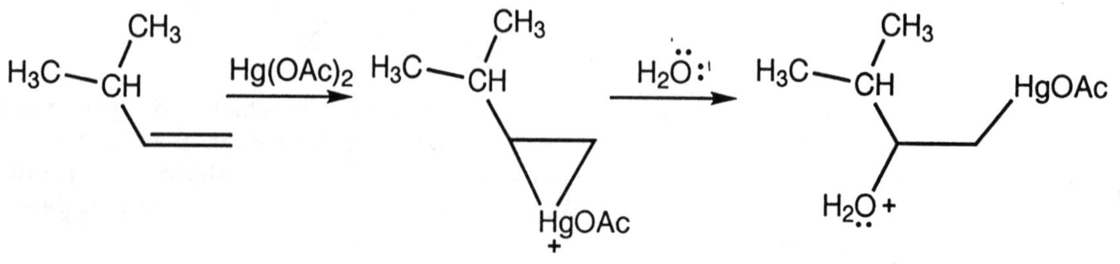

(continued next page)

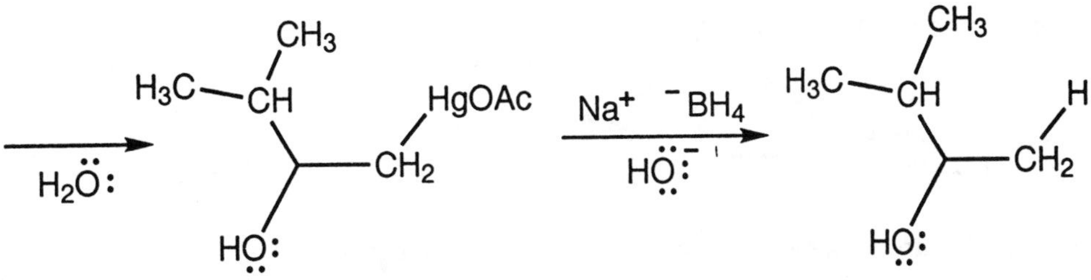

H3C—CH(CH3)—CH(OH)—CH2HgOAc → (Na+ −BH4, HO−) H3C—CH(CH3)—CH(OH)—CH2—H

Hydroboration: Anti-Markovnikov addition without rearrangements

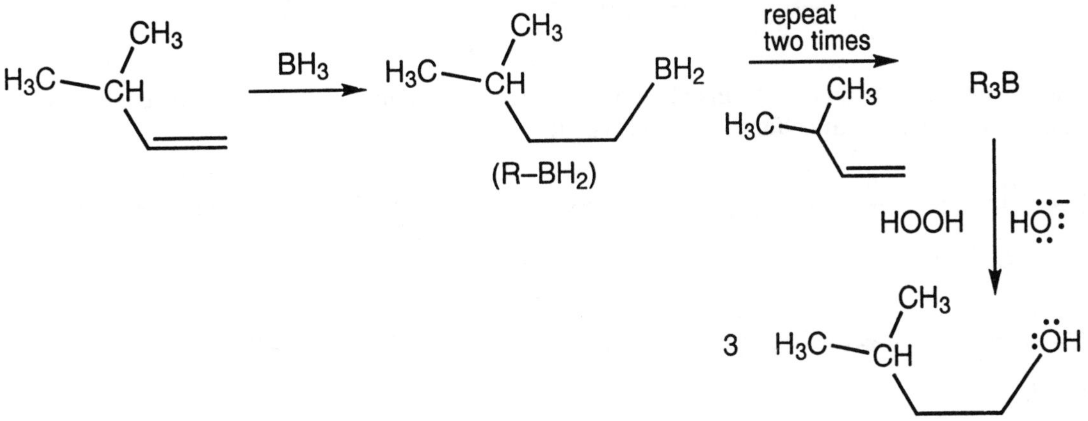

BH3; (R–BH2); repeat two times; R3B; HOOH, HO−; 3 H3C—CH(CH3)—CH2CH2OH

Problem 10.12

N_2

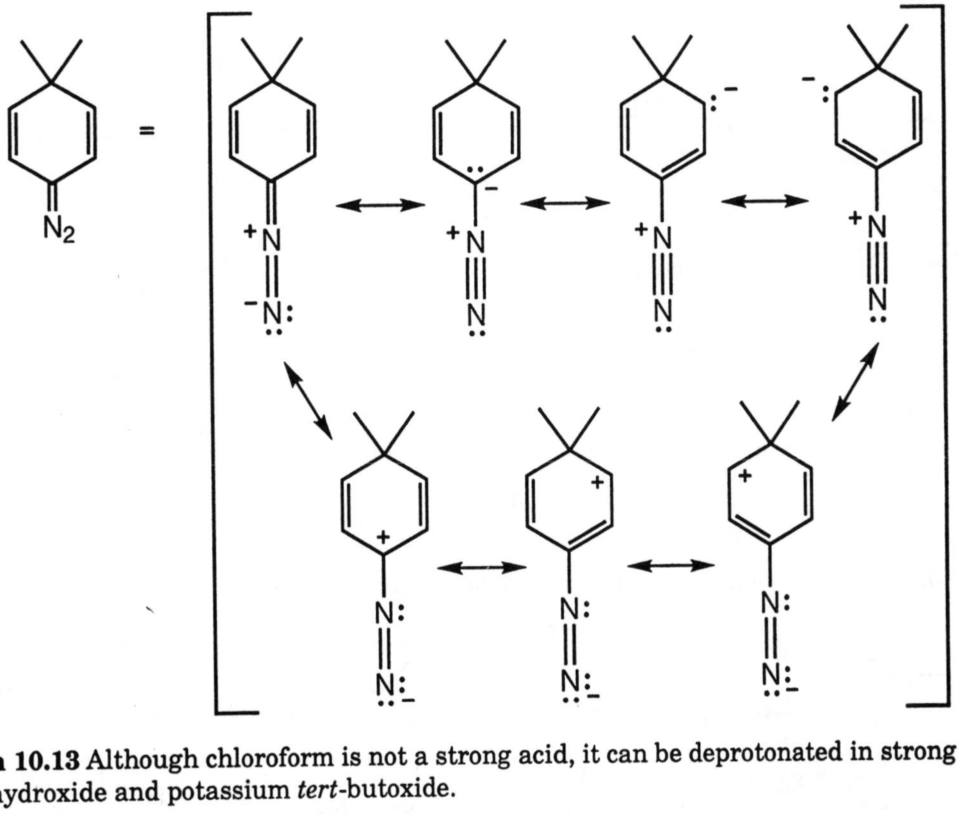

Problem 10.13 Although chloroform is not a strong acid, it can be deprotonated in strong bases such as hydroxide and potassium *tert*-butoxide.

$$HO^- + H{-}CCl_3 \rightleftharpoons {}^-{:}CCl_3 + H_2O$$

In D_2O/DO^- exchange can take place.

$$DO^- + H{-}CCl_3 \rightleftharpoons {}^-:CCl_3 + D{-}OD \rightleftharpoons D{-}CCl_3 + DO^-$$

Another thing $^-CCl_3$ can do is eject a chloride ion to give the carbene:

$$^-:CCl_2{-}Cl \longrightarrow :Cl^- + :CCl_2$$

Problem 10.14 The resonance forms shown in Figure 10.50 emphasize the 1,3 nature of the dipole. Other, often better resonance forms can be drawn.

$$R_2\bar{C}{-}N{=}\overset{+}{N}: \leftrightarrow R_2\bar{C}{-}\overset{+}{N}{\equiv}N: \leftrightarrow R_2C{=}\overset{+}{N}{=}\bar{N}: \leftrightarrow R_2\overset{+}{C}{-}N{=}\bar{N}:$$

diazo compounds

$$R\overset{+}{C}{=}N{-}O^- \leftrightarrow RC{\equiv}\overset{+}{N}{-}O^- \leftrightarrow R\bar{C}{=}\overset{+}{N}{=}O:$$

nitrile oxides

$$R\bar{N}{-}N{=}\overset{+}{N}: \leftrightarrow R\bar{N}{-}\overset{+}{N}{\equiv}N: \leftrightarrow RN{=}\overset{+}{N}{=}\bar{N}: \leftrightarrow R\overset{+}{N}{-}N{=}\bar{N}:$$

azides

$$:\overset{+}{N}{=}N{-}O^- \leftrightarrow :N{\equiv}\overset{+}{N}{-}O^- \leftrightarrow :\bar{N}{=}\overset{+}{N}{=}O:$$

nitrous oxide

$$R_2\overset{+}{C}{-}O{-}O^- \leftrightarrow R_2C{=}\overset{+}{O}{-}O^- \leftrightarrow R_2\bar{C}{-}\overset{+}{O}{=}O:$$

carbonyl oxides

$$R_2\overset{+}{C}{-}N(R){-}O^- \leftrightarrow R_2C{=}\overset{+}{N}(R){-}O^- \leftrightarrow R_2\bar{C}{-}\overset{+}{N}(R){=}O:$$

nitrones

Problem 10.15 This reaction is a simple, "just push the arrows" problem.

$$RCH{=}CHR + R\overset{+}{C}{=}N{-}O^- \longrightarrow \text{isoxazoline (R-C=N-O ring, 4,5-di-R)}$$

$$RCH{=}CHR + R\bar{N}{-}N{=}\overset{+}{N}: \longrightarrow \text{triazoline (R-N-N=N ring, 4,5-di-R)}$$

Problem 10.16 Some approximations must be made here. In this answer, we use the bond energy of ethane for the carbon-carbon bonds, the bond energy of a simple alcohol for the carbon-oxygen bonds, and the bond energy for di-*tert*-butyl peroxide for the oxygen-oxygen bonds. The final ozonide is much more stable than the primary ozonide.

Less stable (weaker overall bonding)

1 C—C	90
2 C—O	184
2 O—O	76
	350 kcal/mol

More stable (stronger overall bonding)

4 C—O	368
1 O—O	38
	406 kcal/mol

Problem 10.17 There is, of course electrostatic repulsion. Many electrons are in proximity to each other, and like charges will repel. However, there is another reason. Like F_2 (fluorine), ROOR contains two adjacent atoms with many electrons. We show the electrons in $2p$ orbitals for clarity but regardless of what orbitals are occupied, orbital overlap will cause some electrons to occupy antibonding orbitals, and this will be strongly destabilizing.

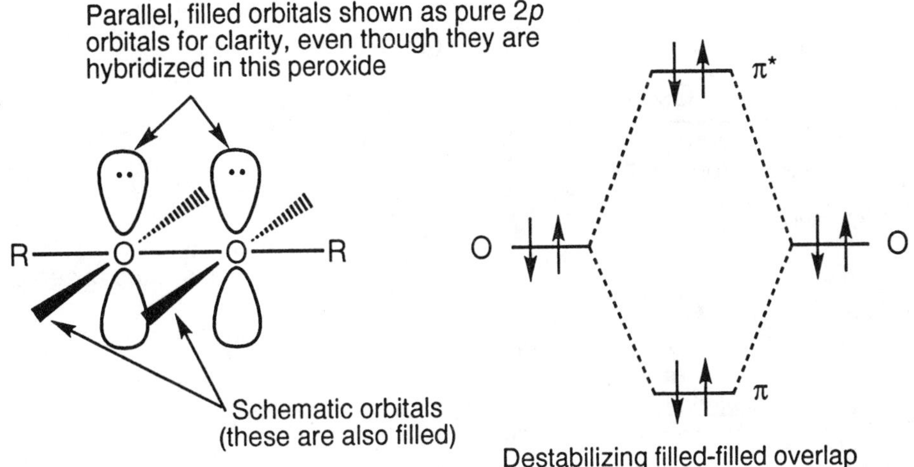
Parallel, filled orbitals shown as pure 2*p* orbitals for clarity, even though they are hybridized in this peroxide
R
O
O
R
Schematic orbitals (these are also filled)
π*
O
O
π
Destabilizing filled-filled overlap

Problem 10.18 A full diagram includes the intermediates as well as starting material and products.

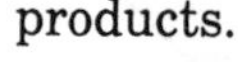

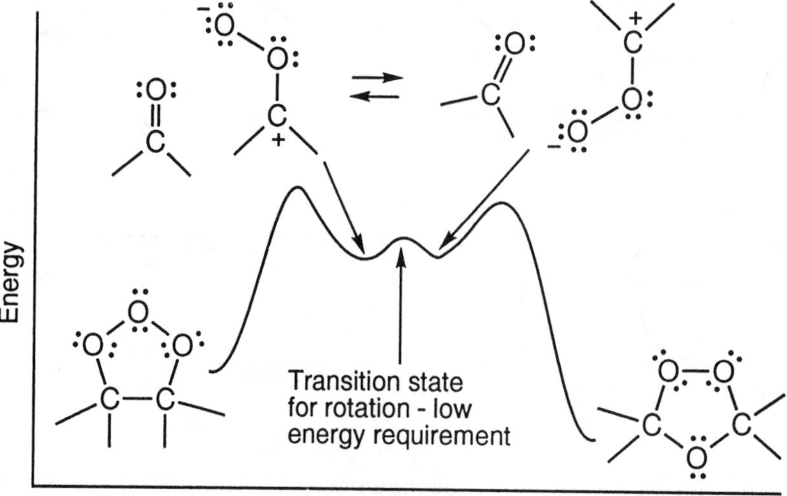

Problem 10.19

(a)

or

Either oxidative or reductive workup can be used

(b)

H

or

H

Since an aldehyde is one of the products, a reductive workup must be used

(d)

The products here are carboxylic acids, and so an oxidative workup must be used

Problem 10.20 The stereochemical relationship of the methyl groups does not change. The cis alkene gives a cis-substituted five-membered ring. There can be no steps in the reaction that allow rotation. If there were, a different diol would also be isolated. The initial reaction of OsO_4 with *cis*-2-butene must be one-step (concerted):

1. OsO_4 concerted (one step)

2. H_2O Na_2SO_3

stepwise

This is the observed product

rotation

Not observed

Problem 10.21 Protonation gives an intermediate, probably a cyclic cation of some kind, that can be captured by water to give the enol after deprotonation. Continued reaction of the enol will eventually produce the ketone.

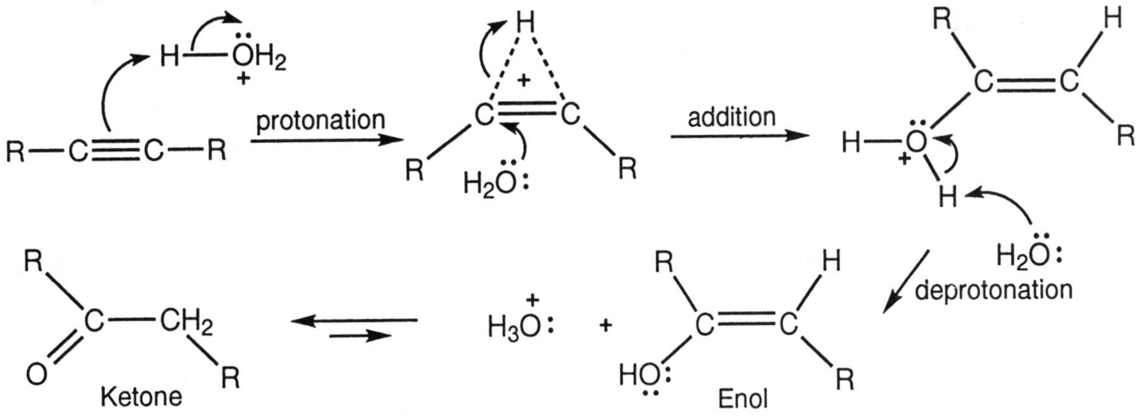

protonation

addition

deprotonation

Ketone

Enol

Problem 10.23 In the cyclic intermediate for this reaction, both carbons will share the positive charge approximately equally, as both are secondary. Accordingly, addition of the nucleophile will occur at both carbons to give, ultimately, two enols in roughly equal amounts.

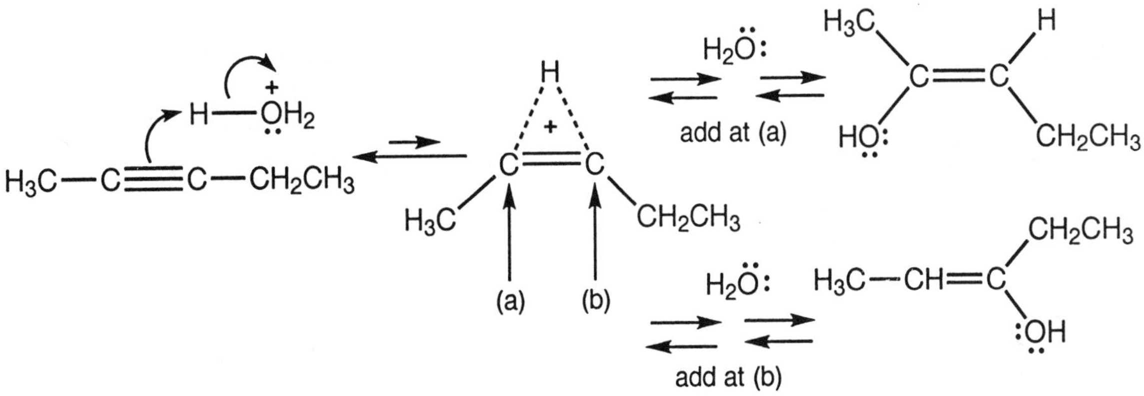

add at (a)

(a) (b)

add at (b)

Conversion of the enols into ketones will give *two* ketones (for a mechanism see Fig. 10.75). This is *not* a useful synthetic process.

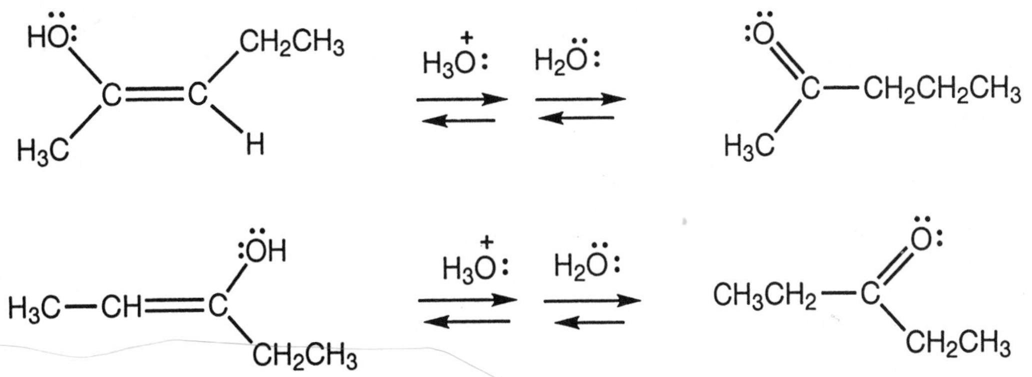

Problem 10.25 The best way to learn mechanisms is to write them backward. This problem presents an opportunity to do that for the mechanism of ketone formation from enols, by asking you to do the reverse reaction. Remember, if you have written the mechanism in one direction, you have automatically written it in the other. All you need to do is to read the first one backward. So you could take the mechanism of Figure 10.75 as your answer.

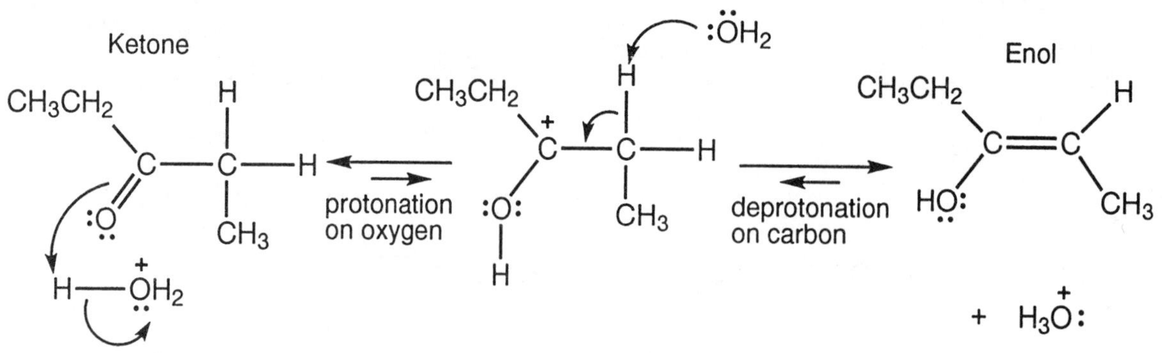

Ketone

CH_3CH_2 H C—C—H :O: CH_3 H—$\overset{+}{O}H_2$

protonation on oxygen

CH_3CH_2 $\overset{+}{C}$—C—H :O: H H CH_3 :$\ddot{O}H_2$

deprotonation on carbon

Enol

CH_3CH_2 H C=C HO: CH_3

+ $H_3\overset{+}{O}$:

Additional Problem Answers

Problem 10.27 These are all one- or two-step syntheses. There is nothing tricky here, only remembering the reactions. If you forget, fall back on a mechanistic analysis and this should lead you to the product in most cases. Remember, we start with achiral materials, and in *all* cases racemic mixtures must result. In the answers, only one enantiomer is shown.

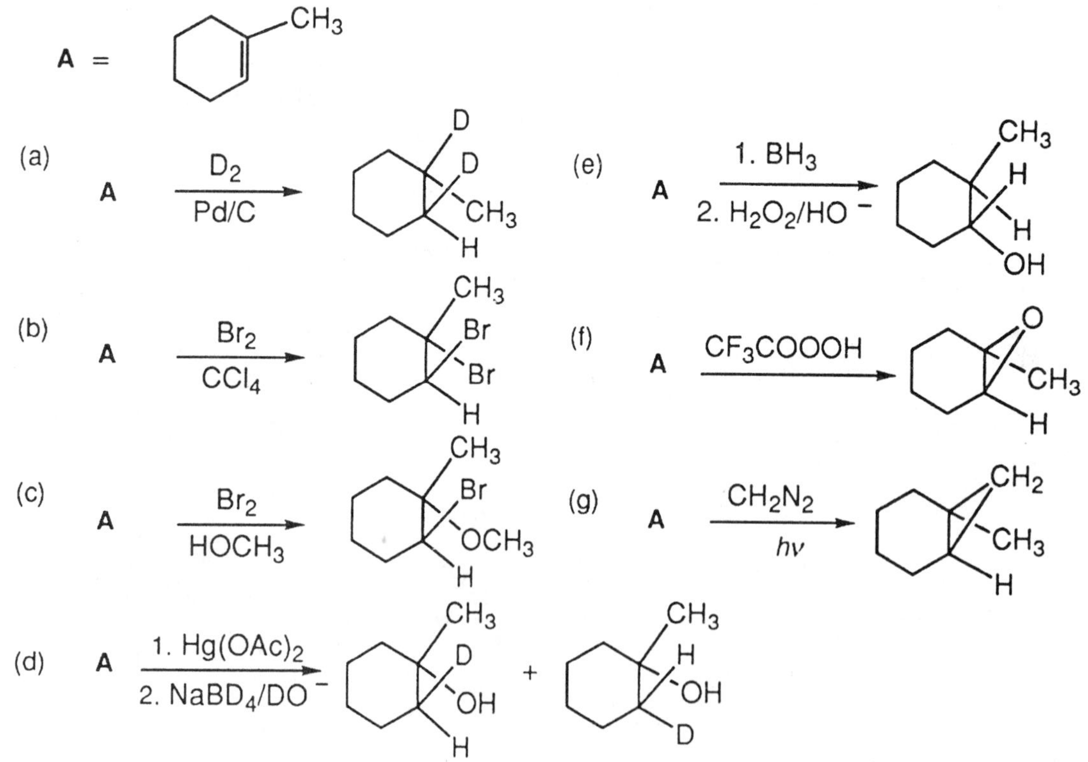

A = CH_3

(a) A $\xrightarrow[Pd/C]{D_2}$ D, D, CH_3, H

(b) A $\xrightarrow[CCl_4]{Br_2}$ CH_3, Br, Br, H

(c) A $\xrightarrow[HOCH_3]{Br_2}$ CH_3, Br, OCH_3, H

(d) A $\xrightarrow[\text{2. NaBD}_4\text{/DO}^-]{\text{1. Hg(OAc)}_2}$ CH_3, D, OH, H + CH_3, H, OH, D

(e) A $\xrightarrow[\text{2. H}_2\text{O}_2\text{/HO}^-]{\text{1. BH}_3}$ CH_3, H, H, OH

(f) A $\xrightarrow{CF_3COOOH}$ O, CH_3, H

(g) A $\xrightarrow[h\nu]{CH_2N_2}$ CH_2, CH_3, H

(h) A $\xrightarrow[2.\ (CH_3)_2S]{1.\ O_3}$ H_3C–CO–$CH_2CH_2CH_2CH_2$–CHO

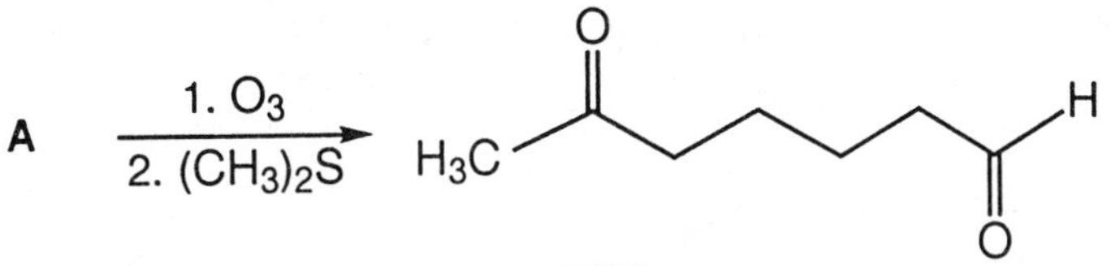

(i) A $\xrightarrow[2.\ NaHSO_3,\ H_2O]{1.\ OsO_4}$ (cis-diol: CH_3, H, OH, OH on cyclohexane)

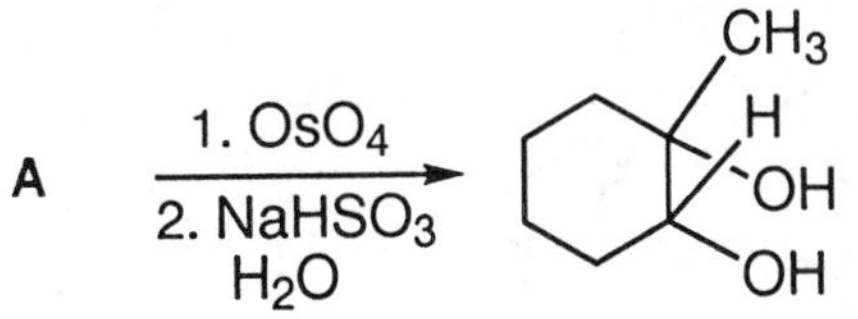

(j) A $\xrightarrow{HN{=}NH}$ (CH_3, H, H, H on cyclohexane)

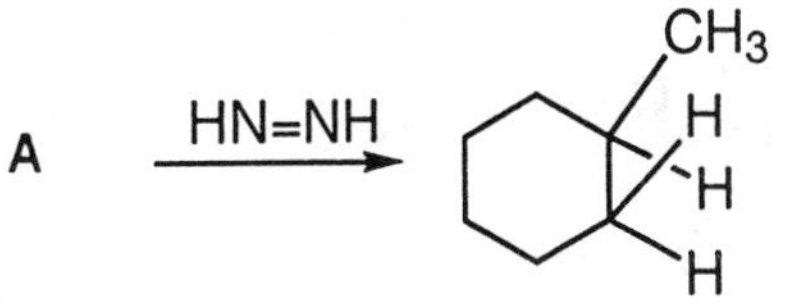

Problem 10.28 Deuterium can be delivered from either side of the planar double bond. The two enantiomers of the methylcyclohexane product result.

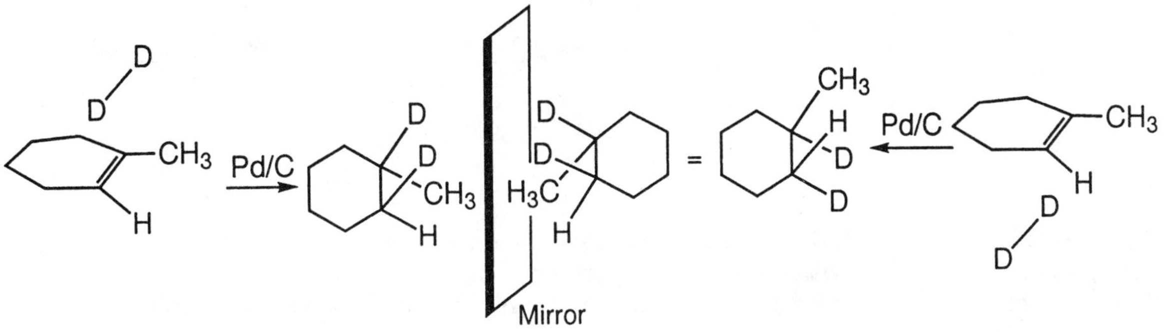

D–D + 1-methylcyclohexene $\xrightarrow{Pd/C}$ (D, D, CH_3, H) | Mirror | (D, D, H_3C, H) = (CH_3, H, D, D) $\xleftarrow{Pd/C}$ 1-methylcyclohexene + D–D

Problem 10.29

(a) $\xrightarrow{HBr}$ cyclohexyl–C(Br)=CH_2 + cyclohexyl–CBr_2–CH_3

(b) $\xrightarrow{Cl_2}$ cyclohexyl–C(Cl)=CHCl [(*Z*/*E*) mix] + cyclohexyl–CCl_2–$CHCl_2$

(c) $\xrightarrow[H_2O/H_3O^+]{Hg^{2+}}$ cyclohexyl–C(=O)–CH_3

(d) $\xrightarrow[2.\ H_2O_2/NaOH]{1.\ (\text{disiamyl})_2BH}$ cyclohexyl–CH_2–C(=O)–H

(e) $\xrightarrow[\text{Lindlar catalyst}]{H_2}$ (*Z*)-cyclohexyl–CH=CH–cyclohexyl (H, H cis)

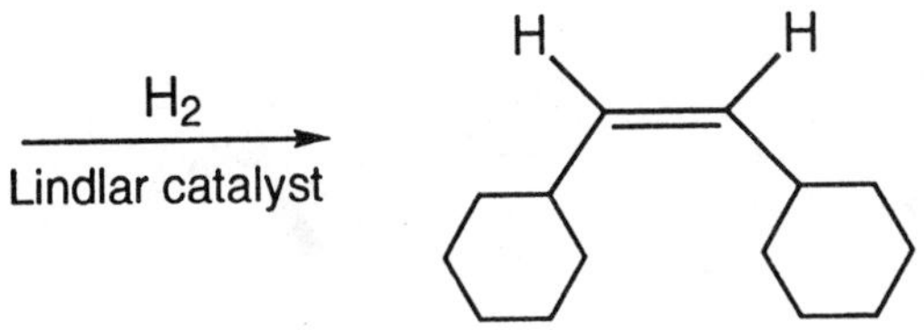

(f) $\xrightarrow[NH_3]{Na}$ (*E*)-cyclohexyl–CH=CH–cyclohexyl (H, H trans)

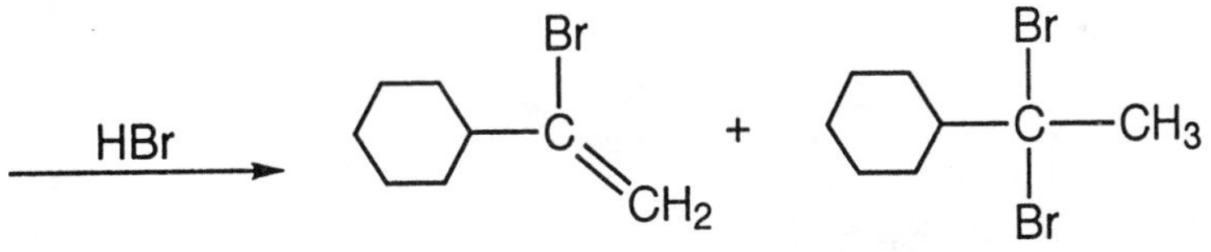

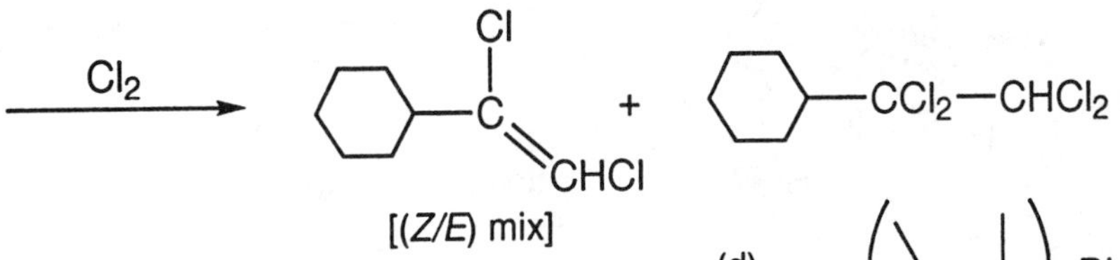

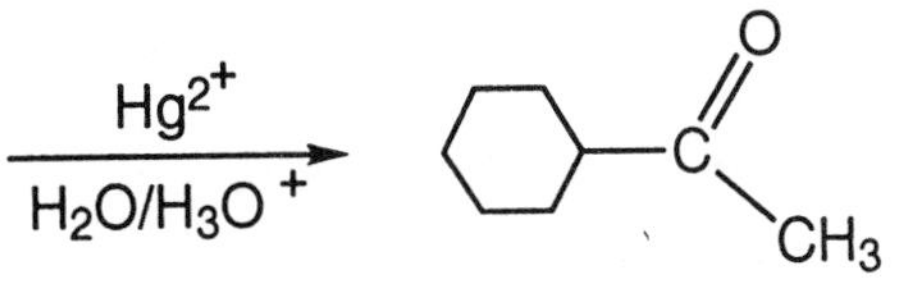

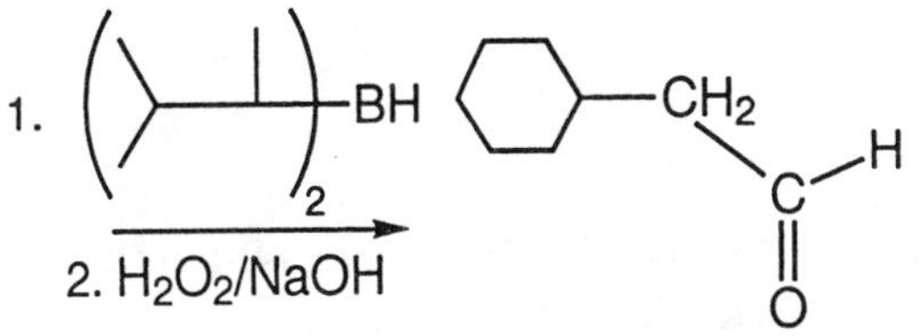

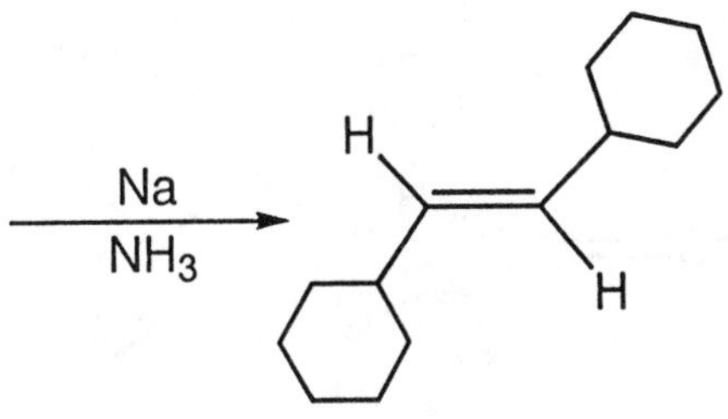

Problem 10.30 In the addition of Br_2 to *cis*-2-butene, an open carbocation must give two diastereomeric products.

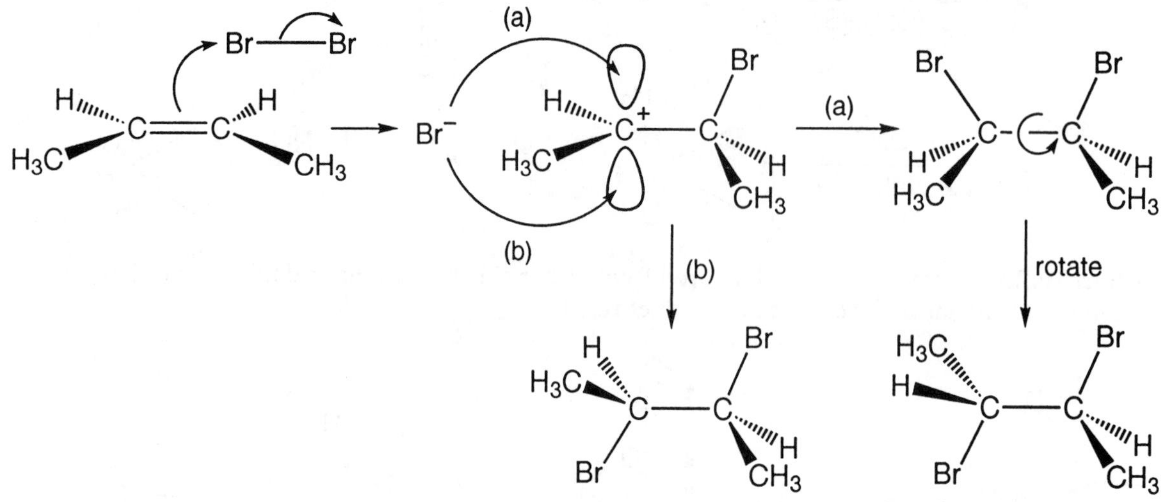

Diastereomers

As this result is not observed, this mechanism must be wrong. Formation of a bromonium ion that opens up in an S_N2 reaction with bromide solves the problem. The bromonium ion must open by addition of bromide from the rear, and this leads to racemic dibromide when *cis*-2-butene is the starting material.

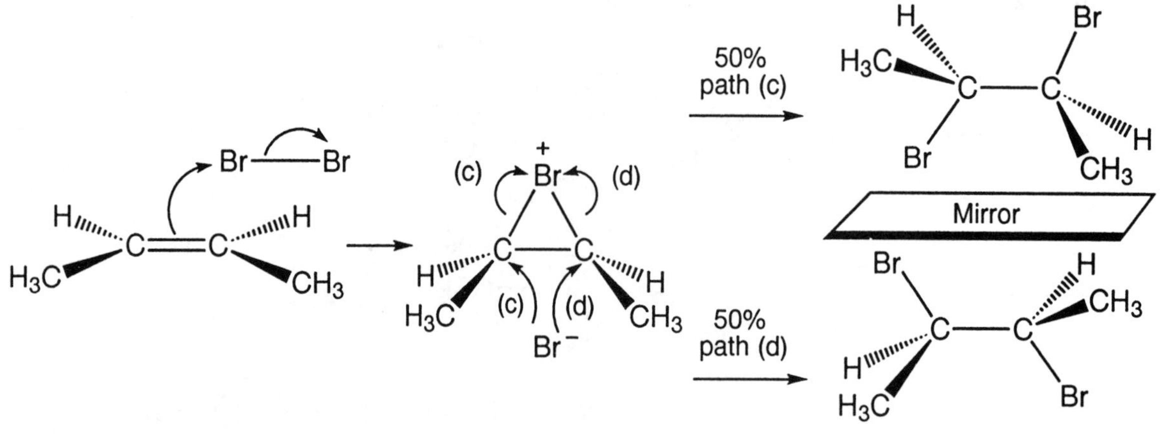

Racemic mixture

Similarly, in the addition to *trans*-2-butene, an open carbocation must lead to two products, a result not observed experimentally:

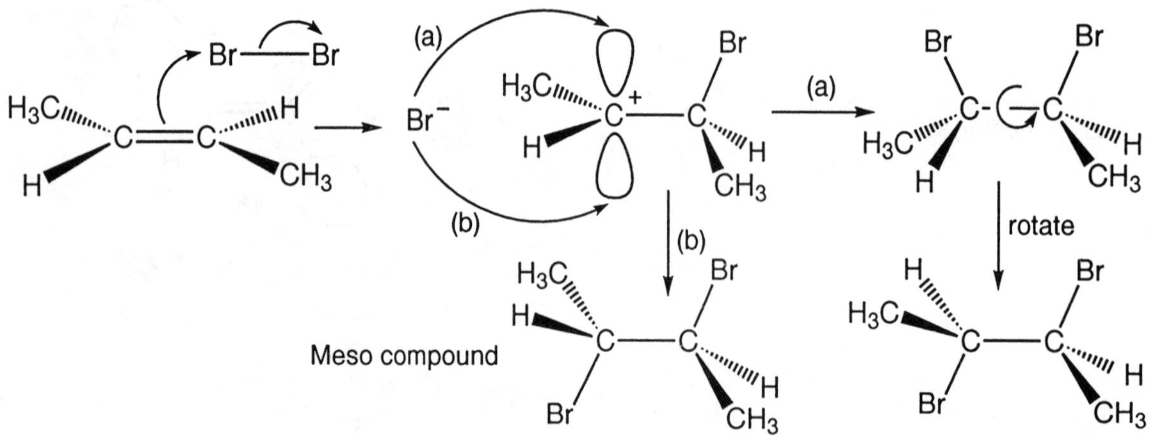

Meso compound

Diastereomers

A bromonium ion shows how the experimentally observed single product, a meso compound, is formed.

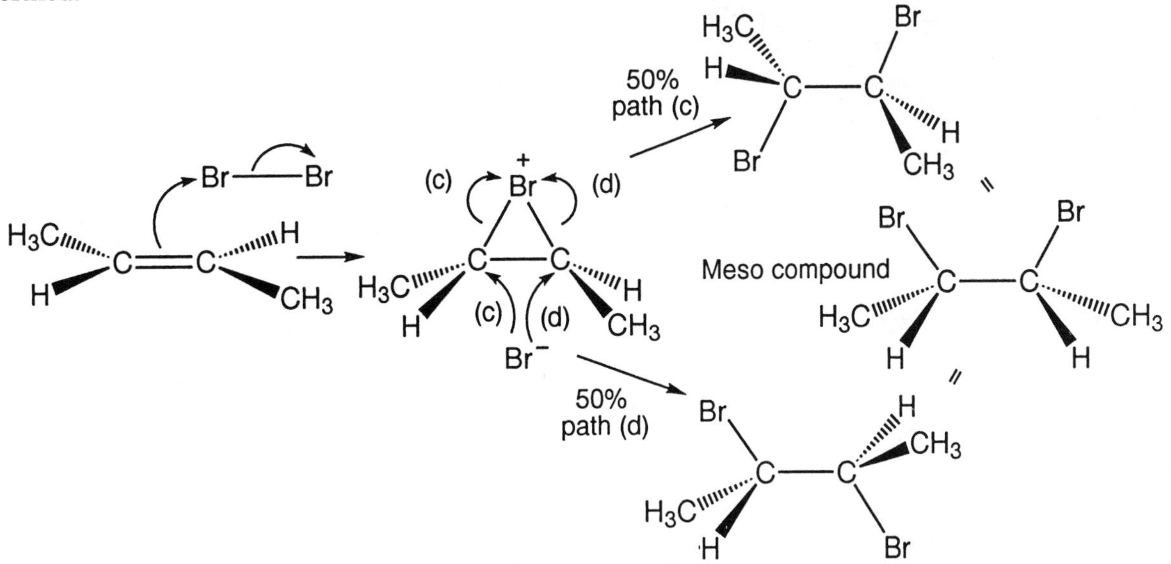

50% path (c)

Meso compound

50% path (d)

Problem 10.31 In the acid-catalyzed reaction, the first step is the protonation of the epoxide oxygen to give a resonance-stabilized cation. The two bonds from carbon to oxygen are different as the tertiary carbon will have a greater share of the positive charge than will the primary carbon.

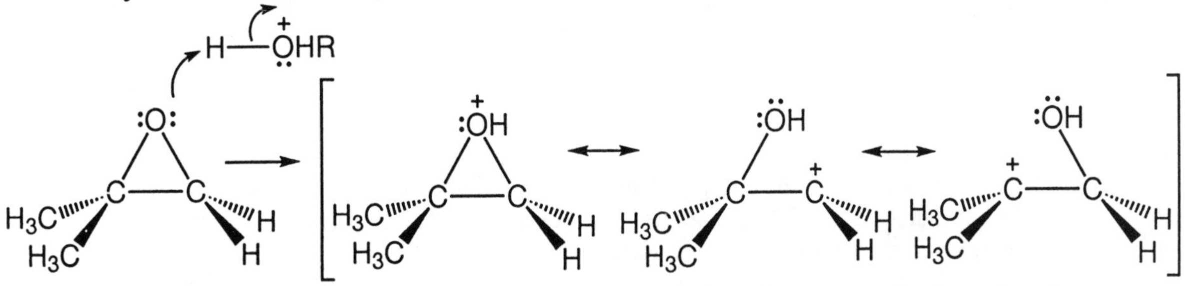

The alcohol will add predominately to break the longer and weaker more substituted carbon-oxygen bond. Deprotonation of the intermediate oxonium ion leads to the product and regenerates the catalyst, the protonated alcohol.

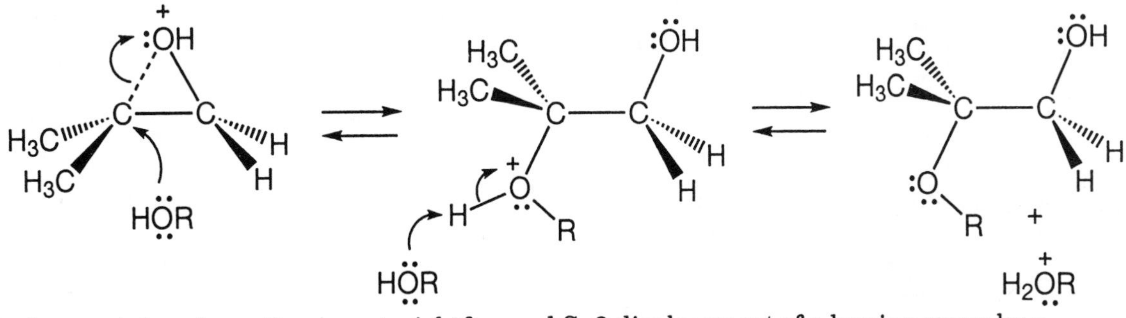

The base-catalyzed reaction is a straightforward S_N2 displacement of a leaving group by a negatively charged nucleophile. The epoxide opens from the less hindered side. Proton transfer completes the reaction and regenerates a molecule of alkoxide.

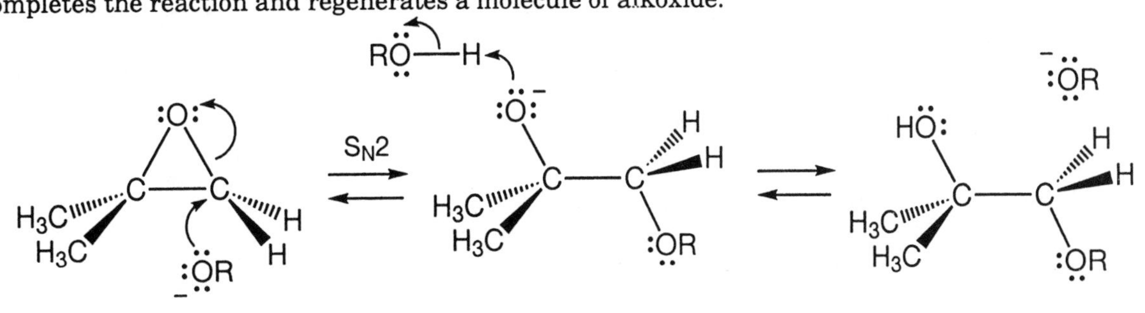

S_N2

Problem 10.32 First of all, what are the basic mechanisms of bromohydrin formation and epoxide formation from the bromohydrin? The first part is a straightforward addition reaction covered many times already. In the second, and more difficult part of the problem, alkoxide formation, followed by intramolecular displacement of bromide in S_N2 fashion. Thus, this process involves *two* inversions of configuration. Two inversions results in net retention of configuration as cis alkene becomes the cis epoxide. A mechanistic analysis should make this clear.

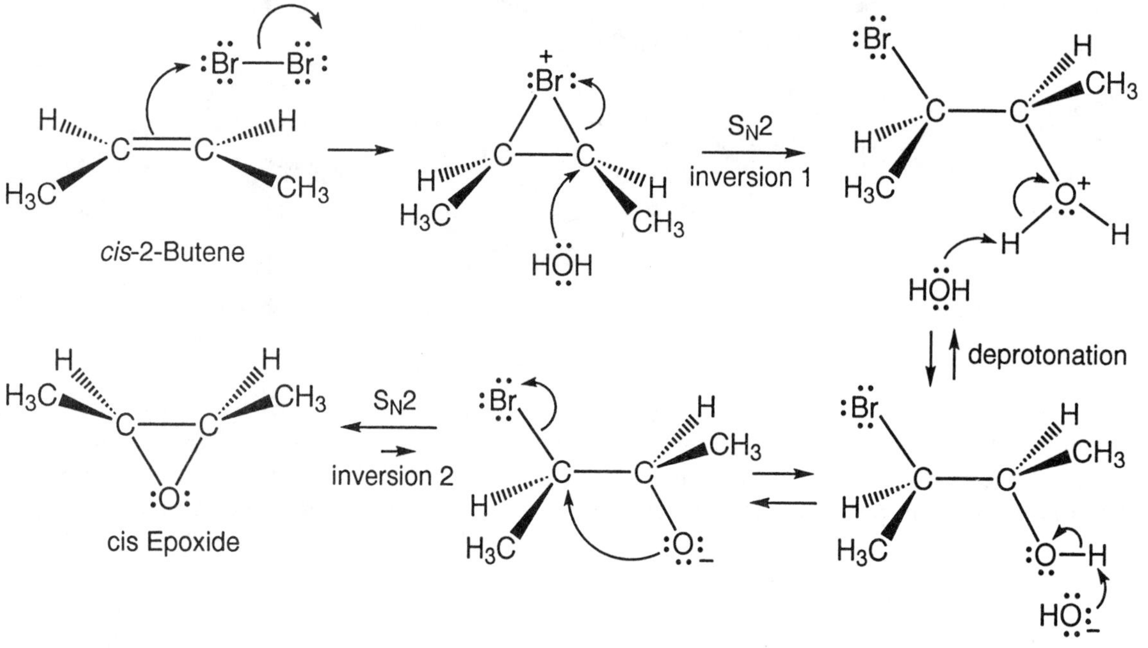

cis-2-Butene

S_N2

inversion 1

deprotonation

S_N2

inversion 2

cis Epoxide

Problem 10.33

(a) Epoxidation is followed by S_N2 opening of the three-membered ring from the rear with hydroxide acting as nucleophile. Notice that in this reaction the epoxide is opened from the sterically less-encumbered side to give a trans diol.

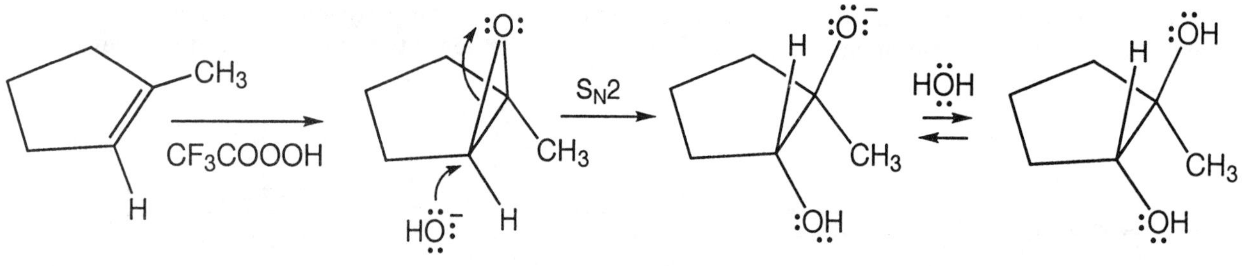

CF_3COOOH

S_N2

(b) By contrast, in this reaction a five-membered ring intermediate is formed and both OH groups originate in this step. Hydrolysis leads to the cis diol.

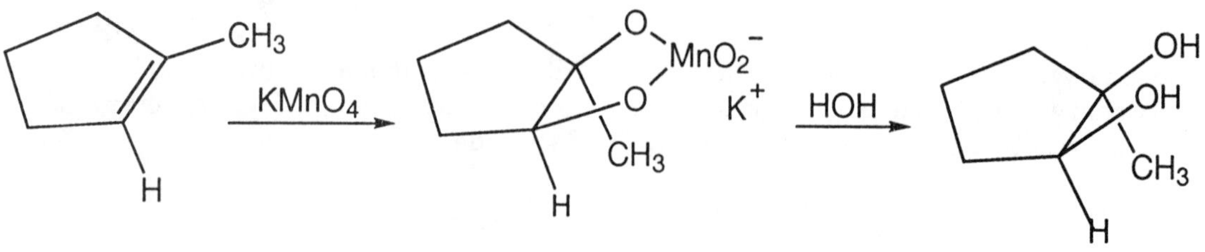

$KMnO_4$

HOH

cis Diol

Problem 10.34 In this intramolecuar version of the periodate cleavage reaction, both new carbonyl groups appear in the same molecule.

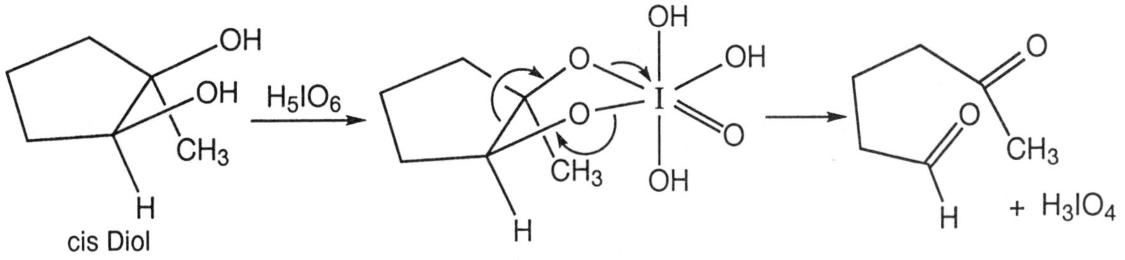

OH
OH H_5IO_6
CH_3
H
cis Diol
OH
O
OH
I
O
O
CH_3
OH
H
O
O
CH_3
H + H_3IO_4

Ozonolysis of 1-methylcyclopentene with a reductive workup [$(CH_3)_2S$ or H_2/Pd] would lead to the same product.

Problem 10.35 In the trans diol, the two hydroxy groups are too far apart to form the cyclic intermediate easily. Reaction of trans diols with periodate is slow at best.

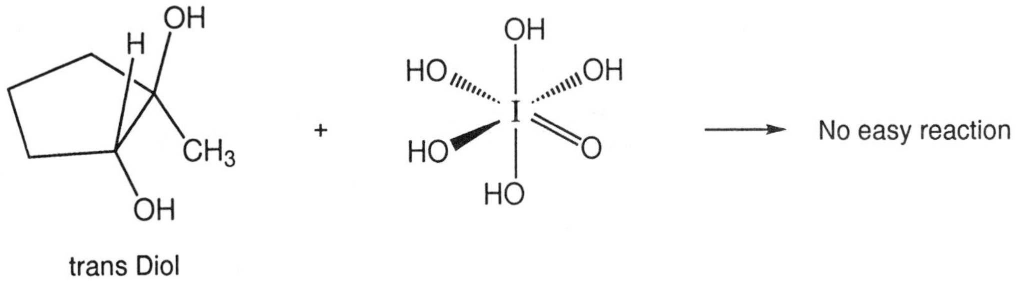

OH
H
CH_3
OH
trans Diol
+
OH
HO
OH
I
HO
O
HO
No easy reaction

Problem 10.36 The ozonolysis results let us piece together the structures of **3** and **4**. Ozonolysis results in the conversion of a carbon-carbon double bond into a pair of carbon-oxygen double bonds. Accordingly, the structures of the products tell us the structures of **3** and **4**.

CH_2
3
1. O_3
2. $(CH_3)_2S$
CH_2
O
O
C
Formation of these two compounds implies the structure of **3**

CH_2
4 H
1. O_3
2. $(CH_3)_2S$
CH_2
O
O
C
H
Formation of these two compounds implies the structure of **4**

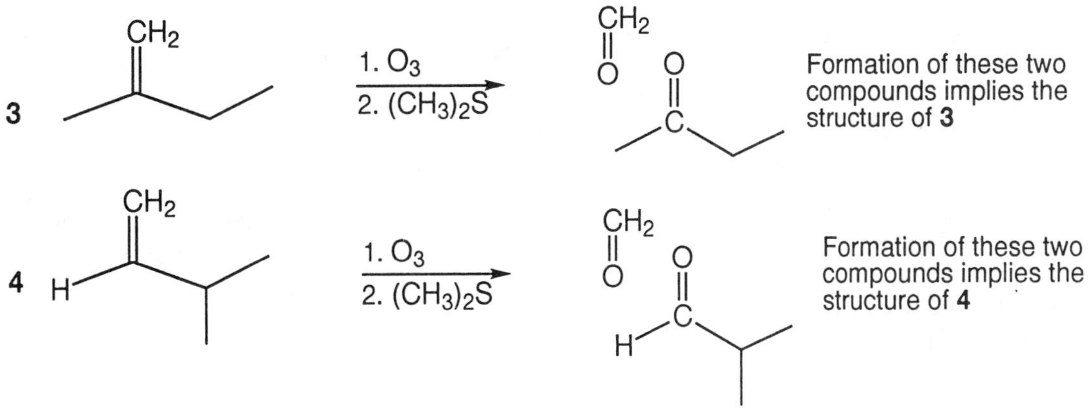

Given that **3** and **4** are formed by partial hydrogenation of **1**, we now know **1**. Full hydrogenation of **1** gives **2**.

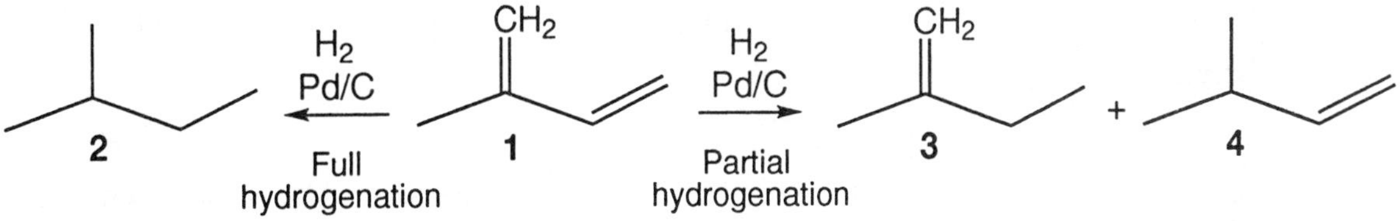

2
H_2
Pd/C
Full hydrogenation
CH_2
1
H_2
Pd/C
Partial hydrogenation
CH_2
3
+
4

Problem 10.37 For α-terpinene (**1**), there is only one way to put the pieces formed on ozonolysis back together. (Remember: with an oxidative workup carboxylic acids, not aldehydes, are formed).

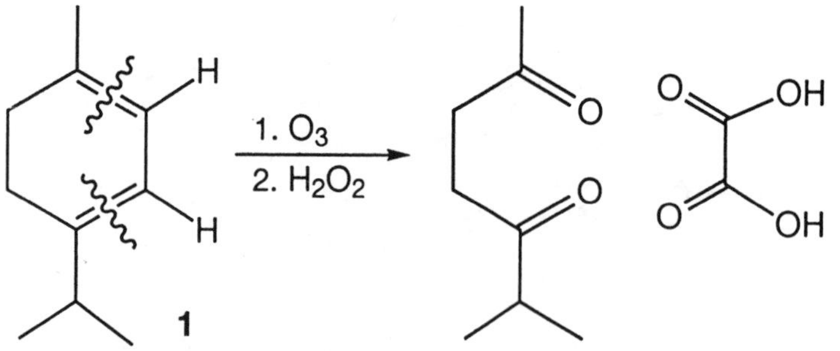

1. O_3
2. H_2O_2

These structures allow us to piece together the structure of α-terpinene **1**

For γ-terpinene (**2**), there are two ways to put the pieces back together. However, only one of the possible cyclohexadienes would yield 1-isopropyl-4-methylcyclohexane on hydrogenation.

1. O_3
2. H_2O_2

H_2
Pd/C

or

But only this possibility hydrogenates to give 1-isopropyl-4-methyl-cyclohexane

1. O_3
2. H_2O_2

This structure would not give 1-isopropyl-4-methylcyclohexane on hydrogenation

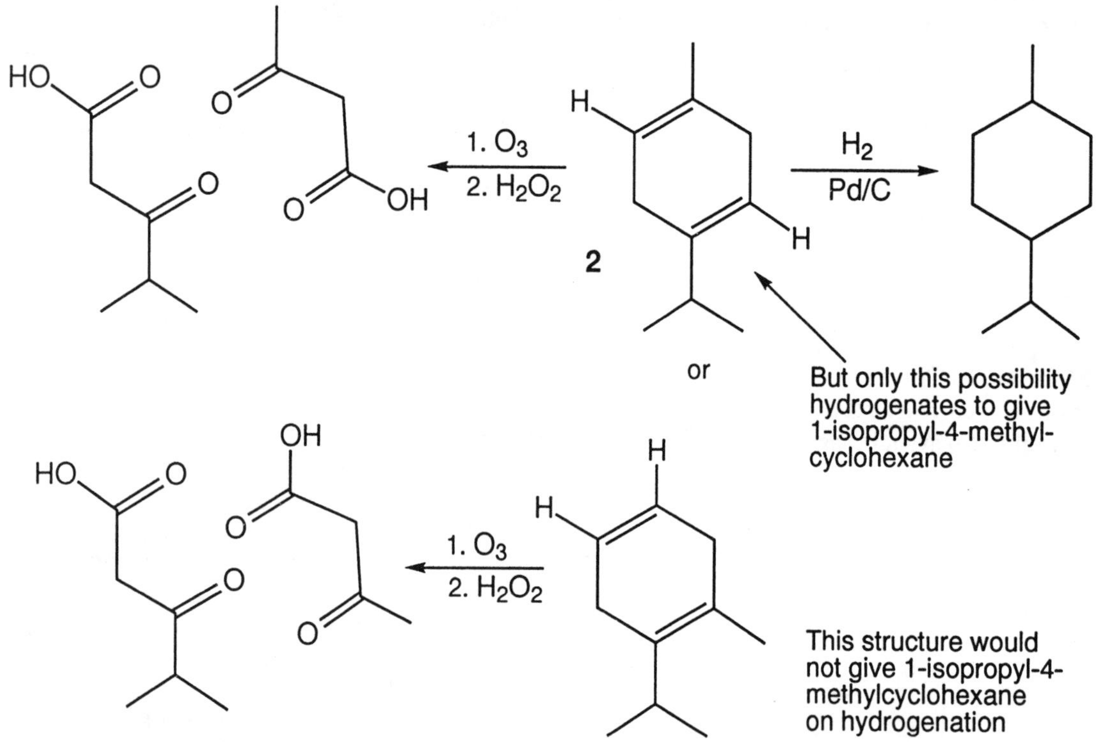

Either set of structures allows us to piece together a possible structure for γ-terpinene

Problem 10.38 First, form the primary ozonide and allow it to break down to give the carbonyl oxide and, in this case, acetone.

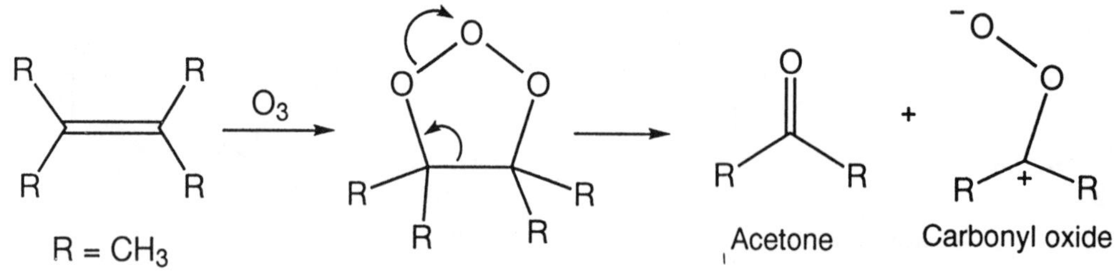

O_3

R = CH_3

Acetone

Carbonyl oxide

If the carbonyl oxide diffuses away from the acetone, it can collapse to the dioxirane.

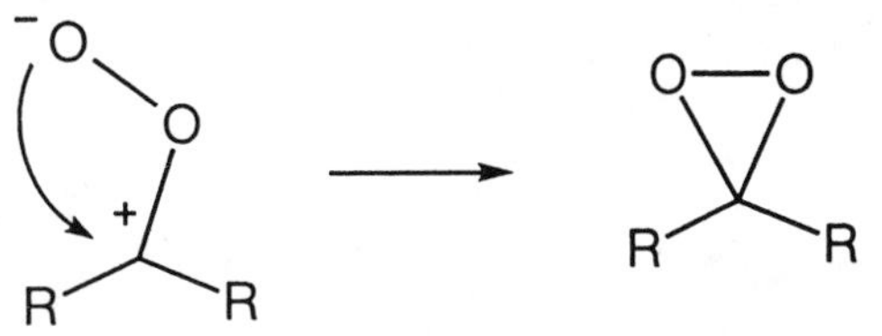

Alternatively, it might be captured by another carbonyl oxide molecule. This eventually leads to the six- and nine-membered rings.

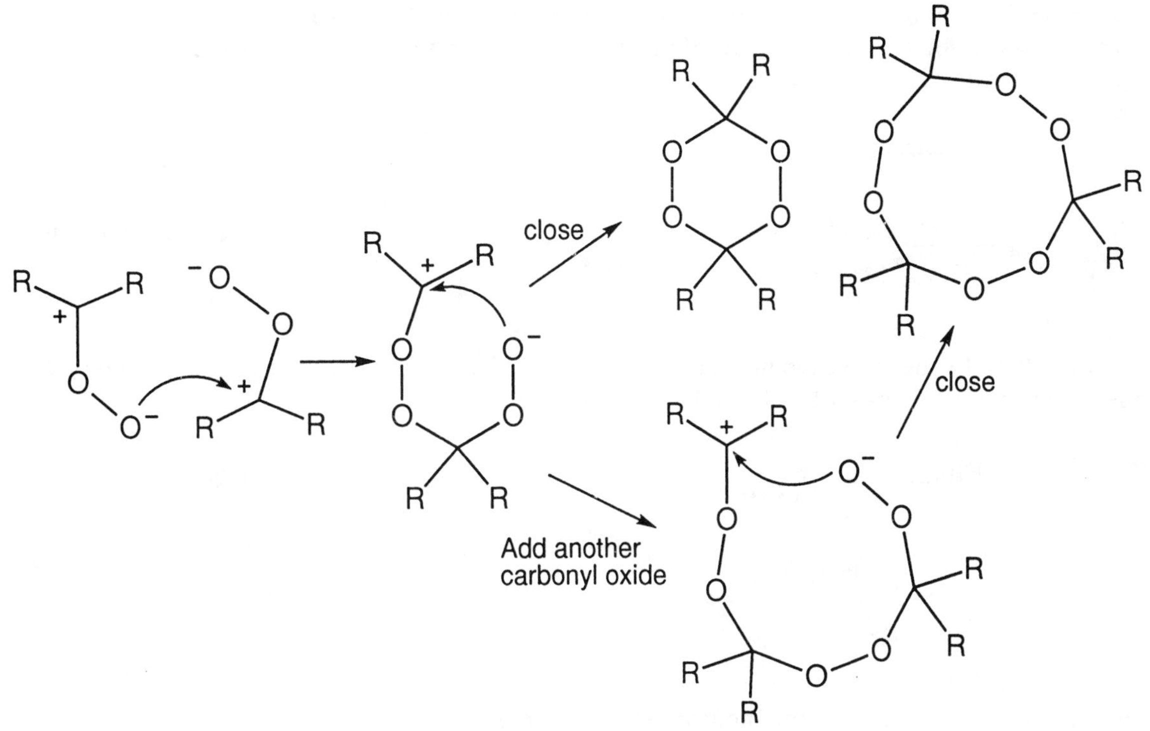

close

Add another carbonyl oxide

close

Problem 10.39 Once formed, the carbonyl oxide can be captured by *any* carbonyl group present. If acetone is present as solvent, it will have a great advantage in the capture process over the carbonyl group produced from the original alkene.

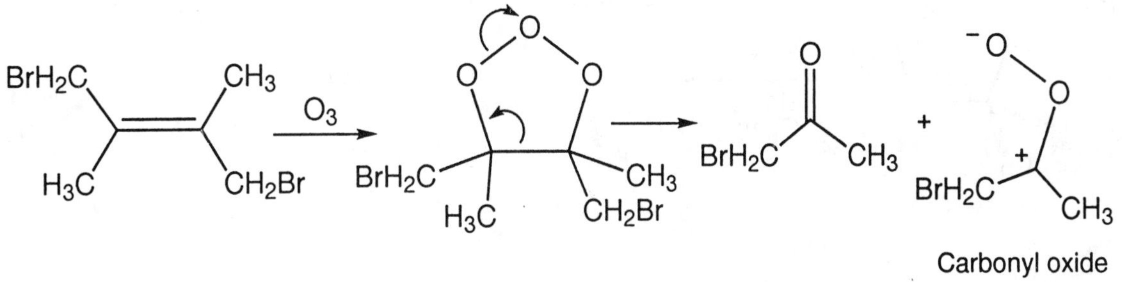

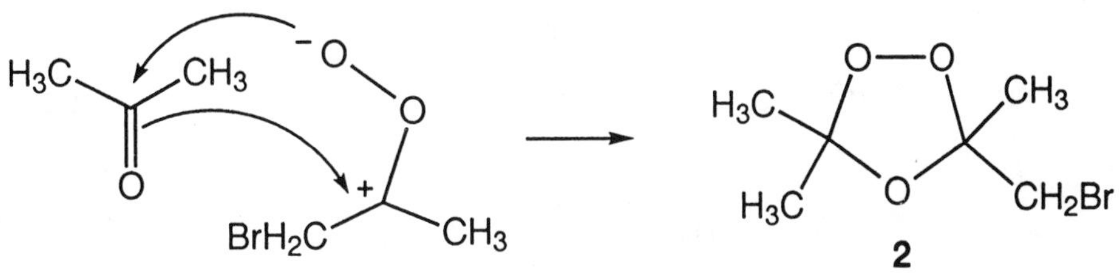

O_3

Carbonyl oxide

2

Problem 10.40 The arrow formalism is easy.

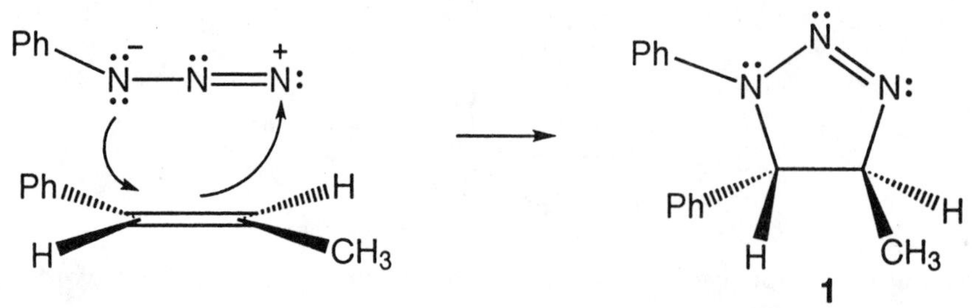

The formation of the single stereoisomer **1** with retention of stereochemistry (the trans alkene gives a trans five-membered ring) means that the mechanism of formation of **1** must involve no intermediates (**2**, for example) capable of rotation that would scramble the stereochemistry originally present in the alkene.

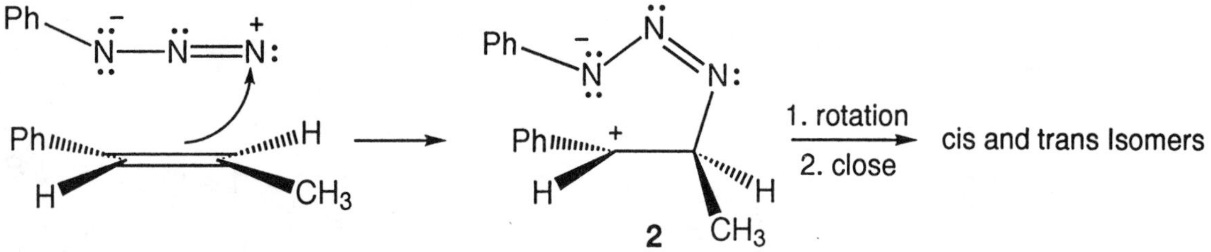

Problem 10.41 In the concerted mechanism nitrogen is lost in a single step to give aziridine **2**. The trans stereochemistry present in **1** will be preserved in **2**.

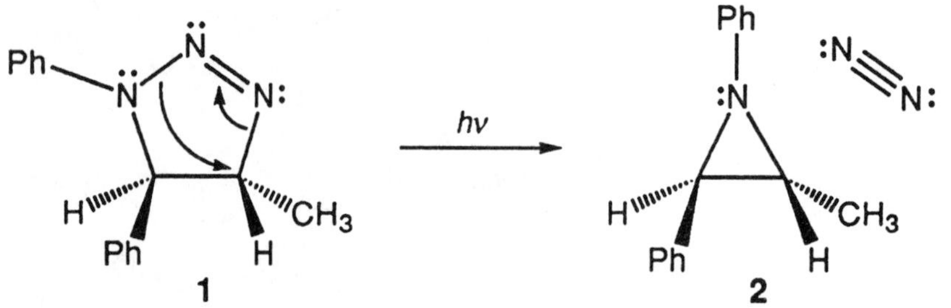

There are many possible two-step (nonconcerted) mechanisms. All must go through an intermediate in which the stereochemical relationships originally present in the alkene can be lost. Both **2** and the stereoisomeric **3** will be formed. Here is one possibility.

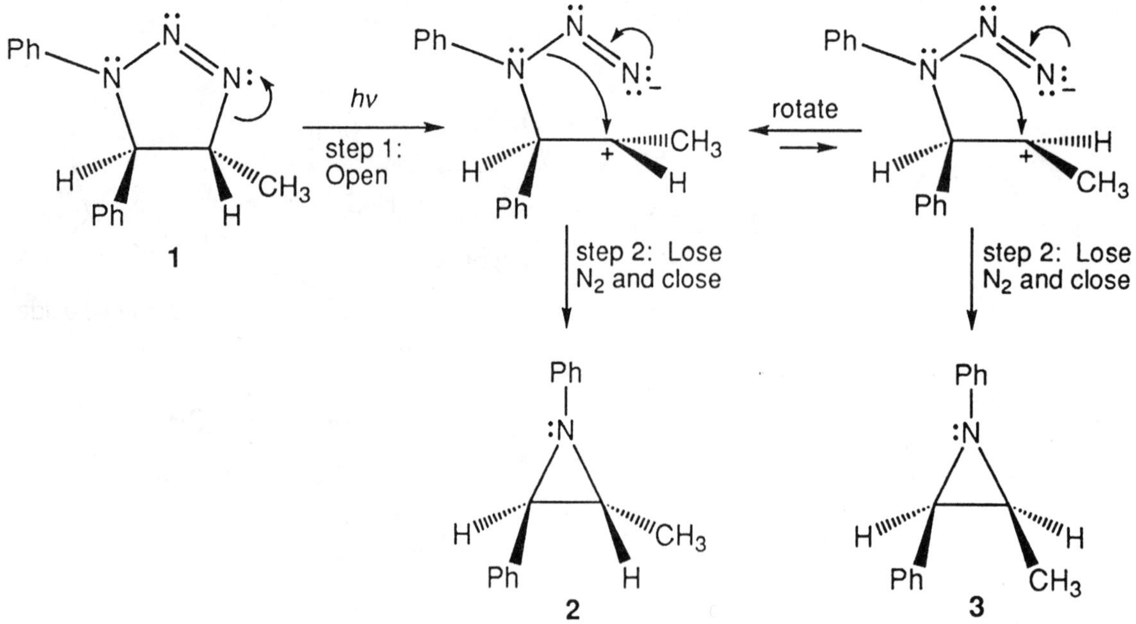

So the formation of **3** becomes diagnostic for the nature of the mechanism of aziridine formation. A one-step mechanism cannot form **3**; a two-step mechanism must form **3**.

Problem 10.42 Azides behave much like diazo compounds. Just as diazo compounds can lose nitrogen to give a carbene (divalent carbon), so azides can lose nitrogen to give "nitrenes" (monovalent nitrogen).

H_3COOC–$\ddot{C}^-$(H)–$N^+{\equiv}N$: → N_2 + H_3COOC–C:(H)

Diazo compound | A carbene

H_3COOC–$\ddot{N}^-$–$N^+{\equiv}N$: → N_2 + H_3COOC–$\ddot{N}$:

Azide | A nitrene

Carbenes add to alkenes to give cyclopropanes. Nitrenes behave in similar fashion, forming the three-membered rings called aziridines.

A carbene + $R_2C{=}CR_2$ → Cyclopropane (ring carbon bearing H and $COOCH_3$)

A nitrene + $R_2C{=}CR_2$ → Aziridine (:N–$COOCH_3$)

2

Problem 10.43 Retention of stereochemistry means that aziridine formation must take place in a single step.

H_3COOC—$\ddot{N}$: + cis-H_3C–CH=CH–CH_3 → (one-step addition) → cis-2,3-dimethyl-N-($COOCH_3$)aziridine **2**

A two-step addition would give both **2** and the stereoisomeric **3**. As only **2** is formed, the reaction must occur in a single step. In turn, that implies a singlet, all paired electron structure for the nitrene reacting species.

H_3COOC—$\ddot{N}$: + cis-H_3C–CH=CH–CH_3 → (step 1: Addition) → diradical intermediate ⇌ (rotation) → rotated diradical intermediate; step 2: Close → **2** + **3**

Problem 10.44

(a) In a one-step reaction, there can be no intermediate; the product must be formed directly. There really isn't much to draw in this case.

—C—H + :CR_2 → —C—C(R)(R)—H

A two step mechanism must involve hydrogen abstraction followed by recombination:

—C—H + :CR_2 → (step 1 (abstraction)) → —C· + H—ĊR_2 → (step 2 (recombine)) → —C—C(R)(R)—H

(b) The Doering and Prinzbach experiment shows that the insertion reaction must be a one-step reaction. A two-step process would generate a resonance-stabilized intermediate allyl radical, and the position of the label would be scrambled. As this does not happen, the reaction must be one step.

An allyl radical

$H_2\overset{*}{C}{=}C(CH_3)_2$ + :CH_2 —step 1 (abstraction)→ [$H_2\overset{*}{C}{=}C(CH_3)\dot{C}H_2$ ⟷ $H_2\overset{*}{\dot{C}}{-}C(=CH_2)CH_3$] + $H{-}\dot{C}H_2$ = $\cdot CH_3$

$\cdot CH_3$ step 2 (recombination) $\cdot CH_3$

$H_2\overset{*}{C}{=}C(CH_2CH_3)CH_3$ $CH_3\overset{*}{C}H_2{-}C(=CH_2)CH_3$

These are the same alkene, but the labeled carbon is in a different position

Problem 10.45 Both hydroboration/oxidation and mercury-catalyzed hydration of 2-pentyne must give a mixture of two ketones, 2-pentanone and 3-pentanone. Neither process is a useful preparative method. By contrast, either procedure applied to the symmetrical alkyne 3-hexyne would give the same product, 3-hexanone. Either would be a reasonable preparative method.

$H_3C{-}C{\equiv}C{-}CH_2CH_3$ (2-Pentyne) —$Hg(OAc)_2$ / H_2O, H_3O^+ or 1. HBR_2 / 2. HOOH, HO^-→ $H_3C{-}C(=O){-}CH_2CH_2CH_3$ + $CH_3CH_2{-}C(=O){-}CH_2CH_3$

$CH_3CH_2{-}C{\equiv}C{-}CH_2CH_3$ (3-Hexyne) —$Hg(OAc)_2$ / H_2O, H_3O^+ or 1. HBR_2 / 2. HOOH, HO^-→ $CH_3CH_2{-}C(=O){-}CH_2CH_2CH_3$

Problem 10.46 In all synthesis problems it is imperative that you work backwards. This is called "retrosynthetic analysis" by those who do it for a living, and *every* successful synthetic chemist analyzes problems this way. There is even a special, "retrosynthetic" arrow that points to the immediate precursors to the target. Thus;

Target ⇒ Precursor molecules
(Retrosynthetic arrow)

Who are we to disagree? We won't. In practice, the rather fancy term "retrosynthetic analysis" simply means, "search for the *immediate* precursor for the target molecule," DO NOT attempt to see all the way back to the ultimate starting material. At this point, when we have relatively few synthetic methods in our arsenal, it may be possible to do synthetic problems "forward". But this is bad technique, and it is best to practice doing these problems the right way.

(a) In this case, we can see that the target, 2-hexanone, could be made from the mercury-catalyzed hydration of 1-hexyne (2-hexyne would also give some 2-hexanone, but this would not be a practical route to the product. Why?).

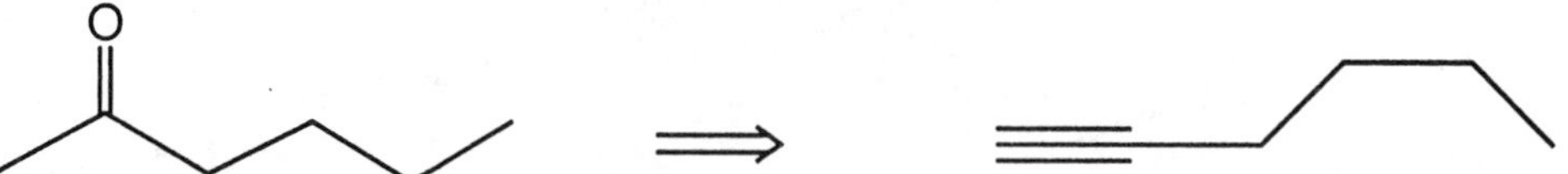

However, 1-hexyne has two more carbons than our allowable starting materials, so we need a synthesis of this molecule. The problem has been reduced to finding a synthesis for 1-hexyne. Here is a suggestion: An S_N2 reaction between an acetylide anion and bromobutane would do the trick.

The acetylide can be formed from the reaction of a large excess of acetylene, an allowed starting material, and sodium amide (the excess acetylene reduces the formation of the diacetylide).

$$HC{\equiv}C:^- \Longrightarrow HC{\equiv}CH + NaNH_2/NH_3$$

We still need to make bromobutane, however, because this material isn't an allowed starting material. We might reduce 1-butyne to 1-butene, hydroborate, and then form the bromide.

Br ⟹ OH ⟹ ⟹

So the synthesis would look like this

H_2 Lindlar catalyst; 1. BH_3 2. H_2O_2/HO^-; OH; HBr; Br; $HC{\equiv}C:^-$; O C; Hg^{2+} $H_2OH_3O^+$

(b) The target molecule, pentanal, should be available from 1-pentyne through a hydroboration-oxidation sequence with diisoamylborane, $HB[CHCH_3CH(CH_3)_2]_2$. 1-Pentyne can be made through a sequence similar to that used for 1-hexyne in (a).

O C H ⟹

Here is the synthesis.

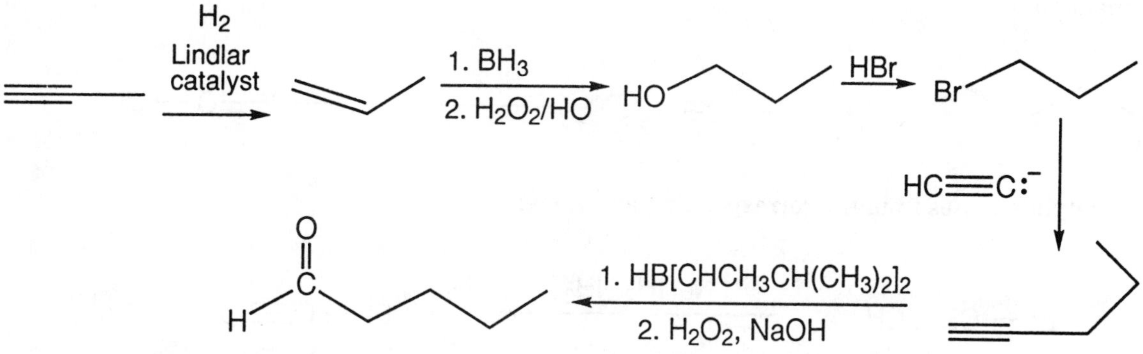

H_2, Lindlar catalyst; 1. BH_3, 2. H_2O_2/HO^-; HO; HBr; Br; $HC{\equiv}C:^-$; 1. $HB[CHCH_3CH(CH_3)_2]_2$, 2. H_2O_2, NaOH; O, C, H

There are other possibilities. For example, we could use the 1-hexyne we made in part (a). The retrosynthetic analysis looks like this

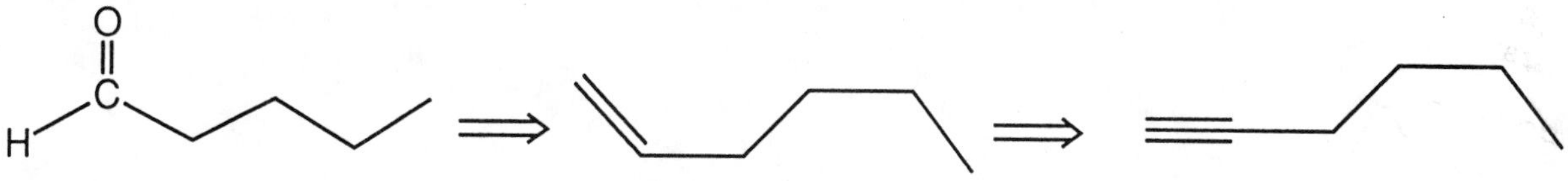

O, C, H

In other words

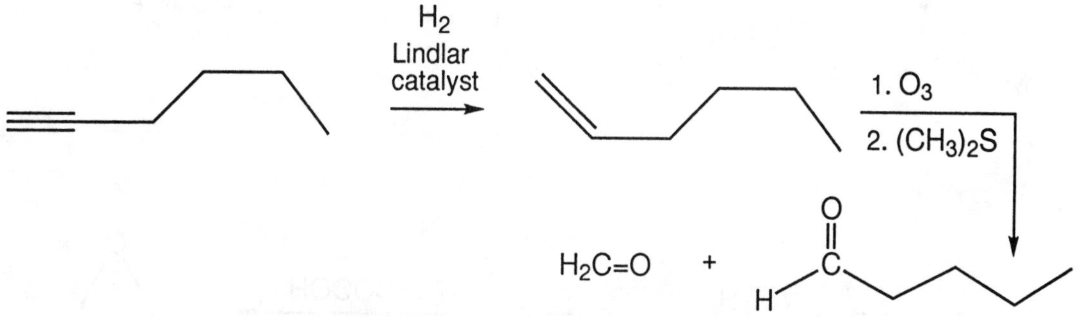

H_2, Lindlar catalyst; 1. O_3, 2. $(CH_3)_2S$

$H_2C{=}O$ + O, C, H

(c) Retrosynthetic analysis suggests that our target epoxide should be available from 3-octyne.

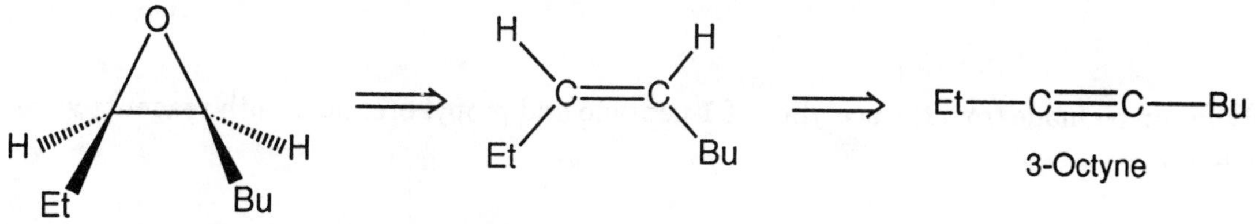

O, H, H, Et, Bu; H, H, C=C, Et, Bu; Et—C≡C—Bu

3-Octyne

Bu = butyl, Et = ethyl

In conventional terms:

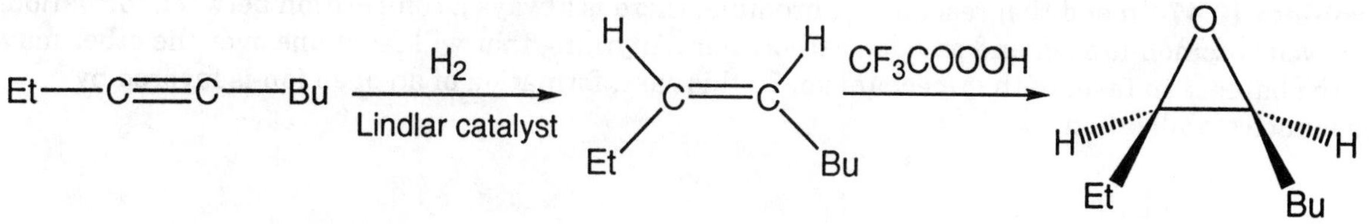

Et—C≡C—Bu; H_2, Lindlar catalyst; H, H, C=C, Et, Bu; CF_3COOOH; O, H, H, Et, Bu

3-Octyne can be prepared by alkylation of the acetylide of 1-butyne with bromobutane, a molecule we made in (a).

$$\text{BrCH}_2\text{CH}_2\text{CH}_2\text{CH}_3 + \text{Et—C}\equiv\text{C:}^- \xrightarrow{S_N2} \text{Et—C}\equiv\text{C—Bu}$$

The acetylide comes from the corresponding acetylene:

$$\text{Et—C}\equiv\text{CH} \xrightarrow{NaNH_2/NH_3} \text{Et—C}\equiv\text{C:}^- \quad Na^+$$

(d) The trans epoxide must come from epoxidation of a trans alkene. In turn, the trans alkene comes from a stereospecific reduction of an alkyne, in this case, 4-octyne.

trans-epoxide ⟹ trans-PrCH=CHPr ⟹ Pr—C≡C—Pr

4-Octyne

Pr = propyl

or

$$\text{Pr—C}\equiv\text{C—Pr} \xrightarrow{Na/NH_3} \textit{trans}\text{-PrCH=CHPr} \xrightarrow{CF_3COOOH} \text{trans epoxide}$$

4-Octyne can be made from the acetylide of 1-pentyne and propyl bromide. Both reagents were made in (b).

$$\text{Pr—C}\equiv\text{C:}^- + \text{BrCH}_2\text{CH}_2\text{CH}_3 \xrightarrow{S_N2} \text{Pr—C}\equiv\text{C—Pr}$$

Problem 10.47 In addition reactions of bromine, there is always a competition between formation of a cyclic bromonium ion and an open carbocation. Anything that will favor one over the other may tip the balance in favor of that mechanism. In this case, formation of an open ion is favored by resonance stabilization.

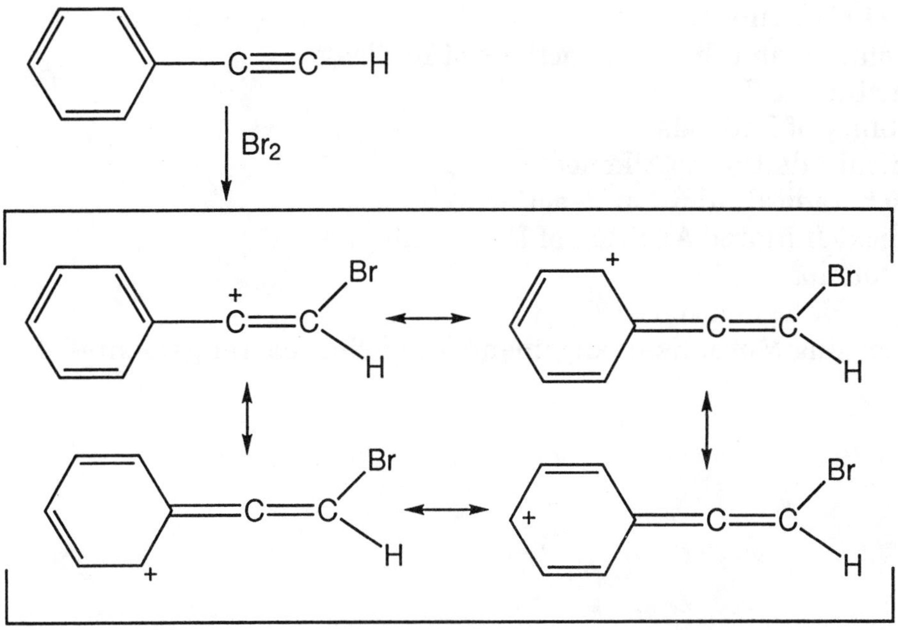

Now, addition of the bromide ion can be to either lobe of the empty 2*p* orbital, thus giving both stereoisomers.

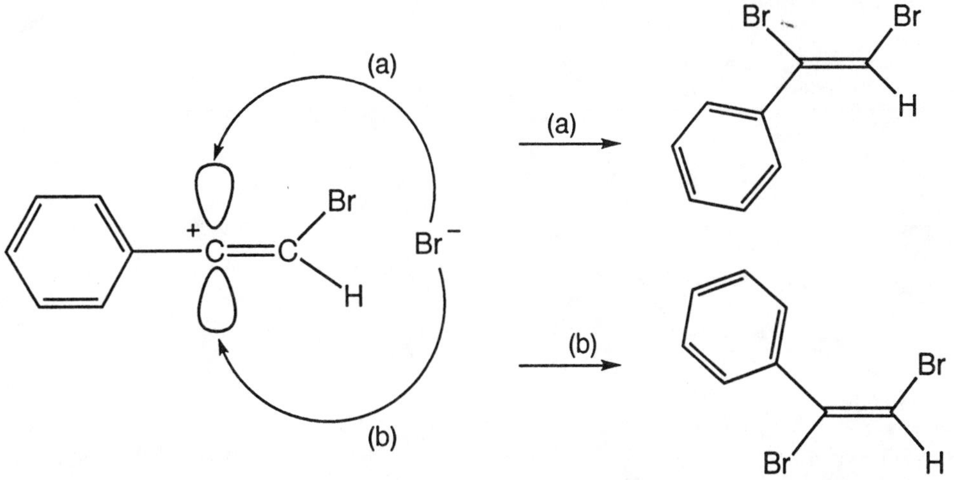

If a cyclic bromonium ion had been the intermediate, only the trans dibromide could have formed. As both stereoisomers are formed, the open cation must be involved.

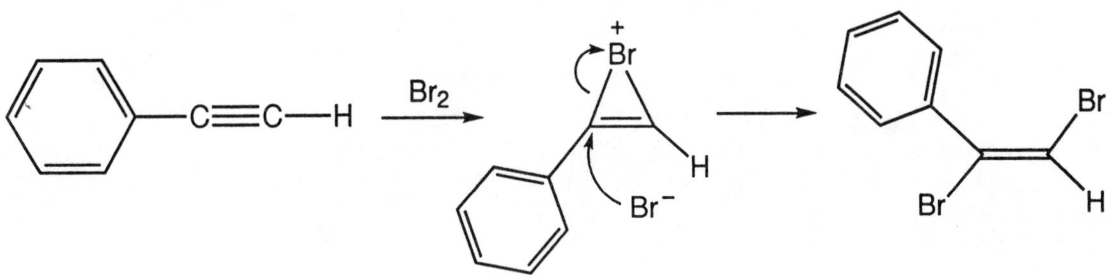

Chapter 11 Outline

11

Radical Reactions

Radical reactions provide new opportunities and new difficulties. Radical reactions allow you to reverse the regiochemistry of addition of hydrogen bromide to alkenes, for example, or to introduce functionality at the position adjacent to a double bond. The following problems allow you to practice such things. Many more radical reactions are known than are discussed in this chapter, and these problems allow you to find a few. Radical chemistry, though different from the polar chemistry we have emphasized so far, nonetheless can be understood through an analysis of structure and orbitals. Keep in mind that chain processes abound in radical chemistry. In such reactions a chain carrying radical is produced in one of the propagation steps. In a sense, this is analogous to the regeneration of a catalyst at the end of a polar reaction. There are clues to the presence of a radical reaction: The presence of peroxides or *N*-bromosuccinimide (NBS) is one clue, for example.

Problem 11.02 Any single step, single barrier, reaction between polar species will do. Protonation of an alkene is only one of many, many possibilities.

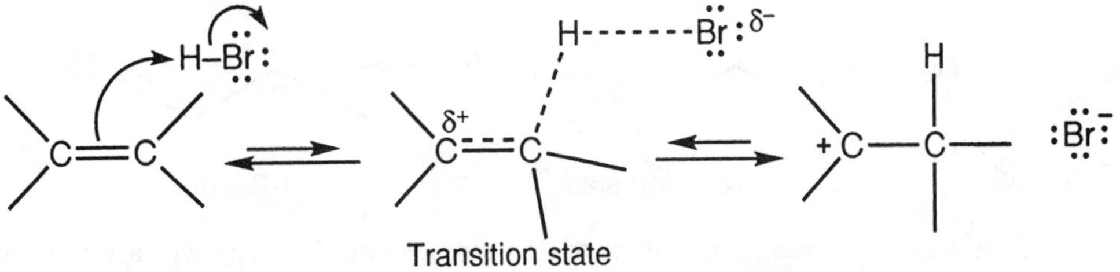

H–Br: δ+ δ− +C :Br:−

Transition state

Problem 11.04 The carbon-carbon bond is weaker (90 kcal/mol) than the carbon-hydrogen bond (100.3 kcal/mol), and therefore is easier to break.

Problem 11.05 In the transition state for abstraction of hydrogen from butane by a methyl radical, the carbon-hydrogen bond of butane is partially broken and the methyl-hydrogen bond is partially made (continued on next page).

Methyl radical $\dot{C}H_3$ H CH_2 → $\delta\cdot$ CH_3 H CH_2 $\delta\cdot$ Transition state → $\dot{C}H_2$ + CH_4 A primary radical

Methyl radical $\dot{C}H_3$ H CH → $\delta\cdot$ CH_3 H CH $\delta\cdot$ Transition state → $\dot{C}H$ + CH_4 A secondary radical

Each of these two reactions is somewhat exothermic, as the carbon-hydrogen bond in methane is stronger than either carbon-hydrogen bond in butane. Accordingly, in each transition state the hydrogen atom is slightly less than halfway transferred (the transition state is slightly "starting material-like"). Recall the Hammond postulate. Do you see why in the transition state for breaking of a secondary carbon-hydrogen bond the hydrogen is slightly less transferred than in the transition state for breaking a primary carbon-hydrogen bond?

Problem 11.06 In a disproportionation reaction, the carbon-hydrogen bond will be partially broken and the double bond partially developed, as one radical abstracts a hydrogen atom from another. In the example shown, a methyl radical plucks a hydrogen atom from the *sec*-butyl radical to give 1-butene and methane.

Methyl radical $\dot{C}H_3$ H CH_2 → $\delta\cdot$ CH_3 H CH_2 $\delta\cdot$ → CH_2 + CH_4

sec-Butyl radical Transition state 1-Butene Methane

Problem 11.07 In a β-cleavage reaction, one particle becomes two. Entropy will always favor such a process. In the expression for the energy change, $\Delta G° = \Delta H° - T\Delta S°$, the entropy term becomes more important as the temperature increases. Accordingly, as the temperature becomes higher, the impact on $\Delta G°$ of the formation of two particles from one becomes more strongly felt, and β-cleavage becomes easier. Here is an example in which the butyl radical fragments to ethylene and an ethyl radical.

H_3C CH_2 β CH_2 α $\dot{C}H_2$ —β-cleavage→ H_3C $\dot{C}H_2$ $H_2C{=}CH_2$

One particle, the butyl radical

Two particles, the ethyl radical and ethylene

Problem 11.08 The hydrogen simply provides a source of abstractable hydrogen from which the initially produced alkyl radicals can pluck hydrogen atoms and become alkanes.

$$CH_3CH_2CH_2CH_3 \longrightarrow CH_3CH_2\dot{C}H_2 + \cdot CH_3 \xrightarrow{H_2} CH_3CH_2CH_3 + CH_4$$

Problem 11.10 There are, of course, numerous possibilities, and the text will go on to discuss a number of them. The stability of radicals increases with substitution, so one might imagine that the bond dissociation energy of carbon-carbon bonds would decrease as substitution increases. One answer to this problem is as simple as "propane!"

	Bond dissociation energy
$H_3C-CH_3 \longrightarrow 2\ \cdot CH_3$	90 kcal/mol
$H_3C-CH_2CH_3 \longrightarrow \cdot CH_3 + \cdot CH_2CH_3$	85 kcal/mol

Problem 11.11 Lowering of the product energy is accompanied by a lowering of the energy of the transition state leading to product, and therefore, of the activation energy for the reaction.

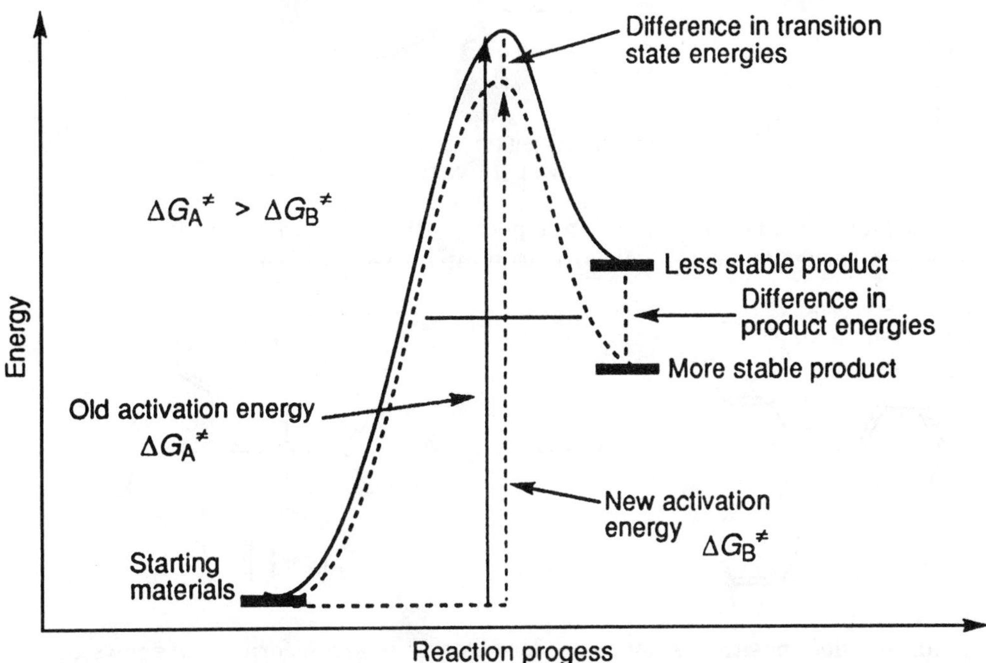

Product 11.12 The arrow formalism for this reaction is easy, although little else about this reaction is. The arrows run from the breaking bonds to the new (forming) bonds. We will meet this molecule and this reaction again in exquisite detail in Chapter 22, p. 1158.

1,1,6,6-Tetradeuterio-1,5-hexadiene $\underset{\Delta}{\rightleftharpoons}$ 3,3,4,4-Tetradeuterio-1,5-hexadiene

Problem 11.13 Azoisobutyronitrile (AIBN) decomposes to give two resonance-stabilized radicals, whereas simple azo compounds do not. This stabilization of the products by resonance makes it easier to break the bonds in the starting materials.

$$(CH_3)_3C-N{=}N-C(CH_3)_3 \xrightarrow{\Delta} 2\,(CH_3)_3C\cdot + N_2 \quad \Delta H^\circ = 42.2\text{ kcal/mol}$$

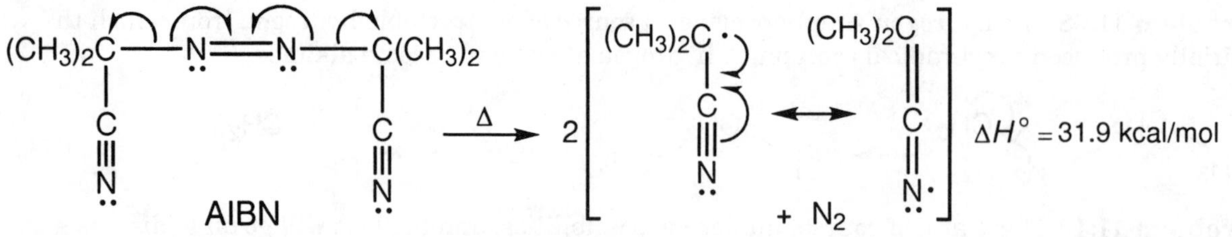

$$(CH_3)_2C{-}N{=}N{-}C(CH_3)_2 \text{ (AIBN)} \xrightarrow{\Delta} 2\left[(CH_3)_2\dot{C}{-}C{\equiv}N \longleftrightarrow (CH_3)_2C{=}C{=}\dot{N}\right] + N_2 \qquad \Delta H^\circ = 31.9 \text{ kcal/mol}$$

Problem 11.14 It is already drawn in the figure. The transition state for the inversion process is the planar radical.

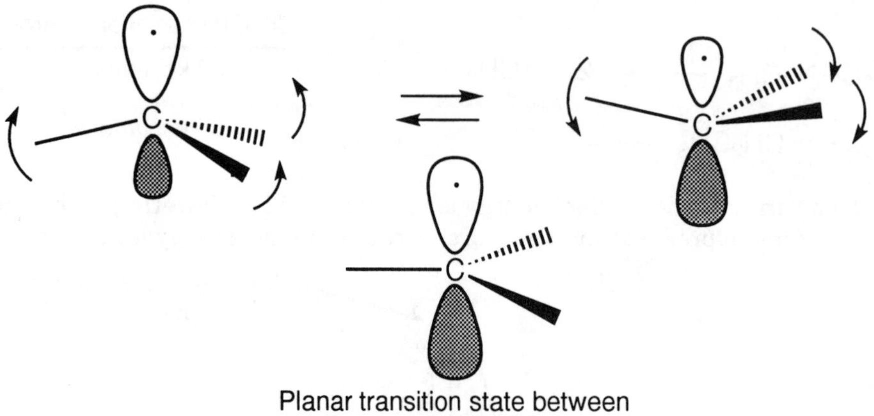

Planar transition state between the two pyramidal forms

Problem 11.15 The "obvious," "just push 'em together" dimer suffers from severe steric problems. The large groups attached to the central carbons bump into each other.

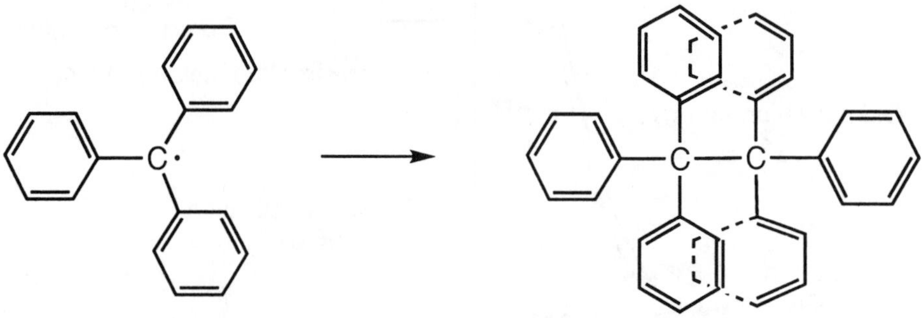

Approach to an "outside" position is not so hindered, and the arrow formalism shows the formation of the real dimer.

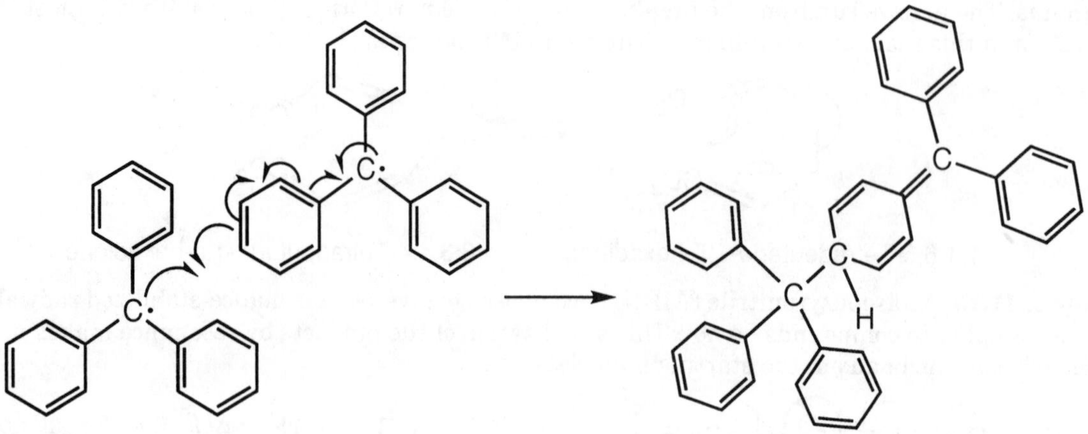

Problem 11.16 The bond formed by abstraction of a H atom is an O-H bond, with an approximate bond energy of 104 kcal/mol (Table 8.2, p. 307). Abstraction of Br would lead to an O-Br bond, which is much weaker, about 56 kcal/mol. Formation of the O-H bond is exothermic, whereas formation of the O-Br bond is endothermic.

$$R{-}\ddot{O}\cdot \; + \; H{-}Br \longrightarrow R{-}O{-}H \; + \; \cdot Br$$

87.6 kcal/mol ~ 104 kcal/mol Exothermic by 16.4 kcal/mol

$$R{-}\ddot{O}\cdot \; + \; Br{-}H \longrightarrow R{-}O{-}Br \; + \; H\cdot$$

87.6 kcal/mol ~56 kcal/mol Endothermic by 31.6 kcal/mol

Problem 11.17 Polar addition of hydrogen bromide to 2-methyl-2-butene will give the more substituted bromide. Peroxide-mediated hydrogen bromide addition leads to the less substituted bromide.

H–Br, no peroxides (← more substituted bromide, Br); H–Br, RO–OR (→ less substituted bromide, Br)

The more substituted alcohol can be made through a hydration reaction, and the less substituted alcohol though a hydroboration procedure.

1. BH_3; 2. H_2O_2/ HO^- → OH

H_3O^+, H_2O → OH

Alternatively, the corresponding bromides can be made and subsequently transformed into the alcohols. Can you see what problems these indirect syntheses will encounter? (see Chapter 7, p. 273).

H–Br, RO–OR → Br; KOH/H_2O, S_N2 → OH

H–Br, no peroxides → Br; H_2O, S_N1 → OH

Problem 11.18 Many exist. The destruction of any chain-carrying radical will terminate the chain reaction. Here are three examples.

$$\cdot CCl_3 + \cdot CCl_3 \longrightarrow Cl_3C{-}CCl_3$$

$$\cdot CCl_3 + \cdot CH_2CH_2CCl_3 \longrightarrow Cl_3C{-}CH_2CH_2CCl_3$$

$$\cdot CH_2CH_2CCl_3 + \cdot CH_2CH_2CCl_3 \longrightarrow Cl_3CCH_2CH_2{-}CH_2CH_2CCl_3$$

Problem 11.19 Addition of a radical to styrene gives an exceptionally well resonance-stabilized, and, hence, stable radical.

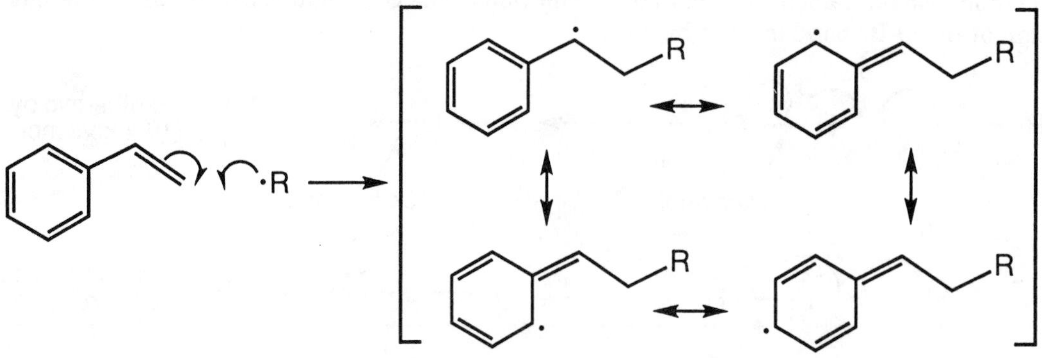

Problem 11.21 Addition of a radical to the vinyl halide can lead to either of two new radicals. In the more stable radical, the free electron is adjacent to a halogen, and is resonance stabilized. The other possible radical is not resonance stabilized, and is less stable.

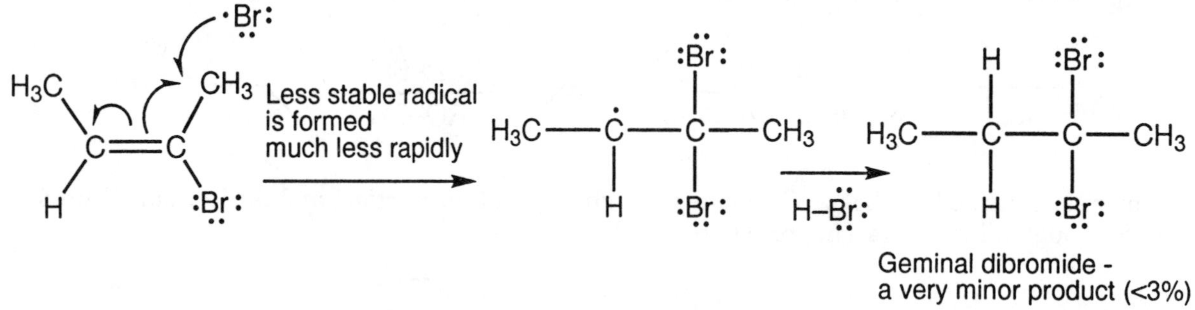

Less stable radical is formed much less rapidly

Geminal dibromide - a very minor product (<3%)

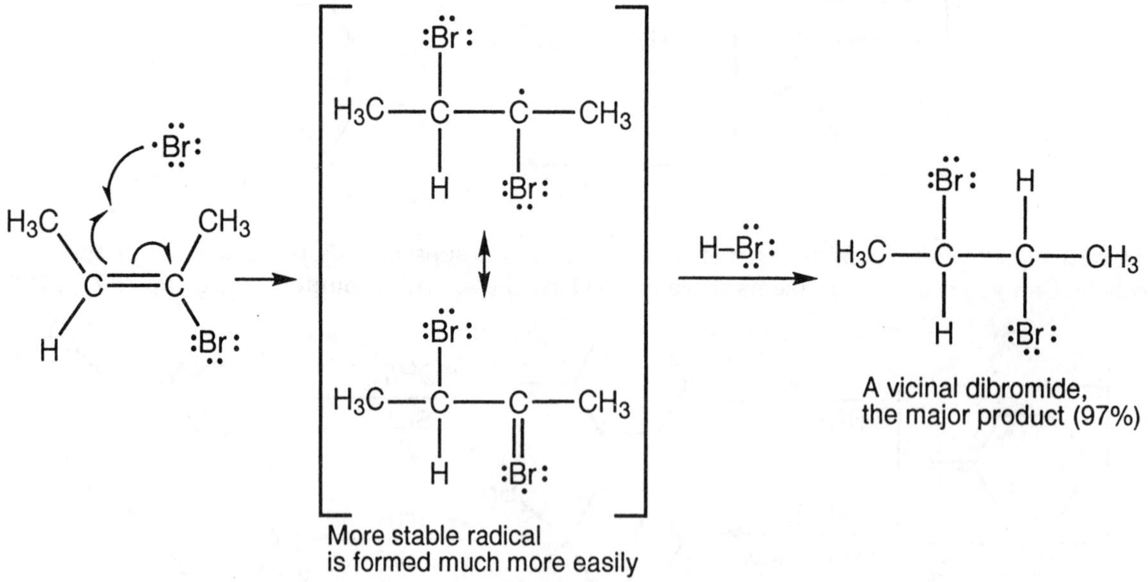

More stable radical is formed much more easily

A vicinal dibromide, the major product (97%)

This resonance stabilization is special, however. In this three-electron system, two electrons are well stabilized in a low energy, bonding molecular orbital, but one electron must occupy an antibonding molecular orbital. This system is stabilized overall, but the situation is not as straightforward as many two-electron cases.

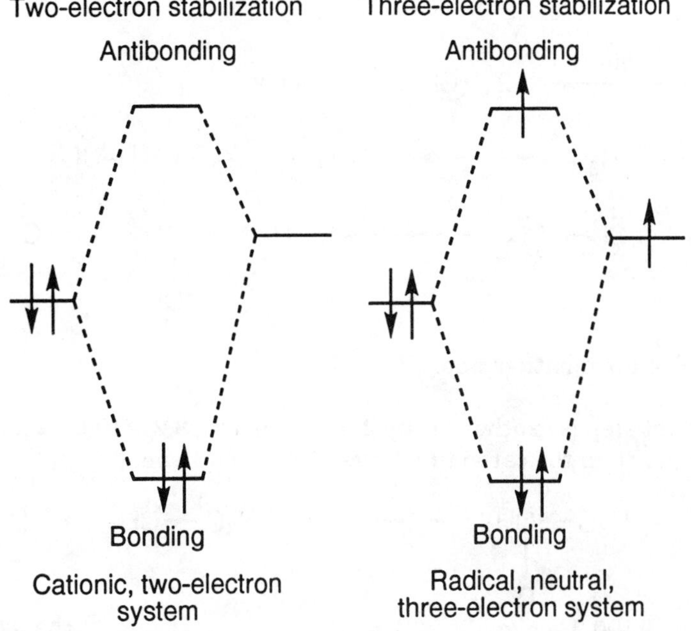

Problem 11.22 Abstraction of the methyl hydrogen leads to an unstabilized radical, whereas abstraction from the methylene position leads to a resonance-stabilized radical. Again, this resonance stabilization is in a three-electron system.

R·

\+ RH

Less stable – unstabilized

H

H

R·

\+ RH

Resonance stabilized – more stable

Problem 11.23 When the carbon-hydrogen bond in methane breaks, a methyl radical and a hydrogen atom are produced. Methyl chloride gives a hydrogen atom and a chloromethyl radical, which is stabilized by resonance, and, therefore, more stable than the unstabilized methyl radical. Accordingly, the bond is easier to break. Are you getting the idea that one of the generic answers to questions in organic chemistry is, It's resonance stabilized and therefore more stable? Well, you're right.

$H_3C—H \longrightarrow \cdot CH_3 + \cdot H$

$H_2C(Cl)—H \longrightarrow [\dot{C}H_2—\ddot{C}l: \longleftrightarrow CH_2=\dot{C}l:] + \cdot H$

Resonance stabilized – more stable – easier to form

Problem 11.24

$$:\ddot{Cl}-\ddot{Cl}: \xrightarrow{h\nu} 2\ \cdot\ddot{Cl}: \quad \text{Initiation}$$

$$\cdot\ddot{Cl}: + H_3C-CH_3 \longrightarrow H_3C-\dot{C}H_2 + H-\ddot{Cl}:$$

$$H_3C-\dot{C}H_2 + :\ddot{Cl}-\ddot{Cl}: \longrightarrow H_3C-CH_2(:\ddot{Cl}:) + \cdot\ddot{Cl}: \quad \text{Propagation}$$

There are many possible termination steps.

Problem 11.25 The first step is exothermic by 2.9 kcal/mol (103.2 – 100.3 = 2.9). The hydrogen-chlorine bond is stronger than the carbon-hydrogen bond in ethane.

$$\cdot\ddot{Cl}: + H_3C-CH_2(H) \longrightarrow H_3C-\dot{C}H_2 + H-\ddot{Cl}:$$

100.3 kcal/mol 103.2 kcal/mol

The second propagation step is also exothermic, this time by 22.5 kcal/mol (81.5 - 59 = 22.5).

$$H_3C-\dot{C}H_2 + :\ddot{Cl}-\ddot{Cl}: \longrightarrow H_3C-CH_2(:\ddot{Cl}:) + \cdot\ddot{Cl}:$$

59 kcal/mol 81.5 kcal/mol

Problem 11.27 The carbonyl oxygen is a base, and can be protonated by hydrogen bromide.

:N—Br: H—Br: → :N—Br: :Br:⁻

Now bromide can form Br_2 by attacking the protonated NBS at bromine:

:N—Br: :Br:⁻ → N: + :Br—Br:

Succinimide itself is produced by a series of proton transfers. Can you write a mechanism for this reaction? (See Section 10.7, p. 430.)

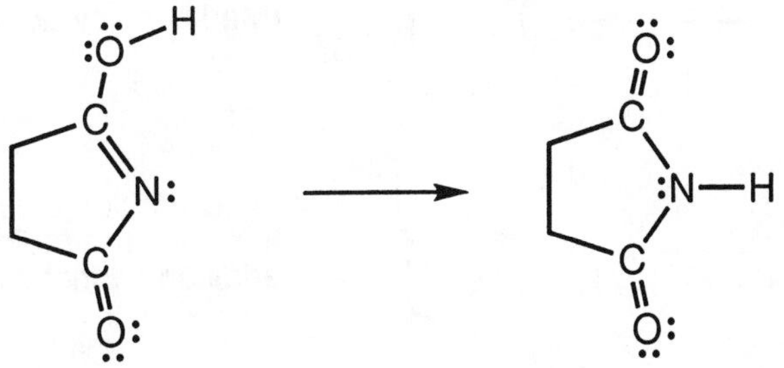

Problem 11.29 In the anionic migration, two electrons must occupy antibonding molecular orbitals; a most unfavorable process.

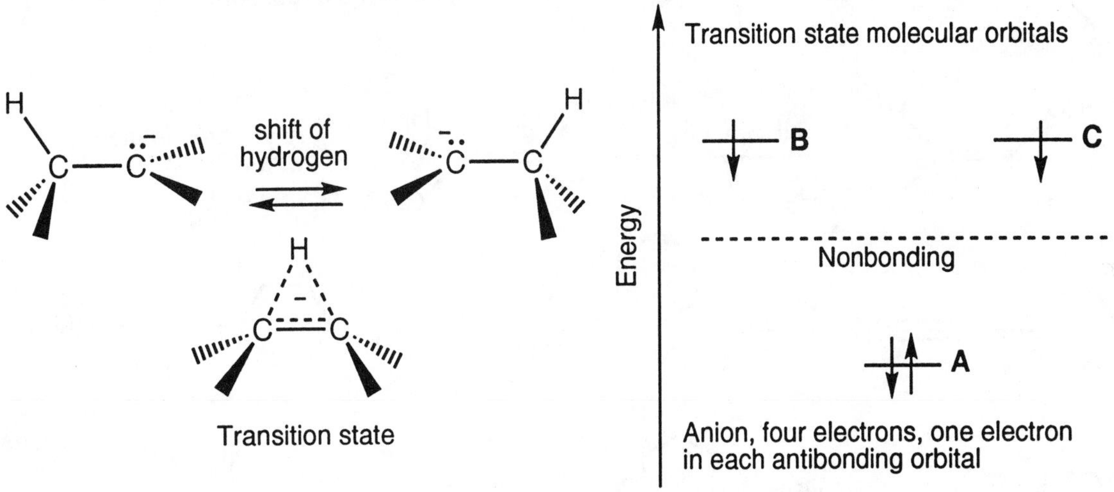

shift of hydrogen

Transition state

Transition state molecular orbitals

Energy

B

C

Nonbonding

A

Anion, four electrons, one electron in each antibonding orbital

Additional Problem Answers

Problem 11.32

(a)

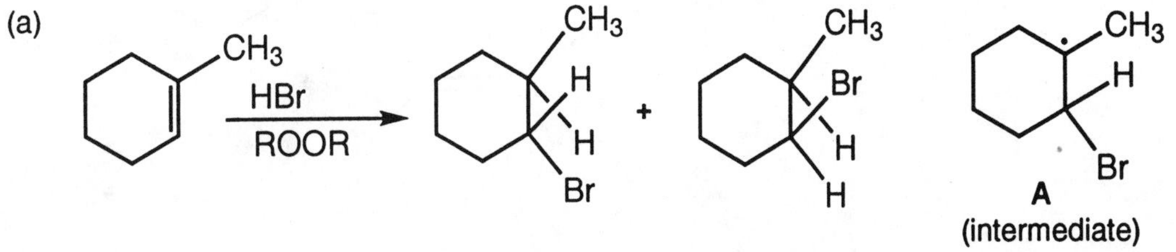

HBr

ROOR

A (intermediate)

This reaction is regiospecific but not stereospecific. The intermediate radical (**A**) can abstract hydrogen from either side.

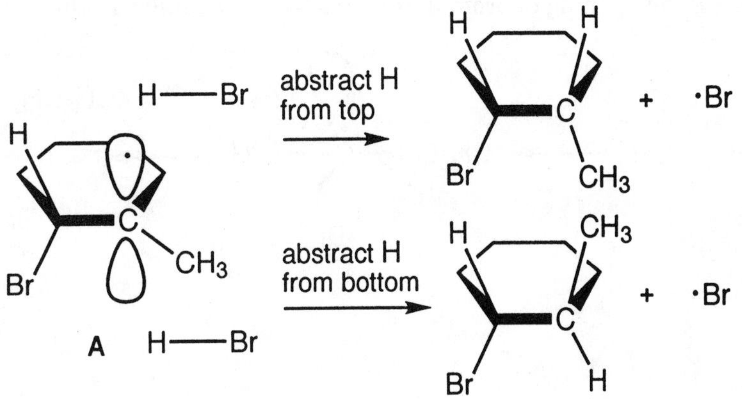

abstract H from top

abstract H from bottom

(b) 1-methylcyclohexene $\xrightarrow{HBr}$ 1-bromo-1-methylcyclohexane — Ionic addition of HBr (Markovnikov addition)

(c) 1-methylcyclohexene $\xrightarrow[ROOR]{HCl}$ 1-chloro-1-methylcyclohexane — Ionic addition of HCl; in this case the radical addition cannot compete

(d) $(CH_3)_2C{=}CH_2$ + CCl_4 $\xrightarrow{h\nu}$ $Cl{-}C(CH_3)_2{-}CH_2CCl_3$ — Radical-chain reaction

(e) $H_3C(H)C{=}C(H)COOCH_3$ $\xrightarrow[h\nu]{NBS,\ CCl_4}$ $BrCH_2(H)C{=}C(H)COOCH_3$ — Allylic bromination with NBS

(f) $C_6H_{11}{-}C{\equiv}CH$ $\xrightarrow{HBr}$ $C_6H_{11}(Br)C{=}CH_2$ + $C_6H_{11}{-}CBr_2{-}CH_3$ (from two additions of HBr)

(g) $C_6H_{11}{-}C{\equiv}CH$ $\xrightarrow[ROOR]{HBr}$ $C_6H_{11}(H)C{=}C(H)Br$ + $C_6H_{11}(H)C{=}C(H)Br$ (other stereoisomer) + $Br{-}CH(C_6H_{11}){-}CH_2Br$

Problem 11.33

(a) Addition of bromide to the more substituted position of the cyclic ion leads to the intital product, a vinyl bromide. Recall the opening of bromonium ions by nucleophiles in the same fashion.

$$CH_3CH_2{-}C{\equiv}CH \longrightarrow CH_3CH_2{-}\overset{+}{C}{=}CH\ (\text{H-bridged}) \xrightarrow{Br^-} CH_3CH_2(Br)C{=}CH_2$$

Now a second molecule of hydrogen bromide can add. Addition of bromide to the more stable, resonance-stabilized, carbocation gives the "double addition" product, the geminal dibromide.

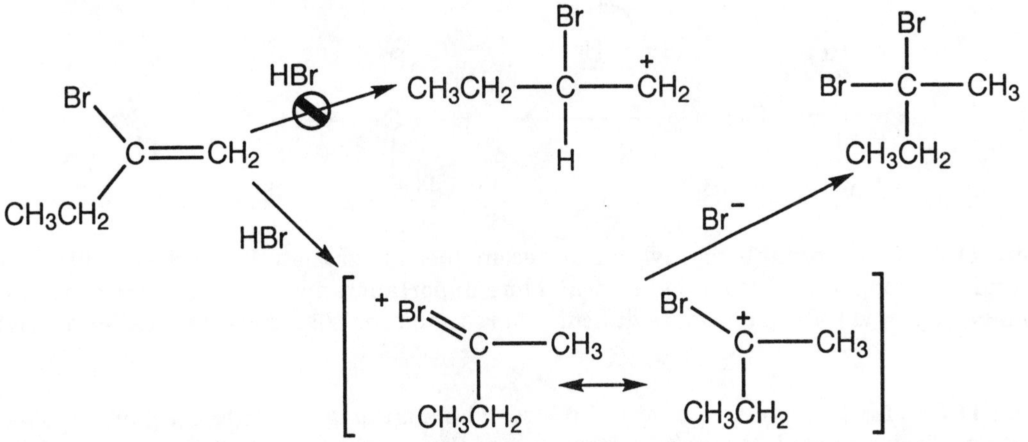

(b) The mechanism of the radical-induced reaction is different, and leads to a different product. Initiation steps lead to bromine atoms. Addition to the alkyne gives the more stable, secondary vinyl radical, not the less stable primary vinyl radical.

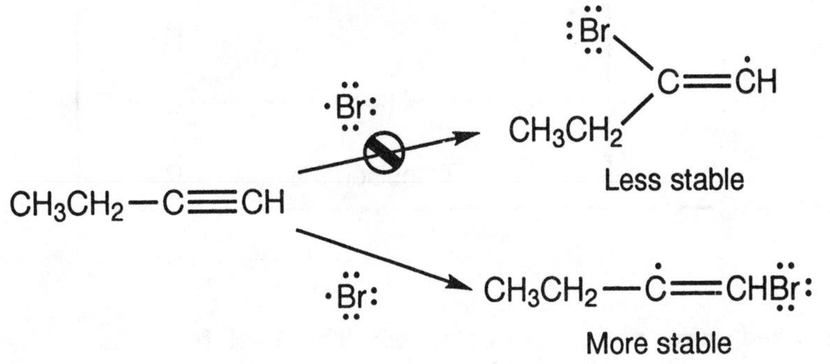

Less stable

More stable

Abstraction of hydrogen leads to (*E*)- and (*Z*)-1-bromo-1-butene and regenerates a bromine atom to carry the chain.

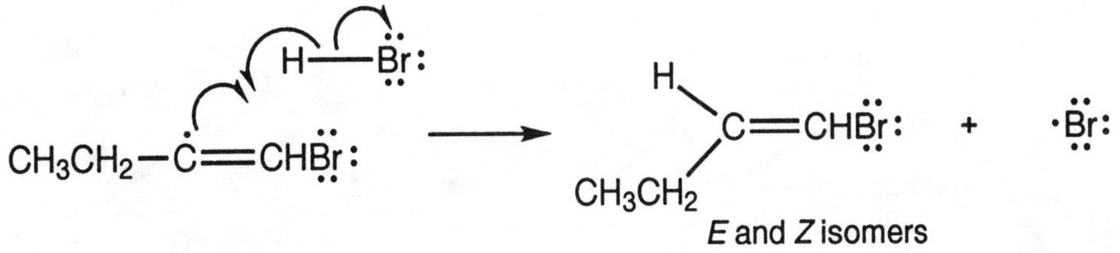

E and *Z* isomers

Now a second molecule of hydrogen bromide can add. A bromine atom first adds to give the resonance-stabilized radical.

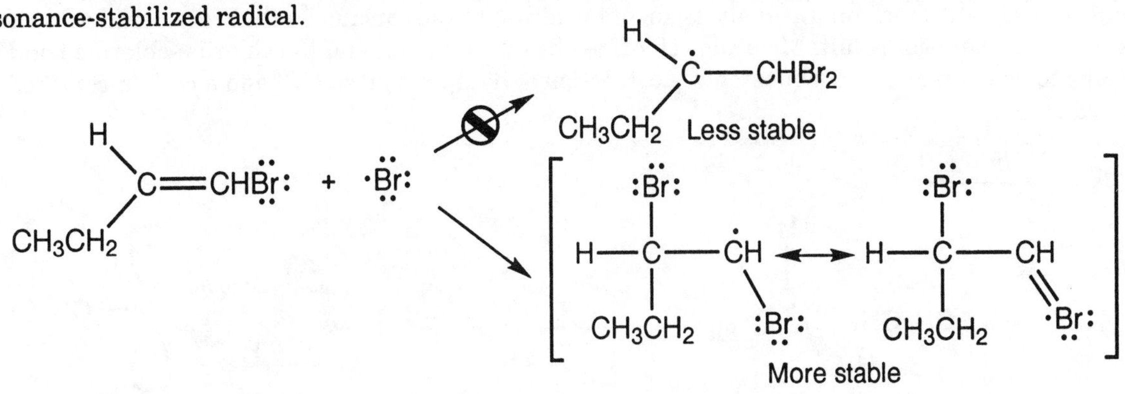

Less stable

More stable

Hydrogen abstraction to give the vicinal dibromide follows, and a new bromine atom is regenerated.

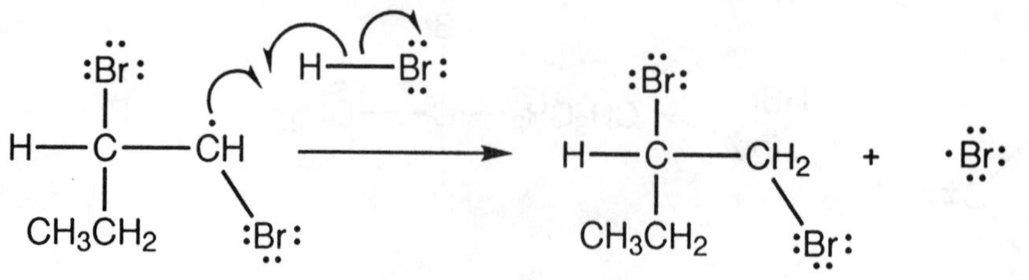

Problem 11.34 The less stable methyl radical recombines faster than the more substituted, more stable, and thus longer lived isopropyl radical. More important, apparently, are steric factors. The more bulky isopropyl radical has more difficulty in recombining than does the smaller methyl radical.

Problem 11.35 This problem is similar to Problem 11.34, although the steric argument is a little more subtle. First, one might argue that formation of the more substituted, more stable radicals should be faster than formation of the less substituted, less stable radicals. That would be right, as the transition state for radical formation will have partially formed radicals and will benefit energetically from substitution.

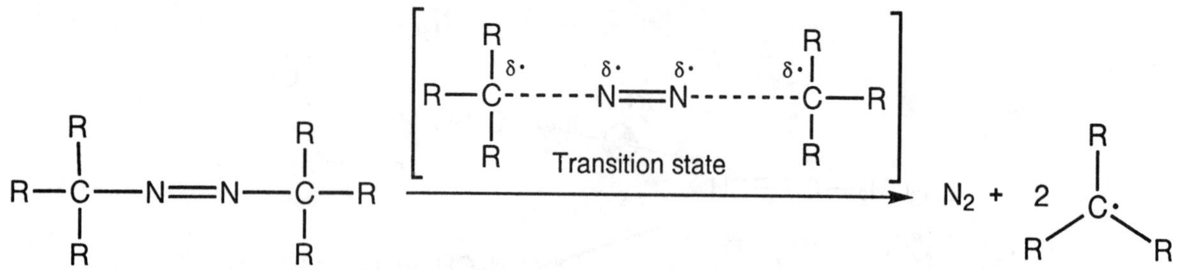

Transition state

But this ignores an even more subtle steric argument. There will be an energy incentive for the larger R groups to go from sp^3 to sp^2 hybridization as the azo compound is transformed into nitrogen and a pair of radicals. The R-C-R angle will change from about 109° to about 120°. The more relatively large methyl groups, the stronger this effect, and the faster the rate of radical formation.

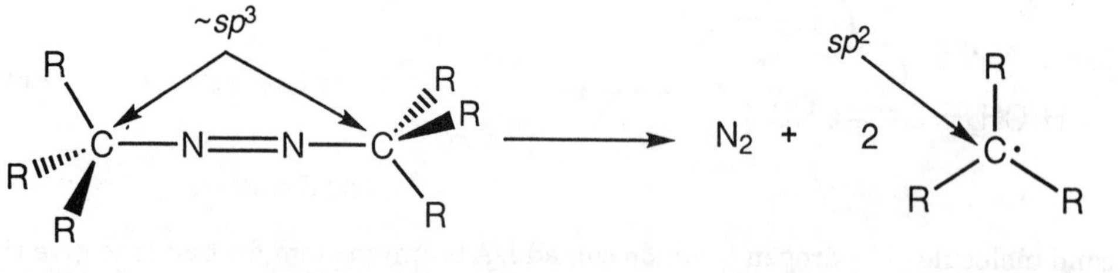

Problem 11.36 A mechanistic analysis should lead you to the conclusion that racemization is the likely stereochemical result. Molecular chlorine absorbs a photon and the chlorine-chlorine bond breaks to give a pair of chlorine atoms. Abstraction of hydrogen gives HCl and a carbon-centered radical.

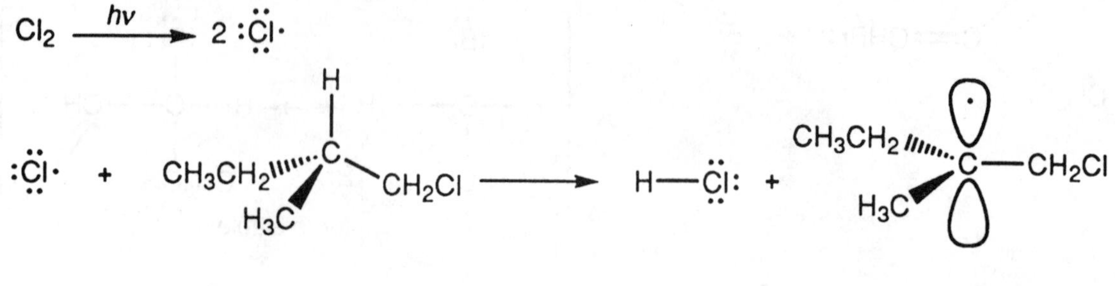

$$Cl_2 \xrightarrow{h\nu} 2\ :\ddot{Cl}\cdot$$

The key point is that the new radical is almost planar, hybridized approximately sp^2. Abstraction of a chlorine atom must take place equally from either side to give a pair of enantiomeric chlorides.

:Cl—Cl:

CH_3CH_2 C—CH_2Cl: ⟶ CH_3CH_2 C(Cl) CH_2Cl: + :Cl·

H_3C H_3C

Mirror

A racemic (50:50) mixture of enantiomers

CH_3CH_2 C—CH_2Cl: ⟶ CH_3CH_2 H_3C C(Cl) CH_2Cl: + :Cl·

H_3C

:Cl—Cl:

Problem 11.37 There are only two possibilities for hydrogen atom abstraction from ethyl chloride. A simple primary radical would be produced from abstraction of a methyl hydrogen, whereas a resonance-stabilized radical appears when a hydrogen atom is removed from the carbon already attached to one chlorine.

Abstraction here

$H_3C—CH_2Cl$: ⟶ $H_2\dot{C}—CH_2Cl$:

$H_3C—CH_2Cl$: ⟶ [$H_3C—\dot{C}H—Cl$: ⟷ $H_3C—CH{=}\dot{C}l$:]

Abstraction here

The more stable radical will be formed preferentially, and will abstract a chlorine atom to yield a geminal dichloroethane, 1,1-dichloroethane, not the vicinal dichloride, 1,2-dichloroethane.

:Cl—Cl:

$H_3C—\dot{C}H—Cl$: ⟶ $H_3C—C(Cl)_2—H$ + :Cl·

Problem 11.38 Decomposition of NBS generates a bromine atom. The bromine atom then abstracts a hydrogen atom from the methylene group adjacent to both the ring and the carbonyl group. This radical is very well resonance stabilized. The positions sharing the free electron are indicated by (•).

:Br·

H

C_6H_5—CH—C(=O)—OCH_2CH_3 ⟶ :Br—H + C_6H_5—$\dot{C}H$—C(=O)—OCH_2CH_3

1

(•) (•) (•) O (•)

The hydrogen bromide formed then reacts with NBS to form the low steady-state concentration of bromine necessary to carry the reaction on page 487. Bromine reacts with the newly formed radical to give the brominated product and regenerate a chain-carrying bromine atom.

:Br—Br:

CH—C(=O)OCH$_2$CH$_3$ ⟶ :Br: CH—C(=O)OCH$_2$CH$_3$ + :Br·

Problem 11.39 The key to this problem is the recognition that the strained three-membered ring will react with a chlorine atom much as will a double bond. An addition reaction gives the radical **B** that then abstracts a chlorine atom from Cl_2. This reaction makes a product molecule and generates a new chlorine atom to carry the chain.

:Cl—Cl: $\xrightarrow{h\nu}$ 2 :Cl·

·Cl: ⟶ :Cl—Cl: :Cl: **B** ⟶ :Cl :Cl· :Cl:

Problem 11.40 Chlorospiropentane is formed by a straightforward photochlorination chain reaction.

:Cl—Cl: $\xrightarrow{h\nu}$ 2 :Cl·

+ ·Cl: ⟶ H—Cl: + :Cl—Cl: ⟶ :Cl: + :Cl·

The strained three-membered ring can also be opened by a chlorine atom (see Problem 11.39). Abstraction of a chlorine atom from chlorine leads to one product and a new chlorine atom.

·Cl: ⟶ :Cl: :Cl—Cl: ⟶ :Cl: :Cl: + :Cl·

The final product is by far the hardest to rationalize. It comes from opening of the second three-membered ring (a β-cleavage), followed by chlorine abstraction.

:Cl: $\xrightarrow{\text{β-cleavage}}$:Cl: :Cl—Cl: ⟶ :Cl: :Cl: + :Cl·

Problem 11.41 The initial addition of an iodine atom to *cis*-2-butene is endothermic and reversible. Addition leads to a radical in which rotation about a carbon-carbon single bond is fast. Reversal of the addition leads to both *cis*- and *trans*-2-butene.

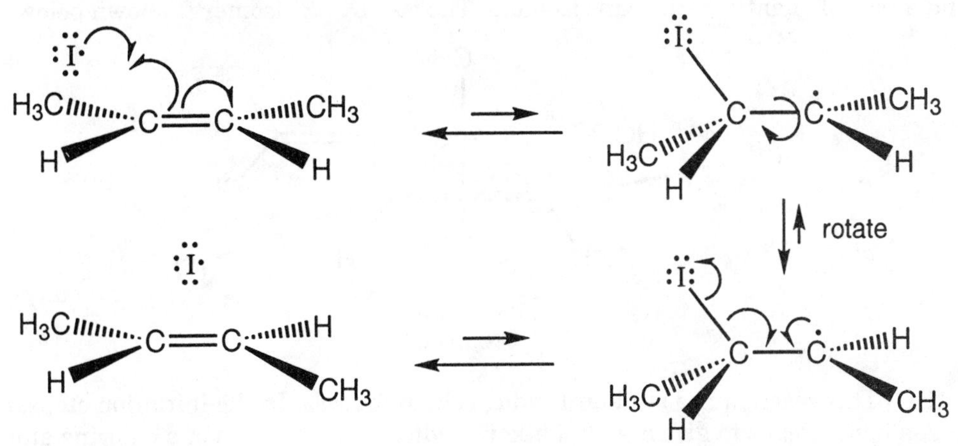

rotate

Problem 11.42
(a) Somehow, the radical $F_3CCl_2C^{\bullet}$ must be formed, as the only reasonable mechanism for formation of **2** involves addition of this radical to the alkene, followed by chlorine abstraction.

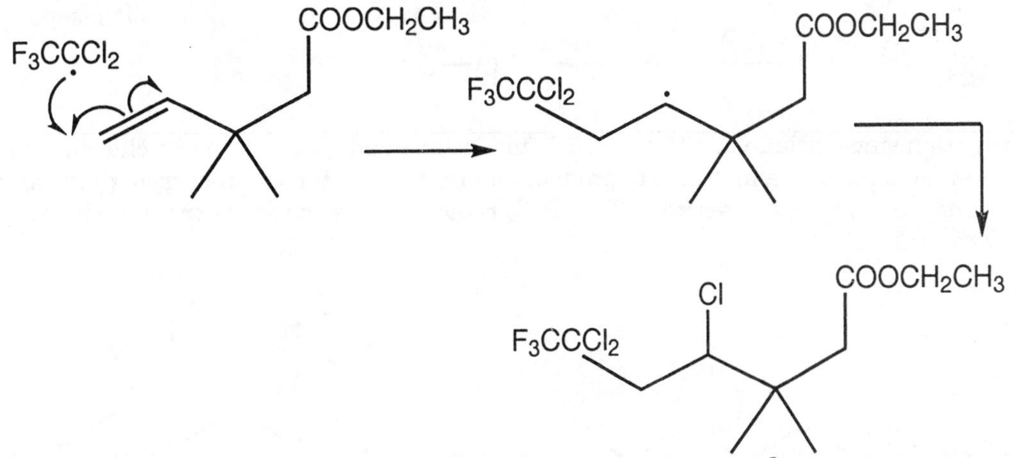

2

The problems now are to find a suitable source of $F_3CCl_2C^{\bullet}$ and the chlorine atom. An examination of Table 8.2 (p. 307) reveals that carbon-chlorine bonds are weaker than carbon-fluorine bonds. Accordingly, one mechanism that has been proposed is a Cu-catalyzed redox-transfer chain mechanism. This initiation step involves abstraction of a chlorine atom from the polyhalide by CuCl.

$$\text{Cu(I)Cl} + F_3CCCl_3 \longrightarrow \text{Cu(II)}Cl_2 + F_3C\dot{C}Cl_2$$

Cu(II)Cl_2 is presumably a more reactive chlorine donor than the polyhalide. Abstraction of chlorine from Cu(II)Cl_2 gives **2** and Cu(I)Cl. The Cu(I)Cl is then recycled to begin the process anew.

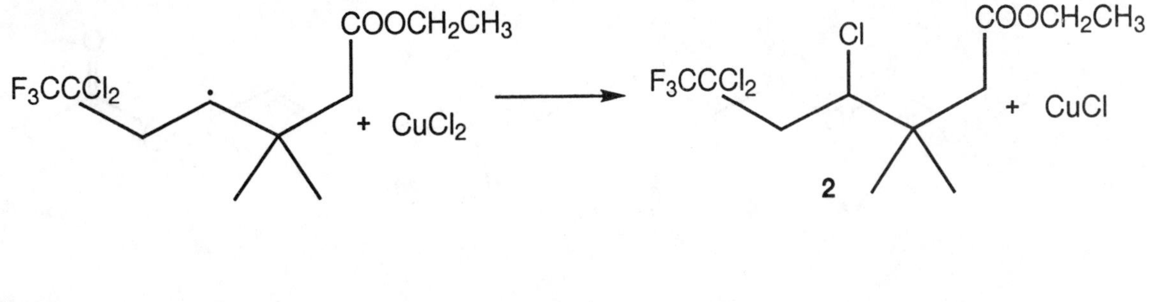

+ $CuCl_2$ → **2** + CuCl

(b) Three steps need to occur in the conversion of **2** into **1**: (1) cyclopropane formation, (2) loss of hydrogen chloride and (3) hydrolysis of the ester to the carboxylic acid.
(c) There are four possible stereoisomers of carboxylic acid **1**. There is a pair of enantiomeric cis isomers and a pair of enantiomeric trans isomers. The cis (1*R*,3*S*) isomer is shown below:

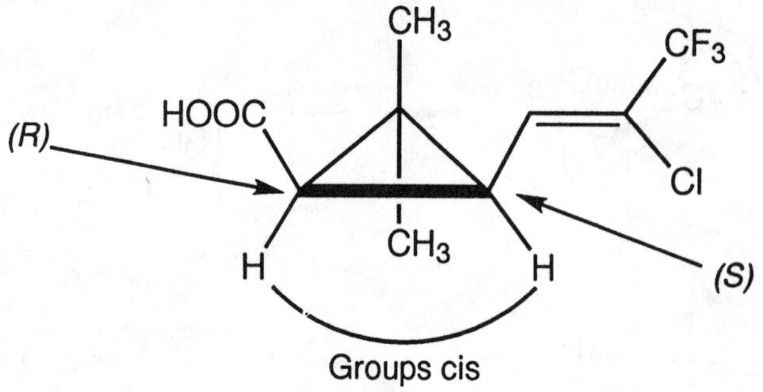

CH3 CF3 HOOC (R) Cl CH3 H H (S)

Groups cis

Problem 11.43 This reaction is a standard radical chain process. In the initiation steps, the weak oxygen-oxygen bond breaks to give a pair of alkoxy radicals. Abstraction of a chlorine atom from PCl_3 gives a dichlorophosphorous radical, $\cdot PCl_2$.

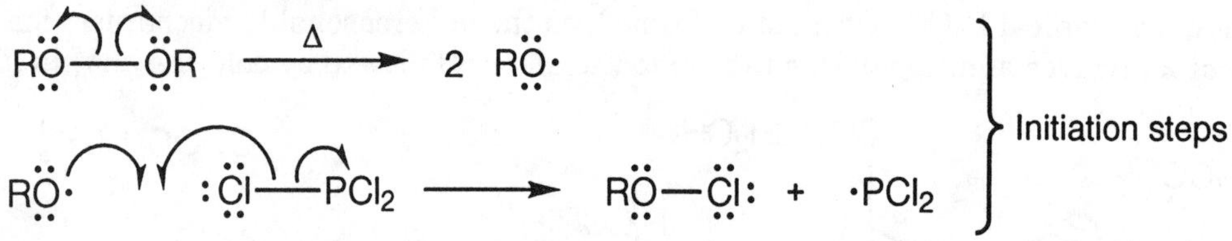

$$RO—OR \xrightarrow{\Delta} 2\ RO\cdot$$

$$RO\cdot + :Cl—PCl_2 \longrightarrow RO—Cl: + \cdot PCl_2$$

Initiation steps

The propagation steps include addition of $\cdot PCl_2$ to 1-octene and abstraction of a chlorine atom from PCl_3. These reactions form a molecule of product and regenerate the chain-carrying radical, $\cdot PCl_2$. Note the regiochemistry of the reaction. The $\cdot PCl_2$ radical adds so as to generate the more stable secondary radical.

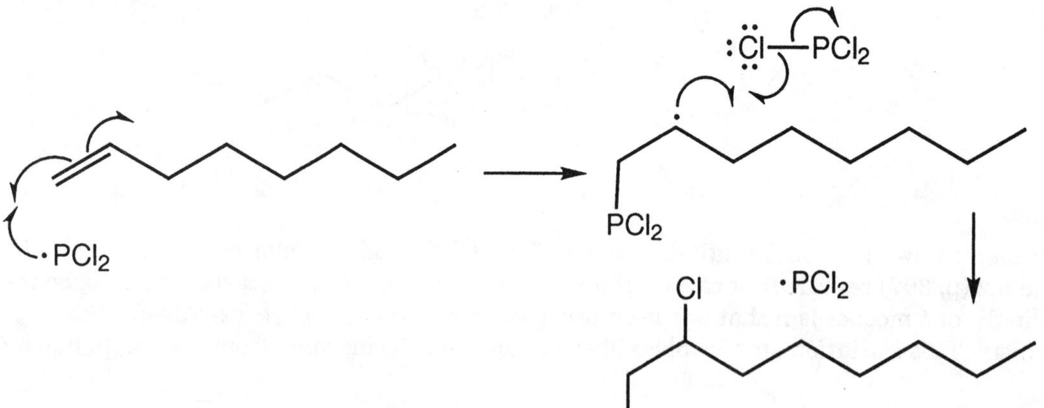

:Cl—PCl2

•PCl2 PCl2

Cl •PCl2

PCl2

Problem 11.44 Decomposition of the peroxide occurs by breaking the weak oxygen-oxygen bond to give a pair of carboxy radicals:

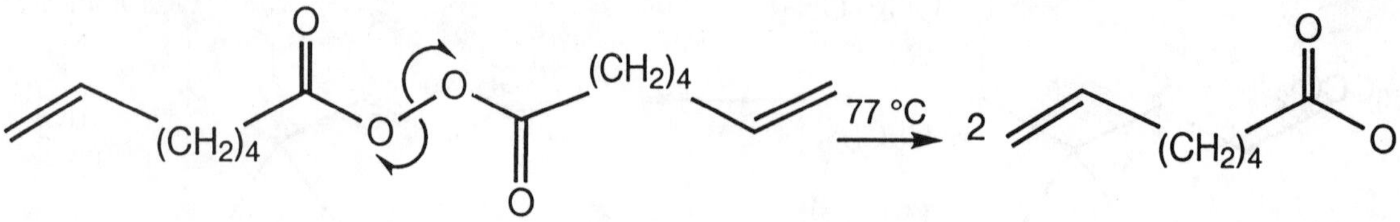

O (CH2)4 O O (CH2)4 O 77 °C 2 O (CH2)4 O•

One product comes from simple abstraction of hydrogen from toluene ($PhCH_3$) by this radical.

$$CH_2{=}CH(CH_2)_4C(=O)O\cdot \xrightarrow{Ph\text{-}CH_3} CH_2{=}CH(CH_2)_4C(=O)OH + Ph\dot{C}H_2$$

The carboxy radical can also lose carbon dioxide to give radical **A**, the source of most of the other products.

$$CH_2{=}CHCH_2CH_2CH_2CH_2C(=O)O\cdot \longrightarrow CO_2 + \underset{\mathbf{A}}{CH_2{=}CHCH_2CH_2CH_2\dot{C}H_2}$$

Disproportionation of **A** produces 1-hexene and 1,5-hexadiene, and dimerization of **A** gives the C_{12} product.

$$\mathbf{A} + \mathbf{A} \xrightarrow{\text{disproportionation}} \text{1,5-hexadiene} + \text{1-hexene}$$

$$\mathbf{A} + \mathbf{A} \xrightarrow{\text{dimerize}} CH_2{=}CH(CH_2)_8CH{=}CH_2$$

The radical **A** can also undergo an intramolecular cyclization. In fact, this can occur in two ways to give either a five- or six-membered ring. Abstraction of hydrogen yields the cycloalkanes.

$$\mathbf{A} \longrightarrow \text{cyclohexyl radical} \xrightarrow[\text{abstraction}]{Ph\text{-}CH_3} \text{cyclohexane} + Ph\dot{C}H_2$$

$$\mathbf{A} \longrightarrow \text{cyclopentylmethyl radical} \xrightarrow[\text{abstraction}]{Ph\text{-}CH_3} \text{methylcyclopentane} + Ph\dot{C}H_2$$

In several of these reactions the benzyl group, $PhCH_2$, appears. If it dimerizes, the final product, $PhCH_2CH_2Ph$, is formed.

$$2\ Ph\dot{C}H_2 \longrightarrow PhCH_2CH_2Ph$$

Problem 11.45 The acyclic product comes from a straightforward radical chain addition reaction. Note that in the last step the chain-carrying trichloromethyl radical is regenerated.

$$RO-OR \xrightarrow{\Delta} 2\ RO\cdot$$

$$RO\cdot + CCl_4 \longrightarrow ROCl\ \ \cdot CCl_3$$

Initiation steps

$$CH_3(CH_2)_4-C\equiv CH\ \ \cdot CCl_3 \longrightarrow CH_3(CH_2)_4-\dot{C}=CH-CCl_3$$

$$Cl-CCl_3 + CH_3(CH_2)_4-\dot{C}=CH-CCl_3 \longrightarrow CH_3(CH_2)_4(Cl)C=CHCCl_3 + \cdot CCl_3$$

The cyclic product is more difficult. What do we have to do? The two major things that must be accomplished in this reaction are formation of a five-membered ring and loss of a chlorine atom. Formation of the cyclic product starts with an intramolecular hydrogen abstraction to give **A**. Radical **A** then undergoes an intramolecular addition to produce **B**. This reaction completes one major goal: the formation of the five-membered ring.

$$CH_3(CH_2)_4-\dot{C}=CH-CCl_3 = \text{H}_3\text{C–HC–H} \ldots \longrightarrow \text{A}$$

A

$$\text{A} \longrightarrow \text{B}$$

A B

Loss of a chlorine atom finishes the mechanism.

$$\text{B} \longrightarrow \text{(2-methylcyclopentyl)–CH=CCl}_2 + \cdot Cl$$

B

Problem 11.46 This is another free radical chain reaction. An alkoxy radical formed in the first initiation step abstracts a hydrogen atom from isobutene to generate a resonance-stabililized methallyl radical.

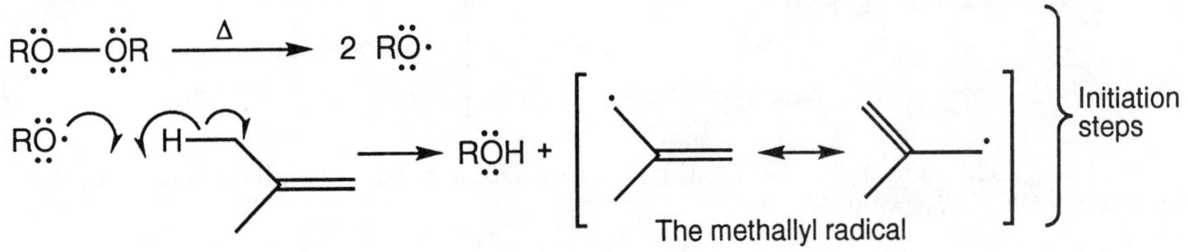

The methallyl radical

Initiation steps

The methallyl radical adds to trichloroethylene to give either radical **A** or **B**. Formation of **A** will be preferred both because it is more stable than **B** (resonance stabilization by two chlorines instead of one) and on the grounds that approach of the methallyl radical to the end of trichloroethylene bearing only one large chlorine will be preferred on steric grounds.

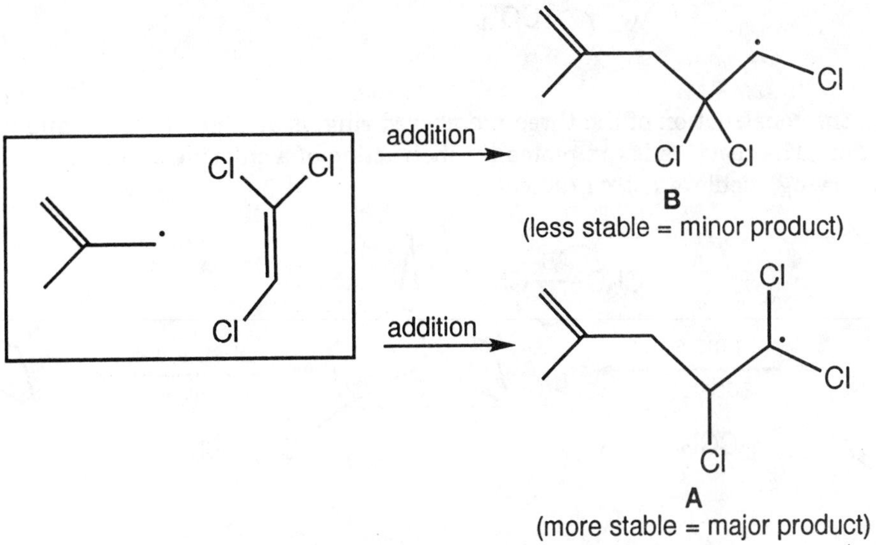

addition

B
(less stable = minor product)

addition

A
(more stable = major product)

Loss of a chlorine atom from **A**, the predominant radical intermediate, leads to **1**, the major product of the reaction. The minor products come from chlorine loss from the less favored radical intermediate **B**.

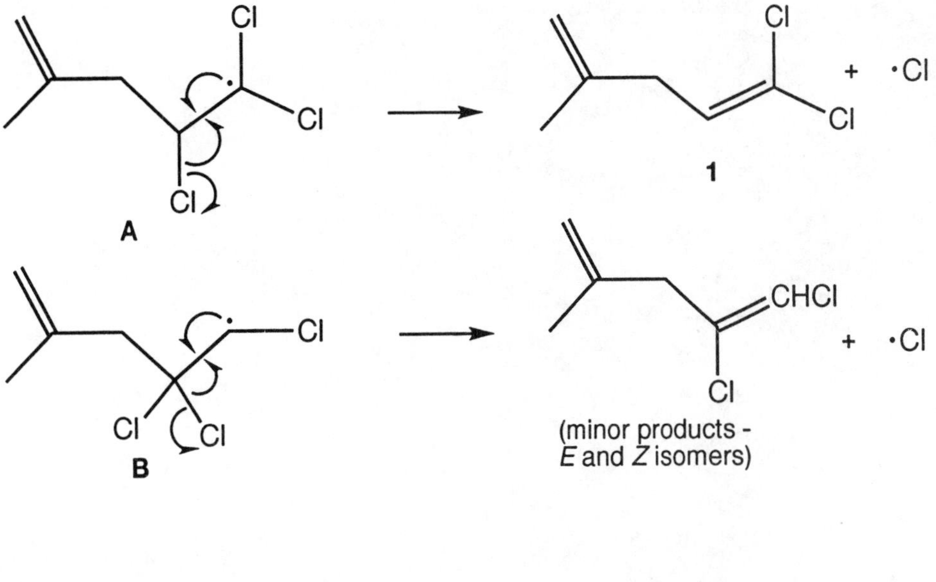

A

1

B

(minor products - *E* and *Z* isomers)

Problem 11.47 The initiation steps are easy.

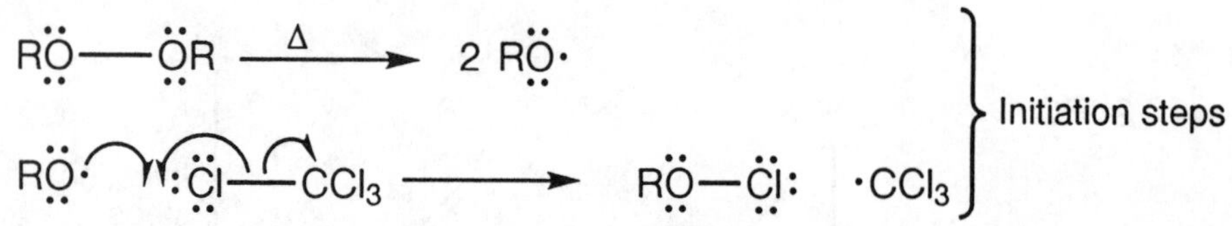

$$R\ddot{O}-\ddot{O}R \xrightarrow{\Delta} 2\ R\ddot{O}\cdot$$

$$R\ddot{O}\cdot \ :\ddot{Cl}-CCl_3 \longrightarrow R\ddot{O}-\ddot{Cl}: \quad \cdot CCl_3$$

Initiation steps

Addition of the trichloromethyl radical to the exocyclic double bond accomplishes one goal, the formation of a CH_2CCl_3 group.

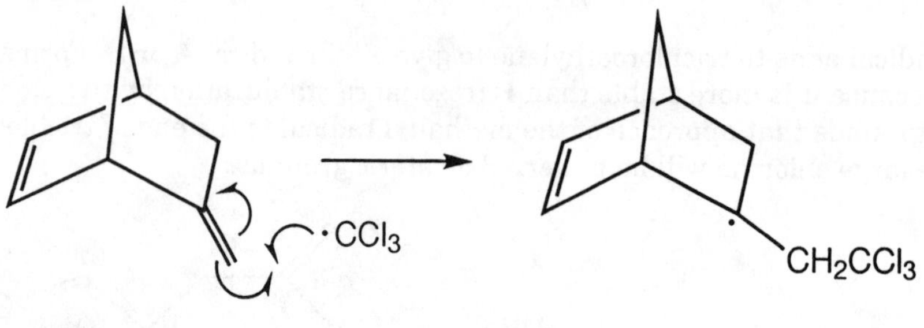

$\cdot CCl_3$

CH_2CCl_3

A second goal, the construction of the three-membered ring, is attained through an intramolecular addition reaction. The reaction is completed by abstraction of a chlorine atom from CCl_4 to give **1** and a chain-carrying trichloromethyl radical.

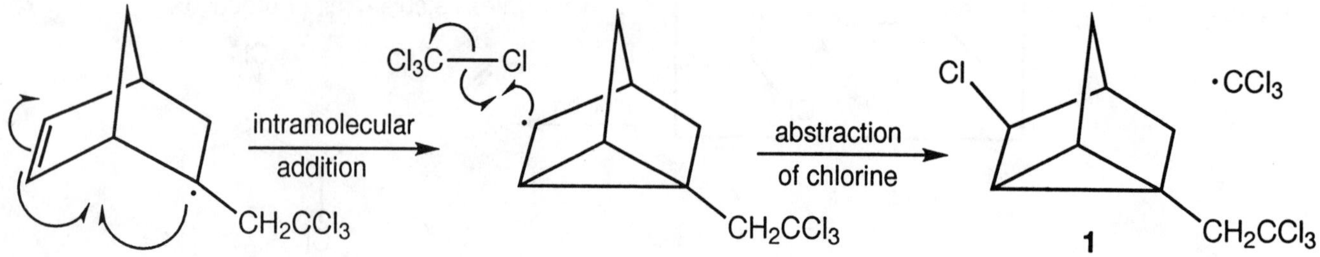

CH_2CCl_3 —intramolecular addition→ (Cl_3C-Cl) CH_2CCl_3 —abstraction of chlorine→ Cl, CH_2CCl_3 **1** $\cdot CCl_3$

Chapter 12 Outline

12.1 Allenes: 1,2-Dienes
12.2 Related Systems: Ketenes and Cumulenes
12.3 Allenes as Intermediates in the Isomerization of Acetylenes
12.4 1,3-Dienes
12.5 The Physical Consequences of Conjugation
12.6 Molecular Orbitals and Ultraviolet Spectroscopy
12.7 Polyenes and Vision
12.8 The Chemical Consequences of Conjugation: Addition Reactions of Dienes
12.9 Thermodynamic versus Kinetic Control of Addition Reactions
12.10 The Allyl System: Three Overlapping 2*p* Orbitals
12.11 The Allyl Cation: S_N1 Solvolysis of Allylic Halides
12.12 S_N2 Reactions of Allylic Halides
12.13 The Allyl Radical
12.14 The Allyl Anion
12.15 The Diels–Alder Reaction of Dienes
12.16 Isoprene: Biosynthesis of Terpenes
12.17 Steroid Biosynthesis
12.18 Something More: The S_N2' Reaction

12

Dienes and the Allyl System; 2*p* Orbitals in Conjugation

This chapter opens a three-chapter sequence in which you will explore the consequences of overlap of more than two 2*p* orbitals, "conjugation". There are both structural and chemical consequences, and the following problems deal with both areas. Two exceptionally important reactions emerge: the formation of allyl cations through protonation of 1,3-dienes and the Diels-Alder reaction. The former opens up synthetic possibilities and leads to a discussion of thermodynamic and kinetic control of reactions. The latter is arguably the most important synthetic reaction in the chemist's arsenal, and leads in a single step to all manner of complex compounds containing six-membered rings. The following problems give lots of practice in both areas.

Problem 12.3 All of these molecules are related to allenes.

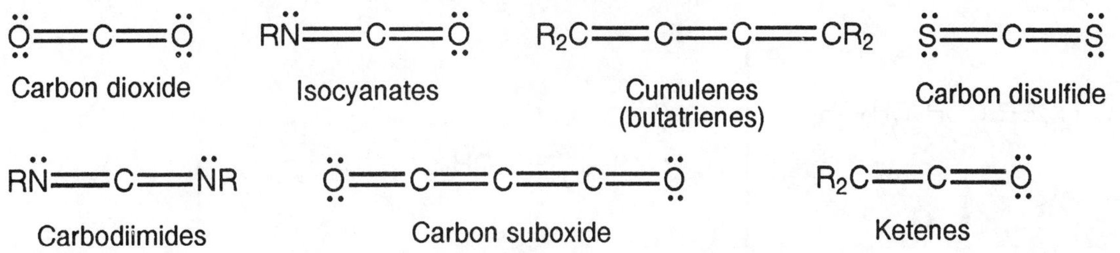

$O{=}C{=}O$ Carbon dioxide

$RN{=}C{=}O$ Isocyanates

$R_2C{=}C{=}C{=}CR_2$ Cumulenes (butatrienes)

$S{=}C{=}S$ Carbon disulfide

$RN{=}C{=}NR$ Carbodiimides

$O{=}C{=}C{=}C{=}O$ Carbon suboxide

$R_2C{=}C{=}O$ Ketenes

Problem 12.4 Just as in ethylene (but not in allene), the end groups of this molecule are coplanar. Therefore, there can be cis and trans stereoisomers.

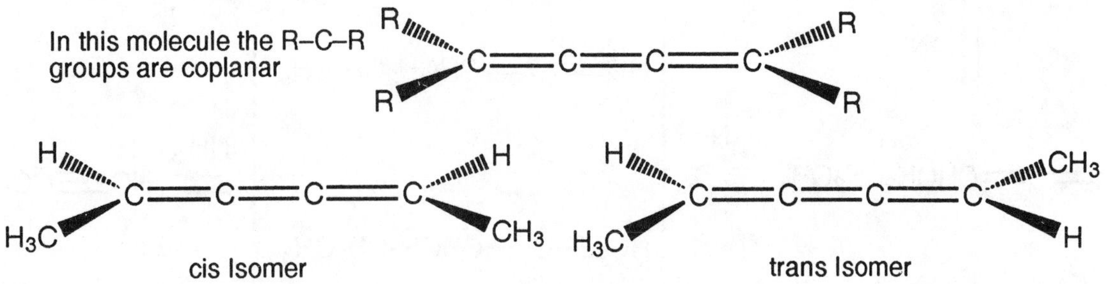

In this molecule the R–C–R groups are coplanar

$R_2C{=}C{=}C{=}CR_2$

$(H)(H_3C)C{=}C{=}C{=}C(H)(CH_3)$ cis Isomer

$(H)(H_3C)C{=}C{=}C{=}C(CH_3)(H)$ trans Isomer

Problem 12.5 These flat molecules are all achiral, just like ethylene. Their mirror images are always superimposable on the originals.

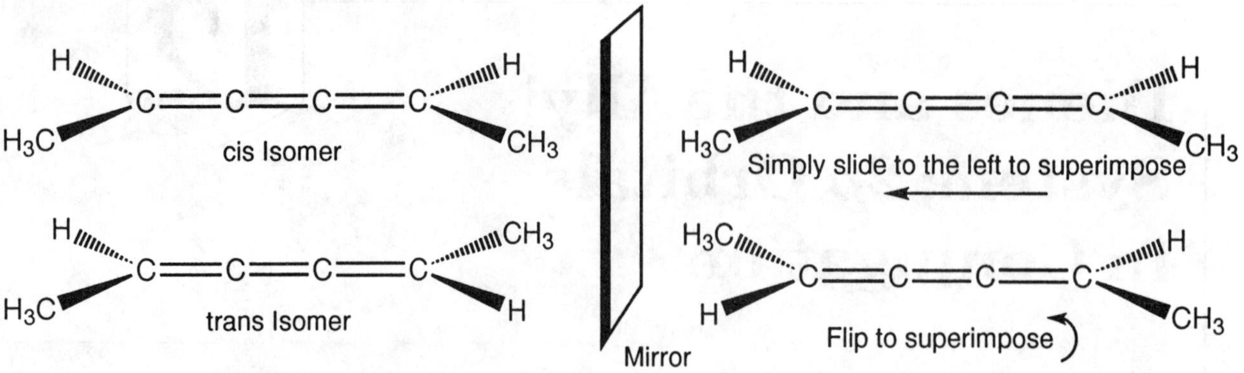

cis Isomer

Simply slide to the left to superimpose

trans Isomer

Mirror

Flip to superimpose

Problem 12.6 This is just a two-stage version of the mechanism outlined in Figures 12.8-12.14. Strong base removes a proton from the carbon adjacent to the triple bond to give resonance-stabilized anion **A**. Reprotonation can occur in two places, one of which gives an intermediate allene. The allene is also acidic, and strong base can remove a proton to produce a second resonance-stabilized anion, **B**. Reprotonation, followed by a third removal of a proton gives **C**, which protonates to give a second allene. One more cycle of deprotonation to anion **D**, followed by reprotonation, generates 1-hexyne.

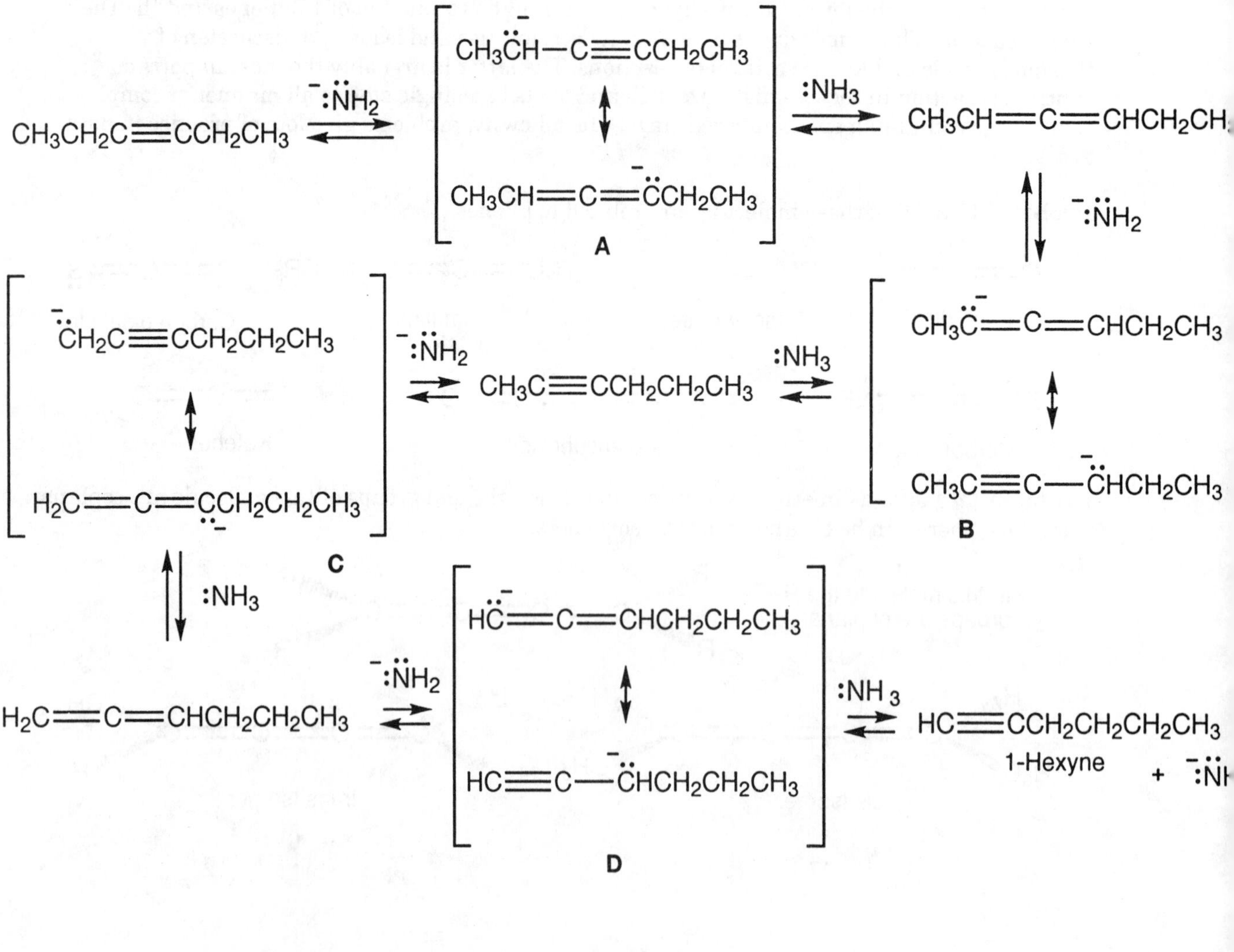

A

B

C

D

1-Hexyne

Now the terminal acetylene can lose its acetylenic hydrogen to generate an acetylide, and the reaction stops at this critical point. The 1-alkyne is a *much* stronger acid than any of the other species in equilibrium. Removal of the acetylenic proton is greatly favored thermodynamically. When the solution is quenched through addition of water, the acetylide protonates to produce 1-hexyne under conditions (neutral) to which it is stable.

$$HC{\equiv}CCH_2CH_2CH_2CH_3 \underset{:NH_3}{\overset{^{-}:\ddot{N}H_2}{\rightleftharpoons}} {}^{-}:C{\equiv}CCH_2CH_2CH_2CH_3$$

$$^{-}:C{\equiv}CCH_2CH_2CH_2CH_3 \overset{H_2O}{\rightleftharpoons} HC{\equiv}CCH_2CH_2CH_2CH_3$$

Problem 12.8 Each carbon in 1,3-butadiene is attached to three other atoms, and will be hybridized sp^2. The carbon-carbon bonds are made from sp^2-sp^2 overlap and the carbon-hydrogen bonds from sp^2-$1s$ overlap.

sp^2-sp^2 sp^2-sp^2 C–H bonds sp^2-$1s$

$H_2C{=}CH{-}CH{=}CH_2$

sp^2-$1s$ sp^2-sp^2

Problem 12.9

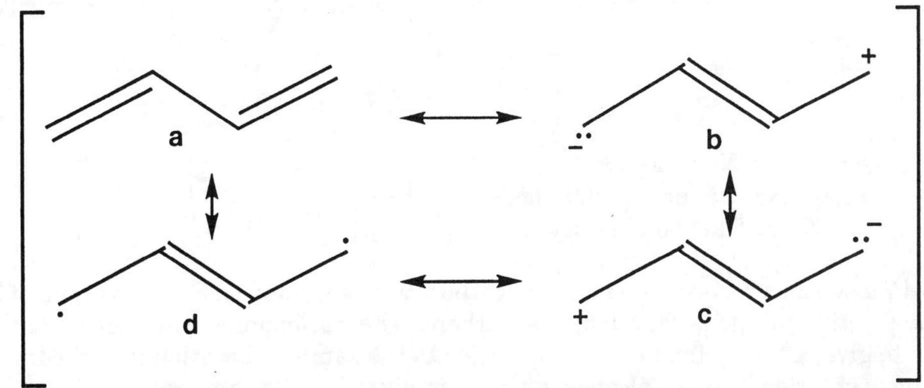

Forms **b**, **c**, and **d** all have one fewer bond than form **a**. In addition, forms **b** and **c** have separated charges. Form **a** represents by far the lowest energy electronic distribution.

Problem 12.10 It is only necessary to interact the orbitals closest in energy; $\pi \pm \pi$ and $\pi^* \pm \pi^*$ in this case. The same four molecular orbitals result as in the procedure outlined in Figure 12.17, but they are not formed in the same order. The figure below shows the orbitals along with the nodes used to order them in energy.

= nodes

π^* ± π^* → $\pi^* + \pi^*$ Φ_2 $\pi^* - \pi^*$ Φ_4

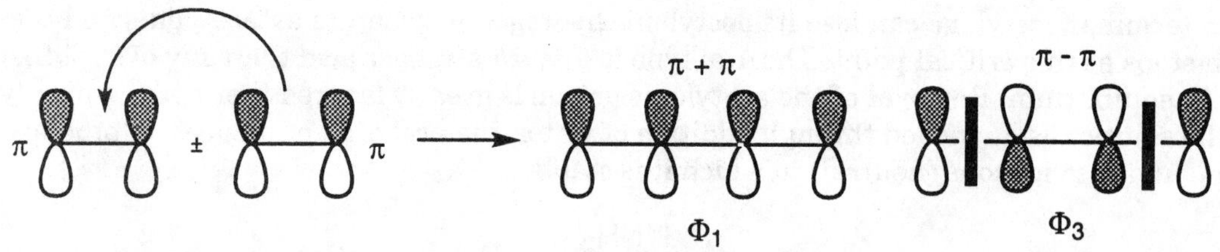

π + π

π - π

π ± π

Φ_1 Φ_3

Problem 12.12 The equation to use is: $\Delta G = -2.3RT\log K$
So, $-2.5 = -1.364 \log K$ [at 25 °C, $2.3RT = 1.364$ kcal/mol (Chapter 8, p. 305)]
$\log K = 1.83$
$K = 68$, and there is only 1.4% of the s-cis compound present at equilibrium.

Problem 12.14 Any of a vast variety of labeling experiments will do. Here is just one, where we use deuterium and detect the two different molecules shown.

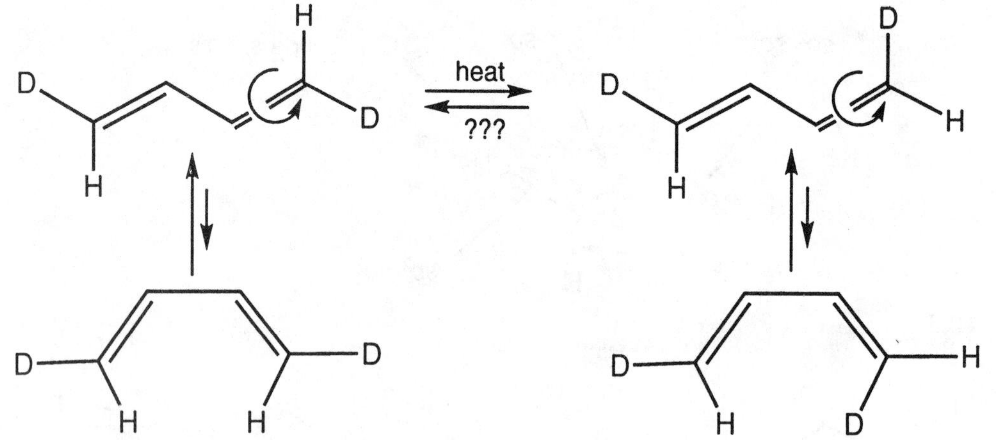

heat

???

Problem 12.15
$E = \text{Nhc}/\lambda$ and the quantity $\text{Nhc} = 28.6 \times 10^3$ (p. 516)
so, $E = (28.6 \times 10^3)/200 = 143$ kcal/mol at 200 nm
and $E = (28.6 \times 10^3)/800 = 35.8$ kcal/mol at 800 nm

Problem 12.17 The same resonance-stabilized carbocation is formed on S_N1 solvolysis of both chlorides. The positive charge is shared by two carbons. The nucleophile, water, can add to each of these carbons to give, after proton loss, two alcohols. As the same carbocation intermediate is formed from each chloride, the two alcohols must be produced in the same ratio.

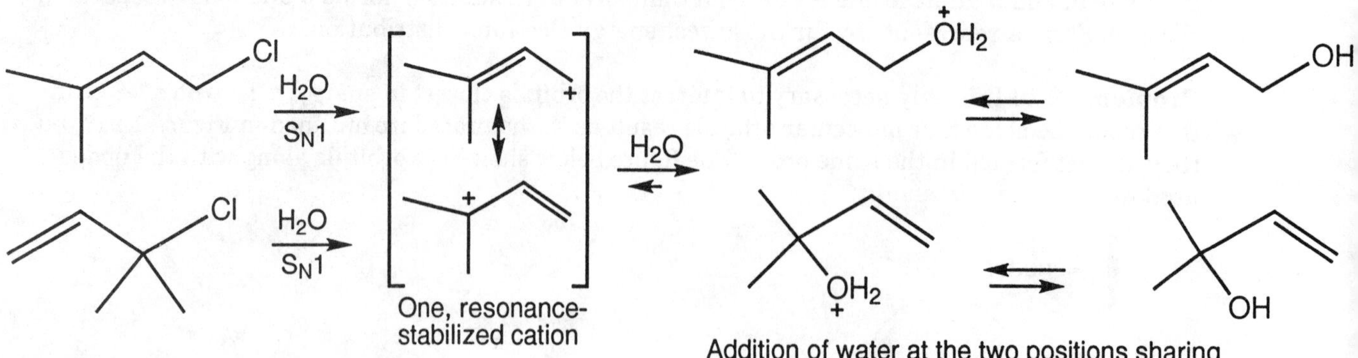

H_2O / S_N1

H_2O

One, resonance-stabilized cation

Addition of water at the two positions sharing the positive charge leads to the two oxonium ions, and, after deprotonation, the two alcohols

Problem 12.19 Write out the two ions formed by protonation at C(1) and C(4) and compare their stabilities. Protonation at C(1) (here using D^+ for clarity) gives a resonance-stabilized allyl cation in which the positive charge is shared by two secondary carbons.

1 D–Cl D + D + Secondary C^+ Secondary C^+

Protonation (D^+) at C(4) also gives a resonance-stabilized allyl cation, but the charge is shared between a secondary and *primary* carbon, and this ion will be less stable.

4 D–Cl D + D + Secondary C^+ Primary C^+

Problem 12.20
The resonance picture shows the individual electronic descriptions contributing to the real, delocalized structures.

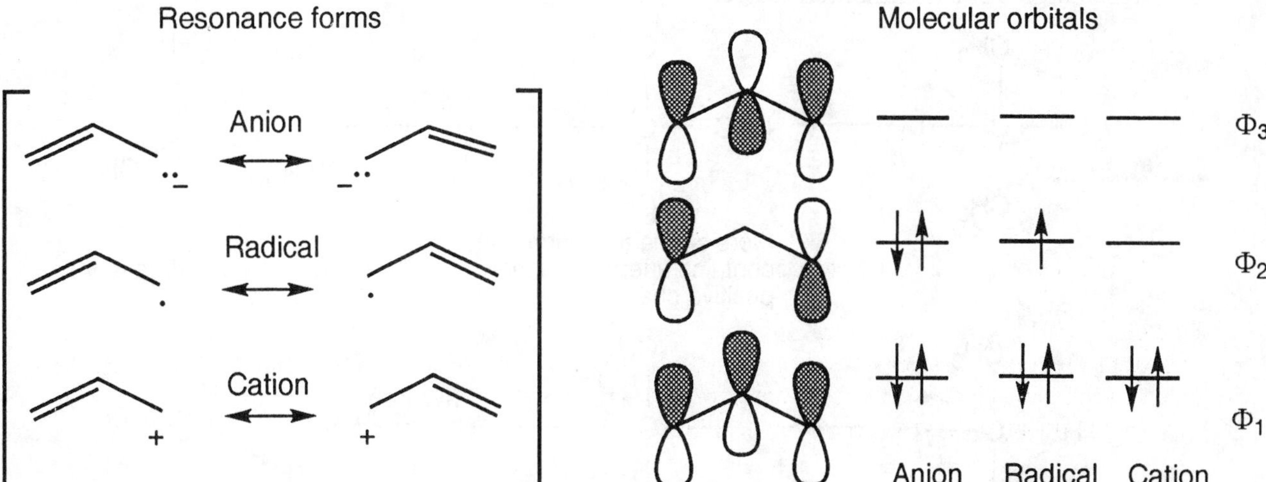

Problem 12.22 The problem is to explain why there should be a connection between the stability of the products (thermodynamics) and the energies of the transition states leading to the products (kinetics). Why should thermodynamics and kinetics be related? Ionization is surely an endothermic process, and, as noted by the Hammond postulate, the more endothermic the reaction, the more the transition state will look like the product. Accordingly, the thermodynamic stability of the allyl cation formed is related to the ease of formation, a kinetic property.

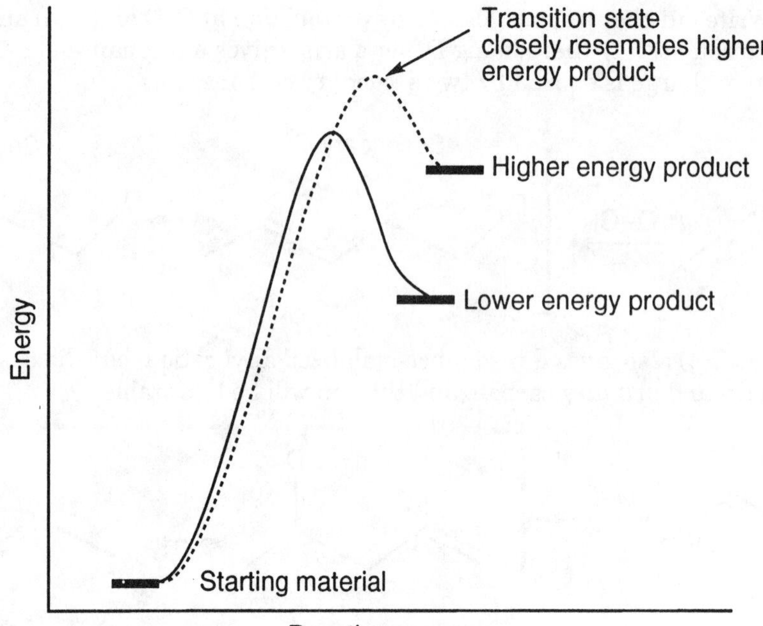

There is another way to put it. The transition states for ionizations to give carbocations will contain partially developed positive charges. A partially developed tertiary positive charge will be more stable than a partially developed secondary positive charge, and this will be reflected in the relative energies of the transition states.

More stable transition state contains a tertiary partial positive charge

Product (a tertiary carbocation)

Less stable transition state contains a secondary partial positive charge

Product (a secondary carbocation)

Problem 12.23 There are two HOMO-LUMO interactions we might consider, HOMO (dienophile), LUMO (diene) or HOMO (diene), LUMO (dienophile). Either will yield the same result, so let's examine the first possibility. Look at the interactions between HOMO (dienophile) and LUMO (diene) where the two new σ bonds are being formed.

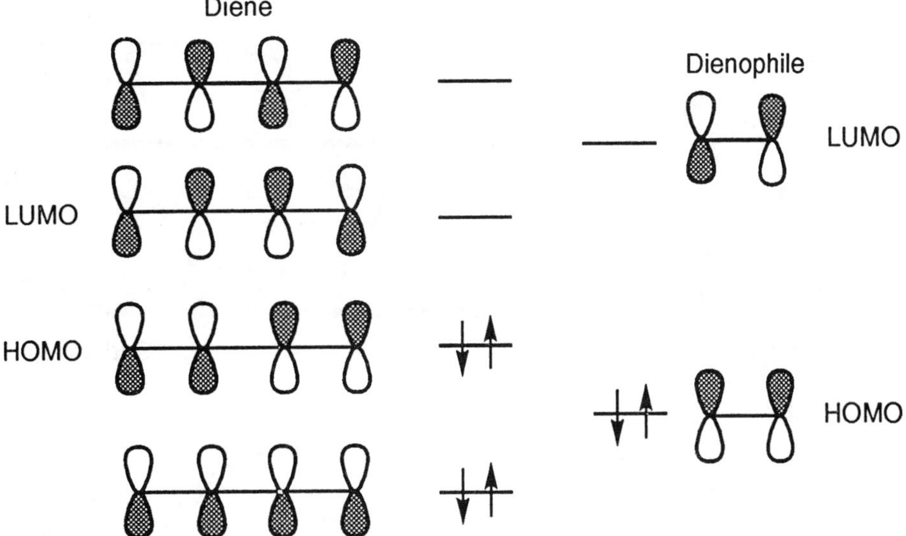

Two new bonding interactions are created in the transition state for the reaction.

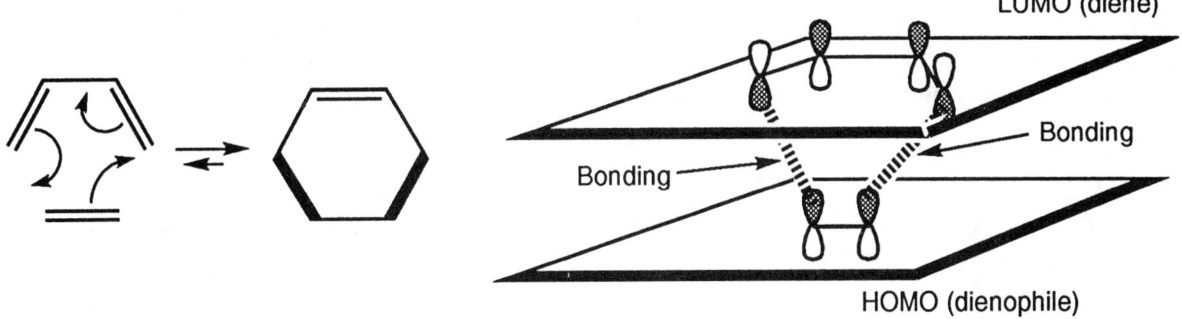

However, this will not be the case if an electron is promoted from the HOMO to the LUMO, which is exactly what occurs on absorption of a photon of light. Here are the molecular orbitals after an electron has been promoted from the HOMO of the diene to the LUMO of the diene.

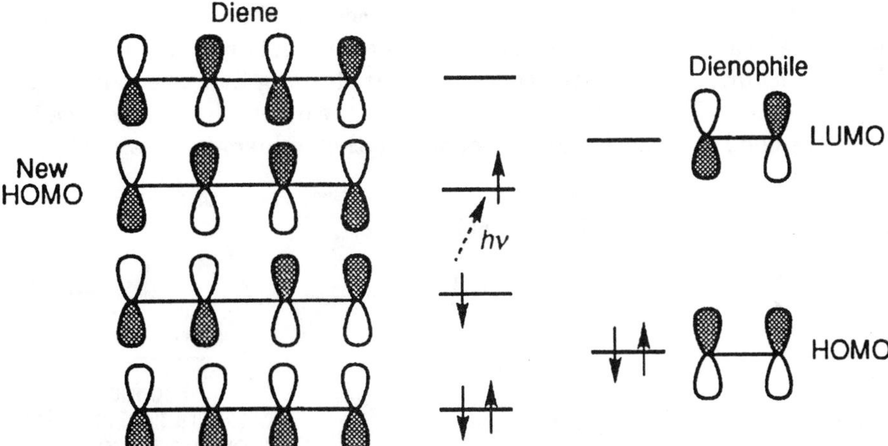

Now look at the lower energy HOMO-LUMO interaction [HOMO (diene) LUMO (dienophile)]. It is not possible to form two bonding interactions in the transition state! The success of the Diels-Alder reaction depends on how the energy is transferred to the molecules. Heat works, but light, which changes the HOMO-LUMO relationships, does not.

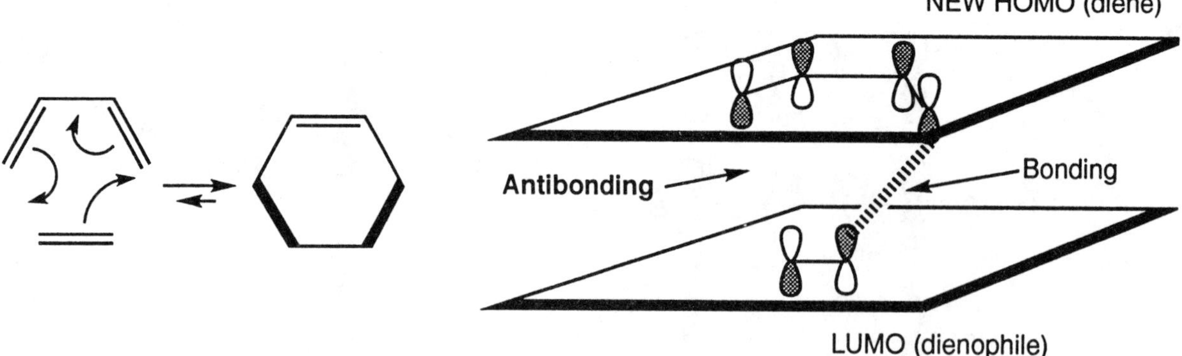

Problem 12.24 Nothing to it.

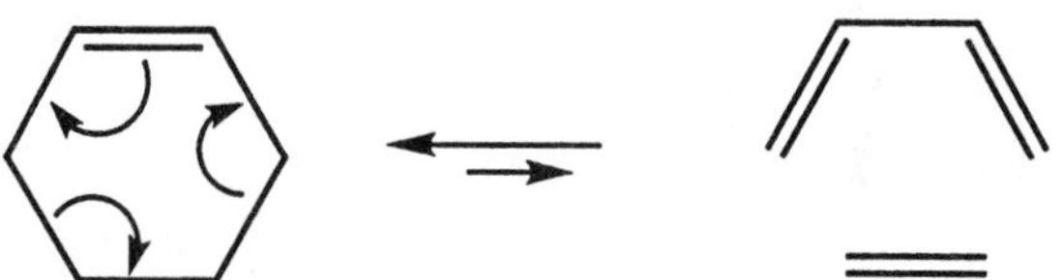

Problem 12.25 In the transition state for bond breaking, the bond is stretching and radical centers are developing on the bonded carbons (δ•). This process is illustrated for carbon-carbon bond breaking in cyclohexane, a reasonable model for ethane.

H_2C-⌇-CH_2 → δ• δ• H_2C------CH_2 → $H_2\dot{C}$—$\dot{C}H_2$

Cyclohexane

Transition state [radicals are developing on two carbons (δ•)]

Product (a diradical)

In cyclohexene, one of the developing radicals is a resonance-stabilized allyl radical and is more stable than the simple localized radicals formed in carbon-carbon bond breaking in cyclohexane. Accordingly, the product diradical, and the transition state leading to it, are more stable than those in the cyclohexane reaction. The energy required to break the σ bond in cyclohexene is lower, and this is reflected in the bond dissociation energy quoted in the problem (85 kcal/mol).

H_2C-⌇-CH_2 → δ• δ• H_2C------CH_2 (δ•) → CH_2 $\dot{C}H_2$

Cyclohexene

Transition state [radicals are developing on two carbons (δ•); one end is delocalized]

Product (a diradical in which one end is resonance stabilized)

Problem 12.27 Vinylcyclobutane lies higher in energy than cyclohexene, largely because of the strain energy in the molecule. In the transition state for formation of vinylcyclobutane (see dashed lines), some of this strain will be present. Accordingly, it, too, lies relatively high in energy.

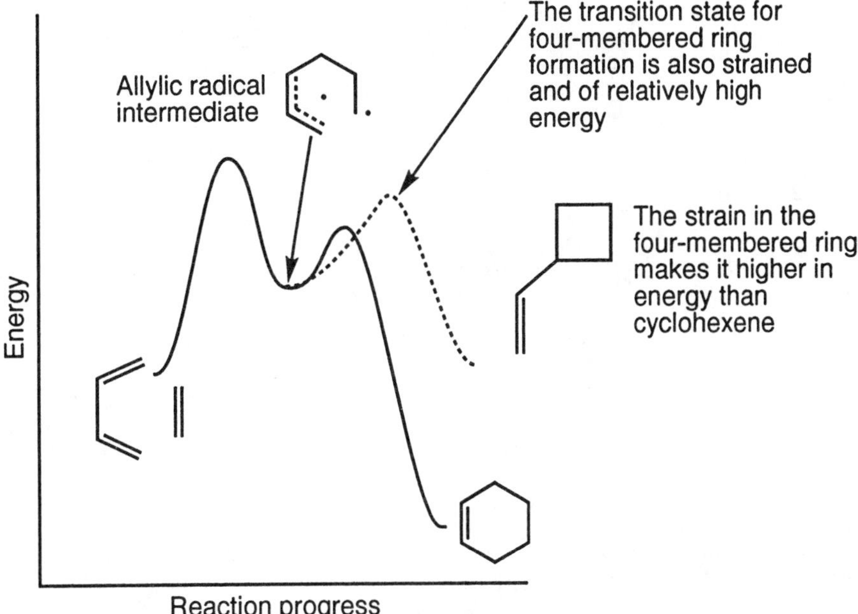
The transition state for four-membered ring formation is also strained and of relatively high energy
Allylic radical intermediate
The strain in the four-membered ring makes it higher in energy than cyclohexene
Energy
Reaction progress

Problem 12.28 A one-step addition to the trans,trans diene must give a single product, the one shown in Figure 12.69 in which the R groups are cis to each other.

trans,trans Diene

Dienophile

Cyclohexene

cis Cycloalkene

This product is exactly what is seen experimentally, so the one-step, concerted mechanism is in accord with experiment. Let's now see if the two-step process works out. If one bond is made before the other, there must be an intermediate diradical in which the second bond can be formed from the same or opposite side as the first. Two different stereoisomers will result. That there should be two isomers depending on the direction of the second bond formation is relatively easy to understand; what's hard to see is just how the groups on the diene appear in the cyclohexene product. Follow the diagram, but do look at models as well.

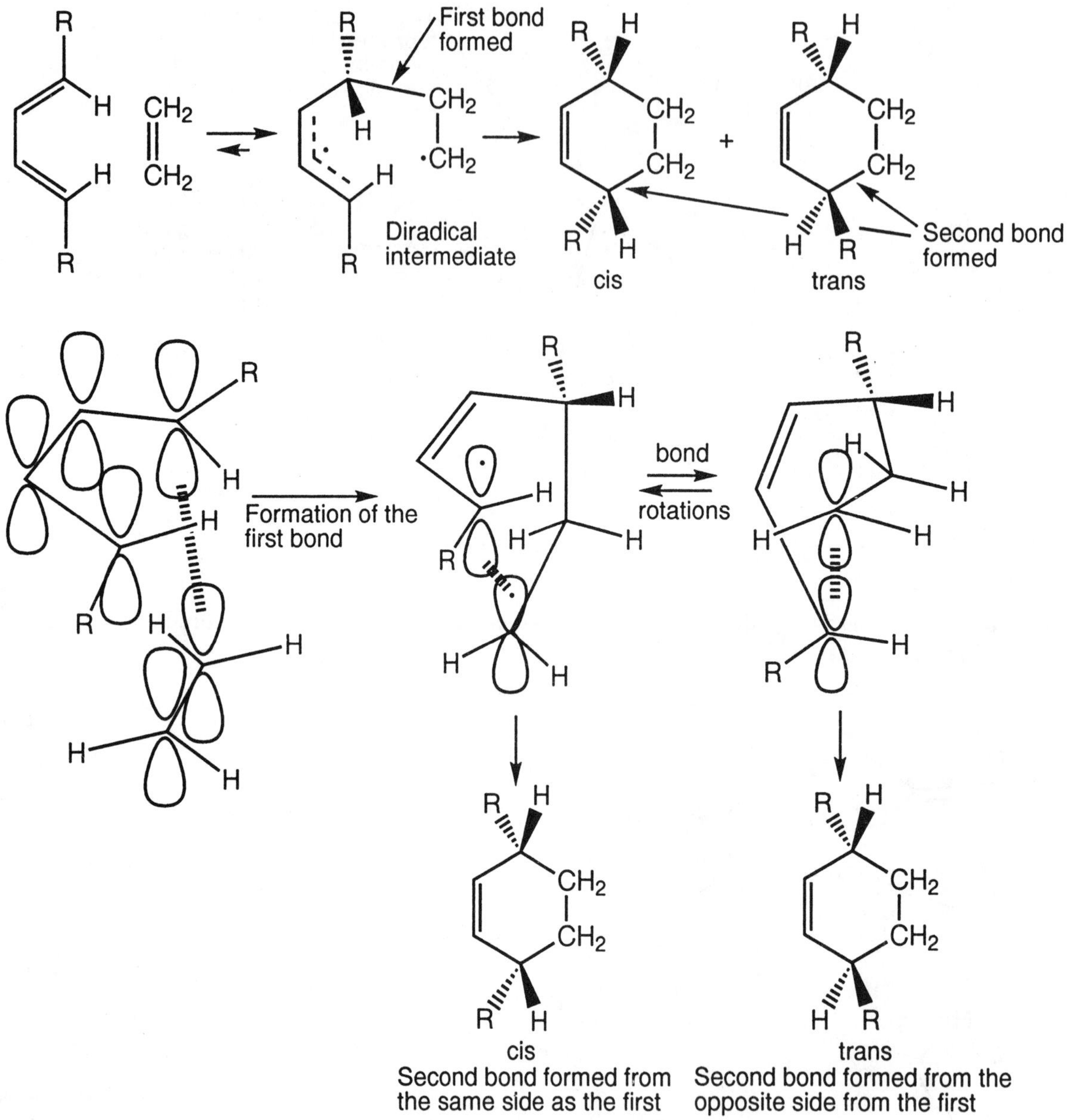

cis — Second bond formed from the same side as the first

trans — Second bond formed from the opposite side from the first

Once again, the experimental results show that the Diels-Alder reaction is concerted; the two new bonds are formed at the same time, and there is no intermediate (Fig. 12.68, p. 541).

Problem 12.31 An initiating radical (R•) first adds to butadiene so as to form a resonance-stabilized allyl radical. This allyl radical then plays the role of the initiating radical and adds to another butadiene to give a new, larger, resonance-stabilized allyl radical. This process is repeated many times to form the polymer.

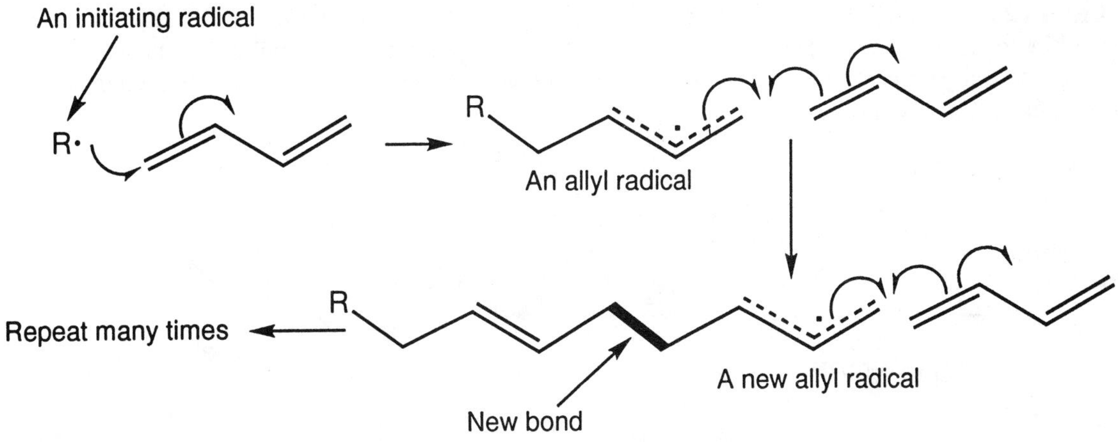

An initiating radical

R·

R

An allyl radical

R

Repeat many times

New bond

A new allyl radical

Problem 12.32 In the rubber molecule shown in Figure 12.78 all the double bonds are *(Z)*. There is another form with all the double bonds *(E)*, that is called gutta percha. Of course there could also be all sorts of polymers with some double bonds *(E)* and others *(Z)*. These molecules may be reasonable answers to the problem as set, but they do not occur in Nature.

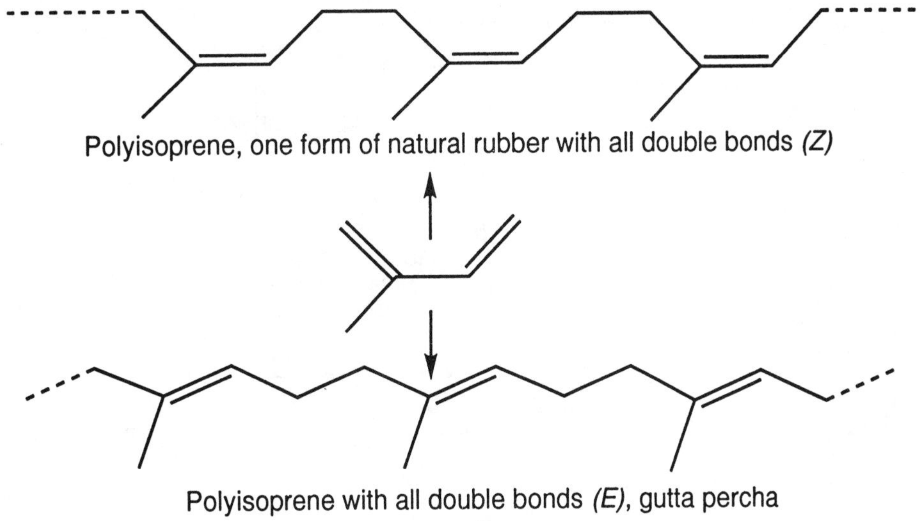

Polyisoprene, one form of natural rubber with all double bonds *(Z)*

Polyisoprene with all double bonds *(E)*, gutta percha

Problem 12.33 Loss of the pyrophosphate ion, shown here as an S_N1 reaction, leads to a resonance-stabilized pyrophosphate anion. The resonance stabilization of the anion makes pyrophosphate a good leaving group.

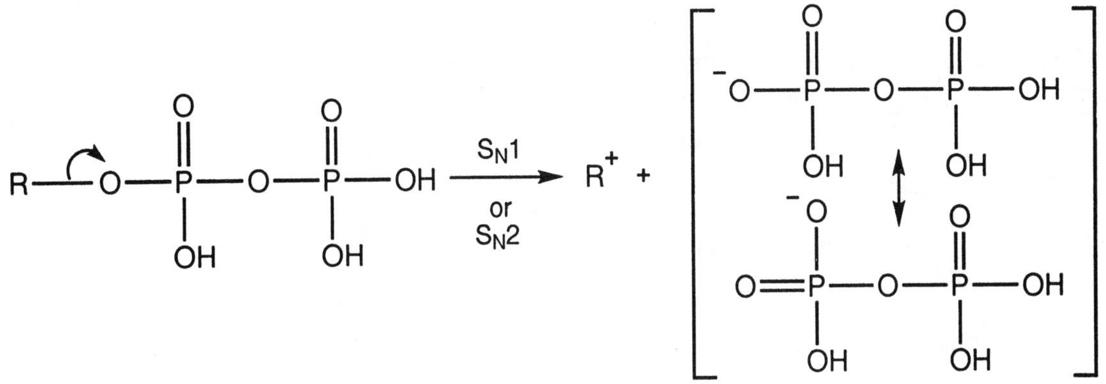

R—O—P(=O)(OH)—O—P(=O)(OH)—OH $\xrightarrow[S_N2]{S_N1 \text{ or}}$ R^+ + [$^-$O—P(=O)(OH)—O—P(=O)(OH)—OH ↔ O=P(O$^-$)(OH)—O—P(=O)(OH)—OH]

Problem 12.35 An S_N1 ionization to give an allyl cation is followed by intramolecular addition to the double bond. Although a tertiary carbocation results, this reaction is not really possible without a prior cis-trans isomerization. The double bond and the carbocation cannot reach each other until this isomerization occurs.

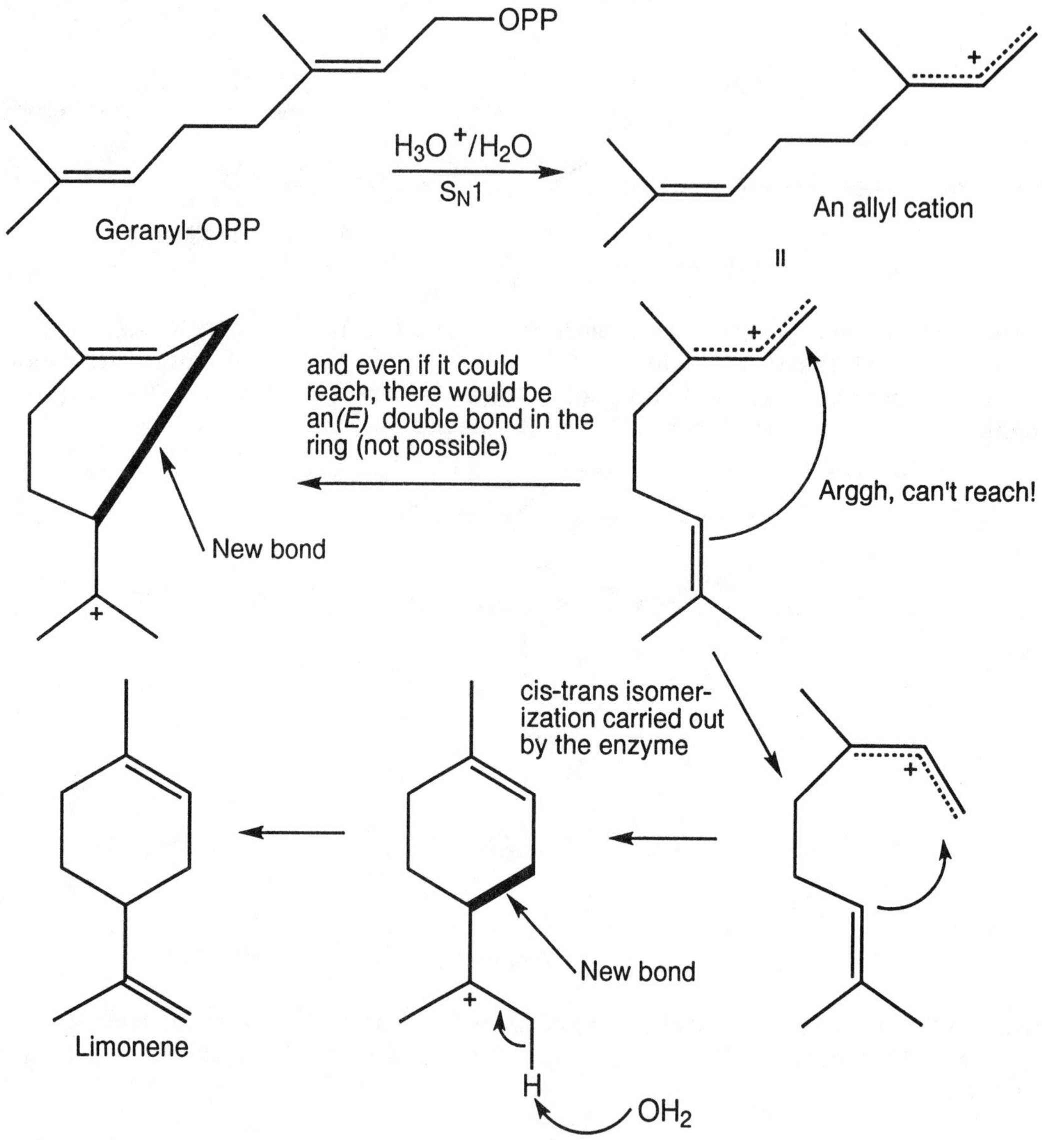

Problem 12.37 This problem is an easy one for a change. Any of a large number of possible labeling experiments will work. Here is one suggestion.

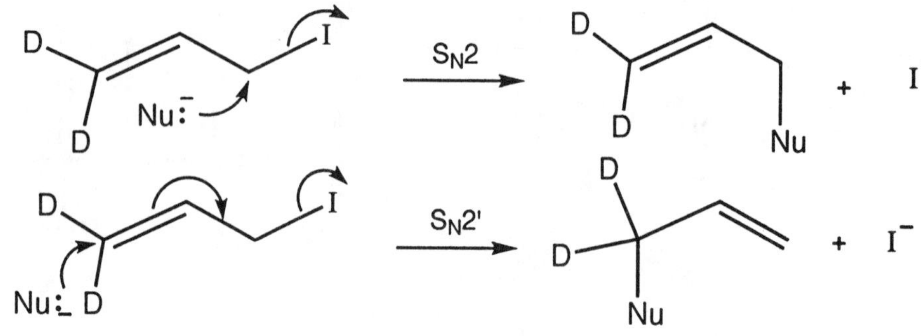

Problem 12.38 An adding nucleophile can enter from the same side as the departing chloride or from the opposite side.

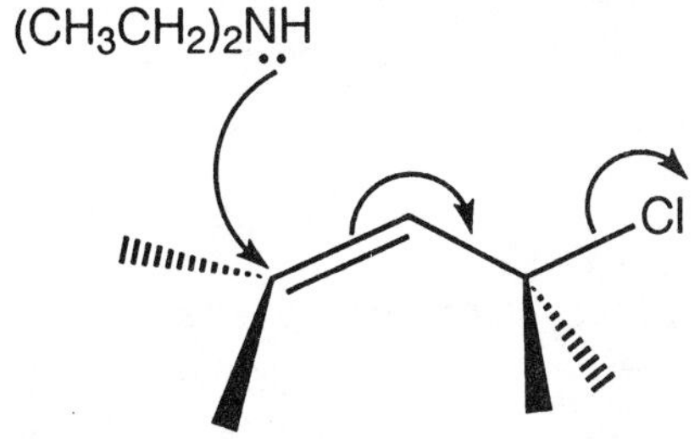

$(CH_3CH_2)_2NH$ Cl

Nucleophile adds from the same side as chlorine

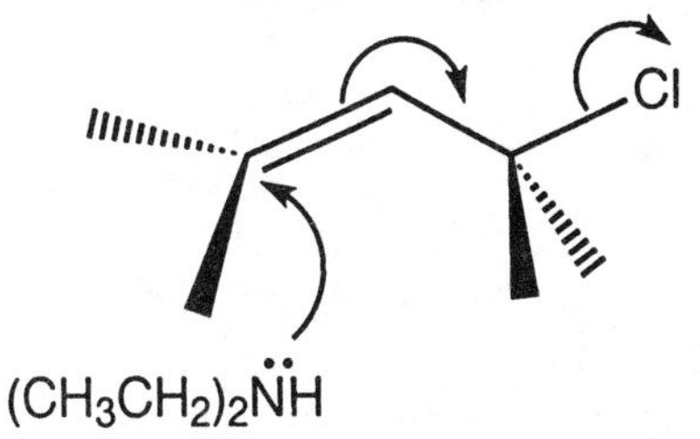

Cl $(CH_3CH_2)_2NH$

Nucleophile adds from the opposite side from chlorine

The data in the problem show that addition of the nucleophile is from the same side as the chlorine. The two products result from the presence of two rotational isomers of the starting material. In each case, diethylamine adds from the same side as chlorine.

$(CH_3CH_2)_2NH$ H Cl H D H CH_3

rotate about C–C single bond

1. add from same side as the chlorine
2. deprotonate ammonium ion

$(CH_3CH_2)_2N$ H C C CH_3 H D H

Major

H H H CH_3 D Cl $(CH_3CH_2)_2NH$

1. add from same side as the chlorine
2. deprotonate ammonium ion

H H D C C H $(CH_3CH_2)_2N$ CH_3

Minor

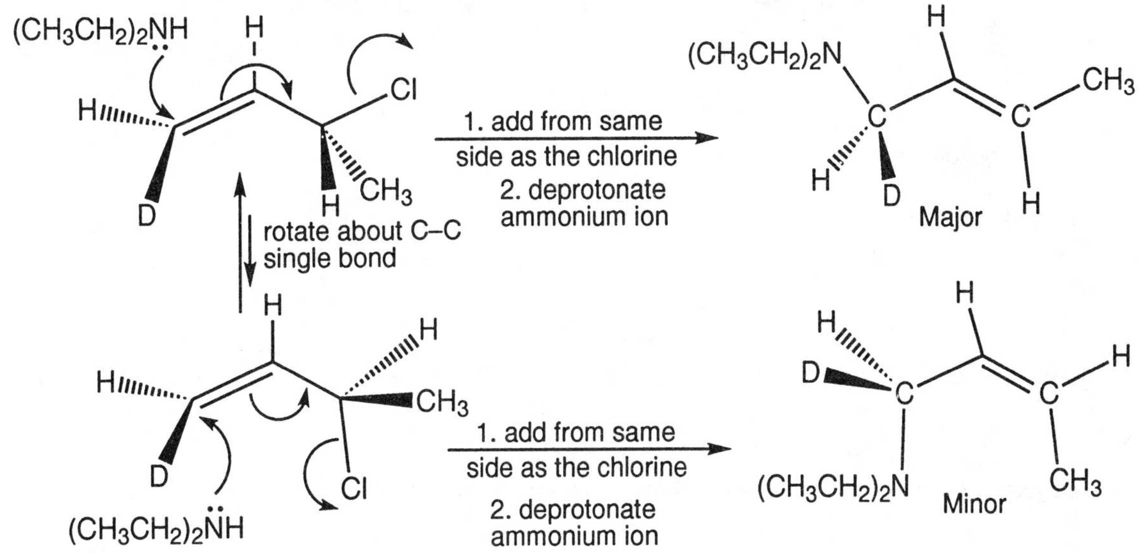

If addition had been from the opposite side, different products would have been observed.

Additional Problem Answers

Problem 12.39 Just follow the rules.

(a)

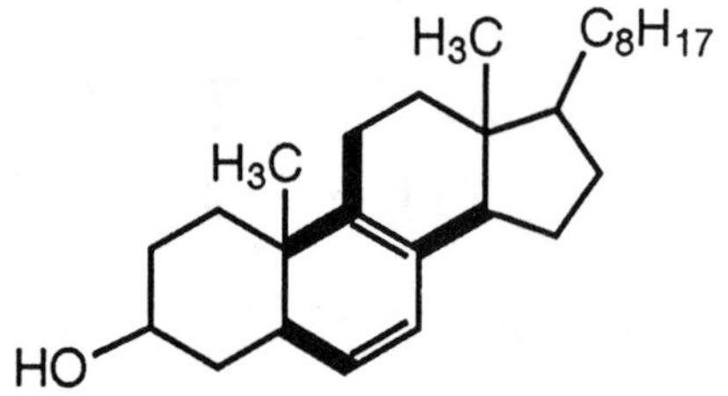

Base 1,3-cyclohexadiene	253
Four alkyl substituents (boldface lines)	20
λ_{max} (calculated)	273 nm
λ_{max} (actual)	275 nm

(b)

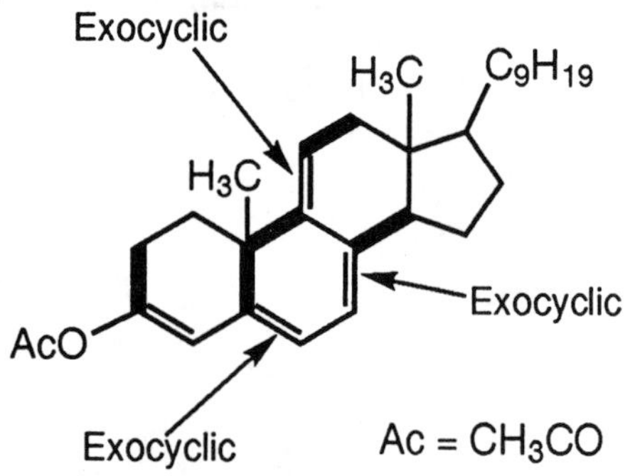

Exocyclic

H_3C C_9H_{19}

Exocyclic

Exocyclic $Ac = CH_3CO$

Base 1,3-cyclohexadiene	253
Two additional conjugated double bonds	60
Five alkyl substituents (boldface lines)	25
Three exocyclic double bonds	15
λ_{max} (calculated)	353 nm
λ_{max} (actual)	356 nm

Problem 12.40 Tricky, tricky, tricky; (a) and (b) are themselves resonance forms. One other carbon shares the negative charge.
(a and b)

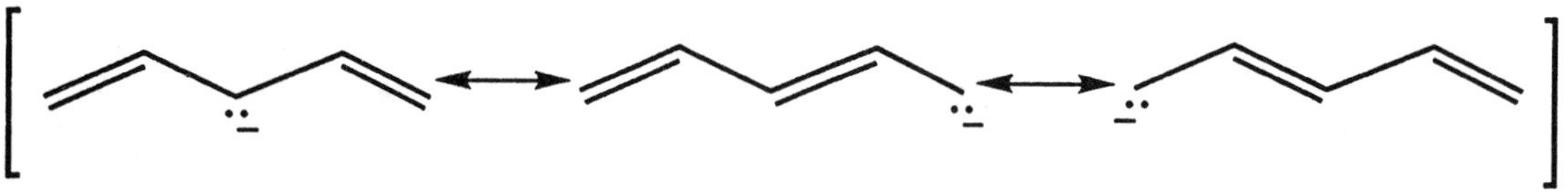

(c) In the cyclohexadienyl cation three carbons, but only three, share the charge.

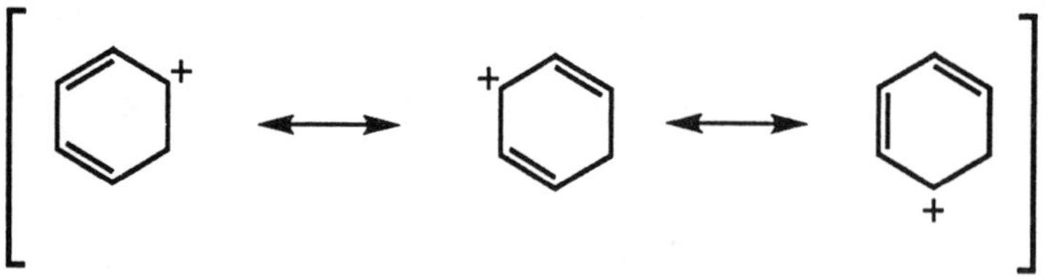

(d) In the cyclopentadienyl anion *all five* carbons share the charge:

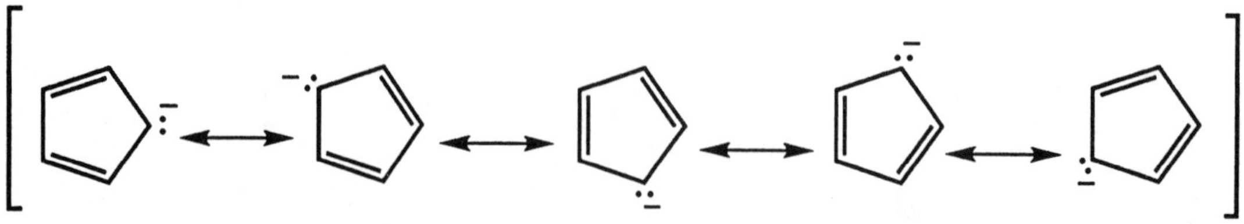

Problem 12.41 In each of the pairs, it is the second, right-hand compound that is chiral. In the figure at the top of the opposite page, to see that the two compounds of the upper pair are identical (superimposable), and the lower pair is nonsuperimposable, rotate 90° along a longitudinal axis. By all means make models! This one is not easy to see.

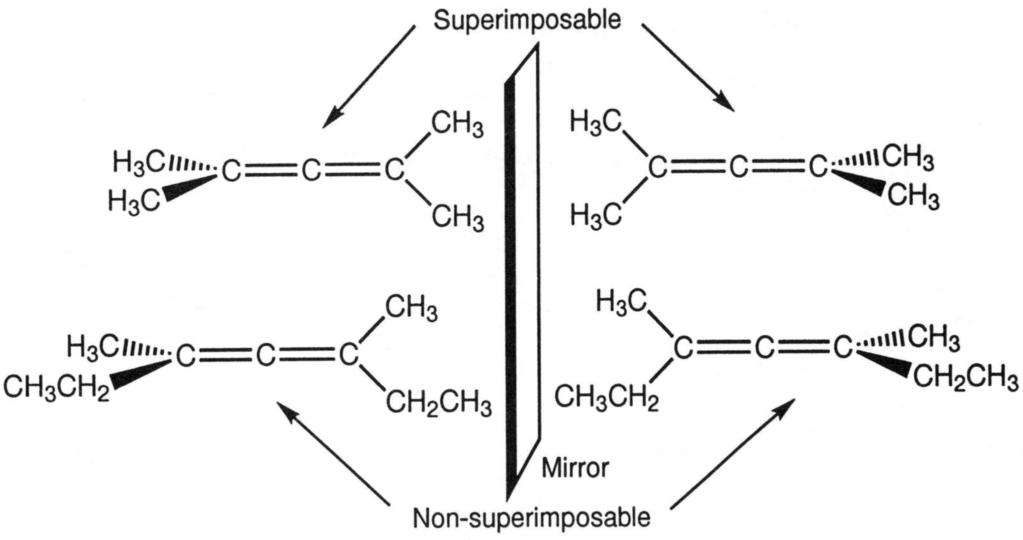

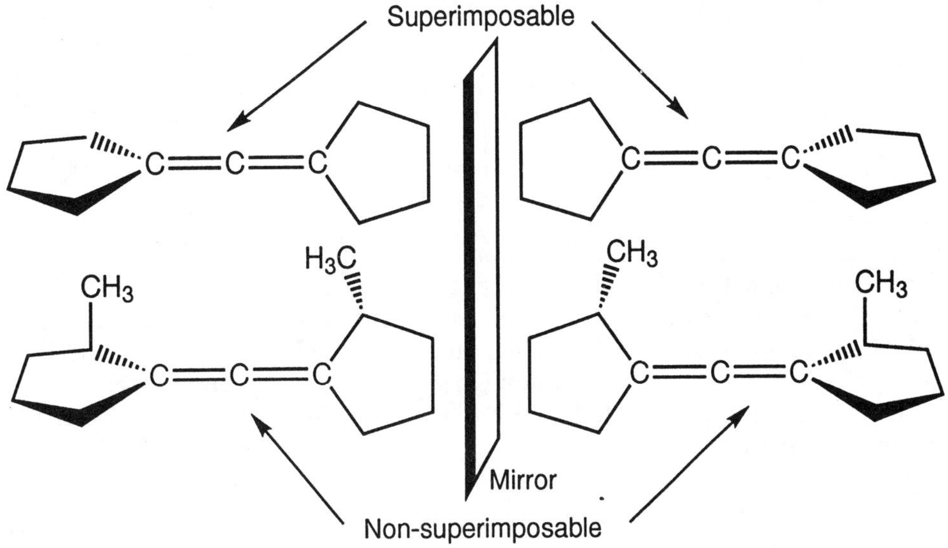

Superimposable

Mirror

Non-superimposable

Superimposable

Mirror

Non-superimposable

Problem 12.42

(a) Protonation gives the resonance-stabilized allyl cation.

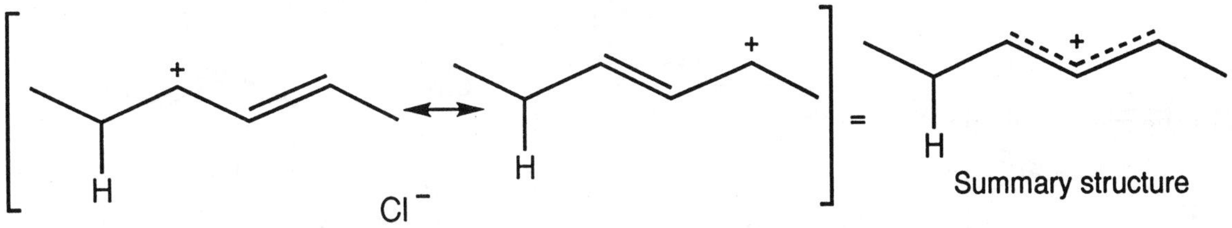

Summary structure

Addition of chloride to the two carbons sharing the positive charge gives the products of 1,2- and 1,4-addition. The major product depends upon the reaction conditions. Kinetic control yields mainly the 1,2-product, whereas thermodynamic control produces the products of both 1,2- and 1,4-addition.

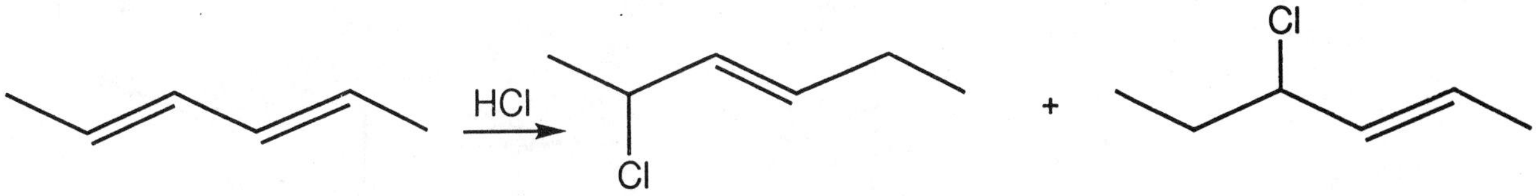

HCl

(b) Hydrogenation gives the saturated alkane, hexane.

H_2/Pd

(c) As in (a), kinetic control gives the product of 1,2-addition, and thermodynamic control gives the product of 1,4-addition as well.

Cl_2 Cl Cl + Cl Cl

(d) The Diels-Alder reaction is a one-step process proceeding through the s-cis form of the diene.

s-trans s-cis $COOCH_3$ C ≡ C $COOCH_3$ $COOCH_3$ $COOCH_3$

Problem 12.43 Compounds **2** and **3** are the 1,2- and 1,4-addition products of 1,3-butadiene and bromine. Compounds **1** and **4** involve addition of one bromine atom and the methoxy group of methyl alcohol. In principle, two mechanisms can be written for these additions, depending on how the intermediate cation is envisioned. If a bromonium ion is involved, direct S_N2 additions of the bromide ion or methyl alcohol to the bromonium ion give the 1,2-addition products **2** and **1**. The 1,4-addition products **3** and **4** could be formed by an S_N2' reaction with bromide ion or methyl alcohol acting as nucleophile.

:Br—Br: → :Br:$^-$ +Br: S_N2 → :Br: :Br: **2**

:Br:$^-$ +Br: S_N2' → :Br: :Br: **3**

$CH_3\ddot{O}H$ +Br: S_N2 → deprotonation → :$\ddot{O}CH_3$:Br: **1**

$CH_3\ddot{O}H$ +Br: S_N2' → deprotonation → :$\ddot{O}CH_3$:Br: **4**

Alternatively, the intermediate could be seen as a resonance-stabilized allylic carbocation. Addition of the two nucleophiles, bromide and methyl alcohol, leads to the observed four products.

:Br: :Br: 2 :Br: 3 :Br: :Br: 1 :Br: :Br: 4

CH3OH deproton-ation :OCH3

It is not possible to distinguish between these two possibilities without further experiments. It is worth noting that the ratios of the two types of 1,2- and 1,4-addition products are appreciably different; structure **2**/structure **3** = 2.2, structure **1**/structure **4** = 15. This result must be ascribed to the greater nucleophilicity of bromide ion relative to methyl alcohol.

Problem 12.44 As with 1,3-butadiene, the kinetic (more easily) formed product will be that from 1,2-addition. Under thermodynamic conditions, the more stable 1,4-adduct prevails. The problem is that there are two possible 1,2-addition products. The issue can be decided by examining the two possible allyl cations formed by protonation at the different ends of the 1,3-diene.

Methyl group does not help stabilize the positive charge

Methyl group does help stabilize the positive charge

Addition of chloride at the relatively nearby position sharing the positive charge leads to the product of 1,2-addition (kinetic control). Addition at the relatively remote position sharing the positive charge gives the more stable, but kinetically disfavored, product of 1,4-addition (thermodynamic control).

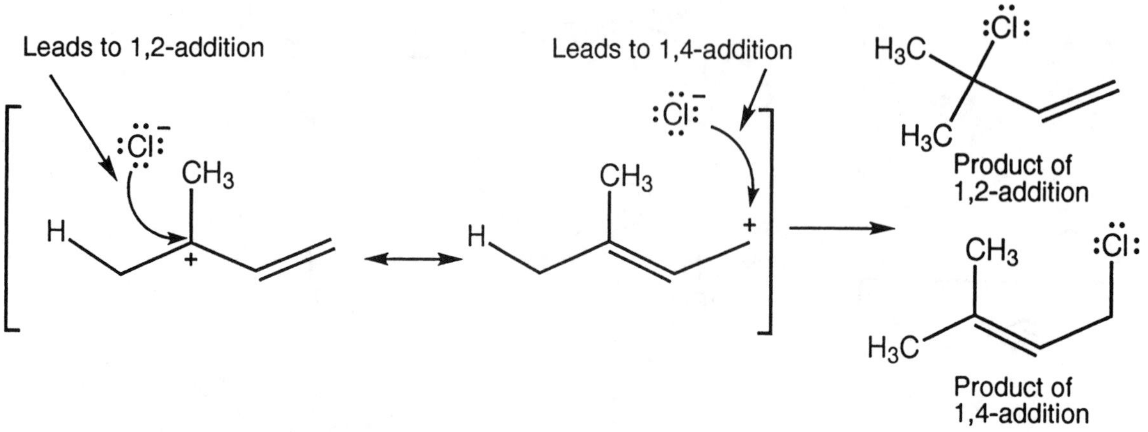

Leads to 1,2-addition

Leads to 1,4-addition

Product of 1,2-addition

Product of 1,4-addition

Problem 12.45 This synthesis problem should be relatively easy. The only real difficulty might be that it is a cascade problem; you have to be able to make one of the products in order to get the others. In this case, the critical compound is 3-hexyne. It comes from formation of the acetylide from 1-butyne followed by an S_N2 reaction on ethyl iodide.

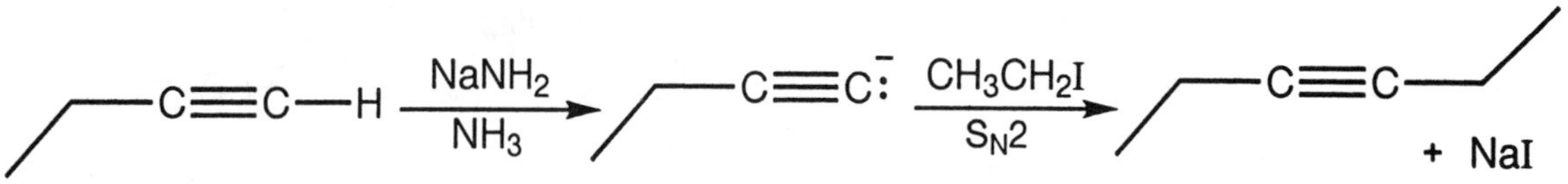

$NaNH_2$ / NH_3; CH_3CH_2I / S_N2; + NaI

Once 3-hexyne is made, the others all follow in straightforward fashion.

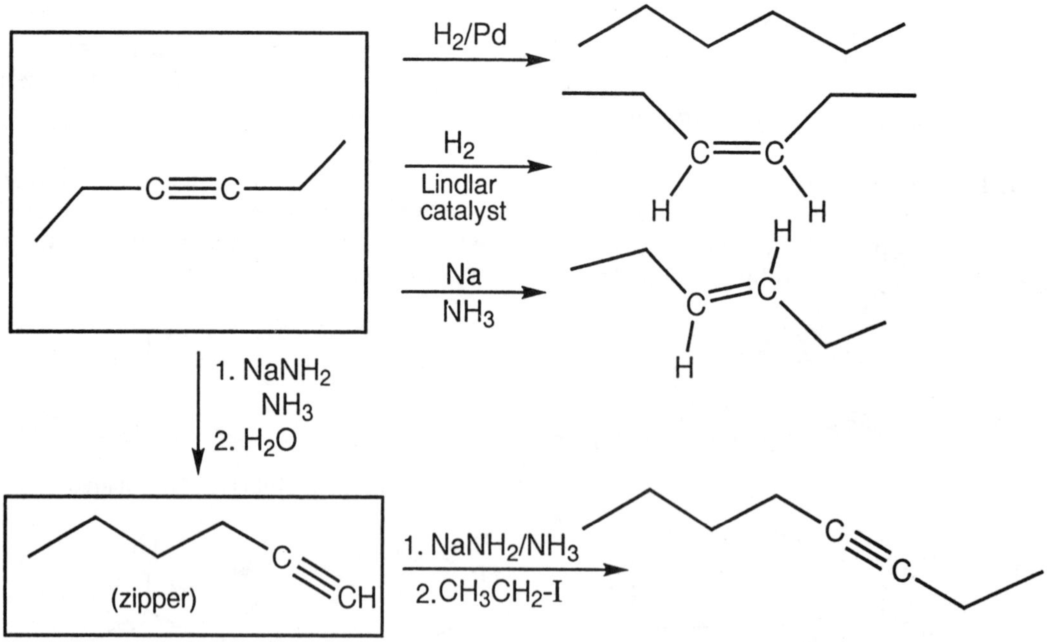

H_2/Pd

H_2 / Lindlar catalyst

Na / NH_3

1. $NaNH_2$ NH_3 2. H_2O

(zipper)

1. $NaNH_2/NH_3$ 2. CH_3CH_2-I

Problem 12.46 Converting 2-butyne into 1-butyne involves a zipper reaction.

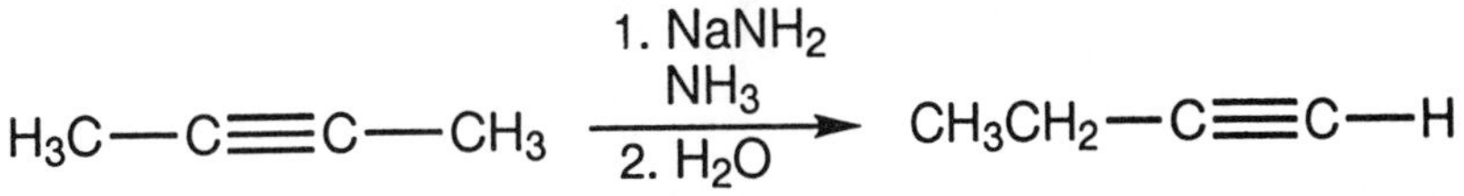

$$H_3C—C≡C—CH_3 \xrightarrow[2.\ H_2O]{1.\ NaNH_2,\ NH_3} CH_3CH_2—C≡C—H$$

As the problem suggests, going the other way poses more problems. Here is one suggestion. First, add two molecules of HBr in polar, Markovnikov fashion to give 2,2-dibromobutane.

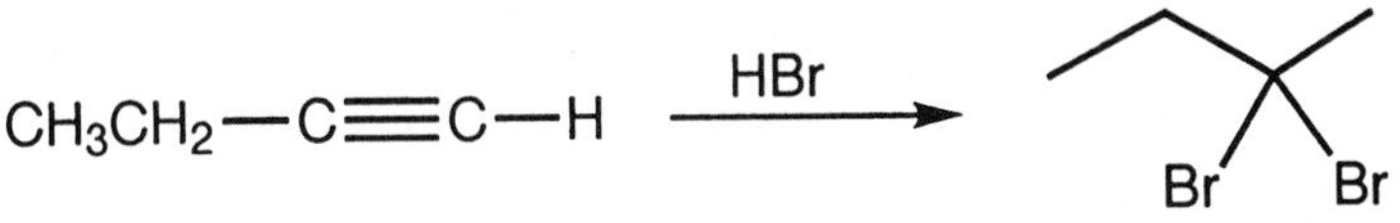

$$CH_3CH_2—C≡C—H \xrightarrow{HBr}$$ Br Br

Now a pair of E2 reactions will give mostly 2-butyne.

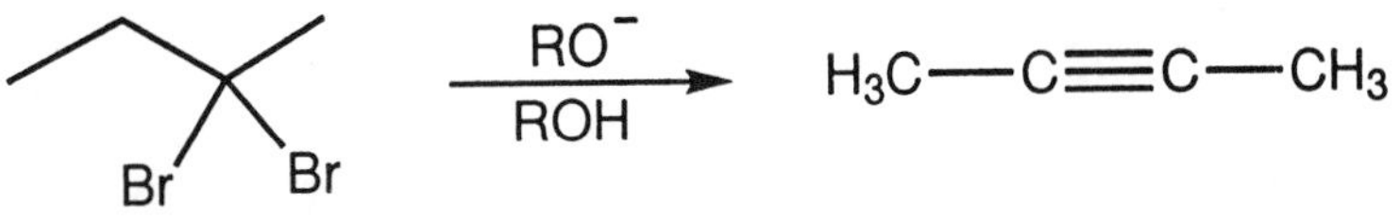

Br Br $$\xrightarrow[ROH]{RO^-} H_3C—C≡C—CH_3$$

Problem 12.47 This problem was designed to illustrate the scope and variety of compounds available from the Diels-Alder reaction.

(a) In this reaction, a 76:24 mixture of endo and exo products is obtained. Be sure you see how these two products arise from a relatively more stable endo transition state and a relatively less stable exo transition state. In doubt? See Figure 12.75 (p. 545).

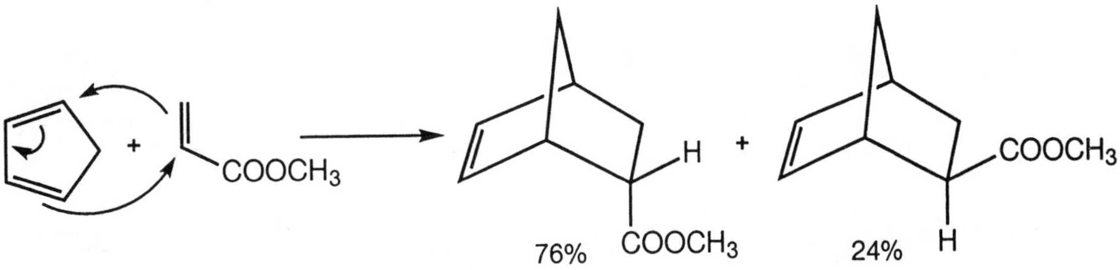

+ COOCH$_3$ → H, COOCH$_3$ 76% + COOCH$_3$, H 24%

(b) The only unusual thing about this variation of the Diels-Alder reaction is the use of an alkyne as a dienophile rather than the more usual alkene. The product is a 1,4-cyclohexadiene rather than a cyclohexene. Note that only one π bond of the alkyne is involved in this Diels-Alder reaction.

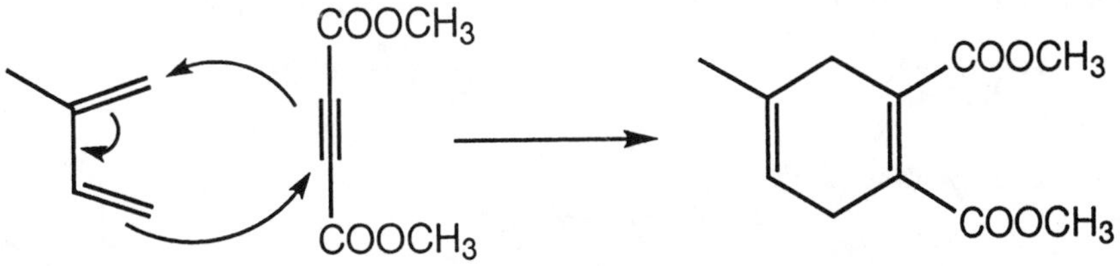

COOCH$_3$, COOCH$_3$ → COOCH$_3$, COOCH$_3$

(c) In this variation, both the diene and the dienophile are cyclic compounds. The first complication is to decide whether the reaction will pass over an endo or exo transition state. In fact, the cycloadduct has the endo stereochemistry, and this reaction is normal in that respect.

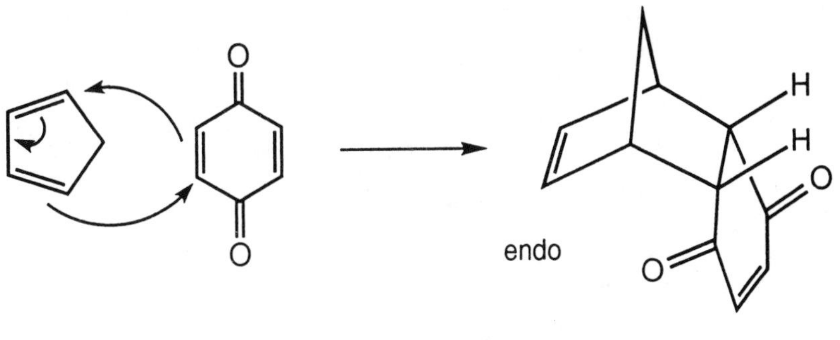

O, O → H, H, O, O endo

You may have noticed that the product of this Diels-Alder reaction still has a double bond with two attached carbonyl groups. This carbon-carbon double bond is also a quite reactive dienophile. When this reaction is run in the presence of two equivalents of 1,3-cyclopentadiene, a 2:1 adduct is readily formed.

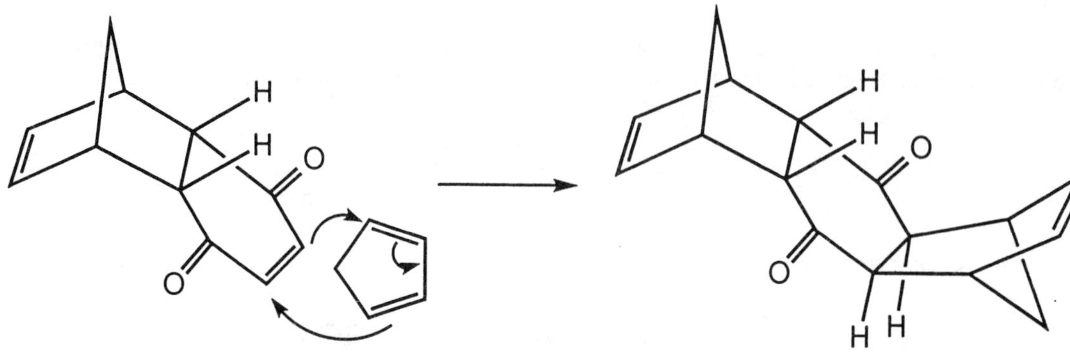

H H O O H H O O H H

(d) This reaction illustrates another useful variation of the Diels-Alder reaction in which the primary adduct loses a small molecule such as carbon monoxide, carbon dioxide, or nitrogen. In this problem the loss of carbon monoxide is indicated by the molecular formula. In problems such as these watch for losses such as this one. The formula will usually tell you, but if it is not given, you must be alert for the possibility, as it is often encountered both in the real world and the world of examination problems.

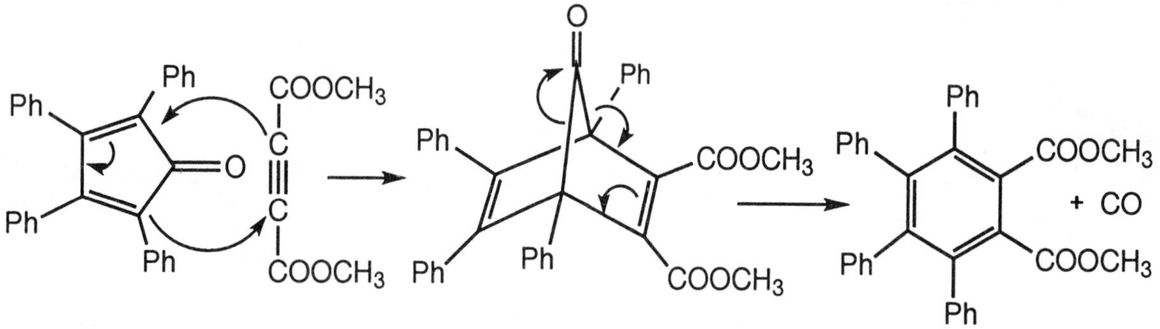

Ph Ph Ph Ph O $COOCH_3$ C C $COOCH_3$ O Ph Ph $COOCH_3$ Ph Ph $COOCH_3$ Ph Ph $COOCH_3$ + CO Ph Ph $COOCH_3$

(e,f) The Diels-Alder reaction has been used to deduce the position of double bonds in steroids. In example (e), the diene is locked into an s-cis arrangement, and the Diels- Alder reaction occurs. In (f), the diene is s-trans, and the Diels-Alder reaction is not possible. Be sure you see why.

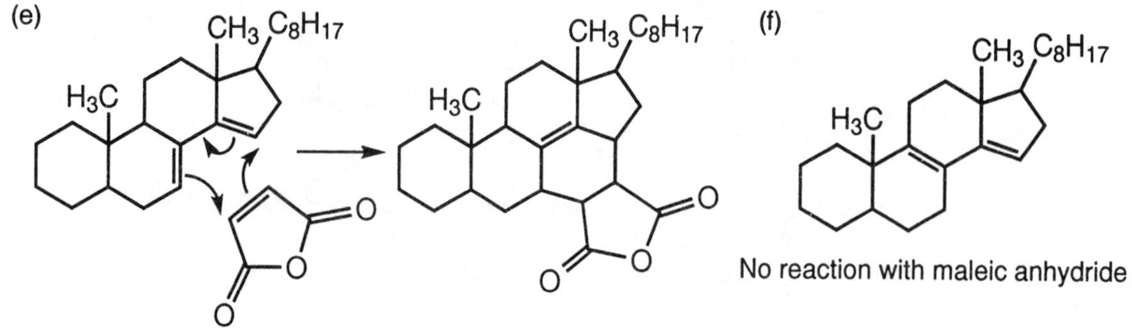

(e) CH_3 C_8H_{17} H_3C O O O CH_3 C_8H_{17} H_3C O O O

(f) CH_3 C_8H_{17} H_3C

No reaction with maleic anhydride

(g) In this example, we first have to realize that a "1,3-diene" is present. True, it contains two nitrogen atoms, but it remains a 1,3-diene nonetheless. Otherwise the first reaction is quite simple.

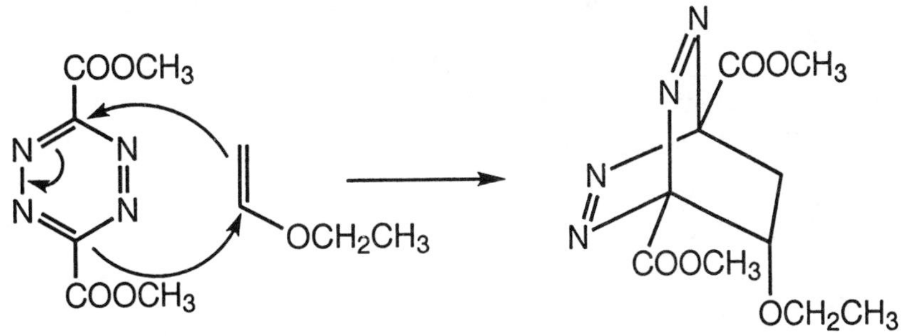

Now, as in (d), a small molecule is lost, this time N_2. Elimination of another molecule, ethyl alcohol, also occurs to give the final product.

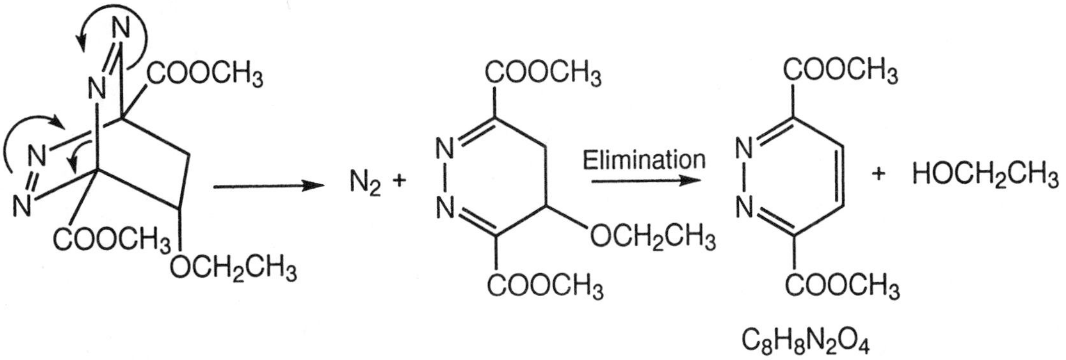

$C_8H_8N_2O_4$

Problem 12.48 The arrow formalism is easy, once you realize that the six-membered ring can bend underneath the remainder of the structure. Closing a four-membered ring gives the cage structure.

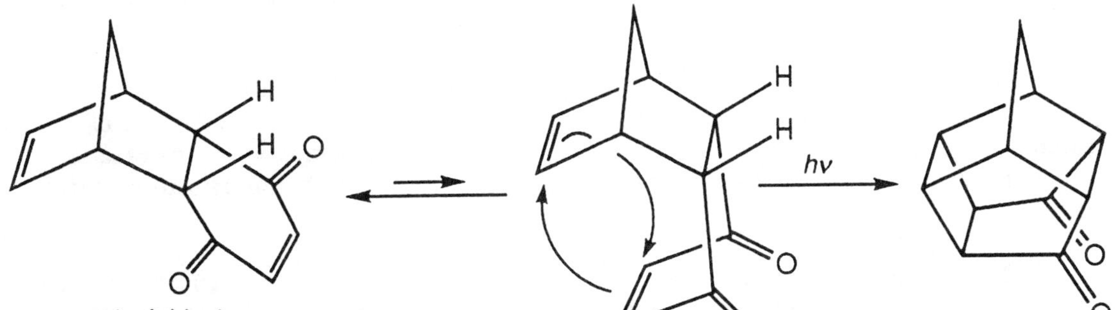

endo Adduct
(the two double bonds can approach each other)

Of course, if the adduct had been exo, no cage formation would have been possible; the double bonds are nowhere near each other.

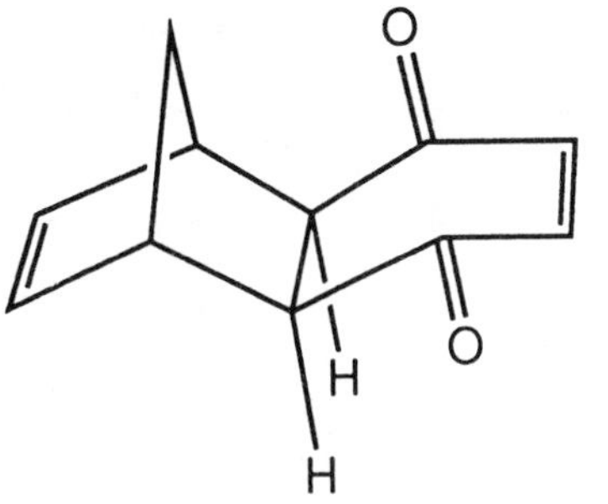

exo Adduct
(no reaction between the two double bonds is possible)

Problem 12.49 This problem is a classic case of thermodynamic versus kinetic control, quite similar to the situation encountered in Section 12.9 (Fig. 12.48, p. 528) for the addition of chlorine to 1,3-butadiene. The exo compound is more stable than the endo compound because the old dienophile points towards the smaller bridge, not towards the larger bridge. However, the transition state leading to the less stable, endo product benefits from extra orbital overlap (p. 545) and is lower in energy than the transition state for formation of the more stable exo product. The Energy versus Reaction progress diagram tells all.

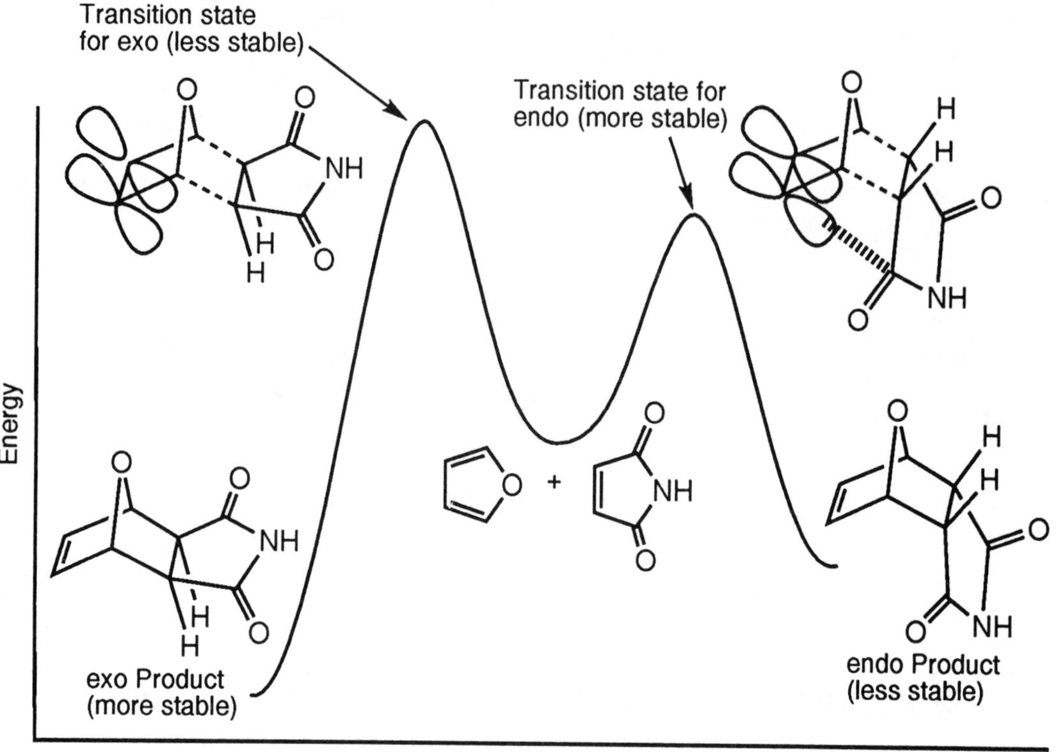

At high temperature all barriers are passable, and the product ratio is determined by the ΔG between the exo and endo *products* (thermodynamic control). The lower steric demands of the exo compound put it at a lower energy than the endo compound. At low temperature the reaction is not reversible and the product ratio is determined by the relative energies of the transition states (kinetic control).

Problem 12.50 First of all, draw an arrow formalisn for the reaction of an allyl species with 1,3-butadiene. Remember: the two participants in the reaction approach each other in parallel planes. There is no attempt in this schematic figure to show the orbital phases; this is a mapping exercise only.

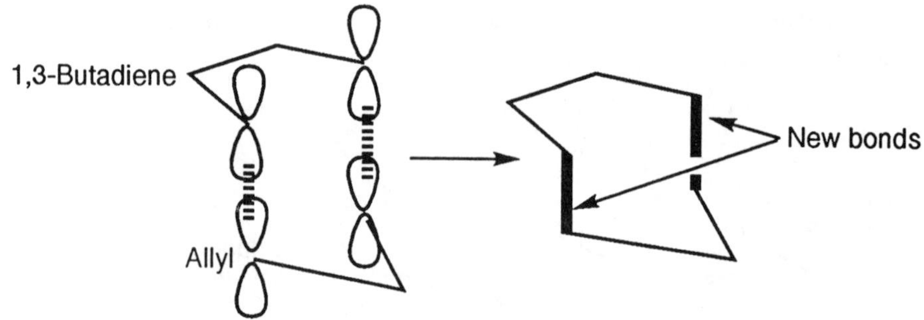

In each reaction there are two HOMO-LUMO interactions possible, HOMO (diene) - LUMO (dienophile), and HOMO (dienophile) - LUMO (diene). Look at both interactions in each case. For the participants in the reactions, 1,3-butadiene, the allyl cation, and the allyl anion, the molecular orbitals are

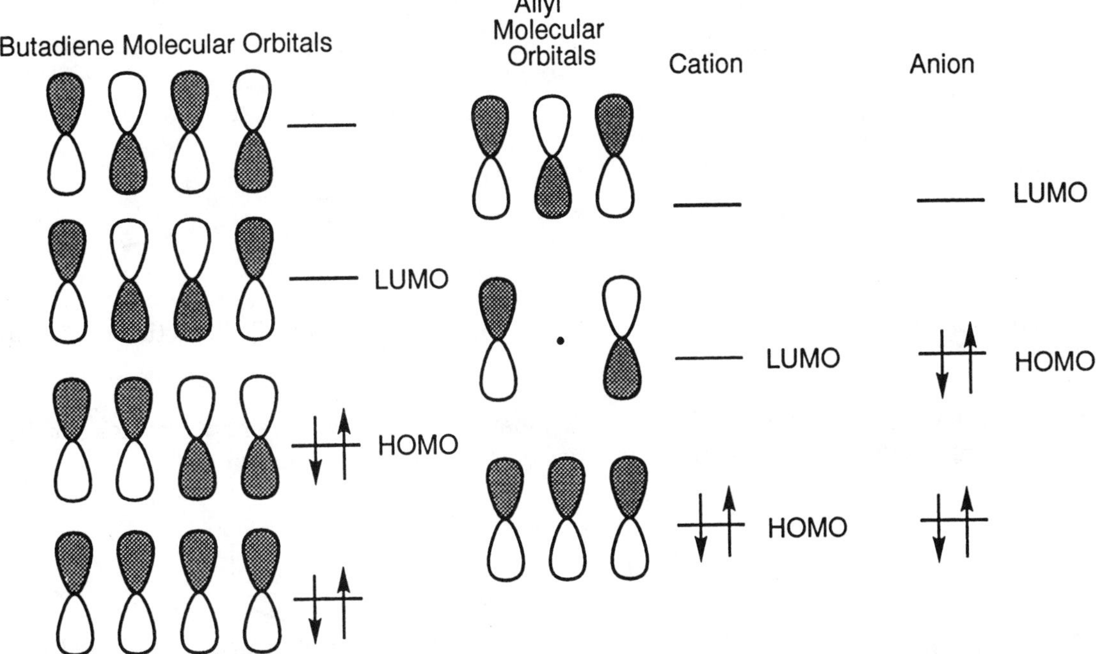

For the reaction to be successful, there must be bonding interactions at the points of formation of both new bonds. Here are the HOMO-LUMO interactions for the reaction of the allyl cation with 1,3-butadiene showing only the points at which new bonds are made.

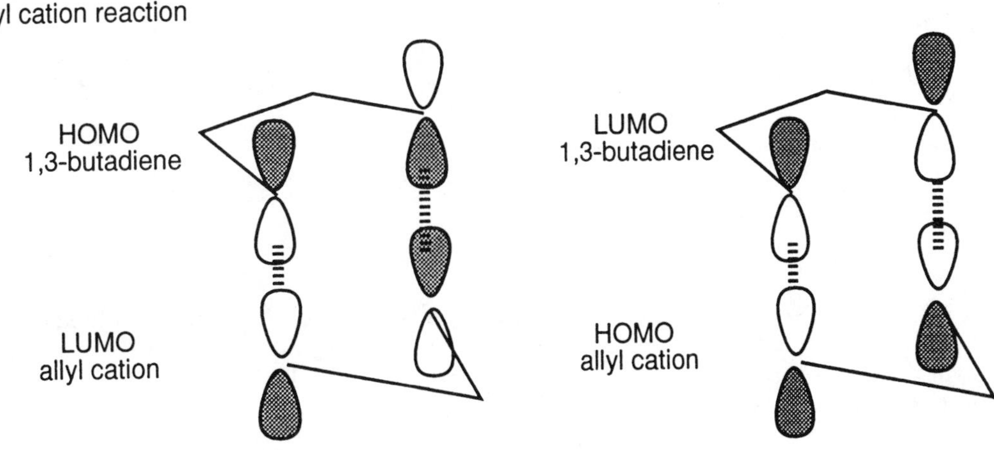

In each case both new interactions are bonding. The new bonds can be made.
The situation turns out quite otherwise in cycloadditions of the allyl anion. The two HOMO - LUMO interactions are on the next page.

Allyl anion reaction

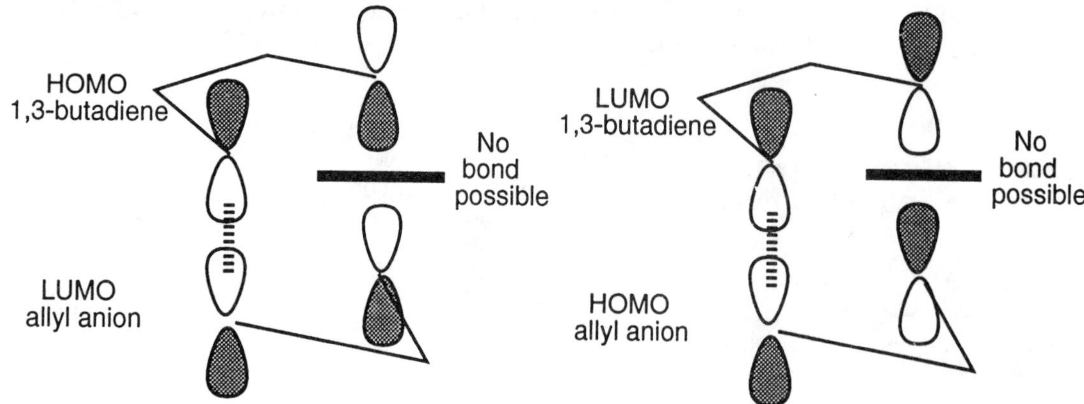

It is not possible to make two bonds in either HOMO-LUMO interaction. The reaction of the allyl anion fails.

Problem 12.51 Work through what the products must be and you will see how it comes out. In each case, there are two Diels-Alder products possible: One endo and one exo.

However, when the products are hydrogenated, each of the adducts from 1,3-cyclohexadiene gives the same product. If the drawing does not make this obvious to you, be sure to make models and convince yourself that the "two" hydrogenated compounds are really the same thing. The hydrogenated adducts from reaction of 1,3-cyclopentadiene are still different.

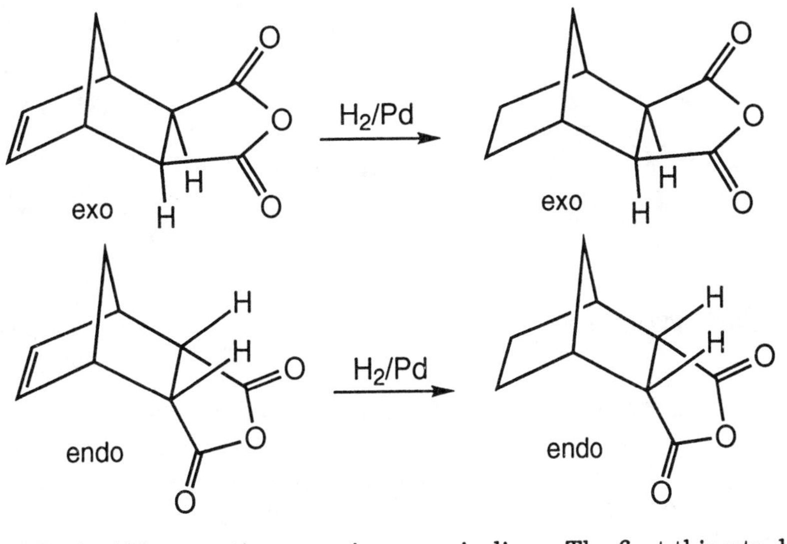

exo H_2/Pd exo

endo H_2/Pd endo

Problem 12.52 All Diels-Alder reactions require an s-cis diene. The first thing to do in solving this problem is to draw the s-cis forms of the diene in each case.

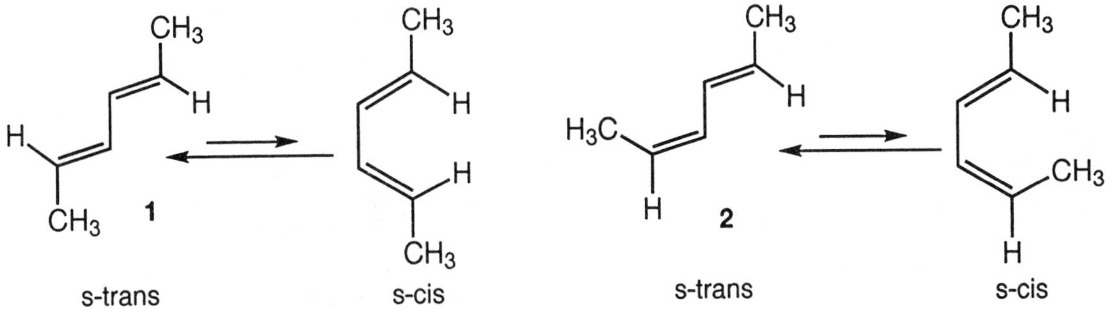

1 s-trans ⇌ s-cis **2** s-trans ⇌ s-cis

In **2**, a relatively large methyl group must be "inside," and this is surely destabilizing. So, at 35 °C there are no significant amounts of the required s-cis diene for **2**. Reaction of maleic anhydride takes place with the s-cis form of **1** to give product. Even this reaction is more complicated than it might seem at first. The stereochemical relationships of the substituents on the diene must be preserved in this one-step reaction, so the methyl groups must be cis in the product. However, addition can take place through either an endo or exo transition state to give endo or exo product. As usual (Chapter 12, p. 545), the endo transition state is lower in energy than the exo transition state and the endo product is formed.

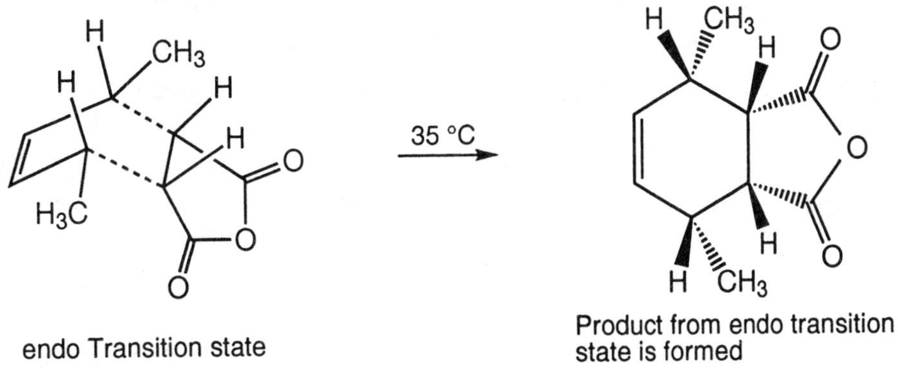

35 °C

endo Transition state

Product from endo transition state is formed

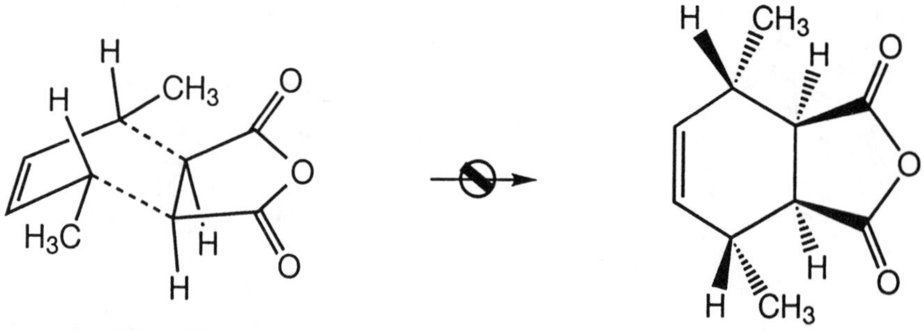

exo Transition state

Product from exo transition state is *not* formed

For **2**, higher temperature is required to form significant amounts of the s-cis form. When reaction occurs, once again the stereochemical relationship of the methyl groups in the s-cis conformation of the diene will be preserved in the adducts. In this case, they must be trans. As the methyl groups are trans, with one "up" and the other "down" there is only one product possible. The exo and endo transition states lead to the same product in this case.

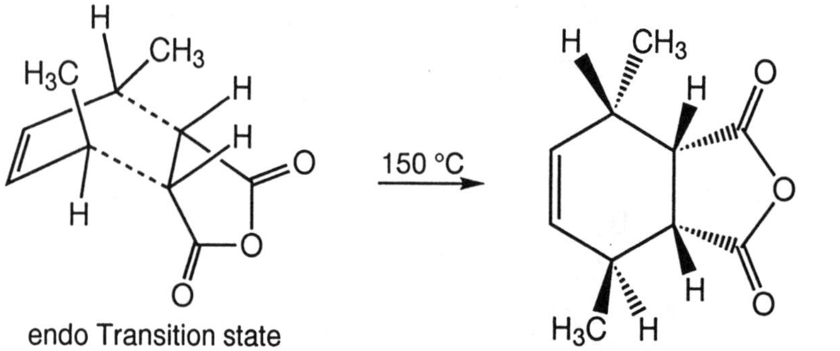

150 °C

endo Transition state

Product from endo transition state is formed

In **3**, both methyl groups must be "inside." We predict that this is too destabilizing, and that no significant amount of Diels-Alder product would be formed at 150 °C. Though this diene was not examined by Alder, there is ample evidence from other examples that this notion is correct.

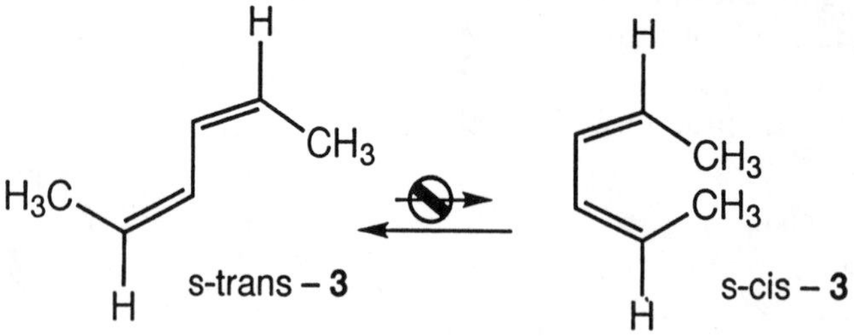

s-trans – **3**

s-cis – **3**

Problem 12.53 If you proposed a Diels-Alder reaction of furan with dimethylmaleic anhydride, followed by catalytic hydrogenation of the carbon-carbon double bond, you'd be in very good company. This route was proposed as early as 1928.

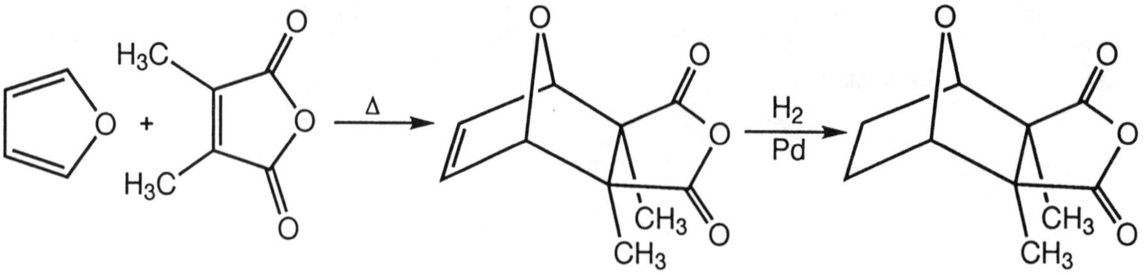

Δ

H_2 / Pd

Unfortunately, the required Diels-Alder reaction was not successful. Dimethylmaleic anhydride is a poor dienophile, presumably because of steric destabilization introduced by the two methyl groups. In addition, cycloadducts of furan are prone to undergo retro Diels-Alder reactions, and it may be that product is formed, but is not thermodynamically stable relative to starting material.

Reactions such as the Diels-Alder reaction, which proceed with a net decrease in volume (two molecules are made into one), can be accelerated under high pressure. This Diels-Alder reaction fails even at pressures up to 40 kbar. However, this synthetic problem was finally solved by William Dauben and his co-workers at Berkeley. Dauben's group used both a dienophile with fewer steric problems, along with high pressure. This is *not* a route you were expected to think up! If you came up with the reasonable ideas shown earlier, you did just fine.

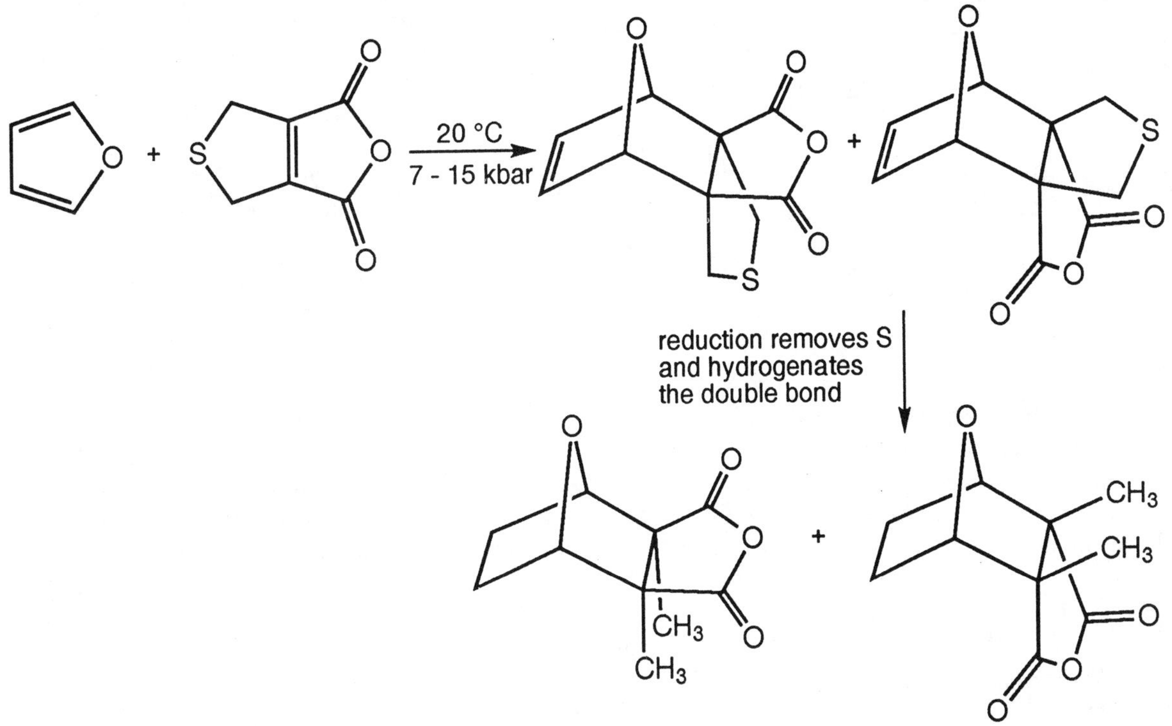

20 °C
7 - 15 kbar

reduction removes S and hydrogenates the double bond

CH_3 CH_3 CH_3 CH_3

The desired exo compound could be crystallized from the mixture and isolated in 51–63% yield.

Problem 12.54(a) This is a simple reverse Diels-Alder reaction brilliantly camouflaged by the molecular architecture. The lesson here is that every time you see a cyclohexene, THINK REVERSE DIELS-ALDER! *All* cyclohexenes are conceptually related to a 1,3-diene and a dienophile. The thermodynamic driving force for this reaction is the strain relief in opening one three-membered ring. In the answer, the product is first drawn without moving any atoms, then "relaxed" to the real structure.

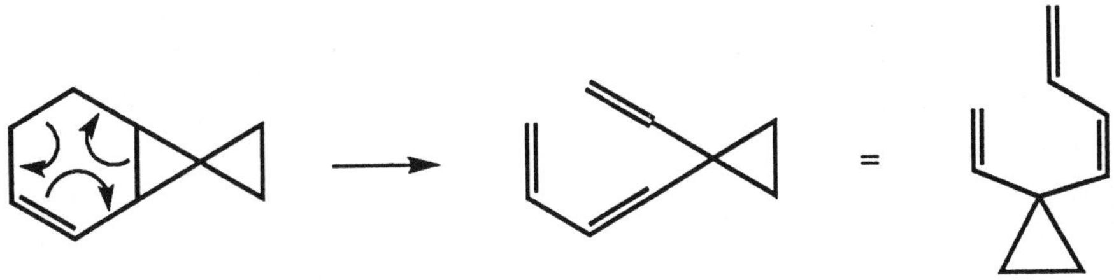

(b) This part also involves a reverse Diels-Alder reaction. In this case, the anthracene formed by the reverse reaction is captured in a "forward" Diels-Alder by maleic anhydride. A good clue here is the structure of the second product. Ask yourself how it can arise from starting material, and the reverse Diels-Alder reaction should appear. Alternatively, work backward. Ask what compounds can react to give the final adduct. The answer is maleic anhydride and anthracene. Again the need to make anthracene from starting material should suggest a reverse Diels-Alder reaction.

H_3COOC $COOCH_3$ — 200 - 250 °C, reverse Diels-Alder → Anthracene + H_3COOC–C(=CH_2)–$COOCH_3$

Anthracene + maleic anhydride — forward Diels-Alder →

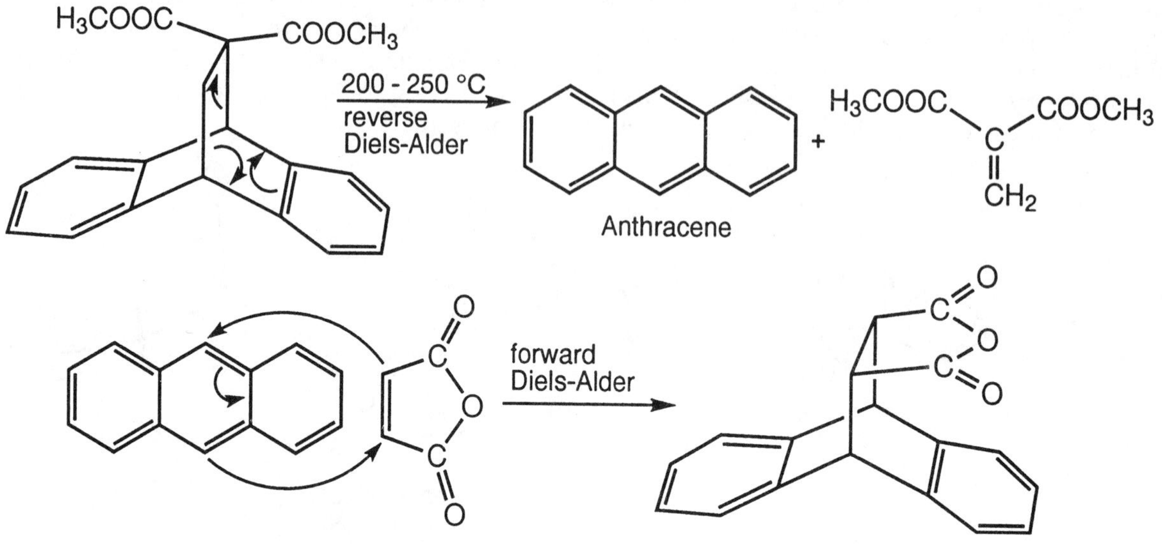

(c) Here we have an intramolecular Diels-Alder reaction. These are notoriously hard to see, especially at the beginning. The reaction is certainly easier to see if the starting material is drawn in a "suggestive" way (as it almost never is in problems!). Aside: Notice that the one-step nature of the Diels-Alder reactions allows us to fix the stereochemistry at four different atoms (the termini where the new σ bonds are made) in this reaction. That's quite a synthetic advantage.

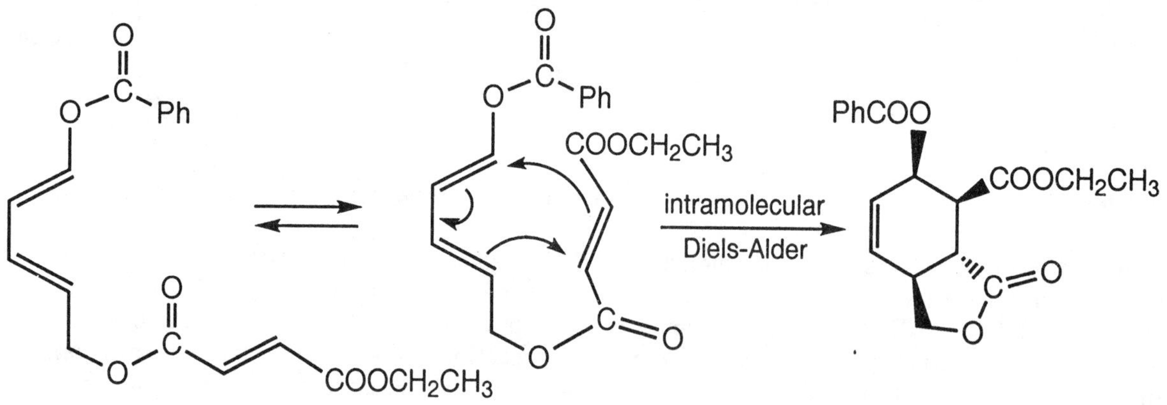

Ph, $COOCH_2CH_3$ ⇌ Ph, $COOCH_2CH_3$ — intramolecular Diels-Alder → PhCOO, $COOCH_2CH_3$

(d) This final transformation is also an example of the intramolecular Diels-Alder reaction, with the furan ring acting as the diene. Once again, it is necessary to draw the molecule in an arrangement that shows the proximity of the diene and dienophile. The product of the Diels-Alder reaction is first drawn without moving any atoms, then relaxed to a more realistic picture. This is a most useful technique.

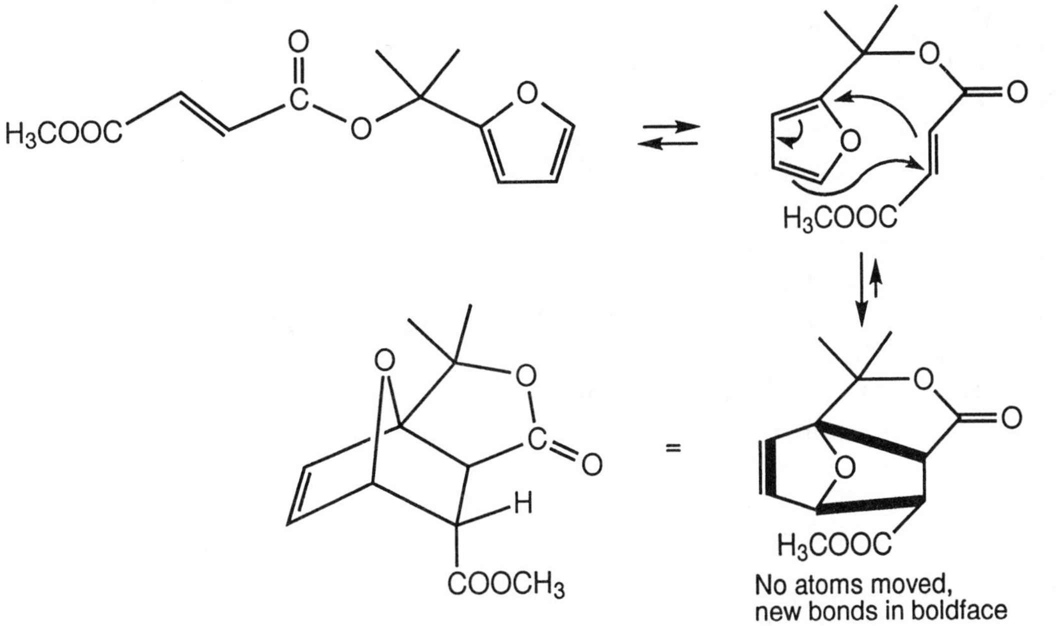

H_3COOC H_3COOC H_3COOC $COOCH_3$

No atoms moved, new bonds in boldface

Problem 12.55 This problem is an example of the S_N2' reaction. The amine is the nucleophile and "OCOPh" is the leaving group. It is easy to see that this is so because the amine finds its way into the product molecule and the "OCOPh" group is lost. The usually more common S_N2 reaction is slow because the rather large isopropyl group guards the rear of the leaving group. Note also, as we saw in Problem 12.38 (p. 559), that the entering nucleophile approaches the double bond from the same side as the leaving group departs, fixing the observed stereochemistry.

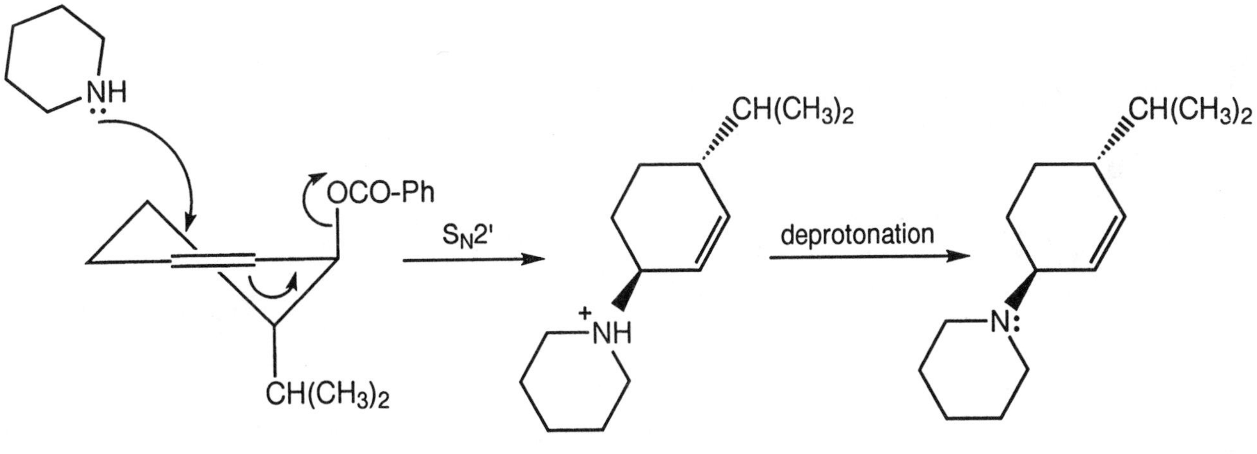

NH OCO-Ph S_N2' $CH(CH_3)_2$ deprotonation $CH(CH_3)_2$ NH N: $CH(CH_3)_2$

Chapter 13 Outline

Conjugation and Aromaticity

13

This chapter consists almost entirely of an exploration of the structural implications of the stabilizing effects of aromaticity. Reactivity does creep in, but only in a small way, as an introduction to the elaborate substitution reactions of Chapter 14. The following problems allow you to explore resonance and molecular orbital theory in order to find examples of molecules that possess the stabilizing quality called aromaticity. You will also get plenty of practice in drawing strange molecules of quite remarkable and varied structures. In addition, you will have a chance to work a bit on reactions. You will see the classic aromatic substitution process, as well as reactions that destroy the usually robust aromatic ring, and reactions at the position adjacent to an aromatic ring, the "benzyl" position.

Problem 13.1 A Kekulé representation predicts two isomers of any 1,2-disubstituted benzene. In one, the two substituents, here methyl groups, are on the same double bond, in the other, they are on different double bonds.

CH_3 CH_3 and CH_3 CH_3

As there is only one real molecule, this poses severe problems for a 1,3,5-cyclohexatriene picture of benzene.

Problem 13.5 In the "tub-flip" process, the first step is conversion of one tub into the other by passing over a planar, bond-localized transition state. This transition state is not a regular octagon as the relatively long single bonds and relatively short double bonds are maintained. This process has an activation energy of 14.6 kcal/mol. The double bonds do not change position (see figure next page).

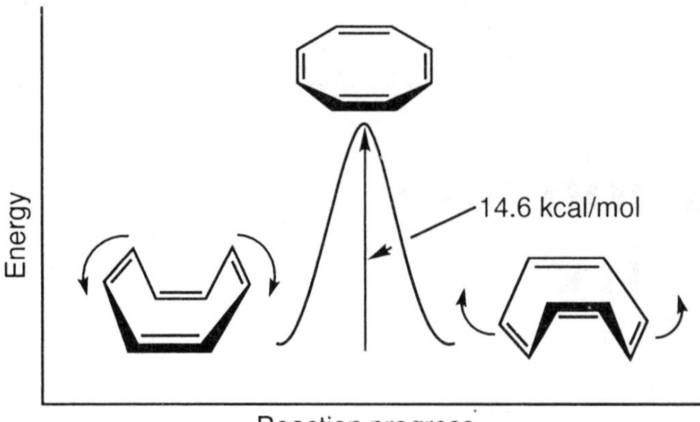

In the second process, there is a "bond switching" as well. The transition state in this reaction is the symmetrical, delocalized regular octagon. The activation energy is a bit higher, 17.0 kcal/mol. The two processes differ in that in the upper reaction the eight-membered ring contains alternating single and double bonds. In the lower reaction, the transition state is the symmetrical, regular octagon.

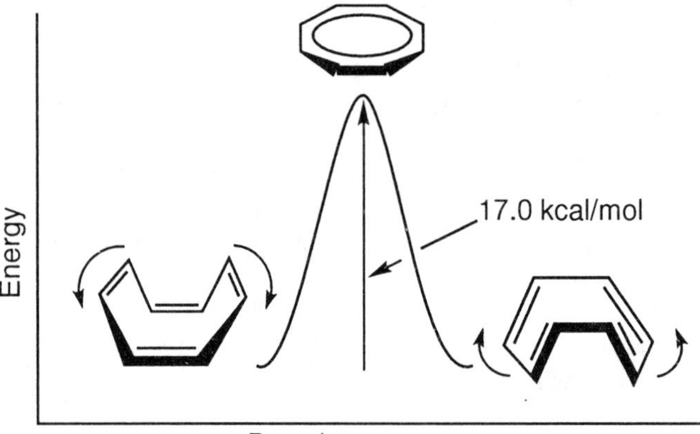

Problem 13.6 Remember the criteria (and why they are important). The molecule in question must be cyclic, planar, fully conjugated, and have $4n + 2$ π electrons.

(a) This compound is not fully conjugated, and cannot be aromatic. The methylene (CH_2) group interrupts orbital overlap.

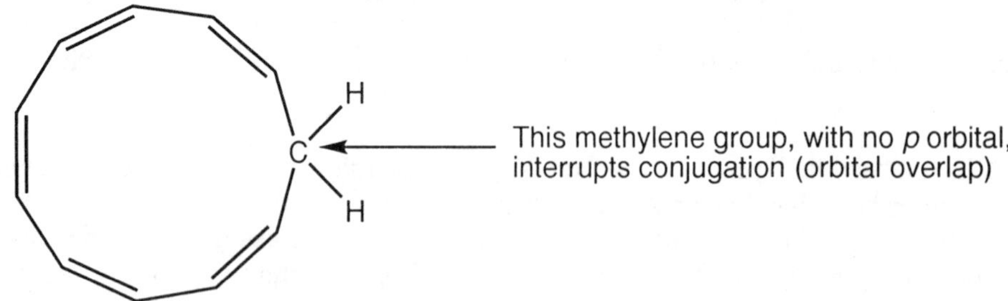

(b and c) These molecules fit all the criteria. They are cyclic, fully conjugated, can be planar, and have $4n + 2$ π electrons [b: 14 = (4 x 3) + 2; c: 18 = (4 x 4) + 2]. Each molecule will have a π molecular orbital system in which all bonding orbitals are full and with no nonbonding or antibonding orbitals occupied.

(d) The periphery of this molecule can be treated as a ring. If so, there is a cycle of 10 π electrons in overlapping (conjugated) orbitals. This molecule, too, will be aromatic.

(e) This time the periphery contains only 8 π electrons. This molecule should not be aromatic (and isn't).

Problem 13.7 The benzene rings in the triphenylmethyl or trityl cation allow the positive charge to be shared by nine other carbons. This delocalization is strongly stabilizing. In the figure, resonance stabilization using only one benzene ring is shown, but the other two rings can act in the same way.

Problem 13.8 The cyclopropenyl cation is an aromatic system (planar, cyclic, fully conjugated), with $4n + 2$ ($n = 0$) π electrons. Accordingly, despite the obvious strain and the presence of a positive charge, it is easy to form. For a cation, it is a very stable species.

A Frost circle shows the molecular orbital system. For the cyclopropenyl cation there are only two π electrons. Accordingly, the bonding molecular orbital is full and there are no electrons in the antibonding orbitals.

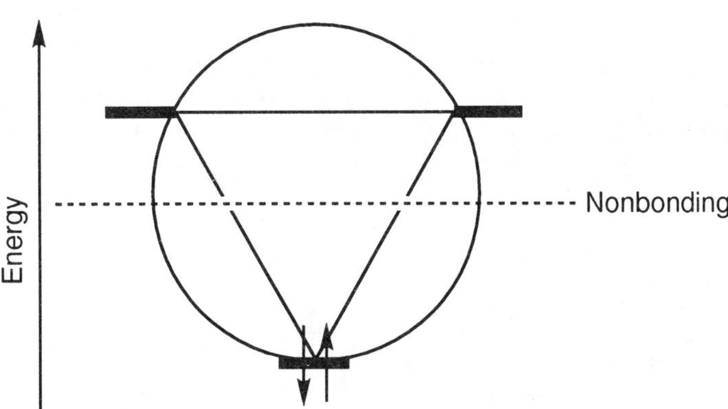

Problem 13.9 The cyclopropenyl anion presents a very different situation. Now there are four ($4n$, $n = 1$) π electrons and antibonding molecular orbitals must be occupied. This system would not be expected to be stable (and it most definitely is not).

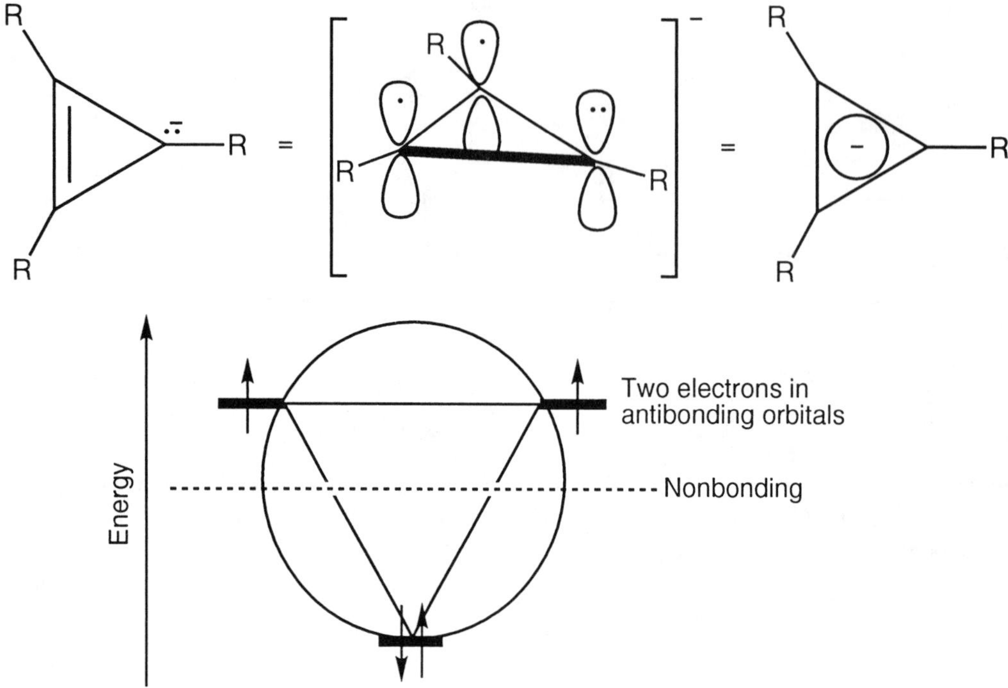
R
R
R
Two electrons in antibonding orbitals
Nonbonding
Energy

Problem 13.10 A trans double bond in a six-membered ring is too strained to be stable. Not only is one hydrogen "inside" and bumping into the other atoms, but the trans double bond is hideously twisted and thus most unstable. Try to make a model.

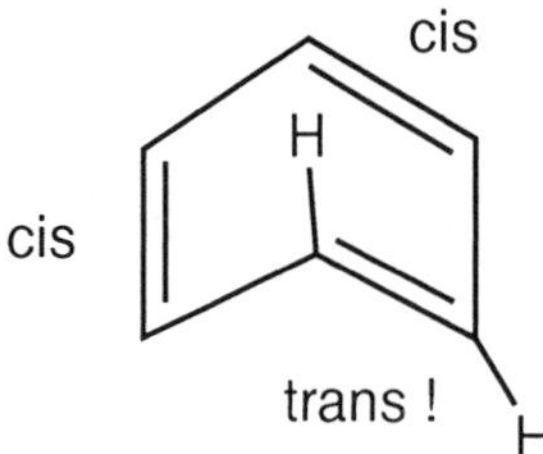
cis
cis
H
trans !
H

Problem 13.11 Here is the reaction shown for only one cyclodecapentaene. This picture is only a crude approximation as the cyclodecapentaenes are not really planar. Although it looks as if the "reach" across the ring to make the new bond is a long one, in reality the nonplanar stuctures bring the orbitals quite close together and make the bond making process easy.

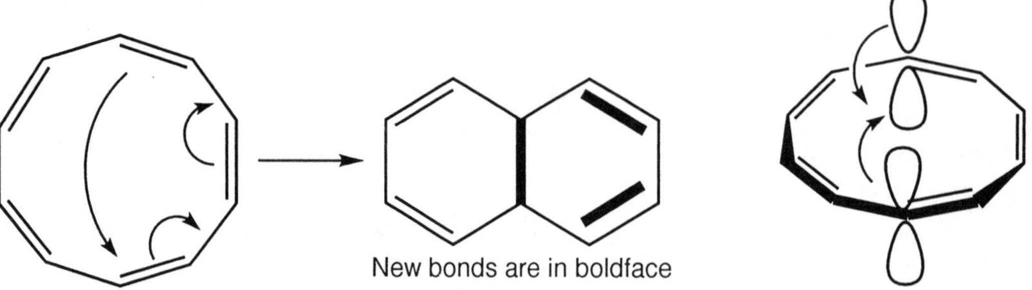
New bonds are in boldface
Side view showing two orbitals bending in to make the cross-ring bond

Problem 13.12 If the working definition of the molecular orbitals of an aromatic molecule includes the words, "with all bonding molecular orbitals filled and no electrons in antibonding or nonbonding orbitals," the cyclooctatetraene dianion fails the test. No less than four electrons occupy nonbonding orbitals. This dianion is remarkably stable, however, and is cyclic, planar, fully conjugated, and does have 10 π electrons ($4n + 2$, $n = 2$).

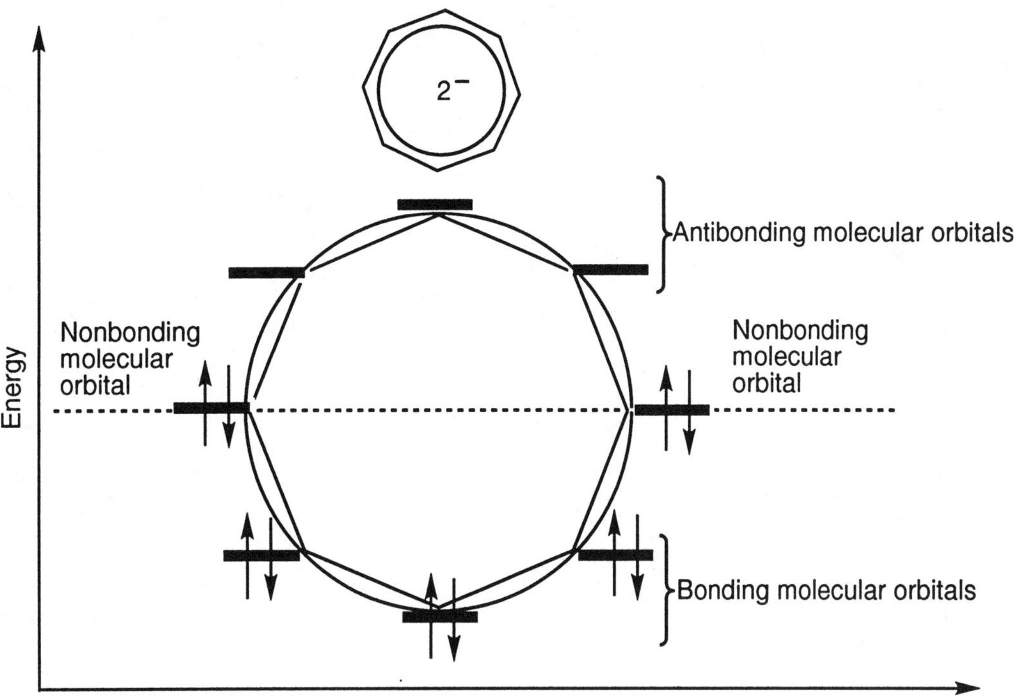

Problem 13.14 There is a singly occupied $2p$ orbital on every carbon of naphthalene and on the nine carbons and one nitrogen of isoquinoline. In isoquinoline, there is also a lone pair of electrons on nitrogen, but these are in an sp^2 orbital, perpendicular to the π system.

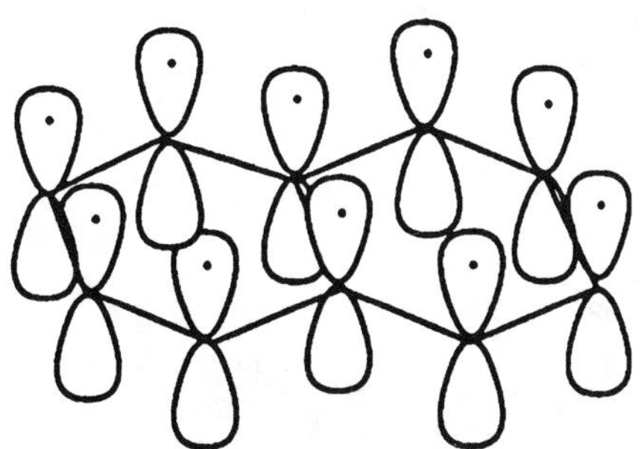

There is a total of 10 π electrons ($4n + 2$, $n = 2$) in naphthalene

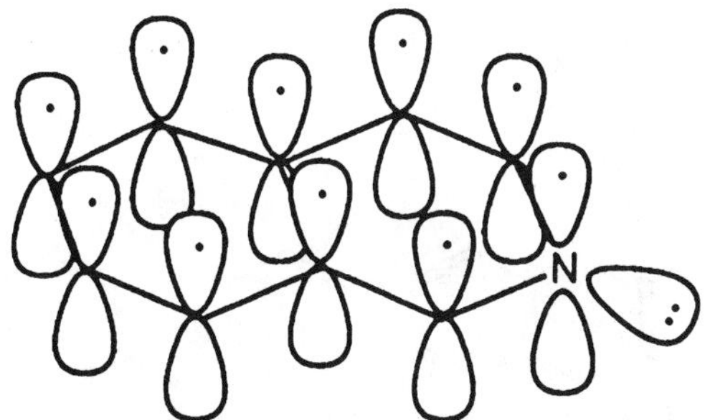

Notice that the lone-pair electrons on nitrogen are not in the π system; there are only 10 π electrons in isoquinoline

Problem 13.15 Certainly not. The spiro carbon cannot be sp^2, and therefore interrupts the π orbital connectivity between the two rings.

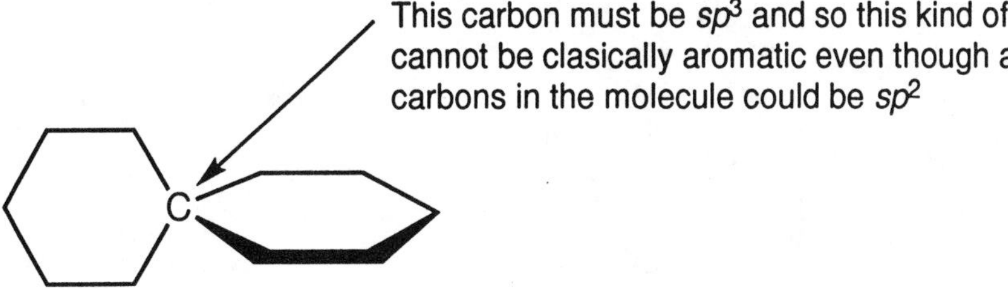

This carbon must be sp^3 and so this kind of molecule cannot be clasically aromatic even though all other carbons in the molecule could be sp^2
C

Problem 13.17 A tricky one. If you just draw circles, you are in big trouble. First of all, this is an isomer problem much like the ones in the early chapters, and you must evolve a system to be sure you have all the possible ring systems. I like to start with a linear array, and then increasingly complicate things.

Four-in-a-row

Three-in-a-row

Two-in-a-row (but watch out for repeats of three-in-a-row isomers); in this scheme, one new ring is added at the upper right corner (boldface) and the fourth ring is moved around the frame

Now add double bonds. If you draw them as circles, all of the above frames will work, but this isn't right! As in Problem 13.16, there is one frame for which it is impossible to make all carbons sp^2 and therefore it cannot be aromatic.

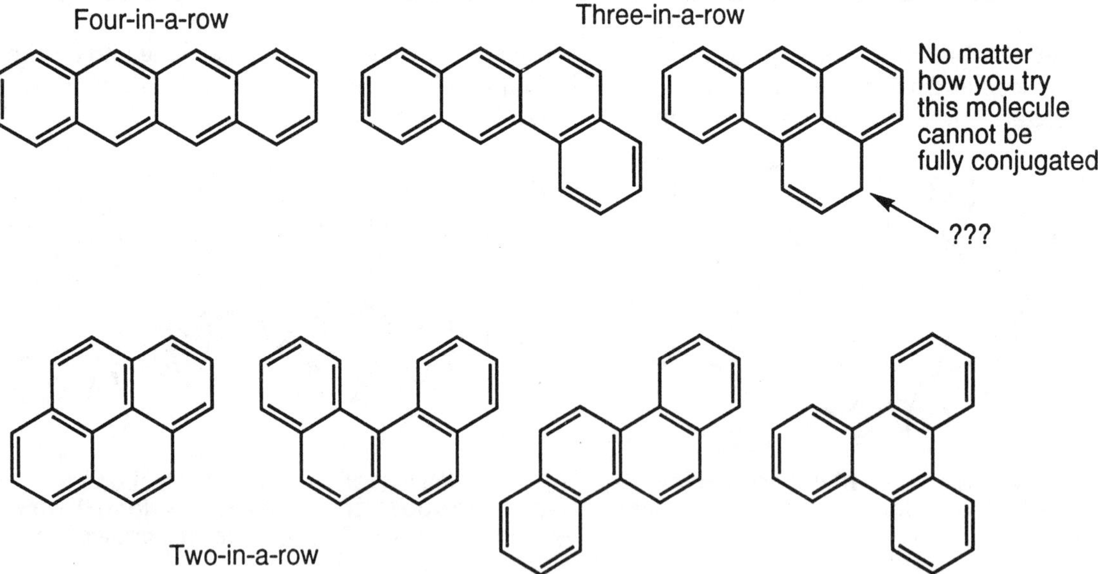
Four-in-a-row
Three-in-a-row
No matter how you try this molecule cannot be fully conjugated
???
Two-in-a-row

Problem 13.18 Hexahelicene is drawn by adding a sixth ring to the five-ring aromatic at the indicated points.

Attach new ring here

A

The problem is that the new ring overlaps with ring **A**. Hexihelicene must twist so that the new ring and ring **A** lie on top of each other. The molecule must be curled in either a right- or left-handed screw form.

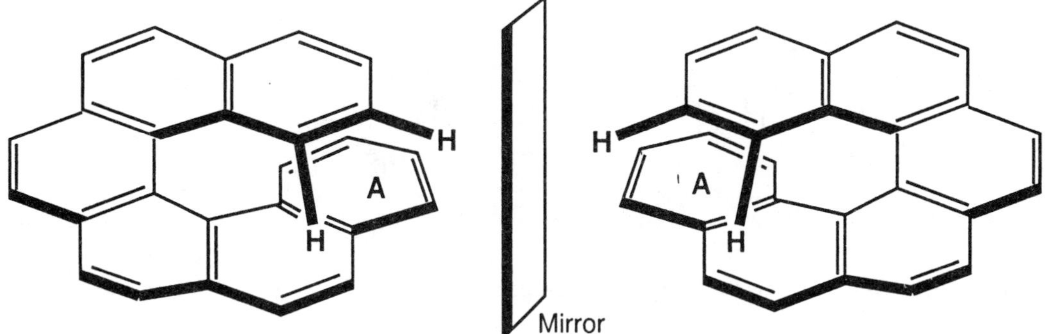

"Twistoflex" is also a molecule that adopts a screw shape, but this time it is along a longitudinal axis, not the vertical direction as with hexihelicene.

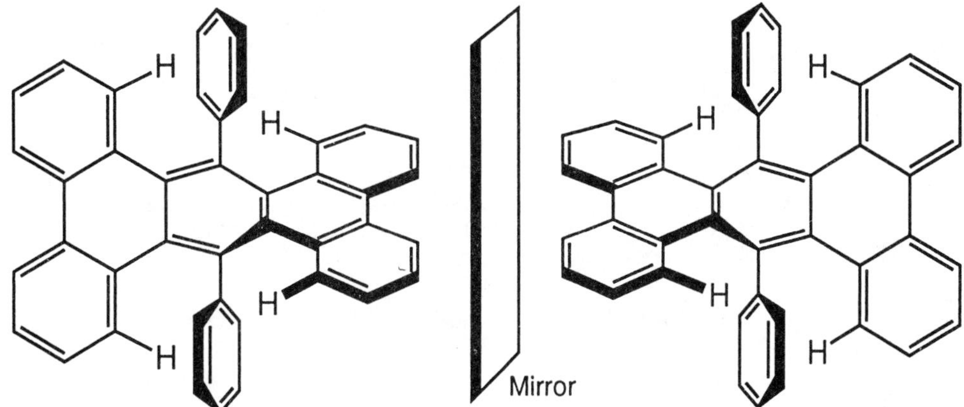

Problem 13.21 In both molecules we can imagine an ionization to give a carbocation. For an adjacent benzene ring to stabilize a positive charge, the orbitals of the ring must overlap with the empty orbital on the benzylic carbon adjacent to the ring. For the trityl system this is quite easy:

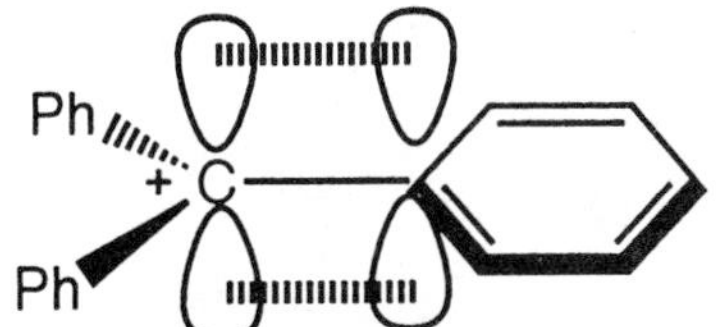

In the trityl cation, overlap between the ring orbitals and the 2*p* orbital on the benzyl position is easy; in this figure only one benzene ring is shown in detail, and only one of the six 2*p* orbitals making up the benzene π orbitals is drawn in

The situation is quite different for tripticenyl. Although the positively charged carbon is flanked by three benzene rings, just as is the case in trityl, the orbitals do not overlap well. They are essentially perpendicular, and the benzene rings offer no stabilization to the adjacent empty orbital.

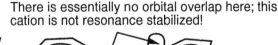

There is essentially no orbital overlap here; this cation is not resonance stabilized!

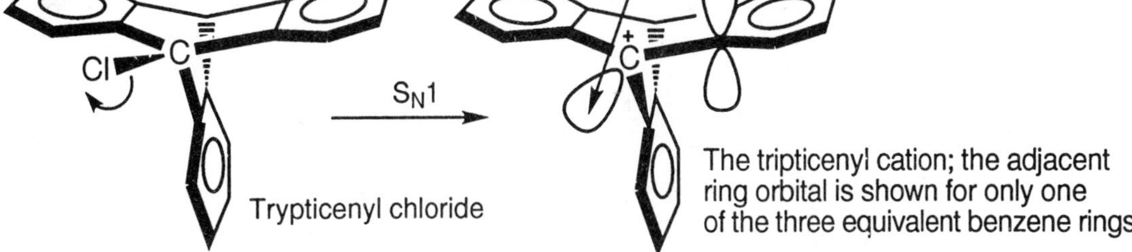

Problem 13.22 The mechanism is exactly parallel to that for allylic bromination using NBS. The first step is formation of a bromine atom from homolytic breaking of the weak nitrogen-bromine bond.

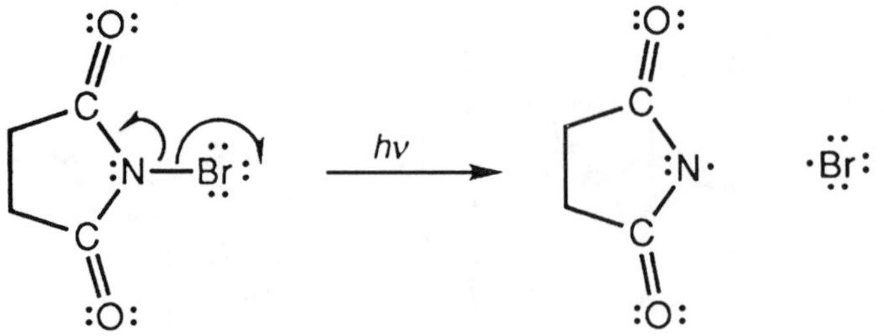

hv

Next, the bromine atom abstracts a benzylic hydrogen from the methyl group of toluene to produce hydrogen bromide and a resonance-stabilized benzyl radical.

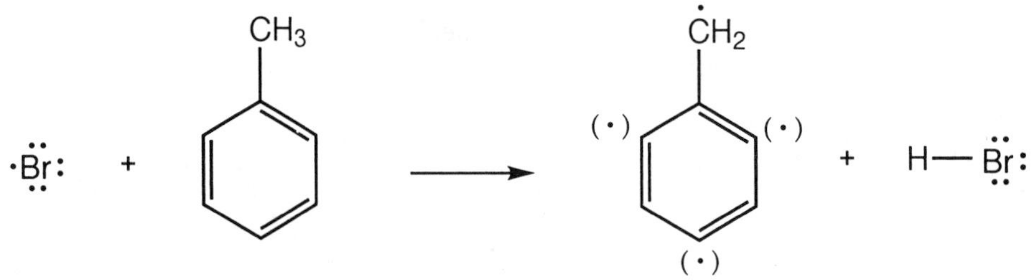

Toluene

The resonance-stabilized benzyl radical; the (•) shows the positions sharing the free electron

N-Bromosuccinimide reacts with the hydrogen bromide formed in this step to give a small amount of bromine.

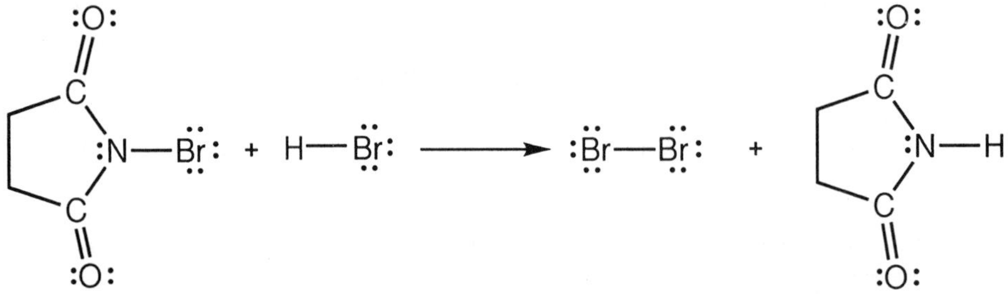

The benzyl radical now abstracts a bromine atom from bromine to give benzyl bromide and a bromine atom that goes on to propagate the chain reaction by abstracting a hydrogen from toluene.

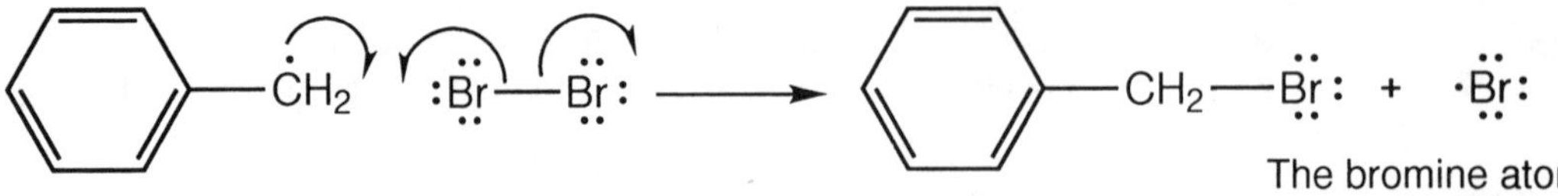

Benzyl bromide

The bromine atom recycles to abstract another hydrogen from the benzylic position of toluene

Additional Problem Answers

Problem 13.24
(a) 1-Bromo-3-fluorobenzene, *m*-bromofluorobenzene
(b) 1-Bromo-4-fluorobenzene, *p*-bromofluorobenzene
(c) 1-Bromo-2-fluorobenzene, *o*-bromofluorobenzene
(d) 4-Methylbenzoic acid, *p*-methylbenzoic acid, *p*-toluic acid
(e) 1,3,5-Trimethylbenzene, mesitylene
(f) 1-Bromo-2-chloro-3-iodobenzene

Problem 13.25 Nothing tricky here - just a few repeats as (b) and (c) are the same, as are (d), (e), and (f).

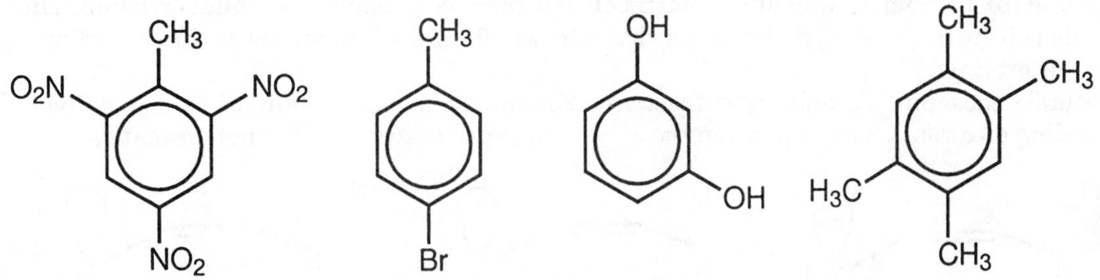

CH_3 O_2N NO_2 NO_2 (a)

CH_3 Br (b and c)

OH OH (d, e and f)

CH_3 CH_3 H_3C CH_3 (g)

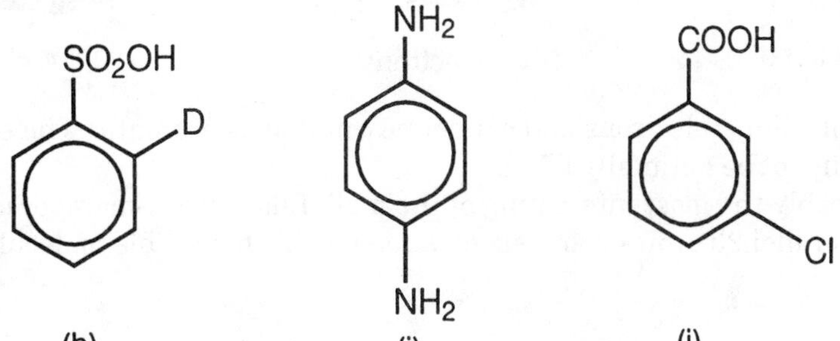

SO_2OH D (h)

NH_2 NH_2 (i)

COOH Cl (j)

Problem 13.26 Dewar benzene is easy, and benzvalene and prismane aren't much harder. In the last two cases, the molecule is redrawn with no atoms moved, then relaxed to benzene.

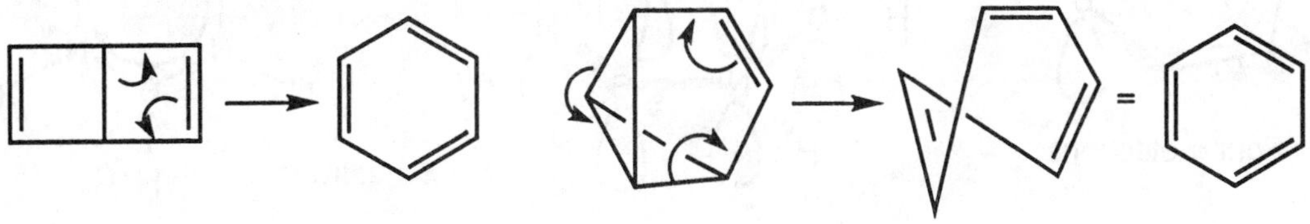

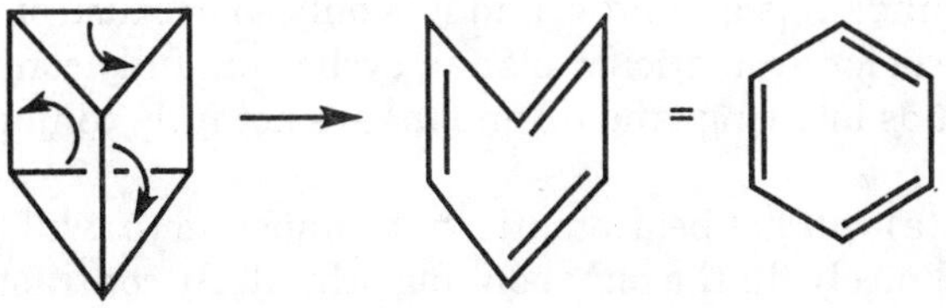

The last molecule, 3,3'-bicyclopropenyl, usually presents a tougher problem. Here is one suggestion.

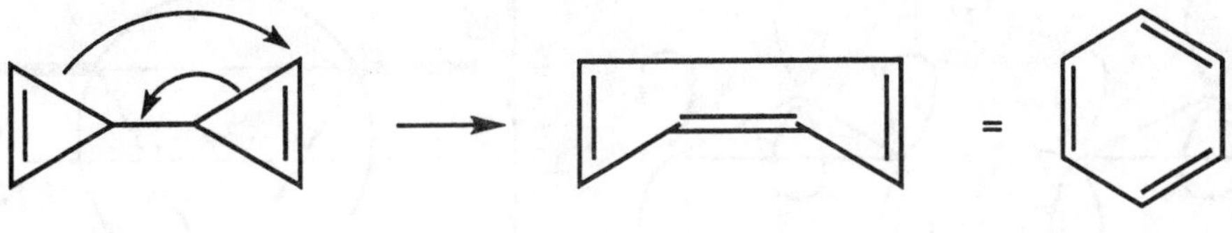

Problem 13.27 Here are the Lewis structures.

Compound (a), azepin, has eight π electrons and is not aromatic.
Compound (b), borepin, is aromatic. Remember that there is an empty $2p$ orbital on boron. This molecule is fully conjugated, and its six π electrons will fit nicely into the set of three bonding molecular orbitals.
Compound (c), oxepin, has only eight π electrons despite the apparent count of 10. Two of the nonbonding electrons on oxygen are in the σ system, not the π system. It is not aromatic.

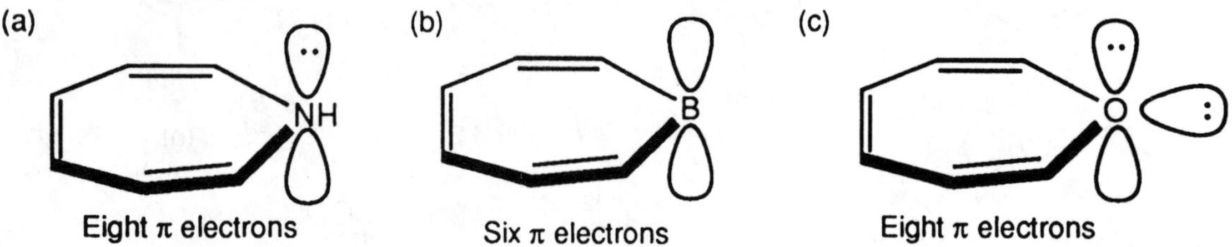

Compound (d) has only four π electrons and cannot be counted as aromatic. The set of three bonding molecular orbitals will not be optimally filled.
Compound (e) is probably the most interesting of them all. Like the six-membered ring benzene, there is a set of six parallel $2p$ orbitals and six electrons to fill them. This molecule, borazine, is aromatic.

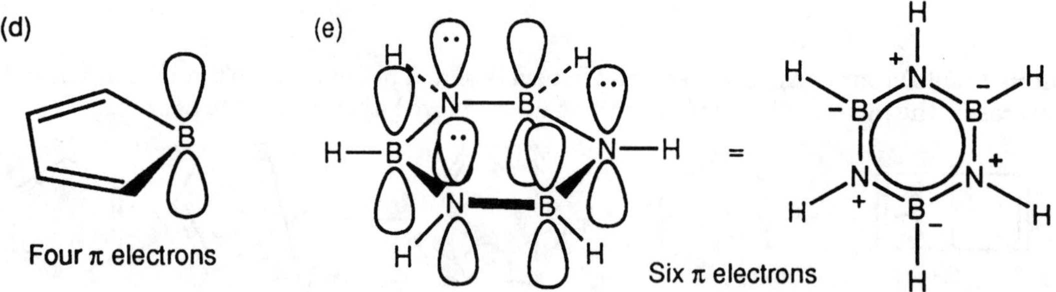

Problem 13.28 Only compound (a) is aromatic, and it is only so because of the benzene ring. None of the others meets the criteria for aromaticity: planar, cyclic, and fully conjugated. Neither (b) nor (e) contains all its double bonds in a ring, and (c) and (d) are not fully conjugated.

Problem 13.29. Compound (a) will not be aromatic, but compound (b) will be. In (b) there are only two π electrons and they will nicely fill the only bonding orbital. By contrast, compound (a) has four π electrons and two of them must occupy antibonding molecular orbitals.

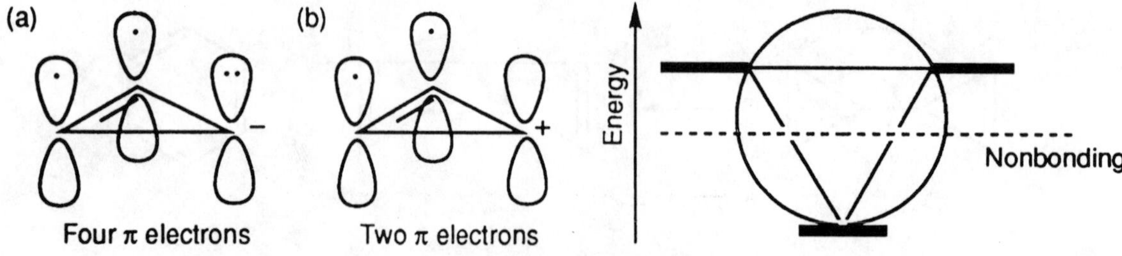

Compound (c) also qualifies as aromatic. Neutral cyclobutadiene has four electrons, and there will be two fewer in the dication. There are only two π electrons, and one bonding molecular orbital. The two positive charges will make this molecule very unstable, however.

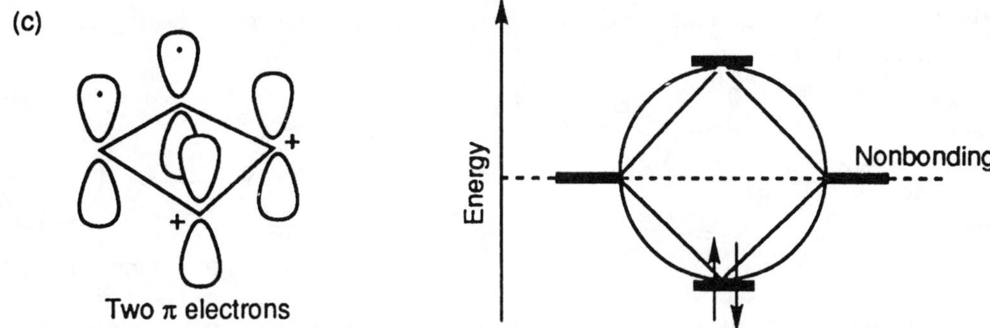

Neither (d) nor (e) is fully conjugated, and thus neither can be aromatic. There is a methylene (CH_2) group interrupting the overlap of $2p$ orbitals.

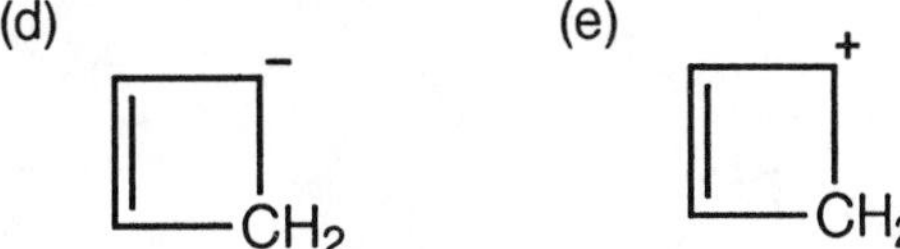

The final two compounds, (f) and (g), are a nonaromatic and aromatic pair, like (a) and (b). Anion (g) is a 10 π electron molecule, but cation (f) is an 8 π electron species.

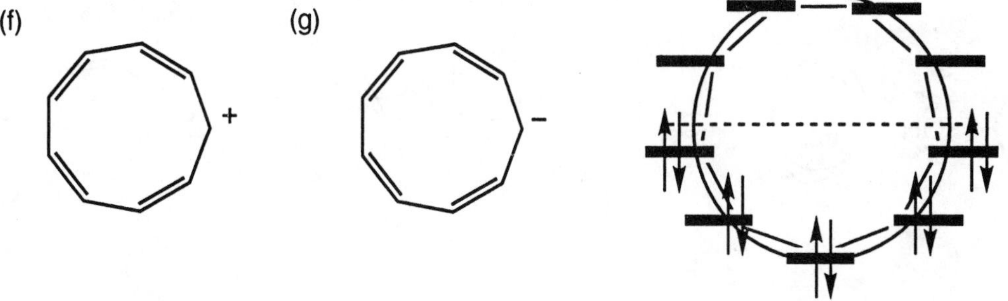

Problem 13.30 If we estimate the heat of hydrogenation for cyclooctatetraene as four times that of cyclooctene, then we predict 4 x 23 = 92 kcal/mol, not much below the experimental number of 101 kcal/mol. No case can be made that cyclooctatetraene is especially stable. By contrast, this molecule appears to be less stable than anticipated.

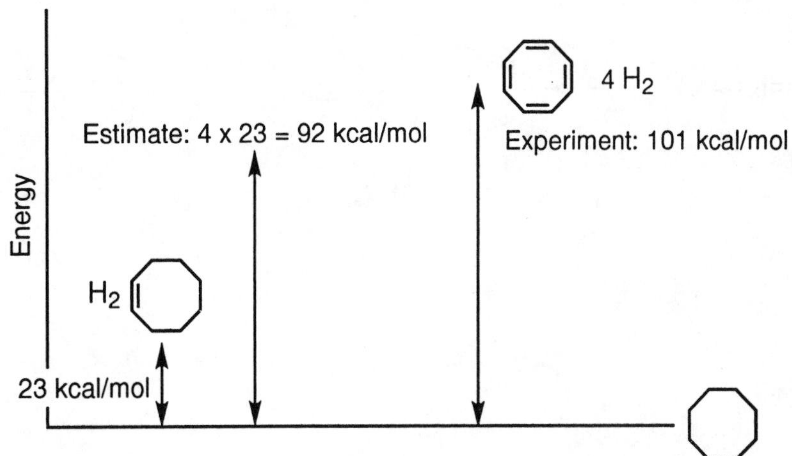

Of course, experiment bears this out: Cyclooctatetraene is a tub-shaped, decidedly nonplanar molecule and reacts as a simple polyene.

Problem 13.31 There are several clues as to what is going on. The presence of hydrogen bromide might evoke thoughts of an elimination reaction once bromine has added, for example.

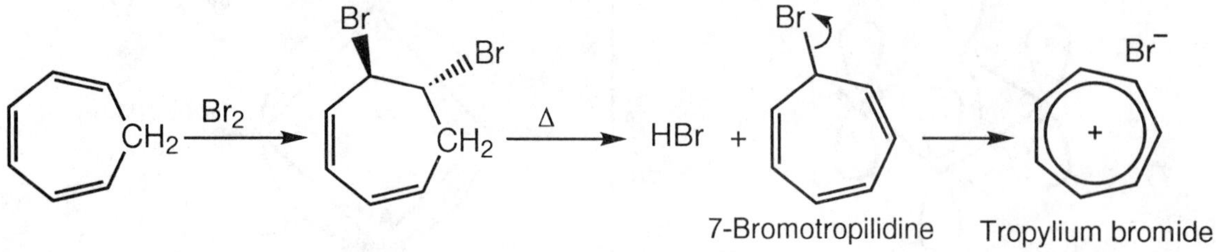

7-Bromotropilidine Tropylium bromide

7-Bromotropilidene is the result, and this molecule easily ionizes to give the very stable tropylium bromide, the yellow solid. Reaction with water gives tropyl alcohol, a molecule that can react again with tropylium bromide to give ditropyl ether.

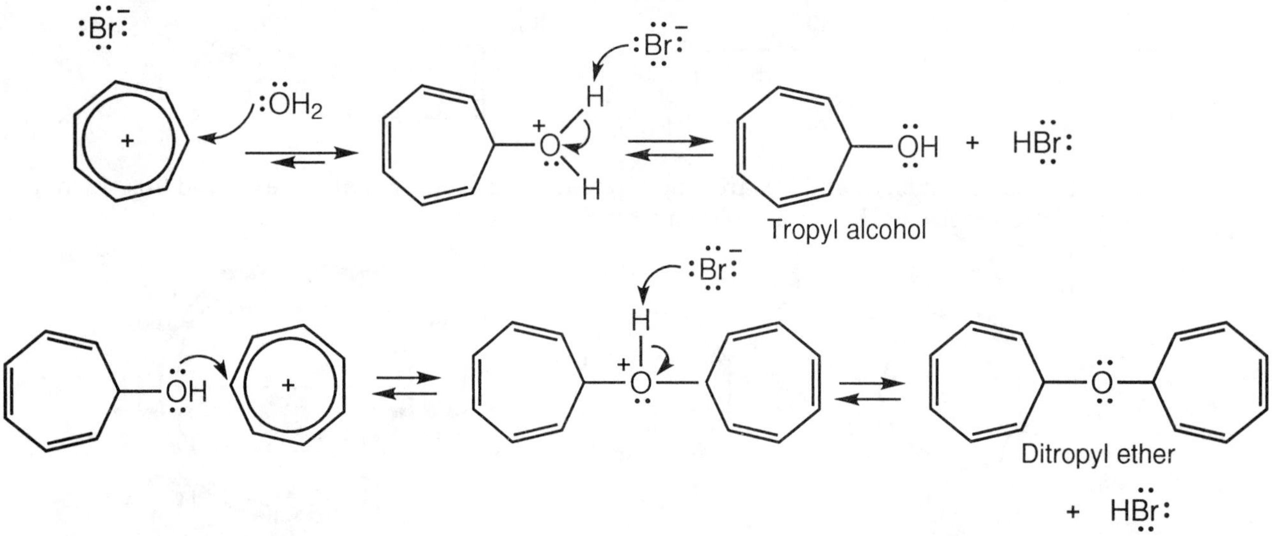

Tropyl alcohol

Ditropyl ether

Problem 13.32 Look at the resonance description of diazocyclopentadiene. This molecule is sharply stabilized by delocalization of the negative charge on carbon. There is an aromatic cyclopentadienide anion within this molecule. This compound is still a *potential* explosive, however, and there are several famous detonations involving this beautiful material. Very unstable diazo compounds, diazomethane (CH_2N_2), for example, are handled in solution and with great caution. Compounds such as diazocyclopentadiene, because of their very stability, can be distilled and handled in bulk. When the occasional explosion does occur, the results can be disastrous.

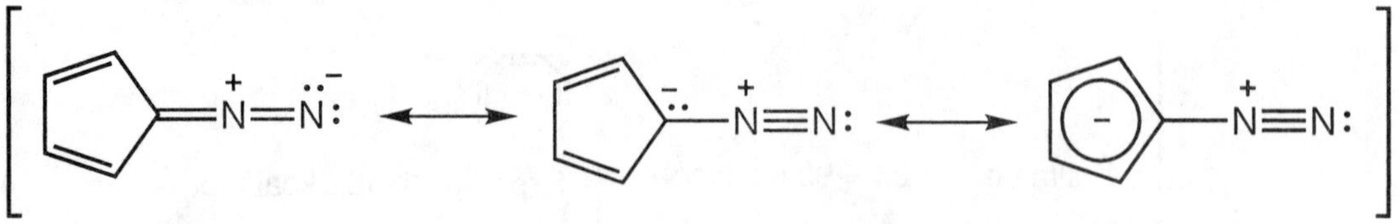

Problem 13.33 The benzene molecular orbitals, **1 - 6**, fall right out if we take combinations of the three allyl orbitals (bonding, Φ_B, nonbonding, Φ_N, and antibonding, Φ_A) in the following ways: Φ_B with Φ_B, Φ_N with Φ_N, and Φ_A with Φ_A.

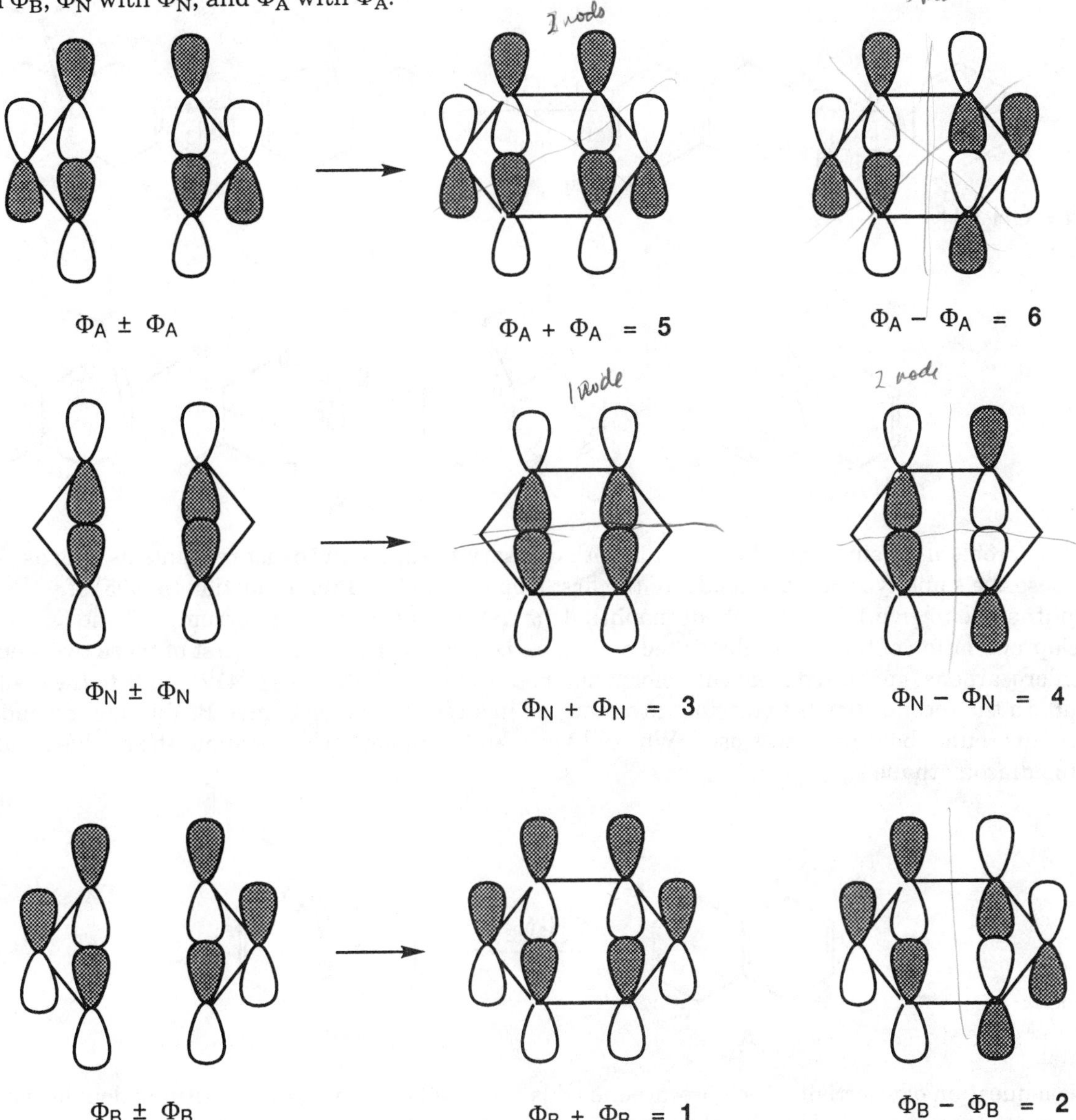

They can be ordered in energy by counting the nodes.

Orbitals	New nodes (ignoring the node present in any p or π orbital)
6	3
4, 5	2
---- Nonbonding ----	
2, 3	1
1	0

Problem 13.34

Na / NH_3 / EtOH → **A** → $CHCl_3$ / $(CH_3)_3COK$ → Na / NH_3 → **B** → Br_2/CCl_4 → KOH / EtOH → **C** → **1**

Et = CH_2CH_3

In this problem (or similar problems), it is not necessary to supply the exact reagents as long as you propose the same type of reagents. Here the first step is a double Birch reduction (p. 605) of naphthalene to give 1,4,5,8-tetrahydronaphthalene (**A**). The next two steps accomplish the cyclopropanation of the more substituted carbon-carbon double bond. In the first of these two steps, dichlorocarbene, generated from chloroform and potassium *tert*-butoxide (p. 412), adds to the double bond. In the second step, the chlorines are removed in a reduction step to give **B**. You might wonder why this roundabout route was used. Why did Vogel and Roth not cyclopropanate the double bond using diazomethane?

A → CH_2N_2 / $h\nu$ → **B**

It is a question of selectivity. Dichlorocarbene adds selectively to the more substituted double bond, but methylene, generated by photolysis of diazomethane, is far more reactive and does not select one double bond over the other. Rather, it reacts indiscriminately to give two addition products.

With cyclopropane **B** in hand, it is necessary to introduce additional unsaturation. This transformation is conveniently accomplished through a bromination-dehydrobromination sequence to give **C**. Finally, tetraene **C** undergoes a smooth transformation (arrows) to the desired bridged cyclodecapentaene, **1**.

Problem 13.35 Here are the Lewis structures.

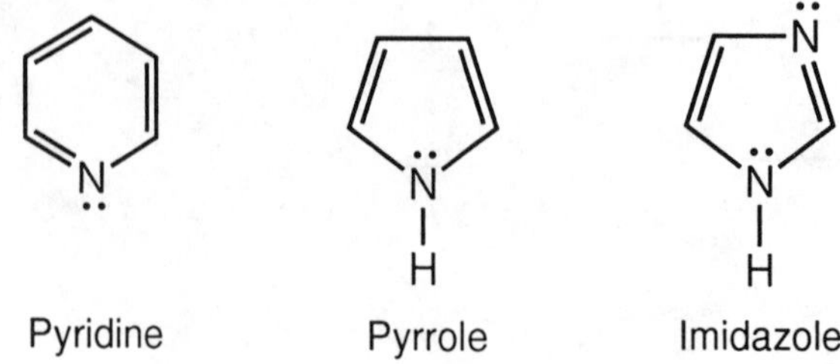

All three of these molecules are heteroaromatic compounds. To find a sensible answer, it is important to examine closely the lone pair electrons on the nitrogen atoms in these compounds. In pyridine, the nitrogen lone pair electrons are not part of the $(4n + 2)$ π system, but lie in the plane of the ring in an sp^2 orbital that is part of the σ system. Accordingly, aromaticity is not disrupted when pyridine is protonated.

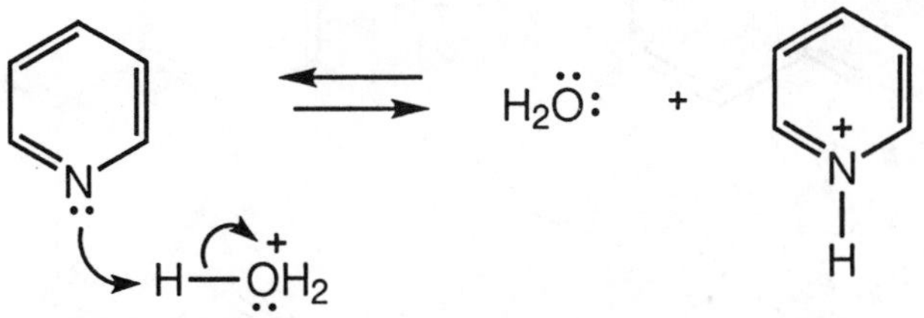

By contrast, the nitrogen lone pair electrons are part of the aromatic sextet in pyrrole. Protonation on nitrogen would result in a loss of aromaticity, and is likely to be a very endothermic process.

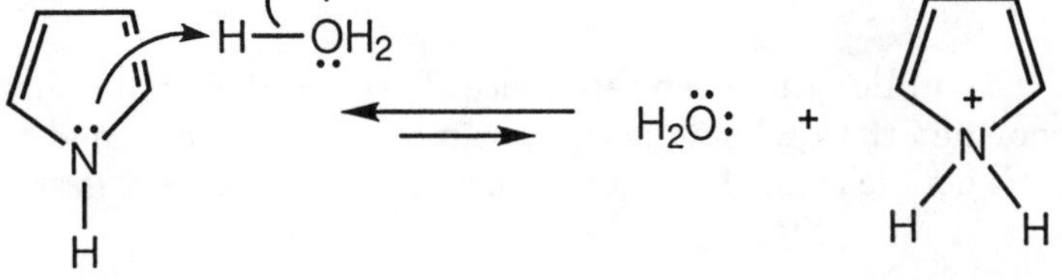

When pyrrole does protonate, it does so on carbon. Although aromaticity is still lost, at least a resonance-stabilized cation is formed.

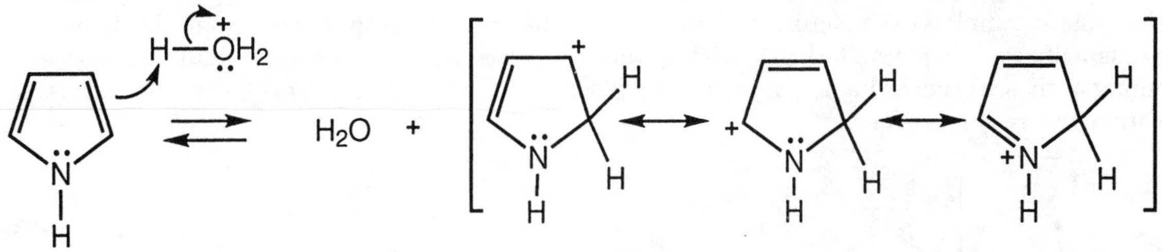

The two nitrogen lone pairs of imidazole are quite different. One nitrogen lone pair is part of the six π electron system and the other is not, as it lies in the σ system. Accordingly, protonation occurs on the latter lone pair as aromaticity is not lost.

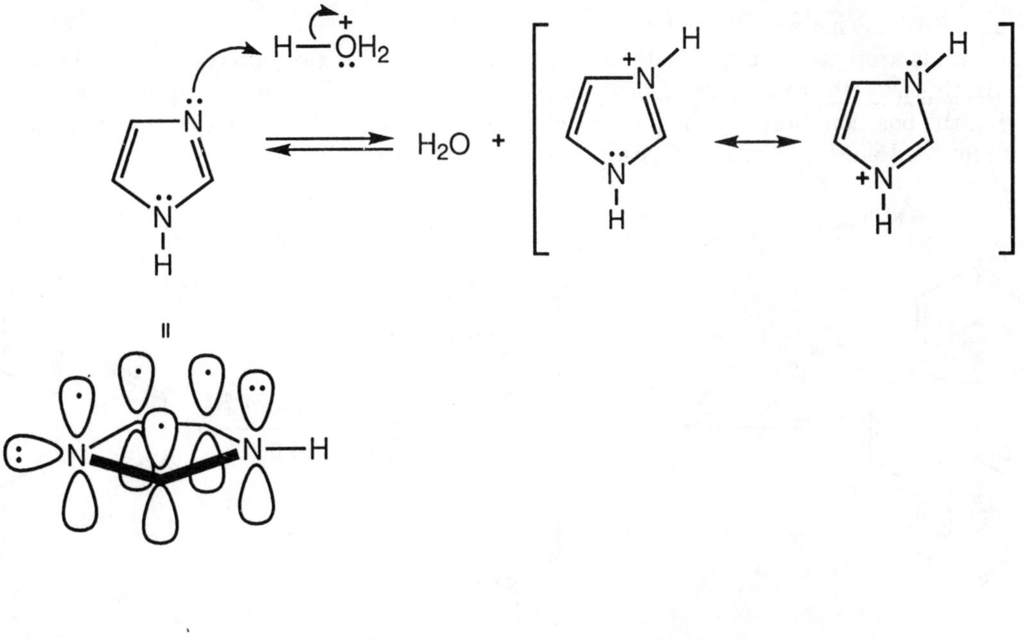

Problem 13.36 This problem is another case where circles are misleading, and Kekulé forms more appropriate. Anthracene can be represented by four resonance structures.

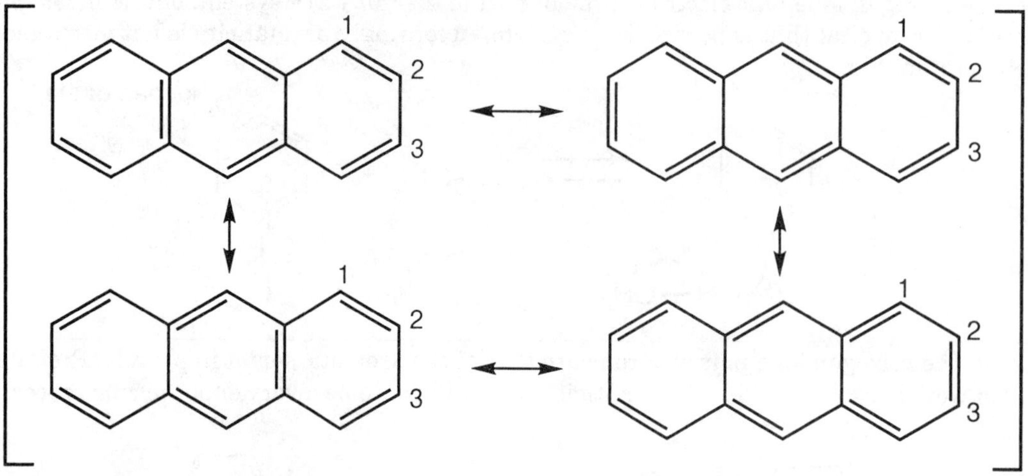

If we make the useful assumption that each resonance structure contributes equally to the hybrid structure of anthracene, then the 1,2-bond has approximately 3/4 double bond character, whereas the 2,3-bond has only 1/4 double bond character. Accordingly, the 1,2-bond is shorter than the 2,3-bond.

Problem 13.37 This problem is just a specific example of the general aromatic substitution process described in the chapter (p. 603). The nitronium ion, ($^+NO_2$), is a Lewis acid and first adds to the benzene ring to give a resonance-stabilized cyclohexadienyl cation. Removal of a proton by any base in the system completes the picture and regenerates the aromatic ring. Even though the arrow formalism shows this proton loss as deriving from one resonance form, do not fall into the trap of thinking of these forms as having separate existence. Write arrow formalisms for proton loss using the other two resonance forms.

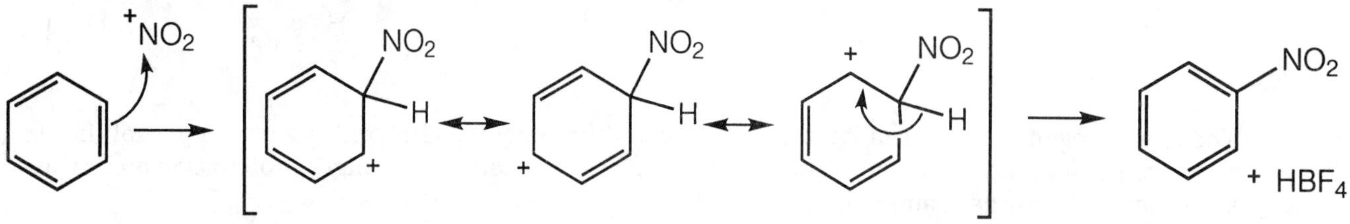

+ HBF_4

Problem 13.38. This molecule is just a fancy benzene. It has a 14 π electron ($4n + 2, n = 3$) perimeter and is aromatic. Aromatic substitution with $^+NO_2$ (nitration) proceeds as in the previous problem. In this case the resonance structures of the intermediate are indicated in shorthand with (+) showing the positions sharing the positive charge. It is worth taking the time to write out the resonance forms. Removal of a proton regenerates the aromatic system.

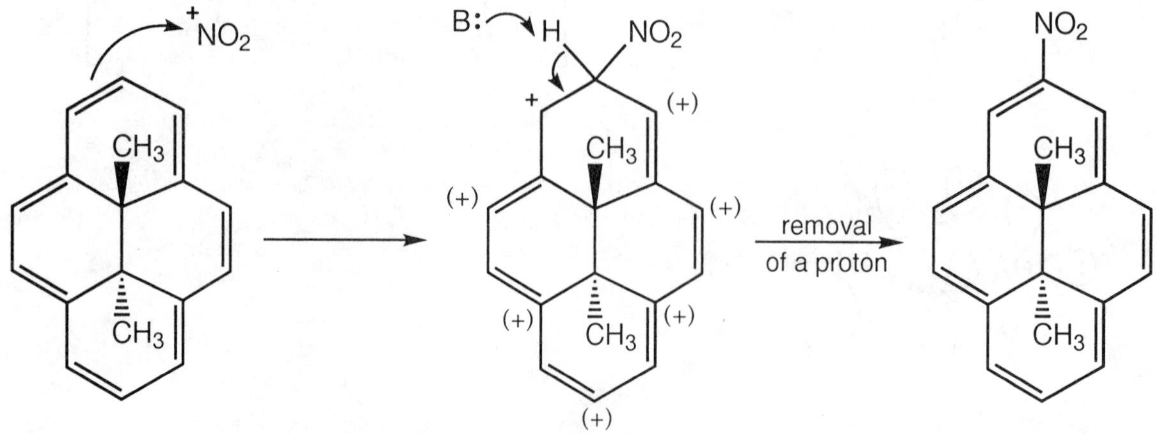

removal of a proton

Problem 13.39 The poor Professor can count, but he doesn't know much about π systems. Dimm had simply added up the electrons in the π bonds, counting the acetylene as four electrons. Thus, he reached a total of 16, and anticipated no aromatic stabilization. What he didn't realize was that two of the acetylene electrons are not in the π system and should not be counted.

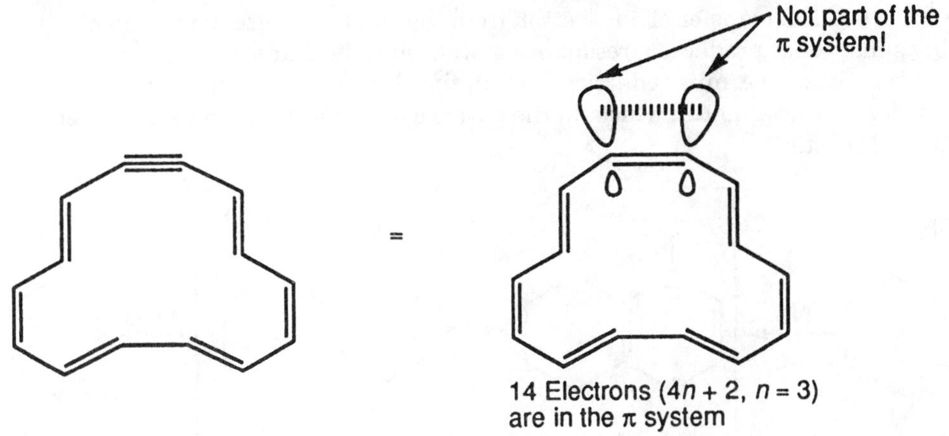

Not part of the π system!

14 Electrons ($4n + 2$, $n = 3$) are in the π system

Problem 13.40

(a) Anthracene adds across the 9,10-positions to form a single adduct. If you drew the product of addition across the 1,4-positions, that's not a bad answer, but it is the symmetrical compound that is really formed. See (b) for an analysis of why this occurs. Naphthalene adds across the 1,4-positions to give exo and endo adducts.

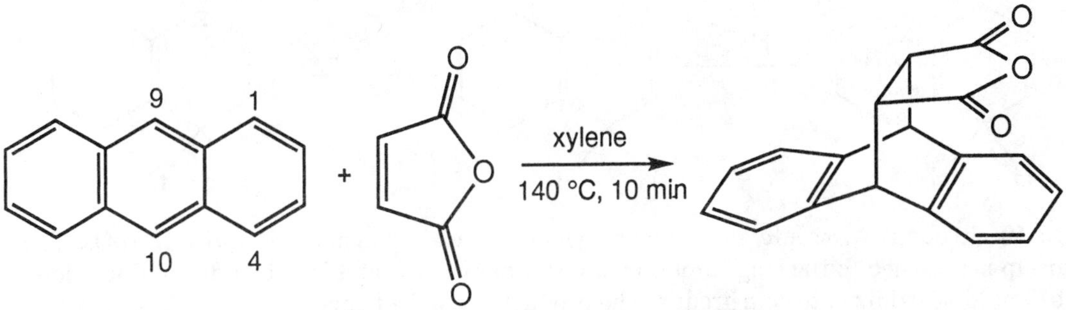

9 1 10 4 + xylene 140 °C, 10 min

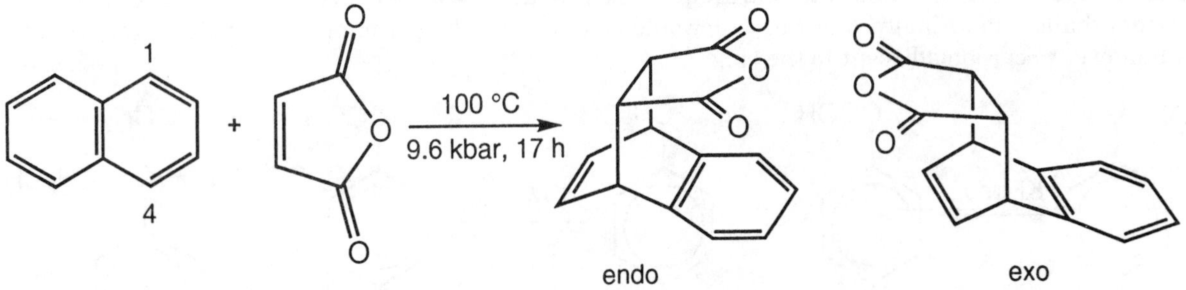

1 4 + 100 °C 9.6 kbar, 17 h

endo exo

(b) The resonance energies of naphthalene and anthracene are 61 and 84 kcal/mol, respectively, as calculated from heats of combustion data. When anthracene undergoes the Diels-Alder reaction across the 9,10-positions, two intact benzene rings remain in the product. In the process, anthracene loses only about 25% of its resonance energy as one anthracene (84 kcal/mol resonance energy) becomes two benzene rings (2 x 32 = 64 kcal/mol resonance energy). Addition across the 1,4-positions would leave behind a naphthalene, with a resonance energy (61 kcal/mol) less than that of two benzenes (about 64 kcal/mol). On the other hand, when naphthalene undergoes the Diels-Alder reaction, about 45% of its resonance energy is lost as naphthalene (61 kcal/mol resonance energy) becomes one benzene ring (32 kcal/mol resonance energy). Although this is simply a statement about the thermodynamics of product energies, and thus not directly relevant to questions of rates (kinetics), loss of aromatic resonance may already be considerable in the transition states leading to the products. Thus, a higher activation energy would be expected for the reaction with naphthalene.

Problem 13.41 This problem is an example of the Birch reduction with a slight, added twist. As implied in Figure 13.80, transfer of an electron from sodium to a benzene with an electron-withdrawing substituent produces a resonance-stabilized radical anion **A**. Apparently, activation by the carbonyl substituent permits reduction even in the absence of a proton donor alcohol. Further electron transfer affords anion **B**, which, in the absence of an alcohol, can be alkylated by methyl iodide in an S_N2 reaction.

COOH ·Na → [COO⁻ Na⁺ H ↔ COO⁻ Na⁺ H] A → NH_3 COO⁻ Na⁺ H H → Na COO⁻ Na⁺ H H B → I—CH_3 S_N2 → H_3C COO⁻ Na⁺ H H → H_3O^+ → H_3C COOH H H

Problem 13.42. Structure (a) would give *p*-*tert*-butylbenzoic acid upon oxidation with $KMnO_4$. The *tert*-butyl group is attached to the ring through a quaternary carbon and is not oxidized. Both side chains in (b) would be oxidized to acid groups. The product would be benzene-1,4-dicarboxylic acid (terephthalic acid). Finally, compound (c) would be untouched by permanganate, as it contains only a quaternary carbon adjacent to the ring.

(a) CH_3 / $C(CH_3)_3$ —$KMnO_4$→ COOH / $C(CH_3)_3$

(b) CH_2CH_3 / $CH(CH_3)_2$ —$KMnO_4$→ COOH / COOH

(c) CH_3 / $H_3C—C—CH_2CH_3$ / phenyl — No reaction

Chapter 14 Outline

14

Substitution Reactions of Aromatic Compounds

In this chapter, we see many reactions in which the aromatic ring is preserved and a few in which it is destroyed. The emphasis in the following problems is upon electrophilic aromatic substitution and the interplay between substituents on the ring. Synthesis appears in an important way as well, and retrosynthetic analysis emerges as an effective way to solve problems of synthesis.

Mechanism problems are becoming increasingly important. In the answers to these problems, we emphasize the importance of analyzing problems before setting out to write structures and arrows.

Problem 14.1 Hydrogenation of benzene leads first to cyclohexadiene, which is hydrogenated further to cyclohexene. Cyclohexene, in turn finally, gives cyclohexane. However, the rates of hydrogenation of the nonaromatic cycloalkenes are much faster than that of the much more stable, and therefore much less reactive, benzene. In practice, as soon as a molecule of cyclohexadiene is formed, it is converted into cyclohexene, and then into cyclohexane much faster than more cyclohexadiene is produced. The nonaromatic intermediates are present, but their concentrations can never build up.

$H_2/Pd/C$ SLOW → ; $H_2/Pd/C$ fast → ; $H_2/Pd/C$ fast →

Problem 14.2 Benzene depends for its great stability on perfect overlap of the six 2*p* orbitals in the ring. When the ring is distorted from planarity, orbital overlap decreases between the *p* orbitals at the "prow" and "stern" positions and those at the sides of the boat. This loss of overlap is strongly destabilizing.

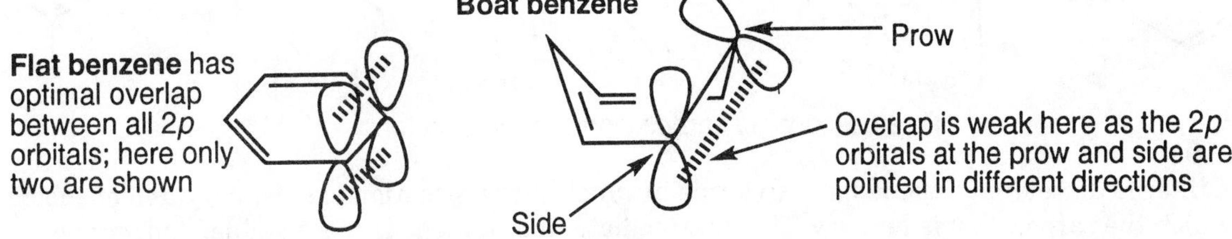
Boat benzene
Prow
Flat benzene has optimal overlap between all 2p orbitals; here only two are shown
Overlap is weak here as the 2p orbitals at the prow and side are pointed in different directions
Side

Problem 14.3 Here is another "just follow the arrows" problem. The structure of the product shows that addition does not take place at the positions to which the bridge is attached. That leaves only one possible Diels–Alder reaction. In a problem this complicated, be sure to draw the initial adduct without moving any atoms. Then, and only then, "relax" it into a better structure.

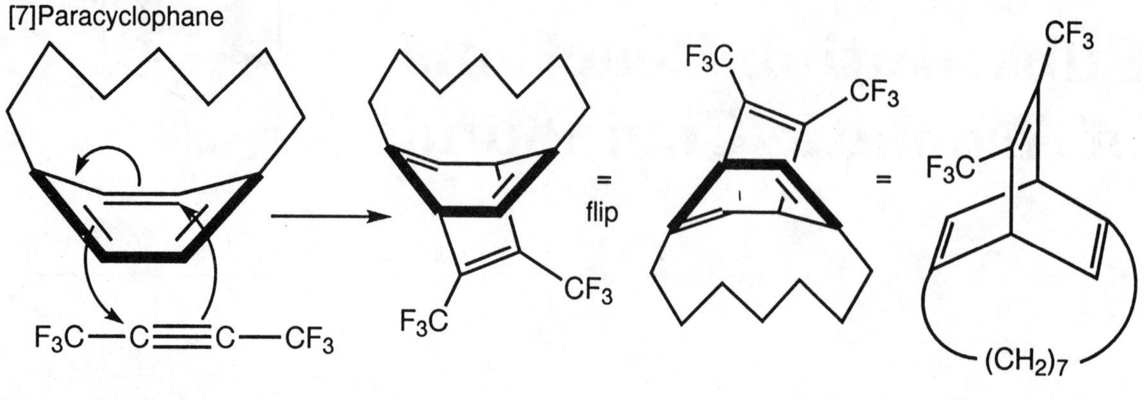

[7]Paracyclophane

$F_3C{-}C{\equiv}C{-}CF_3$

= flip

=

$(CH_2)_7$

Problem 14.5 The mechanism exactly parallels that of Figure 14.32. You need only substitute Br for Cl.

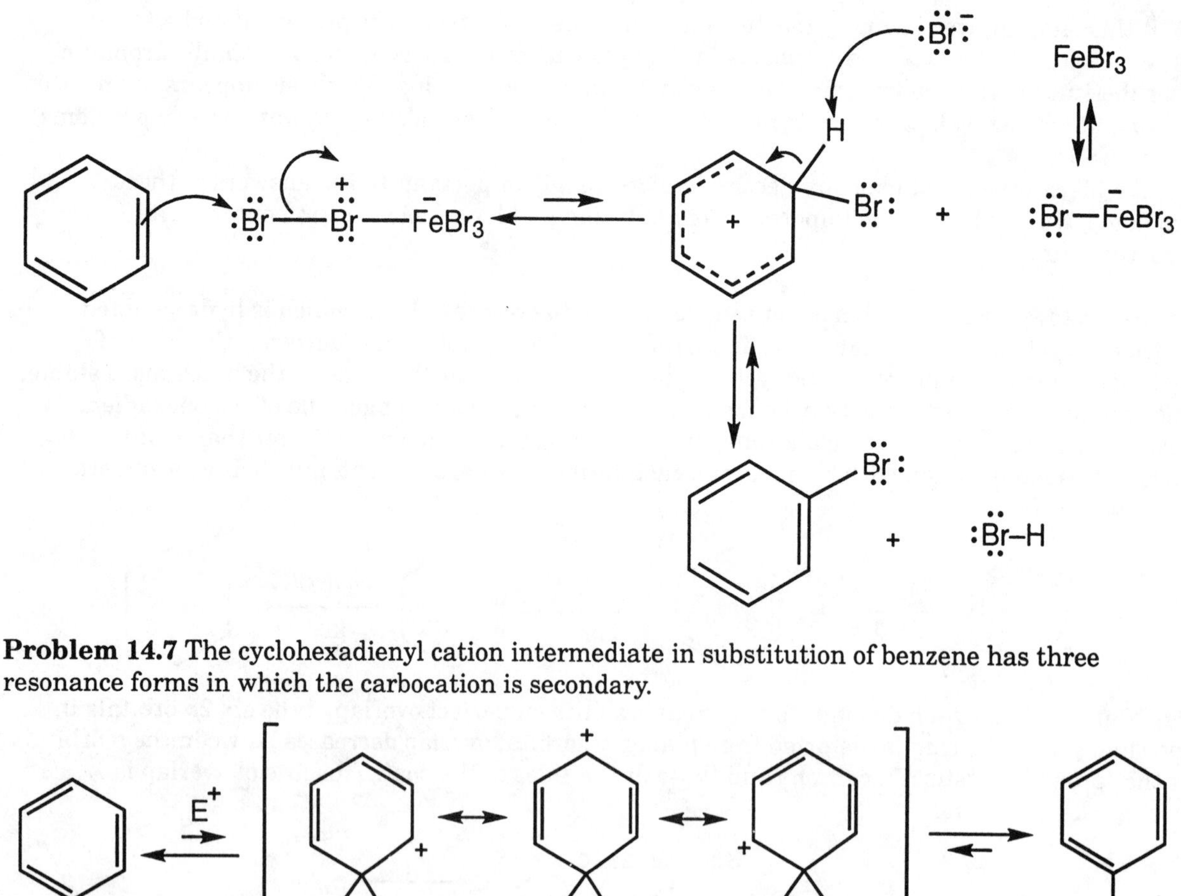

Problem 14.7 The cyclohexadienyl cation intermediate in substitution of benzene has three resonance forms in which the carbocation is secondary.

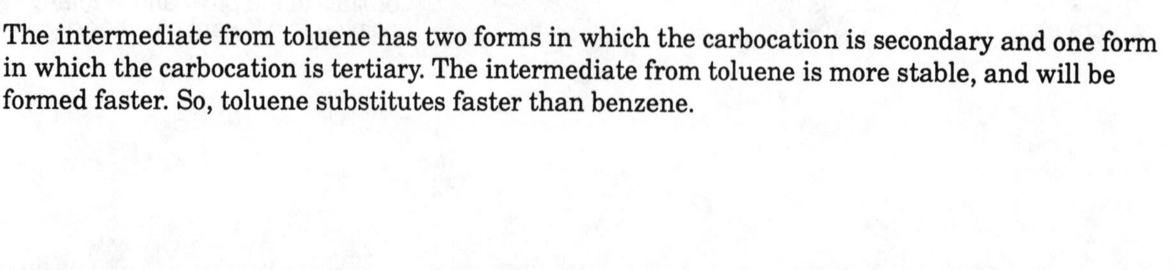

Three secondary carbocationic resonance forms

The intermediate from toluene has two forms in which the carbocation is secondary and one form in which the carbocation is tertiary. The intermediate from toluene is more stable, and will be formed faster. So, toluene substitutes faster than benzene.

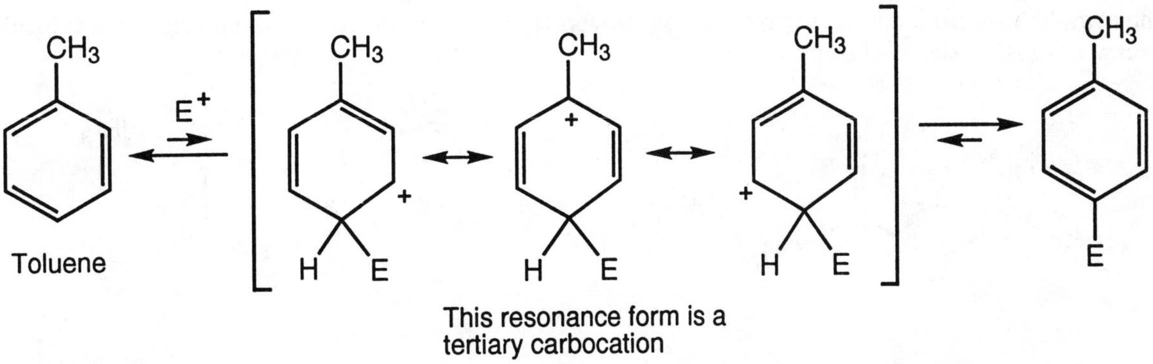

CH3
E+
Toluene
CH3
+
H E
CH3
+
H E
CH3
+
H E
CH3
E
This resonance form is a tertiary carbocation

Problem 14.8 The solution might be to use a vast excess of benzene. That way the advantage of toluene would literally be diluted. The electrophile, E^+, would be far more likely to encounter, and react with, a benzene molecule than a toluene molecule.

Problem 14.9 This reaction is just the reverse of formation of *tert*-butylbenzene from benzene, *tert*-butyl chloride, and $AlCl_3$. The first step is protonation of *tert*-butylbenzene to give a cyclohexadienyl cation. Loss of the *tert*-butyl cation generates benzene.

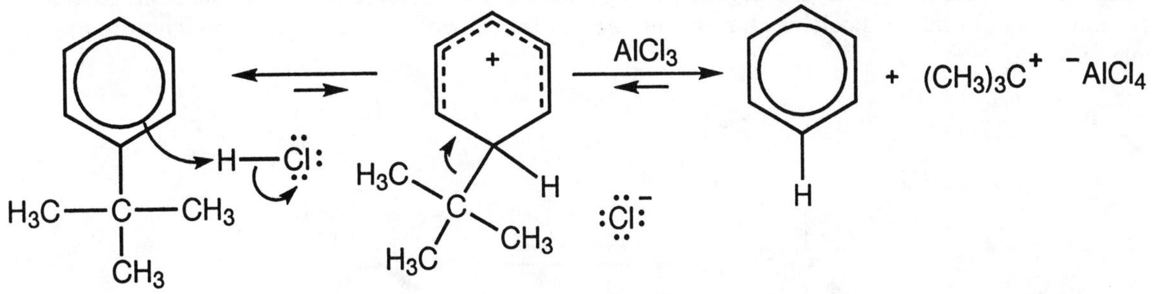

H3C—C—CH3
CH3
H—Cl:
H3C
C
H
CH3
H3C
:Cl:
AlCl3
H
+ (CH3)3C+ −AlCl4

An alternative mechanism uses an E2 elimination reaction to remove the *tert*-butyl group.

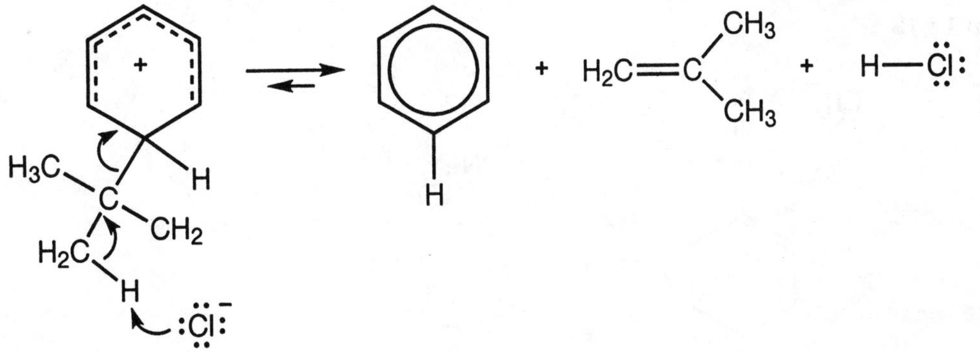

H3C
C
H
CH2
H2C
H
:Cl:
H
+ H2C=C(CH3)CH3 + H—Cl:

Problem 14.11 The first step is formation of a complex as the Lewis acid $AlBr_3$ reacts with acetic anhydride.

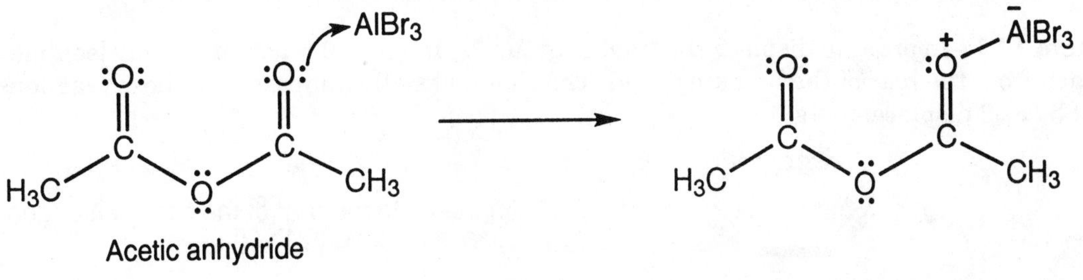

:O: :O: AlBr3
H3C C O C CH3
Acetic anhydride
:O: +:O— −AlBr3
H3C C O C CH3

The complex can break down to give an acylium ion that goes on to react with benzene in a typical aromatic substitution reaction.

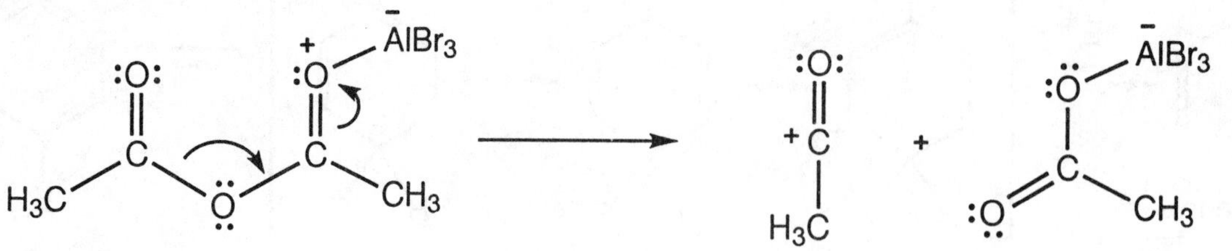

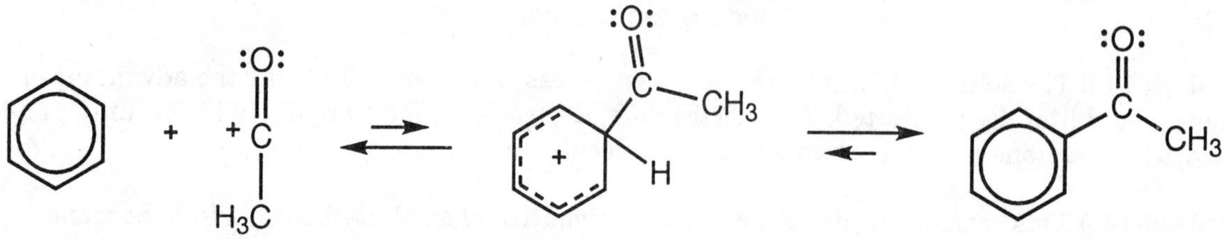

Problem 14.12 There is no pair of electrons on Sb in $^-SbF_6$. It cannot be a nucleophile. If fluorine acts as a nucleophile, a positive charge is placed on fluorine. Fluorine is a most electronegative atom and does not bear a positive charge well. Accordingly, $^-SbF_6$ is not a good nucleophile at fluorine.

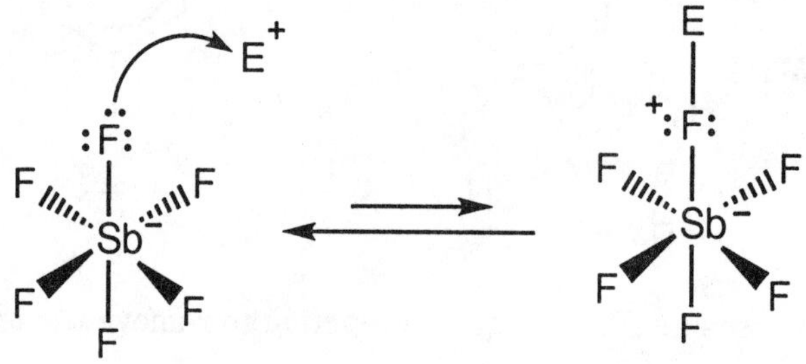

Problem 14.13

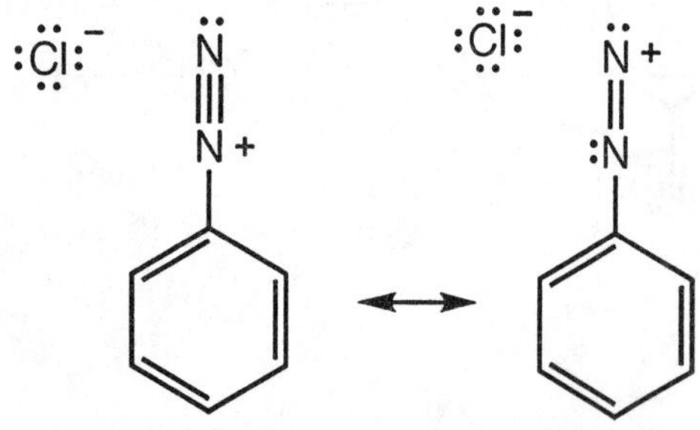

There are also less important resonance forms in which the ring carbons share the positive charge. However, the aromatic sextet is disturbed and such resonance forms are relatively unimportant. We have drawn only one of the two Kekulé forms.

Problem 14.14 In order to displace the leaving group N_2 in an S_N2 reaction, the nucleophile must approach from the rear of the breaking bond. The ring makes this impossible. These reactions cannot be S_N2 displacements.

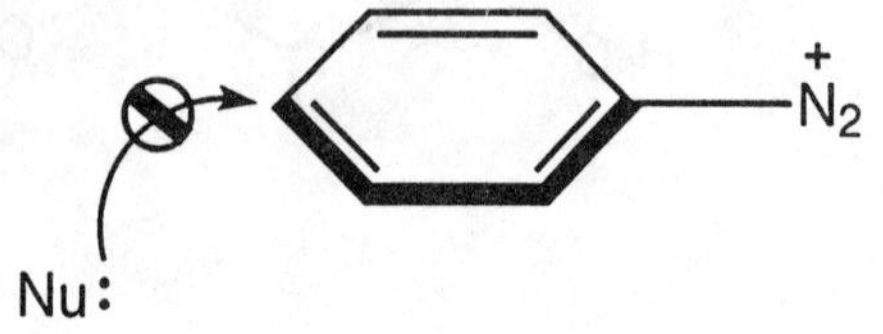

Approach to the rear of the carbon-nitrogen bond is blocked by the ring

Problem 14.15 The orbital containing the odd electron in Ph$^{\bullet}$ does not overlap with the π system. There are no resonance forms in which this electron is delocalized.

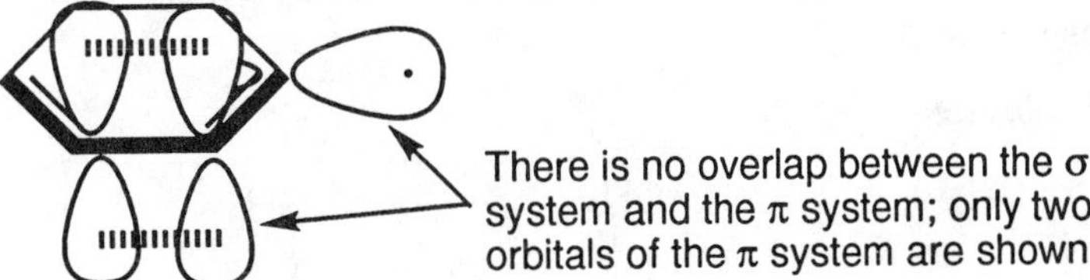

There is no overlap between the σ system and the π system; only two orbitals of the π system are shown

Problem 14.20 The potential problem here is in translation from the thermodynamic notion that the intermediate formed from anisole is lower in energy than the one formed from benzene to the kinetic statement that the lower energy intermediate (from anisole) will be formed faster than the higher energy intermediate (from benzene). We dealt with this situation in some detail in Chapter 8 (p. 320). This connection between thermodynamics and kinetics is usually justified because those factors that stabilize the intermediate are also present to some extent in the transition state leading to the intermediate. Accordingly, the lower energy intermediate is likely to have the lower energy transition state leading to it. The following diagram summarizes the situation:

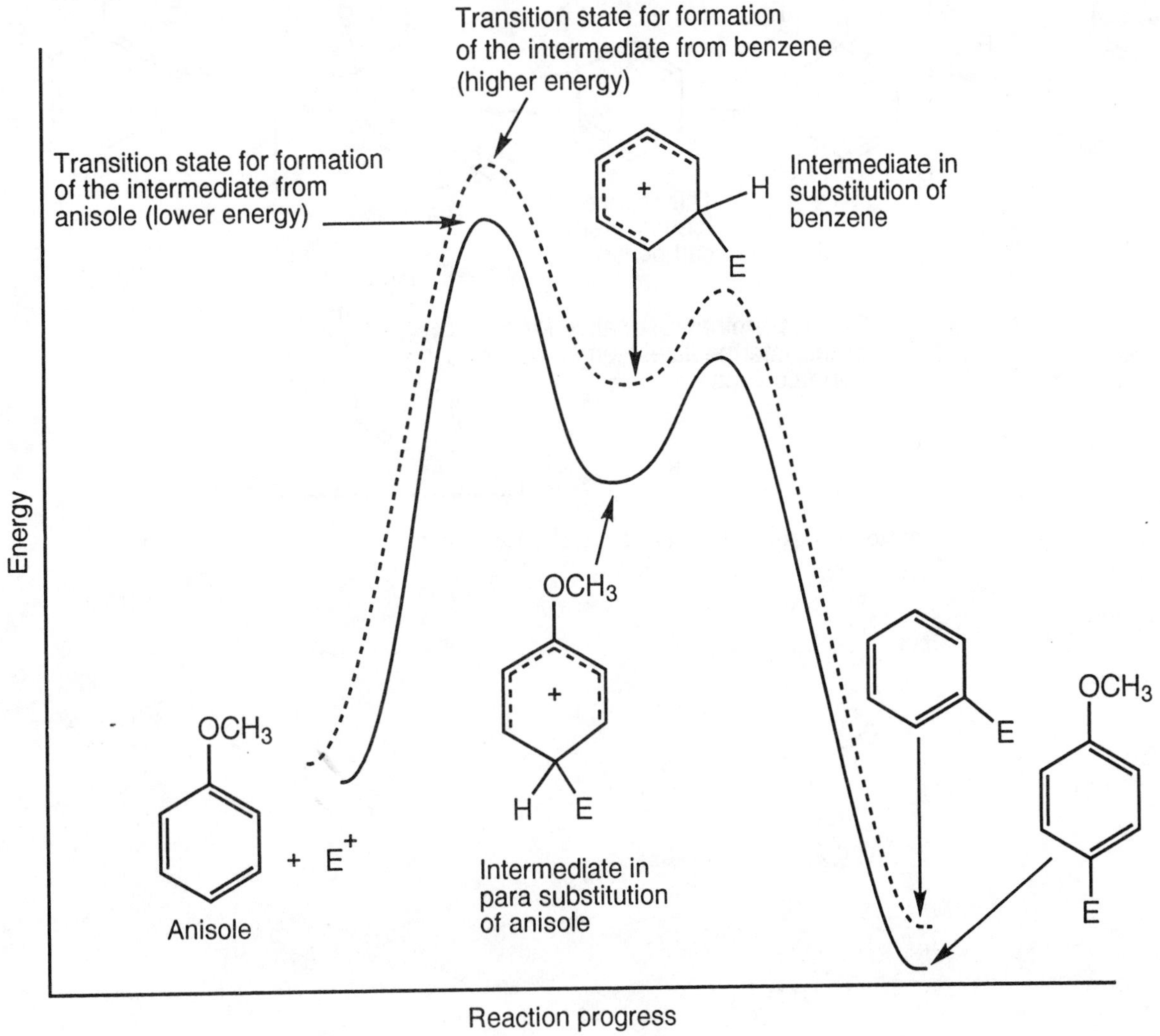

Transition state for formation of the intermediate from benzene (higher energy)

Transition state for formation of the intermediate from anisole (lower energy)

Intermediate in substitution of benzene

H

E

Energy

OCH_3

+

H E

Intermediate in para substitution of anisole

OCH_3

Anisole

+ E^+

E

OCH_3

E

Reaction progress

Problem 14.21 The situation is completely analogous to that for para versus meta substitution. In the cyclohexadienyl cation intermediate formed through ortho substitution, the methyl group operates to stabilize the positive charge on the ring. In the intermediate from meta substitution, the methyl group does not help.

Meta substitution of toluene

CH_3 E^+ CH_3 H E CH_3 H E CH_3 H E CH_3 E

Toluene

All resonance forms are secondary carbocations

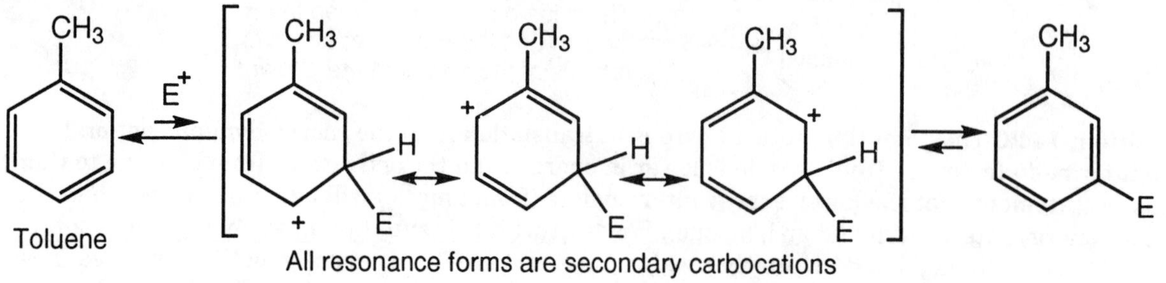

Ortho substitution of toluene

CH_3 E^+ CH_3 E H CH_3 E H CH_3 E H CH_3 E

This resonance form is a tertiary carbocation

H^+ CH_2 E H

There *is* another resonance form for this intermediate! Remember "hyperconjugation"

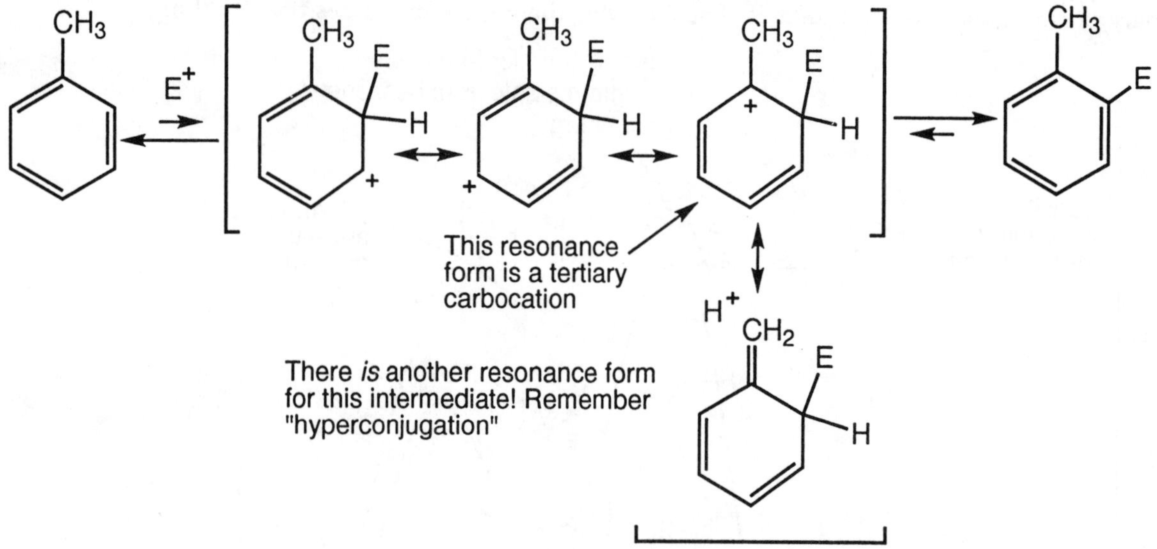

An Energy versus Reaction progress diagram is on the next page.

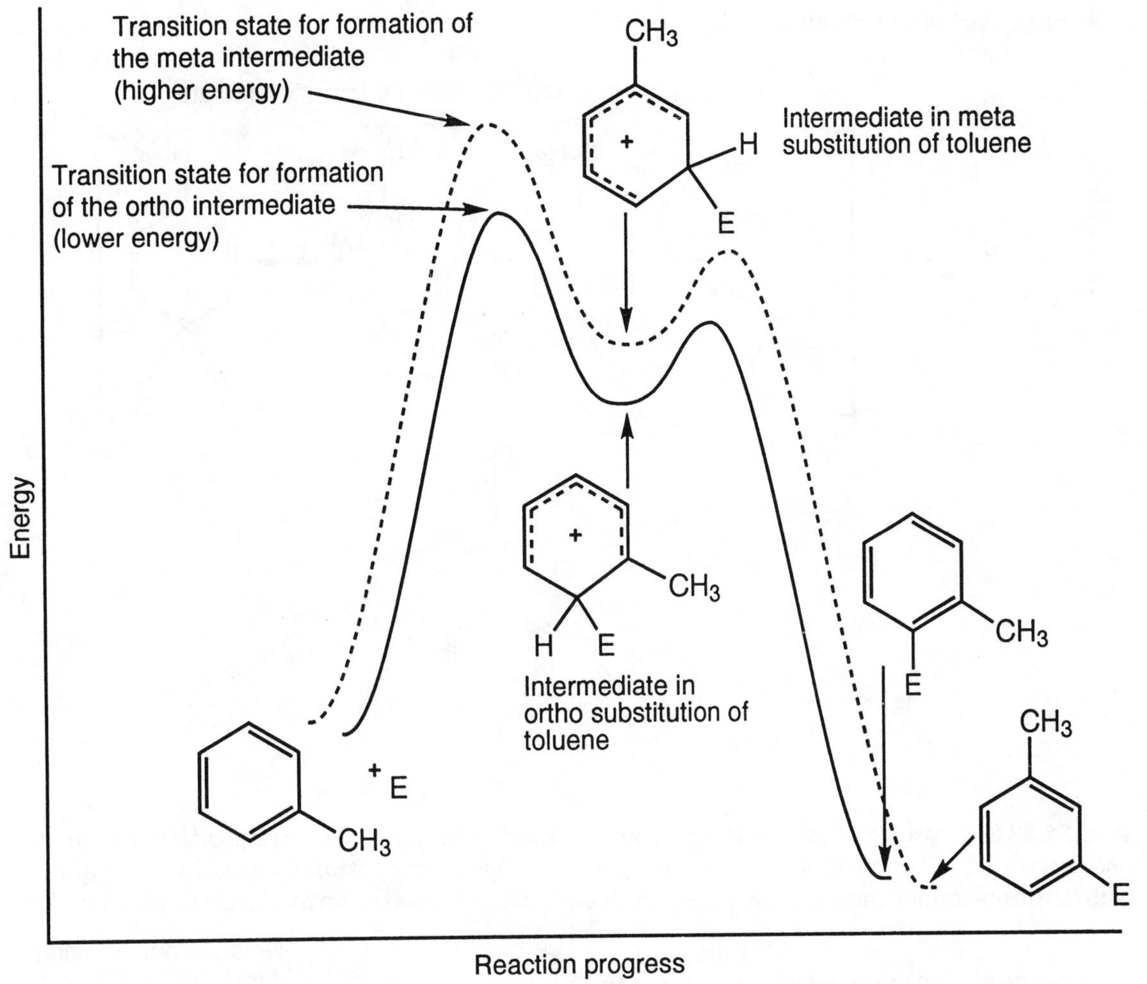

Transition state for formation of the meta intermediate (higher energy)

Transition state for formation of the ortho intermediate (lower energy)

Intermediate in meta substitution of toluene

Intermediate in ortho substitution of toluene

Energy

Reaction progress

Problem 14.22 In each case, the heteroatom (non-carbon heavy atom) can share the positive charge in the carbocation intermediate only if substitution is in the ortho or para position. These intermediates will be of lower energy, and therefore formed faster (careful! see the answer to Problem 14.20). In the figure, only ortho substitution is shown for bromobenzene and only para substitution for aniline.

Ortho substitution of bromobenzene

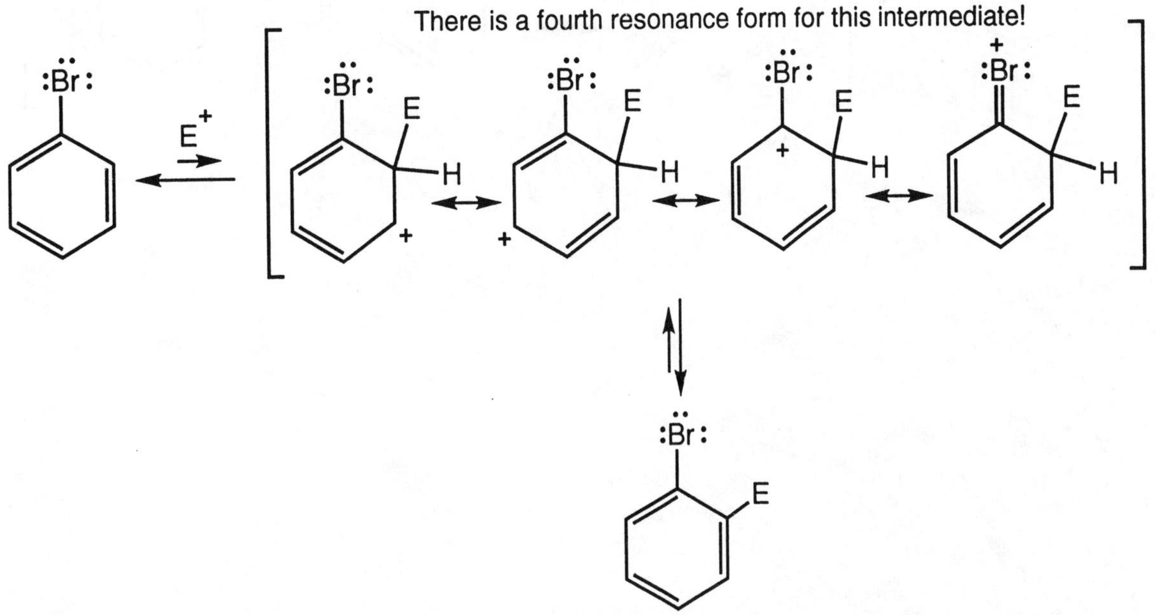

There is a fourth resonance form for this intermediate!

Para substitution of aniline

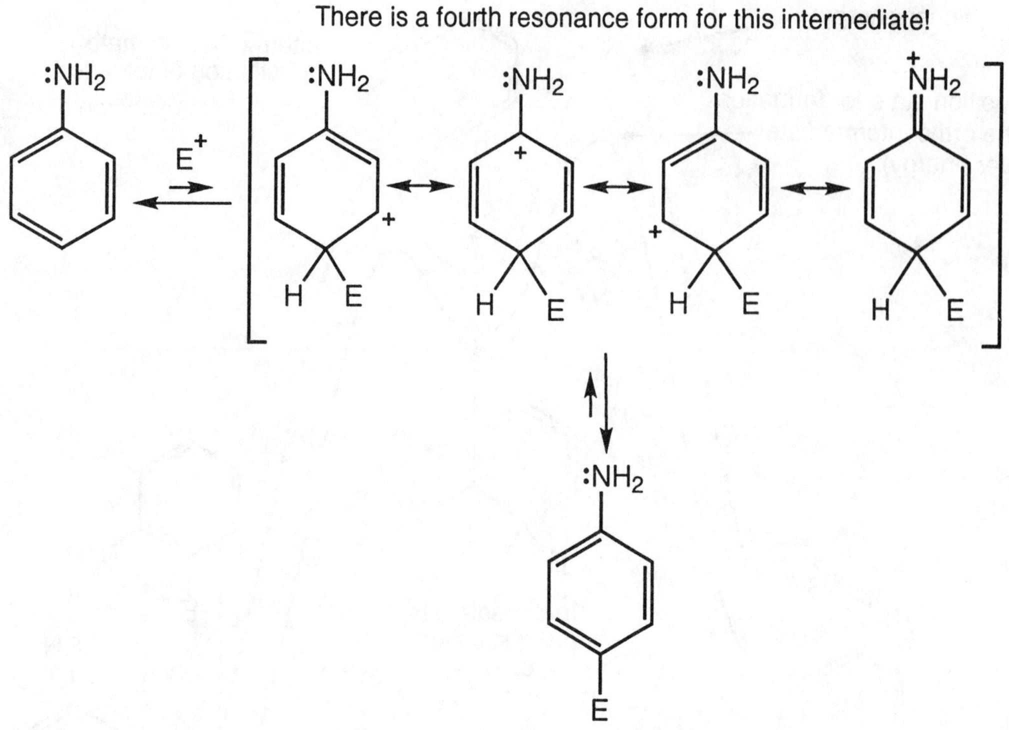

There is a fourth resonance form for this intermediate!

:NH2 E+ :NH2 :NH2 :NH2 $\overset{+}{N}H_2$ H E H E H E H E :NH2 E

Problem 14.23 Substitution in the ortho position, like substitution in the para position, requires the close opposition of two positive charges. This charge opposition is strongly destabilizing, and ortho substitution cannot compete with the less destabilizing substitution at the meta position.

Ortho substitution of the trimethylanilinium ion

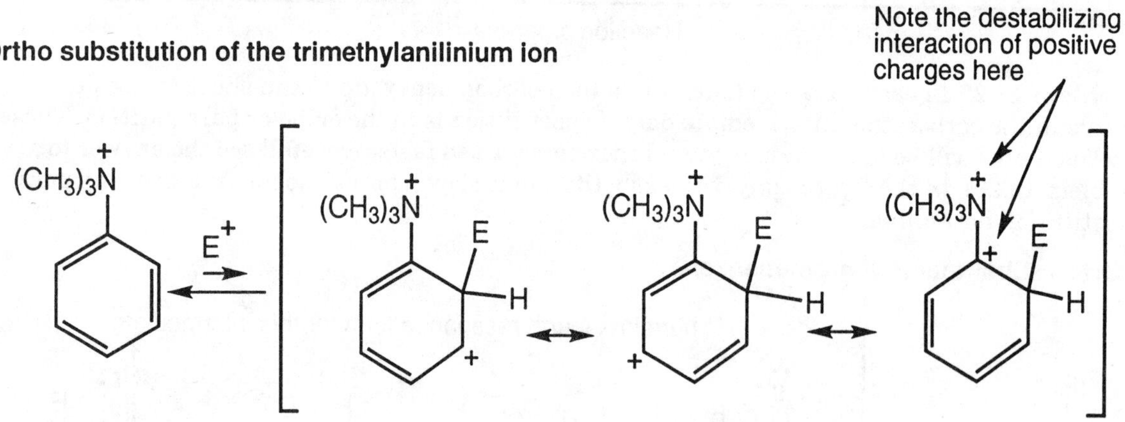

Note the destabilizing interaction of positive charges here

$(CH_3)_3\overset{+}{N}$ E+ $(CH_3)_3\overset{+}{N}$ E H $(CH_3)_3\overset{+}{N}$ E H $(CH_3)_3\overset{+}{N}$ E H

Problem 14.24

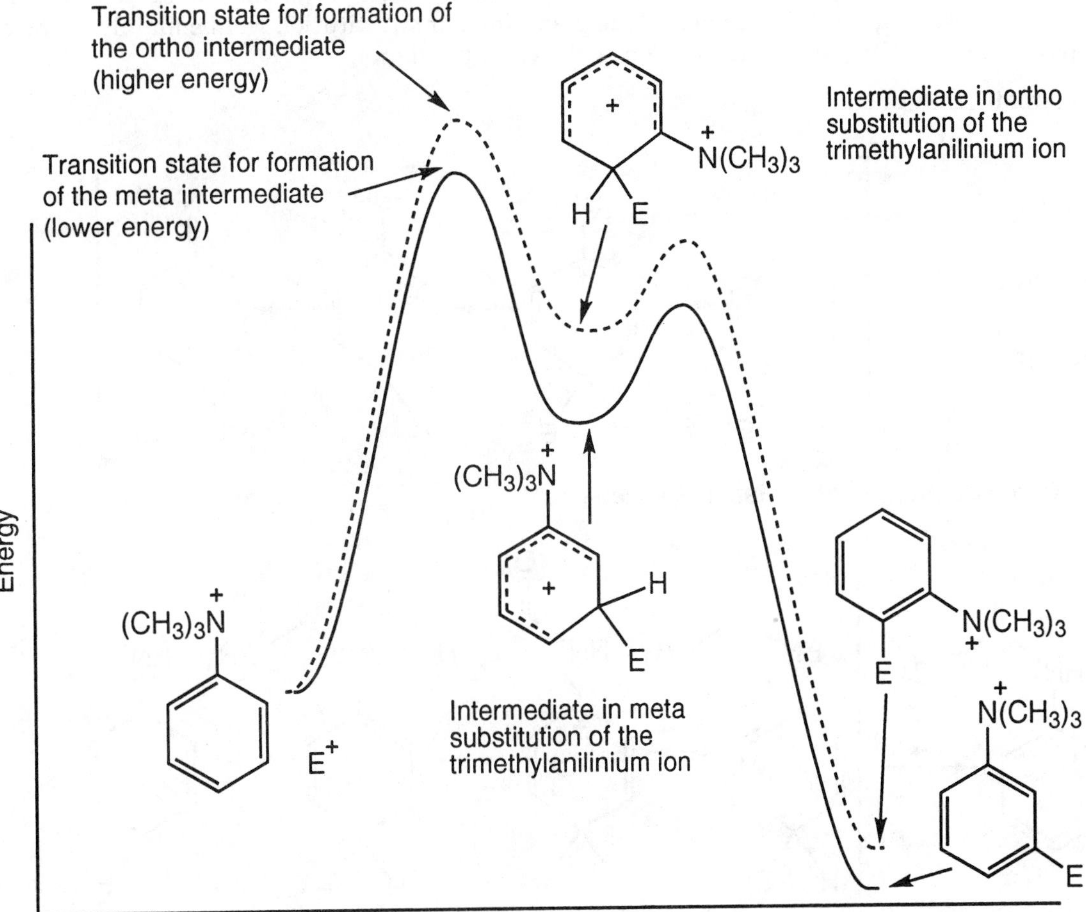

Problem 14.25 Like the carbon–oxygen double bond in ethyl benzoate (p. 661), the carbon-nitrogen triple bond in benzonitrile is strongly polarized δ^+ on carbon, δ^- on nitrogen, thus placing a partial positive charge adjacent to the ring. Substitution in the ortho or para position, again as in ethyl benzoate, opposes two like charges and is destabilizing. Substitution in the meta position is favored energetically.

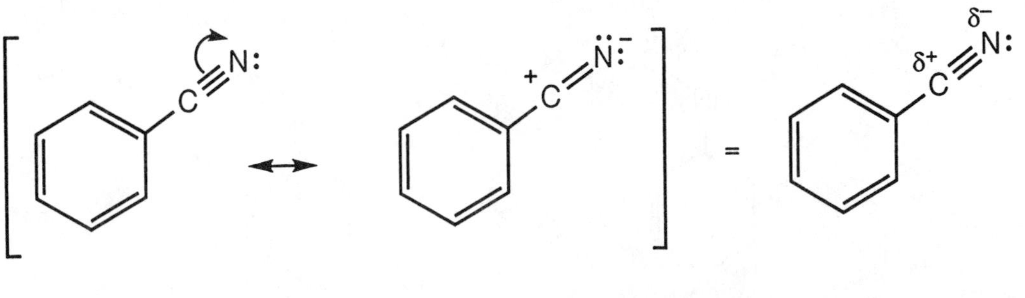

Problem 14.26 As is the case for aniline itself, ortho or para substitution of acetamidobenzene leads to a cyclohexadienyl cation intermediate in which the nitrogen can share the positive charge. In meta substitution, the nitrogen doesn't help and further substitution will be mainly in the ortho and para positions. Only para substitution is shown in the figure.

Para substitution of aniline

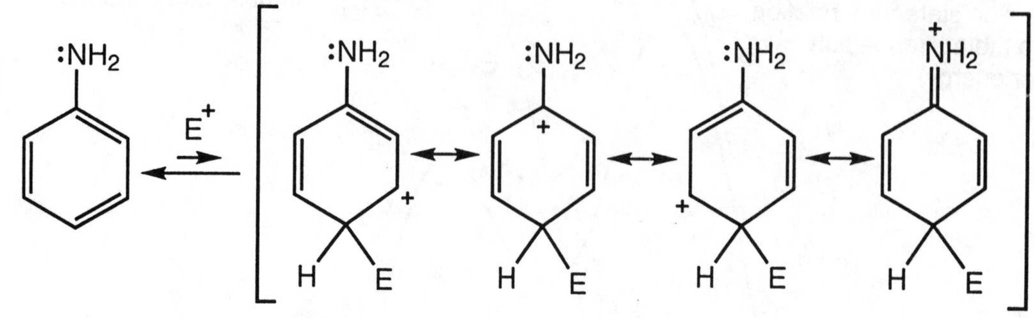

Para substitution of acetamidobenzene

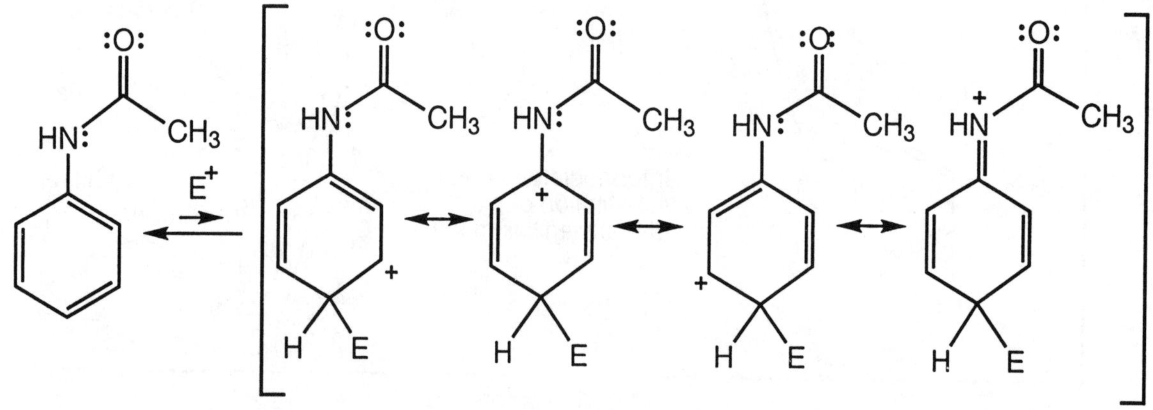

However, in the acyl compound there is a polarized carbon–oxygen double bond in which the carbon bears a partial positive charge. Accordingly, the resonance form in which the adjacent nitrogen is positive will contribute less than in the intermediate formed from aniline. The acyl compound will be less strongly activating than the free amine.

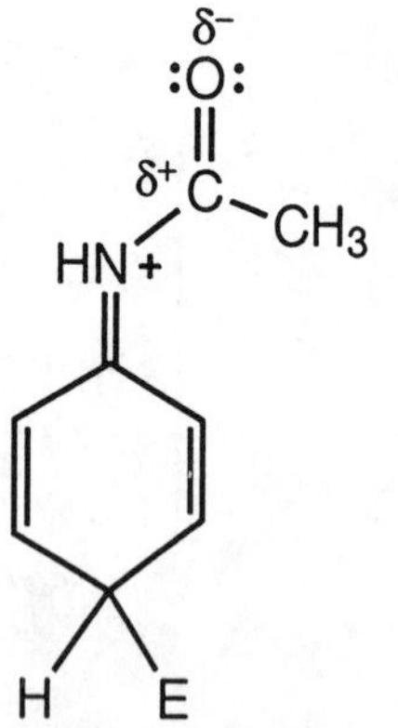

This resonance form contains a destabilizing (+)/ (δ+) interaction

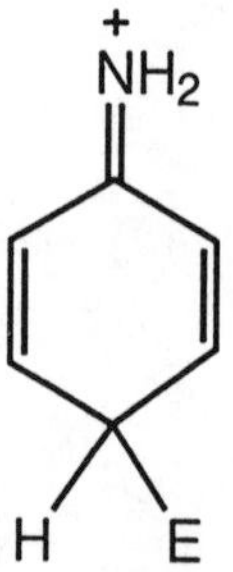

This resonance form does not

Problem 14.28

(a) This problem is easy. The question asks that you find a way to construct a benzene ring with two ortho/para directing groups located meta to each other. There is no way to do this easily using direct substitution of an aminobenzene. However, there is a simple route if you remember that it is possible to reduce nitro groups (meta directors) to amino groups. Let's use the device called "retrosynthetic analysis," introduced first in Chapter 10 (Problem 10.46, p. 448). Recall that in retrosynthetic analysis we ask only for the immediate precursor of our target molecule, working backward to an allowed starting material, in this case, benzene. An arrow is used to point to the immediate precursor. In this case, our target is *meta*-diaminobenzene, which is available from reduction of *meta*-dinitrobenzene. *meta*-Dinitrobenzene comes from nitration of nitrobenzene. Here is the retrosynthetic analysis, followed by the actual synthesis with reagents included.

Retrosynthetic analysis

NH_2, NH_2 ⟹ NO_2, NO_2 ⟹ NO_2 ⟹ (benzene)

Actual synthesis

(benzene) $\xrightarrow[H_2SO_4]{HNO_3}$ NO_2 $\xrightarrow[H_2SO_4]{HNO_3}$ NO_2, NO_2 $\xrightarrow{H_2/Pd/C}$ NH_2, NH_2

(b) This problem seems easy, but is actually a little tricky. Here there are two meta directing groups substituted meta to each other. A retrosynthetic analysis indicates that the target can be made by nitration of methyl phenyl ketone. In principle, there is another way to solve the problem: Why not nitrate first and then acylate? The tricky part of this problem is that this approach will not work. Remember that nitrobenzene is too deactivated to undergo the Friedel–Crafts reaction (p. 660).

Retrosynthetic analysis

H_3C–C=O, NO_2 ⟹ H_3C–C=O ⟹ (benzene)

Actual synthesis

(benzene) $\xrightarrow[AlCl_3]{H_3C-C(=O)-Cl}$ H_3C–C=O $\xrightarrow[H_2SO_4]{HNO_3}$ H_3C–C=O, NO_2

A failed synthetic approach

Benzene $\xrightarrow[H_2SO_4]{HNO_3}$ nitrobenzene $\xrightarrow[AlCl_3]{H_3C-C(=O)-Cl}$ No reaction!

(c) In this part, our task is to find a way to put two meta-directing groups para to each other. The answer takes advantage of your ability to interconvert meta-directing nitro groups and ortho/para directing amino groups. The target can be made by oxidation of the corresponding amino compound, *p*-aminobenzenesulfonic acid. Aniline seems a good starting material for this compound, but it will be necessary to protect the amino group of aniline to avoid substantial formation of the meta isomer (p. 670). The rest of the synthesis is routine.

Retrosynthetic analysis

p-NO_2-C_6H_4-SO_2OH ⟹ p-NH_2-C_6H_4-SO_2OH ⟹ $C_6H_5NH_2$ ⟹ $C_6H_5NO_2$ ⟹ C_6H_6

Actual synthesis

Benzene $\xrightarrow[H_2SO_4]{HNO_3}$ $C_6H_5NO_2$ $\xrightarrow{H_2/Pd/C}$ $C_6H_5NH_2$ $\xrightarrow[\text{(AcCl) pyridine}]{H_3C-C(=O)-Cl}$ C_6H_5NHAc $\xrightarrow{H_2SO_4}$ p-NHAc-C_6H_4-SO_2OH

Separate ortho and para isomers

p-NHAc-C_6H_4-SO_2OH $\xrightarrow[2. H_3O^+]{1. KOH}$ p-NH_2-C_6H_4-SO_2OH $\xrightarrow{CF_3COOOH}$ p-NO_2-C_6H_4-SO_2OH

(d) Here we find an ortho/para director (ethyl) substituted meta to a meta directing group (nitro). Our target can easily be made from the reduction of *m*-nitroacetophenone, a molecule we already made in part (b) of this problem. The suggested answer uses the Wolff-Kishner reduction procedure.

Nitrobenzene is not a good starting material because, even though the nitro group is a meta director, and could be later reduced to an amino group, it deactivates the ring so much that the Friedel–Crafts reaction required to attach the alkyl group will not work.

Retrosynthetic analysis

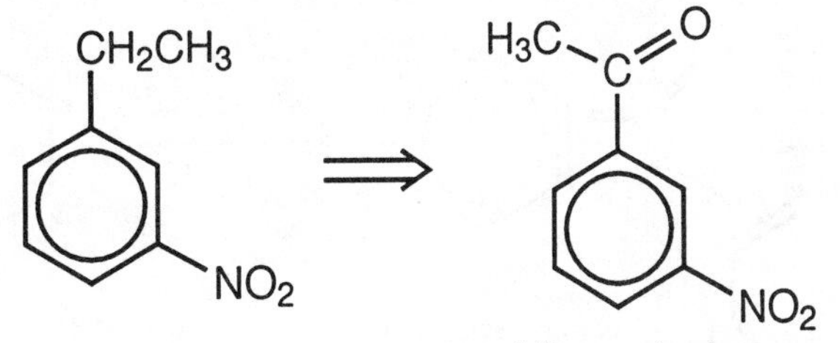

CH_2CH_3 H_3C C=O

NO_2 NO_2

m-Nitroacetophenone
[from (b)]

Actual synthesis

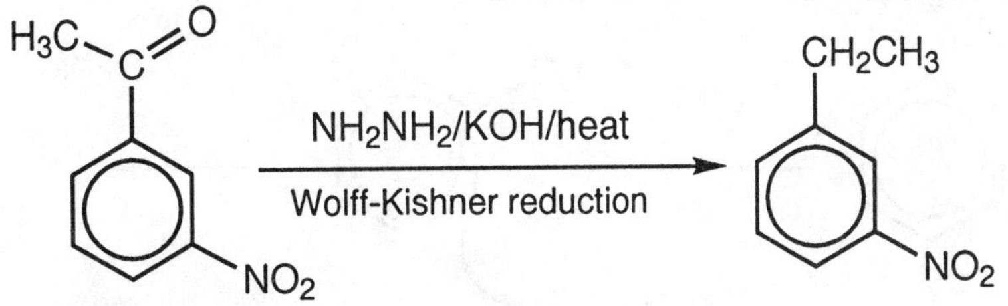

H_3C C=O

NH_2NH_2/KOH/heat

Wolff-Kishner reduction

CH_2CH_3

NO_2 NO_2

m-Nitroacetophenone
[from (b)]

A failed synthetic approach

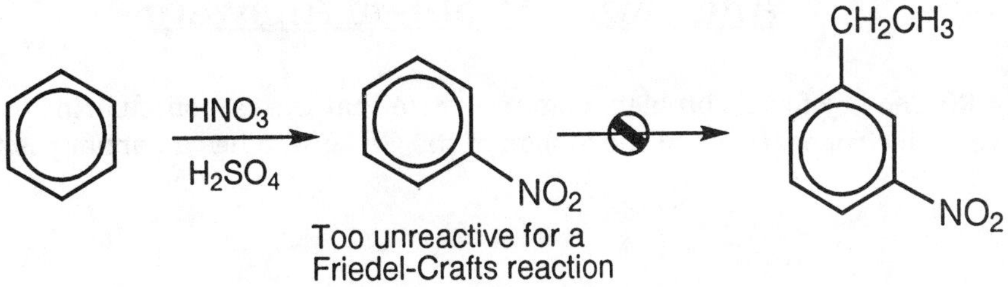

HNO_3

H_2SO_4

CH_2CH_3

NO_2 NO_2

Too unreactive for a
Friedel-Crafts reaction

Problem 14.29

(a) The first reaction is a simple dimerization of benzyne.

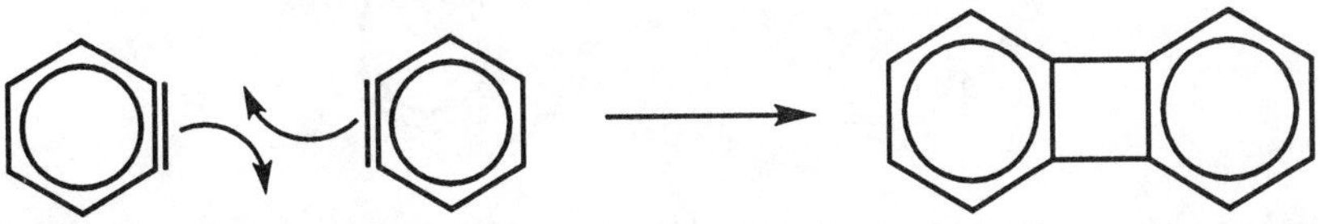

(b and c) These two processes are Diels–Alder reactions. Note that the stereochemical relationships of the diene in (c) are maintained in the product. The reaction must be a concerted one (Chapter 12, p. 542).

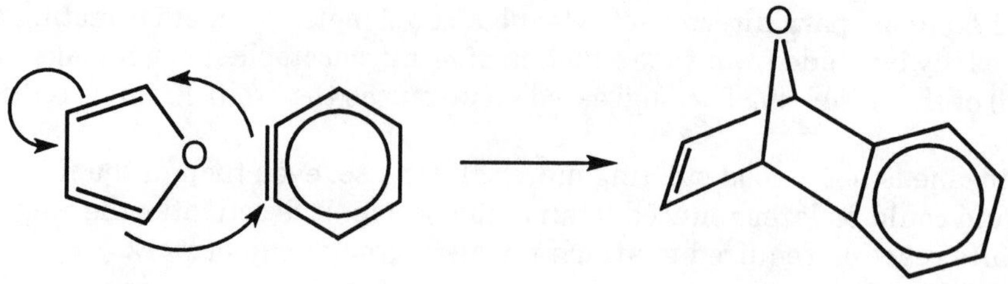

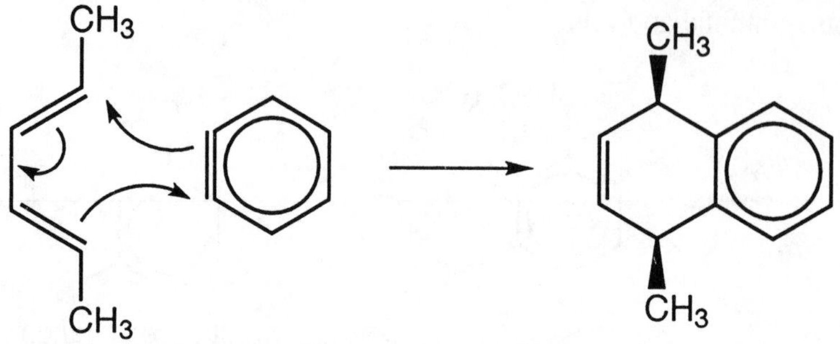

(d) Methoxide ion adds to the very reactive benzyne to give a phenyl anion. The anion is protonated by methyl alcohol to give anisole and methoxide ion.

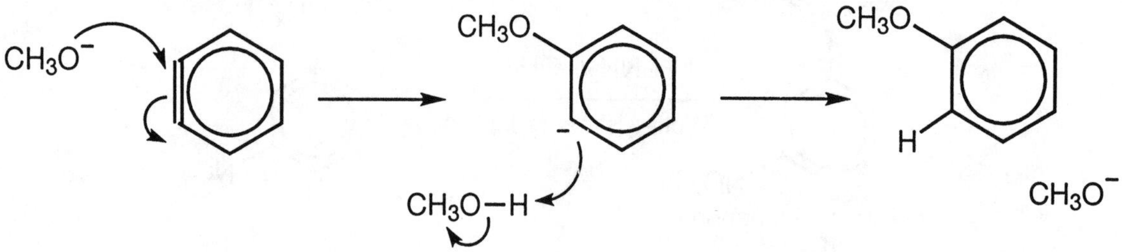

Additional Problem Answers

Problem 14.30 The key to this problem is to recognize that the carbon–fluorine bond is very polar. Fluorine is very electronegative and the carbon of the CF_3 group has a partial positive charge.

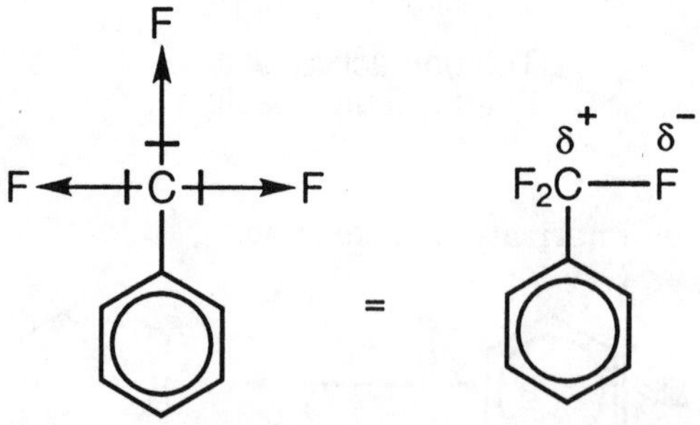

Substitution at either the ortho or para position leads to an opposition of positive charges on adjacent atoms in the cyclohexadienyl cation intermediates (and, most important, in the transition states leading to these intermediates). Substitution in the meta position does not induce this strongly destabilizing opposition of like charges, and is favored.

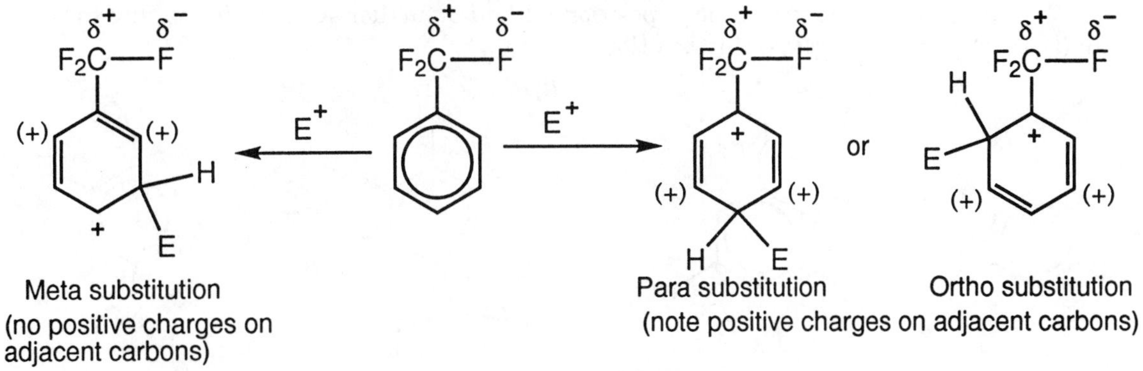

Meta substitution (no positive charges on adjacent carbons)

Para substitution Ortho substitution (note positive charges on adjacent carbons)

As for nitrobenzene (p. 660) and trialkylammonium ions (p. 659), *any* further substitution of trifluoromethylbenzene will be slow because there is charge–charge opposition in all possible intermediates. Trifluoromethylbenzene will be substituted more slowly than benzene itself.

Problem 14.31 The first three parts are easy. The substituents and the unsubstituted positions to which those substituents direct are shown with little arrows. In each case, both substituents direct further substitution to the same position. In (c), the position between the two methyl groups will be disfavored by steric effects.

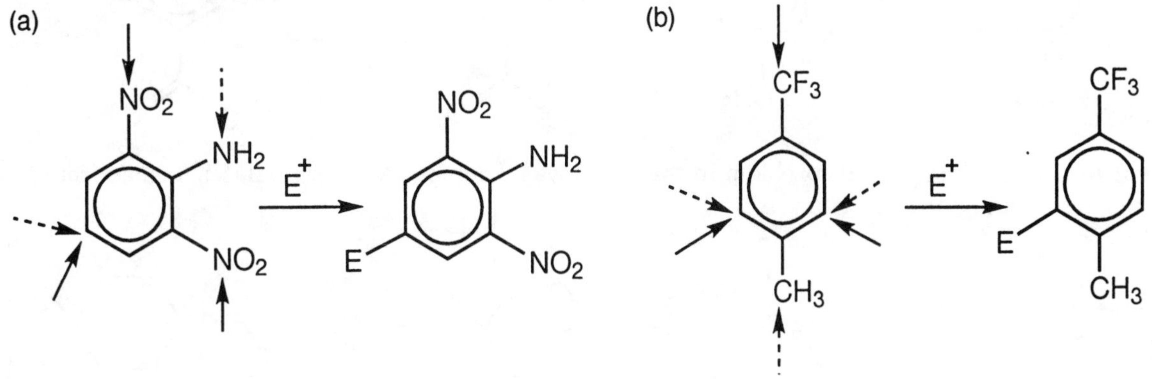

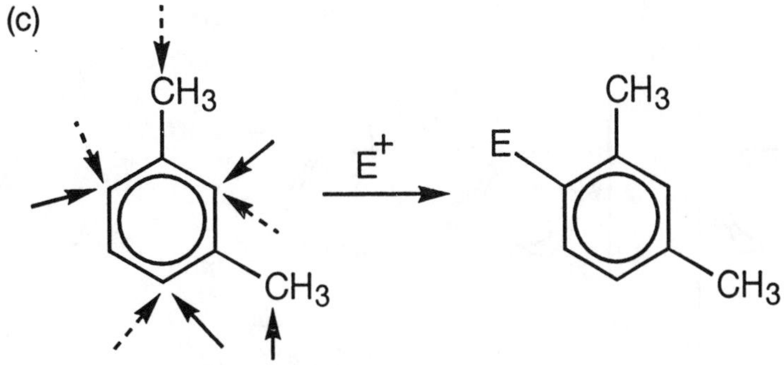

(a)

(b)

(c)

In (d), the two substituents direct further substitution differently. It will be the more powerfully ortho/para directing amino group that directs future substitution.

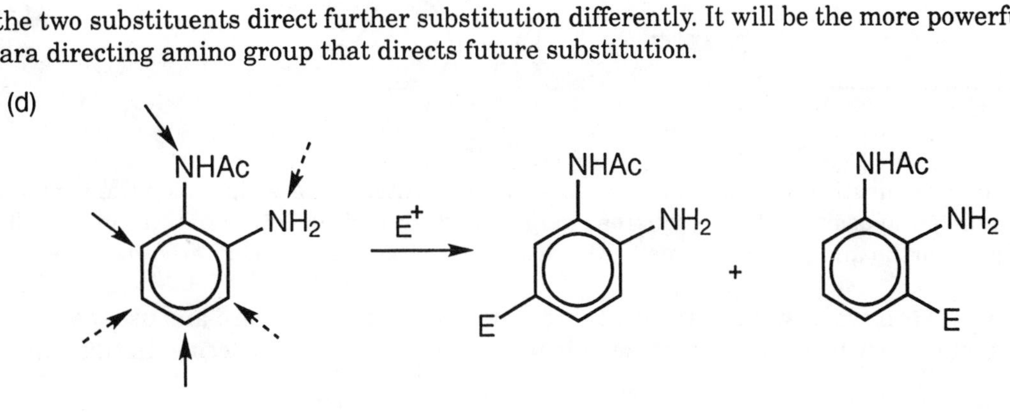

(d)

In (e), both nitro groups direct to the same position, although further substitution is sure to be *very* slow. In (f), there is only one remaining position!

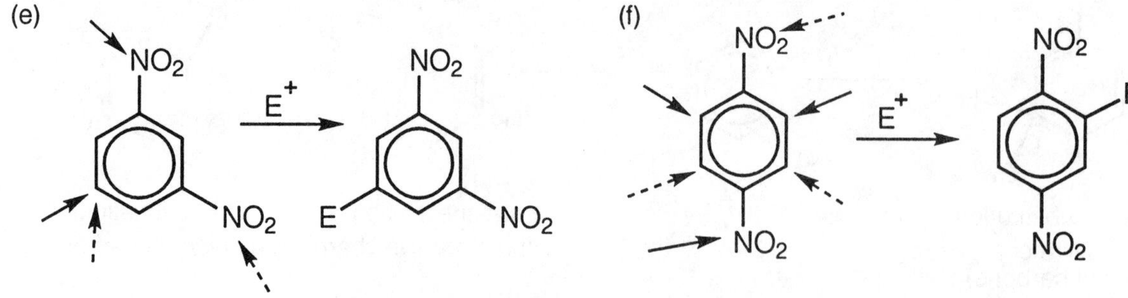

(e) NO_2 E^+ NO_2 NO_2 E NO_2 (f) NO_2 E^+ NO_2 E NO_2 NO_2

In (g), there are two ortho/para directing groups. The stronger of these, OH, will "win" and further substitution will be ortho to the OH. In (h), all three groups direct to only one open position.

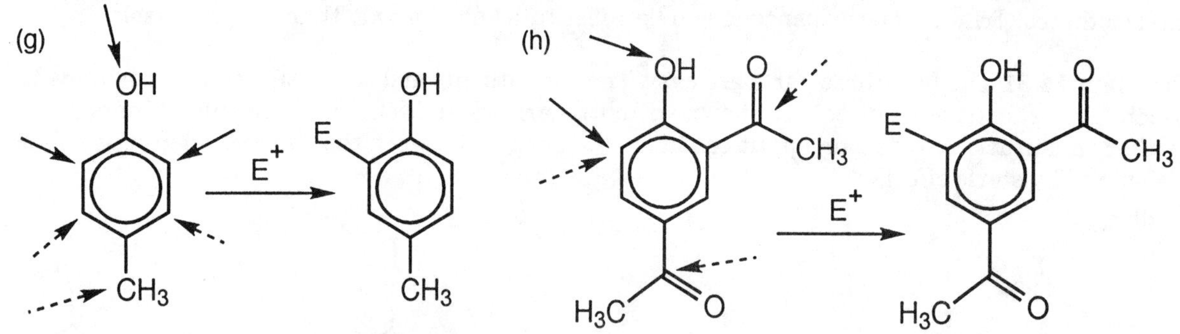

(g) OH E^+ OH E CH_3 CH_3 (h) OH O CH_3 E^+ H_3C O E OH O CH_3 H_3C O

Problem 14.32 Analyze this problem in the usual way. Look at the intermediates formed from ortho, meta, and para substitution.

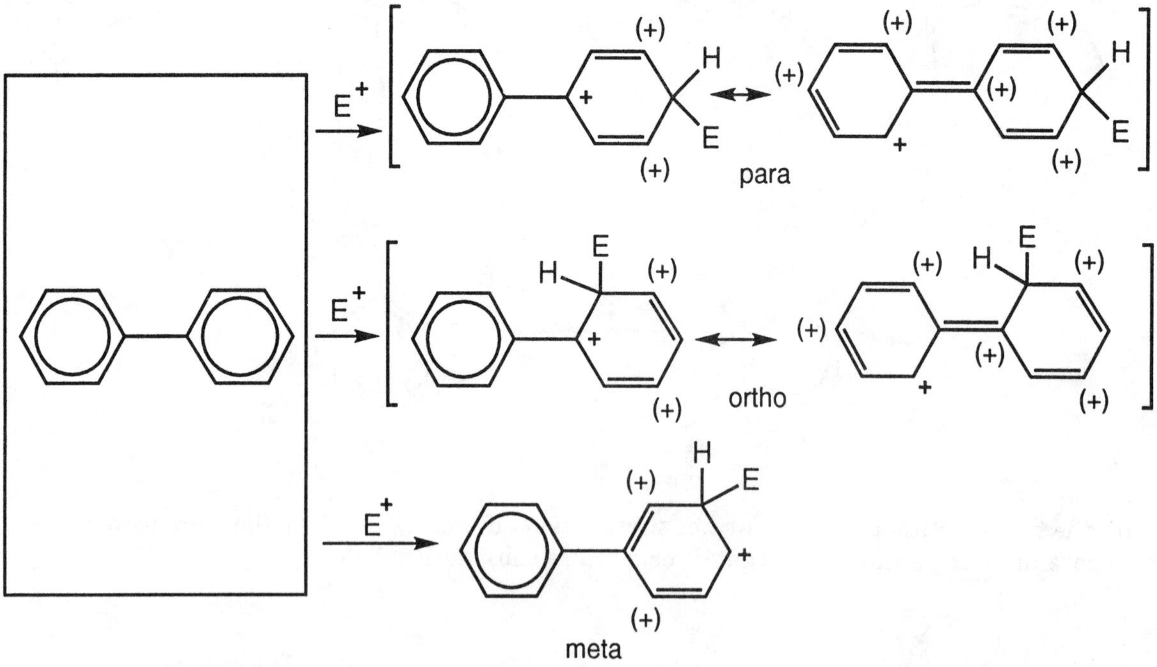

E^+ (+) H E (+) (+) (+) H E (+) (+) para

E^+ H E (+) (+) (+) ortho (+) H E (+) (+) (+)

E^+ H E (+) (+) meta

Ortho and para substitution lead to resonance-stabilized intermediates in which the second ring helps stabilize the positive charge, whereas meta substitution does not. Substitution of biphenyl will be in the ortho and para positions, with the ortho position destabilized by steric effects.

Problem 14.33 In the answers to the next three questions, work backward and use the retrosynthetic arrow to point to the immediate precursor of the target molecule. In this way, you can always work through a successful synthetic route, generating a new target molecule at each step. The overall route will be shown at the end of each problem after an analysis of what must be done.

(a) This warm-up problem is very simple, but there is a small trap. A Friedel–Crafts alkylation reaction will fail, as rearrangements will inevitably occur. Butylbenzene cannot be made directly from benzene; instead, indirection must be used. Friedel–Crafts acylation involves no rearrangements. So, the four-carbon fragment can be introduced this way. Either the Clemmensen or Wolff–Kischner reaction reduces the carbon–oxygen double bond to a methylene group.

$C_6H_5CH_2CH_2CH_2CH_3$ ⇏ C_6H_6 $C_6H_5CH_2CH_2CH_2CH_3$ ⇒ $C_6H_5C(=O)CH_2CH_2CH_3$ ⇒ C_6H_6

Here is the synthesis

C_6H_6 —($CH_3CH_2CH_2C(=O)Cl$, $AlCl_3$)→ $C_6H_5C(=O)CH_2CH_2CH_3$ —(Zn/Hg, HCl)→ $C_6H_5CH_2CH_2CH_2CH_3$

(b) The butyl group is an ortho/para directing group, so nitration of butylbenzene, made in (a) will fail. The nitro group of nitrobenzene is a meta director, but nitrobenzene is too deactivated to take part in a Friedel–Crafts alkylation. We need a meta directing group that can be transformed into the butyl group. The acyl compound we used in (a) will do the job.

m-$O_2NC_6H_4CH_2CH_2CH_2CH_3$ ⇒ m-$O_2NC_6H_4C(=O)CH_2CH_2CH_3$ ⇒ $C_6H_5C(=O)CH_2CH_2CH_3$

Here is the synthesis in detail

$C_6H_5C(=O)CH_2CH_2CH_3$ (Made in (a)) —(HNO_3, H_2SO_4)→ m-$O_2NC_6H_4C(=O)CH_2CH_2CH_3$ —(NH_2NH_2, KOH, ethylene glycol, Δ)→ m-$O_2NC_6H_4CH_2CH_2CH_2CH_3$

(c) Here we have the classic problem of producing an ortho compound without, at the same time, making a para compound. The solution is to use a sulfonic acid group to block the para position ("protect" the para position) temporarily, and to remove it when needed. The retrosynthetic analysis is as follows:

$CH_2CH_2CH_2CH_3$ (o-Br) ⇒ $CH_2CH_2CH_2CH_3$ (Br, SO_2OH) ⇒ $CH_2CH_2CH_2CH_3$ (SO_2OH) ⇒ $CH_2CH_2CH_2CH_3$

We can start with the butylbenzene we made in (a). The details follow:

$CH_2CH_2CH_2CH_3$ (made in (a)) —H_2SO_4→ $CH_2CH_2CH_2CH_3$ (SO_2OH) —Br_2, $FeBr_3$→ $CH_2CH_2CH_2CH_3$ (Br, SO_2OH) —H_2O, H_3O^+→ $CH_2CH_2CH_2CH_3$ (Br)

(d) We know only one way to make a carboxylic acid; the oxidation of a side chain on a benzene ring (Chapter 13). Any side chain in which the carbon attached to the ring bears at least one hydrogen will do. The question wants para substituted product, so a large alkyl group that will shield the nearby ortho positions seems appropriate.

COOH (SO_2OH) ⇒ $CH(CH_3)_2$ (SO_2OH) ⇒ $CH(CH_3)_2$ ⇒ benzene

or:

$CH_2CH_2CH_2CH_3$ (SO_2OH) made in (c)

The detailed synthesis follows:

benzene —$(CH_3)_2CHCl$, $AlCl_3$→ $CH(CH_3)_2$ —H_2SO_4→ $CH(CH_3)_2$ (SO_2OH) —$KMnO_4$→ COOH (SO_2OH)

Problem 14.34 These syntheses all use transformations of diazonium ions.
(a) As both Br and I are ortho/para directors, we might envision some type of electrophilic aromatic halogenation.

I (o-Br) ⇒ I or Br

However, para substitution would clearly be a serious problem with such a procedure as an "isomer free" synthesis was demanded by the question. We still know only one way to make an iodobenzene, the reaction of a diazonium ion with KI. We need a route to *o*-bromoaniline, and this route must somehow avoid the para isomer.

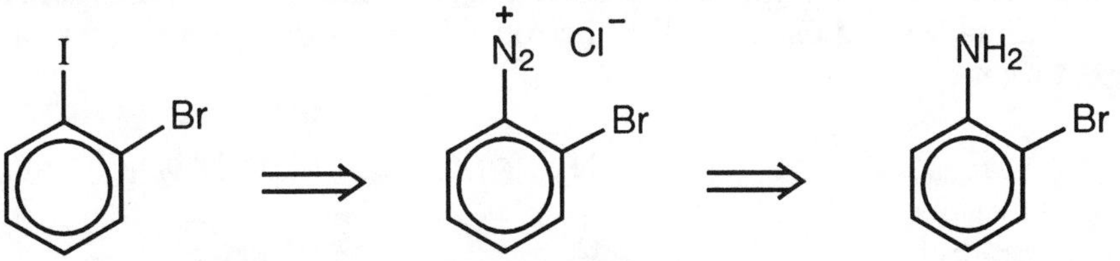

o-Bromoaniline cannot be prepared directly from aniline because the powerfully activating amino group will lead to polybromination. We need both to deactivate the amino group and to block the para position. Note that both the deactivated amine, the acetamido group, and the blocking group, a sulfonic acid, direct further substitution to the position ortho to the acetamido group.

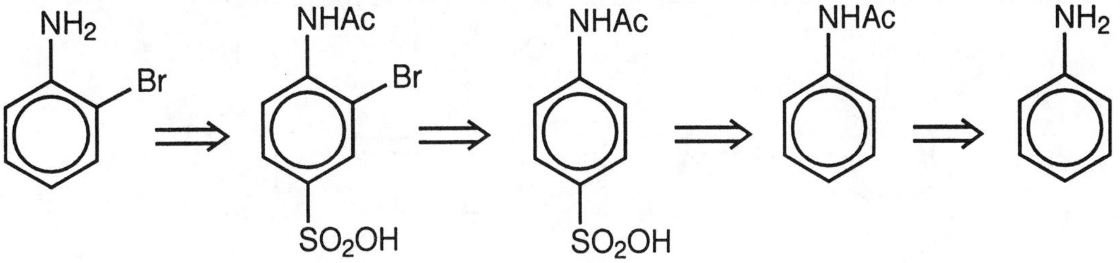

The required aniline is easily available from nitrobenzene and, ultimately, benzene.

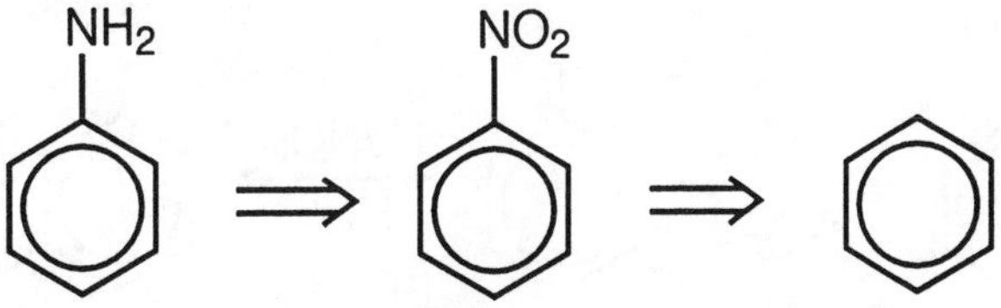

So, our rather complex synthetic procedure, starting from benzene, is shown below.

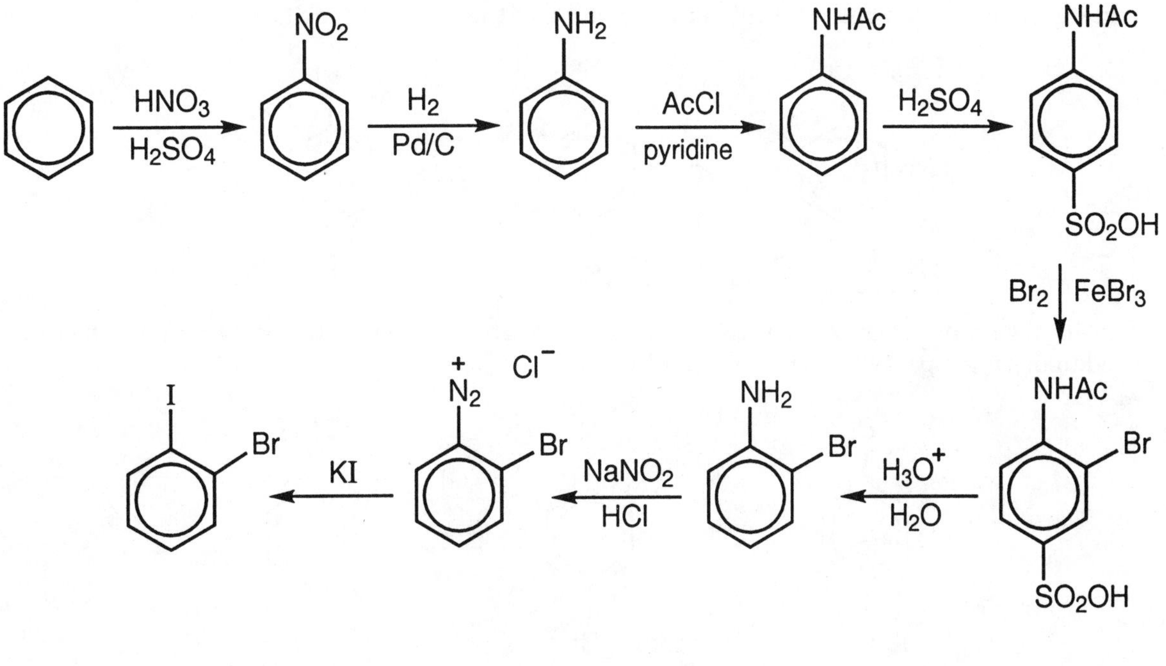

Note the hydrolysis step in which the acetylated amine is "deprotected" and the sulfonic acid group is removed at the same time.

(b) Here we are faced with the classical problem of ortho/para-directing groups placed meta to each other. We only know one way to prepare a fluorobenzene, and that involves the Schiemann reaction of a diazonium ion. Thus, we can use the following route from *m*-bromoaniline, a molecule that becomes our next target.

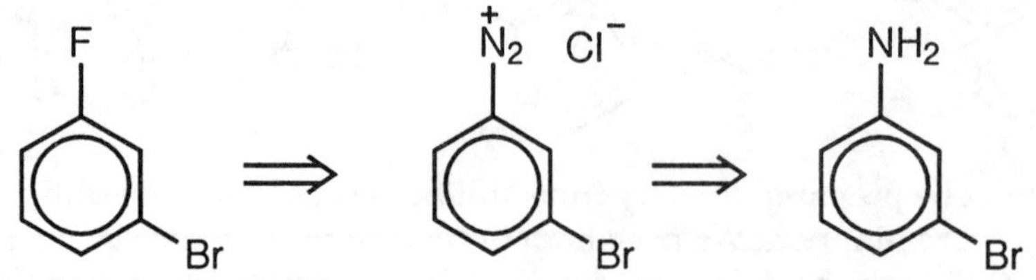

Our ability to interconvert nitro and amino groups suggests the following solution to this problem. This pathway takes advantage of the meta-directing character of the nitro group.

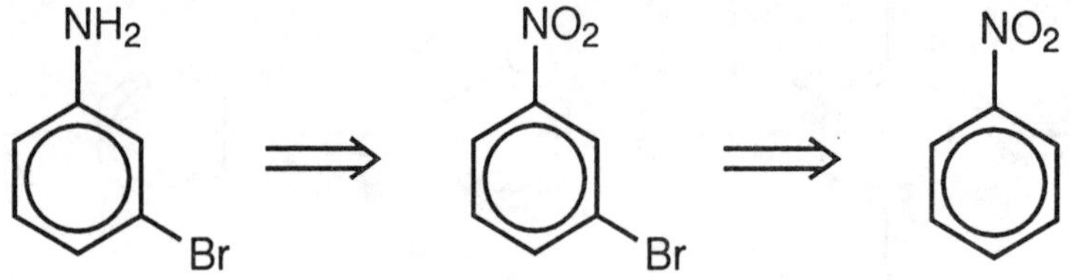

So, starting with nitrobenzene as prepared in (a), our synthesis is

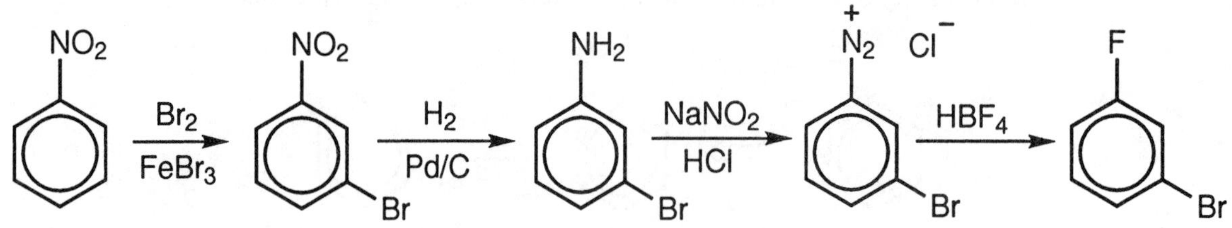

(c) In (c), we are faced with similar problems to those encountered in (a). An electrophilic aromatic bromination route would be complicated by formation of the undesired ortho isomer.

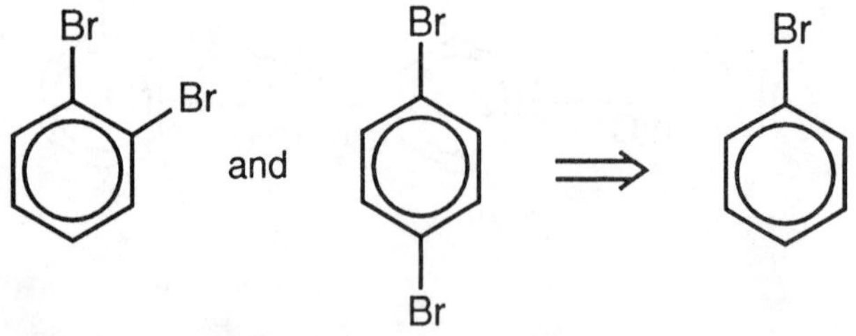

Fortunately, we can introduce the second bromine through Sandmeyer chemistry. Our synthesis is reduced to finding a pathway to *p*-bromoaniline.

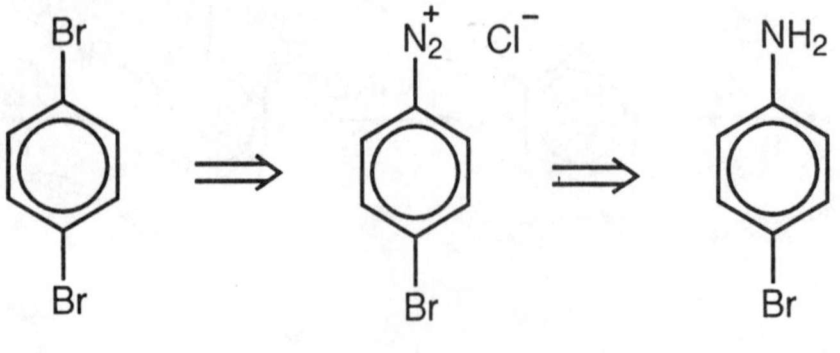

We also know that *p*-bromoaniline is readily available from acetanilide (Figure 14.94, p. 669).

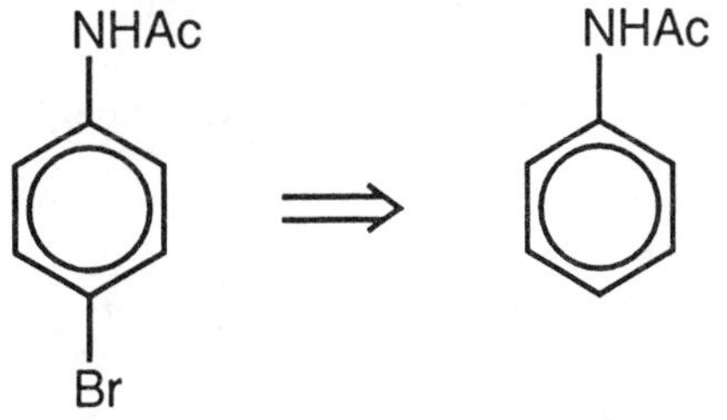

NHAc NHAc

Br

The synthesis of *p*-dibromobenzene, starting from acetanilide prepared in (a), is shown below.

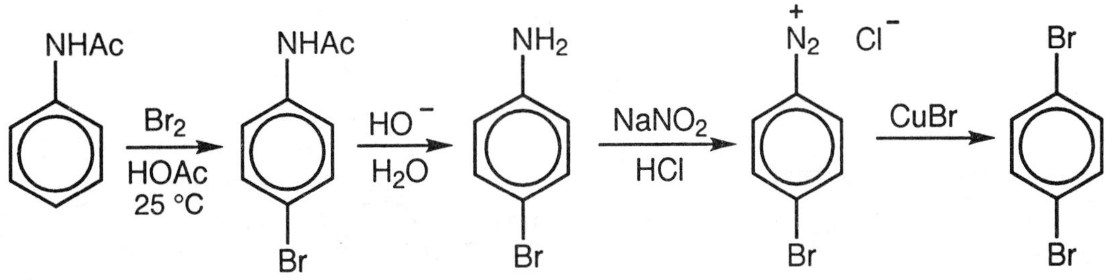

NHAc NHAc NH$_2$ $\overset{+}{N}_2$ Cl$^-$ Br

Br_2 / HOAc 25 °C; HO^- / H_2O; $NaNO_2$ / HCl; CuBr

Br Br Br Br

Problem 14.35

(a) In this synthesis, we are faced with two immediate problems. First, we must introduce two ortho/para-directing groups meta to each other, and second, we need to introduce an ethoxy group. The second problem can be solved by taking advantage of the substantial acidity of phenols, and alkylating the phenoxide ion in an S_N2 reaction.

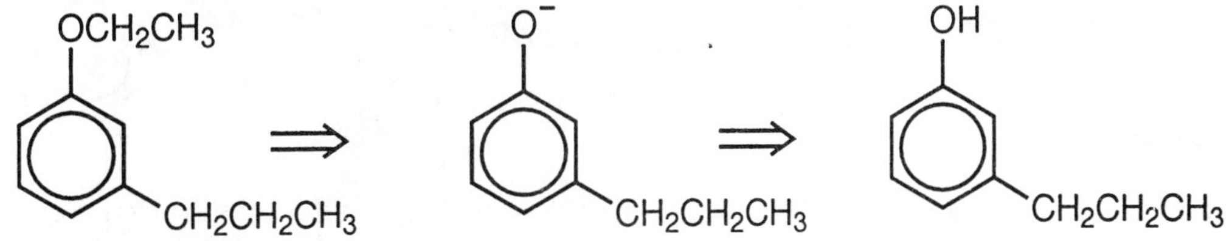

OCH$_2$CH$_3$ O$^-$ OH

CH$_2$CH$_2$CH$_3$ CH$_2$CH$_2$CH$_3$ CH$_2$CH$_2$CH$_3$

We know only one general method of making phenols; the treatment of a diazonium ion with water. Now we need a synthesis of *m*-propylaniline. Meta substitution immediately suggests *m*-nitropropylbenzene as an appropriate precursor.

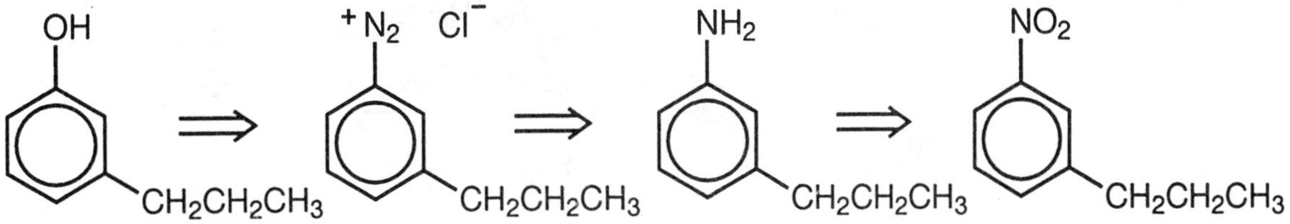

OH $^+N_2$ Cl$^-$ NH$_2$ NO$_2$

CH$_2$CH$_2$CH$_3$ CH$_2$CH$_2$CH$_3$ CH$_2$CH$_2$CH$_3$ CH$_2$CH$_2$CH$_3$

However, we need to be alert for a trap here. *m*-Nitropropylbenzene is not available from the Friedel–Crafts alkylation of nitrobenzene because nitrobenzene is too deactivated to undergo the Friedel–Crafts reaction. Even if the alkylation worked, rearrangements would be likely.

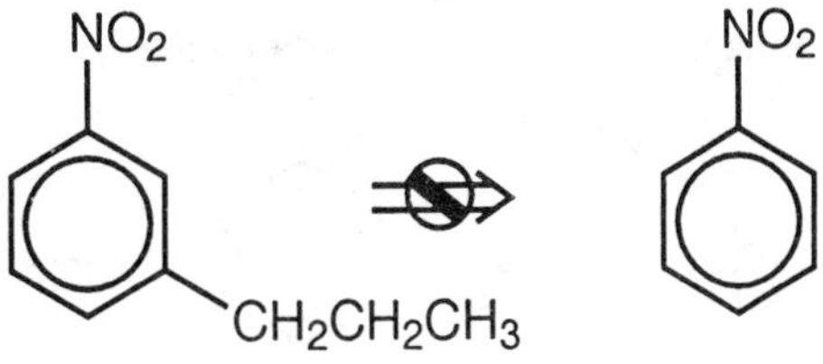

NO$_2$ NO$_2$

CH$_2$CH$_2$CH$_3$

These problems can be circumvented, and the traps avoided, by nitrating propiophenone, also a meta director, and available from benzene through a Friedel–Crafts acylation. Reduction of the ketone then gives the requisite propyl group. In fact, under Clemmensen conditions it may be possibile to reduce the ketone and the nitro group simultaneously.

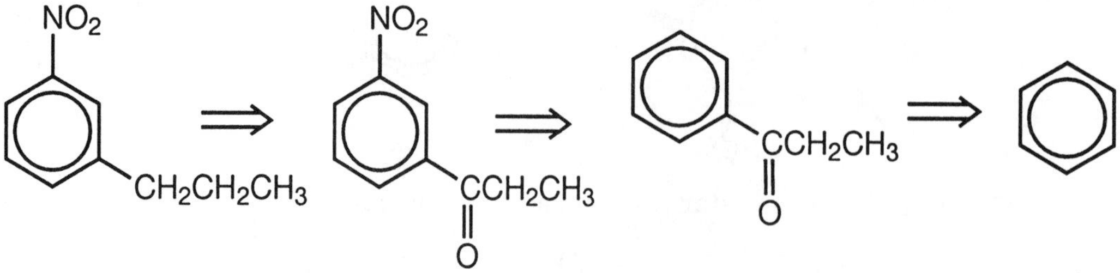

NO_2 NO_2 $CH_2CH_2CH_3$ ⇒ CCH_2CH_3 (C=O) ⇒ CCH_2CH_3 (C=O) ⇒

So, our synthesis, written in a forward fashion, is

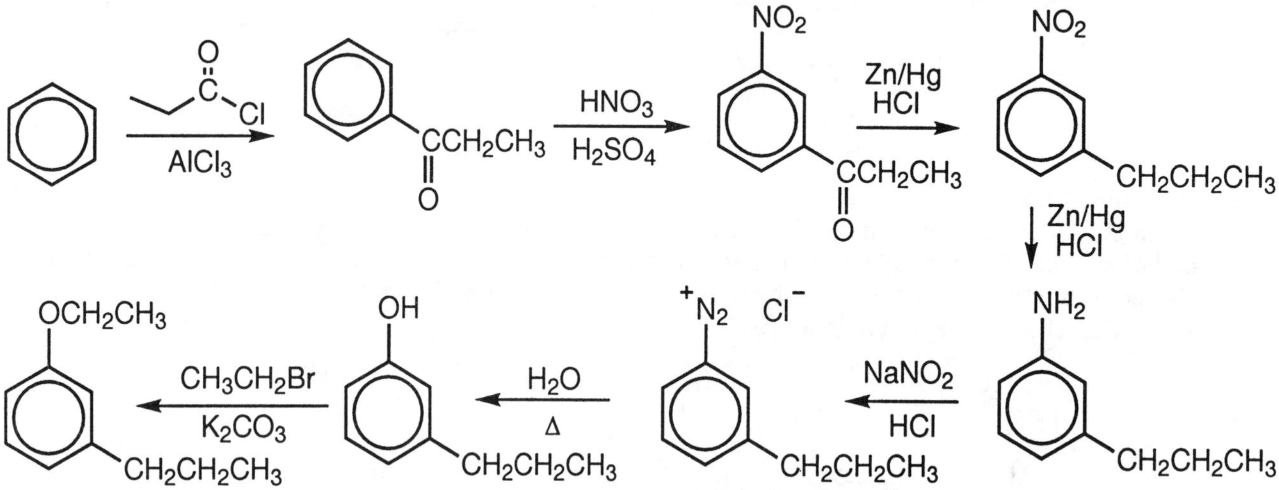

$\xrightarrow[AlCl_3]{CH_3CH_2COCl}$ CCH_2CH_3 (C=O) $\xrightarrow[H_2SO_4]{HNO_3}$ NO_2, CCH_2CH_3 (C=O) $\xrightarrow[HCl]{Zn/Hg}$ NO_2, $CH_2CH_2CH_3$ $\xrightarrow[HCl]{Zn/Hg}$ NH_2, $CH_2CH_2CH_3$ $\xrightarrow[HCl]{NaNO_2}$ $N_2^+\ Cl^-$, $CH_2CH_2CH_3$ $\xrightarrow[\Delta]{H_2O}$ OH, $CH_2CH_2CH_3$ $\xrightarrow[K_2CO_3]{CH_3CH_2Br}$ OCH_2CH_3, $CH_2CH_2CH_3$

(b) Once more, we see two problems. First, we need to do a selective ortho substitution. Second, we need to do a selective introduction of a single deuterium. The latter problem can be neatly solved through a replacement of a diazonium ion with deuterated hypophosphorus acid, D_3PO_2. So, our synthesis is reduced to a preparation of *o*-ethylaniline, a molecule available from *o*-ethylnitrobenzene.

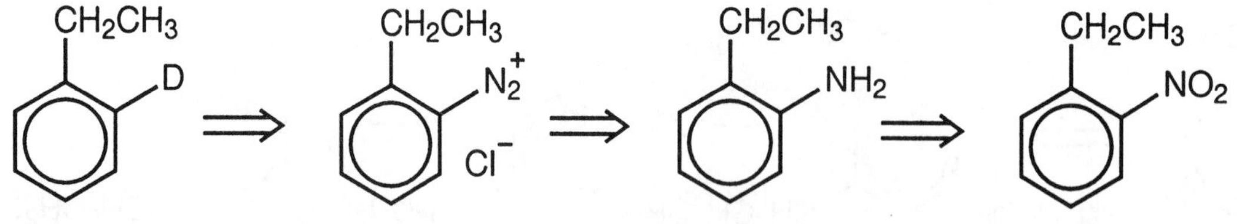

CH_2CH_3, D ⇒ CH_2CH_3, N_2^+ Cl^- ⇒ CH_2CH_3, NH_2 ⇒ CH_2CH_3, NO_2

The selective synthesis of *o*-ethylnitrobenzene suggests a blocking–deblocking protocol employing a sulfonic acid group. Both the ethyl and sulfonic acid groups will direct further substitution to the desired position ortho to the ethyl group. Ethylbenzene is, of course, easily available from a Friedel–Crafts alkylation of benzene.

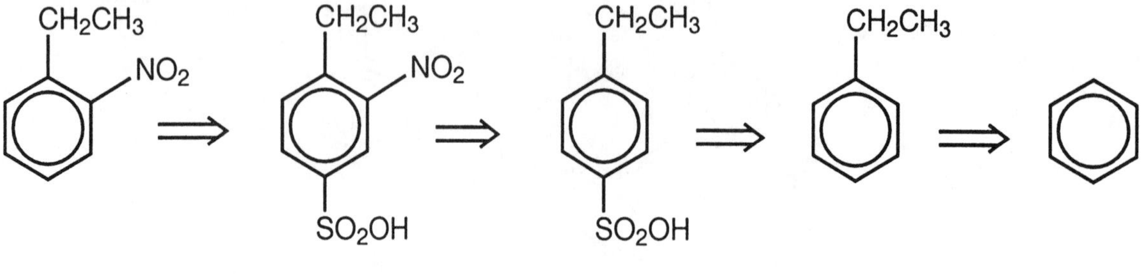

CH_2CH_3, NO_2 ⇒ CH_2CH_3, NO_2, SO_2OH ⇒ CH_2CH_3, SO_2OH ⇒ CH_2CH_3 ⇒

The proposed synthesis is as shown

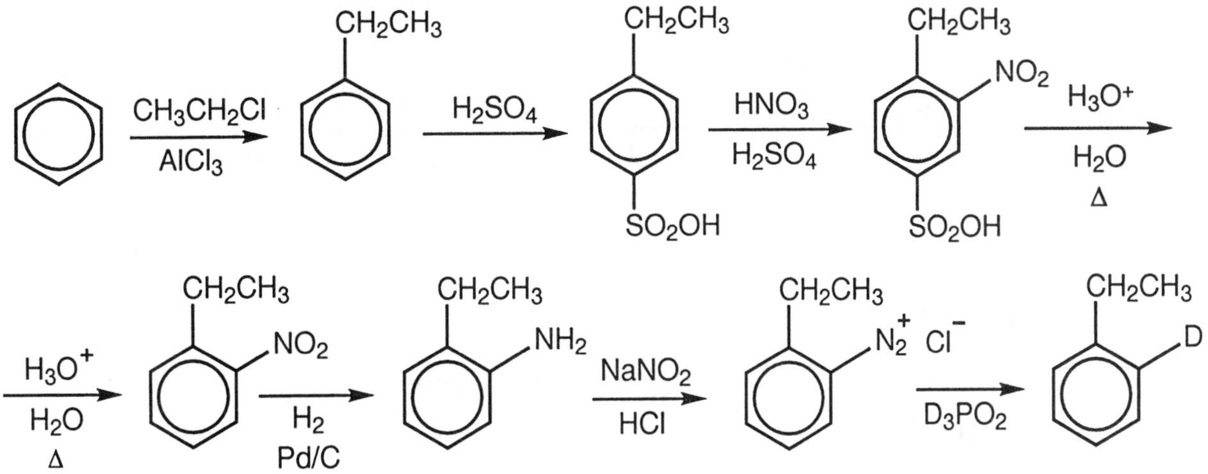

(c) In this last synthesis, we are faced with the classical problem of placing two meta-directing groups para to each other. At this time, we know only one way to introduce a cyano group, and that involves Sandmeyer chemistry. Our task is reduced to a synthesis of *p*-nitroaniline.

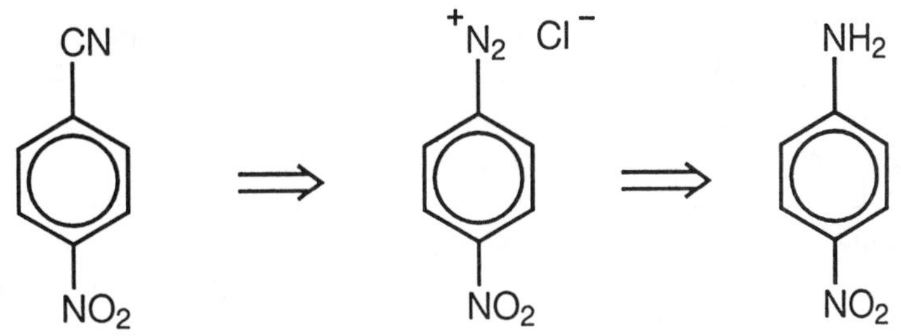

We know that aniline cannot be nitrated directly as a mixture of para and meta isomers will result (see Problem 14.27, p. 670). The solution to this problem is shown in Figure 14.90 (p. 668). It involves the use of a protection–deprotection strategy. The nitration of acetanilide is the weak step in our proposed synthesis as it produces about 20% of the undesired ortho isomer.

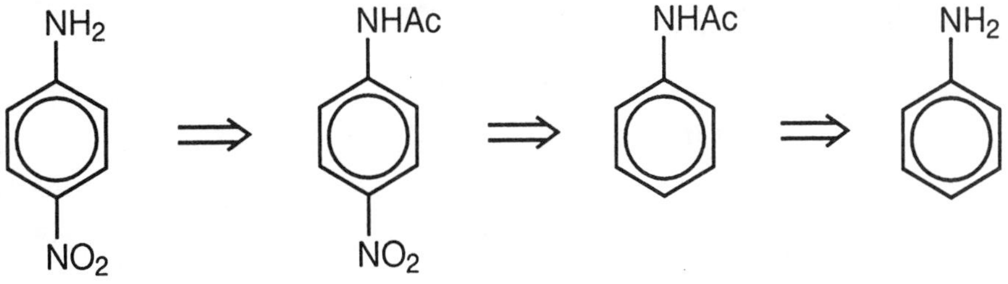

Aniline is easily available in two steps from benzene.

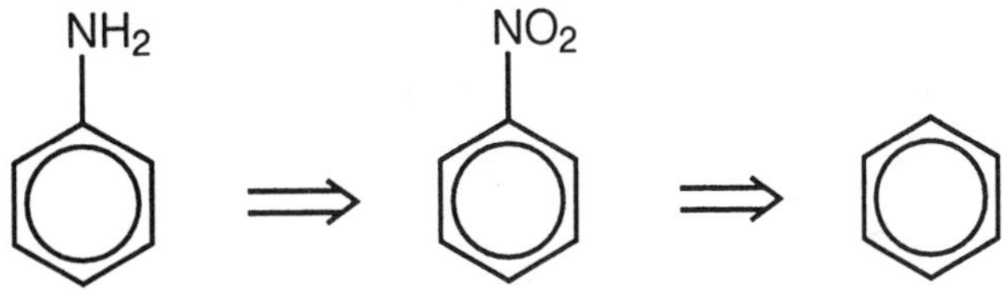

So, here is our proposed synthesis:

$$\text{C}_6\text{H}_6 \xrightarrow[\text{H}_2\text{SO}_4]{\text{HNO}_3} \text{C}_6\text{H}_5\text{NO}_2 \xrightarrow[\text{Pd/C}]{\text{H}_2} \text{C}_6\text{H}_5\text{NH}_2 \xrightarrow[\text{pyridine}]{\text{AcCl}} \text{C}_6\text{H}_5\text{NHAc} \xrightarrow[\text{H}_2\text{SO}_4]{\text{HNO}_3} p\text{-NO}_2\text{C}_6\text{H}_4\text{NHAc} \xrightarrow[\text{H}_2\text{O}]{\text{KOH}} p\text{-NO}_2\text{C}_6\text{H}_4\text{NH}_2 \xrightarrow[\text{HCl}]{\text{NaNO}_2} p\text{-NO}_2\text{C}_6\text{H}_4\text{N}_2^+\ \text{Cl}^- \xrightarrow{\text{CuCN}} p\text{-NO}_2\text{C}_6\text{H}_4\text{CN}$$

Problem 14.36 Up to compound **D**, this reaction sequence is called the Haworth reaction. The initial step is an intermolecular Friedel–Crafts acylation. Aluminum chloride complexes one carbonyl group, and the anhydride opens to give the acylium ion (see Fig. 14.49, for an analogous reaction). The product of the acylation reaction is the *p*-ketocarboxylic acid **A**. Clemmensen reduction of **A** gives the carboxylic acid **B**. Conversion of **B** into the acid chloride **C**, followed by an intramolecular Friedel–Crafts acylation yields **D**, even though the new substituent appears meta to the ortho/para directing methoxy group. That's the only position that can be reached in this intramolecular reaction. The chain is not long enough to reach the other free position on the benzene ring. Finally, reduction of the ketone by a second Clemmensen reduction gives the product, **E** (7-methoxytetralin).

Succinic anhydride + $AlCl_3$ → complex ($:\overset{+}{O}—\overset{-}{A}lCl_3$) → ring-opened acylium ion (C^+) $\xrightarrow[\text{2. } H_2O]{\text{1. anisole } (OCH_3)}$ **A** (p-$CH_3OC_6H_4$–C(=O)–CH_2CH_2–COOH)

A $\xrightarrow[\text{HCl}]{\text{Zn/Hg}}$ **B** (p-$CH_3OC_6H_4$–$CH_2CH_2CH_2$–COOH) $\xrightarrow{SOCl_2}$ **C** (p-$CH_3OC_6H_4$–$CH_2CH_2CH_2$–COCl) $\xrightarrow[\text{2. } H_2O,\ H_3O^+]{\text{1. } AlCl_3}$ **D** (methoxytetralone, C=O) $\xrightarrow[\text{HCl}]{\text{Zn/Hg}}$ **E** (OCH₃-tetralin)

Problem 14.37 Protonation of carbon monoxide leads to an acylium ion. Standard electrophilic aromatic substitution then gives the product.

Cl—H
:C=Ö:
$AlCl_3$
$^{-}AlCl_4$
H
C=Ö:
+
An E⁺ reagent
CH_3
CH_3
(+)
(+)
+
H CHO
proton removal
CH_3
CHO

Problem 14.38 The first step is a straightforward Friedel-Crafts acylation to give acetophenone, **A**. Clemmensen reduction gives ethylbenzene, **B**, a molecule that could also be made directly by Friedel–Crafts alkylation.

O
Cl–C–CH_3
$AlCl_3$
O=C–CH_3
Zn/Hg
HCl
CH_2CH_3
$AlCl_3$
CH_3CH_2Cl
A
B

Ethylbenzene can be photochlorinated exclusively in the benzylic position to give **C**. The E2 reaction gives vinylbenzene, better known as styrene, **D**. Ozonolysis with a reductive workup gives benzaldehyde.

CH_2CH_3
hν
Cl_2
Cl
$CHCH_3$
$(CH_3)_3CO^-$
Na^+
CH=CH_2
1. O_3
2. $(CH_3)_2S$
O=C–H
+ $H_2C=O$
B
C
D

Problem 14.39 This problem presents a pair of "How do we get from here to there?" problems. Neither is particularly hard, and you will probably be able to do them without severe difficulty. Each of these reactions involves protonation of the carbonyl group to generate a carbocation in proximity to a benzene ring. The carbocation is certainly an E⁺ reagent, and reacts with the ring in a typical electrophilic aromatic substitution to create a new ring. It is well worth your time right now to get into the habit of analyzing this kind of problem before you begin to write structures and push arrows. Analysis is always useful, and when we progress to more difficult problems it will be essential for success. Much of what follows may seem obvious, but do not be insulted. Stating the obvious is often useful—it focuses the mind on the objective at hand. You may eventually come to the point at which you do this kind of analysis automatically, but for now making yourself look at the problem in an analytical way may take some effort. We promise you that it is well worth it.

The first obvious, but important, point is that a ring must be closed in this reaction. There are two rings in the product, and only one in the starting material. Obvious? You bet. Did you think of this when you first looked at the problem? We bet not. The thought that a ring must be closed now leads to the second point: We know exactly where the ring must be closed.

Starting material (one ring)

Product (two rings)

Bond that must be closed somehow

This oxygen is lost in the reaction; a bond must somehow be formed between the two dotted carbons

It is also clear that an oxygen atom must be lost in this reaction. It is present in the starting material but absent in the product.

So now we are able to see quite specific goals in this problem. A ring must be closed between the two dotted carbons, and an oxygen atom must eventually be lost. How many ways do we know to form a bond to a benzene ring? Not many, that's for certain. This chapter has been primarily devoted to the reaction of E^+ reagents with benzene rings. Surely, that is the first thing to think of when considering ways to close the second ring by forming a new bond to benzene.

standard aromatic substitution and elimination

This position must become an E^+ reagent to close the second ring.

How to form the E^+ reagent? Simple: protonate the oxygen. Addition to the ring generates the usual cyclohexadienyl cation intermediate. Note again the convention by which we show the positions sharing the positive charge with (+).

$HClO_4$ protonate oxygen

aromatic substitution

proton loss

Now the alcohol is protonated and elimination reactions (E1 as shown, or E2) lead to the final products.

$HClO_4$

The second example in this problem can be analyzed in a very similar fashion. Do this. Find the position at which the new ring must be formed, and find a way to accomplish this, using the first example as a model. The following drawings outline the process:

75-80% H_2SO_4

E^+ Reagent

proton loss

Ph = benzene ring

Now here is the loss of water (elimination) phase:

75-80% H_2SO_4

Problem 14.40 These reactions should ring a bell — the higher energy product is formed preferentially at low temperature, and the lower energy product is formed at higher temperature. This problem is another example of kinetic versus thermodynamic control. Recall our discussion of the additions of Cl_2 and HBr to 1,3-butadiene (Chapter 12, p. 527) and the Diels–Alder reactions of furan and maleimide (Problem 12.49, p. 564). Also recall that aromatic sulfonation is a reversible reaction, especially at high temperature (Section 14.5b, p. 632). The Energy versus Reaction progress diagram tells almost all.

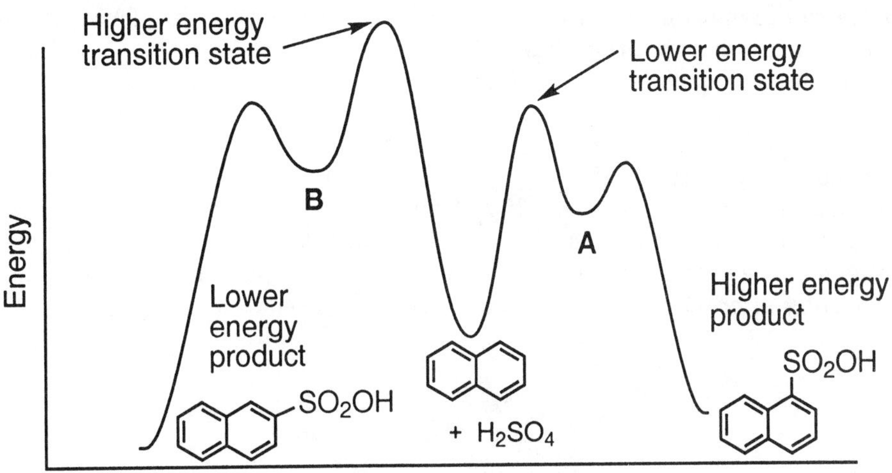

Now we have two questions. First, why is naphthalene-1-sulfonic acid the kinetically preferred (first formed) product, and second, why is naphthalene-2-sulfonic acid the thermodynamic product? As the first question concerns kinetics, we must look at the transition states for the two reactions. The transition states for electrophilic aromatic substitution reactions resemble the cationic intermediates for these endothermic reactions. Recall the Hammond postulate (Chapter 8, p. 320 and Problem 14.20). Therefore, an examination of these intermediates, **A** and **B**, in the diagram above, should prove useful.

For addition at position 1

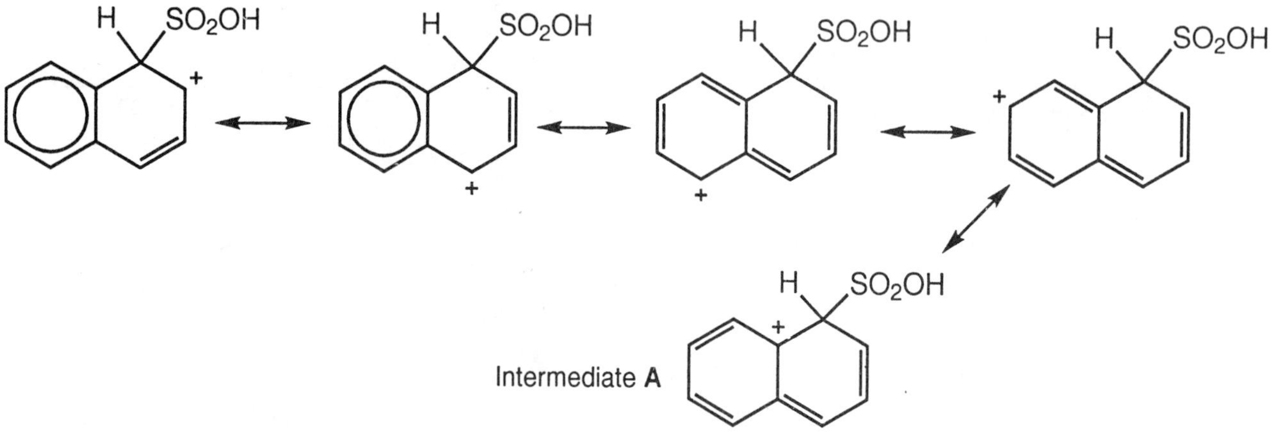

H SO_2OH + ⟷ H SO_2OH + ⟷ H SO_2OH + ⟷ + H SO_2OH ⟷ H SO_2OH +

Intermediate **A**

For addition at position 2

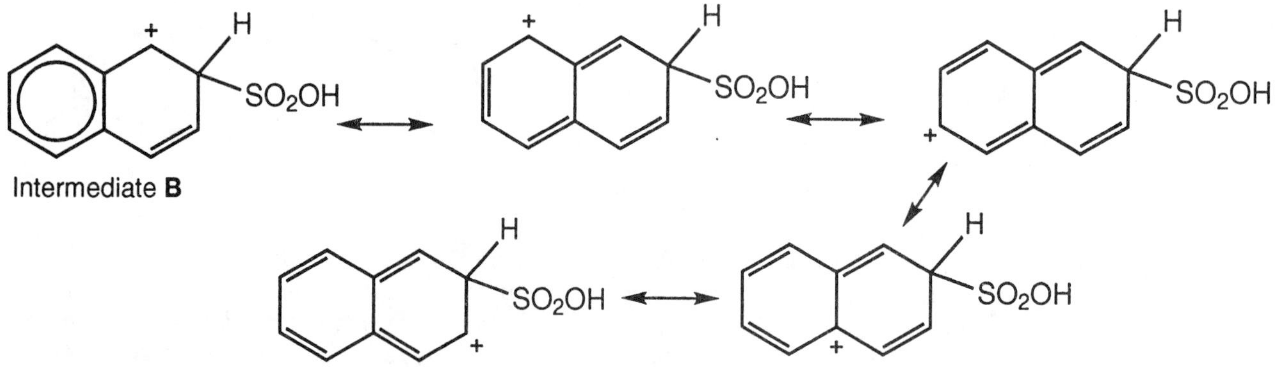

+ H SO_2OH ⟷ + H SO_2OH ⟷ H SO_2OH + ⟷ H SO_2OH + ⟷ H SO_2OH +

Intermediate **B**

Two of the resonance form for intermediate **A,** the cation produced from addition to position-1, retain an intact benzene ring. By contrast, addition to position-2 involves an intermediate (**B**) in which only one form retains an intact benzene ring. Resonance contributors in which a benzene ring is intact are more important contributors (more stable) than the other forms. Accordingly, the intermediate from addition at position 1, *and the transition state leading to it,* are lower in energy than the comparable intermediate, *and the transition state leading to it,* for substitution at the 2-position.

The reason for the increased thermodynamic stability of naphthalene-2-sulfonic acid over naphthalene-1-sulfonic acid is not so clear. If you guessed steric effects, you were right. In the 1-isomer, the sulfonic acid group has an unfavorable steric interaction (called a *peri* interaction) with the hydrogen at the 8-position. The substituents at the 1- and 8-positions are parallel to each other and apparently interact more strongly than other, ortho substituted groups.

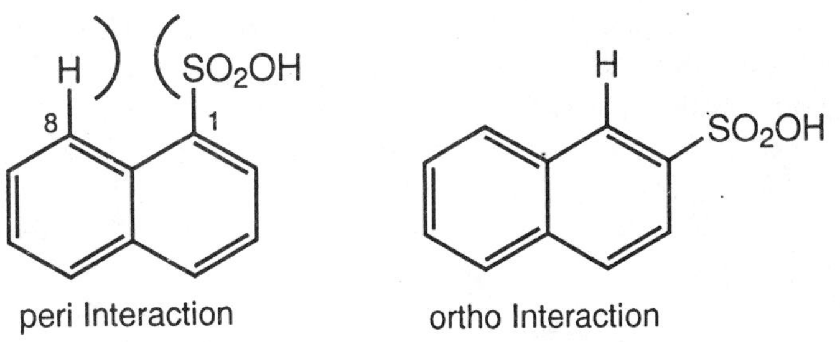

peri Interaction

ortho Interaction

Accordingly, the product of substitution in the 2-position is the more stable compound, and will be favored at equilibrium (thermodynamic control).

Problem 14.41 The only difficult thing about this problem is that there is an intermediate alcohol that is transformed into the chloride. Protonation of formaldehyde ($H_2C{=}O$) leads to an E^+ reagent that undergoes electrophilic aromatic substitution at the 1-position of naphthalene to give the alcohol **A**.

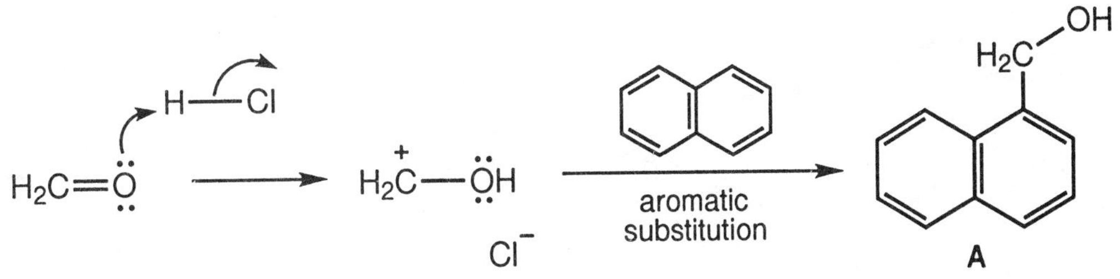

aromatic substitution

A

Now, don't forget earlier chemistry. The alcohol is protonated in acid (converting the poor leaving group OH into the good leaving group OH_2). Displacement by chloride gives the final product.

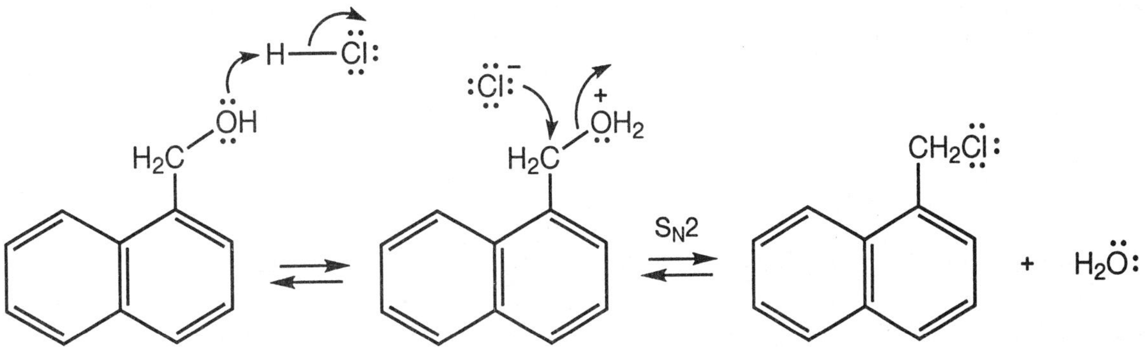

Problem 14.42 This problem is much easier than it looks. It is difficult only for psychological reasons. In this wonderful reaction, investigated by a former student of MJ's, Professor L. T. Scott (b. 1944), then at the University of Nevada, Reno, now at Boston College, the ring at the upper right seems to switch its orientation. And that is exactly what happens. The process is triggered by formation of a carbocation through protonation of the naphthalene part of the molecule.

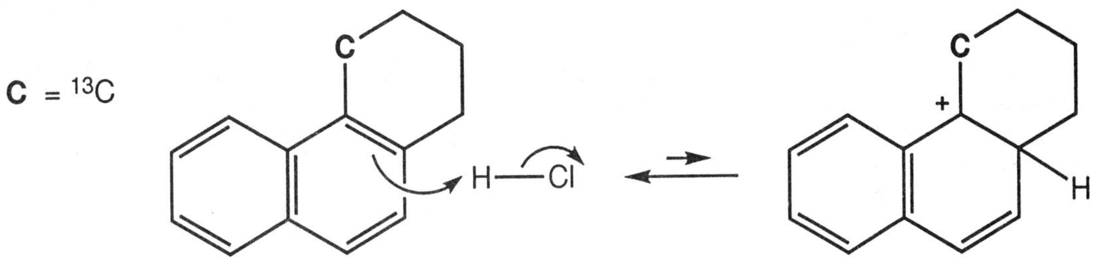

C = ^{13}C

Now, what can happen to this carbocation? It can, of course, rearomatize through proton loss, a reversal of the original protonation, and this must be a very easy reaction. However, in addition, a carbon–carbon bond might rearrange to give an intermediate spiro carbocation.

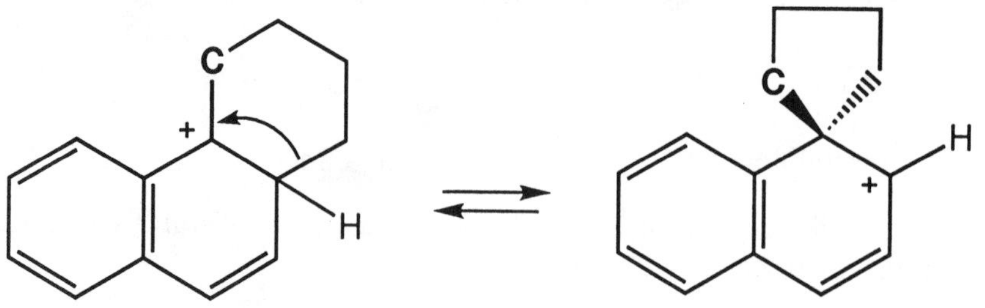

Notice that the spiro carbocation is symmetrical in the sense that there are two equivalent carbon–carbon bonds that can migrate back. If the same bond moves back as originally rearranged, we of course see no change. However, if the other, equivalent carbon–carbon bond moves back, we get a new carbocation. Deprotonation gives the product.

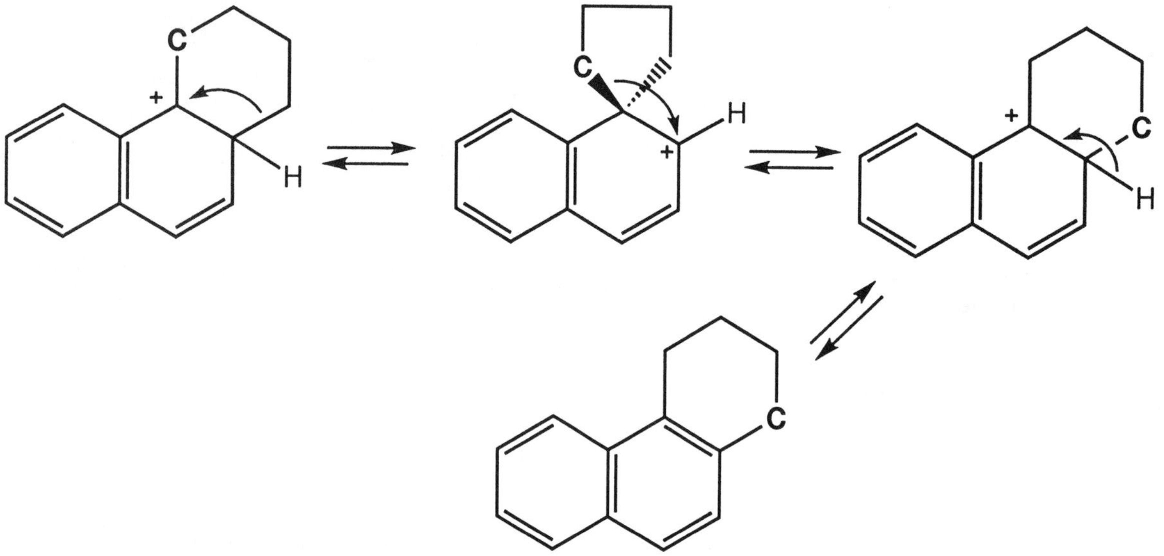

Problem 14.43 Compound **A** is formed through a straightforward Friedel–Crafts alkylation reaction.

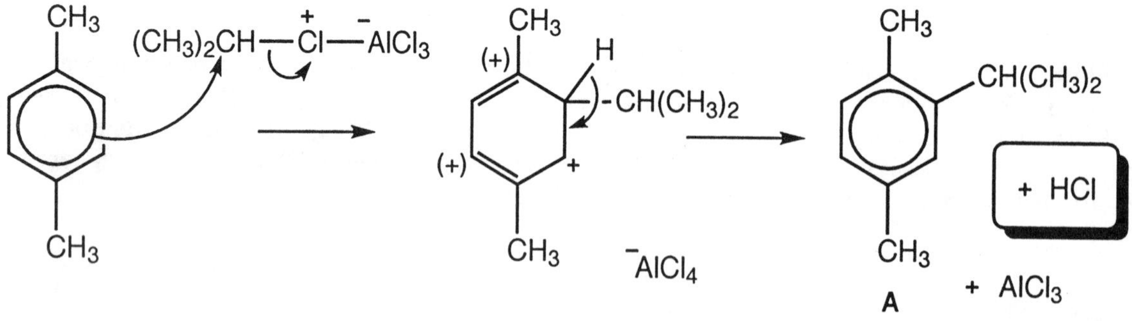

What can HCl do? The hint tells you to focus on this product. If the product **A** is protonated, we can generate an intermediate in which an alkyl shift moves the relatively large alkyl groups apart. That's the key to why this rearrangement happens—the relief of steric strain when the thermodynamically more stable 1,3,5-trisubstituted benzene is formed from a less stable 1,2,4-trialkylbenzene.

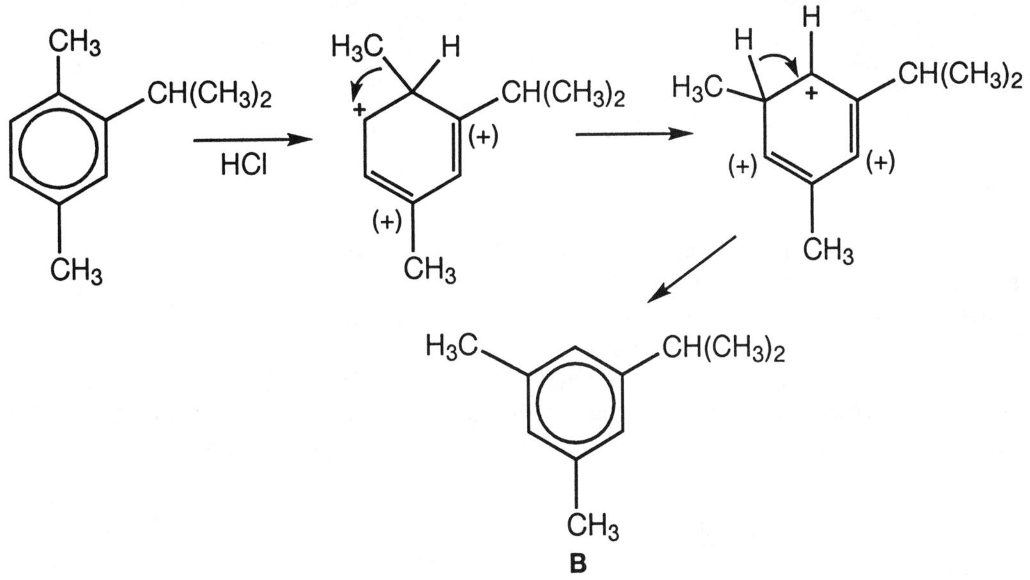

B

Problem 14.44

(a) This problem involves a straightforward nucleophilic aromatic substitution reaction. Like the nitro group, the carbonyl group can stabilize an anion.

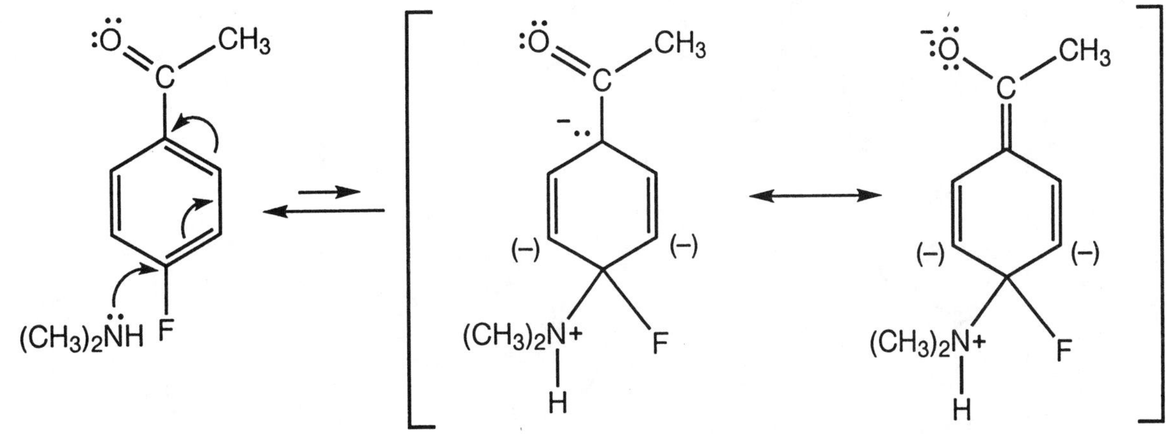

Deprotonation of nitrogen and rearomatization through loss of fluoride completes the reaction.

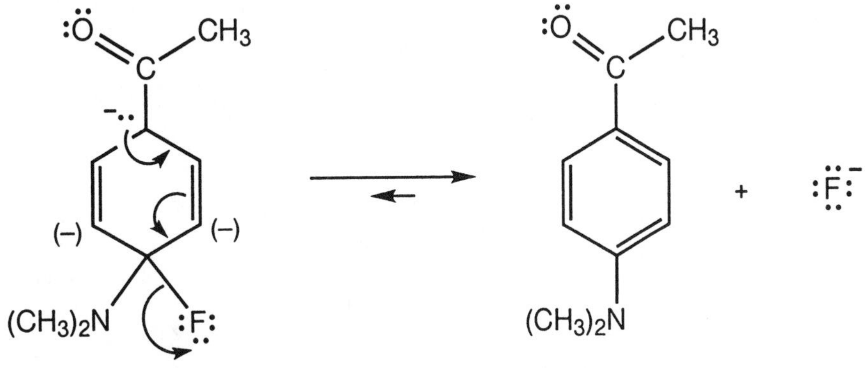

(b) Phthaloyl peroxide decomposes to benzyne when irradiated. Benzyne undergoes a Diels–Alder reaction with the diene to give the product.

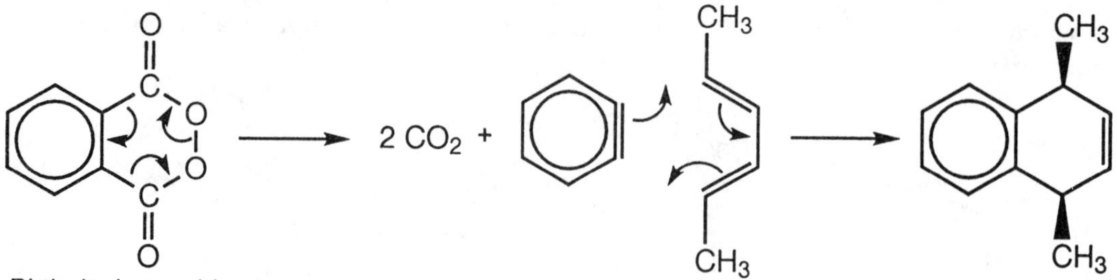

Phthaloyl peroxide $\longrightarrow$ $2\ CO_2$ + benzyne + diene (CH_3, CH_3) $\longrightarrow$ product (CH_3, CH_3)

Note the stereochemistry of the product in the addition of benzyne to *trans,trans*-2,4-hexadiene. The cis stereochemistry shows that the addition takes place in a single step. There can be no intermediates capable of allowing rotation around a single bond.

Problem 14.45 Endothermic addition of methoxide gives an intermediate cyclohexadienyl anion. The question is whether the first step or the second step is rate determining. The initial addition to the ring disrupts the aromatic system and is likely to be the slow step in the reaction.

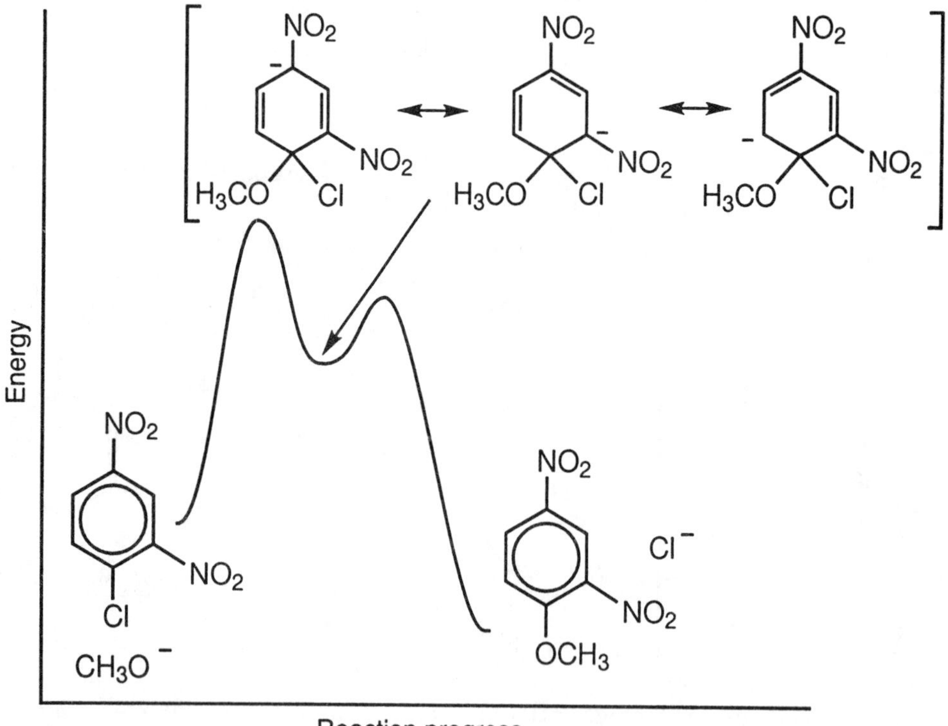

NO_2 NO_2 H_3CO Cl $\longleftrightarrow$ NO_2 NO_2 H_3CO Cl $\longleftrightarrow$ NO_2 NO_2 H_3CO Cl

Energy

NO_2 NO_2 Cl CH_3O^-

NO_2 NO_2 OCH_3 Cl^-

Reaction progress

Problem 14.46 The only possible leaving group is chloride.
(a) The three products are the two possible monosubstitution products and the one possible disubstitution product, **A**, **B**, and **C**.

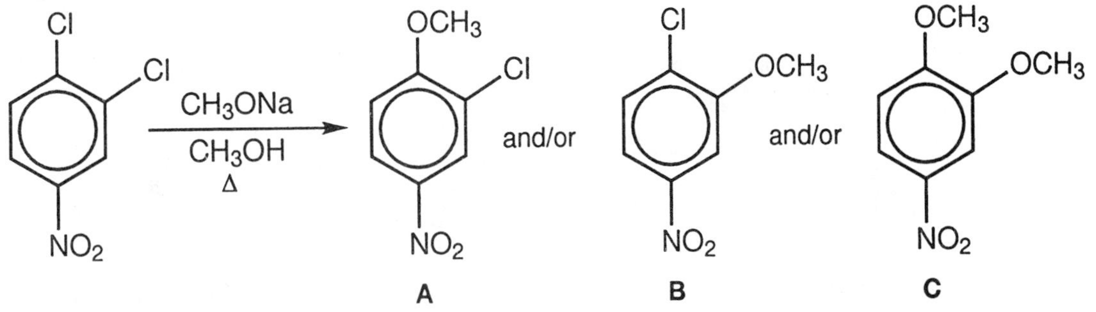

CH_3ONa, CH_3OH, Δ

and/or

and/or

A **B** **C**

(b) The only product that can be formed in this reaction is **A**, 2-chloro-4-nitroanisole. If we examine the mechanism for this nucleophilic aromatic substitution reaction, we can see why.

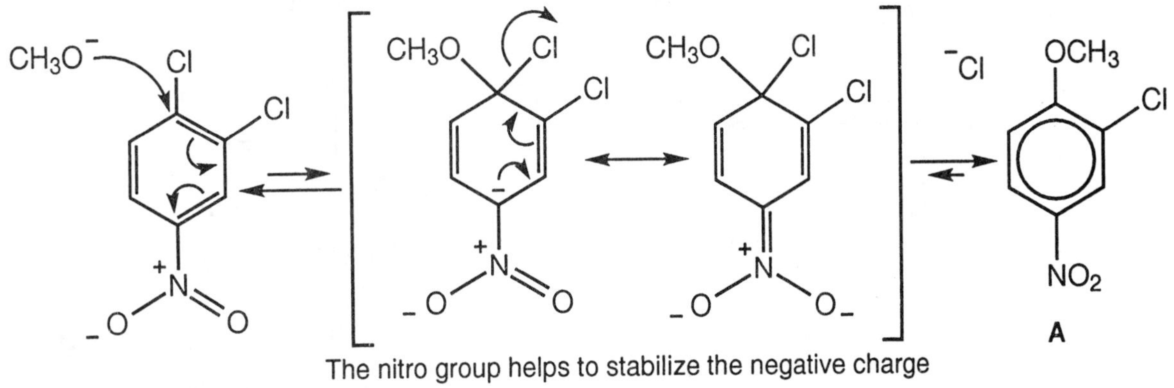

The nitro group helps to stabilize the negative charge

A

Nucleophilic addition at the position para to the nitro group affords an intermediate in which the nitro group can help stabilize the negative charge.

However, addition of methoxide at the other position bearing a chlorine, the position meta to the nitro group, does not give an intermediate in which the nitro group is stabilizing. Accordingly, no substitution at this position is observed.

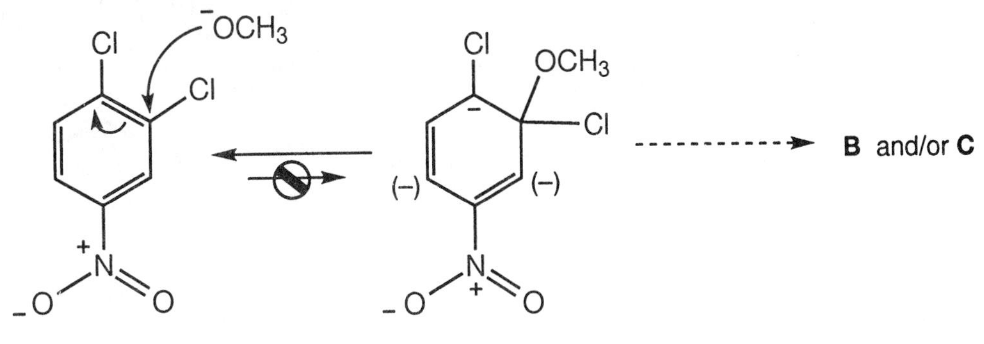

B and/or **C**

Problem 14.47

(a) It helps to draw a good Lewis structure for benzenediazonium-2-carboxylate. Then the arrow formalism becomes easy to draw.

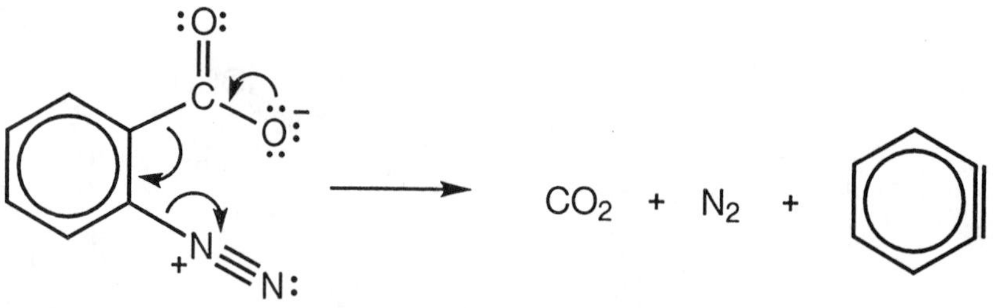

CO_2 + N_2 +

(b) Triptycene is formed through a Diels–Alder reaction across the 9,10-positions of anthracene (see Problem 13.40, p. 619).

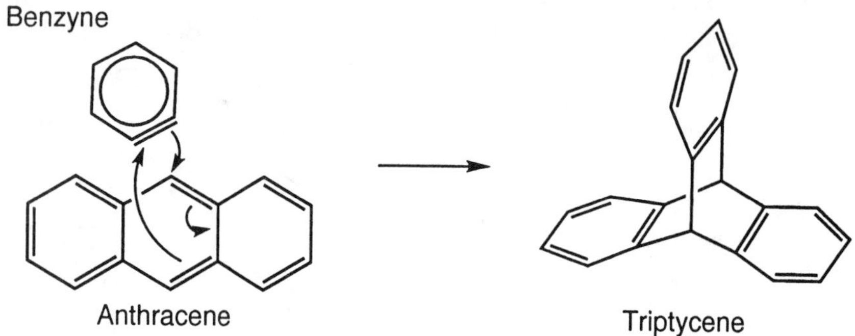

Benzyne

Anthracene

Triptycene
2

Problem 14.48 In the first reaction, amide ion can add in two ways to the unsymmetrical benzyne intermediate to give the two observed products.

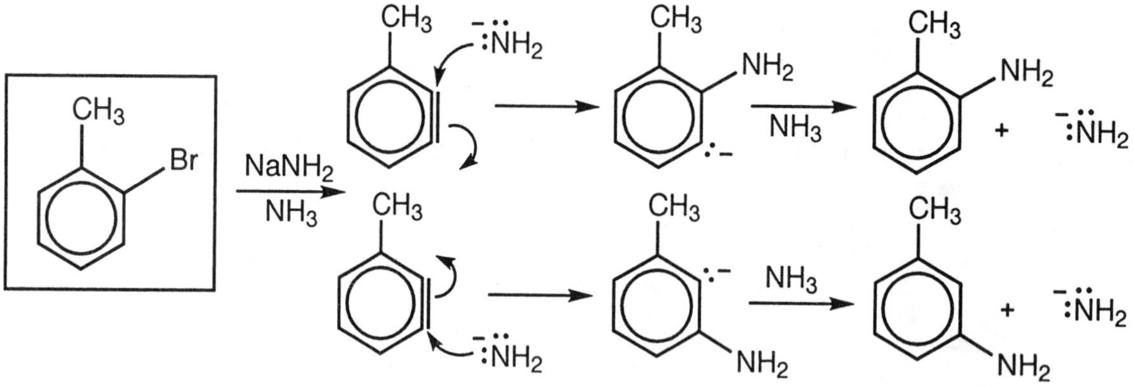

$NaNH_2$ / NH_3

In the second reaction, the same two modes of reaction are theoretically possible, but the presence of a methoxy group strongly influences the situation. Formation of anion **A** is greatly favored over formation of anion **B** because in **A** the adjacent methoxy group inductively stabilizes the negative charge. This effect directs the reaction towards the observed product.

OCH3 Br OCH3 :NH2 OCH3 NH2 The methoxy group is not stabilizing this intermediate much B OCH3 δ− δ+ OCH3 NH3 NH2 :NH2 A OCH3 NH2

This intermediate is strongly stabilized by the methoxy group

Chapter 15 Outline

15.1 Chromatography
15.2 Mass Spectrometry
15.3 Infrared Spectroscopy
15.4 Nuclear Magnetic Resonance Spectroscopy
15.5 Survey of NMR Spectra of Organic Molecules
15.6 Spin–Spin Coupling: The Coupling Constant, J
15.7 More Complicated Spectra
15.8 Decoupled Spectra
15.9 Nuclear Magnetic Resonance Spectra of Other Nuclei
15.10 Something More: Dynamic NMR

15 Analytical Chemistry

Solving spectral problems involves having a reasonable familiarity with the spectral techniques themselves, and then lots and lots of practice. You need to know what kinds of things each spectral technique can tell you: Mass spectrometry is different from nuclear magnetic resonance spectroscopy, for example. These early, in-chapter problems try to highlight what the techniques can do. The Additional Problems (Section 15.13) will go on to start you off on the actual solving of problems involving interpretation of spectra.

Problem 15.1 In Chapter 5, we discussed "resolution," which involves converting a pair of enantiomers (same physical properties) into a pair of diastereomers (different physical properties) by reaction with a single enantiomer of another molecule. The diastereomers can be separated and then the pair of enantiomers regenerated.

Pair of enantiomers —reaction→ Pair of diastereomers —separate→

- One diastereomer —reverse the original reaction→ One enantiomer
- The other diastereomer —reverse the original reaction→ The other enantiomer

In Chapter 5, the technique of column chromatography was also introduced (p. 179). The chromatographic technique for separating enantiomers is a variation on the general theme. Diastereomeric *compounds* are not formed, but diastereomeric *complexes* are. The pair of enantiomers to be separated is passed though a column in which the stationary phase is constructed from a single enantiomer. As the pair to be separated passes through the column, the components will be adsorbed differently on the stationary phase because the complexes formed are diastereomeric. In this technique the strong, covalent bonding of resolution is replaced with weaker partial bonding, or complexing, as the enantiomers are adsorbed on the material of the column.

Problem 15.3 Of course there are many such possibilities. Here are a few.

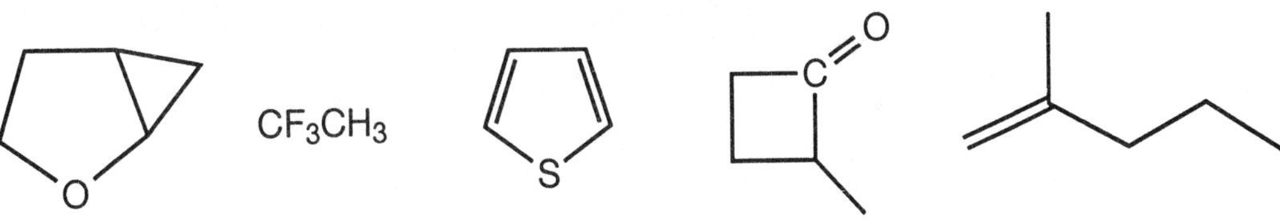

Problem 15.4 The tropylium ion is $C_7H_7^+$, and its aromaticity makes it exceptionally stable (for a carbocation). Here is a possible arrow formalism for its formation from the benzyl cation. The formation of the vinyl cation looks bad, but remember, very high energies are involved in mass spectrometry.

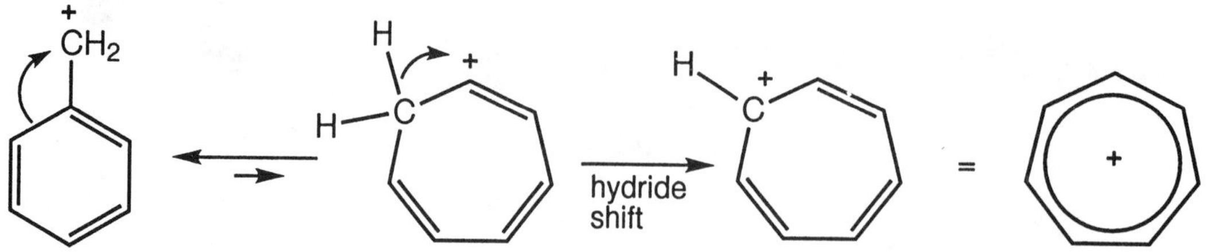

hydride shift

The tropylium ion

Problem 15.5 In any alkene the bonding π molecular orbital is filled with two electrons and the antibonding π^* orbital is empty. When a π electron is ejected in the mass spectrometer, it can only come from the filled π orbital.

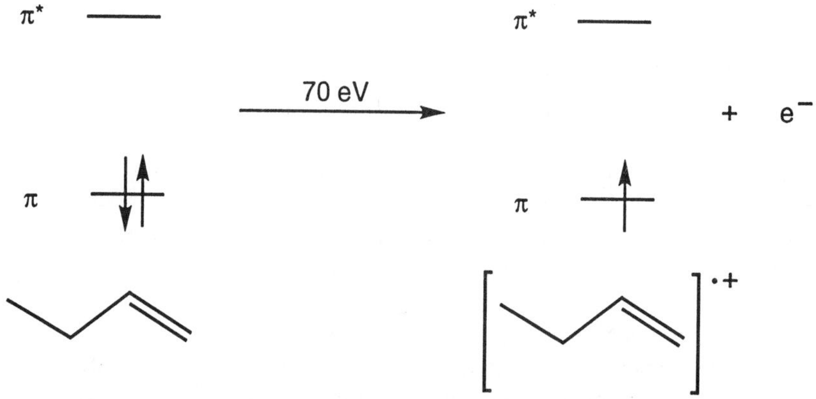

π^* π^*

70 eV

$+ \; e^-$

π π

Problem 15.7 In Figure 15.19 (b), the giveaway is the broad band above 3000 cm^{-1}. This can only result from an OH. There is a strong band at 1720 cm^{-1}; a carbonyl group is present. The combination of these two bands means that you should think of a carboxylic acid, RCOOH. The appropriate strong C-O stretching vibrations are present between 1400 and 1200 cm^{-1}. As there are only two carbons in this molecule, the structure must be acetic acid, CH_3COOH.

In Figure 15.19 (c), the band at 2220 cm^{-1} is in a region of few absorptions in the IR. As the compound contains nitrogen, it is very likely that the compound is a cyanide. There are C-H stretching bands above 3000 cm^{-1}. These bands reveal the presence of either "olefinic" or "aromatic" hydrogens. There is a strong band at 1710 cm^{-1}, indicating that a carbonyl group is present. The band at about ~2740 cm^{-1} tells us that the carbonyl group is an aldehyde. The sequence of medium-to-weak bands at about 1600 cm^{-1} indicates that the molecule is aromatic. A reasonable guess would be a cyanobenzaldehyde, and this one turns out to be the para isomer, as seen by the strong band at about 830 cm^{-1}.

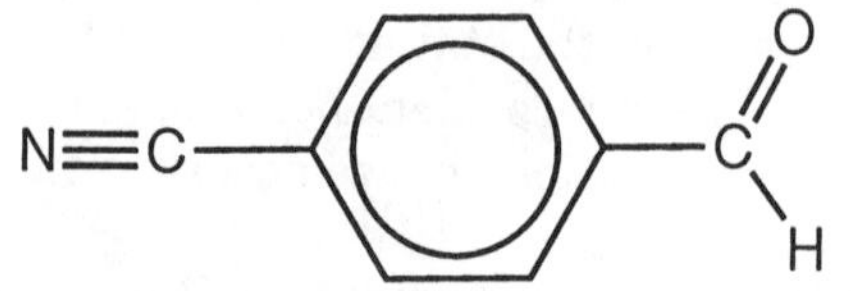

Problem 15.11
(a) The two hydrogens of cyclopropane are homotopic because replacement of each with X gives the same thing.

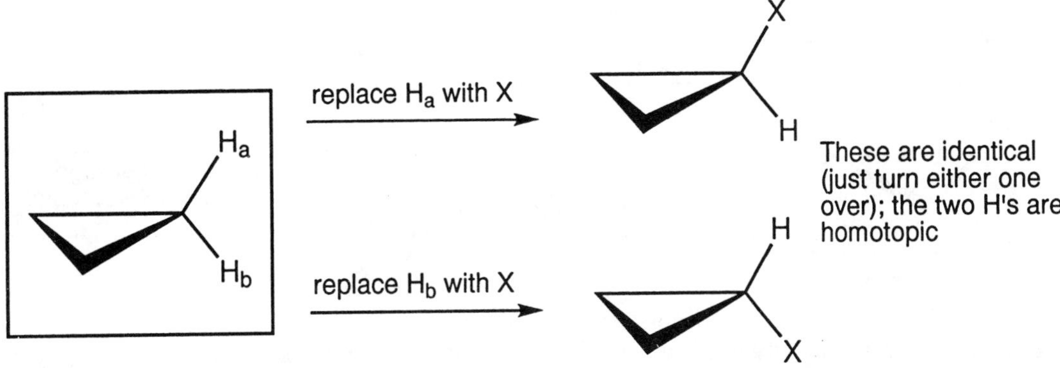

(b) Replacement of the two indicated hydrogens of *tert*-butylcyclohexane gives two diastereomers. The two hydrogens are diastereotopic.

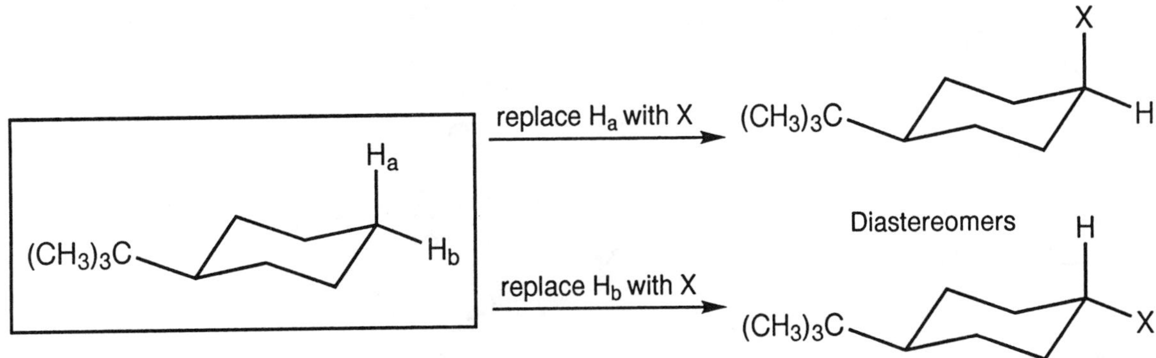

(c) These two hydrogens are homotopic. Replacement of each with X leads to the same thing.

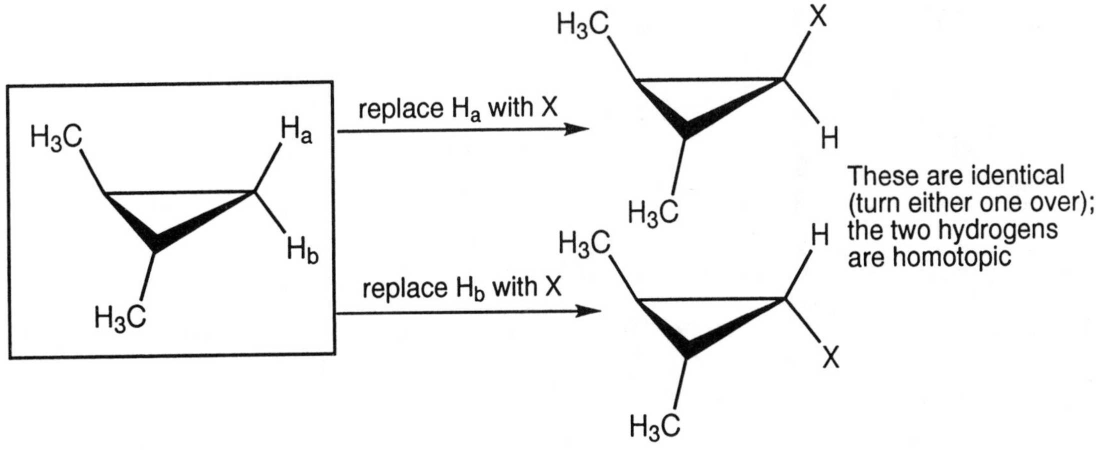

(d) Replacement of the indicated hydrogens with X leads to diastereomers. The two hydrogens are diastereotopic.

replace H_a with X

replace H_b with X

These are diastereomers; the two hydrogens are diastereotopic

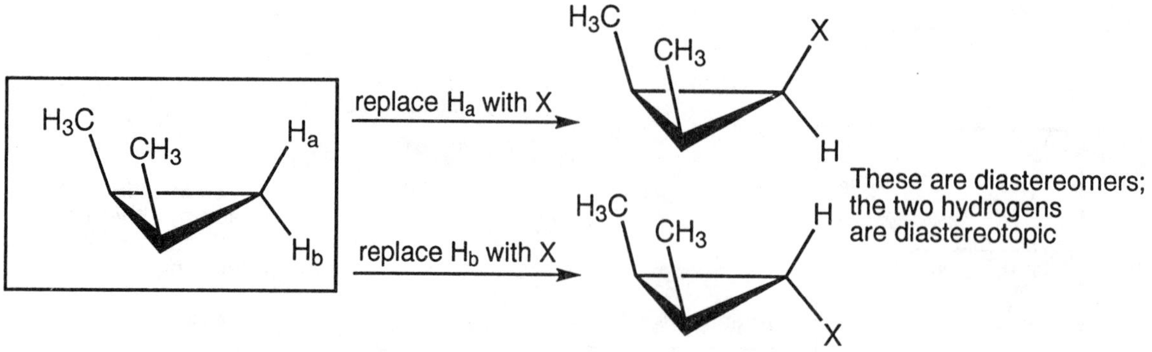

(e) The two indicated hydrogens in propyl alcohol are enantiotopic. Replacement of the two with X leads to a pair of enantiomers. The drawing uses Newman projections to show this.

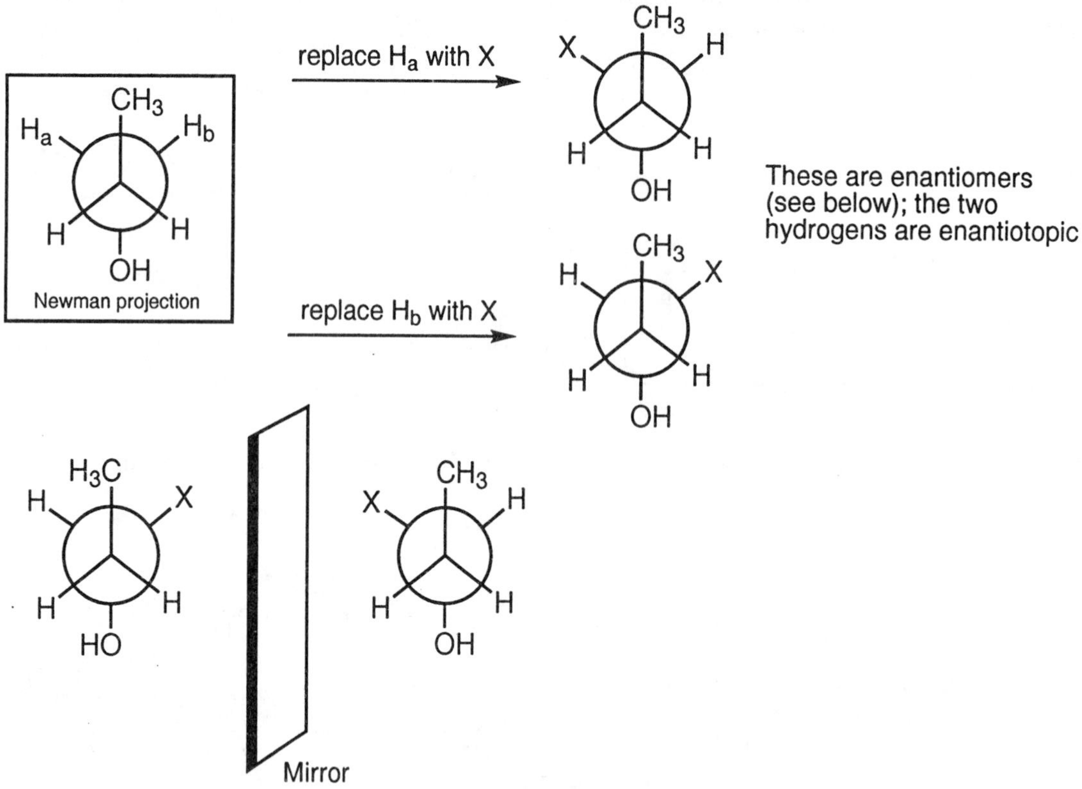

Newman projection

replace H_a with X

replace H_b with X

These are enantiomers (see below); the two hydrogens are enantiotopic

Mirror

Problem 15.13 A look at the resonance forms for an α,β-unsaturated ketone tells the story. The β-position is electron-deficient. Accordingly, shielding by electrons at the β-position will be weak and a relatively weak applied magnetic field will be required to bring hydrogens there into resonance.

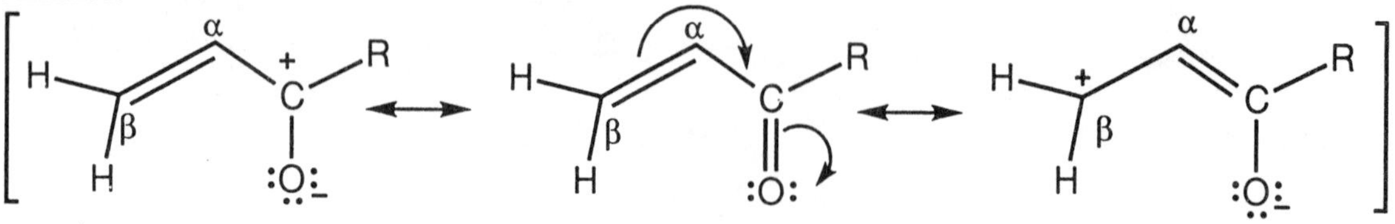

Problem 15.15 The answer is the same for all three compounds. Each compound is aromatic. As discussed in the text (p. 719), in the presence of an applied magnetic field (B_o) there will be a circulation of the π electrons (a ring current) producing an induced magnetic field (B_i) in opposition to B_o. In each molecule, the indicated hydrogens are at a position over or in the center of the aromatic ring. This point is exactly where B_i opposes B_o. A hydrogen in this position will "feel" a net magnetic field, $B_{net} = B_o - B_i$. Accordingly, the applied magnetic field will have to be increased in order to bring the hydrogens into resonance. The presence of high field signals such as the ones in these three molecules is a widely used diagnostic for aromaticity.

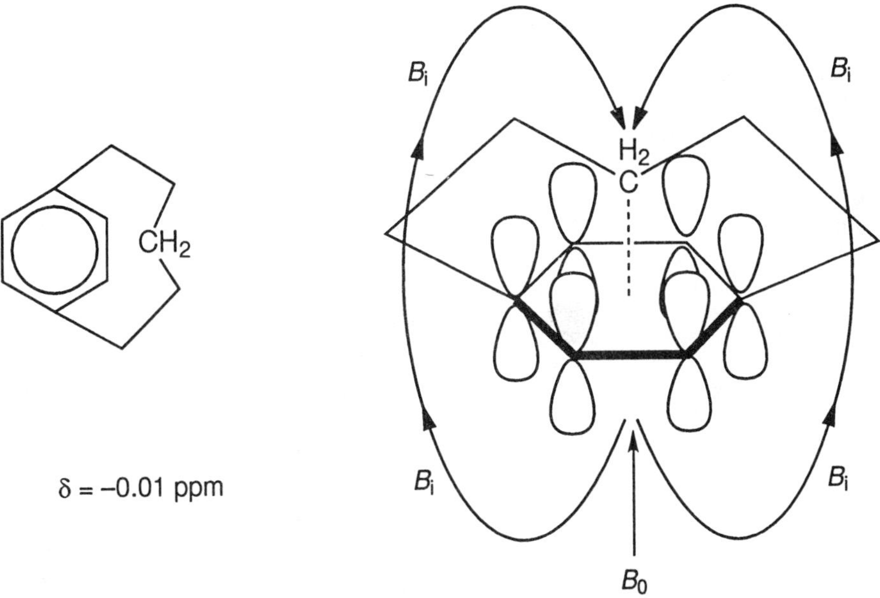

One CH_2 lies over the center of the aromatic ring. The dotted line shows the projection of this CH_2 onto the ring. At this point B_i opposes B_o. The applied field (B_o) will have to be increased to bring these hydrogens into resonance.

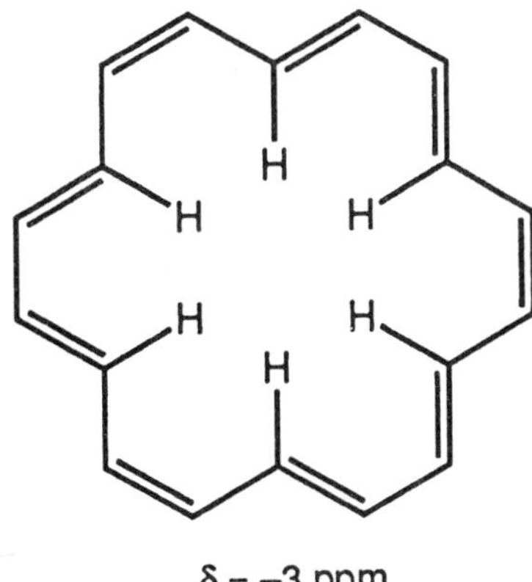

δ = −3 ppm

(4*n* + 2) π electrons (*n* = 4); the "inside" hydrogens will be shifted far upfield as they lie in a region in which B_i strongly opposes B_o

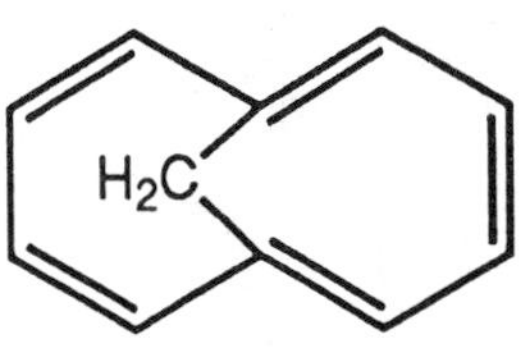

δ = −0.5 ppm

(4*n* + 2) π electrons (*n* = 2); the "bridging" methylene hydrogens will be shifted far upfield as they lie in a region in which B_i strongly opposes B_o

Problem 15.16 There are four equivalent methylene hydrogens, and therefore there will be five lines (n + 1, n = 4). The ratio of intensities is given by Pascal's triangle: 1:4:6:4:1.

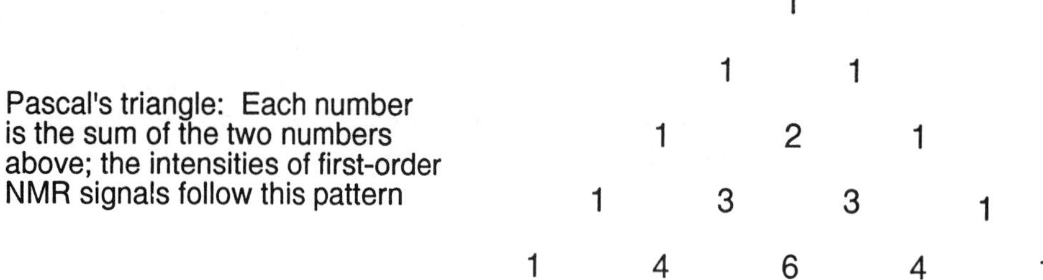

Pascal's triangle: Each number is the sum of the two numbers above; the intensities of first-order NMR signals follow this pattern
1
1 1
1 2 1
1 3 3 1
1 4 6 4 1

Problem 15.17

The dihedral angle between the two cyclopropane carbon-hydrogen bonds is 0°

The dihedral angle between the two cyclopropane carbon-hydrogen bonds is about 120°

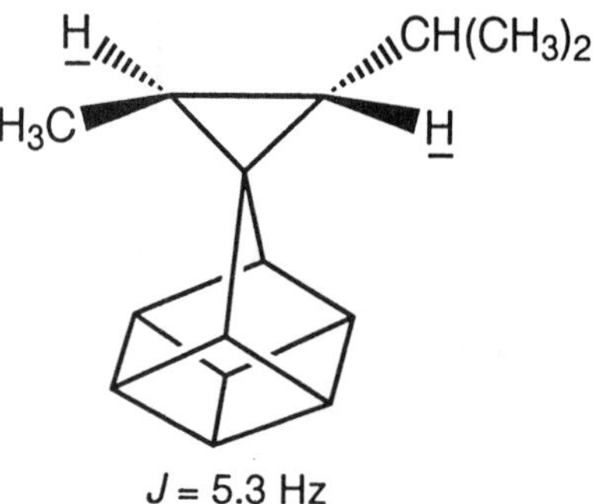

J = 8.4 Hz

J = 5.3 Hz

A look at the Karplus curve (Fig 15.48) shows that the cis (0°) compound should have a larger coupling constant than the trans (120°) compound. This is exactly how the stereochemical assignment was made in the research paper describing these two compounds.

Problem 15.18 Here is an "arrow pushing" problem deep in the middle of a spectroscopy chapter. It is a reminder that you never outgrow your need to push arrows.

$CH_3CH_2\text{—}\ddot{O}\text{—}H + D\text{—}\overset{+}{O}D_2 \rightleftharpoons CH_3CH_2\text{—}\overset{+}{O}(D)(H) + \ddot{O}D_2$

$CH_3CH_2\text{—}\overset{+}{O}(D)(H) + \ddot{O}D_2 \rightleftharpoons CH_3CH_2\text{—}\ddot{O}\text{—}D + H\overset{+}{O}D_2$

$(CH_3CH_2)_2\ddot{N}\text{—}H + {}^{-}\text{:B} \rightleftharpoons (CH_3CH_2)_2\ddot{N}^{-} + B\text{–}H$

$(CH_3CH_2)_2\ddot{N}^{-} + D\text{—}B \rightleftharpoons (CH_3CH_2)_2\text{:N—D} + {}^{-}\text{:B}$

Problem 15.19 Imagine the approach of a generic single enantiomer, C(ABDE) to one of the methylene hydrogens of ethyl alcohol, called H_a. We will pick an approach in which the methylene hydrogen H_a lies in between groups D and E of the enantiomer. In this interaction, D is opposite the methyl group of ethyl alcohol, E is opposite the OH of ethyl alcohol, A is next to the other methylene hydrogen of ethyl alcohol (dotted line), and B is aimed away from ethyl alcohol.

Now let the same enantiomer of C(ABDE) approach the other methylene hydrogen of ethyl alcohol, H_b, in the same way. Now D and the methyl are still opposed, E and OH are still opposed, but it is B which is next to the other methylene hydrogen of ethyl alcohol (dotted line), and A that is aimed away from ethyl alcohol. The two interactions are different; in the presence of this single enantiomer, the two methylene hydrogens of ethyl alcohol, called H_a and H_b here, are different and must give rise to different signals in the NMR spectrum.

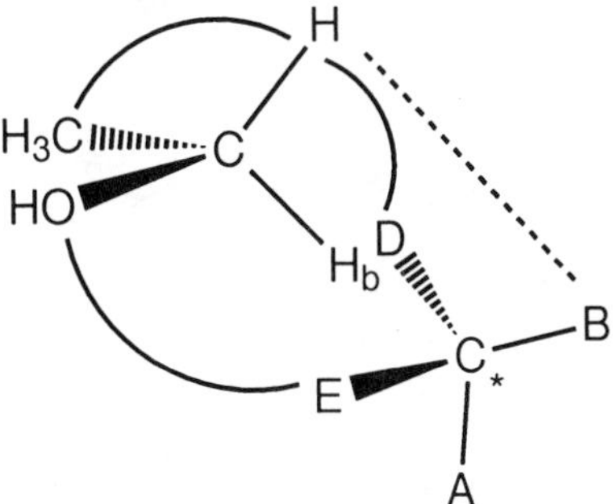

It is very important to realize that this analysis does not depend on these particular interactions between ethyl alcohol and the single enantiomer. In reality, the interaction between ethyl alcohol and the solvent will be the sum of all possible approaches. But the analysis summarized in this answer will be true for *all* approaches. You might try to work through one other approach in order to convince yourself that the analysis does not depend on the particular approach chosen.

Problem 15.21 The first compound is easy. The *tert*-butyl and methyl groups will be largely in equatorial positions. There is only a single hydrogen on the carbon adjacent to the methyl group, and by the $n + 1$ rule the methyl hydrogens will apear as a doublet.

The second compound is more interesting. The methyl groups are diastereotopic and must give rise to different signals. Each methyl group will appear as a doublet in the NMR. Use the technique of replacing each methyl group with X to see this.

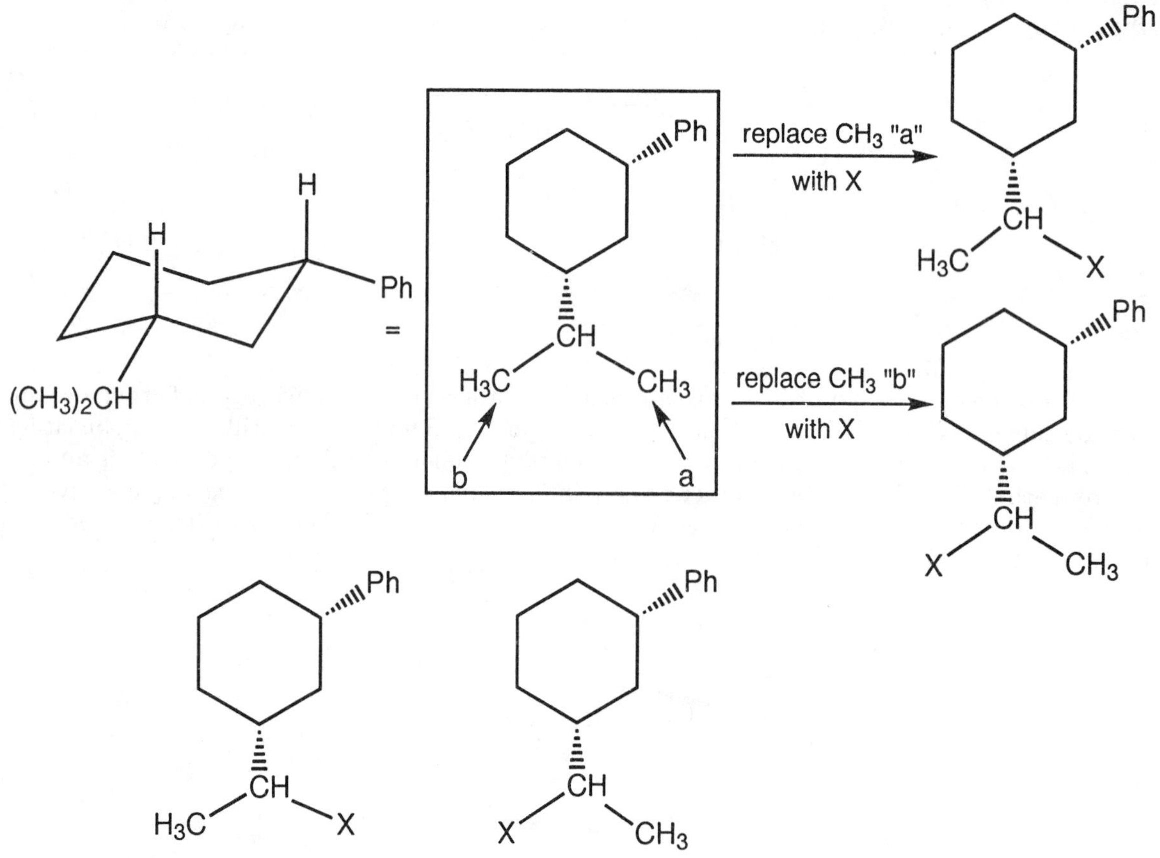

H
H
Ph
$(CH_3)_2CH$
=
Ph
CH
H_3C CH_3
b a
replace CH_3 "a"
with X
Ph
CH
H_3C X
replace CH_3 "b"
with X
Ph
CH
X CH_3

Ph
CH
H_3C X
Ph
CH
X CH_3

These are not the same! They are diastereomers and the methyl groups are diastereotopic

Additional Problem Answers

Problem 15.22 One micron (1μ) is 10^{-4} cm. To convert microns into reciprocal centimeters (cm^{-1}) simply divide 1 by the value in microns and multiply by 10^4. The answers are 3333, 2000, 1429, 1111, and 909 cm^{-1}.

Problem 15.23 The number of degrees of unsaturation is easily calculated either through use of the formula, or, in many cases, by inspection.
(a) Benzene contains one ring and three double bonds. Surely there must be four degrees of unsaturation. There are.

$$\Omega = \frac{2n + 2 - (\text{No. of hydrogens})}{2} = \frac{2(6) + 2 - 6}{2} = 4$$

n = No. of carbons in the related saturated alkane

(b) For pyridine we need to use the formula as adjusted for nitrogen. Count nitrogen as a carbon, and subtract 1 for each nitrogen.

$$\Omega = \frac{2n + 2 - (\text{No. of nitrogens}) - (\text{No. of hydrogens})}{2} = \frac{2(5+1) + 2 - 1 - 5}{2} = 4$$

(c) $\Omega = \dfrac{2n + 2 - (\text{No. of hydrogens})}{2} = \dfrac{2(3) + 2 - 6}{2} = 1$

(d) Don't forget that the chlorine is treated as a hydrogen.

$$\Omega = \frac{2n + 2 - (\text{No. of hydrogens and chlorines})}{2} = \frac{2(5) + 2 - (7 + 1)}{2} = 2$$

(e) $\Omega = \dfrac{2n + 2 - (\text{No. of hydrogens})}{2} = \dfrac{2(3) + 2 - 8}{2} = 0$

(f) $\Omega = \dfrac{2n + 2 - (\text{No. of hydrogens})}{2} = \dfrac{2(3) + 2 - 6}{2} = 1$

(g) $\Omega = \dfrac{2n + 2 - (\text{No. of nitrogens}) - (\text{No. of hydrogens})}{2} = \dfrac{2(3 + 1) + 2 - 1 - 7}{2} = 1$

Problem 15.24 All the structures are consistent with the O—H or N—H stretching band centered roughly at 3235 cm^{-1}, so no distinction is possible through use of this band. (Actually, the IR maven might object that primary amines and amides should show two bands in this region, but this seems a fine point). Structures **A** and **C** can be eliminated because of the absence of a C=O stretch in the 1700 cm^{-1} region. So, the choice is between **B** and **D**. The band at 2240 cm^{-1} is consistent with a cyanide group (C≡N triple bond stretch), and so the structure must be **B**.

Problem 15.25 Compound **1**. The IR spectrum of compound **1** exhibits an intense carbonyl stretch at 1750 cm^{-1}. By itself, this does not help much; however, the presence of a C—O stretch at 1200 cm^{-1} suggests the possibility of an **ester**. Anhydrides and carboxylic acids also show C—O stretches, but these can be ruled out by the absence of a second carbonyl stretch or the absence of an O—H stretch, respectively. Furthermore, the frequency of the carbonyl stretch for compound **1** suggests that it is probably not a conjugated ester (i.e., not an α,β-unsaturated or aryl ester).

R–C(=O)–OR'

Ester

Compound **2**. The IR spectrum of compound **2** displays a strong carbonyl stretch at 1710 cm^{-1}. The important additional bands in this spectrum are the weak C—H stretching absorptions at 2810 and 2715 cm^{-1}, which are characteristic for **aldehydes**. (The higher frequency band is often obscured by overlapping aliphatic C—H stretches.) In addition, the frequency of the carbonyl stretch for compound **2** suggests that this is probably not a conjugated aldehyde.

R–C(=O)–H

Aldehyde

Compound **3**. The IR spectrum of compound **3** shows two carbonyl stretches at 1855 and 1785 cm^{-1}. This pair of bands suggest that the compound is an **anhydride**. Anhydrides should also exhibit a C—O stretching band. Does this one?

R–C(=O)–O–C(=O)–R'

Anhydride

Compound **4**. The IR spectrum of compound **4** exhibits a strong carbonyl stretch at 1720 cm^{-1}. The diagnostic absorption in this spectrum is the very broad, intense absorption at 3600-2500 cm^{-1}, characteristic of the O—H stretch of **carboxylic acid** dimers.

$$R-C(=O)-OH$$

Carboxylic acid

Compound **5**. The IR spectrum of compound **5** displays a strong carbonyl stretch at 1710 cm^{-1}. This spectrum must belong to the **ketone**. Note the absence of a second carbonyl stretch, the aldehyde C—H stretching doublet, an O—H stretch, or a C—O stretch. The frequency of the carbonyl stretch suggests that compound **5** is probably not a conjugated ketone.

$$R-C(=O)-R'$$

Ketone

Problem 15.26 Spectrum **1**. The choices for spectrum **1** are alcohol, carboxylic acid, or phenol. All three classes would be expected to show an O—H stretch. For the alcohol and phenol, this absorption occurs in the 3400-3200 cm^{-1} region of the IR spectrum; whereas for the carboxylic acid, a broad, intense O—H stretch occurs in the 3300-2400 cm^{-1} region. The IR spectrum of compound **1** is most consistent with the latter values. Even stronger evidence for the carboxylic acid is the strong carbonyl stretch at 1720 cm^{-1}. This band would not be expected for the alcohol or phenol. Therefore, the IR spectrum of compound **1** is most likely that of a **carboxylic acid.**

$$R-C(=O)-OH$$

Carboxylic acid

Spectrum **2**. The carbonyl stretch at 1715 cm^{-1} could be diagnostic for an aldehyde, an ester, or a ketone. So, we will have to look elsewhere for a means of differentiating between these classes. An ester would be expected to display an intense C—O stretch in the 1300-1000 cm^{-1} region of the IR spectrum. Such a band is absent in the spectrum of **2**. The important additional bands in the spectrum of **2** are the weak C—H stretching absorptions at 2810 and 2700 cm^{-1}, characteristic of **aldehydes**. (The higher frequency band is often obscured by overlapping aliphatic C—H stretches.)

$$R-C(=O)-H$$

Aldehyde

Spectrum **3**. The weak band at 2220 cm^{-1} could be from the C≡C stretch of a 1-alkyne or the C≡N stretch of a cyanide. However, symmetrically disubstituted alkynes do not exhibit a C≡C stretch because this vibration does not result in a dipole moment change. 1-Alkynes also display a characteristic ≡C-H stretch at about 3300 cm^{-1}; absent in the IR spectrum of compound **3**. Accordingly, this leaves the **cyanide** as the best fit for spectrum **3**.

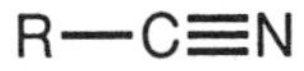

R—C≡N

Cyanide

Spectrum **4**. The intense bands at 3350 and 3180 cm^{-1} are consistent with the N—H stretches of either a primary amine or primary amide, but not with a nitro compound. To differentiate between the amine and the amide, look for the presence or absence of a carbonyl stretch. The IR spectrum **4** shows an absorption centered at 1650 cm^{-1}, the result of a combination of a carbonyl stretch and an N—H bend. Therefore, the IR spectrum **4** is probably that of a **primary amide**.

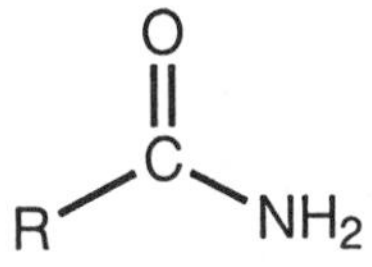

$R-C(=O)-NH_2$

Primary amide

Spectrum **5**. The carboxylic acid can be eliminated immediately because of the absence of the diagnostic broad, intense O—H stretch in the 3300-2400 cm^{-1} region of the IR spectrum. Both anhydrides and esters display a strong C—O stretch in the 1300-1000 cm^{-1} region (present in this case at 1040 cm^{-1}). However, esters normally exhibit only a single carbonyl stretch, whereas anhydrides display two carbonyl stretches. The presence of two carbonyl absorptions at 1810 and 1750 cm^{-1} in the IR spectrum **5** is strong evidence for an **anhydride**.

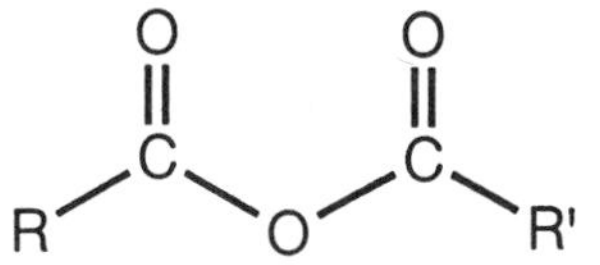

$R-C(=O)-O-C(=O)-R'$

Anhydride

Problem 15.27 Alkene isomers are traditionally distinguished through the use of the =C—H out-of-plane bending vibrations. The third spectrum has a band at 697 cm^{-1}, the position typical of cis alkenes. This molecule is *cis*-2-octene.

Spectrum **2** has a strong band at 966 cm^{-1}, a typical position for a trans alkene. This molecule is *trans*-2-octene. Notice that in the IR spectrum for this relatively symmetrical molecule, there is no visible band for the C=C stretch. By contrast, in the IR spectrum of the less symmetrical cis isomer **3**, this band appears at ~1650 cm^{-1}.

Spectrum **1** must belong to 1-octene by a process of elimination, but let's look at the spectrum anyway to make certain. The two bands for a vinyl group, ($HC{=}CH_2$) appear at 991 and 909 cm^{-1}, and there is a peak for the C=C stretch at about 1640 cm^{-1}. All seems well with the assignment.

Problem 15.28 The keys to this problem are the bands in spectrum **2** at 3333 cm^{-1} and 2128 cm^{-1} for the stretching of the alkyne terminal carbon-hydrogen bond and the triple bond respectively. In the first spectrum, there is no visible band for either the alkyne carbon-hydrogen bond or the carbon-carbon triple bond. The first molecule is 6-phenyl-2-hexyne, a disubstituted acetylene too symmetrical to show a C≡C stretch. The second is the 1-alkyne, 5-phenyl-1-pentyne.

Problem 15.29 To distinguish these positional isomers of a disubstituted benzene, we can use the aromatic out-of-plane C—H bending frequencies.

Spectrum **2** shows bands at 770 and 680 cm^{-1}, and must be the meta isomer. The band at 855 cm^{-1} is also taken by some as diagnostic for meta substitution.

Spectrum **3** shows only one band in the appropriate region, at 803 cm^{-1}, and must be the para isomer.

Spectrum **1** should be the remaining ortho isomer, and the band at 746 cm^{-1} confirms the assignment.

Problem 15.30 As Table 15.1 shows, natural bromine has two common isotopes, ^{79}Br and ^{81}Br, in nearly equal amounts. The two ions come from the two parents, $C_8H_7{}^{79}BrO_2$ ($m/z = 214$) and $C_8H_7{}^{81}BrO_2$ ($m/z = 216$).

Aldehydes generally show very intense (p – 1) peaks as a hydrogen atom is easily lost to give the resonance-stabilized acylium ion.

$$\text{RCHO} \xrightarrow{e^-} [\text{RCHO}]^{\cdot+} \longrightarrow [\text{R–C}^+\text{=O:} \longleftrightarrow \text{R–C}\equiv\text{O:}^+]$$

Of course, in this case the acylium ion still contains bromine, and two peaks will be observed, $C_8H_6{}^{79}BrO_2$ ($m/z = 213$) and $C_8H_6{}^{81}BrO_2$ ($m/z = 215$).

Problem 15.31 Our first task is to derive the empirical formula from the elemental analysis data. If you are unsure of these calculations, be sure to review the problems in Chapter 4 (p. 152). We can find the percentage of oxygen easily, 100 – (80.00 + 6.70) = 13.30.

C $\frac{80.00}{12.01} = 6.66$ H $\frac{6.70}{1.01} = 6.63$ O $\frac{13.30}{16.00} = 0.831$

Now we divide by the smallest value, 0.831:

C $\frac{6.66}{0.831} = 8.01$ H $\frac{6.63}{0.831} = 7.98$ O $\frac{0.831}{0.831} = 1.00$

So, the empirical formula is C_8H_8O, formula weight = 120 g/mol

A quick look at the mass spectrum shows that the highest peak, likely to be the parent ion, is at $m/z = 120$. Accordingly, the empirical formula and the molecular formula are the same, C_8H_8O.

Next, it is useful to work out the degrees of unsaturation (Ω) in this molecule. Remember, $\Omega = [(2n + 2) - (\text{No. of hydrogens})]/2$.

Here, $\Omega = [18 - 8]/2 = 5$. There is a total of five rings and/or π bonds in this molecule. For this small a molecule, this many degrees of unsaturation almost certainly means there will be a benzene ring (four unsaturations) in the molecule.

Now it is finally time to analyze the IR spectrum. The presence of a strong band at 1690 cm^{-1} suggests a carbonyl group. Taken together with the formula and the notion that a benzene ring is present, this limits the choices to the following five possibilities:

CHO / CH_3 — Three possible isomers of tolualdehyde

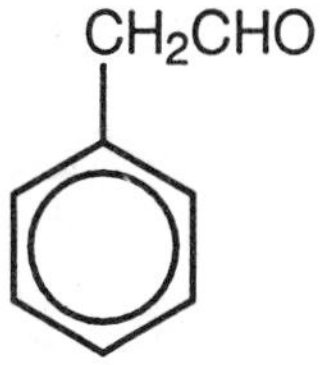

CH_2CHO — Phenylacetaldehyde

$O{=}C{-}CH_3$ — Acetophenone

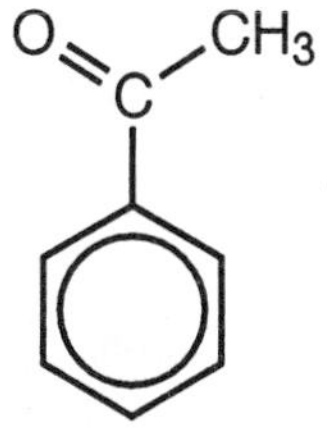

The strong bands at 758 and 690 cm^{-1} indicate a monosubstituted benzene, eliminating the three isomers of tolualdehyde. All aldehydes are eliminated by the lack of the two characteristic bands for the aldehyde C—H stretch at 2900-2700 cm^{-1}. Only acetophenone, the last possibility, remains.

Problem 15.32 First, we need to know the amount of oxygen in this molecule. A little addition and subtraction does the job.

C 70.60
H 5.90
76.50

100 −76.50 = 23.50 = % oxygen in the compound

The calculation of the empirical formula is now as follows:

C $\frac{70.60}{12.01} = 5.88$ H $\frac{5.90}{1.01} = 5.84$ O $\frac{23.50}{16.00} = 1.47$

Now we divide by the smallest value, 1.47

C $\frac{5.88}{1.47} = 4.00$ C $\frac{5.84}{1.47} = 3.97$ O $\frac{1.47}{1.47} = 1.00$

So, the empirical formula is C_4H_4O, formula weight = 68 g/mol

The mass spectrum shows that the molecular formula is twice the empirical formula as the parent ion is 136, (2 x 68). So the molecular formula is $C_8H_8O_2$.

The degrees of unsaturation can now be calculated: $\Omega = [18 - 8]/2 = 5$. A benzene ring is suggested, and would account for four degrees of unsaturation. The IR spectrum reveals a strong band at 1685 cm^{-1}. This absorption shows that a carbonyl group is present as the fifth unsaturation. The pair of peaks at 2820 and 2720 cm^{-1} is diagnostic for aldehydes. The low-frequency position of the carbonyl band suggests that the carbonyl group is conjugated, and in combination with the presence of a benzene ring indicates a benzaldehyde. An atom inventory shows that we are now short CH_3O, probably a methoxy group.

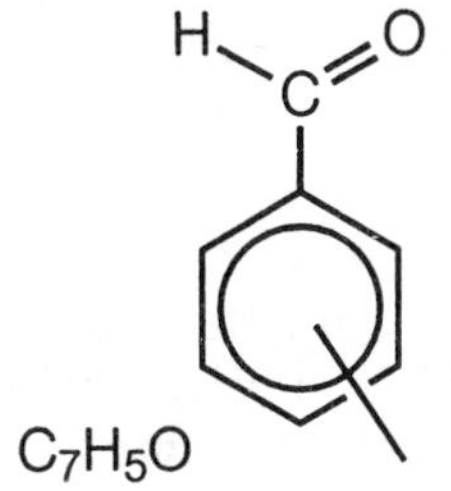

H–C=O

C_7H_5O

$C_8H_8O_2$ less C_7H_5O leaves CH_3O

It is tempting at this point to imagine that compound **A** is a methoxybenzaldehyde, and the only real question left is the position of substitution on the ring. The single band at 833 cm^{-1} is diagnostic for para substitution, and, indeed, this compound is *p*-methoxybenzaldehyde.

H–C(=O)–C₆H₄–OCH_3

Problem 15.33 It is our first task to determine the molecular formulas for isomers **A**, **B**, and **C**. Is there any oxygen? No, the percentages of C, H, and N add up to 100.

C 71.17
H 5.12
N 23.71
100.00

There is no oxygen present!

To determine the molecular formula, apply the usual calculation:

$$C \quad \frac{71.17}{12.01} = 5.92 \qquad H \quad \frac{5.12}{1.01} = 5.07 \qquad N \quad \frac{23.71}{14.01} = 1.69$$

$$C \quad \frac{5.92}{1.69} = 3.50 \qquad H \quad \frac{5.07}{1.69} = 3.00 \qquad N \quad \frac{1.69}{1.69} = 1.00$$

Therefore, the empirical formula = $2(C_{3.50}H_{3.00}N_{1.00}) = C_7H_6N_2$, and the formula weight is: $(C_7H_6N_2)$ = 118 g/mol.

So, the empirical formula = the molecular formula, as the parent ion in the mass spectrum is 118.

Now let's determine the degrees of unsaturation, Ω.

$$\Omega = \frac{2(7+2)+2-2-6}{2} = 6$$

Six degrees of unsaturation suggests the possibility of a benzene ring.

Now it is time to see where we are by doing an atom inventory:

$C_7H_6N_2$	Formula
$-\ C_6$	Benzene ring
CH_6N_2	Remaining

The IR spectra of isomers **A**, **B**, and **C** all show an absorption at about 2220 cm^{-1}. This band could be the result of a C≡C stretch of an alkyne or the C≡N stretch of a nitrile. Assuming that our conjecture that these isomers contain a benzene ring is correct, the alkyne C≡C stretch can be ruled out since the atom inventory shows only one residual carbon atom. Also note that the nitrile C≡N accounts for the remaining two degrees of unsaturation.

—C≡N Two degrees of unsaturation

The IR spectra of isomers **A**, **B**, and **C** also exhibit doublets at about 3490 and 3390 cm^{-1}. These are N—H stretches indicative of a primary amine, which shows two N—H stretches.

R—NH_2 primary amine

All of the available information suggests that isomers **A**, **B**, and **C** are the three possible aminobenzonitriles; that is,

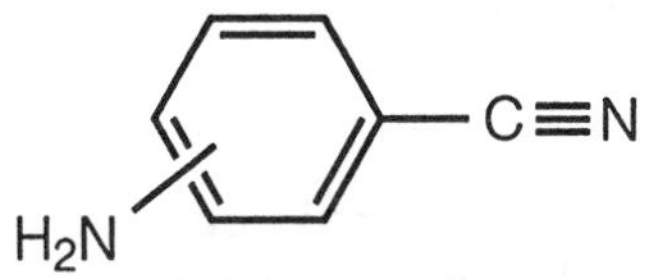

C≡N

H_2N

Isomer **A**. The IR spectrum of isomer **A** exhibits three peaks at 860, 780, and 680 cm^{-1}, which indicates 1,3- or meta-disubstitution. Thus isomer **A** is 3-aminobenzonitrile.

Isomer **B**. The IR spectrum of isomer **B** displays a band at 835 cm^{-1}, which suggests 1,4- or para-disubstitution. Isomer **B** is 4-aminobenzonitrile.

Isomer **C**. Finally, the IR spectrum of isomer **C** shows only an intense peak at 750 cm^{-1}, indicative of 1,2- or ortho-disubstitution. Isomer **C** is 2-aminobenzonitrile (anthranilonitrile).

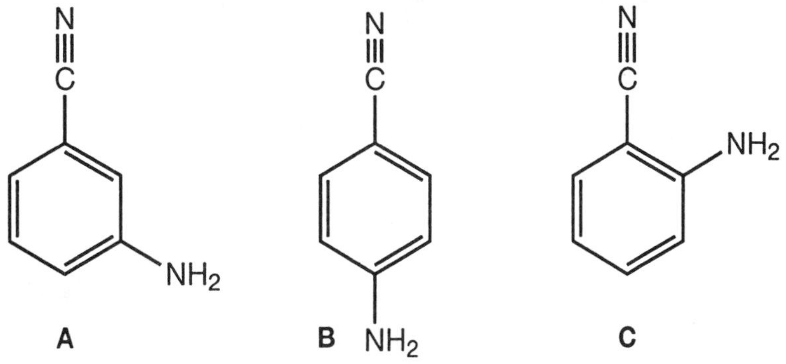

N≡C, NH_2 — **A**

N≡C, NH_2 — **B**

N≡C, NH_2 — **C**

Problem 15.34 This problem presents so much information that it is hard to know where to start. Let's begin by calculating the degrees of unsaturation for $C_7H_{12}O$.

$$\Omega = [2(7) + 2 - 12]/2 = 2$$

A cursory examination of the IR spectral data for compound **A** suggests the presence of some type of carbonyl group as indicated by the strong absorption at 1725 cm^{-1}. What type of carbonyl group? As **A** contains only one oxygen and no nitrogen, we only need to differentiate between an aldehyde and a ketone. An aldehyde should display the characteristic C—H stretching doublet, which is in fact observed at 2833 and 2725 cm^{-1}. Thus, the aldehyde carbonyl group accounts for one of the two degrees of unsaturation.

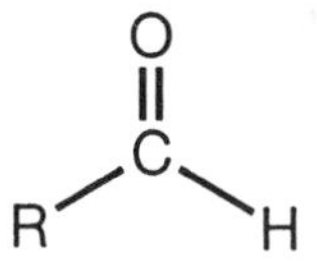

R—C(=O)—H One degree of unsaturation

The other informative peak in the IR spectrum of **A** is the absorption at 975 cm^{-1}. This absorption could be the result of the =C—H bend of a trans-disubstituted alkene. This surmise is strongly supported by the hydrogenation data; that is, the C=C disappears upon hydrogenation, and hence the =C—H bend would also vanish. Note that the trans-disubstituted alkene accounts for the second and remaining degree of unsaturation.

$$\text{HC=CH (trans)} \xrightarrow[\text{Pd/C}]{H_2} \text{—}CH_2\text{—}CH_2\text{—}$$

In addition, the frequency of the aldehyde carbonyl stretch (1725 cm^{-1}) provides us with one more subtle piece of information. The aldehyde and the alkene are **not** directly attached to each other. α,β-Unsaturated aldehydes display a carbonyl stretch at a lower frequency (1685-1705 cm^{-1}) than normal aldehydes (1720-1740 cm^{-1}).

$$\text{—}\underset{\beta}{C}\text{=}\underset{\alpha}{C}\text{—}CHO$$

Finally, the ozonolysis data allow us to make a definitive structural assignment for compound **A**. Thus compound **A** is *trans*-4-heptenal. (Note that the aldehyde group in **A** is also oxidized to a carboxylic acid during the oxidative workup, following the ozonolysis.)

$$\text{trans-4-Heptenal (A)} \xrightarrow[\text{2. HOOH}]{\text{1. } O_3} \text{Propionic acid} + \text{Succinic acid}$$

trans-4-Heptenal (**A**) — Propionic acid — Succinic acid

Problem 15.35 The 300 MHz spectrum is nearly first order. The 60 MHz spectrum is not, and is therefore much more complicated and difficult to interpret. Remember, for a first order spectrum to appear for two different coupled hydrogens such as H_a and H_b, the difference in hertz in the chemical shifts (δ_{ab}) must be about 10 times greater than their coupling constant (J_{ab}). At 60 MHz this condition is not satisfied but at 300 MHz it is.

H_a → $CH_3\underline{CH_2}CH_2CH_2Br$, H_b → $CH_3CH_2\underline{CH_2}CH_2Br$

$$CH_3CH_2CH_2CH_2Br$$

Problem 15.36 We will use the technique outlined in the chapter (p. 711). Simply replace the hydrogens in question one by one with X and determine the relationship between the two "new compounds". If they are identical, the two hydrogens are homotopic, if the "new compounds" are enantiomers, the two hydrogens are enantiotopic, and if they are different compounds entirely, the two hydrogens are diastereotopic. The first three cases are quite easy:

(a)

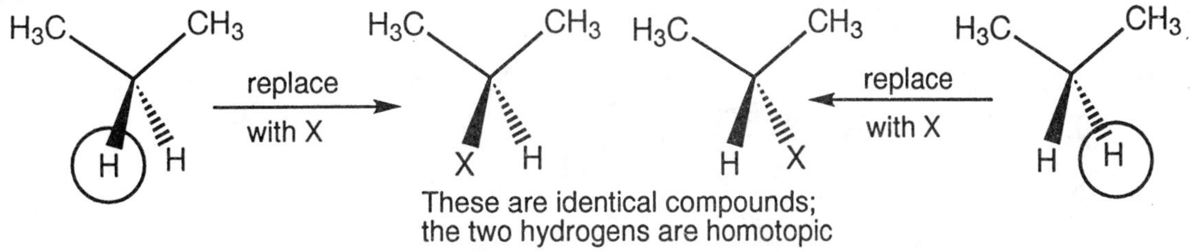

replace with X → ← replace with X

These are identical compounds; the two hydrogens are homotopic

(b)

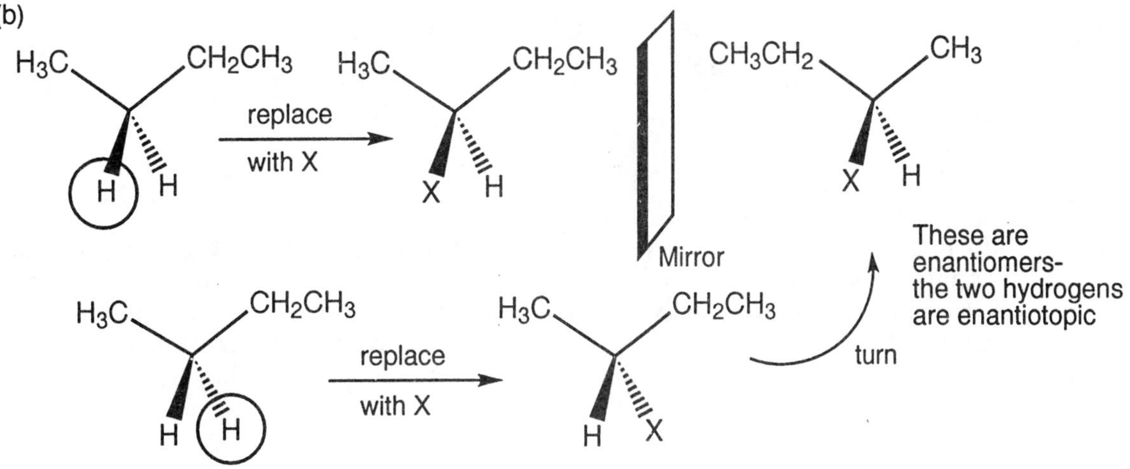

replace with X → ; Mirror

replace with X → ; turn

These are enantiomers- the two hydrogens are enantiotopic

(c)

replace with X → ; Mirror

replace with X → ; turn

These are enantiomers- the two hydrogens are enantiotopic

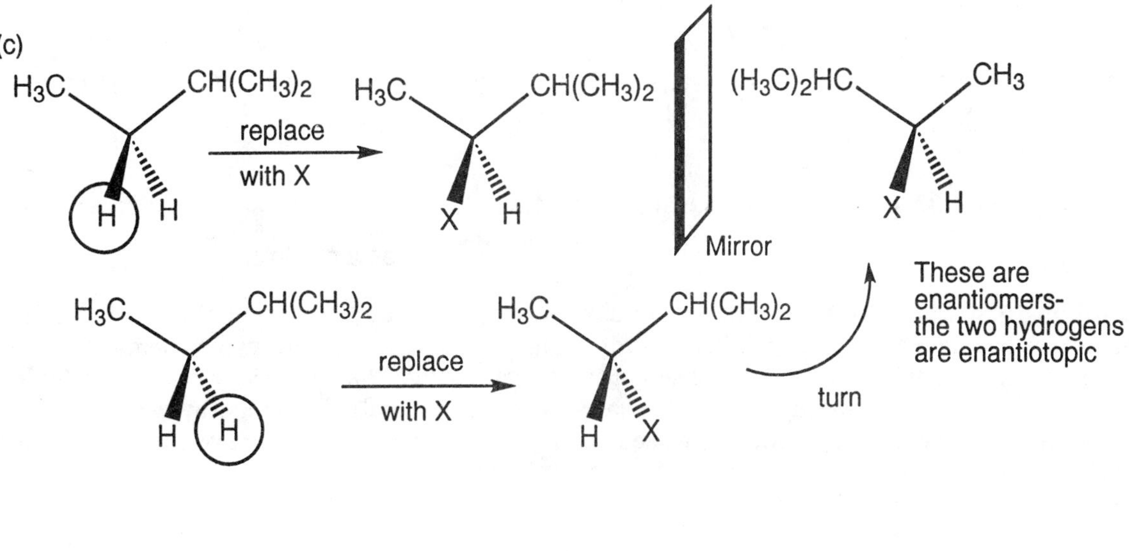

Part (d) is the most interesting case. The "replacement technique" leads to diastereomers. The two hydrogens in question are diastereotopic.

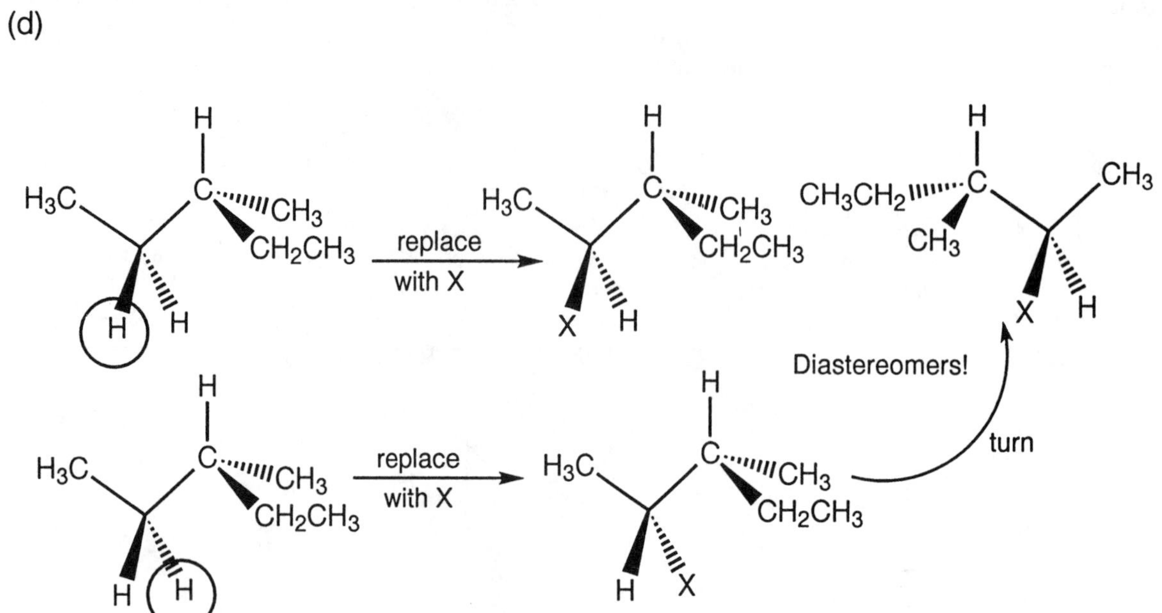

(d)

H H H
H_3C C CH_3 CH_2CH_3 replace with X → H_3C C CH_3 CH_2CH_3 CH_3CH_2 C CH_3 CH_3
H H X H X H
Diastereomers!
H H
H_3C C CH_3 CH_2CH_3 replace with X → H_3C C CH_3 CH_2CH_3 turn
H H H X

Problem 15.37 Parts (a), (b), (c), and (d) are straightforward exercises in looking up approximate chemical shifts and using the $(n + 1)$ rule. In working real problems one generally does not expect to get the observed chemical shifts exactly. A chemist is generally satisfied if he or she gets the approximate value. These examples are not by any means exhaustive. There is no point is making you look up every possible example of a chemical shift! (s) = singlet, (d) = doublet, (t) = triplet, (q) = quartet.

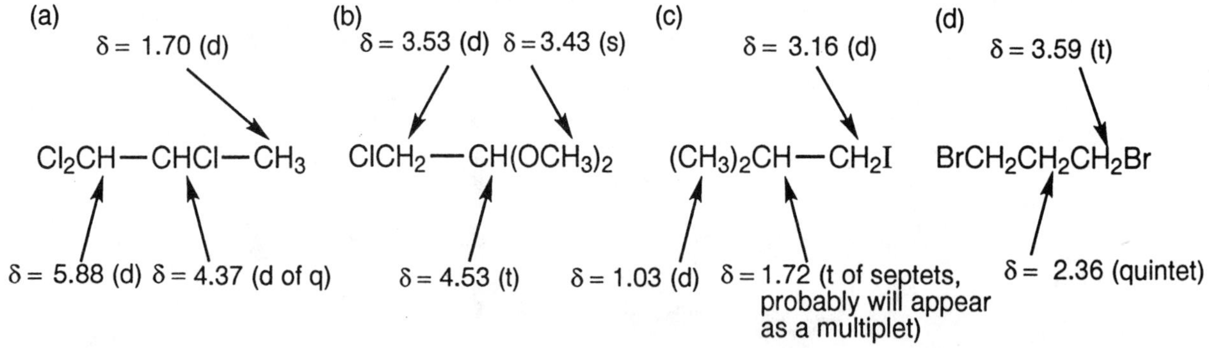

(a) $Cl_2CH—CHCl—CH_3$: $\delta = 5.88$ (d) (Cl_2CH), $\delta = 4.37$ (d of q) (CHCl), $\delta = 1.70$ (d) (CH_3)

(b) $ClCH_2—CH(OCH_3)_2$: $\delta = 3.53$ (d) ($ClCH_2$), $\delta = 4.53$ (t) (CH), $\delta = 3.43$ (s) (OCH_3)

(c) $(CH_3)_2CH—CH_2I$: $\delta = 1.03$ (d) (CH_3), $\delta = 1.72$ (t of septets, probably will appear as a multiplet) (CH), $\delta = 3.16$ (d) (CH_2I)

(d) $BrCH_2CH_2CH_2Br$: $\delta = 3.59$ (t) (CH_2Br), $\delta = 2.36$ (quintet) (central CH_2)

Problem 15.38 Parts (a) and (b) show symmetrical compounds and there will be a single methyl doublet for each molecule. The electron-withdrawing benzene ring will lower the chemical shift of nearby methyl groups somewhat relative to a cyclohexane ring. The only tricky example is (c) in which the diastereotopic methyl groups appear as a pair of doublets. These methyl groups are different! Try replacing each with X and see how the resulting compounds are related—they are always different.

(a) Cl, H_3C–CH–CH_3, δ = 1.25 (d)

(b) Cl, H_3C–CH–CH_3, δ = 0.85 (d)

(c) Cl, H_3C–CH–CH_3, δ ~0.85 (d), δ ~0.85 (d)

Two different doublets

Problem 15.39 The figure shows the spectrum of ultrapure ethyl alcohol. The methylene group appears as a doublet of quartets (or, equivalently, a quartet of doublets).

In ultrapure ethyl alcohol, these hydrogens are split into a quartet by the adjacent methyl hydrogens, and into a doublet by the hydroxyl H

H_3C—CH_2—OH

CH_2

Coupling to OH

Further coupling to CH_3

Problem 15.40 The clue in this problem is the words "scrupulously dry." Alcohols generally show no coupling of the OH hydrogens because of the rapid exchange catalyzed by small amounts of water. If essentially all the water is removed, one sees the first order spectrum (p. 727). In **A** we see a doublet, and there is only one hydrogen in either molecule that can appear as a doublet, the OH of 1-phenylethanol. The OH of the other molecule, 2-phenylethanol, appears as a triplet as it is split by the two equivalent adjacent methylene hydrogens as in **B**.

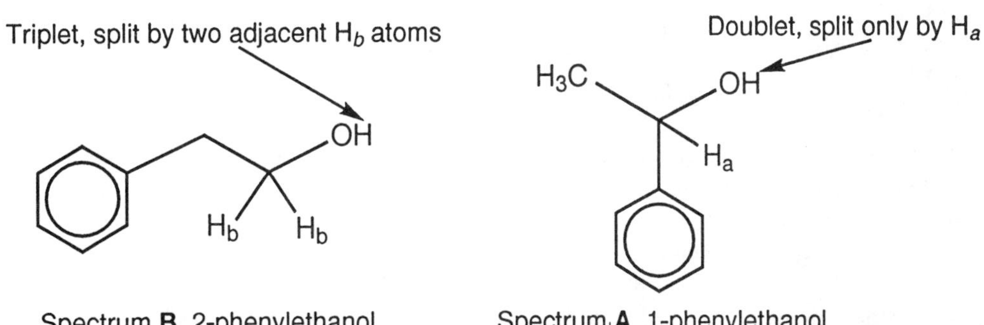

Spectrum **B**, 2-phenylethanol Spectrum **A**, 1-phenylethanol

Problem 15.41 When water is added, the acid-catalyzed exchange reaction effectively decouples the OH hydrogen from all adjacent hydrogens. The OH appears as an average signal, a singlet.

Problem 15.42 We use Eq. 15.4 (p. 706), $\nu = \gamma B_o/2\pi$

for ^{1}H
$\nu = (2.7 \times 10^8 \text{ rad-T}^{-1}\text{s}^{-1})(4.7 \text{ T})/2(3.14 \text{ rad})$
$\nu = 2.02 \times 10^8 \text{ s}^{-1} = 200 \text{ MHz}$

for ^{2}H:
$\nu = (.41 \times 10^8 \text{ rad-T}^{-1}\text{s}^{-1})(4.7 \text{ T})/2(3.14 \text{ rad})$
$\nu = 3.07 \times 10^7 \text{ s}^{-1} = 31 \text{ MHz}$

for ^{13}C:
$\nu = (.67 \times 10^8 \text{ rad-T}^{-1}\text{s}^{-1})(4.7 \text{ T})/2(3.14 \text{ rad})$
$\nu = 5.01 \times 10^7 \text{ s}^{-1} = 50 \text{ MHz}$

Problem 15.43 This problem should be especially easy.

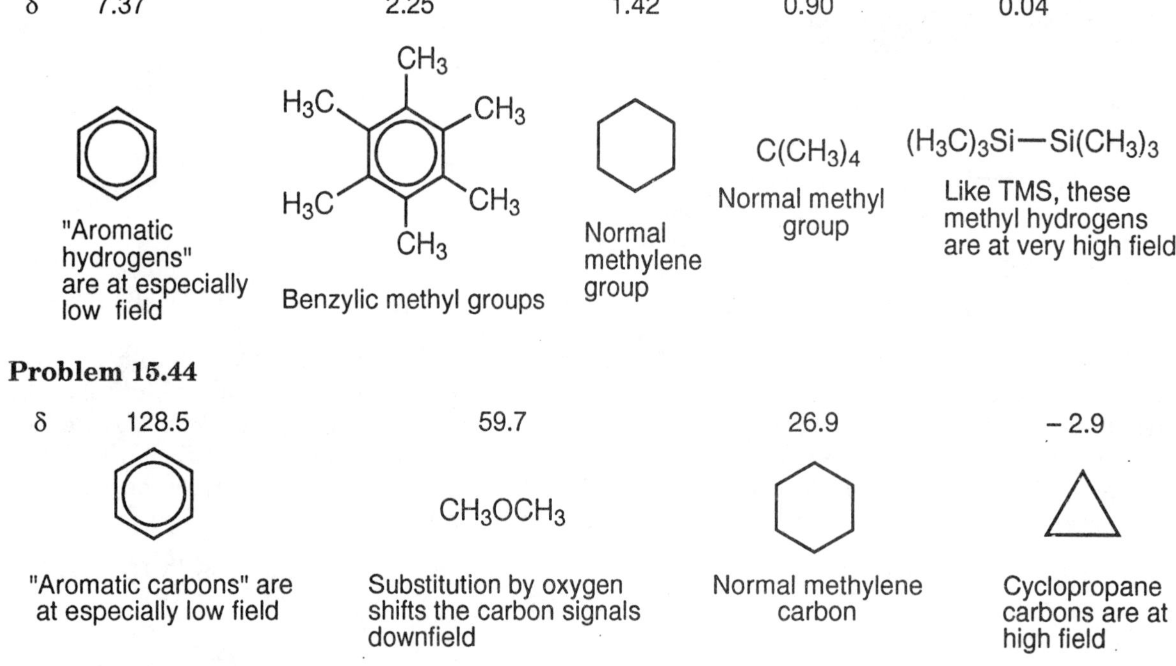

Problem 15.45 The assigned structure can't be correct. There is no hydrogen in the proposed structure that should absorb above about δ ~ 7ppm. This spectrum shows no fewer than three hydrogens at higher field than this. Indeed, one could also have pointed out that in the assigned structure there is no possible absorption for a single hydrogen. It's wrong.

Problem 15.46 Perhaps hydrogen chloride adds to the double bond of the proposed structure to give a chloride. If so, there should be a low-field dd exactly as appears in the spectrum. The lone hydrogen adjacent to the chlorine is split into doublets by the two different (diastereotopic) adjacent hydrogens of the methylene group. The new methylene group would be a complex multiplet at 60 MHz, and the signal centered roughly at δ = 3.4 ppm certainly qualifies for that description.

Ph Ph O= =O H H —HCl→ Ph Ph O= =O Cl H H H

δ ~ 5.5 (dd, 1H)

δ ~ 3.4 (m, 2H)

Problem 15.47 Hydration gives the more substituted alcohol, in this case *tert*-butyl alcohol. The hydroboration-oxidation sequence gives the less substituted alcohol, in this case isobutyl alcohol.

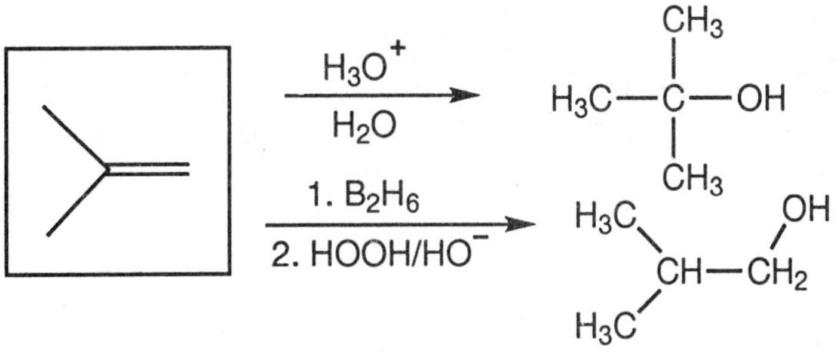

H_3O^+ / H_2O → H_3C—C(CH_3)(CH_3)—OH

1. B_2H_6 2. HOOH/HO^- → $(H_3C)_2CH$—CH_2OH

Spectrum **1** must be the more symmetrical *tert*-butyl alcohol. The hydroxyl hydrogen appears at δ = **1.61** ppm and the nine methyl hydrogens at δ = 1.28 ppm. The more complicated spectrum is that of isobutyl alcohol. The assignments are as follows:

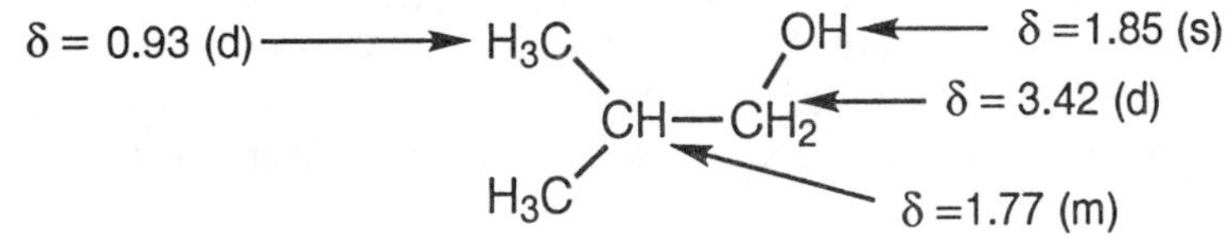

$(H_3C)_2CH$—CH_2—OH: δ = 0.93 (d) (H_3C); δ =1.77 (m) (CH); δ = 3.42 (d) (CH_2); δ =1.85 (s) (OH)

Addition of D_2O will replace the OH hydrogens with D, and their signals will vanish.

Problem 15.48 Hydration gives the more substituted alcohol, in this case 2-methyl-2-butanol. The hydroboration/oxidation sequence gives the less substituted alcohol, in this case 3-methyl-2-butanol.

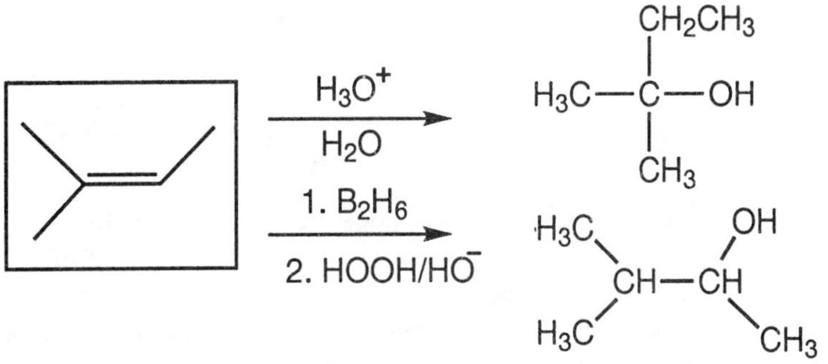

H_3O^+ / H_2O → H_3C—C(CH_2CH_3)(CH_3)—OH

1. B_2H_6 2. HOOH/HO^- → $(H_3C)_2CH$—CH(OH)—CH_3

The more symmetrical compound, 2-methyl-2-butanol gives the simpler spectrum **1**. A critical point to notice is the six-hydrogen <u>singlet</u> for the two equivalent methyl groups. Note especially in spectrum **2** the low-field multiplet for the hydrogen adjacent to the oxygen atom.

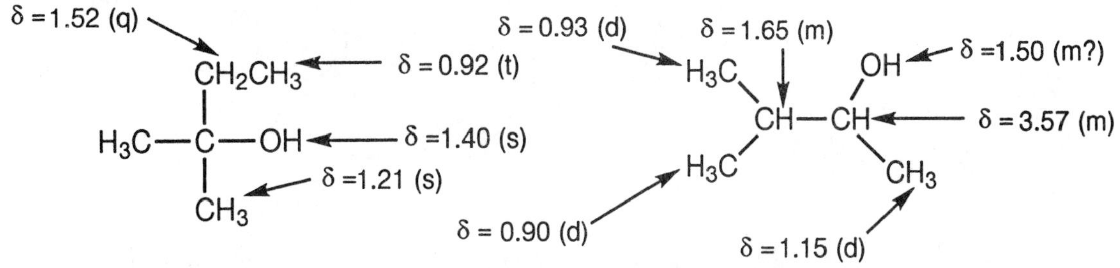

Spectrum **1**: H_3C—C(CH_2CH_3)(CH_3)—OH: δ =1.52 (q) (CH_2); δ = 0.92 (t) (CH_3 of ethyl); δ =1.40 (s) (OH); δ =1.21 (s) (CH_3)

Spectrum **2**: $(H_3C)_2CH$—CH(OH)—CH_3: δ = 0.93 (d) (H_3C); δ = 0.90 (d) (H_3C); δ =1.65 (m) (CH); δ =1.50 (m?) (OH); δ = 3.57 (m) (CH); δ =1.15 (d) (CH_3)

Problem 15.49 These methyl groups are diastereotopic. Use the "replacement technique" to see this. The two methyl groups are different and each will give rise to a separate doublet signal.

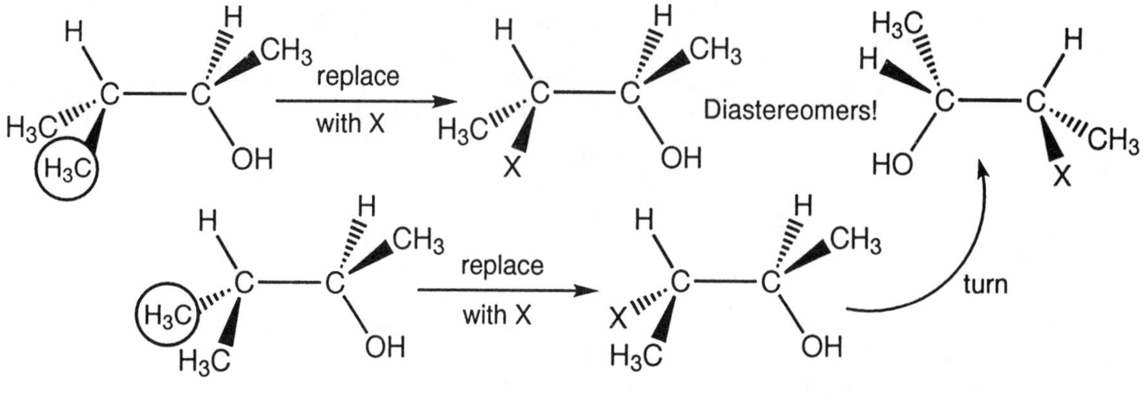

Problem 15.50 Compounds **A** and **B** should be easy. There are only single peaks, one, **B**, in the range for hydrogens attached to a double bond, and the other, **A**, in the range for saturated methylene groups. Compound **C** shows only two signals, one for the hydrogens attached to the double bonds, and the other for the allylic methylene groups. The last, most complicated spectrum must be **D**. The assignments follow:

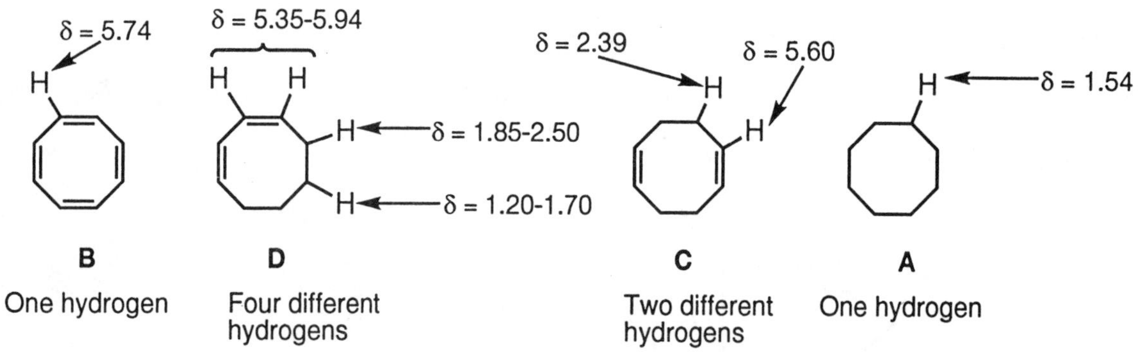

Problem 15.51 This problem should also be easy. There is no need to think about complete assignments of the signals. Instead one takes advantage of symmetry. *p*-Difluorobenzene has only two different carbons and must have spectrum **C**. Similarly, *o*-dichlorobenzene has three different carbon atoms and must have spectrum **B**. The remaining isomer, *m*-dichlorobenzene has four different carbons and must have spectrum **A**. The complete assignments are shown below, but symmetry alone is enough to do the structural assignments.

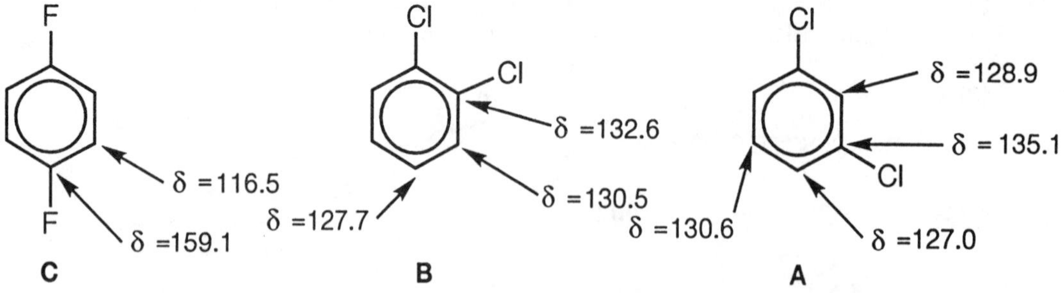

Problem 15.52 How far can we get using symmetry? The answer is "somewhere, but not all the way". We can tell which isomer is 4-methylcyclohexanone, but that's all. 4-Methylcyclohexanone has only five different carbons, whereas both 2-methylcyclohexanone and 2,3-dimethylcyclopentanone have seven different carbons. We can tell which isomer is 4-methylcyclohexanone, but we can go no further than this.

Five signals Seven signals Seven signals

An examination of the coupled ^{13}C NMR spectra should allow us to distinguish the last two compounds. For example, 2-methylcyclohexanone will show one ^{13}C signal as a quartet, the methyl carbon attached to three hydrogens, whereas 2,3-dimethylcyclopentanone will show two such quartets.

Problem 15.53 The three isomers of pentane are pentane, isopentane, and neopentane. Neopentane is easy to assign, as it is the only isomer with only two different carbons. Isopentane has four different carbons and pentane three different carbons. So pentane is **A**, and isopentane is **B**. Note that it is not necessary in this case to be able to assign all the signals, just to count the numbers of different carbons. This practice is common for ^{13}C NMR spectra.

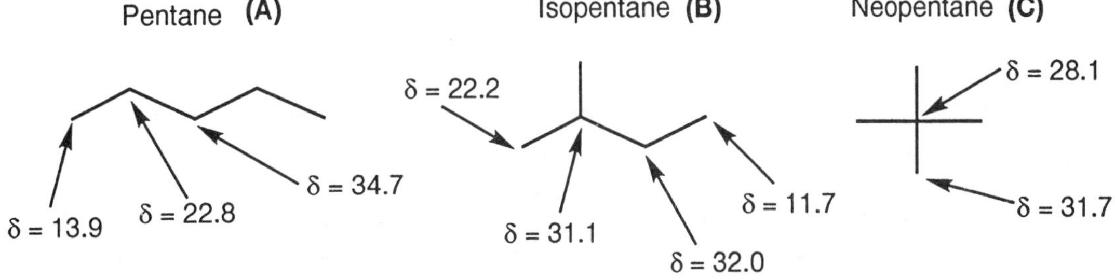

Problem 15.54. Isomer (b) shows a singlet for the two equivalent "aromatic hydrogens," and must be spectrum **1**. Isomer (a) shows two singlets for the two non-equivalent "aromatic hydrogens," and must be spectrum **2**. The other two isomers each show a typical AB quartet, but the spectrum for (c) is much more narrowly split than that for (d). Four-bond coupling is typically much smaller than adjacent, three-bond coupling. Compound (c) must have spectrum **3**, and compound (d) has spectrum **4**.

Problem 15.55 We have the formula $C_{11}H_{12}O_4$, so the first task is to work out the number of degrees of unsaturation, (Ω), for **A** and **B**.
In this case,

$$\Omega = [2(11)+2-12]/2 = 6$$

Six degrees of unsaturation strongly suggests the possibility of a benzene ring.

This surmise is supported by the ^1H NMR spectra. The ^1H NMR spectra for both compounds **A** and **B** are similar in several respects. Both spectra exhibit three aromatic hydrogens and three uncoupled methyl signals. Two of the methyl groups appear at about δ 3.9 ppm, indicating that they are attached to an electronegative substituent (for example, oxygen). The other methyl group appears at about δ 2.5 ppm, suggesting attachment to a carbonyl group or the benzene ring. At the very least, we might surmise that compounds **A** and **B** are trisubstituted benzene derivatives; that is,

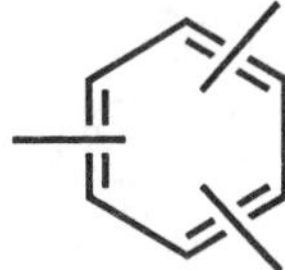

The IR spectra of compounds **A** and **B** reveal the presence of an ester group as indicated by a strong carbonyl stretch at 1720 cm^{-1} and an intense C—O stretch at 1240-1245 cm^{-1}. The relatively low frequency position of the carbonyl absorptions suggests the possibility of conjugation.

Now let's do an atom inventory to see where we are

$C_{11}H_{12}O_4$		Formula
$-\ C_6\ H_3$		Trisubstituted benzene ring
$-\ C_3\ H_9$		Three methyl groups
$-\ C$	O_2	One ester
C	O_2	Remaining

The presence of a residual CO_2 suggests the presence of a second ester group, accounting for the last degree of unsaturation. This result is also consistent with the presence of two low-field methyl groups in the ^{1}H NMR spectra of **A** and **B**. Putting all of the structural fragments together gives trisubstituted benzenes, in which the three substituents are two methyl esters and a methyl group. The ^{1}H NMR aromatic resonances of compounds **A** and **B** can be used to deduce the exact substitution patterns.

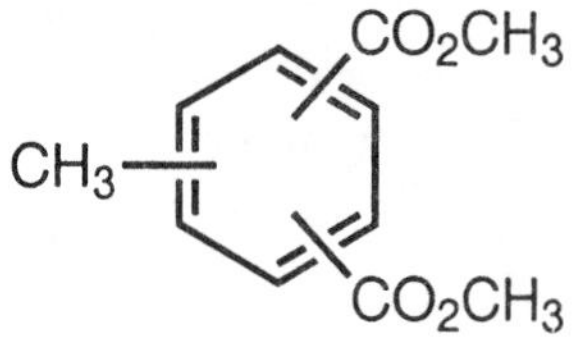

CO_2CH_3

CH_3

CO_2CH_3

Compound **A** Two of the aromatic protons are identical and have a chemical shift of δ 8.05 ppm, whereas the other aromatic proton appears at δ 8.49 ppm. The small 2 Hz coupling constant is indicative of meta coupling. The only possible arrangement of substituents with only meta coupling is dimethyl 5-methylisophthalate. Note that H_4 and H_6 are identical and are each adjacent to one ester group. The H_2 is adjacent to two ester groups and accordingly appears at the lower field.

^{1}H NMR ($CDCl_3$): δ = 2.46 (s, 3H)
3.94 (s, 6H)
8.05 (d, J = 2 Hz, 2H)
8.49 (t, J = 2 Hz, 1H)

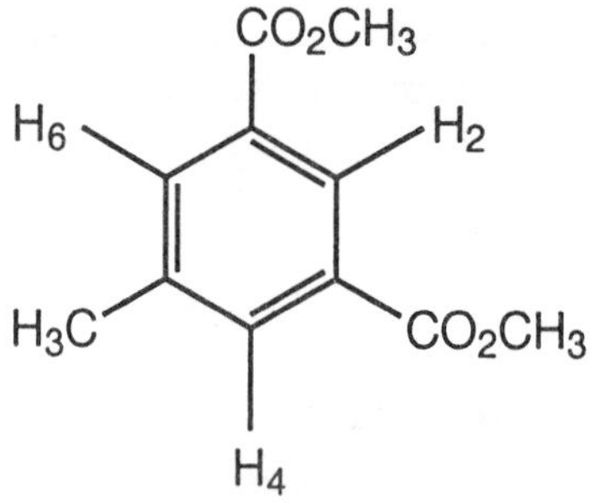

CO_2CH_3 H_6 H_2 H_3C CO_2CH_3 H_4

Dimethyl 5-methylisophthalate

A

Compound **B** The 8 Hz coupling constant is indicative of ortho coupling, whereas the 2 Hz coupling constant is appropriate for meta coupling. This coupling pattern suggests a 1,2,4-trisubstituted benzene ring; that is,

^{1}H NMR ($CDCl_3$): δ = 7.28 (d, J = 8 Hz, 1H) H_a
8.00 (dd, J = 8 Hz, J = 2 Hz, 1H) H_m
8.52 (d, J = 2 Hz, 1H) H_x

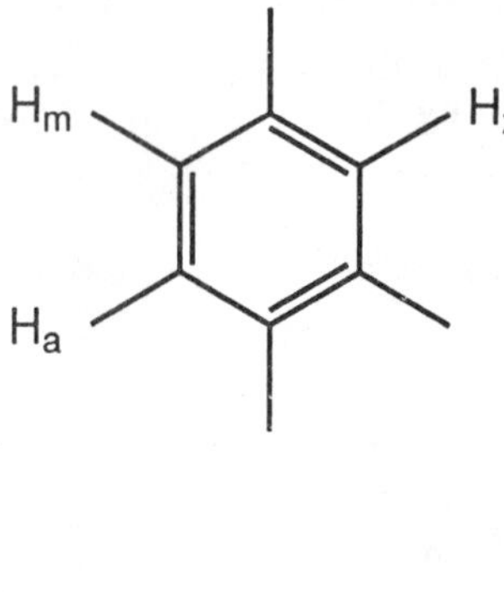

H_m H_x H_a

The aromatic hydrogen H_x, which exhibits only meta coupling, is furthest downfield and should be flanked by both ester groups. Dimethyl 4-methylisophthalate is the only reasonable stucture for compound **B**.

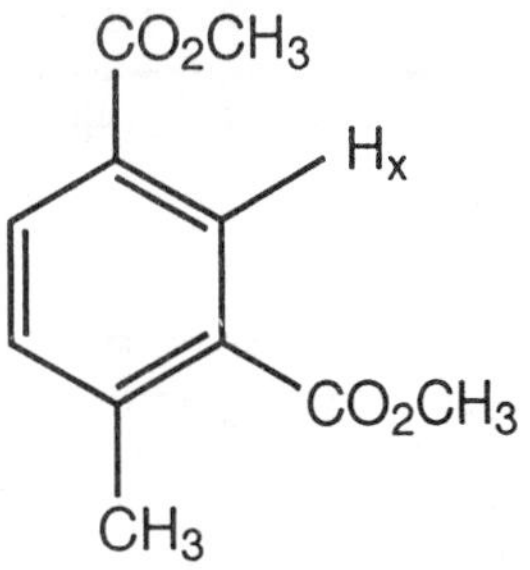

Dimethyl 4-methylisophthalate

B

Problem 15.56 Our first task is to determine the molecular formulas of **A** and **B**. We can get the percentage of oxygen in the compounds by subtraction, and then work out the formula in the usual way.

$$\begin{array}{lr} C & 63.15 \\ \underline{H} & \underline{5.30} \\ & 68.45 \end{array} \qquad \begin{array}{r} 100.00 \\ \underline{-68.45} \\ 31.55\%\ O \end{array}$$

$$\begin{array}{ll} C & 63.15/12.01 = 5.26 \\ H & 5.30/1.01 = 5.25 \\ O & 31.55/16.00 = 1.97 \end{array} \qquad \begin{array}{ll} C & 5.26/1.97 = 2.67 \\ H & 5.25/1.97 = 2.66 \\ O & 1.97/1.97 = 1.00 \end{array}$$

$$(C_{2.67}H_{2.66}O_{1.00}) \times 3 = C_{8.01}H_{7.98}O_{3.00}$$

$$\text{Empirical formula} = C_8H_8O_3$$

As we know the molecular weight from freezing point depression to be about 150 g/mol, we know that the empirical formula = molecular formula.

Second, let's work out the number of degrees of unsaturation, Ω. In this case,

$$\Omega = [2(8)+2-8]/2 = 5$$

Although the presence of five degrees of unsaturation might suggest the presence of a benzene ring, the IR and NMR spectra of compounds **A** and **B** do not support this hypothesis, particularly in the case of compound **B**.

The presence of three oxygen atoms and two high frequency carbonyl bands in the IR spectra of compounds **A** and **B** suggest the possibility of an anhydride; that is,

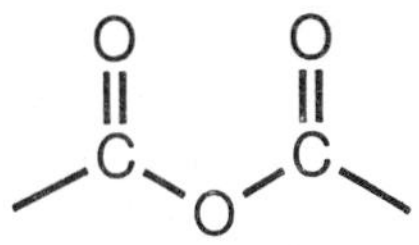

Two degrees of unsaturation

(A true IR maven would also recognize that the anhydrides are cyclic because the lower frequency carbonyl band is more intense than the higher frequency band.)

Now let's do an atom inventory:

$C_8H_8O_3$	Formula
$-\ C_2\ \ O_3$	Anhydride
C_6H_8	Remaining

There must be at least one carbon-carbon double bond (from the hydrogenation data), leaving two degrees of unsaturation unaccounted, and suggesting the possibility of one or two rings. (Alternatively, a disubstituted alkyne and one ring could be present).

The ^{1}H NMR spectra of compounds **A** and **B** should help with the structural assignments.

Compound **A** The ^{1}H NMR spectrum of compound **A** suggests the presence of two vinylic hydrogens (δ 5.65-6.25 ppm), two deshielded hydrogens (δ 3.15-3.70 ppm), possibly adjacent to the carbonyl groups, and four slightly deshielded hydrogens (δ 2.03-2.90 ppm), probably allylic. One possible way to join these fragments is shown below. Note that the relative numbers of hydrogens given by the integral need to be multiplied by two to bring the hydrogen count to eight.

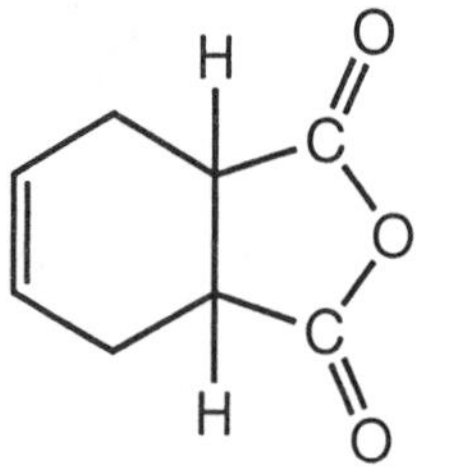

A

The five degrees of unsaturation are two carbonyl groups, one carbon-carbon double bond, and two rings. The type of ring juncture (cis or trans) is not obvious at this time.

Compound **B** The ^{1}H NMR spectrum of compound **B** shows the absence of vinylic hydrogens and the presence of four slightly deshielded hydrogens (δ 2.15-2.70 ppm) - possibly allylic - and four hydrogens with a "normal" aliphatic chemical shift (δ 1.50-2.10 ppm). It should also be noted that the carbonyl absorptions in the IR spectrum of compound **B** are about 15-20 cm^{-1} lower in frequency than for compound **A**, indicating the possibility of conjugation. Putting all this information together suggests the following structure for compound **B**.

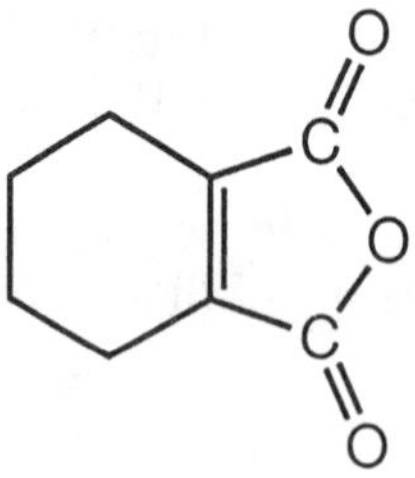

B

Compound **C** The hydrogenation experiments imply that compound **C** is *cis*-1,2-cyclohexane-dicarboxylic anhydride. This structure is consistent with the ^{1}H NMR spectrum, which exhibits

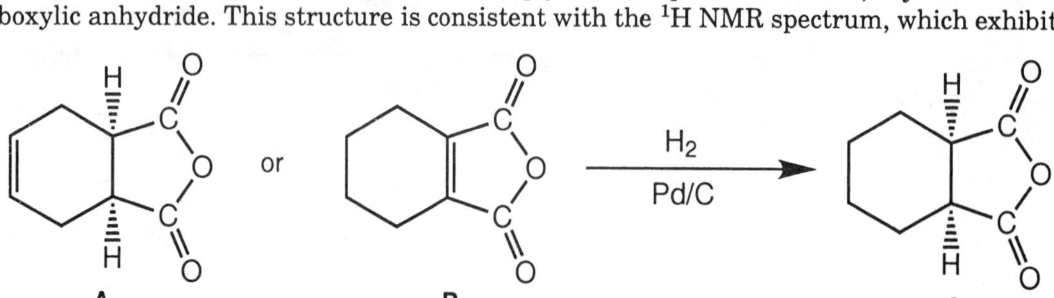

A or **B** $\xrightarrow[\text{Pd/C}]{H_2}$ **C**

eight hydrogens with a "normal" aliphatic chemical shift (δ 1.10-2.30 ppm) and two deshielded hydrogens (δ 3.05-3.55 ppm) adjacent to the carbonyl groups. Notice also that as **C** must have a cis ring juncture (from the hydrogenation of **B**), then compound **A'**, which also gives **C** upon hydrogenation, must also have a cis ring junction.

Problem 15.57 Compound A. In this problem, we are given no information with which to determine the molecular formula. Although we have no elemental analysis data, we do have a mass spectrum, and this tells us the molecular weight of the compound, as the parent ion is 148. We can also gain information from looking at the fragment ions, 105 and 77. These result from sequential losses of masses 43 (148 – 105), and 28 (105 – 77). At this point we can't be certain, but masses of 43 and 28 suggest C_3H_7 and CO, respectively.

The ^1H NMR spectrum shows five aromatic hydrogens (δ 7-8 ppm), and this, in turn will surely make us think of a monosubstituted benzene. The septet integrating for a single proton looks very much like an "isopropyl" hydrogen [C<u>H</u>$(CH_3)_2$], and its chemical shift implies that it is adjacent to an electronegative atom.

The IR spectrum shows a strong band at 1675 cm^{-1}, possibly a conjugated ketone, and is consistent with the loss of CO shown by the mass spectrum. Collecting the pieces leads to the following parts:

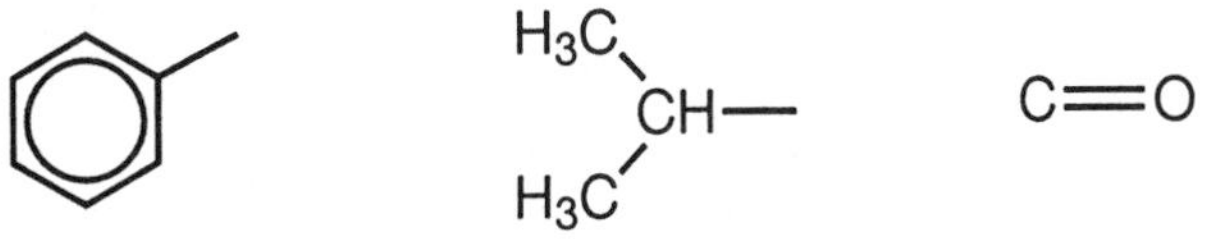

H_3C / CH— / H_3C C=O

Do these parts add up to a molecule with a molecular weight of 148? Yes, just zip them together to give isobutyrophenone:

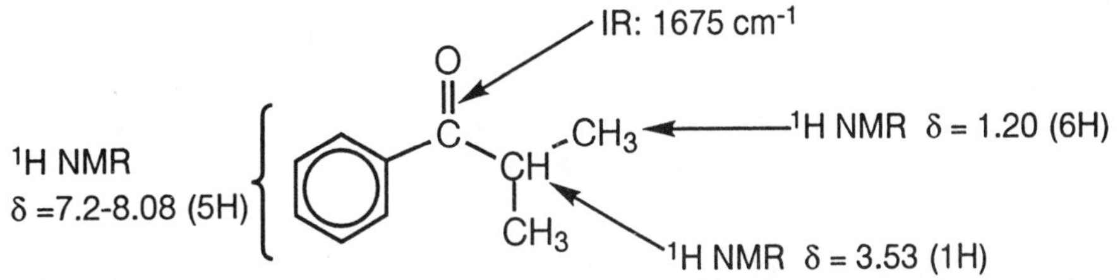

IR: 1675 cm^{-1}
^1H NMR δ = 1.20 (6H)
^1H NMR δ =7.2-8.08 (5H)
^1H NMR δ = 3.53 (1H)

Note also that the mass spectrum becomes quite logical. If the isopropyl group is lost to give an acylium ion, we see loss of C_3H_7. Subsequent loss of CO leads to the fragment of molecular weight 77.

Compound B. We know this molecule is isomeric with compound **A**, isobutyrophenone, so the formula is $C_{10}H_{12}O$. The IR spectrum shows a strong band at 1705 cm^{-1}, at substantially higher frequency than the carbonyl stretch of compound **A**, and indicative of an unconjugated carbon-oxygen double bond. The ^1H NMR spectrum shows five "aromatic" hydrogens, and so a monosubstituted benzene ring is apparently present, as in compound **A**. The combination of an upfield triplet (3H) and a somewhat downfield quartet (2H) is always strongly suggestive of an ethyl group (CH_2CH_3). There is also an uncoupled two-hydrogen signal.

Now we can put the pieces together:

CH_2 C=O CH_3CH_2

The combination is benzyl ethyl ketone:

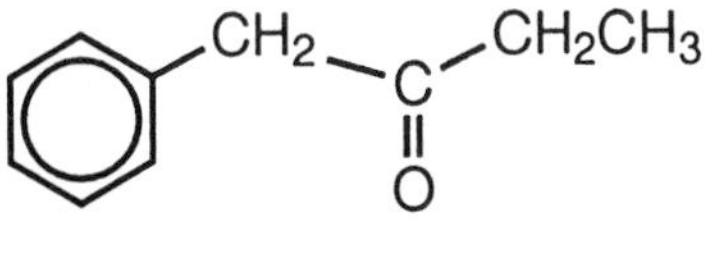

$C_6H_5CH_2C(=O)CH_2CH_3$

Problem 15.58 We start by determining the molecular formula of **B**. The cyclopentadienone dimer **A**, $C_{38}H_{32}O_2$, has a molecular weight of 520 g/mol. From the mass spectral data, the molecular weight of compound **B** is 492 g/mol, which means that 28 g/mol has been lost upon photolysis or heating of **A**. This loss corresponds to the elimination of carbon monoxide (CO) and implies a molecular formula of $C_{37}H_{32}O$ for compound **B**.

A reasonable expectation is that the dimer of cyclopentadienone **A** is a simple Diels-Alder adduct that can readily eliminate CO in a reverse cycloaddition reaction (sometimes called a cheletropic extrusion) to give **B**. Now we need to check to see if the rest of the spectral data are consistent with the proposed structure of compound **B**.

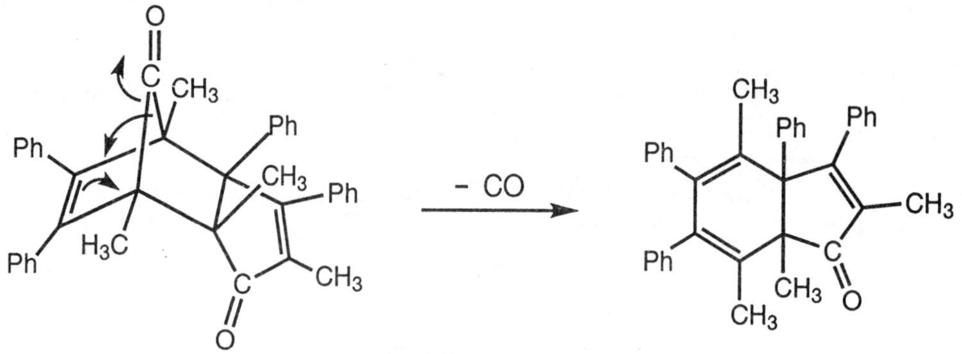

- CO

Ph = C_6H_5 = benzene ring **B**

<u>IR and NMR Spectra of Compound B</u> The presence of a band at 1704 cm^{-1} in the IR spectrum of compound **B** is consistent with the C=O stretch of an α,β-unsaturated ketone.

The ^{1}H NMR spectrum of compound **B** is also consistent with the proposed structure. For example, there are 20 aromatic hydrogens and 4 different methyl singlets.

Finally, what is the stereochemistry of **B**? The answer to this question is not obvious from the available spectral data. If one makes the "educated" assumptions that the dimer of cyclopentadienone **A** has cis stereochemistry (exo or endo) and that the reverse cycloaddition reaction is concerted, then **B** should also have cis stereochemistry.

Problem 15.59

(a) The ^{13}C chemical shift of the β-carbon of α,β-unsaturated ketones such as **A** can be rationalized on the basis of the lower electron density at this position. The resonance formulation shows it well.

A

Similarly, the ^{13}C chemical shift of the β carbon of vinyl ethers such as **B** occurs further upfield because of the greater electron density at this position.

$CH_3\ddot{O}-CH=CH_2 \longleftrightarrow CH_3\overset{+}{O}=CH-\overset{-}{C}H_2$

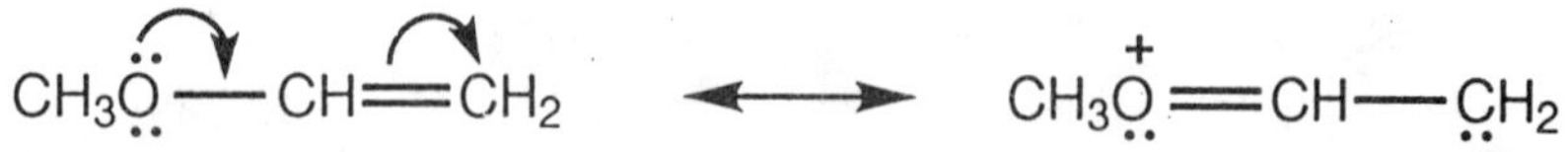

B

(b) The ^1H NMR chemical shifts of the ortho, meta, and para protons of aniline (**C**) can also be correlated by considering the appropriate resonance contributors.

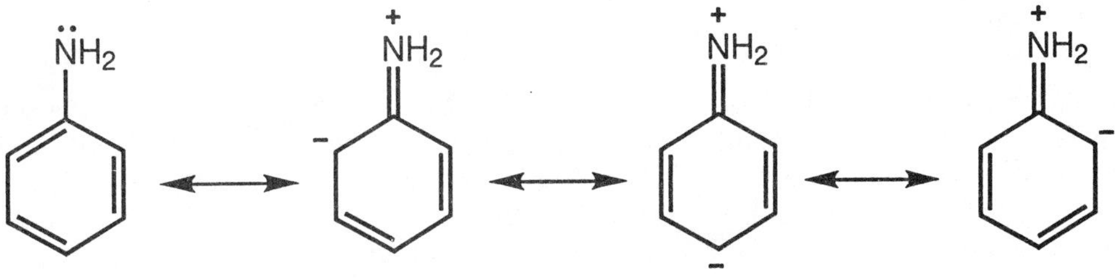

C

Therefore, the ortho and para hydrogens of **C** appear at a higher field than the *meta* hydrogens because of the greater electron density at these positions.

A similar treatment of acetophenone (**D**) would predict that the ortho and para hydrogens would be deshielded because of the lower electron density at these positions.

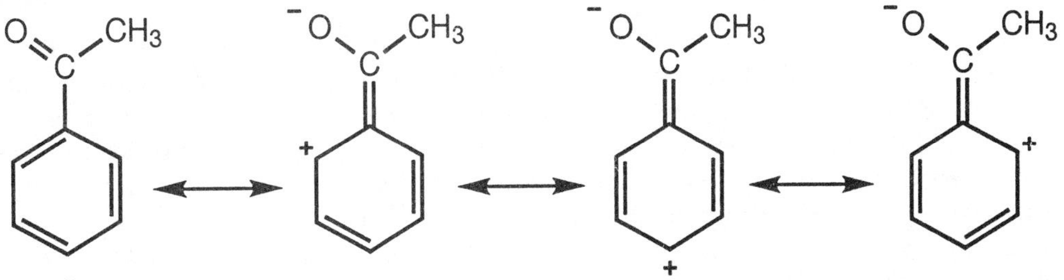

D

However, **D** shows only two deshielded aromatic hydrogens, not three, as predicted by this resonance argument. Clearly, there must be another factor (or factors) operating here. We already know that the hydrogens on a benzene ring are deshielded because of the induced field of the π ring current. The carbon-oxygen double bond in **D** also has an induced field of circulating π electrons. Because the carbonyl bond and the benzene ring of **D** are coplanar, the ortho protons lie in the deshielding zone of the adjacent carbonyl group. As a result, the two ortho protons are shifted further downfield than the other three ring protons. The ortho protons of **D** may also be inductively deshielded by the adjacent carbonyl group.

For chlorobenzene (**E**), the ^1H NMR chemical shifts of the aromatic protons can be rationalized by a consideration of inductive and resonance effects. In this case, these two effects, which oppose each other, are apparently roughly in balance. By contrast, in alkyl chlorides, the chlorine substituent is deshielding because only the inductive effect is operative, as there can be no resonance effect. This situation is similar to electrophilic aromatic substitution, in which a chlorine substituent is an ortho/para director thanks to resonance, but is a deactivating substituent because of the inductive effect.

The ^{13}C NMR chemical shifts of **C** and **E** are, in general, consistent with the proton chemical shifts for these compounds. For example, note the upfield chemical shifts of C(2) and C(4) in aniline (**C**). Finally, note that the ^{13}C NMR chemical shifts for C(1) of compounds **C** and **E** are good indications of the inductive effect of the attached substituent.

Problem 15.60 First, let's determine the molecular formula of compound **A**. From the mass spectral data, we know that compound **A** has a molecular weight of 152 g/mol. From the ^{13}C NMR spectrum, there must be a minimum of eight carbons. In addition, we know from the ^1H NMR spectrum that there are at least eight hydrogens. Any molecule of the formula C_8H_8 has a molecular weight of 104 g/mol, which leaves 48 g/mol unaccounted. The simplest fragments that could account for this missing mass are C_4 or O_3. As the IR spectrum of compound **A** suggests the presence of at least two oxygen atoms (the peak at 3205 cm^{-1} could be an O—H stretch and the peak at 1675 cm^{-1} could be a C=O stretch), let's assume that the missing fragment is O_3 and see where this leads. Therefore, the tentative molecular formula of compound **A** is $C_8H_8O_3$. Now let's work out the number of degrees of unsaturation:

$$\Omega = [2(8)+2-8]/2 = 5$$

Five degrees of unsaturation suggests the possibility of a benzene ring. Compound **A** is soluble in 5% aqueous NaOH solution but not in 5% aqueous $NaHCO_3$ solution. Carboxylic acids and phenols are strong enough acids to react with NaOH to yield water-soluble salts. However, unlike carboxylic acids, most phenols are too weakly acidic to undergo reaction with $NaHCO_3$. Thus this solubility data suggests that compound **A** is a substituted phenol.

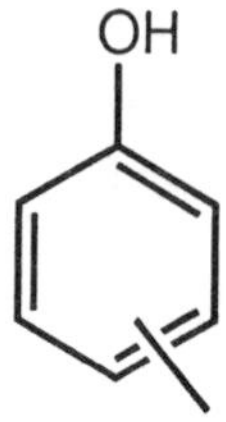

OH

The presence of the broad peak at 3205 cm^{-1} in the IR spectrum of compound **A** is consistent with the O—H stretch of the proposed phenol. In addition, the presence of the strong band at 1675 cm^{-1} suggests the possibility of a C=O stretch, accounting for the last degree of unsaturation. The frequency of the carbonyl stretch further indicates that the carbonyl group is probably conjugated. What type of carbonyl group? An anhydride can be ruled out because of the absence of a second carbonyl stretch in the infrared; an aldehyde can be eliminated because of the absence of the C—H stretching doublet; and a carboxylic acid can be ruled out from the solubility data. These eliminations leave the possibility of either an ester or a ketone. There are several bands in the 1300-1100 cm^{-1} region of the IR spectrum that could be an ester C—O stretch. However, the ketone cannot be ruled out on this basis. (The ^{13}C NMR spectrum of compound **A** is very helpful in this regard, as we will see shortly.)

Now is a good time for an atom inventory.

$C_8H_8O_3$	Formula
$-\ C_6$	Benzene ring
$-\ \ H\ O$	OH of phenol
$-\ C\ \ \ O$	Carbonyl group
$C\ H_7O$	Remaining

Now let's look at the proton and carbon NMR spectra of compound **A**. First of all, let's deal with the ketone versus ester problem. The carbonyl carbon of ketones (and aldehydes) has a ^{13}C NMR chemical shift in the δ 190-220 ppm range, whereas the carbonyl carbon of esters appears in the δ 150-180 ppm range. The ^{13}C NMR spectrum of compound **A** fails to show a signal for a ketone, but does exhibit a singlet in the ester range (δ 170.7 or 162.0 ppm).

It is obvious from both the ^1H and ^{13}C NMR spectra of compound **A** that the remaining unassigned carbon (from the atom inventory) is a deshielded CH_3 group (the singlet at δ 3.92 ppm in the proton spectrum and the quartet at δ 52.1 ppm in the carbon spectrum). If we put all of the structural fragments together, we obtain a methyl hydroxybenzoate; that is,

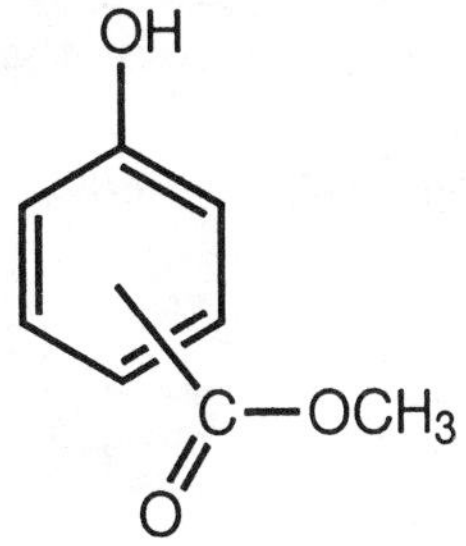

Note that this preliminary structure also accounts for the four aromatic protons in the δ 6.85-7.83 ppm region of the ^{1}H NMR spectrum of compound **A**. Is compound **A** the ortho, meta, or para isomer? There are several ways to approach this question; happily, all of them give the same answer. First, the IR spectrum of compound **A** shows a single aromatic C—H bending absorption at 757 cm^{-1}, which is indicative of an ortho-disubstituted benzene. Second, note that all four aromatic hydrogens in the δ 6.85-7.83 ppm region of the ^{1}H NMR spectrum show ortho coupling (J = 8 Hz). The only disubstituted isomer in which four different aromatic hydrogens all have at least one adjacent ortho proton is the ortho isomer. Finally and more esoterically, the chemical shift of the phenolic O—H hydrogen in the ^{1}H NMR spectrum of compound **A** is very informative. The O—H hydrogen of phenols normally has a chemical shift of δ 4.0-7.5 ppm. However, an ortho carbonyl group shifts the phenolic hydrogen downfield to δ 10.0-12.0 ppm because of intramolecular hydrogen bonding. The phenolic hydrogen of compound **A** appears at δ 10.8 ppm. Thus compound **A** is methyl salicylate (methyl 2-hydroxybenzoate).

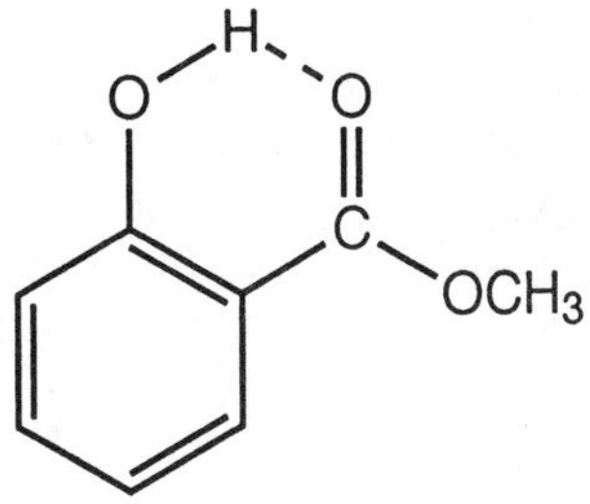

A

The ^{1}H NMR chemical shifts and coupling constants for compound **A** are summarized below.

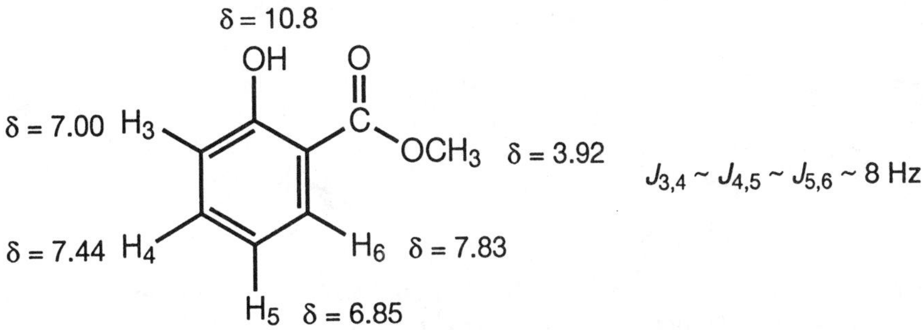

δ = 10.8 OH; δ = 7.00 H_3; δ = 7.44 H_4; H_5 δ = 6.85; H_6 δ = 7.83; OCH_3 δ = 3.92

$J_{3,4} \sim J_{4,5} \sim J_{5,6} \sim 8$ Hz

A

The H(3) is shielded by the adjacent OH group and is split into a doublet by H(4). (Remember that J_{meta} and J_{para} were not observed at 300 MHz.) H(4) has a "normal" aromatic chemical shift and is split into a "triplet" by H(3) and H(5). H(5) is shielded by the para OH group and is split into a "triplet" by H(4) and H(6). Finally, H(6) is deshielded by the adjacent ester carbonyl bond and is split into a doublet by H(5).

A tentative assignment of the ^{13}C NMR resonances of compound **A** is as follows:

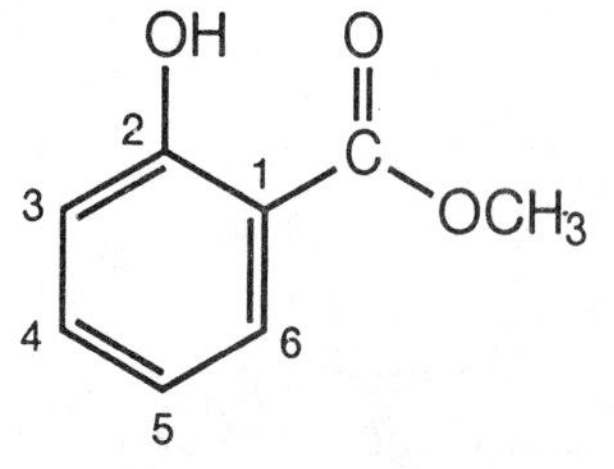

OH O C OCH3 1 2 3 4 5 6

A

δ 52.1(q) CH_3
112.7(s) C-1
117.7(d) C-3 (or C-5)
119.2(d) C-5 (or C-3)
130.1(d) C-6 (or C-4)
135.7(d) C-4 (or C-6)
162.0(s) C-2 (or C=O)
170.7(s) C=O (or C-2)

Problem 15.61 First, we will determine the molecular formulas of compounds **B** and **C**. Compound **B** is a dehydration product of α-terpineol (**A**) (i.e., $C_{10}H_{18}O - H_2O = C_{10}H_{16}$). If compound **B** reacts with maleic anhydride to give a 1:1 adduct, compound **C** has a molecular formula of ($C_{10}H_{16}$ + $C_4H_2O_3$) = $C_{14}H_{18}O_3$. The molecular weight for this formula is 234 g/mol. Thus, our surmise that compound **C** is a 1:1 adduct of **B** and maleic anhydride is consistent with the mass spectrum of **C**. This proposal is also supported by the presence of 18 hydrogens in the ^1H NMR spectrum of **C**.

Now we need to work out the number of degrees of unsaturation in **C**.

$$\Omega = [2(14) + 2 - 18]/2 = 6$$

So far in these problems, six degrees of unsaturation has generally suggested the possibility of a benzene ring, thus accounting for four degrees of unsaturation. Could this problem be an exception to this expectation? There seems no easy way to get an aromatic compound in this reaction.

As was already noted, compound **C** is a 1:1 adduct of **B** and maleic anhydride. Futhermore, **B** is a dehydration product of α-terpineol (**A**). It is worth speculating at this point about the structure of compound **B**. Simple dehydration of compound **A** could afford dienes **D** and **E**; perhaps one of these is compound **B**. This seems unlikely, as neither **D** nor **E** is a conjugated diene, and therefore, neither can react in Diels-Alder fashion with maleic anhydride.

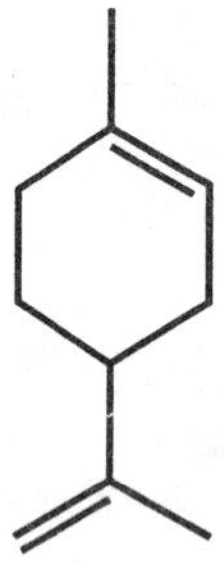

D 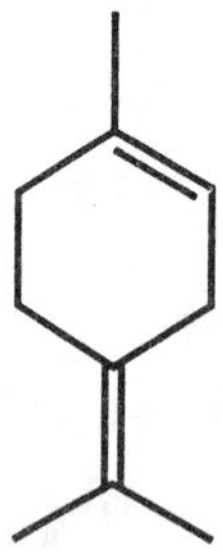 E

However, a more complicated dehydration of **A** could afford the conjugated diene, α-terpinene. You should be able to write a mechanism for this dehydration.

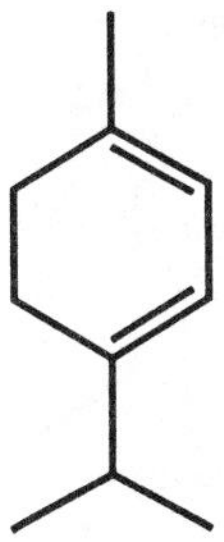

α-Terpinene

Clearly, α-terpinene and maleic anhydride could undergo a Diels-Alder reaction to give a 1:1 cycloadduct, and this could be compound **C**. Let's see if the rest of the spectral data are consistent with this proposal.

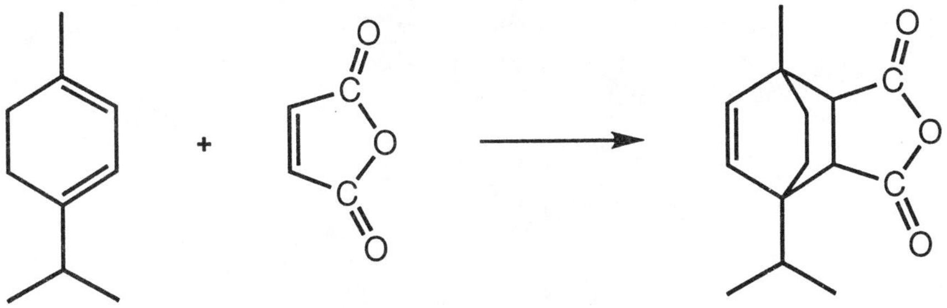

C

The two carbonyl stretches at 1840 and 1780 cm^{-1} in the IR spectrum of **C** are consistent with the presence of a cyclic anhydride.

The NMR spectral data are also consonant with the proposed structure for **C**. The chemical shift assignments are summarized below. Note that the isopropyl methyl groups are diastereotopic as required by the intrinsic asymmetry of the proposed structure. Also note that the six degrees of unsaturation for compound **C** are three rings, 1 C=C, and 2 C=O bonds.

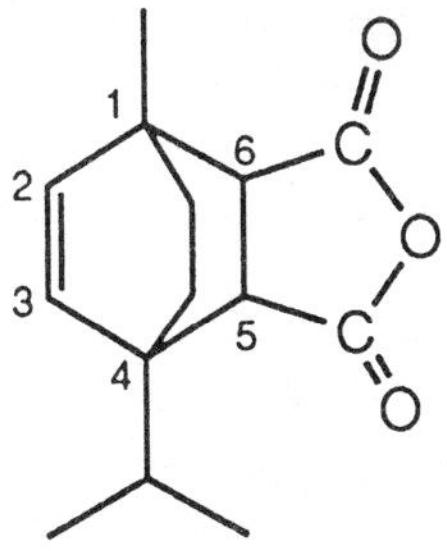

δ	Assignment
0.98, 1.08	$(C\underline{H}_3)_2CH$
1.36	Bridge CH_2's
1.46	C(1) CH_3
2.60	$(CH_3)_2C\underline{H}$
2.80, 3.20	H(5) and H(6)
5.96	H(2) and H(3)

C

Two questions still remain concerning the stereochemistry of cycloadduct **C**.
(1) Are H(5) and H(6) cis or trans to each other? (2) Is the cycloadduct endo or exo? One of these questions can be answered with the available spectral data, the other cannot. As you might expect from what you already know about the Diels-Alder reaction, the cis stereochemistry present in maleic anhydride is preserved in the cycloadduct; i.e., H(5) and H(6) have a cis relationship. This assignment is supported by the magnitude of their coupling constant (J = 9 Hz). From the Karplus curve, it is known that vicinal coupling constants are dependent on the dihedral angle (ϕ) between the vicinal protons — with large coupling constants for dihedral angles of 0° and 180° and smaller coupling constants for dihedral angles approaching 90°.

Typical values for cis ($J_{exo,exo}$ and $J_{endo,endo}$; $\phi \sim 0°$) and trans ($J_{exo,endo}$; $\phi \sim 120°$) vicinal coupling constants for bicyclo[2.2.2]octenes are shown below.

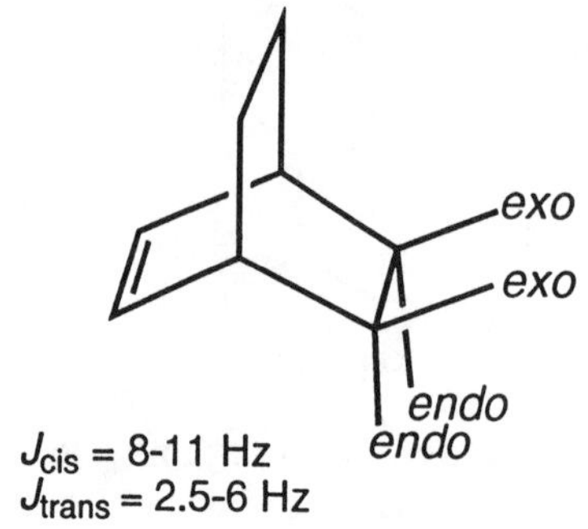

exo
exo
endo
endo

J_{cis} = 8-11 Hz
J_{trans} = 2.5-6 Hz

Although it is not possible to deduce the exo versus endo stereochemistry of cycloadduct **C** from the available spectral data, **C** has endo stereochemistry. The actual stereochemistry has been determined by using more complex spectral techniques, by chemical transformations, and by independent synthesis of the exo isomer.

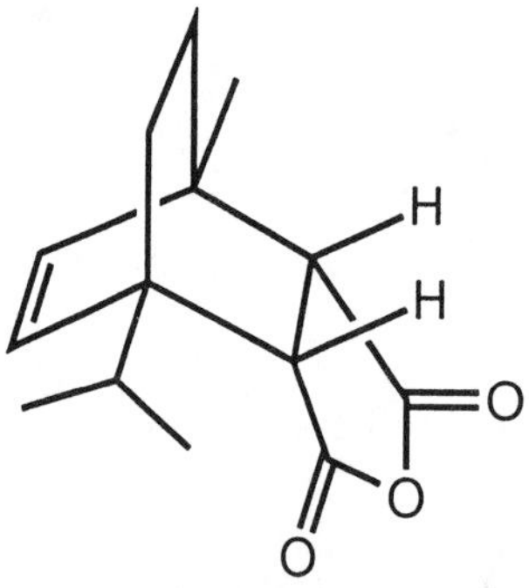

H
H
O
O
O

endo-**C**

Problem 15.62 The spectrum of 2,2,2-trichloroethanol is easy to analyze. The hydroxyl hydrogen appears as a broad singlet at δ = 3.4 ppm, and the methylene hydrogens are a singlet at δ = 4.2 ppm, a position quite appropriate for hydrogens adjacent to an oxygen atom. The chlorines do not couple. The real question here is why the spectrum of 2,2,2-trifluoroethanol is so different. The critical question is why the signal for the methylene hydrogens appears as a quartet. Fluorine, unlike chlorine, has a nuclear spin of 1/2. The methylene hydrogens are therefore split by the three adjacent fluorines into a 1:3:3:1 quartet (apply the "*n*+1 rule"). Why are the fluorines invisible? Should there not be a triplet in the spectrum? No. The gyromagnetic ratio (γ) for fluorine is very different from that for hydrogen and this triplet will be shifted far from the hydrogen spectrum. There is a triplet, but we can't see it unless we tune the spectrometer for fluorine and look for it.

Chapter 16 Outline

Carbonyl Chemistry 1: Addition Reactions

16

Problem 16.1 There are three different bonds in formaldehyde, the carbon-hydrogen bonds and the σ and π bonds between carbon and oxygen (Fig. 16.2). The carbon-hydrogen bonds are constructed through $1s$ - sp^2 overlap, the carbon-oxygen σ bond by sp^2(C) - sp^2(O) overlap, and the carbon-oxygen π bond from $2p$(C) - $2p$(O) overlap.

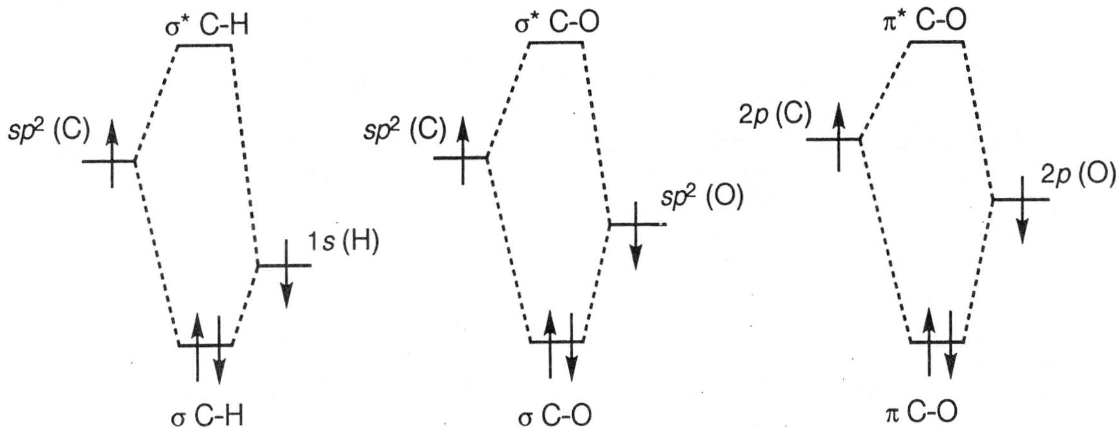

Problem 16.2 Just as the π bond "leans" toward the more electronegative oxygen atom, so will the sp^2-sp^2 σ bond. Similarly, as π^* leans toward carbon, so will σ^*. The σ bond is made of more than 50% of the oxygen orbital and the σ^* bond of more than 50% of the carbon orbital.

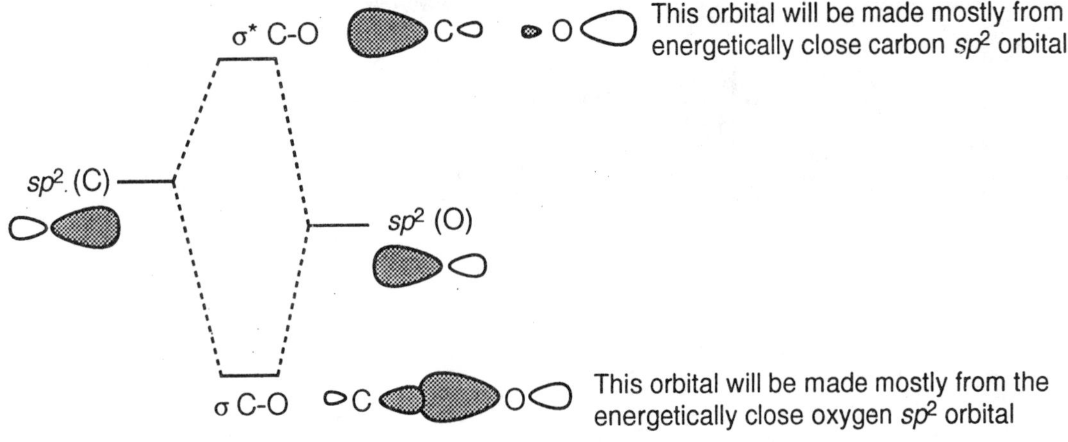

Problem 16.3 The only difference between this model and the one in Figure 16.2 is that the oxygen is hybridized *sp*. The carbon is still sp^2.

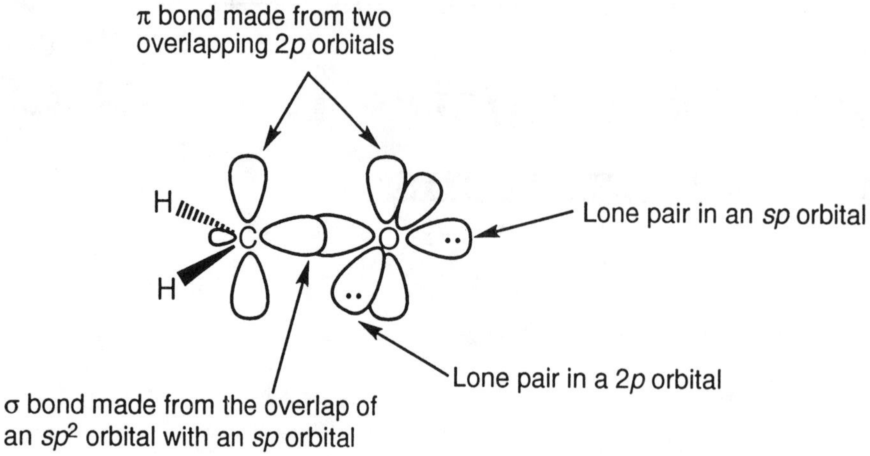

π bond made from two overlapping 2*p* orbitals

Lone pair in an *sp* orbital

Lone pair in a 2*p* orbital

σ bond made from the overlap of an sp^2 orbital with an *sp* orbital

Problem 16.6 The name 2-propanone is redundant. The suffix "one" means ketone, and there is only a single position, C(2), where there can be a ketone. If the carbonyl were at either of the other carbons, the compound would be an aldehyde.

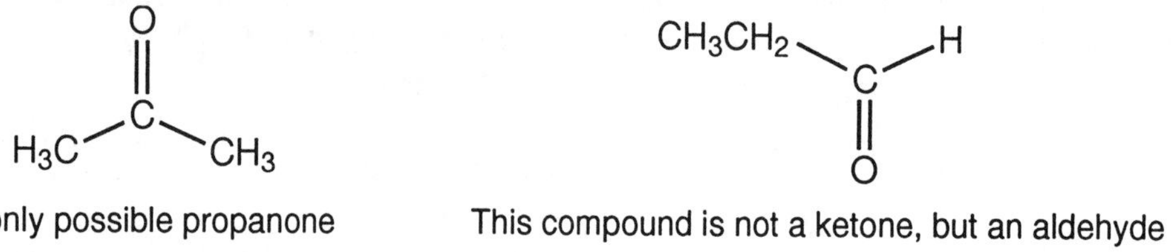

The only possible propanone

This compound is not a ketone, but an aldehyde

Problem 16.8 Addition to the oxygen end of the carbonyl, or to the alkene would transfer the negative charge to carbon. It is energetically favorable to keep the negative charge on the relatively electronegative oxygen of the hydroxide ion.

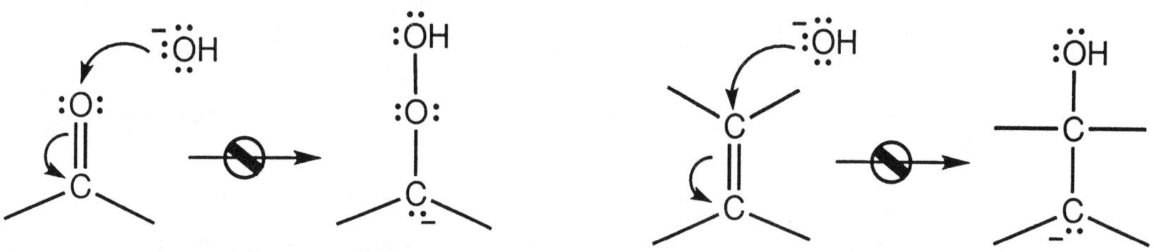

Problem 16.9 As long as there is *some* acetaldehyde present *at equilibrium*, it can react to form a small amount of cyanohydrin. As the aldehyde is used up, more will be regenerated. Eventually, all the acetaldehyde can be converted into cyanohydrin in this way.

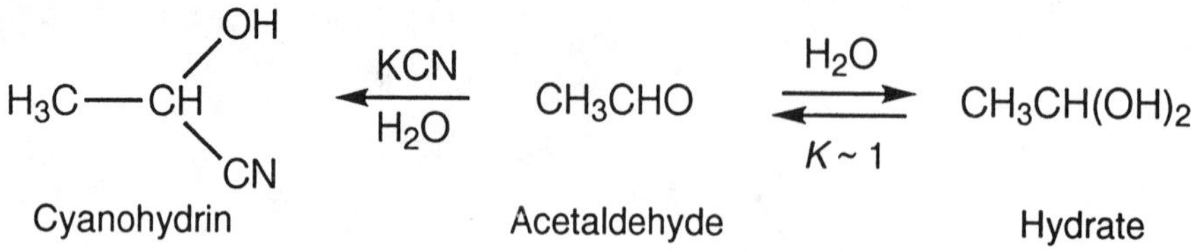

$$H_3C{-}CH(OH)(CN) \xleftarrow[H_2O]{KCN} CH_3CHO \underset{K\sim 1}{\overset{H_2O}{\rightleftharpoons}} CH_3CH(OH)_2$$

Cyanohydrin　Acetaldehyde　Hydrate

Problem 16.10 Addition reactions are generally less successful with ketones than aldehydes. Ketones are more substituted than aldehydes, and therefore more stable. This stability results in the carbonyl form being relatively favored at equilibrium compared to the addition product.

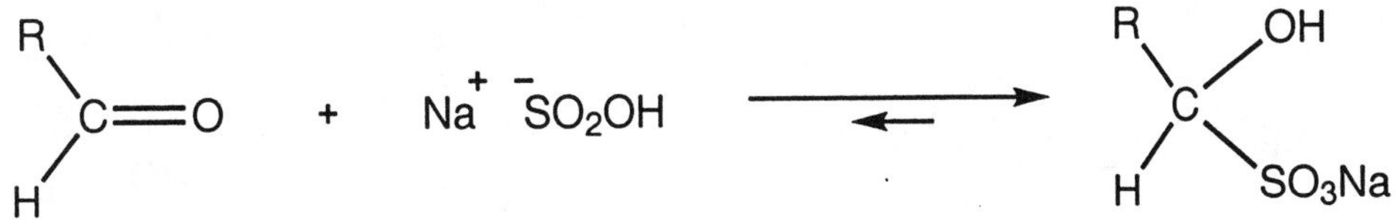

For the less stable aldehydes, it is the addition product that is usually favored

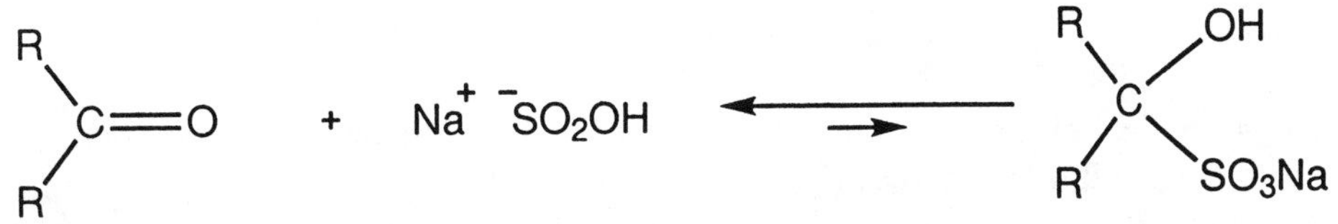

For the more stable ketones, it is the carbonyl compound that is usually favored

Problem 16.11 The nucleophilic hydroxyl oxygen adds to the Lewis acid carbonyl group. In the uncatalyzed addition, this generates an ionic structure. Deprotonation of one oxygen and protonation of the other leads to the final product. These protonations and deprotonations can be accomplished by any bases and acids present. The order of these reactions is not certain, and they are not shown in the figure.

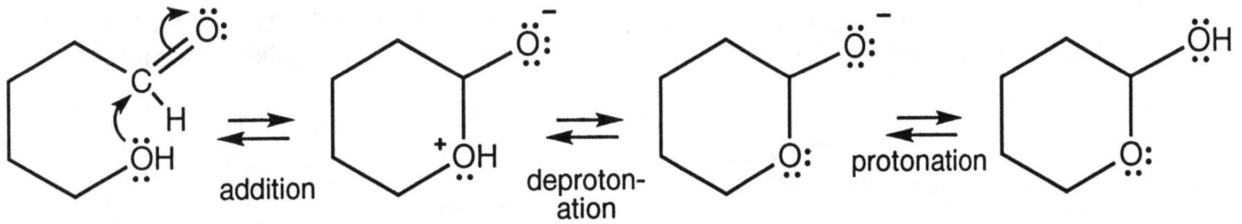

addition

deproton-
ation

protonation

Problem 16.12 This problem is a tough one. The OH at the indicated carbon (see arrow) can be in either the axial or equatorial position. This OH is shown as the more stable equatorial form.

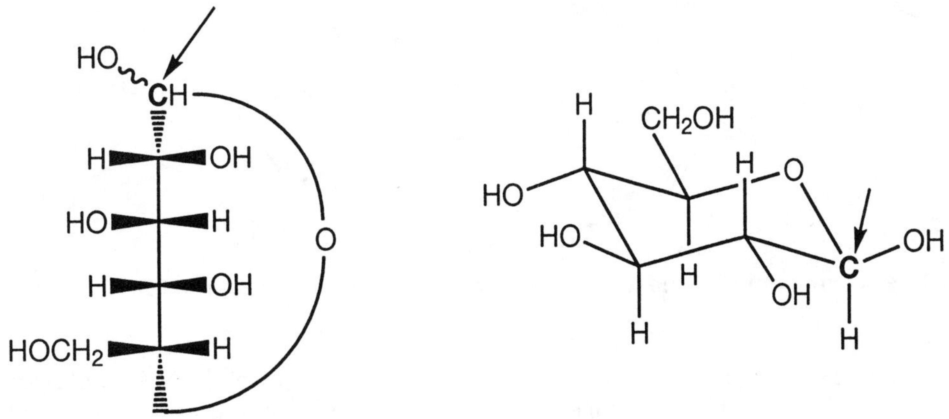

Problem 6.13 As in any acid-catalyzed acetal formation, the first step is protonation of the carbonyl group to give a resonance-stabilized intermediate.

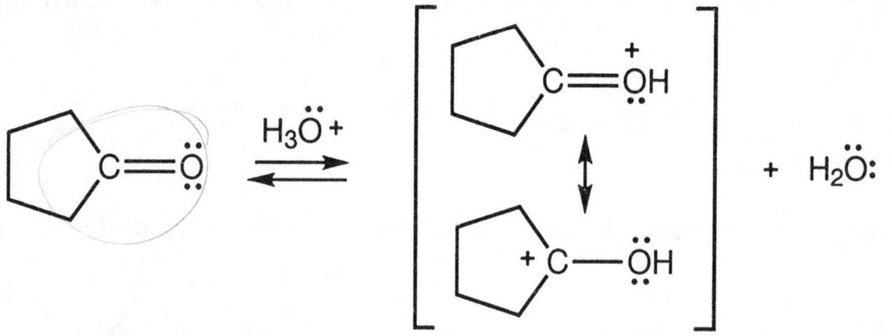

Ethylene glycol is an alcohol, and adds to the strongly Lewis acidic protonated carbonyl. Deprotonation and reprotonation gives **A**.

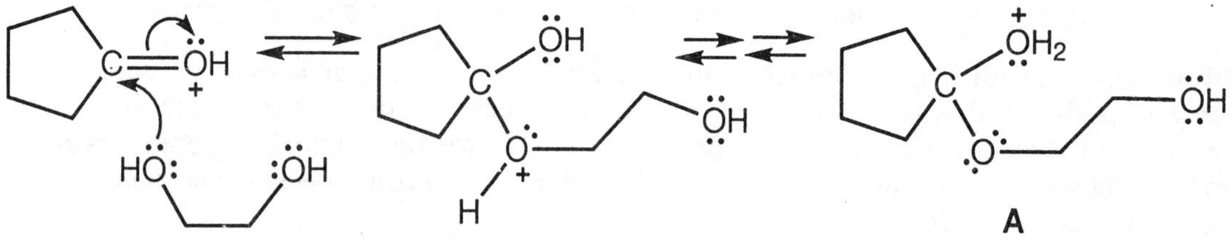

Intermediate **A** loses water to give the resonance-stabilized **B**. Now, in the critical step, the second OH group adds in intramolecular fashion to give **C**. A final deprotonation leads to the acetal.

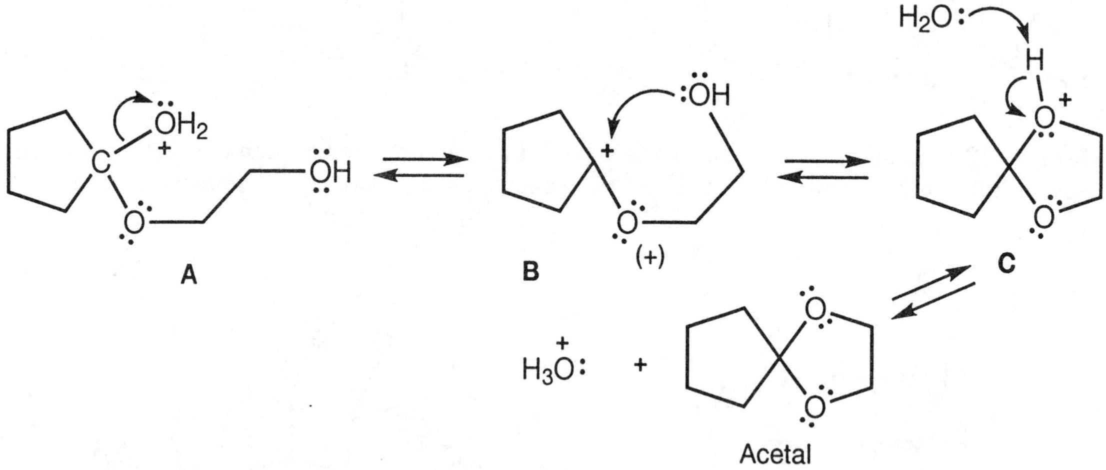

Problem 16.14 Argggh. A trick question. Simply run the mechanism in the previous problem backward!

Problem 16.16 Hydrate formation is reversible. In labeled water, a small amount of hydrate **A** is formed. When **A** re-forms acetone, the label will be incorporated half of the time. If enough labeled water is present, eventually all of the acetone will acquire the label.

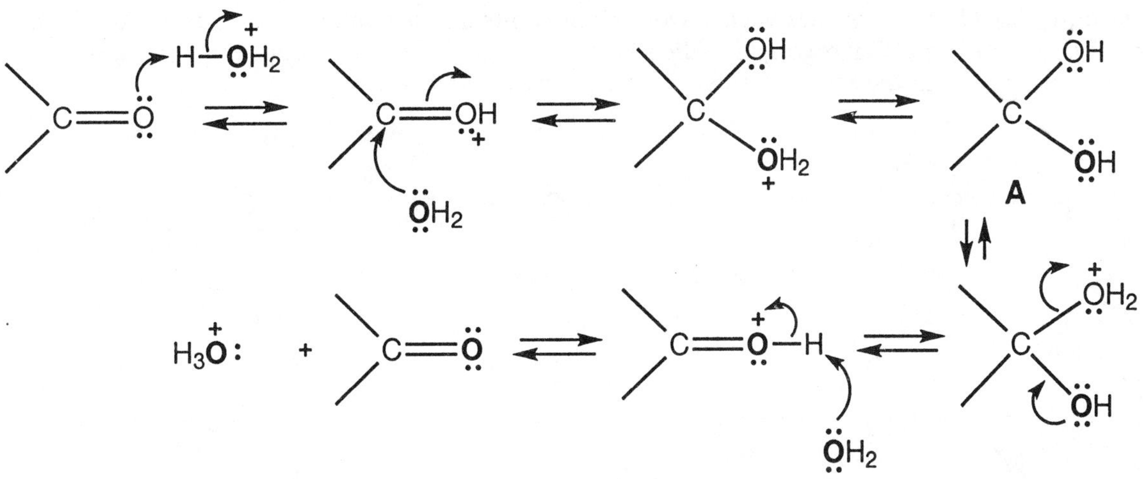

A

Hemiacetal

Carbinolamine

Addition steps

Protonation steps with reformation of the base

Hydrate

Problem 16.17 The bases involved are RO^-, RHN^-, and HO^-, the conjugate bases of ROH, RNH_2, and H_2O.

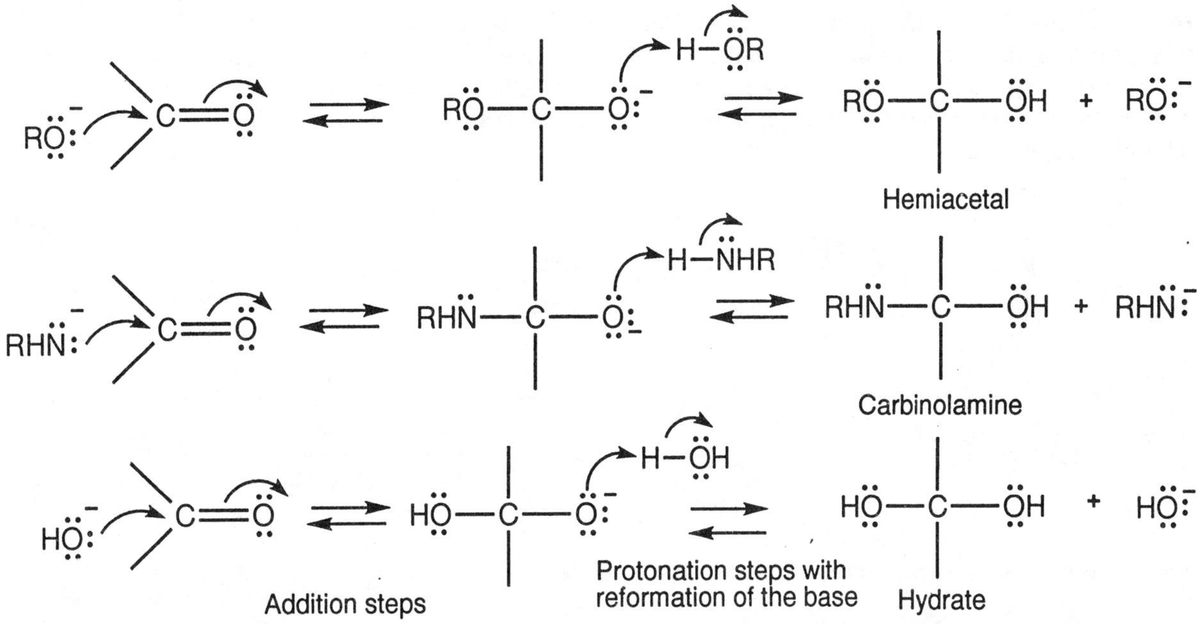

Problem 16.18 In the first step, the carbonyl group is protonated. The nucleophilic amine then adds to give intermediate **A**. Deprotonation of nitrogen and protonation of oxygen leads to **B**, from which water can be lost to give **C**. A final deprotonation leads to the imine product.

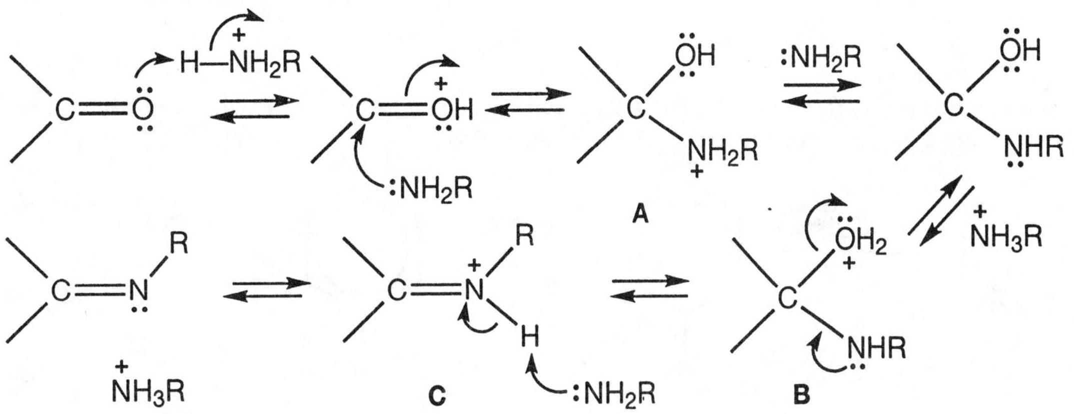

A

B

C

Problem 16.20 Let's work backward. Given the previous paragraph in the text, certainly the final step is the quenching of an organometallic reagent with D_2O. The organolithium or Grignard reagent is made from the bromide and either lithium or magnesium.

$$(CH_3)_3C{-}D \xleftarrow{D_2O} (CH_3)_3C{-}Li \xleftarrow{Li} (CH_3)_3C{-}Br$$

tert-Butyl bromide

$$(H_3C)_2HC{-}CH_2D \xleftarrow{D_2O} (H_3C)_2HC{-}CH_2Li \xleftarrow{Li} (H_3C)_2HC{-}CH_2Br$$

Isobutyl bromide

Now, the only remaining question is how to convert the alkene into the bromide. For this you must remember material in Chapter 9. Simple addition of HBr to isobutene gives *tert*-butyl bromide (Chapter 9, p. 350), but, in order to make isobutyl bromide, a hydroboration sequence is necessary to produce isobutyl alcohol, which can then can be converted into the bromide with HBr. Alternatively, isobutene can be brominated in the anti-Markovnikov sense with HBr/ROOR (Chapter 11, p. 462).

$$(CH_3)_3C{-}Br \xleftarrow{HBr} (H_3C)_2C{=}CH_2$$

$$(H_3C)_2HC{-}CH_2Br \xleftarrow{HBr} (H_3C)_2HC{-}CH_2OH \xleftarrow[2.\ ^{-}OH/H_2O_2]{1.\ B_2H_6} (H_3C)_2C{=}CH_2$$

The third part of the problem is similar in its final steps:

$$C_6H_5D \xleftarrow{D_2O} C_6H_5MgBr \xleftarrow{Mg} C_6H_5Br$$

Bromobenzene is made from nitrobenzene by a sequence of conversions encountered in Chapter 14. The last step is a Sandmeyer reaction (Chapter 14, p. 647).

$$C_6H_5Br \xleftarrow{CuBr} C_6H_5N_2^+ \xleftarrow{HONO} C_6H_5NH_2 \xleftarrow{H_2/Pd} C_6H_5NO_2$$

Problem 16.21 One conceptually simple test would be to measure the stereochemistry of the reaction. If the mechanism involves an S_N2 displacement, the reaction must proceed with complete inversion of configuration.

R_2CuLi + C—I $\xrightarrow{S_N2}$ R—C

When this experiment is performed, it is found that the stereochemical outcome *is not* pure inversion, but a mixture of inversion, retention, and/or racemization depending on the nature of the R group and the leaving group. The mechanism of this reaction is more complicated than a simple S_N2 displacement.

Problem 16.23 This reaction is nothing more than an E1 elimination (Chapter 7, p. 273). The oxygen atom is protonated and water lost to give the tertiary carbocation, **A**.

HÖ: CH₃ — H_3O^+ / protonation ⇌ $H_2\overset{+}{O}$: CH₃ — E1 ⇌ CH₃ +C **A** + :ÖH₂

There are two different protons that can be lost in the second step. The one leading to formation of the more stable trisubstituted double bond will be preferred to the one leading to the less stable, disubstituted double bond.

$H_2\ddot{O}$: H CH₃ C+ ⟶ $H_3\overset{+}{O}$: + CH₃ C (more stable product)

$H_2\ddot{O}$: H CH₂ C+ ⟶ $H_3\overset{+}{O}$: + CH₂ C (less stable product)

Additional Problem Answers

Problem 16.26
(a) Heptanal, (b) 3,3-dichlorobutanal, (c) butanedial (d) 2,2-dimethylpropanal (pivaldehyde) (e) 3-chloro-2,2-dimethylpropanal, (f) *m*-fluorobenzaldehyde (3-fluorobenzaldehyde).

Problem 16.27

(a) (b) 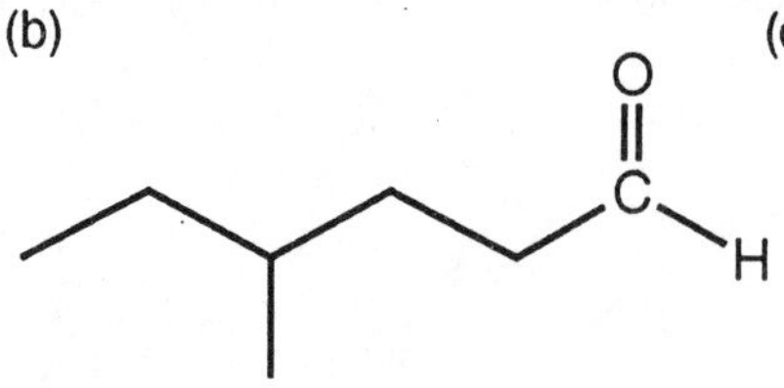 (c) 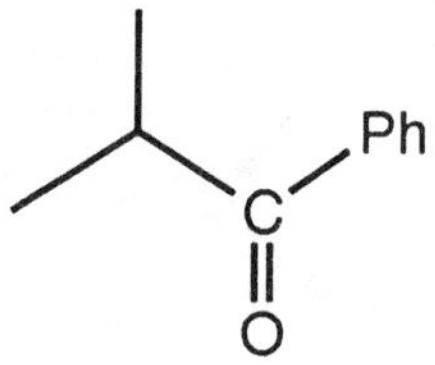 (d)

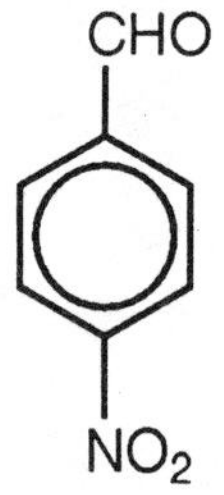

Problem 16.28

(a) 3-Methylcyclobutanone, (b) Cyclohexyl phenyl ketone, (c) 4,6-Diamino-2-octanone

Problem 16.29

(a) (b) 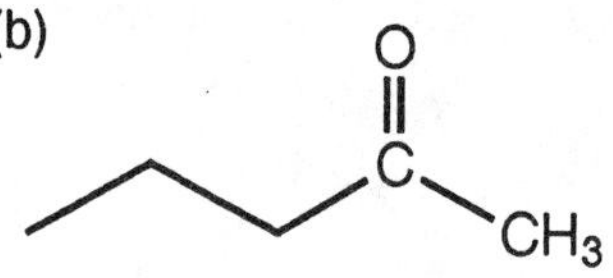 (c)

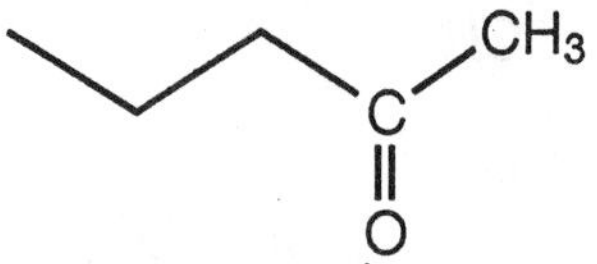

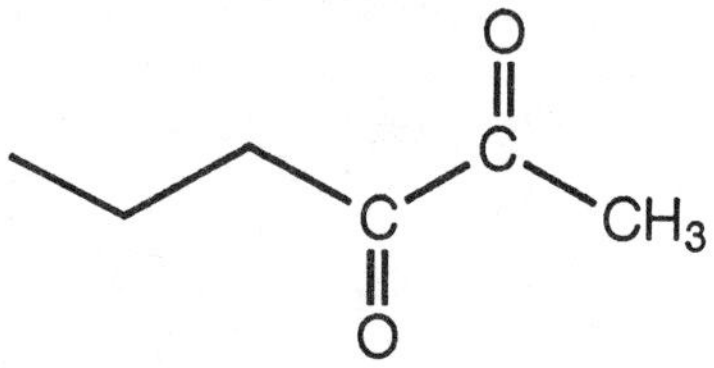

Note that (a) and (b) are the same!

(d) (e)

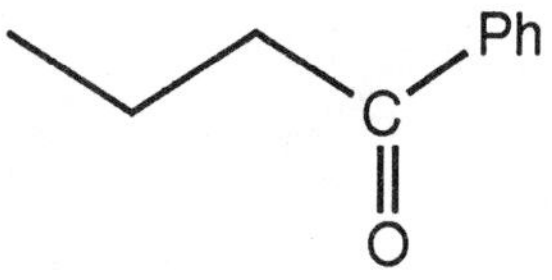

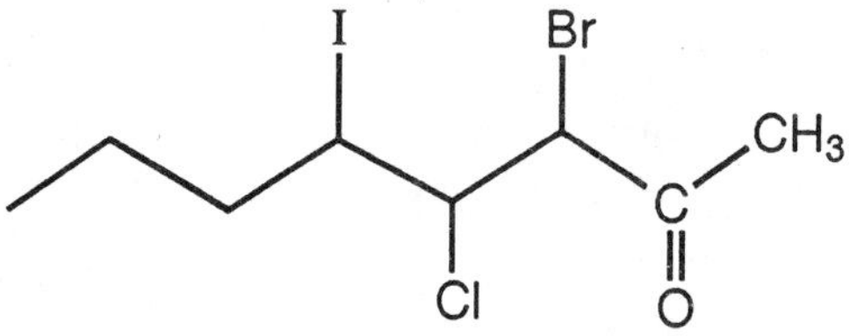

Problem 16.30

(a) (b) 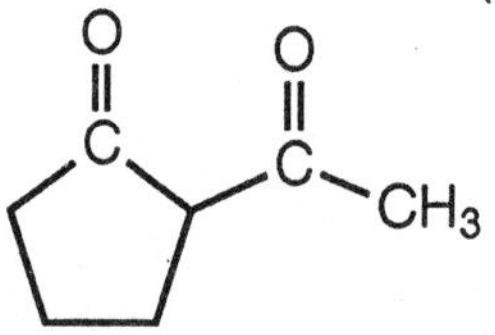 (c)

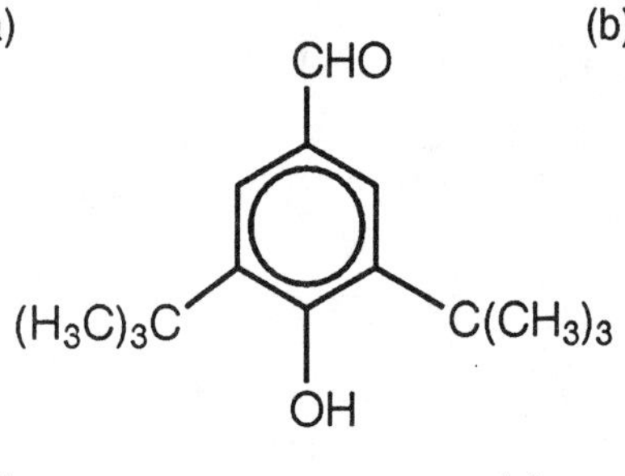

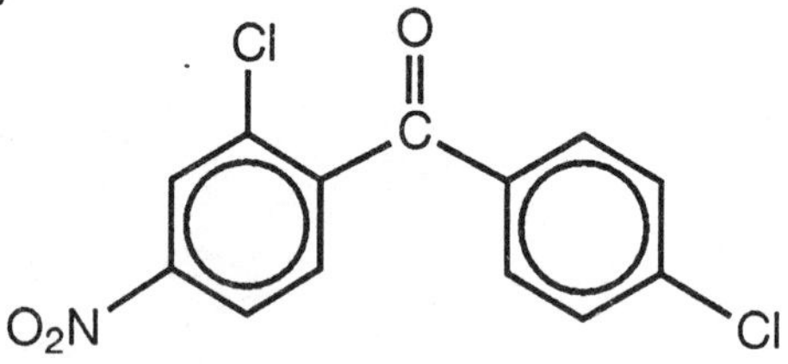

(d) 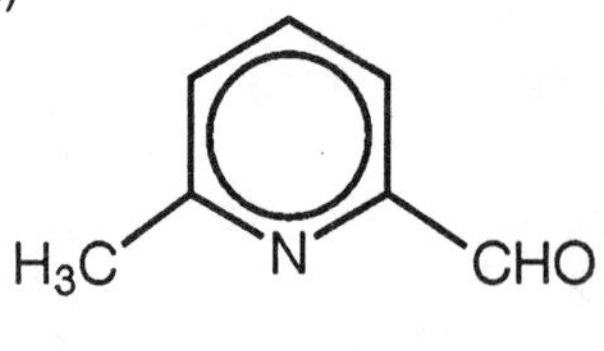 (e) 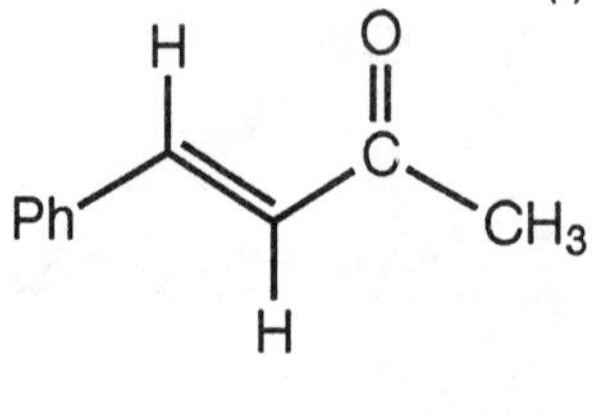 (f) $H_2NCH_2CH(OCH_3)_2$

Problem 16.31

(a) Cyclopentanones should absorb in the IR at about 1750 cm^{-1}. Acyclic carbonyls should absorb at about 1720 cm^{-1}. In fact, cyclopentanone itself absorbs at 1751 cm^{-1} and methyl propenyl ketone at 1723 cm^{-1}.

(b) Cyclobutanones absorb about 40 cm^{-1} higher than the less strained cyclopentanones. 3-Methyl-cyclobutanone absorbs at 1789 cm^{-1}.

(c) The appearance of a low-field signal for the carbonyl carbon will be a dead giveaway. Cyclohexanone absorbs at δ 207.9 ppm, and no signal in the spectrum of diallyl ether will be close to that position. Moreover, the ^{13}C NMR spectrum of cyclohexanone must show signals for four different carbons, whereas the spectrum for diallyl ether will show only three.

(d) There are many ways to make the distinction. One way that definitely won't work is to use the strong IR band for the carbonyl stretch. Small acyclic aldehydes and ketones are too close together for that kind of distinction. These two molecules absorb at 1715 and 1719 cm^{-1}, for example. However, you should be able to pick out the two bands for the C—H stretch in the aldehyde spectrum. These typically appear at about 2850 and 2750 cm^{-1}, and this aldehyde shows them at 2840 and 2732 cm^{-1}. The ^{1}H NMR spectra should also be definitive. First of all, the aldehyde should show a distinctive low field signal at about δ 9-10 ppm. This one appears at δ 9.77 ppm. The detailed spectra will also be different, and these are summarized below.

δ 0.98 (CH₃), δ 2.19 (HC), δ 2.29 (CH_2), δ 9.77 (CHO): $(CH_3)_2CH-CH_2-CHO$

δ 1.11 (CH₃), δ 2.6 (HC), δ 2.15 (CH_3 of C=O): $(CH_3)_2CH-C(=O)-CH_3$

Problem 16.32

(a) Cyclohexanone + HO~~~OH (benzene, H_3O^+) → spiro 1,3-dioxane acetal — Acetal formation

(b) Cyclohexanone + Ph-NH_2 → C=N–Ph — Imine formation from a primary amine

(c) Cyclohexanone + pyrrolidine (NH) → 1-(cyclohexenyl)pyrrolidine — Enamine formation from a secondary amine

(d) $PhC(=O)CH_3$ $\xrightarrow[\text{2. } H_2O/H_3O^+]{\text{1. PhMgBr}}$ $Ph{-}C(OH)(Ph){-}CH_3$ Grignard addition followed by protonation

(e) $PhC(=O)H$ $\xrightarrow[H_2O]{NaHSO_3}$ $\xrightarrow[H_2O]{NaCN}$ $Ph{-}C(OH)(CN){-}H$ Bisulfite addition followed by cyanide addition

(f) $PhC(=O)Ph$ $\xrightarrow[\text{2. } H_2O/H_3O^+]{\text{1. } CH_3Li}$ $Ph{-}C(OH)(Ph){-}CH_3$ Alkyllithium addition followed by protonation

(g) $PhC(=O)CH_3$ $\xrightarrow[CH_3OH]{NaBH_4}$ $Ph{-}C(OH)(H){-}CH_3$ Reduction by hydride followed by protonation

(h) $PhC(=O)Ph$ $\xrightarrow[\text{2. } H_2O/H_3O^+]{\text{1. } LiAlH_4}$ $Ph{-}C(OH)(Ph){-}H$ Reduction by hydride followed by protonation

(i) 2-methylcyclohexanol (OH, CH_3) $\xrightarrow[H_2SO_4]{Na_2Cr_2O_7}$ 2-methylcyclohexanone (O, CH_3) Oxidation of a secondary alcohol to a ketone

(j) $PhCH_2OH$ $\xrightarrow{H_2CrO_4}$ $PhCOOH$ Oxidation of a primary alcohol to a carboxylic acid

(k) $PhCH{=}CHCH_2OH$ $\xrightarrow[\text{pyridine}]{CrO_3}$ $PhCH{=}CHCHO$ Oxidation of a primary alcohol to an aldehyde

(l) $PhC(=O)(CH_2)_4CH_2I$ $\xrightarrow[\text{2. BuLi; 3. } \Delta]{\text{1. } Ph_3P}$ 1-phenylcyclohexene (Ph) Intramolecular Wittig reaction

Part (l) is the only tricky one. It is hard because it involves a multistep intramolecular reaction. The following intermediates are involved:

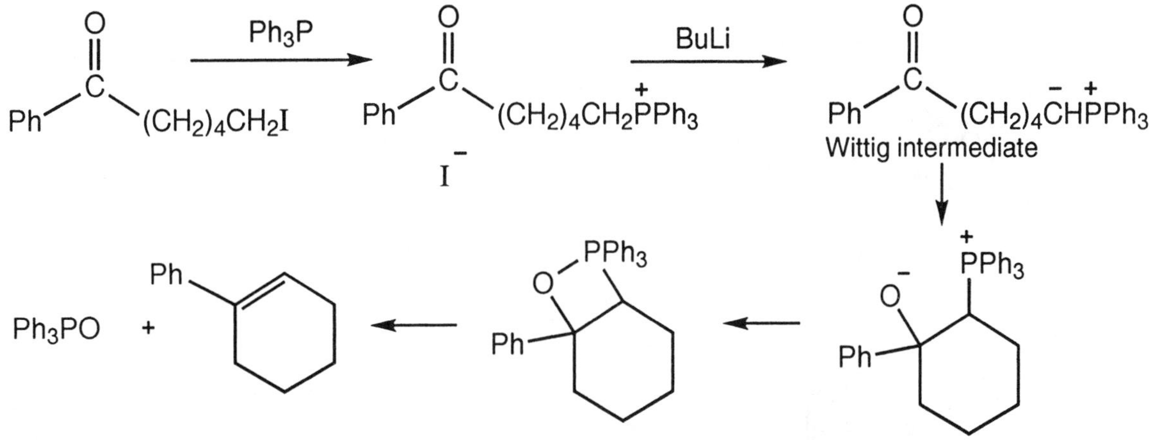

Ph–C(=O)–$(CH_2)_4CH_2I$ —(Ph_3P)→ Ph–C(=O)–$(CH_2)_4CH_2\overset{+}{P}Ph_3$ I^- —(BuLi)→ Ph–C(=O)–$(CH_2)_4\overset{-}{C}H\overset{+}{P}Ph_3$
Wittig intermediate

Ph_3PO + 1-phenylcyclohexene ← oxaphosphetane ← betaine (O^-, $\overset{+}{P}Ph_3$)

Problem 16.33 We might argue that the transition state for protonation on oxygen places a partial positive charge on carbon, whereas the transition state for protonation on carbon places the positive charge on oxygen.

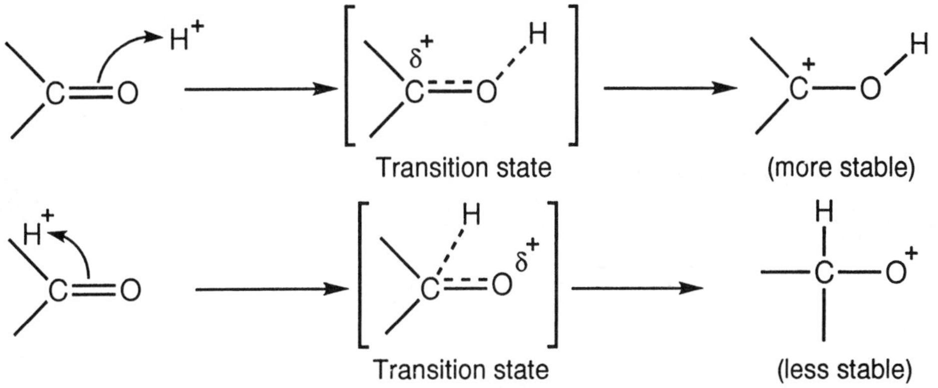

C=O + H^+ → [C$^{\delta+}$···O···H] → $\overset{+}{C}$–O–H
Transition state (more stable)

H^+ + C=O → [H···C···O$^{\delta+}$] → H–C–O^+
Transition state (less stable)

It is far better energetically to place the positive charge on the relatively electropositive carbon than on the relatively electronegative oxygen. Protonation will be favored on oxygen.

Problem 16.34
(a) This problem is just a reverse acetal formation.

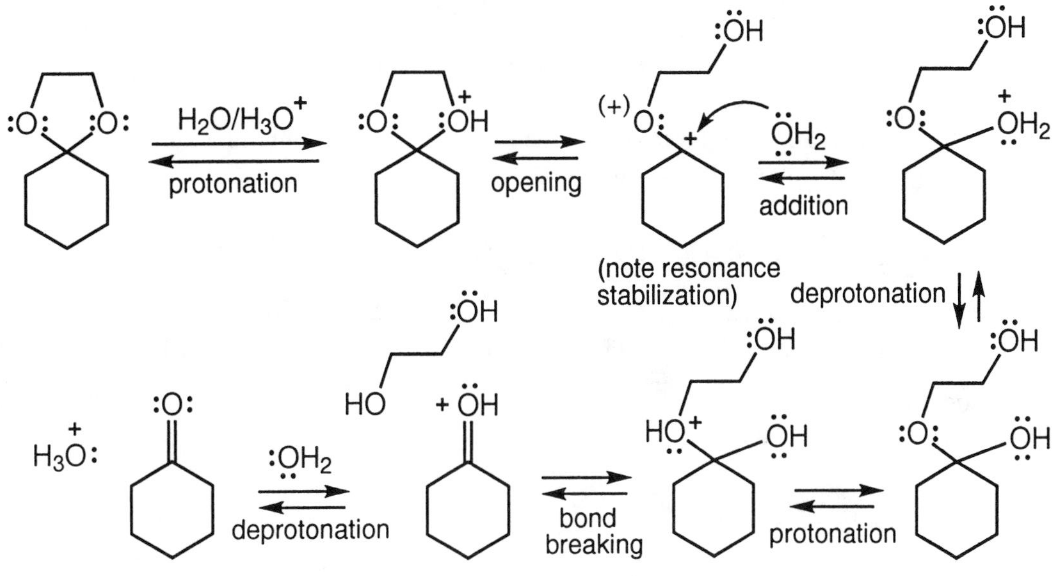

H_2O/H_3O^+ protonation; opening; addition; (note resonance stabilization); deprotonation; protonation; bond breaking; deprotonation

$\ddot{O}H_2$; $:\ddot{O}H$; HO; $+\ddot{O}H$; $:O:$; $H_3\overset{+}{O}:$

(b) In (b), you are asked to hydrolyze an immonium ion. The analogy here is to carbonyl chemistry, with the immonium ion playing the part of a protonated carbonyl.

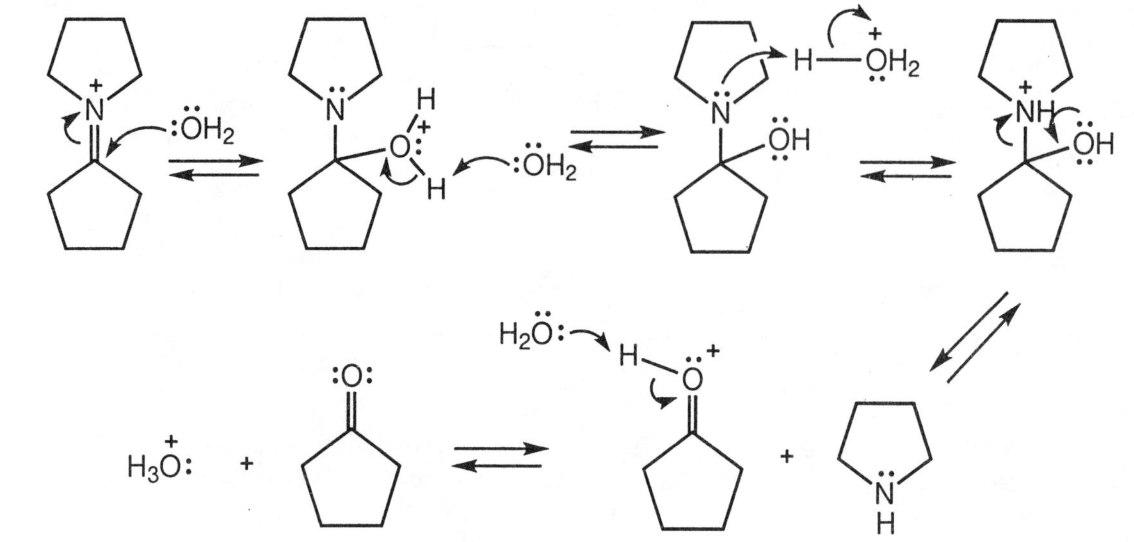

(c) This reaction involves the conversion of a hemiacetal into a full acetal. The steps are essentially those of (a).

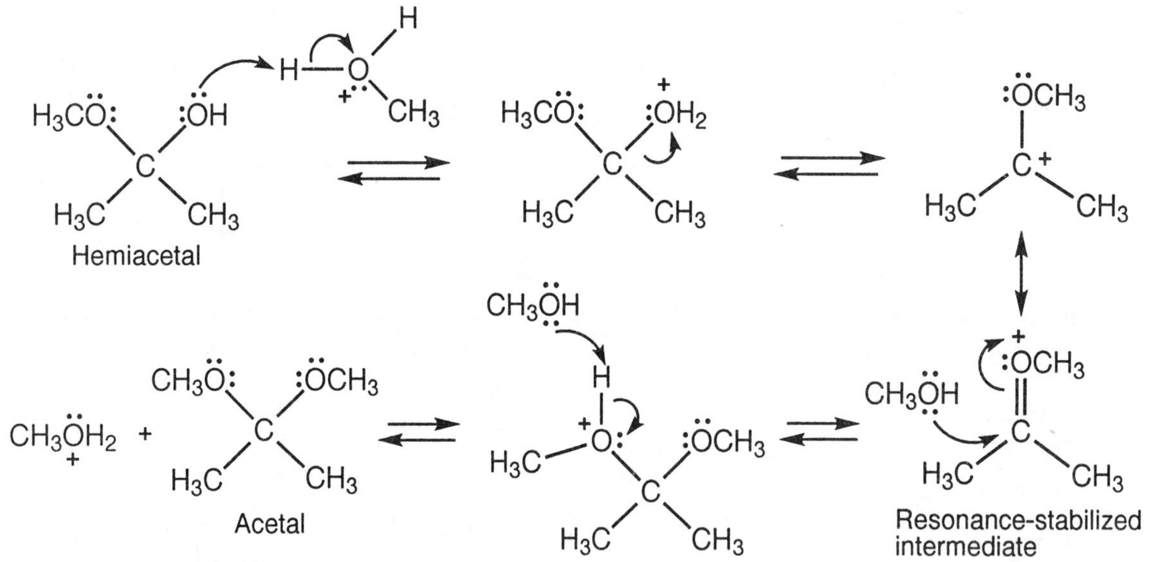

Hemiacetal

Resonance-stabilized intermediate

Acetal

(d) In base, a hemiacetal cannot go on to the full acetal. The only possible reaction here is reversal to the carbonyl compound, generally an exothermic process.

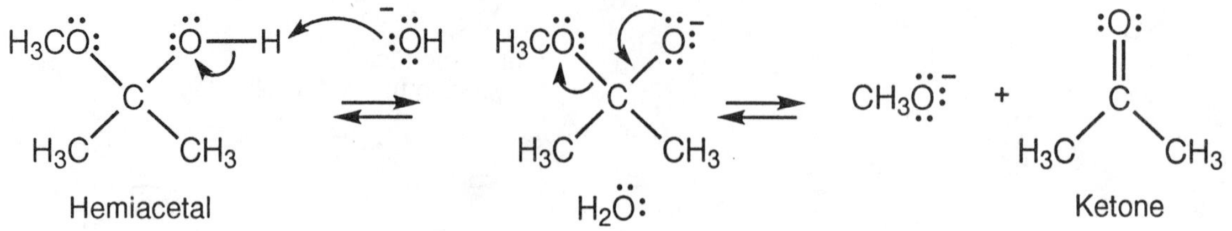

Hemiacetal

Ketone

(e) Full acetals are stable in base. There is no reaction here.
(f) This reaction is a version of imine formation in which a carbonyl compound reacts with a primary amine. In this case, the imine has the special name of "oxime."

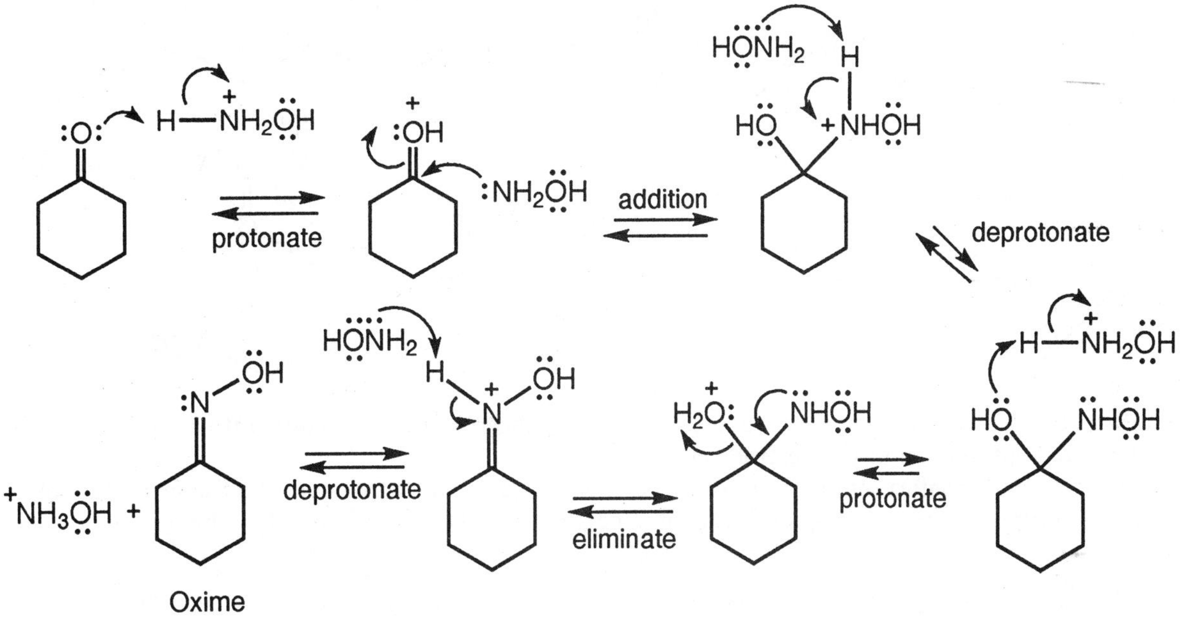

protonate addition deprotonate protonate eliminate deprotonate

Oxime

Problem 16.35 There are two possible directions from which the carbonyl group can be reduced. The product is a mixture of diastereomeric alcohols.

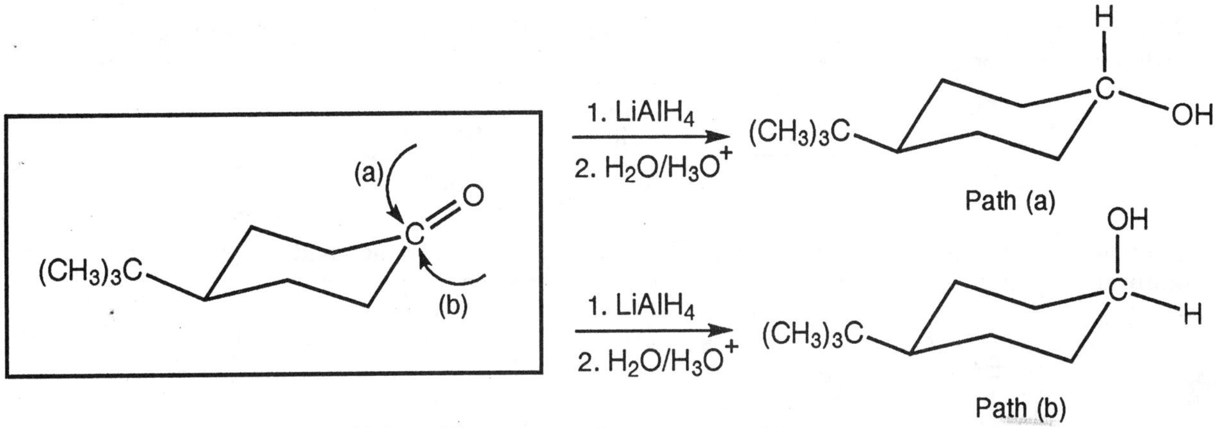

1. $LiAlH_4$
2. H_2O/H_3O^+

Path (a)

Path (b)

When the environment on each side of the carbonyl group is comparable, as in this ketone, the more stable product is normally the major product. In this case, the more stable product is the alcohol with the OH equatorial. In more complicated metal hydride reductions, the stereochemical outcome is more difficult to rationalize. Both the ease of approach of the reducing agent (steric approach control) and the stability of the product (product development control) are important.

Problem 16.36 In the first step the Grignard reagent is formed. Next an intramolecular addition takes place to give alkoxide **A**. Protonation gives the product.

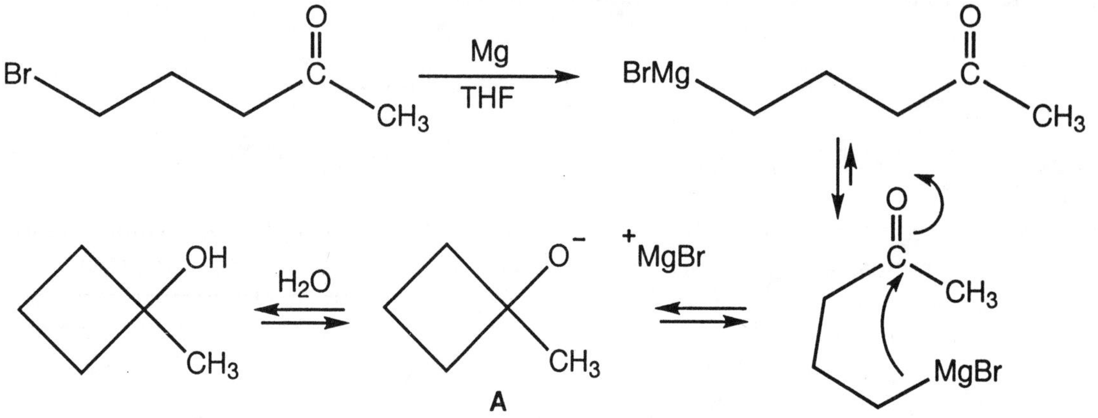

Problem 16.37 Most of these parts require you to transform isopropyl alcohol into a molecule that can incorporate the label in the desired way.
(a) For example, transformation of isopropyl alcohol into 2-bromopropane allows the formation of the alkyllithium reagent or Grignard reagent. Reaction with D_2O completes this part of the question.

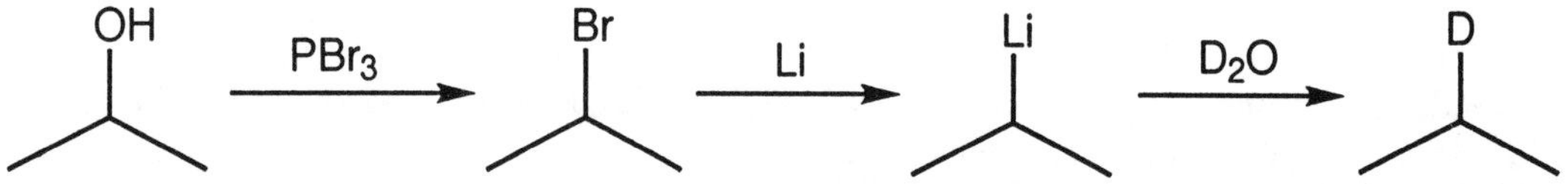

(b) Oxidation of isopropyl alcohol to acetone allows reduction in deuterated medium to give 2,2-dideuteriopropane:

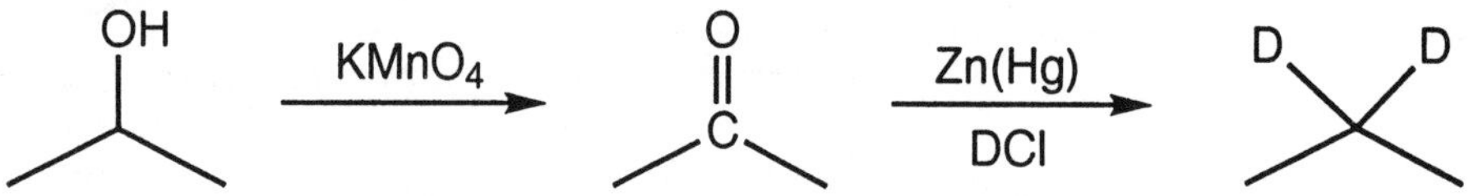

(c) For this part we need to make propene. We can use 2-bromopropane, made in part (a), and do an elimination reaction. "Hydrogenation" using D_2 makes the desired product.

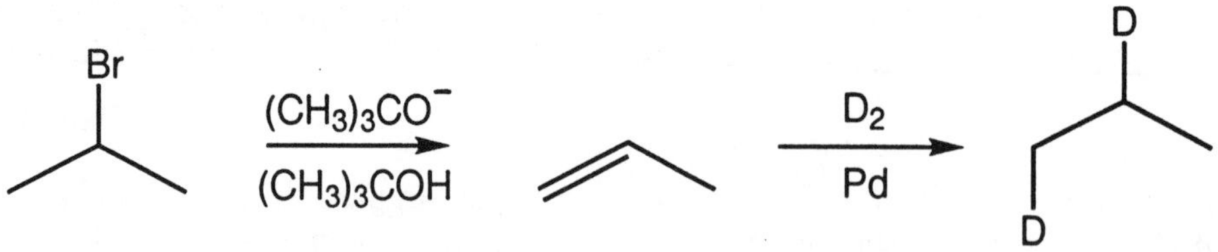

(d) Reduction of the acetone, made in part (b), with $NaBD_4$ or $LiAlD_4$, followed by hydrolysis will do the job:

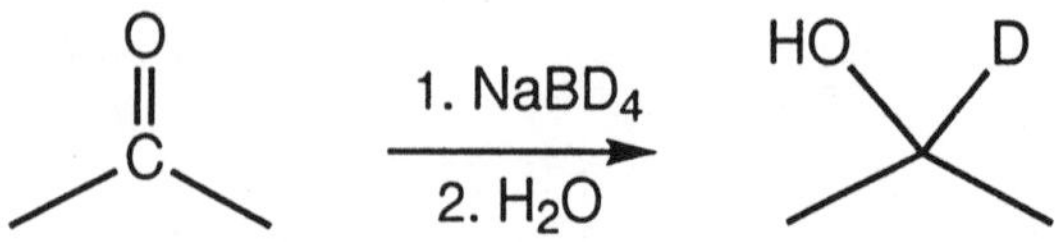

(e) Two propane backbones need to be sewn together here. A nice way is to allow 2-propyllithium (part a) to react with acetone (part b). The resulting alcohol can be transformed as in part a to the deuterated material.

1. Li (2-propyllithium); 2. H_2O → OH; 1. PBr_3; 2. Li; 3. D_2O → D

Problem 16.38 Here is another problem requiring sequential transformations of a simple starting material. In each case at least one carbon atom must be added to the three-carbon starting material. One carbon can be added by transforming propyl alcohol into an organometallic reagent, allowing it to react with formaldehyde to give a new alcohol. Manipulation of this alcohol (butyl alcohol) can lead to all the desired products.

(a) Reaction of butyl alcohol with PCl_3 (or many other reagents) will give the desired chloride.

OH —PBr_3→ Br —Li→ Li —1. $H_2C{=}O$ 2. H_2O→ CH_2OH —PCl_3→ Cl

(b) Butyl chloride can also be used as starting material for 1-cyanobutane.

Cl —^-CN (S_N2)→ CN + ^-Cl

(c) Butyl chloride can be made into the organolithium reagent and treated with water to make butane.

Cl —Li→ Li —H_2O→

(d) This slightly more complex transformation requires that we find a way to combine a three-carbon fragment with a four-carbon fragment. One way to do this is to allow propyllithium (a) to react with oxidized butyl alcohol, butanal.

OH —CrO_3 pyridine→ H–C=O —1. Li (propyllithium) 2. H_2O→ OH

Problem 16.39 This problem demands a series of relative complex alcohol syntheses. The last two parts require the conversion of an intermediate alcohol into a new material. In questions of this complexity, it is definitely worthwhile to work backward, starting with the desired product and asking, What is the immediate precursor of this molecule? (retrosynthetic analysis). In this way, we can always work back to a set of four-carbon alcohols. It is very likely that there will be more than one way to make each target. Notice that in this problem we make use of the special retrosynthetic arrow that points from the desired product to the immediate reactants needed to make it.

(a) In this problem, the target secondary alcohol is potentially available from two different sets of precursors, both involving addition of a Grignard reagent to an aldehyde.

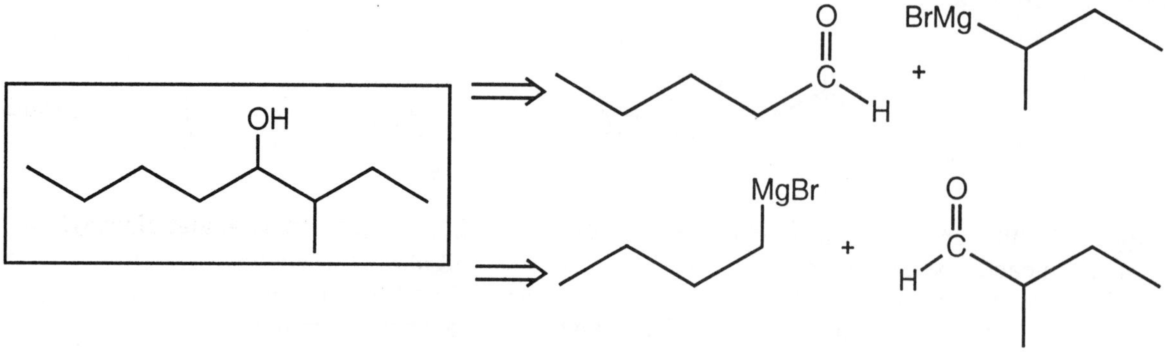

OH

⟹ O=C–H + BrMg

⟹ MgBr + H–C=O

Although both of these routes are feasible, we will concentrate on the first possibility. (You should work out the other synthesis). The required Grignard reagent for the first route is available from *sec*-butyl alcohol. The reaction partner, pentanal, can be made from butyl magnesium bromide and formaldehyde.

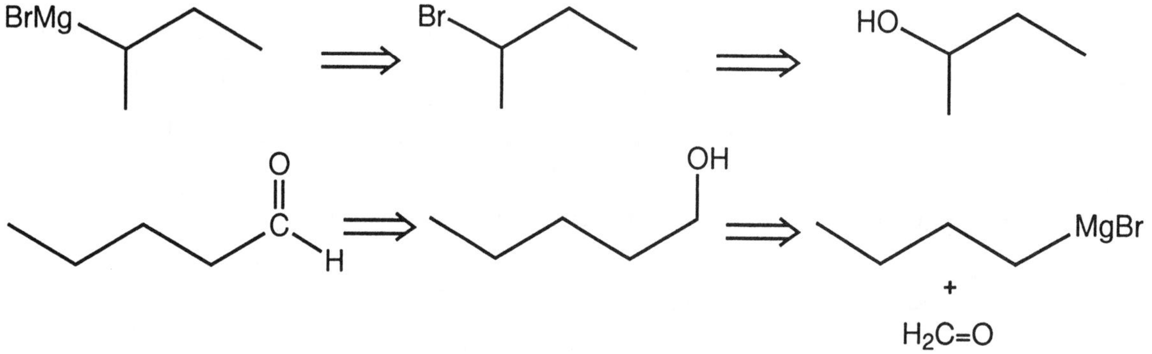

BrMg ⟹ Br ⟹ HO

O=C–H ⟹ OH ⟹ MgBr + $H_2C{=}O$

Butyl magnesium bromide is available from butyl alcohol, and formaldehyde can be made through oxidation of methyl alcohol. An outline of the full synthesis follows:

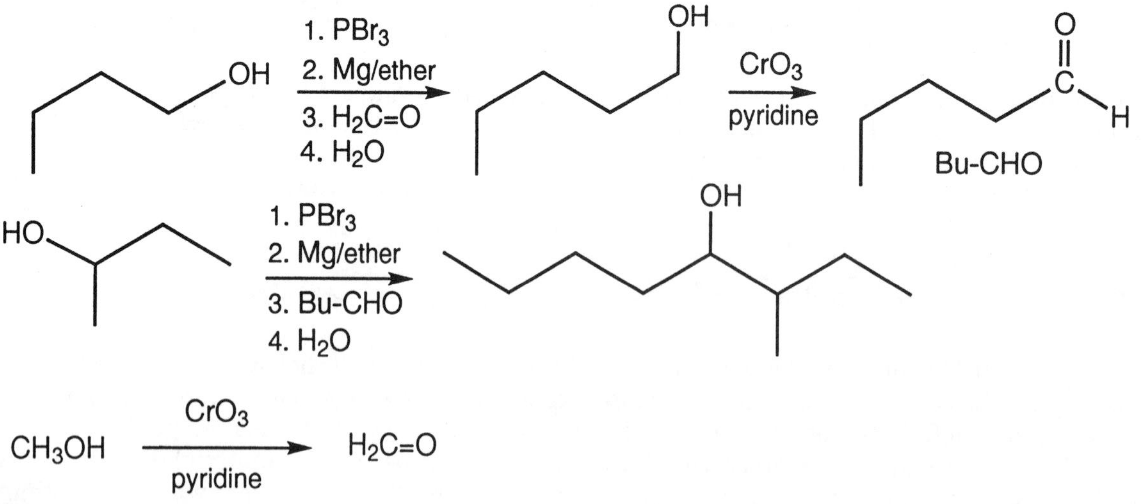

OH —(1. PBr_3; 2. Mg/ether; 3. $H_2C{=}O$; 4. H_2O)→ OH —(CrO_3, pyridine)→ O=C–H (Bu-CHO)

HO —(1. PBr_3; 2. Mg/ether; 3. Bu-CHO; 4. H_2O)→ OH

CH_3OH —(CrO_3, pyridine)→ $H_2C{=}O$

(b) This tertiary alcohol "deconstructs" in only one way. The immediate precursors must be dipropyl ketone and propyl magnesium bromide (propyl Grignard).

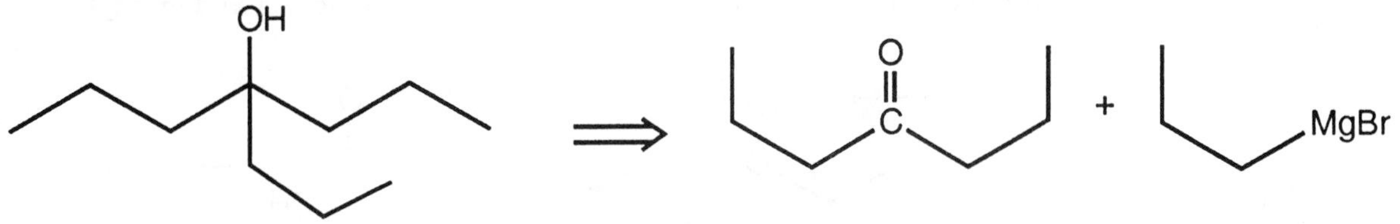

Propyl Grignard can be made the same way butyl Grignard was made in part a, and dipropyl ketone comes from 4-heptanol, itself formed from butanal and propyl Grignard.

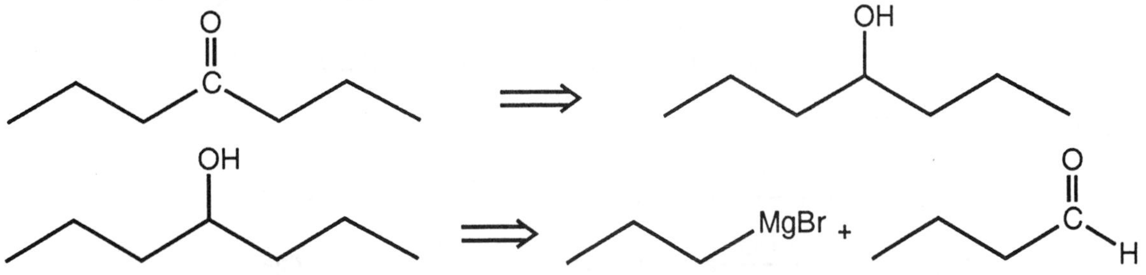

Butanal can be made from oxidation of butyl alcohol, so the synthesis looks as follows:

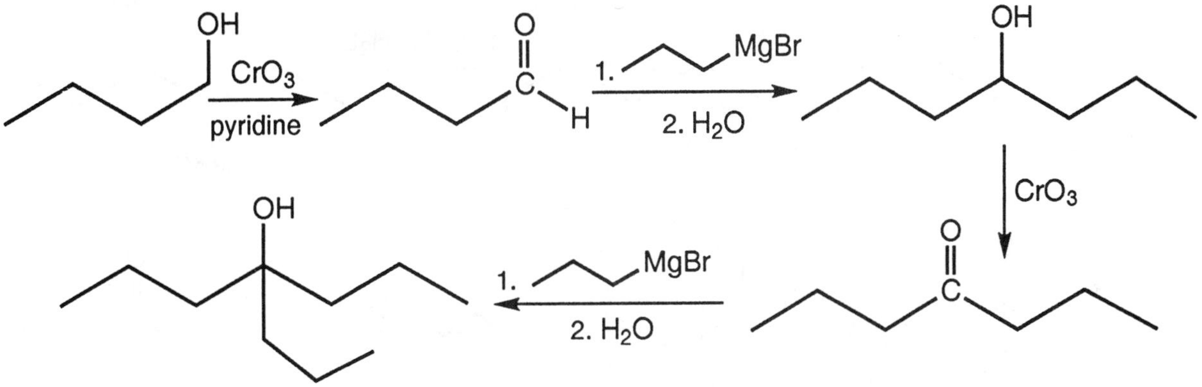

(c) The target tertiary bromide is available from the corresponding alcohol.

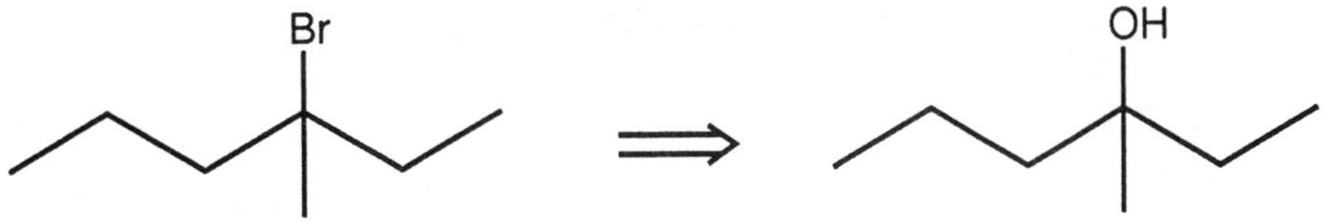

The alcohol, in turn, is available from three different sets of precursors, all combinations of organometallic reagents and ketones:

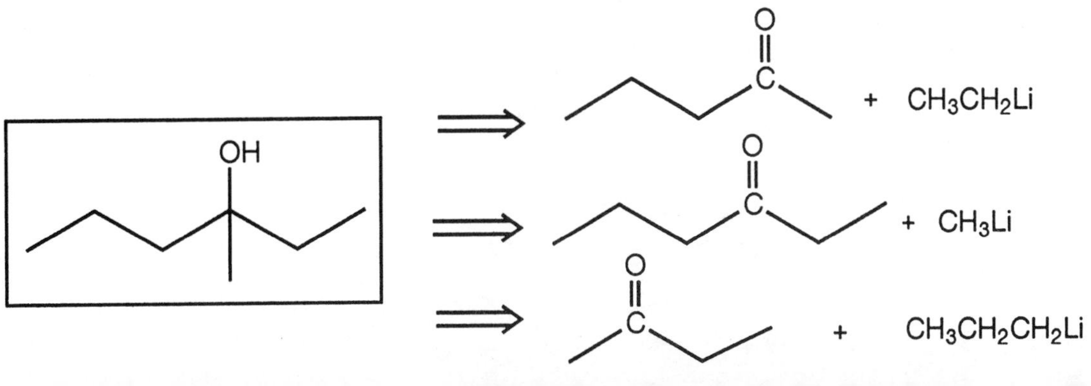

We will follow the second path, but you might work out the other two for practice. The required ketone, 3-hexanone, is available from butanal and ethyllithium. Butanal can be made from butyl alcohol, and ethyllithium and methyllithium derive, ultimately, from ethyl alcohol and methyl alcohol.

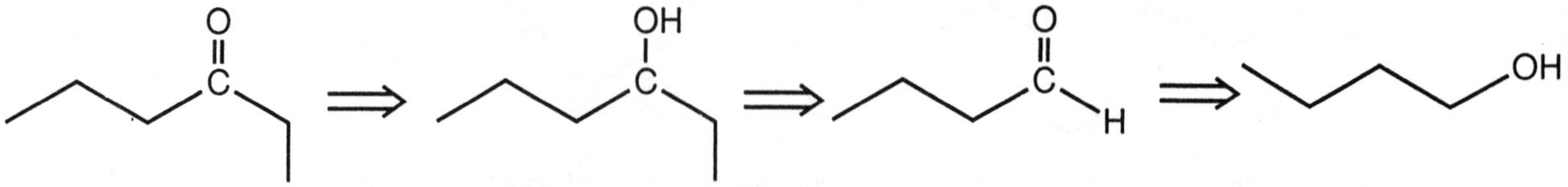

$CH_3CH_2Li \Rightarrow CH_3CH_2Br \Rightarrow CH_3CH_2OH$

$CH_3Li \Rightarrow CH_3Br \Rightarrow CH_3OH$

So, here is the synthesis:

$CH_3OH \xrightarrow{PBr_3} CH_3Br \xrightarrow{Li} CH_3Li$

$CH_3CH_2OH \xrightarrow{PBr_3} CH_3CH_2Br \xrightarrow{Li} CH_3CH_2Li$

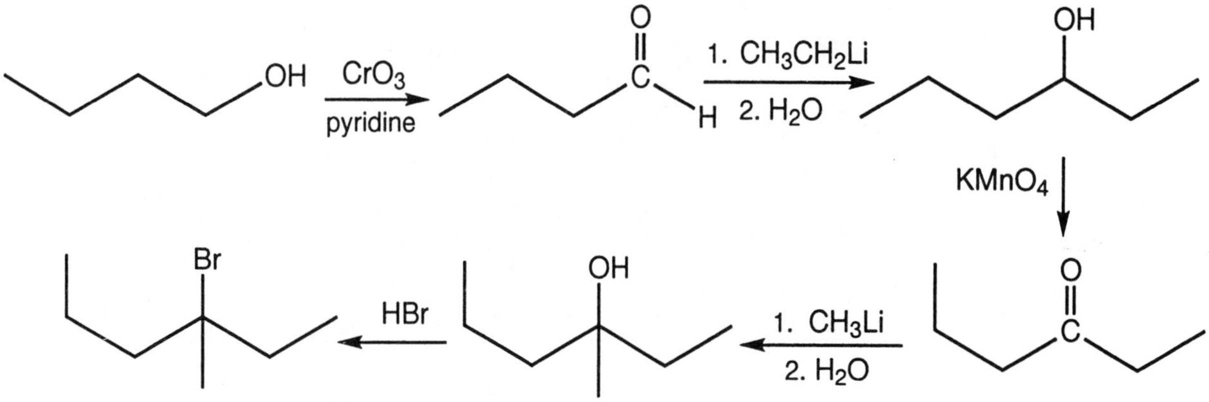

CrO3 / pyridine; 1. CH_3CH_2Li 2. H_2O; $KMnO_4$; 1. CH_3Li 2. H_2O; HBr

(d) This last problem is easy. The target molecule comes from dipropyl ketone, a molecule we made in (b).

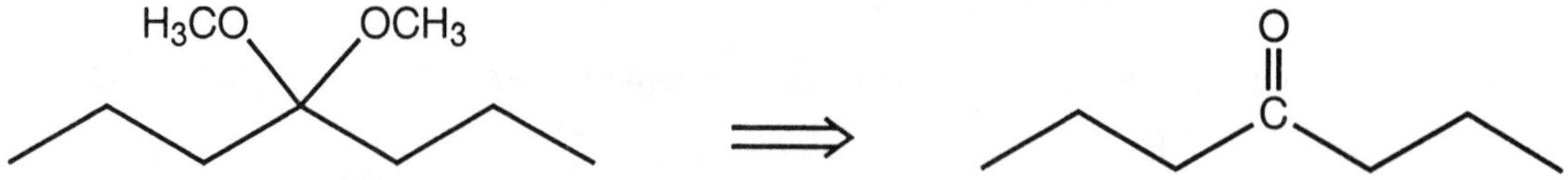

So, this is a one-step synthesis.

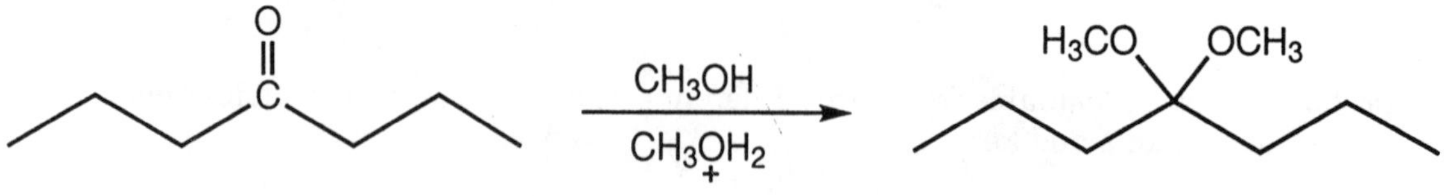

$\xrightarrow[CH_3\overset{+}{O}H_2]{CH_3OH}$

Problem 16.40 The difficulty here is recognition. The starting material is an acetal, and must reverse in acid to give the corresponding aldehyde. However, the standard hydrolysis can be short-circuited in this case because there in a nucleophile lurking within the molecule in the form of the nitrogen atom. This nitrogen can capture the Lewis acid formed by loss of ethyl alcohol. A series of proton transfers and loss of ethyl alcohol completes the reaction.

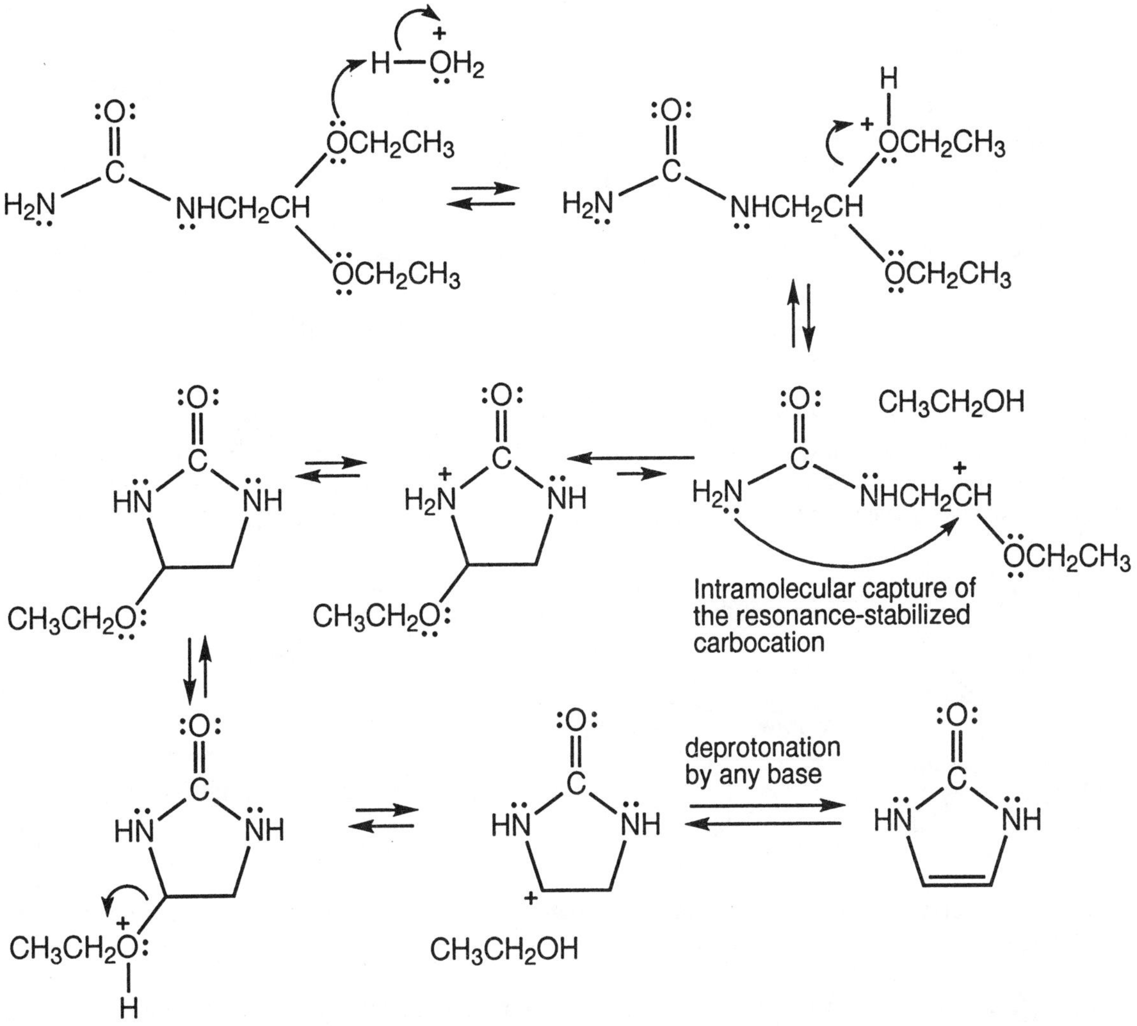

Intramolecular capture of the resonance-stabilized carbocation

deprotonation by any base

The process shown above looks complicated, but it is really not so bad. One critical insight is necessary, and the rest is nuts and bolts protonation-deprotonation and addition-elimination chemistry. How to see the crucial ring forming step? Analyze before you begin. Look at the product and compare it to the starting material. What needs to be accomplished? In this case, it is very likely that the series, N—CO—N in the five-membered ring product (arrows) is formed from the same N—CO—N series in the starting material. If this is so, it becomes easy to see where the connection must be made. Then the task is to see how to do the connecting.

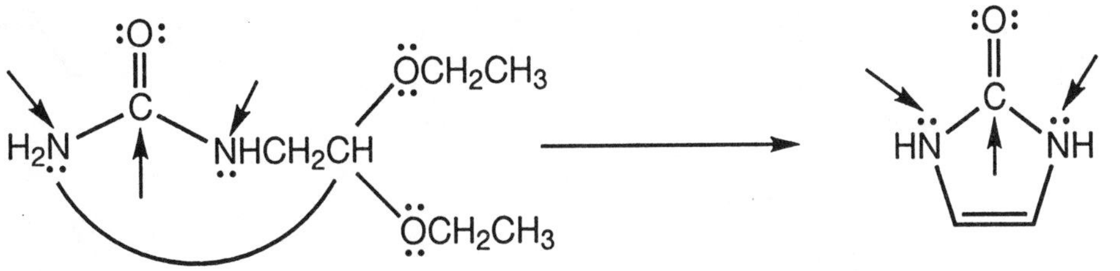

If this N-CO-N sequence becomes the N-CO-N sequence in the product (very likely), the points of ring formation are identified. That makes the problem much easier.

Problem 16.41 Here we have a seven-membered acetal converted into a five-membered acetal. This problem is tricky because the O—C$(CH_3)_2$—O sequence in the starting material is *not* the O—C$(CH_3)_2$—O sequence in the product. The seven-membered acetal opens and an intermediate is captured by the OH within the same molecule. In this sense, this problem is like the last one. Start by working out the mechanism for opening the acetal. This reaction involves only protonation of a ring O and opening to give a resonance-stabilized carbocation.

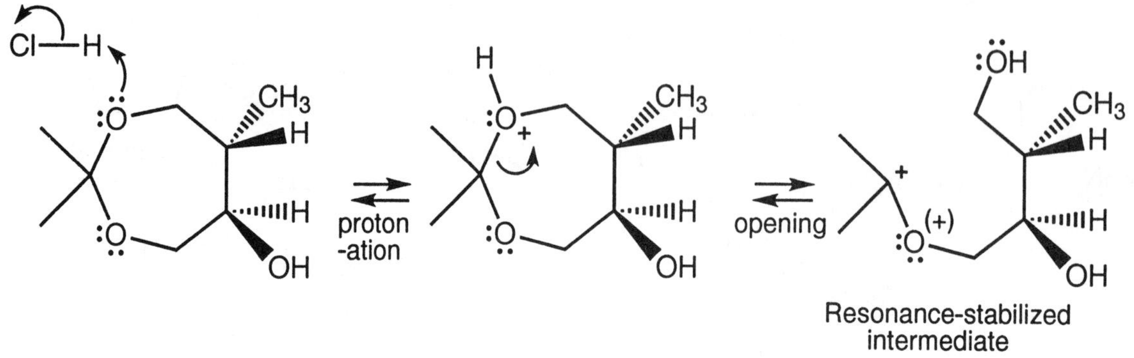

Cl—H CH3 H H OH proton -ation H CH3 H H OH opening :OH CH3 H H OH

Resonance-stabilized intermediate

Now the "upper right" of the product molecule is in place. That's the kind of thing that makes a problem solver feel he or she is on the right track. If parts of the molecule begin to fall into place as you work through a potential answer, that's a good sign. In this case, we need only close up the five-membered ring using the other OH, deprotonate, and we are done. It helps greatly to redraw the molecule first:

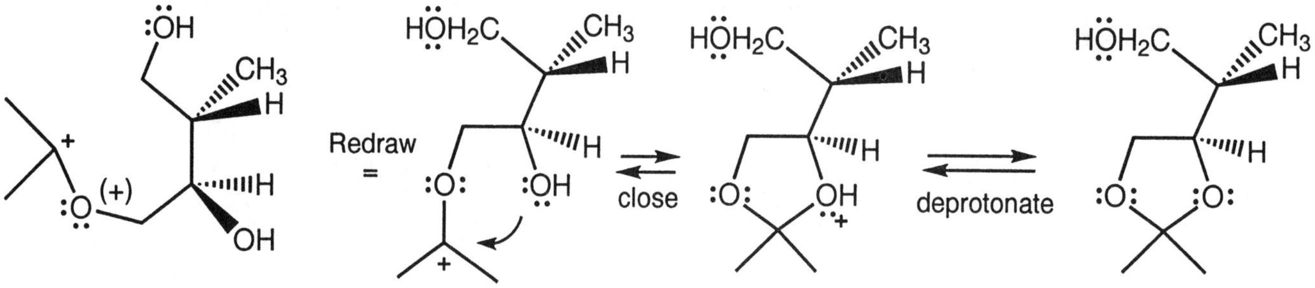

:OH CH3 H H OH Redraw = HOH2C CH3 H H :O: :OH close HOH2C CH3 H H :O: OH deprotonate HOH2C CH3 H H :O: :O:

Why does the reaction go? There is substantially more strain in a seven-membered ring than in a five-membered ring.

Problem 16.42 Here is yet another problem involving an acetal. This problem becomes much easier if we do a little analysis before we start. The question to ask is, "What must be the precursor to the final acetal?" This is a "retrosynthetic" question. Acetals are produced from carbonyl compounds and alcohols under acidic conditions. The alcohol is there as one starting material, and the reaction conditions are surely acidic, so this seems a promising line of thought. The final acetal could be formed from the starting diol and the aldehyde shown (**A**).

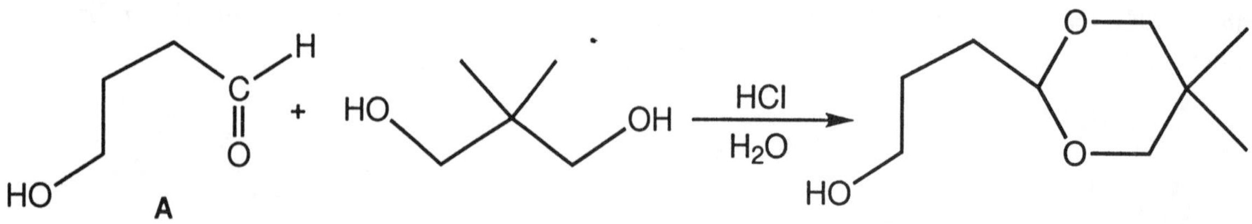

HO C=O H **A** + HO OH $\xrightarrow[H_2O]{HCl}$ HO O O

Now the problem is to find a way to make aldehyde **A** from the five-membered ring in the problem. That's not so hard. Protonation to give the resonance-stabilized carbocation is followed by addition of water to give a hemiacetal. Opening of the hemiacetal gives exactly the aldehyde we need.

:ÖH2 H3Ö: (+) :ÖH2 + addition deproton-ation Hemiacetal H—Cl B H (resonance-stabilized carbocation)

H—Cl H :ÖH2 HÖ :O: H3Ö: ÖH protonation deprotonation and ring opening C H Hemiacetal Desired aldehyde

Now, straightforward acetal formation makes the product (mechanism not shown).

H C O + HO OH HCl H2O O O HO HO

There is another, more imaginative (and probably correct) version of the mechanism. Suppose the resonance-stabilized cation **B** is captured, not by water, but by the alcohol? Now a sequence of protonations, deprotonations, and ring openings and closings leads directly to the product.

(+) :ÖH :ÖH + add H :ÖH deprotonate :ÖH B H

(continued next page)

protonate

(note resonance stabilization)

open

close

deprotonate

How is one to chose between these two reasonable mechanisms? The answer is that we don't really have to. Both are probably taking place, and the mix will depend on reaction conditions and, especially, on the relative concentrations of water and the diol.

Problem 16.43 The hint tells you to think first about the reaction of the alcohol with the oxidizing agent. Oxidation of this primary alcohol proceeds first to the aldehyde. We know that aldehydes react in acid with alcohols to form hemiacetals, *that are themselves substituted alcohols.* Further oxidation of the hemiacetal gives the ester.

$Na_2Cr_2O_7$, H_2SO_4/H_2O, 35 °C

H_2SO_4/H_2O

$Na_2Cr_2O_7$, H_2SO_4/H_2O

Ester

Hemiacetal (an alcohol)

Problem 16.44 The formula shows that two nitrogens are incorporated into this product. Like hemiacetals, carbinolamines can react further with amines to give diamino compounds called aminals. These do not generally predominate at equilibrium, but they are partners in the interconverting mixture of compounds that eventually leads to the imine, the thermodynamically favored structure.

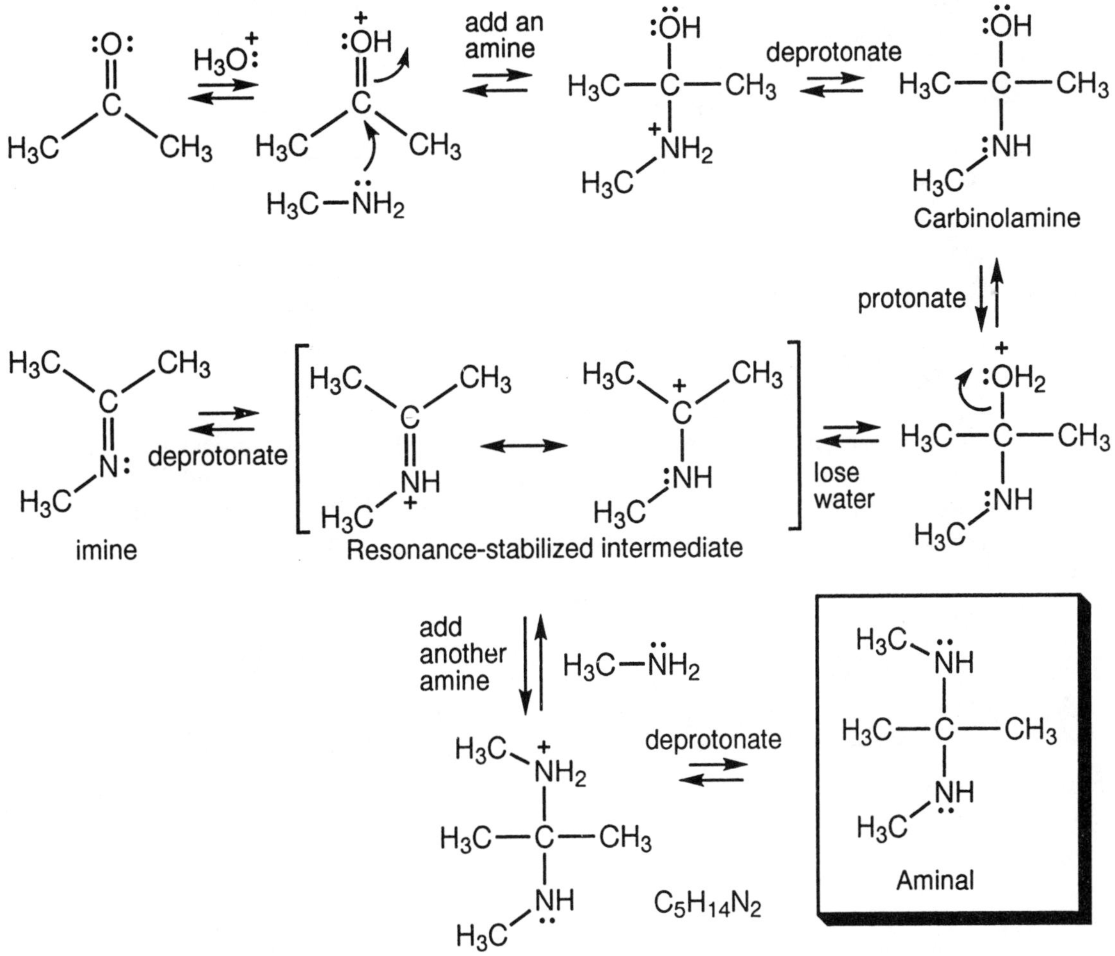

H3C, CH3, :O:, C, H3O+, :OH+, H3C—NH2, add an amine, :OH, H3C—C—CH3, NH2+, deprotonate, :NH, Carbinolamine, protonate, :OH2+, lose water, Resonance-stabilized intermediate, deprotonate, N:, imine, NH+, add another amine, $C_5H_{14}N_2$, Aminal

Problem 16.45 The reaction of propionaldehyde and morpholine sets up a complex equilibrium in which the enamine, a carbinolamine, and an aminal are all involved. The analysis by ^1H NMR spectroscopy shows that little enamine is present. Nonetheless, as enamine is used up in the Diels-Alder addition, more is formed. Eventually, all of the components in the equilibrium can be converted into enamine and then to Diels-Alder adducts.

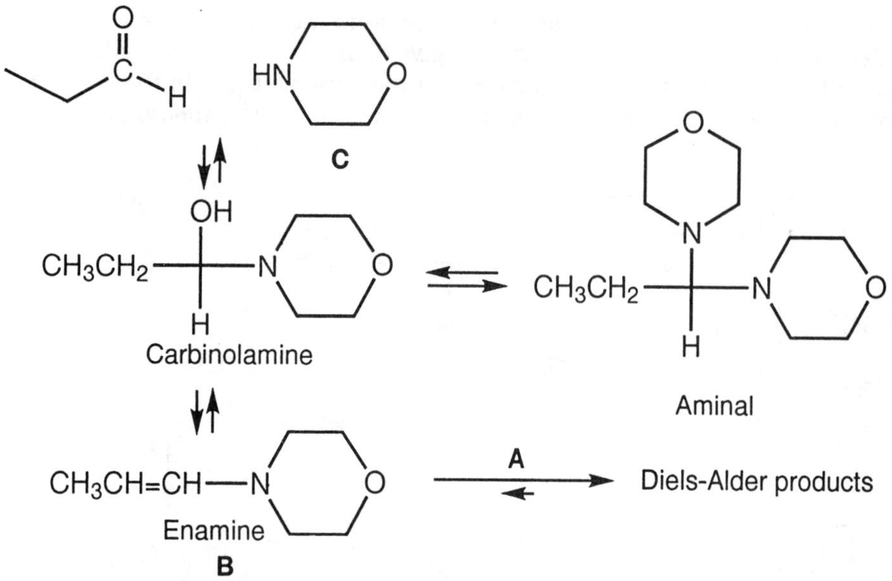

Carbinolamine

Aminal

Enamine

Diels-Alder products

Problem 16.46 This reaction sequence is an example of a synthesis that requires the use of a protecting group. If the Grignard reagent of *p*-bromoacetophenone were prepared directly, it would add to the carbonyl group of another molecule of starting material. To avoid this complication, the carbonyl group of *p*-bromoacetophenone is first protected as a cyclic acetal to give **A**. The Grignard reagent **B** can then be prepared safely.

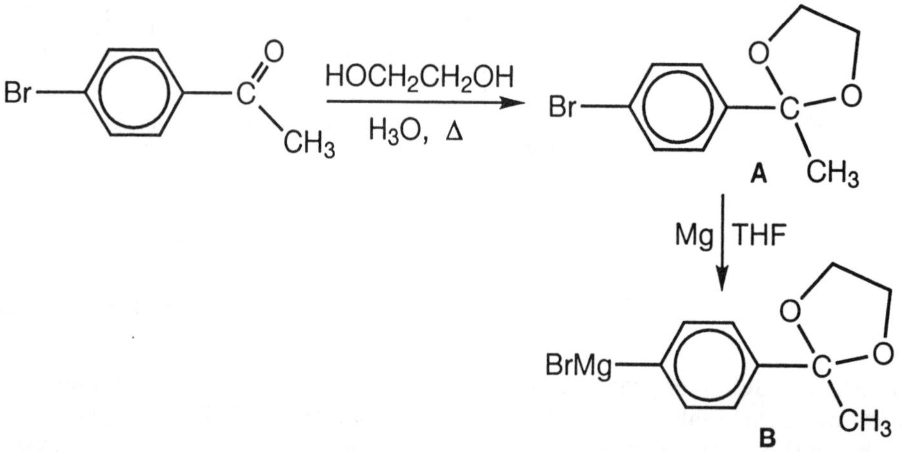

Now reaction with *p*-methoxybenzophenone affords **C** after mild hydrolysis. Finally, the carbonyl group can be regenerated upon more vigorous hydrolysis to yield ketone **D**.

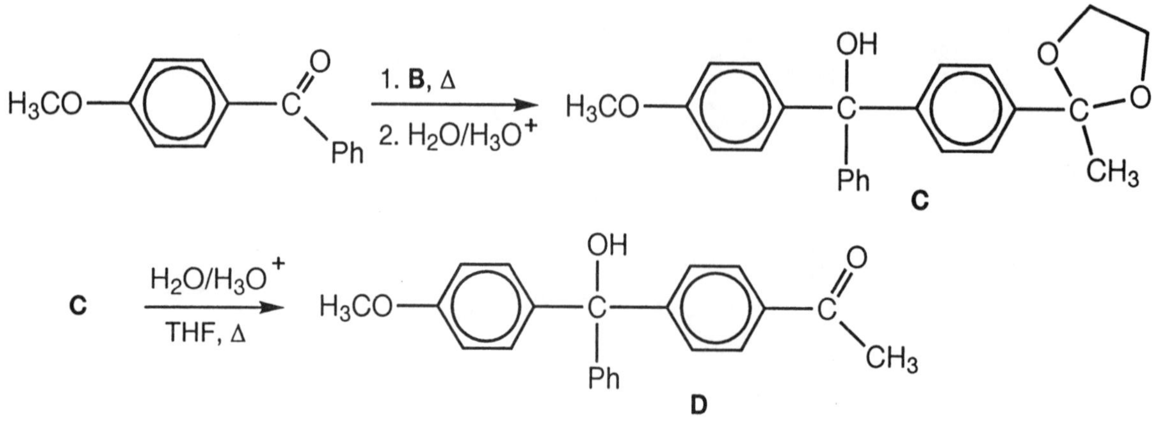

The structure of **D** is supported by the spectral data. The parent ion in the mass spectrum at m/z = 332 is consistent with a molecular formula of $C_{22}H_{20}O_3$. The presence of an O–H stretch at 3455 cm^{-1} and a conjugated carbonyl band at 1655 cm^{-1} are also consistent with the proposed structure. The ^1H and ^{13}C NMR data are also consonant with the assignment. Pertinent assignments are indicated below with the ^{13}C NMR chemical shifts in parentheses.

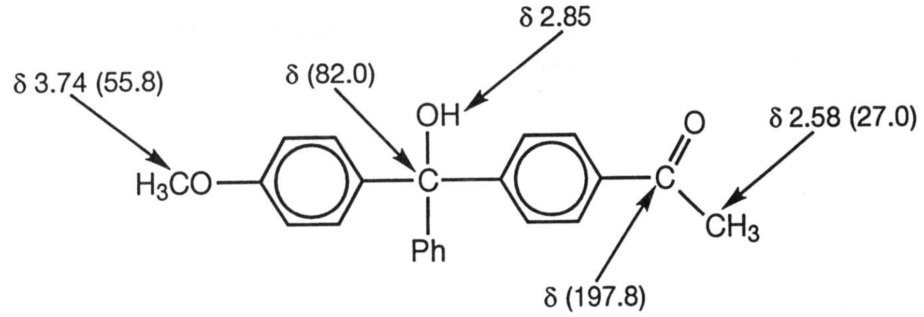

δ 2.85
δ 3.74 (55.8)
δ (82.0)
OH
O
δ 2.58 (27.0)
H_3CO
C
C
CH_3
Ph
δ (197.8)

Problem 16.47 This reaction sequence is an example of the Wittig reaction, followed by hydrolysis of the enol ether (**C**), to give 2-ethylbutanal (**D**). This sequence represents a convenient method for the following conversion: R—CO—R′ ⟶ RR′CH—CHO.

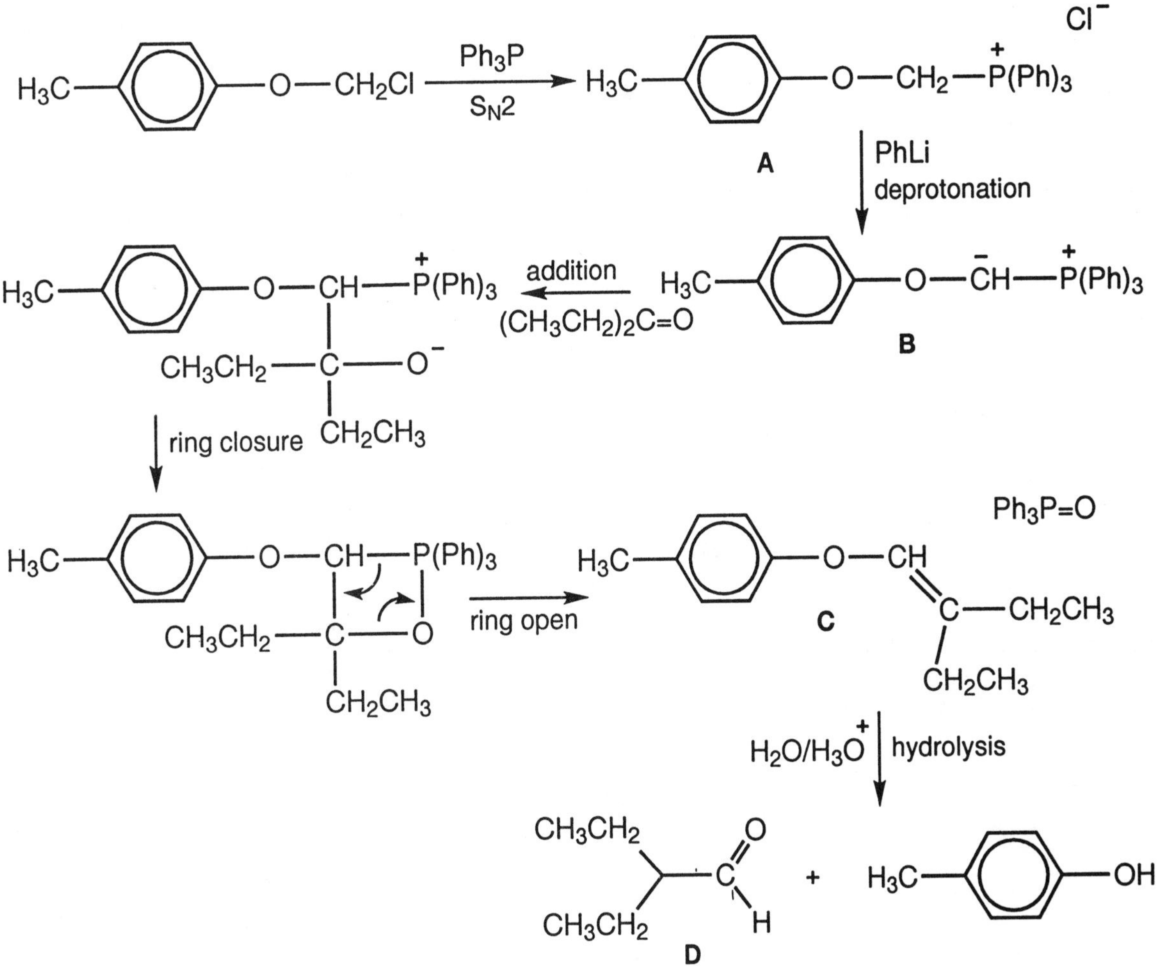

H_3C–C_6H_4–O—CH_2Cl —(Ph_3P, S_N2)→ H_3C–C_6H_4–O—CH_2—$\overset{+}{P}(Ph)_3$ Cl^- **A**

A —(PhLi, deprotonation)→ H_3C–C_6H_4–O—$\overset{-}{C}H$—$\overset{+}{P}(Ph)_3$ **B**

B —(addition, $(CH_3CH_2)_2C{=}O$)→ H_3C–C_6H_4–O—CH—$\overset{+}{P}(Ph)_3$ / CH_3CH_2—C—O^- / CH_2CH_3

ring closure

H_3C–C_6H_4–O—CH—P(Ph)$_3$ / CH_3CH_2—C—O / CH_2CH_3 —(ring open)→ H_3C–C_6H_4–O—CH=C—CH_2CH_3 / CH_2CH_3 **C** + $Ph_3P{=}O$

C —(H_2O/H_3O^+, hydrolysis)→ $(CH_3CH_2)_2CH$—CHO **D** + H_3C–C_6H_4–OH

The spectral data for **D** are consistent with the proposed structure. The IR spectrum displays the characteristic aldehyde C—H stretching doublet at 2817 and 2717 cm^{-1}, as well as the intense C=O stretch at 1730 cm^{-1}. The presence of the low-field doublet at δ 9.51 ppm in the ^{1}H NMR spectrum is also indicative of an aldehyde C—H. Note the small coupling to the adjacent methine hydrogen. The methyl groups appear as a triplet centered at δ 0.92, whereas the methylene and methine hydrogens appear as an overlapping multiplet at δ 1.2-2.3 ppm. The ^{13}C NMR spectrum is also consistent with the proposed structure. Assignments are shown below. Note especially the low-field chemical shift for the aldehyde carbonyl carbon.

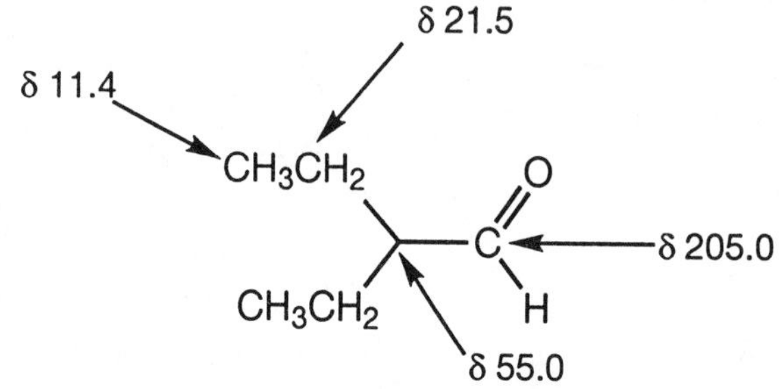

δ 21.5
δ 11.4
CH_3CH_2
O
C ← δ 205.0
CH_3CH_2
H
δ 55.0

Chapter 17 Outline

17

The Chemistry of Alcohols Revisited and Extended: Glycols, Ethers, and Related Sulfur Compounds

This chapter provides breathing space between the concept-heavy chapters on carbonyl chemistry on either side of it. It summarizes what we know of alcohols and makes a few extensions into the chemistry of related diols, ethers, and sulfur compounds. The following problems are largely "nuts and bolts" exercises, and give you a chance to practice both old and new syntheses and reactions of alcohols and their relatives. It also provides an important chance to reacquaint yourself with the fundamental building block reactions from earlier chapters, the S_N1, S_N2, E1, and E2 reactions. As usual, there are a few taxing problems toward the end of the series.

Problem 17.01

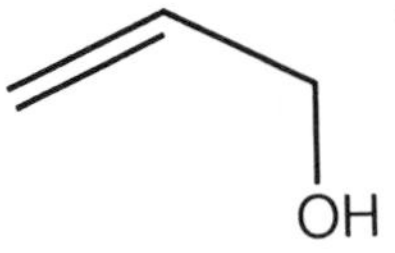

OH
2-Propen-1-ol
(allyl alcohol)

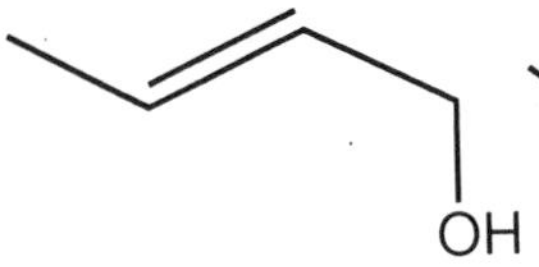

OH
trans-2-Buten-1-ol
(*E*)-2-buten-1-ol

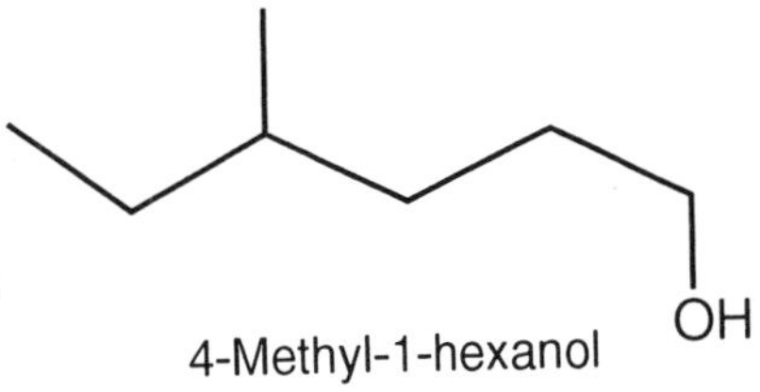

OH
4-Methyl-1-hexanol

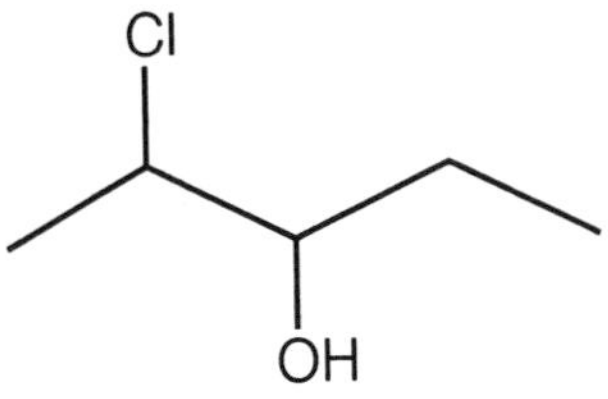

Cl
OH
2-Chloro-3-pentanol

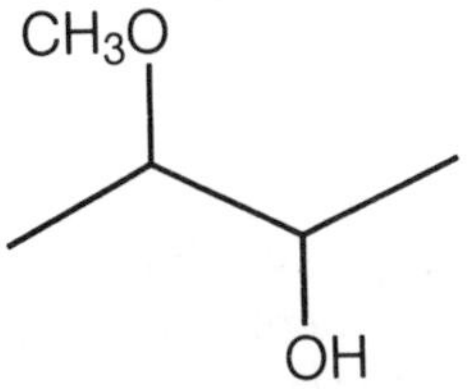

CH_3O
OH
3-Methoxy-2-butanol
(OH gets precedence)

HS
OH
2-Mercaptoethanol
(OH gets precedence
over SH)

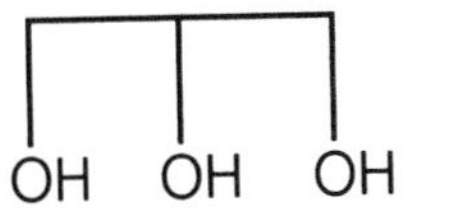

OH OH OH
1,2,3-Propanetriol
(1,2,3-trihydroxypropane)

Problem 17.2 Angle and hybridization are intimately connected; if you know one, you know the other. The angle in methyl alcohol is almost exactly the tetrahedral angle, and oxygen must be almost exactly hybridized sp^3.

Problem 17.4 It is more difficult for *tert*-butyl alcohol to form hydrogen-bonded dimers and oligomers (collections of several molecules) than it is for methyl alcohol. The three methyl groups take up much more space than do three hydrogens and this steric effect makes hydrogen bonding between molecules more difficult. Accordingly, the sharp peak of the monomeric alcohol is more apparent in the spectrum of *tert*-butyl alcohol than it is in the spectrum of methyl alcohol.

Problem 17.5 In the "syn-unsaturated" compound, *intramolecular* hydrogen bonding is possible between the double bond (acting as Brønsted base) and the hydroxyl hydrogen. In neither of the other two compounds is this possible. These two compounds, the "anti-unsaturated" and "saturated," participate in the usual *intermolecular* hydrogen bonding, but cannot form *intramolecular* hydrogen bonds. As usual, dilution makes intermolecular hydrogen bonding more difficult, but has no effect on intramolecular hydrogen bonding.

In this syn-unsaturated compound, intramolecular hydrogen bonding is possible

anti-Unsaturated

Saturated

Only intermolecular hydrogen bonding is possible for these molecules

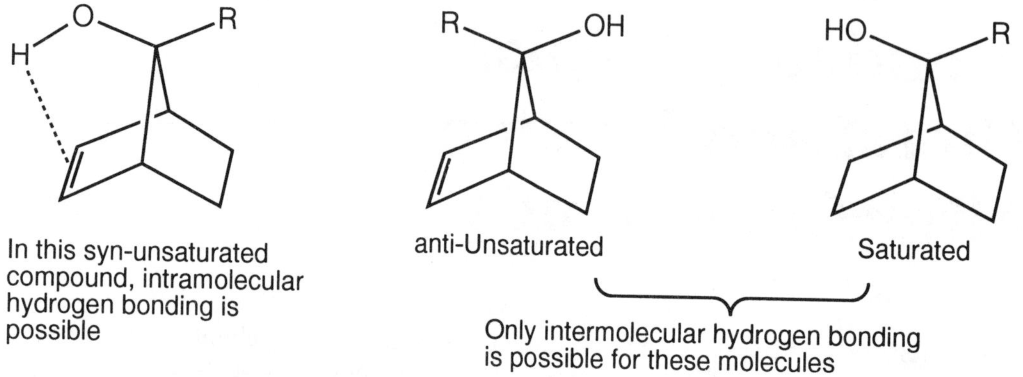

Problem 17.6 This reaction is a standard S_N1 solvolysis. The oxonium ion appears as the second intermediate in the process.

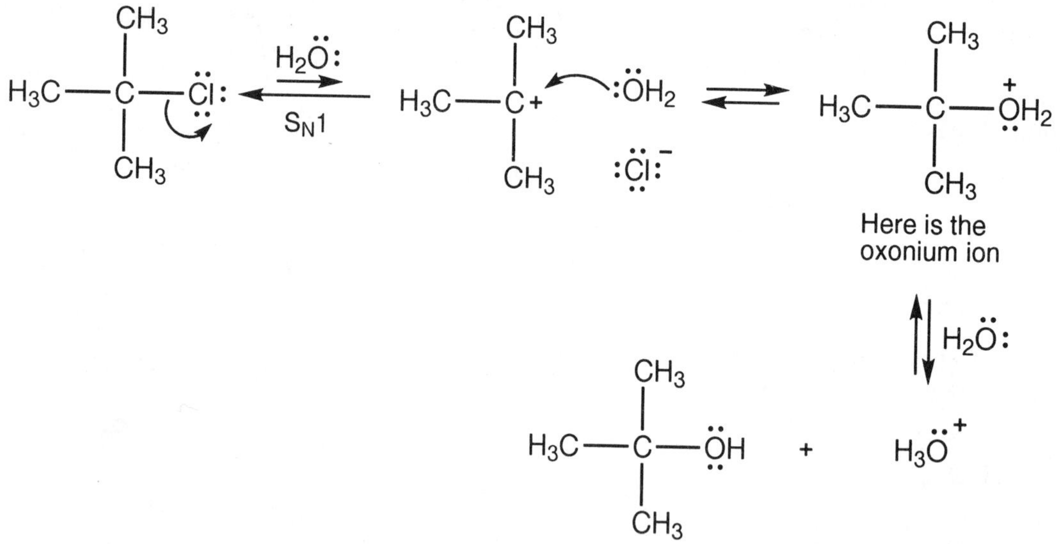

Here is the oxonium ion

Problem 17.10 The sulfonate anions are well resonance stabilized as the negative charge is shared by three oxygen atoms. The transition state for S_N1 ionization or S_N2 displacement will have a partial negative charge (δ^-) developed on the sulfonate oxygen and the transition state will be stabilized by delocalization.

R—O—SO_2—Ph ⟶ $R^{\delta+}$·········$O^{\delta-}$—SO_2—Ph ⟶ R^+ + [resonance forms of $PhSO_3^-$]

Transition state for this S_N1 reaction

R—O—SO_2—Ph ⟶ $Nu^{\delta-}$:·····$R^{\delta+}$·········$O^{\delta-}$—SO_2—Ph ⟶ Nu–R + [resonance forms of $PhSO_3^-$]

Transition state for this S_N2 reaction

Problem 17.11 The first of the reactions in Figure 17.24 is a simple S_N2 displacement on a methyl group. If there are no nucleophiles other than ethyl alcohol there can be no side products. In the second reaction, the E2 reaction can compete, and one would expect some ethylene to be formed. In the third reaction, both S_N1 and S_N2 mechanisms are possible. Regardless of mechanism, one would expect some propylene to be formed by either an E1 or E2 reaction. In the fourth reaction the mechanism will be S_N1/E1 and substantial amounts of isobutylene must accompany the product of S_N1 reaction, *tert*-butyl ethyl ether.

$$CH_3OTs \xrightarrow{\text{EtOH, 75 °C}} CH_3OCH_2CH_3 + HOTs$$

$$CH_3CH_2OTs \xrightarrow{\text{EtOH, 75 °C}} CH_3CH_2OCH_2CH_3 + CH_2{=}CH_2 + HOTs$$

$$(CH_3)_2CHOTs \xrightarrow{\text{EtOH, 75 °C}} (CH_3)_2CHOCH_2CH_3 + CH_3CH{=}CH_2 + HOTs$$

$$(CH_3)_3C\text{–}Cl \xrightarrow{\text{EtOH, 75 °C}} \underset{(56\%)}{(CH_3)_3COCH_2CH_3} + \underset{(44\%)}{(CH_3)_2C{=}CH_2} + HCl$$

Problem 17.12 This problem is nothing more than a simple S_N2 displacement of a good leaving group by an alkoxide.

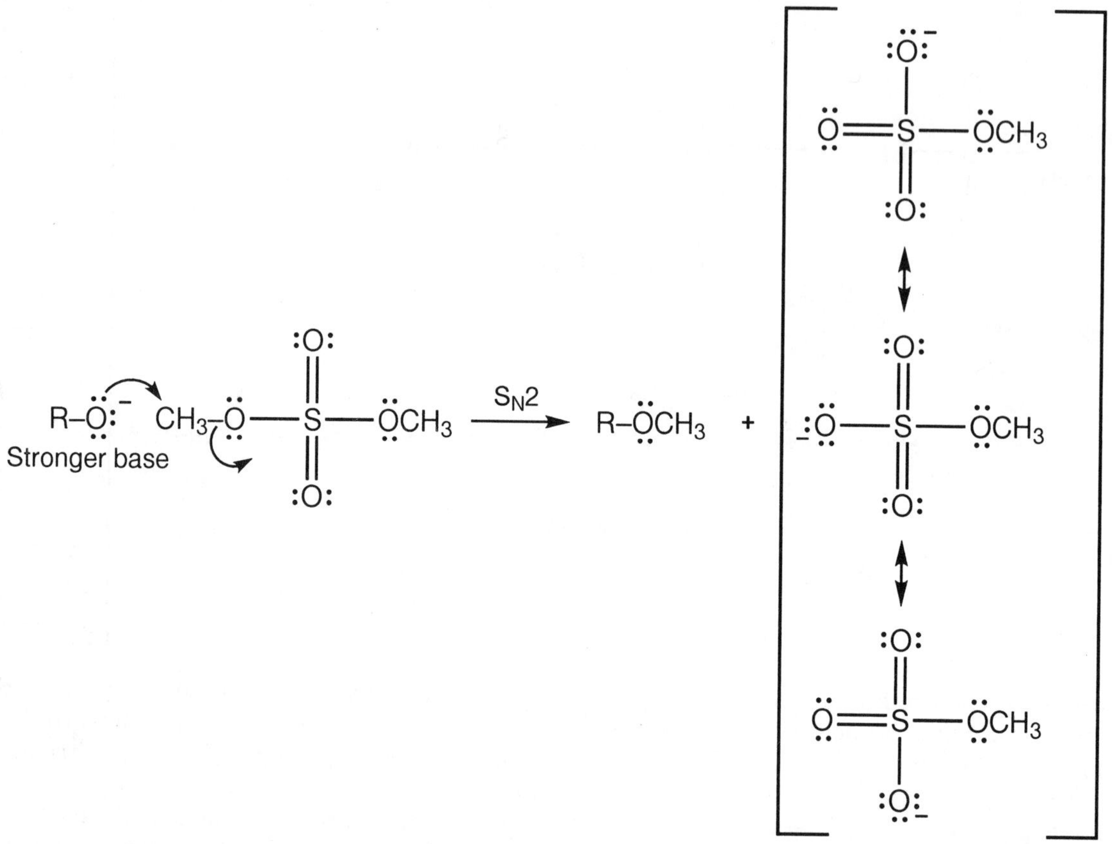

R–O:$^-$ CH$_3$–O–SO$_2$–OCH$_3$ $\xrightarrow{S_N2}$ R–OCH$_3$ + $^-$O–SO$_2$–OCH$_3$

Stronger base

Problem 17.13 Like the carbon–oxygen double bond, the sulfur–oxygen double bond is attacked by nucleophiles to give an intermediate **A**. Intermediate **A** is deprotonated and the good leaving group chloride is lost. The result is a chlorosulfite ester.

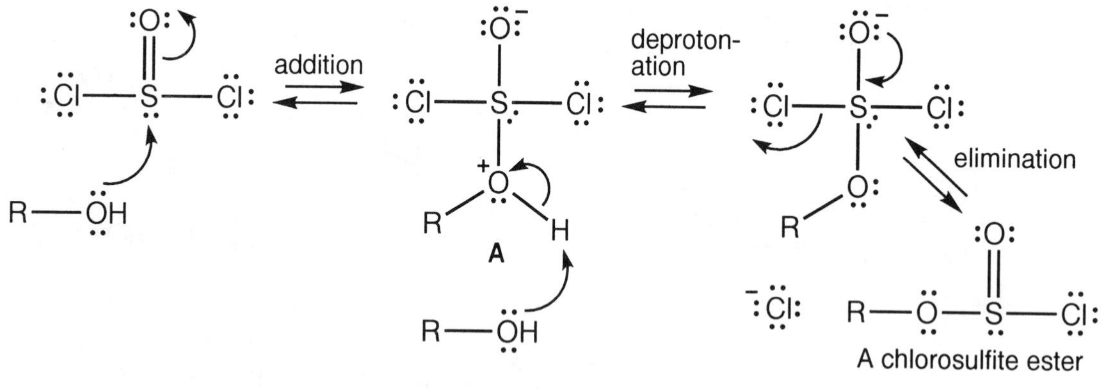

addition

deproton-
ation

elimination

A

A chlorosulfite ester

Problem 17.14 Once again, the reaction proceeds by transforming the poor leaving group OH into a better one. Reaction of the alcohol with PBr_3 leads to an intermediate that can either go on to a bromophosphite ester, or undergo S_N2 displacement to give the alkyl bromide and $HOPBr_2$.

R—OH + :PBr_3 → $R-\overset{+}{O}H-PBr_2$ ⇌ $R-O-PBr_2$ + HBr

A bromophosphite ester

:Br:⁻ ; S_N2

R–Br: + $HOPBr_2$

Problem 17.15 Review! In hydroboration–oxidation, the OH enters at the position occupied by boron. In hydroboration, the transition state for the addition of BH_3 to the double bond has a partial positive charge on carbon, and the boron inevitably becomes attached to the less substituted position. The end result is anti-Markovnikov addition.

H—BH_2

δ^- H······BH_2 δ^+

Transition state

H BH_2 (= RBH_2)

repeat twice

R_3B

NaOH / HOOH

3 H OH

The OH becomes attached to the carbon bearing the boron; the result is anti-Markovnikov addition of water to the alkene

In oxymercuration–reduction, the intermediate mercurinium ion is opened by water by addition at the more substituted carbon to break the weaker carbon–mercury bond. The water enters at the more substituted position. Reduction by $NaBH_4$ results in the more substituted alcohol.

$Hg(OAc)_2$

$\overset{+}{H}gOAc$:$\ddot{O}H_2$

HgOAc $^+OH_2$

HgOAc :OH

$NaBH_4$ / NaOH

Markovnikov addition :OH

Problem 17.16 These substitution reactions will be complicated by eliminations to give alkenes. No elimination is possible in the methyl case, but alkenes can be formed in the second two examples. Both *cis*- and *trans*-2-octene will be formed, and, of course, no stereoisomers are possible for 1-octene or styrene.

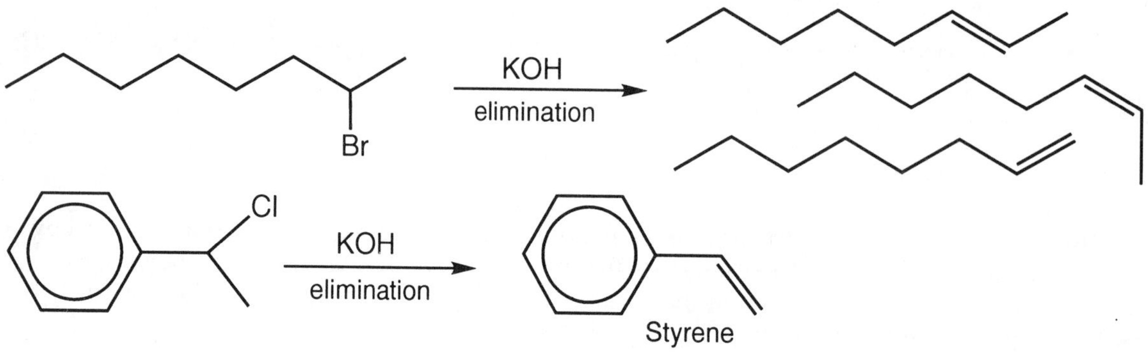

KOH
elimination

Br

Cl

KOH
elimination

Styrene

Problem 17.17 Several ways are possible. In any case, it is necessary to change the leaving group from OH to something that can be displaced by a nucleophile. The traditional ways are protonation or formation of a sulfonate ester (a tosylate, OTs = OSO_2PhCH_3).

$$CH_3CH_2{-}OH \xrightarrow{\text{H–I}} CH_3CH_2{-}\overset{+}{O}H_2 \xrightarrow[S_N2]{I^-} I{-}CH_2CH_3 + OH_2$$

$$CH_3CH_2{-}OH \xrightarrow[\text{pyridine}]{\text{Ts–Cl}} CH_3CH_2{-}OTs \xrightarrow[S_N2]{KI} I{-}CH_2CH_3 + KOTs$$

Problem 17.18 Epoxides contain three-membered rings, and are therefore severely strained. This strain raises their energy and increases reactivity.

Problem 17.19 Oxidation by basic permanganate leads to 1,2-diols through an overall syn addition. *cis*-2-Butene leads to a meso diol, and *trans*-2-butene to a racemic mixture of 1,2-diols. The figure sketches out the formation of the cyclic ester from both the top and bottom of the alkene, and the subsequent opening to the diol.

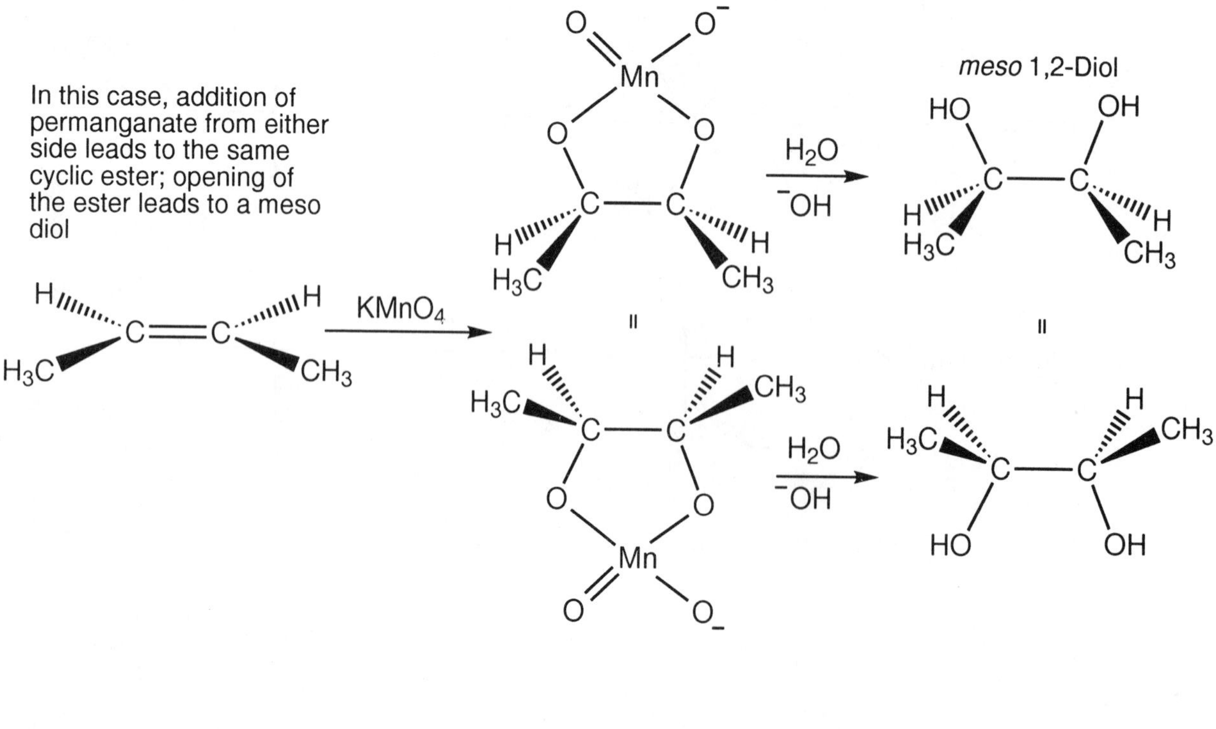

In this case, addition of permanganate from either side leads to the same cyclic ester; opening of the ester leads to a meso diol

$KMnO_4$

H_2O
^-OH

meso 1,2-Diol

The products are shown for simplicity in eclipsed forms. Of course, these are not the energy minimum forms. The staggered form, the real energy minimum of the meso compound, is shown here.

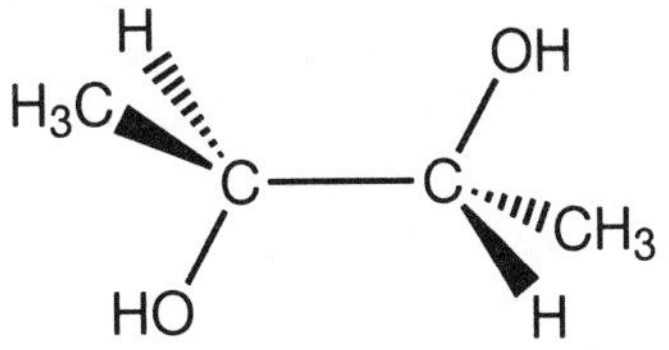

Here is the reaction with *trans*-2-butene:

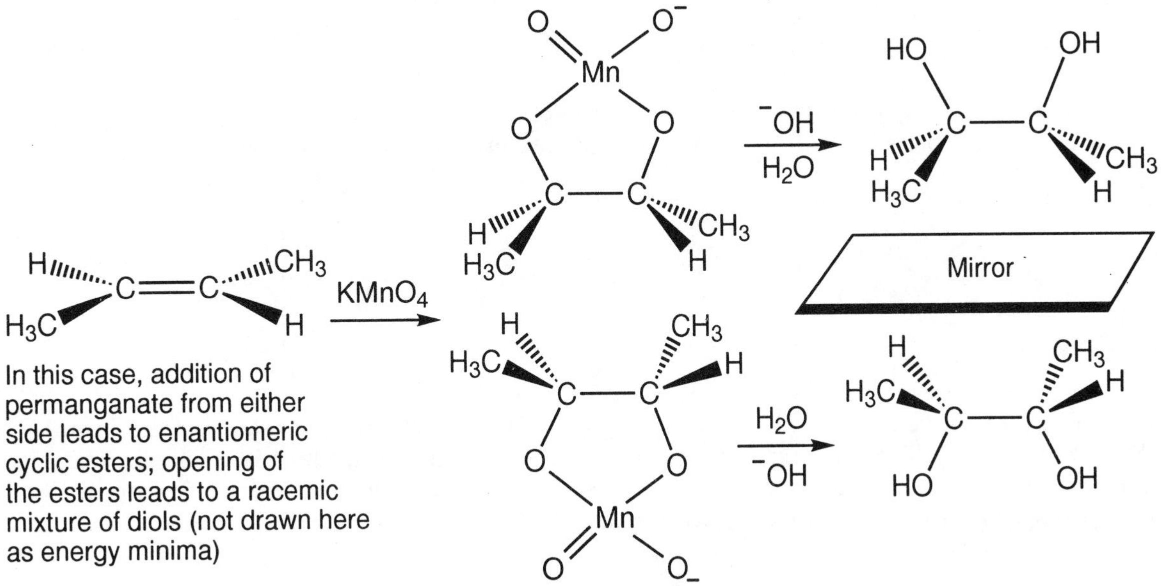

In this case, addition of permanganate from either side leads to enantiomeric cyclic esters; opening of the esters leads to a racemic mixture of diols (not drawn here as energy minima)

$KMnO_4$; ^-OH / H_2O ; Mirror

Problem 17.22 In order to do the periodate cleavage of 1,2-diols, it is necessary to make a cyclic ester of periodic acid.

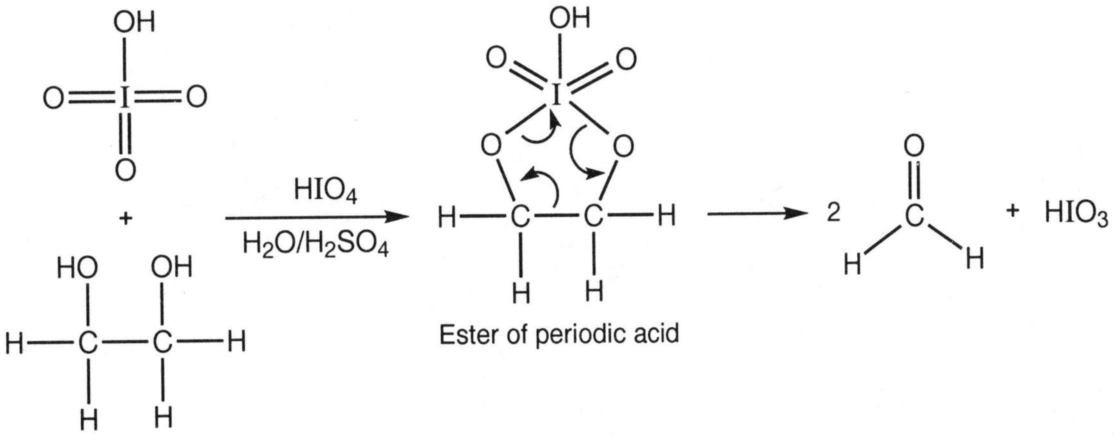

HIO_4 / H_2O/H_2SO_4 → 2 $H_2C{=}O$ + HIO_3

Ester of periodic acid

Ester formation is much easier to do in *trans*-1,2-cyclohexanediol than in the cis isomer. In the trans isomer, both bonds can be to equatorial positions, whereas in the cis isomer the bonds to iodine must span the axial and equatorial positions.

trans-1,2-Cyclohexanediol $\xrightarrow[H_2O/H_2SO_4]{HIO_4}$ (Equatorial, Equatorial; IO_3H)

cis-1,2-Cyclohexanediol $\xrightarrow[H_2O/H_2SO_4]{HIO_4}$ (Axial, Equatorial; IO_3H)

Problem 17.23 Simply brominate an alkene in the presence of solvent water. The initially formed bromonium ion is opened by water to give a protonated bromohydrin. Deprotonation gives the final product.

Br_2/H_2O

Problem 17.25 There is a straightforward analogy to the oxymercuration reactions run in water (Chapter 10, p. 402). The first step is formation of a mercurinium ion, which is opened by breaking the more substituted, and therefore weaker, carbon-mercury bond to give an isolable intermediate. In this case, the opening nucleophile is alcohol, not water.

$Hg(OAc)_2$ → Mercurinium ion ($HOCH_2CH_3$, HgOAc) → ring opening (S_N2) → deprotonation ($HOCH_2CH_3$) → HgOAc, OCH_2CH_3 + $H_2\overset{+}{O}CH_2CH_3$

In the second step, the product is reduced with $NaBH_4$ to remove the mercury. The result is an ether with the ether oxygen attached to the more substituted carbon of the original alkene.

HgOAc, OCH_2CH_3 $\xrightarrow[NaOH]{NaBH_4}$ OCH_2CH_3

Problem 17.26 Phenol ionizes to give the resonance-stabilized phenoxide ion. By contrast, cyclohexanol gives a localized, much higher energy alkoxide.

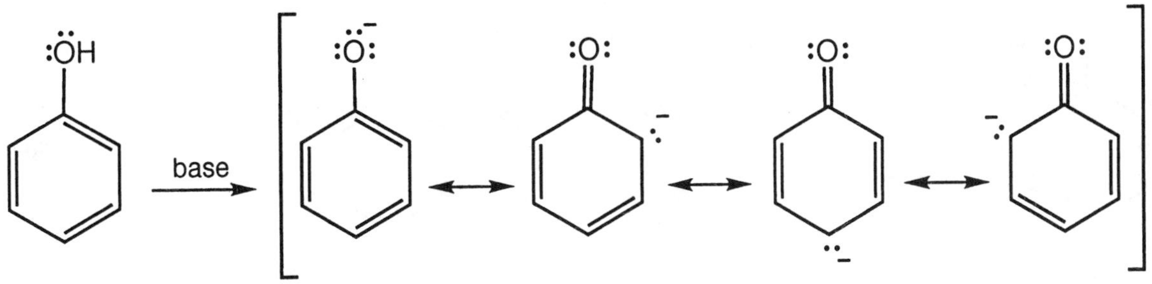

base

Resonance stabilized

base

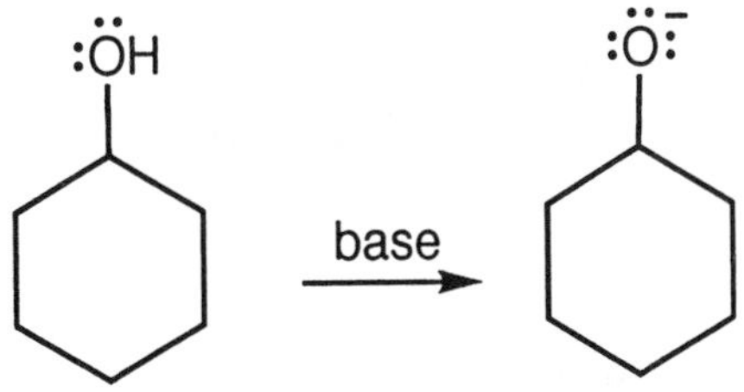

A localized anion

Problem 17.27 As Problem 17.26 shows, there is electron density and negative charge at the ortho and para carbons of the six-membered ring of the phenoxide ion. Although most of the charge is on oxygen, and most alkylation occurs at that position, there can be alkylation at carbon as well. In the drawing below, don't make the mistake of thinking that the resonance forms undergo individual reactions. The arrows are drawn from individual resonance forms only for bookkeeping purposes.

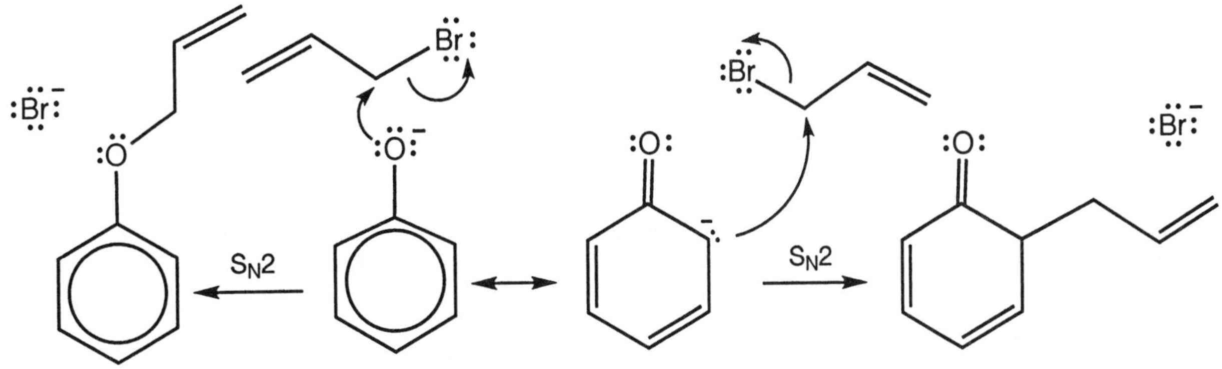

S_N2

S_N2

The initial product is a ketone, but enolization leads to rearomatization, and therefore is a very easy process.

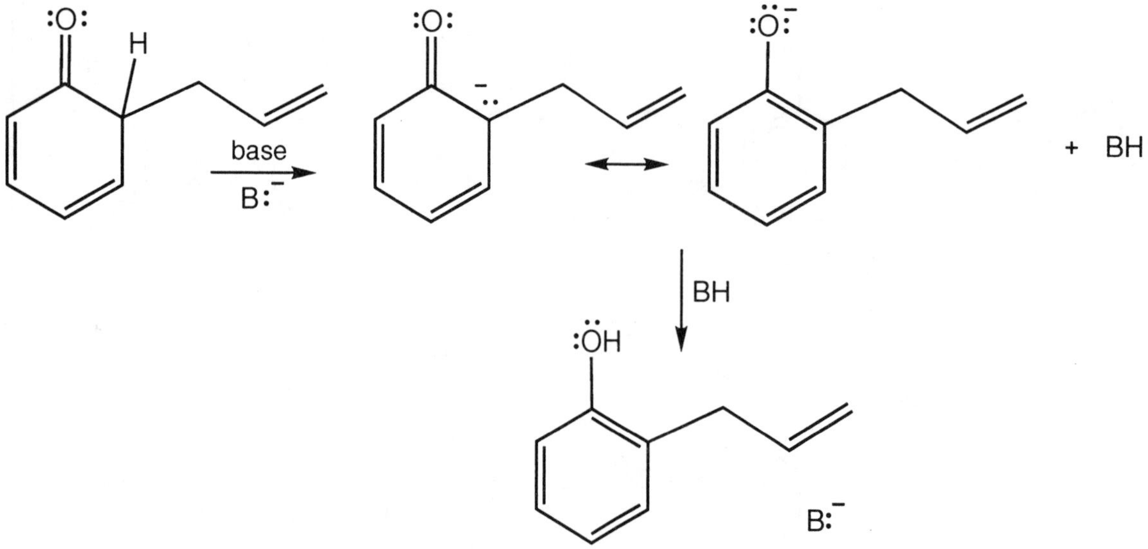

Problem 17.28 Protonation by hydrogen iodide yields an intermediate in which only one S_N2 reaction is possible. There can be no S_N2 reaction at the carbon of the benzene ring—the ring blocks access for the iodide ion.

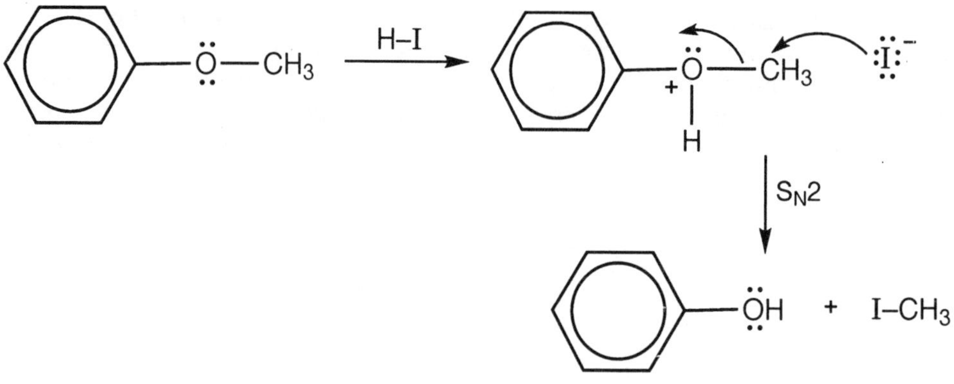

Problem 17.29 Below is a standard ether cleavage reaction applied in a cyclic system. It is complicated by the transformation of the initial product, an iodo alcohol, into a diiodide.

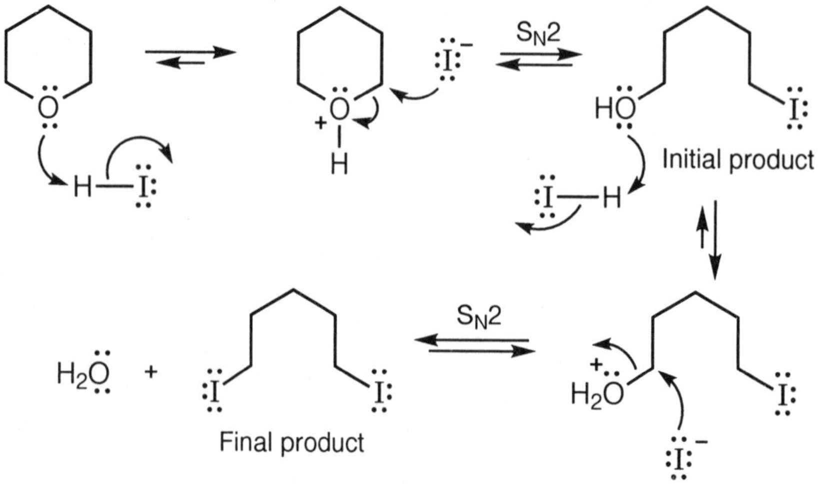

Problem 17.31 Lithium aluminum hydride can reduce the weak oxygen–oxygen bond to generate a pair of harmless alkoxides.

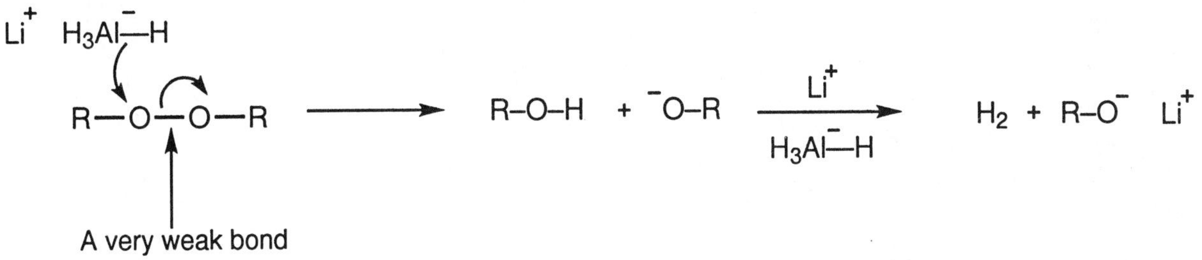

$$Li^+ \; H_3\bar{Al}{-}H \quad R{-}O{-}O{-}R \longrightarrow R{-}O{-}H + {}^-O{-}R \xrightarrow[H_3\bar{Al}{-}H]{Li^+} H_2 + R{-}O^- \; Li^+$$

A very weak bond

Problem 17.32 The epoxide oxygen is first protonated to give a cyclic oxonium ion (notice the similarity to a bromonium ion). Like bromonium ions, this cyclic oxonium ion can be opened by S_N2 reaction with any nucleophiles present. Here the only nucleophile is water. Deprotonation of the new oxonium ion gives the 1,2-glycol.

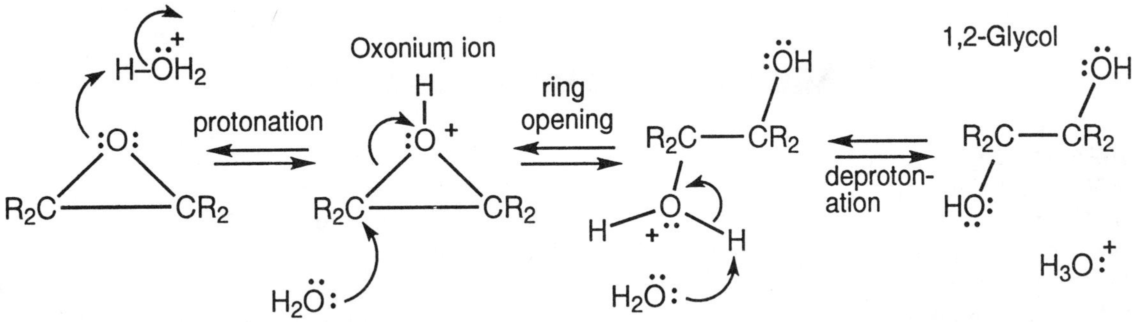

Oxonium ion

protonation

ring opening

deprotonation

1,2-Glycol

Problem 17.33 There is nothing difficult or profound here, despite the apparent complexity of the molecule. This substitution pattern allows an experimenter to determine the stereochemical course of the reaction. If this methyl transfer is an S_N2 reaction, inversion must take place. The rest of the molecule is really no more that a complicated R group.

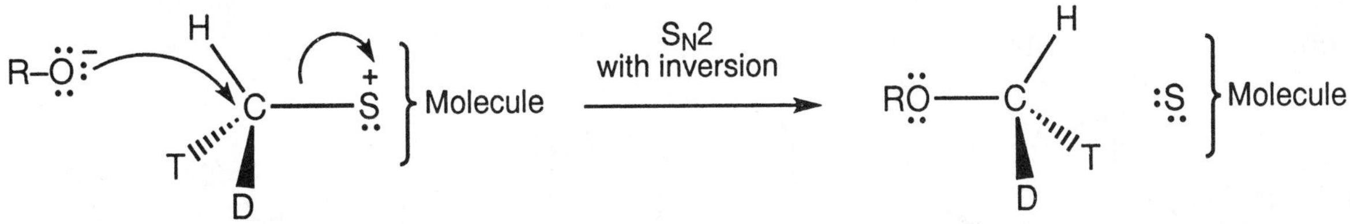

Molecule

S_N2 with inversion

Molecule

Problem 17.35 Remember, boron trifluoride etherate is a source of BF_3, and BF_3 is a strong Lewis acid (trivalent boron has an empty 2*p* orbital). The BF_3 plays the role of proton here—of Lewis acid catalyst for the initial addition to the carbonyl group.

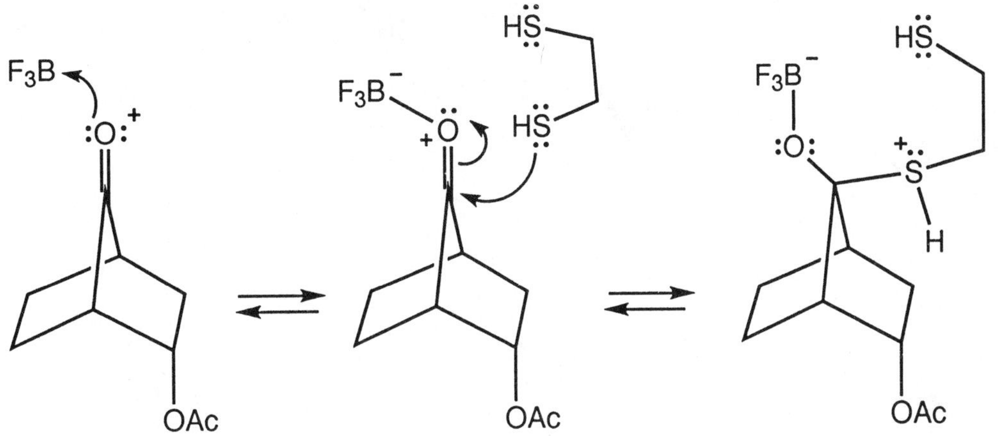

OAc

OAc

OAc

The hydroxide ion in the last step hydrolyzes the acetate, so conveniently abbreviated in the problem as "OAc." The shorthand hides the carbonyl group, but you must be able to see past OAc to the real structure. Then, the question should be easy, at least if you worked out the mechanism of Figure 14.90 (p. 667).

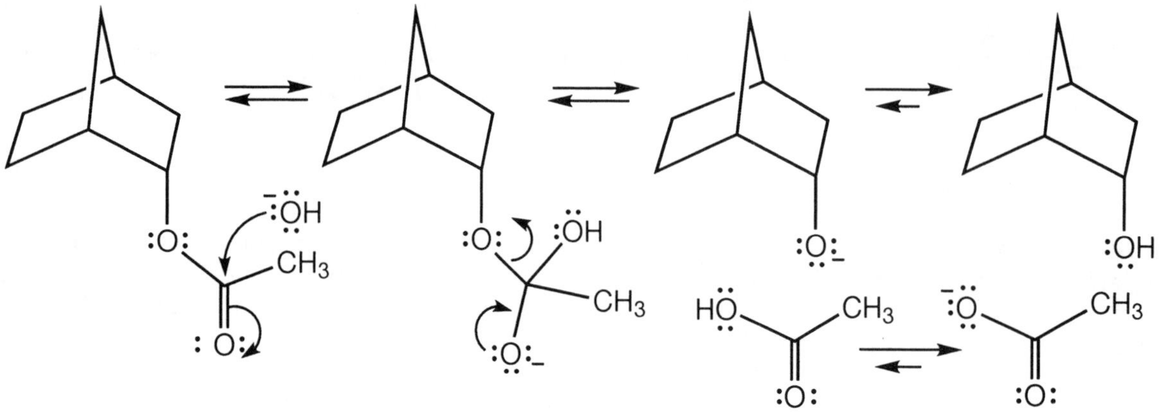

Additional Problem Answers

Problem 17.36

(a) 3-Methyl-1-butanol
(b) 3-Methyl-2-butanol
(c) 2-Methyl-2-butanol
(d) 2-Methyl-1-butanol
(e) 2-Fluoro-1-pentanol
(f) 4-Phenyl-1-pentanol
(g) *trans*-1,2-Cyclobutanediol
(h) 4-Penten-1-ol
(i) *trans*-3-Penten-1-ol
(j) *m*-Ethoxyphenol
(k) 1,3-Propanediol
(l) Propyl mercaptan (propanethiol)
(m) Dipropyl sulfide
(n) Dipropyl sulfone
(o) Dipropyl sulfoxide

Problem 17.37

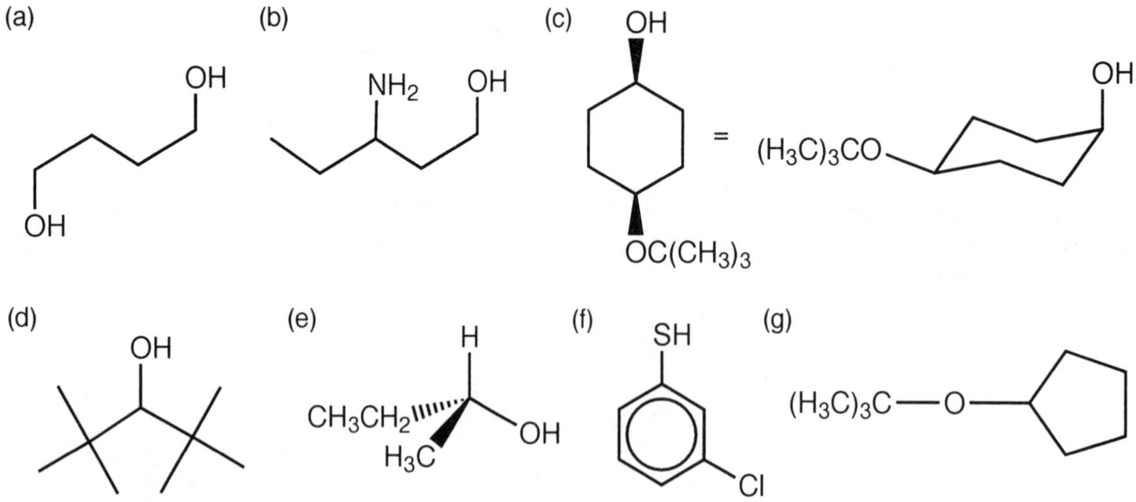

(a) OH OH

(b) NH_2 OH

(c) OH OC(CH$_3$)$_3$ = (H$_3$C)$_3$CO OH

(d) OH

(e) H CH_3CH_2 OH H_3C

(f) SH Cl

(g) (H$_3$C)$_3$C—O

Problem 17.38 This problem gives you a series of common or "trivial" names. There is not much you can do if you do not know these names, other than look them up. Each of these compounds could be named in a more complicated way, but nearly all organic chemists would use these common names.

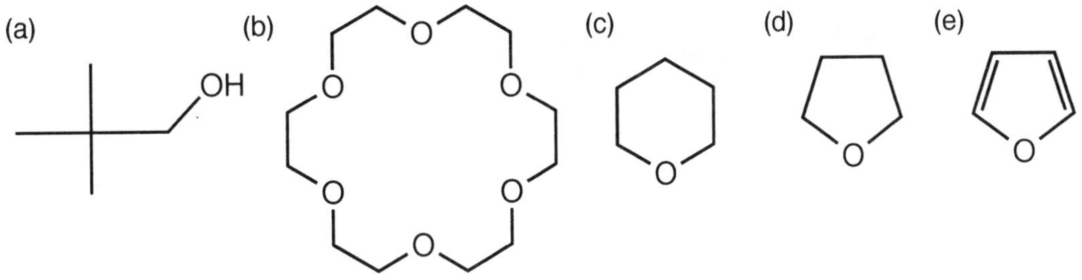

(a) (b) (c) (d) (e)

Problem 17.39
(a) In solution, at least, ethyl alcohol is a stronger acid than isopropyl alcohol (Table 17.3, p. 824). So we would expect that the primary alcohol would be the stronger acid here.

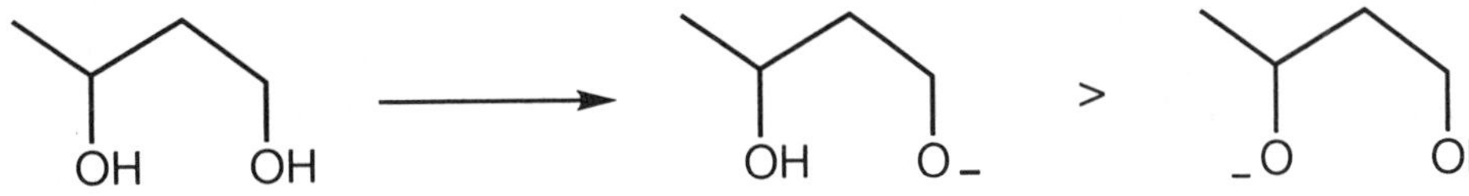

(b) In the gas phase, the acidity order for alcohols is the reverse of what obtains in solution. So in this case, it will be the tertiary alkoxide that will be preferentially formed.

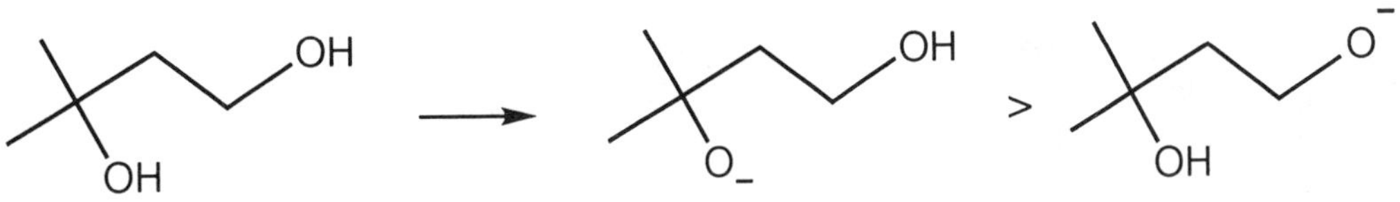

(c) Thiols are stronger acids than alcohols, so the SH will be preferentially ionized.

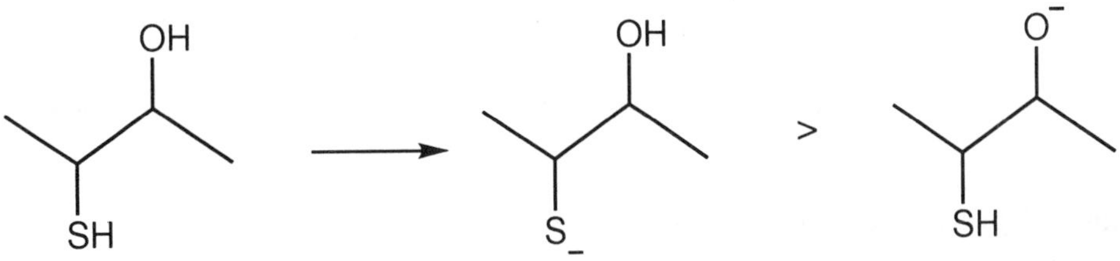

Problem 17.40 The carbon–chlorine bond is strongly polarized, with the negative end of the dipole on chlorine. There is a partial positive charge on carbon. That δ^+ will be stabilizing to a neighboring anion. The closer it is, the greater the stabilization. Accordingly, the stability of the related alkoxide, and thus the acidity order of the alcohols, is in the following order:

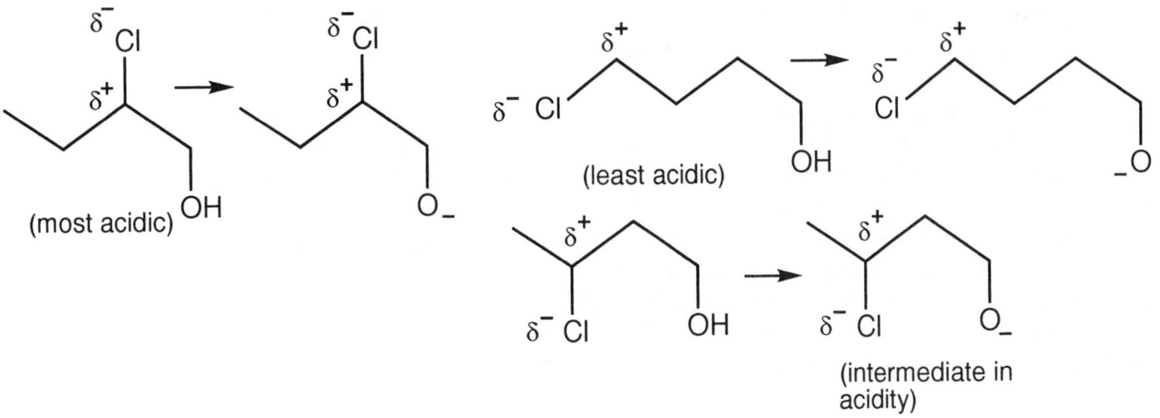

(most acidic)

(least acidic)

(intermediate in acidity)

Problem 17.41

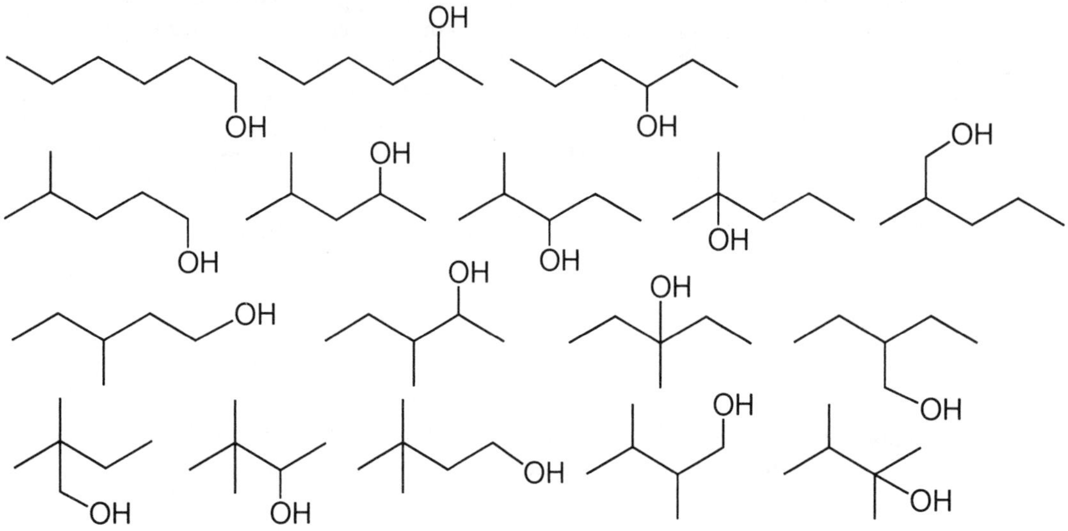

Problem 17.42 First, let's determine the molecular formula for Compound **1**.

C 81.76
H 10.98
92.74

100.00
-92.74
7.26 %O

C 81.76/12.011 = 6.807
H 10.98/ 1.008 = 10.89
O 7.26/15.999 = 0.454

C 6.807/0.454 = 14.99
H 10.89/0.454 = 23.99
O 0.454/0.454 = 1.00

$$\text{Empirical formula} = C_{15}H_{24}O$$

$$\text{Formula weight } (C_{15}H_{24}O) = 220 \text{ g/mol}$$

Because the mass spectrum of compound **1** displays a parent ion with $m/z = 220$, the empirical formula equals the molecular formula.

Now, how many degrees of unsaturation are present?

$$\Omega = [2(15) + 2 - 24]/2 = 4$$

Four degrees of unsaturation suggests the possibility of a benzene ring, which accounts for all of the degrees of unsaturation.

Next, examine the spectral data. The IR spectrum of compound **1** shows a sharp, medium intensity band at 3660 cm^{-1}. The high frequency and sharpness of this absorption suggest the possibility of a "free" O—H stretch; that is, no hydrogen bonding. This information will prove to be particularly useful later on.

All 24 hydrogens are apparent in the ^1H NMR spectrum of compound **1**. However, there are only seven signals in the ^{13}C NMR spectrum, which indicates a reasonably high degree of symmetry. The 18H singlet at δ 1.43 ppm in the ^1H NMR spectrum of **1** is highly suggestive of two identical *tert*-butyl groups. This supposition is further supported by the singlet at δ 34.2 ppm and the quartet at δ 30.4 ppm in the ^{13}C NMR spectrum. The 3H singlet at δ 2.27 ppm in the ^1H NMR spectrum is indicative of a methyl group attached to the benzene ring. This methyl carbon dutifully appears as a quartet at δ 21.2 ppm in the ^{13}C NMR spectrum. The 1H singlet at δ 5.00 ppm in the ^1H NMR spectrum could be due to a phenolic hydrogen. Finally, the 2H singlet at δ 6.98 ppm in the ^1H NMR spectrum is consistent with two identical phenyl hydrogens. The doublet at δ 125.5 ppm in the ^{13}C NMR spectrum supports this assignment. It is also worth noting that there are only three other signals for the phenyl carbons, all of which are singlets. Also note the deshielded signal at δ 151.5 ppm, which corresponds to the phenyl carbon directly attached to the oxygen. Thus, compound **1** must be a tetrasubstituted benzene derivative with a high degree of symmetry.

Given the deduced structural fragments and the high degree of symmetry, there are really only two structures, **A** and **B**, that we need to consider for compound **1**.

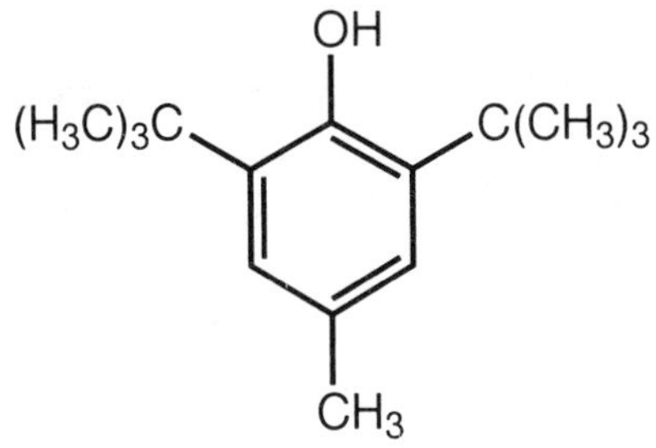

OH, $(H_3C)_3C$, $C(CH_3)_3$, CH_3 — **A**

CH_3, $(H_3C)_3C$, $C(CH_3)_3$, OH — **B**

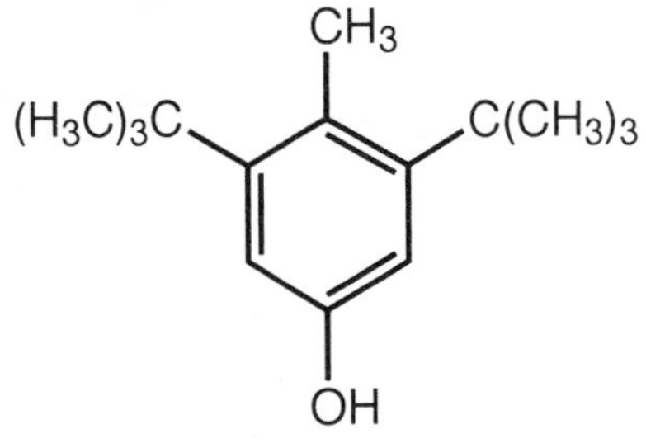

Both of these structures would exhibit only seven signals in the ^{13}C NMR spectrum. Structure **A** appears to be the better candidate for at least two reasons. First of all, the phenyl hydrogens ortho to the OH group in structure **B** would be appreciably shielded in the ^1H NMR spectrum, as would the corresponding carbons in the ^{13}C NMR spectrum. (Why?) This is clearly not the case. Second, we return to the IR spectrum of compound **1**. Recall that the O—H stretch of **1** is quite sharp and appears at a very high frequency, suggesting the absence of hydrogen bonding. It is obvious that intermolecular hydrogen bonding involving the OH group would be sterically inhibited by the large *tert*-butyl groups in structure **A** but not in structure **B**. Thus compound **1** is 2,6-di-*tert*-butyl-4-methylphenol, **A**, commonly called butylated hydroxytoluene (BHT); which is employed as a stabilizer (antioxidant) at concentrations of 0.025% in solvents such as THF.

Problem 17.43 There is quite a number of products possible! First of all, what's likely? Protonation of the alcohol, followed by loss of water, to give the carbocation starts us off on an $S_N1/E1$ process.

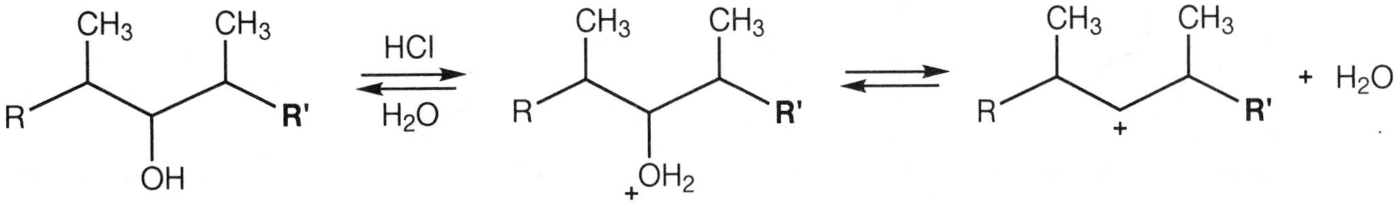

CH_3 CH_3 R R' OH ⇌ (HCl / H_2O) CH_3 CH_3 R R' $^+OH_2$ ⇌ CH_3 CH_3 R + R' + H_2O

There are two nucleophiles present, chloride ion and water. So, two possible products are the starting alcohol and the corresponding chloride, **A**. This is the S_N1 part of this reaction.

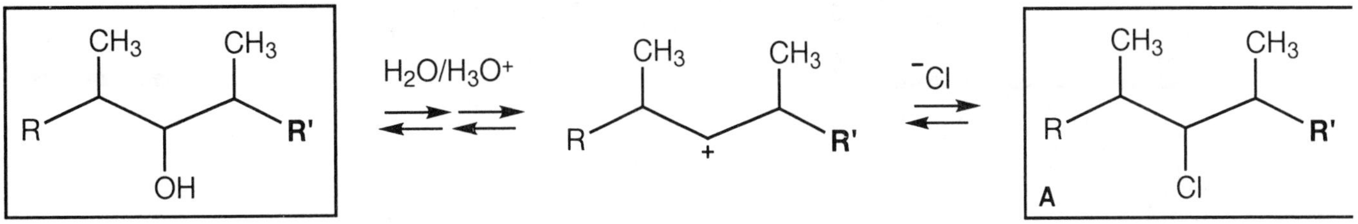

CH_3 CH_3 R R' OH ⇌ (H_2O/H_3O^+) CH_3 CH_3 R + R' ⇌ (^-Cl) CH_3 CH_3 R R' Cl **A**

The E1 part involves deprotonation of the carbocation to give an alkene. Deprotonation can occur in two directions, and there are two stereoisomers of each alkene. Now we have products **B–E**.

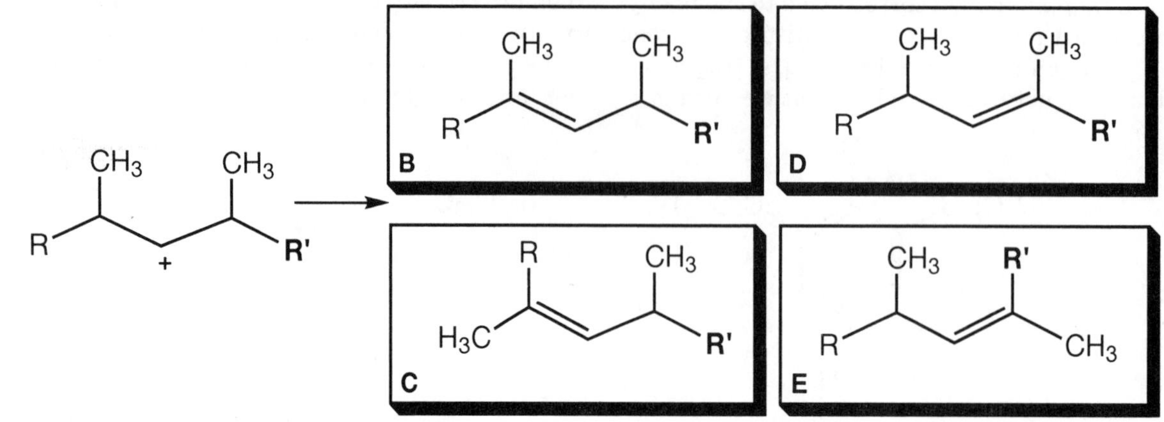

But remember — if a hydride shift to give a more stable carbocation can occur, it will. In this case, there are two such reactions, each of which generates a tertiary carbocation from a secondary carbocation.

Secondary carbocation

Both tertiary carbocations

Migrate hydride H_a

Migrate hydride H_b

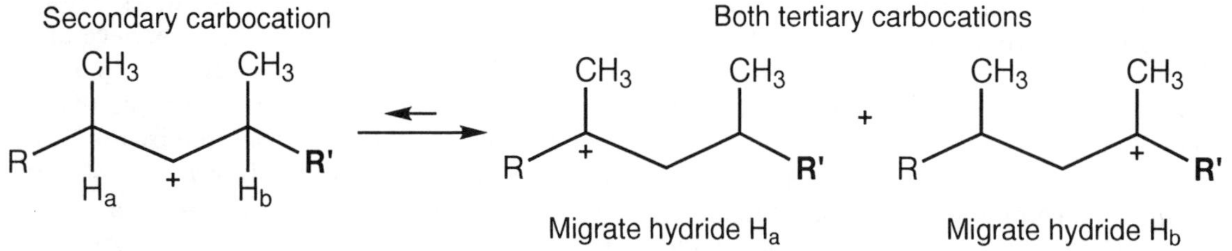

Each new carbocation can react with the two nucleophiles present, water and chloride ion. Four new products, **F–I**, are formed.

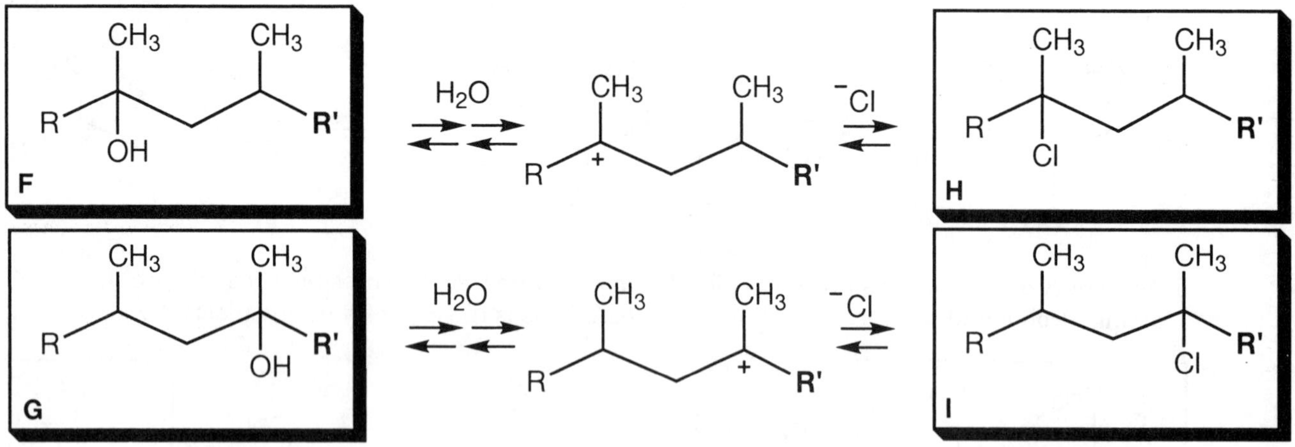

Finally, each of the tertiary carbocations can lose a proton in an E1 reaction to give two more new products, **J** and **K**, as well as alkenes **B**, **C**, **D**, and **E**.

D + E ⇌ R–C⁺(CH₃)–CH₂–CH(CH₃)–R' ⇌ **J**: R–CH(CH₃)–CH₂–C(=CH₂)–R'

B + C ⇌ R–C⁺(CH₃)–CH₂–CH(CH₃)–R' ⇌ **K**: R–C(=CH₂)–CH₂–CH(CH₃)–R'

Problem 17.44 Parts (a) and (b) require oxidation under different conditions. To get the aldehyde, we must be careful to exclude water. The traditional route is to use CrO_3 in pyridine. Many oxidizing agents will suffice to generate the acid; a common one for primary alcohols is potassium permanganate.

1-butanol —(CrO_3, pyridine)→ butanal (a)

1-butanol —($KMnO_4$)→ butanoic acid (b)

Parts (c) and (d) require that we make butyl bromide first. The S_N2 displacement by cyanide gives (c), and formation of the Grignard (or organolithium) reagent, followed by treatment with D_2O leads to (d).

1-butanol —(PBr_3)→ butyl bromide —(KCN)→ butyl–CN (c)

butyl bromide —(1. Li; 2. D_2O)→ butyl–D (d)

Problem 17.45 The alcohol is 2-butanol, and here are the transformations:

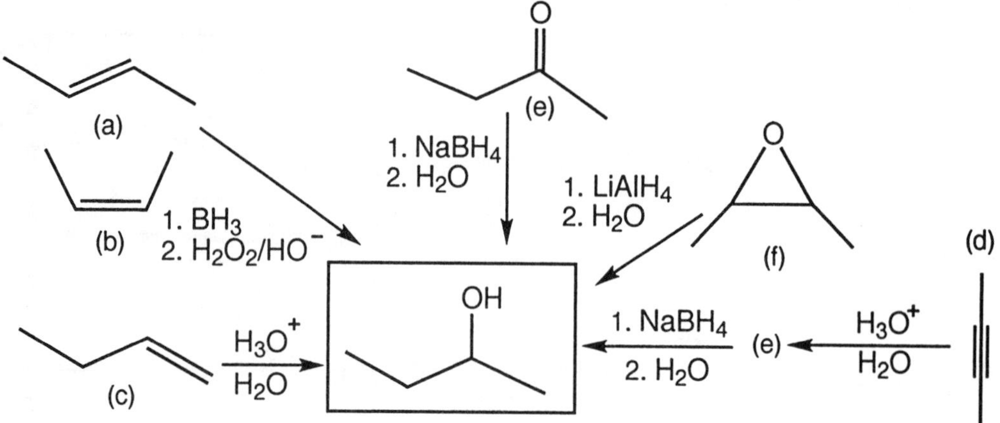

Problem 17.46 The ether cleavage reaction proceeds through protonation of the ether oxygen and displacement by halide ion. In the first example, this can happen in two ways.

$CH_3CH_2CH_2CH_2\text{—}\ddot{O}\text{—}CH_2CH_3 \xrightleftharpoons{HI} CH_3CH_2CH_2CH_2\text{—}\overset{H}{O^+}\text{—}CH_2CH_3$ (attack by $:\ddot{I}:^-$ at (a) or (b))

$\xrightarrow[(a)]{S_N2}$ $CH_3CH_2CH_2CH_2\text{-}\ddot{I}:$ + $CH_3CH_2\ddot{O}H \xrightarrow{HI} CH_3CH_2\text{-}\ddot{I}:$

$\xrightarrow[(b)]{S_N2}$ $CH_3CH_2\text{-}\ddot{I}:$ + $CH_3CH_2CH_2CH_2\ddot{O}H \xrightarrow{HI} CH_3CH_2CH_2CH_2\text{-}\ddot{I}:$

In the second reaction, there is only one possible S_N2 cleavage—one of the two possible displacements involves a tertiary carbon, and the S_N2 reaction never succeeds under such circumstances.

$(CH_3)_3C\text{—}\ddot{O}\text{—}CH_2CH_3 \xrightleftharpoons{HI} (CH_3)_3C\text{—}\overset{H}{O^+}\text{—}CH_2CH_3$ (attack by $:\ddot{I}:^-$ at (a): S_N2 blocked!; or at (b))

$\xrightarrow[(a)]{S_N2}$ (blocked) no reaction

$\xrightarrow[(b)]{S_N2}$ $CH_3CH_2\text{-}\ddot{I}:$ + $(CH_3)_3C\text{-}\ddot{O}H$

Problem 17.47 The inversion accompanying any S_N2 reaction will turn the (*R*) compound into the (*S*) compound in this case. The only problem is that OH is not a good leaving group. Reaction of the alcohol with "tosyl chloride" (*p*-toluenesulfonyl chloride) accomplishes the transformation of OH into the good leaving group, OTs. Displacement by cyanide ion does the trick.

Tosyl chloride, Ts-Cl; pyridine; (R); (R) Tosylate; S_N2; (S); ^-OTs

Problem 17.48 This problem simply involves "remember the reaction" questions. It involves much review, as well as a few new reactions from this chapter.

(a) Ether cleavage leads to CH_3OH and CH_3I (see Problem 17.46).
(b) No reaction. Ethers do not react with bases such as alkoxides.
(c) Iodide is a good enough leaving group for the S_N2 reaction to succeed. The products are iodide ion and $(CH_3)_3COCH_3$.
(d) This reaction looks the same as (c) but it's not. This time, S_N2 displacement is impossible because of the tertiary nature of the substrate. No displacement reaction occurs. Instead, an E2 reaction is likely, and isobutene, methyl alcohol and iodide ion are the products.

$$CH_3O^- + H{-}CH_2{-}C(CH_3)_2{-}I \xrightarrow{E2} CH_3OH + H_2C{=}C(CH_3)_2 + I^-$$

(e) The tosylate of methyl alcohol is formed, CH_3OTs (see Problem 17.47).
(f) Periodate cleaves 1,2-diols to give a pair of carbonyl compounds. In this case, the products will be benzaldehyde (PhCHO) and acetaldehyde (CH_3CHO).
(g) To be fair, a number of possible reactions can occur in this part. The product is generally the epoxide, formed through alkoxide formation and intramolecular S_N2 displacement of chloride.

NaOH; Alkoxide; S_N2; Epoxide; Cl^-

However, epoxides are sensitive to opening by nucleophiles such as hydroxide, and, as no formula is given for the product, a reasonable answer might be the 1,2-diol.

NaOH, S_N2; H_2O

(h) This problem is old stuff. The bromonium ion is formed and then opened in S_N2 fashion by bromide ion or methyl alcohol to give a dibromide and a bromo ether, after deprotonation. Be careful to note that the opening is regiospecific; the bromide and methyl alcohol add at the more substituted carbon of the bromonium ion.

Br_2 Br (+) Br Br Br OCH$_3$

(opened by bromide) (opened by methyl alcohol)

(i) This problem is simple Markovnikov hydration to give the tertiary alcohol. The tertiary carbocation is an intermediate.

H_3O^+ H H_2O H OH

(j) The combination of oxymercuration–reduction in alcohol also leads to Markovnikov addition. But in alcohol the product is the ethyl ether.

$Hg(OAc)_2$ CH_3CH_2OH CH_3CH_2O HgOAc NaBH$_4$ NaOH CH_3CH_2O H

(k) S_N2 displacement gives methyl phenyl ether (anisole).

$$PhO^- \; H_3C{-}I \xrightarrow{S_N2} PhO{-}CH_3 + I^-$$

(l) This time S_N2 is not possible, as there is no access to the rear of the carbon–iodine bond. There is no reaction.

(m) Alcohols react with dihydropyran (DHP) to form stable acetals that serve as protecting groups for the alcohol. Protonation is followed by addition of the alcohol and deprotonation.

DHP $\xrightarrow{H_3O^+}$ (+) Note resonance stabilization $\xrightarrow{CH_3OH}$ $CH_3\overset{+}{O}H$ $\xrightarrow{CH_3OH}$ CH_3O Acetal + $CH_3\overset{+}{O}H_2$

(n) Trimethylsilyl groups also serve as protecting groups for alcohols. This reaction is the deprotection of the alcohol as the fluoride ion reacts to form trimethylsilyl fluoride and liberate the alkoxide. Protonation of the alkoxide gives the alcohol.

$$CH_3CH_2O{-}Si(CH_3)_3 \xrightarrow{F^-} CH_3CH_2O^- + F{-}Si(CH_3)_3 \overset{H_2O}{\rightleftharpoons} CH_3CH_2OH + HO^-$$

(o) This reaction is straightforward acid-catalyzed opening of an epoxide. The trans diol is the final product. The protonated epoxide is opened in an S_N2 reaction, so the stereochemistry of the diol must be trans.

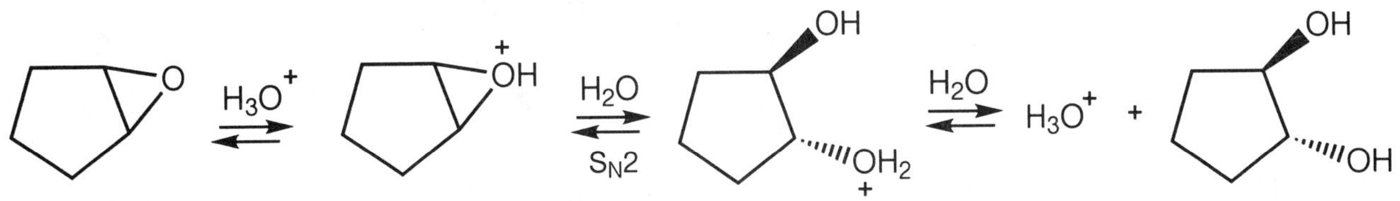

(p) Epoxides also open in base. Addition of deuteride leads to the alkoxide. A second protonation step gives the alcohol. Once again, the product must be trans.

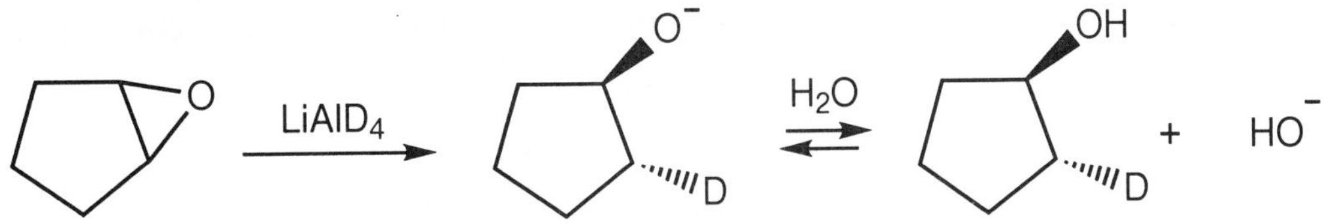

(q) Trimethyl sulfonium ions are methylating agents. An S_N2 reaction with phenoxide acting as nucleophile and dimethyl sulfide as leaving group leads to methyl phenyl ether (anisole).

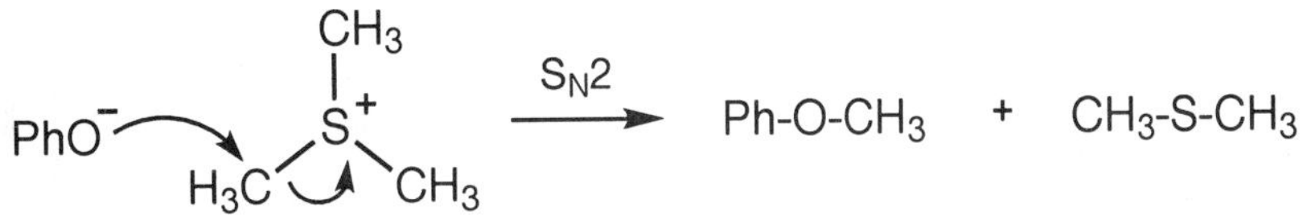

(r) This part involves the Raney nickel desulfurization reaction to form alkanes. In this case, the products will be propane and hexane.

(s) Oxidation of thiols with nitric acid leads to sulfonic acids.

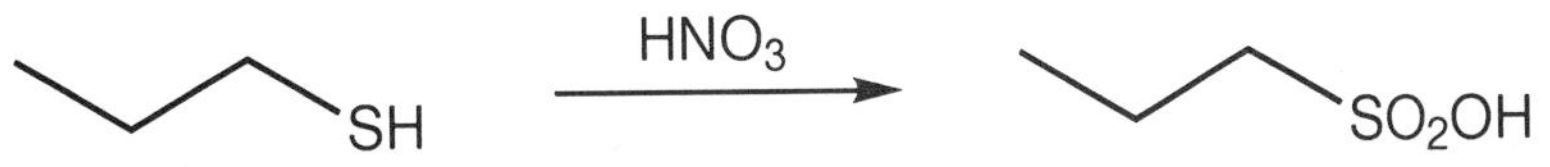

(t) This reaction is a simple reduction of the aldehyde to a primary alcohol.

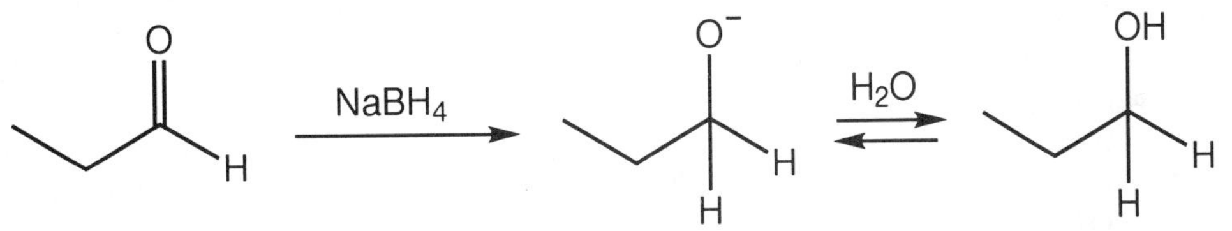

Problem 17.49 Vicinal (1,2) diols react in acid to give carbonyl compounds in what is called the pinacol rearrangement. In this chapter, we saw the prototype of this reaction, the conversion of 2,3-dihydroxy-2,3-dimethylbutane into pinacolone. When the 1,2-diol is cyclic, this reaction becomes a ring contraction process. The steps are exactly the same as for pinacol itself, however. Be sure to make the comparison (see p. 841). The problem gives clues that an aldehyde is the product. Note the IR stretching frequency at 1729 cm^{-1} and the signal in the ^1H NMR spectrum at $\delta = 9.5$ ppm, a position diagnostic for aldehydes.

Rearrangement to give a resonance-stabilized carbocation

H_3O^+ protonation; $+\ OH_2$; OH_2 deprotonation; H_3O^+

Problem 17.50 This item should be a simple roadmap problem, as there are only straightforward steps. The epoxide is first opened by hydride to give the alkoxide, then protonated by water to give **A**. Treatment with PCl_3 gives the corresponding chloride, **B**. Lithium generates the organolithium reagent, and this adds to acetone to give the alcohol **C**, after protonation of the alkoxide. Acid-catalyzed elimination in the Saytzeff (more substituted double bond is formed) sense produces the final product.

$LiAlH_4$; H_2O; PCl_3; Li; $(CH_3)_2C{=}O$; H_2O; H_3O^+ / H_2O

$C_6H_{12}O$ **A**

$C_6H_{11}Cl$ **B**

$C_9H_{18}O$ **C**

Problem 17.51 This roadmap problem is slightly more challenging than the last one. The first step is easy enough; bromine forms a bromonium ion that is opened in water to give the bromo alcohol, **A**. Treatment with base gives a molecule that does not contain bromine. So, we clearly must find some way to get rid of a bromine. The alkoxide is formed and undergoes an intramolecular S_N2 displacement of bromide to give the epoxide, **B**. Compound **B** opens in acid to give diol **C**. 1,2-Diols are cleaved by periodic acid to give a pair of carbonyl compounds. Notice that you could have used that fact to reconstruct **C** from the two carbonyl compounds shown at the end of the problem. That's an important technique. If you get stuck, work a roadmap problem from any point of information, especially the end. For example, if you couldn't remember the transformation from **B** to **C**, you could have worked backward to **C** from the structures of the two aldehydes.

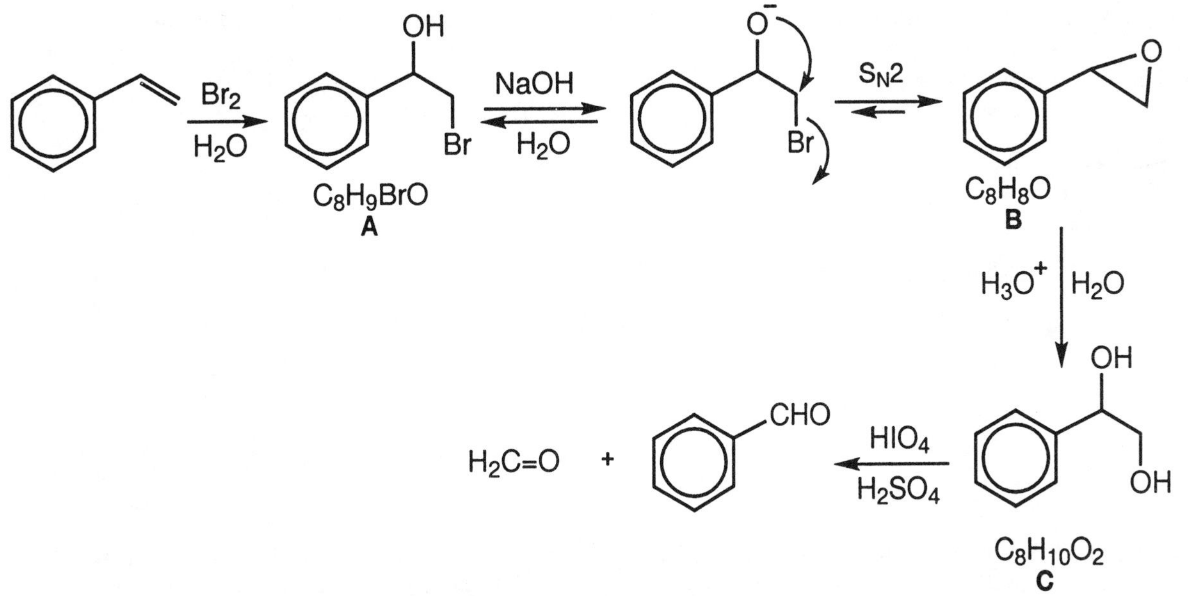

C_8H_9BrO
A

C_8H_8O
B

$C_8H_{10}O_2$
C

Problem 17.52 Surely this problem begins with protonation of the alcohol. There is only one possible elimination, and this generates one of the two products. The only difficulty in this problem is the second structure. Remember the pinacol rearrangement in Problem 17.49? There, too, we saw a ring contraction reaction. This process is related. Migration of a carbon–carbon bond gives a tertiary carbocation **A**, and now a proton can be removed to give the ring-contracted product.

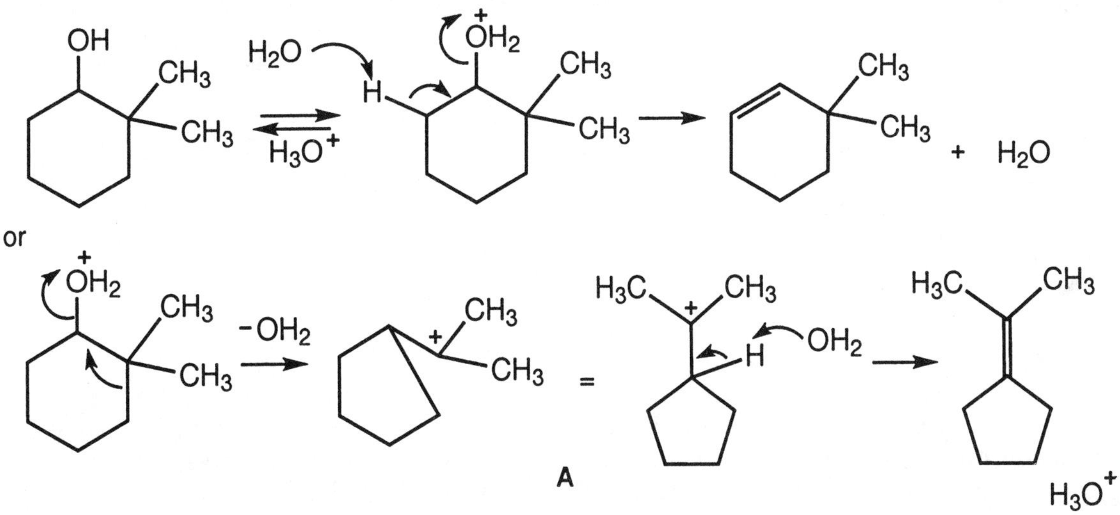

Problem 17.53 Sodium hydride forms the alkoxide by deprotonating the alcohol, and S_N2 displacement of chloride introduces the protecting group.

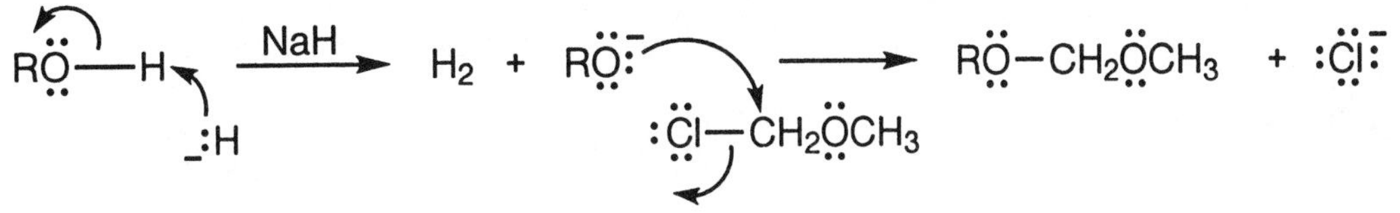

As the MOM-protected alcohol is an acetal, it is not surprising that it is labile in aqueous acid. Treatment with acid regenerates the original alcohol. Protonation of the oxygen atom, followed by S_N1 loss of alcohol, produces the alcohol and a resonance-stabilized carbocation.

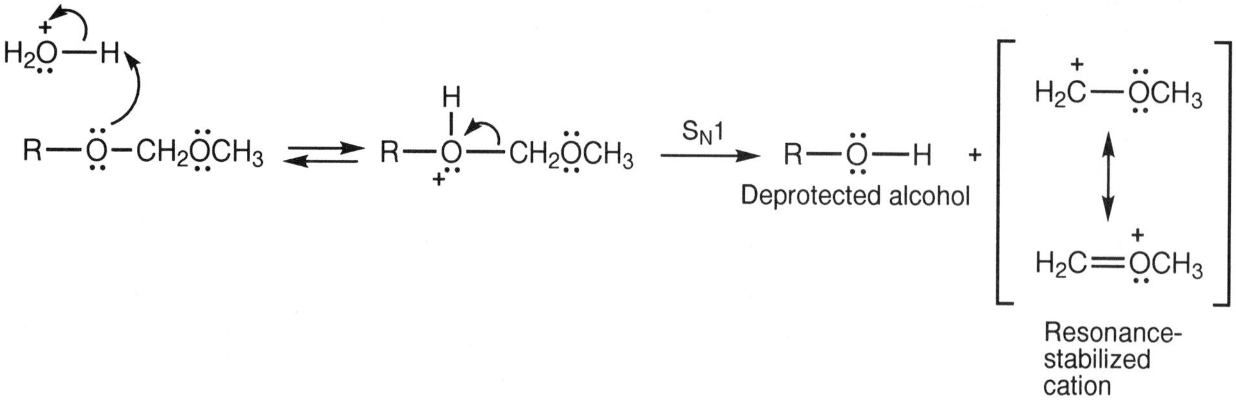

Deprotected alcohol

Resonance-stabilized cation

Problem 17.54 Here we have a pinacol rearrangement with choices. The product is formed by protonation of one OH, loss of water, and migration of hydride to give a resonance-stabilized cation. Deprotonation gives the final product.

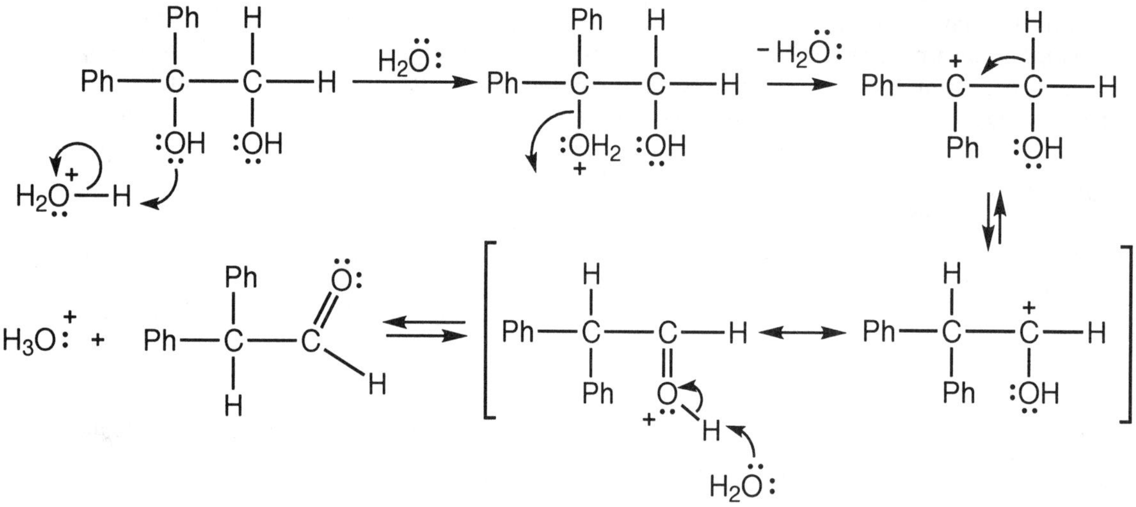

Why doesn't the same reaction take place at the other OH group? The difficulty is that the initial loss of water would give an unstabilized primary carbocation. Such an energetically unfavorable process cannot compete with the other protonation and loss of water to give a relatively stable tertiary carbocation that is well resonance stabilized by the two adjacent benzene rings. The product not formed would result from phenyl migration and deprotonation, but these steps cannot take place if the primary carbocation is never formed in the first place.

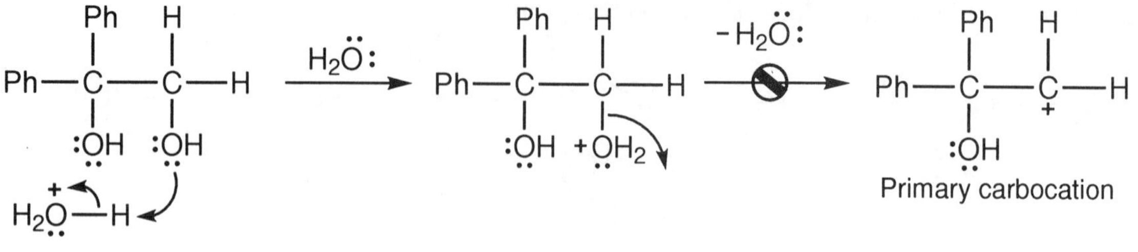

Primary carbocation

Problem 17.55 It will be far easier to form the resonance-stabilized phenoxide (pK_a of a phenol is ~ 10) than it will be to form the simple, unstabilized alkoxide (pK_a of ethyl alcohol is 15.9).

pK_a ~ 10

OH

OH

pK_a ~ 16

NaOH/ H_2O

CH_3-O-SO_2OCH_3

O^-

OH

OCH_3

OH

NaOH/H_2O

OH

O^-

Problem 17.56 It is not necessary to supply the exact reagents used in the actual synthesis, as long as you give the same type of reagent that will accomplish the purpose. The first step in this synthesis is the reduction of a ketone to an alcohol. Although this reduction could be accomplished by $NaBH_4$ as shown, the actual reducing agent used was aluminum isopropoxide, the critical ingredient in the Meerwein–Ponndorf–Verley–Oppenauer reaction, to be discussed in Chapter 18 (p. 931).

CH_3O

O

C

CH_2Br

CH_3O

$NaBH_4$

CH_3OH

CH_3O

OH

CH

CH_2Br

CH_3O

The second step in this reaction sequence involves the formation of an epoxide from a bromohydrin, a reaction we have now seen several times. The base forms a small amount of the alkoxide, which displaces bromide in an intramolecular S_N2 reaction.

CH_3O

OH

CH

CH_2Br

CH_3O

HO^-

CH_3O

O^-

CH

CH_2—Br

CH_3O

S_N2

CH_3O

O

CH_3O

The epoxide is then reopened by isopropylamine in an intermolecular S_N2 reaction that takes place at the less substituted end of the epoxide. Amines are good Brønsted bases and are often conveniently isolated as the salts of mineral acids. In this case, the hydrochloride salt was isolated when the amine was treated with HCl.

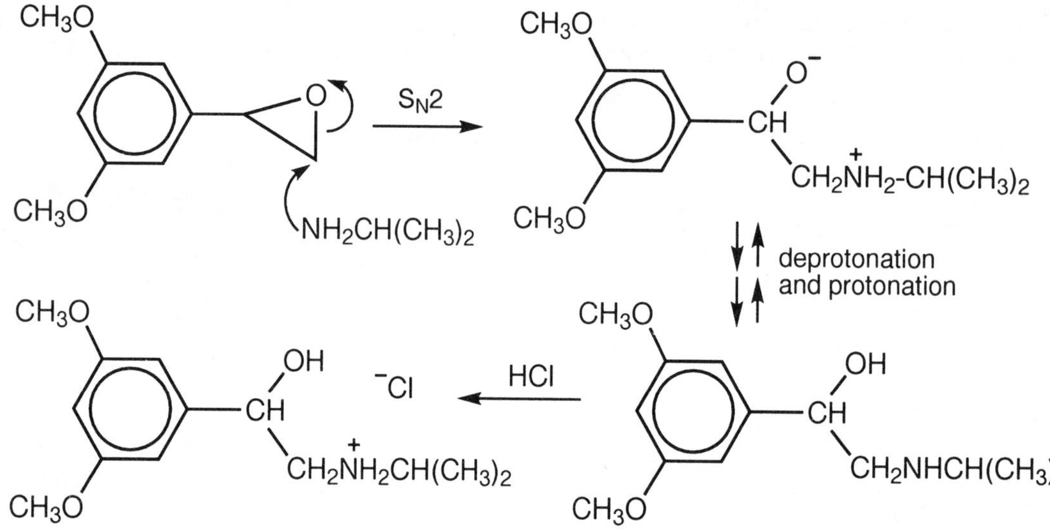

Finally, the methyl protecting groups were removed by treatment with HBr. Be sure you understand the mechanism of this ether-cleavage reaction (p. 850).

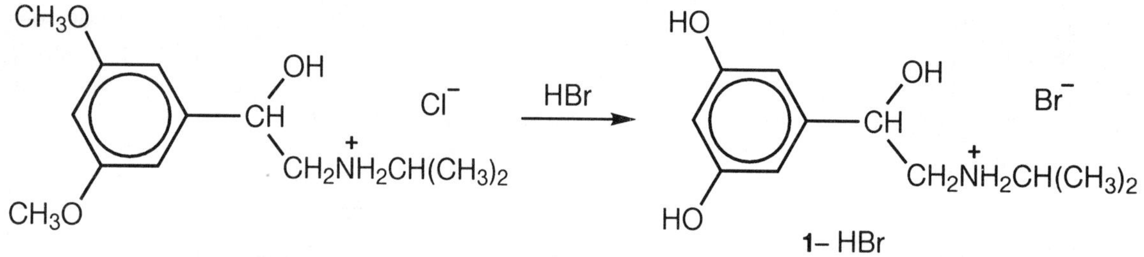

Problem 17.57 From Section 17.13d (p. 861), we know that sulfides such as **1** can be oxidized to sulfoxides or sulfones, depending on the type and amount of oxidizing agent used. The only trick here is to appreciate that sulfoxides are pyramidal. Accordingly, diastereomeric (exo and endo) sulfoxides **2** and **3** are produced upon mild oxidation of **1**. Further oxidation gives the same sulfone, **4**, in each case.

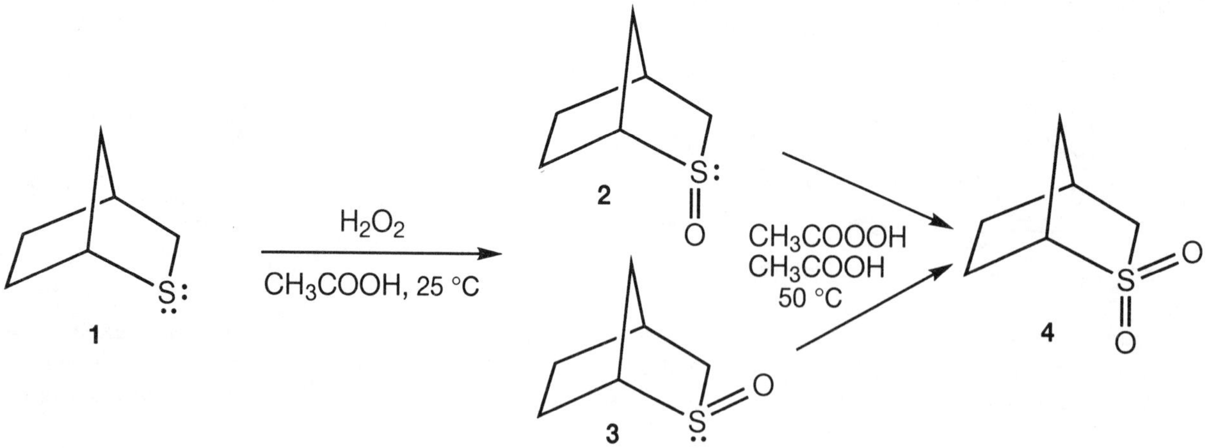

Problem 17.58 Treatment with thionyl chloride is a standard method for converting alcohols into chlorides. Compound **2** is the expected chloride.

Ph–CO–CH(OH)–Ph (**1**) $\xrightarrow{SOCl_2}$ [Ph–CO–CH(Ph)–O–S(=O)–Cl] $\xrightarrow{\Delta}$ Ph–CO–CH(Cl)–Ph (**2**, $C_{14}H_{11}ClO$) + SO_2

Intermediate chlorosulfite

The IR stretch at 1689 cm^{-1} is appropriate for the carbonyl group, and the ^{1}H NMR spectrum is consistent with the proposed structure. Notice especially the methine singlet at δ 6.33 ppm and the two ortho hydrogens of the phenyl ring at δ 7.75–8.05 ppm, deshielded by the adjacent carbonyl group.

Compound **3** is benzil. The molecular weight of 210 g/mol is consistent with the parent ion seen in the mass spectrum. In addition, the base peak in the mass spectrum is m/z = 105, which corresponds to the resonance–stabilized benzoyl ion, [Ph—CO]$^+$. That is, the compound "splits in half."

Ph–CO–CO–Ph (**3**, **Benzil**) $\xrightarrow{e^-}$ [Ph–$\overset{+}{C}$=O ⟷ Ph–C≡O^+] + 2e$^-$

m/z = 210 g/mol m/z = 105 g/mol

The IR spectrum of **3** shows a carbonyl stretch at 1645 cm^{-1}, consistent with a 1,2-diketone (conjugated). The ^{1}H NMR spectrum shows only aromatic hydrogens, and the four ortho hydrogens are deshielded by the adjacent carbonyl groups. Finally, the ^{13}C NMR spectrum shows only five different carbons, as required for benzil. Note the very low field position of the carbonyl carbon (δ 194.3 ppm).

Now the problem is to see how benzil is formed. Although there are several possibilities, the most plausible mechanism involves formation of the cyclic enediol sulfite **A** from the chlorosulfite intermediate involved in chloride formation.

[Mechanism scheme: chlorosulfite intermediate (Ph–C(=O)–CH(Ph)–O–S(=O)Cl) → cyclization with carbonyl oxygen attacking S → loss of Cl^- → Cl^- removes the methine H → cyclic enediol sulfite **A** (Ph–C=C–Ph with O–S(=O)–O ring)]

A

Compound **A** then decomposes thermally to benzil and sulfur monoxide. As the problem states, SO disproportionates to give SO_2 and S.

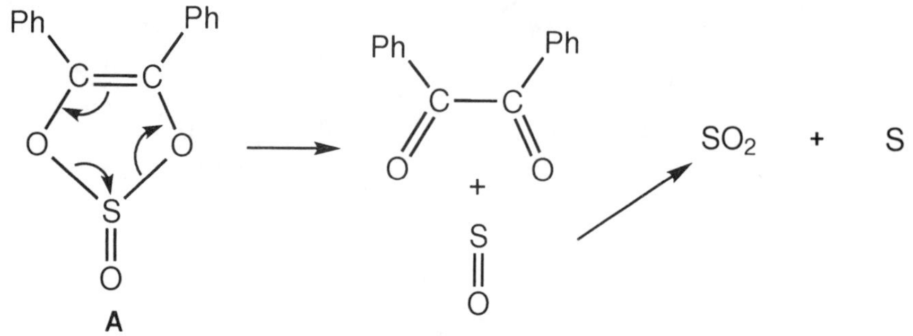

Although attempts to isolate **A** have not been successful in reactions of benzoin itself, related intermediates with different aryl groups can be isolated.

Problem 17.59 The decidedly nonpolar toluene is not a particularly good solvent for cations. The simplest explanation for the inversion of configuration found in solvent toluene is that chloride ion is blocked from addition to the cation from the same side from which it departed by SO_2. It is even possible that chloride ion, once formed, will do a direct displacement on the chlorosulfite intermediate.

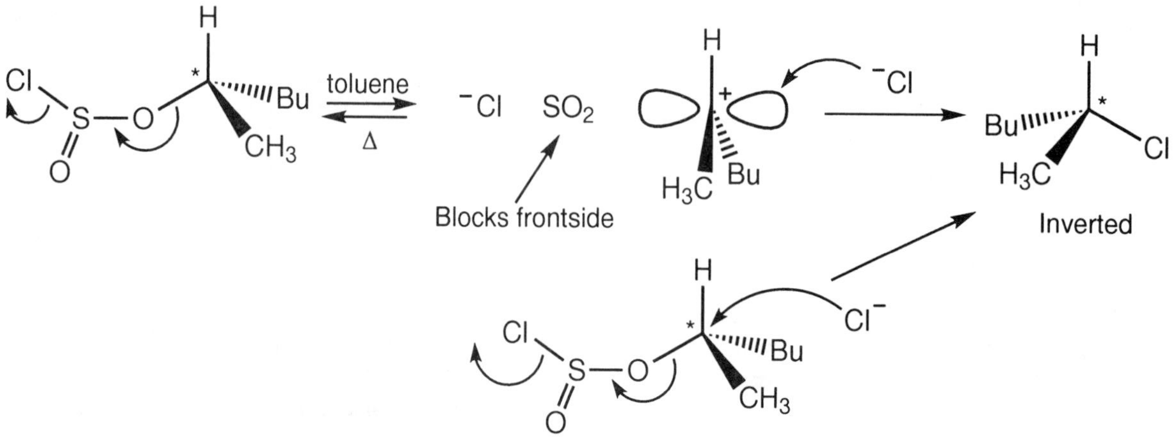

By contrast, the more polar molecule dioxane is a much better solvent for cations than toluene. It is also a nucleophile, and it is this nucleophilicity that is crucial to the observed retention of configuration. Dioxane apparently participates in the ionization of chloride (as shown for chloride itself in the figure above). Although dioxane is not as strong a nucleophile as chloride, the highly polar solvent makes this reaction possible. The result is an oxonium ion.

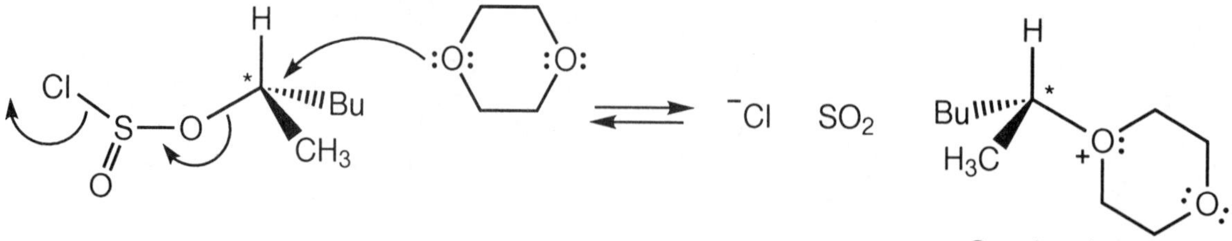

After the SO_2 diffuses, the oxonium ion can be attacked by chloride to displace dioxane and generate the chloride. The observed retention of configuration is actually the result of a pair of inversions, the first by dioxane, and the second by chloride.

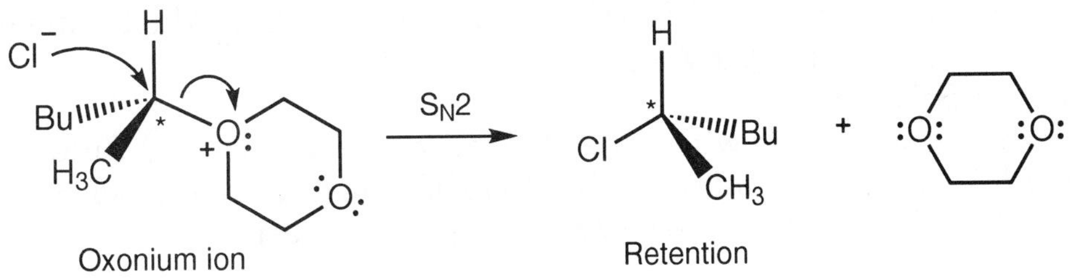

Oxonium ion $\xrightarrow{S_N2}$ Retention

Chapter 18 Outline

18

Carbonyl Chemistry 2: Reactions at the α-Position

In this chapter we reach a point we have been approaching for some time. We now know so much of the grammar and vocabulary of the discipline, and have done so many simple problems (in this analogy, "written sentences"), that we can now go on to do quite complicated problems, that is, to write some of those "paragraphs" we have been warning you about. Now we have a full arsenal at our command: a great deal of structure, most basic reactions, and an ability to determine structure through spectroscopy. We can now deal with problems that resemble the questions chemists deal with in the real world of chemistry. It will be very important to analyze each hard problem. These exercises become much, much easier when you have an idea of where you are going, of what bonds you are trying to make or break. That sounds simple, but it is amazing how few people really start problems with that simple question to oneself: What happens in this reaction?

There are still drill problems (some are review) in this chapter, and their solutions should come relatively simply. But don't worry if the hard problems do not come so easily; they are meant to tax you, to demand some hard work and careful thought. Some hard problems will be dealt with best over time. If a problem resists solution, and some will, come back to it after a while; let your subconscious work on it for a time. Most research chemists carry unsolved problems around in their heads, sometimes for years, returning to them now and then. There is nothing wrong with emulating that process, at least with hard, real world, problems. People think at vastly different rates, and it is rare moment indeed that requires a *rapid* solution of a problem (unfortunately, hour exams tend to be such moments).

Why put such "hard stuff" in an introductory book? That's a good question, and deserves a thoughtful answer. Is it possible to do a reasonable job in a course in organic chemistry without having a great deal of success with problems at the level of difficulty of some of the problems in this chapter? You bet. If you can do the bread and butter exercises well, that certainly constitutes a respectable performance. But you should at least be exposed to more complicated material. The fun in this business comes in solving problems, and for some of you it will be the pleasure of working out complex problems that leads you on in chemistry. In this chapter, we also lapse further into schematic drawings. Full Lewis "dot" structures are not always drawn, for instance. Be careful to fill in electron dots where necessary.

Problem 18.3 The drawing below shows exchange through formation of the anion and reprotonation using D_2O to give the deuterioaldehyde. However, the anion formed by removal of the aldehydic hydrogen is *not* resonance-stabilized and cannot be formed easily. If the anion is not formed, exchange cannot take place.

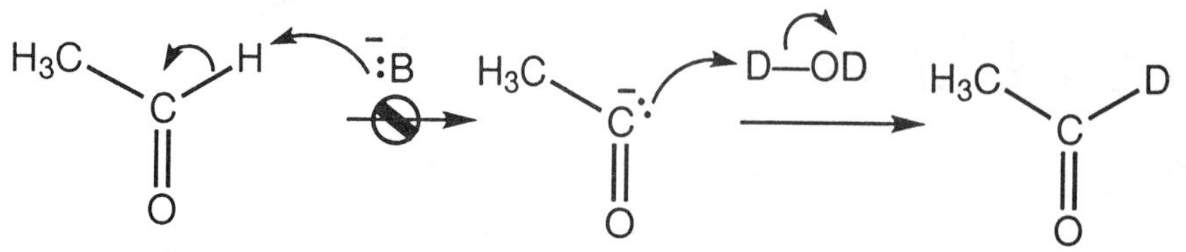

Not resonance-stabilized

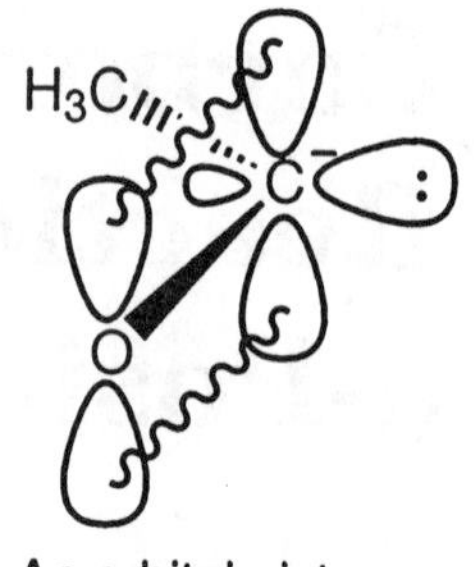

An orbital picture of the putative anion

Problem 18.4 This problem sets up an apparent contradiction: α-hydrogens exchange in base - but this particular α-hydrogen doesn't. Why? The way to start on this question is to remind yourself why α-hydrogens exchange - the formation of resonance-stabilized enolates. For example, the two other α-hydrogens exchange in the usual way.

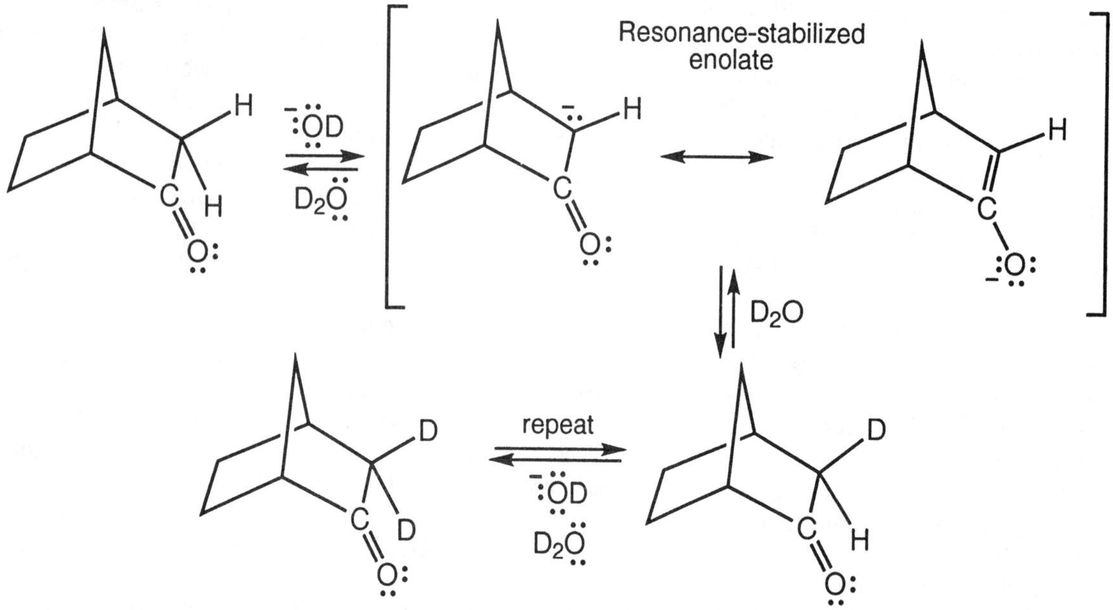

Resonance-stabilized enolate

repeat

Why might the enolate formed through removal of the bridgehead α-hydrogen not be resonance-stabilized? It seems on paper that a similar mechanism can be written for exchange of the bridgehead hydrogen, but this is an illusion.

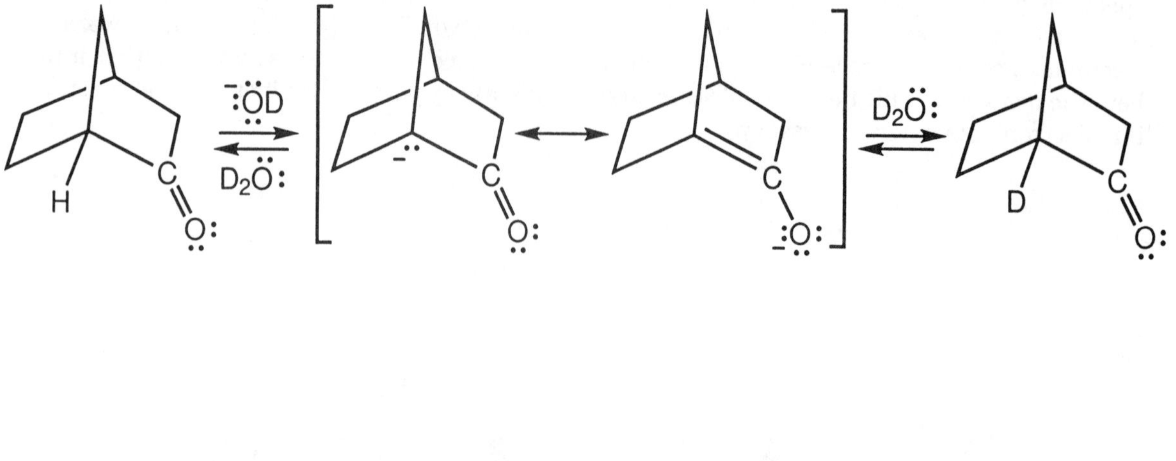

In reality, there is no resonance stabilization. Although it is easy to draw the delocalization on paper, there really is no resonance. The problem is that the orbital at the bridgehead does not overlap with the orbitals making up the π system of the carbon-oxygen double bond. The paper is lying to you: There is no resonance stabilization because the non-overlapping orbitals do not allow the electrons to be delocalized. If the anion is not resonance stabilized, it cannot be formed. If the anion is never formed, there can be no exchange.

No overlap here!

Problem 18.6 Vinyl alcohol is itself both an acid and a base! This compound can catalyze its own conversion to acetaldehyde. Can you write a mechanism?

Problem 18.7 The carbonyl compound appears to be somewhat more stable than the enol form. In doing this ΔH calculation, we are making many (unwarranted) assumptions including, for example, the notion that all carbon-hydrogen and all carbon-carbon bonds are equal in energy, and that the carbon-oxygen double bond can be approximated by a pair of carbon-oxygen single bonds. Alternatively, we can use the value for the carbon-oxygen double bond from Table 2.1 (p. 51). In either case we find the enolization reaction to be somewhat endothermic in the general case.

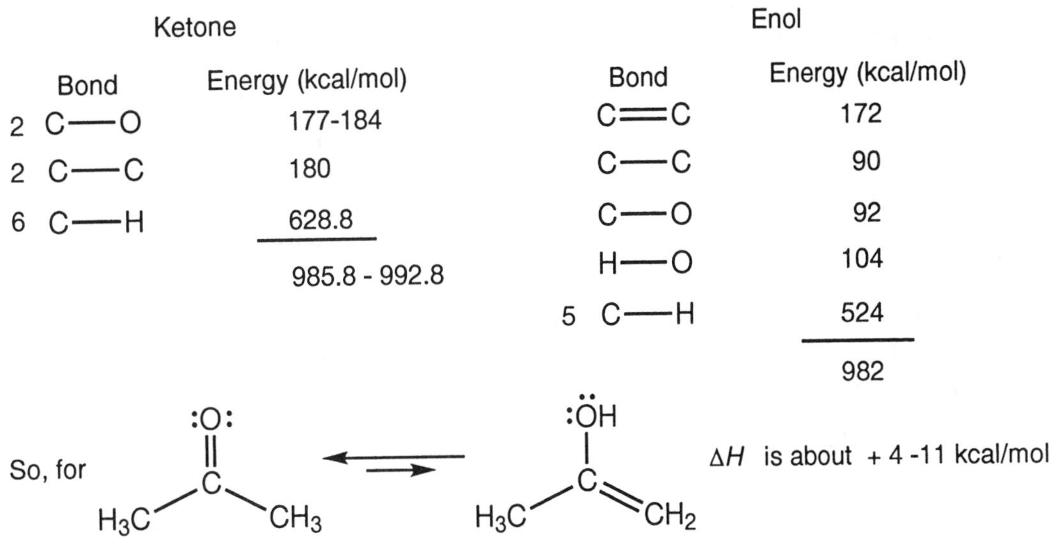

Ketone

Bond	Energy (kcal/mol)
2 C—O	177-184
2 C—C	180
6 C—H	628.8
	985.8 - 992.8

Enol

Bond	Energy (kcal/mol)
C=C	172
C—C	90
C—O	92
H—O	104
5 C—H	524
	982

Problem 18.8 In any S_N2 reaction, the nucleophile must displace the leaving group from the rear of the departing bond. The groups attached to the sp^2 hybridized carbon of the double bond effectively prevent access of the nucleophile.

Bump!

Problem 18.9 This problem recalls a similar situation involving the S_N1 reaction. The rate-determining step of the S_N1 reaction is the slow ionization to give a carbocation. The faster product-determining steps in which nucleophiles capture the carbocation follow. In each of the three seemingly different reactions of this problem, the rate-determining step is the same, enolate formation. Only after the endothermic formation of the enolate do the subsequent, product-determining steps take place.

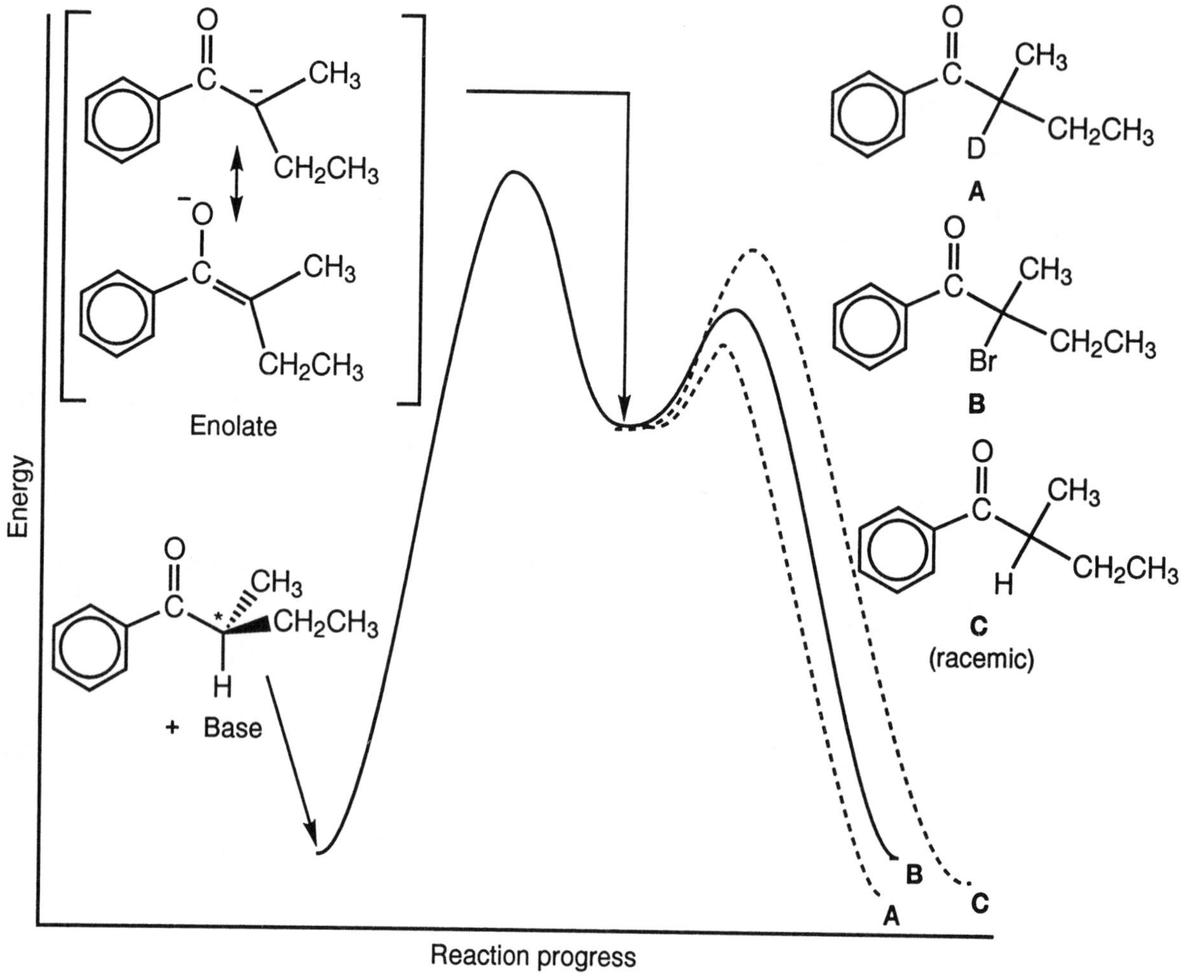

Enolate

+ Base

A

B

C (racemic)

Energy

Reaction progress

Problem 18.10 In nucleophilic additions such as S_N2 reactions or additions to carbonyl groups, steric effects are important. The nitrogen atom in LDA is effectively guarded by two large isopropyl groups that get in the way during addition reactions.

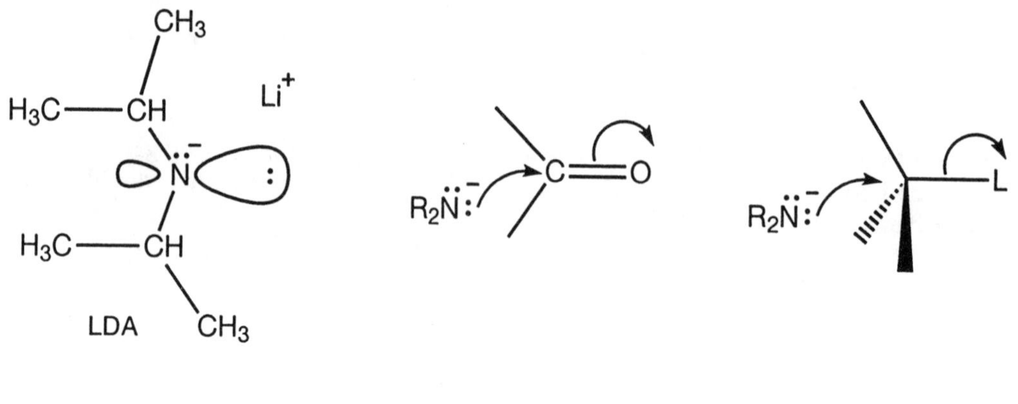

LDA

Problem 18.11 In the acid-catalyzed reaction of carbonyl compounds with amines, the first steps produce a protonated carbinolamine, **A**. As long as the nitrogen has at least one attached hydrogen, deprotonation to a carbinolamine can occur.

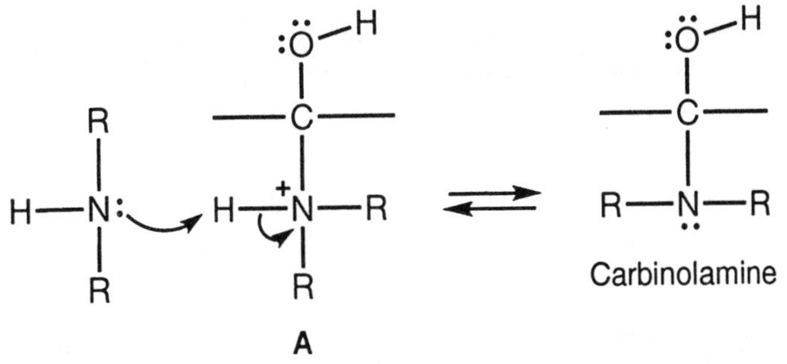

Carbinolamine

A

When the amine is tertiary, there is no hydrogen that can be removed from the nitrogen atom of **A**. The only possible reaction is reversal to starting material.

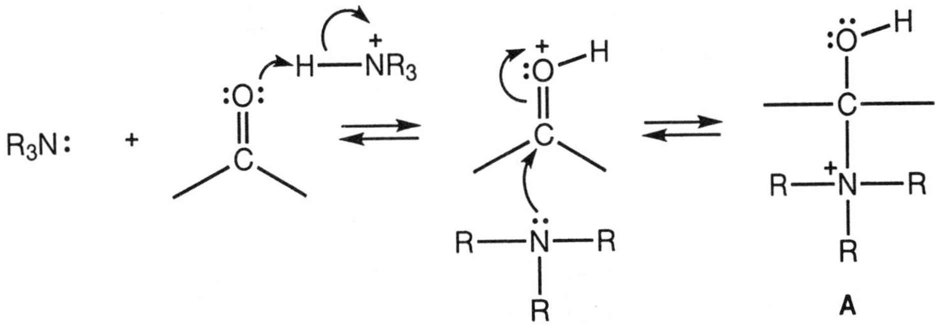

A

When the amine is primary or secondary, proton loss from **A** *can* occur, and the carbinolamine can be formed. If the amine is primary, protonation of the oxygen can be followed by elimination to give an imine.

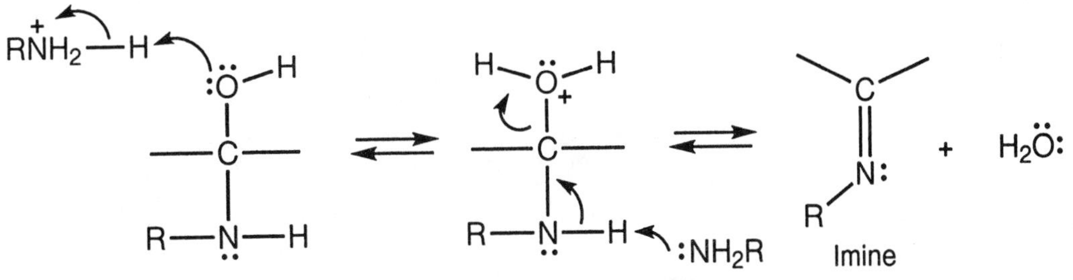

Imine

If the amine is secondary, no imine is possible. Instead, water loss leads to an immonium ion. If there is an available hydrogen on carbon, it can be lost to give the enamine.

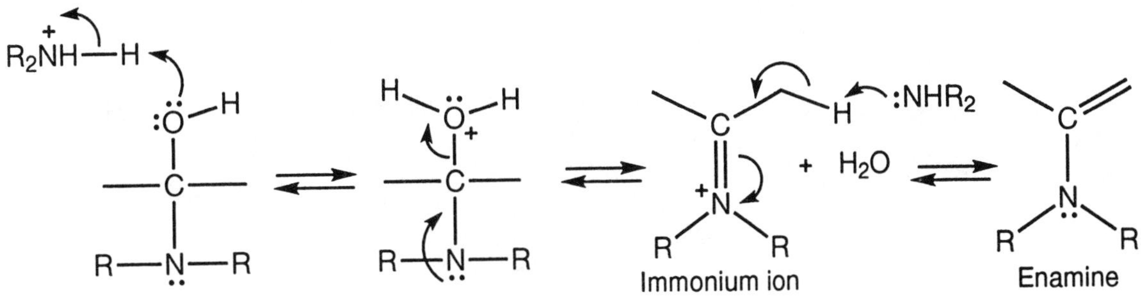

Immonium ion

Enamine

Only secondary amines can give enamines because only secondary amines can give carbinolamines (tertiary amines cannot) but are blocked from imine formation. Primary amines can give imines or enamines, but the imines are usually favored.

Problem 18.12 This problem is simply the reverse of the mechanism for the formation of the immonium ion from the ketone in acid. Just write the mechanism backward.

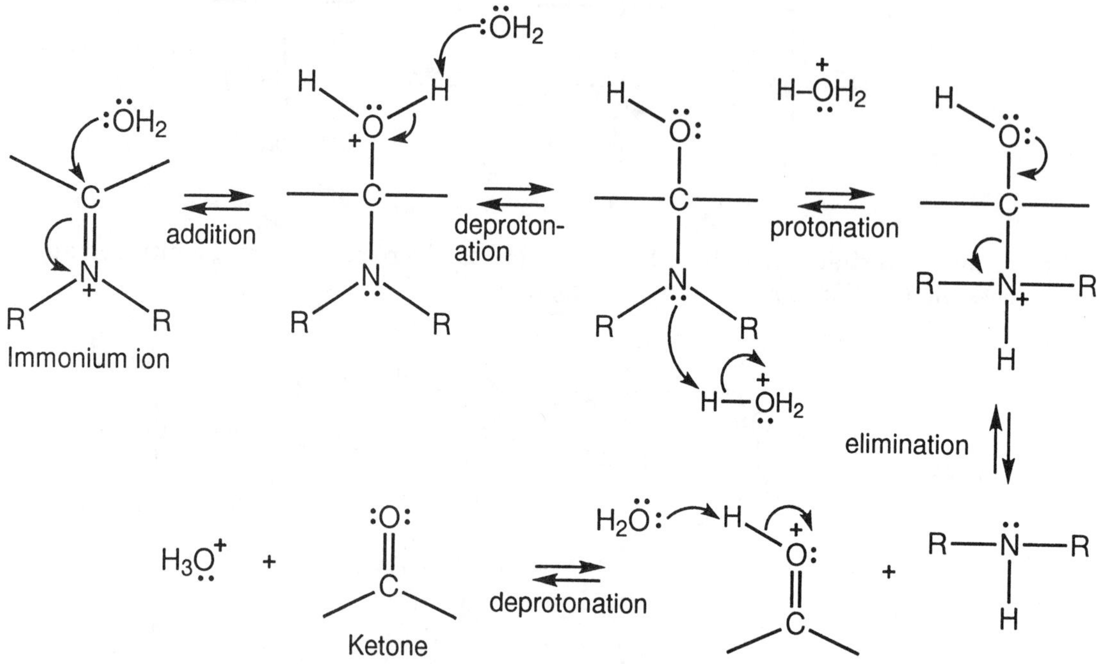

:ÖH₂ addition deproton- ation $H-\ddot{O}H_2^+$ protonation elimination Immonium ion H_3O^+ + Ketone deprotonation $H_2\ddot{O}$: + R—N—R H

Problem 18.14 This problem is simple. All the products are α,β-unsaturated carbonyl compounds formed by loss of water from the β-hydroxy carbonyl compounds.

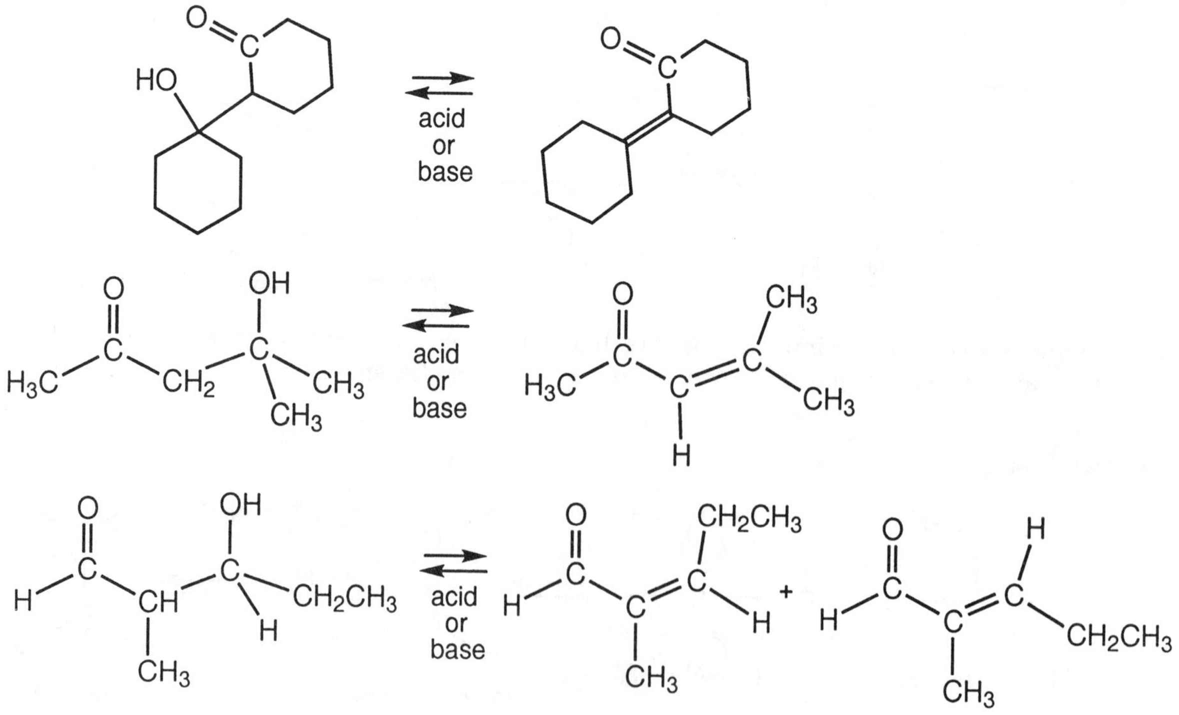

HO acid or base

H_3C CH₂ OH CH_3 CH_3 acid or base H_3C CH_3 CH_3 H

H CH OH CH_2CH_3 H CH_3 acid or base H CH_2CH_3 H CH_3 + H H CH_2CH_3 CH_3

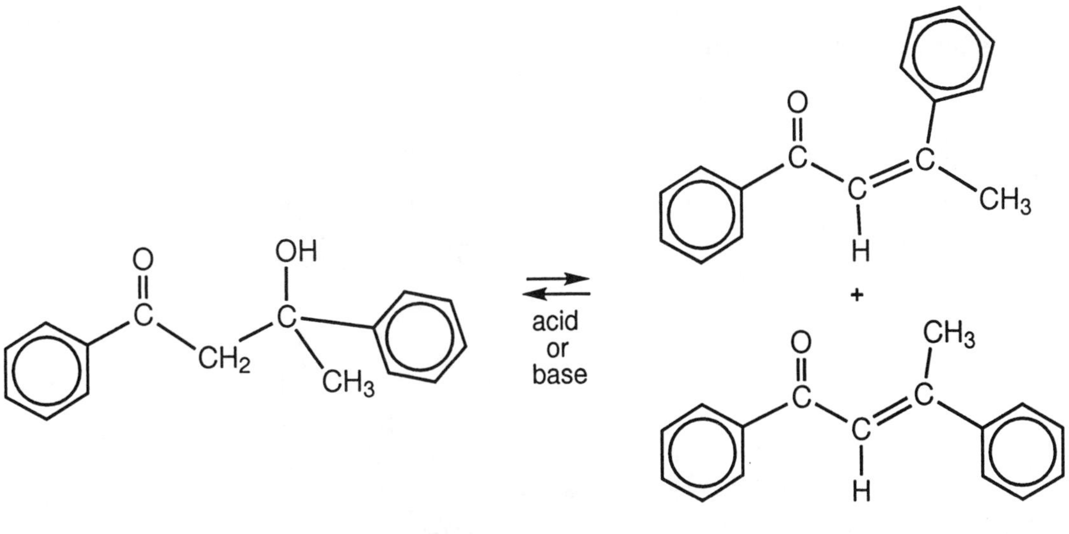

Now ask yourself a harder question: What carbonyl compounds could lead to these β-hydroxy ketones?

Problem 18.16 More drill here. As we have often said, it is important to know reactions backward and forward - to be able to write mechanisms in both directions. These mechanisms are exactly the reverse of the "forward" processes. In acid, the first step is protonation of the carbonyl oxygen. This molecule decomposes to an enol and protonated carbonyl compound. Equilibration leads to two molecules of acetone and regenerates the catalyst for the reaction, H_3O^+.

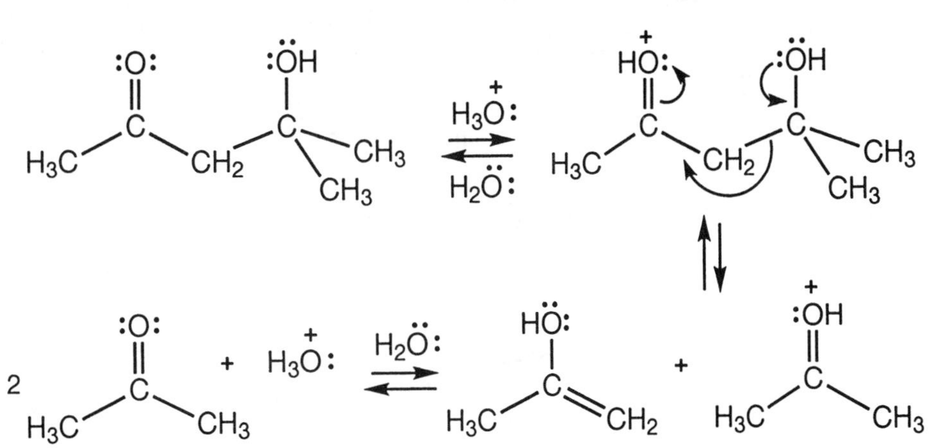

In base, the first step is formation of the alkoxide ion. Now decomposition leads to a molecule of acetone and one of the enolate. Protonation by water gives a second molecule of acetone and regenerates the catalyst, hydroxide ion.

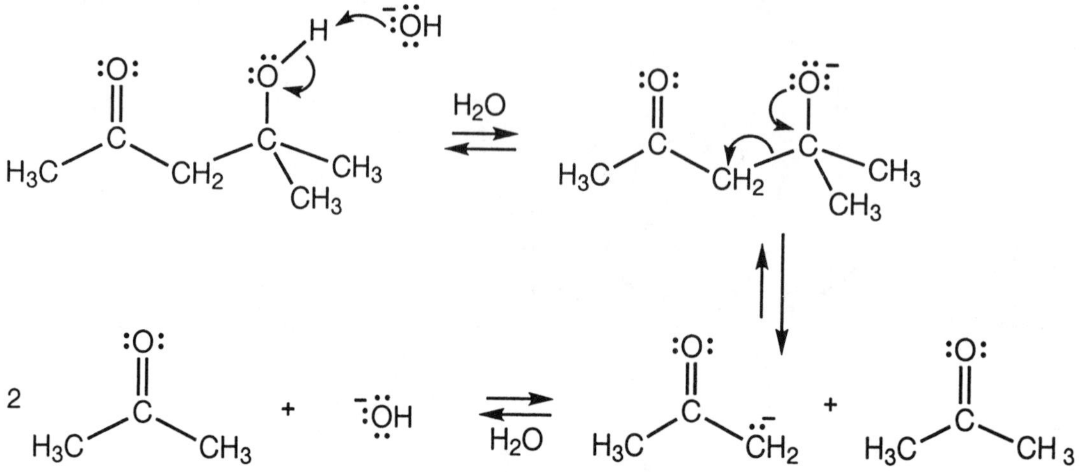

Problem 18.17 This problem provides practice in seeing backward from products to potential starting materials. First just deconstruct the double bond of the α,β-unsaturated carbonyl compound to give the β-hydroxy ketone from which it must have been formed. Then, further deconstruct the β-hydroxy ketone into the two molecules [or two pieces of one molecule in (c)] from which it must have been constructed by breaking the critical bond attaching the two molecules (boldface bond).

(a)

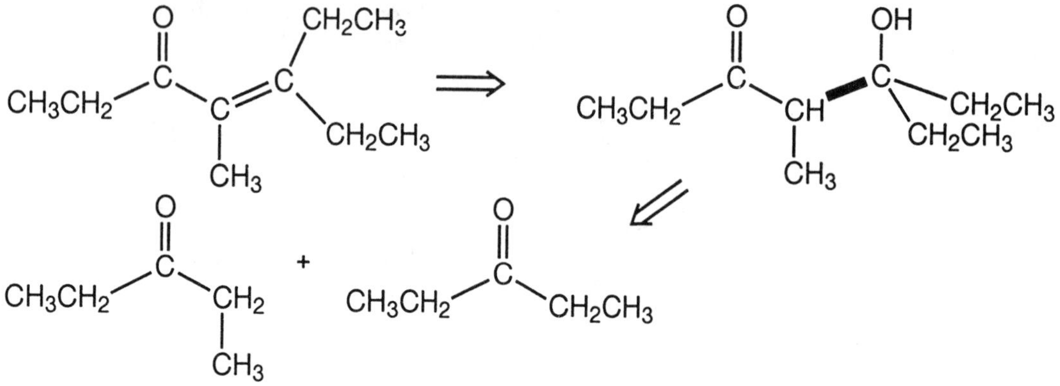

(b)

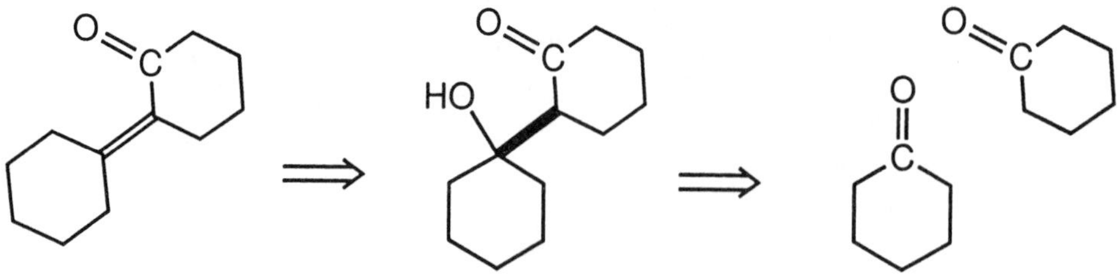

(c)

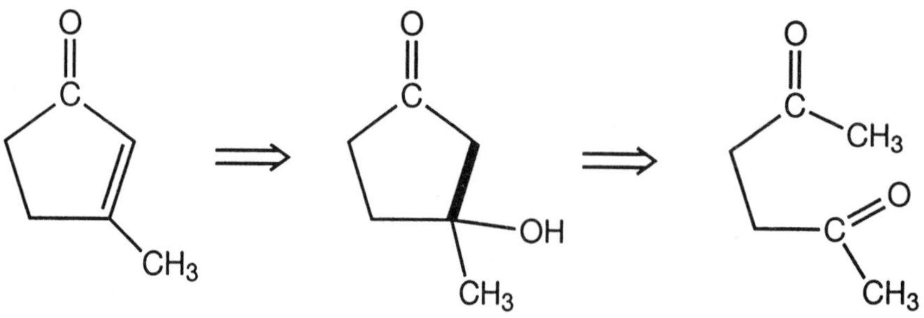

Problem 18.18 This problem involves still more drill on aldol condensations! These are intramolecular reactions, and are a little harder to do than their intermolecular versions. Nothing fundamental has changed, however. In base, the first step is formation of an enolate.

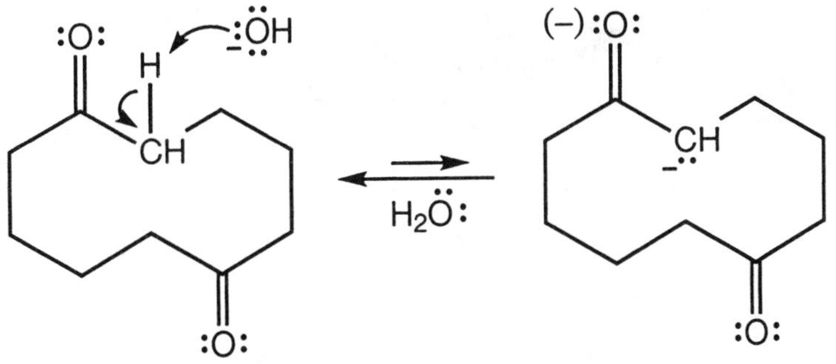

:O: H :ÖH CH H_2O: (−) :O: CH :O: :O:

Note again the use of the shorthand convention for drawing resonance forms; the (–) indicates that the negative charge is shared by oxygen

Now the enolate could add to another molecule of the starting dione in a standard, base-catalyzed aldol condensation. But, a Lewis acid carbonyl compound lurks within the same molecule, well within reach of the anion. Addition leads to a β-hydroxy ketone that can dehydrate to product.

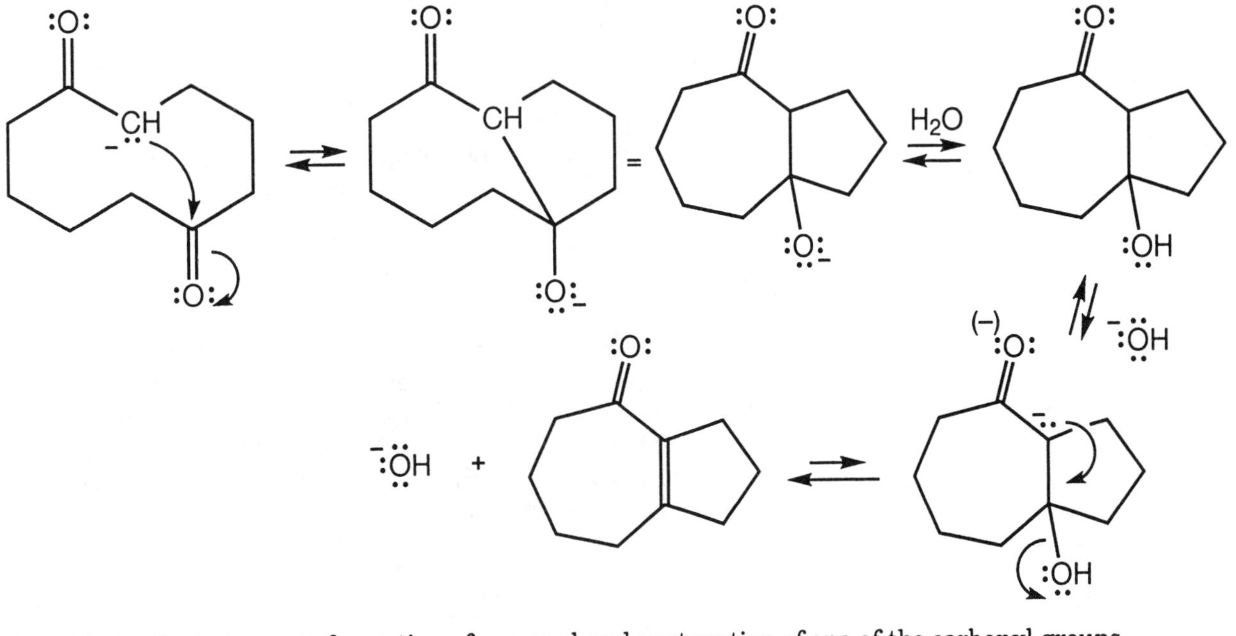

:O: CH :O: :O: CH :O: = :O: :O: H_2O :O: :OH (−) :O: :ÖH :O: :ÖH + :OH

In acid, the first steps are formation of one enol and protonation of one of the carbonyl groups.

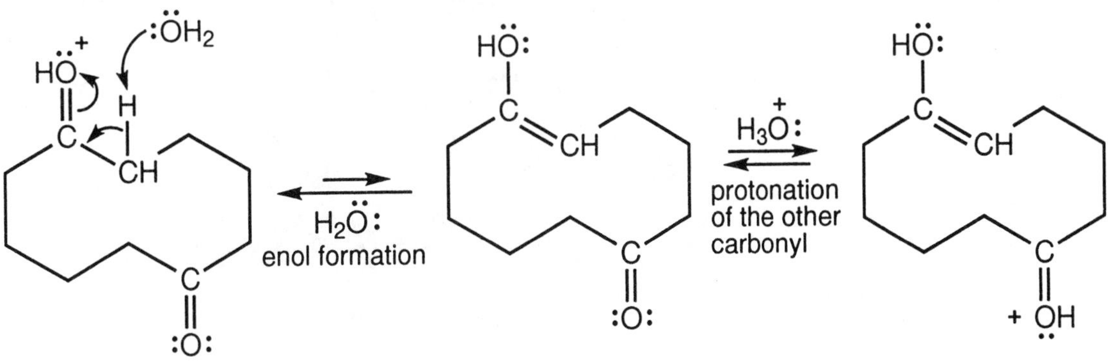

HO + :ÖH₂ H C CH C :O: H_2O: enol formation HÖ: C CH C :O: H_3O^+: protonation of the other carbonyl HÖ: C CH C + OH

Now the enol, a modest nucleophile, adds to the protonated carbonyl group, a strong Lewis acid. As in the base-catalyzed reaction, intramolecular reaction is more favorable than intermolecular reaction. A sequence of deprotonation, protonation, and elimination steps completes the reaction and regenerates the catalyst, H_3O^+.

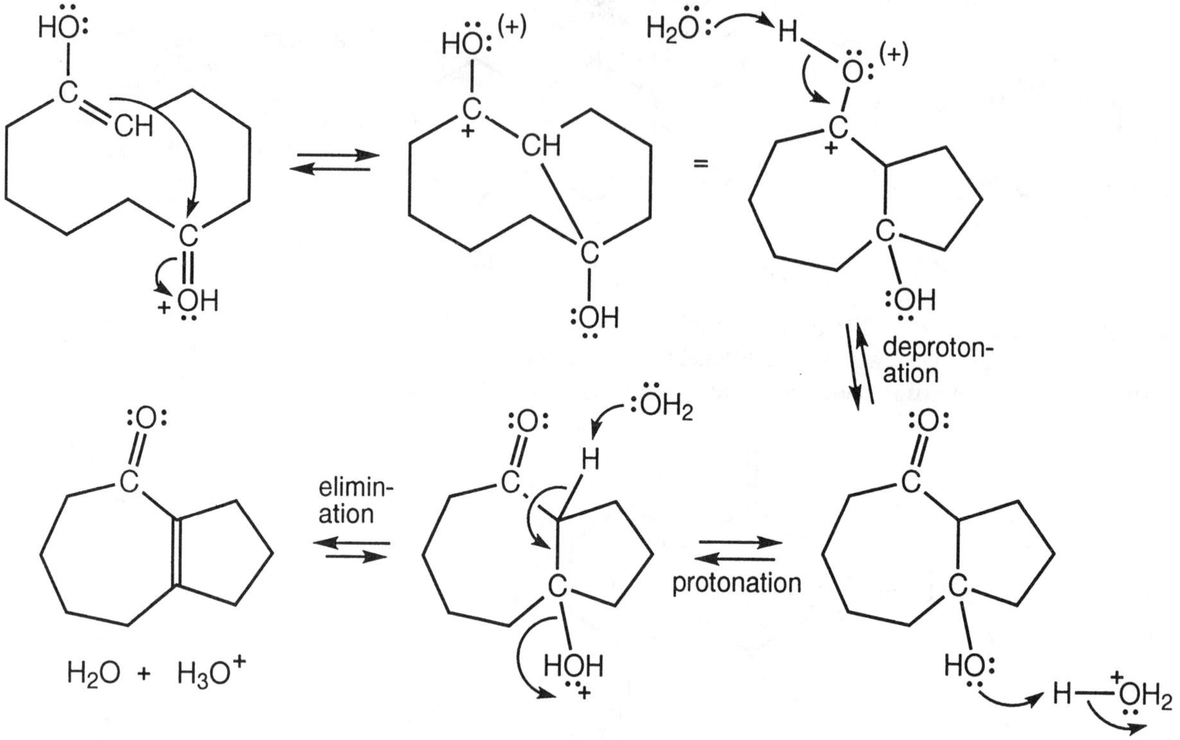

deproton-ation

elimin-ation

protonation

H_2O + H_3O^+

Problem 18.19 More drill, this time on the reverse intramolecular aldol condensation. The mechanistic steps of the previous problem are repeated in the reverse order, starting at the β-hydroxy ketone stage. This problem is very similar to Problem 18.16, except that this time the reaction is intramolecular. In acid, protonation of the carbonyl is followed by the reverse aldol step in which the ring is opened. Deprotonation leads to an enol that regenerates the diketone. If formation of the ketone from the enol in acid is the least bit obscure, by all means write out the detailed mechanism.

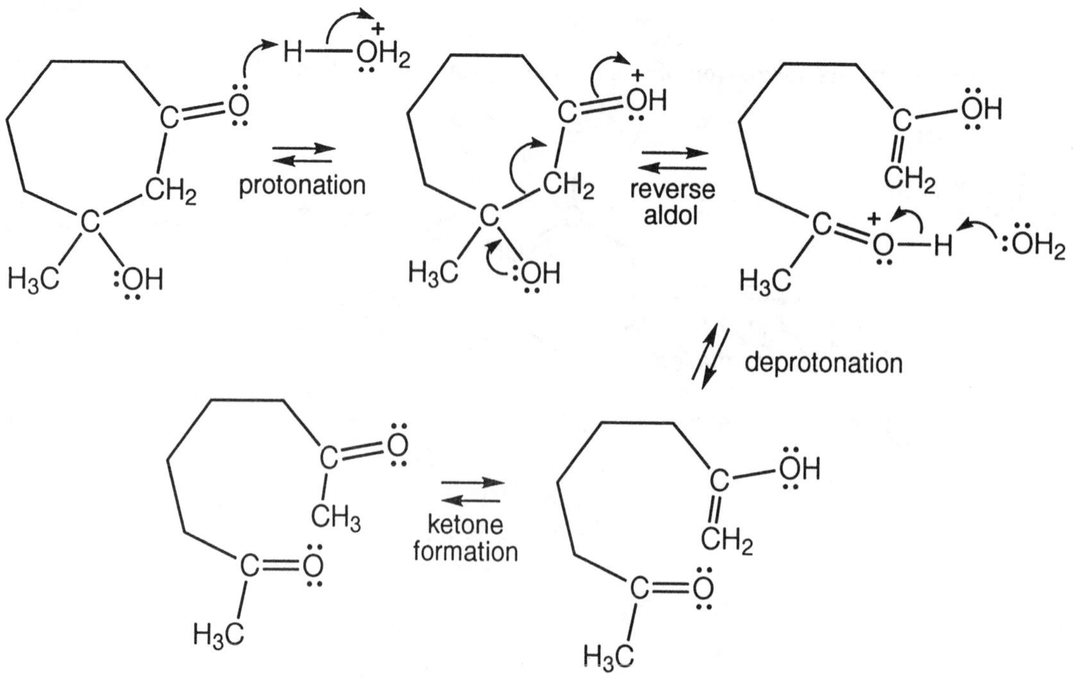

protonation

reverse aldol

deprotonation

ketone formation

In base, the alkoxide is formed and the ring opened to give an enolate. The enolate is protonated at carbon to give the diketone and regenerate hydroxide ion catalyst. Protonation may well occur faster at oxygen to give the enol. The enol, of course, will equilibrate with the diketone, and the diketone will be very strongly favored.

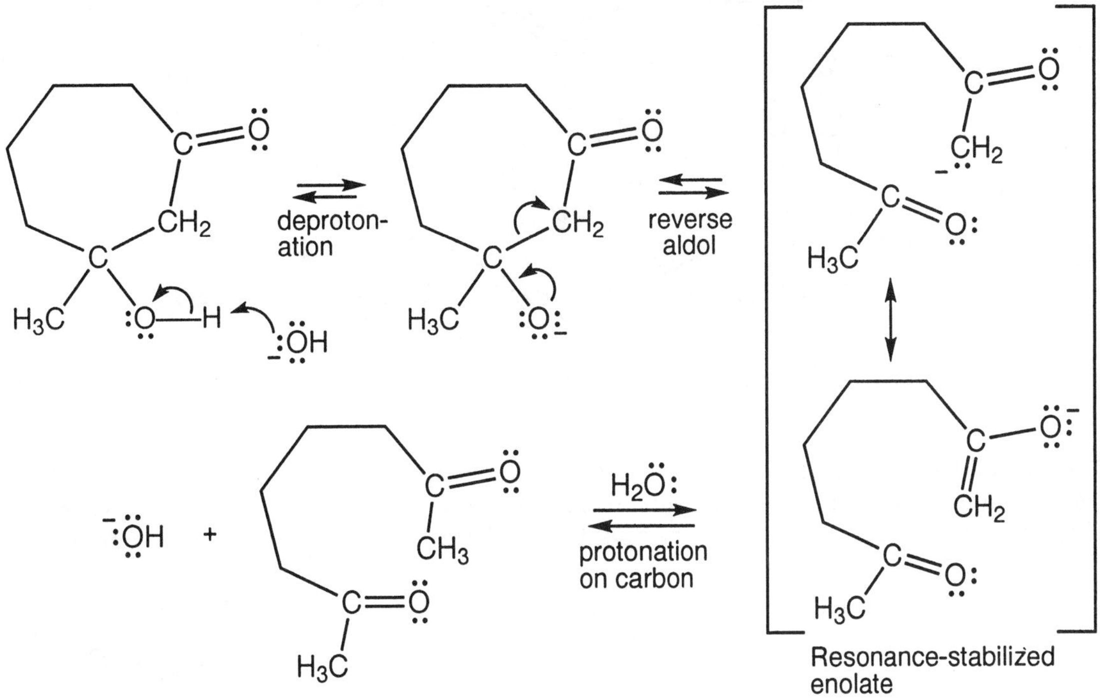

deproton-ation

reverse aldol

Resonance-stabilized enolate

protonation on carbon

Problem 18.20 In base, any aldehyde will form a hydrate. However, this is an equilibrium reaction and there will almost always be enough of the carbonyl compound available for other reactions. Hydrate formation does not interfere because it is reversible.

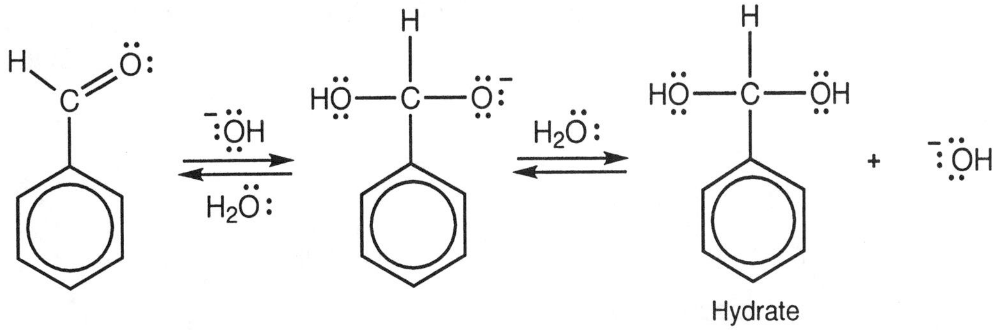

Hydrate

Problem 18.21

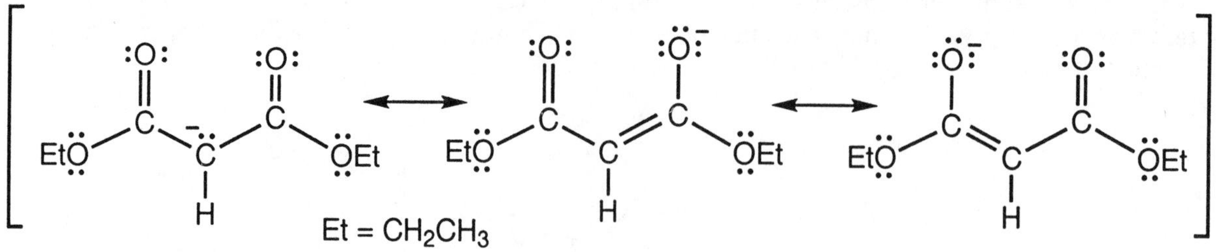

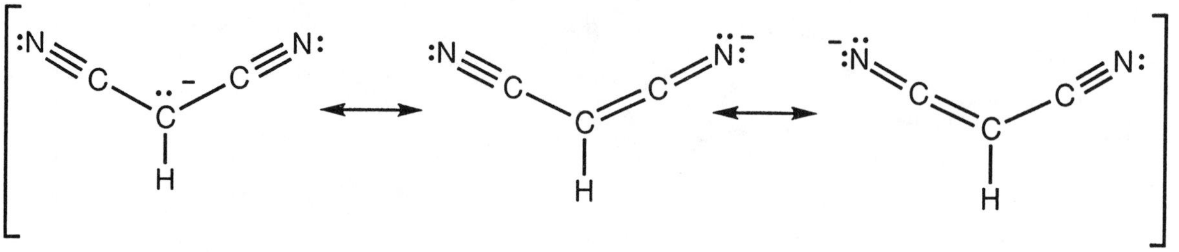

Et = CH_2CH_3

Problem 18.22 Retrosynthetic analysis leads to the following ideas. In each case various basic catalysts could be used. In particular, NaOEt/HOEt should suffice in each reaction. Dehydration probably will take place under the conditions of the condensation reaction or on subsequent acidification.

(a)

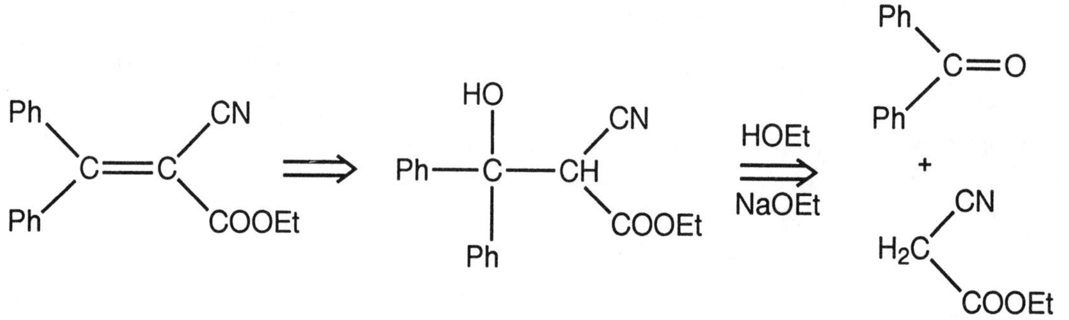

Ph₂C=C(CN)COOEt ⟹ Ph—C(OH)(Ph)—CH(CN)COOEt ⟹ (HOEt, NaOEt) Ph₂C=O + H_2C(CN)COOEt

(b)

$(CH_3)_3C$(H)C=C(COOEt)COOEt ⟹ $(CH_3)_3C$—C(OH)(H)—CH(COOEt)COOEt ⟹ (HOEt, NaOEt) $(CH_3)_3C$(H)C=O + H_2C(COOEt)COOEt

(c)

Ph(H)C=C(cyclopentadiene) ⟹ H—C(Ph)(OH)—cyclopentadienyl ⟹ (HOEt, NaOEt) Ph(H)C=O + cyclopentadiene

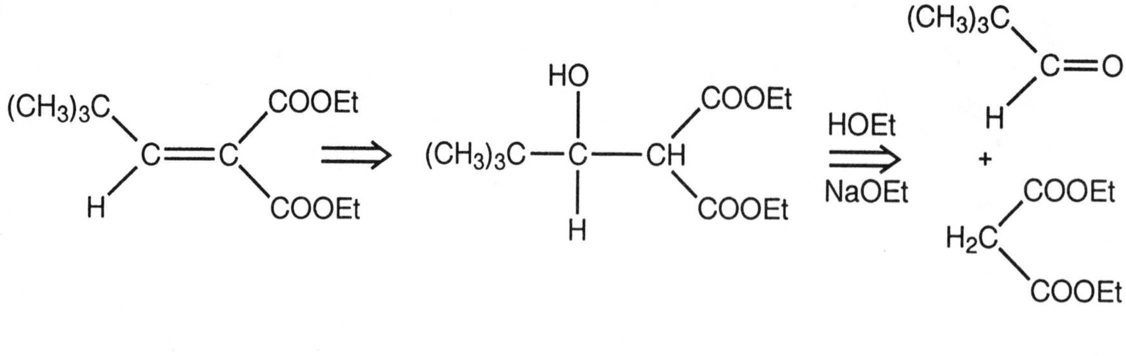

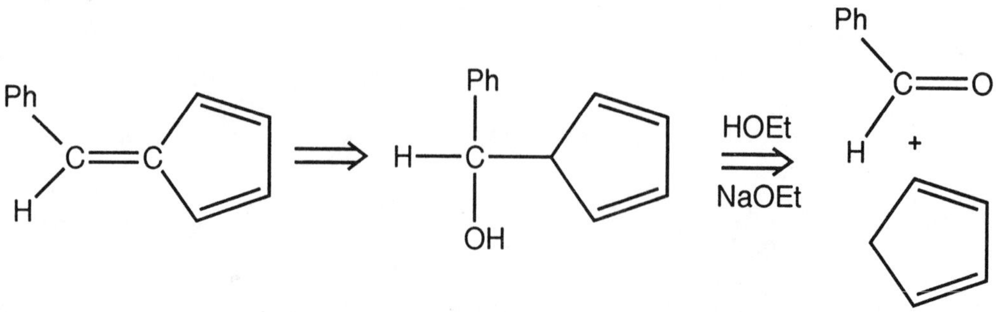

Problem 18.24 This problem insists that we do the aldol first, so there is no doubt about where to start. Notice that an acidic hydrogen has been generated after dehydration. This is the difficult part of the problem - seeing that a resonance-stabilized carbanion can be generated after the initial aldol/dehydration sequence.

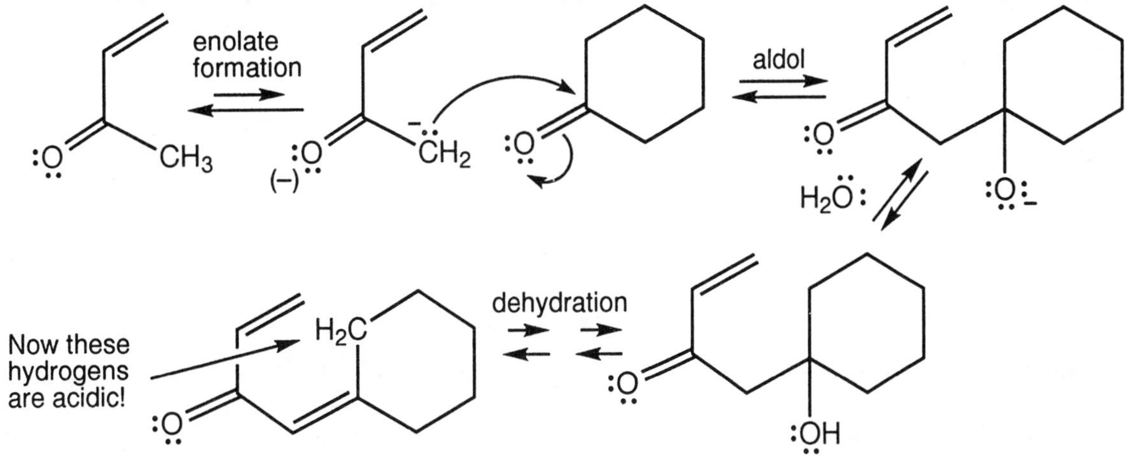

enolate formation

aldol

dehydration

Now these hydrogens are acidic!

Once the new enolate is formed, a Michael reaction can take place to close the second ring. Protonation at carbon gives the product.

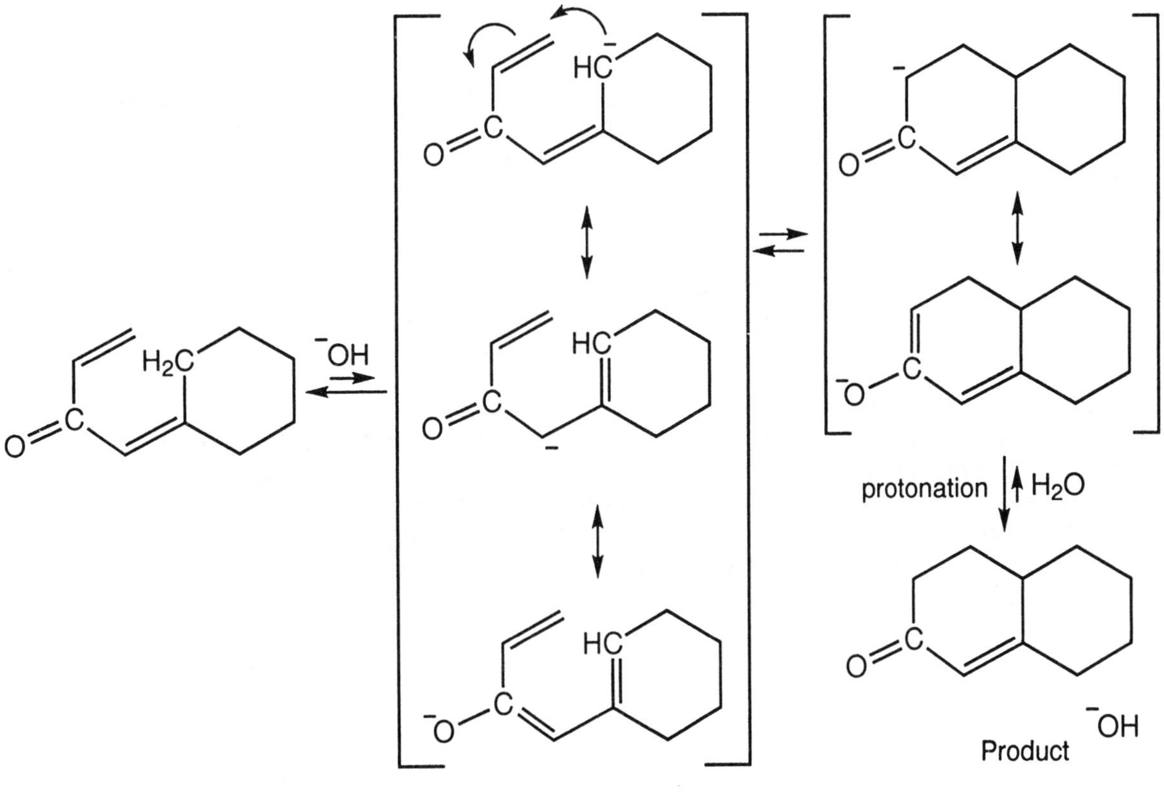

protonation

Product

Resonance-stabilized enolate

Problem 18.25

(a) The first step is a straightforward base-catalyzed enolization. If this sequence is not obvious, be certain to go back and review keto-enol equilibrations. Once the enolate is formed in base, re-formation of the ketone can take place in two ways. If the proton is added from the bottom side, starting material is regenerated. If, however, the proton is added from the top side, the product is formed.

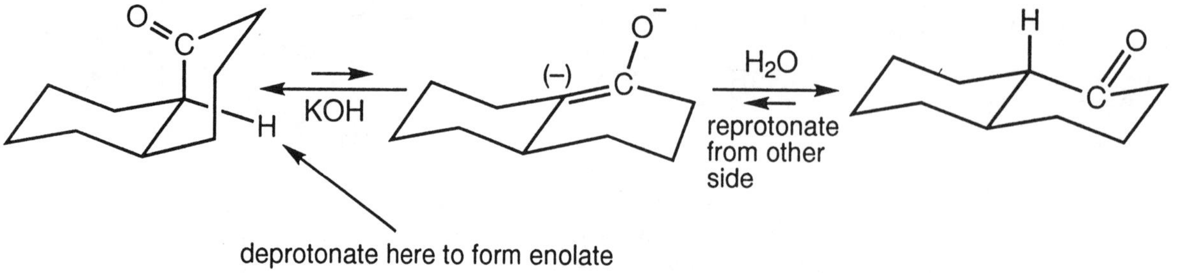

KOH

H_2O

reprotonate from other side

deprotonate here to form enolate

(b) It never hurts to start with a little analysis. We have to close a ring here, and the bromine atom is lost in the process. The new carbon-carbon bond that attaches the two pieces seems likely to be formed through an addition to the carbonyl group. We start with enolate formation. As there is only one type of α-hydrogen, there is no problem in finding the proper enolate. Addition of the enolate to the carbon-oxygen double bond yields intermediate **A**, and intramolecular S_N2 displacement of bromide gives the product.

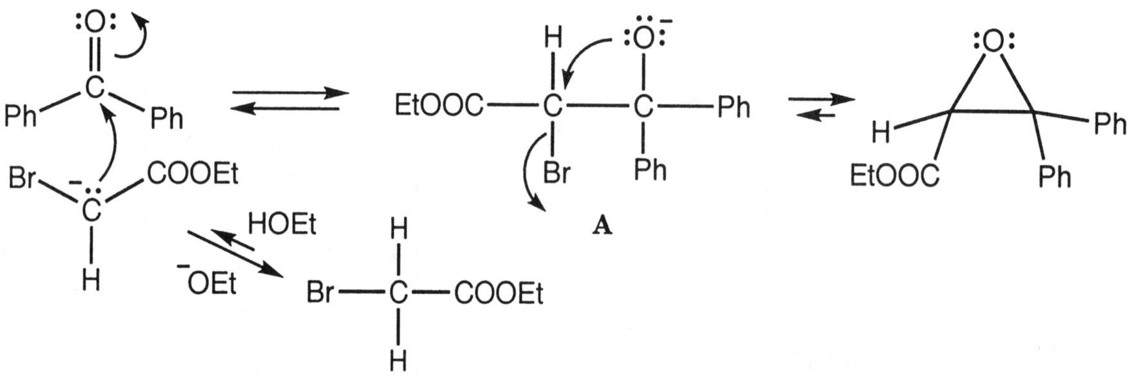

HOEt

$^-$OEt

A

(c) This problem clearly asks you to find a way to open a ring; to break a carbon-carbon bond. We are used to seeing things in the other direction - to finding ways to make carbon-carbon bonds. Success in this problem requires that we see the aldol condensation in both directions, forward and reverse. Here we have an intramolecular base-catalyzed reverse aldol condensation. The key is to recognize the starting material as a β-hydroxy ketone. *All* β-hydroxy ketones are potentially the products of aldol condensations. In this case, the first step is formation of alkoxide **A**. The reverse aldol opens the four-membered ring and forms the resonance-stabilized enolate **B**. Addition of a proton at carbon yields the diketone product.

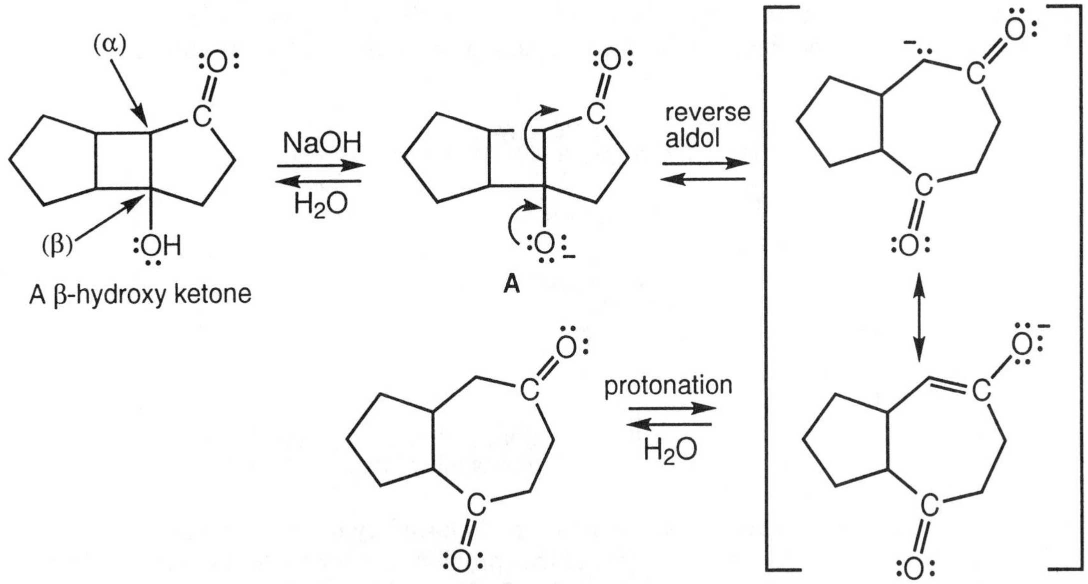

(α)
:O:
C
NaOH
H₂O
(β)
:OH
A β-hydroxy ketone
:O:
C
:O:
A
reverse aldol
:O:
C
C
:O:
:O:
C
:O:
C
protonation
H₂O
:O:
C
C
:O:

(d) The recognition point in this problem is the formation of a cyclohexenone. Cyclohexenones can be made by the combination of Michael and aldol condensations known as the Robinson annulation. In this case, two enolates could be formed; one from methyl vinyl ketone and one from cyclohexanecarboxaldehyde. We'll follow Magid's second rule, and use the second enolate to do a Michael addition to methyl vinyl ketone. Protonation and a second enolate formation leads to **A**. The second step in the Robinson annulation is an aldol condensation, which in this case gives **B**. A sequence of protonation, formation of a third enolate, and elimination gives the product.

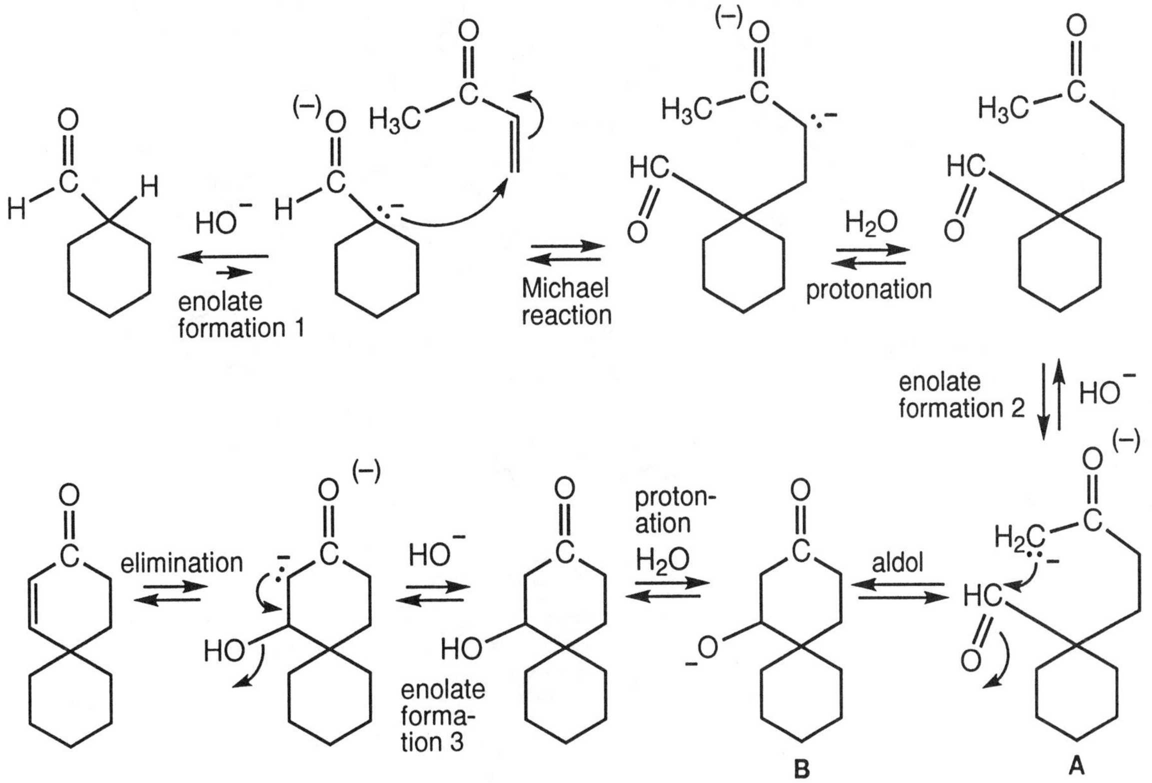

O
C
H H
HO⁻
enolate formation 1
(–)O
C
H
O
C
H_3C
Michael reaction
(–)O
C
H_3C
HC
O
H_2O
protonation
O
C
H_3C
HC
O
enolate formation 2
HO⁻
O (–)
C
H_2C
HC
O
A
aldol
O
C
⁻O
B
proton-ation
H_2O
O
C
HO
enolate forma-tion 3
HO⁻
O (–)
C
HO
elimination
O
C

(e) Now we come to tougher problems, and analysis before you begin will become even more important. In this case, we can see the probable final resting place of benzaldehyde within the product.

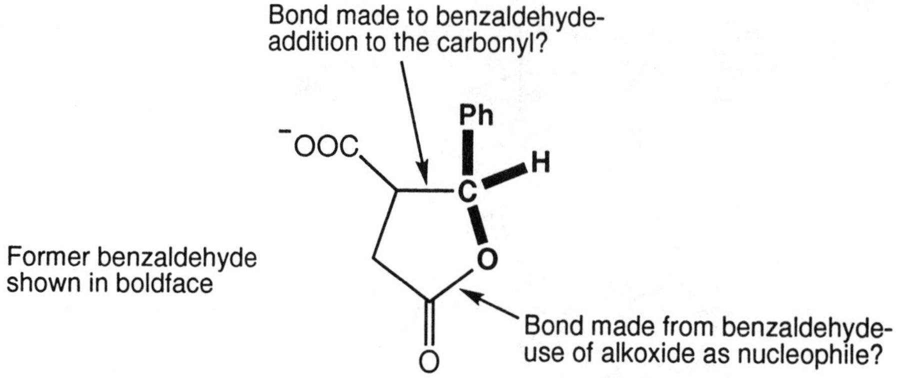

Bond made to benzaldehyde-addition to the carbonyl?

Former benzaldehyde shown in boldface

Bond made from benzaldehyde-use of alkoxide as nucleophile?

It seems that an addition reaction to benzaldehyde has been followed by incorporation of the resulting alkoxide into a new ring. At least the start of this problem is easy since there is only one possible enolate. Addition to the carbon-oxygen double bond of benzaldehyde leads to **A**, and accomplishes one of our goals, the formation of the bond to benzaldehyde. The hard part of this problem is the next step. The nucleophilic alkoxide ion adds to one carbonyl group to give **B.** How do we choose between the two different carbonyl groups? Addition to the other carbonyl would generate **C**, a strained four-membered ring, and this reaction will be slower than the one shown. Opening of **B** gives the product.

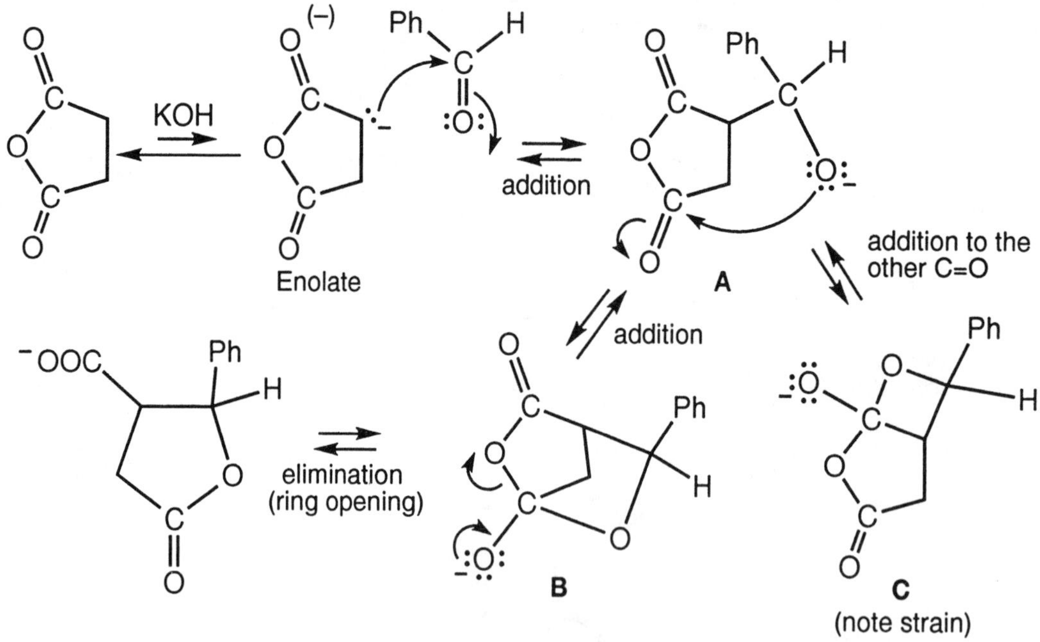

KOH

Enolate

addition

A

addition to the other C=O

addition

elimination (ring opening)

B

C (note strain)

(f) This one is really hard. Many things can happen when the starting material reacts with hydroxide, but we will first try the Michael reaction (Magid's second rule) to give **A**. The hint tells us that a 10-membered ring is involved, so the ring junction must be broken. Let's do a reverse Michael reaction to give **B**. Protonation of enolate **B**, followed by ketonization of the enol gives **C**, a molecule with two "doubly α" hydrogens. Removal of one of them by hydroxide leads to **D**. In **D**, an intramolecular aldol reaction to give **E** is possible. Compound **E** also contains a "doubly α" hydrogen, and elimination of hydroxide leads to the final product.

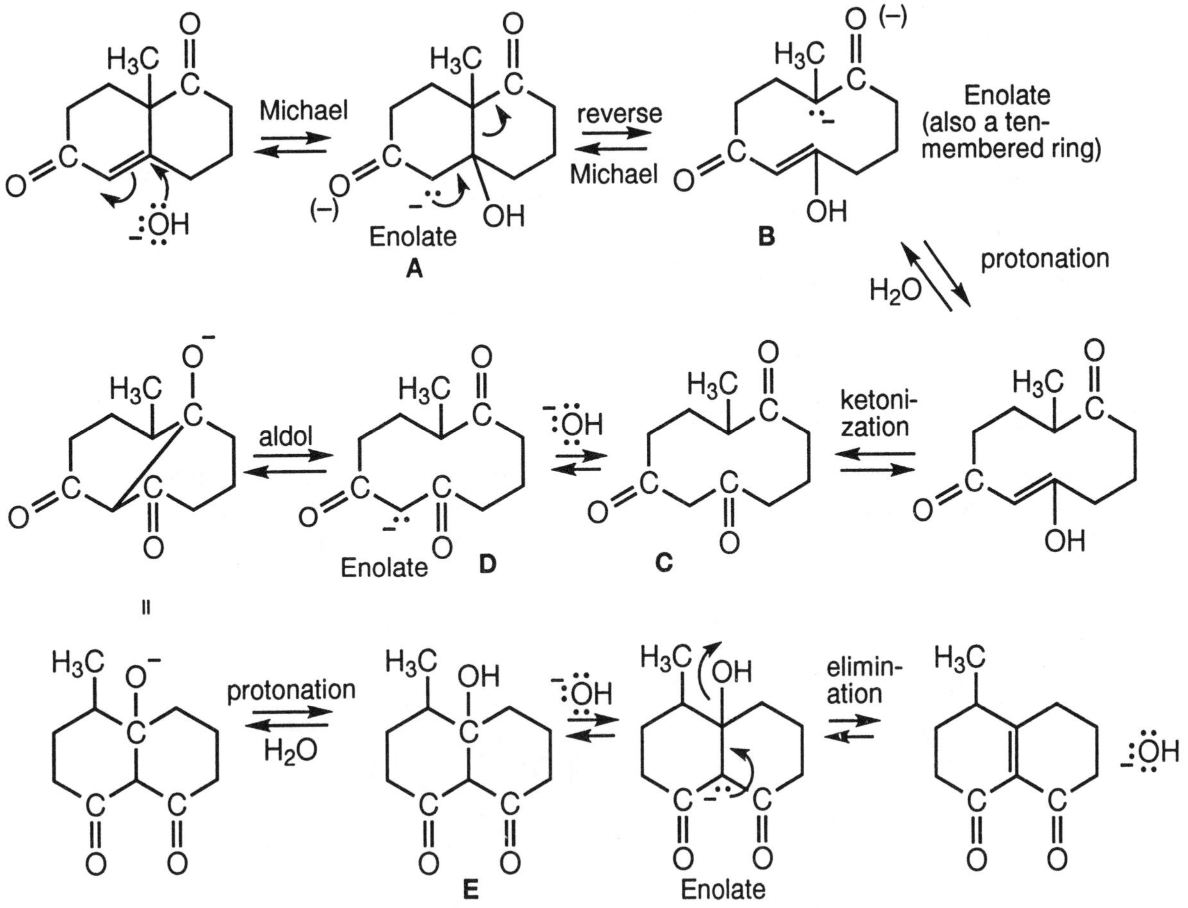

Michael
reverse Michael
Enolate A
B
Enolate (also a ten-membered ring)
protonation H_2O
ketonization
C
Enolate D
aldol
protonation H_2O
E
Enolate
elimination

Problem 18.28 In the hydride shift step of the reaction the carbon-hydrogen bond of one aldehyde is breaking, and the new hydrogen-carbon bond is being made. The dashed lines show the partially made and partially broken bonds.

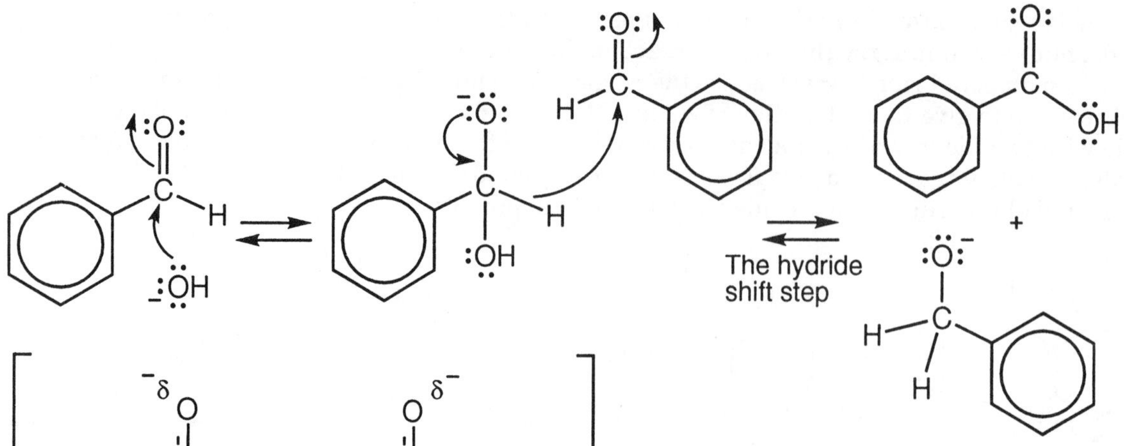

The hydride shift step

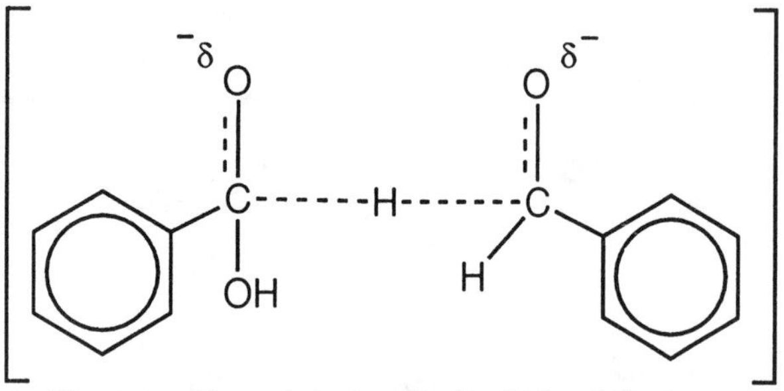

The transition state for the hydride shift step

Empty 2*p* orbital

Empty 3*p* orbital

Problem 18.29 Aluminum sits right below boron in the periodic table. Trisubstituted boron compounds should be familiar. Boron is hybridized sp^2 and there is an empty 2*p* orbital on boron. The situation for aluminum is the same except that it is a 3*p*, not a 2*p*, empty orbital.

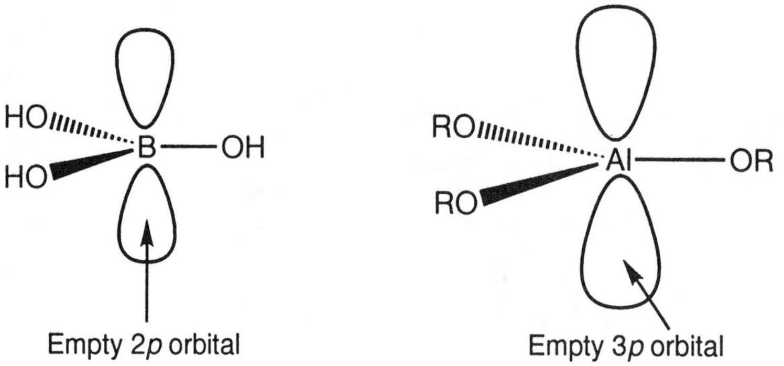

Problem 18.30 A "complex" between the carbonyl compound and the aluminum alkoxide is formed, just as it is in the first step of the reaction.

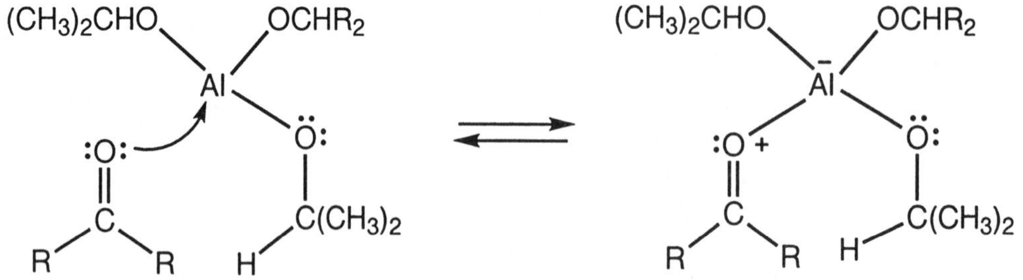

Next, the hydride migrates, reducing the old carbonyl and oxidizing one of the alkoxide groups. This generates a molecule of acetone and $(R_2CHO)_2AlOCH(CH_3)_2$.

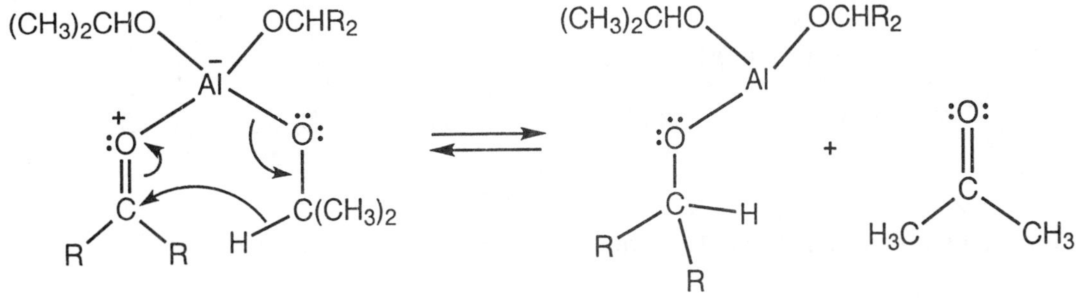

Problem 18.31 This alkoxide, $Al(OR)_3$ has an empty $3p$ orbital (see Problem 18.29), and is a good Lewis acid. The nucleophile H_2O first adds to this Lewis acid forming a tetrahedral intermediate.

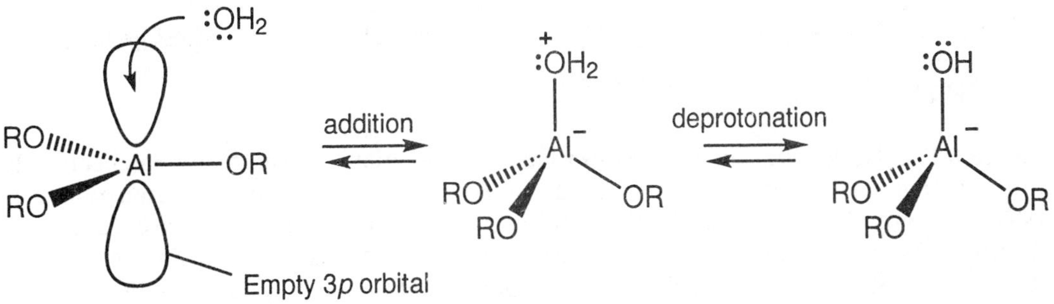

This intermediate can reverse to regenerate the $Al(OR)_3$ and H_2O, or can lose RO^-, and produce $HOAl(OR)_2$.

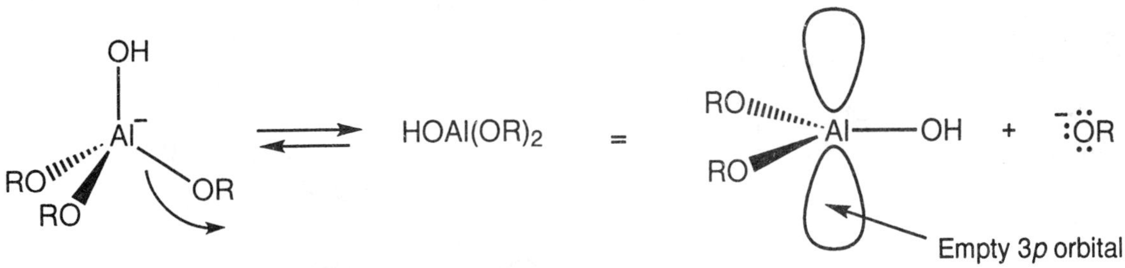

The overall conversion is $Al(OR)_3 \rightleftharpoons HOAl(OR)_2$. One "OR" has been replaced with an "OH". Repetition two times leads to the final product, $Al(OH)_3$.

Problem 18.32 Not much! This reaction simply regenerates undeuterated isobutene.

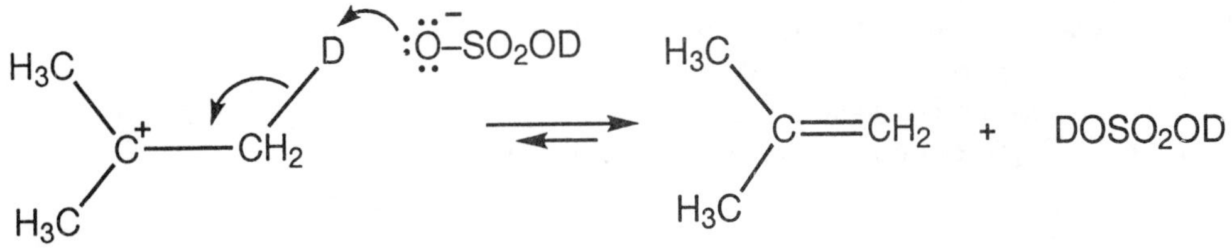

The reaction can now start over again.

Problem 18.33 Any monodeuterio isobutene will do. The first step is addition of a deuteron from deuteriosulfuric acid to give the tertiary carbocation.

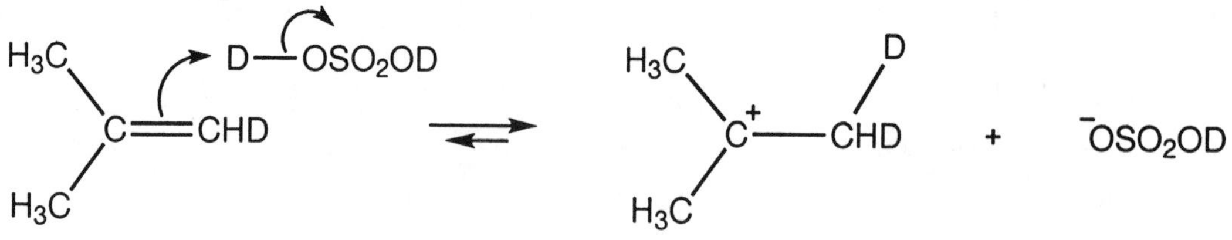

Removal of any proton leads to a dideuterated isobutene.

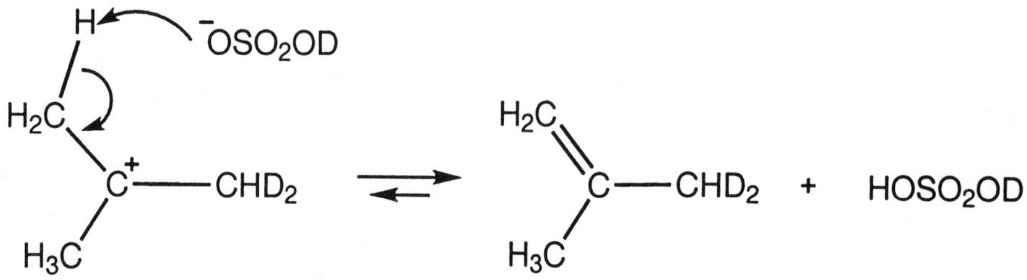

Problem 18.34 (a) Notice first the intramolecular redox reaction. One carbonyl has been reduced to the alcohol stage and the other has been oxidized to the carboxylic acid stage. Reaction (a) is an intramolecular Cannizzaro reaction. Addition of hydroxide is to the more reactive carbonyl group of the molecule, the aldehyde, to give adduct **A**. Hydride transfer takes place within the molecule to generate **B**. Protonation of the alkoxide and deprotonation of the acid leads to the product.

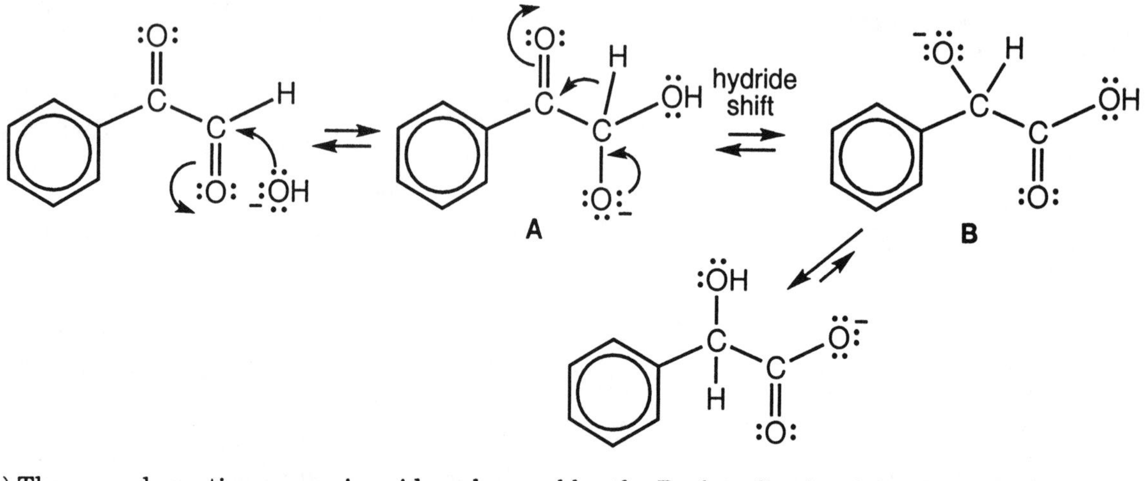

(b) The second reaction occurs in acid, and resembles the Barlett-Condon-Schneider reaction and the pinacol rearrangement. The structure of the product gives the clue to what happens - a way must be found to transfer the deuterium from one side of the molecule to the other. The first steps are protonation of one hydroxyl group and loss of water to generate the tertiary carbocation, **C**. This carbocation is a strong Lewis acid, and the deuteride (D^-) shifts to give the resonance-stabilized cation **D**. Deprotonation by water gives the product.

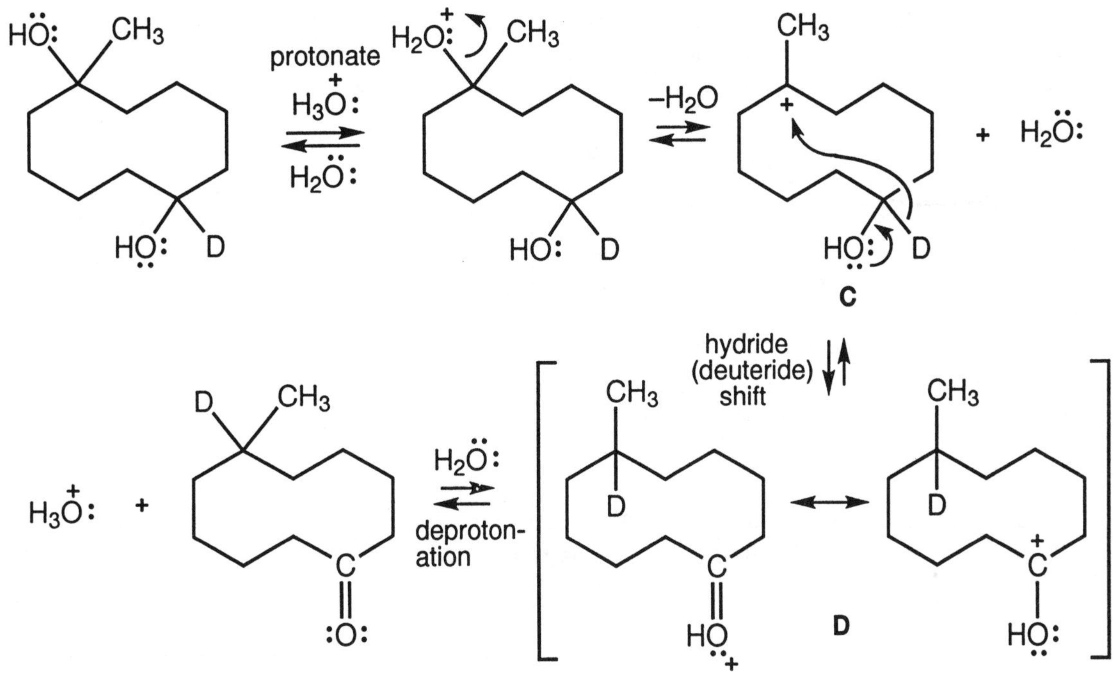

(c) This part advises the construction of a good three-dimensional drawing, so let's start with that. The drawing includes the hydrogen adjacent to the hydroxyl group, drawn in what should be a suggestive way.

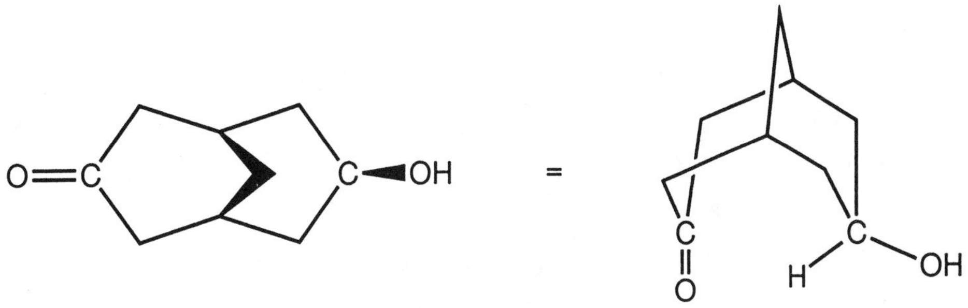

Of course, the four hydrogens adjacent to the carbonyl group will exchange in deuterated base. (If this is the least bit obscure, be certain you can draw the mechanism for this garden-variety α-exchange!) In strong base another reaction will certainly occur, the loss of the hydroxyl proton to give the alkoxide.

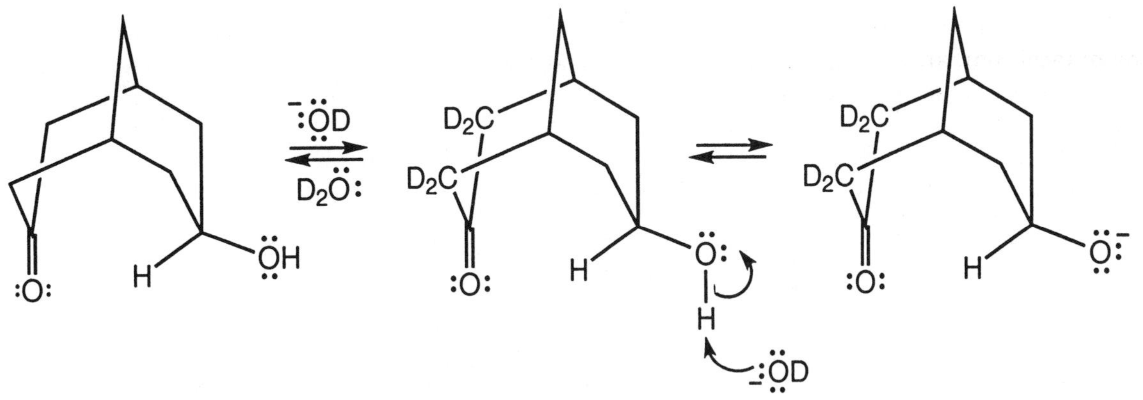

Now an intramolecular Cannizzaro reaction occurs as the hydride is transferred to the strategically placed carbonyl group on the other side of the molecule. The carbonyl group is reduced and the alkoxide is oxidized. The newly formed alkoxide is deuterated by D_2O.

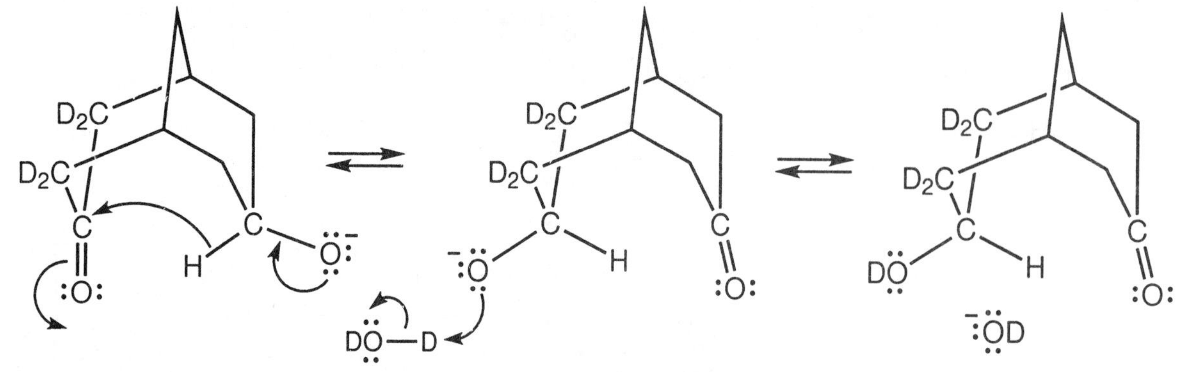

The new carbonyl activates the adjacent hydrogens which exchange to give the final molecule in which eight carbon-hydrogen bonds have been exchanged for deuterium.

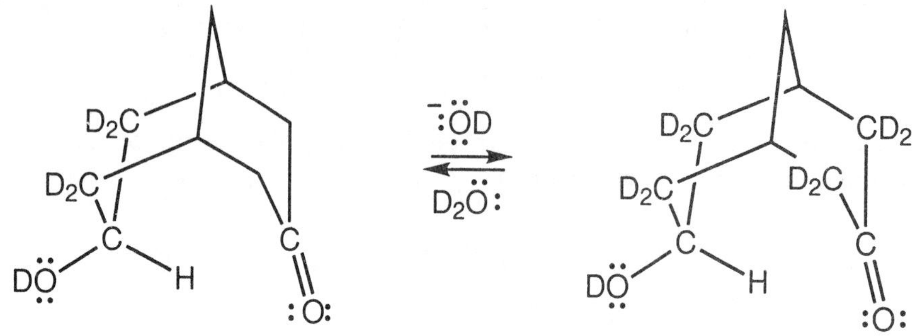

In the other stereoisomer no hydride shift is possible, as the hydrogen points to the outside of the molecule and not towards the acceptor carbon-oxygen double bond.

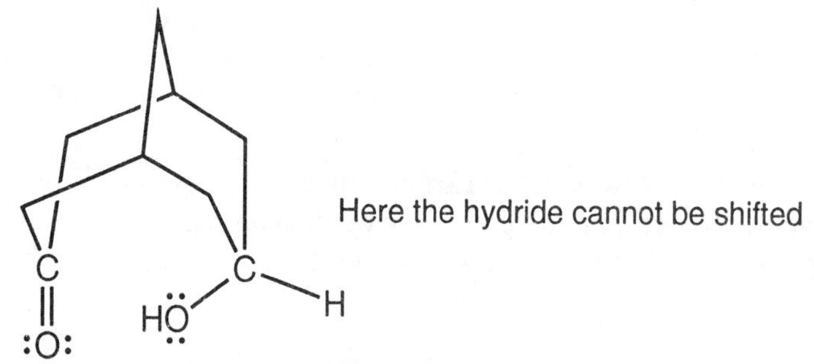

Here the hydride cannot be shifted

The only possible reaction is exchange of the four hydrogens adjacent to the carbonyl, and exchange of the hydroxyl hydrogen .

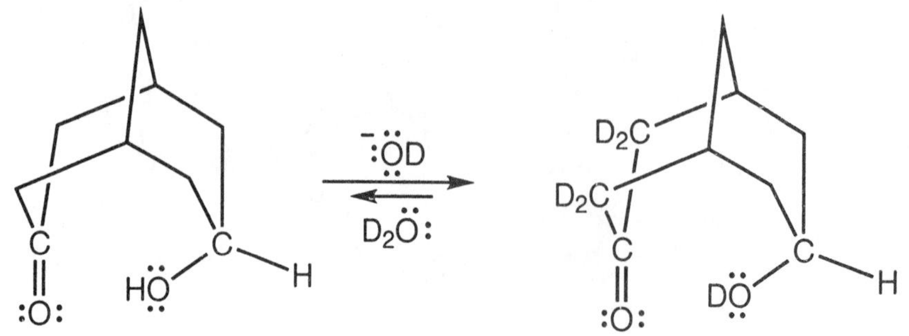

Additional Problem Answers

Problem 18.35(a) This problem involves a straightforward keto-enol equilibration. In base, the active ingredient is the enolate, whereas in acid it is the protonated carbonyl. Both intermediates are resonance-stabilized.

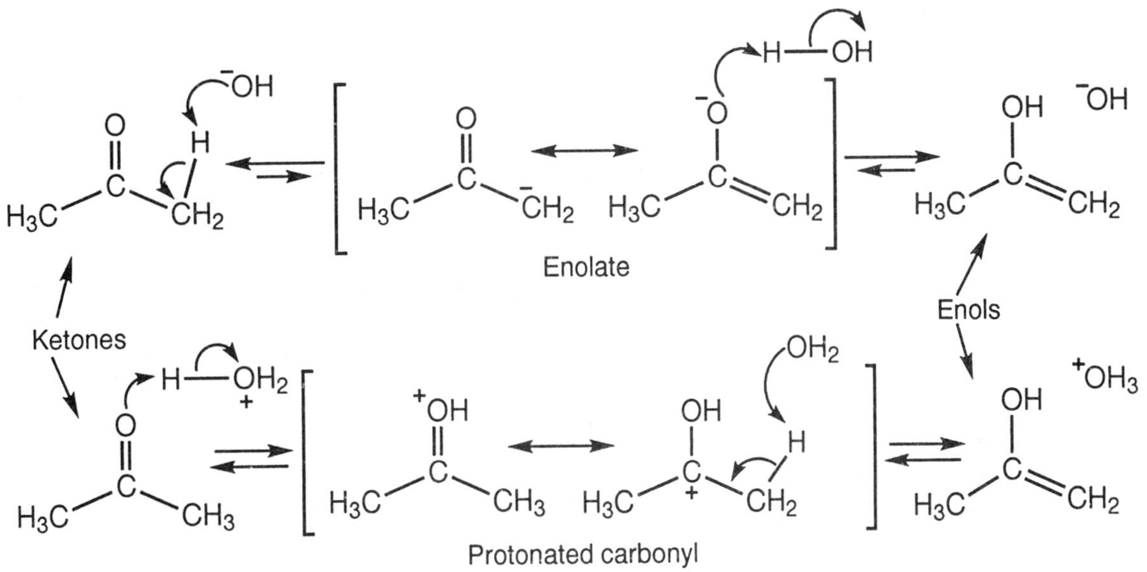

Ketones

Enolate

Enols

Protonated carbonyl

(b) This imine-enamine equilibrium is just the nitrogen counterpart of the keto-enol equilibrium in (a).

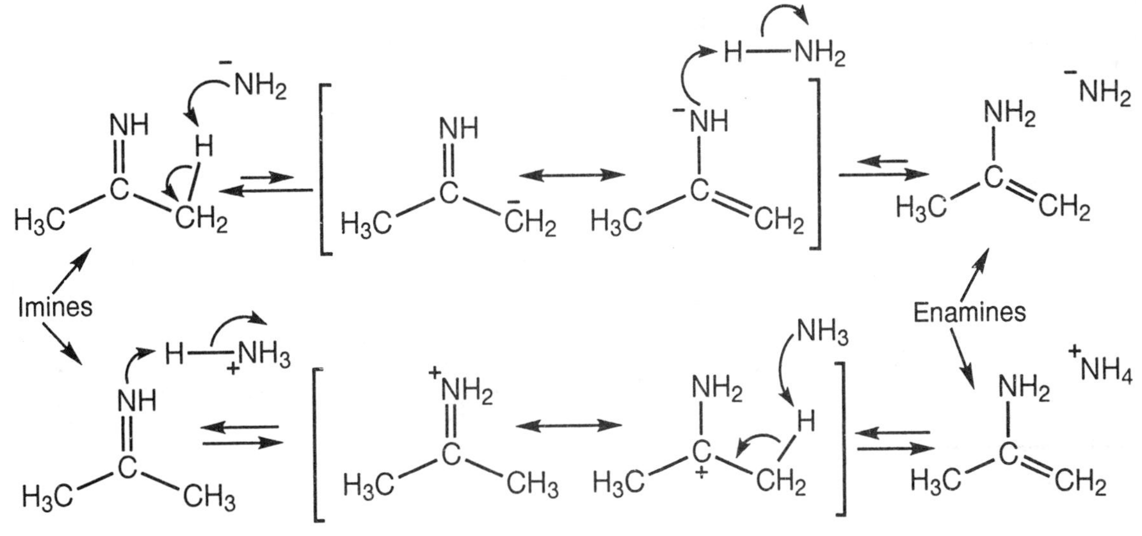

Imines

Enamines

(c) Hydrate formation can be both acid- and base-catalyzed.

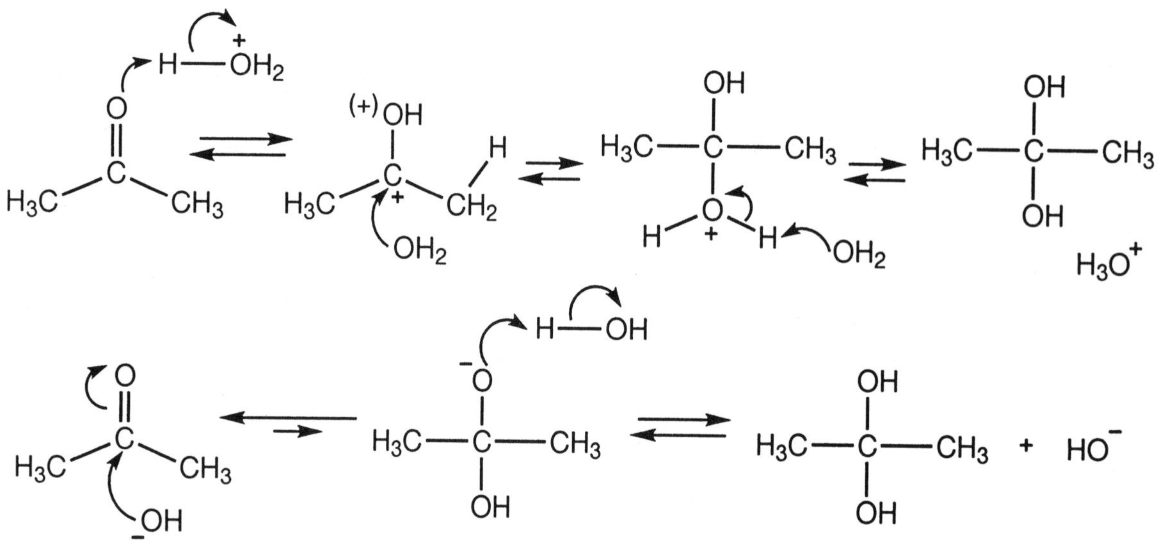

Problem 18.36 Removal of the methylene hydrogen leads to an enolate that is stabilized by delocalization by both the carbonyl group and the aromatic ring. It is surely more stable than the other enolate in which only delocalization by the carbonyl group is possible. Enolate **A** also contains the more substituted double bond, and hence is more stable than enolate **B**, which contains the less substituted double bond.

A, the more stable enolate

B, the less stable enolate

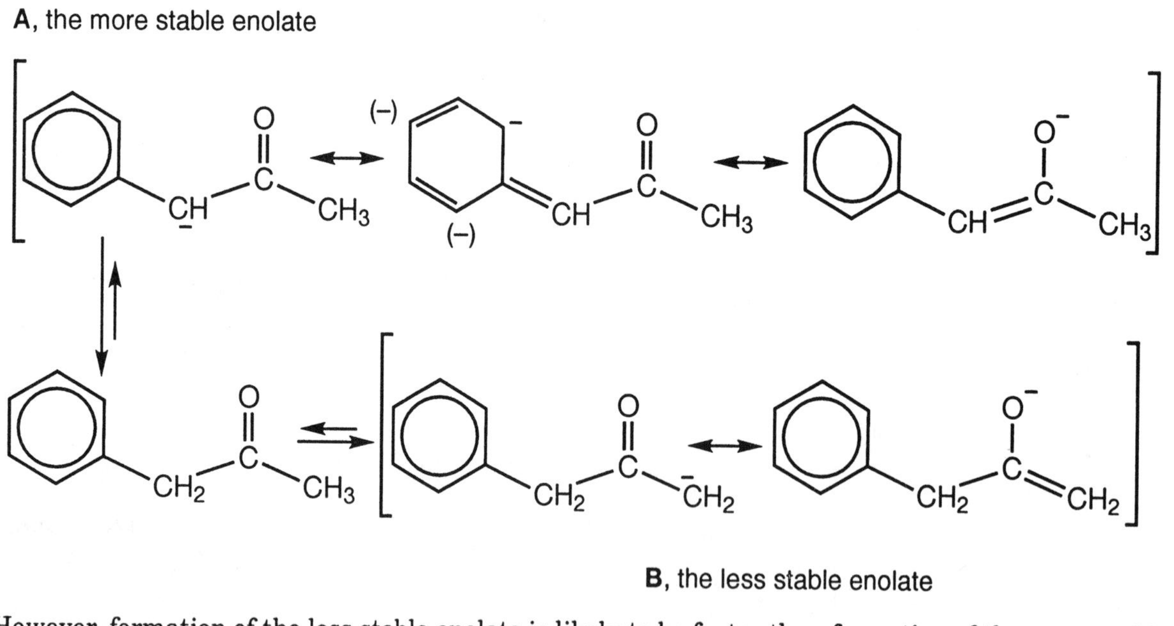

However, formation of the less stable enolate is likely to be faster than formation of the more stable enolate, especially if a large base such as LDA is used. Approach to the relatively open methyl group is sterically easier than is approach to the more "buried" methylene group.

Problem 18.37 Parts (a)-(c), and (e) should be straightforward. The positions α to the carbonyl group can be exchanged through the usual formation of the enolate. The products are shown in their exchanged forms.

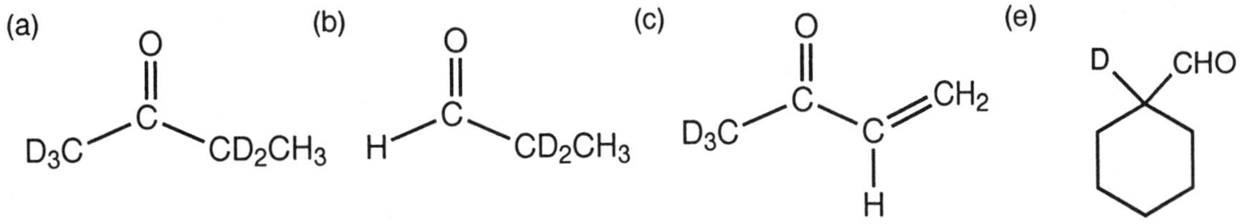

(a) $D_3C-CO-CD_2CH_3$ (b) $H-CO-CD_2CH_3$ (c) $D_3C-CO-CH=CH_2$ (e) 1-deuteriocyclohexanecarbaldehyde (D, CHO)

Part (d) is more troublesome. To get this one completely right you have to recognize two things, and one is tricky. First, the presence of the carbon-carbon double bond in conjugation with the carbonyl group allows the γ-position to exchange. The figure shows one γ-hydrogen exchanging through enolate **A**.

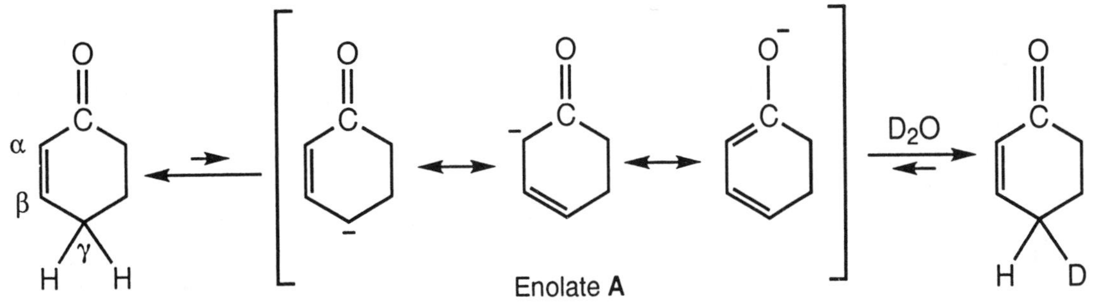

α β γ H H D₂O H D

Enolate **A**

Of course, the other γ-hydrogen and the usual α-hydrogens also exchange. So, one might think that the final result will be as shown for (d).

(d) O C CD₂ D D

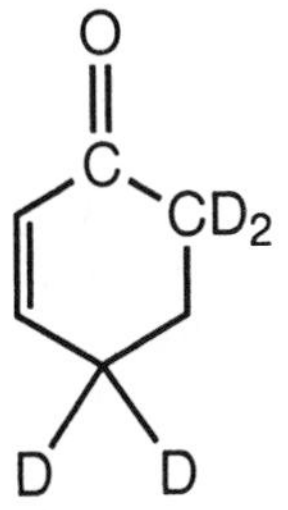

But there is more to it. What if enolate **A** adds deuterium at the α-position? Deuterium is now incorporated into the allylic α-position of **A** to give **B**. Now either H or D can be removed in base. Removal of D simply reverses the reaction, but removal of H leads to an enolate, **C**, in which deuterium is incorporated at one of the vinyl positions.

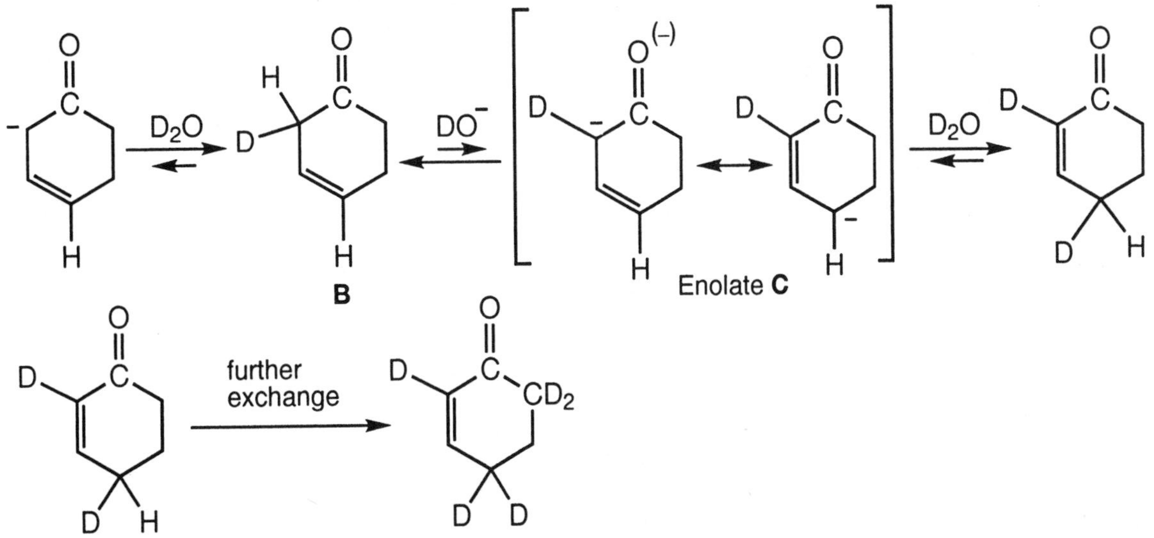

D₂O DO⁻ D₂O

B Enolate **C**

further exchange

Problem 18.38 Enolate formation demands orbital overlap. Stabilization by an adjacent carbonyl group requires that the π orbitals of the carbon-oxygen double bond overlap with an orbital on an adjacent, α-carbon. The problem with the hydrogen in question is that, although formally α, there is no overlap of the appropriate orbitals. Accordingly, there can be no enolate formation and no exchange.

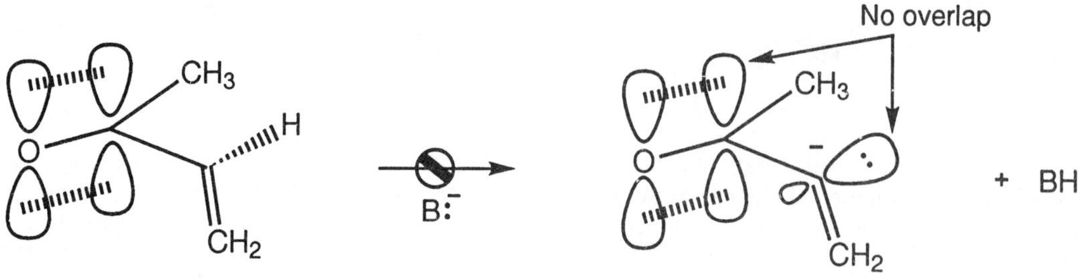

No overlap

CH_3 H CH_2 B:⁻ CH_3 CH_2 + BH

Problem 18.39 Part (a) is conventional; there is only a single exchangable α-hydrogen. Notice that there are two possible exchanged compounds; the entering deuterium can be exo or endo. The figure shows the planar enolate. In (b), there are no exchangable hydrogens at all, so no deuteriums can be incorporated.

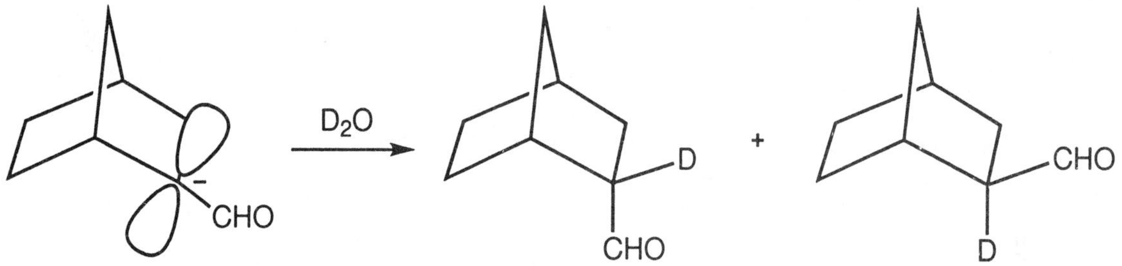

CHO D_2O D CHO + CHO D

It is (c) and (d) that may cause trouble. Compound (c) exchanges the pair of α-methylene hydrogens but not the α-methine hydrogen. Although this hydrogen is formally α, there is no good overlap between an orbital at the bridgehead position and the carbonyl group. The same trouble plagues (d); there is no overlap between the formally α-hydrogens and the carbonyl orbitals, and no exchange can take place.

(c)

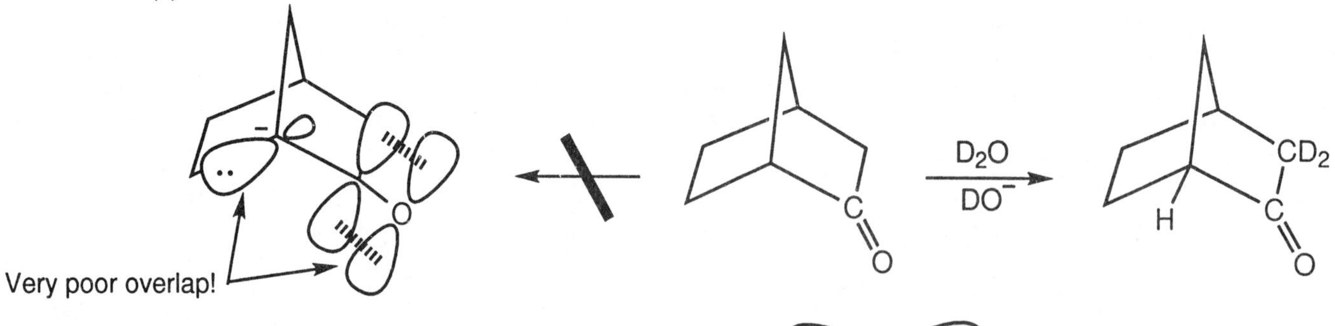

Very poor overlap! D_2O DO^- CD_2 H

(d)

H D_2O DO^- No overlap

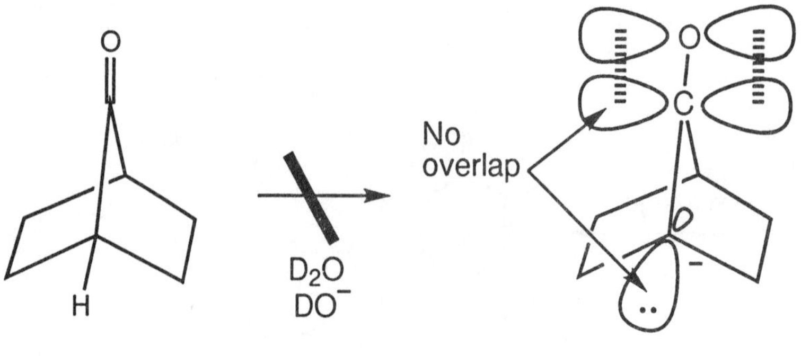

Problem 18.40 Recall from Section 18.1 (Fig. 18.22) that β-dicarbonyl compounds such as 2,4-pentanedione exist largely in their enol forms.

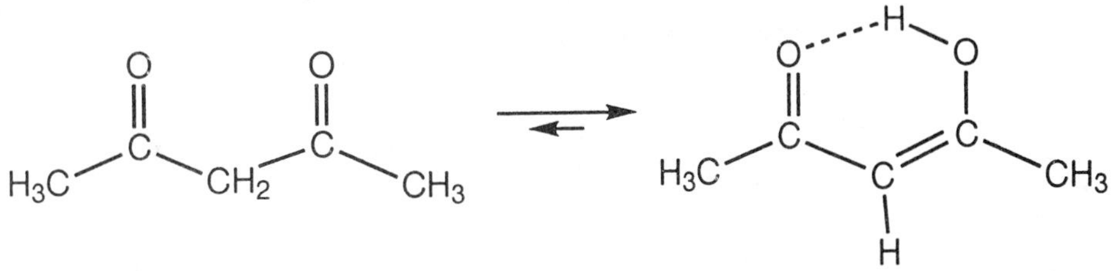

In this case, both the β-diketone and enol form are observed in the ^{13}C NMR spectrum. For the β-diketone form, we would expect to see three signals: one signal for the methyl carbons (δ 24.3 or 30.2 ppm), one signal for the methylene carbon (δ 58.2 ppm), and one signal for the carbonyl carbons (δ 191.4 or 201.9 ppm). Perhaps what is not as easy to see is that the enol form would also display only three different carbons. The trick is that there are two rapidly equilibrating enol forms.

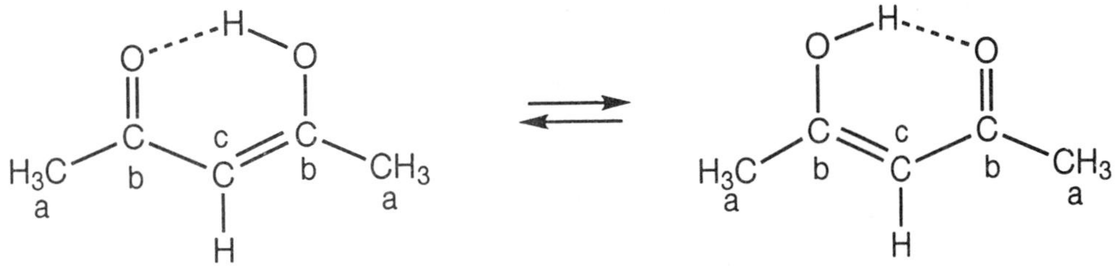

Methyl carbons (a) are equivalent, as are the ketone/enol carbons (b). The actual peak assignments are shown below.

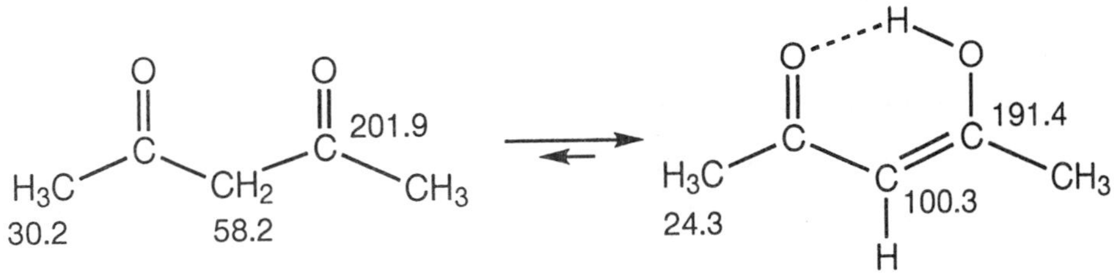

Problem 18.41. These should all be easy transformations to remember. Part (a) is acetal formation, (b) is reduction to a methylene group (two methods are shown), (c) is imine (hydrazone) formation, and (d) is reduction to the alcohol. The only difficulty might come from (e), which requires you to recall the Wittig reaction from Chapter 16 (p. 799).

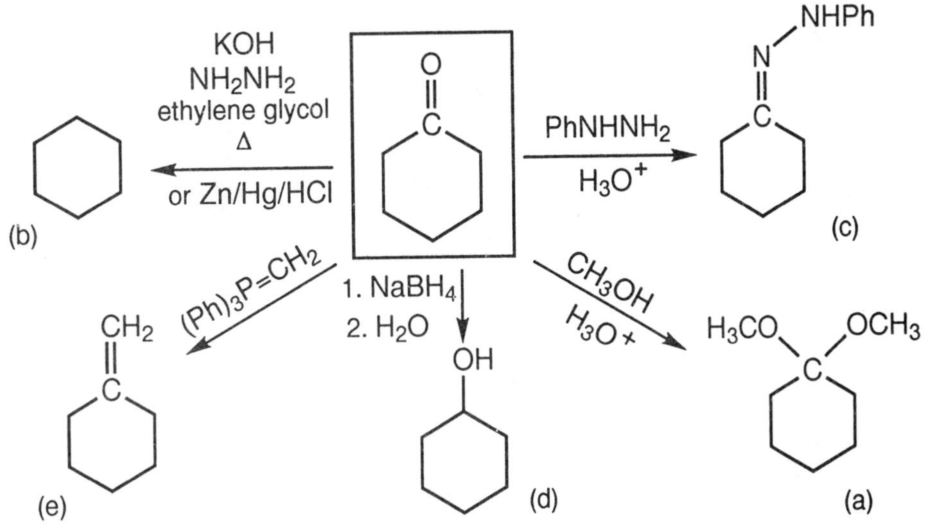

Problem 18.42 The reactions all involve enolate (or enol) chemistry from this chapter. Part (a) requires an alkylation. Enamine chemistry or LDA-produced enolates can be used. Parts (b) and (c) are simple α-bromination and α-exchange reactions. Note that mono α-bromination is best done in acid. Part (d) requires an aldol condensation and dehydration.

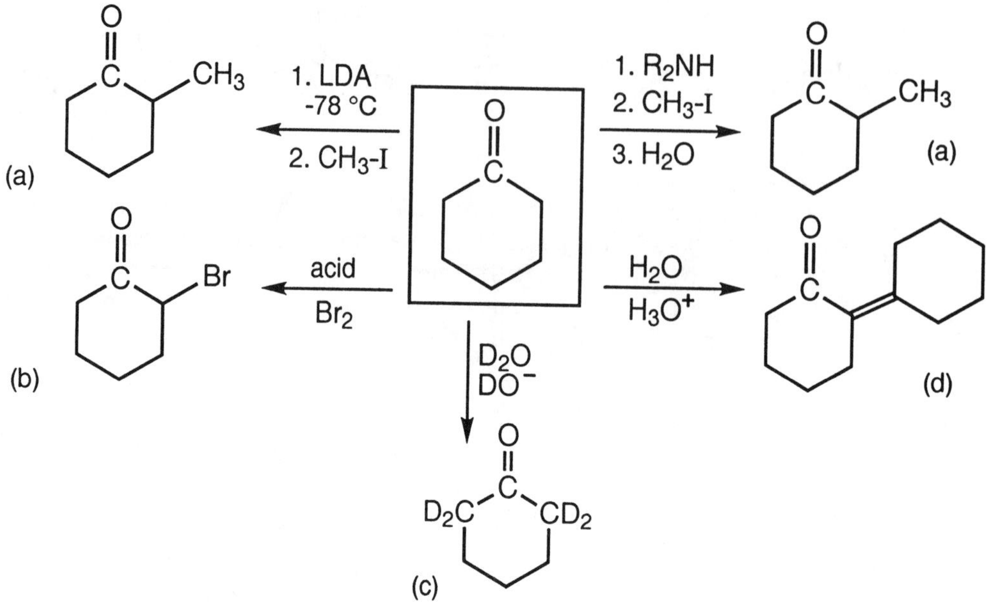

1. LDA -78 °C 2. CH_3-I (a)

1. R_2NH 2. CH_3-I 3. H_2O (a)

acid Br_2 (b)

H_2O H_3O^+ (d)

D_2O DO^- (c)

Problem 18.43 This problem shows two isomerizations. (a) The first of these involves enolate formation and reprotonation. In this case, there are two carbons available for reprotonation. One leads back to starting material, but the other accomplishes the isomerization shown.

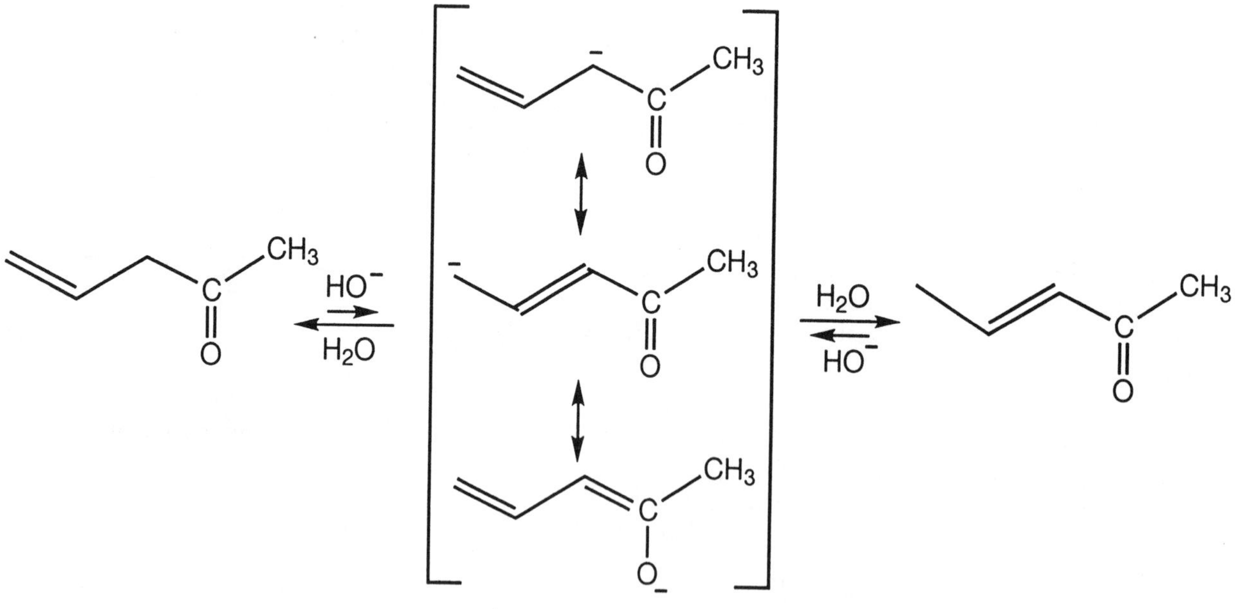

HO^- / H_2O ⇌ [resonance structures of enolate] ⇌ H_2O / HO^-

(b) In the second example a planar enol is produced. Protonation of the enol by water can occur from either side and give the two products.

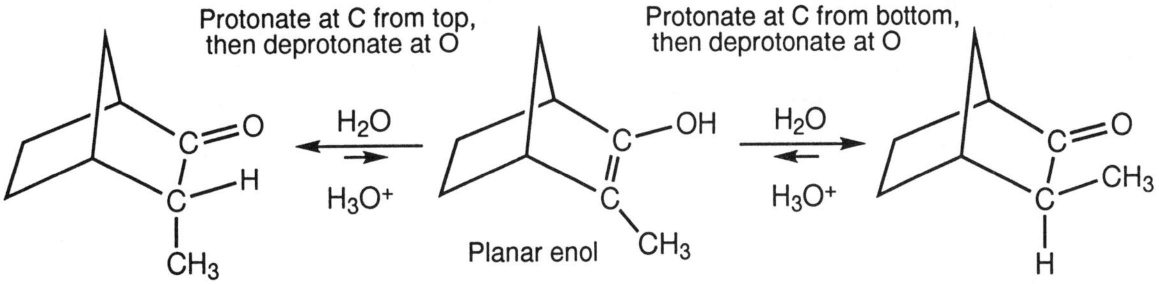

Protonate at C from top, then deprotonate at O

Protonate at C from bottom, then deprotonate at O

H_2O H_3O^+ Planar enol H_2O H_3O^+

Problem 18.44 This problem provides more practice in seeing simple aldol condensations in synthetic terms.

(a)

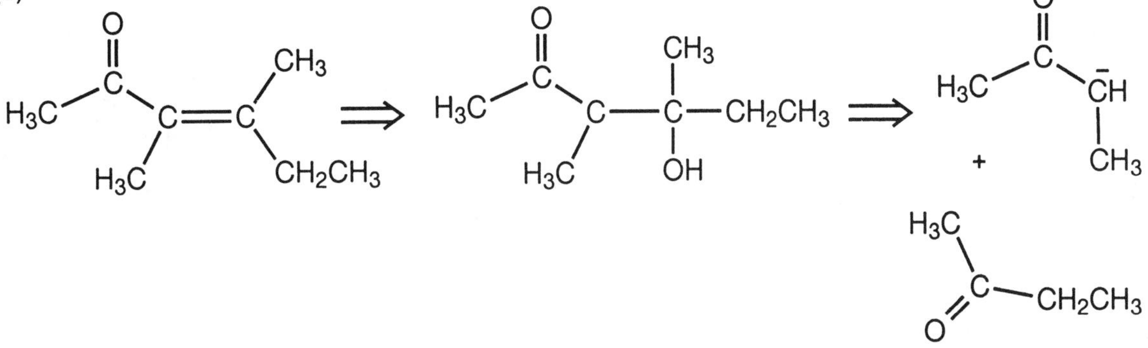

As the retrosynthetic analysis shown above indicates, in principle, the target molecule could be prepared from two molecules of 2-butanone through an aldol condensation and dehydration. However, in practice, a mixture of products would be the likely result. First, the proposed dehydration would probably give an *(E/Z)* mixture of α,β-unsaturated ketones. Second, as 2-butanone has two different sets of α-hydrogens, two enolates are possible. Ultimately, the other enolate would lead to an *(E/Z)* mixture of another α,β-unsaturated ketone, **1** (write a mechanism for its formation).

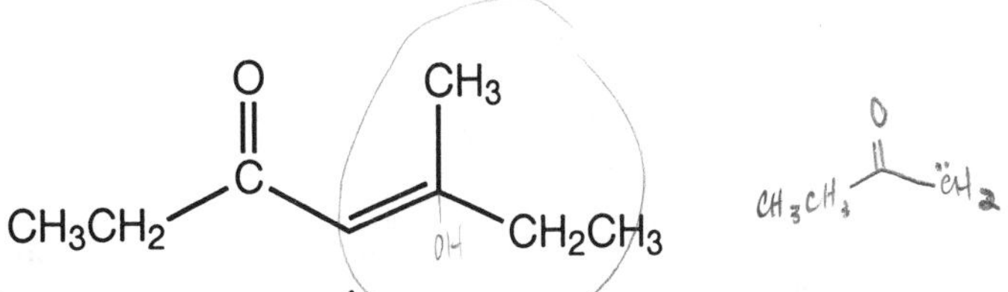

1

In fact, compound **1** appears to be the major product of the base-catalyzed self condensation of 2-butanone, whereas the target *(E)*-3,4-dimethyl-3-hexen-2-one is the favored (although not exclusive) product of the acid-catalyzed self condensation.

(b)

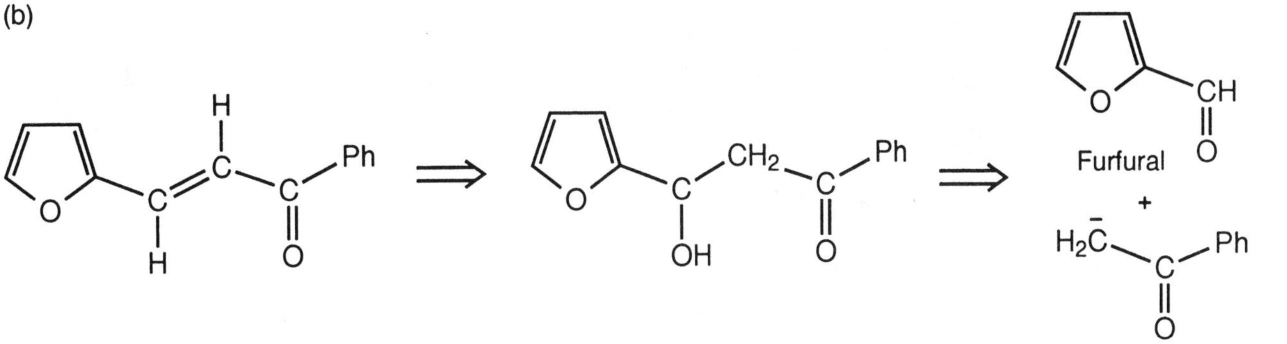

Furfural

Acetophenone enolate

The target molecule should be available from an aldol condensation/dehydration sequence starting from furfural and the enolate of acetophenone. Note that there is only one possible enolate here as furfural has no α-hydrogens. Although this enolate could add to the two different carbonyl groups (those in furfural and acetophenone), addition to the more reactive aldehyde of furfural should be greatly favored over addition to the less reactive ketonic carbonyl of acetophenone. In addition, the target trans isomer should be greatly favored over the sterically more congested cis isomer.

(c)

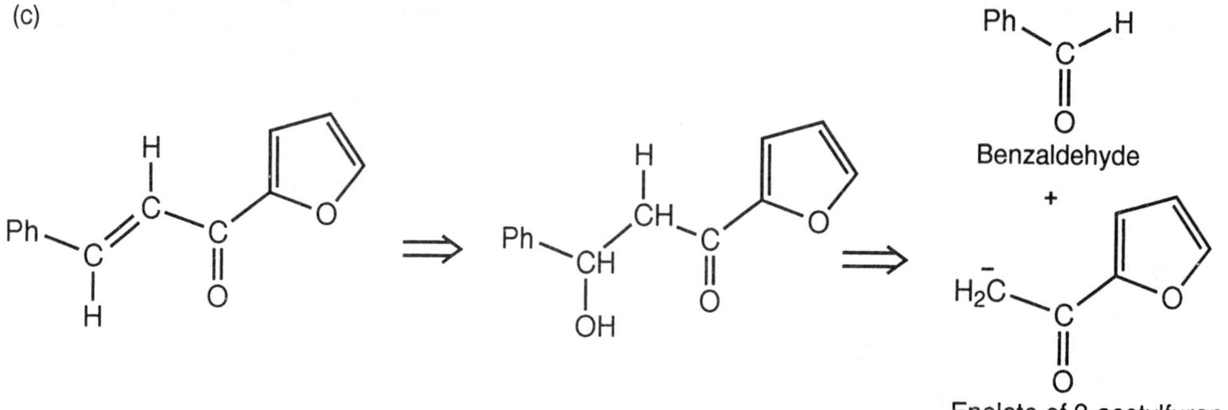

Benzaldehyde

+

Enolate of 2-acetylfuran

This target molecule should be available from an aldol condensation/dehydration sequence from benzaldehyde and 2-acetylfuran. The reasoning is the same as that for (b).

(d)

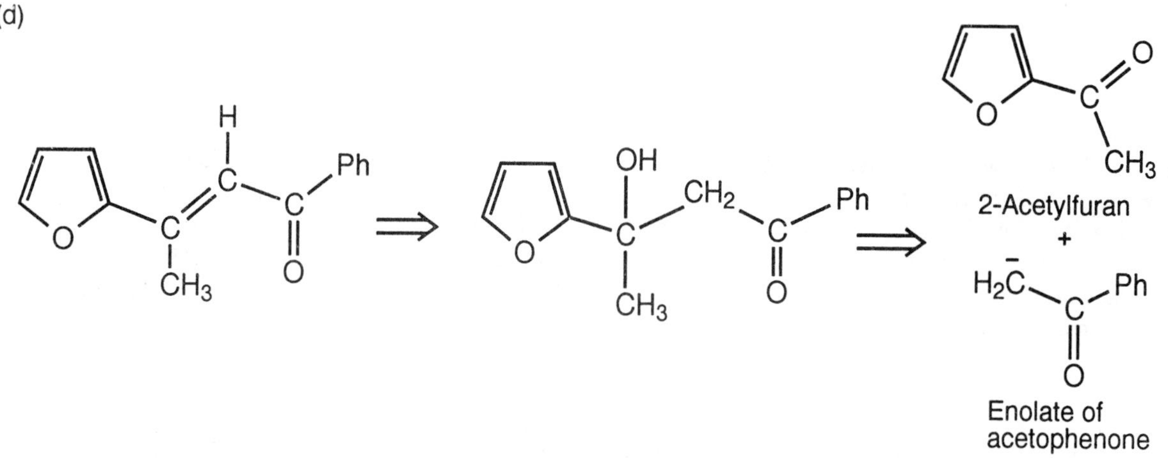

2-Acetylfuran

+

Enolate of acetophenone

In principle, the target molecule is available through an aldol condensation/dehydration sequence starting from acetophenone and 2-acetylfuran. In a practical sense, we have a lot of problems here. Both compounds have α-hydrogens and two enolates can be formed. Addition of the enolates to the two different ketone carbonyl groups will ultimately lead to four possible products after dehydration. There will, of course, also be *(E/Z)* pairs after dehydration. The three additional possible products are shown below. You should write arrow formalism mechanisms for their formations.

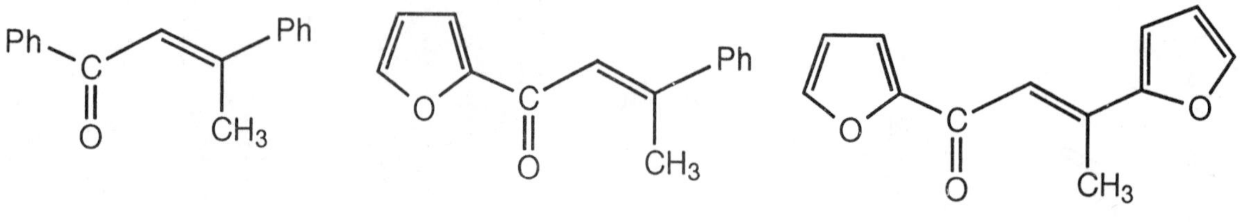

Problem 18.45 The first step is to identify the reaction from the conditions given, and to analyze what happens during that process. Here we have an example of the haloform reaction, in which a methyl ketone is converted into a carboxylic acid containing one fewer carbon atom. Accordingly, compound **1** must be *tert*-butyl methyl ketone (pinacolone - do you recall how to make this compound?). The mechanism for the haloform reaction is sketched below.

$(CH_3)_3C$–C(=O)–CH_2–H (**1**) + ^-OH ⇌ $(CH_3)_3C$–C(=O)–$\bar{C}H_2$ + Br–Br ⇌ $(CH_3)_3C$–C(=O)–CH_2Br

repeat bromination sequence twice more

$(CH_3)_3C$–C(=O)–CBr_3 + ^-OH ⇌ (addition) $(CH_3)_3C$–C(O^-)(OH)–CBr_3 ⇌ (elimination) $(CH_3)_3C$–C(=O)–OH + $^-CBr_3$

⇅

$(CH_3)_3C$–C(=O)–O^- + $CHBr_3$ $\xrightarrow[H_2O]{H_3O^+}$ $(CH_3)_3C$–C(=O)–OH **2**

According to this mechanism, three equivalents of Br_2 and four equivalents of sodium hydroxide are required for the conversion of **1** into **2**.

Problem 18.46 As soon as you see LDA, look for an α-hydrogen that can be removed by this strong base. Both possible enolates are formed from LDA and molecule **1**, but removal of the methylene hydrogen is far easier (faster) than removal of the methine hydrogen (slower), even though this leads to the less stable enolate, **A**. It is easier to approach the relatively unhindered methylene position than the more encumbered tertiary position. As enolate formation by LDA is essentially irreversible, the faster formation of **A** ensures that it will lead to the majority of methylated product.

LDA = $[(CH_3)_2CH]_2N^-$ Li^+

1 → **A** (the less stable, but faster formed enolate) + H_3C–I $\xrightarrow{S_N2}$ 66% (methylated ketone, CH_3) + I^-

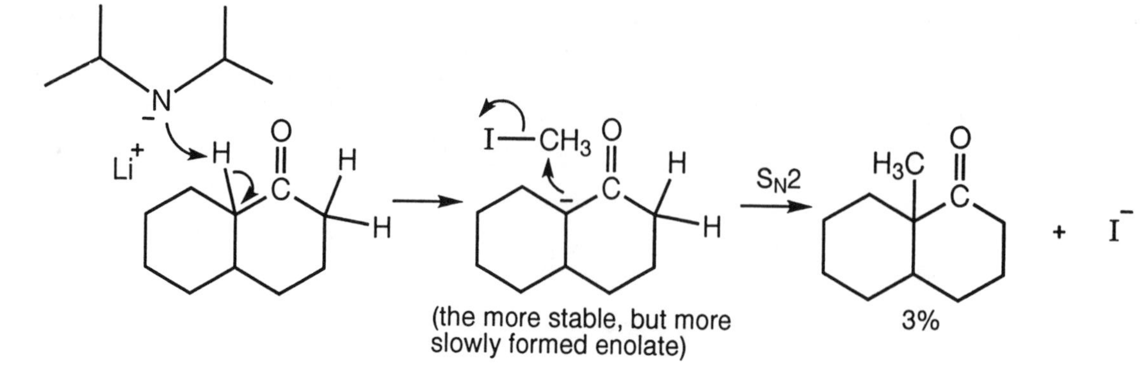

(the more stable, but more slowly formed enolate)

3%

Problem 18.47 The silicon-oxygen bond is enormously strong (~ 109 kcal/mol), whereas the silicon-carbon bond is much weaker (about 69 kcal/mol) and this drives the reaction toward silicon-oxygen bond formation. So, thermodynamics will clearly favor alkylation on oxygen and formation of the more substituted, more stable double bond. The problem shows that the regiochemistry depends on the base used to form the enolate. As we saw in Problem 18.46, the strong, sterically hindered base LDA irreversibly removes a proton from the position of easier access. In this case, the methylene hydrogen is removed and the less substituted, less stable enolate is formed.

Less accessible hydrogen (methyl group in the way)

More accessible hydrogens

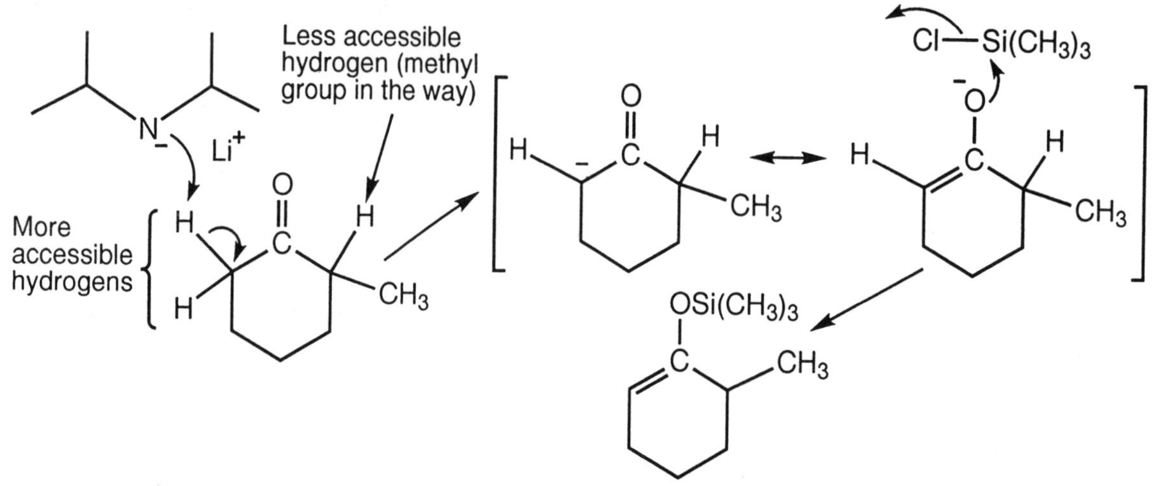

When triethylamine is used as base, enolate formation is reversible. So, even if the more stable enolate is formed more slowly, the reversible nature of the reaction allows it to predominate eventually. Under such conditions the thermodynamically more stable enolate will lead to most of the silated product.

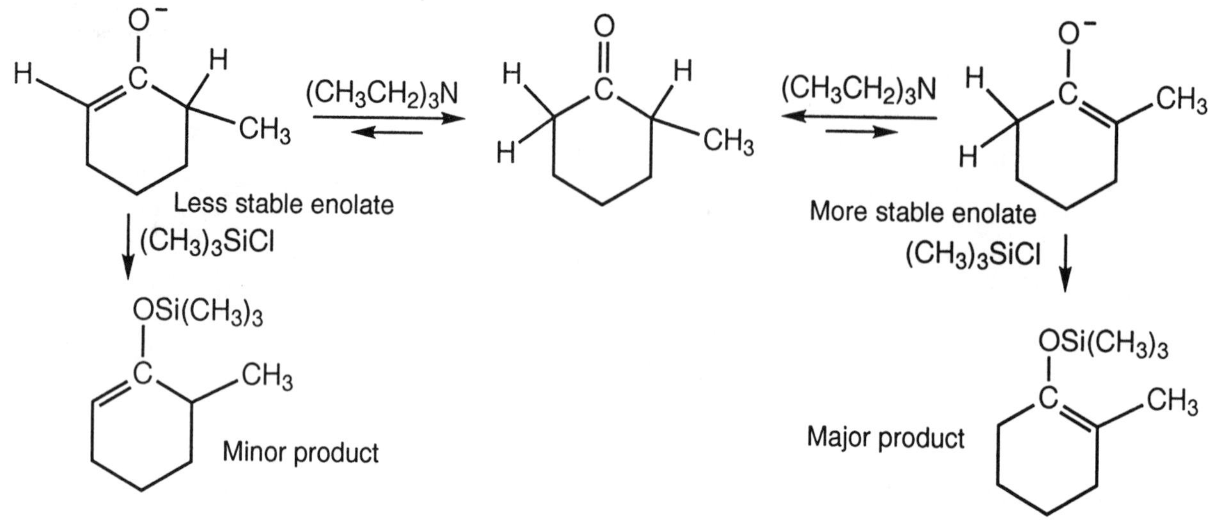

Less stable enolate

More stable enolate

Minor product

Major product

Problem 18.48 Once again, the solution to this problem depends on the irreversible removal of the sterically more accessible hydrogen by LDA. When **1** is deprotonated under equilibrating conditions with potassium *tert*-butoxide in *tert*-butyl alcohol at room temperature, the thermodynamically favored, more substituted enolate dominates, and this ultimately leads to **2**.

$^-$O−C(CH$_3$)$_3$ H C O CH$_3$ **1** Br

(CH$_3$)$_3$COK, (CH$_3$)$_3$COH, 25 °C

O$^{(-)}$ C CH$_3$ Br (more substituted, more stable enolate)

O C CH$_3$ **2**

However, when **1** is deprotonated under kinetic conditions with LDA at −72 °C, the less-substituted enolate is formed by removal of the more easily accessible hydrogen, and this subsequently leads to **3**.

LDA = Li$^+$ N$^-$

H O C CH$_2$ H **1** Br

LDA/THF, − 72 °C

O$^{(-)}$ C CH$_2$ Br (less substituted, less stable enolate)

O C **3**

Problem 18.49 This reaction is an example of a "double" crossed aldol condensation (Claisen-Schmidt condensation). Note that only ketone **1** has α-hydrogens; in fact, it has two sets of α-hydrogens. Also, the diketone **2** is more reactive than an ordinary ketone (why?). The reaction proceeds first to intermediate **A**, and then through dehydration to the α,β-unsaturated ketone **B**.

O C PhH$_2$C CH$_2$Ph **1**

CH$_3$CH$_2$OH HO$^-$

O$^{(-)}$ C PhCH CH$_2$Ph

O C Ph C Ph O **2**

O C PhCH CH$_2$Ph Ph—C C=O O$^-$ Ph

CH$_3$CH$_2$OH

(continued at top left of next page)

(from previous page)

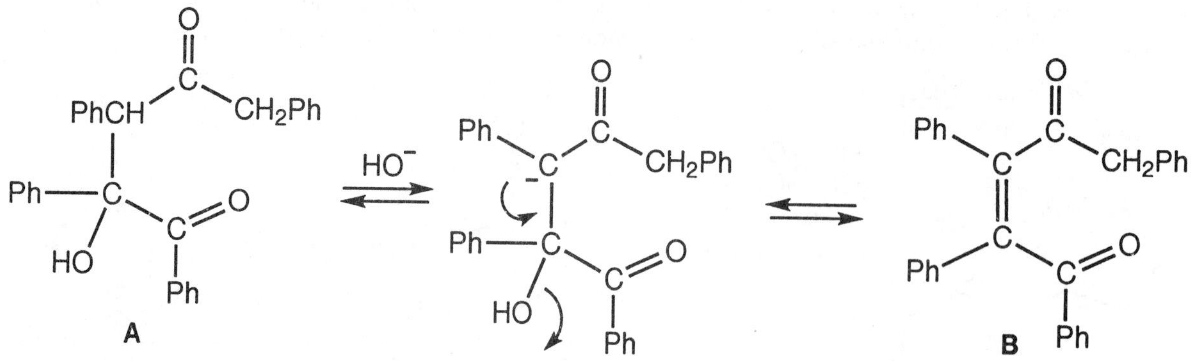

Now the process simply is repeated, this time in an intramolecular sense to give **3**. (It is also possible that the second aldol condensation occurs before the first dehydration from **A** to **B**.)

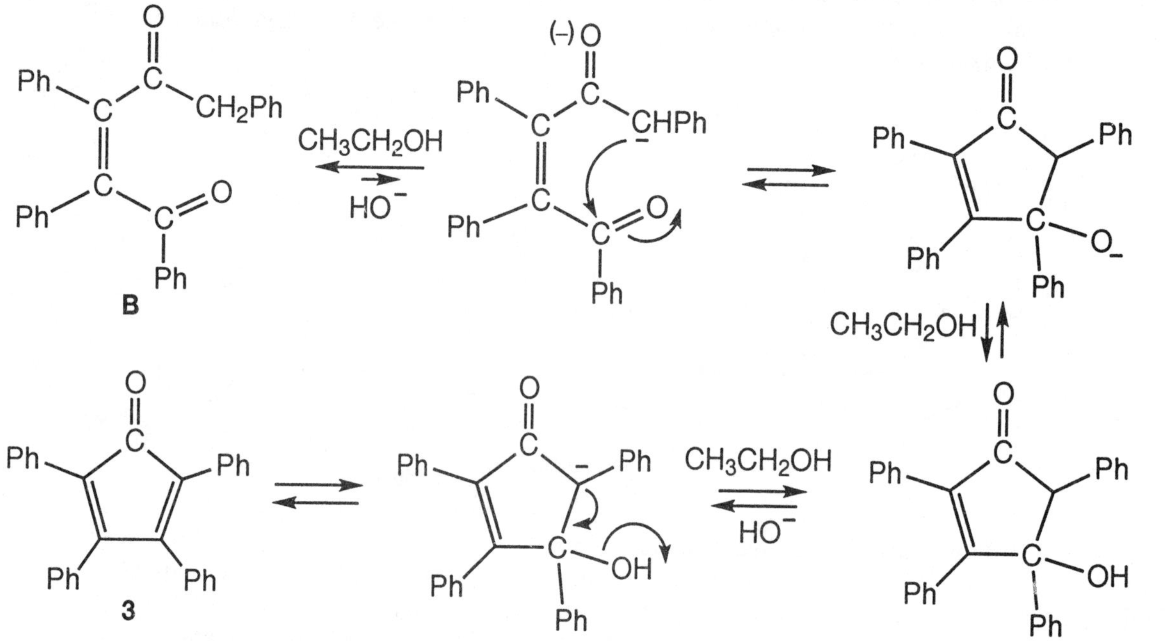

Problem 18.50 Here we have a relatively straightforward "combination" problem. A little analysis shows that we need to find a way to combine two molecules of dimedone with one aldehyde.

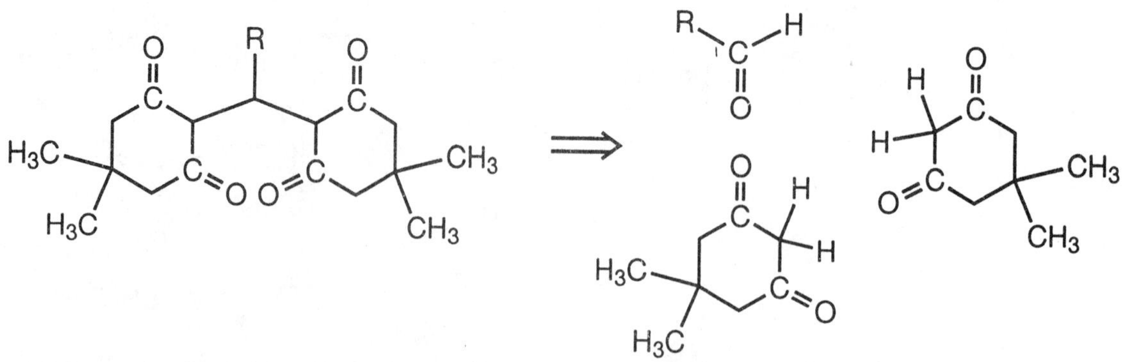

The dimedone derivatives, **2**, are formed by a combination of a Knoevenagel condensation (to give **A**) and a Michael reaction. Enolate formation is followed by addition to the aldehyde and dehydration to give the α,β-unsaturated compound **A**. Michael addition of another enolate leads, after protonation, to **2**.

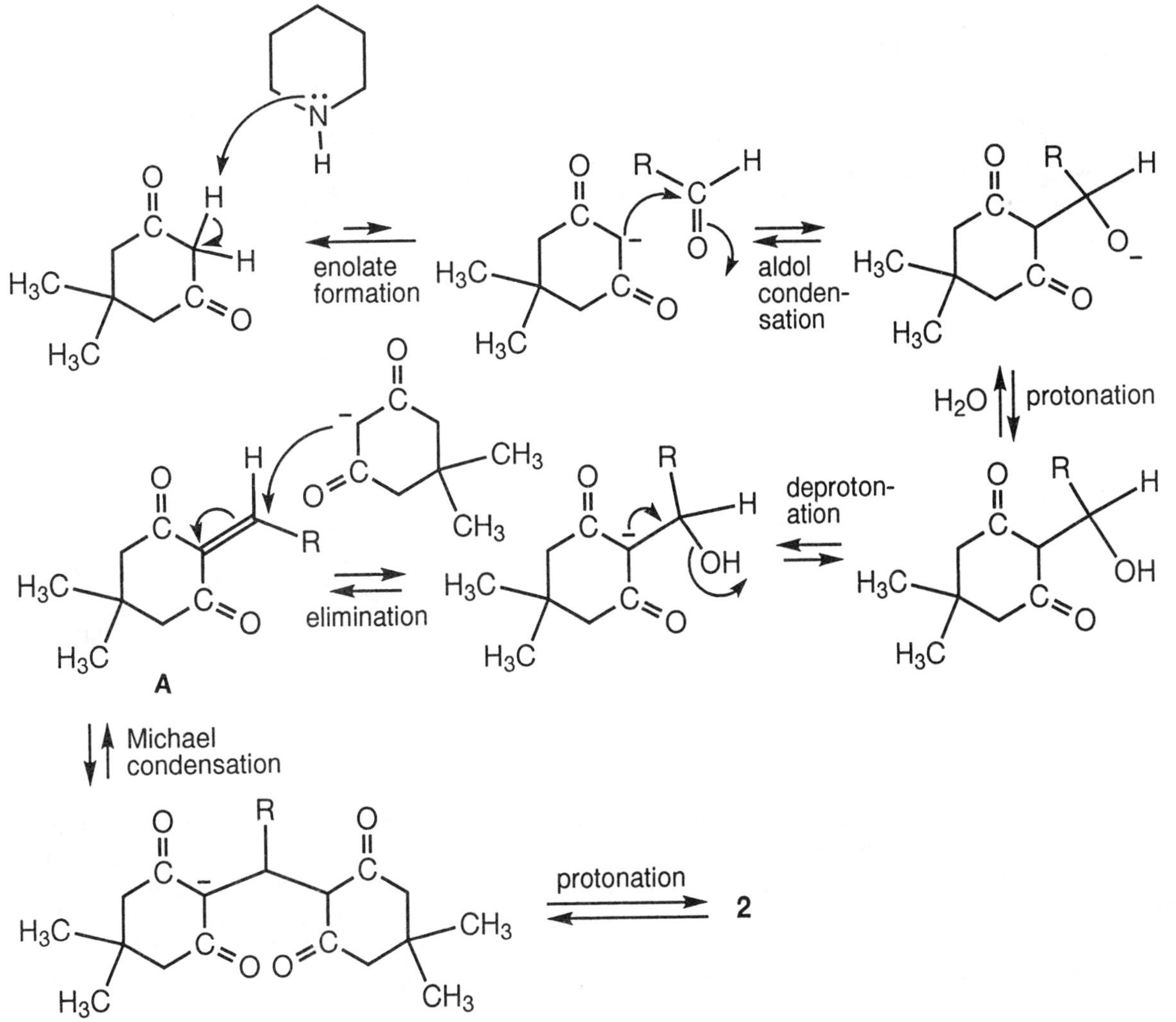

enolate formation

aldol conden- sation

H_2O protonation

deproton- ation

elimination

A

Michael condensation

protonation

2

The conversion of **2** into **3** is reminiscent of an acid-catalyzed intramolecular aldol condensation, except that, because of geometrical constraints in this system, it is the enol oxygen, not carbon, that acts as the nucleophile and adds to the protonated carbonyl.

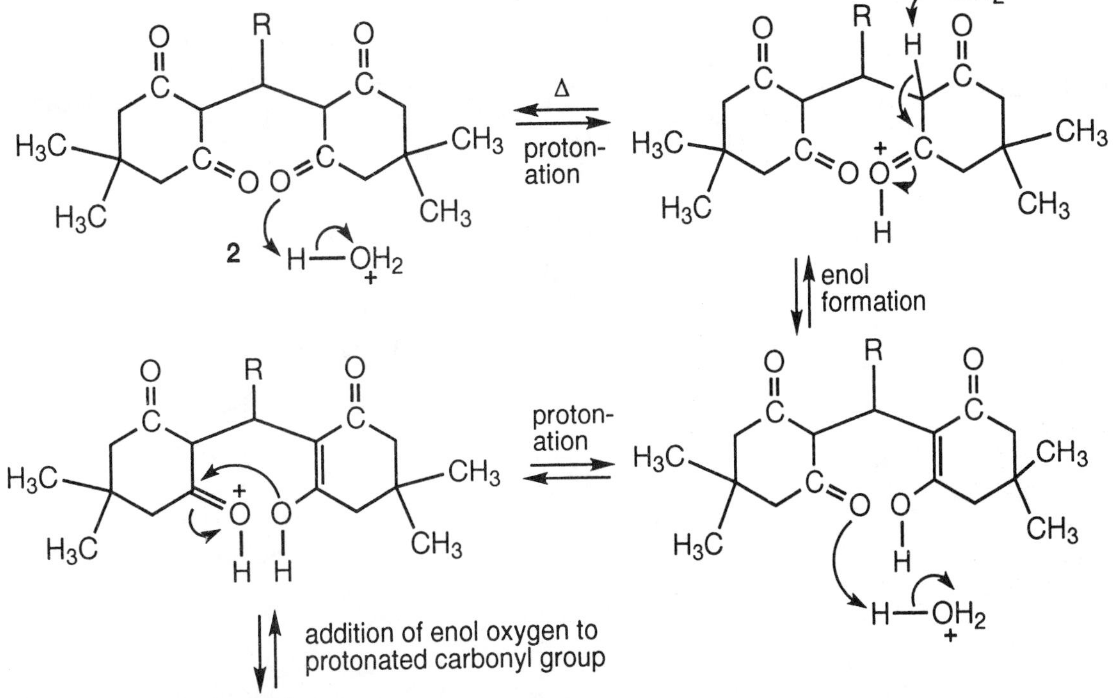

Δ

proton- ation

2

enol formation

proton- ation

addition of enol oxygen to protonated carbonyl group

(continued on next page)

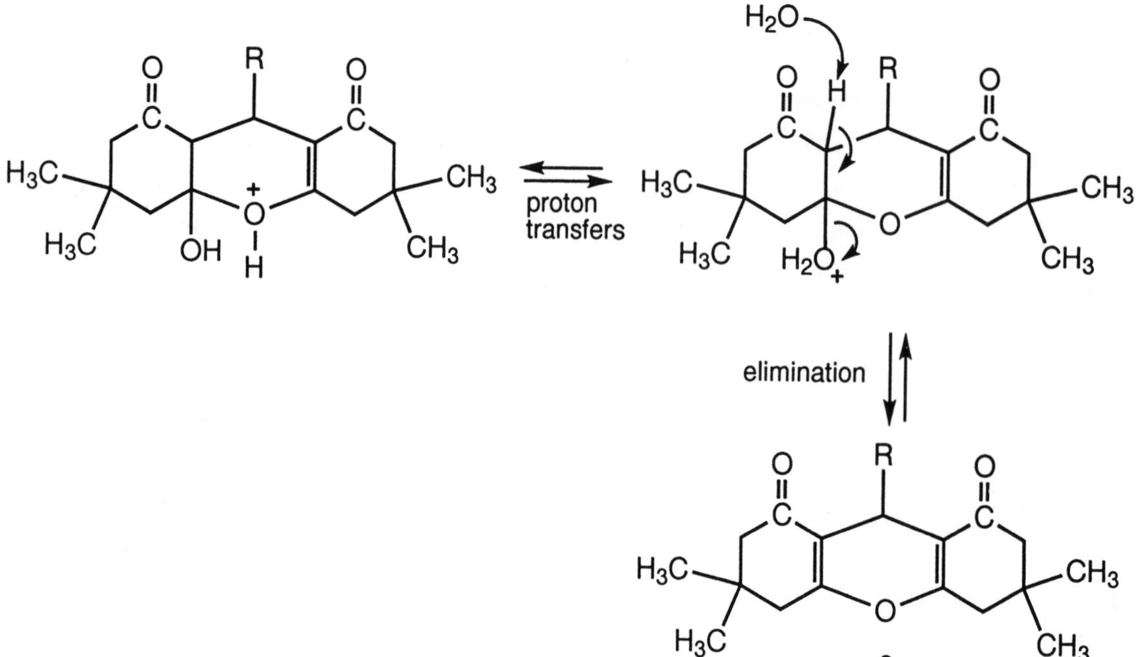

Problem 18.51 Clearly, the chloride must be lost and a three-membered ring containing oxygen created. Perhaps chloride is displaced by an oxyanion formed through addition of an enolate to the carbonyl. There is only one enolate possible, and that starts the reaction with an addition to the carbonyl group of cyclohexanone. Ordinarily, addition reactions are completed by protonation of the newly formed alkoxide, but in this case there is a leaving group, the chloride, poised for displacement. An intramolecular S_N2 reaction finishes the process.

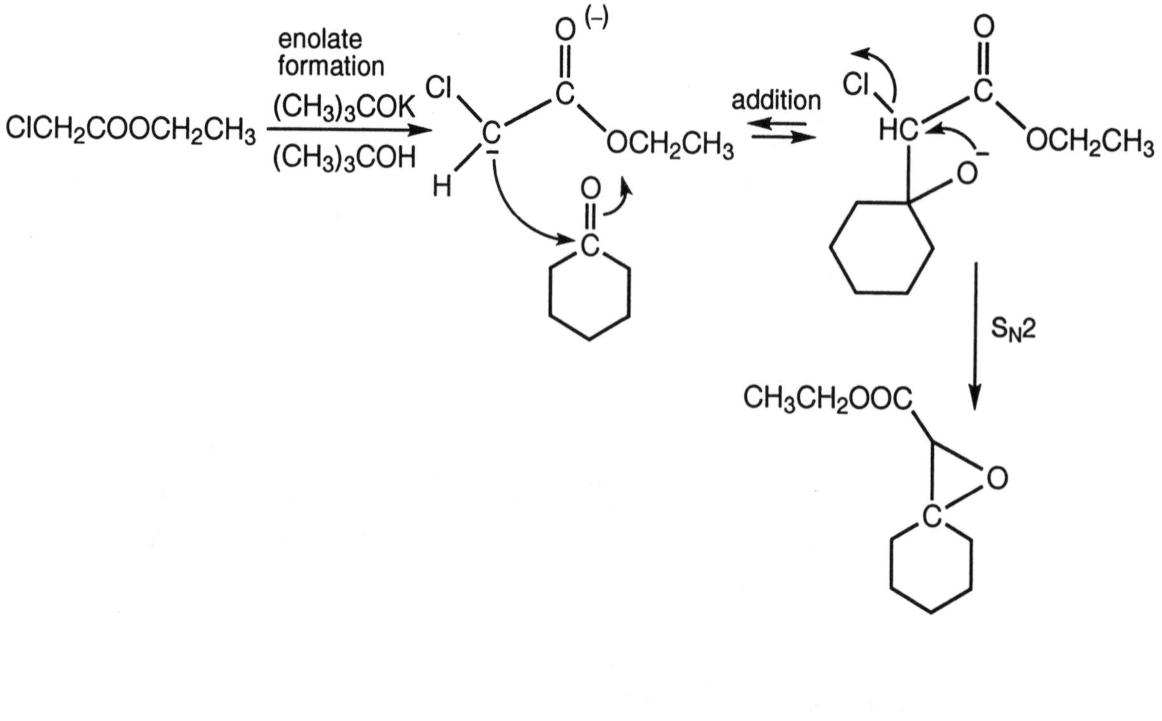

Problem 18.52

(a) This problem involves nothing more than the standard sequence of anion formation, addition, and dehydration. The only trick is seeing that a nitro group can nicely stabilize an adjacent anion. That effect appeared earlier when we described nucleophilic aromatic substitution (Section 14.13, p. 673).

CH3NO2 → NaOH, anion formation → [H2C⁻—N⁺(=O)O⁻ ↔ H2C=N⁺(O⁻)O⁻]; addition to PhCHO; Ph—C(CH2NO2)(H)—O⁻ → H2O, protonation → Ph—C(CH2NO2)(H)—OH → HCl, protonation → Ph—C(CHNO2)(H)—OH2⁺ (Cl⁻) → dehydration → Ph(H)C=C(H)NO2

(b) Here the strong base is able to remove a benzylic hydrogen to give an anion that is stabilized by the adjacent benzene ring. Intramolecular addition and dehydration finish the sequence.

N-formyl-o-toluidine → (CH3)3COK, (CH3)3COH (⁻OC(CH3)3) → benzylic anion adds to CHO → indoline-2-O⁻ → (CH3)3COH → indoline-2-OH → (CH3)3COK → anion, loss of OH → indole

Problem 18.53

(a) We must find a way to combine two molecules of **2** with one of **1**. In this problem, the "standard" sequence of enolate formation, addition, and elimination leads to a molecule that can act as an acceptor in a Michael reaction [see Problem 18.50 for a similar sequence]. This reaction is a tandem Knoevenagel-Michael. Notice that both the enolates in this sequence result from removal of an especially acidic "doubly α" hydrogen.

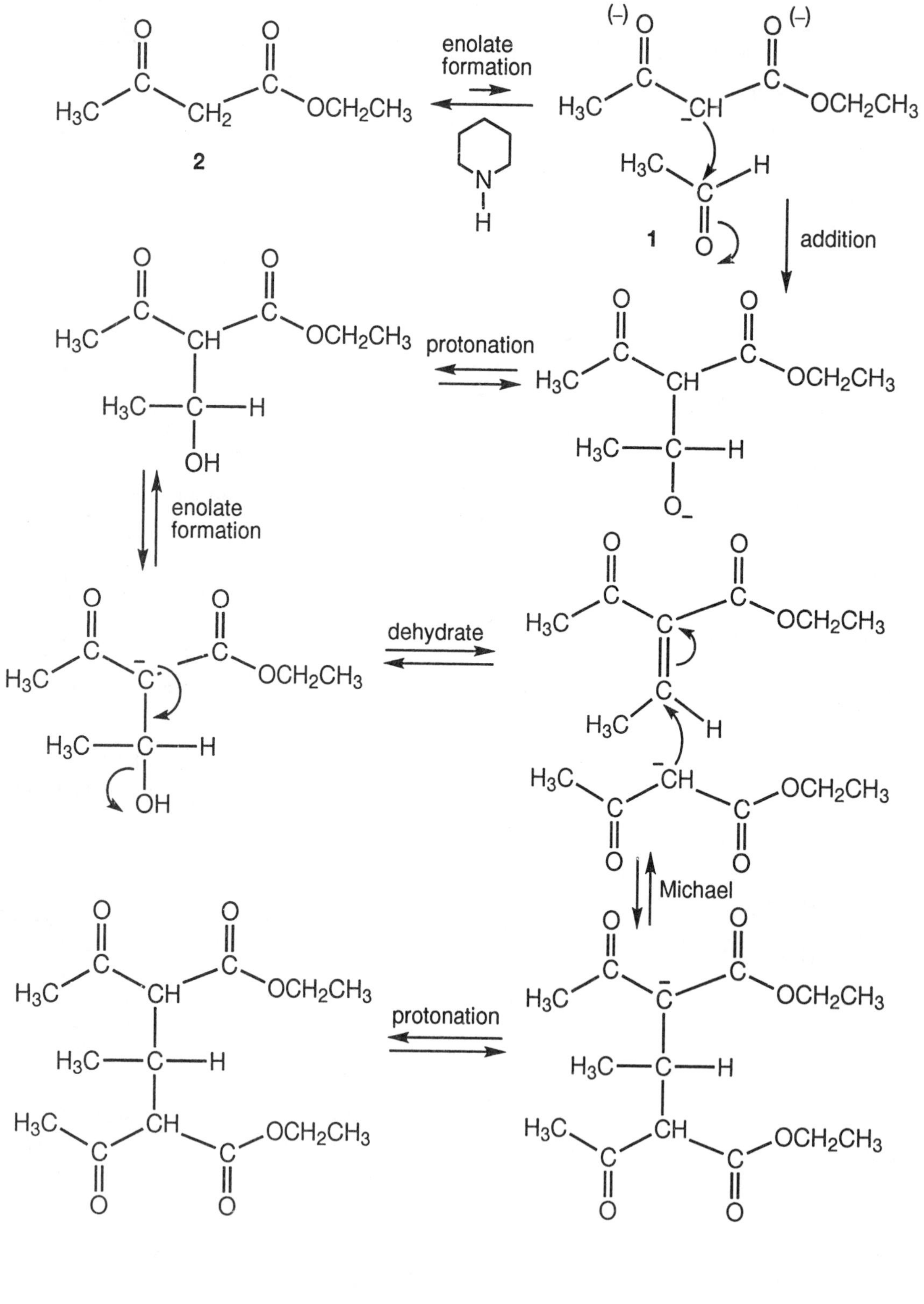

enolate formation

2

1

addition

protonation

enolate formation

dehydrate

Michael

protonation

(b) Here condensation occurs under reducing conditions (note the H_2/Pd), and the α,β-unsaturated carbonyl compound is hydrogenated as it is formed. Start with the formation of the better enolate by removal of the most acidic hydrogen. Then add to the carbonyl group of propionaldehyde and dehydrate. Hydrogenation gives the final product.

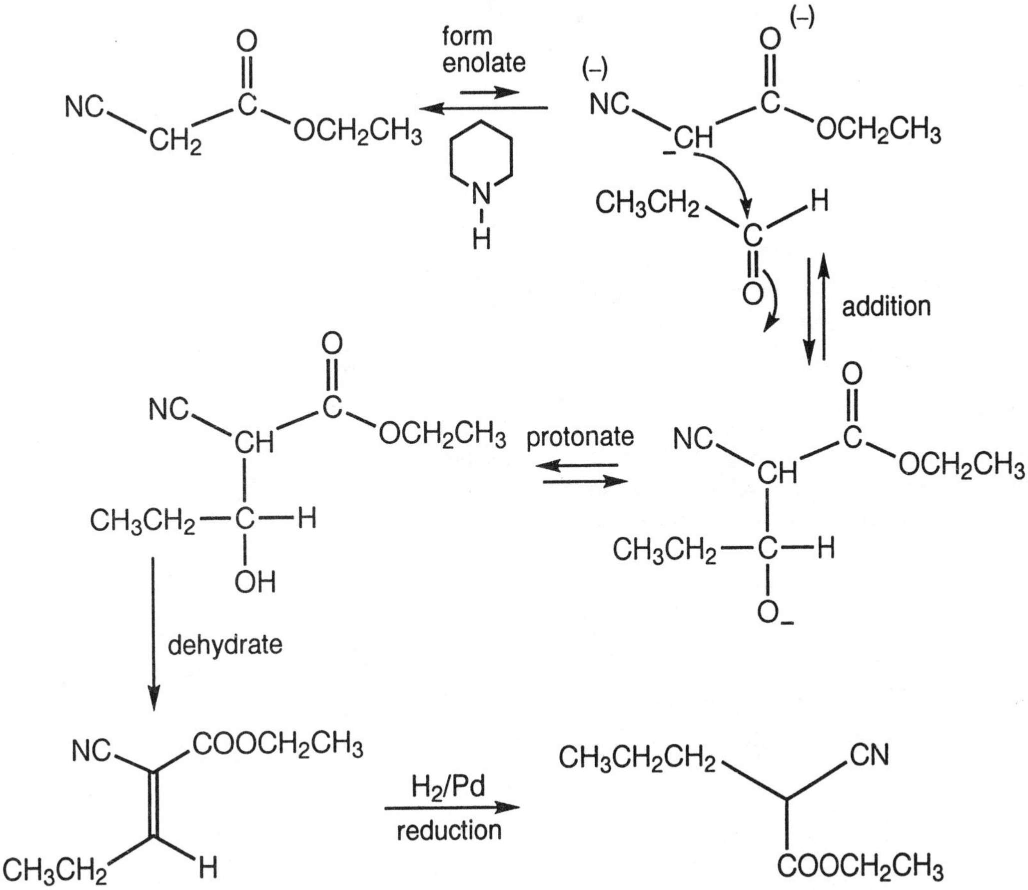

form enolate

addition

protonate

dehydrate

H_2/Pd reduction

Problem 18.54 This reaction is nothing more than an intramolecular base-catalyzed aldol condensation. However, it is complicated because, as you undoubtedly notice after a little analysis, there are four possible enolates (actually more if you consider *(E/Z)* stereochemistry) that can be formed from **1**. In fact, all four are probably formed. The simpler, but closely related compound **3** exchanges nine hydrogens for deuterium when treated with D_2O/DO^-, for example, so all four enolates of **3** must be generated.

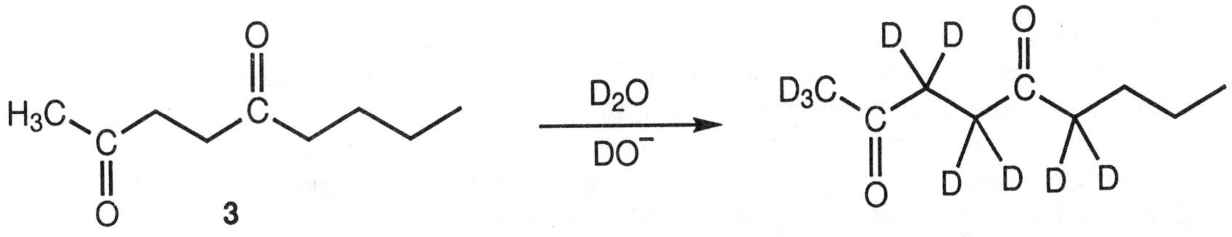

D_2O / DO^-

Of the four possible enolates that can be formed from **1**, two can only form three-membered rings through intramolecular additions to carbon-oxygen double bonds, and these will surely be disfavored thermodynamically. (You should draw these two enolates and the three-membered rings they form). The two remaining possible enolates can each lead to an energetically more favorable five-membered ring. However, it is the more substituted, and thus more stable, enolate that leads to the observed *cis*-jasmone (**2**).

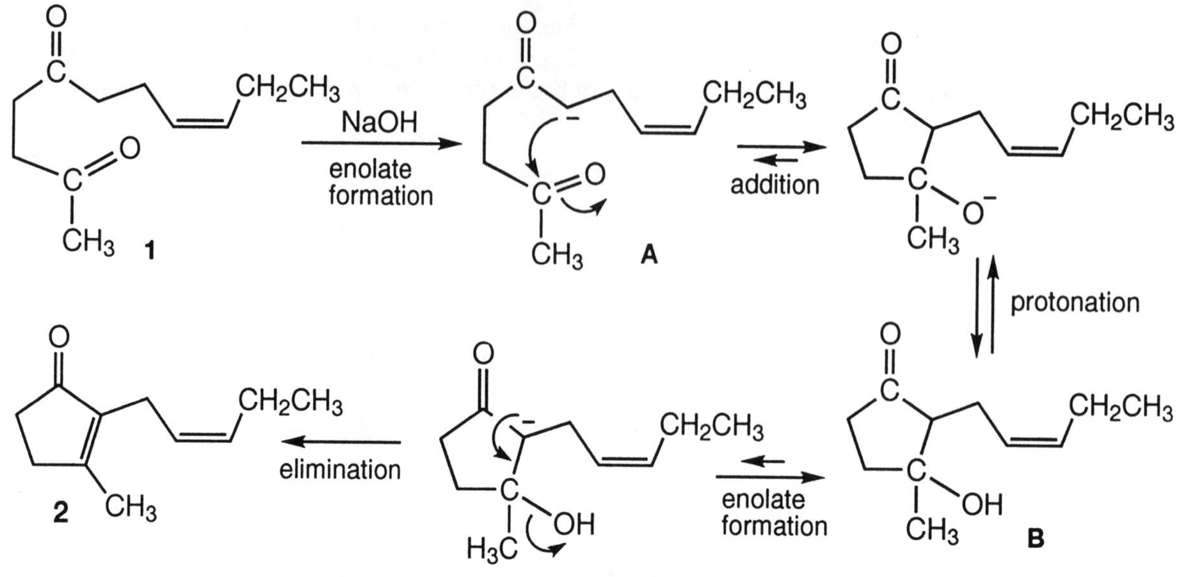

The other cyclopentenone could arise from the less stable, less substituted enolate in the following way. The steps are equivalent to those in the figure above. Enolate formation, intramolecular addition to the carbonyl group, and dehydration would give **4**.

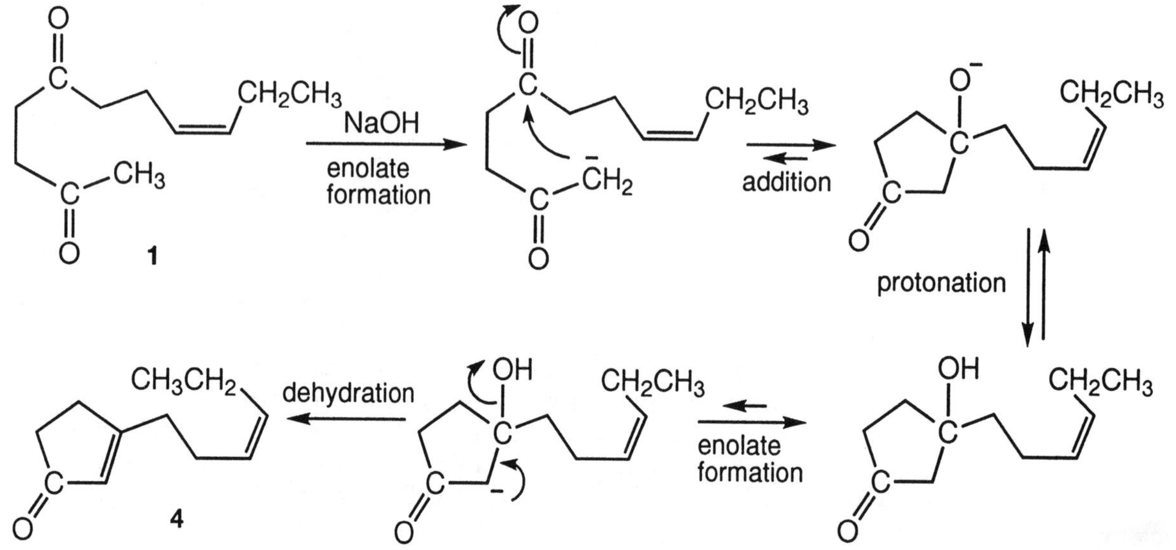

Why is *cis*-jasmone (**2**) formed in this base-catalyzed aldol process? Two immediate possibilities come to mind. Perhaps the tetrasubstituted enone **2** is thermodynamically more stable than the trisubstituted enone **4**, and under the reaction conditions the production of **4** is reversible through a retro-aldol reaction (write a mechanism). Alternately, *cis*-jasmone may be the kinetically favored product. The situation is far from resolved. Notice that we could have worked this problem nicely backward. Enone **2** must have come from dehydration of **B**, and **B** is the inevitable result of aldol condensation of enolate **A**.

Problem 18.55 The parent masses (**1** = 146 g/mol; **2** = 234 g/mol) allow us to see that compound **1** is a 1:1 combination of acetone and benzaldehyde (minus water), and compound **2** incorporated two molecules of benzaldehyde with one of acetone (again, minus two waters). Both **1** and **2** are formed by crossed aldol condensations (Claisen-Schmidt reactions). Notice that acetone has α-hydrogens whereas benzaldehyde does not. The presence of an excess of acetone should favor "single" aldol condensation/dehydration.

$H_3C-CO-CH_3$ (Excess) + $H-CO-Ph$ —(aldol condensation; NaOH, H_2O)→ $H_3C-CO-CH_2-CH(OH)-Ph$ —(dehydration)→ $H_3C-CO-CH=CH-Ph$ (**1**)

With an excess of benzaldehyde, a "double" aldol condensation-dehydration should be favored.

$H_3C-CO-CH_3$ + $H-CO-Ph$ (Excess) —(aldol condensation; NaOH, H_2O)→ $H_3C-CO-CH_2-CH(OH)-Ph$ —(dehydration)→ $H_3C-CO-CH=CH-Ph$ (**1**) —(repeat aldol)→ $Ph-CH(OH)-CH_2-CO-CH=CH-Ph$ —(dehydration)→ $Ph-CH=CH-CO-CH=CH-Ph$ (**2**)

The spectral data for compounds **1** and **2** support these conjectures. The mass spectrum for compound **1** is consistent with the proposed α,β-unsaturated ketone, 4-phenyl-3-buten-2-one ($C_{10}H_{10}O$, MW = 146 g/mol). The IR spectrum of compound **1** shows a strong carbonyl stretch at 1667 cm^{-1}, consistent with a conjugated ketone. Finally, the ^1H NMR spectrum is also consonant with the proposed structure. The methyl hydrogens appear at δ = 2.38 ppm, and the aromatic hydrogens appear as a 5H multiplet at δ = 7.30-7.66 ppm. The olefinic hydrogen resonances, which appear as doublets at δ = 6.71 and 7.54 ppm are quite informative. First, note the low field position of one of the hydrogens. This signal is the β-vinyl hydrogen and is deshielded both by the adjacent benzene ring and the carbonyl resonance:

$H_3C-CO-CH=CH_\beta-Ph$ ⟷ $H_3C-C(O^-)=CH-C^+H_\beta-Ph$ (H_β: δ = 7.54 ppm)

Note also that the magnitude of the coupling constant, J = 16 Hz, is appropriate for trans coupling (Chapter 15, p. 724). The stereochemical assignment is supported by the presence of a =C—H bending (wag) band at 973 cm^{-1} in the IR spectrum of **1**.

The spectral data for **2** are also consistent with the proposed "double" aldol product, 1,5-diphenyl-1,4-pentadien-3-one. The parent ion in the mass spectrum is appropriate for a molecular formula of $C_{17}H_{14}O$ (MW = 234 g/mol). The IR spectrum of **2** displays a strong carbonyl stretch at 1651 cm^{-1}, once again consistent with a conjugated carbonyl group. Notice the absence of a methyl group signal in the ^1H NMR spectrum of **2**. The trans stereochemistry of compound **2** is supported by the magnitude of the coupling constant for the vinyl hydrogens (16 Hz) and the characteristic =C—H bending at 984 cm^{-1} in the IR spectrum of **2**.

Problem 18.56 This transformation involves a rather rare example of an acid-catalyzed intramolecular aldol condensation giving a seven-membered ring. In this case, the necessary components are revealed only after the acid-catalyzed hydrolysis of a diacetal to a dialdehyde. Recall that the acid-catalyzed aldol condensation involves a reaction between a protonated carbonyl group and an enol. The following figure only sketches out the process. By all means fill in the detailed, arrow formalism mechanisms if you are at all uncertain about any of the steps. Note how we cannot escape earlier material. This problem is impossible unless we know that acetals are hydrolyzed to carbonyl compounds in aqueous acid.

H2C CH2
(CH3O)2HC CH(OCH3)2
H_3O^+ / H_2O
Acetal hydrolysis
H2C CH2
O=CH HC=O
enolization
H2C CH
O=CH HC—OH
protonation H_3O^+
H2C CH
HO+=CH HC—OH
aldol
CHO
OH
dehydration
CHO

Problem 18.57 This reaction is the Michael-like addition of an enamine to an alkene activated by a nitro group to give intermediate **A**. The nitro group stabilizes an adjacent negative charge more strongly than does a benzene ring, which explains the regiochemistry of the addition.

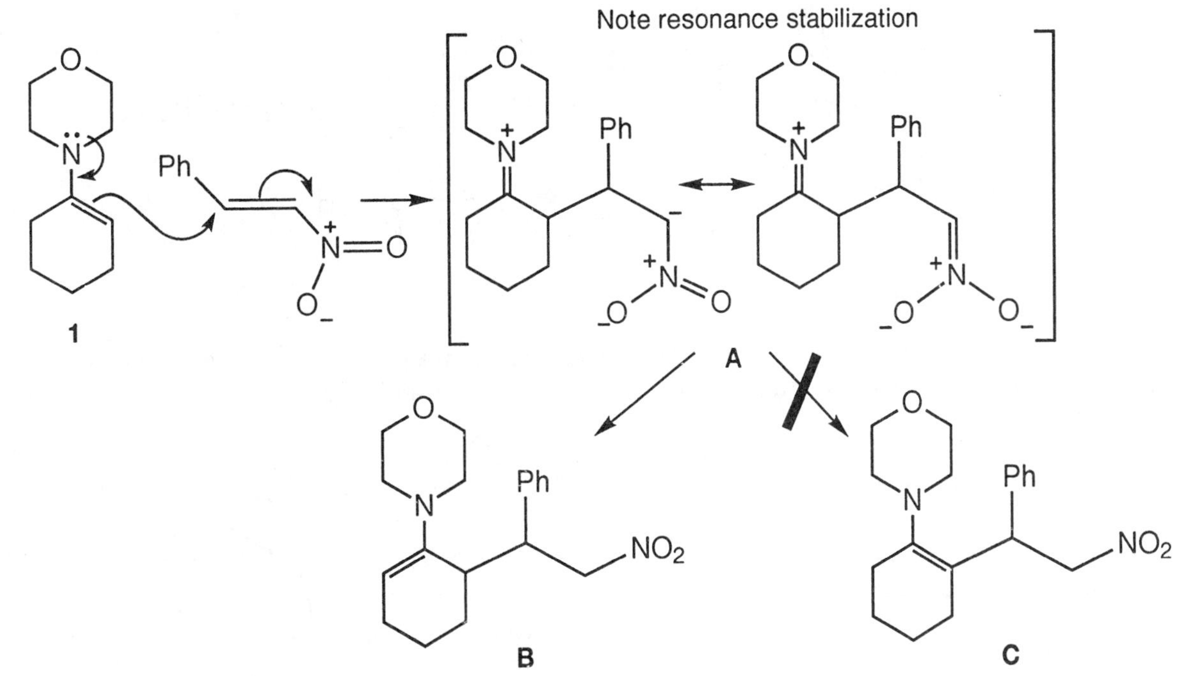

Note resonance stabilization

1

A

B

C

Deprotonation and reprotonation leads to a new enamine, **B**. There is experimental evidence that the structure is as shown, but you may well wonder why deprotonation does not form the other possible enamine, **C**. The answer isn't known, but it may involve nothing more complicated than the relative ease of access to a methylene group compared to that for the more congested methine hydrogen. Acid-catalyzed hydrolysis leads to the observed nitro ketone, **3**.

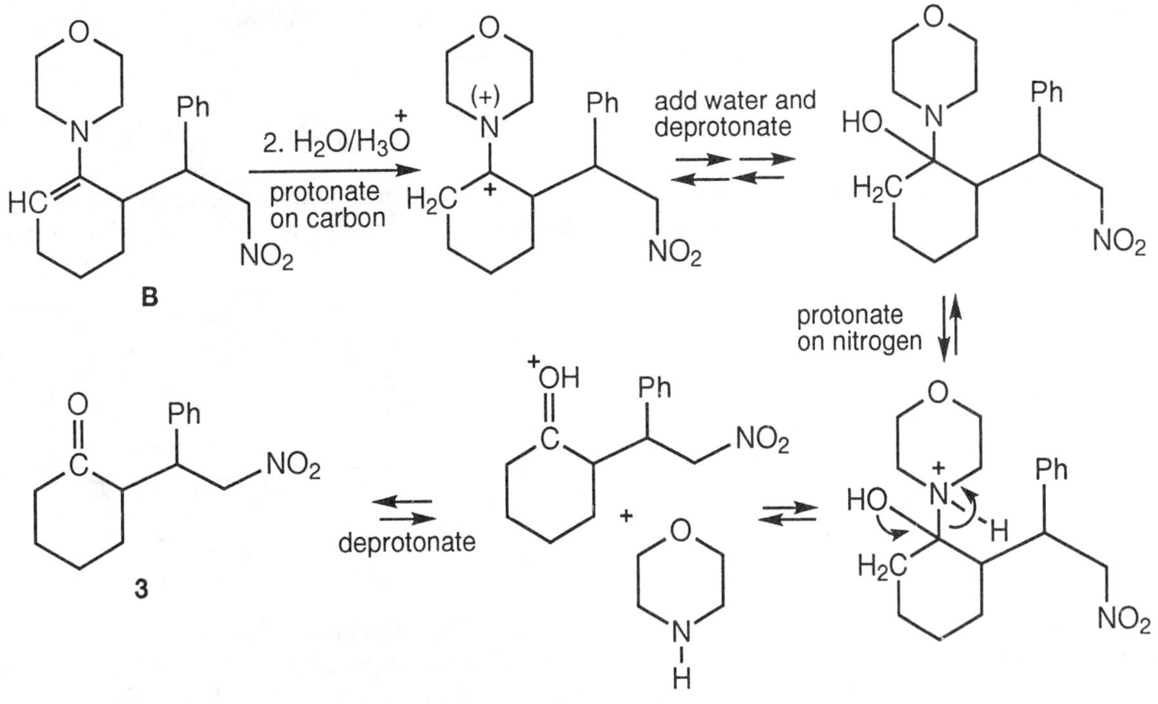

2. H_2O/H_3O^+

protonate on carbon

add water and deprotonate

protonate on nitrogen

deprotonate

B

3

Problem 18.58 Reduction of one equivalent of carbonyl compound by DIBAL-H is simple to explain. The DIBAL-H acts as any hydride reducing reagent ($LiAlH_4$ and $NaBH_4$ are other examples) and delivers a hydride to the Lewis acidic carbon of the carbonyl group.

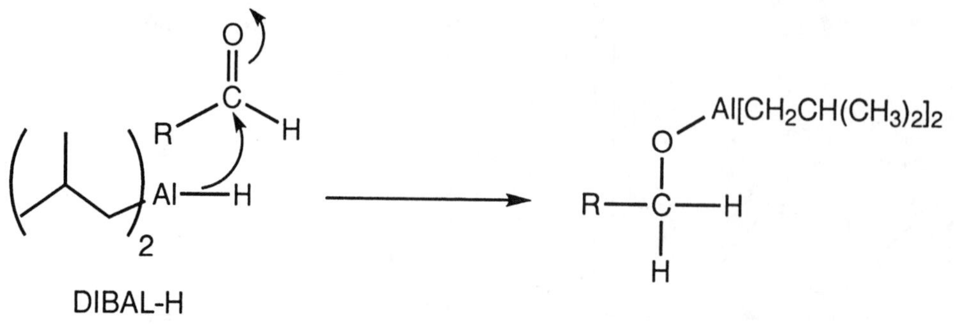

DIBAL-H

$Al[CH_2CH(CH_3)_2]_2$

The second and third equivalents of hydride are available from the isobutyl groups, in a manner reminiscent of the Meerwein-Ponndorf-Verley-Oppenauer reaction (p. 931).

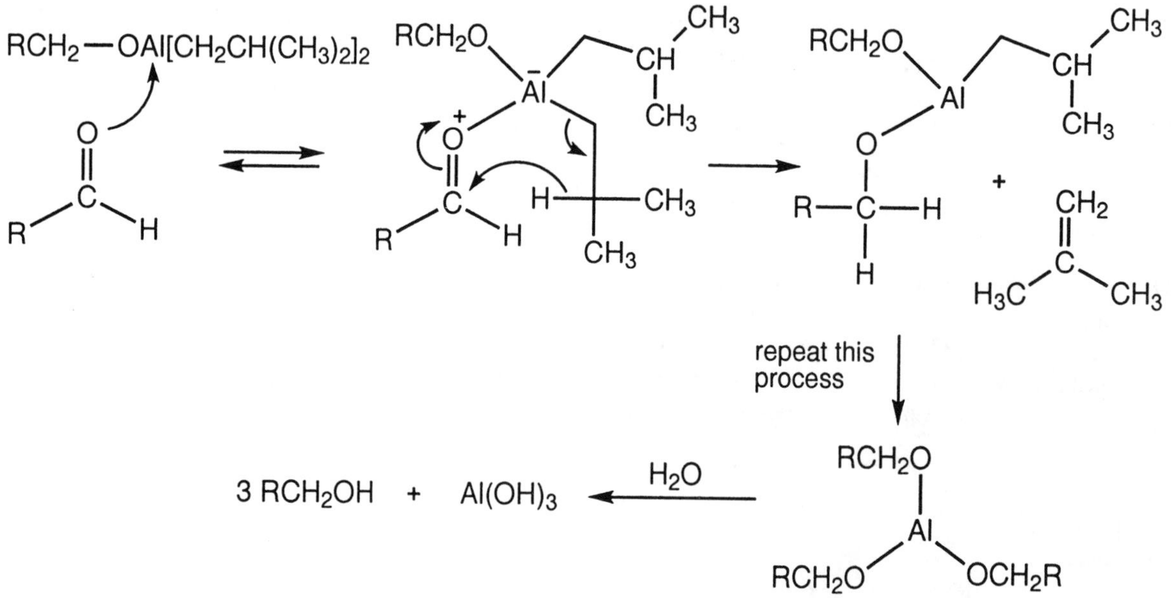

$RCH_2{-}OAl[CH_2CH(CH_3)_2]_2$

repeat this process

3 RCH_2OH + $Al(OH)_3$ ← H_2O

Problem 18.59 This problem starts with an easy reaction. Cyclopentyl methyl ketone (**1**) and pyrrolidine (**2**) must react to give an enamine. In this case, there are two enamines possible, **A** and **A'**. Be sure you can write mechanisms for their formations.

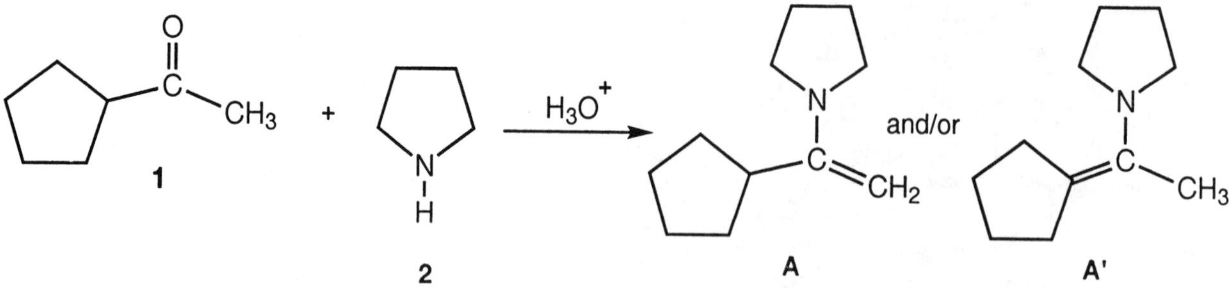

1 + **2** → (H_3O^+) **A** and/or **A'**

In the original literature, this enamine was represented as the less-substituted isomer **A**. Alkylation of the enamine with methyl α-(1-bromomethyl)acrylate (**3**) affords compound **B**. The two enamines should give different alkylation products, **B** and **B'**. Which is formed? Recall that mild hydrolysis of **B** yields compound **E** for which we have spectral data. The ^{1}H NMR spectral data is most compatible with the ketone derived from the less substituted enamine. For example, structure **E'** would be expected to exhibit a 3H singlet at $\delta = 2.0\text{-}2.5$ ppm for the ketone methyl group. Such a signal is clearly not present.

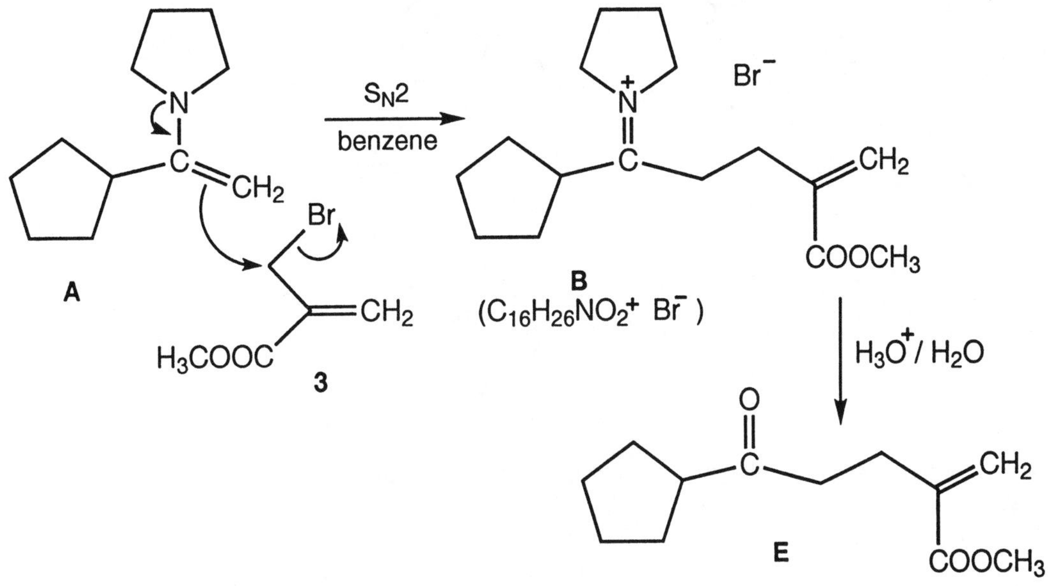

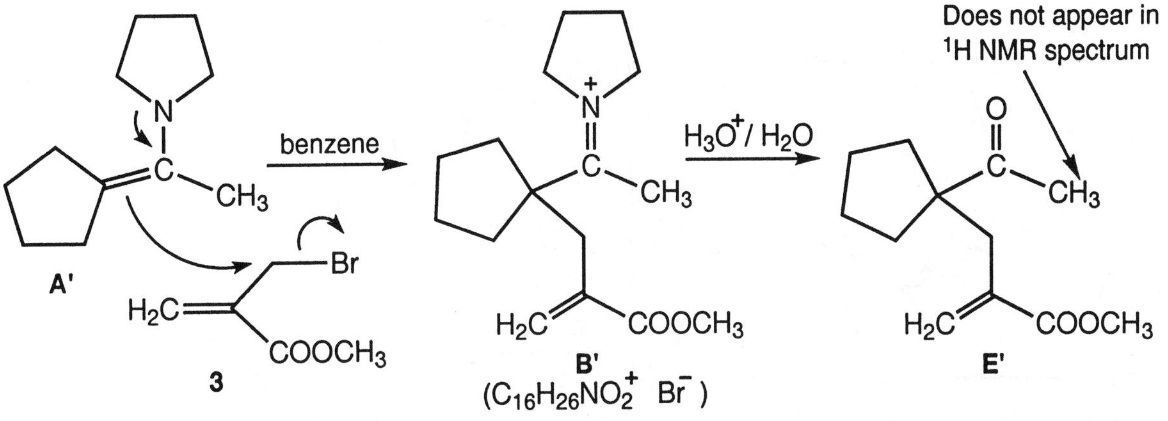

A + 3 —(S_N2, benzene)→ B ($C_{16}H_{26}NO_2^+$ Br^-) —(H_3O^+ / H_2O)→ E

A' + 3 —(benzene)→ B' ($C_{16}H_{26}NO_2^+$ Br^-) —(H_3O^+ / H_2O)→ E'

Does not appear in ^{1}H NMR spectrum

So, apparently the less substituted enamine **A** selectively undergoes alkylation with **3** to give salt **B**. Even if enamine **A** is the minor component of a mixture of enamines **A** and **A'**, it appears to be the more reactive isomer, presumably for steric reasons. Remember that as long as the two possible enamines are in equilibrium, the equilibrium will shift to replace the minor component as it is used up in the alkylation reaction. It is also worth commenting on the alkylation reaction itself, as the methyl α-(1-bromomethyl)acrylate (**3**) is such an unusual electrophile. The alkylation reaction could proceed by three different mechanistic pathways: (1) an S_N2 displacement as shown above, (2) an S_N2' displacement, and (3) a Michael reaction followed by expulsion of bromide. The last possibility is illustrated below. You should write an arrow formalism for the third possibility. Unfortunately, it is not known what the correct mechanistic path(s) is (are).

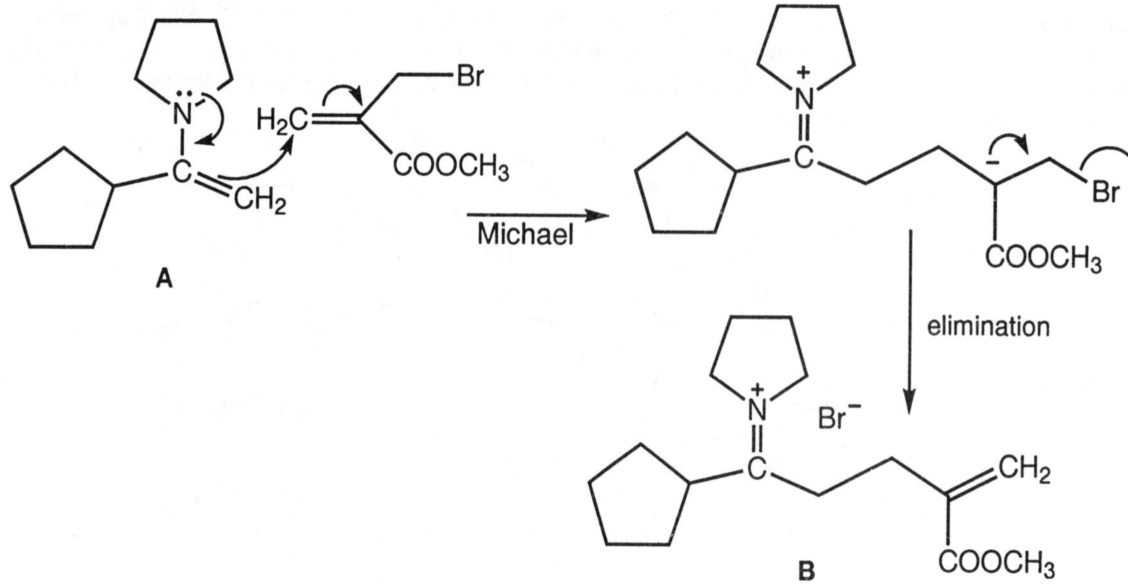

Michael

A

elimination

B

Treatment of **B** with triethylamine generates a new enamine; once again, two enamines, **C** and **C'**, are possible.

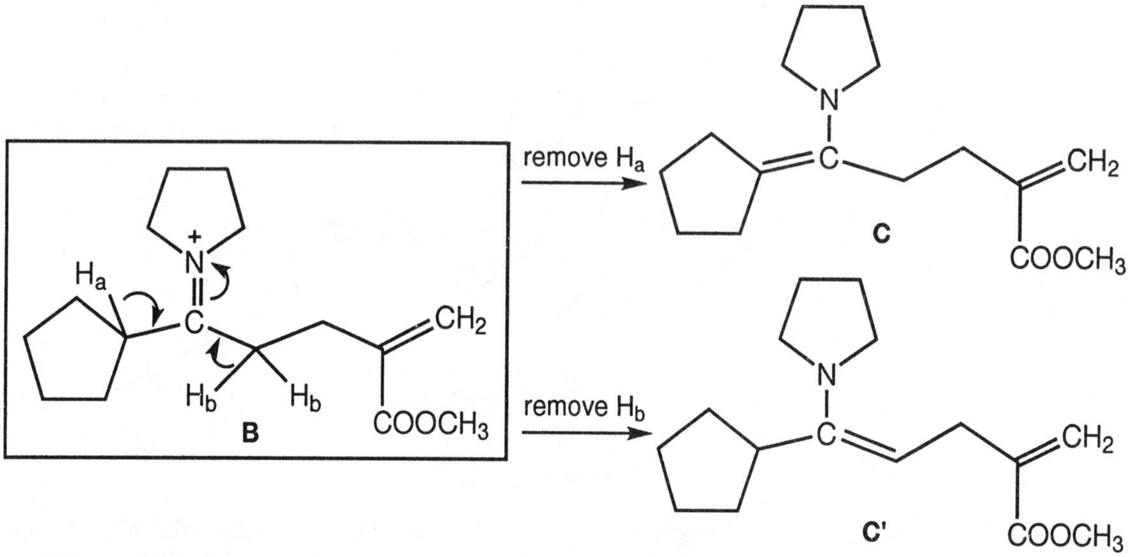

remove H_a

remove H_b

B

C

C'

Even if an equilibrating mixture of the two possible enamines **C** and **C'** forms, the more substituted enamine **C** leads ultimately to the observed spirodecanone **4** through an intramolecular Michael addition and mild hydrolysis.

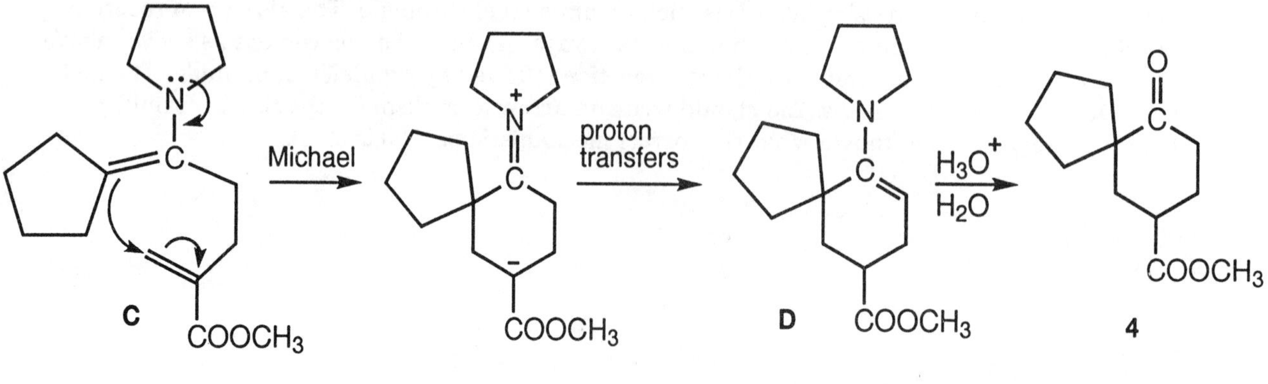

Michael

proton transfers

H_3O^+ / H_2O

C **D** **4**

Note that the less substituted enamine **C'** could also undergo an intramolecular Michael addition. However, in this case, a thermodynamically unfavorable four-membered ring would result.

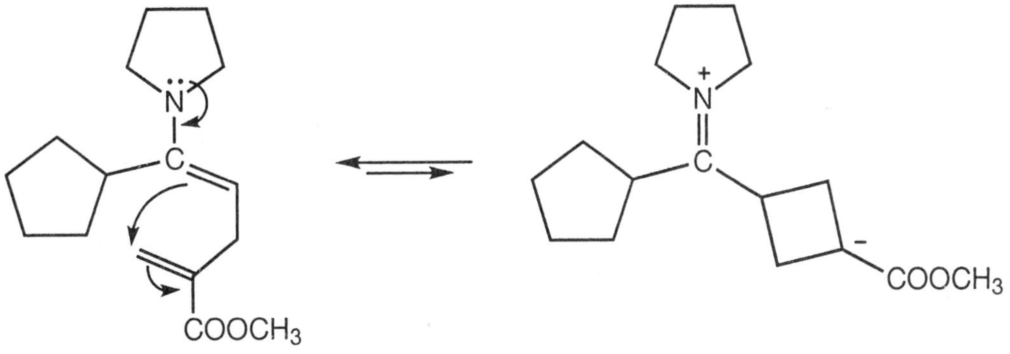

Although there are many mechanistic possibilities in this synthetic sequence, the process is still quite efficient as the spirodecanone **4** was obtained in an overall yield of 78% from **A** and **3**.

Problem 18.60 Recall from Problem 18.27 that deprotonated thiazolium ions can catalyze benzoin-type condensations. That is what is happening here, with the added complication of a decarboxylation. We first deprotonate thiamine pyrophosphate (TPP) with an enzymatic base. The resulting anion adds to the α-keto group of pyruvate **1** to give intermediate **A**. Decarboxylation then occurs to give the resonance-stabilized intermediate **B**.

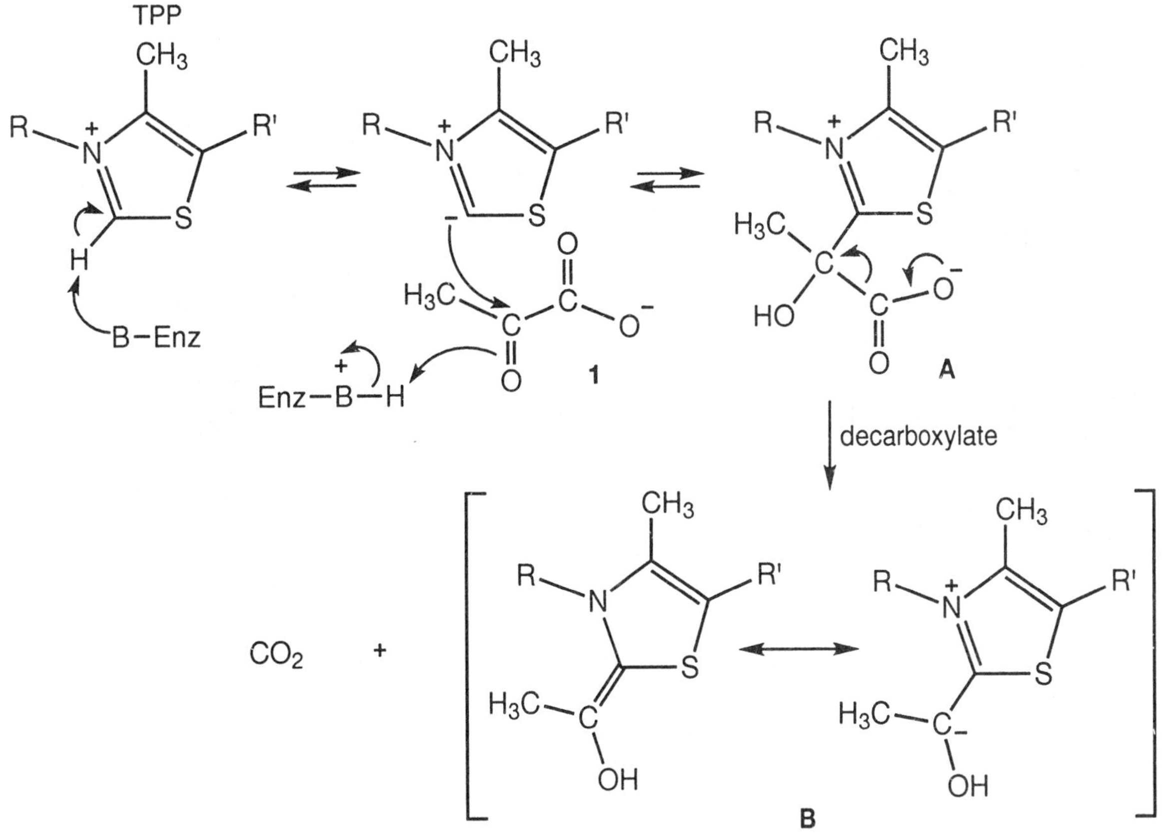

Intermediate **B** now adds to the α-keto group of a second molecule of pyruvate to yield **C**. Intermediate **C** can eliminate the TPP anion to give the product, acetolacetate (**2**).

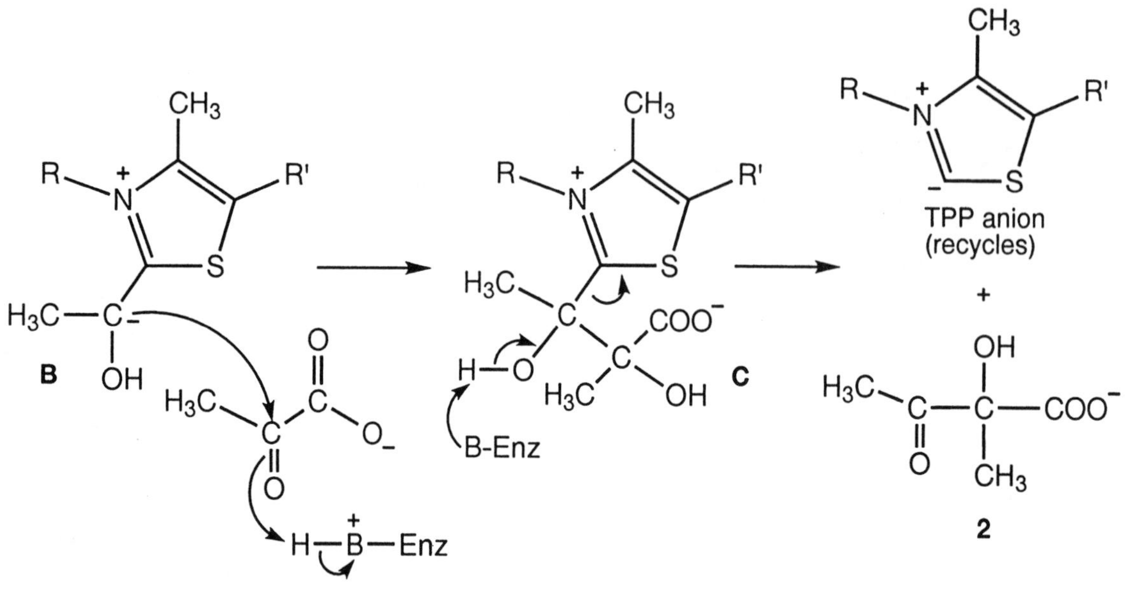

TPP anion (recycles)

B C 2

Problem 18.61 These are both double Michael problems. Part (a) is nothing more than a pair of straightforward addition reactions. Part (b) incorporates the complication of an intramolecular displacement after the second Michael.

(a)

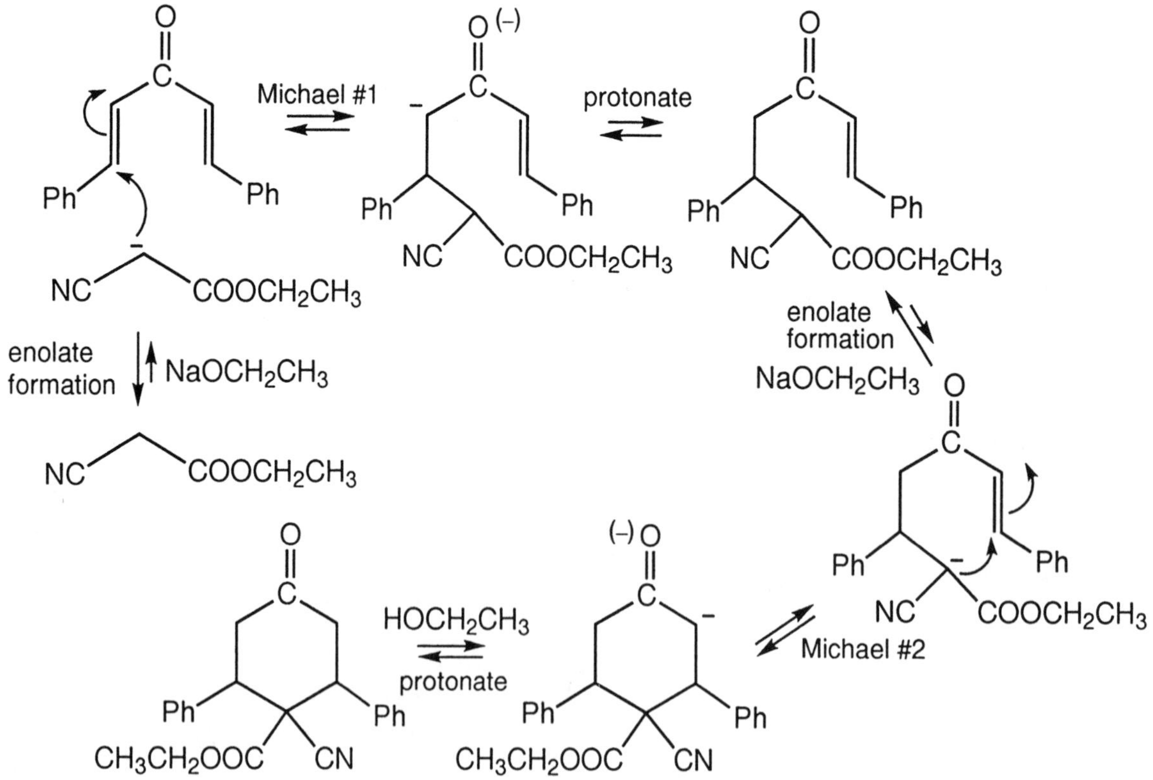

Michael #1 protonate enolate formation $NaOCH_2CH_3$ Michael #2 $HOCH_2CH_3$ protonate

(b)

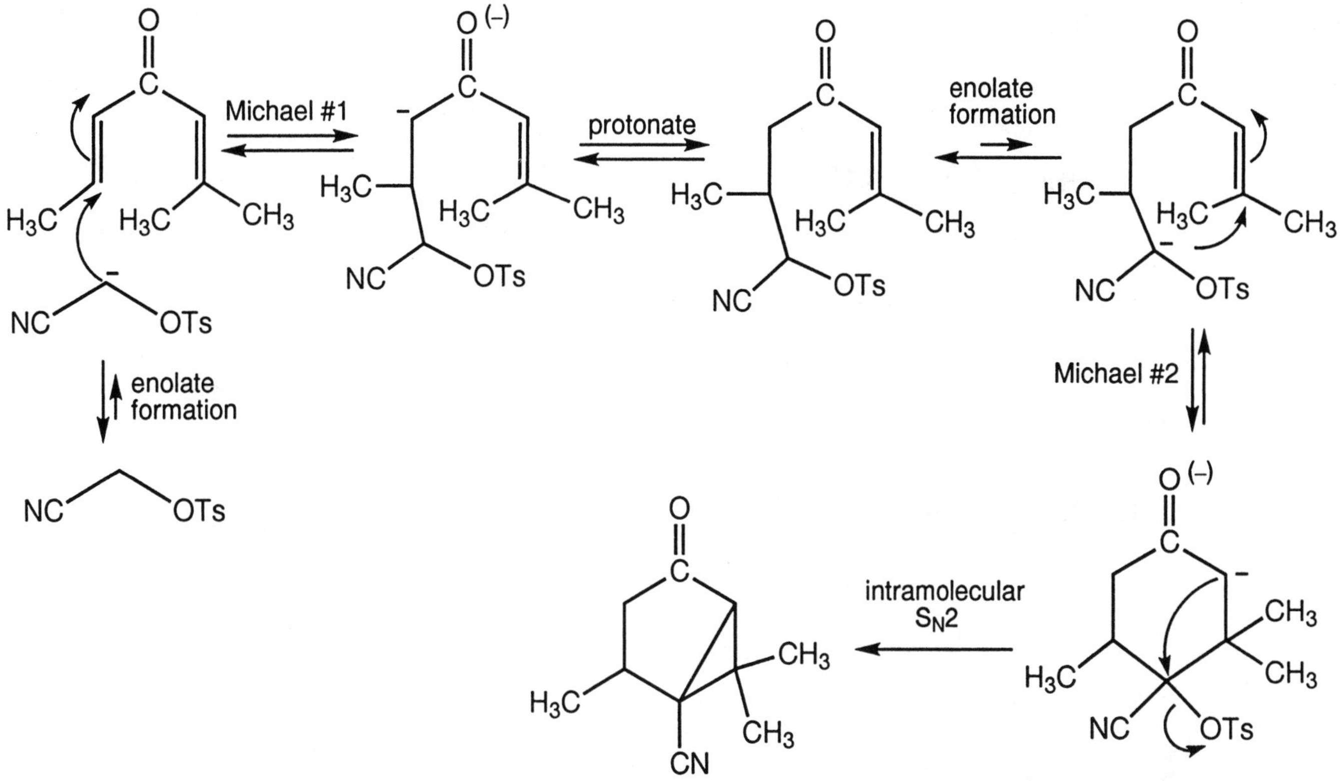

Michael #1
protonate
enolate formation
enolate formation
Michael #2
intramolecular S_N2

Problem 18.62 First analyze what must be done. Try to see the remnants of the starting materials in the product. That's not so hard if we use the methyl groups as markers ($R = CH_2CH_3$).

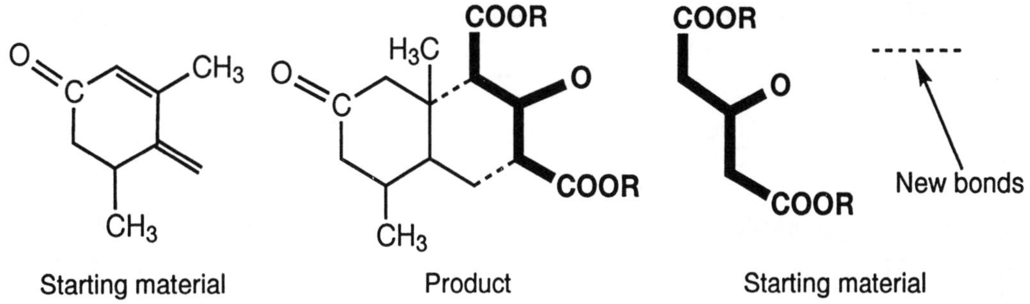

Starting material Product Starting material

New bonds

This reaction is another double Michael, but the two additions must be done in the correct order. If you try the wrong one first, the second cannot occur. There are two choices once an enolate has been formed, "Michael **A**" or "Michael **B**."

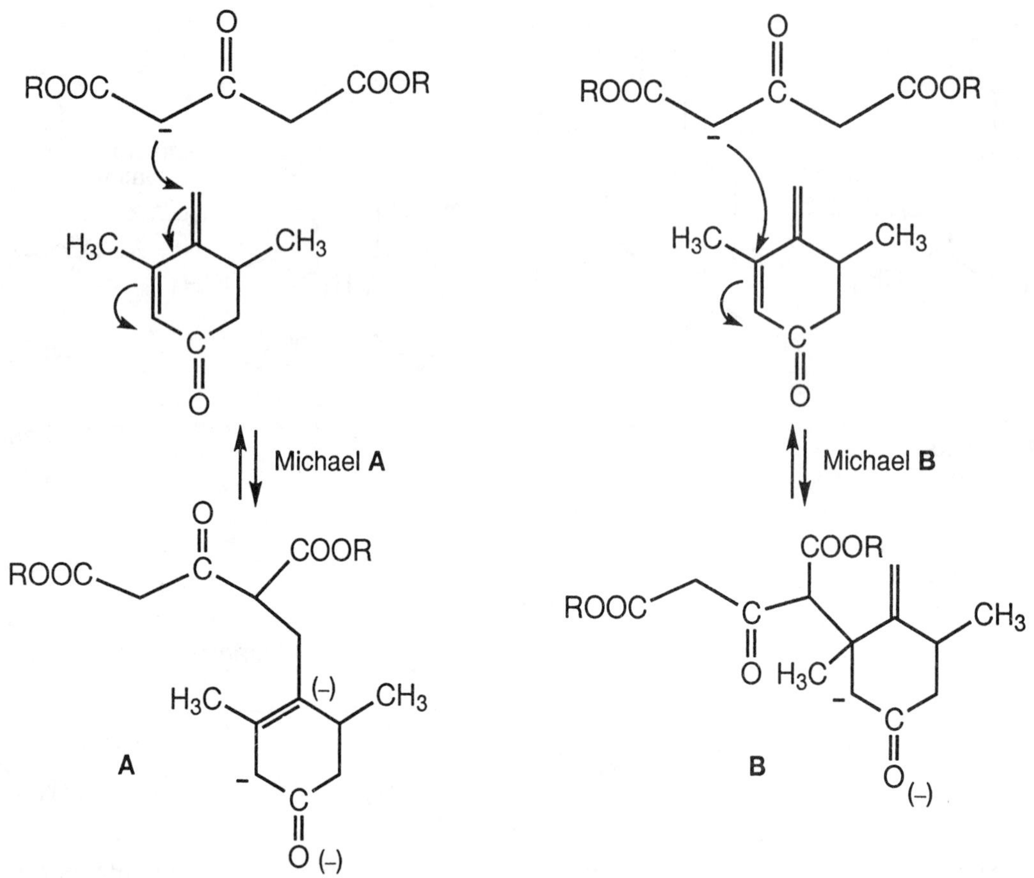

ROOC COOR C O Michael **A** H3C CH3 C O

ROOC COOR C O Michael **B** H3C CH3 C O

A **B**

Intermediate **A** can protonate to give a new α, β-unsaturated ketone, **C**, but **B** cannot. If it is not clear how the protonation of **A** occurs, by all means carefully draw out all the resonance forms for **A**. Enolate formation and a second Michael addition, followed by protonation, gives the product.

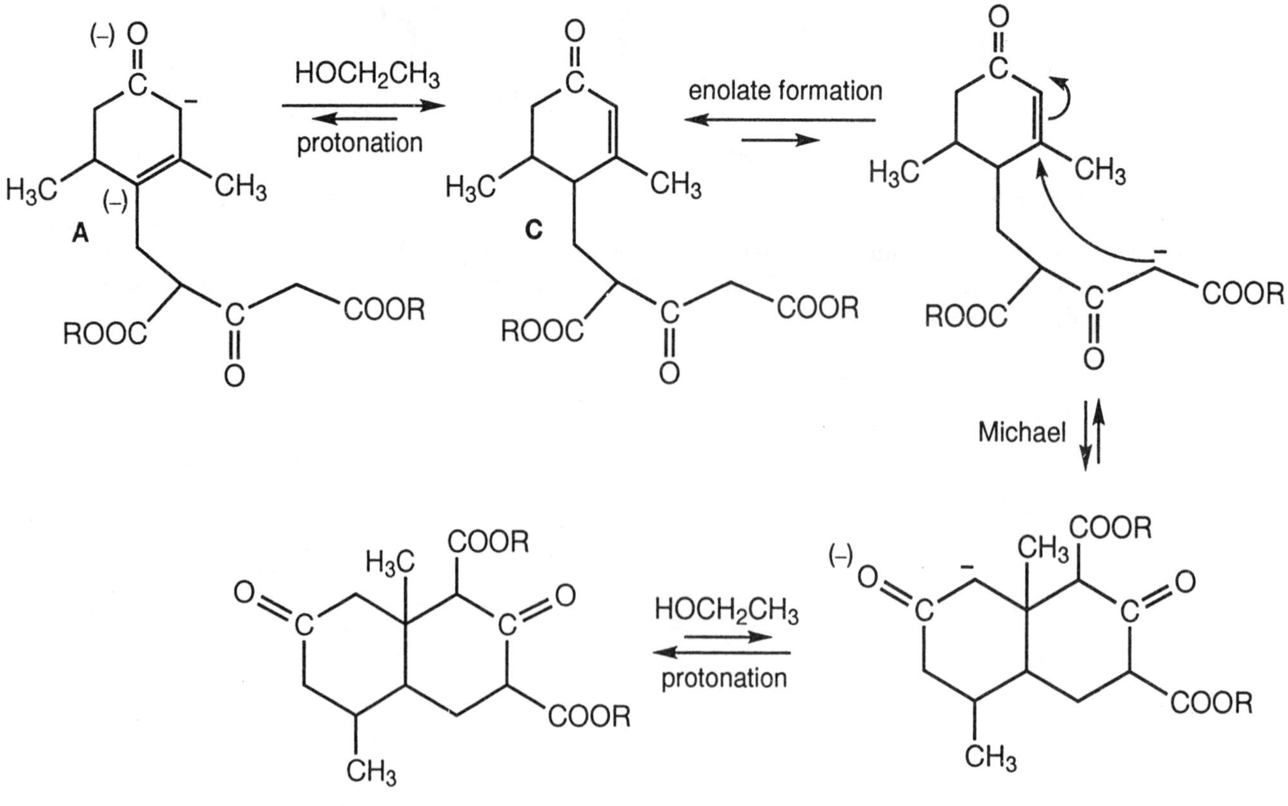

HOCH2CH3 protonation enolate formation Michael HOCH2CH3 protonation

Problem 18.63 On the surface, this is a simple alkoxide formation, followed by alkylation. This reaction is a Williamson ether synthesis.

$$R—OH \xrightarrow{NaH} R—O^- \xrightarrow[(S_N2)]{H_3C—I} R—O—CH_3 + I^-$$

However, there is the problem of the seemingly strange stereoisomerization that takes place. Alcohol **1** is a β-keto alcohol, which means that it can, at least in principle, be made through an aldol condensation. We can find that aldol condensation in many ways, but one good method is to write the reverse aldol condensation starting from **1**.

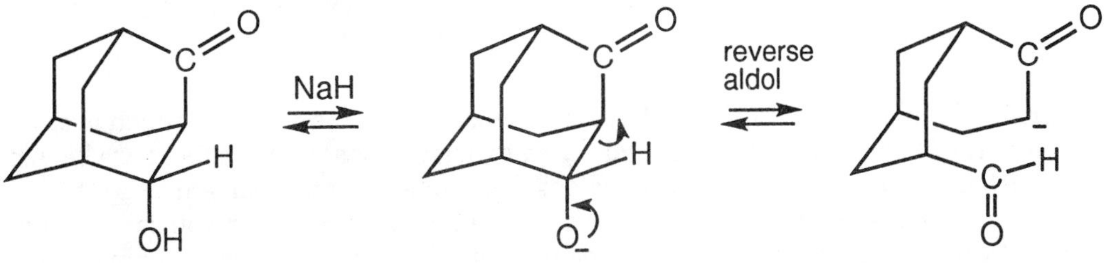

What can the enolate **A** do? Of course it can simply reverse to give **1**, that's merely the forward aldol. But there are two ways this can happen! Opened intermediate **A** can undergo carbon-carbon bond rotations to give **A'**, and aldol condensation of **A'** gives the isomerized alkoxide, **B**. Methylation gives the product.

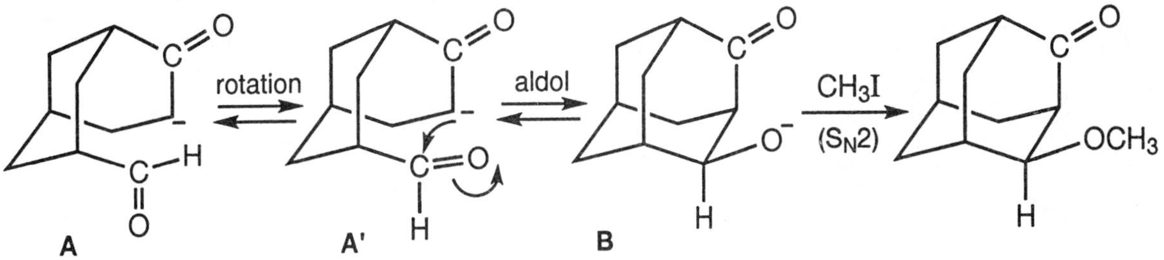

Problem 18.64 Note first that this is a redox reaction. One carbonyl has been reduced and the other oxidized. This redox process should make you consider a hydride shift, but there is one very substantial problem: There is no hydrogen to be shifted! It is not only hydride that can shift. Here there is a migration of a phenyl group, a benzene ring, with its pair of electrons. In the trade, this reaction is called the benzilic acid rearrangement. There really is only one way to start. There are no α-hydrogens, so no enolates can be formed. Hydroxide ion must add to one of the two equivalent carbonyl groups. In the section on hydride shifts, there was a problem (much like this one) in which a hydride migrated in what was essentially an intramolecular Cannizzaro reaction (Fig. 18.111). Here there is no hydride, and it is a benzene ring that plays the role of the migrating group. This reaction generates an alkoxide, and proton transfer to make the far more stable carboxylate anion must be very fast. A final acidification generates the free acid (next page).

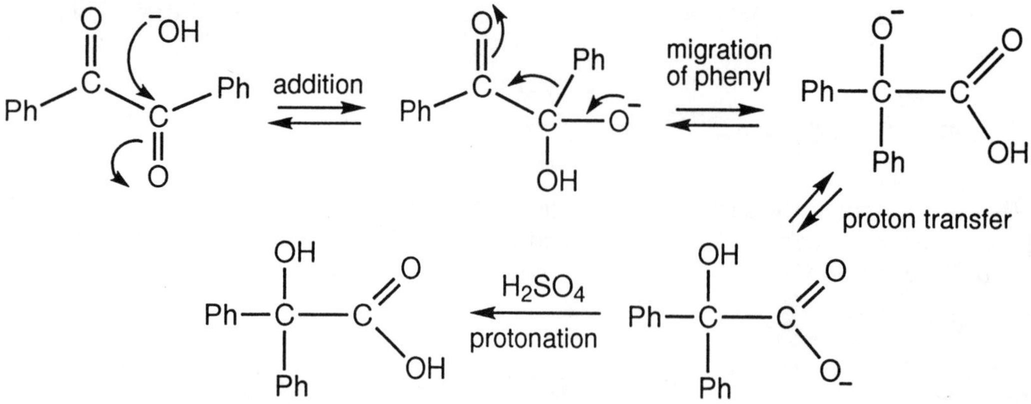

addition
migration of phenyl
proton transfer
H_2SO_4
protonation

The key to this problem is the recognition that a benzene ring has moved. That much is obvious from a glance at the starting material (two separated benzene rings) and product (two benzene rings on one carbon). Although that much *is* obvious, it is nonetheless important to go through the analysis that makes you articulate the thought that a ring must move. Then you can set out to accomplish that task, asking at every point in the problem, Is there a way now to move the ring?

Chapter 19 Outline

19.1 Nomenclature and Properties of Carboxylic Acids
19.2 Structure of Carboxylic Acids
19.3 Infrared and Nuclear Magnetic Resonance Spectra of Carboxylic Acids
19.4 Acidity and Basicity of Carboxylic Acids
19.5 Reactions of Carboxylic Acids
19.6 Reactivity of Carboxylic Acids at the α-Position
19.7 Syntheses of Carboxylic Acids
19.8 Something More: Fatty Acids

19

Carboxylic Acids

Problem 19.2 The only tricky part is making sure you figure out the absolute configurations of the two stereogenic carbons in the last example. Notice that the "cis" designation, necessary when we are talking about the racemic material, becomes unnecessary if we specify the absolute configurations, as the *(R)* and *(S)* designations require the cis stereochemistry in this molecule.

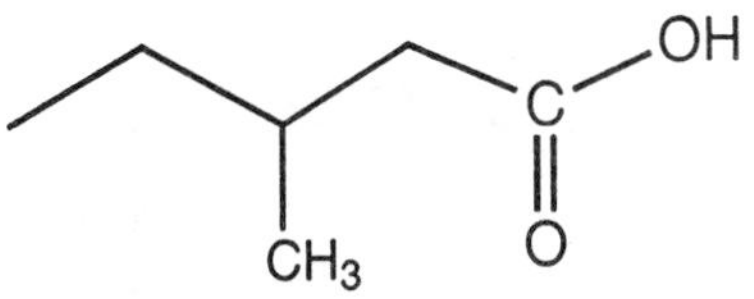

3-Methylpentanoic acid
(3-methylvaleric acid)

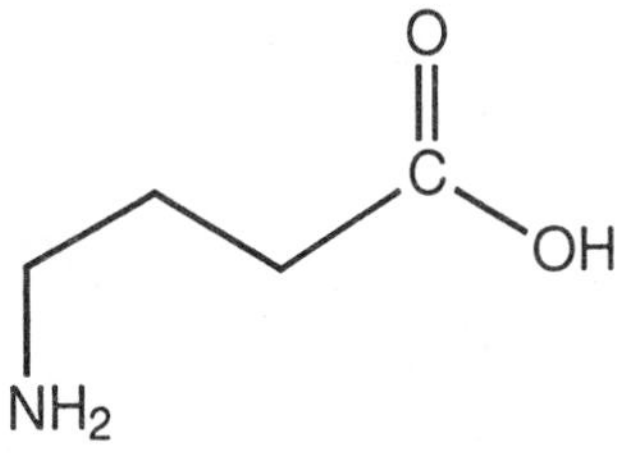

4-Aminobutanoic acid
(4-aminobutyric acid)

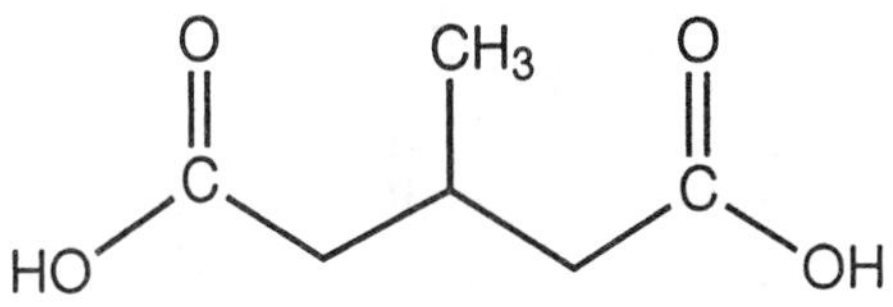

3-Methylpentanedioic acid
(3-methylglutaric acid)

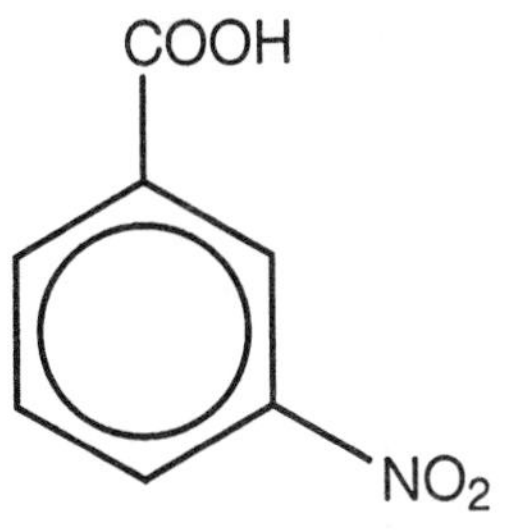

m-Nitrobenzoic acid
(3-nitrobenzenecarboxylic acid)

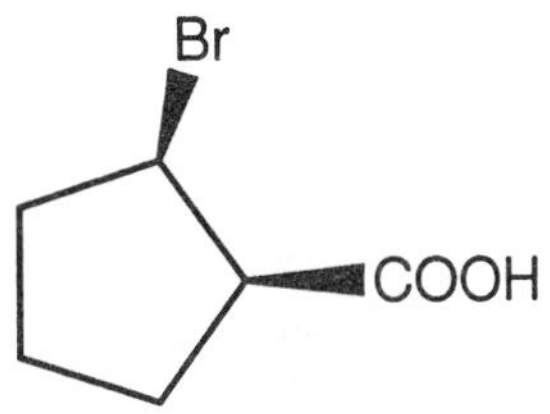

(racemic)
cis-2-Bromocyclopentanecarboxylic acid

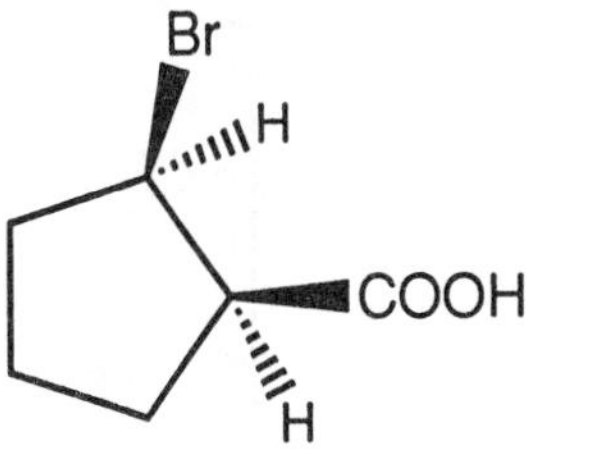

(this enantiomer)
(1*R*),(2*R*)-2-Bromocyclopentanecarboxylic acid

Problem 19.3 In the syn form, an intramolecular hydrogen bond is possible; in the anti form it is not. As bond formation is stabilizing, the syn form will be more stable than the anti form.

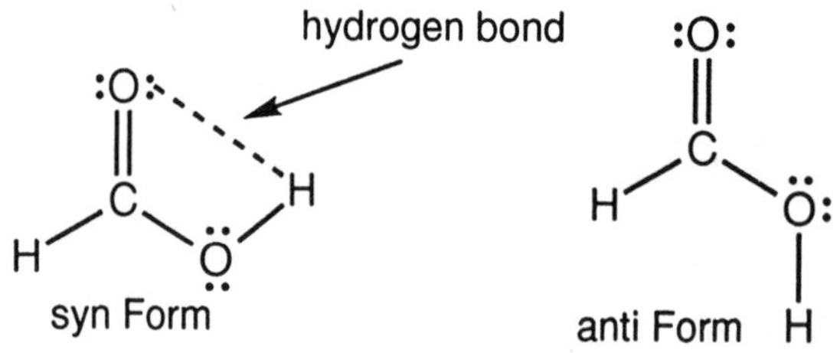

hydrogen bond

syn Form

anti Form

Problem 19.4 Review from Chapter 8:
$\Delta G = -2.3RT \log K$
At 25 °C, $2.3RT$ is 1.364 kcal/mol (Chapter 8, p. 305). So, if the energy difference (ΔG) between syn and anti forms is given as six kcal/mol:
$-6 = -(1.364) \log K$, and: $K \sim 25{,}000$

Problem 19.8 It is the same old story. Although the rate of a reaction is determined not by the energy of the product, but by the energy of the transition state leading to product, very often the factors influencing the energy of the product will also be at work in determining the energy of the transition state. This parallelism is especially true for endothermic reactions in which the transition state will resemble the product. In this case, the transition state for protonation of the carbonyl oxygen will be stabilized by delocalization of the developing positive charge.

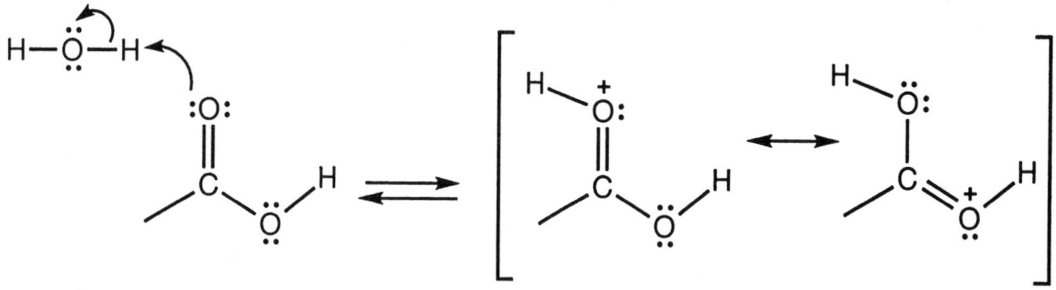

Protonation of the carbonyl oxygen leads to a delocalized intermediate

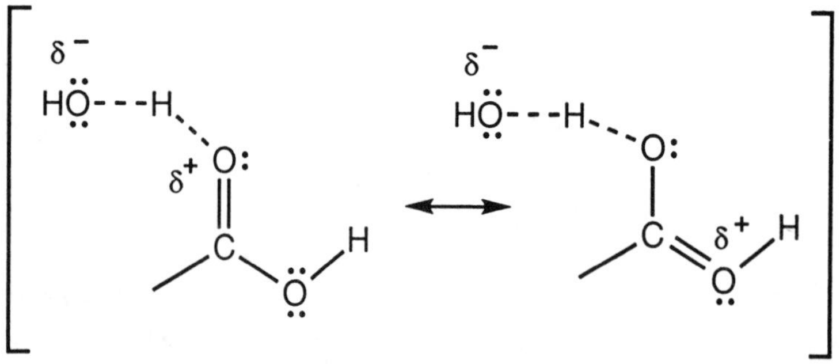

The delocalized transition state for protonation

Problem 19.9 The mechanism involves a tetrahedral intermediate **A** that contains three equivalent oxygens, one ^{18}**O**, the others ^{16}O.

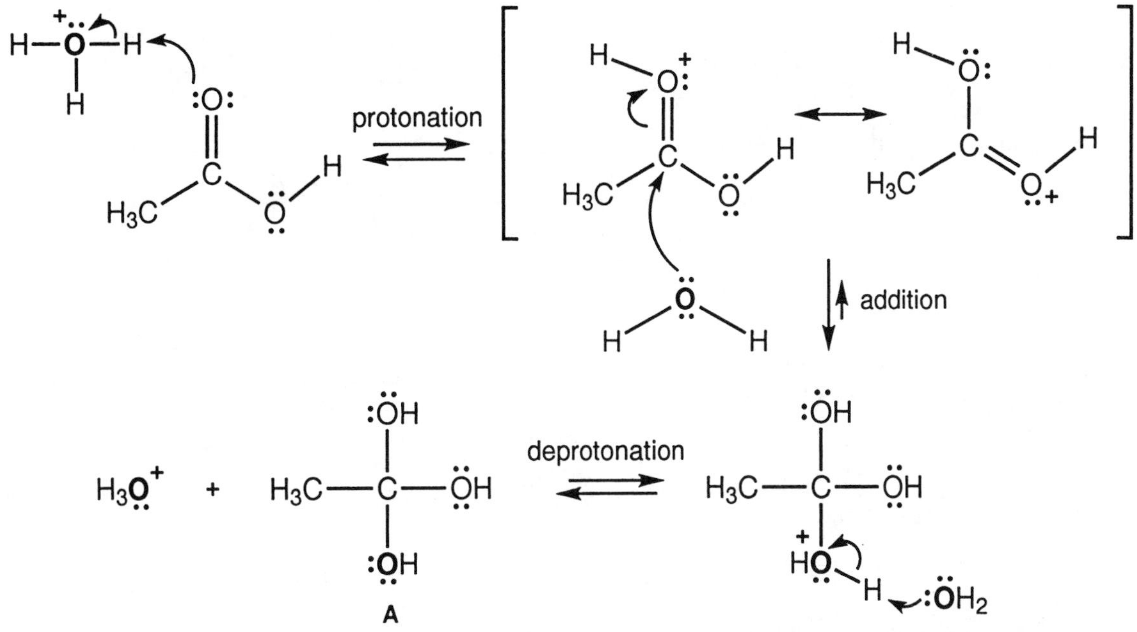

protonation

addition

deprotonation

A

Intermediate **A** can revert to acetic acid to give either carbonyl-labeled acetic acid or hydroxyl-labeled acetic acid. The following drawing only sketches the mechanism (it is, of course, just the reverse of the steps outlined in the first figure of this problem).

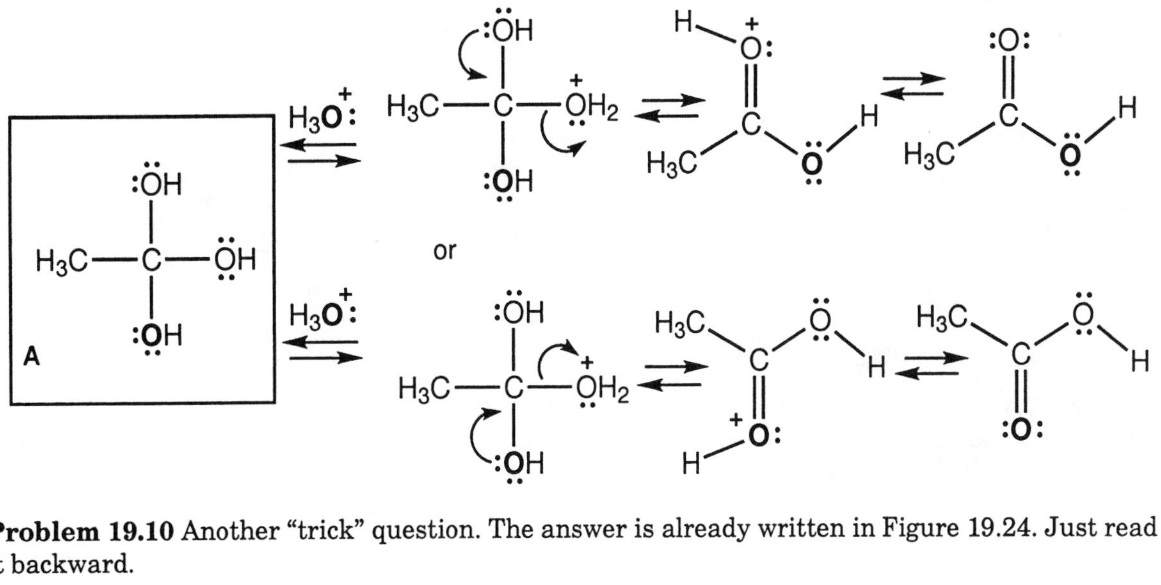

A

or

Problem 19.10 Another "trick" question. The answer is already written in Figure 19.24. Just read it backward.

Problem 19.11. These are both addition-elimination processes. In each case, the alcohol adds to the double bond to oxygen, and a good leaving group is lost to generate the ester:

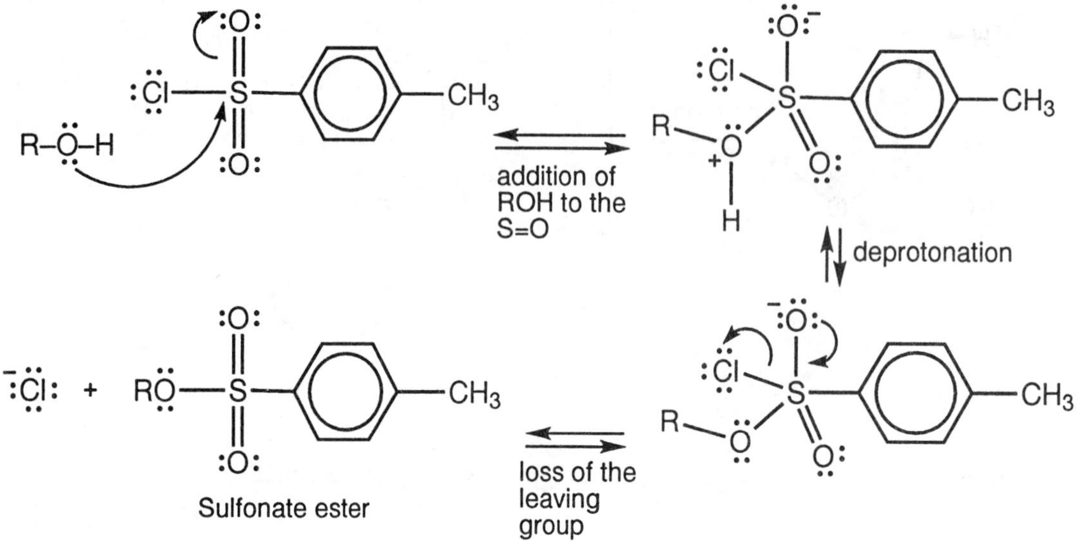

addition of ROH to the S=O

deprotonation

loss of the leaving group

Sulfonate ester

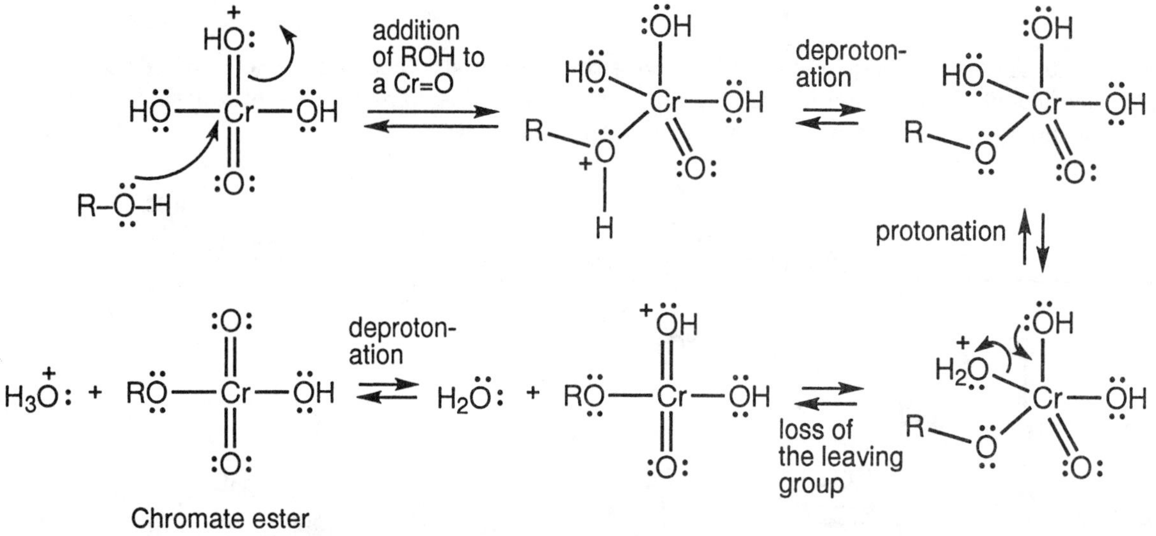

addition of ROH to a Cr=O

deproton-ation

protonation

loss of the leaving group

deproton-ation

Chromate ester

Problem 19.12 The critical insight is to recognize the product as a cyclic ester, a lactone. This reaction is just an intramolecular Fischer esterification. The steps are identical to those of Figure 19.24, except that they all occur within the same molecule.

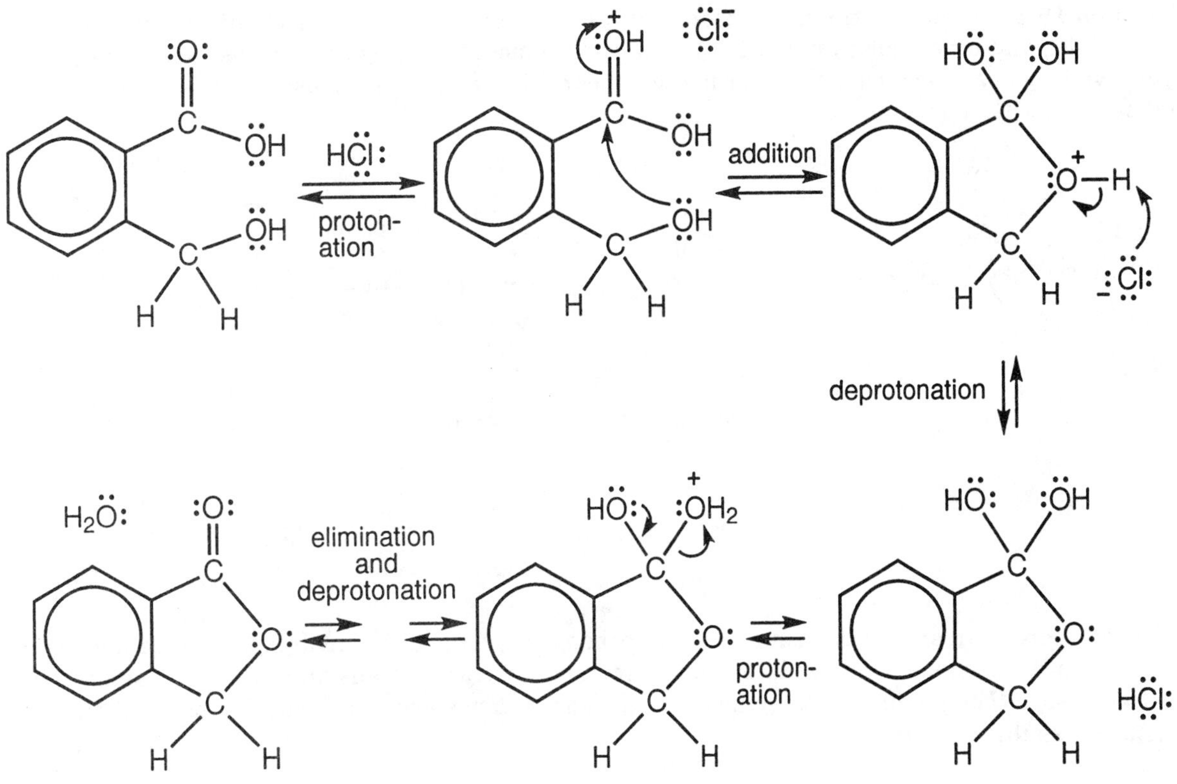

HCl: proton-ation

addition

deprotonation

elimination and deprotonation

proton-ation

HCl:

A cyclic ester - a lactone

Problem 19.13 The phrase "a neutral molecule" in this problem, coming where it does, must surely suggest that the product is a lactone. In base, the carboxylate salt will be formed first. Once iodine reacts with the carbon-carbon double bond to produce the iodonium ion, the perfectly poised carboxylate opens (S_N2 - displacement from the rear) it to give the lactone, the "neutral" product.

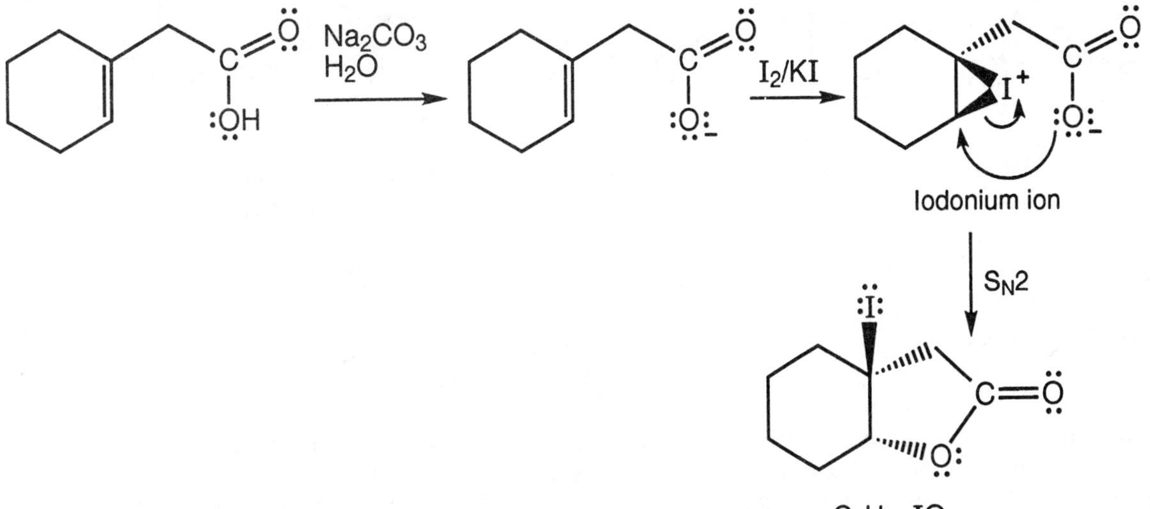

Na_2CO_3 H_2O

I_2/KI

Iodonium ion

S_N2

$C_8H_{11}IO_2$

Problem 19.16 Do you get the impression that the addition-elimination mechanism is important? You're right, and here comes yet another example. The most tempting thing is just to use the hydroxyl oxygen as the nucleophile, form a tetrahedral intermediate, **A**, lose chloride, deprotonate, and be done with it.

eliminate chloride

Tetrahedral intermediate **A**

deprotonation

However, there is an error. Addition using the carbonyl oxygen, not the hydroxyl hydrogen, leads to a more stable, resonance-stabilized, intermediate in which both oxygens and the carbon share the positive charge. The product is the same as in the mechanism shown first, but the mechanistic details are slightly different.

Resonance-stabilized intermediate

loss of chloride and deprotonation

\+ HCl

Problem 19.17 This transformation begins with addition of hydride and loss of chloride, still another example of the addition-elimination sequence. This time, however, the product is an aldehyde, and aldehydes are reduced by $LiAlH_4$ (Chapter 16, p. 794) to give alkoxides that form alcohols on protonation. In this reaction, the initial product, the aldehyde, is not stable to the conditions and is further reduced.

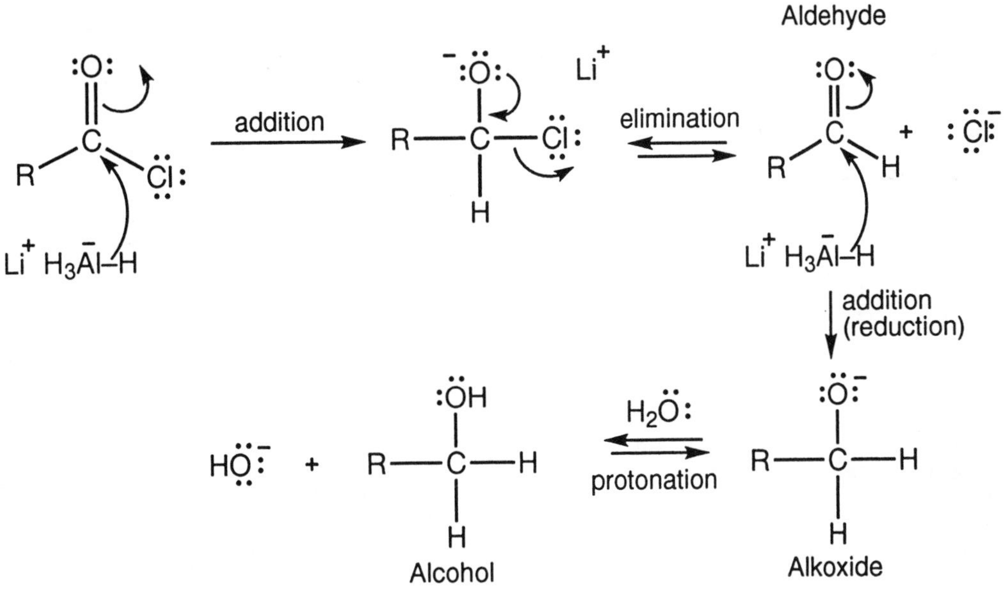

Aldehyde

addition

elimination

Li⁺ H3AlH

addition (reduction)

H_2O:

protonation

Alcohol

Alkoxide

Problem 19.18 Once again, this problem involves a series of addition-elimination steps. Phosgene reacts with an acid in the first addition-elimination sequence to give intermediate **A** and a chloride ion. The chloride ion adds to **A** to give **B** from which carbon dioxide and chloride can be irreversibly lost. This sequence is the second addition-elimination. A final deprotonation leads to the acid chloride and hydrochloric acid. Many variations on this mechanism are reasonable. For example, one might write first an equilibrium in which phosgene deprotonates the acid to give a carboxylate and a protonated carbonyl. The carboxylate could add to the protonated carbonyl to give an **A**-like intermediate. The mechanism(s) of reaction with oxalyl chloride is (are) similar.

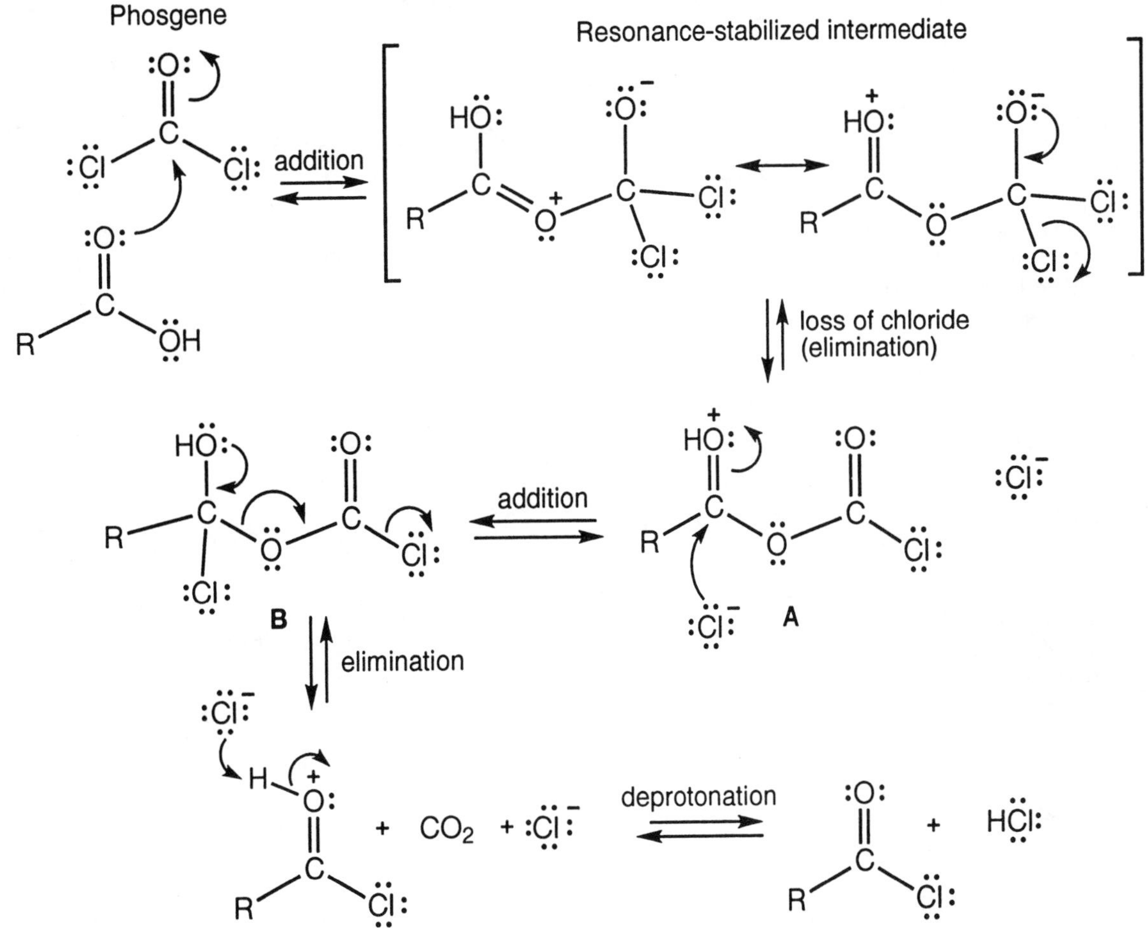

Phosgene

addition

Resonance-stabilized intermediate

loss of chloride (elimination)

addition

B

A

elimination

+ CO_2 + :Cl:⁻

deprotonation

+ HCl:

Problem 19.19 Phosgene, like any acid chloride, reacts rapidly with water in an addition-elimination process. One product is hydrochloric acid, and this acid is a most unpleasant molecule to have in one's lungs.

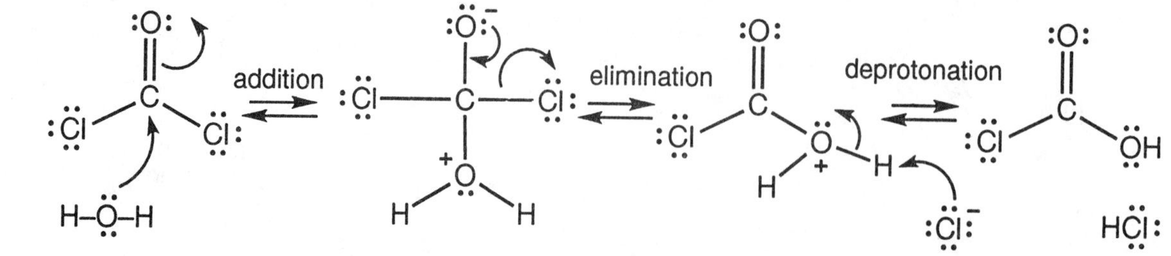

Phosgene

addition

elimination

deprotonation

Problem 19.20 The most important form will be the one with both negative charges on the relatively electronegative oxygen atoms.

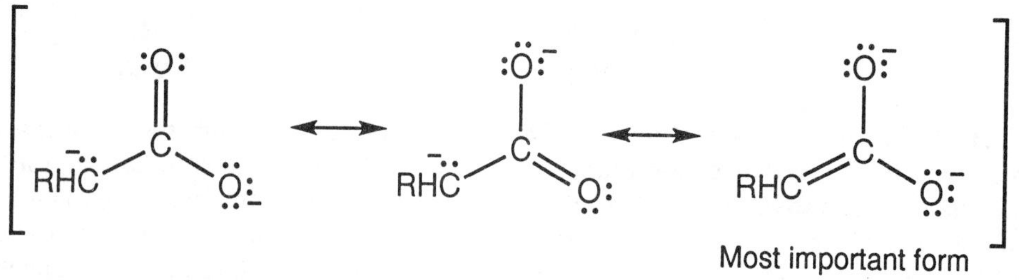

Most important form

Problem 19.21 In each case, formation of the dianion is followed by alkylation at carbon. A final protonation completes the reaction. In acid, the γ-hydroxy acid would probably form the related lactone. Can you write a mechanism?

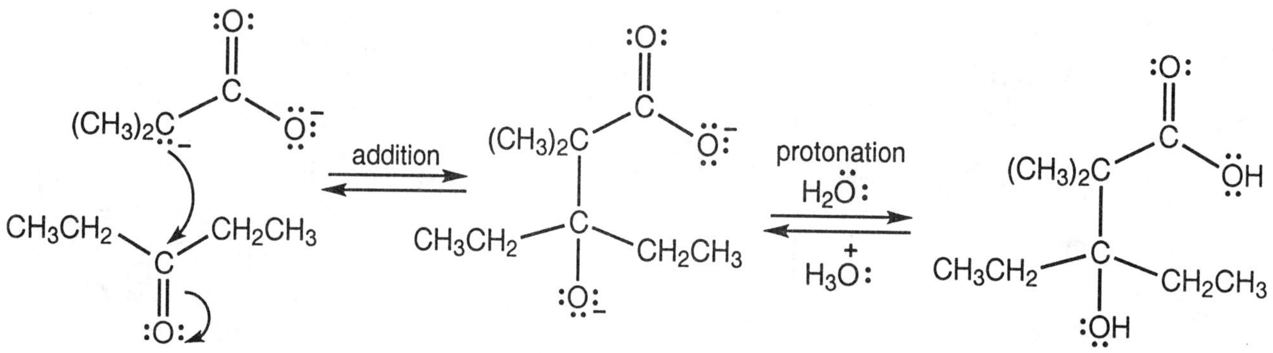

addition

protonation

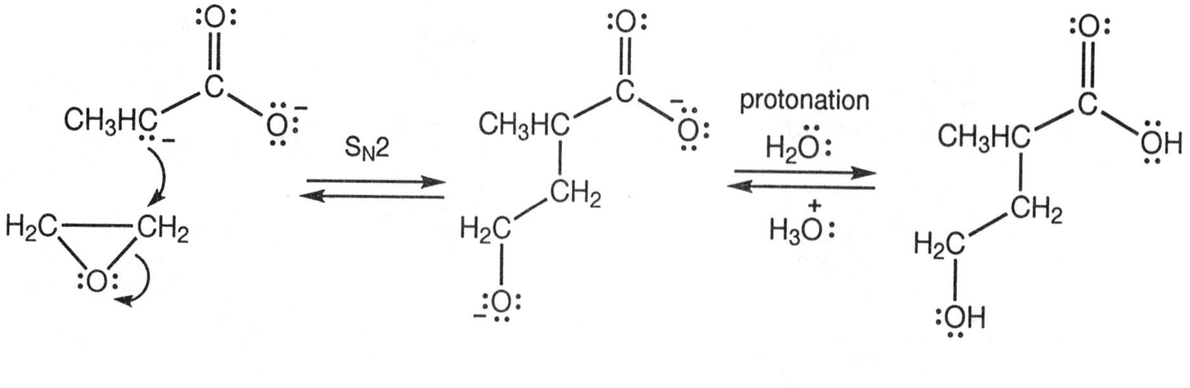

$H_2\ddot{O}$:

$H_3\overset{+}{O}$:

S_N2

protonation

$H_2\ddot{O}$:

$H_3\overset{+}{O}$:

Problem 19.22 More addition-elimination reactions! The same mechanism will suffice for all three reactions. The nucleophile first adds to the carbonyl group to give a tetrahedral intermediate, and the good leaving group bromide (or chloride in the specific example) is then lost. The reaction may be complicated by S_N2 displacement of the α-bromide.

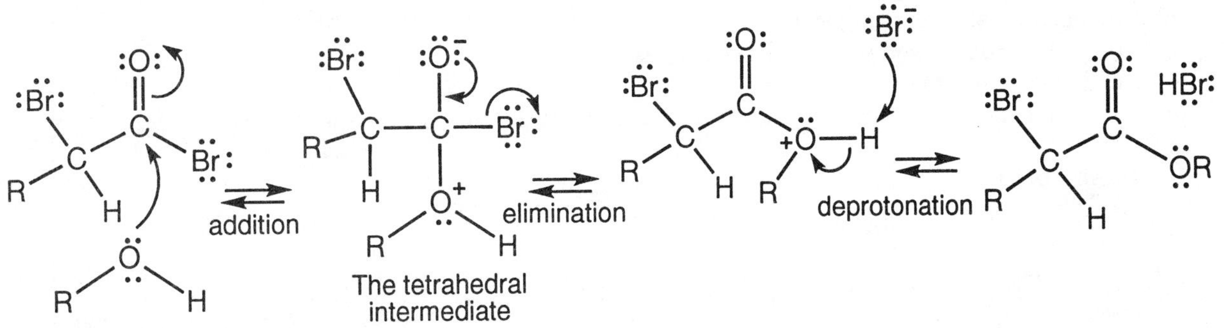

addition
The tetrahedral intermediate
elimination
deprotonation

Problem 19.23 Hydroxide ion adds to the carbon-oxygen double bond to give a tetrahedral intermediate. This time the leaving group is ^-OR. Three such reactions complete the hydrolysis reaction.

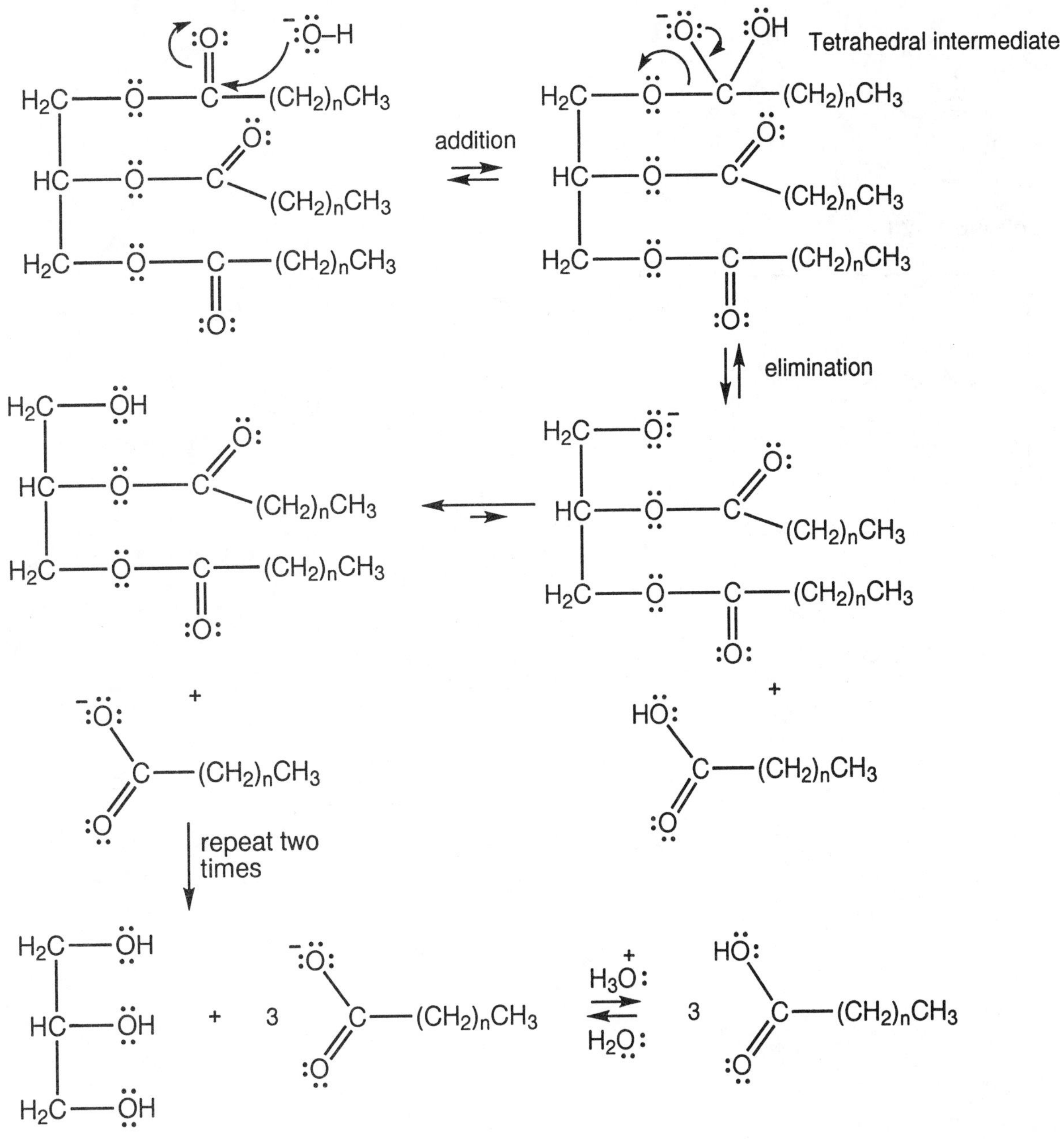

addition
Tetrahedral intermediate
elimination
repeat two times

Additional Problem Answers

Problem 19.24

(a) Cyclopropanecarboxylic acid
(b) 2-Bromo-4-methylpentanoic acid
(c) *trans*-4-Hydroxycyclohexanecarboxylic acid
(d) *p*-Chlorobenzoic acid
(e) 2-Amino-3-methylbutyric acid (2-amino-3-methylbutanoic acid, valine)

Problem 19.25

(a) H_2N–CH(CH_3)–COOH

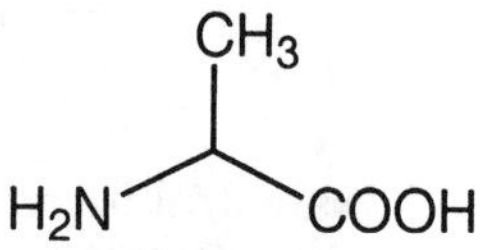

(b) H_3C–CH(OH)–COOH

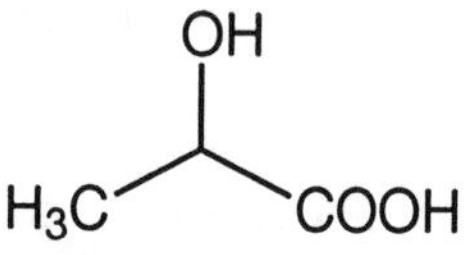

(c) H_3C–CH=C(CH_3)–COOH

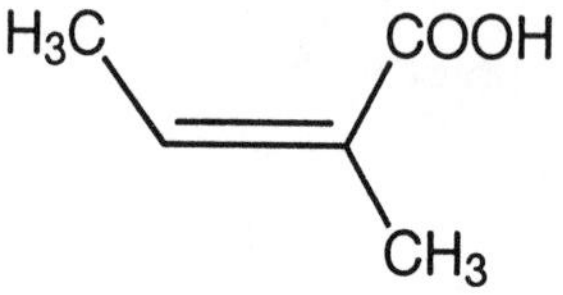

(d) HO–C(COOH)(CH_2COOH)$_2$ (labels: HO, COOH, COOH, COOH)

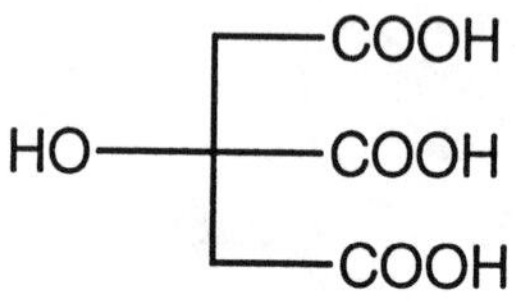

(e) COOH, OH (benzene ring)

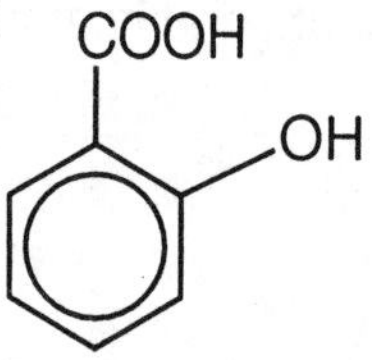

(f) H_3C–C(=O)–COOH

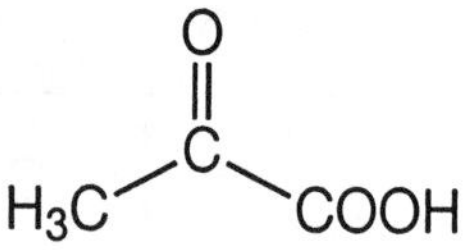

Problem 19.26

OH, C, O

$CH_3\overset{+}{O}H_2$ / CH_3OH → (a) OCH$_3$, C, O — Fischer esterification

DCC / CH_3NH_2 → (b) $NHCH_3$, C, O — Amide formation

$SOCl_2$ → (c) Cl, C, O — Acid chloride formation

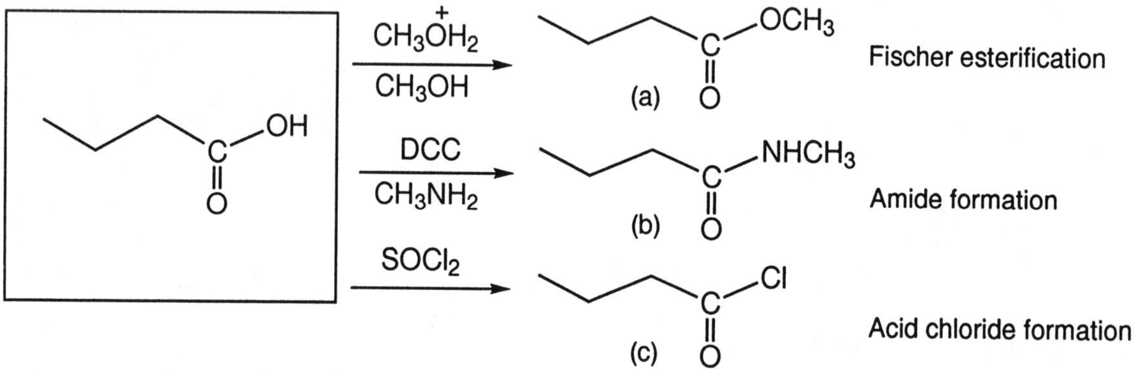

Butanoic acid ($CH_3CH_2CH_2COOH$):

$\xrightarrow[\text{2. } H_2O/H_3O^+]{\text{1. } LiAlH_4}$ (d) 1-butanol — Reduction to the alcohol

$\xrightarrow[\text{2. } CH_3CH_2CH_2COCl]{\text{1. } KOH/H_2O}$ (e) butanoic anhydride — Anhydride formation

$\xrightarrow[\text{2. } H_2O]{\text{1. 2 } CH_3Li}$ (f) 2-pentanone $\xrightarrow[\text{2. } H_2O]{\text{1. } NaBH_4}$ (g) 2-pentanol

$\xrightarrow[Br_2]{PBr_3 \text{ (cat.)}}$ (h) $CH_3CH_2CHBrCOOH$ — α-Bromination with a catalytic amount of PBr_3

$\xrightarrow{PBr_3/Br_2}$ (i) $CH_3CH_2CHBrCOBr$ — A full equivalent of PBr_3 introduces two Br atoms

$\xrightarrow[\text{2. } CH_3I \text{; 3. } H_2O/H_3O^+]{\text{1. 2 LDA}}$ (j) $CH_3CH_2CH(CH_3)COOH$ — Formation of the dianion leads to α-alkylation

Problem 19.27

(a) This reaction is an example of a standard Fischer esterification. The bromide is not displaced under acidic conditions as there is no nucleophile strong enough.

$$BrCH_2CH_2COOH \xrightarrow[CH_3OH,\ \Delta]{CH_3\overset{+}{O}H_2} BrCH_2CH_2COOCH_3$$

(b) Treatment of a carboxylic acid with DCC results in the formation of an anhydride.

$$RCH_2COOH \xrightarrow[CH_3CN]{DCC} RCH_2C(=O)OC(=O)CH_2R$$

(c) This reaction is a classic example of the saponification of a glyceride (a fat) to afford a fatty acid and glycerol.

$$\text{Glyceride (tri-}C_{13}H_{27}CO_2\text{ ester)} \xrightarrow[\text{2. } H_2O/H_3O^+]{\text{1. NaOH}/H_2O,\ \Delta} 3\ C_{13}H_{27}COOH \text{ (Myristic acid)} + \text{Glycerol}$$

(d) In the presence of an amine, the anhydride (formed from the carboxylic acid and DCC) yields an amide.

$$RCH_2COOH \xrightarrow[CH_3CN]{DCC} RCH_2C(O)OC(O)CH_2R + DCU \xrightarrow{R'R''CH\text{–}NH_2} RCH_2C(O)NHCHR'R'' + RCH_2COOH \text{ (recycles)}$$

(e) Carboxylic acids react with bases such as LiH to form carboxylate anions. Acidification of the salt merely regenerates the carboxylic acid.

$$C_6H_{11}COOH \xrightarrow{\text{1. LiH}} C_6H_{11}COO^- \xrightarrow{\text{2. } H_3O^+/H_2O} C_6H_{11}COOH$$

(f) This process is a bit more complicated. It is a variation of the standard ketone synthesis from a carboxylic acid and two equivalents of an alkyllithium reagent. In this case, the lithium salt of the carboxylic acid is formed upon treatment with LiH. Addition of one equivalent of organolithium reagent, followed by hydrolysis, gives the ketone.

$$C_6H_{11}COOH \xrightarrow{\text{1. LiH}} C_6H_{11}COO^- \xrightarrow{\text{2. } CH_3Li} C_6H_{11}C(O^-)_2CH_3 \xrightarrow[H_3O^+]{\text{3. } H_2O} C_6H_{11}C(O)CH_3$$

(g) This part is an example of the Hell-Volhard-Zelinsky reaction. Remember that when only a catalytic amount of PBr_3 is used, the α-halocarboxylic acid is the product. Otherwise it is the α-halo acid halide.

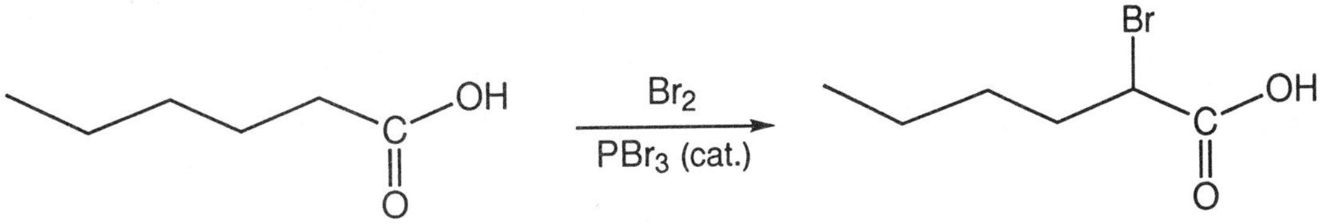

Br_2, PBr_3 (cat.)

(h) This reaction is a classic example of the carboxylation of a Grignard reagent.

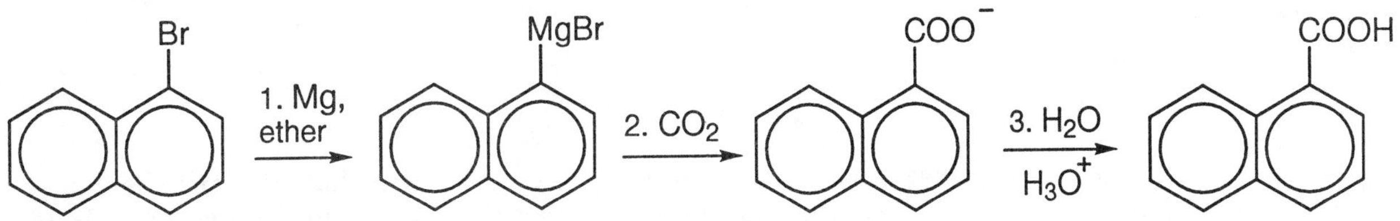

1. Mg, ether; 2. CO_2; 3. H_2O, H_3O^+

Problem 19.28

(a) Methyl groups are electron-releasing when attached to sp^2 hybridized carbons. Formation of the anion will be retarded by such an influence, and therefore acetic acid is weaker than formic acid.

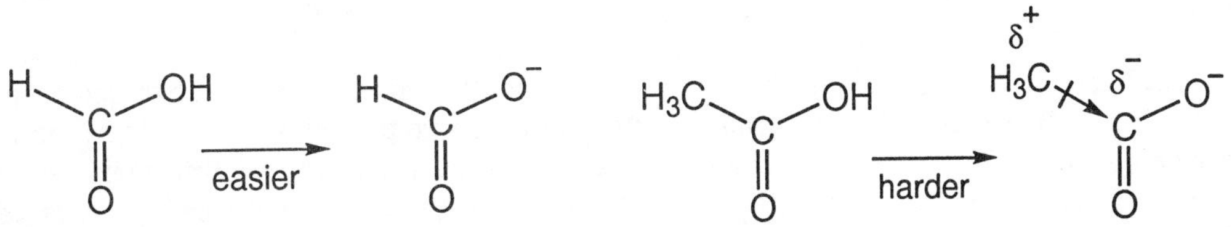

easier; harder

(b) The nitro group is strongly electron-withdrawing (note the positive charge on N) and this will stabilize a nearby negative charge. Accordingly, the nitro compound is a much stronger acid than acetic acid.

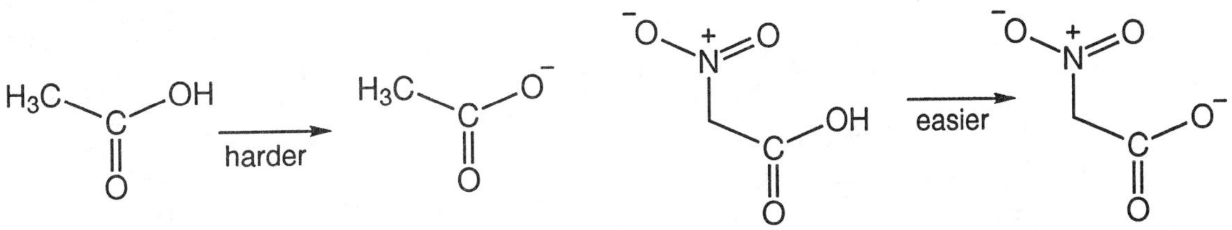

harder; easier

(c) The benzene ring is also electron-withdrawing, and its presence will stabilize the carboxylate anion. Its presence makes benzoic acid a stronger acid than cyclohexanecarboxylic acid.

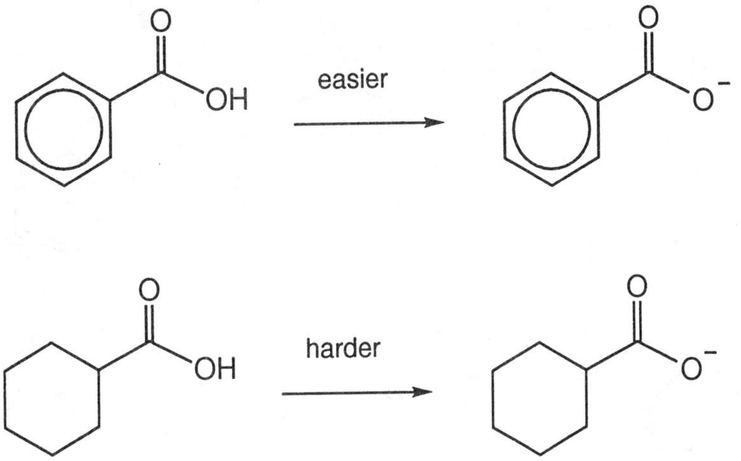

easier

harder

(d) This is a tough one. All carbonyl groups are polar, with the carbon atoms relatively positive. In the cis (Z) diacid, the partially positive carbon helps to stabilize the nearby negative charge, which makes the cis diacid a stronger acid than the trans (E) diacid.

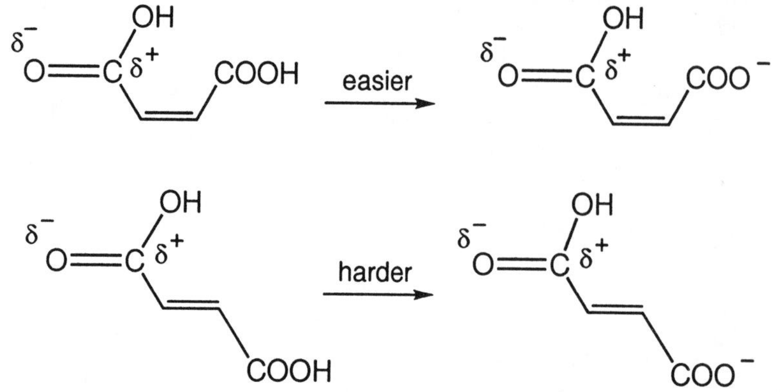

easier

harder

(e) The situation changes dramatically once one proton has been lost. Now dianions are to be formed and it is much more favorable to have the two negative charges on opposite sides of the double bond, and thus quite remote from each other, than it is to have them on the same side of the double bond.

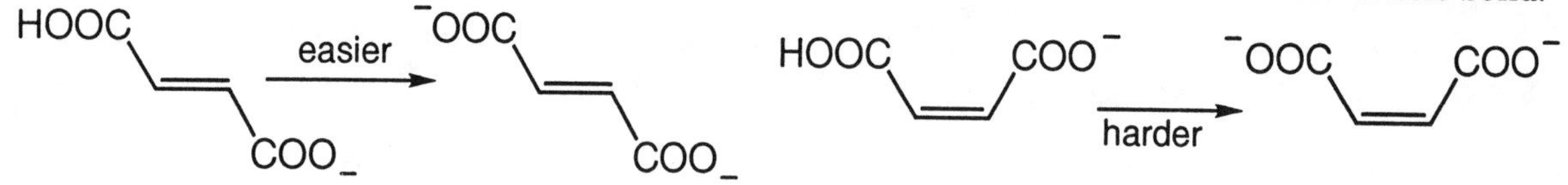

easier

harder

Problem 19.29 Although CDI is a derivative of phosgene, it is more convenient to use because it is a nontoxic solid. *N,N′*-Carbonyldiimidazole behaves similarly to phosgene in its reactions with carboxylic acids (see Problem 19.18). Both reagents react through addition-elimination sequences, but the byproduct of the reaction of CDI is the relatively inert molecule imidazole, rather than HCl as it is for reaction with phosgene.

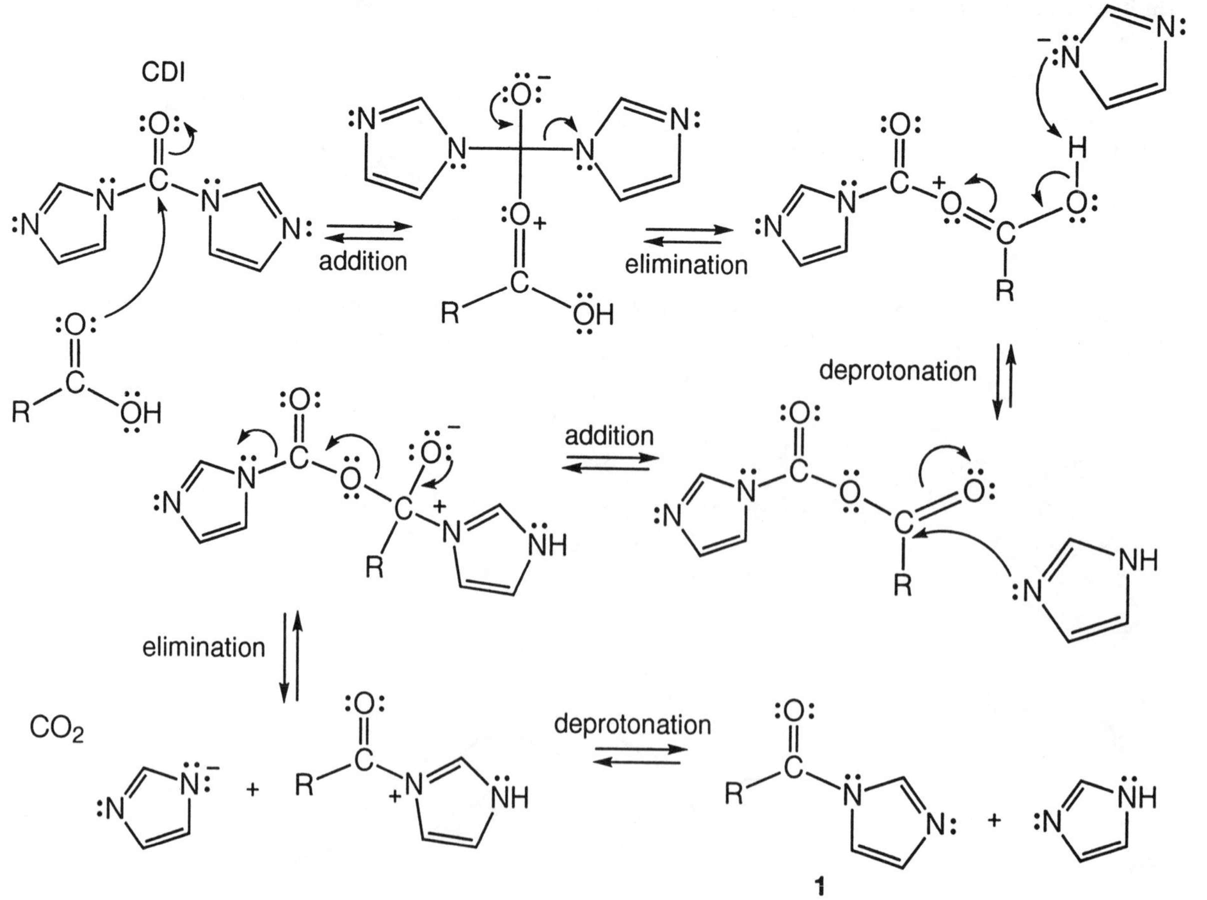

CDI

addition

elimination

deprotonation

addition

elimination

CO_2

deprotonation

1

Imidazolides (**1**) behave very much like acid chlorides in reactions with amines. The nucleophilic nitrogen of the amine adds to the carbonyl group and an excellent leaving group (why?) is then lost in the elimination phase of this addition-elimination process.

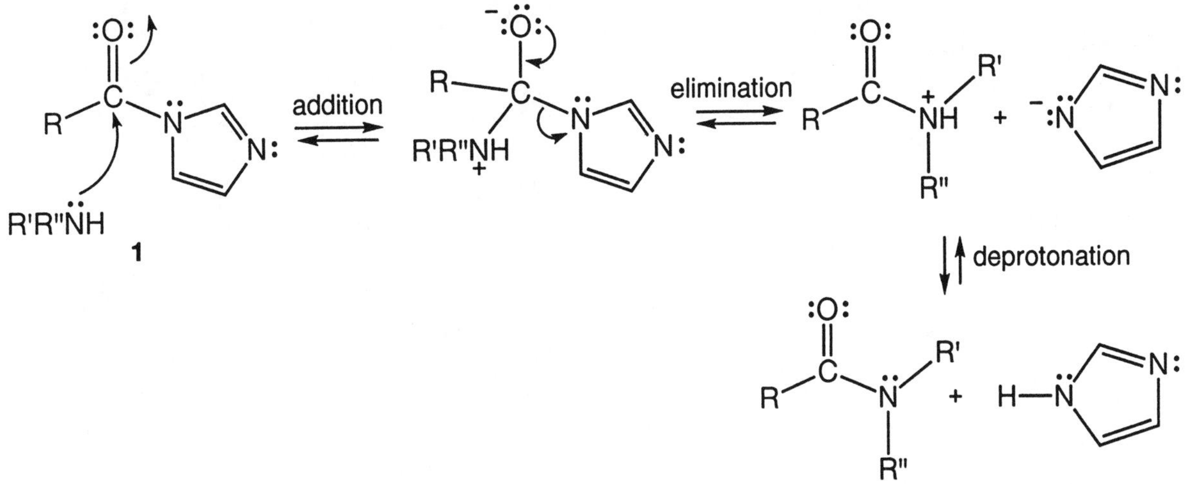

addition

elimination

deprotonation

R'R"NH

1

Problem 19.30 The first step in this process is formation of a "mixed anhydride" through reaction of the carboxylic acid with acetic anhydride. Addition of the acid to the anhydride carbonyl, followed by loss of acetate and deprotonation, gives the mixed-anhydride intermediate.

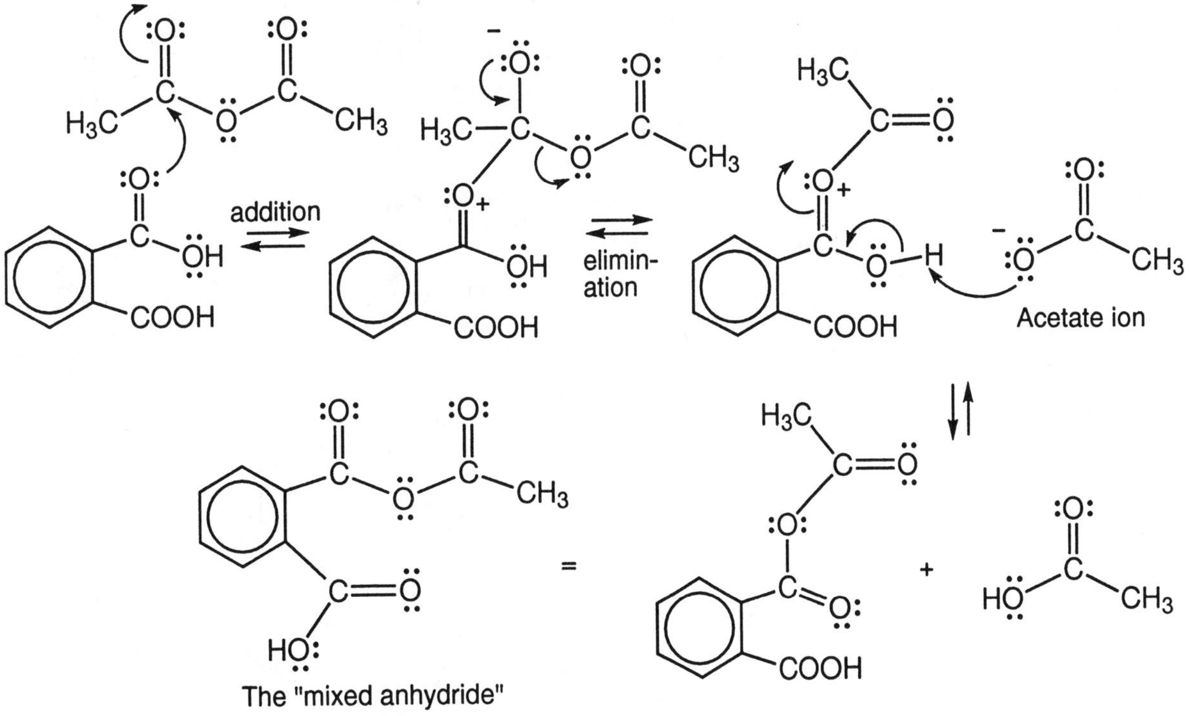

addition

elimin-
ation

Acetate ion

The "mixed anhydride"

The mixed anhydride now undergoes an intramolecular reaction very similar to the intermolecular reaction just shown.

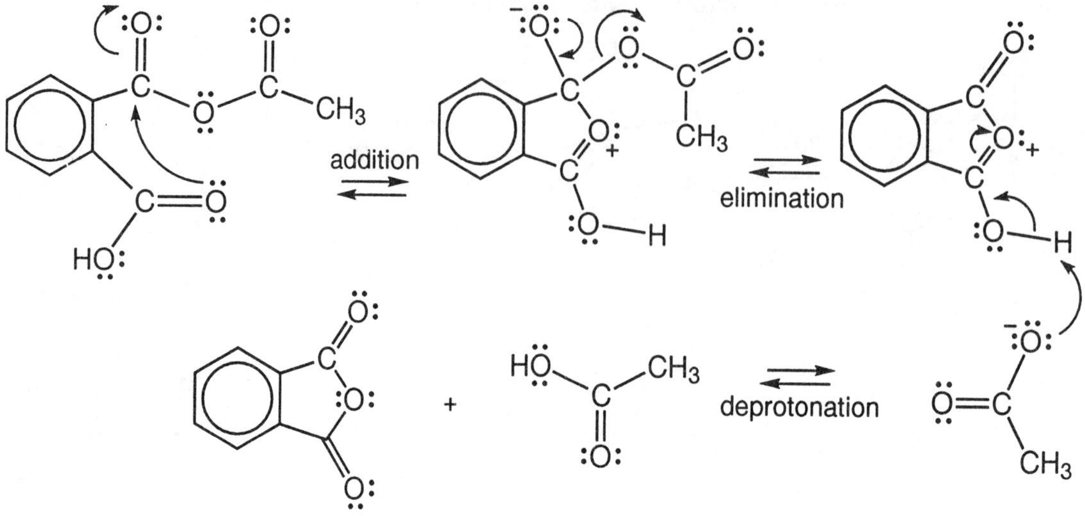

addition

elimination

deprotonation

Problem 19.31 Protonation of the carboxylic acid carbonyl, followed by addition of the intramolecular nitrogen atom, begins the process.

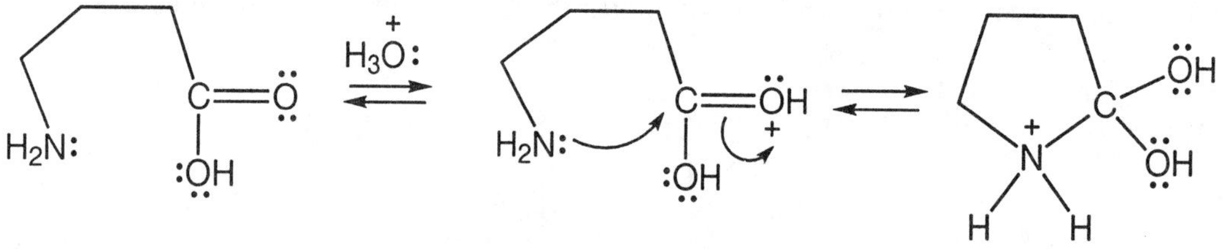

Proton transfers, loss of water, and deprotonation finish off the reaction.

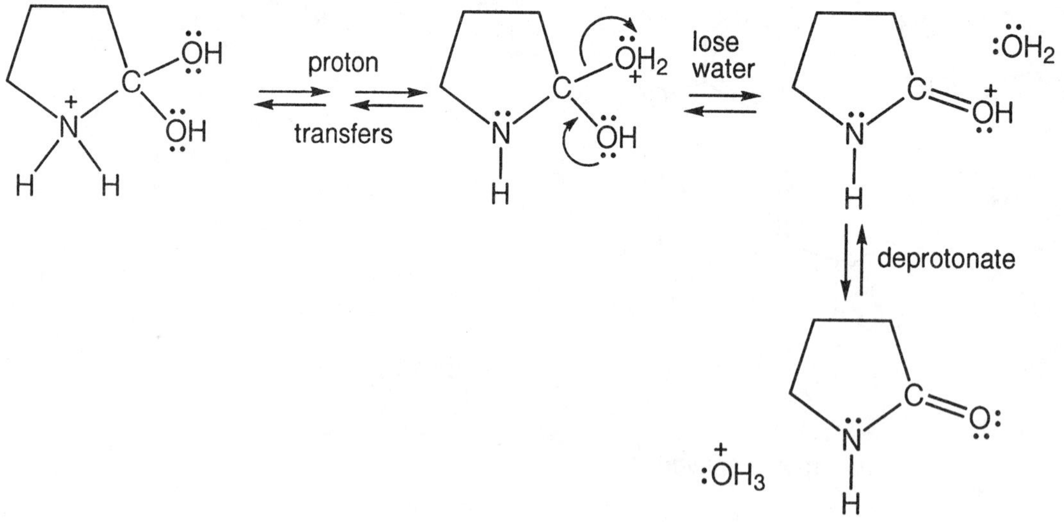

proton transfers

lose water

deprotonate

Lactam

Problem 19.32 The Vilsmeier reagent is a strong Lewis acid and reacts with the carboxylic acid in an addition reaction. Deprotonation and loss of chloride leads to intermediate **A**. Addition of chloride gives an intermediate **B**, which eliminates DMF to give the product.

$$(CH_3)_2\overset{+}{N}{=}CH{-}Cl \ (\text{Vilsmeier reagent}) + Cl^- + RCOOH \xrightleftharpoons{\text{addition}} (CH_3)_2N{-}CH(Cl){-}\overset{+}{O}{=}C(R){-}OH$$

$$\xrightleftharpoons{\text{deprotonate}} (CH_3)_2N{-}CH(Cl){-}O{-}C(R){=}O \xrightleftharpoons{\text{elimination}} (CH_3)_2\overset{+}{N}{=}CH{-}O{-}C(R){=}O \ (\mathbf{A}) + Cl^-$$

$$\xrightleftharpoons{\text{addition}} (CH_3)_2\overset{+}{N}{=}CH{-}O{-}C(Cl)(R){-}O^- \ (\mathbf{B}) \xrightleftharpoons{\text{elimination}} (CH_3)_2N{-}CHO \ (\text{DMF}) + Cl{-}C(=O){-}R$$

Problem 19.33
(a) This conversion requires the formation of a bromide containing the same number of carbons as the starting carboxylic acid. Acids can be converted into alcohols through reduction with lithium aluminum hydride. Alcohols can be made into bromides in may ways, including reaction with PBr_3. The combination of these two steps gives the first product.

$$CH_3(CH_2)_{10}COOH \xrightarrow[\text{2. } H_2O]{\text{1. } LiAlH_4} CH_3(CH_2)_{10}CH_2OH \xrightarrow{PBr_3} CH_3(CH_2)_{11}Br$$

(b) The Hunsdiecker reaction is a convenient method for preparing an alkyl bromide containing one fewer carbon atom than the starting carboxylic acid.

$$CH_3(CH_2)_{10}COOH \xrightarrow[\text{2. } AgNO_3]{\text{1. KOH}} CH_3(CH_2)_{10}COOAg \xrightarrow[CCl_4,\ \Delta]{Br_2} CH_3(CH_2)_{10}Br$$

(c) The third part of this problem requires the addition of a single carbon, which can be easily done through formation of the Grignard reagent, reaction with carbon dioxide, and neutralization with dilute acid. The starting material was made in part (a).

$$CH_3(CH_2)_{11}Br \xrightarrow[\text{2. } CO_2 \text{ 3. } H_2O/H_3O^+]{\text{1. Mg/ether}} CH_3(CH_2)_{11}COOH$$

Problem 19.34 This reaction is an example of the Kolbe electrolysis with a few added twists. First of all, it is a "mixed" Kolbe in which two different carboxylates are employed. Second, in one of the carboxylates there is a carbon-carbon double bond appropriately positioned to trap the initially generated alkyl radical **A** to give the cyclized radical **C**. It is also worth noting that this process is most efficient when hexanoic acid is used in excess. In this way, radical **B** reacts preferentially with radical **C** rather than with itself to give a dimer. When a fourfold excess of hexanoic acid was used, the product tetrahydrofuran was obtained in a 52% yield.

H_3C O CH_3 COOH + COOH

KOH
CH_3OH

H_3C O CH_3 C O⁻ K⁺ O + C O⁻ K⁺ O

electrochemical
oxidation

H_3C O CH_3 C O O

lose
CO_2

H_3C O CH_3 H_2C **A**

radical
addition

B

lose
CO_2

Note odd
electron!

H_3C O CH_3 H_2C **C**

combine **B** and **C**

H_3C O CH_3 C_6H_{13}

Problem 19.35 The reaction of the *(E)* ester is a simple acid-catalyzed reverse acetal formation. This is an important review problem. If formation of the diol seemed mysterious, BE CERTAIN to go back to Section 16.8 and review acetal formation.

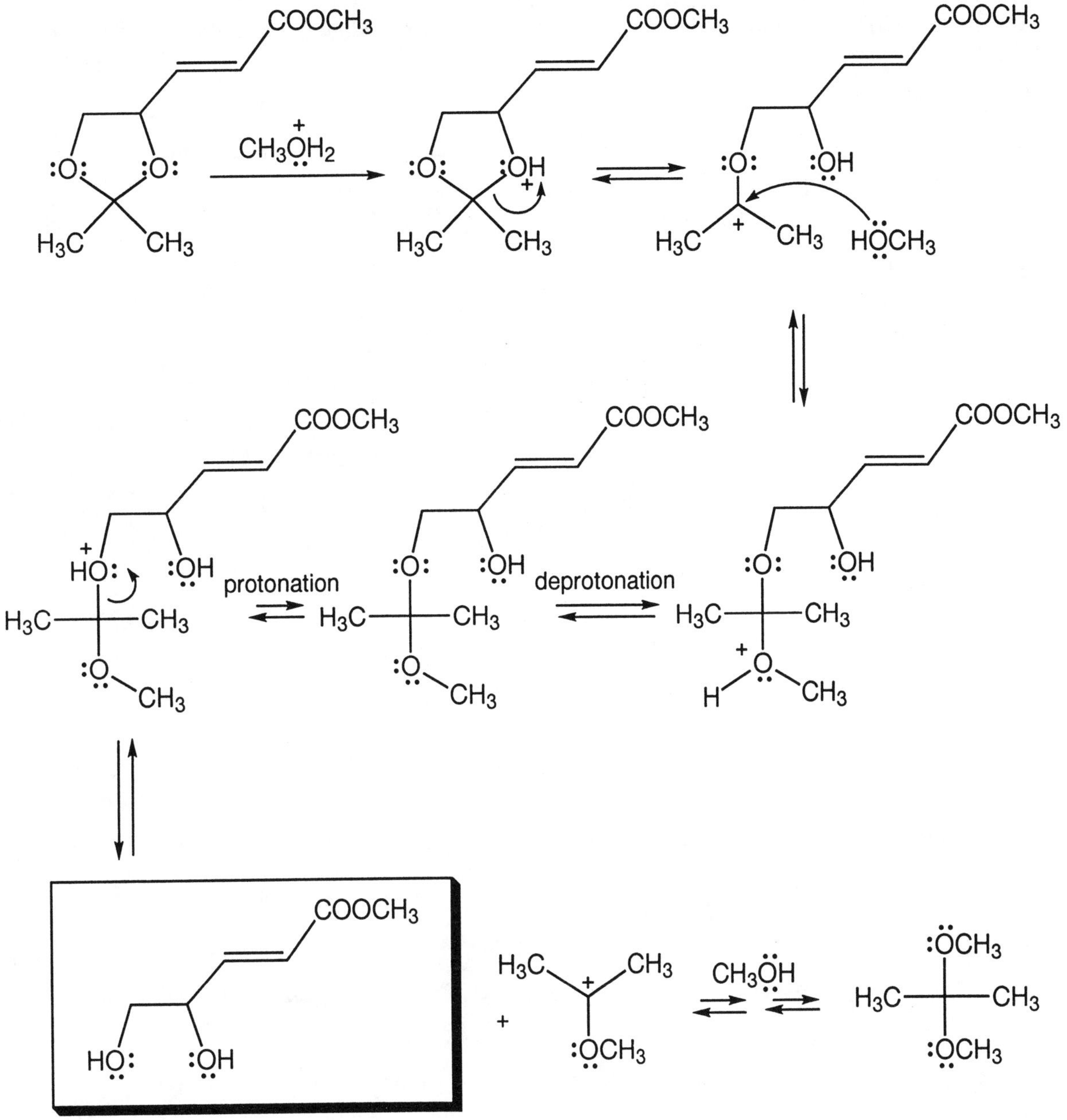

$CH_3\overset{+}{O}H_2$

protonation

deprotonation

CH_3OH

In the *(Z)* isomer there are more options. The ester is in position to react with the product diol. This reaction is really only a Fischer esterification carried out in an intramolecular way. The steps leading to the diol are exactly the same as in the *(E)* isomer. Now new things happen as a hydroxyl group is within range of the ester group. Protonation is followed by addition of the hydroxyl to give a five-membered ring. Proton transfers, loss of methyl alcohol, and deprotonation give the lactone product.

Problem 19.36

(a) This one is simple. A Kolbe electrolysis will do the trick.

(b) This problem is a little harder. The α-bromide is first made and then *tert*-butoxide is used to induce an E2 reaction (remember them?).

(c) In this part, you need to do two things with your starting material. First, use the Hunsdiecker reaction to make bromocyclopentane. This compound is converted into the corresponding organolithium reagent. Two equivalents of this will react with the starting acid to make dicyclopentylketone.

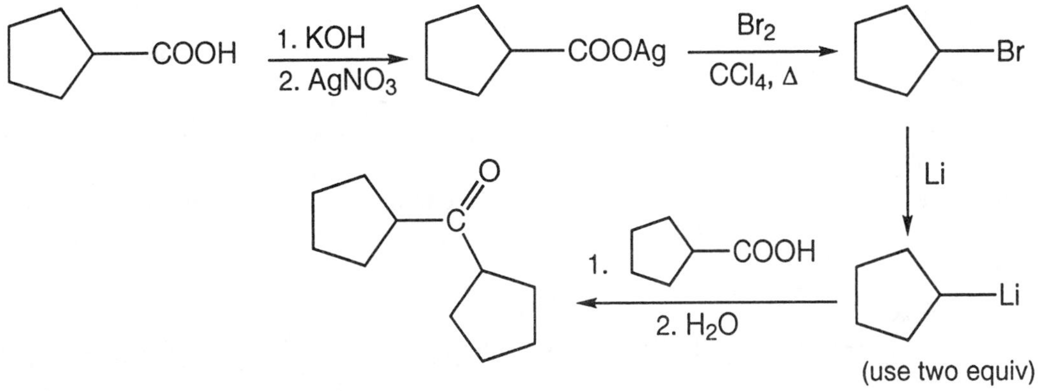

(d) Here, too, a number of things must be done. Reduction of the starting acid with lithium aluminum hydride gives the alcohol. This compound can be used in a Fischer esterification reaction to make the desired ester.

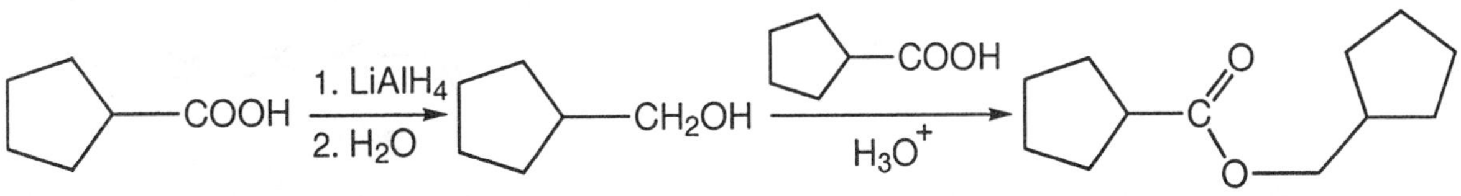

Problem 19.37 The IR spectrum of compound **3** (as well as its appearance in this chapter) suggests the presence of a carboxylic acid. The critical bands are the broad intense OH stretch at 3330-2500 cm^{-1} and the carbonyl stretch at 1696 cm^{-1}. The presence of the carboxylic acid is further supported by the presence of a 1H singlet at δ 11.6 ppm in the ^{1}H NMR spectrum. In addition, the ^{1}H NMR spectrum of **3** implies the presence of three adjacent methylene groups as indicated by the quintet at δ 1.95 ppm and the two triplets at δ 2.34 and 2.59 ppm, each integrating for two hydrogens. The 3H singlet at δ 3.74 ppm suggests the presence of some sort of deshielded methyl group. Finally, the set of doublets at δ 6.75 and 7.05 ppm with J = 8 Hz is most consistent with a para disubstituted benzene ring.

An atom inventory of the suspected structural fragments indicates a mass of 178 g/mol. As the mass spectrum of **3** shows a parent ion at m/z = 194, we are missing a mass of 16, most easily accommodated by an oxygen atom. Thus, the molecular formula is $C_{11}H_{14}O_3$.

	Mass	Fragment(s)
	45	COOH
	42	$(CH_2)_3$
	15	CH_3
	76	C_6H_4
Total	178	$C_{11}H_{14}O_2$ (plus one more O to accommodate the mass spectrum) = $C_{11}H_{14}O_3$

It also appears reasonable that the oxygen atom is attached to the methyl group as in OCH_3, because of the low-field position of the signal for the methyl singlet (δ 3.74 ppm).

There are only two reasonable structures that can be assembled from these structural pieces: 4-(4-methoxyphenyl)butyric acid (**3**), and 4-(3-methoxypropyl)benzoic acid (**3a**).

CH_3O—C₆H₄—$CH_2CH_2CH_2COOH$ (**3**)   HOOC—C₆H₄—$CH_2CH_2CH_2OCH_3$ (**3a**)

Possibility **3a** can be eliminated from consideration because the methylene hydrogens adjacent to the methoxy group would be expected to appear at a lower chemical shift than the observed δ 2.34 or 2.59 ppm.

With a structure for **3** in hand, it is now possible to speculate about the structures of **2** and estragole (**1**) itself. From the method of preparation of **3**, it appears reasonable that compound **2** is 4-(3-bromopropyl)anisole.

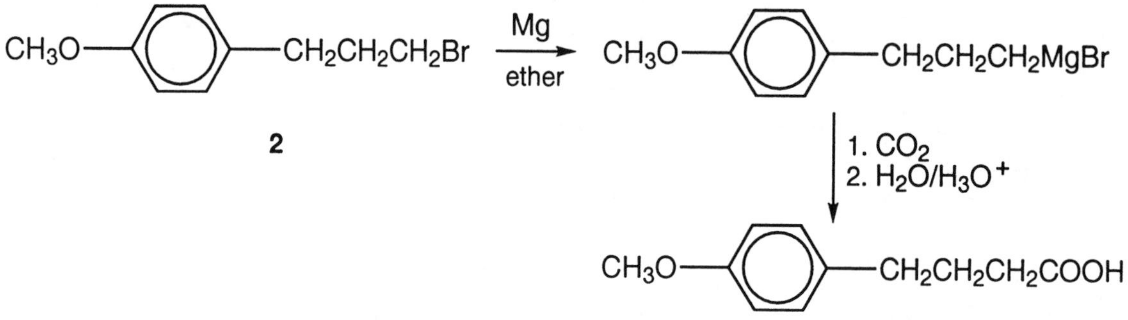

CH_3O—C₆H₄—$CH_2CH_2CH_2Br$ (**2**) —Mg, ether→ CH_3O—C₆H₄—$CH_2CH_2CH_2MgBr$ —1. CO_2 2. H_2O/H_3O^+→ CH_3O—C₆H₄—$CH_2CH_2CH_2COOH$ (**3**)

Finally, compound **2** appears to be the product of a peroxide-induced anti-Markovnikov addition of HBr to an alkene. That alkene, estragole, must be 4-allylanisole.

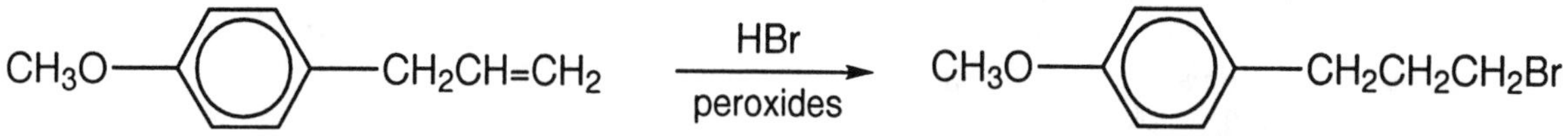

CH_3O—C₆H₄—$CH_2CH{=}CH_2$ (**1**) —HBr, peroxides→ CH_3O—C₆H₄—$CH_2CH_2CH_2Br$ (**2**)

Problem 19.38 The reaction of cyclopentadiene and maleic anhydride should be familiar stuff. This example of the Diels-Alder reaction gives **1**, *endo*-norbornenedicarboxylic acid anhydride.

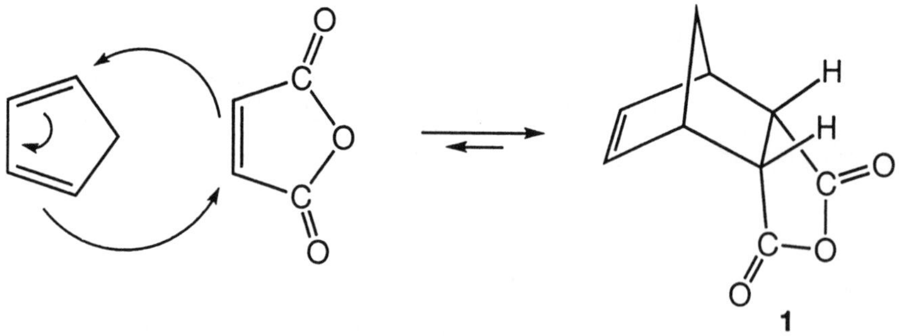

The spectral data for compound **1** are in accord with the proposed structure. The weak parent ion at m/z = 164 is consistent with a molecular formula of $C_9H_8O_3$. The two carbonyl stretches at 1854 and 1774 cm^{-1} are indicative of a five-membered cyclic anhydride. The ^1H NMR spectral assignments are shown below.

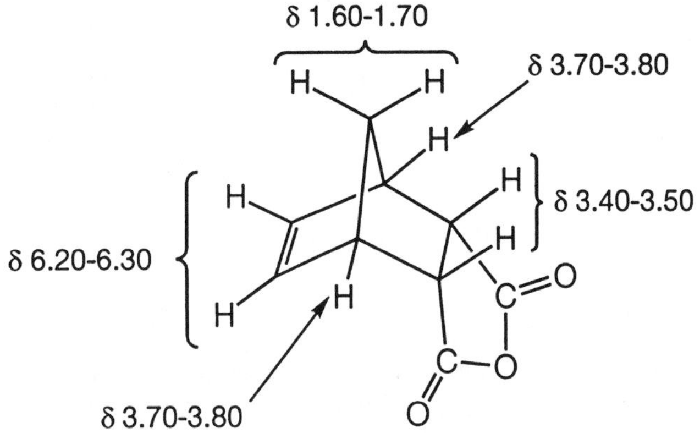

δ 1.60-1.70
δ 3.70-3.80
δ 3.40-3.50
δ 6.20-6.30
δ 3.70-3.80

It must be admitted that, while the endo cycloadduct **1** is formed exclusively, this level of analysis does not rule out the exo isomer.

Hydrolysis of anhydride **1** gives the dicarboxylic acid **2** as shown below:

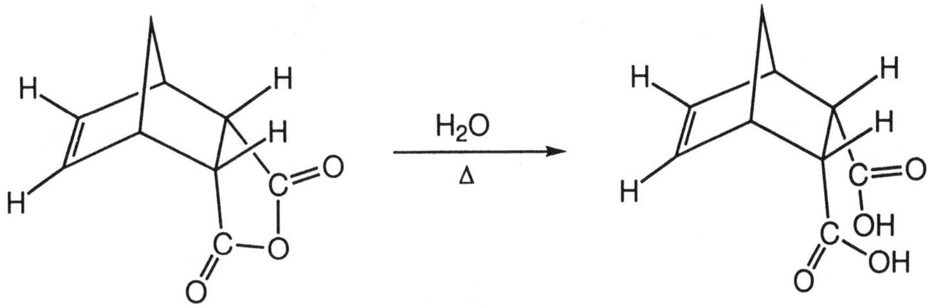

H_2O, Δ

1 **2**

Again, the spectral data for diacid **2** are consistent. Note, in particular, the broad, intense O—H stretch in the IR, as well as the appearance of a single C=O stretch at lower frequency than that for the anhydride. In addition, the appearance of a 2H singlet at δ 11.35 ppm is consistent with the OH hydrogens of a dicarboxylic acid. Note that this is given in the problem as 1H: integrals are relative, and you must adjust the ratios given to match the molecular formula.

So far, so good. Now comes the hardest part of this problem, deducing the structure of **3**. First, we know that compound **3** is isomeric with **2** ($C_9H_{10}O_4$). Second, the IR spectrum of **3** still shows the presence of a carboxylic acid (broad hydroxyl band at 3450-2500 cm^{-1} and the carbonyl stretch at 1690 cm^{-1}). However, there is also a higher frequency carbonyl band at 1770 cm^{-1} so far unexplained. Finally, the ^1H NMR spectrum of **3** is much more complicated than that of either **1** or **2**. Clearly, compound **3** is less symmetric than its precursors. Notice that there is only one carboxylic acid group remaining in **3** as indicated by the 1H singlet in the NMR at δ 12.4 ppm. There is a new 1H signal in the NMR at δ 4.75-4.85 ppm and the 2H signal for the hydrogens attached to the double bond of **2** (δ 5.6 - 5.7 ppm) is gone.

It appears that one carboxyl group and the carbon-carbon double bond in **2** have reacted to form **3**. Now comes the mechanistic analysis. What can compound **2** do in concentrated sulfuric acid? How about protonation of the double bond to give carbocation **A**?

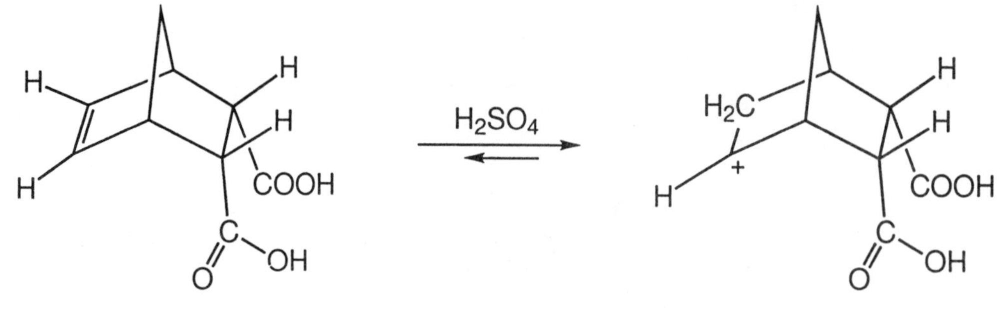

H_2SO_4

A

As carbocation **A** inherits endo stereochemistry from its ancestors **1** and **2**, intramolecular capture of the carbocation by a carbonyl oxygen is possible. Deprotonation then leads to the lactone **3**.

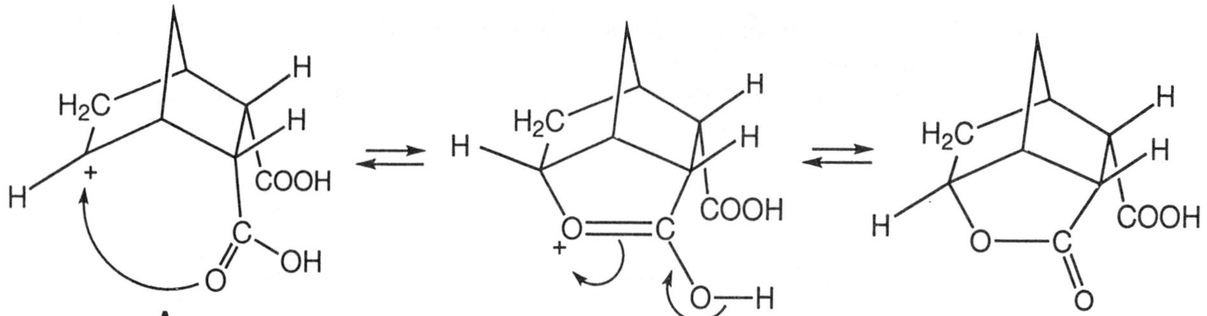

A

3

Lactone **3** is much less symmetrical than either anhydride **1** or diacid **2**, accounting for the far more complex NMR spectrum. The 1770 cm^{-1} band in the IR is appropriate for a five-membered lactone carbonyl group. Note the absence of hydrogens attached to double bonds in **3** and the presence of a single, low-field hydrogen, $\mathbf{H_a}$, deshielded by the adjacent oxygen.

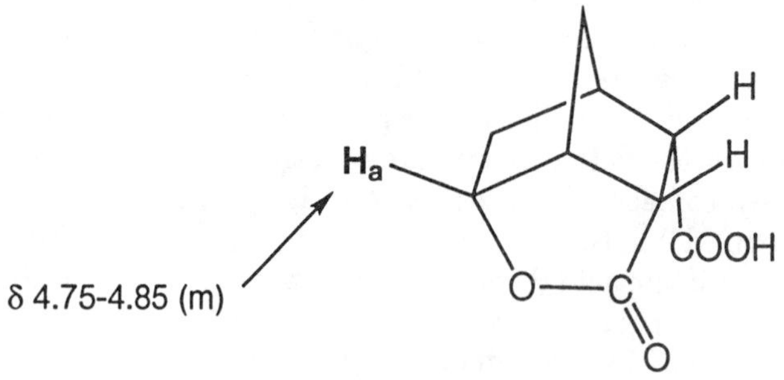

δ 4.75-4.85 (m)

3

Chapter 20 Outline

20

Derivatives of Carboxylic Acids: Acyl Compounds

This chapter continues the exploration of the addition-elimination mechanism begun in Chapter 19. This process dominates the chemistry of the acyl compounds explored here. There are some quite taxing problems in this section. Do try to analyze what needs to be done before you start. It is much easier that way.

Problem 20.1

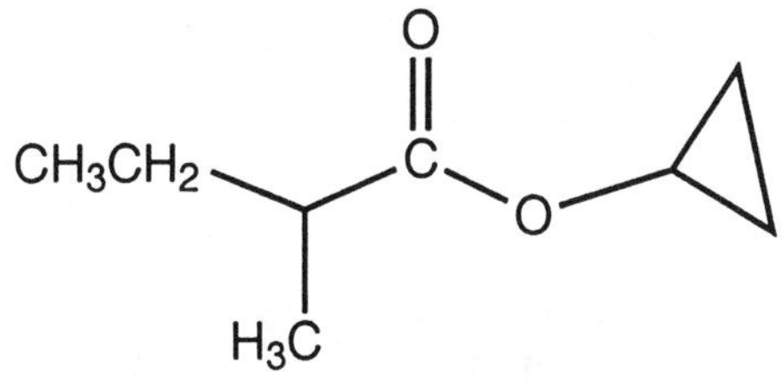

CH_3CH_2–CH(H_3C)–C(=O)–O–cyclopropyl

Cyclopropyl 2-methylbutyrate

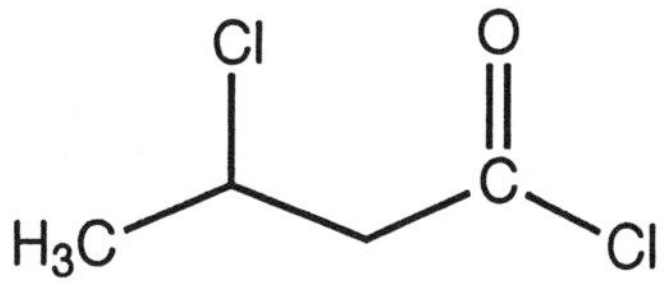

H_3C–CH(Cl)–CH_2–C(=O)–Cl

3-Chlorobutanoyl chloride

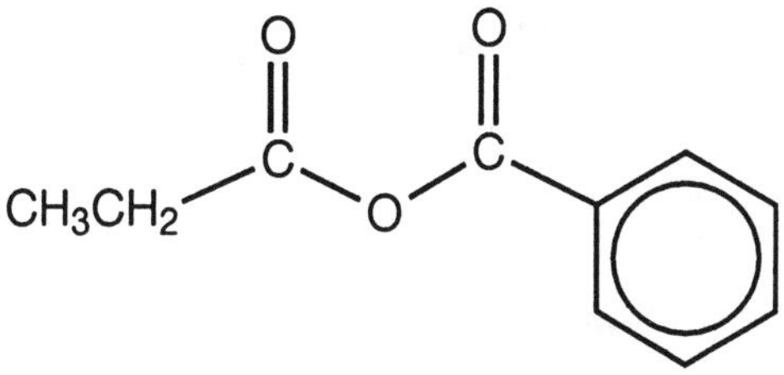

CH_3CH_2–C(=O)–O–C(=O)–C_6H_5

Benzoic propionic anhydride

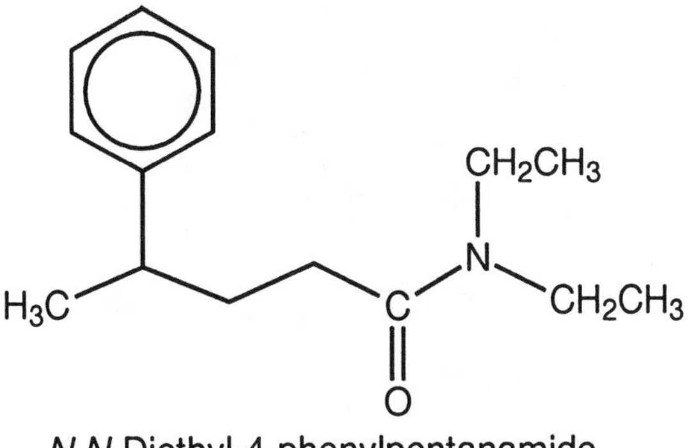

H_3C–CH(C_6H_5)–CH_2CH_2–C(=O)–N(CH_2CH_3)(CH_2CH_3)

N,*N*-Diethyl-4-phenylpentanamide

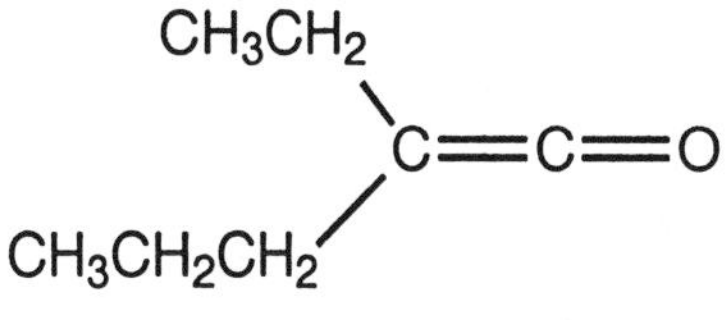

(CH_3CH_2)($CH_3CH_2CH_2$)C=C=O

Ethyl propyl ketene

Problem 20.3 Your nasal passages are moist (or at least they should be). When an inhaled acid chloride encounters that moisture, a hydrolysis reaction takes place and HCl is formed. That is what is detected, usually quite unpleasantly.

addition elimination deprotonation

HCl:

Problem 20.4 Old stuff (Chapter 14). This problem is pure review of Friedel-Crafts acylation. It is a standard, no-frills aromatic substitution problem.

complex formation

$AlCl_3$

An E^+ reagent (acylium ion)

Deprotonation by any base in the system, here B:⁻

Resonance-stabilized cyclohexadienyl cation

Problem 20.5 This problem is somewhat complicated, although no individual step is very hard. The first steps involve opening the anhydride in a typical addition-elimination process, with the nitrogen of aniline acting as the nucleophile.

addition deprotonation

Aniline

elimination

Intermediate **A**

protonation

The reaction sometimes goes on to product without the aid of acetic anhydride, but it is greatly facilitated by the ability of the anhydride to create a good leaving group out of the poor leaving group, OH (a repeating theme if ever there were one).

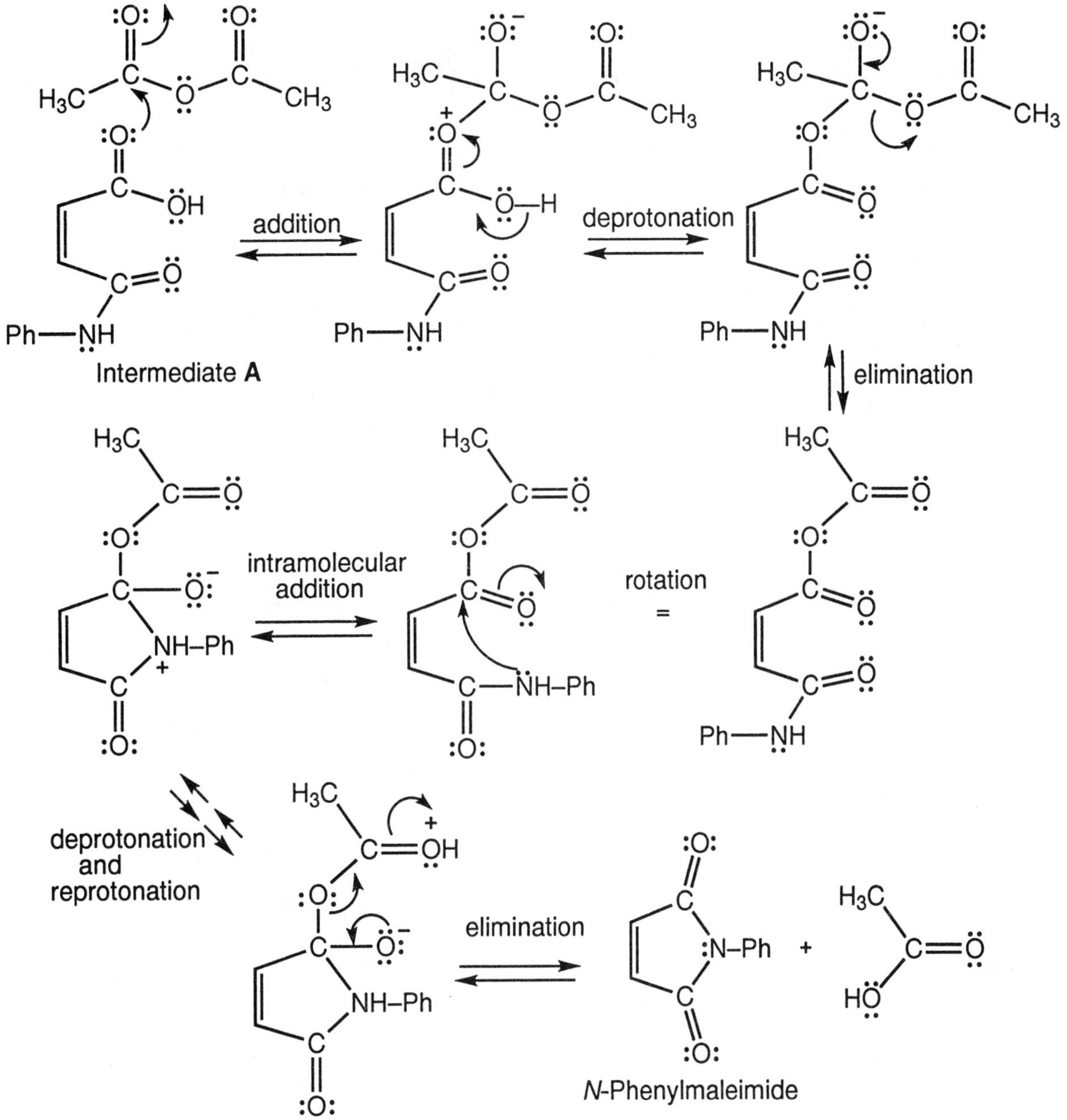

Intermediate **A**

addition

deprotonation

elimination

intramolecular addition

rotation =

deprotonation and reprotonation

elimination

N-Phenylmaleimide

Problem 20.6 The mechanism is the usual one. There is complex formation and generation of an E^+ reagent, here the acylium ion, followed by addition to benzene to give a resonance-stabilized cyclohexadienyl cation. Deprotonation by any base in the system leads to rearomatization and formation of the substituted benzene.

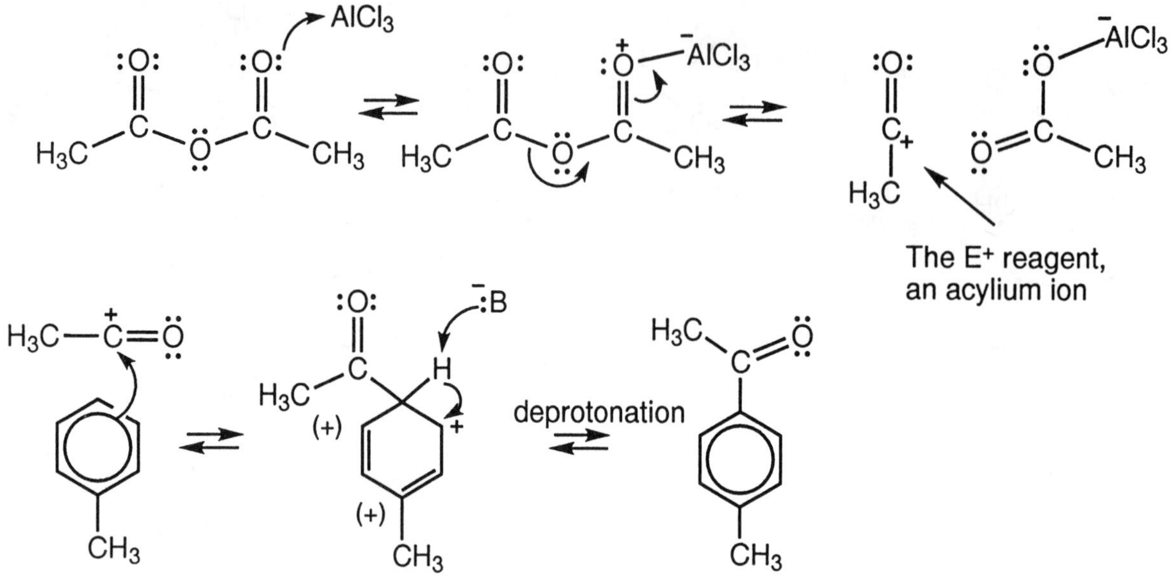

The E^+ reagent, an acylium ion

deprotonation

Problem 20.7 There are two species in this reaction for which resonance stabilization is especially important. These are the protonated ester and the protonated carboxylic acid.

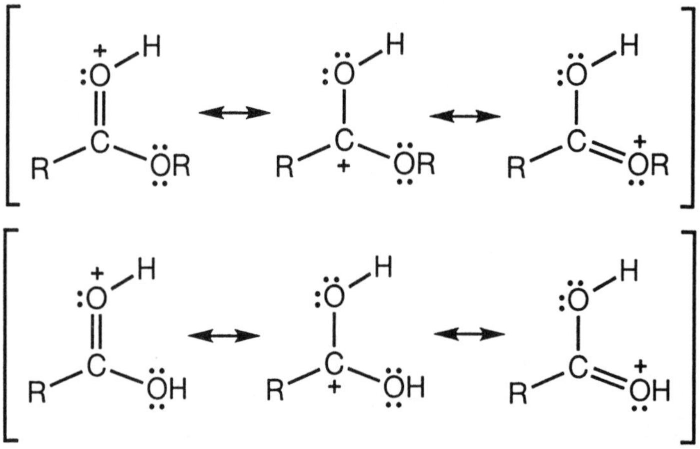

Of course, there is resonance stabilization of the ester and acid molecules themselves, although this is not as important as delocalization in the protonated species (why?).

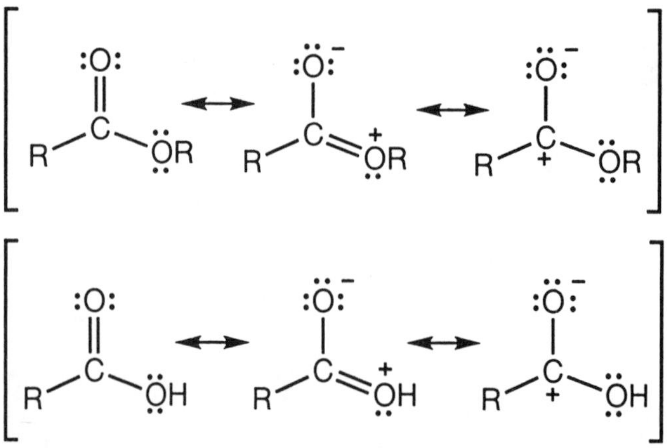

Problem 20.8 The base-catalyzed reaction involves the usual addition-elimination process. Alkoxide adds reversibly, and loss of the original alkoxide leads to the new ester.

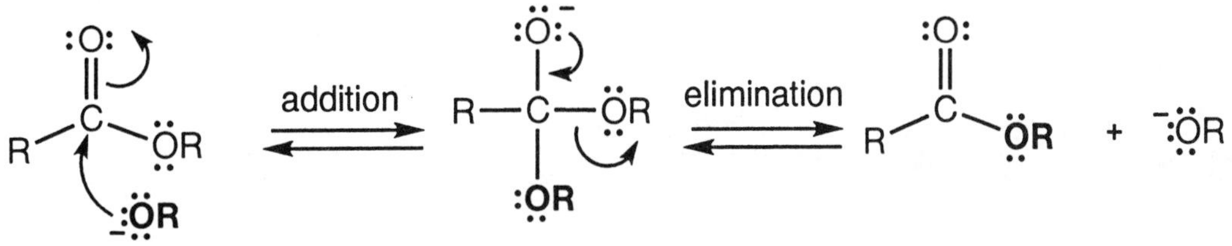

addition
elimination

In acid, protonation leads to a resonance-stabilized intermediate to which alcohol adds. Deprotonation and reprotonation leads to an intermediate that can lose alcohol and generate the new ester.

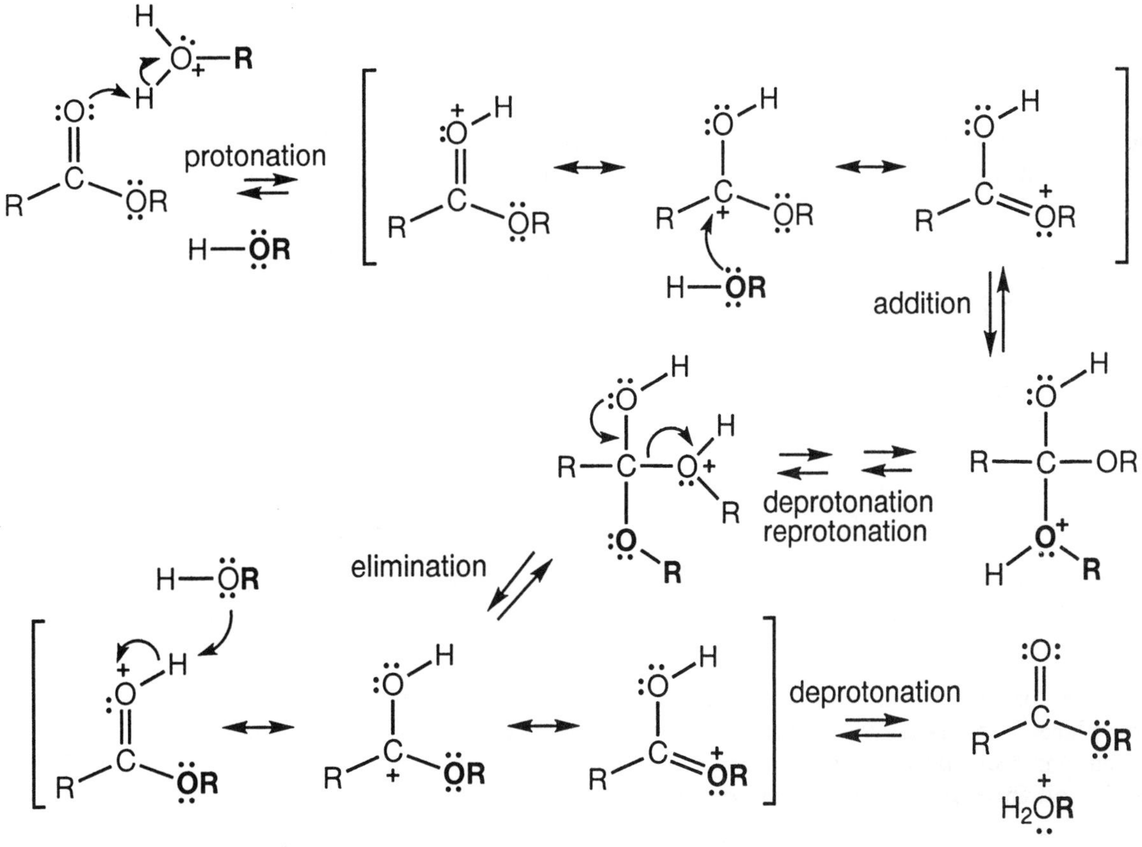

protonation
addition
deprotonation
reprotonation
elimination
deprotonation

Problem 20.11 This problem involves a garden-variety addition-elimination process. Ammonia adds to the ester, protons are transferred, and methyl alcohol is lost.

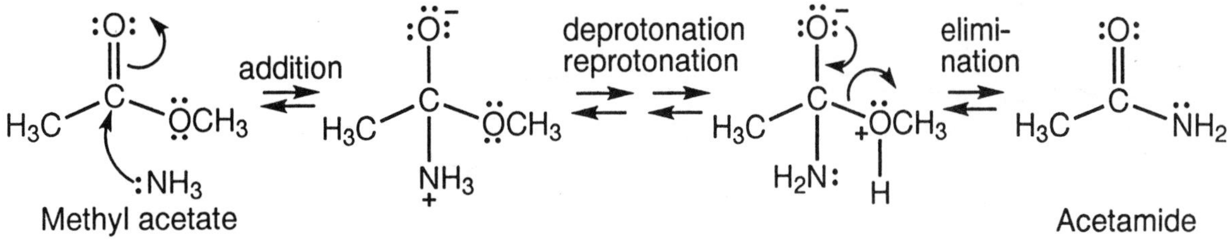

addition
deprotonation
reprotonation
elimi-
nation
Methyl acetate
Acetamide

Problem 20.12 Phosgene reacts with an alcohol as would any acid chloride to make the ester (p. 1016). In this case, both chlorides are replaced with "OR".

Phosgene

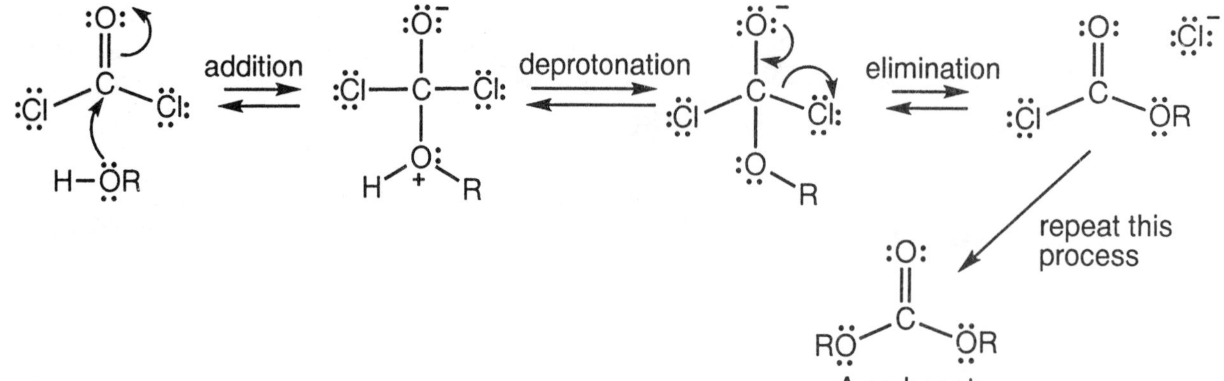

addition

deprotonation

elimination

repeat this process

A carbonate

The carbonate is just a "double ester", and reacts with an organometallic reagent in the ususal addition-elimination fashion. This reaction first generates an ester and then a ketone.

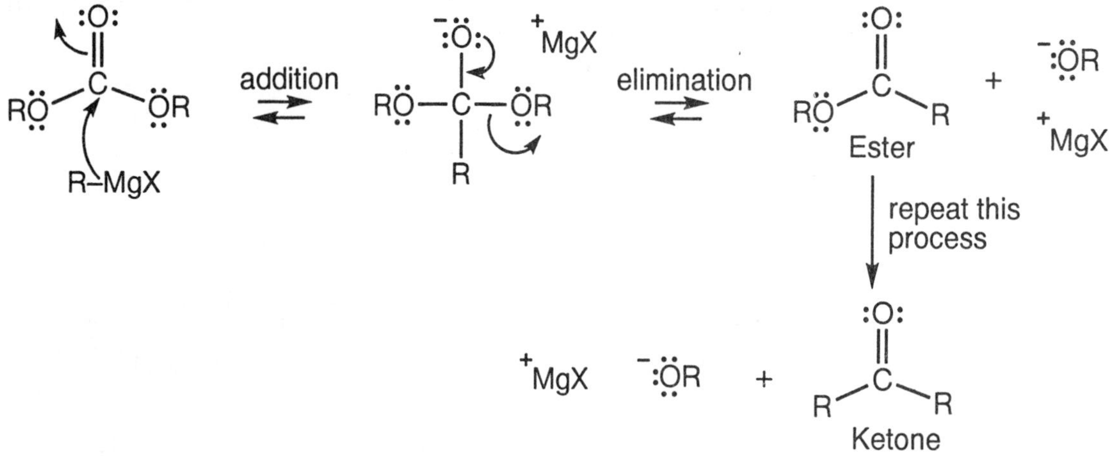

addition

elimination

Ester

repeat this process

Ketone

But the ketone is born in the presence of the organometallic reagent and cannot survive. Addition takes place in the usual way to give an alkoxide. As there is no longer a reasonable leaving group, the reaction stops here. A final protonation generates the tertiary alcohol. Notice that in this reaction all three R groups *must* be the same.

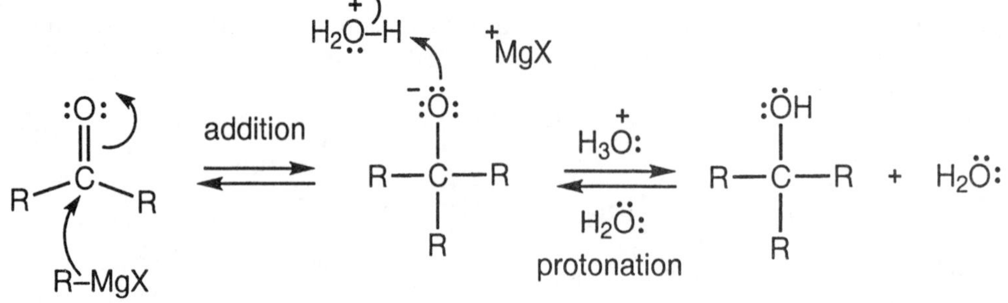

addition

protonation

Problem 20.13 One reason is simple. Diisobutylaluminum hydride is much bigger than lithium aluminum hydride and addition reactions are naturally slower. It's a case of steric inhibition of reactivity.

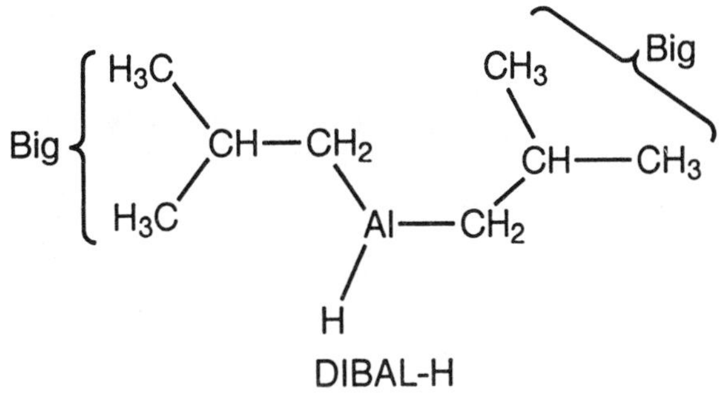

Big
H_3C CH—CH_2 CH_3 Big
H_3C Al—CH_2 CH—CH_3
H
DIBAL-H

Problem 20.15 The acid-catalyzed alcoholysis of an amide takes place through a standard addition-elimination mechanism. The first step is protonation of the amide oxygen. Addition leads to a tetrahedral intermediate, and elimination generates the protonated carboxylic ester. Deprotonation by any base gives the ester itself.

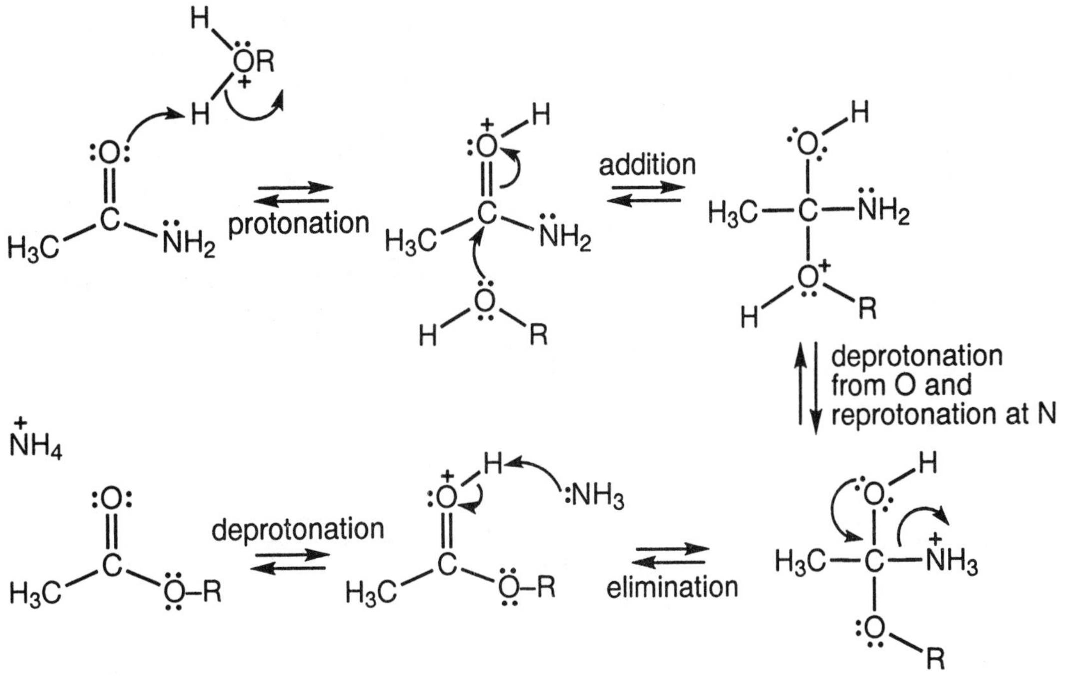

protonation
addition
deprotonation from O and reprotonation at N
$\overset{+}{N}H_4$
deprotonation
:NH_3
elimination

The reaction in base is also addition-elimination. Alkoxide adds to the carbonyl group to give a tetrahedral intermediate that can expel amide ion in the elimination phase of the reaction. Notice that alkoxide is regenerated in the last step of the reaction.

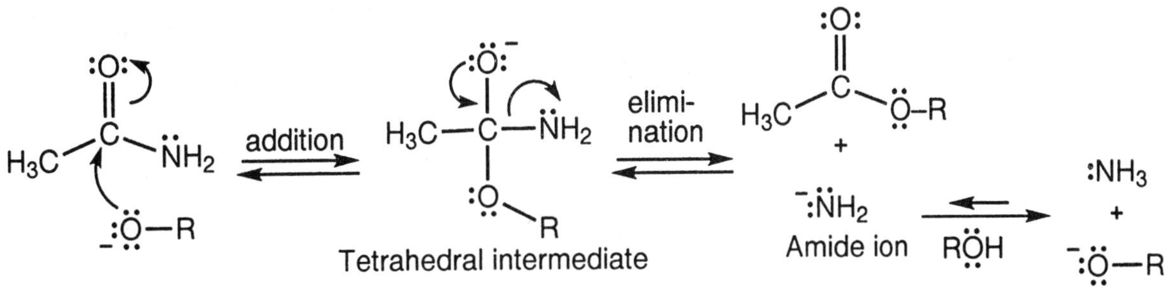

addition
elimi-nation
Tetrahedral intermediate
Amide ion ROH
:NH_3 + $^-$:O—R

Problem 20.16 In each case, a resonance-stabilized intermediate is formed, but the amidate is favored because the charge is partially borne by a relatively electronegative nitrogen rather than a carbon.

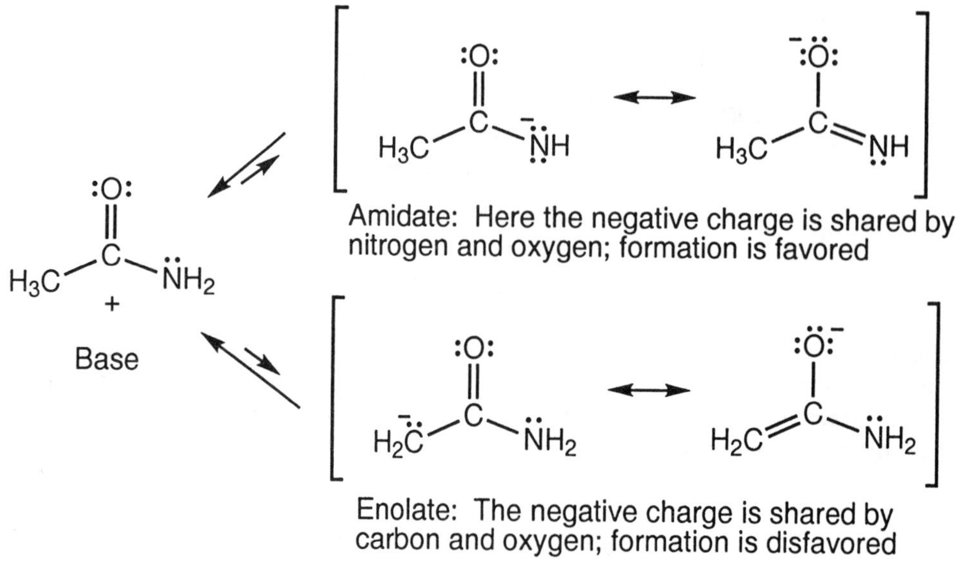

Amidate: Here the negative charge is shared by nitrogen and oxygen; formation is favored

Enolate: The negative charge is shared by carbon and oxygen; formation is disfavored

Problem 20.17 For an *sp* hybridized nitrogen, the lone pair electrons are in an *sp* orbital with 50% *s* character. They are held more tightly than electrons in an sp^2 orbital (33% *s* character), and will therefore be less available for reaction. Electrons in an sp^3 hybrid orbital (25% *s* character) are still less tightly held and will be more easily used in reactions with Lewis acids.

Problem 20.18 This reaction is the reverse of many studied in Chapter 16. There we saw imine formation from the reaction of carbonyl-containing compounds with amines. Here we see the reverse. The reaction begins with addition of water to the Lewis acid imine. Proton shifts follow, and then loss of ammonia completes the reaction.

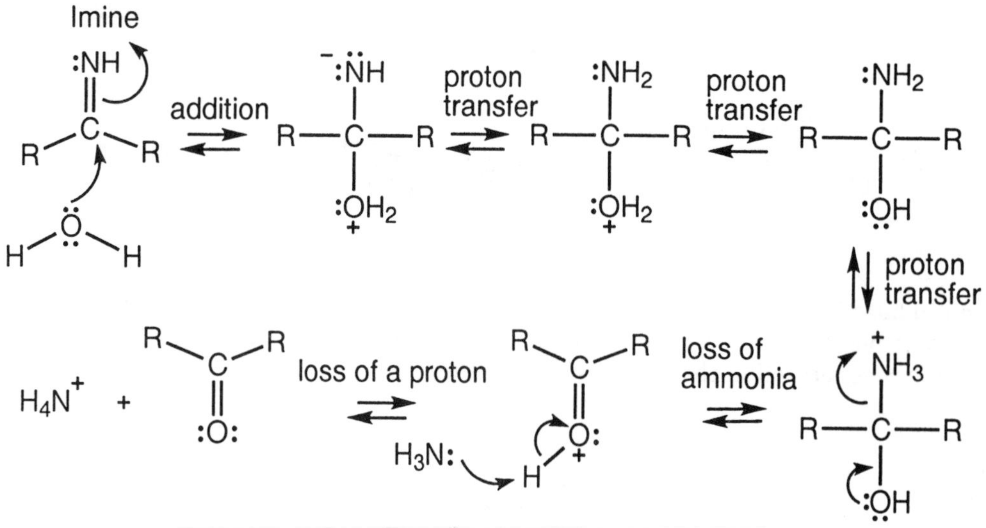

Imine

addition

proton transfer

proton transfer

proton transfer

loss of ammonia

loss of a proton

Problem 20.21 Two enolates are possible and each can add to the two esters. These reactions produce a total of four products.

$RO-C(=O)-CH_3$ + ^{-}OR ⇌ $RO-C(=O)-\overset{-}{C}H_2$

1. react with $RO-C(=O)-CH_3$ 2. acidification → $RO-C(=O)-CH_2-C(=O)-CH_3$ + HOR

1. react with $RO-C(=O)-CH_2CH_3$ 2. acidification → $RO-C(=O)-CH_2-C(=O)-CH_2CH_3$ + HOR

$RO-C(=O)-CH_2CH_3$ + ^{-}OR ⇌ $RO-C(=O)-\overset{-}{C}HCH_3$

1. react with $RO-C(=O)-CH_3$ 2. acidification → $RO-C(=O)-CH(CH_3)-C(=O)-CH_3$ + HOR

1. react with $RO-C(=O)-CH_2CH_3$ 2. acidification → $RO-C(=O)-CH(CH_3)-C(=O)-CH_2CH_3$ + HOR

Here is a detailed mechanism for one product.

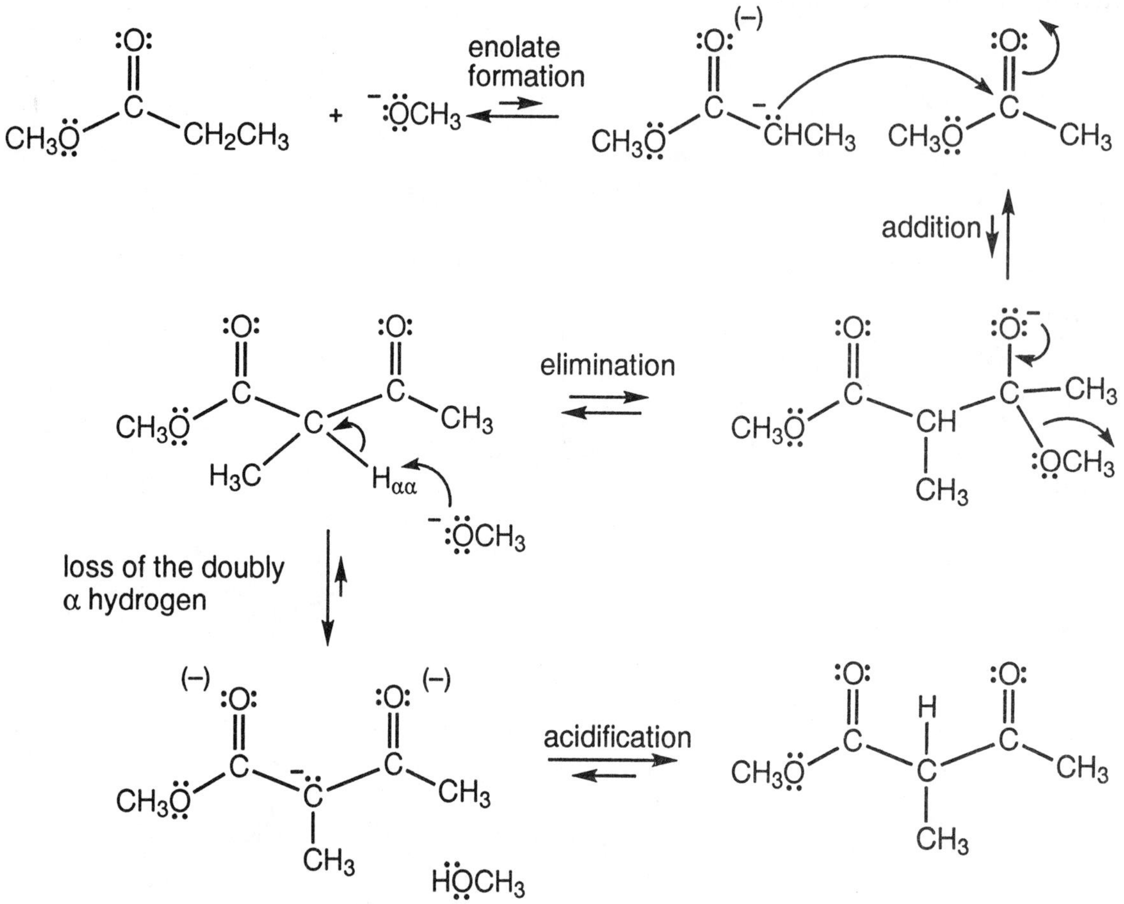

enolate formation

addition

elimination

loss of the doubly α hydrogen

acidification

Problem 20.22 Review again. Esters are more strongly stabilized by resonance than are aldehydes and ketones, and thus are less reactive.

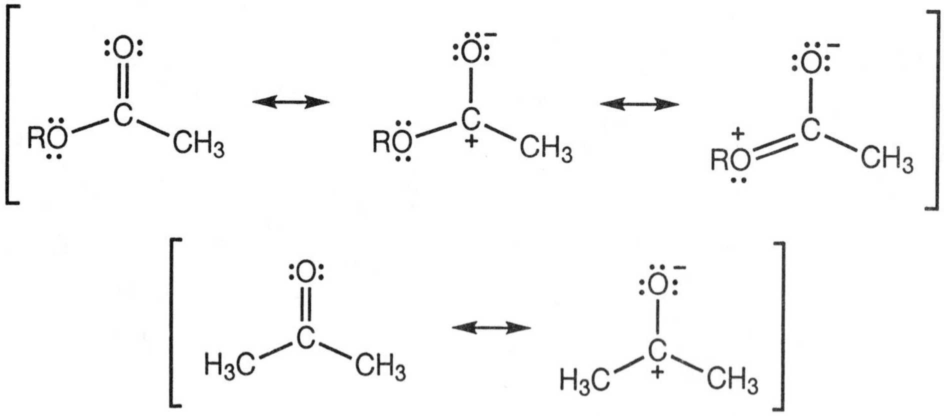

Problem 20.23 Another easy one. Malononitrile is acidic (Table 20.5) because the anion is so well stabilized. Formation of the anion followed by S_N2 displacement on ethyl iodide leads to the once-alkylated product. Repetition can give the dialkylated compound.

Resonance-stabilized anion

$CH_3CH_2—I:$

S_N2

CH_2CH_3

repeat sequence

Problem 20.24

(a) A sequence of Claisen condensation, two separate alkylations and a decarboxylation does the trick. It is actually not necessary to isolate the Claisen product. One could do the first alkylation on the salt initially formed in the condensation step.

1. NaOR/HOR (full equivalent)
2. H_2O/H_3O^+

1. NaOR/HOR
2. CH_3I

1. NaOR/HOR
2. CH_3CH_2I

1. NaOH/H_2O
2. acid, Δ

+ CO_2

(c) This time, a simple Claisen-decarboxylation sequence should work.

$$2\ PhCH_2C(=O)OR \xrightarrow[2.\ H_2O/H_3O^+]{1.\ NaOR/HOR\ \text{(full equivalent)}} PhCH_2C(=O)CH(Ph)C(=O)OR \xrightarrow[2.\ \text{acid},\ \Delta]{1.\ NaOH/H_2O} PhCH_2C(=O)CH_2Ph$$

Problem 20.25

(b) Two alkylations are followed by hydrolysis and decarboxylation.

$$CH_2(C(=O)OR)_2 \xrightarrow[2.\ CH_3CH_2CH_2\text{–}I]{1.\ NaOR/HOR\ \text{(full equiv.)}} CH_3CH_2CH_2CH(C(=O)OR)_2 \xrightarrow[2.\ CH_3\text{–}I]{1.\ NaOR/HOR} (H_3C)(CH_3CH_2CH_2)C(C(=O)OR)_2 \xrightarrow[2.\ \Delta]{1.\ H_2O/H_3O^+} (H_3C)(CH_3CH_2CH_2)CHC(=O)OH$$

(c) This part is very similar to part (b) except that a final esterification of the acid must be included.

$$CH_2(C(=O)OR)_2 \xrightarrow[2.\ CH_3CH_2CH_2\text{–}I]{1.\ NaOR/HOR\ \text{(full equiv)}} CH_3CH_2CH_2CH(C(=O)OR)_2 \xrightarrow[2.\ C_6H_5\text{–}CH_2\text{–}I]{1.\ NaOR/HOR} (PhCH_2)(CH_3CH_2CH_2)C(C(=O)OR)_2 \xrightarrow[2.\ \Delta]{1.\ H_2O/H_3O^+} (PhCH_2)(CH_3CH_2CH_2)CHC(=O)OH \xrightarrow{EtOH,\ H_3O^+} (PhCH_2)(CH_3CH_2CH_2)CHC(=O)OEt$$

Problem 20.26

(a) Acid hydrolysis generates a β-keto acid, that, like almost all β-keto acids, is prone to decarboxylation.

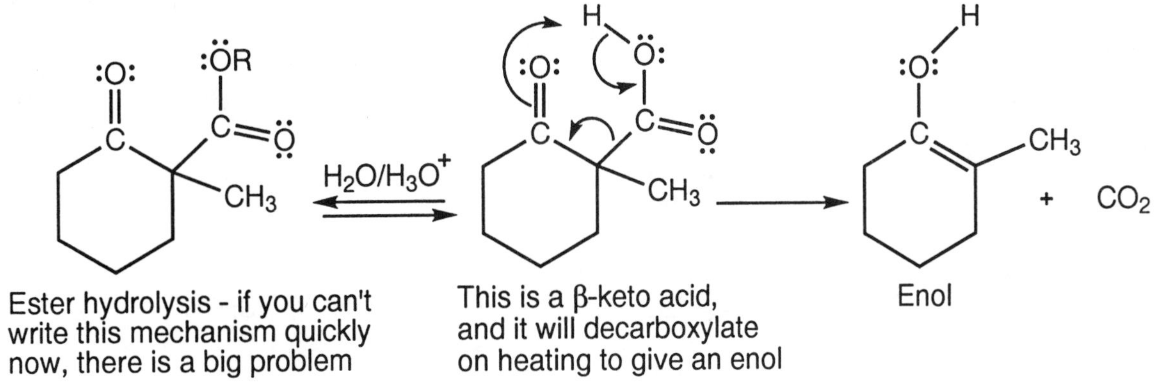

H_2O/H_3O^+

CH_3 + CO_2

Ester hydrolysis - if you can't write this mechanism quickly now, there is a big problem

This is a β-keto acid, and it will decarboxylate on heating to give an enol

Enol

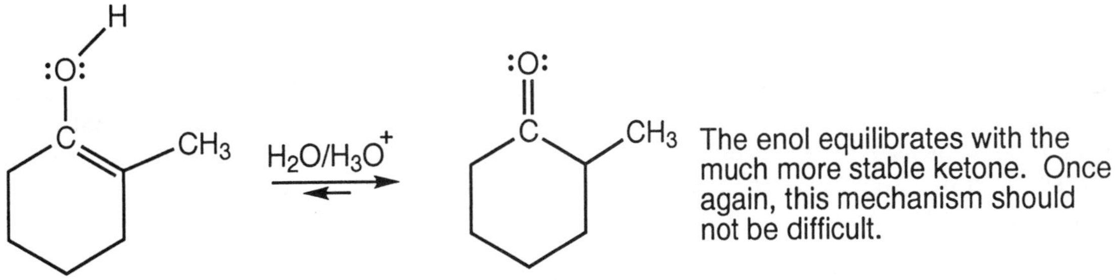

H_2O/H_3O^+

The enol equilibrates with the much more stable ketone. Once again, this mechanism should not be difficult.

In base, a reverse Claisen takes place—note the lack of doubly α hydrogens in the starting material.

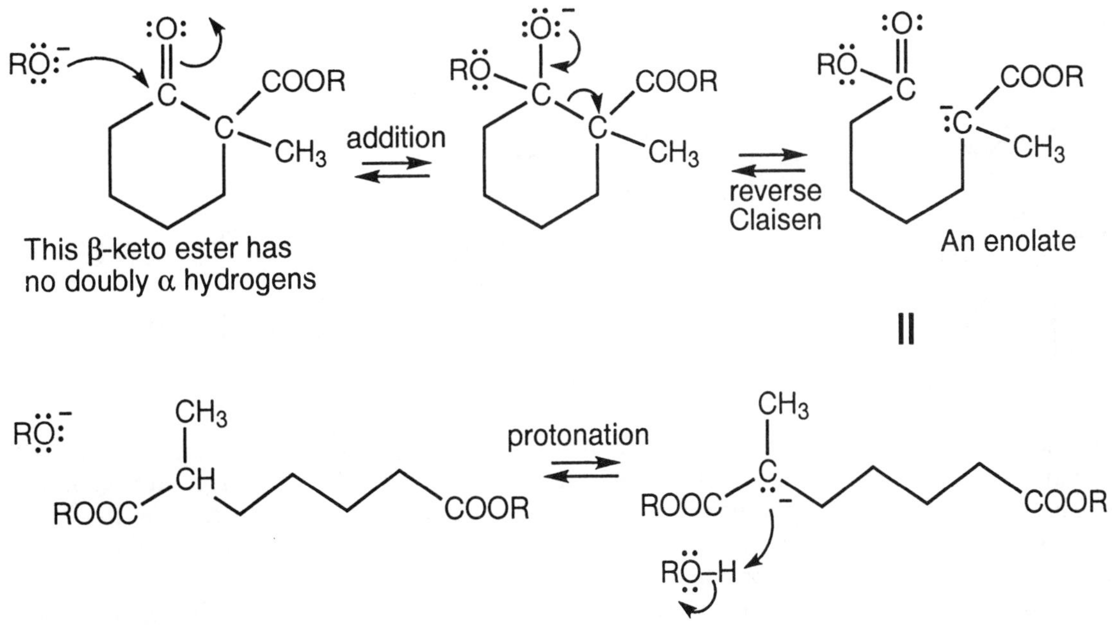

addition

reverse Claisen

This β-keto ester has no doubly α hydrogens

An enolate

protonation

(b) Here we are faced with a common problem: How to choose among several possible modes of reaction. At this point it's hard to make this choice, and you may just have to explore as many possibilities as you can. However, we can always try to use the Wisdom of the Ages, obey Magid's Second Rule (Chapter 18, p. 924), and try the Michael reaction first, using the enolate formed from dimethyl malonate.

Michael addition

$:\bar{C}H(COOCH_3)_2$

NaOCH$_3$

$CH_2(COOCH_3)_2$

$CH(COOCH_3)_2$

Resonance-stabilized enolate

The addition generates an enolate, so the Michael addition to the carbon-carbon double bond is a reasonable reaction. Now what's possible? The reaction can always reverse, but that path only leads back to the starting molecules. Perhaps an examination of the first arrow leading along the reversal path will give you an idea. It is also possible, as the carbon-carbon double bond reforms, to open the ring through generation of a new enolate.

This arrow starts both the reverse reaction and the ring opening reaction that will eventually lead to product.

$CH(COOCH_3)_2$

On to product

Reversal

$CH(COOCH_3)_2$

$CH(COOCH_3)_2$

Another resonance-stabilized enolate

One of the hardest things that you must learn is to analyze the problem before you start to push arrows and electrons around. In this case, the critical thing to see is that the original ring *must* open. There must be a way to get the ether oxygen out of the ring, and ring opening is the only possible way. As the product contains an all-carbon ring, there must also be a way to close the ring up again. This is an "open-close" problem.

Protonation of the enolate at carbon gives an aldehyde. Notice that the ring opening has generated a triply α hydrogen that can easily be removed in base to give a new enolate. This enolate sits poised, ready to add to the new aldehyde. So the ring opening has generated the two species necessary to close the all-carbon ring, the nucleophilic enolate and the Lewis acid aldehyde carbonyl. Protonation of the alkoxide formed through addition gives an alcohol.

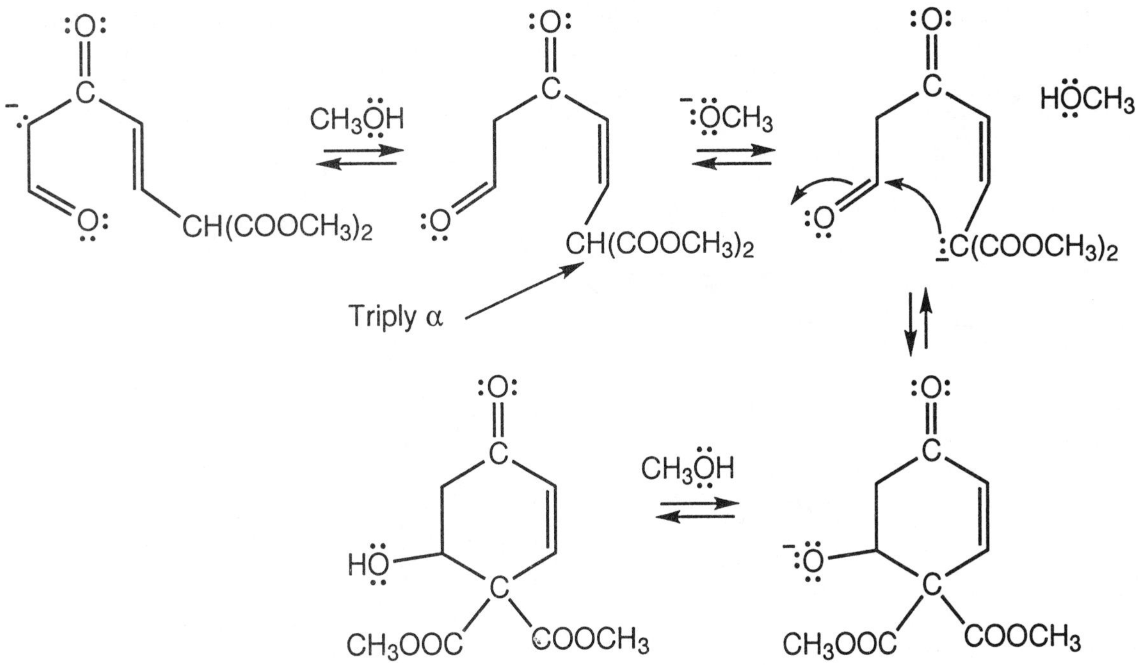

Triply α

Finally, an elimination reaction generates the product diene. It is true that hydroxide is a poor leaving group, but the anion, an enolate, is easy to form, and this E1cB reaction (Chapter 7, p. 288) can take place.

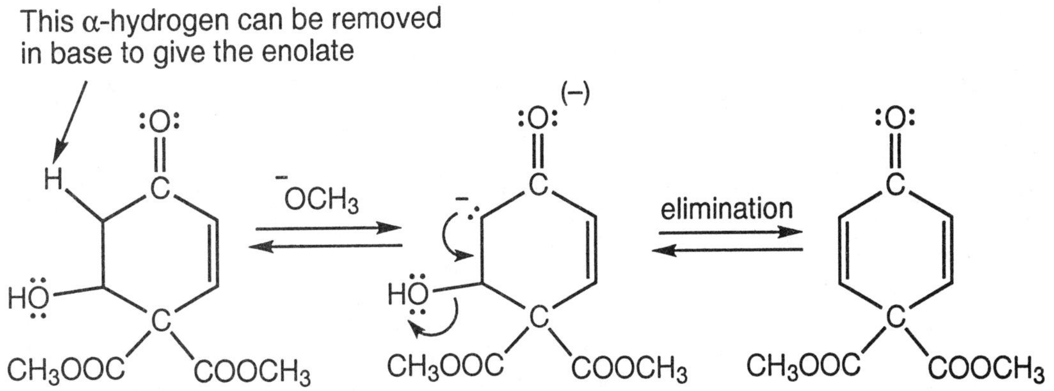

This α-hydrogen can be removed in base to give the enolate

elimination

In the second part of this problem an ester group is removed and an aromatic compound is formed. One can already see *why* this reaction happens; it must be driven by the great stabilization of the aromatic product. Again the tough part is seeing how to start. An analysis of the problem tells us that an ester is lost, so that should focus attention on reaction with an ester group. Addition of alkoxide to the ester gives a tetrahedral intermediate (**A**). This species can either revert to regenerate the starting ester (hardly a productive pathway) or lose dimethyl carbonate to generate the aromatic system. This route surely seems a good way to try. A simple protonation completes the reaction.

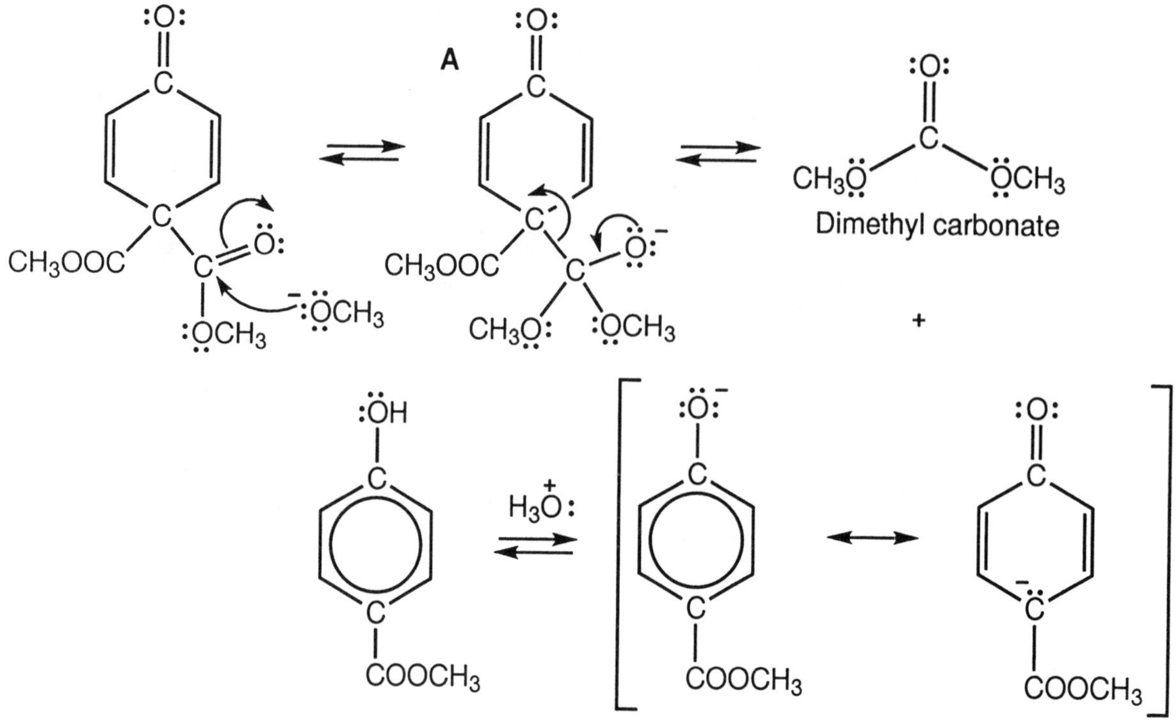

A

Dimethyl carbonate

(c) This problem is easy but it *looks* hard because the starting material and products have such different structures. That can make a problem difficult, especially at this stage. We'll try to solve this problem by exploring all the possible reactions of the starting material in the hopes that one will reveal an obvious path to the product. There are two different α- positions that can lead to two enolates, and two different carbonyl groups that could participate in addition reactions.

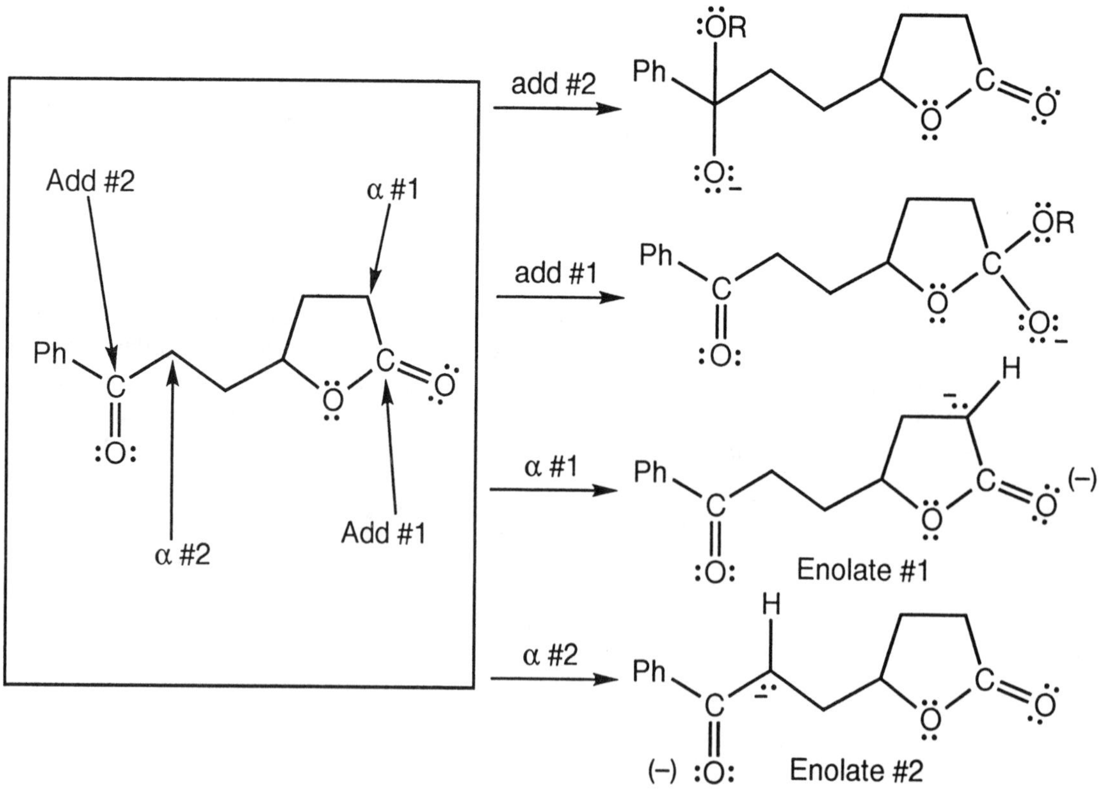

Add #2 α #1 α #2 Add #1

add #2 add #1 α #1 α #2

Enolate #1

Enolate #2

Enolate #2 can open directly to the product by an intramolecular S_N2 reaction! The carboxylate anion acts as leaving group. So here's a good solution; two simple steps, enolate formation and S_N2 ring opening.

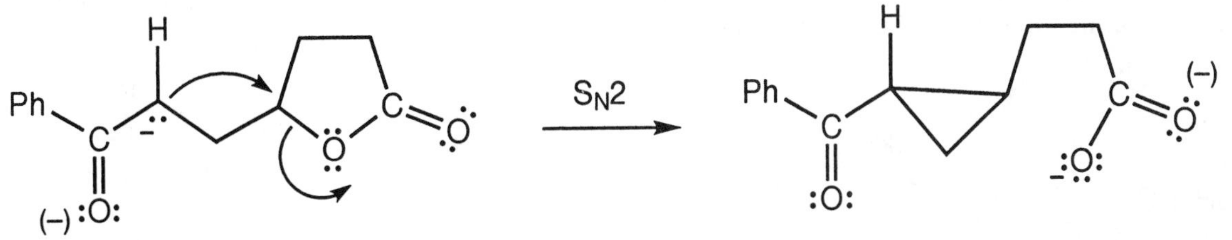

Problem 20.28 There is nothing tricky here. The only problem is to make the connection between carbon disulfide (CS_2), which you have not encountered before, and carbon dioxide (CO_2), which you have (Chapter 19, p. 991). Addition to carbon disulfide is followed by an S_N2 reaction to make the xanthate.

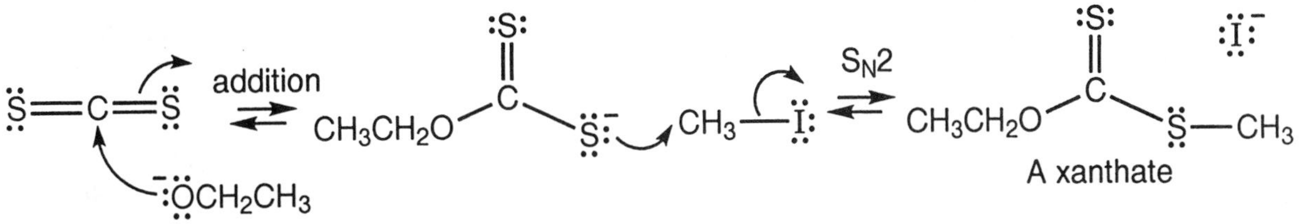

A xanthate

Now a reaction very much like that of Figure 20.93 occurs:

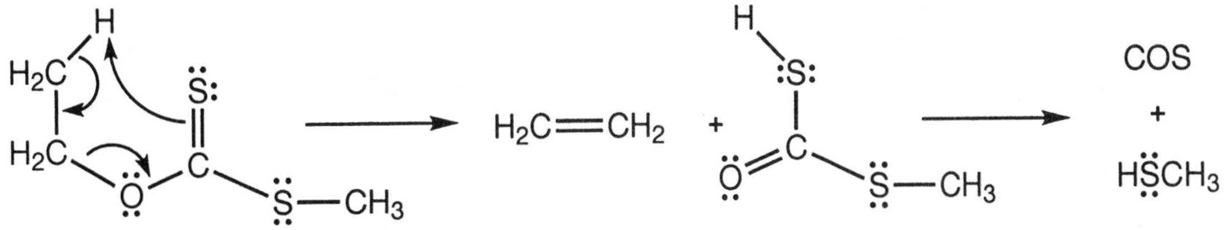

Problem 20.29

(a) These are both examples of thermal elimination reactions. In (a), it is the negative oxygen atom that acts as base and removes a hydrogen.

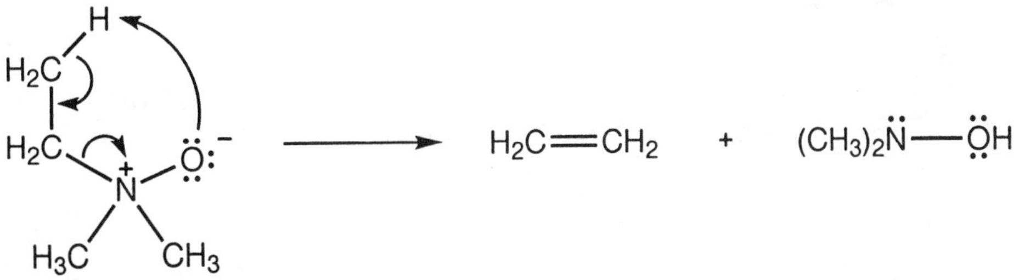

Reaction (b) is just a double ester elimination.

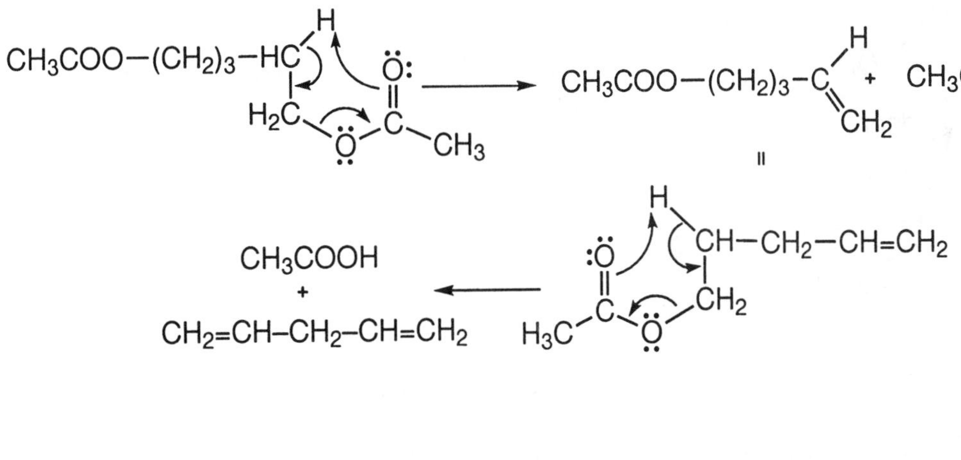

Additional Problem Answers

Problem 20.30 These two problems are straightforward addition-elimination sequences. In these cases, an amide and an acyl azide are the products.

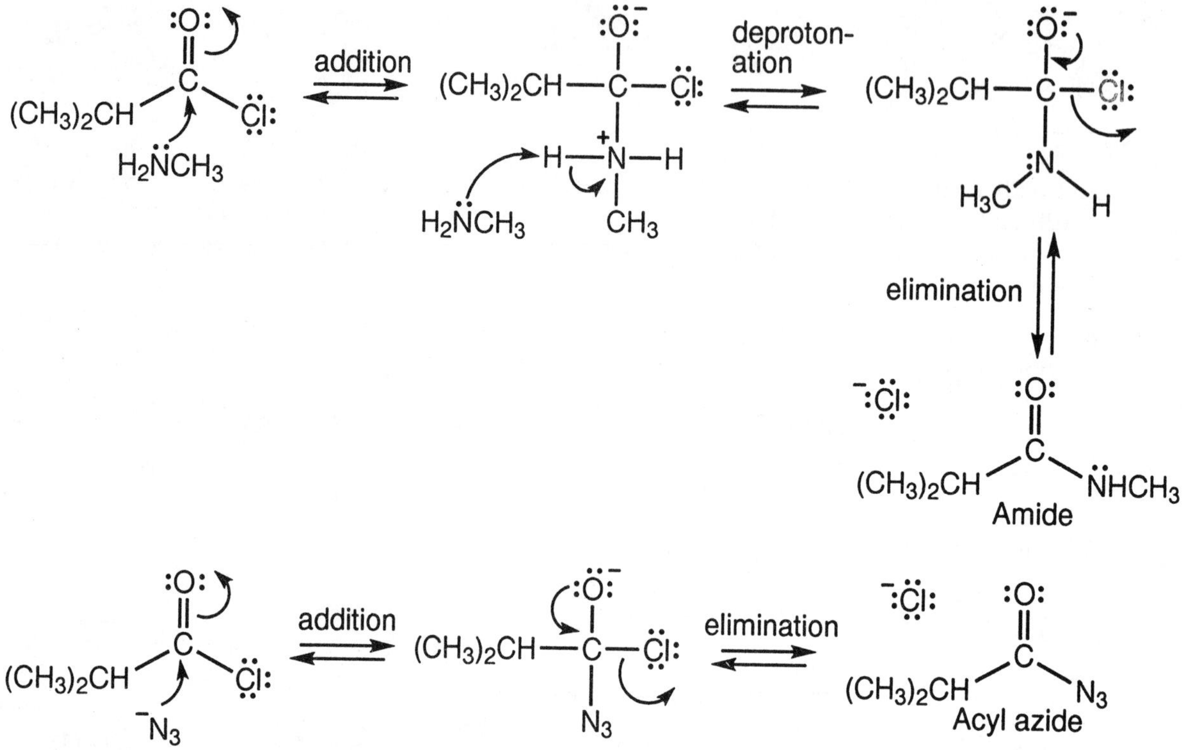

addition
deproton-ation
elimination
Amide

addition
elimination
Acyl azide

Problem 20.31 Here are two addition-elimination sequences done in acid.
(a)

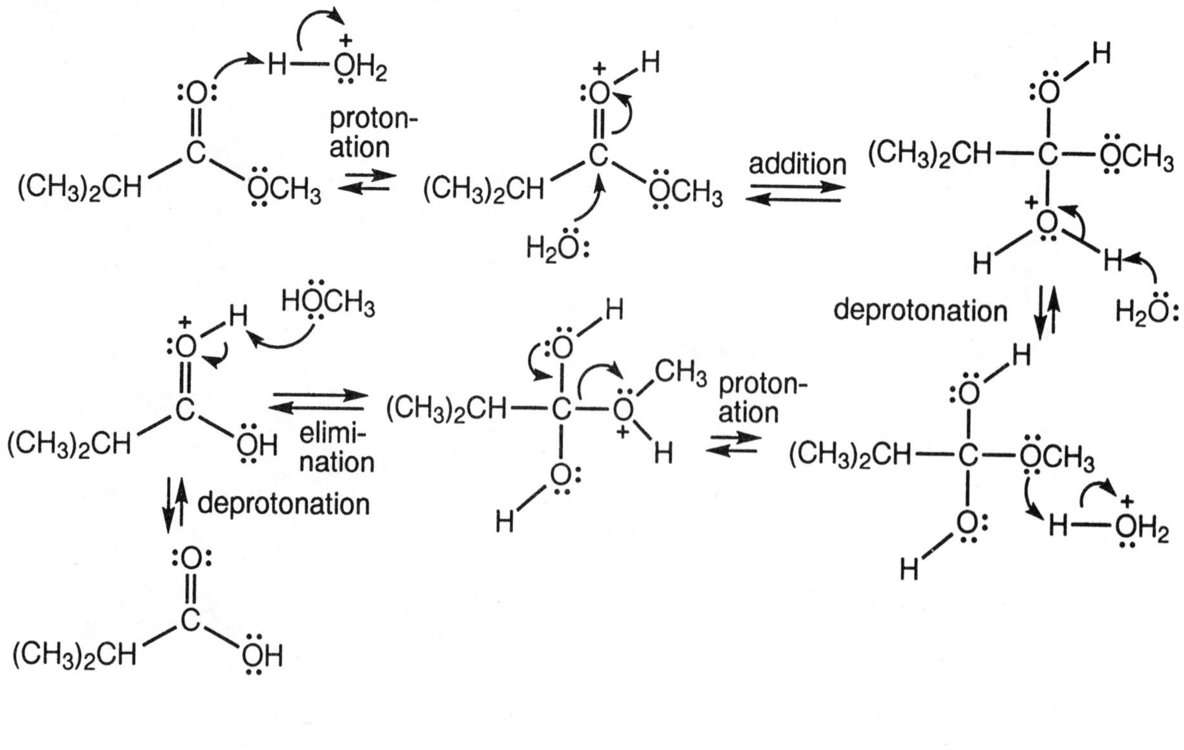

proton-ation
addition
deprotonation
proton-ation
elimi-nation
deprotonation

(b)

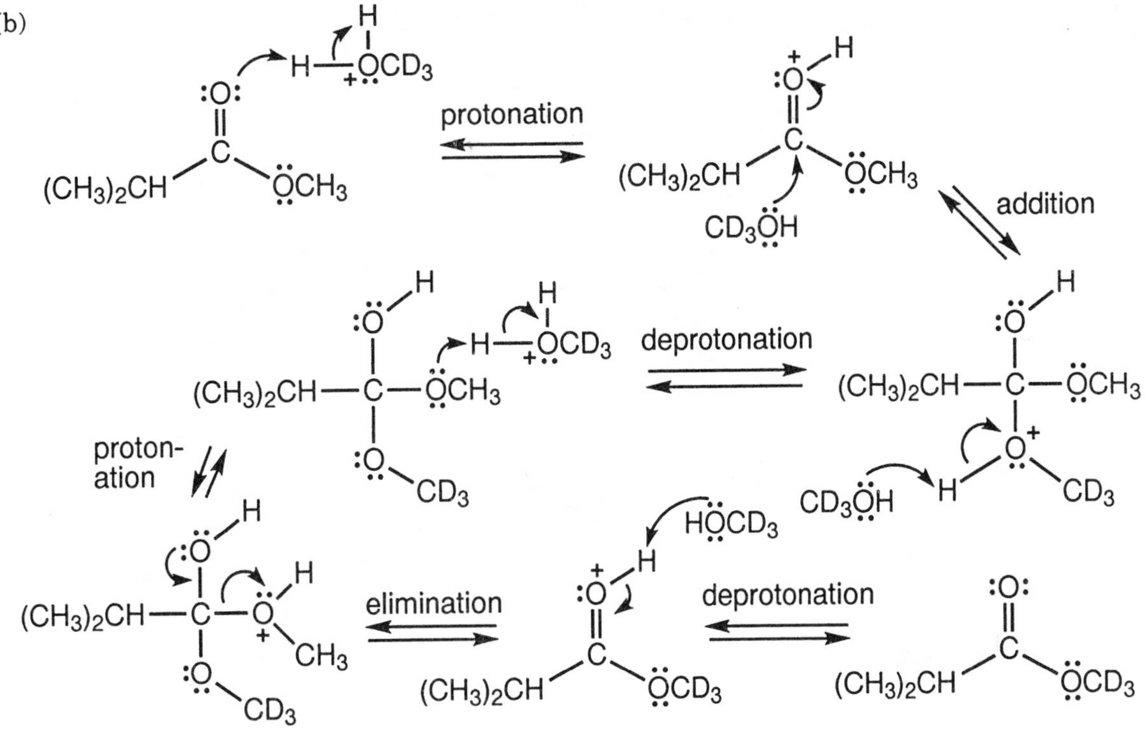

Problem 20.32 Here is another series of protonations, deprotonations, additions, and eliminations. Be careful about the details.

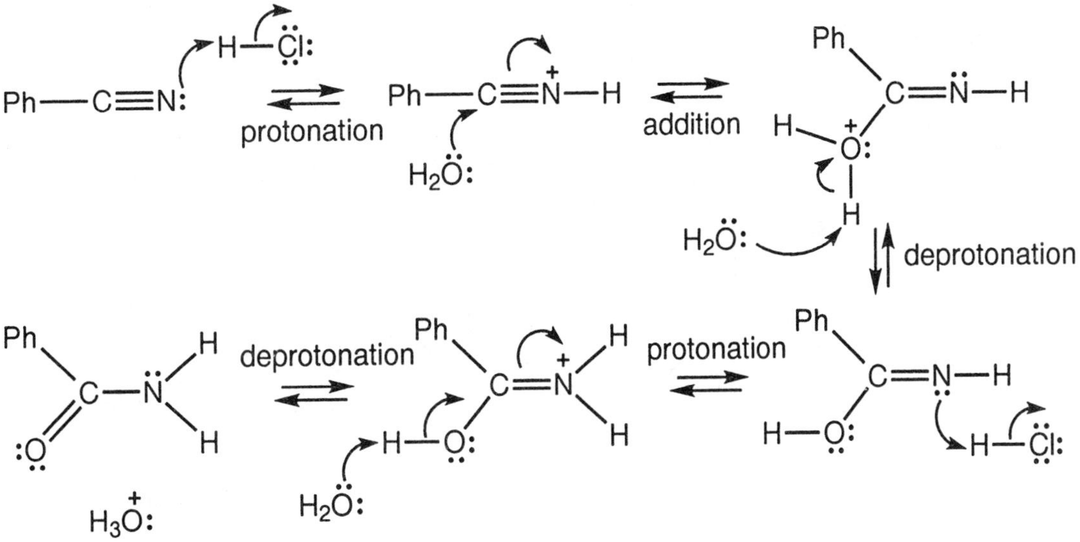

Problem 20.33 This reaction is a straightforward intramolecular Claisen condensation (Dieckmann). The first product is the anion formed by removing the doubly α proton. This enolate is protonated by acetic acid in a second step to give the product.

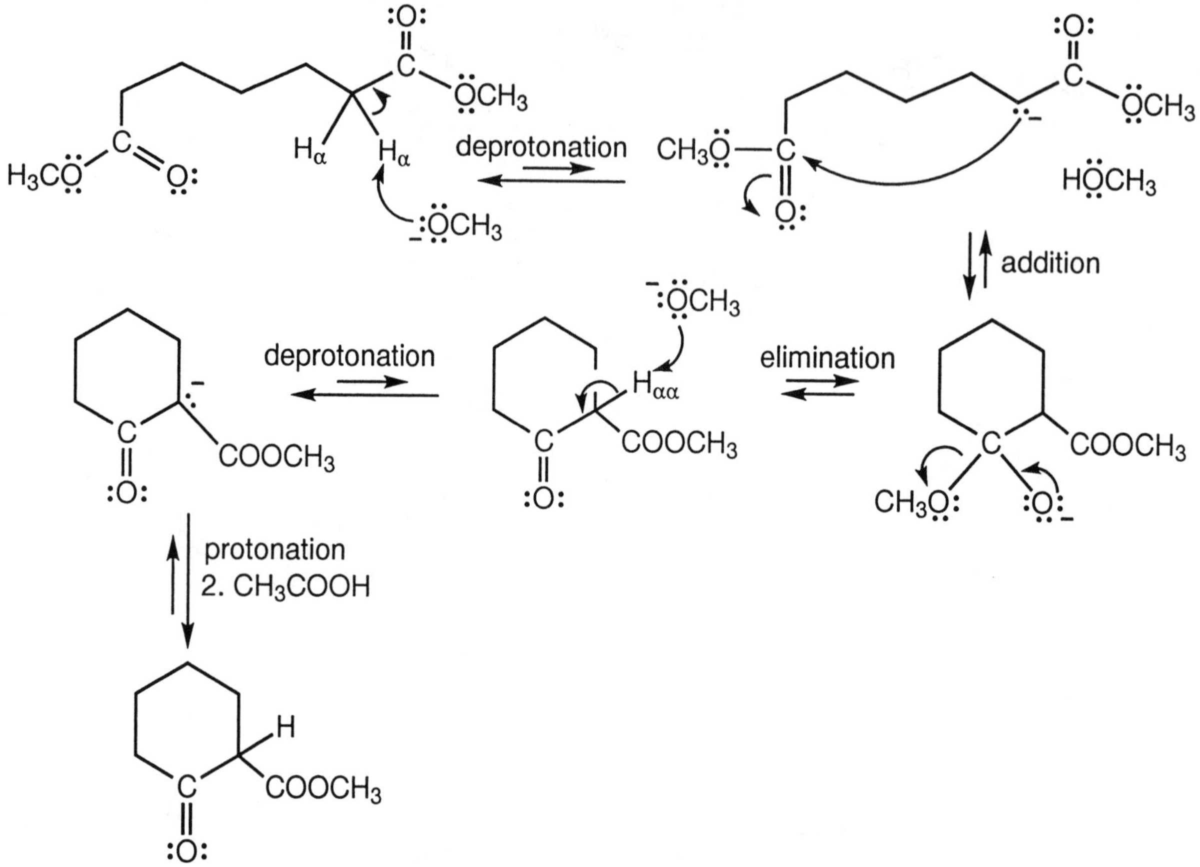

deprotonation

addition

deprotonation

elimination

protonation
2. CH_3COOH

Problem 20.34 In this process, the steps of the forward Dieckmann of Problem 20.33 are reversed.

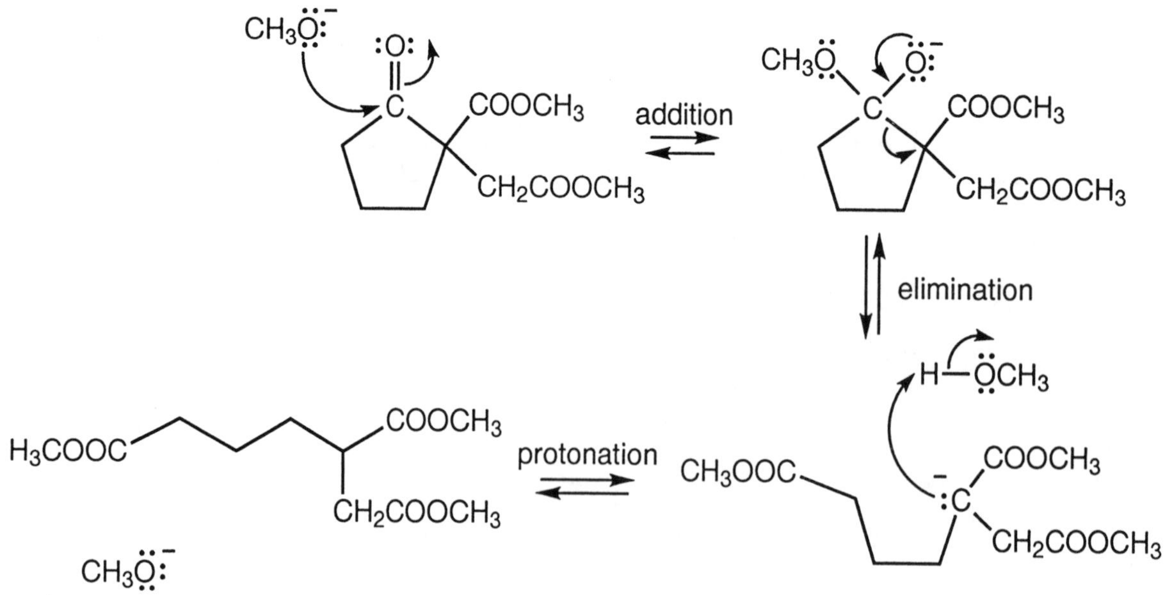

addition

elimination

protonation

Problem 20.35

(a) (b) 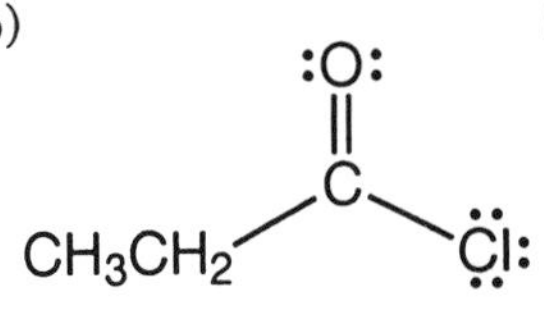 (c)

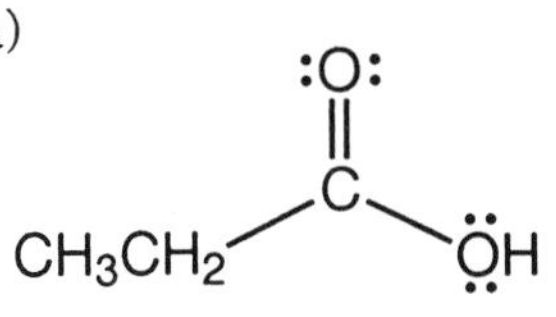

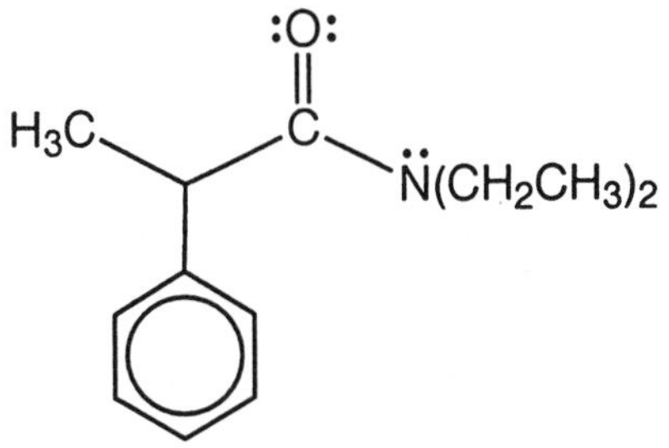

(d) 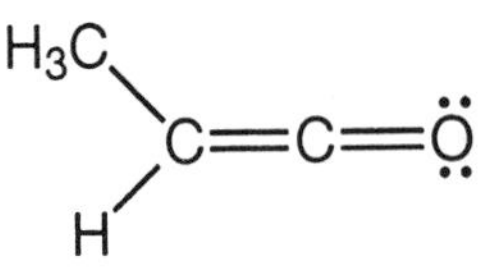 (e) 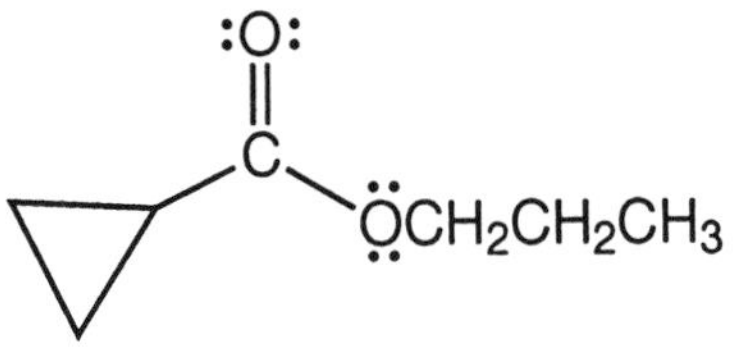 (f)

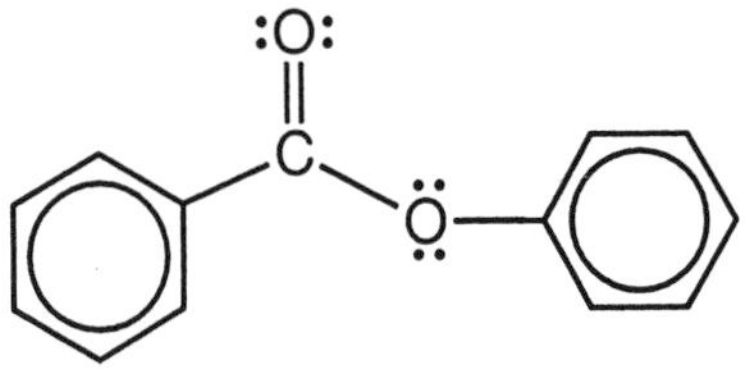

Problem 20.36

(a) Methyl cyclohexanecarboxylate
(b) Ethyl acetate
(c) Methyl propionate
(d) Cyclopropyl hexanoate
(e) Pentanoic acid
(f) Pentanoyl chloride
(g) Pentanamide
(h) *p*-Bromobenzamide
(i) *N,N*-Dimethylpentanamide
(j) 2-Cyanopropane
(k) Methyl phenyl ketene

Problem 20.37(a)
Consider the resonance descriptions of acetone and acetaldehyde shown below.

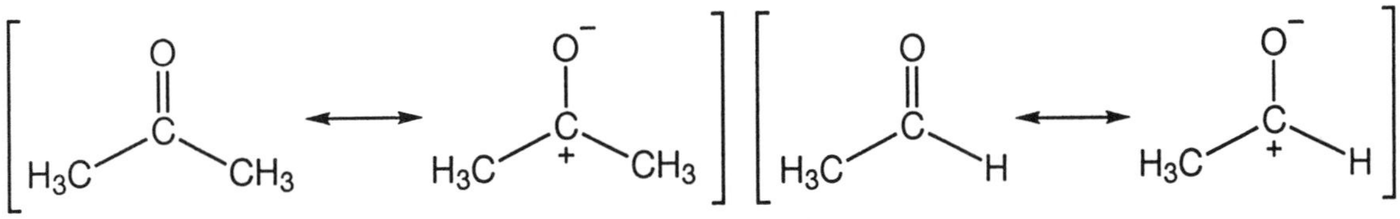

The "extra" methyl group in acetone will stabilize the polar resonance form in acetone relative to that in acetaldehyde. The polar form contributes more to the resonance hybrid in acetone than it does in acetaldehyde. As this form has only a single bond between carbon and oxygen, the carbonyl bond will be weaker in acetone than in acetaldehyde, and therefore the stretching frequency is lower (1719 vs. 1733 cm^{-1}). But wait. Doesn't the "extra" methyl group in acetone also stabilize the carbon-oxygen double bond relative to that in acetaldehyde? Sure, but these hyperconjugative effects are especially strong for charged species, and it is the effect of the methyl group on the polar form that dominates.

(b) Look again at the contributing resonance forms.

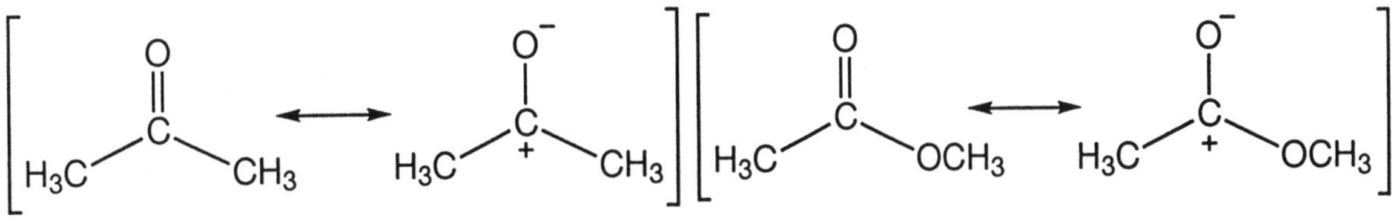

Polarization of the C–O σ bond

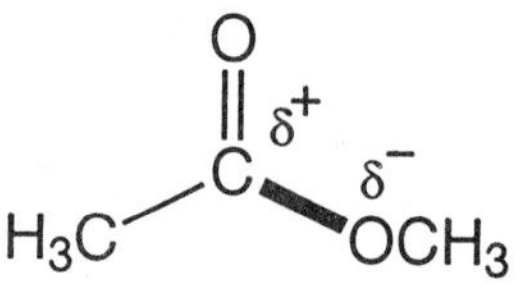

As the figure indicates, now the polarity of the carbon-oxygen σ bond makes the polar form in the ester especially unfavorable. The polar form will contribute less to the resonance hybrid than it does in acetone, and the carbonyl group will be relatively more "double," and therefore stronger. Accordingly, it appears at higher frequency than in acetone (1750 vs. 1719 cm^{-1}).

But wait again. Doesn't the ester oxygen effect the situation in another way? Isn't there another resonance form in which there is a single bond in the original carbonyl group? Shouldn't this make the carbon-oxygen double bond weaker, and the absorption at lower frequency?

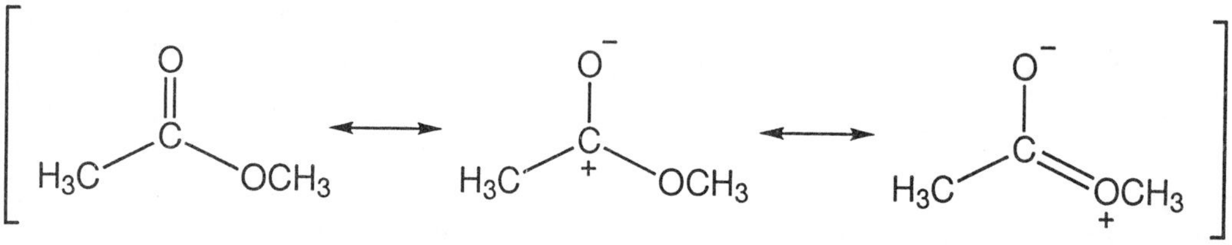

Yes, but the inductive effect mentioned earlier must dominate. Here is where we are so perilously close to circular reasoning. The statement, "The inductive effect must dominate the resonance effect, and therefore the absorption appears at higher frequency" is valid, but, it surely is clear that in the absense of a measured spectrum it would be difficult to predict the relative positions of the absorptions.

(c) Once again, start by looking at the contributing forms.

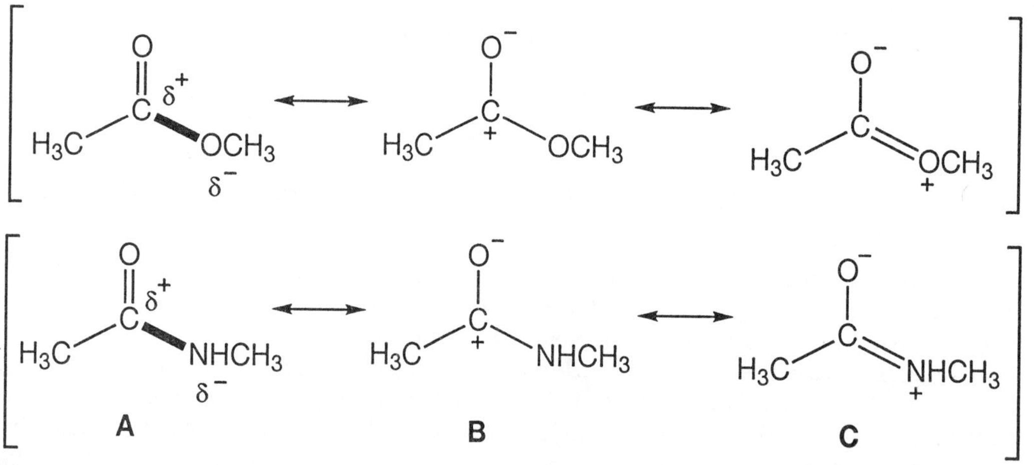

A B C

The problem is to explain why in the amide the dipole shown in form **A** doesn't result, as it does for the ester, in a high frequency absorption. Once again, one can rationalize. First of all, the dipole in form **A** does not have as big an effect in the amide as it does in the ester. Nitrogen is less electronegative than oxygen. Second, form **C** will be more important for the amide than it is for the ester, as a positive charge on nitrogen (less electronegative) is more stable than it is on oxygen (more electronegative). In the amide, the combination of a reduced dipole and increased contribution of form **C** leads to a weaker carbon-oxygen double bond and a lower frequency absorption.

Problem 20.38 Yes. The greater the positive charge on the carbon of the carbonyl group, the lower the position of the carbon in the ^{13}C NMR spectrum. In the ester the dipole in the carbon-oxygen σ bond reduces the contribution of the dipolar resonance form. The carbon is less positive than that in acetone and appears far upfield (δ 169 vs. 205 ppm) of the carbonyl carbon in acetone.

Problem 20.39

(a)

Cyclohexane-COOH → (1. $SOCl_2$) → cyclohexane-COCl → (2. $HN(CH_3)_2$) → cyclohexane-$CON(CH_3)_2$ → (3. $LiAlH_4$; 4. H_2O) → cyclohexane-$CH_2N(CH_3)_2$

(b)

Succinic anhydride → (CH_3OH, Δ) → $HOOC-CH_2CH_2-COOCH_3$

Opening of the anhydride by methyl alcohol leads to one ester and one acid group

(c)

Cage ester $-O-C(=O)-CH_3$ → (CH_3ONa; CH_3OH, Δ) → cage alcohol $-OH$ + CH_3COOCH_3

This part includes a base-catalyzed transesterification. The cage is hiding a simple reaction.

(d)

Caprolactam (cyclic $C(=O)-NH$) → (HCl; H_2O, Δ) → $HOOC(CH_2)_5\overset{+}{N}H_3$ Cl^-

The cyclic amide is opened by water to give the carboxylic and the amine salt. In this case, both products are in the same molecule.

(e)

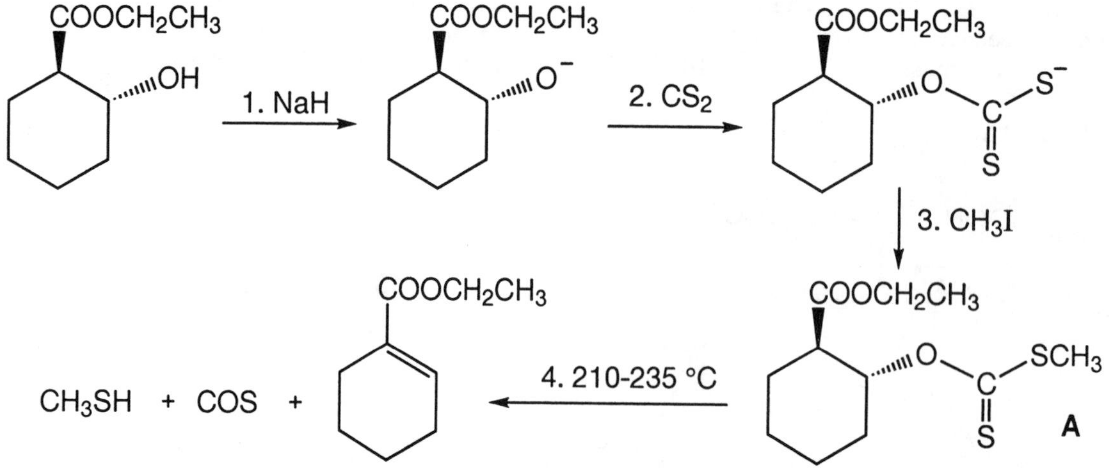

$COOCH_2CH_3$ / OH —1. NaH→ $COOCH_2CH_3$ / O^- —2. CS_2→ $COOCH_2CH_3$ / O–C(=S)–S^-

3. CH_3I

CH_3SH + COS + $COOCH_2CH_3$ (cyclohexene) ←4. 210-235 °C— $COOCH_2CH_3$ / O–C(=S)–SCH_3 **A**

The xanthate ester **A** is first prepared, and then undergoes a thermal elimination reaction to give the conjugated ester.

(f)

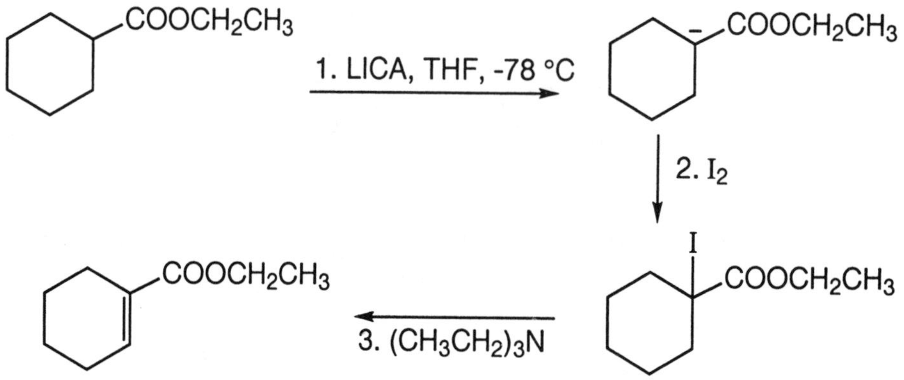

$COOCH_2CH_3$ —1. LICA, THF, -78 °C→ $^-$ $COOCH_2CH_3$

2. I_2

$COOCH_2CH_3$ ←3. $(CH_3CH_2)_3N$— I $COOCH_2CH_3$

There's more than one way to skin a cat. Here is another route that leads to the same ester formed in (e), but does so under milder conditions. The ester enolate is first prepared by using the strong base LICA at low temperature. Iodination at low temperature leads to an iodide that can undergo an E2 reaction when treated with an amine.

Problem 20.40 The conversion of aldoximes to nitriles is formally a dehydration reaction. This reaction is yet another example of the conversion of a poor leaving group (hydroxyl) into a better one (acetic acid).

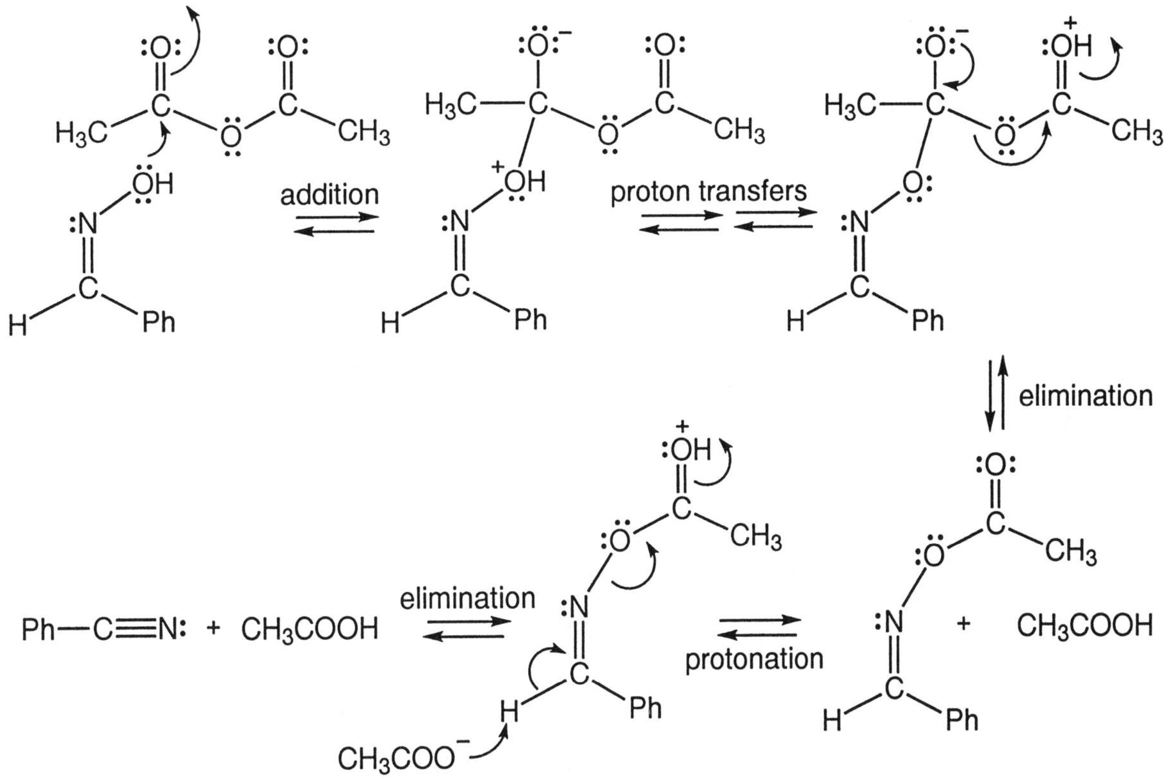

addition

proton transfers

elimination

elimination

protonation

Ph—C≡N: + CH_3COOH

CH_3COO^-

+ CH_3COOH

Oximes (R_2C=N—OH) are formed from the reaction of aldehydes and ketones with hydroxylamines.

Problem 20.41 This process is named the Ritter reaction. The first step in the sequence is the formation of the relatively stable *tert*-butyl cation.

$$(CH_3)_3COH \xrightleftharpoons{H_2SO_4} (CH_3)_3C\overset{+}{O}H_2 \rightleftharpoons (CH_3)_3\overset{+}{C} + H_2O$$

The *tert*-butyl cation is then captured by the nucleophilic nitrile to give the resonance-stabilized cation **A**.

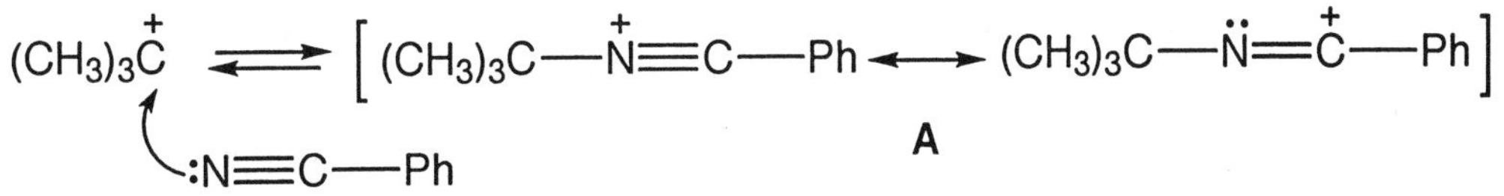

$$(CH_3)_3\overset{+}{C} \rightleftharpoons \left[(CH_3)_3C—\overset{+}{N}≡C—Ph \longleftrightarrow (CH_3)_3C—\ddot{N}=\overset{+}{C}—Ph\right]$$

:N≡C—Ph

A

Addition of water to **A**, followed by a series of proton shifts, gives the observed product. Notice the relation between the last step in this figure and keto-enol equilibria.

$(CH_3)_3C—\ddot{N}=\overset{+}{C}—Ph$ + $:\ddot{O}H_2$ ⇌ $(CH_3)_3C—\ddot{N}=C(Ph)—\overset{+}{O}H_2$; $:\ddot{O}H_2$ ⇌ $(CH_3)_3C—\ddot{N}=C(Ph)—\ddot{O}H$ + $H_3\overset{+}{O}:$ ⇌ $(CH_3)_3C—\overset{+}{N}(H)=C(Ph)—\ddot{O}—H$; $H_2\ddot{O}:$ ⇌ $(H_3C)_3CH\ddot{N}—C(=O)—Ph$

2

Problem 20.42 (a) Tertiary alcohols in which all three R groups are the same are easily made from the reaction of a carbonate with a Grignard reagent.

$H_3CO—C(=O)—OCH_3$ → (1. 3BuMgBr; 2. H_2O/H_3O^+) Bu_3COH

Dimethyl carbonate

The carbonate comes from reaction of phosgene and methyl alcohol, and the Grignard reagent can be made from treatment of butyl bromide with Mg.

$COCl_2 + 2CH_3OH \longrightarrow H_3CO—C(=O)—OCH_3$

BuOH →(HBr) BuBr →(Mg, ether) BuMgBr

b. The reaction of propionic acid with two equivalents of butyllithium, followed by protonation, will do the trick. Butyllithium comes from butyl bromide (part a) and lithium.

2BuBr →(Li) 2BuLi → (1. CH_3CH_2COOH; 2. H_2O/H_3O^+) $CH_3CH_2—C(=O)—Bu$

c. Reaction of propionic acid with ethyl alcohol and an acid catalyst (Fischer esterification) leads to ethyl propionate. Reaction with butyllithium, followed by protonation, completes the synthesis.

$$CH_3CH_2COOH \xrightarrow[CH_3CH_2\overset{+}{O}H_2]{CH_3CH_2OH} CH_3CH_2COOCH_2CH_3 \xrightarrow[2.\ H_2O/H_3O^+]{1.\ 2BuLi} CH_3CH_2{-}C(OH)(Bu){-}Bu$$

d. A Friedel-Crafts reaction between propionyl chloride and benzene will produce the product. The acid chloride is made directly from the carboxylic acid with thionyl chloride.

$$CH_3CH_2COOH \xrightarrow{SOCl_2} CH_3CH_2COCl \xrightarrow[AlCl_3]{C_6H_6} CH_3CH_2{-}C(=O){-}C_6H_5$$

e. Dehydration of propionamide will make the nitrile. The amide can be made from the acid chloride (part d) and ammonia.

$$CH_3CH_2COCl \xrightarrow{NH_3} CH_3CH_2CONH_2 \xrightarrow[heat]{P_2O_5} CH_3CH_2CN$$

f. This product comes from a Claisen condensation of ethyl propionate (part c). Don't forget to use a full equivalent of base in the Claisen condensation (why?).

$$CH_3CH_2COOCH_2CH_3 \xrightarrow[2.\ H_2O/H_3O^+]{1.\ CH_3CH_2OH,\ CH_3CH_2ONa} CH_3CH_2{-}C(=O){-}CH(CH_3){-}C(=O){-}OCH_2CH_3$$

g. This ketone is made from the Claisen product in part f through alkylation, followed by decarboxylation (acetoacetic ester synthesis).

$$CH_3CH_2{-}C(=O){-}CH(CH_3){-}C(=O){-}OCH_2CH_3 \xrightarrow[2.\ BuBr]{1.\ CH_3CH_2OH,\ CH_3CH_2ONa} CH_3CH_2{-}C(=O){-}C(H_3C)(Bu){-}C(=O){-}OCH_2CH_3$$

$$\xrightarrow[H_2O/H_3O^+]{\Delta} CH_3CH_2{-}C(=O){-}CH(CH_3){-}Bu$$

Problem 20.43. All of these target molecules can be prepared by acetoacetic ester or malonic ester syntheses. Recall that acetoacetic esters are sources of ketones and malonic esters are sources of acetic acids. We will present a retrosynthetic analysis for each target molecule then propose a detailed synthesis in the forward sense.

a. The target ketone is available from ethyl acetate via a Claisen condensation, followed by two alkylations, acid hydrolysis, and decarboxylation; a standard acetoacetic ester synthesis.

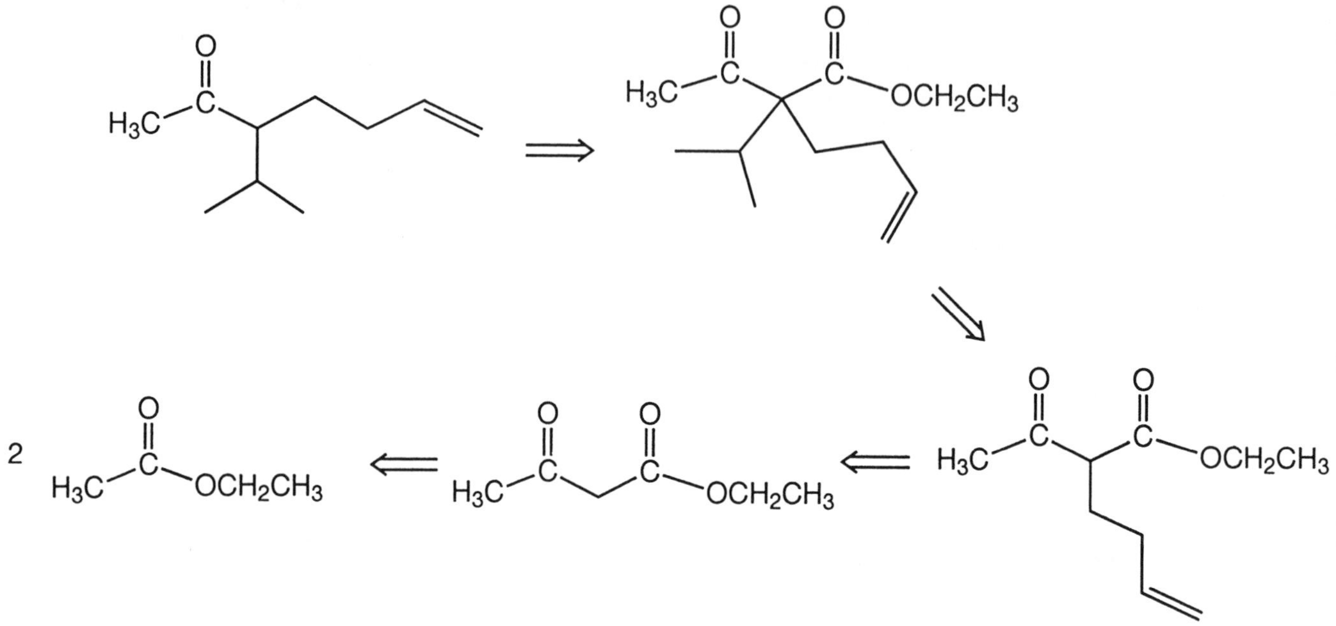

$H_3C-C(=O)-CH(CH(CH_3)_2)CH_2CH_2CH=CH_2 \Longrightarrow H_3C-C(=O)-C(CH(CH_3)_2)(CH_2CH_2CH=CH_2)-C(=O)-OCH_2CH_3$

$\Longrightarrow H_3C-C(=O)-CH(CH_2CH_2CH=CH_2)-C(=O)-OCH_2CH_3 \Longrightarrow H_3C-C(=O)-CH_2-C(=O)-OCH_2CH_3 \Longrightarrow 2\ H_3C-C(=O)-OCH_2CH_3$

In a forward direction, with reagents:

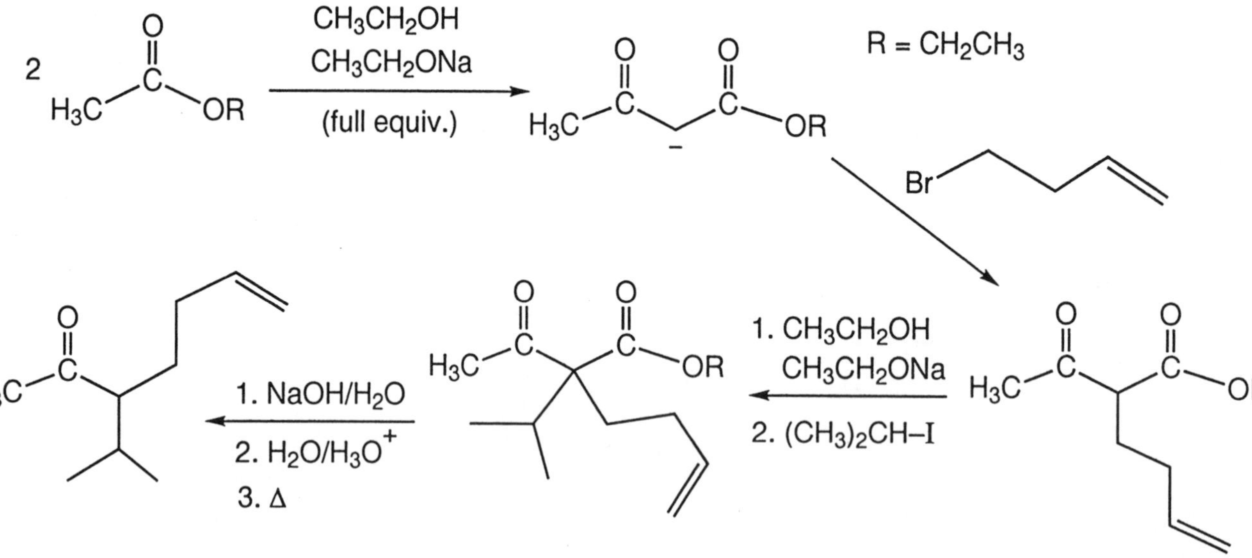

$2\ H_3C-C(=O)-OR \xrightarrow[\text{(full equiv.)}]{CH_3CH_2OH,\ CH_3CH_2ONa} H_3C-C(=O)-\overset{-}{C}H-C(=O)-OR$ $\quad R = CH_2CH_3$

$\xrightarrow{Br-CH_2CH_2CH=CH_2} H_3C-C(=O)-CH(CH_2CH_2CH=CH_2)-C(=O)-OR$

$\xrightarrow[\text{2. } (CH_3)_2CH\text{–I}]{\text{1. } CH_3CH_2OH,\ CH_3CH_2ONa} H_3C-C(=O)-C(CH(CH_3)_2)(CH_2CH_2CH=CH_2)-C(=O)-OR$

$\xrightarrow[\text{2. } H_2O/H_3O^+\text{; 3. } \Delta]{\text{1. } NaOH/H_2O} H_3C-C(=O)-CH(CH(CH_3)_2)CH_2CH_2CH=CH_2$

b. The target is available from dimethyl malonate through a standard malonic ester synthesis. A pair of alkylations is followed by hydrolysis and decarboxylation.

HO–C(=O)–CH(R)–CH₃ R = $C_{10}H_{21}$, decyl

⇓

CH_3O–C(=O)–C(R)(CH_3)–C(=O)–OCH_3 ⟹ CH_3O–C(=O)–CH(R)–C(=O)–OCH_3 ⟹ CH_3O–C(=O)–CH_2–C(=O)–OCH_3

Forward, with reagents:

CH_3O–C(=O)–CH_2–C(=O)–OCH_3 —(1. CH_3OH, CH_3ONa; 2. RI (R = decyl))→ CH_3O–C(=O)–CH(R)–C(=O)–OCH_3 —(1. CH_3OH, CH_3ONa; 2. CH_3Br)→ CH_3O–C(=O)–C(R)(CH_3)–C(=O)–OCH_3 —(1. KOH/H_2O; 2. H_2O/H_3O^+; 3. Δ)→ HO–C(=O)–CH(R)–CH_3

c. The key reaction in this proposed synthesis is a Dieckmann condensation that is followed by an alkylation and decarboxylation. The required precursor, dimethyl adipate, is easily available from cyclohexene through an ozonolysis with oxidative workup, followed by Fischer esterification.

2-benzylcyclopentanone (C=O, CH_2Ph) ⟹ 2-benzyl-2-($COOCH_3$)cyclopentanone (C=O, CH_2–Ph, $COOCH_3$)

⇓

2-($COOCH_3$)cyclopentanone ⟹ Dimethyl adipate (H_3COOC–$(CH_2)_4$–$COOCH_3$) ⟹ HOOC–$(CH_2)_4$–COOH ⟹ cyclohexene

Forward, with reagents:

1. O_3 2. H_2O_2 → HOOC…COOH; CH_3OH, $CH_3\overset{+}{O}H_2$ → H_3COOC…$COOCH_3$; CH_3OH, CH_3ONa → enolate (–, $COOCH_3$); $PhCH_2Cl$ → Ph, $COOCH_3$; 1. KOH/H_2O 2. H_2O/H_3O^+ 3. Δ → Ph

d. The target is available through a standard malonic ester synthesis. The "twist" here is the use of 1-bromo-3-chloropropane to do both alkylation reactions, thus closing the four-membered ring.

HOOC– ⟹ H_3COOC, H_3COOC ⟹ OCH_3, O=C, O=C, OCH_3, Cl ⟹ OCH_3, O=C, O=C, OCH_3

Forward, with reagents:

OCH_3, O=C, O=C, OCH_3 → 1. CH_3OH, CH_3ONa 2. $BrCH_2CH_2CH_2Cl$ → OCH_3, O=C, O=C, OCH_3, Cl → CH_3OH, CH_3ONa → H_3COOC, H_3COOC → 1. KOH/H_2O 2. H_2O/H_3O^+ 3. Δ → HOOC–

Problem 20.44. This reaction is an example of the Perkin reaction, named for Sir William Henry Perkin, Jr. (1848 - 1907), who synthesized the first synthetic dye, mauve. The Perkin reaction involves the condensation of an aromatic aldehyde with an anhydride to form a cinnamic acid. In this example, the first step involves the formation of the enolate of acetic anhydride.

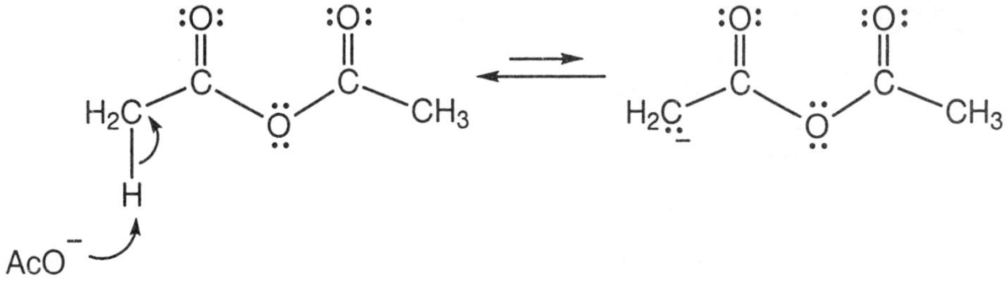

The enolate then adds to the aldehyde in a reaction closely related to the aldol condensation, giving alkoxide **A**. An addition–elimination sequence leads to the intermediate given in the problem, **B**.

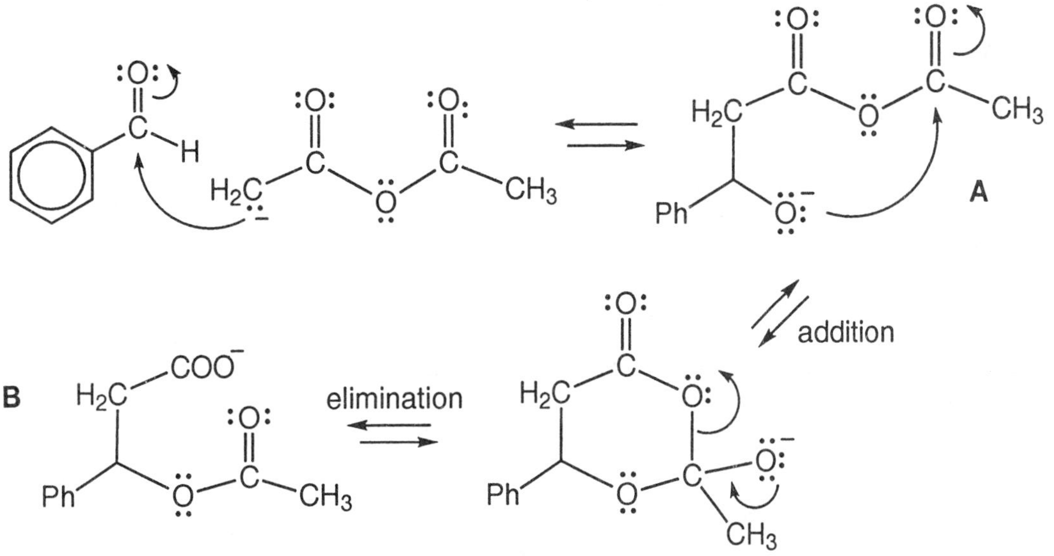

Now, it looks as though we are close. An elimination of acetate does the job. If you wrote such an answer that's certainly a reasonable approach. In fact, the process is a bit more complicated, as acetic anhydride gets into the picture again to form **C**. Compound **C** is hydrolyzed in water to give cinnamic acid.

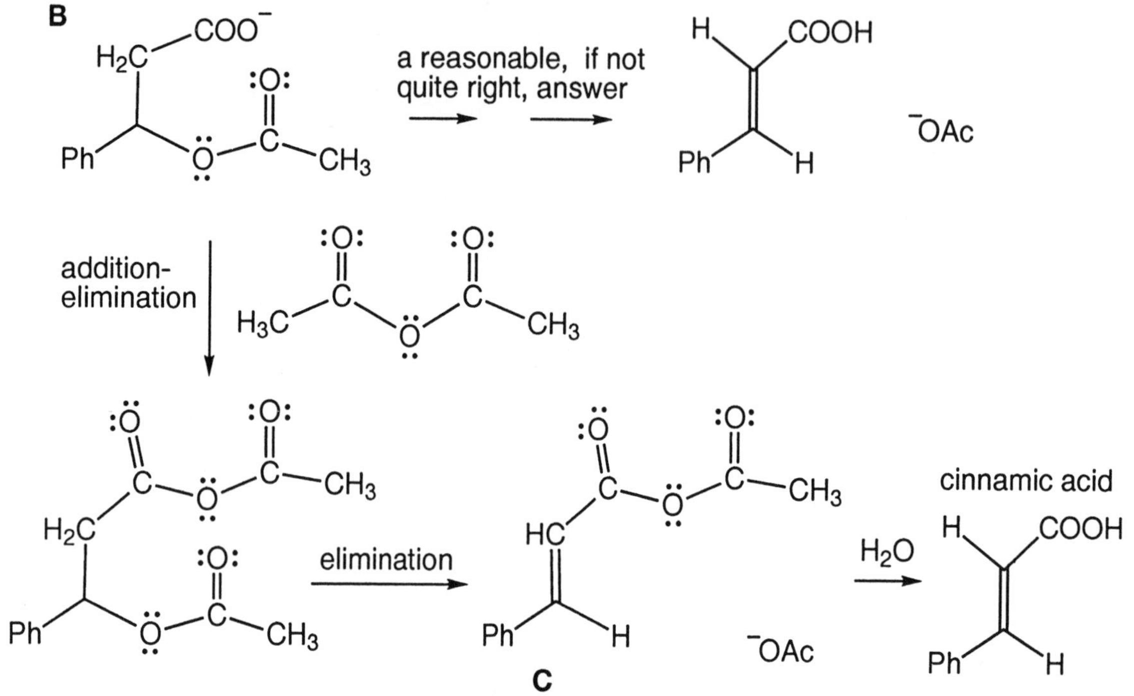

B
a reasonable, if not quite right, answer
addition-elimination
elimination
C
H_2O
cinnamic acid

Problem 20.45. The hint tells you how to start; the nucleophile must be the oxygen of DMF. DMF and oxalyl chloride react first in an addition–elimination process to give **A**, a very strong Lewis acid. Addition of chloride, followed by another elimination, gives the Vilsmeier reagent:

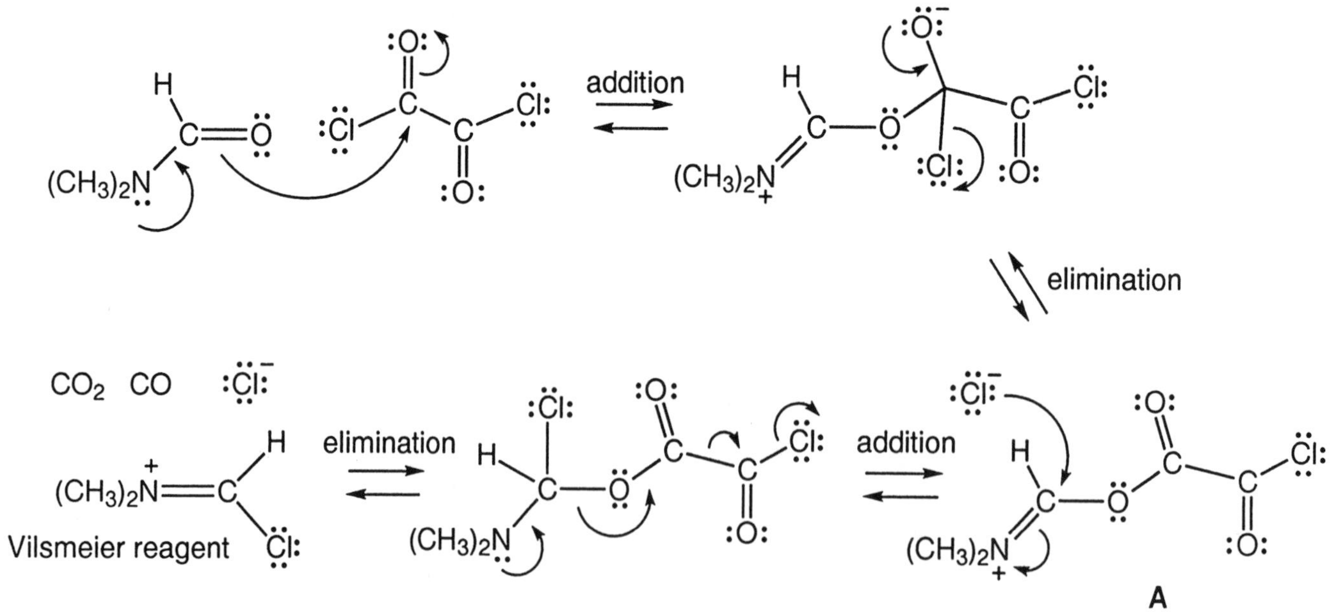

addition
elimination
CO_2 CO
Vilsmeier reagent
elimination
addition
A

Problem 20.46. Clearly, a ring is to be closed. A quick analysis should show where the starting pieces are in the product (bold bonds), and where the new bonds must be formed. Now think about reactions that let you make those new bonds.

new bond
Ph
H3COOC
COOCH3
COOCH3
new bond
C
O
CH3
H3COOC
Ph
H3C
COOCH3
H3COOC
COOCH3

This reaction is an example of a "tandem" reaction. In this case, a Michael addition is followed by a Dieckmann condensation. Removal of a "doubly α" hydrogen from the starting material by methoxide gives enolate **A**. This then adds in Michael fashion to methyl cinnamate to give a new enolate, **B**, and to make one of the required new bonds. Intermediate **B** is poised to close the five–membered ring in a Dieckmann condensation to give **C**, and making the second new bond. A final protonation leads to the observed product.

A
Ph Methyl cinnamate
H3C
COOCH3
H
H3COOC
COOCH3
:ÖCH3
COOCH3
COOCH3
B
(bond in boldface formed above as **A** → **B**)
Haα
C
Salt formation ends Dieckmann
CH3COOH
protonation
Final product

Problem 20.47. The starting material for this problem is a β-keto ester, the potential product (as are *all* β-keto esters) of a Claisen condensation (here in its cyclic version, a Dieckmann condensation). However, remember (or, as the hint says, see Problem 20.26a) that the Claisen and Dieckmann condensations depend for their success on the presence of a "doubly α" hydrogen in the product so that a salt can be formed. This Claisen product does not have such a hydrogen, and so the reverse Claisen, the reversion to the thermodynamically more stable starting materials, is inevitable.

The beginning: a reverse Dieckmann.

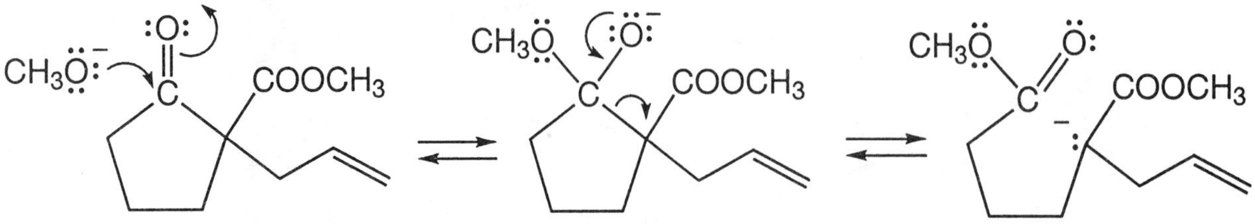

A new enolate is formed through proton transfers and a new Dieckmann condensation takes place to give compound **A**. Compound **A** does contain a "doubly α" hydrogen that can be removed to give **B**.

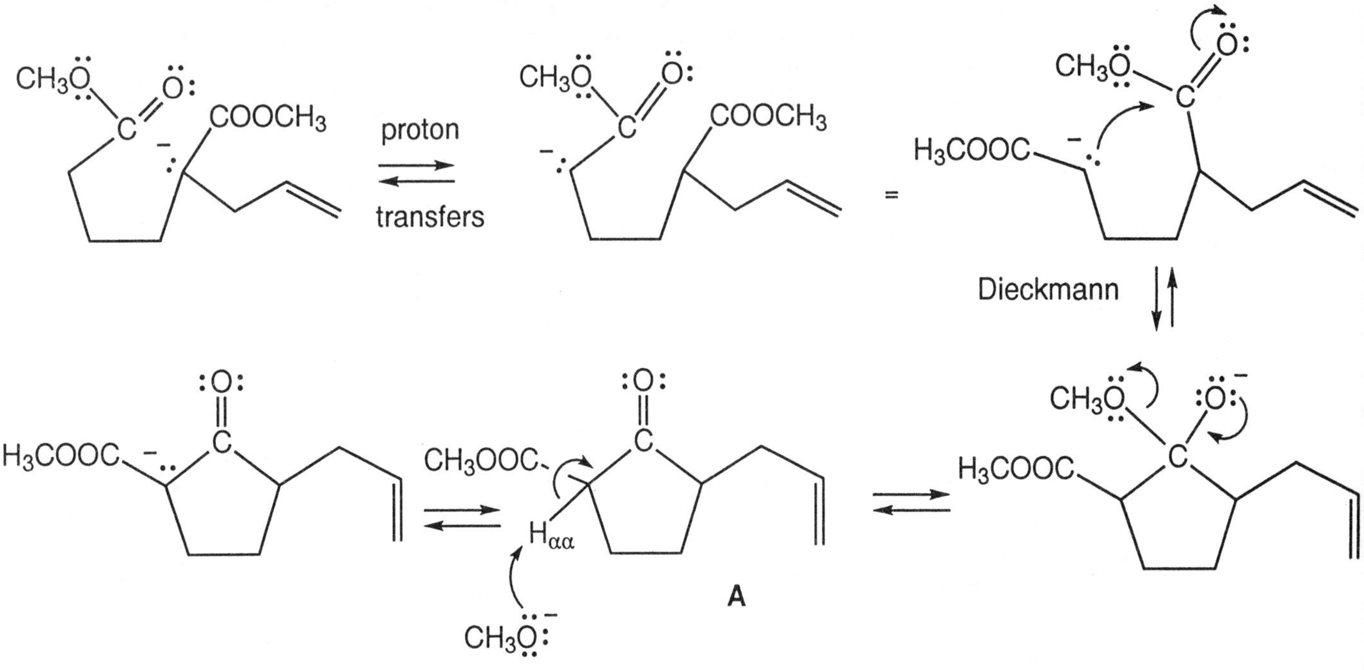

proton transfers

Dieckmann

A

Alkylation of this enolate gives the final product.

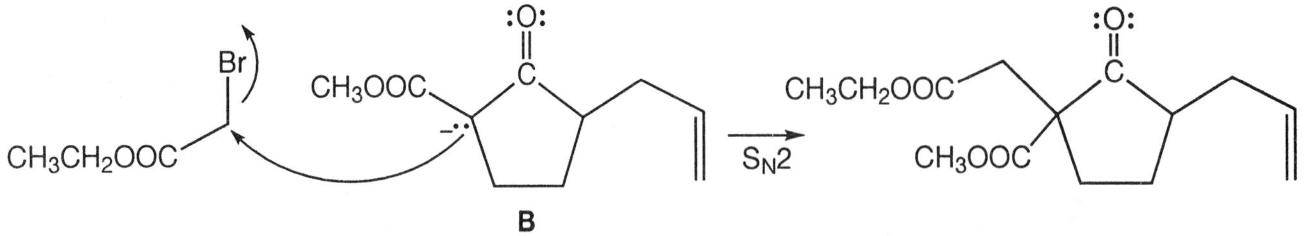

S_N2

B

We could probably have done this problem backwards. It is quite clear that the final product comes from an alkylation of enolate **B**. How can **B** arise? Only though the Dieckmann condensation shown above. Now the problem is to find a way to make the critical enolate **B**. Somehow the ring must open. That's a critical insight. Now you are directed to find a way to open the ring, and that makes starting the problem much easier.

Problem 20.48. Do some analysis first. A new ring must be made. It is likely that the oxygen atom in the ring is the oxygen of the phenol starting material. The other atoms in the ring probably come from the starting materials as shown.

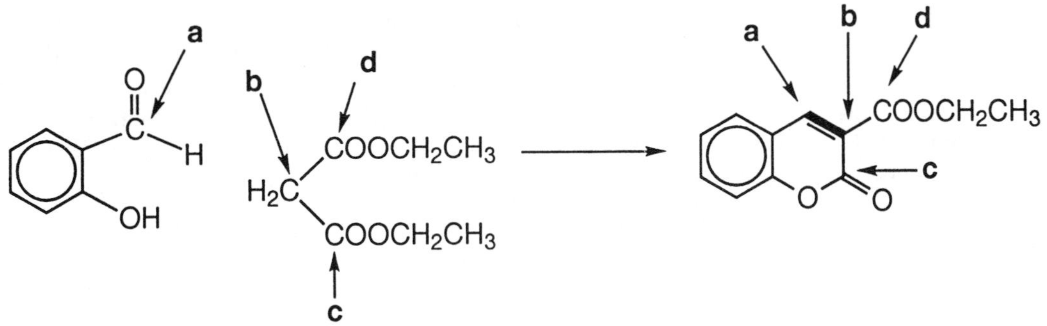

The new bonds to be made are between the oxygen and carbon **c**, and carbons **a** and **b**. We are now directed to find ways to make these bonds. That's a great advantage. Now we can start this problem with confidence.

The most acidic carbon-bound hydrogens are the "doubly α" hydrogens of malonic ester. The base piperidine removes one of these hydrogens, and the resulting enolate adds to the aldehyde to give **A**. One of the new bonds, carbon **a** to carbon **b**, has been made. The alkoxide is protonated and water eliminated to give **B**. At this point we must think about closing the new ring, incorporating the oxygen atom. The quite acidic phenolic hydrogen can be removed to give an alkoxide poised perfectly to do an intramolecular transesterification of one of the ethyl esters and make the oxygen-carbon **c** bond. This closes the ring, and leads to the product.

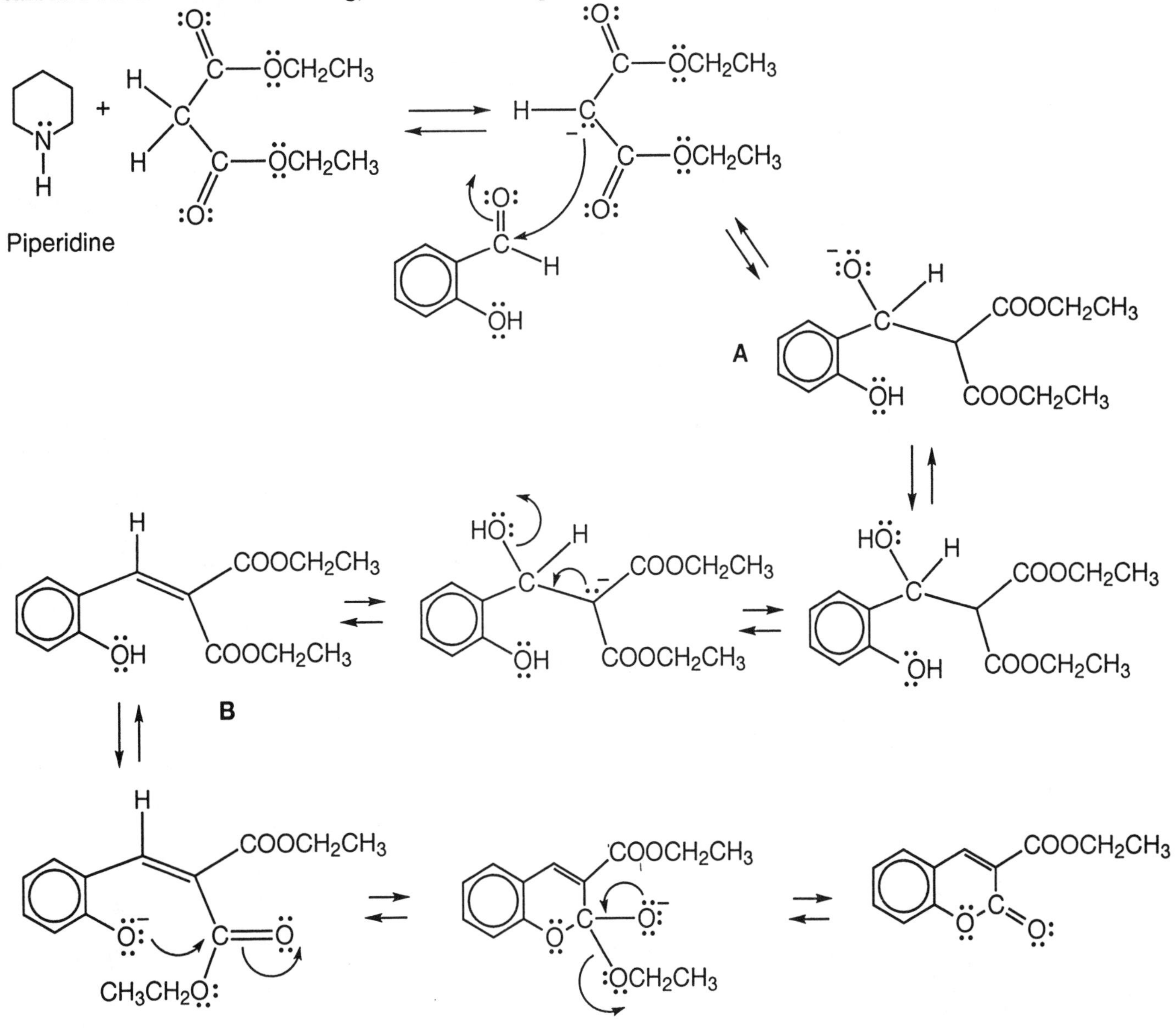

Piperidine

A

B

Problems 20.49, 20.50. The difficulty with the mechanism shown in Problem 20.50 (and in more than one place in the chemical literature) is the increase in strain incurred as the four-membered ring is closed to give the bicyclo[2.1.0]pentane system. One's intuitive feeling is that there must be a better way! The anion shown, **A**, is the most stable enolate possible in this system, so this does seem a good way to start. The hint tells you to think about reducing strain. Opening the three-membered ring surely relieves strain and in this case generates a new enolate, **B**.

COOCH3 COOCH3 COOCH3 OCH3 C O: Reverse Michael COOCH3 OCH3 C O: A B

Here comes the key step. The problem is to find a way to close a five–membered ring so as to create the product we know is formed. This cannot be done from **B** directly. So, either there is a way for **B** to rearrange, or **B** is a dead end. Notice that a proton shift, or more likely a protonation-deprotonation sequence, generates **C**, a resonance–stabilized enolate even more stable than **B**. This anion can close to product, although a final double bond isomerization is also necessary (**D** to **E**).

H3COOC OCH3 C O CH H H B H3COOC H3CO C O CH2 H C H3COOC H3CO C O CH2 H

:O: H3COOC C E H3COOC :O: C H—OCH3 CH3O: H :O: C H3COOC D CH3O :O: C H3COOC

So, how is one to know? That's a tough question. For some time at Princeton we have offered 10 hour test points for a workable way to distinguish the two mechanisms. Although there have been some good suggestions, no one has ever won the full ten points. One of the main difficulties is that isotopic labels seem ineffective in this case, and other, more intrusive labels such as methyl groups have the potential of so changing the molecule that the mechanism changes. You try.

Problem 20.51. The best way to start this problem is to deduce the molecular formula of compound **2**. The ^{13}C NMR spectrum suggests a minimum of nine carbons, and the ^{1}H NMR spectrum indicates a minimum of six hydrogens. C_9H_6 has a mass of 114 g/mol. The mass spectrum of compound **2** shows a parent ion of m/z = 146, so we are short a mass of 32 g/mol. This deficit can be most easily accommodated by two oxygen atoms, and this produces a molecular formula of $C_9H_6O_2$, corresponding to seven degrees of unsaturation. Given the structure of the starting material, salicylaldehyde, it is not unreasonable to attribute four of the seven degrees of unsaturation to the benzene ring.

The IR spectrum of compound **2** displays a carbonyl stretch at 1704 cm^{-1}. The possibility of an aldehyde or carboxylic acid is unlikely because of the absence of the corroborating C—H or O—H stretches. In addition, aldehydes and carboxylic acids can be eliminated by the absence of the expected low–field signal in the ^{1}H NMR spectrum of **2**. This leaves the possibility of either an ester or a ketone. The ^{13}C NMR spectrum serves to make this distinction. The carbonyl carbon of ketones (and aldehydes) has a chemical shift in the 190–220 ppm range, whereas the carbonyl carbon of esters appears in the 160–190 ppm range. In this case, the absence of a singlet in the 190–220 range makes the presence of a ketone most unlikely. Moreover, there is a signal at 160.6 ppm, quite consistent with the presence of an ester. Finally, the frequency of the carbonyl stretch in the infrared spectrum of **2** is appropriate for a *conjugated* ester (1704 cm^{-1}). The ester group accounts for the fifth degree of unsaturation.

The ^{1}H NMR spectrum of **2** shows two 2H multiplets at δ 7.2–7.6 ppm, consistent with a disubstituted benzene (possibly ortho disubstituted, considering the structure of the starting salicylaldehyde). Further, there are two 1H doublets at δ 6.42 and 7.72 ppm, coupled to each other. This coupling constant, 9 Hz, suggests a cis alkene. The alkene double bond accounts for the sixth degree of unsaturation.

All the atoms are now accounted for ($C_6H_4 + CO_2 + C_2H_2 = C_9H_6O_2$), but we have not yet identified the last degree of unsaturation. This must be the result of a ring. All the evidence suggests that **2** is the compound known as coumarin:

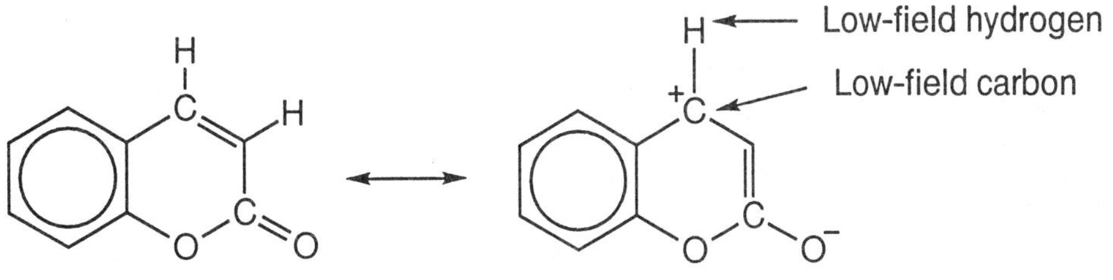

Coumarin

Notice that the low–field position of one of the olefinic hydrogens (δ 7.72 ppm) and one olefinic carbon (δ 143.4 ppm) are nicely accommodated by the coumarin structure, as shown by the resonance form shown above in which the β-carbon bears a positive charge.

Now, how is this molecule formed? In this system, the most acidic hydrogen is the phenolic proton. The first step is O–acylation to give acetylsalicylaldehyde, **3**.

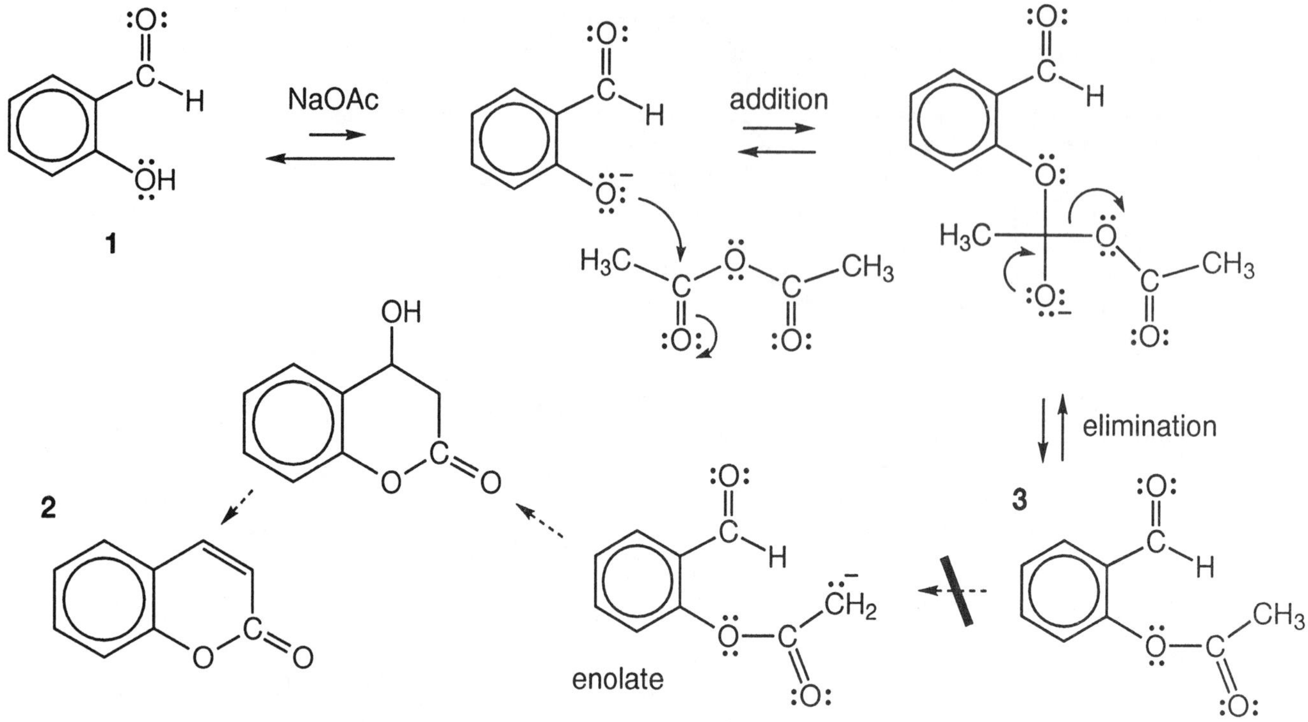

NaOAc

addition

H_3C CH_3

elimination

1 **2** **3**

enolate

The most economical route to **2** would seem to be formation of the enolate, intramolecular addition to the aldehyde group and dehydration, as outlined with dotted arrows in the figure. However, this cannot be correct, as the problem tells us that **3** does not go efficiently to **2** in the absence of acetic anhydride. Acetic anhydride must somehow be involved in the **3** to **2** conversion. If you get stuck at this point, don't worry. It has been proposed that a Perkin reaction now takes place (Problem 20.44) to give **4**. Addition–eliminations lead to **5** (use the indicated oxygen of **4**), and an elimination of acetate gives **2**.

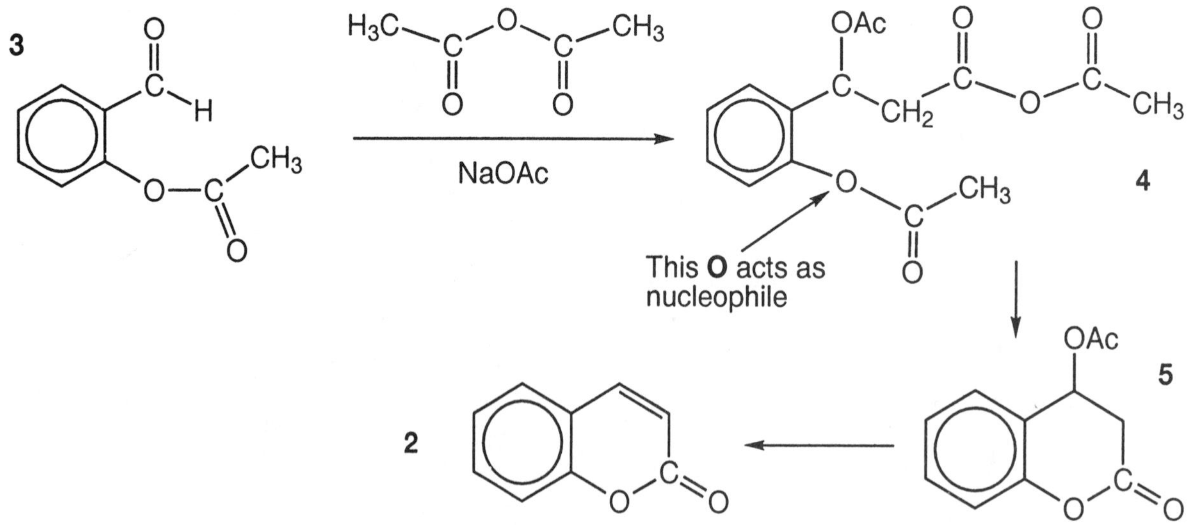

NaOAc

This **O** acts as nucleophile

3 **4** **5** **2**

Problem 20.52. Treatment of diethyl malonate with base gives the enolate. In Section 17.11, we saw that anions can open epoxides in S_N2 reactions. Here, there are two possible modes of opening. Although it was originally thought that only the less hindered route was followed, it is now clear that both openings occur to give a pair of alkoxides. The molecular formulae of **A** and **B** show that ethanol has been lost, and so it is clear that intramolecular transesterifications have occurred through the usual addition–elimination reactions.

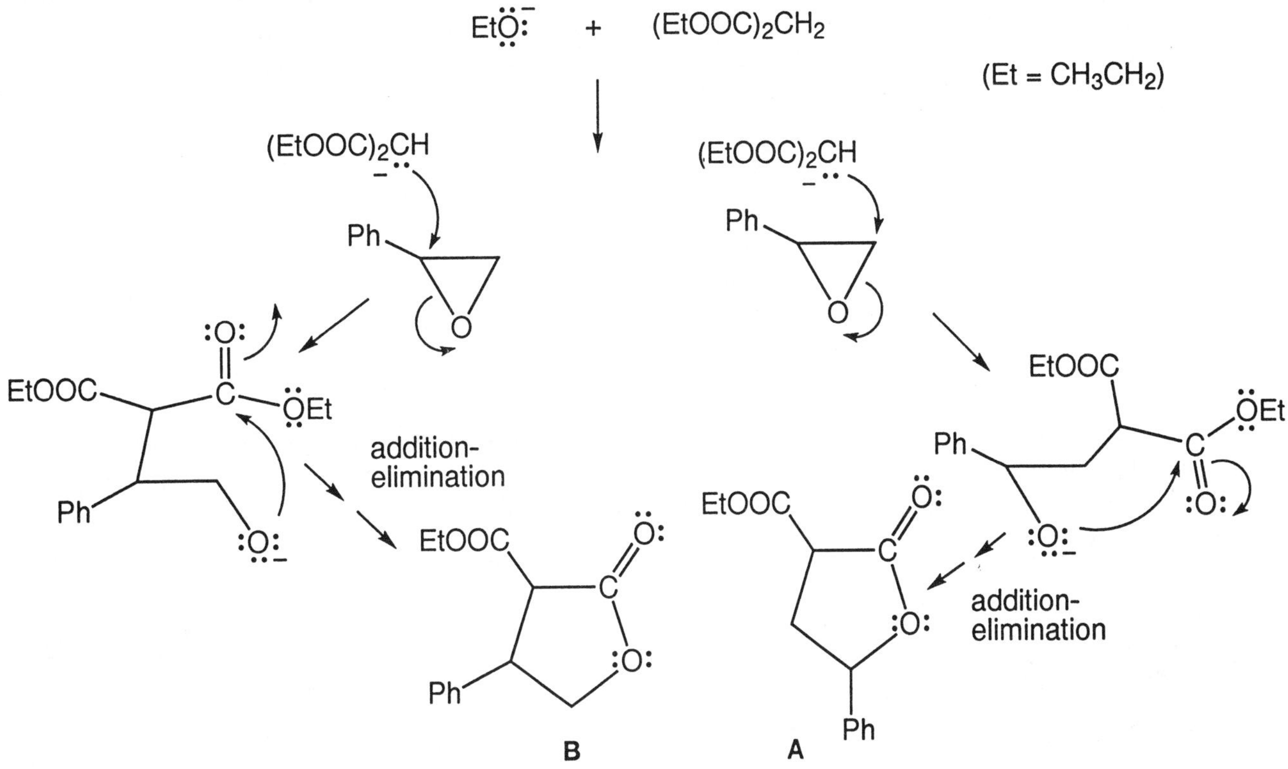

EtÖ:⁻ + $(EtOOC)_2CH_2$

($Et = CH_3CH_2$)

$(EtOOC)_2CH^-$ Ph   O   EtOOC   :O: C   ÖEt   Ph   :O:⁻   addition-elimination   EtOOC   Ph   **B**   **A**   Ph

Saponification (treatment with base, followed by acidification) transforms the esters into acids by straightforward ester hydrolyses. All acids β to carbonyl groups are prone to decarboxylation (acetoacetic acids and malonic acids are examples in this chapter), and **C** and **D** lose CO_2 on heating to give lactones **E** and **F**.

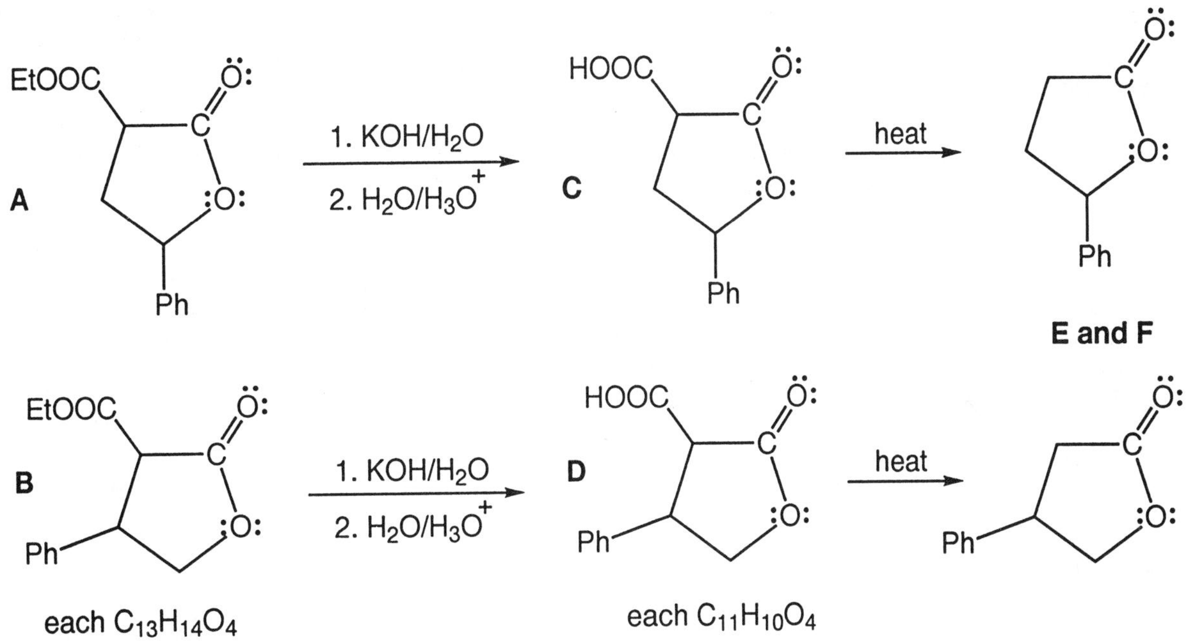

A   EtOOC   Ph   1. KOH/H_2O 2. H_2O/H_3O^+ **C**   HOOC   Ph   heat   Ph

E and F

B   EtOOC   Ph   1. KOH/H_2O 2. H_2O/H_3O^+ **D**   HOOC   Ph   heat   Ph

each $C_{13}H_{14}O_4$ each $C_{11}H_{10}O_4$

Now we are left to decide which lactone is **E** and which is **F**. Both **E** and **F** show a parent ion in the mass spectrum with m/z = 162, consistent with the proposed structures. In addition, the IR spectra of both **E** and **F** exhibit an intense band at about 1780 cm^{-1}, consonant with a carbonyl stretch of a γ-lactone. The crucial spectral distinctions occur in the ^{1}H and ^{13}C NMR spectra. In particular, **E**, γ-phenyl-γ-butyrolactone, shows a single low-field hydrogen (H_x) at δ 5.45-5.55 ppm in the ^{1}H NMR spectrum. This hydrogen is deshielded by both the adjacent phenyl ring and the adjacent oxygen atom. There is no corresponding hydrogen in **F**. The carbon to which H_x is attached in **E** appears as a doublet at δ 81.2 ppm in the ^{13}C NMR spectrum. In **F**, β-phenyl-γ-butyrolactone, two hydrogens (H_x and $H_{x'}$) on the carbon adjacent to the other oxygen appear at δ 4.28 and 4.67 ppm, and the carbon to which they are attached is a triplet at δ 73.8 ppm.

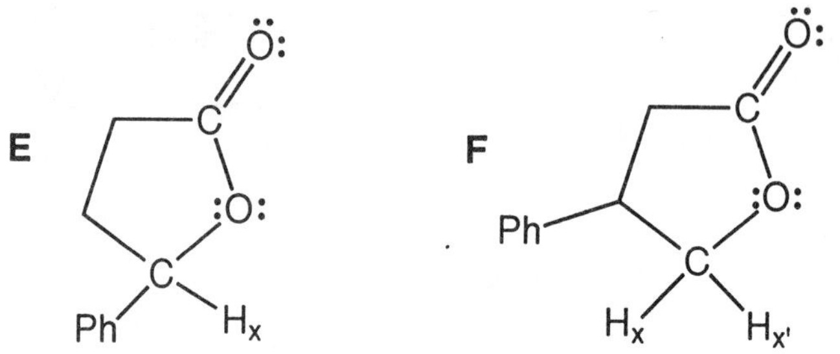

E Ph, H_x **F** Ph, H_x, $H_{x'}$

Problem 20.53. Reaction of ethyl butyrate and DIBAL–H initially affords addition product **A**. Hydrolysis at -70 °C ultimately gives butyraldehyde. The success of this aldehyde synthesis is undoubtedly the result of the stability of the addition product **A** at -70 °C.

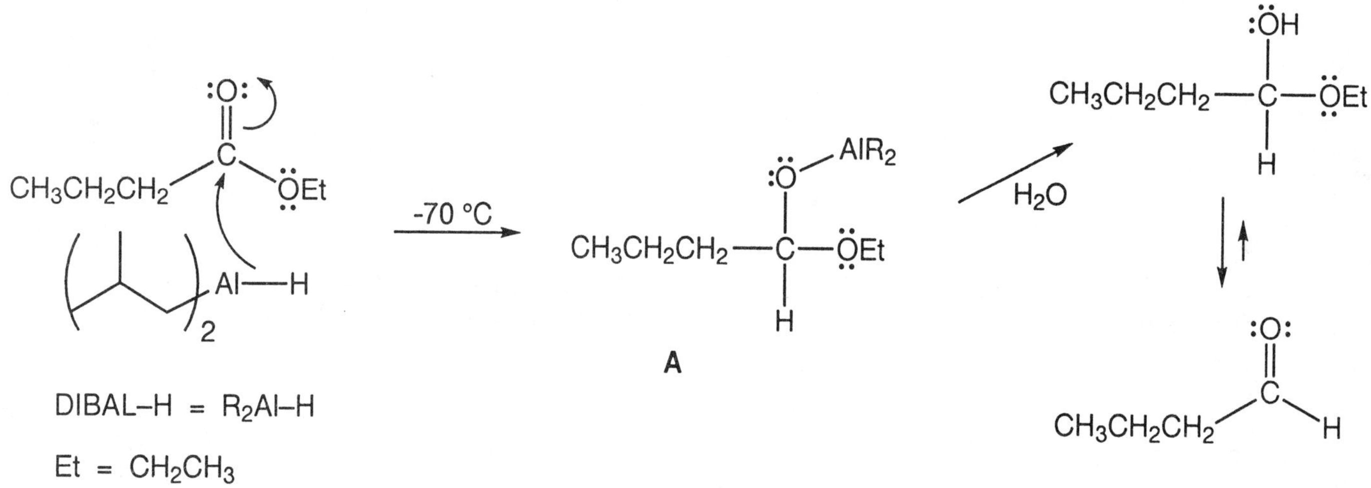

$CH_3CH_2CH_2C(=O)OEt$ + $R_2Al–H$ —(-70 °C)→ $CH_3CH_2CH_2–CH(OAlR_2)–OEt$ (**A**) —(H_2O)→ $CH_3CH_2CH_2–CH(OH)–OEt$ ⇌ $CH_3CH_2CH_2CHO$

DIBAL–H = $R_2Al–H$

Et = CH_2CH_3

However, when the temperature of the reaction mixture is allowed to rise, compound **A** decomposes to butyraldehyde and ethoxydiisobutylaluminum ($EtOAlR_2$).

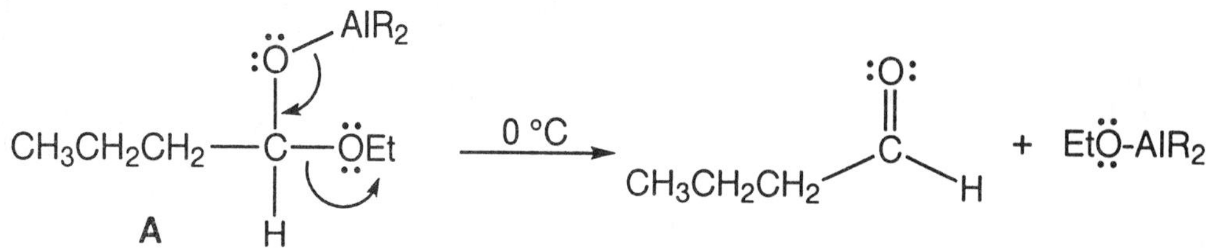

$CH_3CH_2CH_2–CH(OAlR_2)–OEt$ (**A**) —(0 °C)→ $CH_3CH_2CH_2CHO$ + $EtO–AlR_2$

The liberated butyraldehyde can now react with compound **A** in a fashion reminiscent of the Meerwein-Ponndorf-Verley-Oppenauer equilibration to give intermediate **B** that can decompose to ethyl butyrate (starting material) and butoxydiisobutylaluminum (**C**), that is hydrolyzed to butyl alcohol.

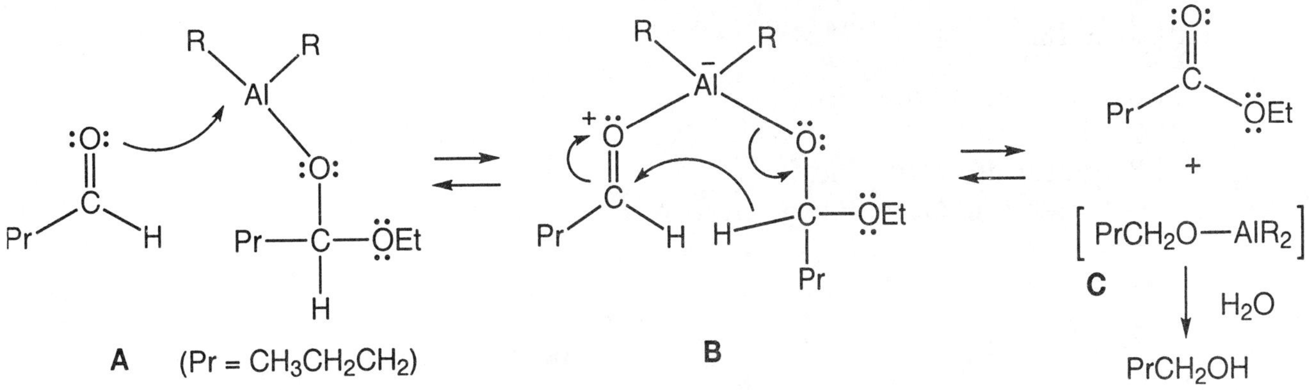

A (Pr = $CH_3CH_2CH_2$)

B

$[PrCH_2O{-}AlR_2]$ **C** $\xrightarrow{H_2O}$ $PrCH_2OH$

Chapter 21 Outline

21.1 Nomenclature
21.2 Structure and Physical Properties of Amines
21.3 Spectroscopic Properties of Amines
21.4 Acid and Base Properties of Amines
21.5 Reactions of Amines
21.6 Syntheses of Amines
21.7 Alkaloids
21.8 Synthesis of Quinine
21.9 Something More: Phase-Transfer Catalysis

21 Introduction to the Chemistry of Nitrogen-Containing Compounds: Amines

Many of the themes of alcohol chemistry are recapitulated in this chapter on amines. Both amines and alcohols are simultaneously Lewis acids and Lewis bases. Displacement reactions and addition reactions dominate the reactivity scene. Differences and complications do arise, largely because of the fact that amines have one more hydrogen attached to the heteroatom than alcohols. The following questions pose structural, synthetic, and mechanistic problems. The problem-solving techniques you already know should lead you through them.

Problem 21.2 Easy. Only perfect tetrahedra will have the exact tetrahedral angle of 109.5°. In amines, the four substituents surrounding the central nitrogen atom are not all the same, and all the angles cannot be the same.

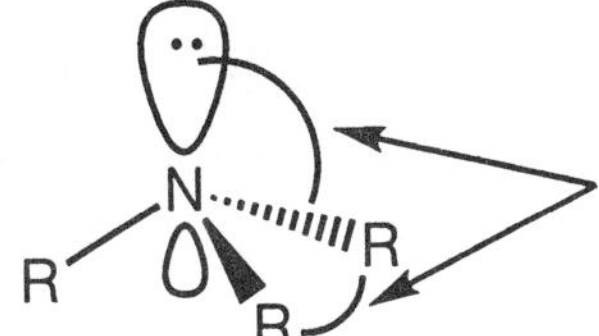

These two angles cannot be exactly the same.

Problem 21.3 This question is so simple as to be hard. If the hybridization is sp^2 at the transition state, and sp^3 at the starting material, it must be between sp^2 and sp^3 at some point between the starting material and transition state.

Problem 21.4 This problem is quite tough. In aziridines, there is more angle strain in the transition state than there is for noncyclic amines. In an amine, the most favorable angle is approximately the tetrahedral angle, about 109°. In the sp^2-hybridized transition state, this angle is expanded from about 109° to 120°. One might say that there is approximately 11° of angle strain. In an aziridine, in which one angle is restricted to 60° through incorporation into the three-membered ring, the transition state is very highly strained. Now there is 60° of angle strain.

Accordingly, it is much more difficult to reach the transition state for inversion in aziridines than it is in acyclic amines.

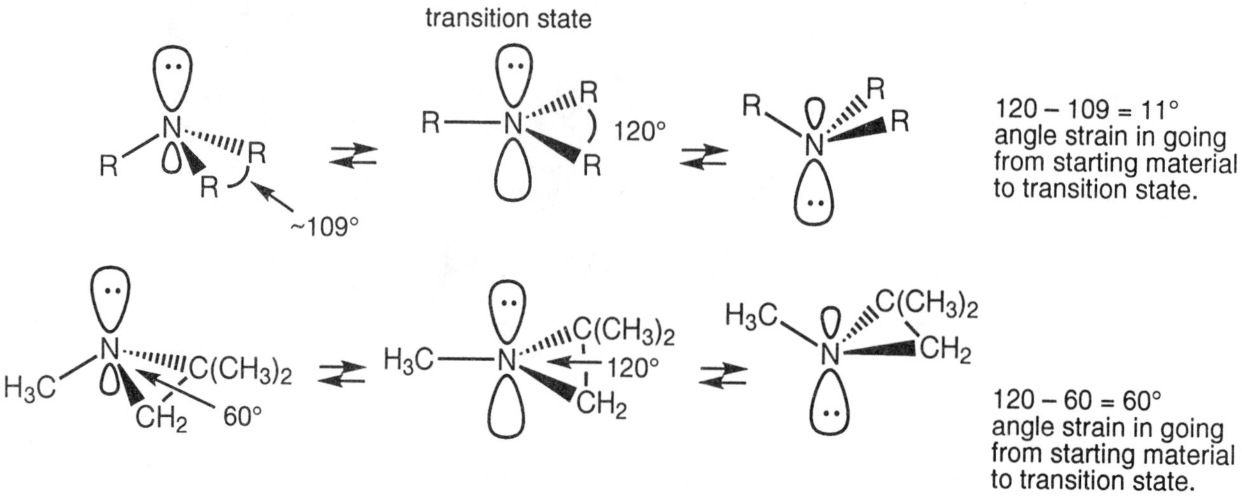

transition state

~109°

120°

120 – 109 = 11° angle strain in going from starting material to transition state.

H_3C $C(CH_3)_2$ CH_2 60°

120°

120 – 60 = 60° angle strain in going from starting material to transition state.

Problem 21.6 Citric acid (lemon) acts as a Brønsted acid toward the Brønsted base, the amine (fish odor), in a proton-transfer reaction to form a salt. Unlike the amine, the salt is not volatile and cannot easily be detected by our sense of smell. The fish has been deodorized by the acid–base reaction.

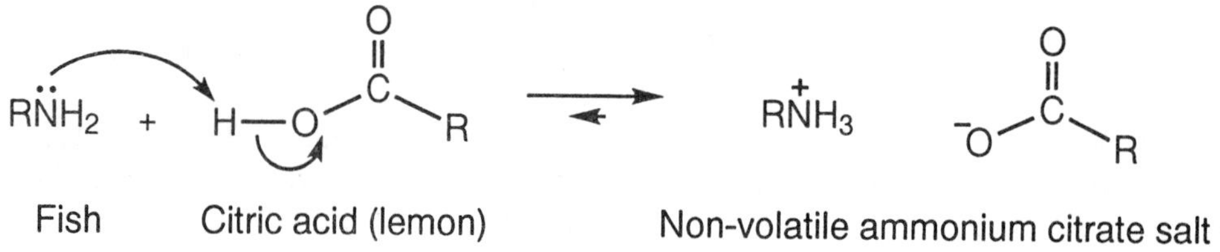

$R\ddot{N}H_2 + H{-}O{-}C(=O){-}R \rightleftharpoons R\overset{+}{N}H_3 \quad {}^{-}O{-}C(=O){-}R$

Fish | Citric acid (lemon) | Non-volatile ammonium citrate salt

Problem 21.8 The anilinium ion ($pK_a = 4.6$) is a much stronger acid than the ethyl ammonium ion ($pK_a = 10.8$) because the conjugate base of the anilinium ion, aniline, is stabilized by resonance whereas the conjugate base of the ethyl ammonium ion, ethylamine, is not. Moreover, the electron-withdrawing benzene ring destabilizes adjacent electron deficiency.

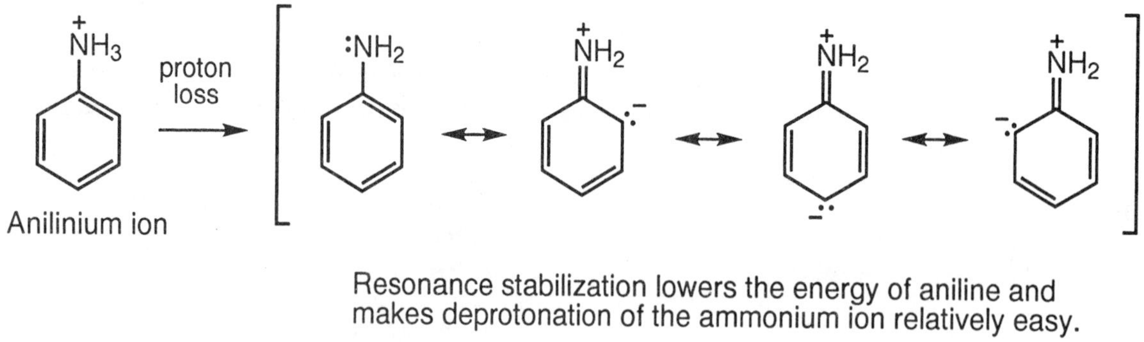

Anilinium ion —proton loss→ [resonance structures of aniline]

Resonance stabilization lowers the energy of aniline and makes deprotonation of the ammonium ion relatively easy.

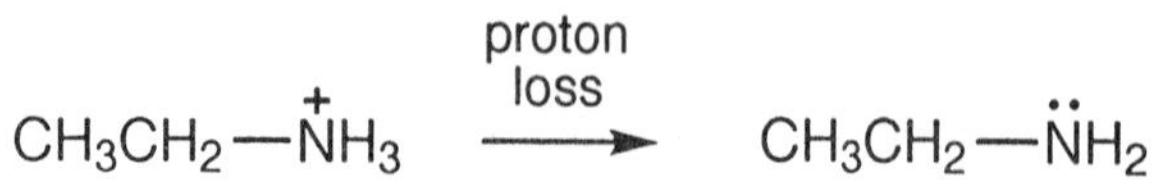

$CH_3CH_2{-}\overset{+}{N}H_3 \xrightarrow{\text{proton loss}} CH_3CH_2{-}\ddot{N}H_2$

In this case, neither reactant nor product is stabilized by resonance.

Problem 21.9 Oxonium ions are destroyed by nucleophiles through an S_N2 alkylation process.

$$R_3O^+ + {}^-Nu \xrightarrow{S_N2} Nu{-}R + R_2O$$

Most negatively charged counterions are nucleophiles, but $^-BF_4$ is an exception. There is no lone pair of electrons on boron and therefore no possible nucleophilicity.

Problem 21.10 No S_N2 displacement is possible with *tert*-butyl bromide, and the conditions of very strong base ($^-NH_2$) do not seem to favor an S_N1 ionization. About the only reaction possible is the bimolecular elimination process (E2), leading, in this case, to isobutene.

$$(CH_3)_3CBr + {}^-NH_2 \xrightarrow{E2} H_3N: + :Br^- + (H_3C)_2C{=}CH_2$$

Problem 21.11 This problem is a review question for general addition–elimination reactions of carbonyl compounds. As discussed in detail in Chapter 16 (p. 780), primary amines lead to imines whereas secondary amines, which cannot give imines, lead to enamines.

In these two reactions, the first steps are the same, addition of the primary or secondary amine and deprotonation to give the carbinolamine.

Carbinolamines

Now, however, the two reaction mechanisms diverge. In the primary amine case, protonation of the carbinolamine oxygen leads to an intermediate from which water can be lost to give a resonance-stabilized carbocation. Now removal of the proton attached to nitrogen leads to the product imine.

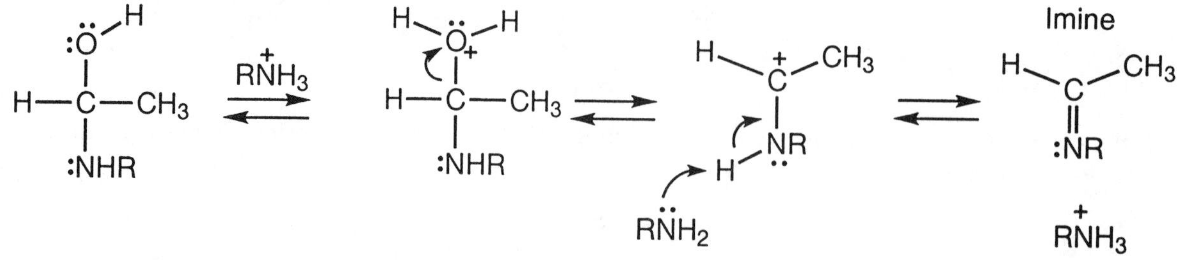

In the secondary amine case, there is no proton attached to nitrogen, and an imine cannot be formed. However, the positive charge can still be removed if there is a proton attached to carbon. Loss of a carbon-attached proton gives the enamine.

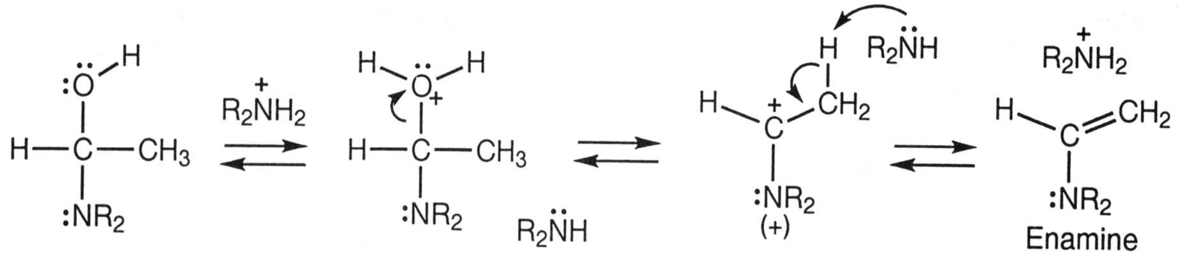

Problem 21.12. These are both Hofmann elimination reactions. In (a), the amine nitrogen is methylated twice, forming the ammonium ion. Elimination gives the product.

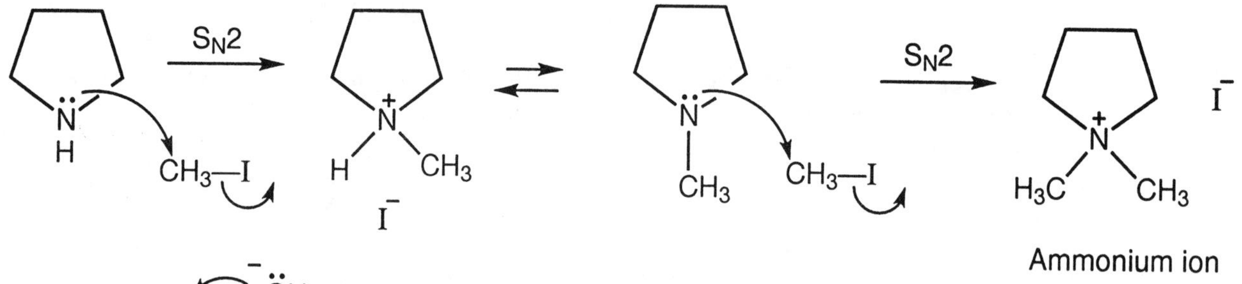

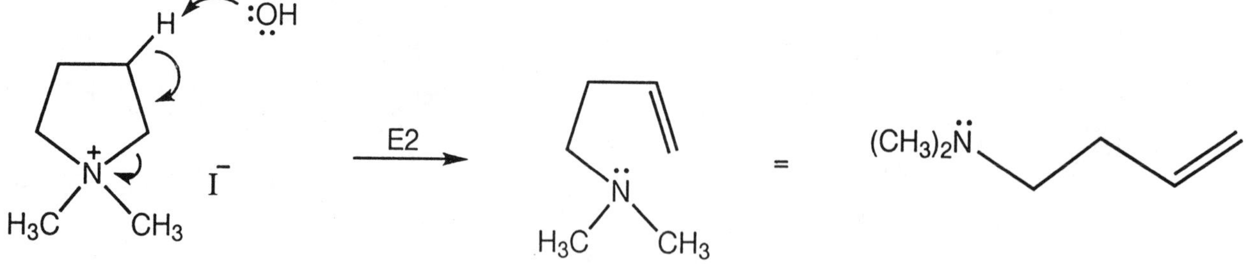

In (b), the Hofmann elimination is carried out twice. The first stages repeat the steps of (a) exactly to give an acyclic amino alkene. Methylation and elimination give the diene.

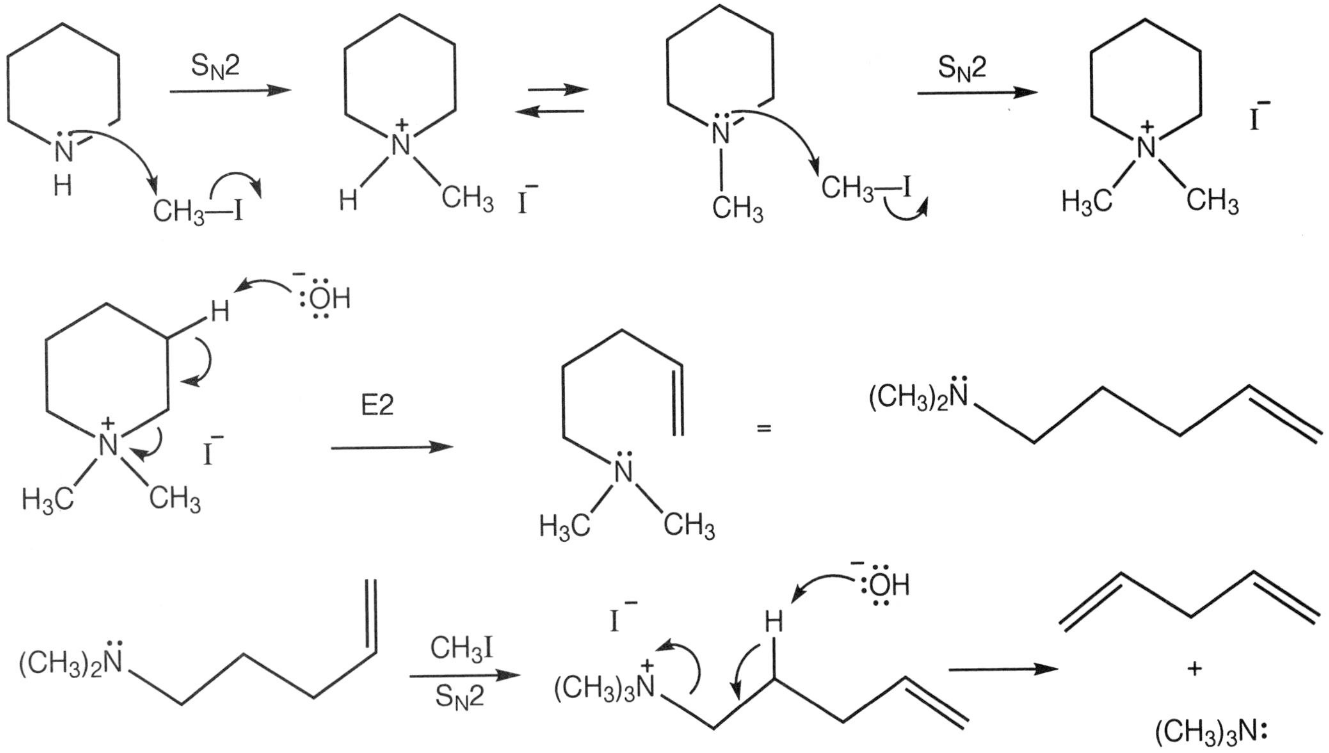

S_N2 CH_3—I S_N2 CH_3—I I^- E2 $(CH_3)_2\ddot{N}$ CH_3I S_N2 $(CH_3)_3\overset{+}{N}$ $^-:\ddot{O}H$ $(CH_3)_3N:$

Problem 21.13 More review here. Substitution of the ammonium ion in the ortho or para position results in the close opposition of two positive charges. If substitution is in the meta position, no such 1,2-destabilizing opposition of two like charges results (boxed structures). The transition state for meta substitution, in which positive charge is building up, will be more stable than that for either ortho or para substitution in which the developing positive charge is relatively close to the positive charge on nitrogen already present.

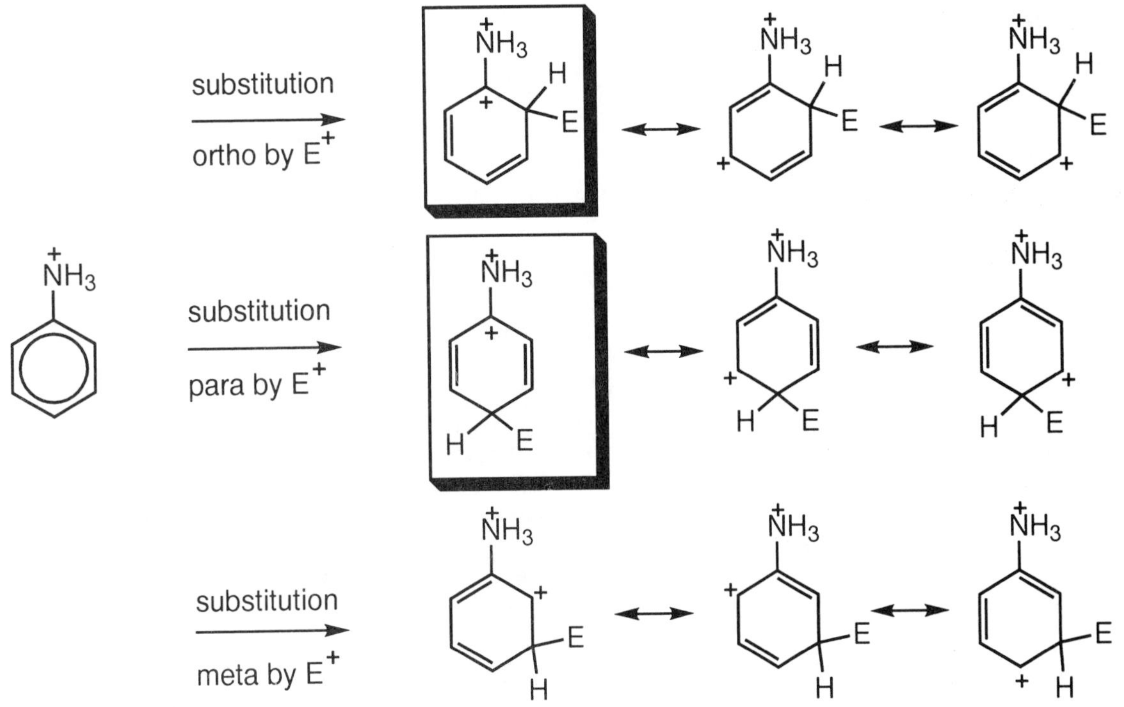

substitution ortho by E^+

substitution para by E^+

substitution meta by E^+

Any substitution reaction that introduces two like charges into the same molecule (any of the reactions shown above) will be slow compared to a reaction that introduces only a single positive charge such as the substitution of benzene.

substitution

of benzene by E^+

H E H E H E

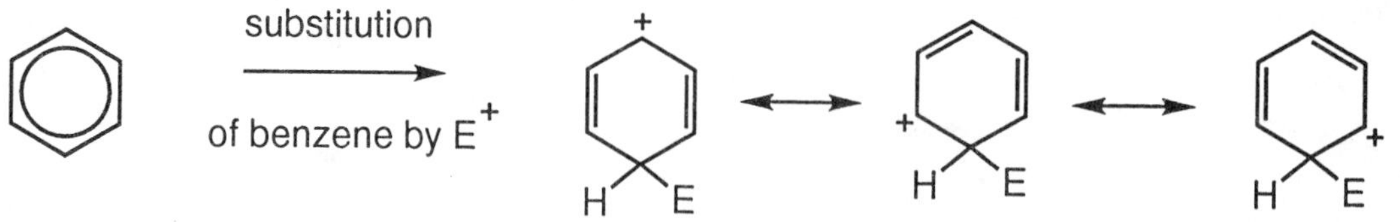

Problem 21.14 The compound N_2O_3 is essentially an anhydride of nitrous acid (HON=O) and could be formed in a standard way by protonation of one nitrous acid, and addition of a second acid molecule. Deprotonation and protonation steps are then followed by loss of water.

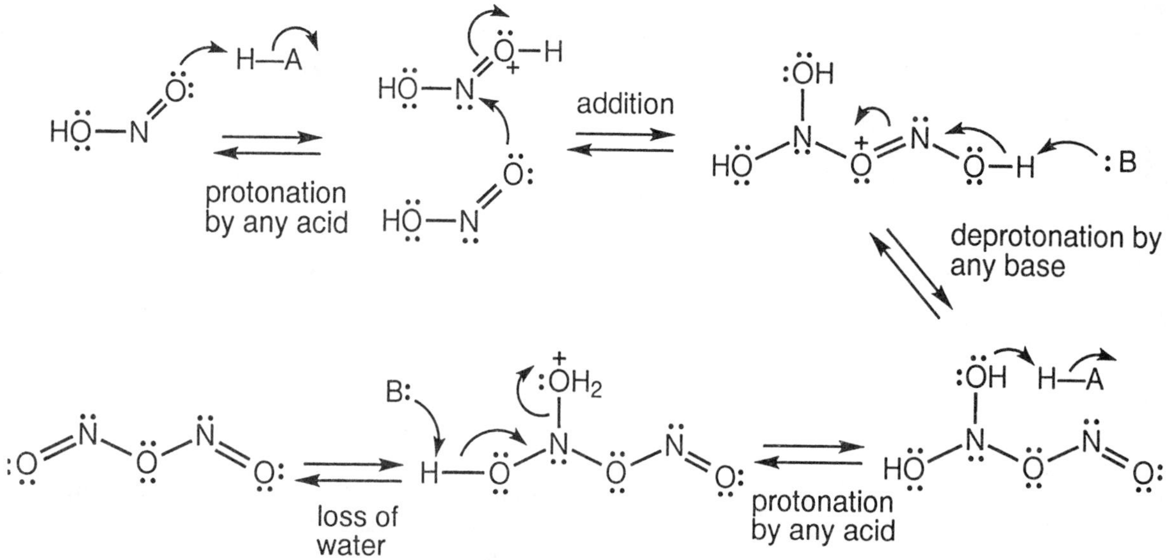

protonation by any acid

addition

deprotonation by any base

loss of water

protonation by any acid

The new anhydride, N_2O_3, is protonated, and a series of addition–elimination reactions similar to the one shown above does the nitrosation.

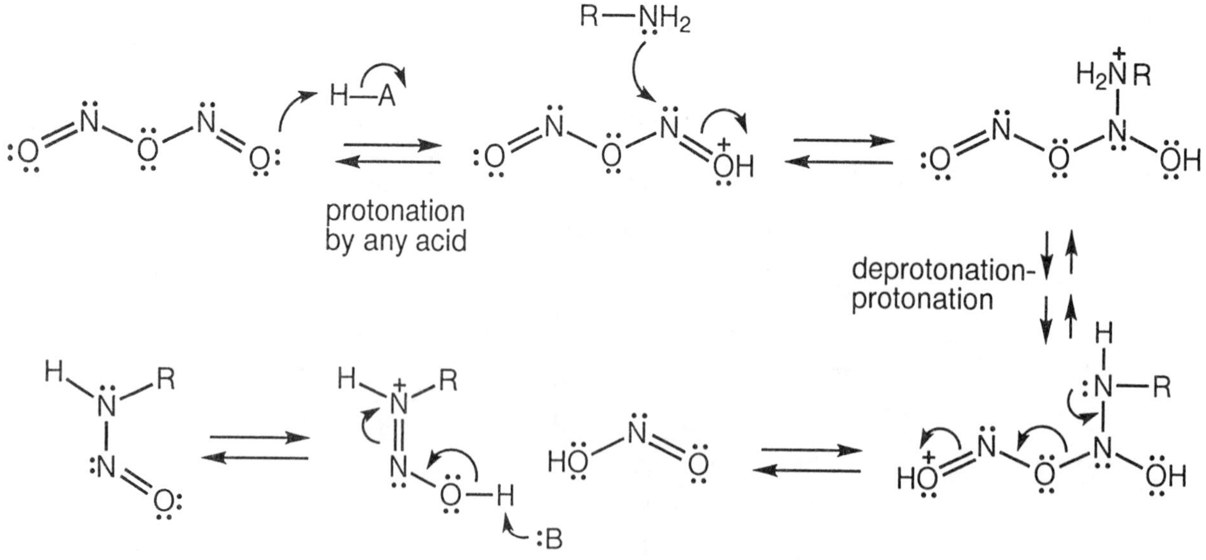

protonation by any acid

deprotonation-protonation

Problem 21.15 Diazomethane is first protonated by acetic acid to give the acetate ion and the methyldiazonium ion. Displacement of the excellent leaving group N_2 by acetate leads to methyl acetate.

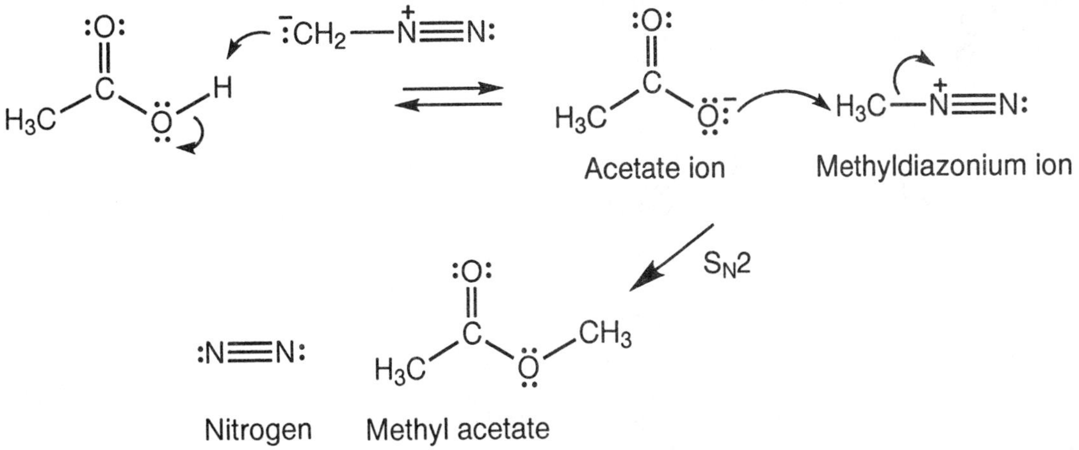

Acetate ion   Methyldiazonium ion

S_N2

Nitrogen   Methyl acetate

Problem 21.16 Still more electron and hydrogen pushing, this time in base. One important part of this tough problem is figuring out where the heavy atoms of diazomethane (C—N—N) come from. The methyl group and the attached two nitrogen atoms of NMU (shown in boldface type) seem likely candidates. The problem now is how to get rid of the "extra" atoms.

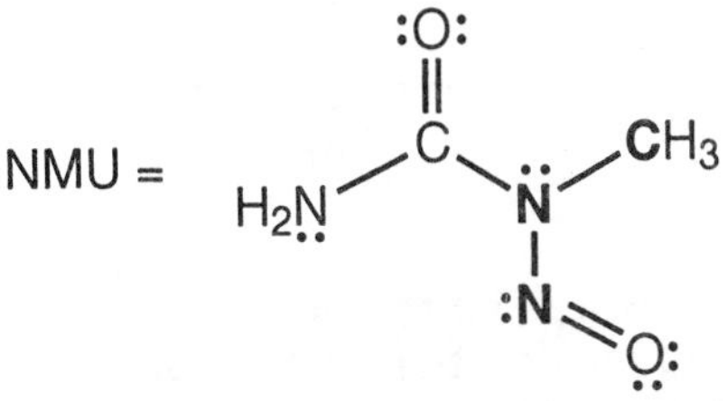

NMU =

Addition of hydroxide to the carbonyl group is followed by loss of CH_3NNO^-, a resonance-stabilized anion, and therefore a decent leaving group. Protonation, loss of hydroxide, and deprotonation complete the reaction.

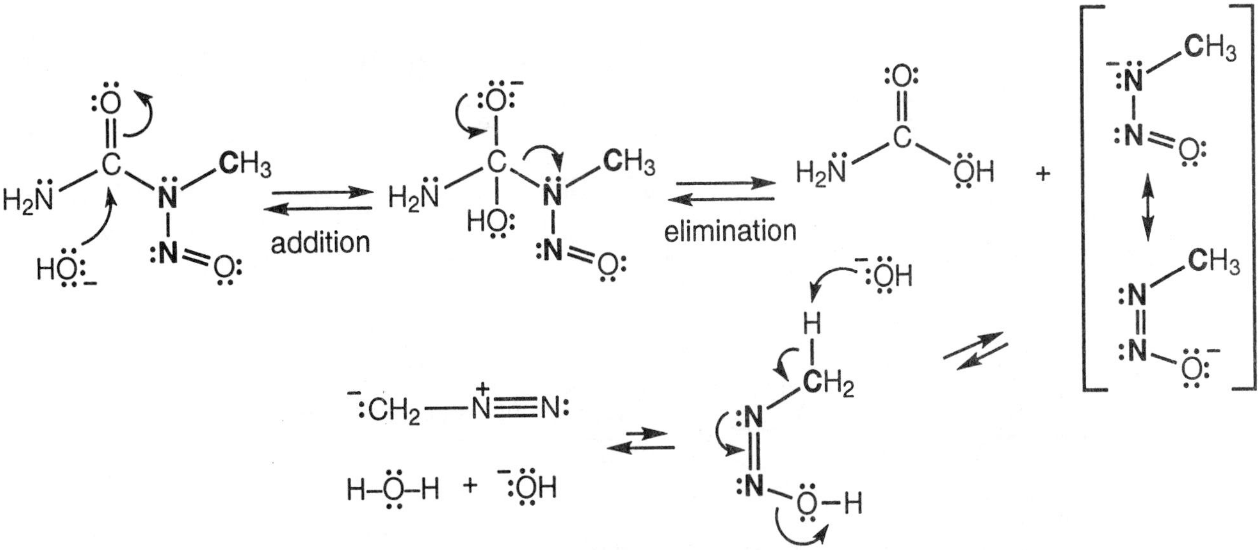

addition   elimination

Problem 21.17 It is not a carbon–hydroxy bond but an oxygen–hydroxy bond that breaks. The oxygen–oxygen bond is very weak (40 kcal/mol) and bond breaking is easier than usual.

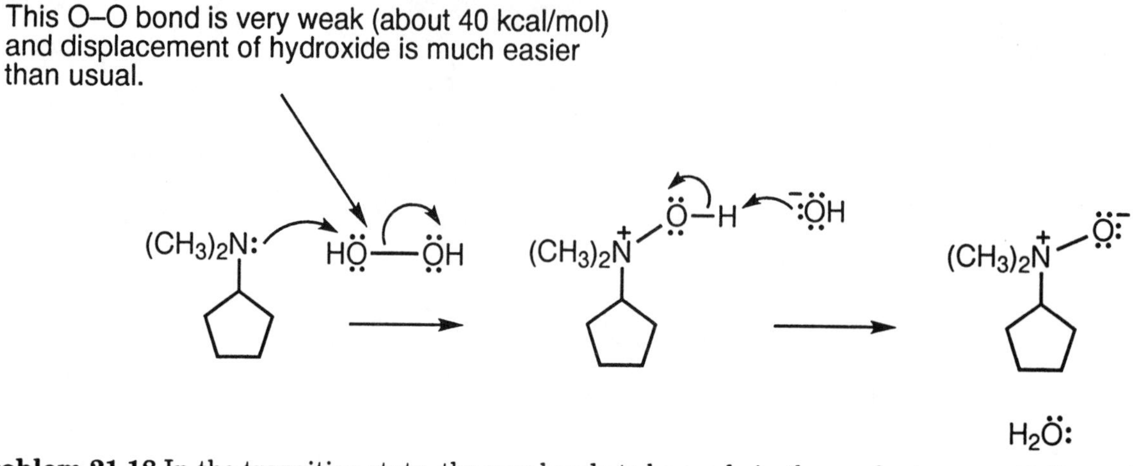

This O–O bond is very weak (about 40 kcal/mol) and displacement of hydroxide is much easier than usual.

$(CH_3)_2N$: HO–OH → $(CH_3)_2\overset{+}{N}$–O–H :OH$^-$ → $(CH_3)_2\overset{+}{N}$–O:$^-$

H_2O:

Problem 21.18 In the transition state, the new bonds to be made in the product are partially made and the bonds to be broken in the starting material are partially broken. The arrows in the arrow formalism point to the breaking and making bonds.

$(CH_3)_2\overset{+}{N}$–O:$^-$ H H H

Starting material

δ^+ $(CH_3)_2N$–O δ^- H H H

Transition state – the breaking and making bonds are dotted to show that they are partial bonds

H H

$(CH_3)_2N$–OH

Products

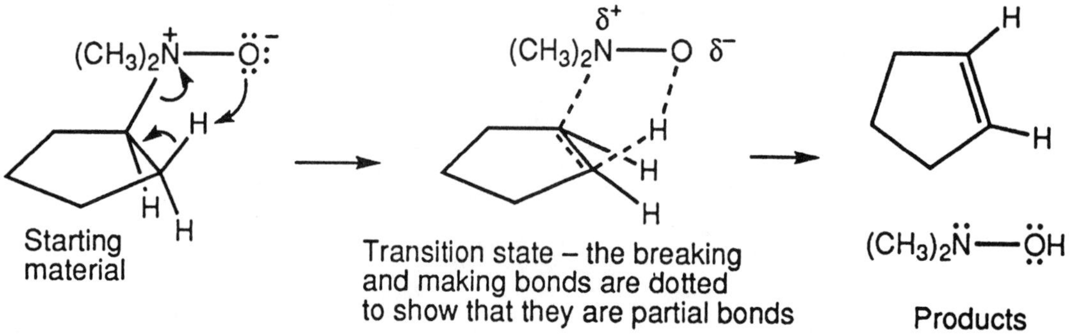

Problem 21.20(a) Epoxides react with nucleophiles to open the three-membered ring (Chapter 17, p. 853). Here ethylamine acts as nucleophile in the S_N2 reaction. Protonation gives the first product.

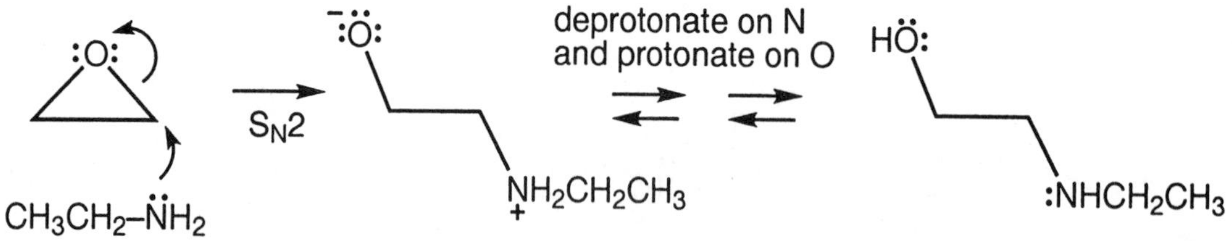

:O: CH_3CH_2–NH_2 $\xrightarrow{S_N2}$ $^-$:O: $\overset{+}{N}H_2CH_2CH_3$ deprotonate on N and protonate on O ⇄ ⇄ HO: :$NHCH_2CH_3$

The first-formed product is still a strong enough nucleophile to react with the epoxide, and so a second step takes place.

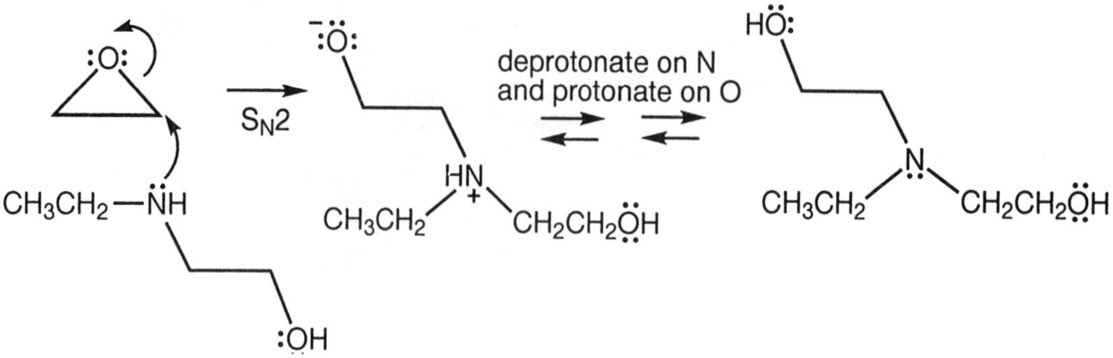

:O: CH_3CH_2–NH :OH $\xrightarrow{S_N2}$ $^-$:O: $\overset{+}{H}N$ CH_3CH_2 CH_2CH_2OH deprotonate on N and protonate on O ⇄ ⇄ HO: N CH_3CH_2 CH_2CH_2OH

Although this product is still a nucleophile, it is apparently too sterically hindered to react further with the epoxide.

(b) This is a simple Michael addition of the nucleophilic amine (Chapter 18, p. 918) to the β-position of the α,β-unsaturated double bond.

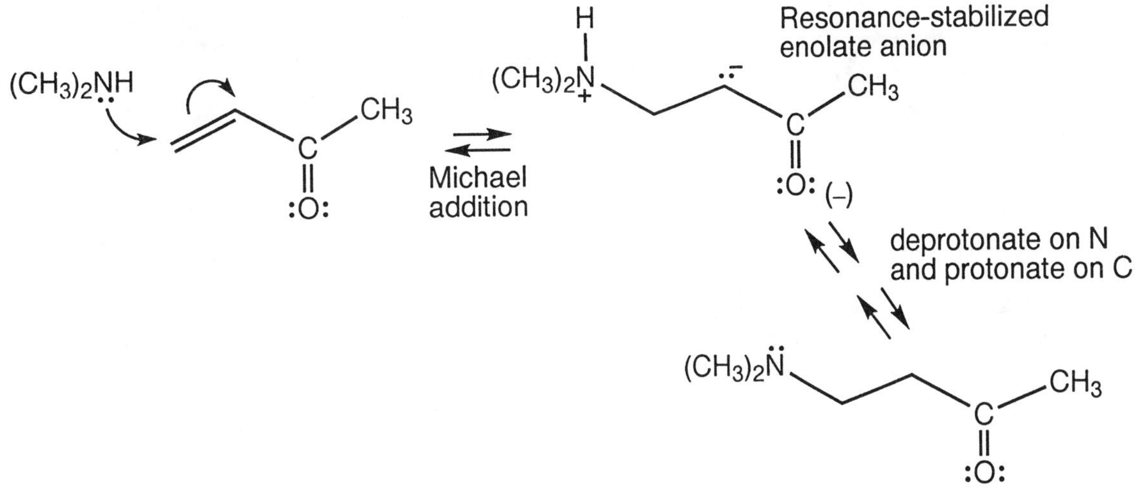

Resonance-stabilized enolate anion

Michael addition

deprotonate on N and protonate on C

Problem 21.22 This reaction is a Hofmann elimination (Chapter 7, p. 284; Chapter 21, p. 1093). In the first step, the amine is alkylated through an S_N2 reaction with methyl iodide to give the

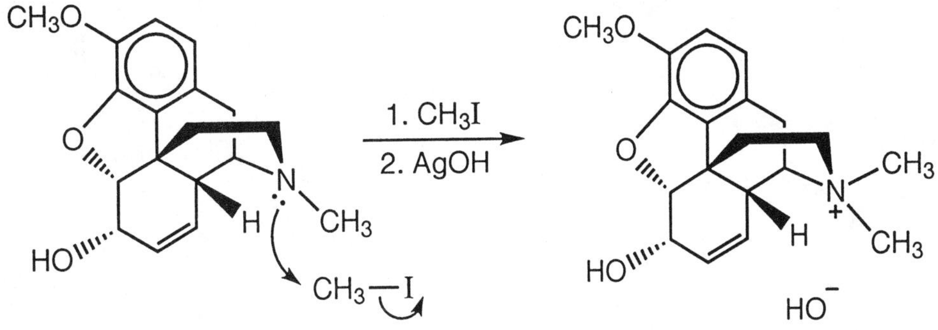

1. CH_3I
2. AgOH

ammonium ion.

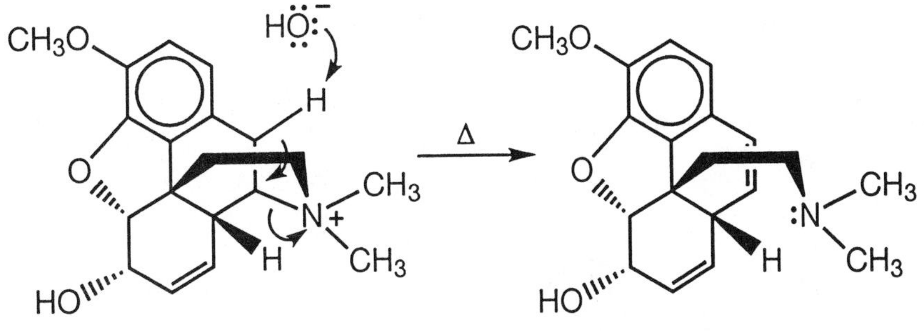

Δ

Problem 21.23 Super-easy, unfortunately. Alcohols are converted into acetates on treatment with acetyl chloride or acetic anhydride.

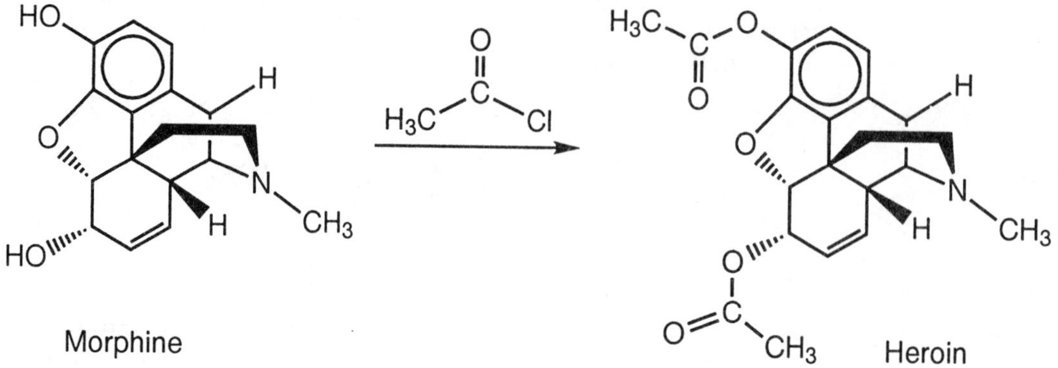

Morphine → (H_3C–CO–Cl) → Heroin

Problem 21.24 Two easy base-induced ester hydrolyses accomplish the conversion into ecognine (Chapter 20, p. 1022). The two reactions may occur in either order, and the one chosen is arbitrary.

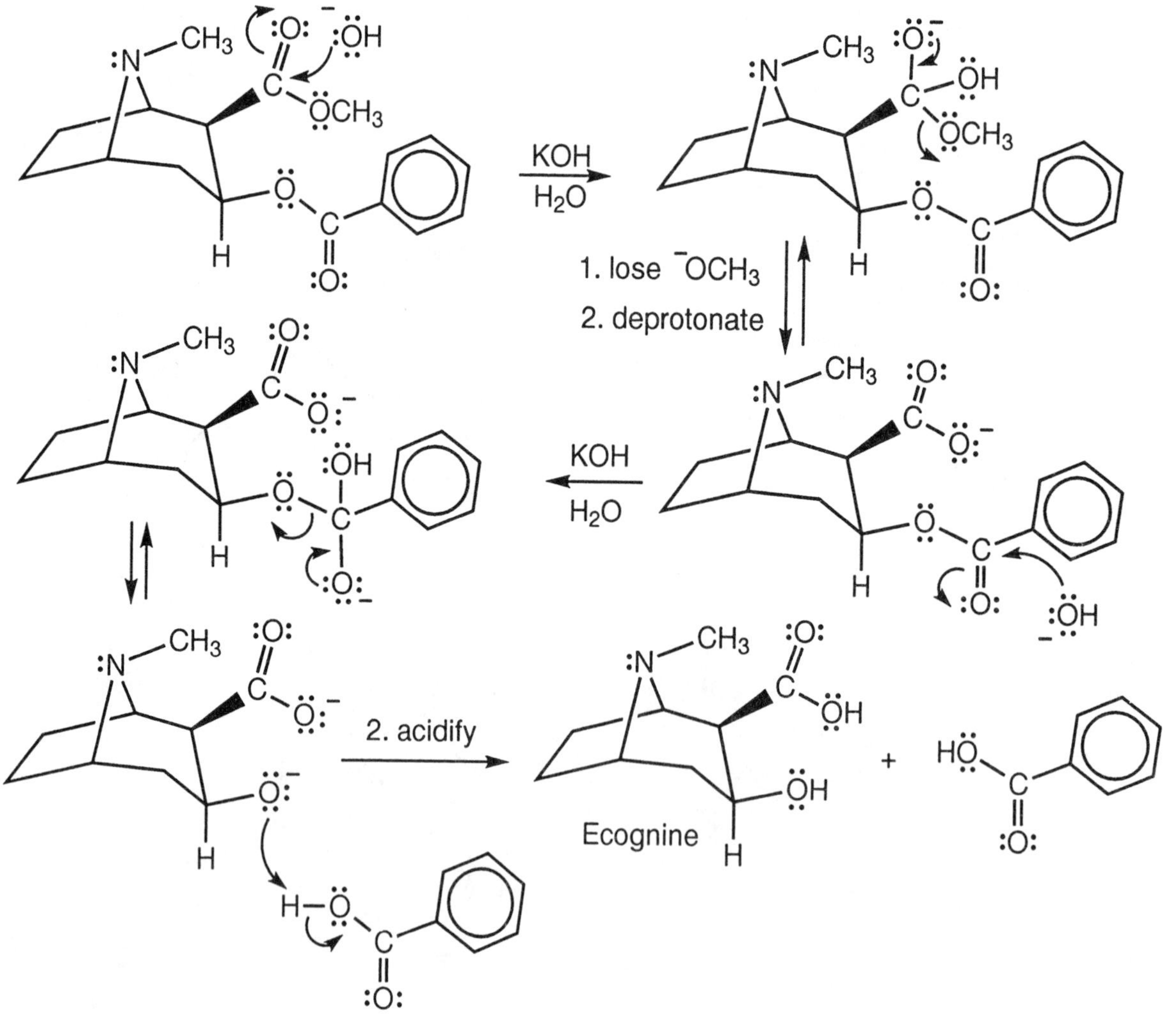

KOH / H_2O

1. lose $^-OCH_3$

2. deprotonate

KOH / H_2O

2. acidify

Ecognine

The oxidation is also straightforward. A chromate ester is first produced. An elimination reaction then gives the β-ketoacid. Decarboxylation gives tropinone.

Problem 21.25(a)

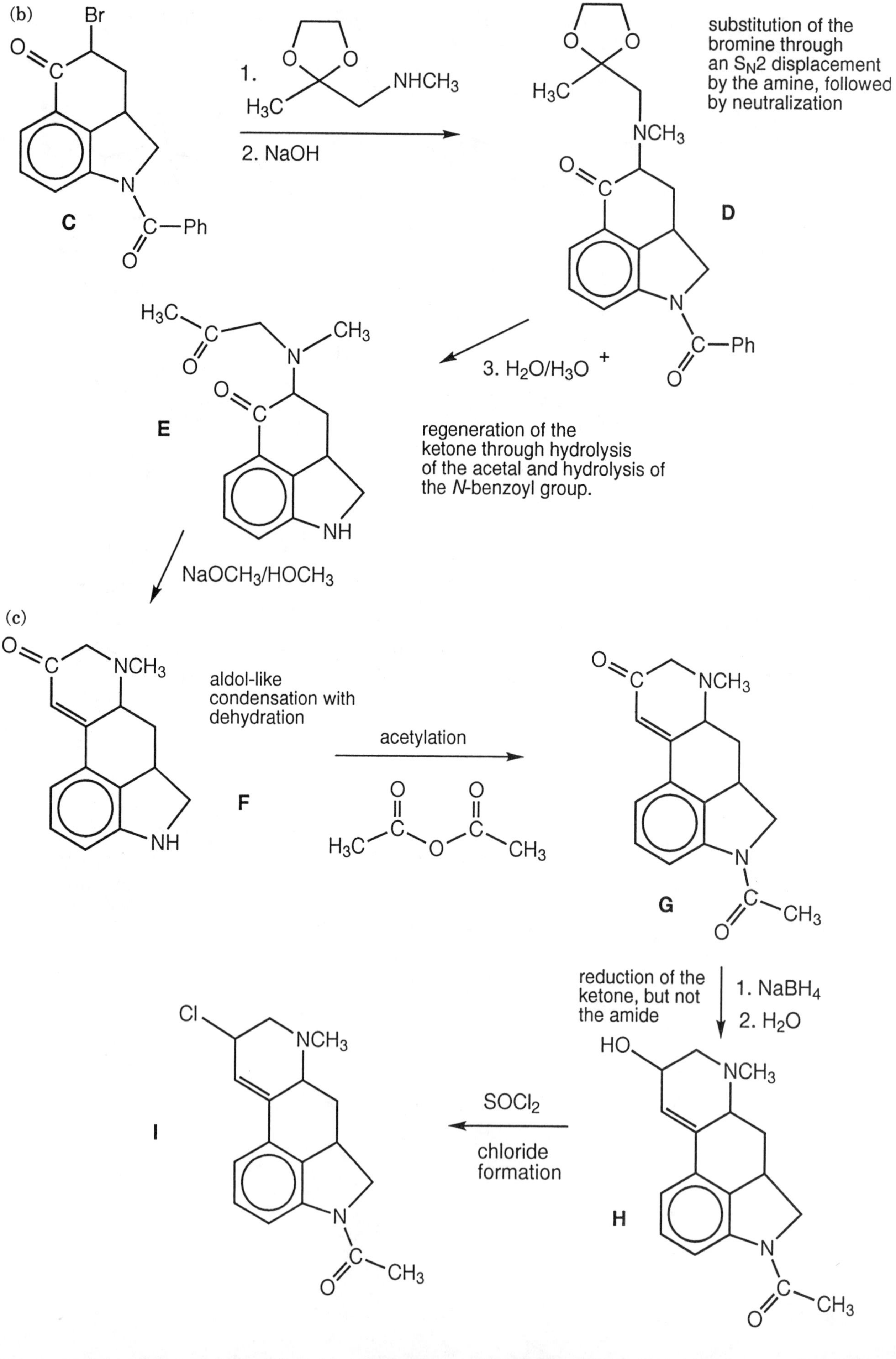

(b)

C

1. (2-methyl-1,3-dioxolan-2-yl)methyl-$NHCH_3$

2. NaOH

substitution of the bromine through an S_N2 displacement by the amine, followed by neutralization

D

3. H_2O/H_3O^+

regeneration of the ketone through hydrolysis of the acetal and hydrolysis of the *N*-benzoyl group.

E

$NaOCH_3/HOCH_3$

(c)

F

aldol-like condensation with dehydration

acetylation

$H_3C-C(=O)-O-C(=O)-CH_3$

G

reduction of the ketone, but not the amide

1. $NaBH_4$

2. H_2O

H

$SOCl_2$

chloride formation

I

(d)

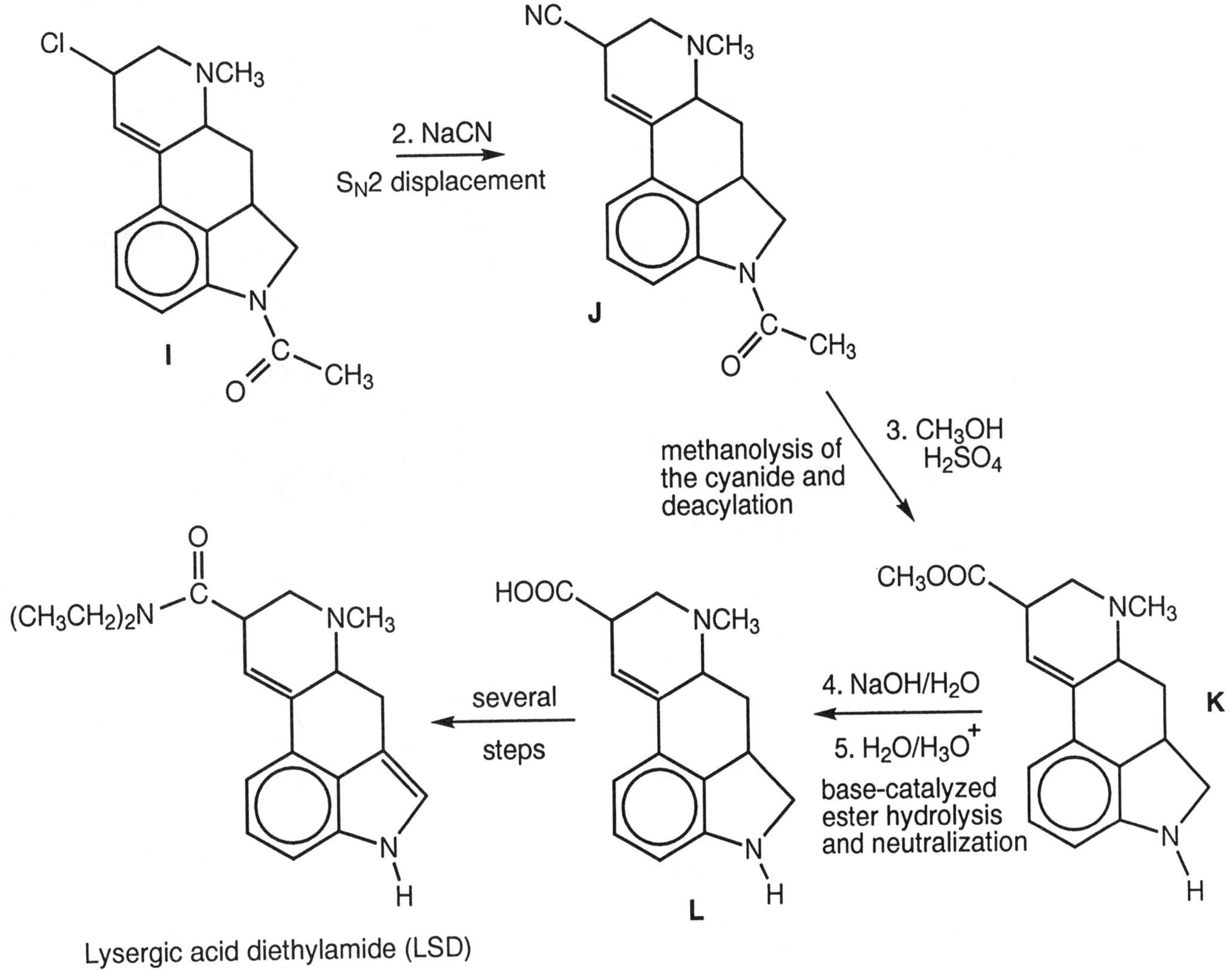

2. NaCN
S_N2 displacement

I

J

3. CH_3OH
H_2SO_4

methanolysis of the cyanide and deacylation

K

4. NaOH/H_2O

5. H_2O/H_3O^+

base-catalyzed ester hydrolysis and neutralization

L

several steps

Lysergic acid diethylamide (LSD)

Problem 21.26(a) The first step is a mixed Claisen condensation. Note that only one partner in the condensation has an α-hydrogen and can form an enolate. As usual in the Claisen condensation, the initial product is a salt. Neutralization gives the β-keto ester.

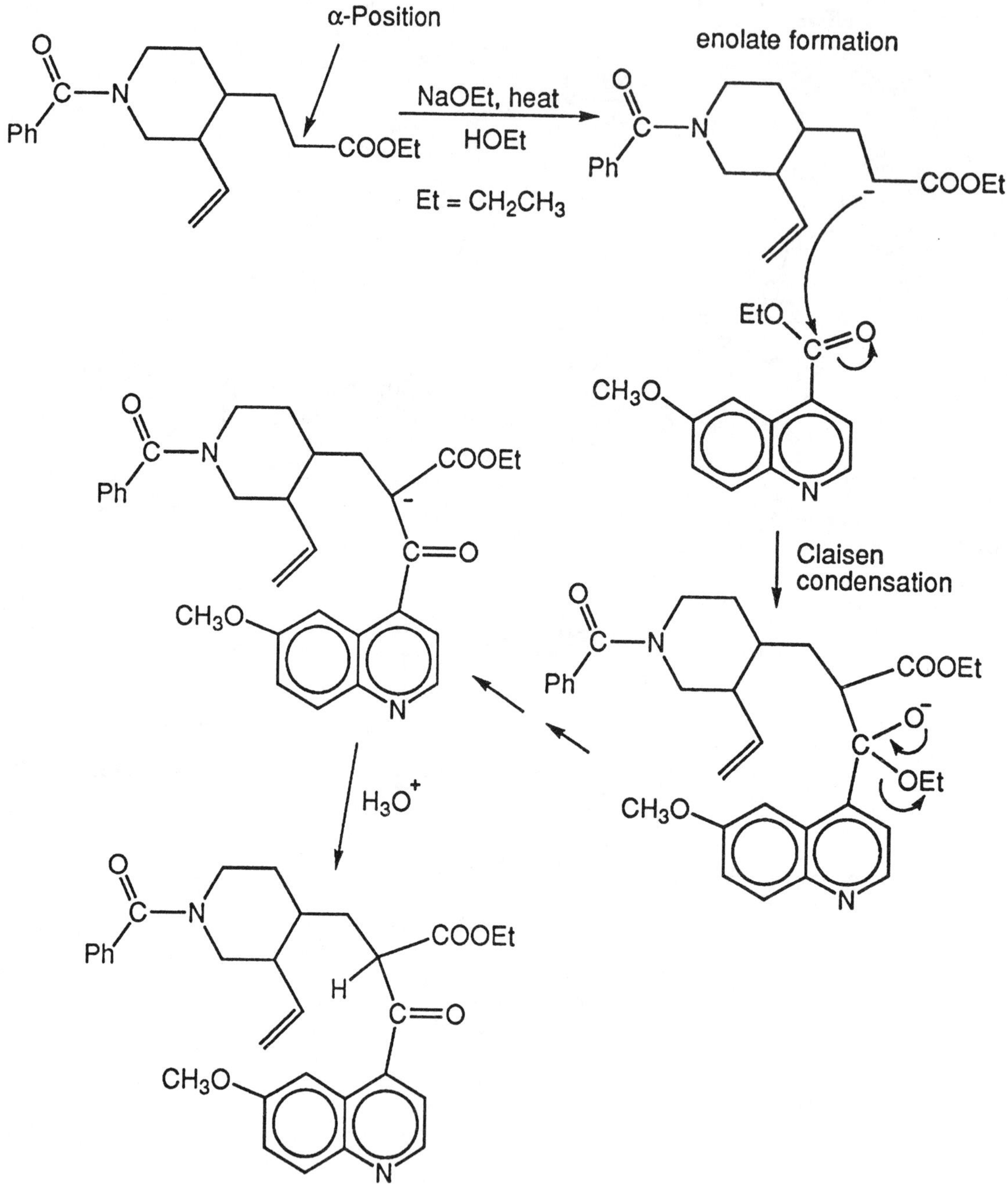

α-Position

NaOEt, heat

HOEt

Et = CH_2CH_3

enolate formation

Claisen condensation

H_3O^+

(b) First, the α-position is brominated.

1. Br_2, NaOH, H_2O

Now, treatment with base generates the amide ion. Intramolecular displacement of the α-bromide produces the bicyclic compound. The arrow looks hideously long, but this is an artifact of the two-dimensional representation of these three-dimensional molecules.

NaOH / H_2O

reduction

Quinine

Answers to Additional Problems

Problem 21.27

(a) Primary amine: 2-pentanamine, 2-aminopentane, 1-methylbutylamine.
(b) Secondary amine: dipentylamine, *N*-pentylpentanamine.
(c) Tertiary amine: tripentylamine, *N,N*-dipentylpentanamine.
(d) Primary amine: 3-fluoro-2-pentanamine, 2-amino-3-fluoropentane.
(e) Quaternary ammonium ion: tetrapentylammonium iodide.
(f) Both amines are primary: 1,3-diaminopentane, 1,3-pentanediamine, 3-aminopentanamine, 3-aminopentylamine.

Problem 21.28

(a) Primary amine: 3-buten-2-amine, 2-amino-3-butene.
(b) Tertiary amine: ethylmethylphenylamine, *N*-ethyl-*N*-methylaniline, *N*-ethyl-*N*- methyl-benzenamine.
(c) Primary amine: 2-propynamine, 2-propynylamine, propargylamine.
(d) Primary amine: 3-amino-2-butanol, 3-amino-2-hydroxybutane.

Problem 21.29

(a) Tetramethyleneamine, azacyclopentane, pyrrolidine.
(b) 3-Methylpyridine, 3-picoline.
(c) 2-Phenylpyrrole.
(d) 5-Aminoquinoline.

Problem 21.30

(a) Amine **1** can displace iodide in S_N2 fashion to give a pair of diastereomeric ammonium ions.

1

CH_3 H_3C CH_3CH_2—I $\xrightarrow{S_N2}$ CH_3 CH_2CH_3 H CH_3 I^- + H CH_2CH_3 CH_3 CH_3 I^-

$C_8H_{18}IN$ $C_8H_{18}IN$

(b) However, when amine **2** undergoes the same kind of displacement reaction, a proton can be removed from the initially formed ammonium ion to give an amine. Amine inversion will interconvert the two possible isomers, and only one set of signals will be observed in the NMR spectrum.

CH_3 H CH_3CH_2—I $\xrightarrow{S_N2}$ CH_3 CH_2CH_3 H H I^- + CH_3 H H CH_2CH_3 I^-

2

$C_7H_{16}IN$ $C_7H_{16}IN$

base base

CH_3 CH_2CH_3 H rapid inversion CH_3 H CH_2CH_3

$C_7H_{15}N$ $C_7H_{15}N$

Problem 21.31 At low temperature, what you see is what you get. There are five different carbons in the molecule and, of course, five different signals in the ^{13}C NMR spectrum. At higher temperature, amine inversion sets in and the two carbon-attached methyl groups and the two ring carbons become equivalent. There are now only three different carbons (shown as a, b, and c) and three signals in the ^{13}C NMR spectrum.

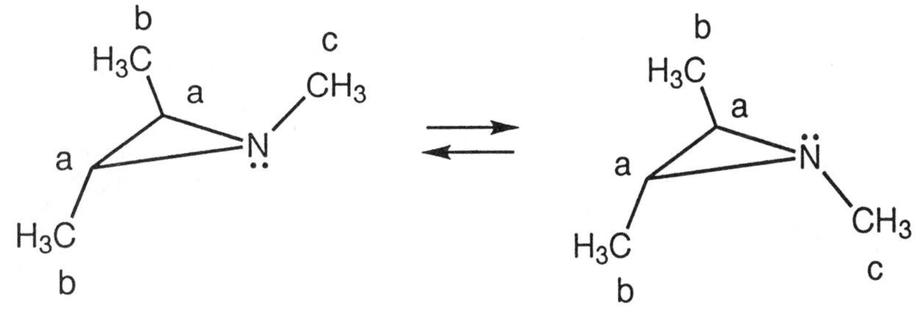

Problem 21.32

For the equilibrium: $H_2O + HA \rightleftharpoons H_3O^+ + A^-$ $\quad K_a = \dfrac{[H_3O^+][A^-]}{[HA]}$

For the equilibrium: $H_2O + A^- \rightleftharpoons HO^- + HA$ $\quad K_b = \dfrac{[HO^-][HA]}{[A^-]}$

$$K_a \times K_b = \frac{[H_3O^+][A^-]}{[HA]} \times \frac{[HO^-][HA]}{[A^-]} = [H_3O^+][HO^-] = K_w = 10^{-14}$$

if $K_a \times K_b = 10^{-14}$, then, $pK_a + pK_b = 14$

Problem 21.33 The lower the pK_b, the greater the base strength. So, methylamine is a stronger base than ammonia, and dimethylamine is a (slightly) stronger base still. Yet trimethylamine is a weaker base than the other methylated species, and almost as weak a base as ammonia itself. Indeed, we might well have been disturbed at the small difference between methylamine and dimethylamine as the two are almost of equal base strength. The problem is one of solvation. Each methyl group increases base strength, but decreases the possibilities for solvation of the charged intermediates (and the transition states leading to them). So, for trimethylamine, each methyl group stabilizes the intermediate by helping to disperse the charge, but, at the same time, it interferes with the stabilizing effects of solvation. It's a balancing act. See the discussion of these effects in terms of the pK_a values of the conjugate acids in this chapter (p. 1086).

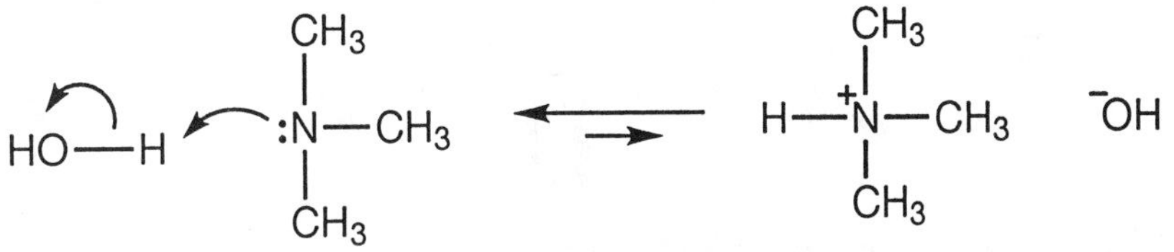

$$HO{-}H + :N(CH_3)_3 \rightleftharpoons H{-}\overset{+}{N}(CH_3)_3 \quad {}^{-}OH$$

Each methyl group helps to disperse the charge, but also interferes with stabilization of the ions by solvent

Problem 21.34 Oxygen is more electronegative than nitrogen, and will withdraw electrons more effectively, making the adjacent carbon more electron deficient (greater δ^+). All other effects being more or less equal, the adjacent hydrogens are more shielded in the amines than they are in the alcohols, and will appear at higher field.

δ^+ ⟶ δ^-

$RCH_2—NH_2$

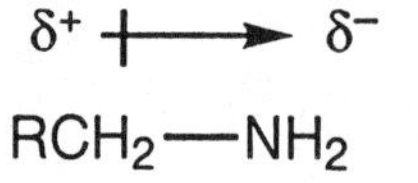

The smaller dipole means that the hydrogens are more shielded than in the alcohols. They will appear at higher field.

δ^+ ⟶ δ^-

$RCH_2—OH$

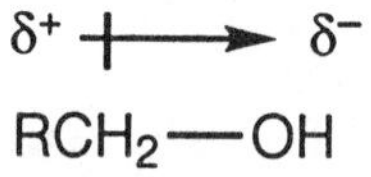

The greater dipole means that the hydrogens are less shielded than in the amines. They will appear at lower field.

Problem 21.35 The two hydrogens are diastereomeric, and thus there should be two signals. Use the technique first outlined in Chapter 15. To see how many signals should appear, replace each in turn with an "X" and see what the relationship is between those two hypothetical compounds.

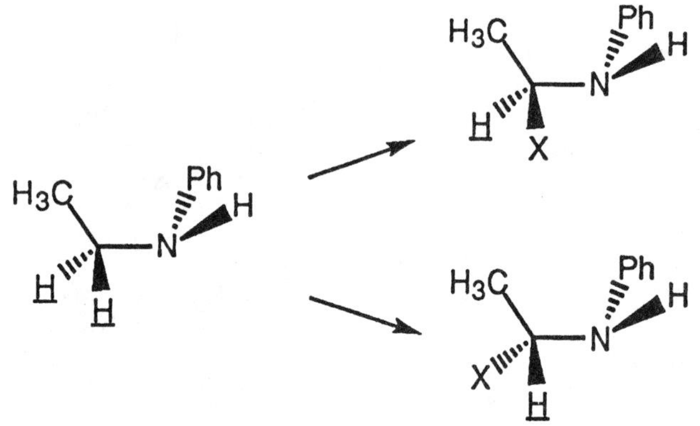

These are stereoisomers, but not mirror images; these two molecules are diastereomers. The two hydrogens replaced with "X" are diastereomeric, and must give different signals in the NMR spectrum.

So, why is only one signal observed? The reason is amine inversion. Amine inversion renders the two methylene hydrogens enantiomeric, and they must give only one signal.

amine inversion

rotation

mirror

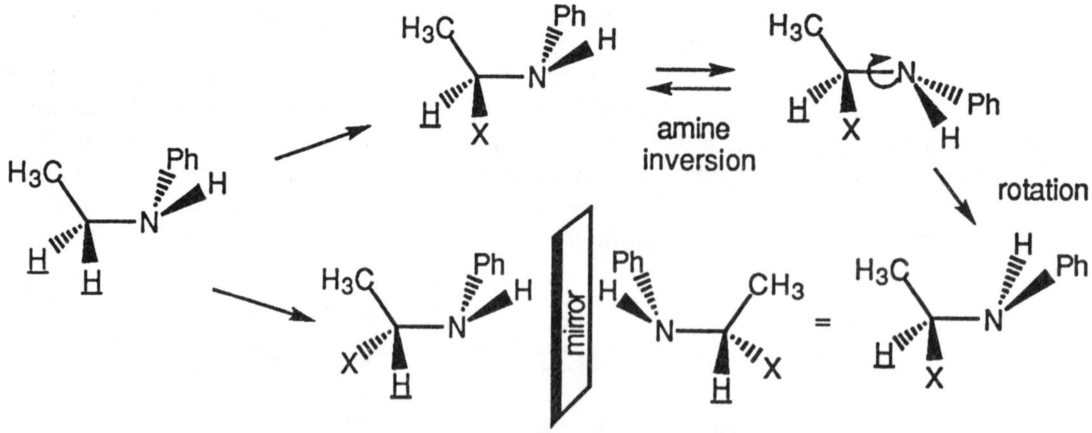

These are mirror images; these two molecules are enantiomers. The two hydrogens replaced with "X" are enantiomeric, and must give one signal in the NMR spectrum.

Problem 21.36

a. $HONO/HCl/H_2O$ — This reaction involves formation of the unstable diazonium ion, followed by displacement by water and deprotonation.

b. $HONO/HCl/CH_3CH_2OH$ — This reaction involves formation of the unstable diazonium ion, followed by displacement by ethyl alcohol and deprotonation.

c. 1) 3 CH_3I, 2) AgO, heat — Hofmann elimination.

d. BuLi — Removal of one of the amine protons by a very strong base.

e. $Cl-C(=O)-CH_3$, pyridine — Acetylation, amide formation by addition-elimination.

f. $HONO/HCl/H_2O$ — Nitrosation of a secondary amine to give a stable nitroso compound.

g. $H_3C-C(=O)-CH_3$ — Enamine formation from a secondary amine.

h. 1) 2 CH_3I, 2) AgO, heat — Hofmann elimination again.

Problem 21.37 This problem is essentially pure review of aromatic substitution reactions as applied to amines.

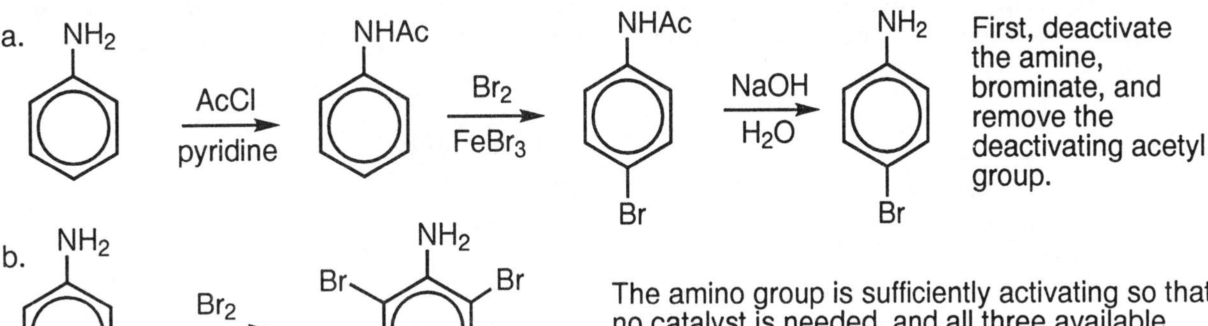

First, deactivate the amine, brominate, and remove the deactivating acetyl group.

The amino group is sufficiently activating so that no catalyst is needed, and all three available ortho/para positions are substituted.

c. Aniline (NH_2) → (HONO, HCl) → benzenediazonium chloride ($\overset{+}{N}_2$ ^-Cl) → (CuBr) → bromobenzene (Br) → (1. Mg; 2. CO_2; 3. H_2O) → benzoic acid (COOH)

d. Benzenediazonium chloride ($\overset{+}{N}_2$ ^-Cl) → (CuCN) → benzonitrile (CN) → (1. $LiAlH_4$; 2. H_3O^+) → benzylamine (CH_2NH_2)

In both parts (c) and (d), Sandmeyer chemistry is used to make an intermediate that is further elaborated.

e. Aniline (NH_2) → (CF_3CO_3H) → nitrobenzene (NO_2) → (Br_2, $FeBr_3$) → m-bromonitrobenzene (NO_2, Br)

The ortho/para directing amino group must be first oxidized to the meta directing nitro group.

Problem 21.38 This problem is hard because the target, **1**, surely looks very little like the required starting material. All that obviously remains of glycine is, apparently, the carboxy group, COO. The alkene, 2,3-dimethyl-2-butene, seems present as six of the carbons of the three-membered ring.

2,3-dimethyl-2-butene becomes: tetramethylcyclopropane bearing $COOCH_3$ } from glycine

As always, it is best to work backward, and that should focus us on ways of making cyclopropanes. As you know very few routes to such molecules, the first choice of steps isn't so bad. Carbenes, compounds containing divalent carbon, add to alkenes to give cyclopropanes. So, a retrosynthetic analysis would suggest carbomethoxycarbene (**A**) and the allowed 2,3-dimethyl-2-butene as reactants.

1 (cyclopropane with $COOCH_3$) ⟹ 2,3-dimethyl-2-butene + $\ddot{C}H$–$COOCH_3$

Carbene **A**

Now the problem is how to make carbene **A**. Carbenes come from diazo compounds, and our problem now becomes finding a synthesis of methyl diazoacetate (**B**).

$\ddot{C}H$–$COOCH_3$ ⟹ CHN_2–$COOCH_3$

Carbene **A** Methyl diazoacetate **B**

In Figure 21.53, we saw the transformation of an amine into a diazo compound, and that suggests the diazotization of glycine methyl ester, itself formed from the esterification of glycine.

CHN_2–$COOCH_3$ ⟹ CH_2NH_2–$COOCH_3$ ⟹ CH_2NH_2–$COOH$

Methyl diazoacetate **B**

So here is the synthesis in the forward direction, with reagents:

$$H_2NCH_2COOH \xrightarrow[\text{2. neutralize}]{\text{1. } CH_3OH,\ HCl} H_2NCH_2COOCH_3 \xrightarrow{HONO} N_2CHCOOCH_3 \quad \textbf{B}$$

$$N_2CHCOOCH_3 + (CH_3)_2C{=}C(CH_3)_2 \xrightarrow{h\nu} \textbf{1}$$

$COOCH_3$

Problem 21.39 This problem is a straightforward Cope elimination preceded by a sequence of reactions designed to make the starting *N*-oxide. Here are the structures (see Fig. 21.56).

COOH $\xrightarrow[150\ °C]{SOCl_2}$ COCl **A** $C_7H_{11}ClO$ $\xrightarrow[\text{benzene}]{(CH_3)_2NH}$ $CON(CH_3)_2$ **B** $C_9H_{17}NO$ $\xrightarrow[\text{ether, 35 °C; 2. } H_2O]{\text{1. } LiAlH_4}$ $CH_2N(CH_3)_2$ **C** $C_9H_{19}N$

acid chloride formation — conversion to the amide through addition-elimination — reduction of the amide

C $\xrightarrow[25\ °C]{30\%\ H_2O_2}$ $CH_2\overset{+}{N}(CH_3)_2\text{–}O^-$ **D** $C_9H_{19}NO$ $\xrightarrow[\text{2 hours}]{100\ °C}$ CH_2 **E** C_7H_{12}

oxidation to the *N*-oxide — Cope elimination

Problem 21.40 This problem involves a Mannich reaction with two possible regiochemical outcomes. The first step involves formation of the immonium ion **B**.

$$(CH_3)_2NH + H_2C{=}O \xrightarrow[145\ °C]{CF_3COOH} H_2C{=}\overset{+}{N}(CH_3)_2 \quad \textbf{B}$$

Now, two enols, **C** and **D**, are possible and each could lead to a product.

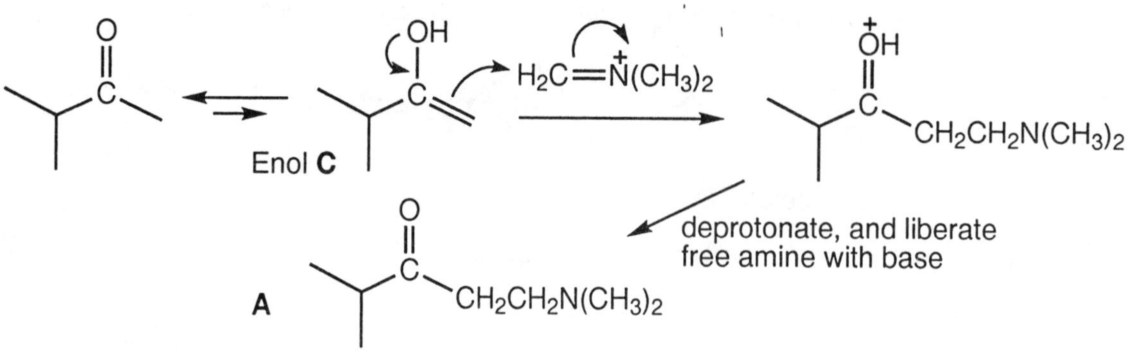

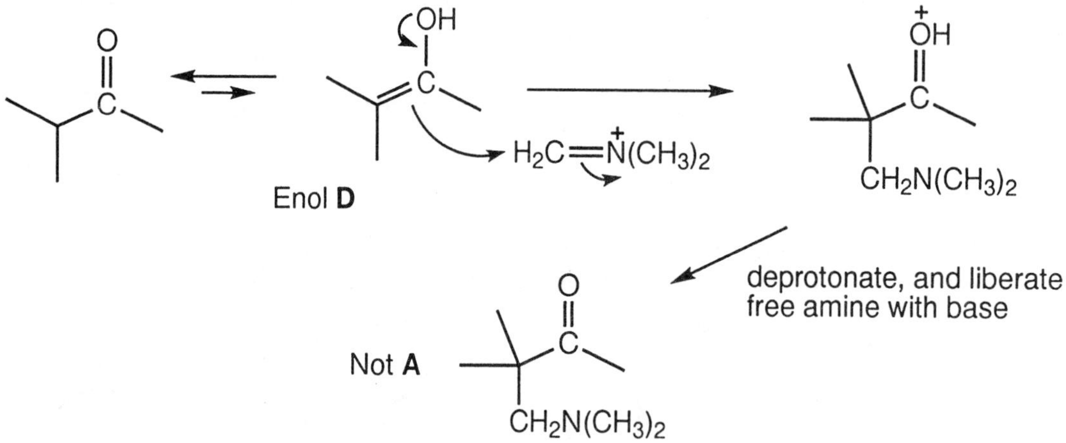

Enol **C**

$H_2C{=}\overset{+}{N}(CH_3)_2$

$CH_2CH_2N(CH_3)_2$

deprotonate, and liberate free amine with base

A $CH_2CH_2N(CH_3)_2$

Enol **D**

$H_2C{=}\overset{+}{N}(CH_3)_2$

$CH_2N(CH_3)_2$

deprotonate, and liberate free amine with base

Not **A** $CH_2N(CH_3)_2$

The ^{1}H NMR spectrum allows a choice to be made between **A** and "not **A**".

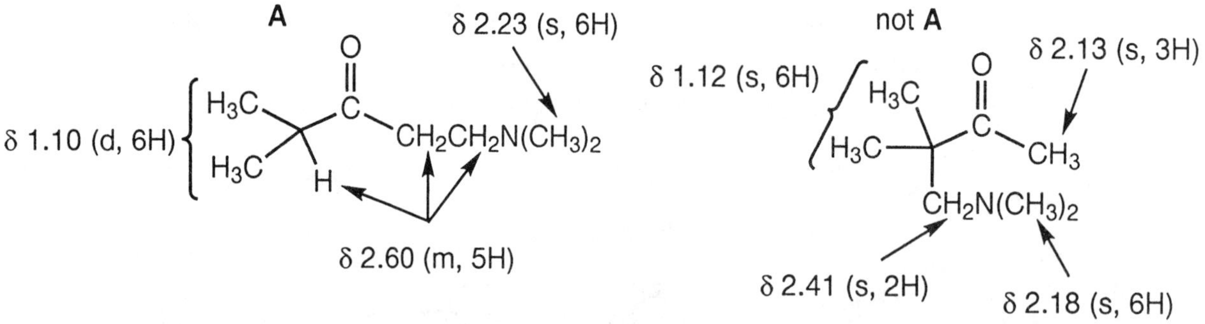

A

δ 1.10 (d, 6H) H_3C, H_3C, H

δ 2.23 (s, 6H)

$CH_2CH_2N(CH_3)_2$

δ 2.60 (m, 5H)

not **A**

δ 1.12 (s, 6H) H_3C, H_3C

δ 2.13 (s, 3H) CH_3

$CH_2N(CH_3)_2$

δ 2.41 (s, 2H)

δ 2.18 (s, 6H)

One critical distinguishing feature is the appearance of the carbon-bound methyl groups as a doublet in **A**. They would be (and are) a singlet in "not **A**".

Problem 21.41 This problem is tough because there is a large number of "small" steps. Some of the basic outlines should be clear, however. Clearly, the ether oxygen will become the second OH of the top benzene ring. The allylic OH will probably be lost in an elimination reaction that helps build the lower aromatic ring in apomorphine. Somehow the bridge in morphine must become the new azacyclohexane ring. The real trick in this problem is to see how to do that transformation, so let's try it first. What can happen in acid? Protonation, that's what. First, protonate the double bond, then move a hydride to give the more stable, tertiary carbocation.

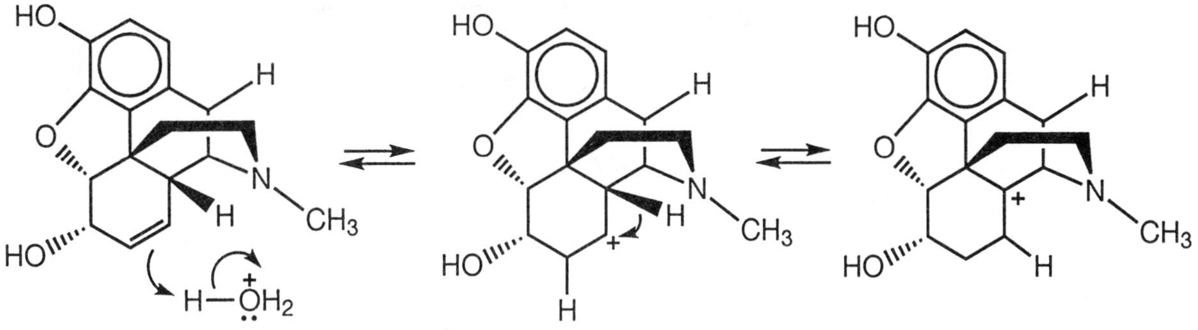

Secondary carbocation

Tertiary carbocation

Now a carbon in the bridge can migrate to give another tertiary carbocation:

The migrating carbon is labelled "CH_2" for clarity.

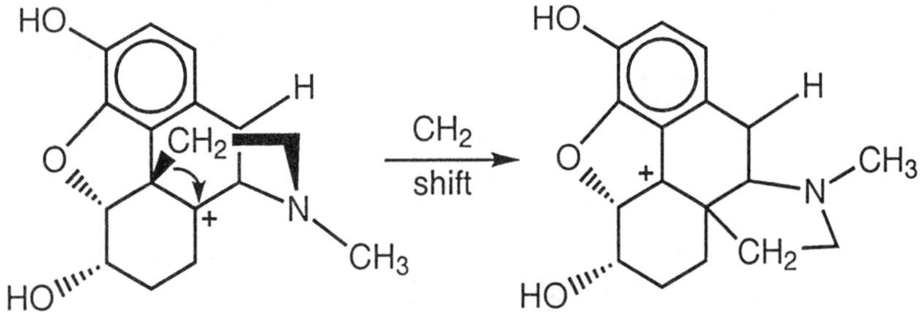

CH_2 shift

Tertiary carbocation

New tertiary carbocation

What has been accomplished? We are on the way to constructing the new azacyclohexane, and we have removed the bridge. Now we need to expand the new five-membered ring to a six-membered ring, and to do that we need an adjacent carbocation so we can do another carbon migration. A series of 1,2-hydride shifts (or elimination–protonation steps) can "walk" the positive charge around the ring to the useful position next to the new ring.

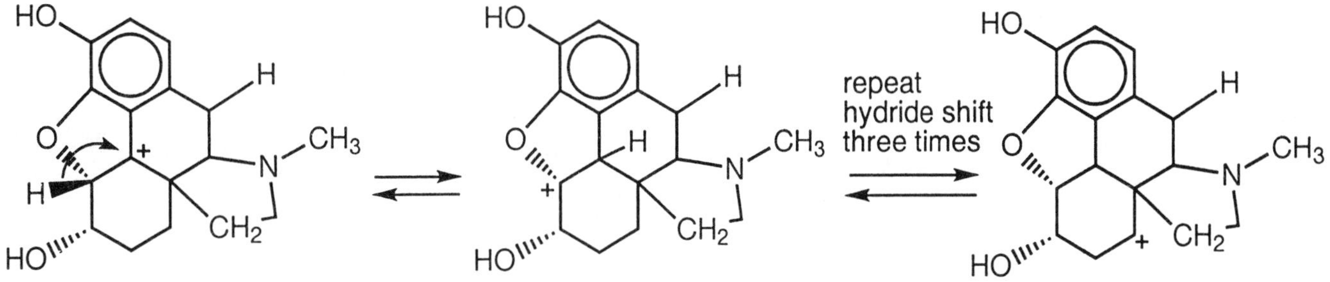

repeat hydride shift three times

Now migrate the ring CH_2 again to give the tertiary carbocation, and lose a proton to give an alkene.

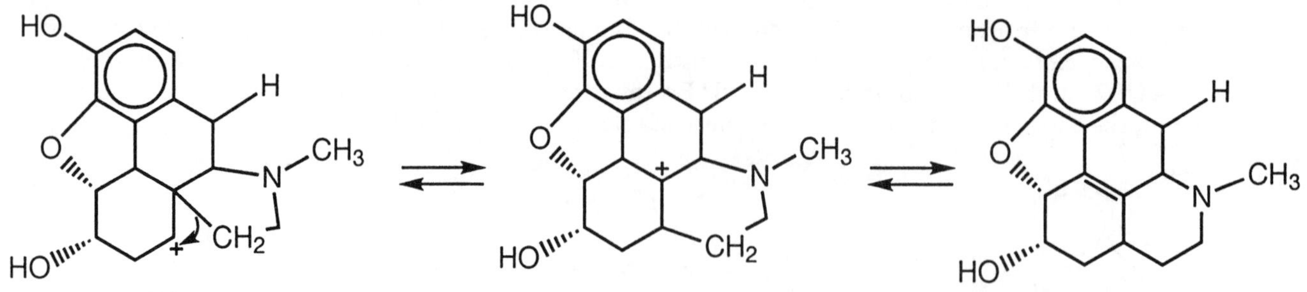

Two protonation–elimination sequences finish off this hard problem. The eliminations are shown as E2, but E1 reactions are also possible.

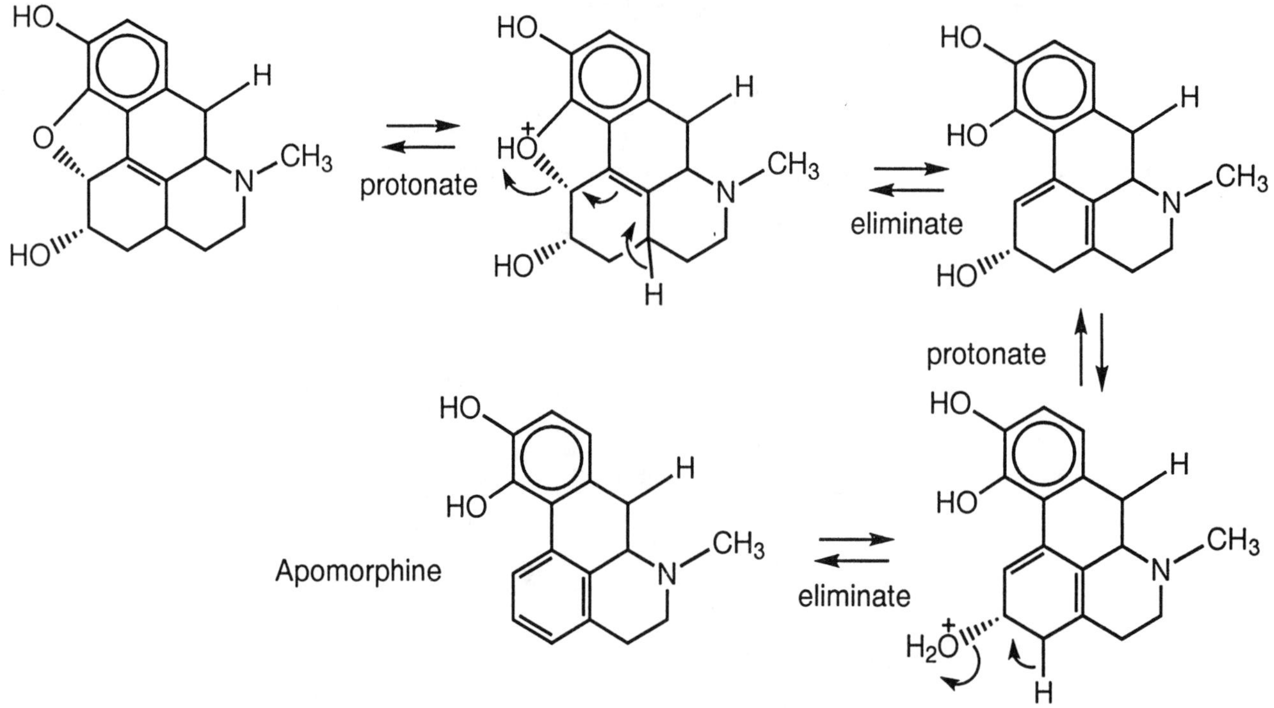

protonate

eliminate

protonate

eliminate

Apomorphine

Problem 21.42 These reactions are examples of the Tiffeneau–Demyanov ring expansion. The first step in each case is the conversion of the amino group into a diazonium ion. We won't write the mechanism for this transformation (see Fig. 21.49), but you should.

(a) Here, displacement of nitrogen, an excellent leaving group, by an adjacent carbon–carbon single bond forms the ring-expanded, resonance-stabilized carbocation **A**. Deprotonation of **A** by any base leads to cyclohexanone. Be careful not to form the primary carbocation **B**. *Remember:* Simple primary carbocations are very high in energy and are almost never formed. They serve as mechanistic "stop signs" in problems.

HO CH$_2$NH$_2$ HONO HO CH$_2$—$\overset{+}{N}_2$ H O H$_2$O H O+ A HO $\overset{+}{C}$H$_2$ B O + H$_3$O$^+$

(b) In the second example, two different modes of ring expansion (a and b) are possible, as two different carbon–carbon bonds are in position to help displace the leaving group.

OH C $\overset{+}{O}$H C deprotonate O C "b" OH CH$_2$NH$_2$ HONO OH CH$_2$–$\overset{+}{N}_2$ "b" or OH CH$_2$–$\overset{+}{N}_2$ "a" "a" OH C=$\overset{+}{O}$H deprotonate C=O

Problem 21.43 In all cases, the amino group is first converted into a diazonium ion. In order to see what happens next, we need better three-dimensional drawings of these diazonium ions. So, first draw out the chair conformations of these molecules, and be careful to keep the large *tert*-butyl group in the lower energy equatorial position. In each case nitrogen will be displaced by a group or bond in the proper position (anti-periplanar) to assist departure by displacement from the rear.

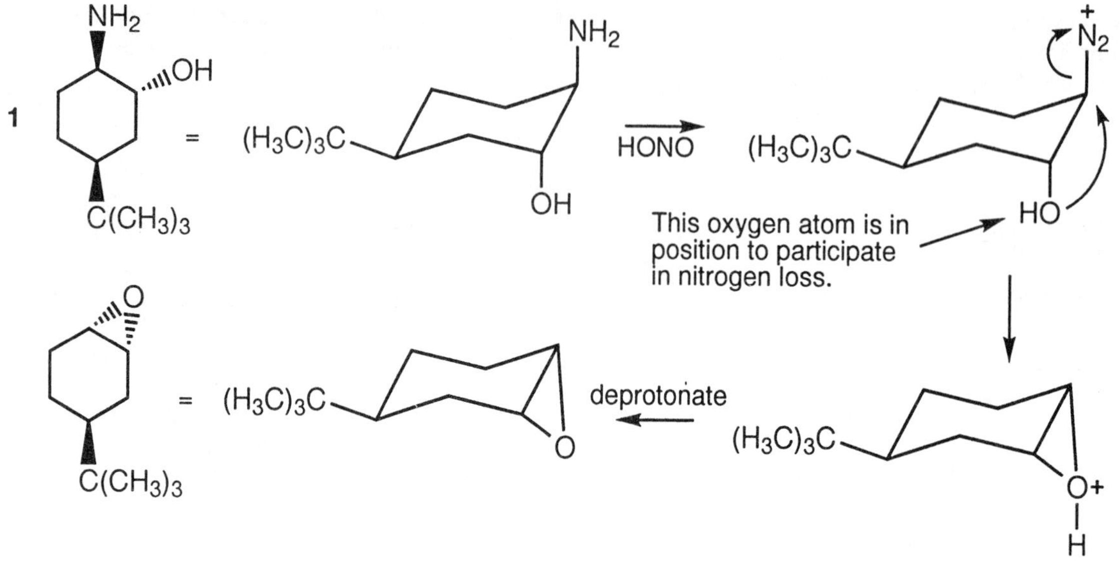

This oxygen atom is in position to participate in nitrogen loss.

In the second case, the hydroxyl oxygen is no longer in a position to displace nitrogen. Instead, the axial carbon–hydrogen bond is in position. A hydride shift takes place to give the resonance-stabilized cation **A**. Deprotonation gives the ketone.

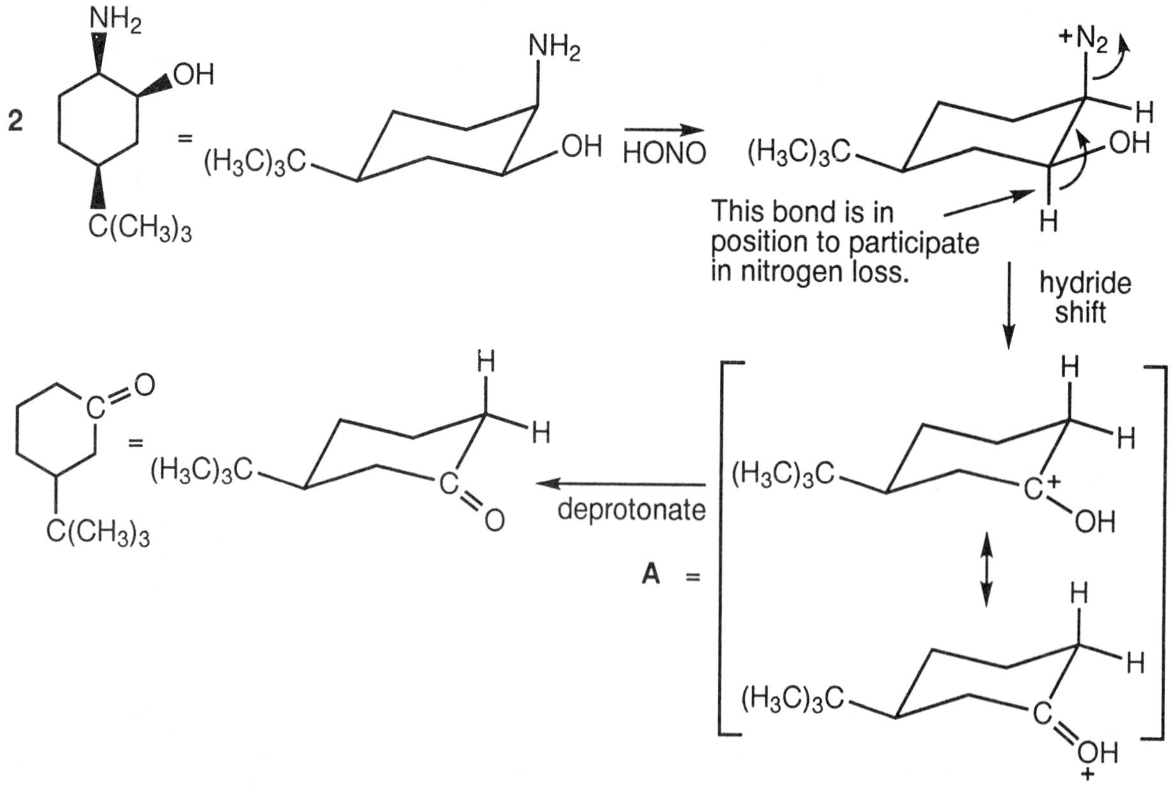

This bond is in position to participate in nitrogen loss.

In the third example, neither adjacent exo-ring bond is in position to participate in displacement of nitrogen. However, there is one ring bond that is. Migration leads to ring contraction, and a new resonance-stabilized carbocation **B**, now attached to a five-membered ring. Deprotonation gives the product.

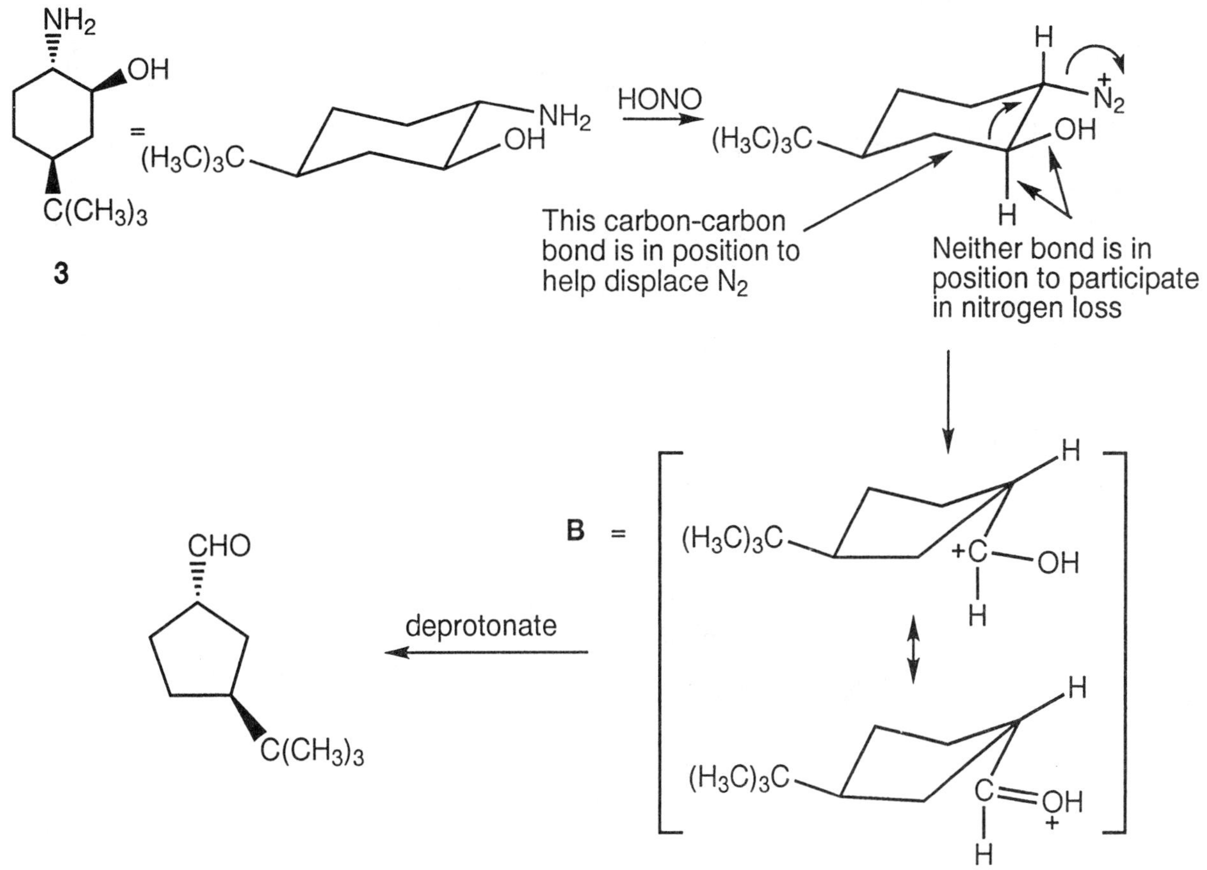

3

HONO

This carbon-carbon bond is in position to help displace N_2

Neither bond is in position to participate in nitrogen loss

B =

deprotonate

The last example is similar; once again no exo-ring bond is in position to overlap correctly with the equatorial leaving group, N_2. A ring bond migrates instead, and the product is a five-membered ring.

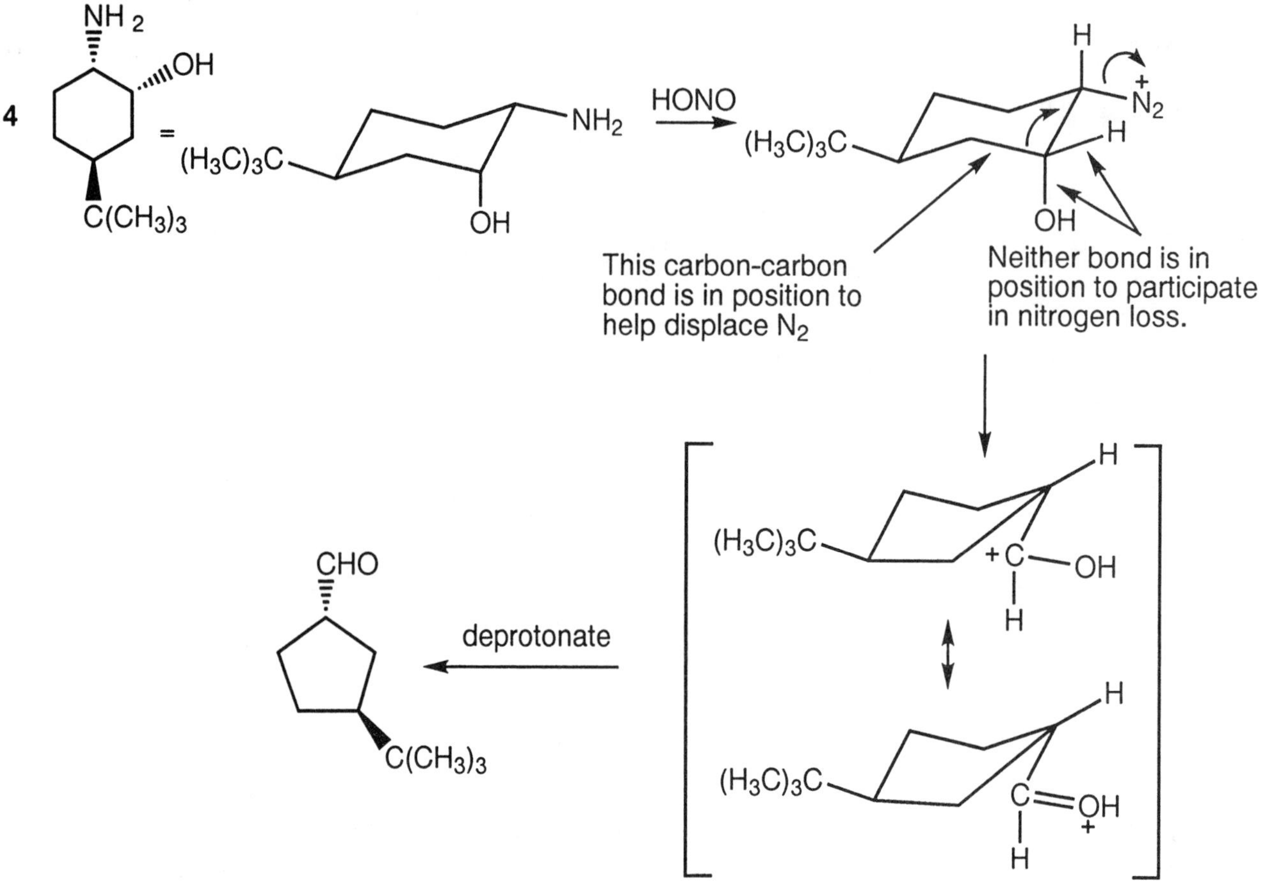

4

This carbon-carbon bond is in position to help displace N_2

Neither bond is in position to participate in nitrogen loss.

deprotonate

Problem 21.44 Just as in the previous problems, the amine is first converted into a diazonium ion (**A**), containing an excellent leaving group, N_2. Rearrangements of adjacent bonds lead to carbocations that give the observed products. In one possibility, the bicyclic system is opened to give a tertiary carbocation. Capture by water gives the product after deprotonation.

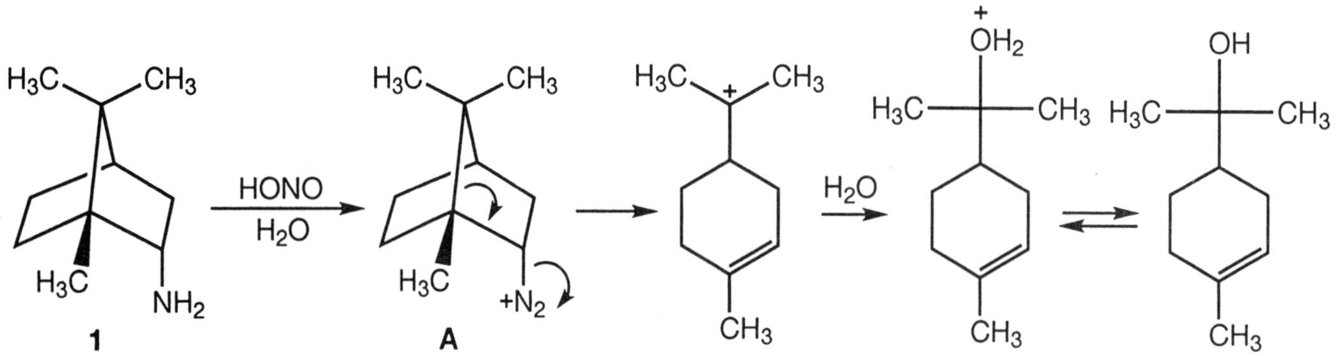

1 HONO / H_2O → **A** → H_2O →

Alternatively, nitrogen can be lost in E1 fashion, and a different cage bond can migrate. This reaction generates a different tertiary carbocation, **B**. Capture by water gives the alcohol, and loss of a proton from a methyl group generates the alkene.

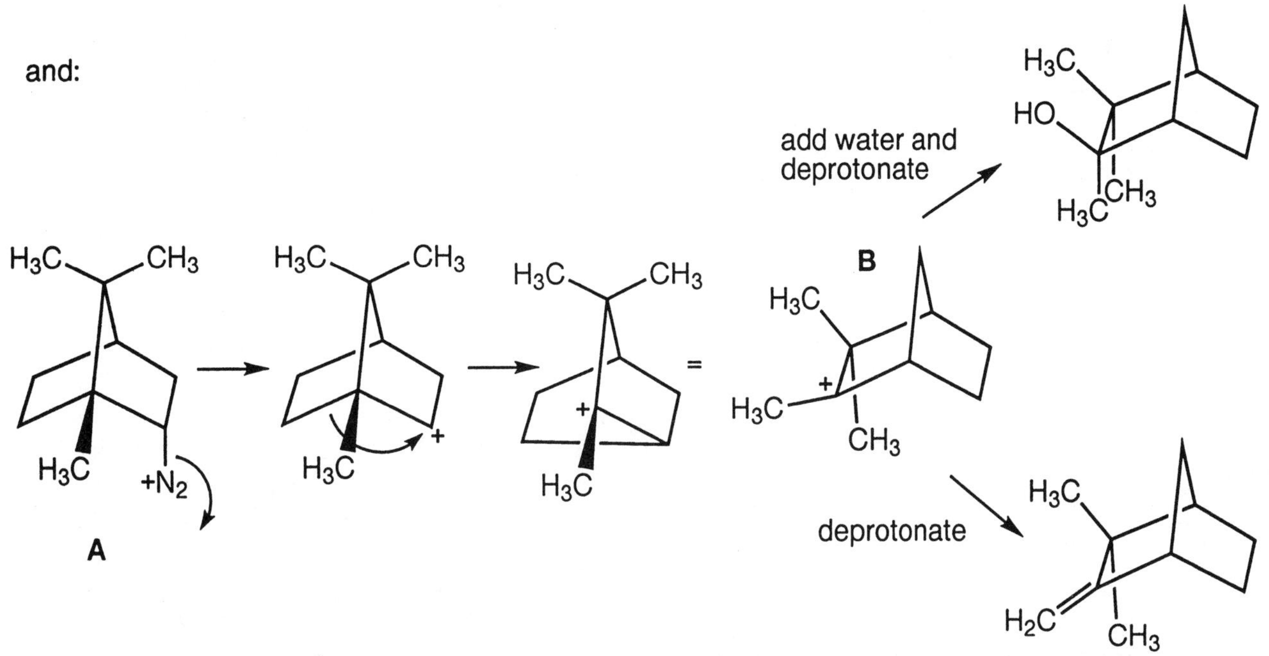

and:

add water and deprotonate

deprotonate

A

B

Problem 21.45 The first step of the reaction sequence involves esterification of the carboxylic acid group of nicotinic acid with diazoethane. Compound **A** is ethyl nicotinate.

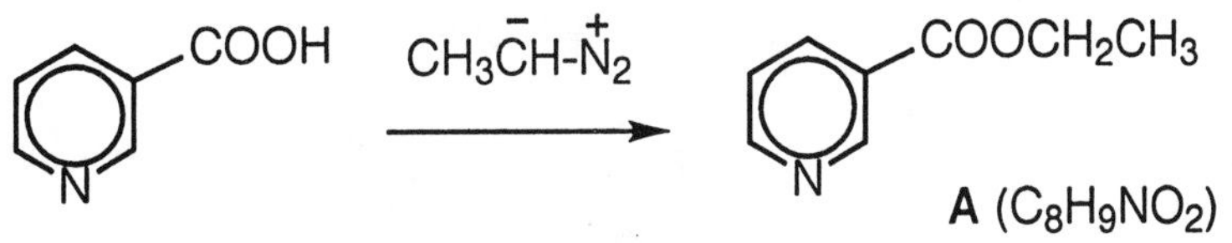

$CH_3\bar{C}H\text{-}\overset{+}{N}_2$

A ($C_8H_9NO_2$)

In a reaction reminiscent of a crossed Claisen condensation, the enolate of *N*-methylpyrrolidinone adds to the ester carbonyl group of **A** to give, after the elimination phase of this addition–elimination process, the ketoamide **B**. This reaction is another process in which the real product of the condensation step is a resonance-stabilized double enolate. A neutralization is required to give stable **B**.

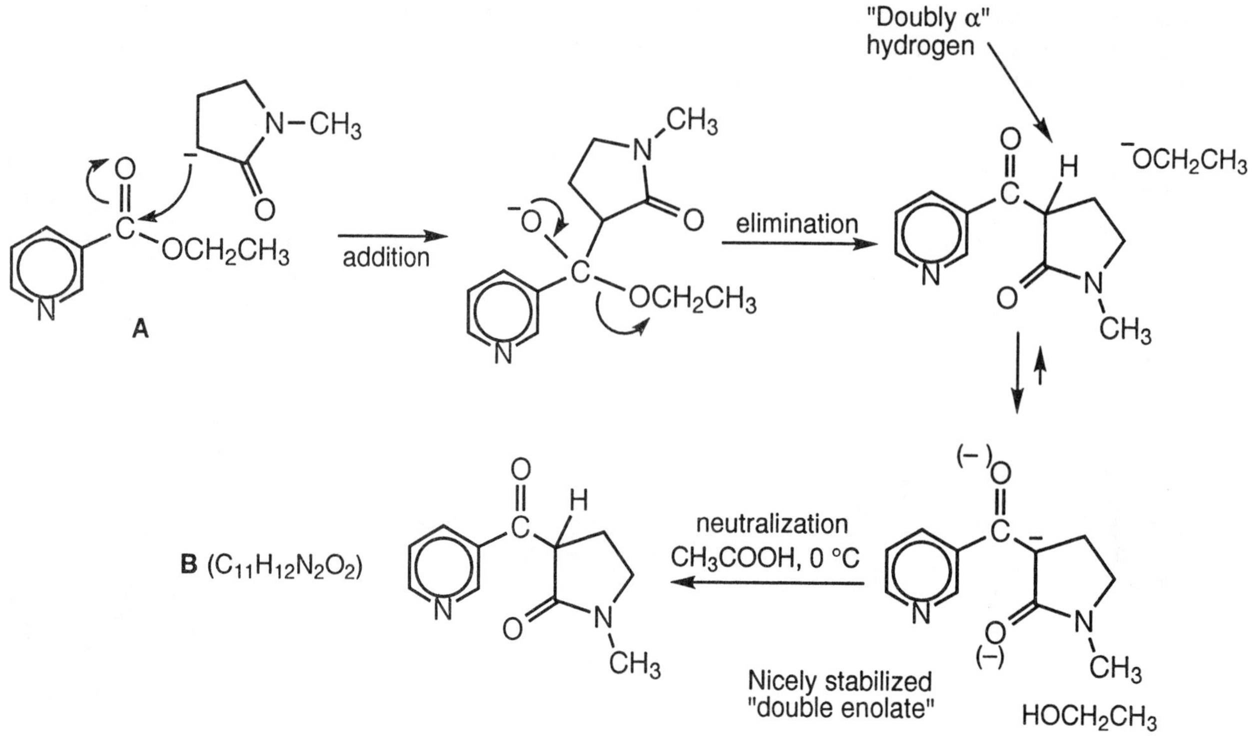

"Doubly α" hydrogen

A

addition

elimination

$^{-}OCH_2CH_3$

neutralization
CH_3COOH, 0 °C

B ($C_{11}H_{12}N_2O_2$)

Nicely stabilized "double enolate"

$HOCH_2CH_3$

The acid-catalyzed hydrolysis of **B** initially yields the β-ketocarboxylic acid **2**, which easily decarboxylates to give the γ-aminoketone salt **C**. The HCl will also protonate the pyridine nitrogen to give a double salt.

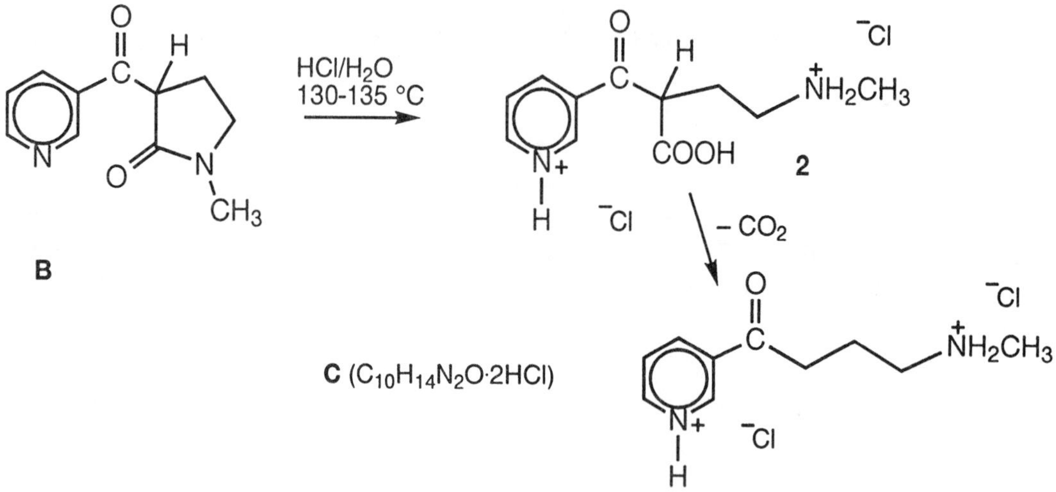

HCl/H_2O
130-135 °C

B

2

$- CO_2$

C ($C_{10}H_{14}N_2O{\cdot}2HCl$)

Finally, neutralization of **C** with KOH/CH_3OH first affords the free aminoketone **3**, which then undergoes an intramolecular reductive amination to the racemic nicotine (**1**).

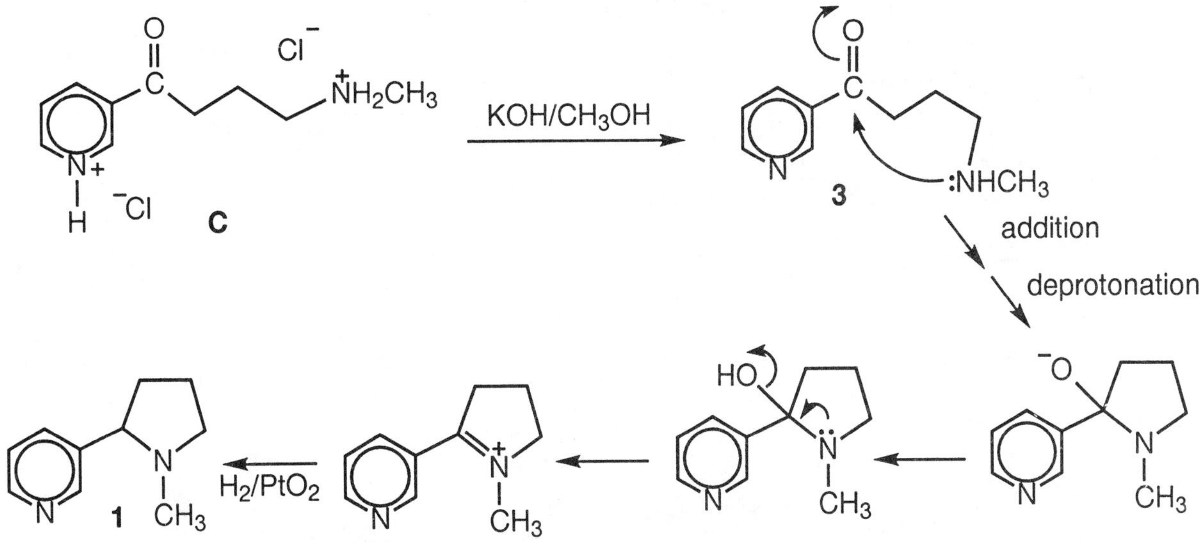

KOH/CH_3OH

addition

deprotonation

H_2/PtO_2

Problem 21.46 This synthesis of tropinone (**3**), aptly called the Robinson–Schöpf reaction, involves a double Mannich condensation, followed by two decarboxylations. Succindialdehyde (**1**) and methylamine react to yield the bis-immonium salt **A**. We won't write the detailed mechanism for this simple transformation, but you should be certain you can do it easily.

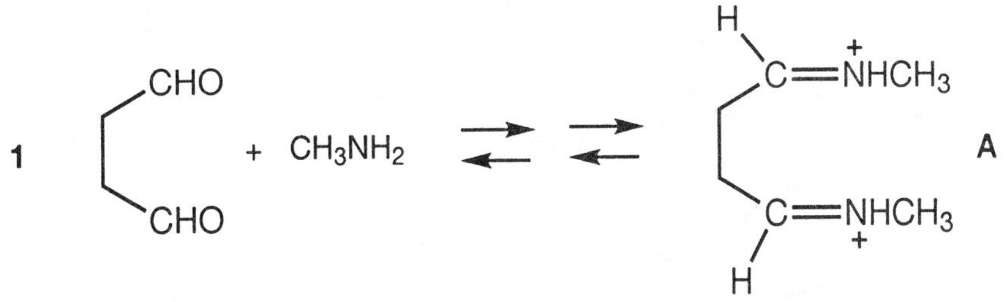

1 + CH_3NH_2 ⇌ ⇌ **A**

The first Mannich condensation now occurs between one immonium salt and the enol form of **2**. The amine **B** is the product. Now intramolecular cyclization takes place to give a new immonium salt **C** (see next page).

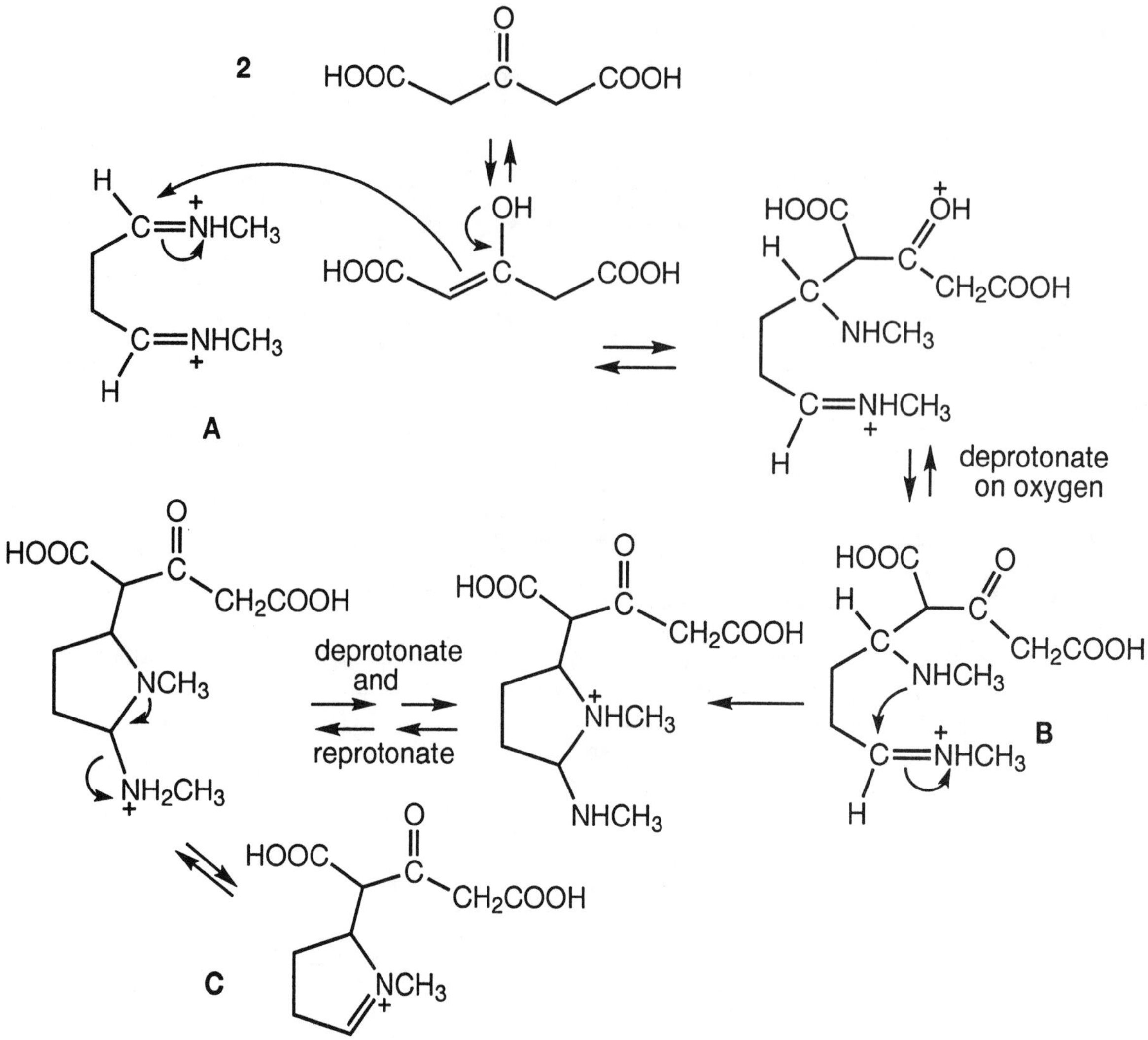

Now a second, intramolecular Mannich reaction takes place to give **D**. As **D** is a β-ketocarboxylic acid, two decarboxylations readily occur to give tropinone (**3**).

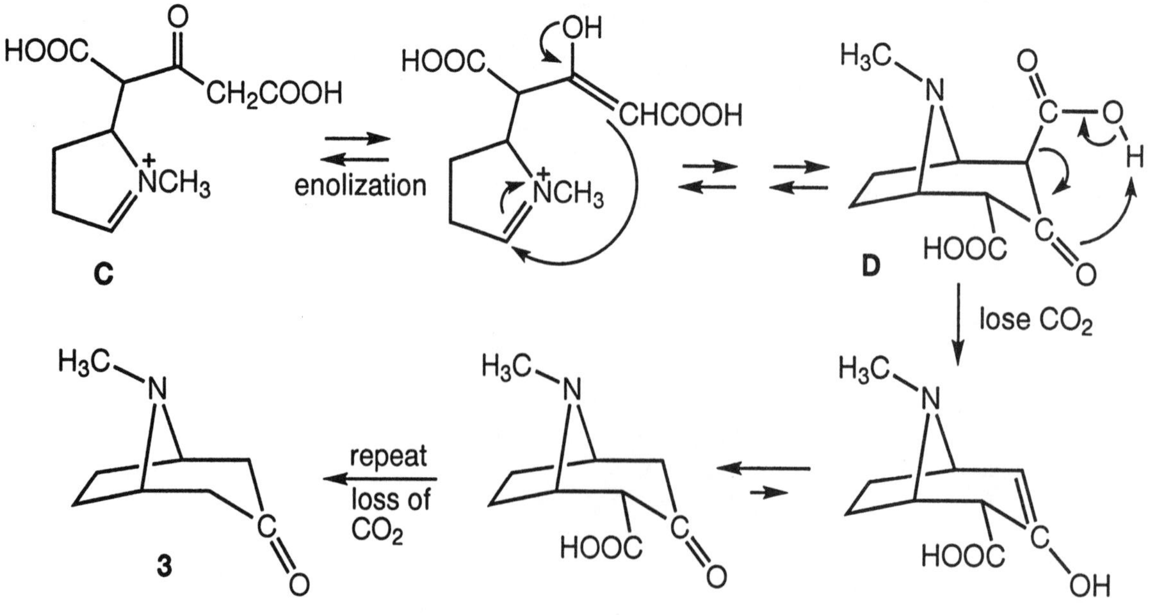

Chapter 22 Outline

22.1 Concerted Reactions
22.2 Electrocyclic Reactions
22.3 Cycloaddition Reactions
22.4 Sigmatropic Shift Reactions
22.5 Sigmatropic Shifts of Carbon
22.6 Other Sigmatropic Shifts
22.7 Something More: A Molecule with a Fluxional Structure

22

Aromatic Transition States: Orbital Symmetry

Problem 22.2 There are always two conrotatory and two disrotatory modes. The two conrotatory and two disrotatory modes will always give the same bonding or antibonding interaction at the newly formed bond. Here are the "missing" two modes from Figure 22.14.

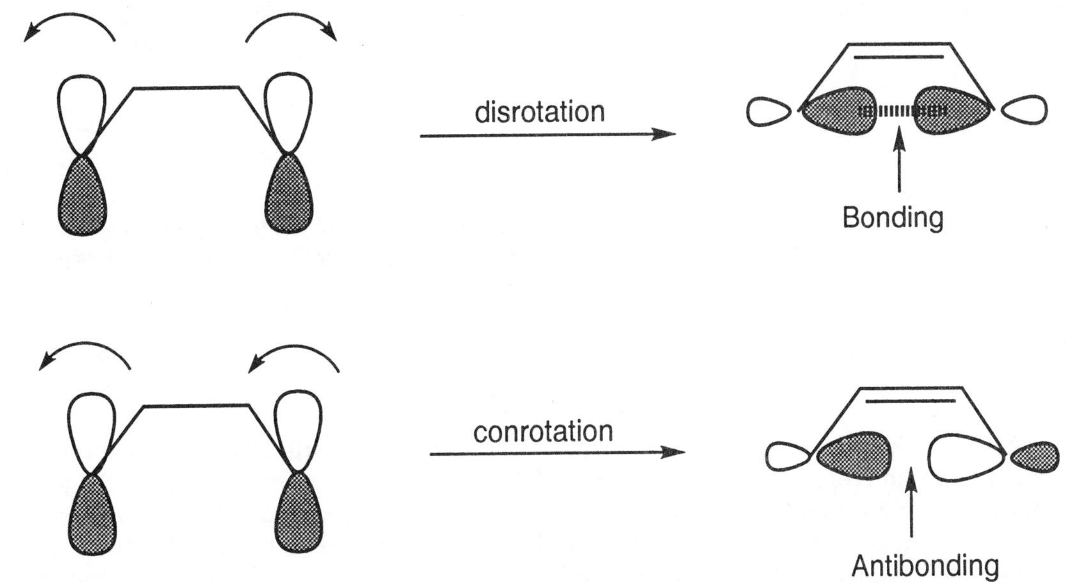

Problem 22.3 The two possible disrotatory modes lead to the same isomer in which there is one cis and one trans double bond. Be certain that you see that these planar molecules are the same.

disrotation $h\nu$

trans,cis (*E*,*Z*)

Same molecule

disrotation $h\nu$

cis,trans (*Z*,*E*)

Problem 22.4 The other disrotatory and conrotatory modes will give the same stereochemical results as the first ones.

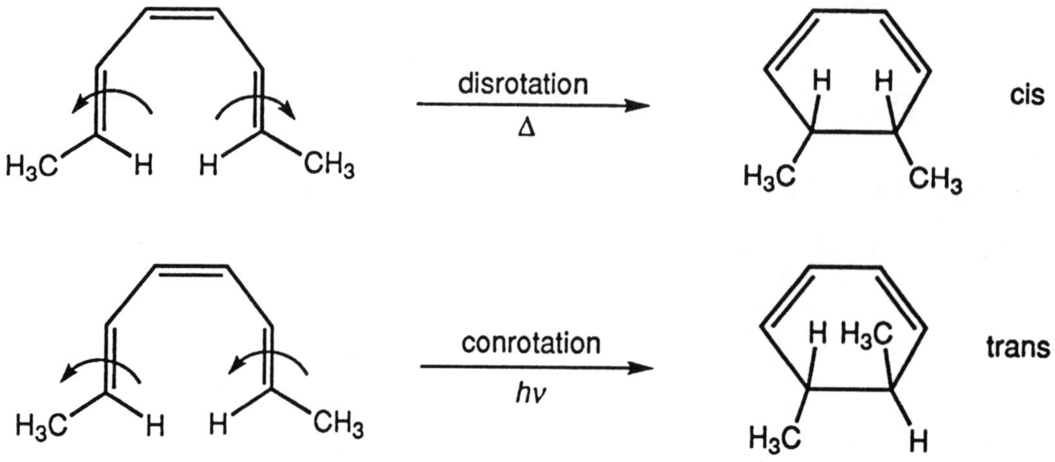

Note that this conrotatory motion produces the enantiomer of the molecule of Figure 22.20. The two conrotatory modes will be followed exactly equally, producing a racemic mixture of enantiomers.

Problem 22.5 This one is easy. In the trans,trans,trans molecule the ends of the double bonds, where the new bond must form, cannot reach each other. As there can be no rotation about the central trans double bond (this costs over 50 kcal/mol), there can be no reaction.

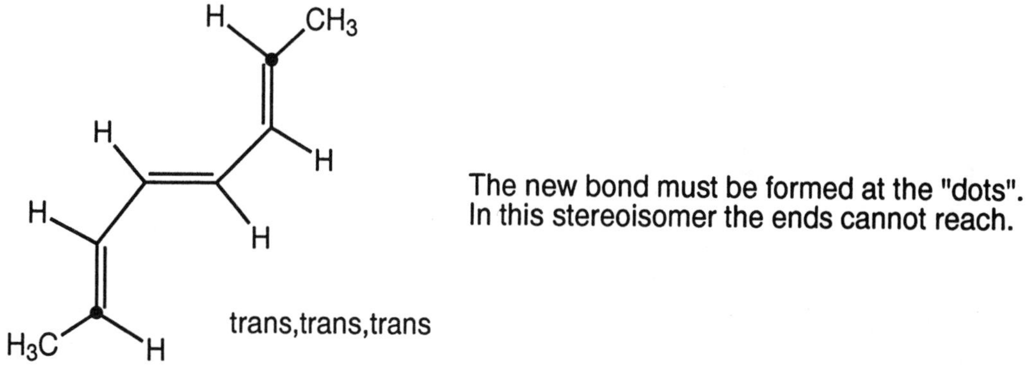

The new bond must be formed at the "dots".
In this stereoisomer the ends cannot reach.
trans,trans,trans

Problem 22.6 For these six-electron processes, the thermal reaction will be disrotatory and the photochemical reaction conrotatory. The following figure shows only one disrotatory and one conrotatory mode for each molecule. The others will give the same results. Note that we are still using the convention of writing only one member of racemic pairs of enantiomers. As the starting materials are achiral, no optical activity can really be induced. The other disrotatory mode for the cis,cis,trans molecule will produce the enantiomer of the trans molecule shown. Similarly, for the cis,cis,cis isomer the other conrotatory mode will produce the enantiomer of the molecule shown.

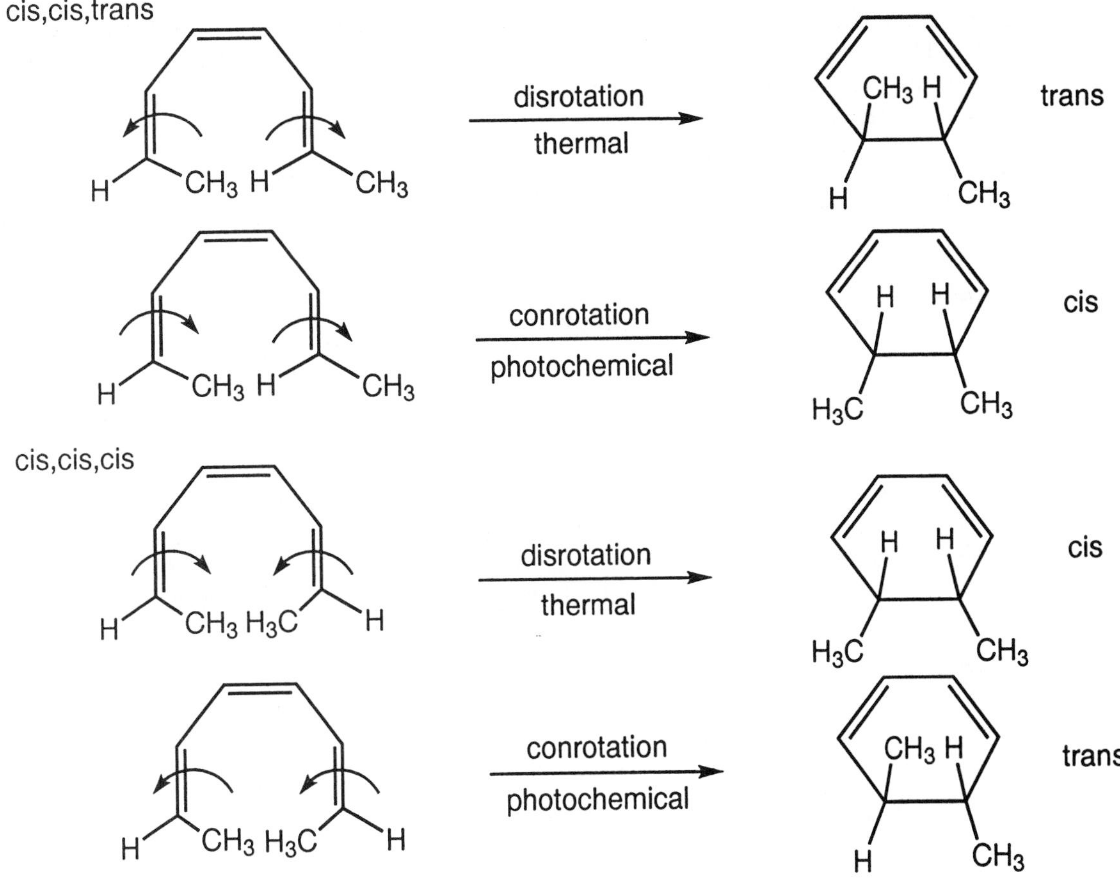

Problem 22.7

(a) This molecule incorporates an eight-electron system, a $4n$ number, so we should expect the thermal reaction to take place in a conrotatory manner (Table 22.2). We can verify this by looking at the molecular orbitals involved. The HOMO for octatetraene is Φ_4. Here are the four lowest molecular orbitals.

Top views of the molecular orbitals of octatetraene		
	Φ_4	+ + − − + + − − HOMO
	Φ_3	+ + + − − + + +
	Φ_2	+ + + + − − − −
	Φ_1	+ + + + + + + +

Now the task is to see how the orbitals must move. The figure shows the ends of the HOMO, Φ_4, as well as one conrotatory motion. If you follow through where the methyl groups must go, you can see why it is the trans stereochemistry that appears in the product.

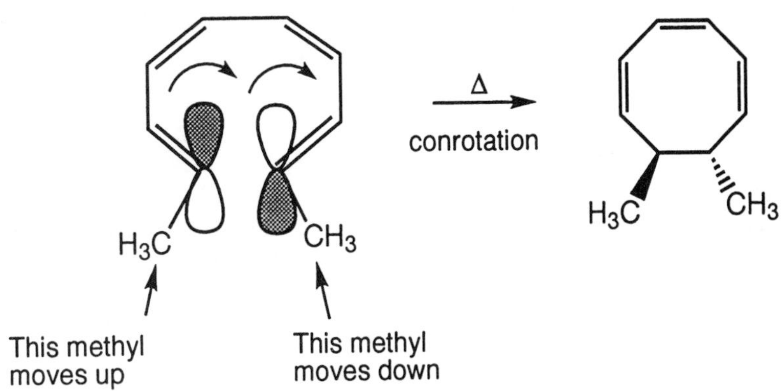

(b) The rearrangement of Dewar benzene to benzene involves the opening of a cyclobutene to a diene. The thermal motion must be conrotatory, and this conrotation must lead to a trans double bond. Of course, a trans double bond cannot be accomodated in a six-membered ring. There is no simple, disrotatory motion possible. Orbital symmetry will not permit it.

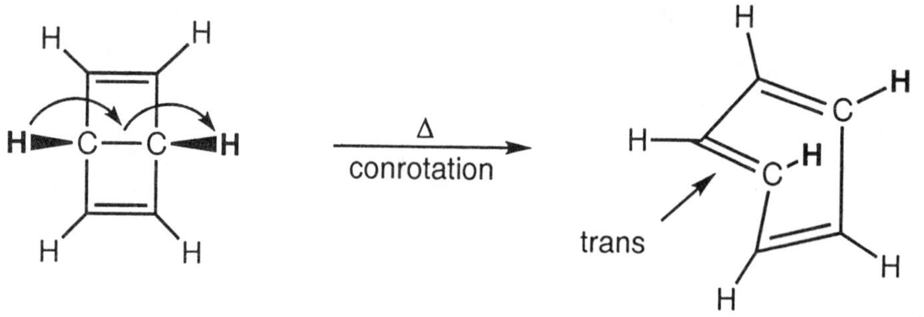

Δ
conrotation
trans

(c) Compound "**X**" is formed by opening the cyclobutene. A thermal opening of a cyclobutene must be conrotatory, so **X** must incorporate one trans double bond. Compound **X** contains a 10-membered ring, so the trans double bond is not impossibly strained, but it is unstable and rearranges quickly to the product shown in the problem, *trans*-9,10-dihydronaphthalene*. The problem now is why it is the trans stereoisomer that is formed in this second step.

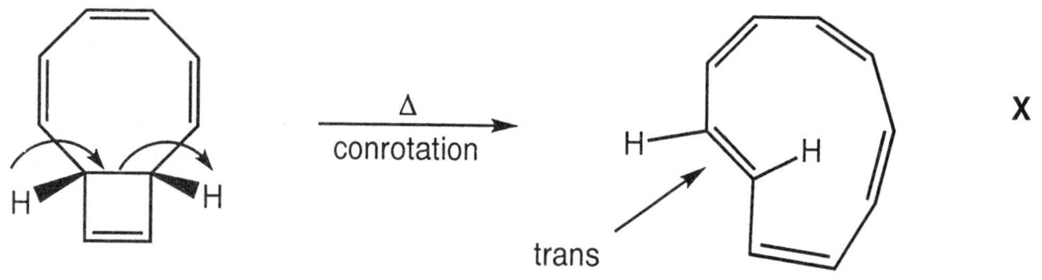

Δ
conrotation
trans
X

trans,cis,cis,cis,cis-Cyclodecapentaene

Further reaction of **X** involves a hexatriene–cyclohexadiene interconversion and must take place thermally in a disrotatory way (Table 22.2). This rotation leads directly to the *trans*-9,10-dihydronaphthalene product.

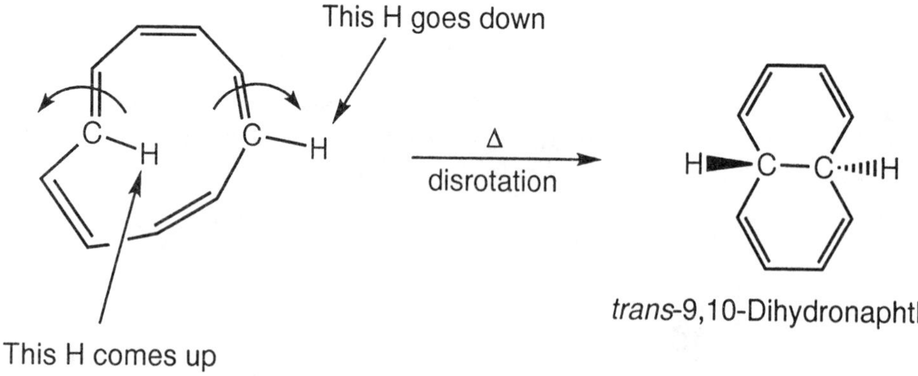

This H goes down
Δ
disrotation
This H comes up

trans-9,10-Dihydronaphthalene

*The authors of this Study Guide have argued long and bitterly over the naming of this compound. One of us argues for the traditional name given to the compound when it was first made (by this same author) in 1967. The other claims that the more modern name, 4a,8a-dihydronaphthalene, should be used. Guess which author had the last turn with the proofs?

(d) Each reaction involves two hexatriene–cyclohexadiene reactions. The thermal reactions must be disrotatory and the photochemical reactions must be conrotatory (Table 22.2).

Δ, two disrotations → cis,cis

The left-hand hydrogens come up

The right-hand hydrogens go down

H up, H down, H down, H up

$h\nu$, two conrotations → trans,trans

Problem 22.8 In this case, the interactions would be HOMO (ethylene = π^*) – LUMO (butadiene = Φ_3). The photochemical Diels–Alder is still forbidden.

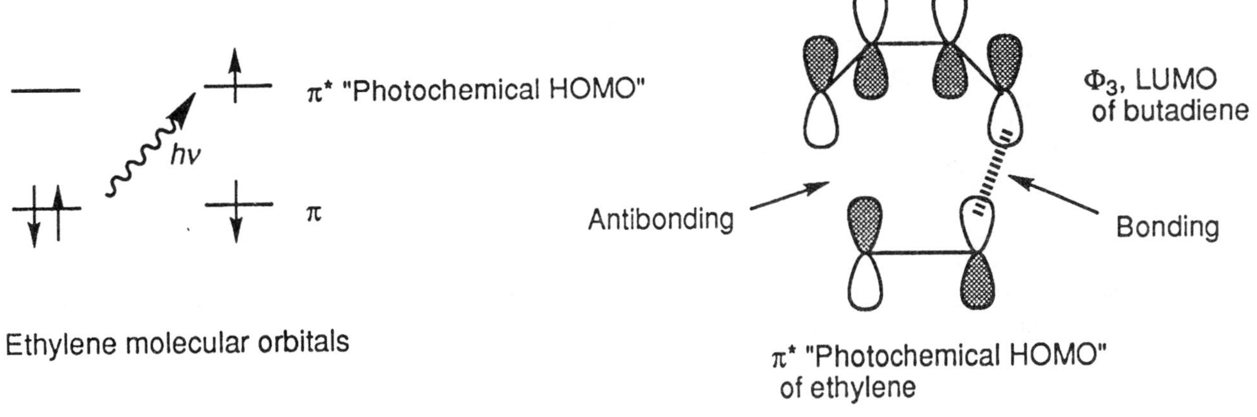

Problem 22.9 A ΔH calculation can easily be performed.

Bonds Broken

2 π bonds =

2 x 66 = 132 kcal/mol

$H_2C{=}CH_2$ + $H_2C{=}CH_2$ → cyclobutane

Bonds Made

2 σ bonds =

2 x 90 = 180 kcal/mol

– 25 kcal/mol strain

Net = 155 kcal/mol

So, the reaction is calculated to be exothermic by approximately 23 kcal/mol (155 – 132 kcal/mol). In this calculation, the carbon-carbon bond strength for ethane was used (Table 8.2, p. 307). More appropriate would be a lower value for the breaking of the central carbon-carbon bond in butane. If this value, ca. 83 kcal/mol, is used, the reaction is calculated to be exothermic by only 9 kcal/mol.

Problem 22.11 The more substituted a double bond, the more stable it is (Chapter 4, p. 134). The isomer on the right is the more stable compound.

Two disubstituted double bonds

Less stable

One disubstituted double bond and one trisubstituted double bond

More stable

Problem 22.13

(a) This problem involves a straightforward [1,9] shift of hydrogen. Don't forget, the shift numbering protocol has nothing to do with the naming protocol.

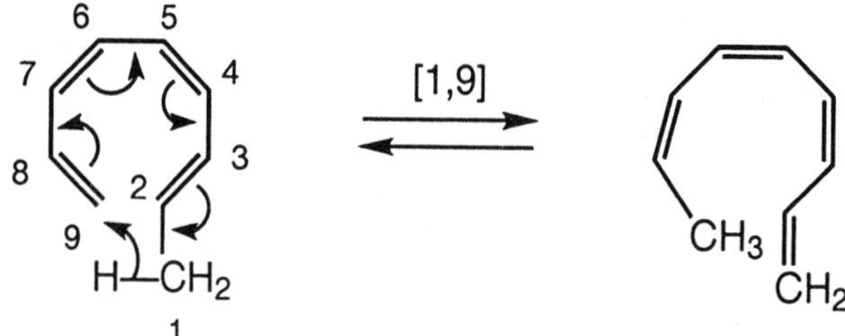

(b) This part is tricky only because of the way the molecule is drawn. It is a [1,3] shift of the pentadienyl group.

(c) The label lets you follow this through as a [3,5] shift of a pentadienyl group. The bond breaks at position 1 and re-forms at position 5, attaching to position 3 of the allyl fragment. It is the 3- and 5-positions that come together to make the new bond.

(d) This reaction is a simple [3,3] shift. The 3-position of the starting migrating group becomes attached to the 3-position of the framework over which the migration takes place.

(e) This problem is another hard one. It is a [1,5] shift of the indicated carbon. The R group is there only as a marker so that you can tell that a reaction has taken place.

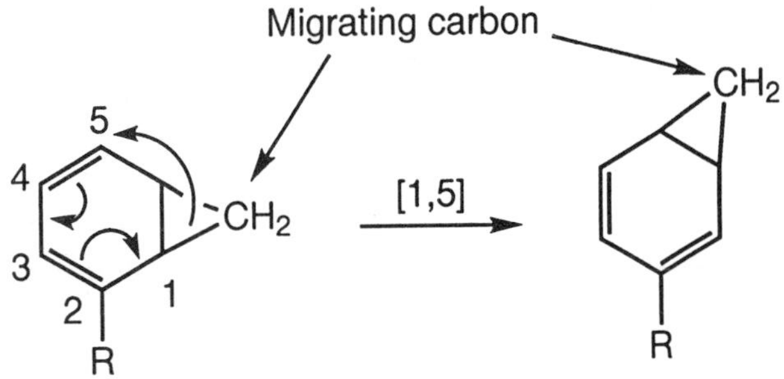

Problem 22.15 Not a chance. The HOMO is Φ_2 of allyl. Therefore, the thermal [1,3] shift requires an antarafacial shift of hydrogen and the 1*s* orbital is not large enough to span the distance.

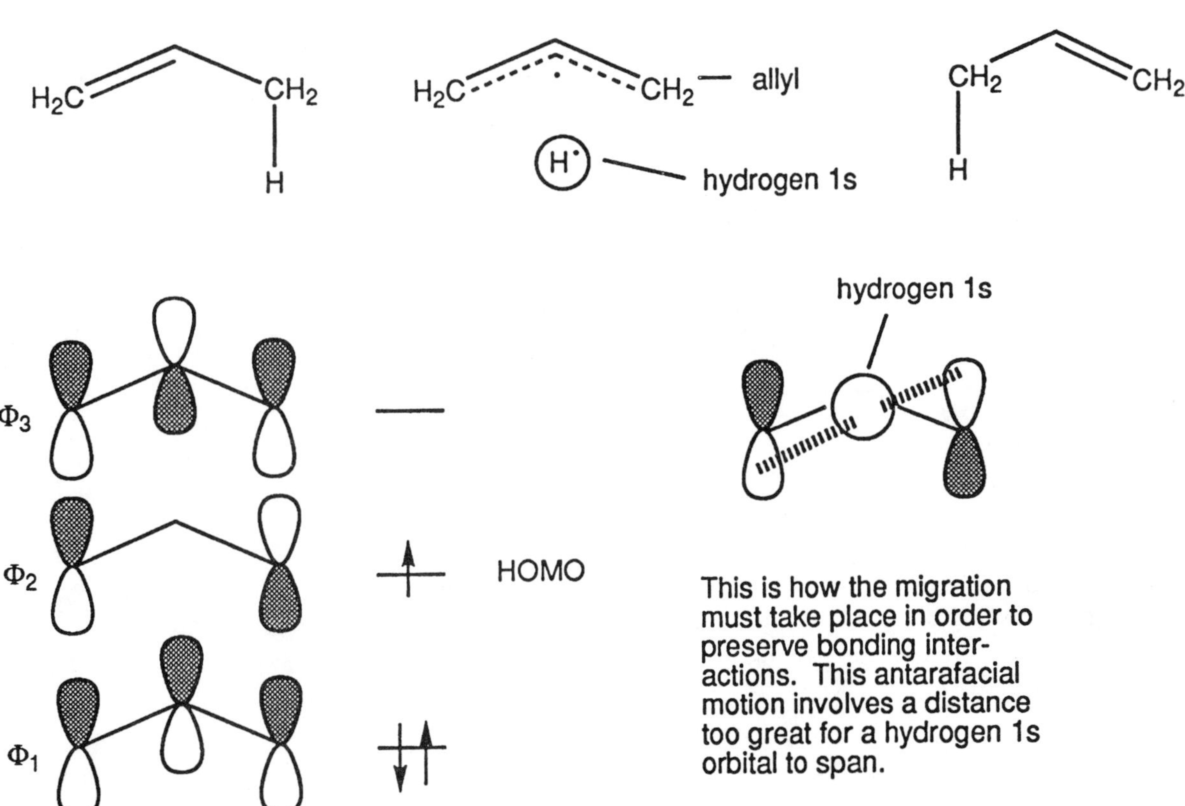

Problem 22.16
(a) The key reaction is a [1,5] shift to give the nonaromatic molecule "isoindene". Isoindene can be formed by a [1,5] shift of either hydrogen or deuterium, and both reactions are shown in the figure. A rapid re-formation of the aromatic ring through another [1,5] shift, again of either hydrogen or deuterium, leads to the isomerized products. The difficulty in this problem lies in seeing the necessity for the very endothermic first step. Notice that the aromatic circle makes the shift harder to see. Write Kekulé forms! The convention is to count the shift as a [1,5] shift, even though it could be called a [1,2] shift just as well. No matter what you call it, it is the same reaction.

H [1,5] H [1,5]

Isoindenes

D [1,5] H [1,5]

(b) Three-membered rings can often play the parts of double bonds. This reaction shows just such a case as the [1,5] shift of hydrogen takes place with opening of the cyclopropane. The reaction is quite analogous to a [1,5] shift in a 1,3-pentadiene.

(c) An arrow formalism shows the reaction to be a [1,5] shift of carbon.

arrow formalism – a [1,5] shift of carbon

As the C–1 bond breaks, the new bond between C and position 5 will form. Notice that the lobes making up the C–1 bond must be of the same sign—they are a bond, not an antibond.

Breaking bond

A pentadienyl radical

Now fill in the phase relationships for the pentadienyl radical.

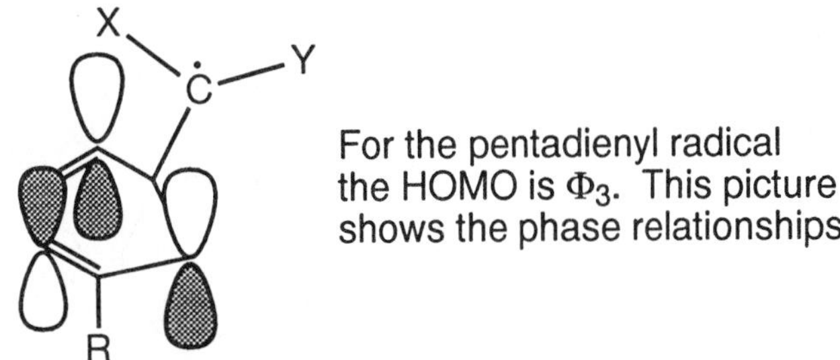

Now look at how the new bond at position 5 must be formed. Two lobes of the same phase must overlap, and in this case that must be two light lobes. Look how the bond must rotate to accomplish this.

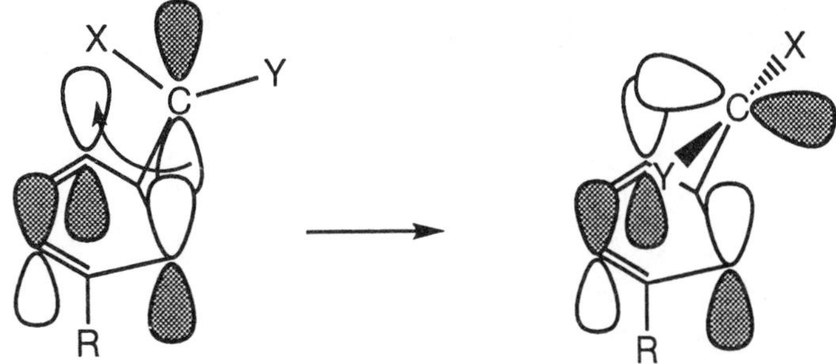

Y must swing *in* as X swings *out*. The two groups will alternate as the *walk* of CXY around the ring goes on.

View of the molecule from the top:

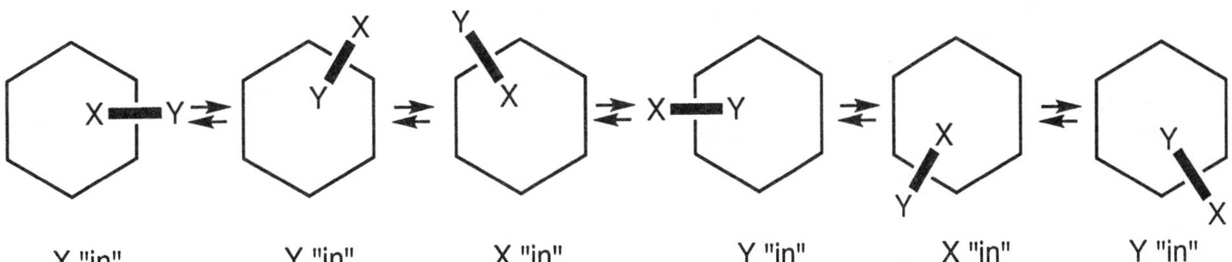

Problem 22.18 In this case, the Cope rearrangement involves a large amount of strain relief. The three-membered ring is opened in a disrotatory fashion to give the cycloheptadiene. Of course, the relatively strain free cycloheptadiene will be strongly favored at equilibrium.

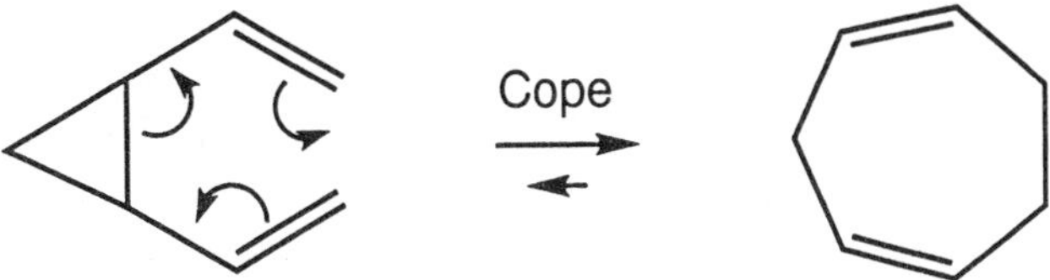

Problem 22.19 At low temperature the Cope rearrangement is stopped, and one sees the ^{1}H NMR spectrum expected of the simple, static structure. For example, there are four olefinic hydrogens at about δ 5.8 ppm.

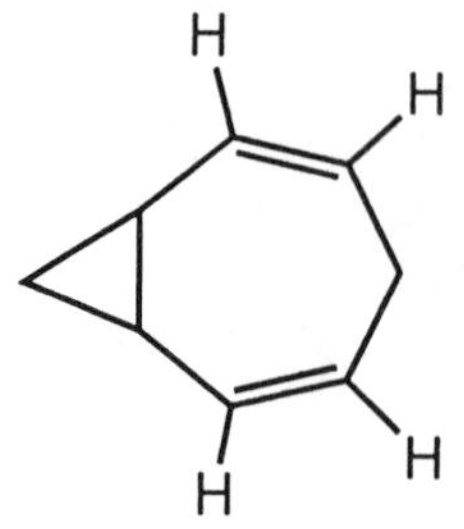

At high temperature, the Cope rearrangement is fast, and a completely averaged spectrum is obtained. There are only two hydrogens that are always olefinic, and these appear at about δ 5.8 ppm (H_a). Four hydrogens are half olefinic and half cyclopropyl, and these appear at about δ 3.7 ppm (H_b).

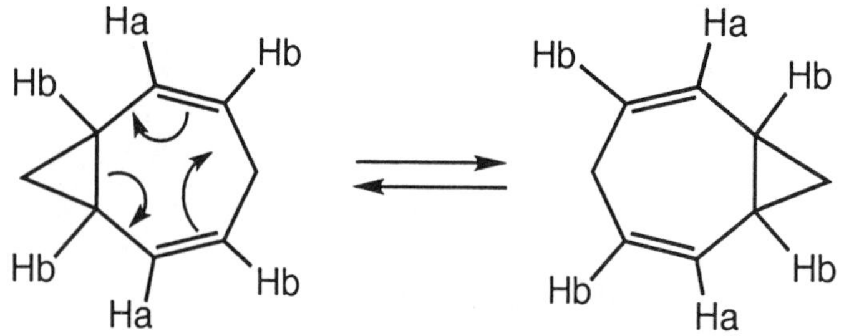

In between these extremes of the frozen out spectrum and the averaged spectrum, we see the 20 °C intermediate spectrum. We see a spectrum of no particular definition as incomplete averaging of the signals is taking place. Near room temperature we are neither at the "frozen out" nor completely averaged regimes, but somewhere in between.

Problem 22.22 Photochemical 2 + 2 reactions are allowed by orbital symmetry, and this reaction is just a retro 2 + 2 addition. The formation of benzene drives the reaction to the right.

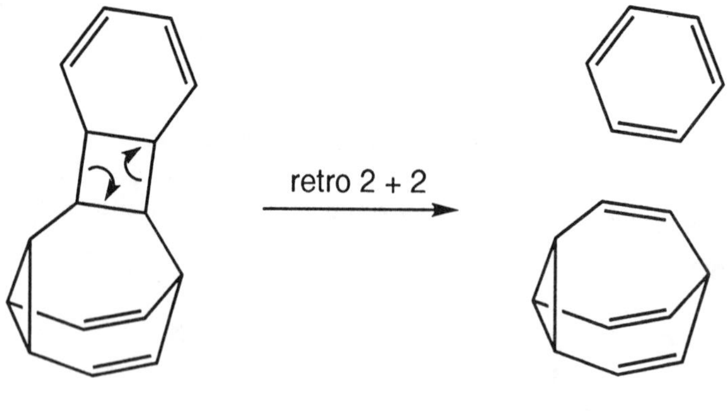

Problem 22.23 This problem shows two ways of disguising the Cope rearrangement.

(a) Whenever you see a complex molecule containing double bonds, think "Cope." This advice is especially useful when the subject of thermal reactions is being discussed, and, of course, that's where we are right now. Count to see if there is a 1,5-diene present.

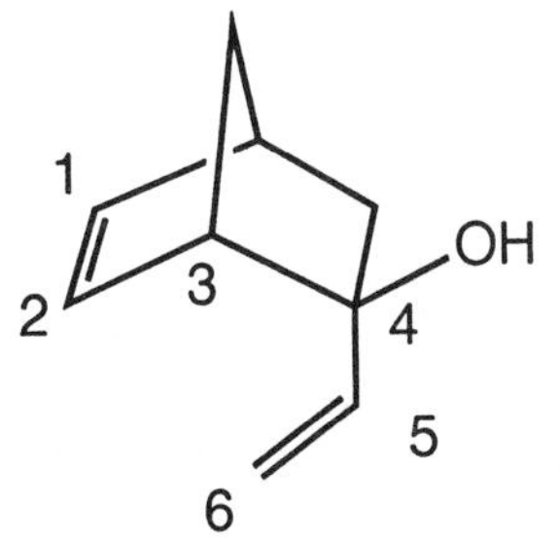

A 1,5-diene, so a Cope is possible.

If there is, try the Cope, no matter how odd the system looks. Write the three "Cope" arrows, forming the bond between carbons 1 and 6 and breaking the bond between carbons 3 and 4. After you draw the arrow formalism be sure to redraw the molecule first without moving anything! Now you can refashion the molecule in a reasonable way. In this case, the product is an enol that is, of course, in equilibrium with the related keto form.

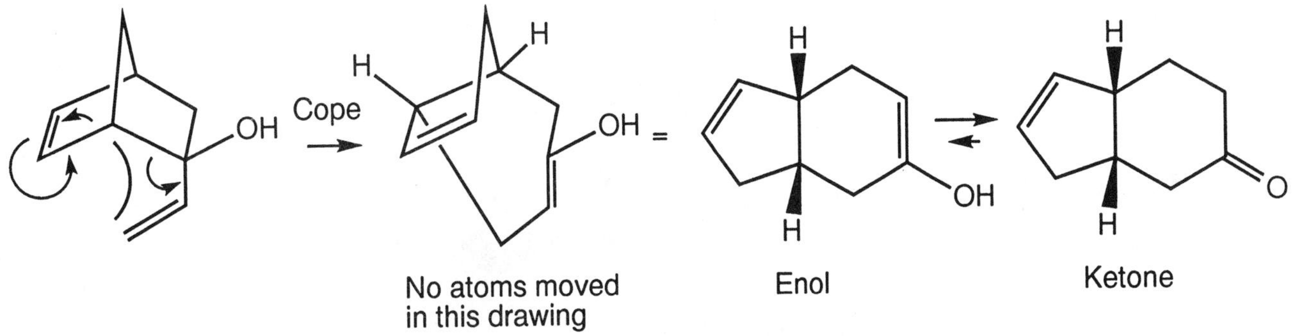

No atoms moved in this drawing

Enol

Ketone

(b) In this example, the aromatic ring helps to hide the 1,5-diene at the heart of the Cope rearrangement. Nonetheless, it is a 1,5-diene, and a Cope-like rearrangement is possible. In this case, it leads to a ketone that is much less stable than it enol form, a phenol. This variation of the Cope is called the Claisen rearrangement. It's the same Claisen of the Claisen condensation (Chapter 20, p. 1037) and the Claisen–Schmidt condensation (Chapter 18, p. 914).

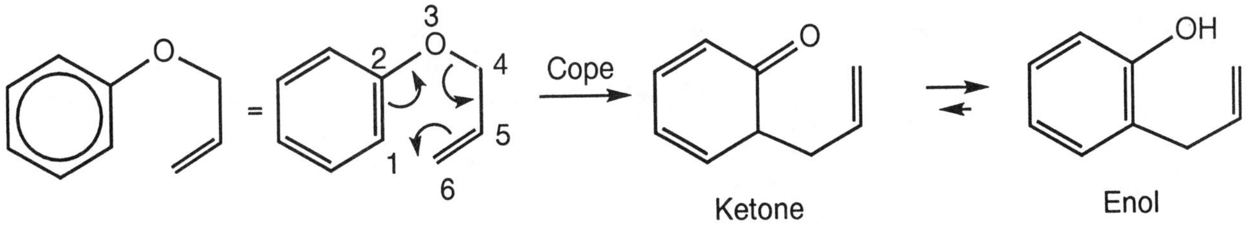

Ketone

Enol

Answers to Additional Problems, Chapter 22.

Problem 22.24 First of all, stereochemistry is very important. If there is a trans double bond in the heptatriene, there is no way that the ends of the system can come together. The reaction must fail; no [1,7] shift is possible.

These three isomers must fail; the ends cannot meet

Only in this stereoisomer can the [1,7] shift take place

For heptatrienyl, there are seven π electrons, and therefore the HOMO is Φ_4. For photochemical transformations, the HOMO will be Φ_5. We need only examine the ends of the π system to see what is possible. In this case, the photochemical [1,7] shift must be suprafacial and the thermal [1,7] shift must be antarafacial. In other words, the [1,7] shift looks just like the [1,3] shift.

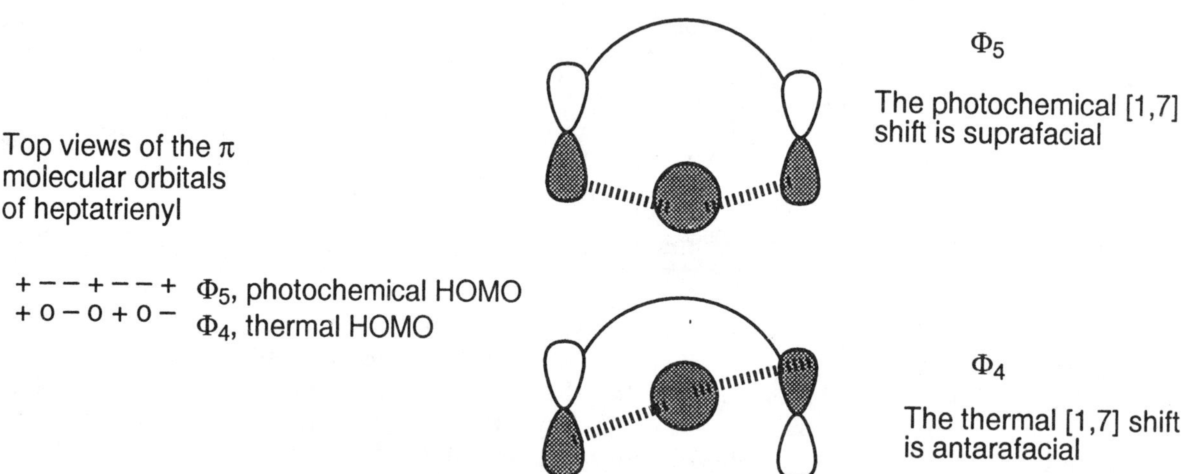
Top views of the π molecular orbitals of heptatrienyl
+ − − + − − + Φ5, photochemical HOMO
+ 0 − 0 + 0 − Φ4, thermal HOMO
Φ5
The photochemical [1,7] shift is suprafacial
Φ4
The thermal [1,7] shift is antarafacial

Problem 22.25 The conversion of 7-dehydrocholesterol (**1**) to previtamin D_3 (**2**) involves an electrocyclic ring opening of a 1,3-cyclohexadiene to a 1,3,5-hexatriene. The figure shows that this change occurs in a conrotatory fashion. The methyl group moves in a clockwise fashion, as does the hydrogen. Be sure you can see this. This process must be photochemical, as the reaction involves six ($4n + 2$) π electrons, and conrotatory, six-electron reactions are photochemically, not thermally allowed. Be certain you can explain why this is so.

In the next reaction, we get to use the answer to the preceding problem. The transformation of previtamin D_3 (**2**) to vitamin D_3 (**3**) itself involves a [1,7] sigmatropic shift. If this shift occurs under thermal conditions, it must be antarafacial. If it is a photochemical process, it must be suprafacial. Although these two processes cannot be differentiated using these molecules, the antarafacial nature of a similar thermal [1,7] shift of hydrogen was recently demonstrated using a deuterium labeling experiment.

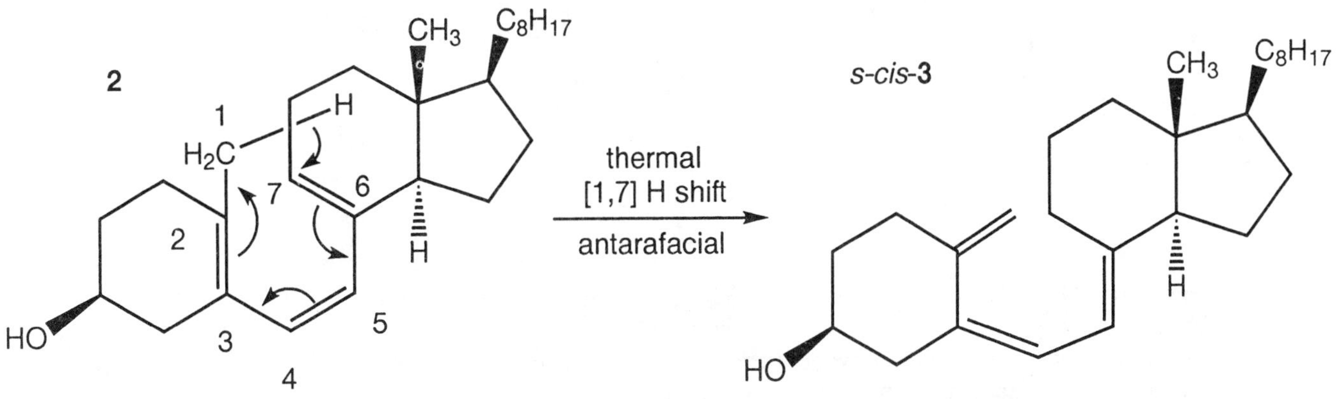

Problem 22.26 The set of arrows in (a) is misleading. The arrow formalism implies a butadiene–cyclobutene interconversion. Under thermal conditions, this requires a conrotatory motion, and this motion must generate not **2**, but a trans version of **2**, too highly strained to be isolable.

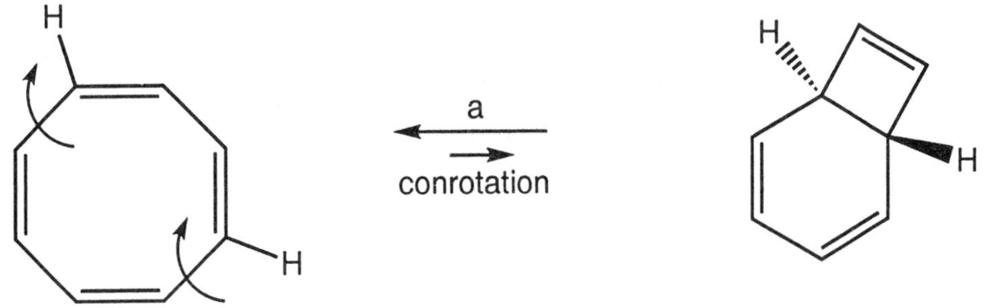

1 *trans*-**2**

If we look at the reaction in the other direction, we can see that a conrotatory opening of **2** must lead not to **1**, but to a version of **1** in which there is a trans double bond. This molecule is too strained to be formed.

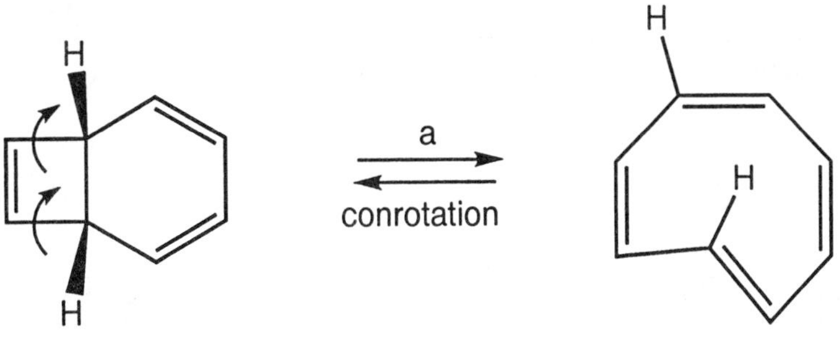

2 *cis,cis,cis,trans*-**1**

By contrast, arrow formalism (b) shows a thermal, disrotatory interconversion of a 1,3,5-hexatriene and a 1,3-cyclohexadiene. There is no difficulty with stereochemistry or strain here. This arrow formalism is much more appropriate.

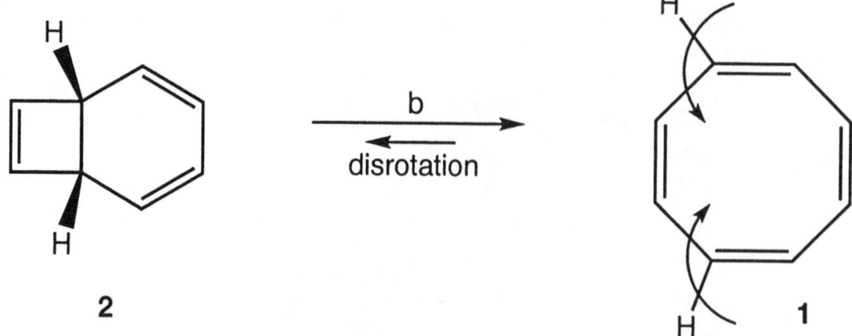

Problem 22.27 The charges disturb many people, but there is no reason to be upset. They become important only when it becomes time to count electrons.
(a) This reaction is an electrocyclic interconversion of the cyclopropyl and allyl cations, and the question to be decided is whether the reaction is conrotatory or disrotatory.

This bond breaks - will it be conrotatory or disrotatory?

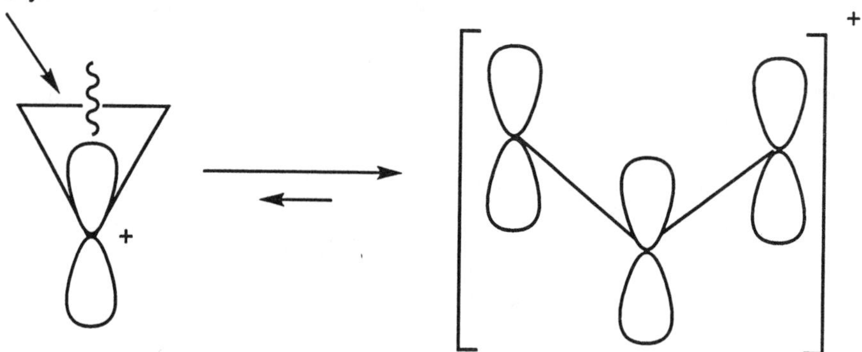

The cyclopropyl cation - note the 2*p* orbital at one carbon

The allyl cation

The number of electrons involved should tell us the answer without looking any further. Now the positive charge is important because it allows us to count electrons properly. The allyl cation has only two π electrons (destined, on closing, to become the two electrons in the new cyclopropane bond). Two electrons is a $(4n + 2)$ number, $n = 0$, and so we can expect this reaction to be disrotatory. So

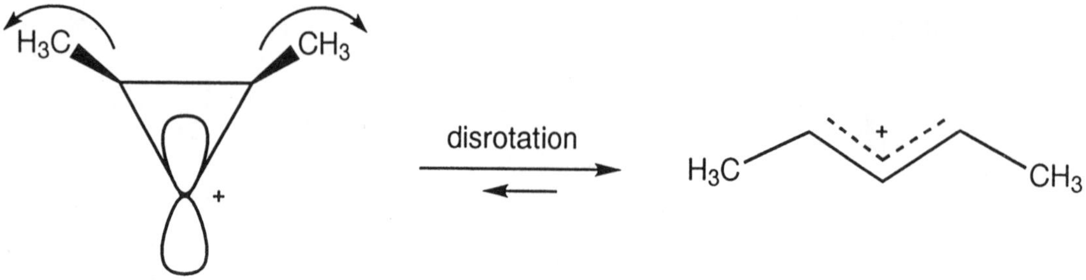

H_3C CH_3 + disrotation H_3C + CH_3

Let's do a better job of analysis. As usual, we will look at the open partner in the reaction, in this case the allyl cation. The HOMO for the cation is Φ_1. In order to make a bonding interaction, rotation must be in the disrotatory sense.

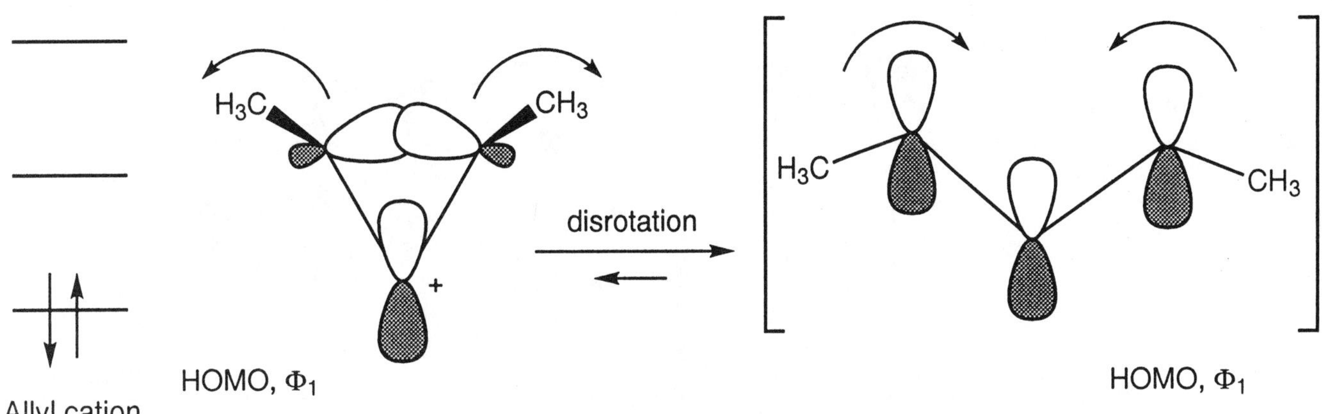

The other possible disrotation leads to the allyl cation with the two methyl groups inside. This process will be disfavored thermodynamically for steric reasons.

(b) For the anion, there are four electrons, and we can expect a thermal, conrotatory process in this $4n$ system. Here is the orbital analysis.

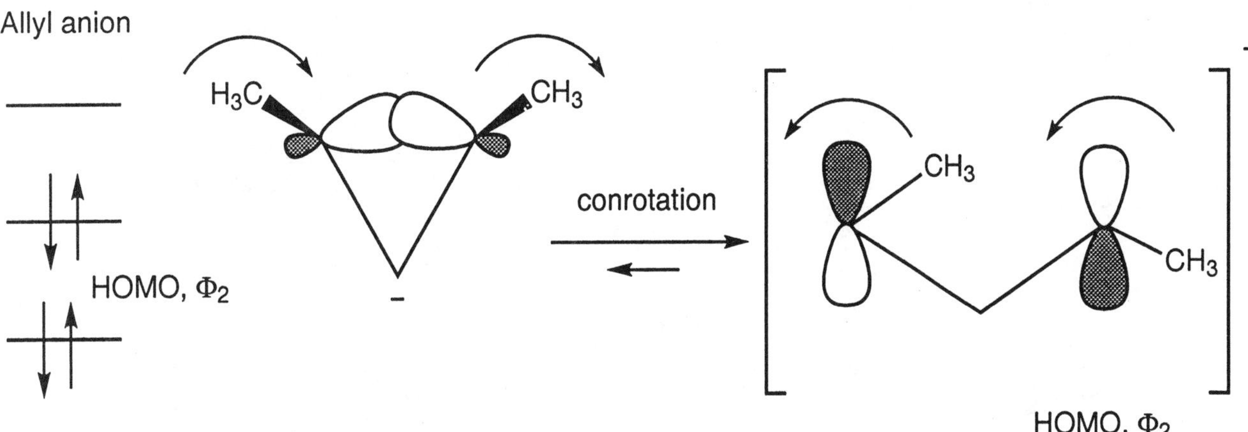

The other possible conrotation leads to the same product.

Problem 22.28 First of all, here are arrow formalisms for the two possible reactions.

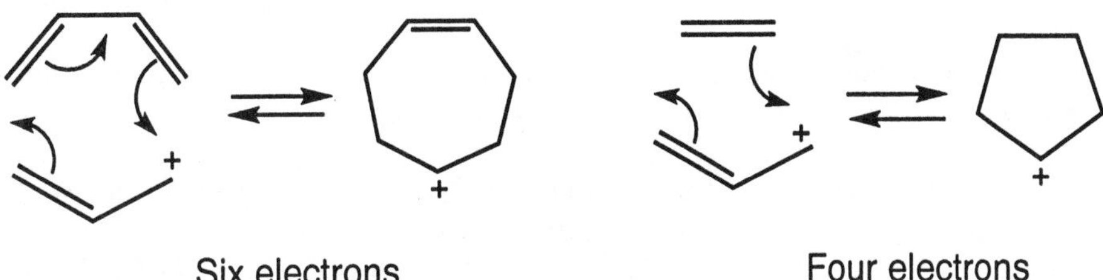

Once again if we count electrons, we can answer the question with very little analysis. The first process is a six-electron cycloaddition, and, like the Diels–Alder reaction, would be expected to be a thermal process. The second reaction is a four-electron cycloaddition and, like the cycloaddition of a pair of ethylenes, should be a photochemical reaction. Now, let's do a better HOMO–LUMO analysis to see which of the two should be concerted photochemically. We will use the "photochemical HOMO" of allyl and the LUMO of the reaction partner (although an analysis that did just the opposite, and used the "photochemical HOMO" of the reaction partner and the LUMO of the allyl cation, would work out exactly the same).

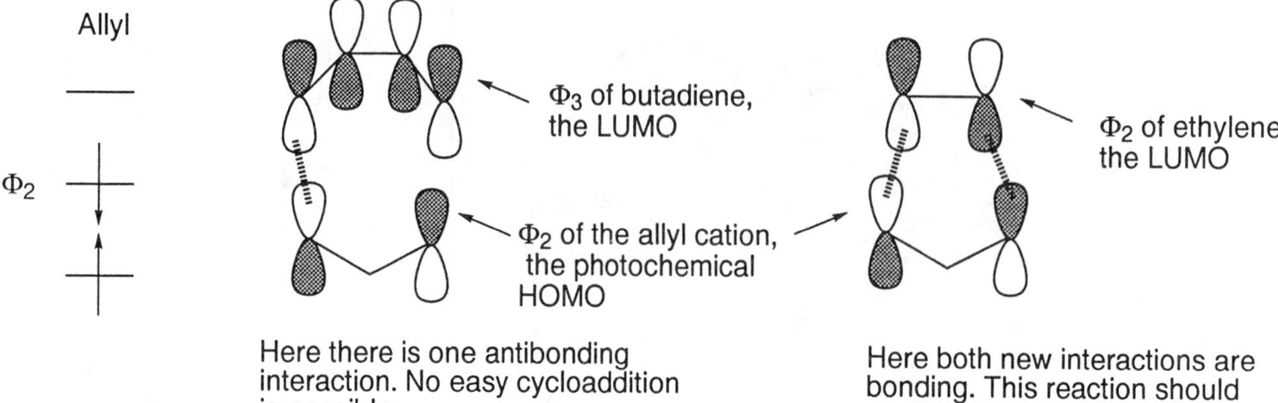
Allyl
Φ_2
Φ_3 of butadiene, the LUMO
Φ_2 of the allyl cation, the photochemical HOMO
Φ_2 of ethylene, the LUMO
Here there is one antibonding interaction. No easy cycloaddition is possible.
Here both new interactions are bonding. This reaction should work well.

Problem 22.29 The thermal opening of a cyclobutene must be a conrotatory process. Follow through the two reactions and look for a difference.

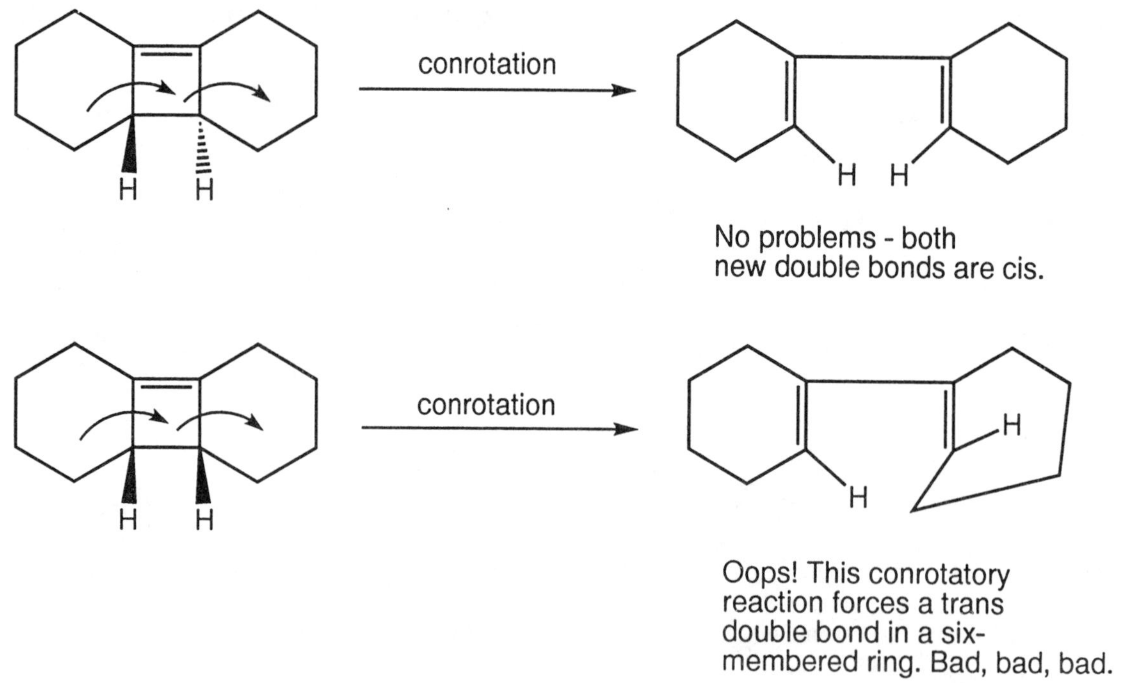
conrotation
H H
H H
No problems - both new double bonds are cis.
conrotation
H H
H
H
Oops! This conrotatory reaction forces a trans double bond in a six-membered ring. Bad, bad, bad.

It's not hard to find. The cis compound must produce a compound containing a trans double bond in a six-membered ring. This reaction will surely be a slow process.

Problem 22.30 Concerted photochemical 2 + 2 cycloadditions and their reversals are allowed by orbital symmetry, and it is a reverse 2 + 2 cycloaddition that takes place here. As the reactions are concerted, the cis cyclobutene leads to the cis alkene and the trans cyclobutene leads to the trans alkene.

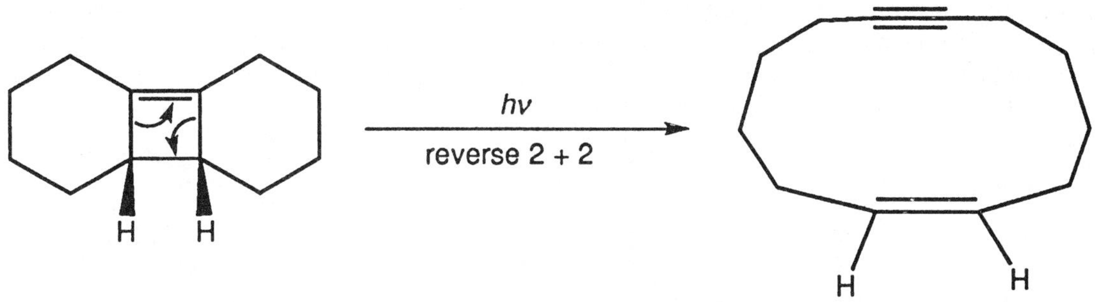

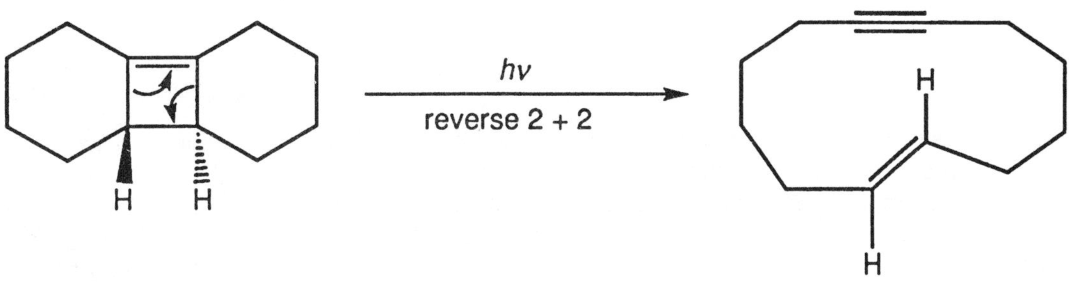

hv

reverse 2 + 2

hv

reverse 2 + 2

Problem 22.31 There are several clues in the drawing. First of all, whenever you see maleic anhydride think of the Diels–Alder reaction. Maleic anhydride is a favorite dienophile. Second, this question immediately follows a problem on the opening of cyclobutenes, and we deliberately drew the starting material without the aromatic circle, so you could see the hidden cyclobutene. Finally, all cyclohexenes are potential Diels–Alder products, and you might be able to take the product apart to its component parts. In any event, this problem involves the opening of the starting material to a diene, followed by capture by maleic anhydride. Notice how the last step recovers the aromaticity lost in the original opening. That is why the endocyclic 1,3-diene does not undergo a Diels–Alder addition.

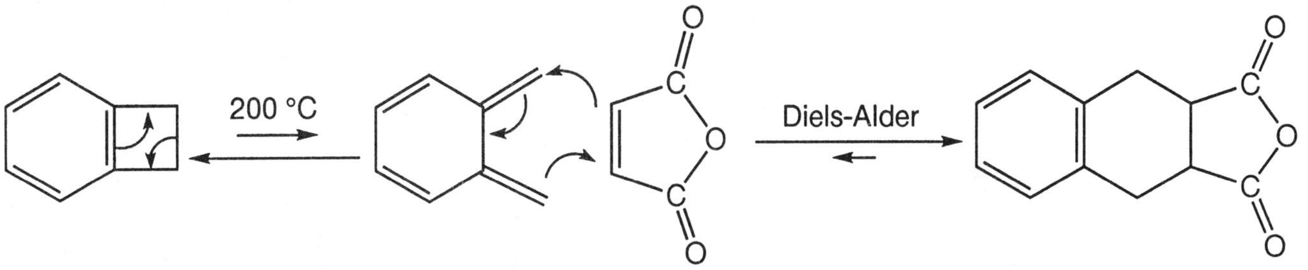

200 °C

Diels-Alder

Problem 22.32 Now we have to worry about stereochemistry. The thermal opening of the cyclobutene to the diene will be conrotatory, and in this case that will lead to intermediate **A**, as shown. The concerted, thermal, Diels–Alder reaction of **A** will lead to the compound in which the phenyl groups are cis.

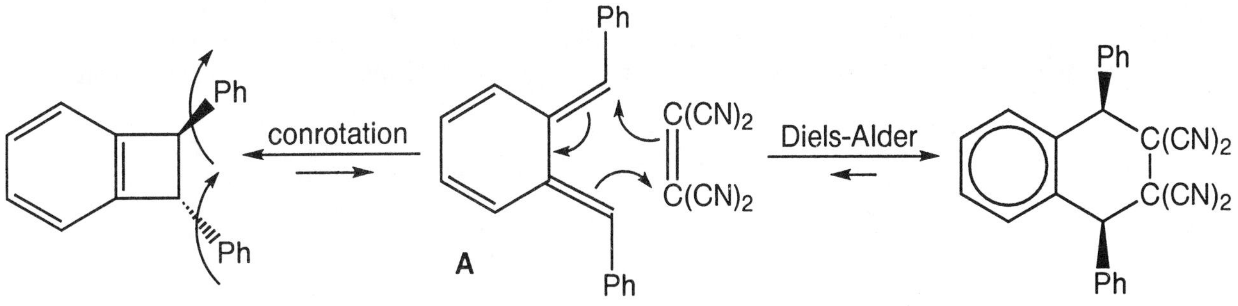

conrotation

Diels-Alder

A

Problem 22.33 Let's do what the problem suggests and work backward from the structure of the product. Let's also be mindful of the presence of maleic anhydride. A Diels–Alder reaction is likely to be involved somehow. So, let's deconstruct the cyclohexene in the product to its constituent parts, the diene **A** and a dienophile (maleic anhydride).

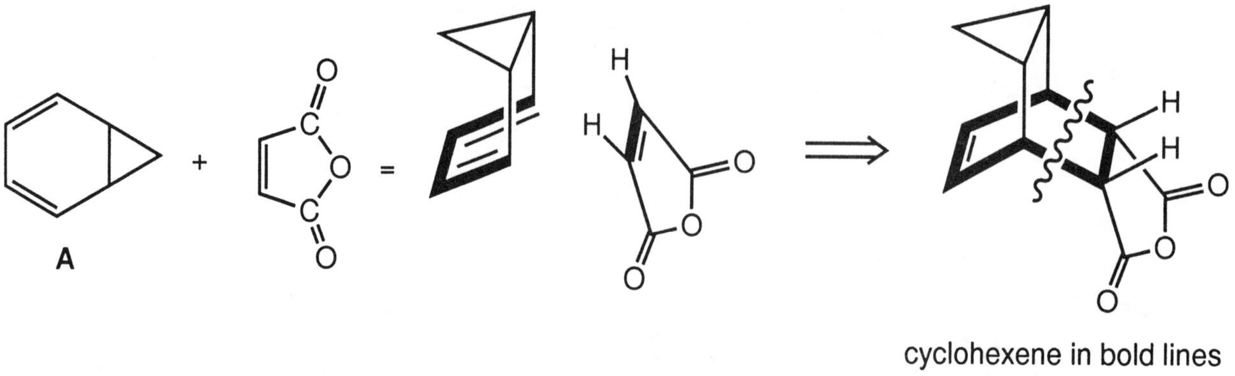

cyclohexene in bold lines

How can **A** arise from our starting material, cycloheptatriene? We saw many cyclohexadiene-hexatriene rearrangements in this chapter, and this reaction involves another. In this reaction, cyclohexatriene first closes to **A** (called norcaradiene), and **A** is then captured by maleic anhydride to give the product. The closing to **A** is sure to be endothermic because of the strain added by the three-membered ring, but a little **A** is formed, and as it is captured by maleic anhydride, more is produced.

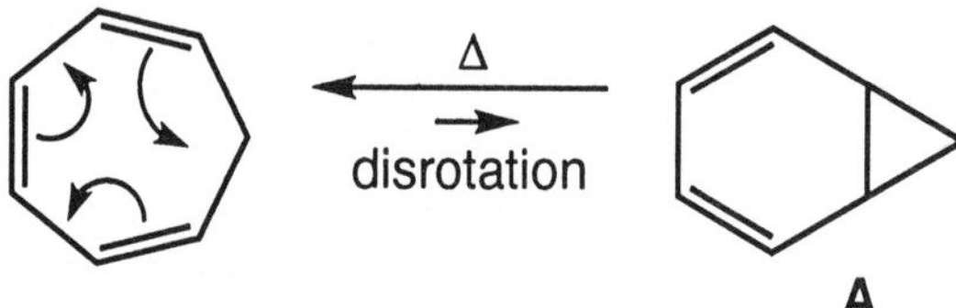

Problem 22.34 The arrow formalism is relatively easy, although it may be difficult to draw the polycyclic product molecule correctly. As usual in such cases, we have first drawn the molecule with only minimal movement of the reaction partners, and then "relaxed" to the product structure.

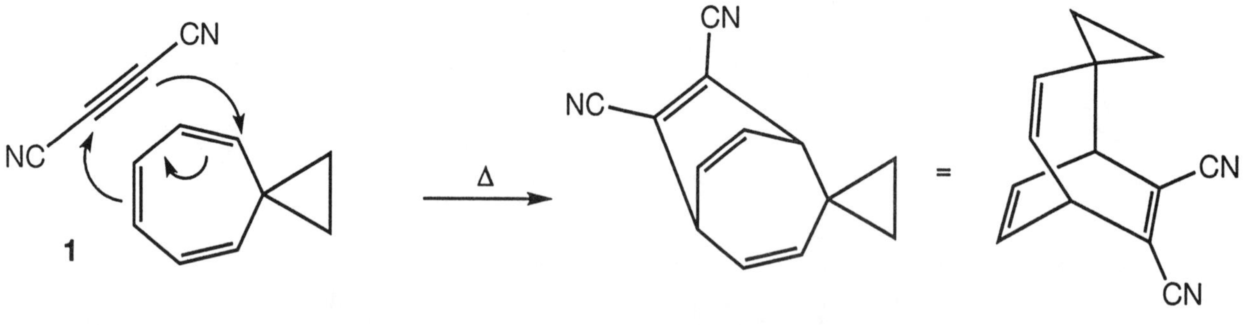

Now the question is why compound **1** reacts in normal Diels–Alder fashion, whereas other cycloheptatrienes do not. As shown in Problem 22.33, they usually first isomerize to a small amount of the "norcaradiene" form, and then react. Look at what happens when **1** isomerizes to a norcaradiene. A compound with two spiro three-membered rings **(A)** is formed. It is this extra strain that prevents **1** from forming enough of the norcaradiene to undergo the "usual" reaction.

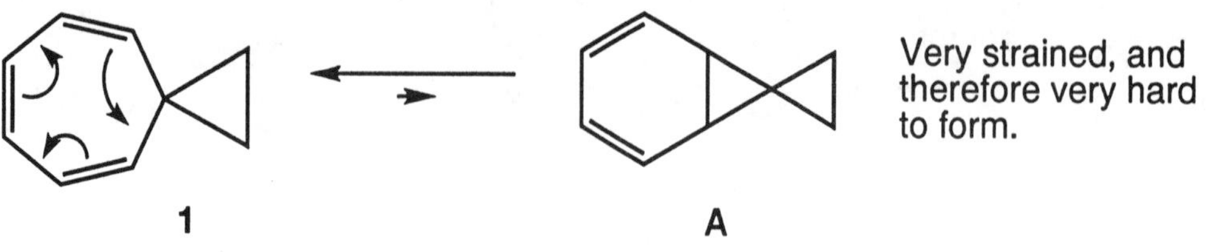

Problem 22.35 In this classic problem, we face two difficulties. First, we need a general mechanism for the formation of the product, and then we need to worry about the stereochemical details. The mention of stereospecificity should make you think about orbital symmetry (not to mention the placement of the problem in this chapter). Protonation of the carbonyl oxygen is the first step (you can't forget Chapter 16, and acid–base chemistry, just because we are in an "orbital" chapter).

H_3PO_4

The figure shows two resonance forms for the protonated carbonyl compound, but there are others. In fact, protonation has yielded a pentadienyl cation. Electrocyclic closure of the cation leads to **A**, and deprotonation and a keto–enol equilibration gives the product.

A

deprotonation and ketonization

Now, what about the stereochemistry? Let's look at the thermal electrocyclic closure of a pentadienyl cation. The system and top views of its π molecular orbitals are shown in the figure.

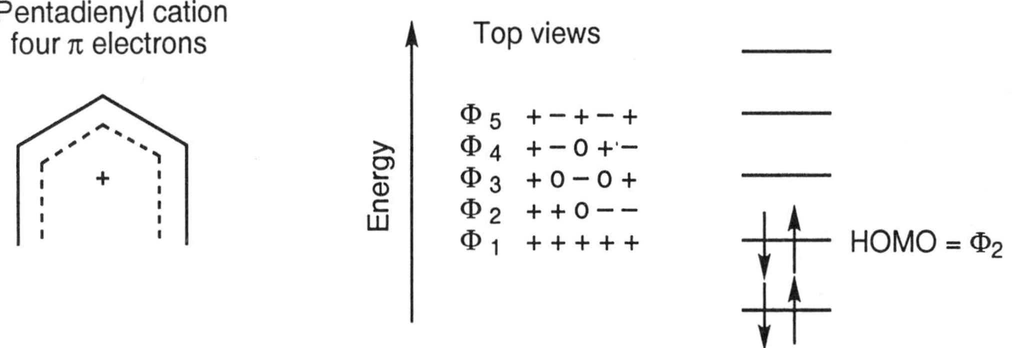

The HOMO is Φ_2 for this four-electron system, and closure must be conrotatory.

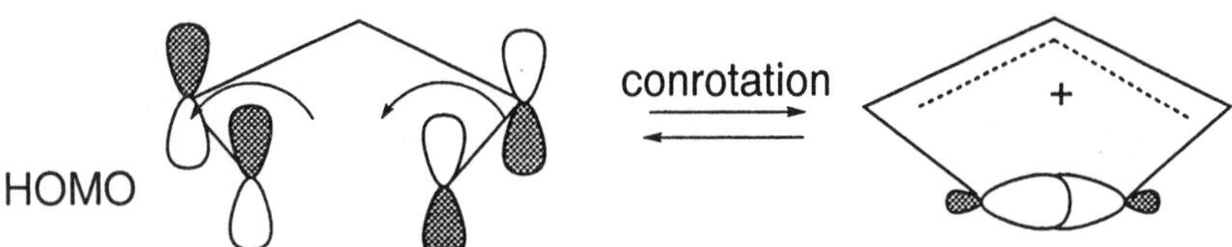

Now look at the more complicated pentadienyl cation in this problem. Conrotatory closure must lead to the observed trans stereochemistry.

conrotation

deprotonation and ketonization

A

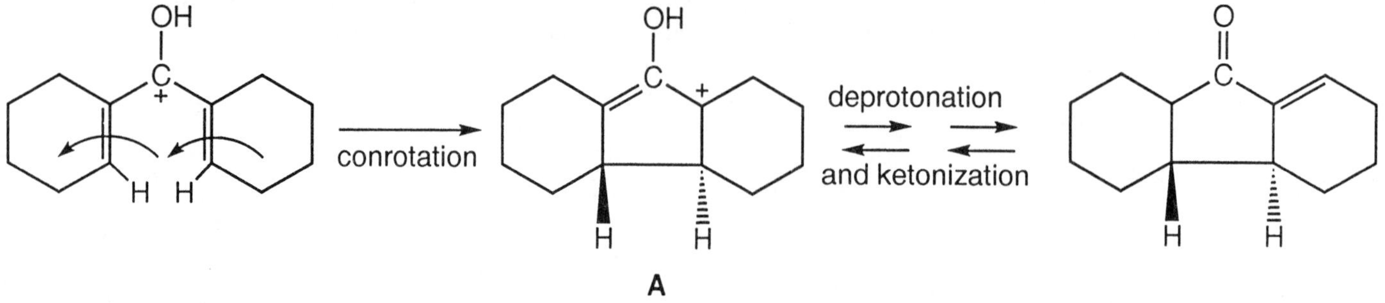

Problem 22.36 This problem is difficult only because there is an intermediate in the reaction and we didn't tell you. There is no way to go directly from starting material to product. Surely, in an orbital symmetry chapter, a cyclobutene must conjure the notion of opening to a butadiene. The thermal opening must be conrotatory, and this leads to intermediate **A**, in which there is one trans double bond.

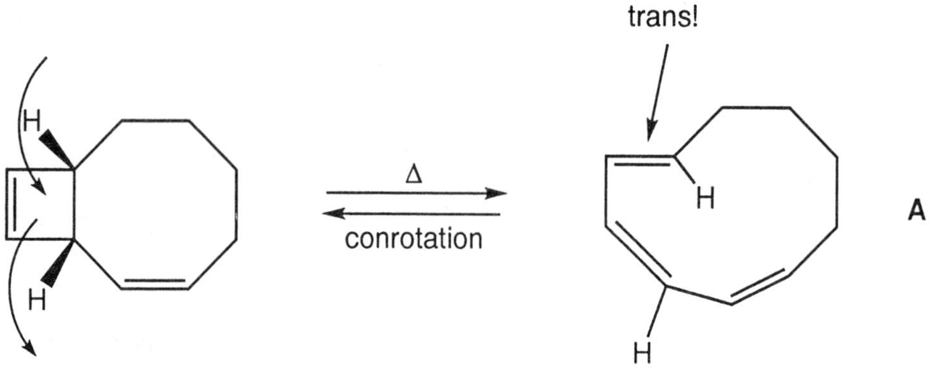

trans!

Δ

conrotation

A

Now **A**, a hexatriene, closes to the final product, a cyclohexadiene. This thermal reaction must be disrotatory, and this leads to the observed trans stereochemistry. The figure first shows the arrow formalism and then the stereochemistry.

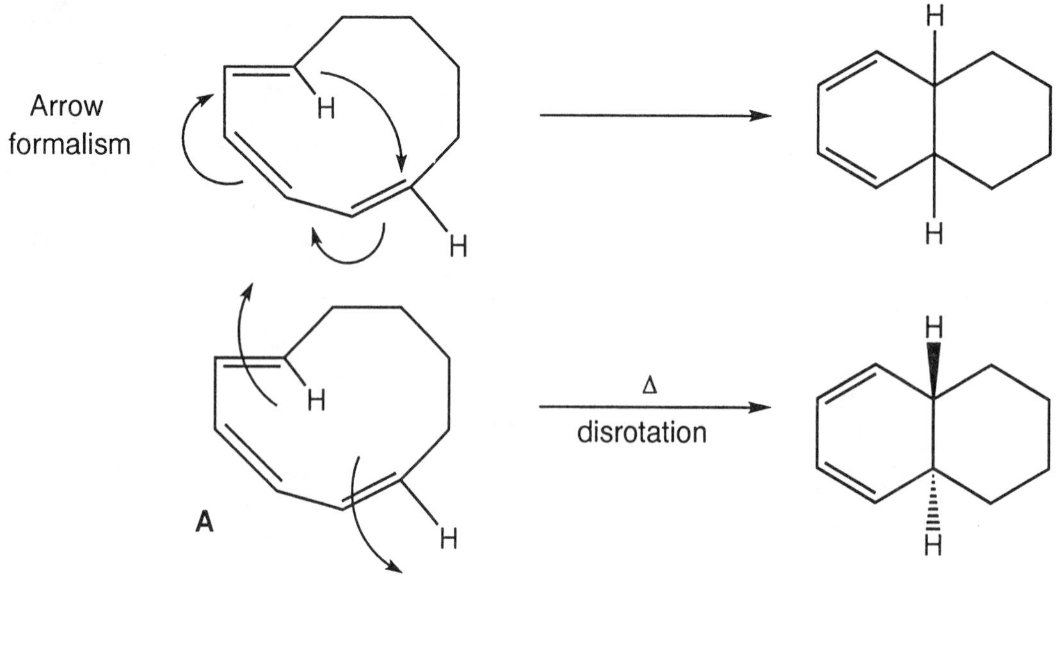

Arrow formalism

Δ

disrotation

A

Problem 22.37 Here we have a problem much like Problem 22.27. Despite the nitrogen atom, this is really nothing more than a cyclopropyl anion–allyl anion equilibrium. Like the cyclopropyl anion, this is a four-electron system. The nonbonding pair of electrons on nitrogen and the pair of electrons in the bond are the ones involved.

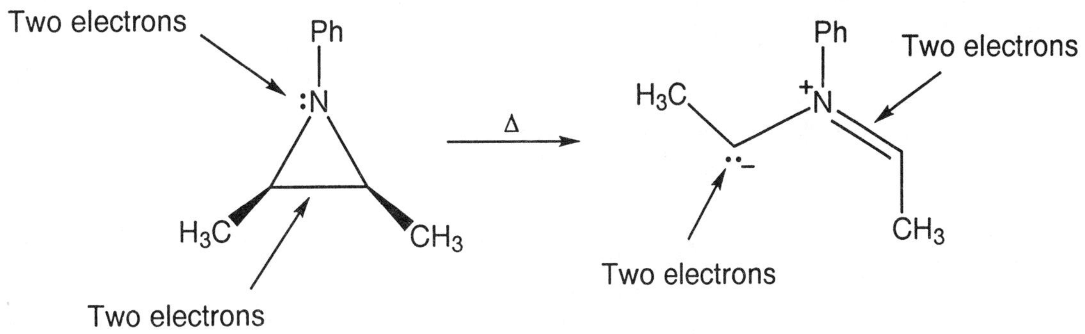

Two electrons
Ph
:N
H_3C
CH_3
Two electrons
Δ
Ph
H_3C
^+N
Two electrons
CH_3
Two electrons

A thermal, four-electron process will be conrotatory (it's the same as the cyclobutene–butadiene four-electron system), and the observed stereochemistry is exactly what conrotation must produce.

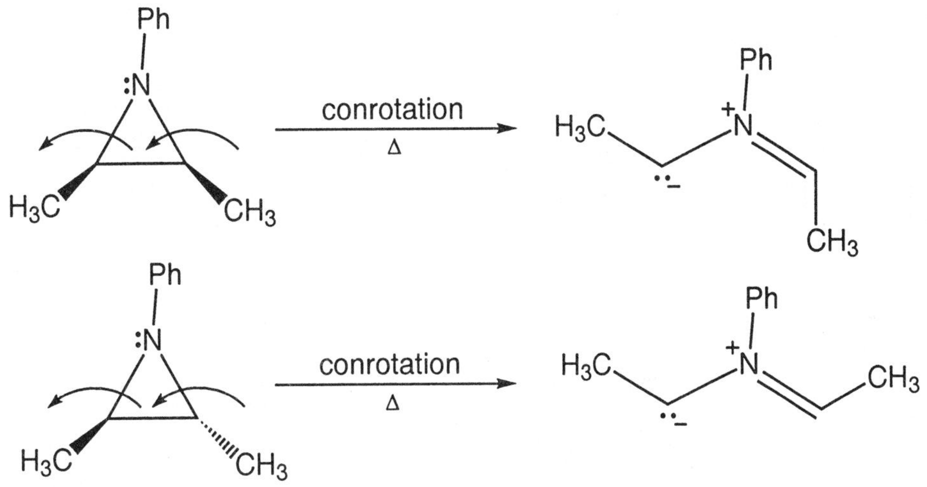

Ph
:N
H_3C
CH_3
conrotation
Δ
Ph
H_3C
^+N
CH_3

Ph
:N
H_3C
CH_3
conrotation
Δ
Ph
H_3C
^+N
CH_3

Problem 22.38 Both parts of this problem involve a Diels–Alder reaction that cannot take place until the starting material undergoes another reaction first.

(a) Watch that cyclobutene. There probably has never been a cyclobutene in an orbital symmetry problem that didn't open, and this one is no exception. Once it opens thermally (conrotation, please) the double bond attached to the carbonyl group captures it to give the product.

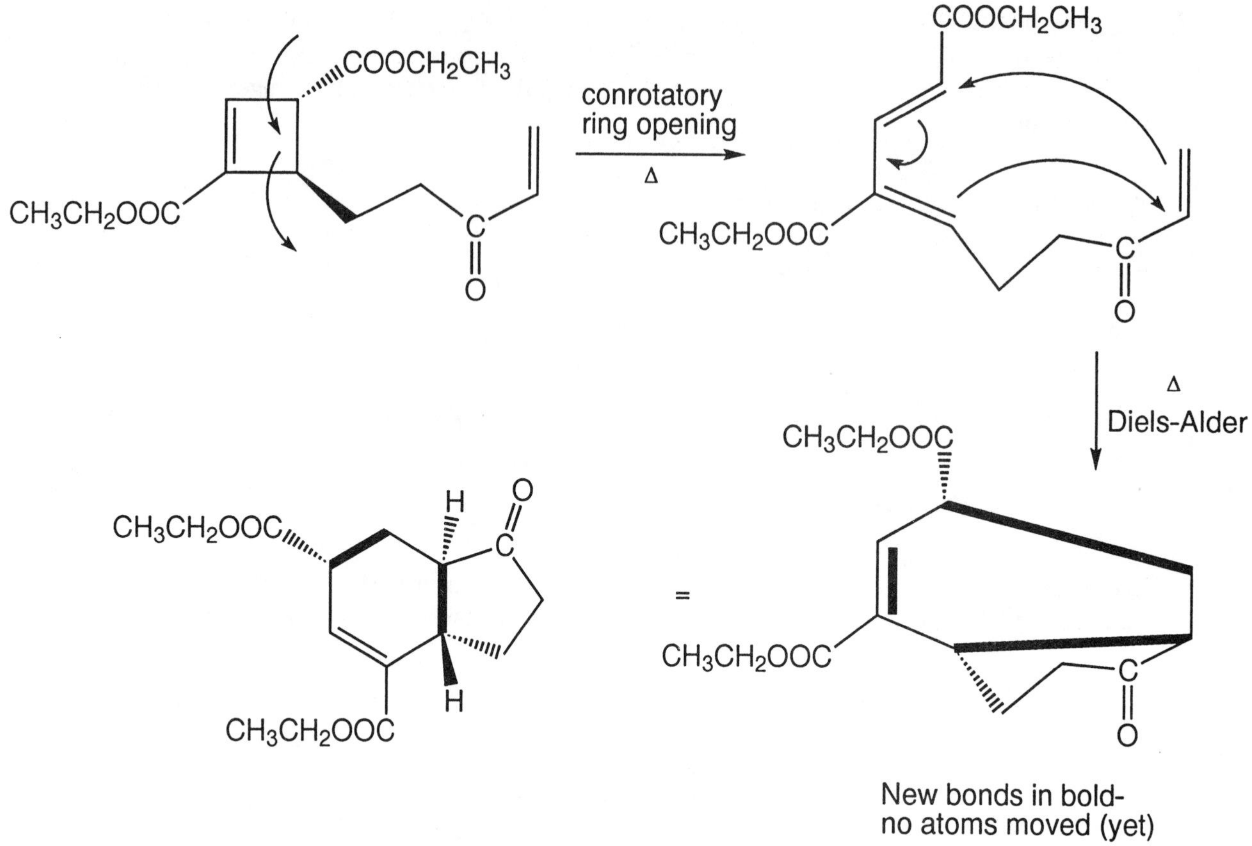
COOCH₂CH₃
CH₃CH₂OOC
conrotatory
ring opening
Δ
COOCH₂CH₃
CH₃CH₂OOC
Δ
Diels-Alder
CH₃CH₂OOC
H
H
CH₃CH₂OOC
CH₃CH₂OOC
CH₃CH₂OOC
New bonds in bold-
no atoms moved (yet)

(b) Here, the first reaction is a disrotatory six-electron thermal closure of a 1,3,5-hexatriene to a 1,3-cyclohexadiene. Once this has happened the lurking double bond (dienophile) captures the 1,3-diene.

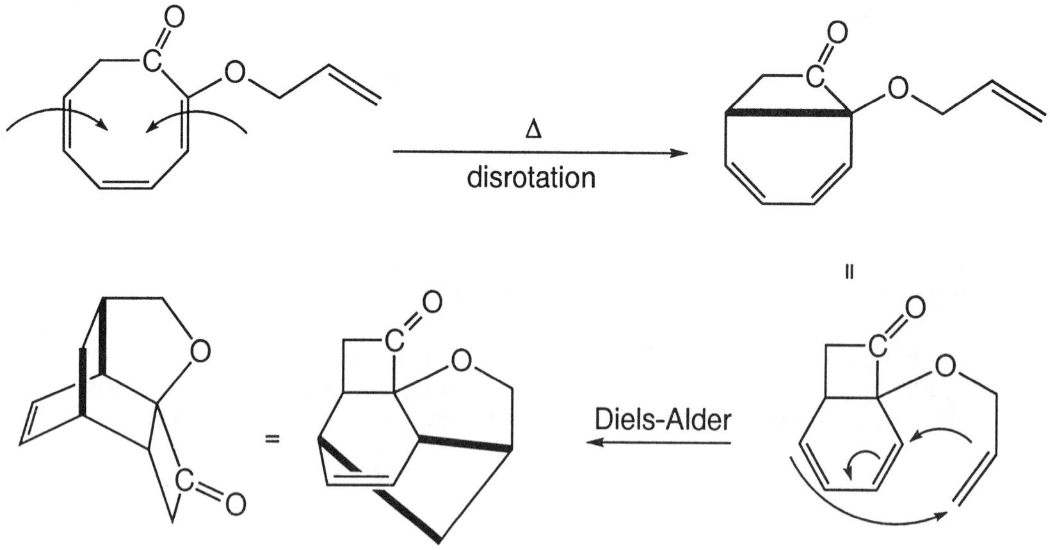
Δ
disrotation
Diels-Alder

Problem 22.39

(a) The hint says that there are two intermediates in which the three-membered ring is gone, so it surely makes sense to focus on this part of the molecule. Presumably, the cyclopropane is destroyed in a first step, and reconstituted later. There are two things to think about whenever you see a cyclopentadiene. The first is a Diels–Alder reaction. There is no visible dienophile to go with the cyclopentadiene, so this doesn't seem a promising line of inquiry. The second is a thermal [1,5] sigmatropic shift. That turns out to be a useful way to think in this case. A [1,5] shift of carbon generates intermediate **A**. There are two things **A** can do. It can go right back to starting material, or it can do another [1,5] shift of hydrogen to produce intermediate **B**. Now a [1,5] shift in **B** regenerates the cyclopropane and makes the product. This problem is hard for two reasons. First, there are two intermediates in between starting material and product, and that level of complexity almost always makes for a difficult problem. Second, there is misdirection here. It looks as if the methyl group is migrating somehow, and this is not the case. It remains steadfast throughout, always attached to the same carbon as the molecule rearranges around it, eventually moving it to another position.

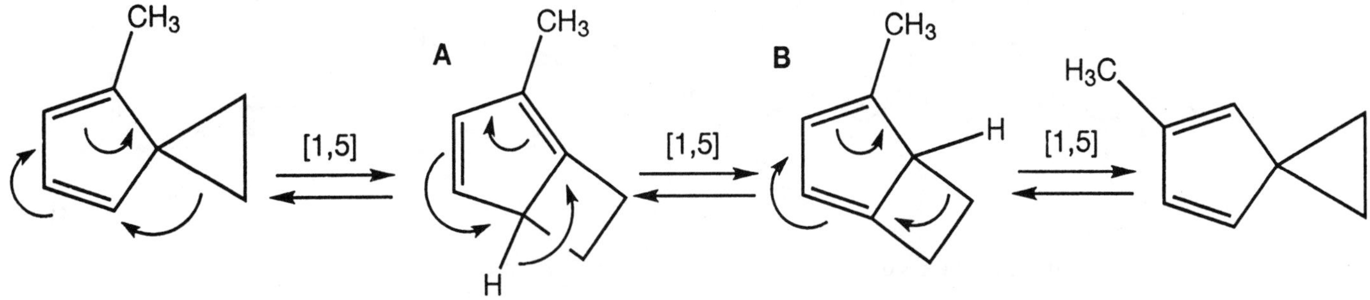

(b) This part is hard because it is difficult to visualize the structures involved. What can happen photochemically? The two most common reactions that we know about are 2 + 2 cycloadditions of alkenes and [1,3] shifts. In this case, it is a simple (?) photochemical [1,3] shift of carbon that leads to product.

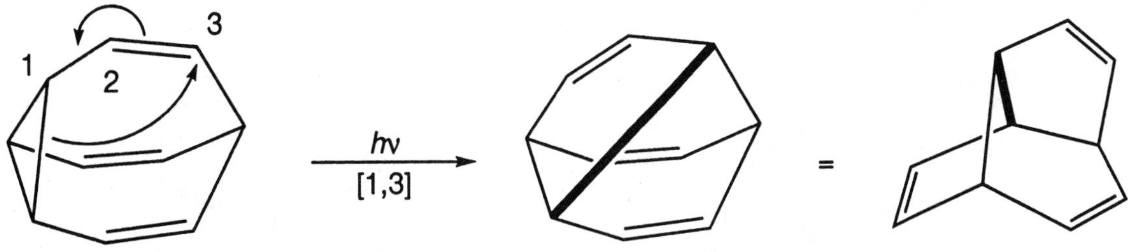

Problem 22.40

a. There are many possible labeling experiments that could be (and have been) used here, incuding isotopic labels. Perhaps the simplest label would be a methyl group or groups attached to either C_1 or C_3 of the allyl group. For example, the methyl labeled allyl phenyl ether **A** would afford only the *o*-allylphenol **B** if the rearrangement occurred by an intramolecular concerted [3,3] shift. Note the so-called "allylic shift" where the methyl label originally α to the oxygen is found γ to the ring.

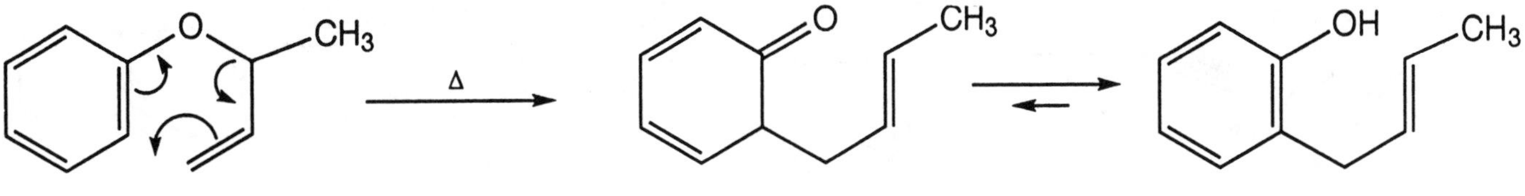

By contrast, if the rearrangement were intermolecular, and involved radical intermediates, both phenol **B** and phenol **C** would be expected. The allyl radical can recombine with the dienone radical in two different ways, and conversion of the ketone to the enol forms would give not only **B**, but **C** as well.

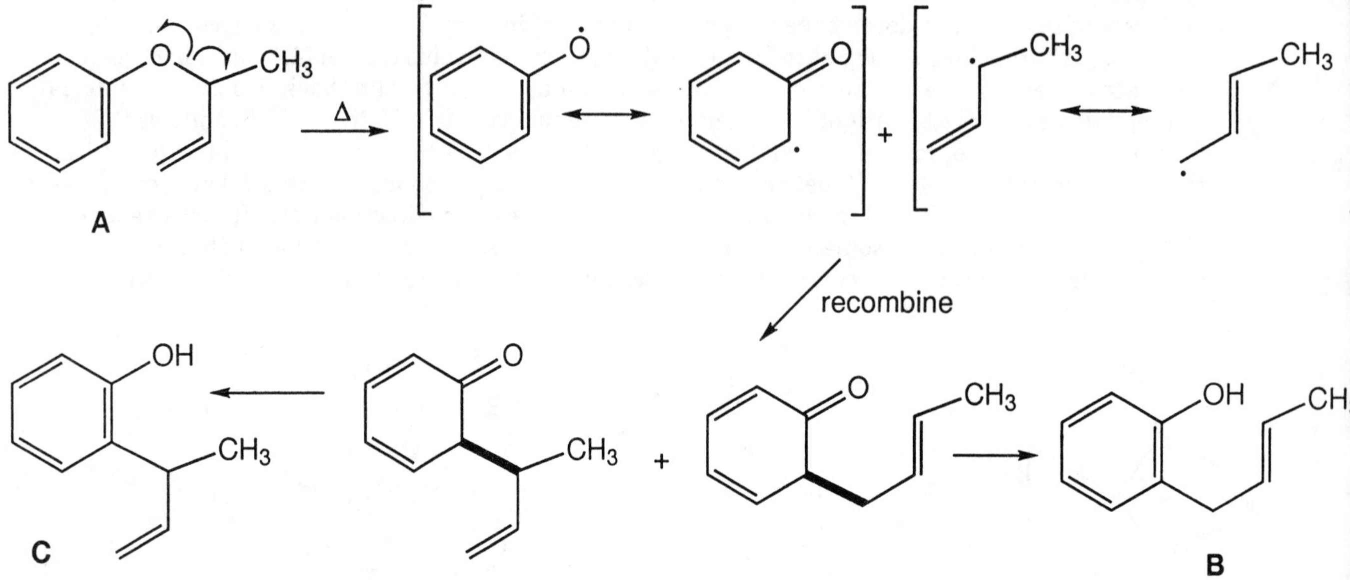

A

Δ

recombine

C

B

In fact, the experiment reveals only **B**, and the mechanism must be concerted.

(b) The mechanism for the para-Claisen rearrangement begins the same way, with an intramolecular [3,3] shift to give dienone **D**. However, in this case, formation of the aromatic phenol cannot take place because there is no hydrogen α to the carbonyl group. The ortho positions are blocked with methyl groups.

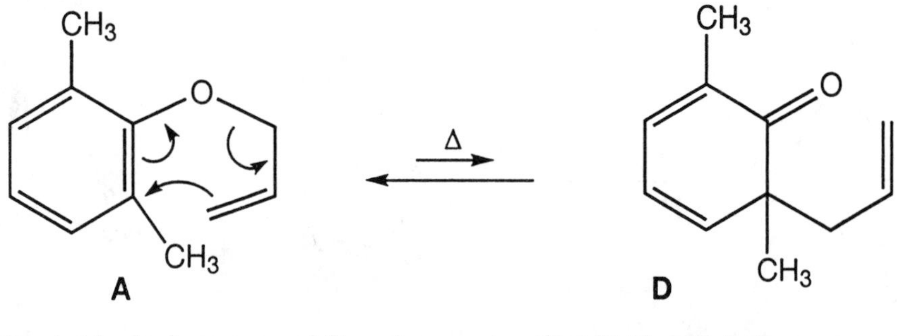

A D

As aromatization is blocked, compound **D** undergoes another [3,3] shift (a Cope rearrangement) to give dienone **E**. Compound **E** easily rearranges to the aromatic compound.

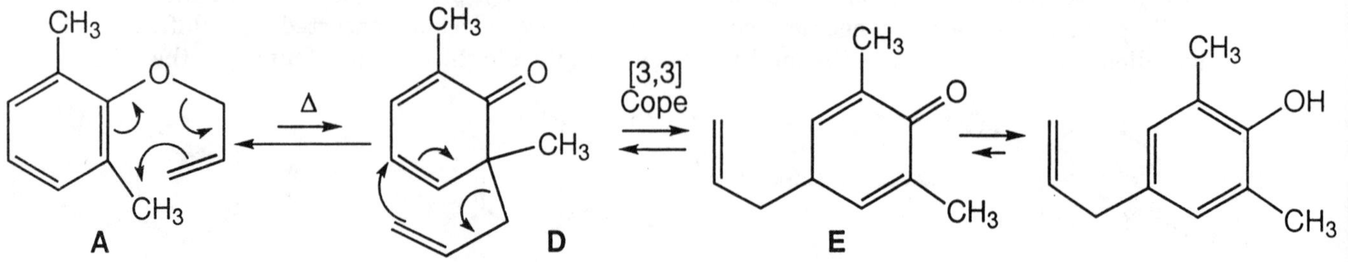

[3,3] Cope

A D E

If we use the α-methyl label here, the intramolecular [3,3] shift mechanism predicts only the formation of **F**. An intermolecular mechanism would give a mixture of **F** and **G**. Experimentally, only **F** is observed.

Intramolecular:

CH_3 CH_3 O CH_3 CH_3 Δ O CH_3 H_3C CH_3 O CH_3 CH_3 CH_3 OH CH_3 CH_3 **F**

Intermolecular:

CH_3 O CH_3 CH_3 Δ CH_3 O CH_3 CH_3 O CH_3 CH_3 CH_3

recombine

OH H_3C CH_3 **F** H_3C O H_3C CH_3 H_3C + O H_3C CH_3 H_3C OH H_3C CH_3 **G** not formed H_3C

Problem 22.41 As the starting material is an allyl aryl ether, you should be thinking Claisen rearrangement (especially considering the two previous questions!). Although it is true that allyl aryl ethers with the para and ortho positions blocked normally don't react, we have a special case here. Begin as before, with a concerted [3,3] shift to give dienone **A**.

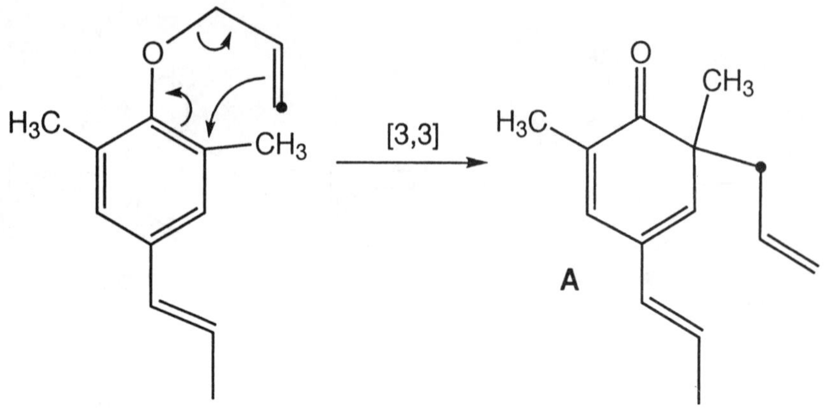

As there is no hydrogen α to the carbonyl group in **A**, this dienone cannot rearomatize to a phenol. What can it do instead? Often students make the interesting observation that a thermally allowed [1,5] shift of carbon would give dienone **B**, which could give the desired phenol. Although this mechanism is certainly economical, it is also incorrect. The proposed [1,5] shift involves passage over a long distance, and cannot occur in an intramolecular fashion. You might make a model to convince yourself that this is so.

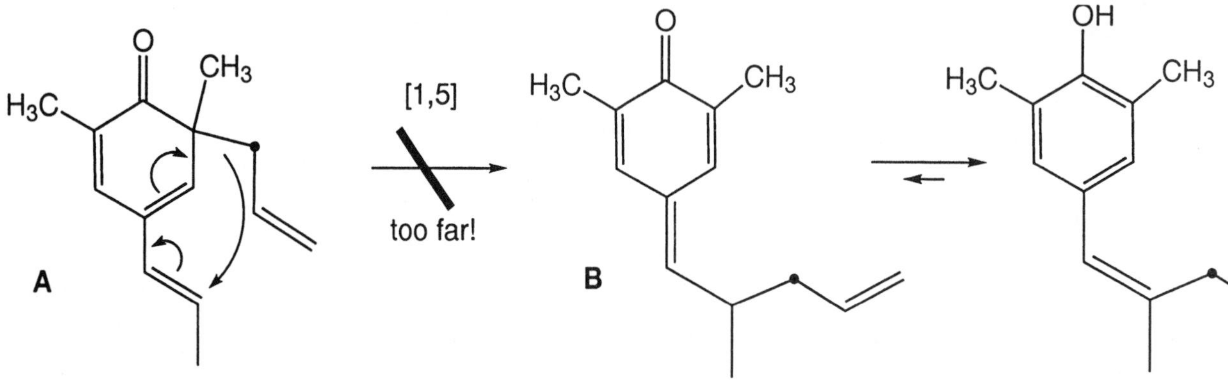

What else can dienone **A** do? How about another [3,3] shift to give dienone **C,** just as we saw in the para-Claisen rearrangement in Problem 22.40? Now dienone **C** can undergo yet another Cope rearrangement to give **B,** the molecule we failed to make earlier.

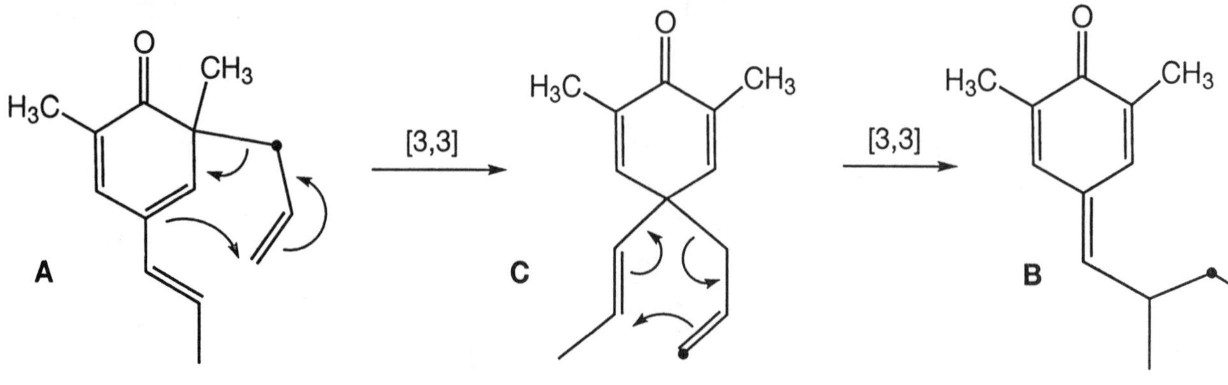

Now, although a [1,5] hydrogen shift, followed by phenol formation, rearomatizes and apparently generates the product, this [1,5] shift too passes over far too long a distance to be reasonable. An intermolecular reaction is far more likely. Simply protonate on oxygen and deprotonate on carbon. Where does the acid come from? In practice, there is usually enough acid on the glass surface to accomplish such reactions.

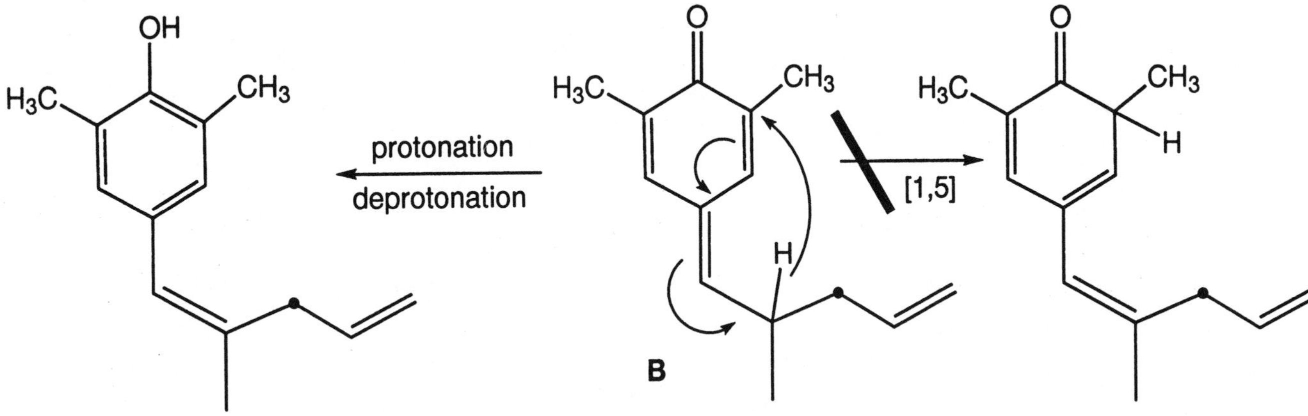

OH
H_3C CH_3
protonation
deprotonation
O
H_3C CH_3
H
B
[1,5]
O
H_3C CH_3
H

Problem 22.42

(a) This answer is simple and shows clearly in the diagrams in the problem and below. In **2**, the ends of the diene system cannot reach each other, whereas in conformation **1** they can.

(b) Conformational preference for **2** is presumably a result of intramolecular hydrogen bonding between the C_4 hydroxyl hydrogen and the side chain carboxylate.

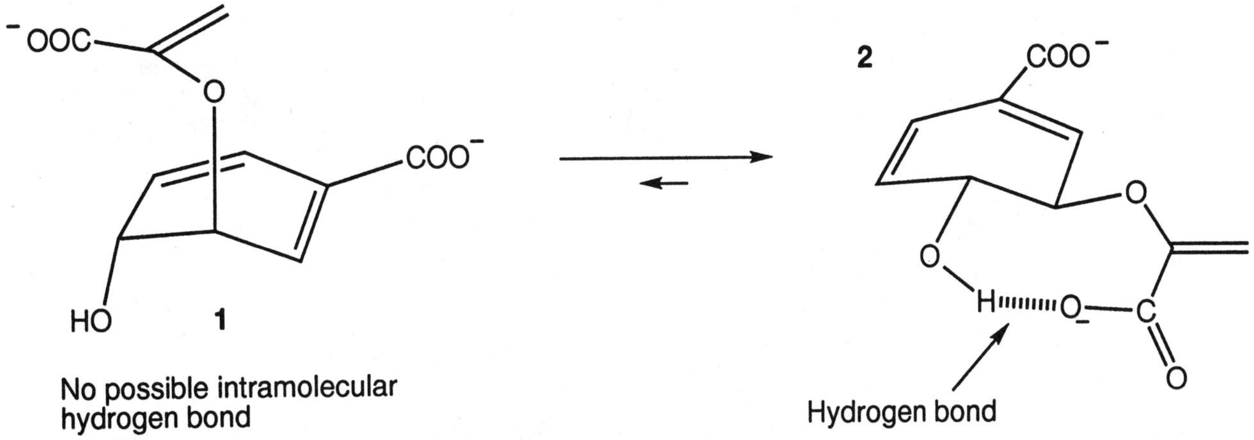

^-OOC
O
COO^-
HO **1**
No possible intramolecular
hydrogen bond
2 COO^-
O
O
H O C
O
Hydrogen bond

(c) The Energy versus Reaction progress diagram for the nonenzymatic reaction is similar to the one we encountered for the Cope rearrangement of *cis*-1,2-divinylcyclopropane (Fig. 22.71) in that it involves an equilibrium between two conformations, and only the higher energy conformation can react. However, it is suggested in this case that the conformational isomerism is the slow (rate-determining) step. So, the postulated role of the enzyme is to stabilize the transition state for this conformational isomerization.

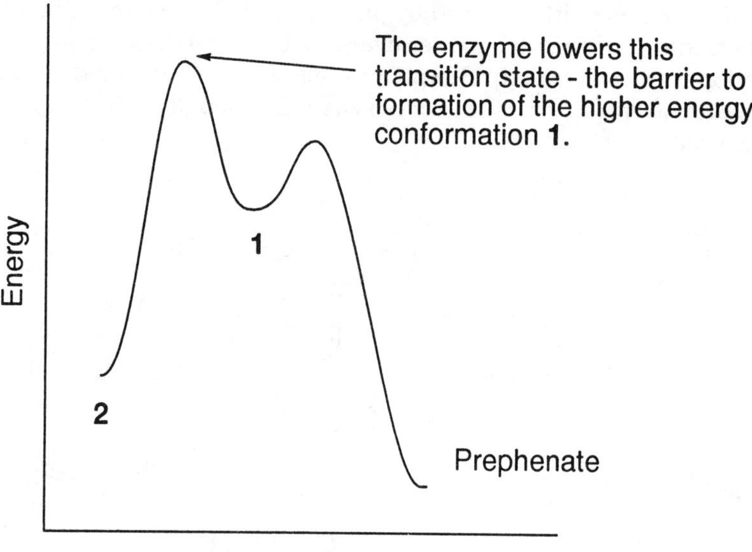
The enzyme lowers this transition state - the barrier to formation of the higher energy conformation **1**.
Energy
1
2
Prephenate
Reaction progress

Unfortunately, this simple picture is not correct. Largely through the efforts of Professor Jeremy Knowles and his co-workers, it has been shown that chorismate mutase selectively binds the 10-20% of the pseudodiaxial conformer of chorismate (**1**) present in solution. As conformational interconversion is fast, there is no need to postulate an enzyme-catalyzed conformational change. Consequently, the enzyme must stabilize the transition state for the Claisen rearrangement itself. In short, the Energy versus Reaction progress diagram is very similar to the one we encountered in Fig. 22.71.

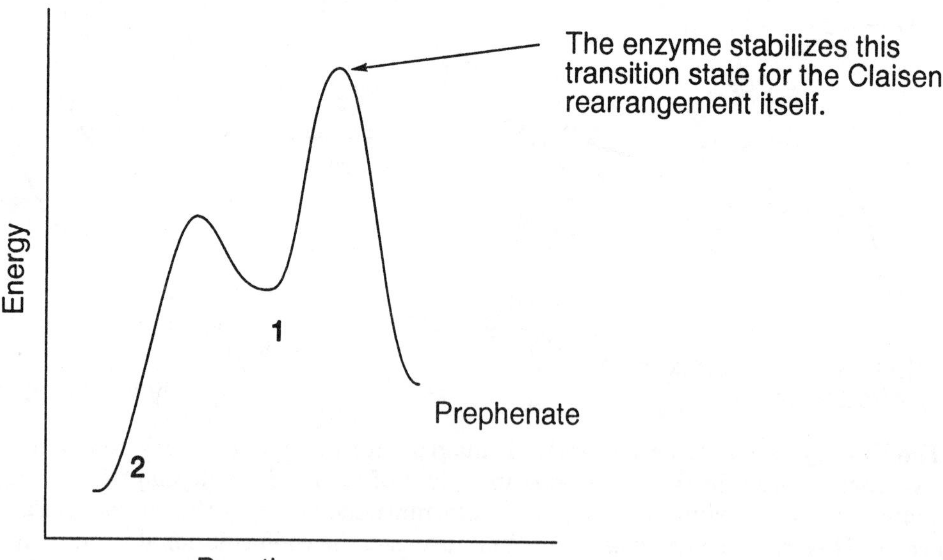
The enzyme stabilizes this transition state for the Claisen rearrangement itself.
Energy
1
2
Prephenate
Reaction progress

Problem 22.43

(a) As we saw in Chapter 20 (Fig. 20.57), a variety of nucleophiles adds to the carbonyl group of ketenes. In this case, the nitrogen nonbonding electrons of the imine **2** add to the carbonyl group of ketene **1** to give the dipolar intermediate **A**. Dipolar intermediate **A** can then undergo cyclization to yield the observed β-lactam **3**.

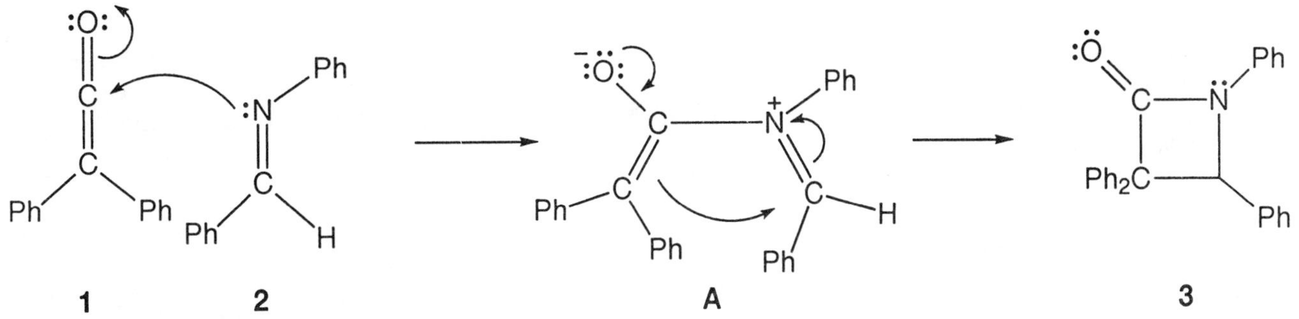

(b) The stereochemistry of these cycloaddition reactions has been rationalized on the basis of a conrotatory ring closure of the dipolar intermediate [**A** above in (a)] in which steric interactions are minimized. This intermediate is a 4 π electron system, and therefore a conrotatory thermal process is expected (Remember: Once again, the cyclobutene-1,3-butadiene interconversion). In this case, conrotatory closure of **B** leads to the observed β-lactam.

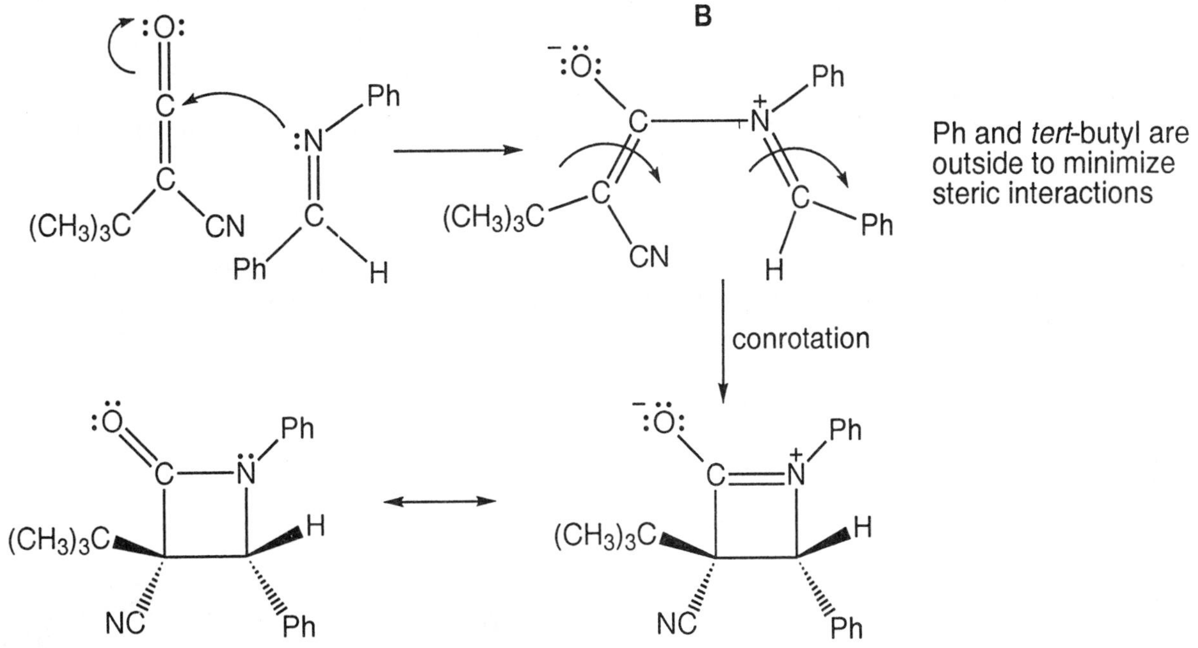

Ph and *tert*-butyl are outside to minimize steric interactions

Problem 22.44 As compound **2** is an isomer of **1**, it must have a molecular formula of $C_{11}H_{12}O_2$. There are six degrees of unsaturation for this molecular formula. It is not unreasonable to assume that four of the six degrees of unsaturation are the result of the presence of the benzene ring, particularly given the presence of a benzene ring in starting material.

The IR spectrum of compound **2** suggests an ester as indicated by a strong carbonyl absorption at 1740 cm^{-1}. The presence of an ester is further supported by the signal at δ 170.6 ppm in the ^{13}C NMR spectrum of **2**. This chemical shift is consistent with a carbonyl carbon of an ester or a carboxylic acid, but not with that of an aldehyde or ketone. The carboxylic acid possibility can be eliminated by the absence of a corroborating O—H stretch in the IR spectrum and a low field carboxylic acid hydrogen in the ^{1}H NMR spectrum of **2**. The frequency of the stretch in the IR spectrum suggests that the ester C=O group is not conjugated. Note that the ester carbonyl group accounts for the fifth degree of unsaturation.

The ^1H NMR spectrum of compound **2** shows the following signals: a methyl singlet at δ 2.08 ppm (possibly attached to a carbonyl or benzene ring), a methylene doublet at δ 4.71 ppm (attached to an electronegative substituent such as O and an adjacent CH), two signals for hydrogens attached to a double bond (accounting for the final degree of unsaturation) at δ 6.27 and 6.63 ppm (the signal at δ 6.27 is coupled to the methylene signal at δ 4.71 ppm), and finally, a 5H aromatic multiplet at δ 7.2–7.4 ppm (clearly a monosubstituted benzene ring).

Putting all these structural fragments together gives cinnamyl acetate as the most reasonable structure for compound **2**. The isomeric **3** can be eliminated from consideration because the methyl group of **3** would have a much lower chemical shift than δ 2.08 and 20.9 ppm in the ^1H and ^{13}C NMR spectra.

$PhCH{=}CHCH_2O{-}C({=}O)CH_3$

2

$PhCH{=}CHCH_2{-}C({=}O)OCH_3$

3

Two other structures might be considered as well, **4** and **5**, even though there is no obvious mechanistic pathway to them.

$CH_3O{-}C({=}O){-}CH{=}CHCH_2Ph$

4

$H_3C{-}C({=}O){-}OCH{=}CHCH_2Ph$

5

Compound **4** can be quickly rejected on the same grounds that served to eliminate **3**, the absence of a signal in the ^1H NMR spectrum for an OCH_3 group. Compound **5** is tougher. There are hints in both the ^1H and ^{13}C NMR spectra that **5** cannot be correct. In compound **5** a resonance form, **5'**, will contribute. That should result in a relatively high-field absorption for the β-carbon and its hydrogen.

$$\left[H_3C{-}C({=}O){-}O{-}CH{=}CHCH_2Ph \longleftrightarrow H_3C{-}C({=}O){-}\overset{+}{O}{=}CH{-}\overset{-}{C}HCH_2Ph \right]$$

5 **5'**

As the data show, there are no unusually high-field vinyl signals, and **5** is probably wrong. Moreover, there is a straightforward mechanism for producing **2**. Let's do a mechanistic analysis now.

When α-phenylallyl acetate (**1**) and cinnamyl acetate (**2**) are drawn in more suggestive conformations, it is easy to see that this isomerization is just a thermally induced [3,3] shift.

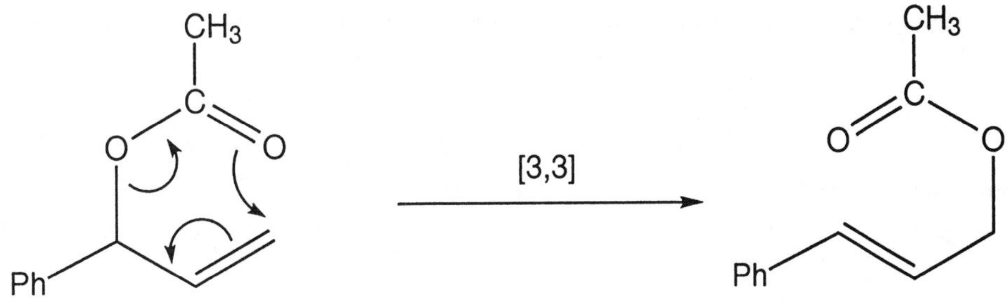

What is the stereochemistry of the double bond in cinnamyl acetate (**2**)? Hydrogens that are substituted cis on a double bond have smaller coupling constants (J = 6–14 Hz) than those substituted trans (J = 11–18 Hz). In the case of compound **2**, the coupling constant is 16 Hz, strongly suggesting a trans double bond. This stereochemical assignment is also supported by the olefinic C—H bending vibration at 960 cm^{-1}, characteristic of trans disubstituted alkenes.

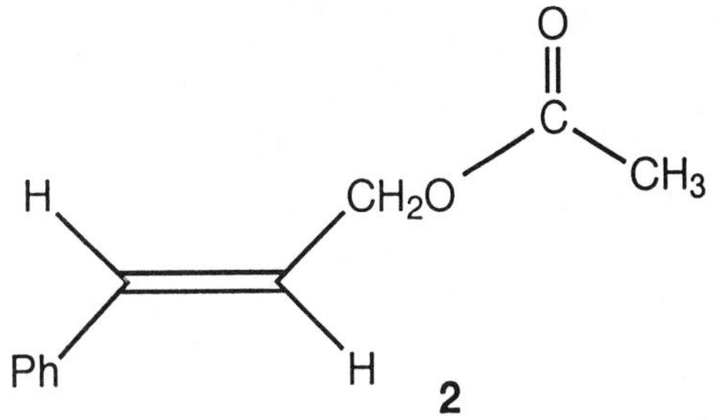

Problem 22.45 A sequence of alternating Diels–Alder and retro Diels–Alder reactions gets the job done.

The semibullvalene product **3** undergoes a fast Cope rearrangement that makes the two methyl groups equivalent (δ 1.13 ppm singlet), and the vinyl and cyclopropyl hydrogens equivalent (δ 4.79 ppm singlet, shown as "H" in figure). The methoxy hydrogens (δ 3.73 ppm singlet) are also equlivalent. The ^{13}C NMR spectrum can be rationalized in a similar way.

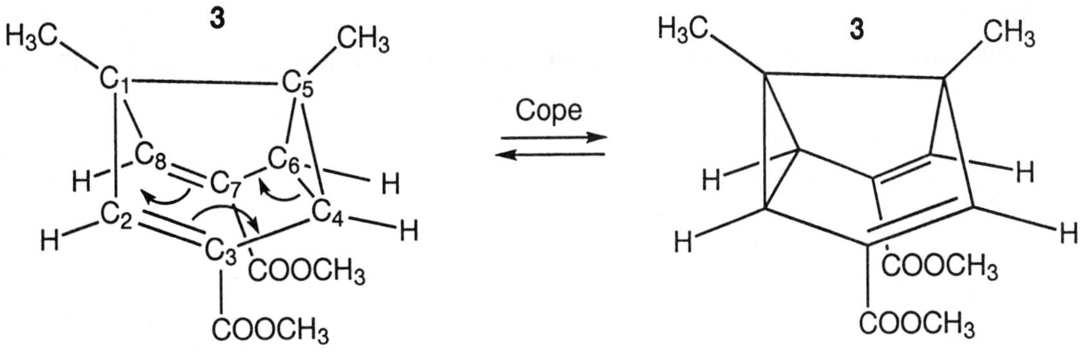

δ 14.9 (q)	Methyls
51.4 (q)	OCH_3 groups
60.6 (s)	C_1 and C_5
93.7 (d)	C_2, C_4, C_6, and C_8
127.2 (s)	C_3 and C_7
164.7 (s)	Carbonyl carbon

Chapter 23 Outline

23.1 Heteroatoms as Neighboring Groups
23.2 Neighboring π Systems
23.3 Single Bonds as Neighboring Groups
23.4 Coates' Cation
23.5 Something More: A Family of Concerted Rearrangements of Acyl Compounds

23

Introduction to Polyfunctional Compounds: Intramolecular Reactions and Neighboring Group Participation

Intramolecular reactions play a very large role in organic chemistry. For example, both product structures and rates of reactions are strongly influenced by the presence of internal nucleophiles. Intramolecular displacements and additions are very common. The following problems show both simple chemistry and more complex reactions in which neighboring group participation, or "anchimeric assistance" appears.

Problem 23.2 This problem is nothing more than a standard, no-frills, hemiacetal formation in an intramolecular setting. Addition of the hydroxyl oxygen atom (a nucleophile) to the carbonyl group within the same molecule is followed by protonation and deprotonation steps.

addition

protonation

deprotonation

Hemiacetal

Problem 23.4 Acetic acid can attack at two "side" carbons to give the major products of ring opening. However, there is a third carbon that can participate in the S_N2 reaction with acetic acid, the methyl carbon. This reaction retains the ring and leads to 2-methyltetrahydrofuran.

This reaction leads to 2-methyltetrahydrofuran and, after deprotonation, methyl acetate.

Addition here leads to one product.

S_N2

Addition here leads to another product.

2-Methyltetrahydrofuran

Problem 23.5 Almost any labeling experiment will do. In one, we might use ^{13}C to produce two products in which the ^{13}C appears in a different position.

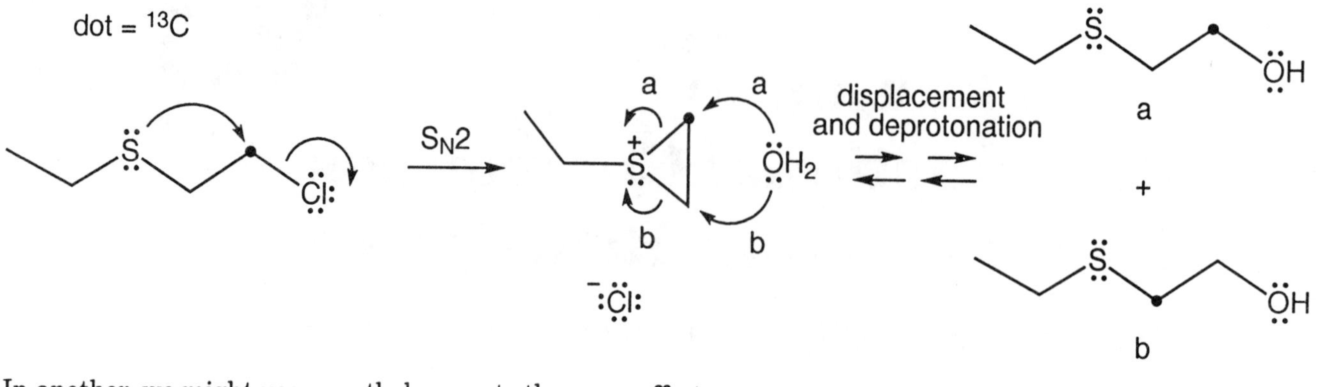

dot = ^{13}C

S_N2

displacement and deprotonation

a

b

In another, we might use a methyl group to the same effect.

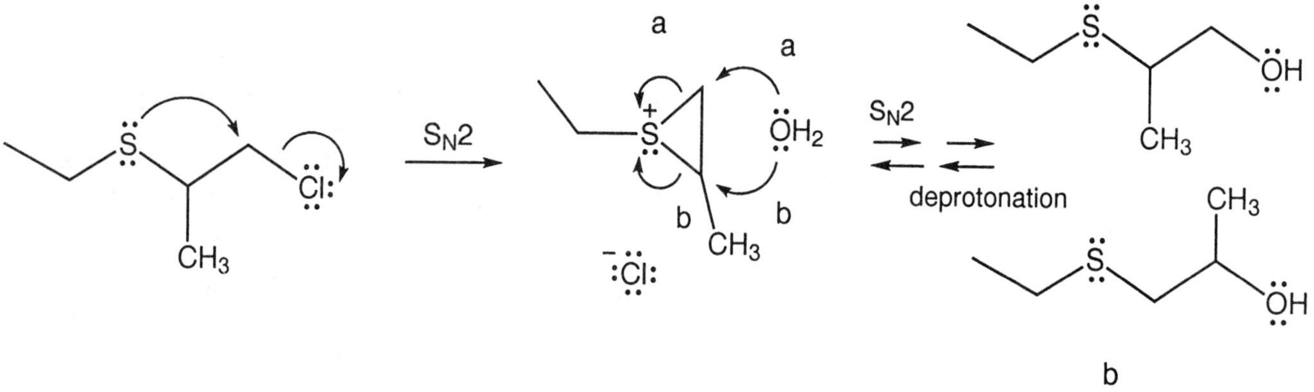

S_N2

deprotonation

a

b

Neighboring group effects would be signaled by the formations of the rearranged products labeled (b) in each case.

Problem 23.6 This one is pretty much up to you. It's all in the text. Figure 23.4 shows the first clue, the observation of an "odd" or "backward" stereochemical result. The second clue, the occurrence of a strange rearrangement, appears in Figure 23.10. Finally, there is clue 3, the observation of an unusually fast rate of reaction as in Figure 23.13.

Problem 23.7 The structure of the product shows that a ring bond must be broken. The five-membered ring containing nitrogen can be spotted in the starting material, and must be retained in the reaction. Although there is no intramolecular S_N2 reaction possible because of the orientation of the leaving group (no frontside S_N2 reactions!), there is an intramolecular elimination that can take place. This reaction leads directly to the product.

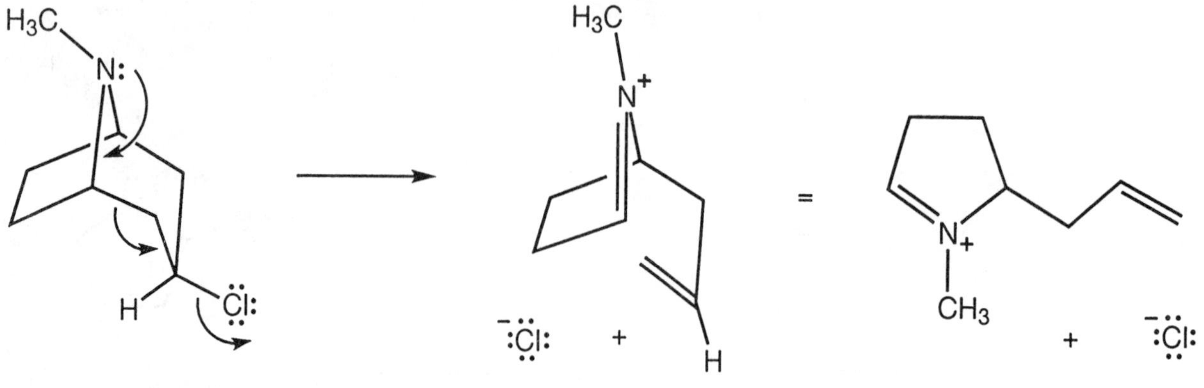

Problem 23.8(b) The two ester groups remain unchanged in the product, and the three carbons in the chain must become the three carbons of the cyclopropane ring. An easy intramolecular S_N2 displacement by the enolate forms the product in a single step.

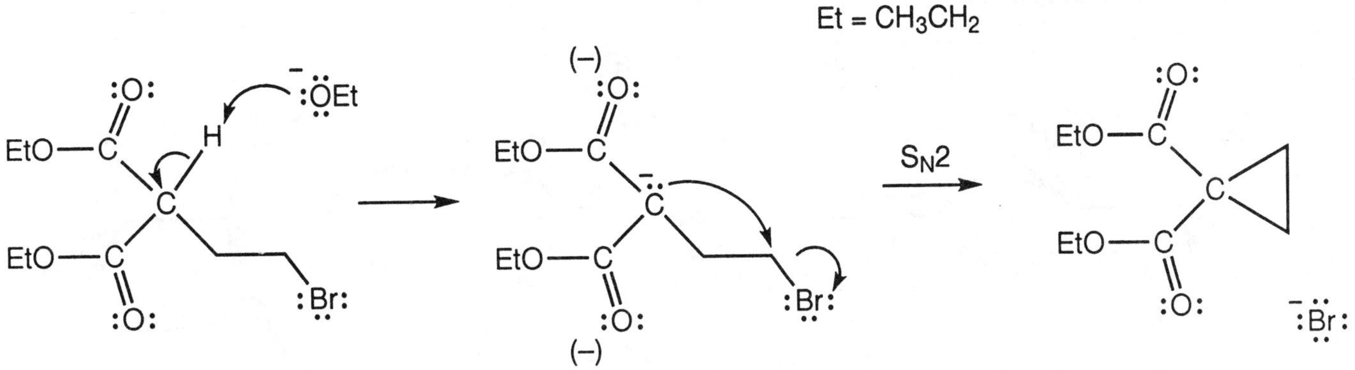

Et = CH_3CH_2

(c) This problem should also be easy. The nitrogen acts as an internal nucleophile, displacing chloride to give the cyclic ammonium ion. Hydroxide ion opens the ion at the less hindered position to give the product. The usual pattern of an intramolecular S_N2 displacement, followed by an intermolecular S_N2 displacement, appears.

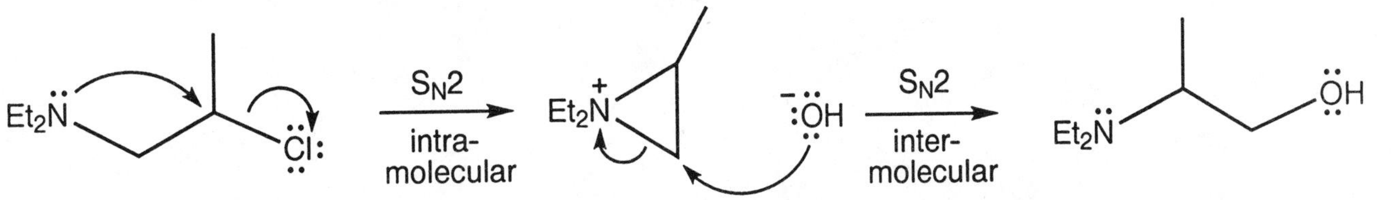

Problem 23.9 If fluorine acts as a neighboring group, a positively charged fluorine atom, a fluoronium ion, would result. Fluorine is very electronegative, and positively charged fluorine is very high in energy. So, such intermediates are only rarely encountered.

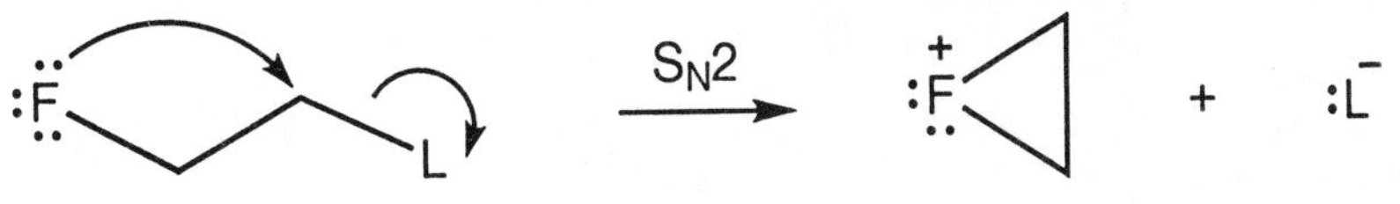

Fluoronium ion

Problem 23.10 A really easy one. Simply add bromine to *cis*-2-butene.

Problem 23.11 All you have to do in this problem is follow the pattern of Figure 23.25. If you do, you can see that the other enantiomer would also give the meso dibromide if a conventional mechanism were followed. Protonation, followed by displacement of water by bromide ion, gives the achiral meso compound.

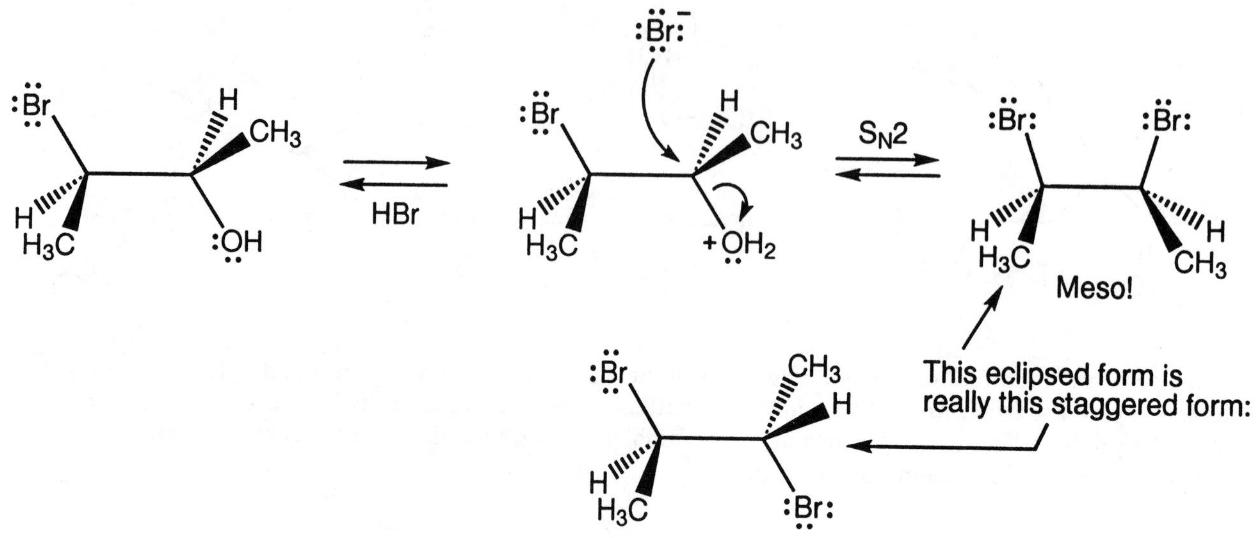

HBr

S_N2

Meso!

This eclipsed form is really this staggered form:

Problem 23.12 Bromination begins with bromonium ion formation. Opening of the bromonium ion by water leads to a bromohydrin. The bromonium ion from *trans*-2-butene can be formed in two ways (addition to the top or bottom of the alkene) to give a pair of enantiomeric bromonium ions.

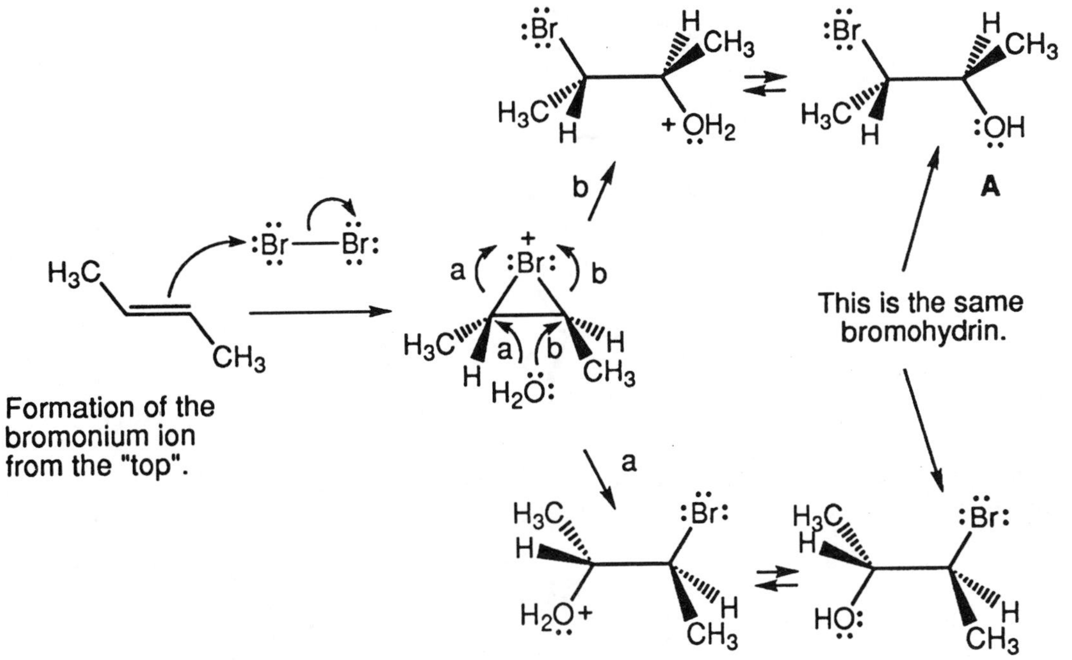

Formation of the bromonium ion from the "top".

This is the same bromohydrin.

A

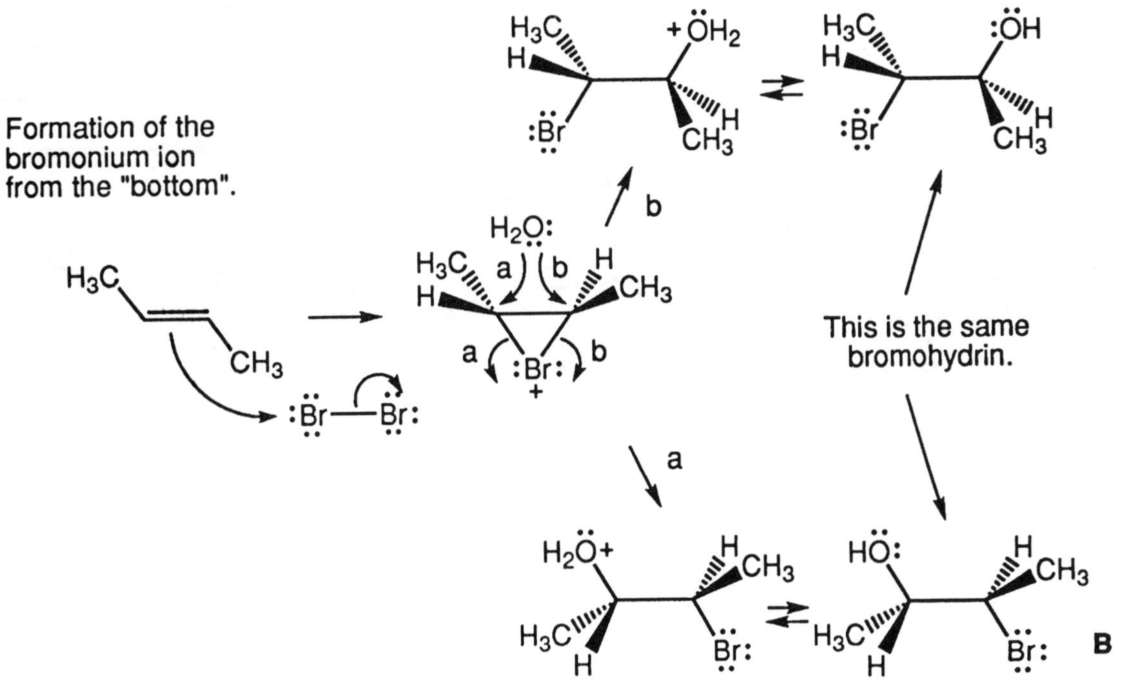

Formation of the bromonium ion from the "bottom".

This is the same bromohydrin.

B

The two possible bromohydrins, **A** and **B**, are enantiomers.

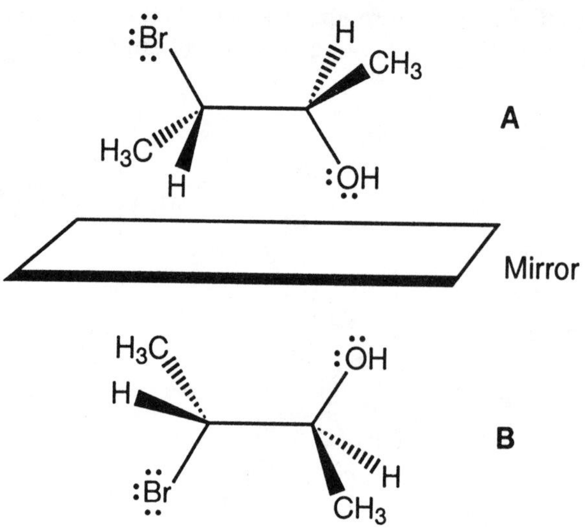

Mirror

A

B

Treatment of the bromohydrin with HBr leads to the protonated alcohol and then, by intermolecular displacement of water by bromide, to the racemic pair of enantiomers shown.

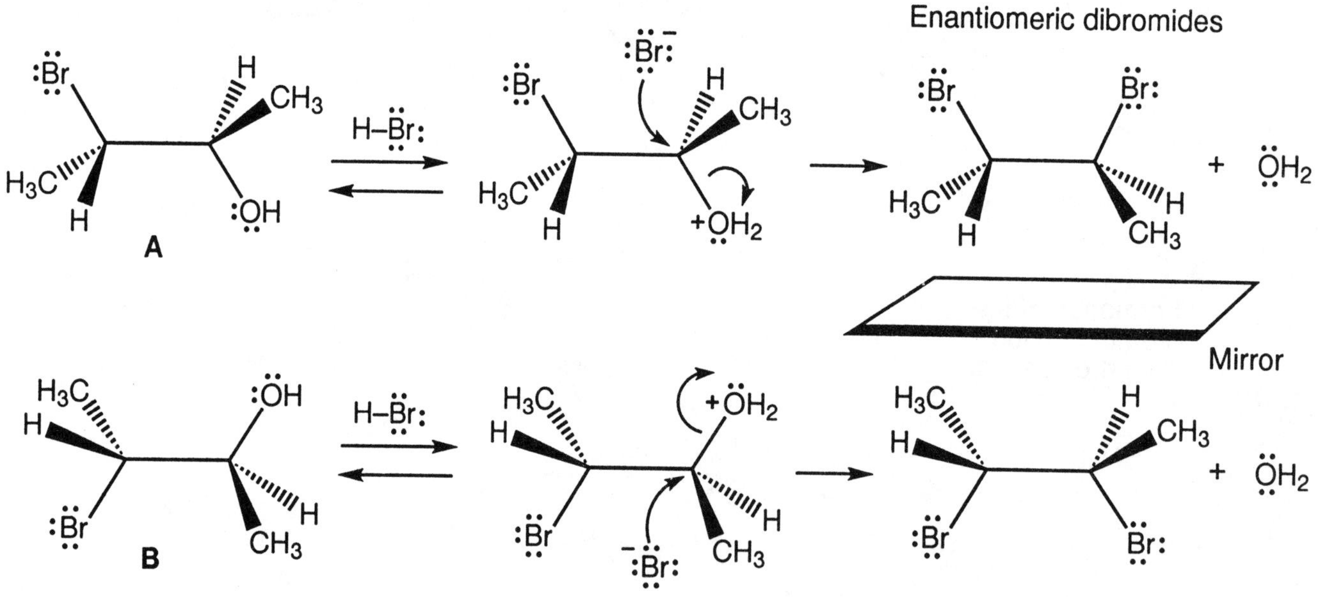

Enantiomeric dibromides

Mirror

However, that's not what happens. The reaction takes another course, formation of *meso*-2,3-dibromobutane, and a neighboring group must be involved. It is a displacement by intramolecular bromine that takes place on the protonated alcohol.

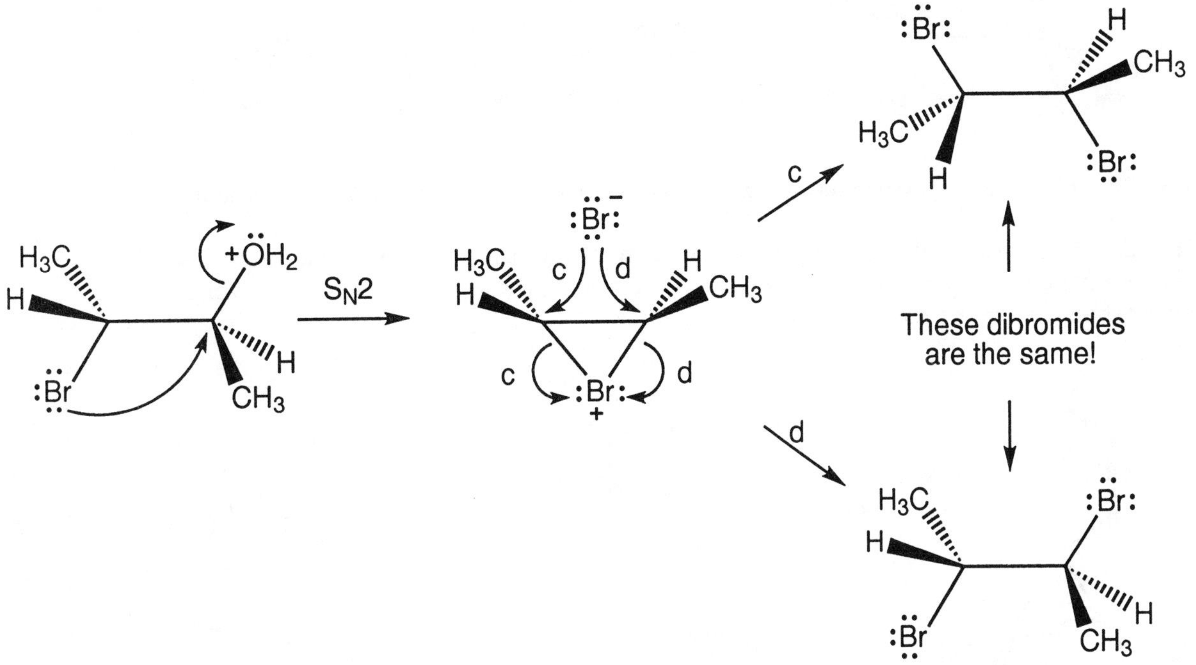

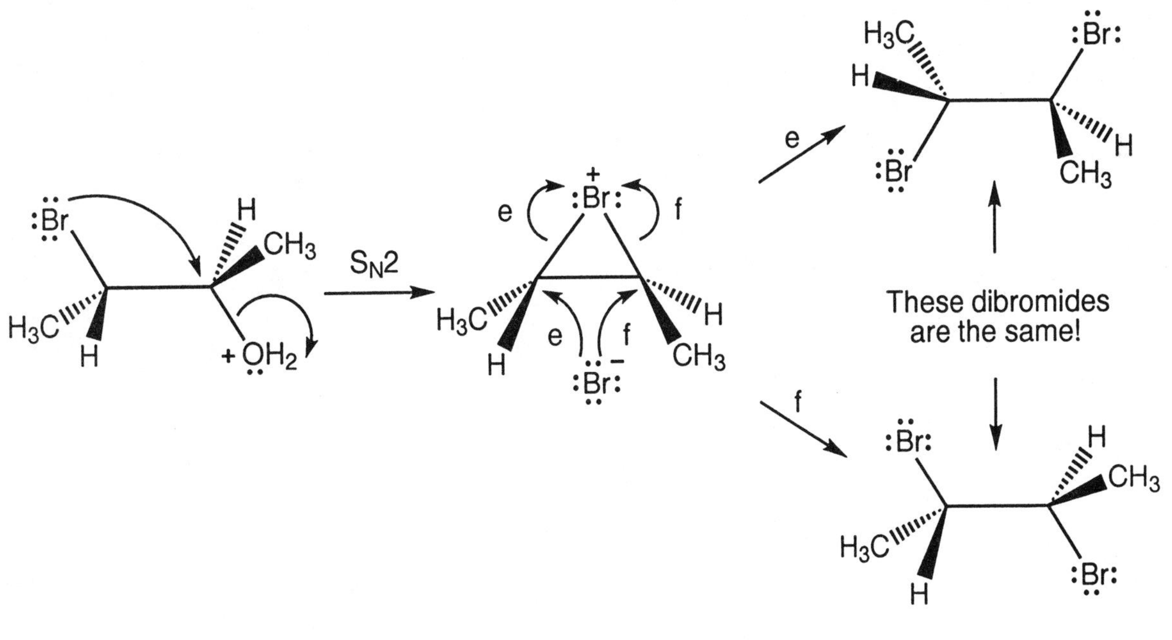

These dibromides are the same!

These dibromides are the same!

In fact, *all* of these dibromides are the same. Each is the meso isomer.

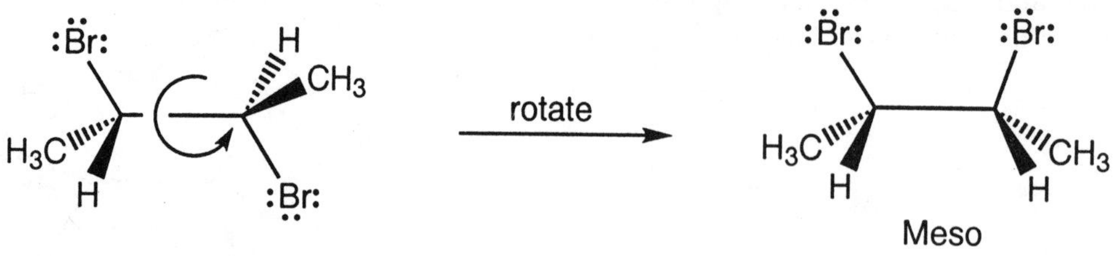

Meso

Problem 23.14 They are stereoisomers, but not mirror images. They must be diastereomers.

Problem 23.15 As for any phenonium ion (see Chapter 14 for countless examples of such species formed as intermediates in aromatic substitution reactions), there are three forms in which different carbons share the positive charge.

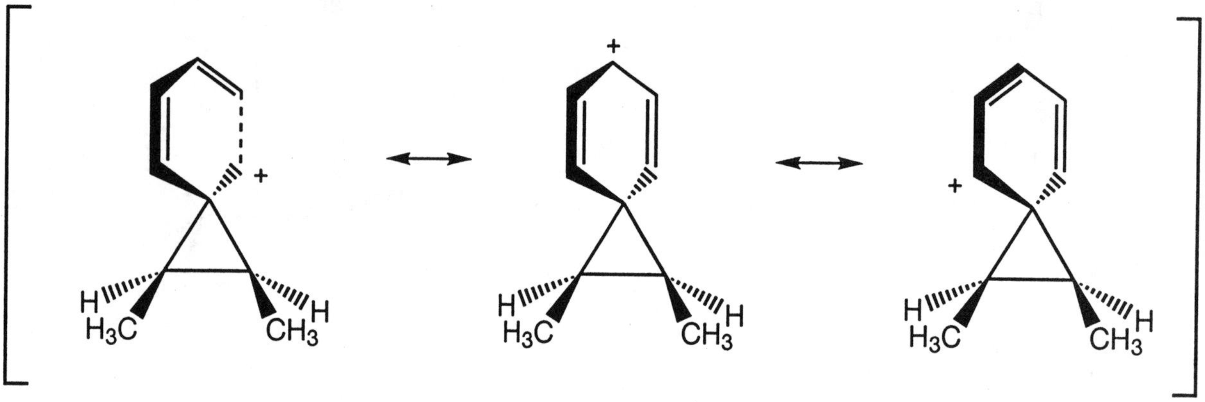

Problem 23.16 The steps are exactly the same as those outlined in the text for the other isomer of 3-phenyl-2-butyl tosylate. Follow Figures 23.31–23.33.

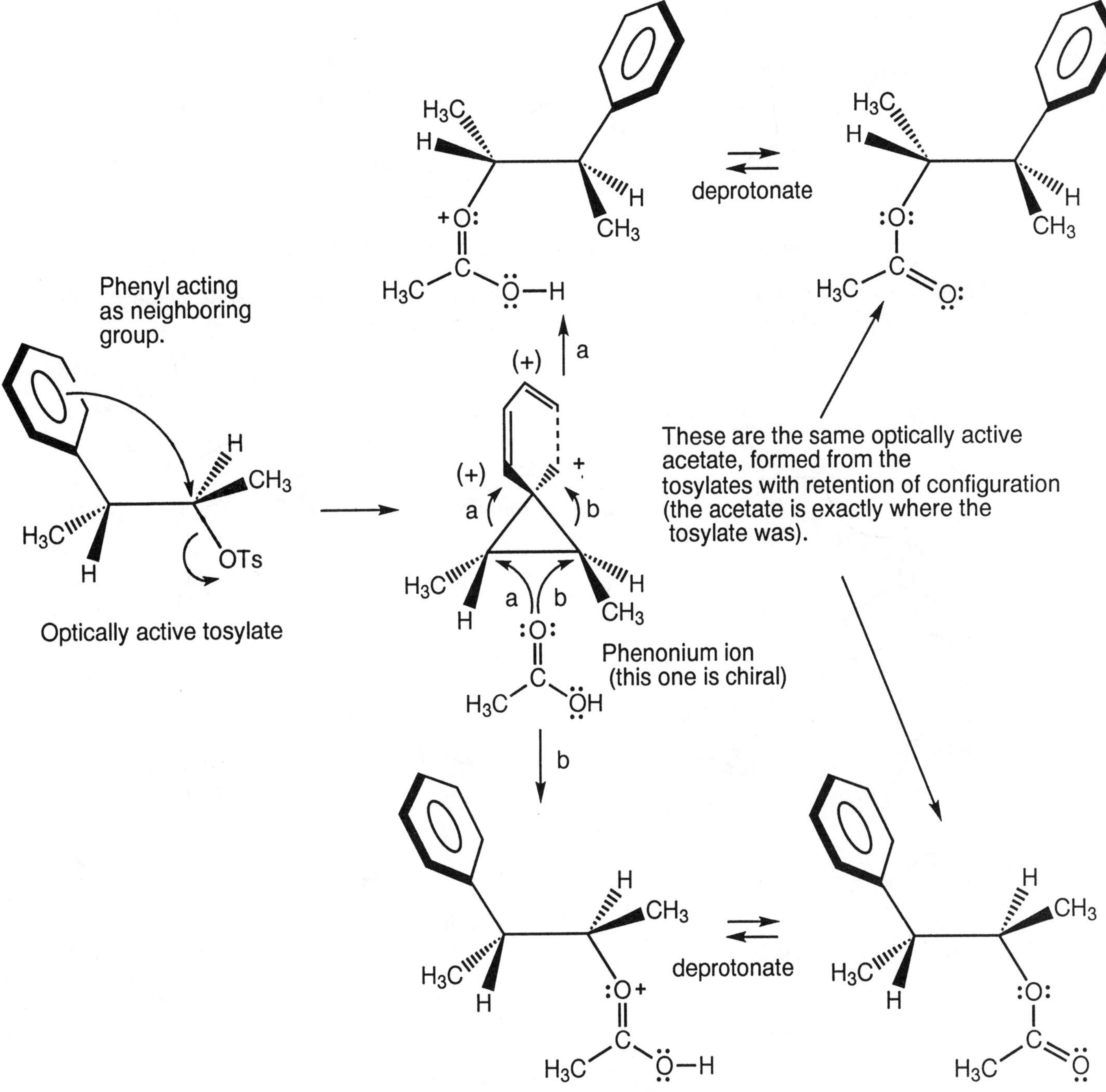
Phenyl acting
as neighboring
group.
OTs
Optically active tosylate
(+)
a
b
:O:
Phenonium ion
(this one is chiral)
deprotonate
These are the same optically active
acetate, formed from the
tosylates with retention of configuration
(the acetate is exactly where the
tosylate was).
deprotonate

Problem 23.17 First of all, there are only three imaginable bicyclo[2.2.1]heptenes. Only one of them avoids having the double bond at a bridgehead. In a bicyclic system this small, the orbitals making up the bridgehead double bond will be severely twisted and therefore the molecule will be of very high energy.

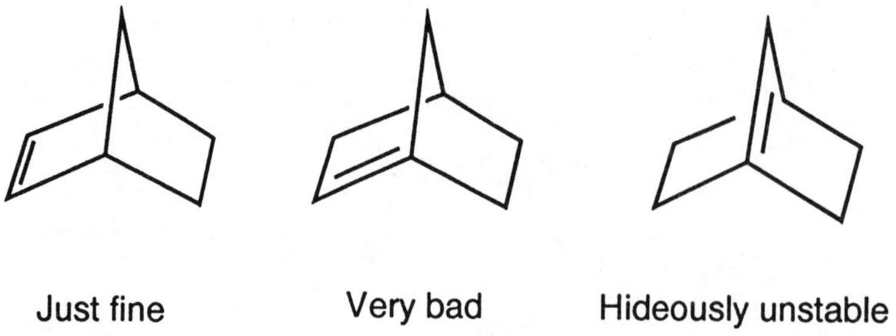

Just fine Very bad Hideously unstable

Problem 23.18 Presumably, it is the strain of the product (and of the transition state leading to product) that makes addition to the other position unfavorable.

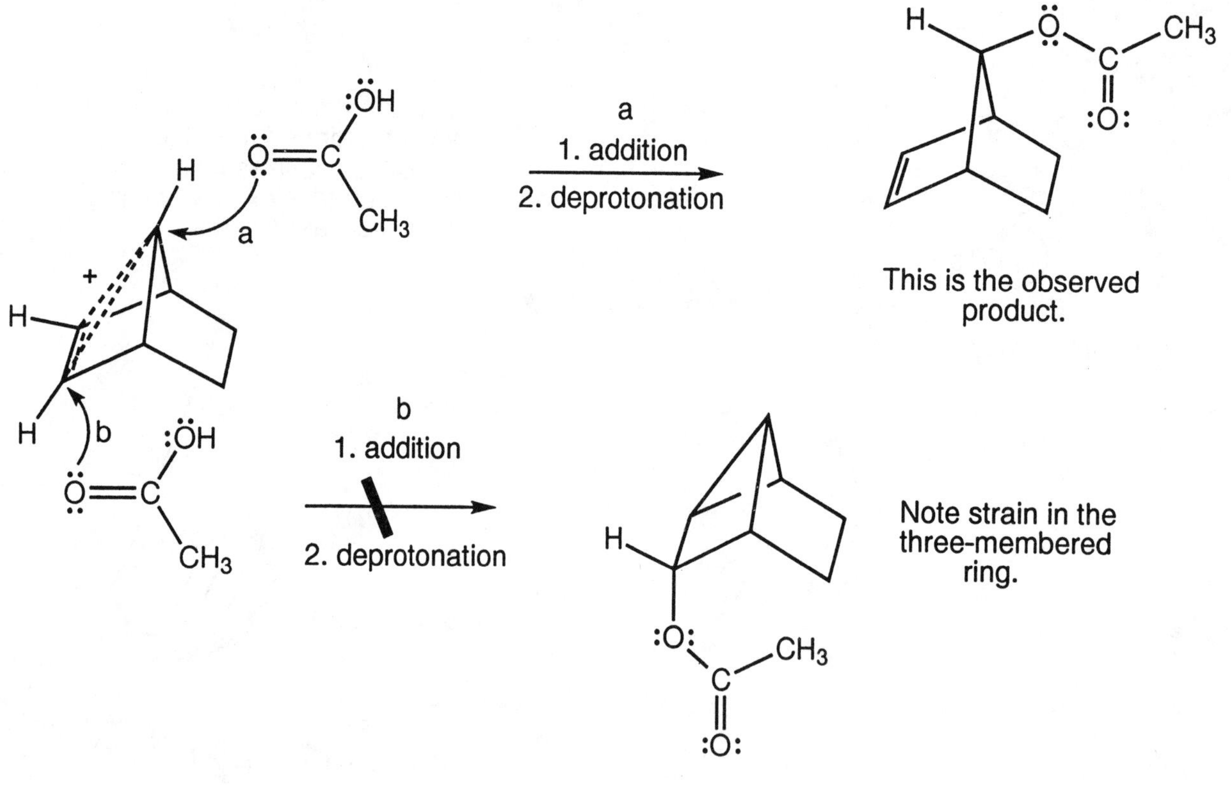

a
1. addition
2. deprotonation

This is the observed product.

b
1. addition
2. deprotonation

Note strain in the three-membered ring.

Problem 23.19 Let's combine the π molecular orbitals of ethylene with a carbon 2*p* orbital. As in the construction of the molecular orbitals of cyclic H_3 (p. 64), there is no interaction between the 2*p* orbital and π*.

π* No interaction with 2*p* → One node

π ± →

π + 2*p* No new nodes π − 2*p* One new node

To order these, simply count nodes. These molecular orbital energies could also be generated by using a Frost circle (Chapter 13, p. 579).

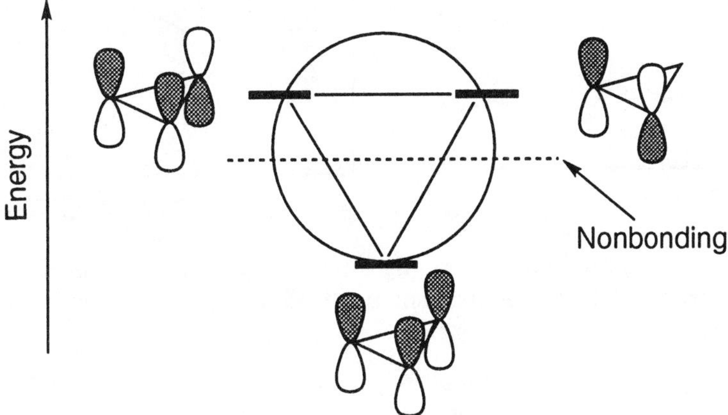

Problem 23.20 The positive charge is shared among the three carbons as shown.

H
H_3C
=
H_3C
H_3C
H_3C

Problem 23.21 There is, by definition only one barrier in the concerted rearrangement (**A**). In the stepwise process (**B**) there must be two transition states, one for formation of the less stable secondary carbocation and the second for rearrangement (hydride shift) to the more stable tertiary carbocation.

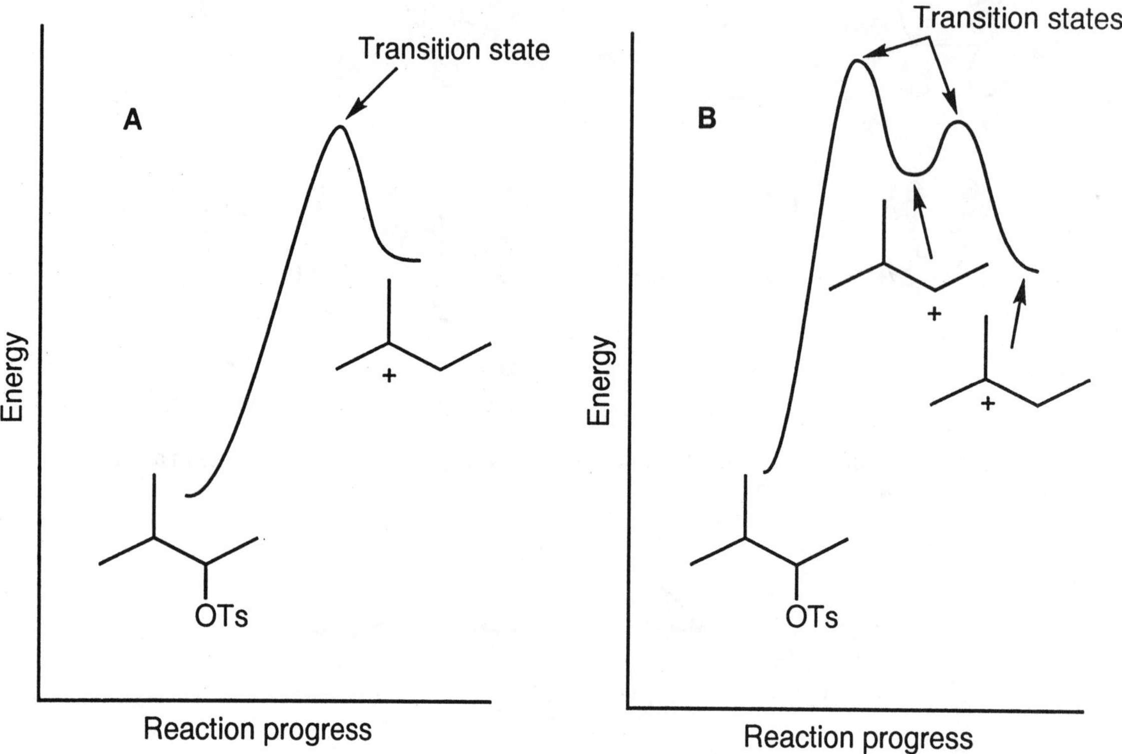

Problem 23.22 There really should not have been any controversy over phenonium ions. They are not substantively different from the intermediate ions involved in aromatic substitution.

+ H E H E H E

Intermediate in aromatic substitution

+

Phenonium ion

Perhaps π participation to form the phenonium ion was confused with σ participation to form something quite different—a three-center, two-electron bonding system in which the positive charge is not delocalized into the benzene ring.

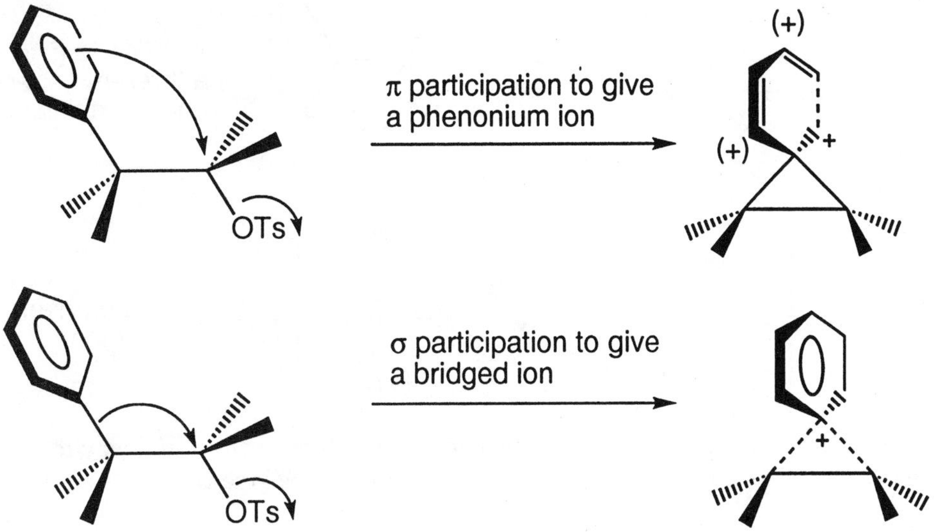

OTs

π participation to give a phenonium ion

(+)

(+)

+

OTs

σ participation to give a bridged ion

+

Problem 23.23 The positive charge is shared among the three carbons as shown.

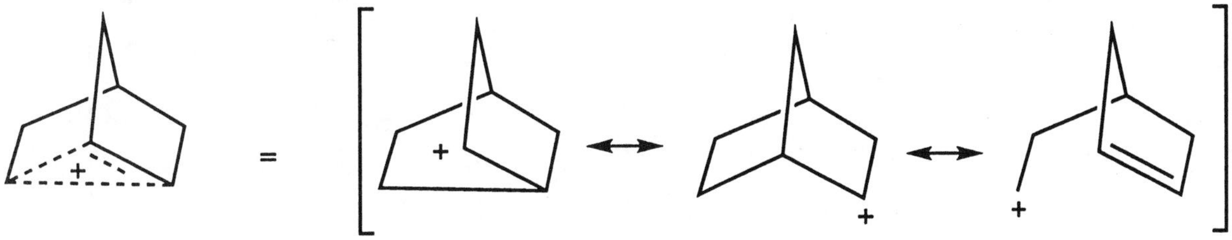

Problem 23.26 These structures are the most important resonance forms.

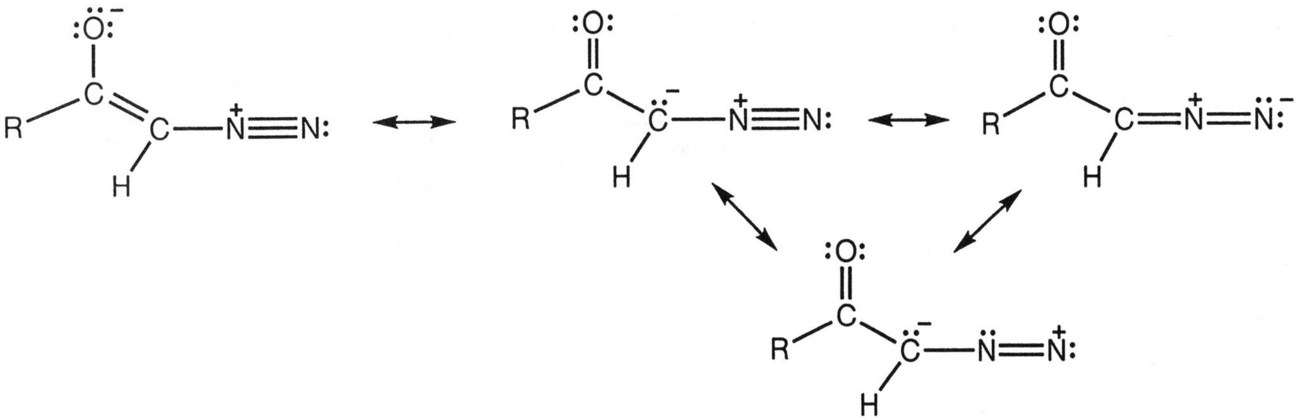

Problem 23.29 If nitrene chemistry is akin to carbene chemistry, there should be addition reactions to alkenes and carbon–hydrogen insertion reactions with alkanes, and there are, as the following two reactions show.

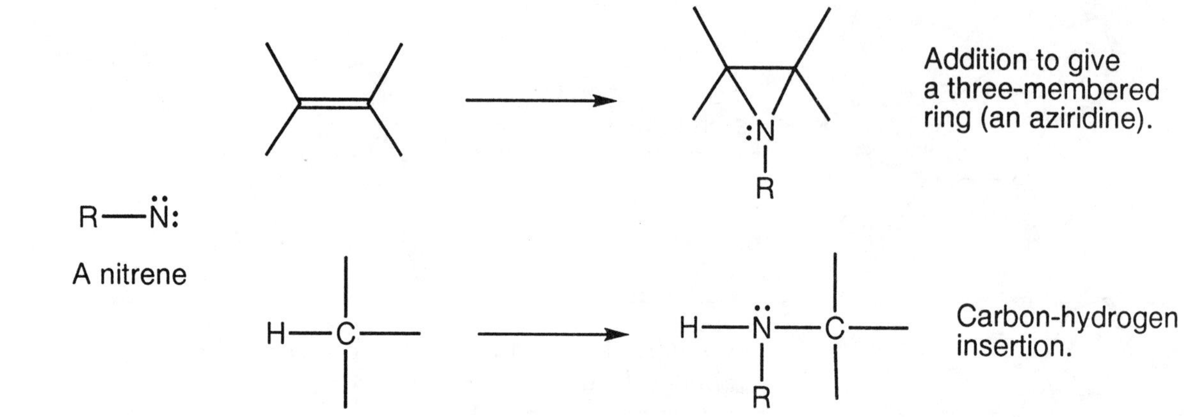

R—N̈:

A nitrene

Addition to give a three-membered ring (an aziridine).

Carbon-hydrogen insertion.

Problem 23.30 Addition to the carbon–oxygen double bond takes place in the usual way to give a resonance-stabilized intermediate. Deprotonation and protonation steps give the carbamate ester.

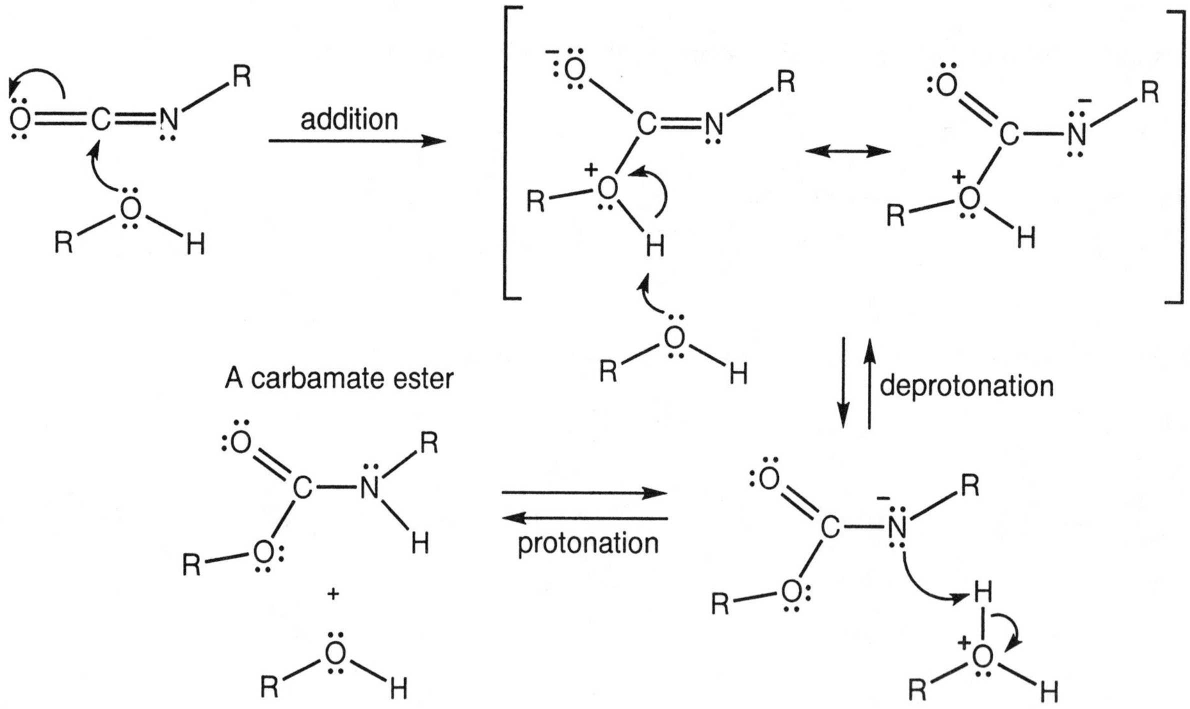

addition

deprotonation

A carbamate ester

protonation

Problem 23.31 The reaction begins exactly as in Problem 23.30 with the substitution of water for the alcohol. This addition leads to a carbamic acid.

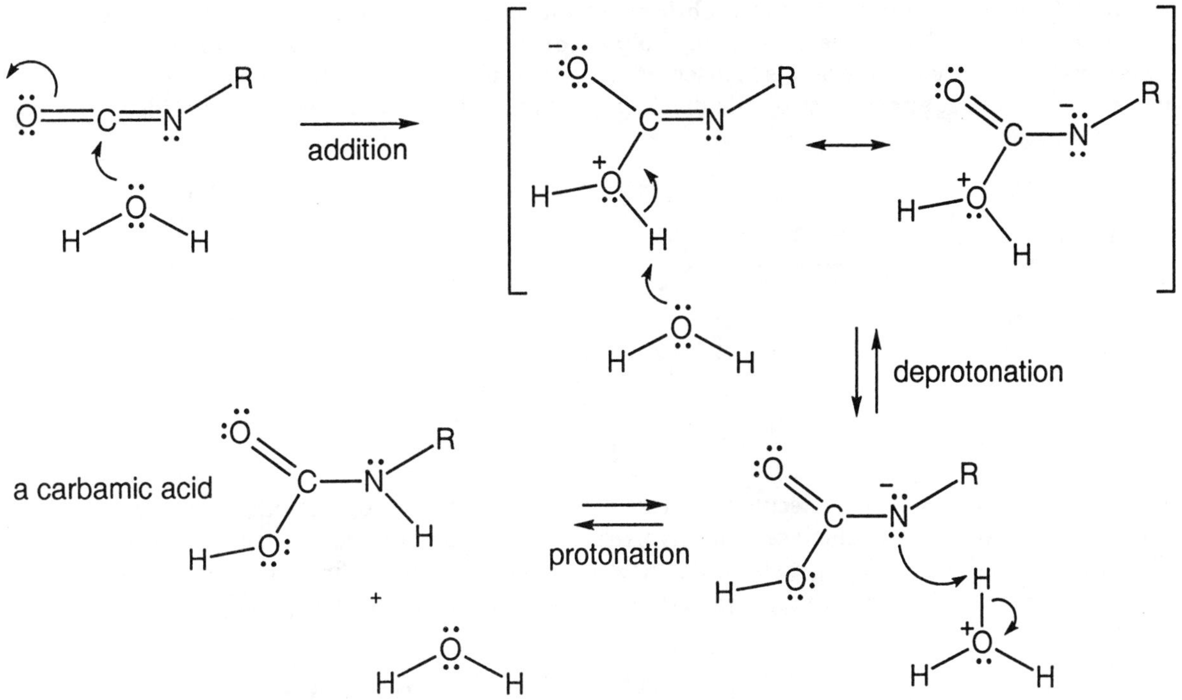

addition

deprotonation

a carbamic acid

protonation

Carbamic acids are not isolable, but decarboxylate to give CO_2 and an amine.

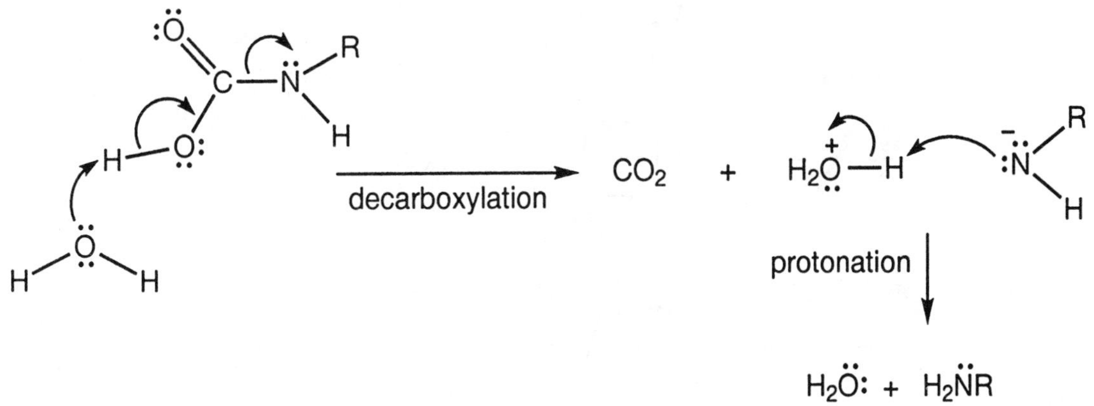

decarboxylation

CO_2 + $H_2\overset{+}{O}$—H

protonation

H_2O: + H_2NR

Answers to Additional Problems, Chapter 23

Problem 23.32 This reaction involves a straightforward intramolecular displacement of bromide by the nucleophilic amine. As always, it probably helps to draw good Lewis structures for the starting materials first. Then attention is focused on the nonbonding pair of electrons on the nitrogen atom that will act as the nucleophile displacing bromide. Deprotonation by hydroxide completes the reaction.

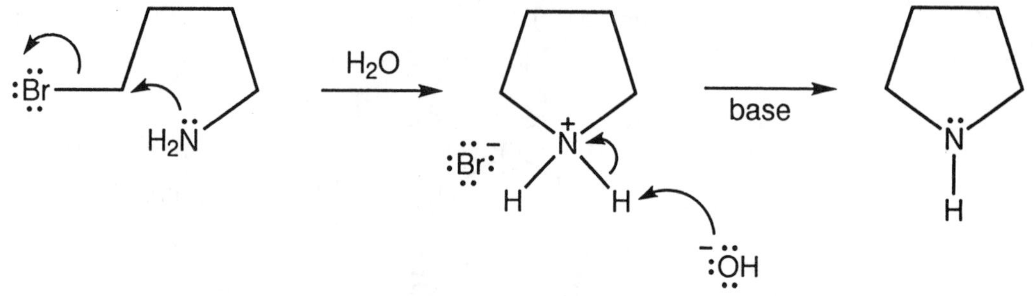

Problem 23.33 Here, too, an intramolecular displacement is followed by deprotonation. In the smaller molecule, intramolecular displacement involves the formation of a highly strained four-membered ring, and is slowed relative to the intermolecular reaction. In the first example, intramolecular formation of the relatively strain free five-membered ring wins the competition.

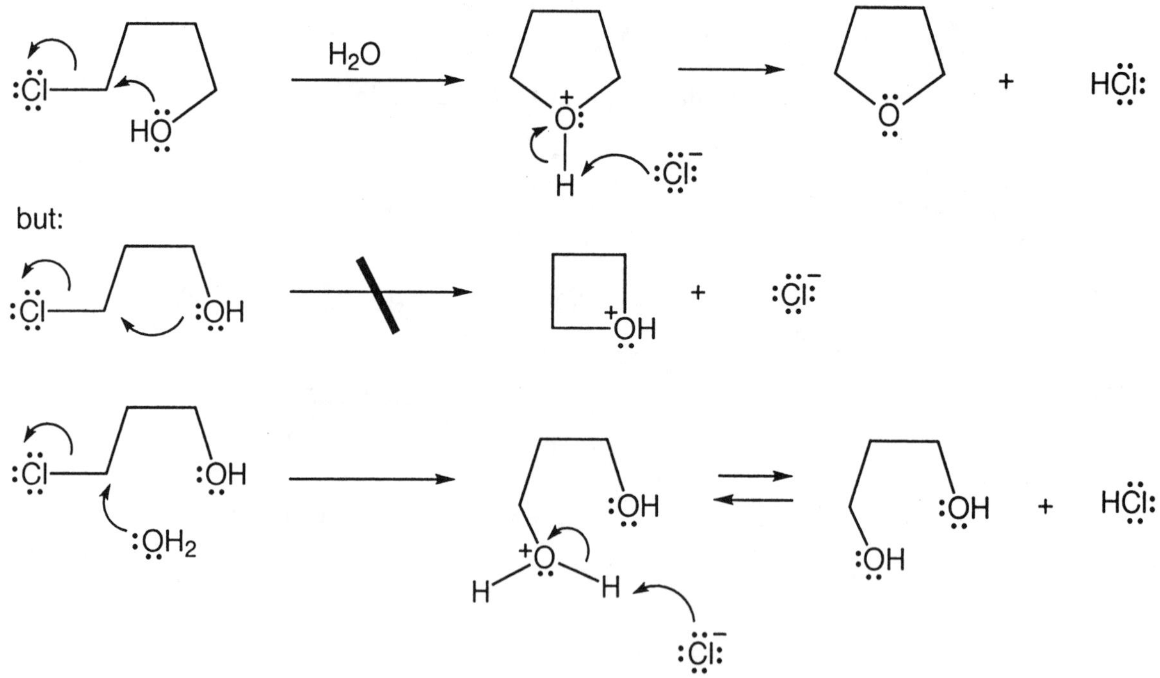

but:

Problem 23.34 Here, intramolecular S_N2 displacement of the secondary chloride wins over intermolecular reactions. There is a cyclic ammonium ion (aziridinium ion) intermediate that is opened in a second, intermolecular S_N2 reaction.

Et = CH_3CH_2
Bz = CH_2Ph

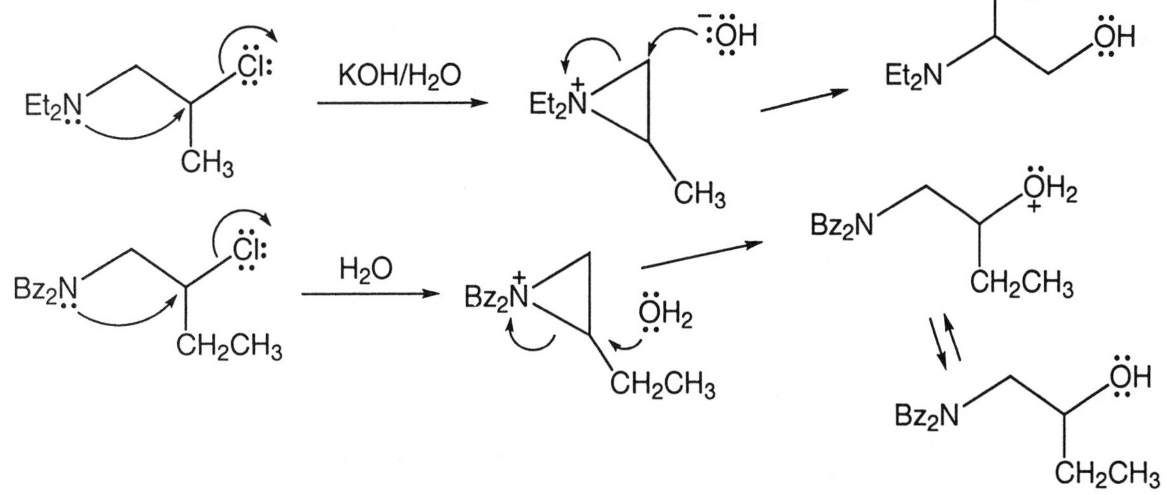

KOH/H_2O

H_2O

The question becomes why hydroxide opens the aziridinium ion at one position and water at the other. Recall that this is exactly the pattern seen in opening of epoxides and other three-membered ring intermediates. Strong nucleophiles (here, hydroxide), react at the sterically less demanding point, whereas openings in acid, in which weaker nucleophiles (here, water) are involved, take place by breaking the weaker (more substituted) bond.

Problem 23.35 Compound **1** can do an intramolecular displacement, whereas neither **2** nor **3** can, which accounts for the increase in rate. However, the data do not allow us to determine whether there is a bridged intermediate or an open intermediate involved in this reaction. The rate increase could come from assistance by the π bond in either a symmetrical fashion to give a bridged ion, or unsymmetrical fashion to give an open, "classical" carbocation.

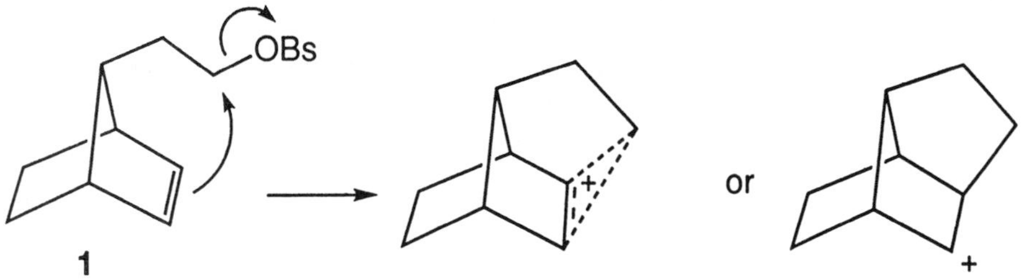

1 or **2** **3**

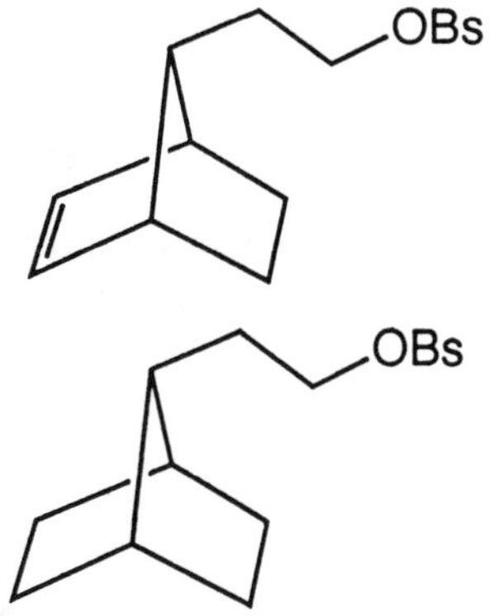

No such reaction possible

Problem 23.36 The product is formed with retention, and there is a rate difference between **1** and **2** of 10^5-10^6. Two clues to neighboring group participation are present. Compound **3** must be formed by neighboring group participation of sulfur. It takes a look at these molecules in three dimensions to see the reason why **1** can undergo an intramolecular displacement and **2** cannot.

chair flip

S_N2

S_N2

1

chair flip

3

But:

chair flip

2

No intramolecular S_N2 possible

No intramolecular S_N2 possible

In **1**, the internal nucleophile, sulfur, can displace the leaving group from the rear (even though this displacement must take place in the less favored, diaxial conformation). In **2**, there is no conformation in which intramolecular S_N2 is possible. A frontside displacement is required, and such reactions do not occur. A relatively slow intermolecular S_N2 can occur.

Problem 23.37 Consider the transition states for intramolecular displacement by the two phenyl groups in question.

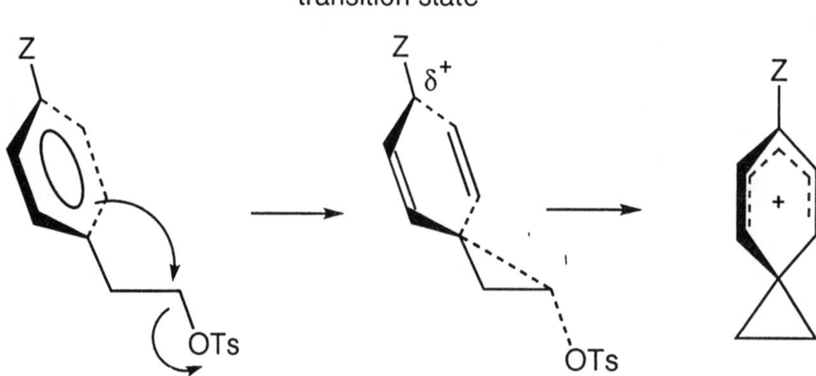

When Z = OCH_3, the transition state is stabilized, as the methoxy group helps to delocalize the adjacent partial positive charge. The energy of the transition state is lowered, and the rate of intramolecular displacement increased. When Z = NO_2, the electron-withdrawing nitro group destabilizes the transition state, and the rate of intramolecular displacement is lessened.

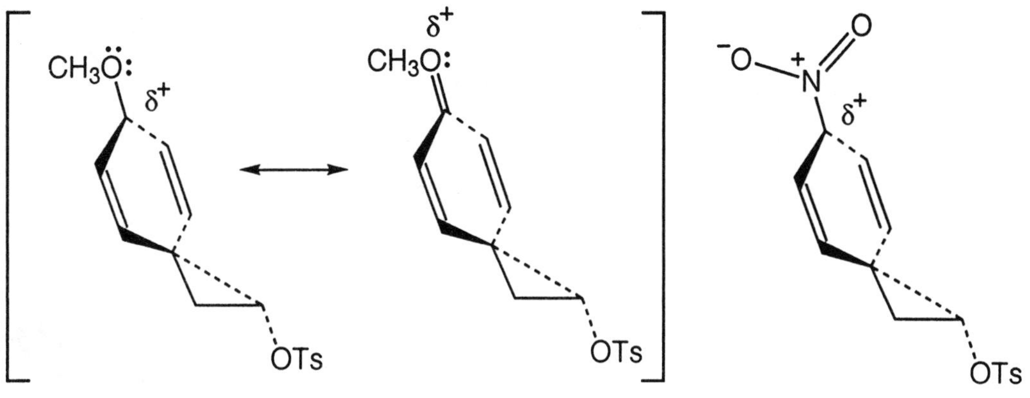

Transition state stabilized

Transition state destabilized

Problem 23.38 The double bond in **1** is not in a position to participate in ionization of the tosylate, but one of the σ bonds is (bold face line). An allylic carbocation (**A**) is formed, and then captured by water to produce **2**.

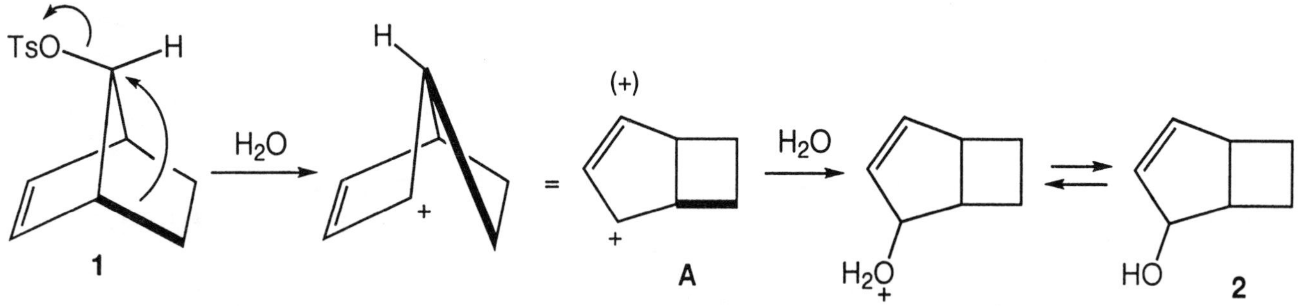

In compound **1,** participation of the σ bond leads to a relatively stable allylic cation. That would not be the case in **3**, and reaction is much slower.

Problem 23.39 Here the clues to neighboring group participation are the "strange rearrangement" in which the chlorine and methoxy groups seem to change places, and the mysterious switch in stereochemistry from cis in the starting material to trans in the product. The methoxy oxygen is the neighboring group. A three-dimensional picture helps a lot.

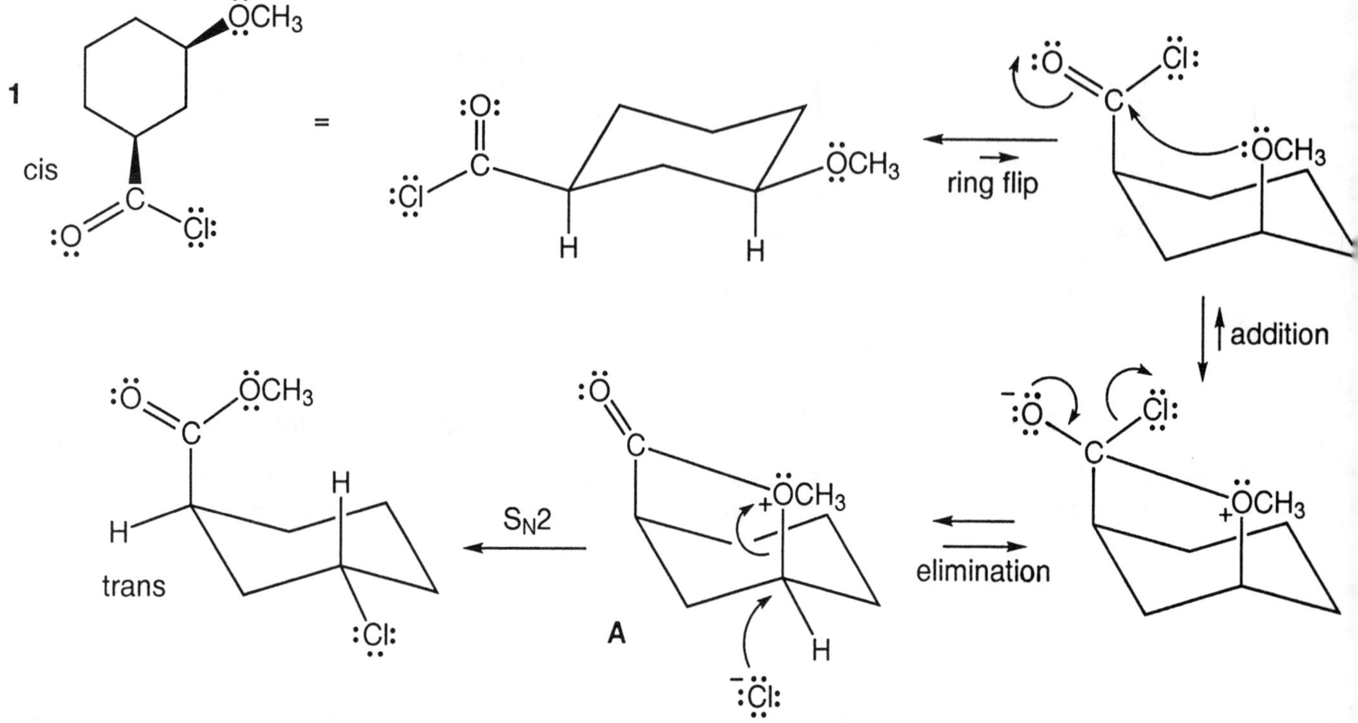

1 cis = ring flip addition elimination S_N2 A trans

In the starting cis isomer (but not in the trans compound **2**), the methoxy group is in position to undergo an addition–elimination reaction generating bicyclic ion **A**. Opening by chloride (S_N2) from the rear yields the product.

Problem 23.40 Remember Magid's First Rule? "All neutral products are lactones." Here's a "neutral product", and I'll bet it's a lactone. Even if Magid's First Rule has lapsed into deserved obscurity, this problem is not too difficult. Bromine and a double bond add up to bromonium ion formation. This intermediate is opened by the nearby acid carbonyl, and deprotonated to the product. There are two possible points of attack on the bromonium ion. One, the route actually followed, leads to lactone **A**. A reasonable answer to this problem might take the other approach, opening the bromonium ion to a six-membered ring, and ultimately giving **B**.

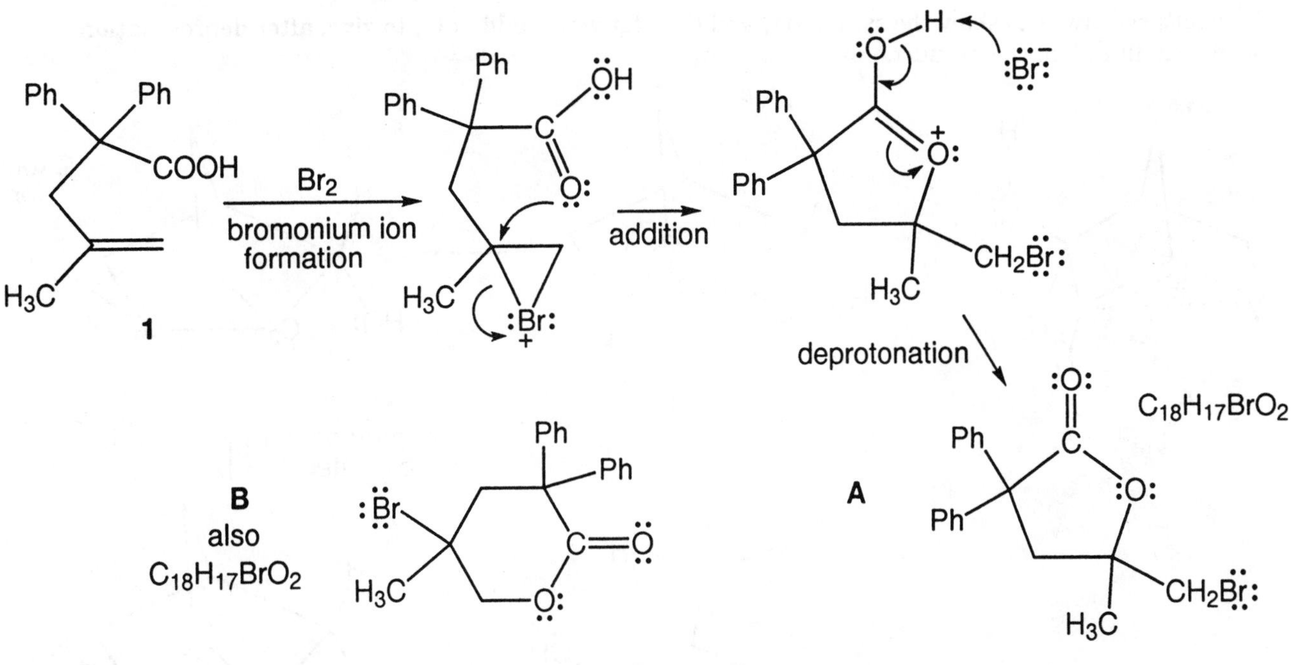

Ph Ph COOH H3C 1 Br_2 bromonium ion formation addition deprotonation A $C_{18}H_{17}BrO_2$ B also $C_{18}H_{17}BrO_2$

Problem 23.41 This reaction is an example of neighboring group participation by the π system of a benzene ring in which the product is sufficiently stable to be isolable. Potassium *tert*-butoxide is sterically hindered and is unlikely to be effective at intermolecular displacements. There is little strain in the five-membered ring product. Deprotonation of the phenol generates a phenoxide that undergoes a rapid intramolecular displacement of the leaving group at the nearby carbon.

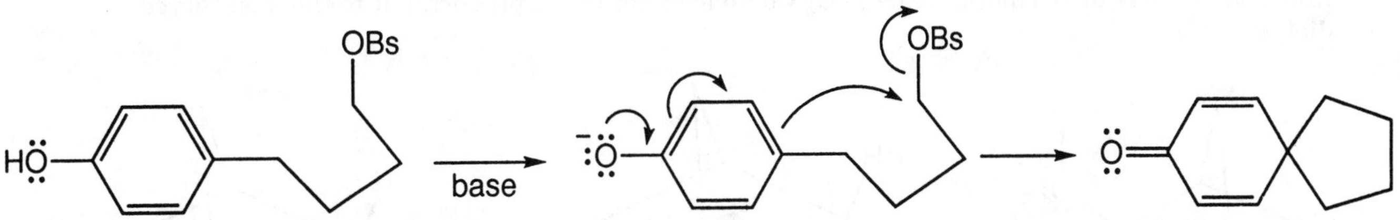

OBs base

Problem 23.42 Surely the first step in acid will be protonation of the epoxide. Normally, a protonated epoxide would be opened from the backside by water. In this case, an adjacent carbon–carbon σ bonds (bold face line) is poised in perfect position to do a faster intramolecular displacement to give bridged intermediate **A**.

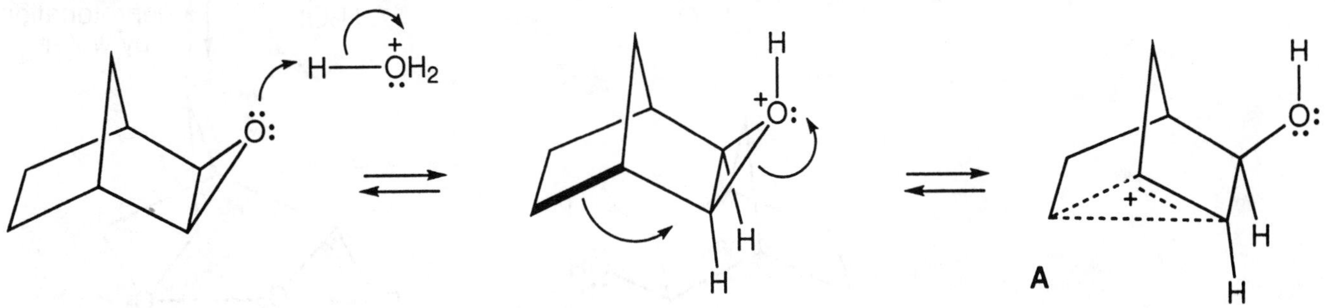

A

Now let's redraw **A** to show the symmetry, and then let water add at C_1 to give, after deprotonation of the oxonium ion, the product.

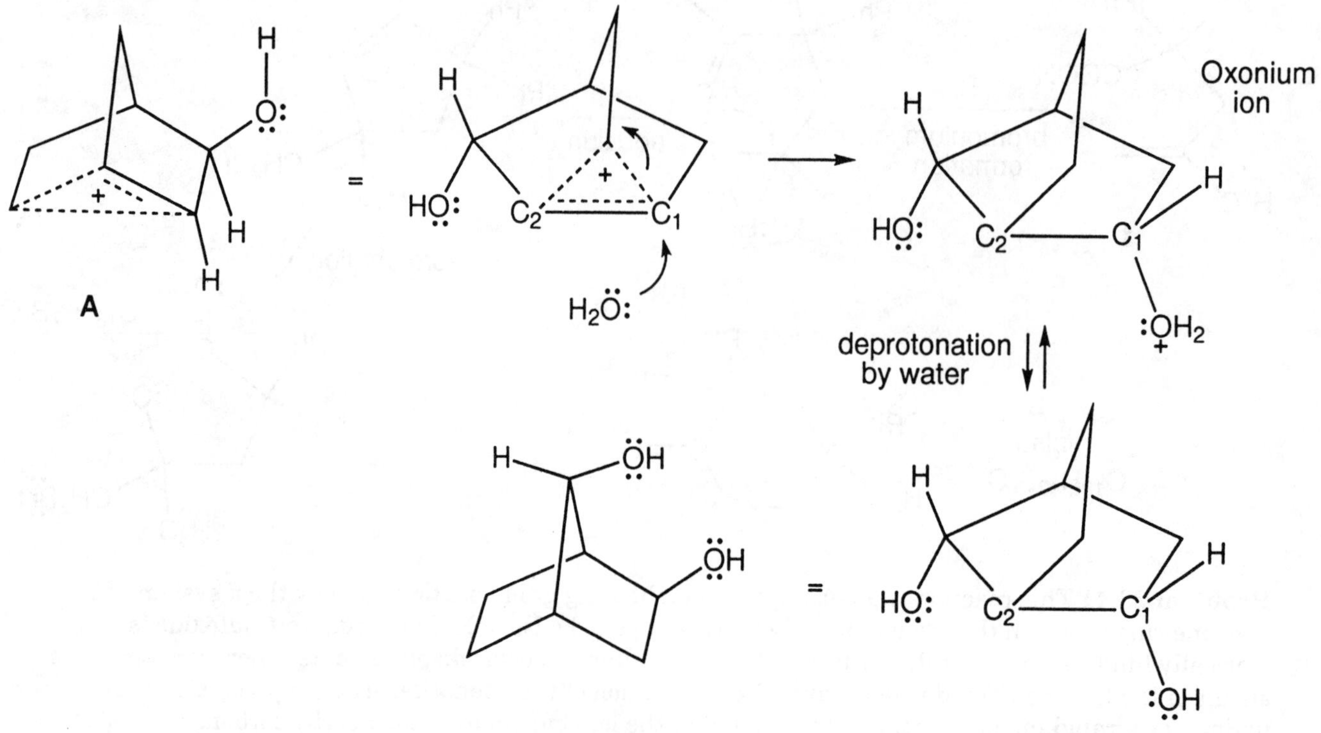

Addition of water at the more hindered C_2 would lead not to the product, but to the unobserved diol, **B**.

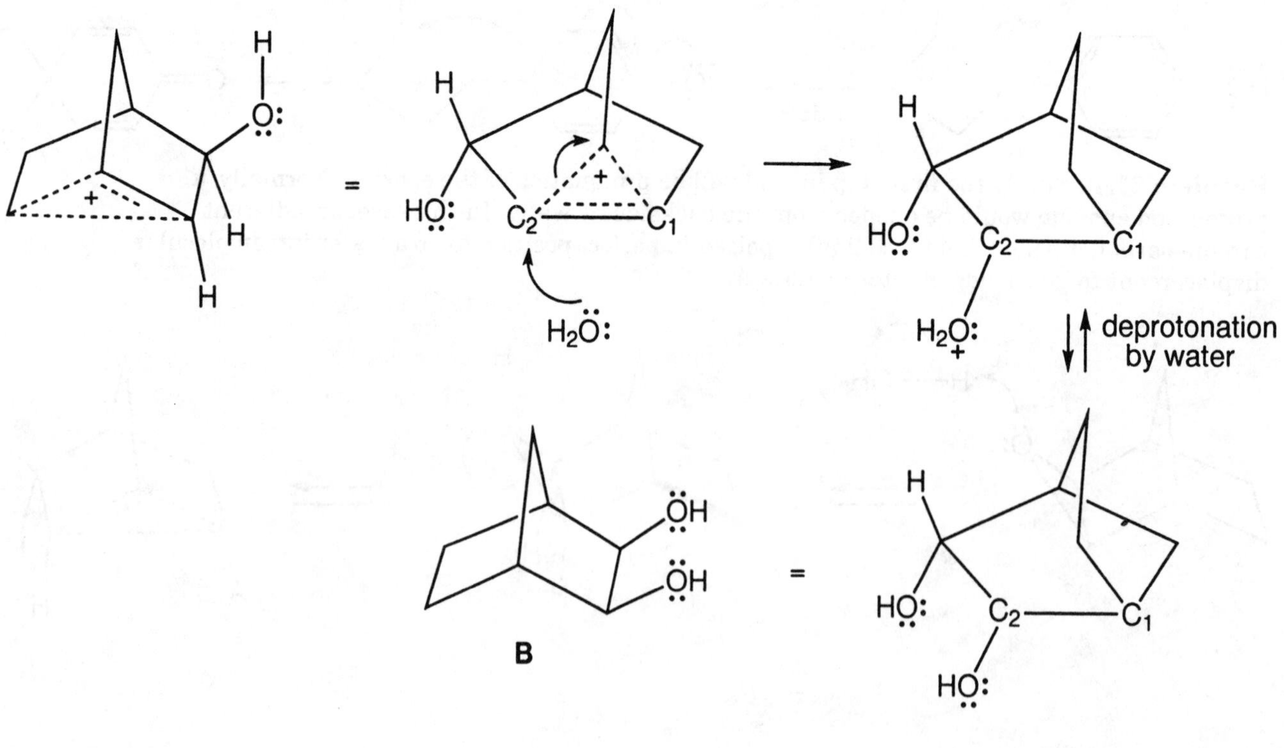

Problem 23.43 As we saw in Problem 23.24, this reaction is just another way of forming the 2-norbornyl cation. Symmetrical displacement of the leaving group by the double bond of the five-membered ring gives the bridged ion **A**. Opening by acetic acid in an intermolecular S_N2 reaction produces the exo acetate. Note that the bridged ion for the monomethyl derivative **A** (R = CH_3, R' = H), is not symmetrical. From studies of the products formed it became apparent that acetic acid addition occurred only at the tertiary carbon of the bridged ion.

R
R'
ONs
NsO
R
R'
R
R'
A
R
R'
:O:
H_3C C ÖH
(AcOH)
deprotonate
R
R'
:O+
C—OH
H_3C
OAc
R'
R

What do the kinetic data tell us about the nature of the carbocation? Although we have drawn them as bridged, it is possible to view the reaction as proceeding through pairs of equilbrating open cations. As before (Section 23.2b), the rate data help us to differentiate bridged from unbridged ions.

The monomethyl derivative reacts seven times faster than the parent. This is, of course, compatible with either mechanistic scheme. In the bridged ion, one of the contributing structures contains a tertiary carbocation. Similarly, for the open ions, one of the equilibrating structures is a tertiary carbocation. However, the two mechanistic schemes make different predictions for the dimethyl compound. In a bridged carbocation, both methyl groups contribute at the same time. If one methyl group produces a rate enhancement of 7, two must produce $7^2 = 49$. On the other hand, if it is the formation of an open, tertiary carbocation that is important, the second methyl group increases this possibility by only a factor of 2. The two methyl groups never exert their influence at the same time. Thus the predicted rate increase would be $7 + 7 = 14$. The experimental rate for the dimethyl derivative is 38.5, much closer to the predicted value for the bridged ion than that for a system of equilibrating open cations.

Problem 23.44

(a) This problem is an example of the Hofmann rearrangement using chlorine instead of bromine.

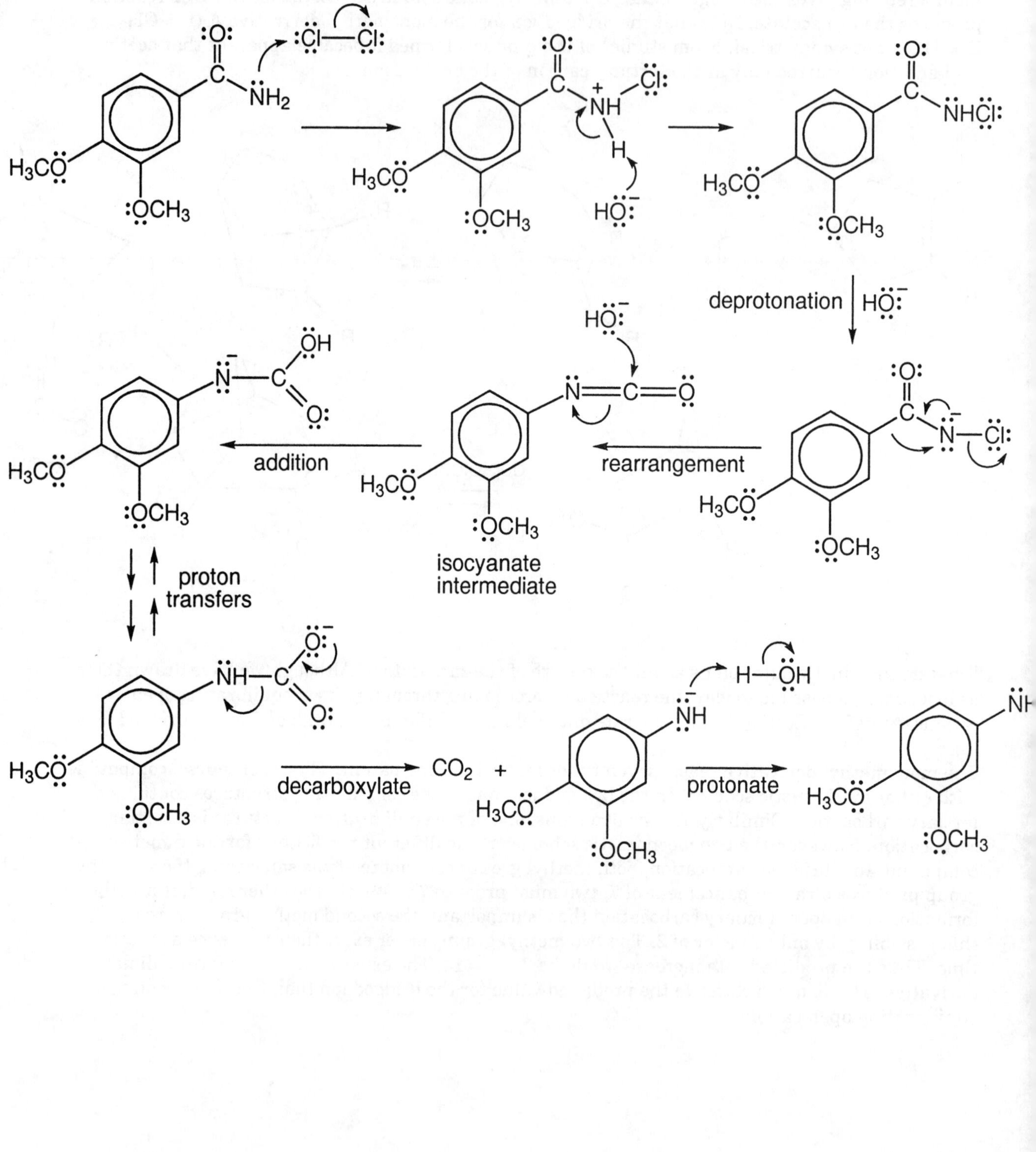

deprotonation

addition

rearrangement

isocyanate intermediate

proton transfers

decarboxylate

protonate

(b) This reaction is a Wolff rearrangement in which an initially formed ketene is trapped by methyl alcohol reaction solvent. Begin by drawing a more detailed and suggestive view of the diazo ketone. Loss of nitrogen with bond migration (bold face line) gives the ketene. Capture by methyl alcohol ultimately gives the ester.

hν
Wolff
rearrangement
N_2 +
$CH_3\ddot{O}H$
addition
deprotonation
protonation
H_3C
H
$COOCH_3$
=

Problem 23.45 The problem here is not so much finding the neighboring group participation, but rather the transformation of the initial product into isobutyraldehyde. It should be obvious now that the rate increase comes from the intramolecular displacement of leaving group by the oxygen of the methoxy group to give oxonium ion **A**.

$CH_3\ddot{O}$:
$(CH_3)_2C$
OBs
1
H_2O
CH_3
$(CH_3)_2C$
A
+ $^-$OBs

Although **1** can do this, the neopentyl compound **2**, of course, cannot. Now, how to get to the final product from **A**? The main question seems to be how to lose the required one carbon atom. The most serious candidate for loss would seem to be the methyl group attached to oxygen.

A stepwise mechanism involves opening **A** to give a tertiary carbocation, **B**. A hydride shift generates the resonance-stabilized **C**, and removal of the methyl group through displacement by water (don't just pop it off as $^{+}CH_3$), gives the product.

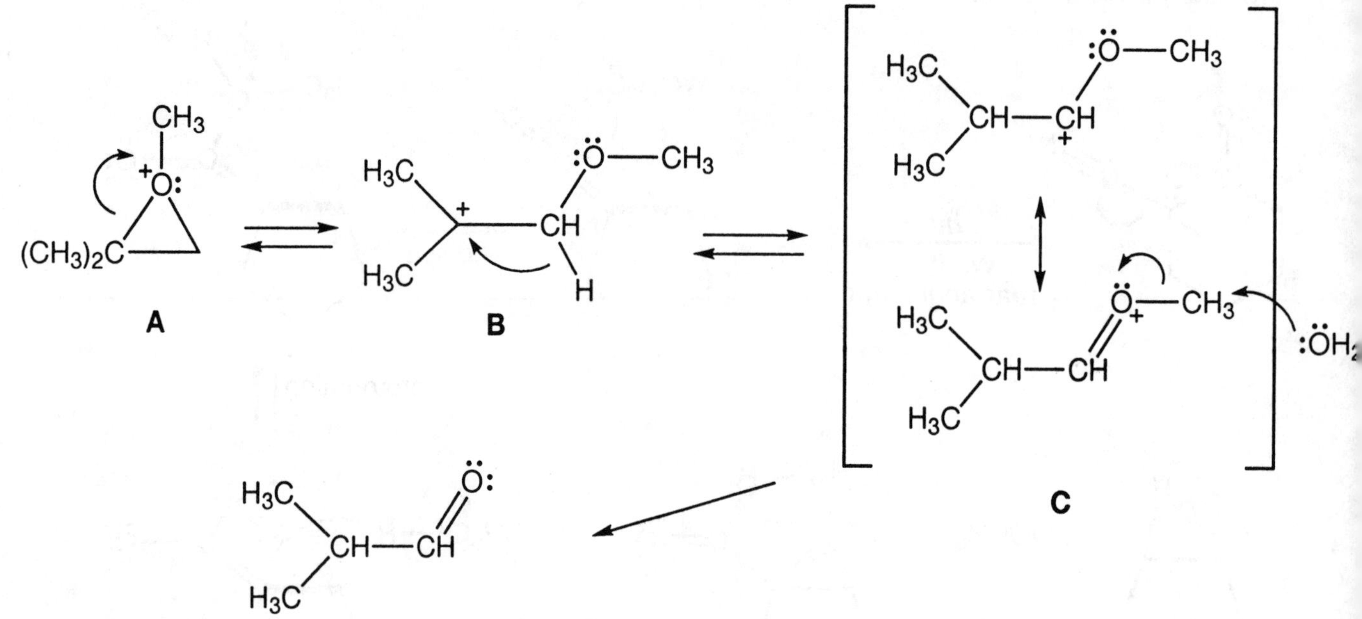

Problem 23.46 In base, removal of the amide hydrogen by acetate takes place to give a resonance-stabilized anion **A**. Displacement of tosylate by the oxygen end of the anion then occurs to give **1**.

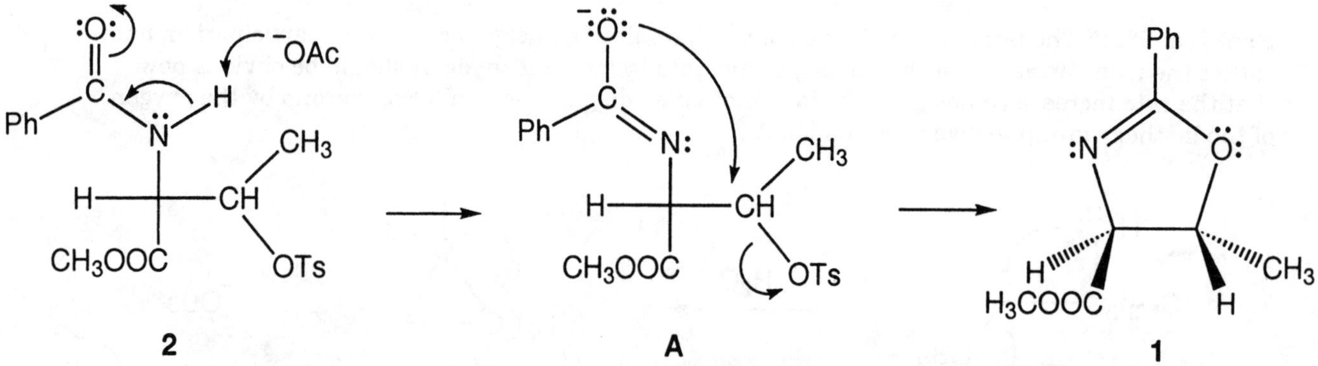

We can use the stereochemistry of **1** to deduce the stereochemistry of **2**. Displacement of tosylate by oxygen is an S_N2 reaction and must occur with inversion. Thus we can specify the geometry of **A**, and of **2**.

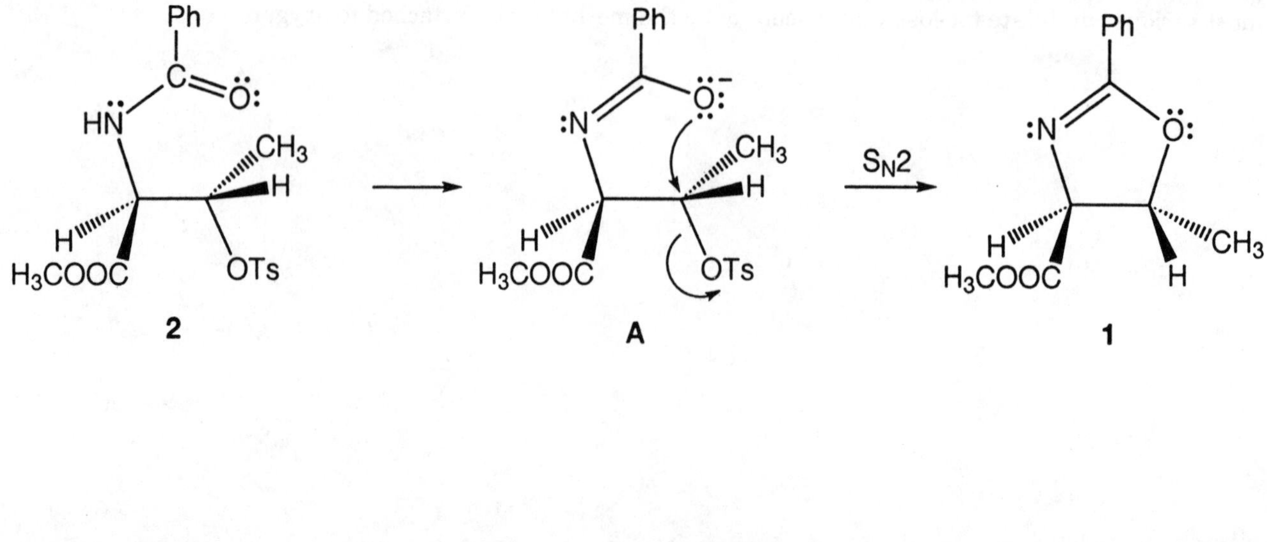

Problem 23.47

(a) Conversion of **1** into **2** could involve the standard alcohol-to-chloride transformation via a chlorosulfite intermediate **A**.

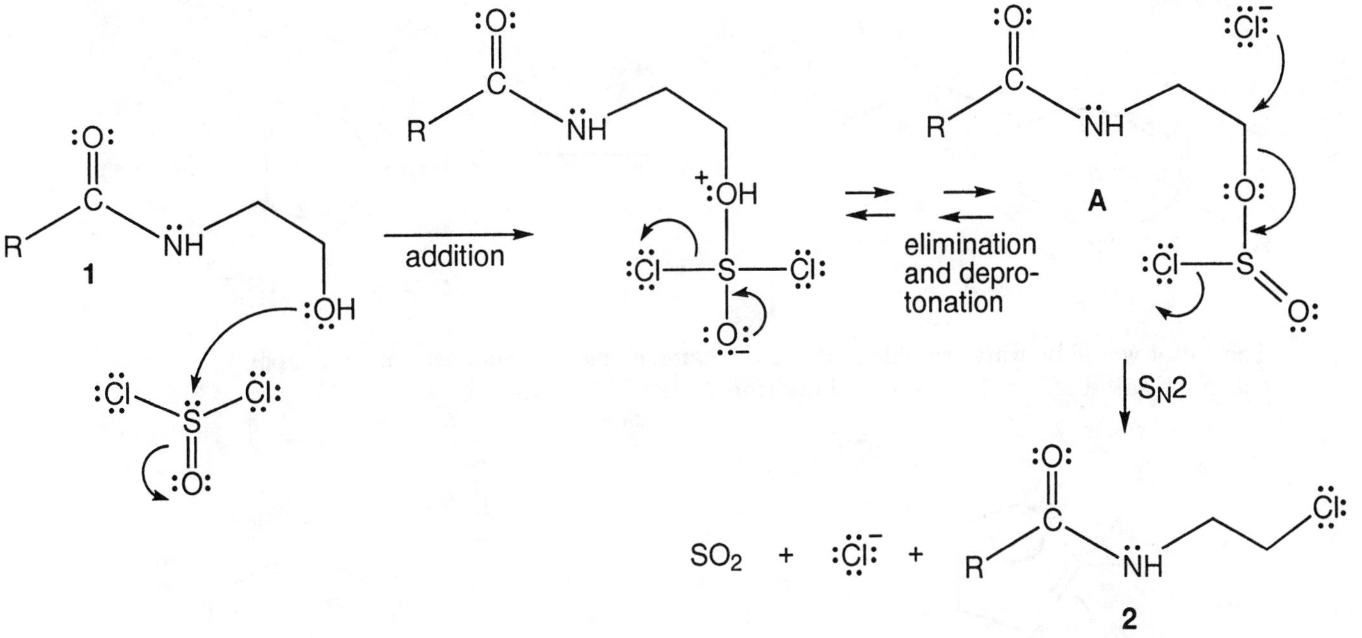

addition

elimination and deprotonation

A

S_N2

SO_2 + :Cl:⁻ + **2**

Treatment of **2** with base leads to proton loss to give the resonance-stabilized anion, **B**. Intramolecular displacement of chloride leads to **3**.

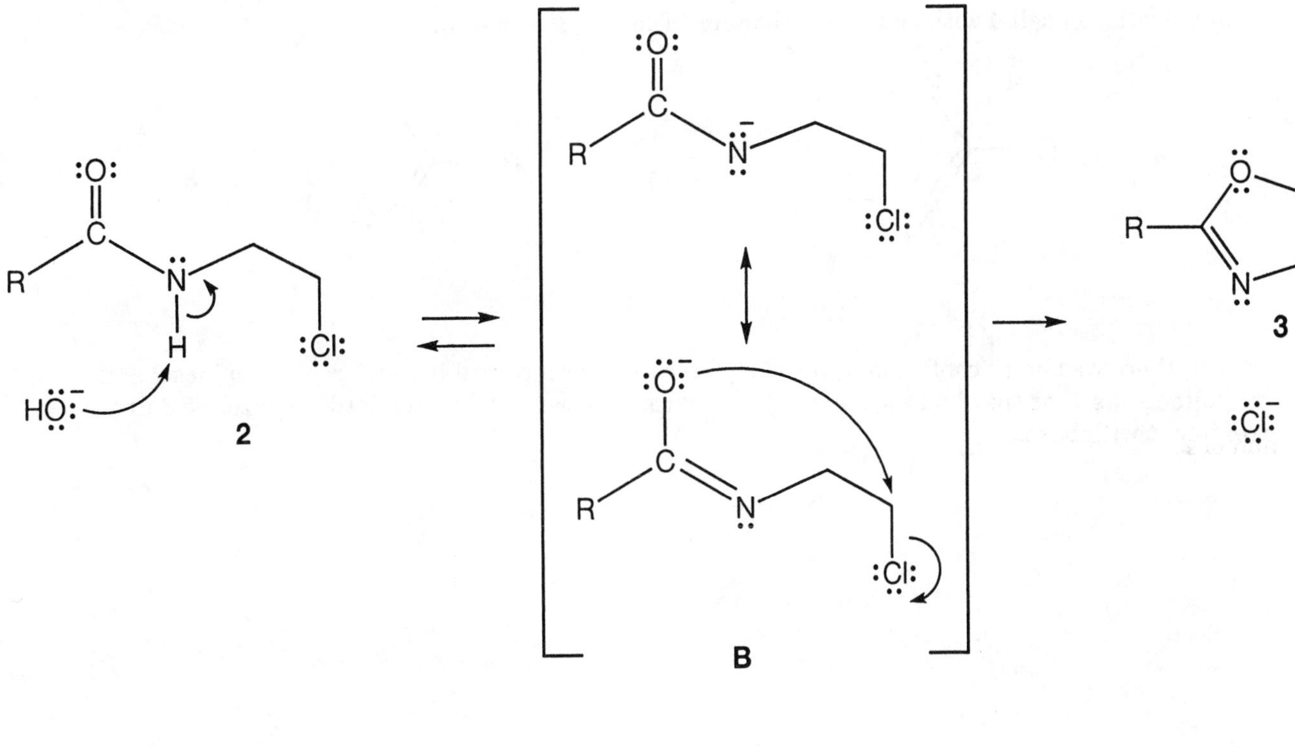

2

B

3

(b) However, when **1** is treated with thionyl chloride at low temperature, compound **4** is isolated rather than **2**. What could compound **4** be? Notice that in chlorosulfite **A** the amide oxygen is poised to displace sulfur dioxide and chloride ion in intramolecular fashion to produce the oxazoline salt, **4**.

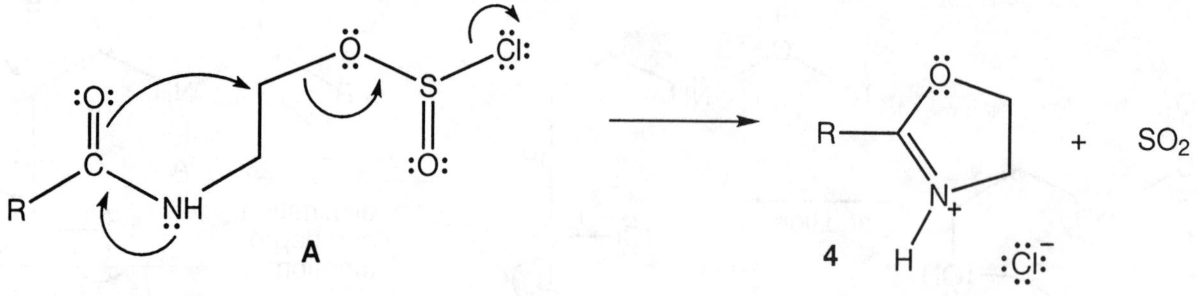

The salt **4** would be water soluble and has the same elemental composition as **2**. Upon heating, oxazoline salt **4** affords **2** by a second intermolecular S_N2 reaction.

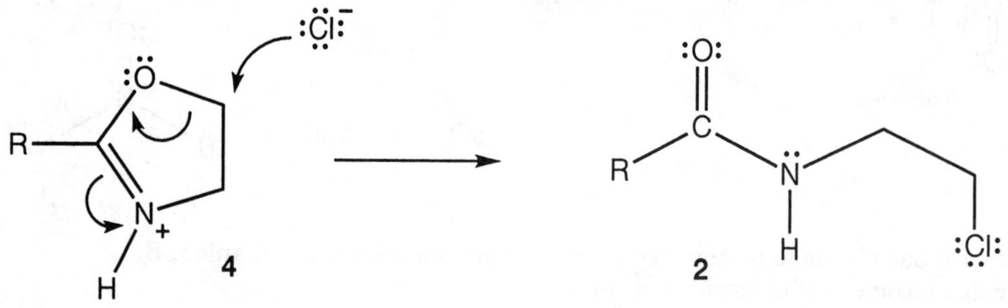

Neutralization of salt **4** with aqueous carbonate, of course, liberates **3**.

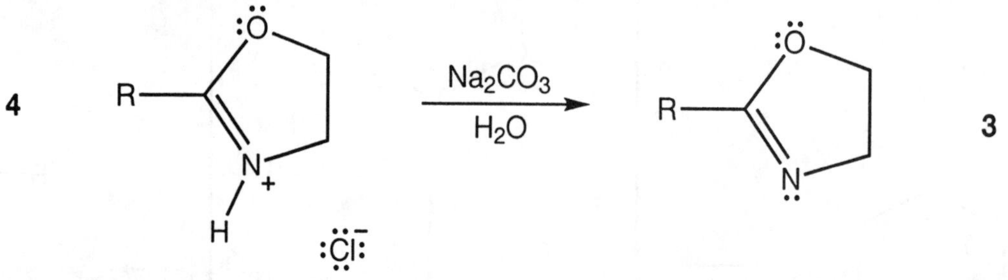

In fact, there was much confusion in the early literature with regard to β-chloroethylamides **2** and oxazoline salts **4**, as they had the same elemental compositions and both afforded oxazoline **3** upon treatment with base.

Problem 23.48 In base, the dicarboxylic acid is first converted into the dicarboxylic acid salt. Bromine then adds to the double bond to give bromonium ion **A**. Addition is from the more accessible, exo face of the molecule. The bromonium ion is now opened from the backside by an intramolecular S_N2 reaction of the perfectly positioned carboxylic acid salt. Neutralization of the carboxylate salt with hydrochloric acid gives the product.

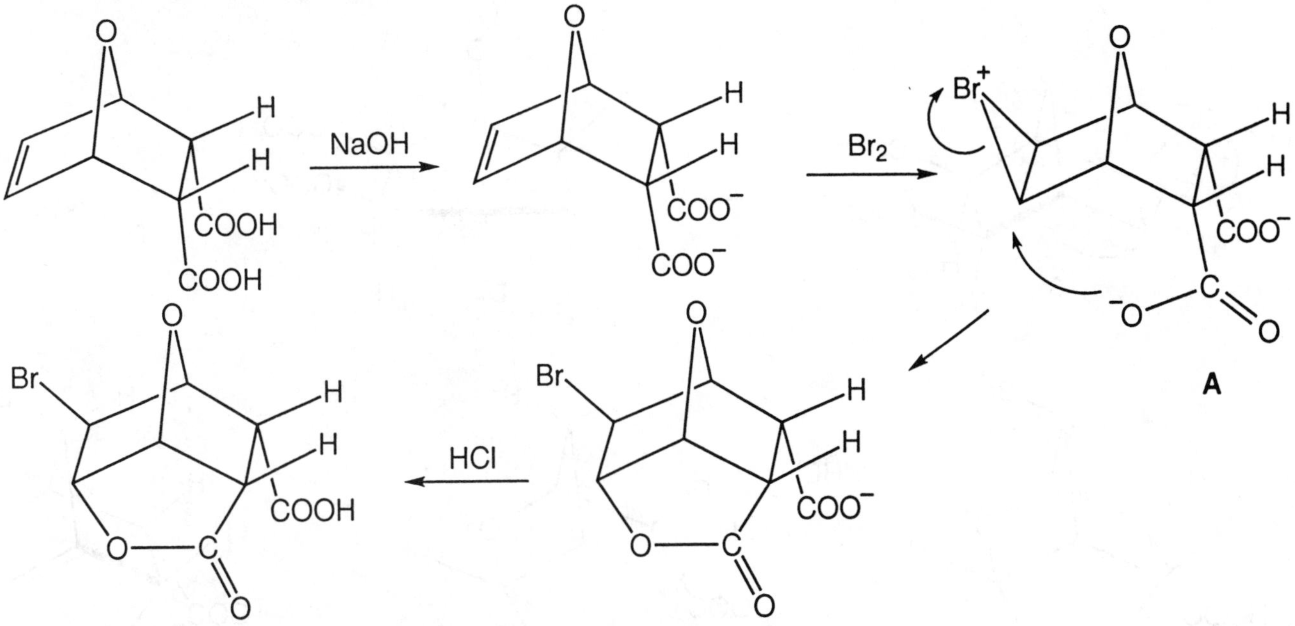

NaOH → ; Br_2 → **A** ; HCl ←

In the second reaction sequence, things are a bit more complicated. Begin by opening the anhydride to give the dicarboxylate **B**, the exo version of **A**.

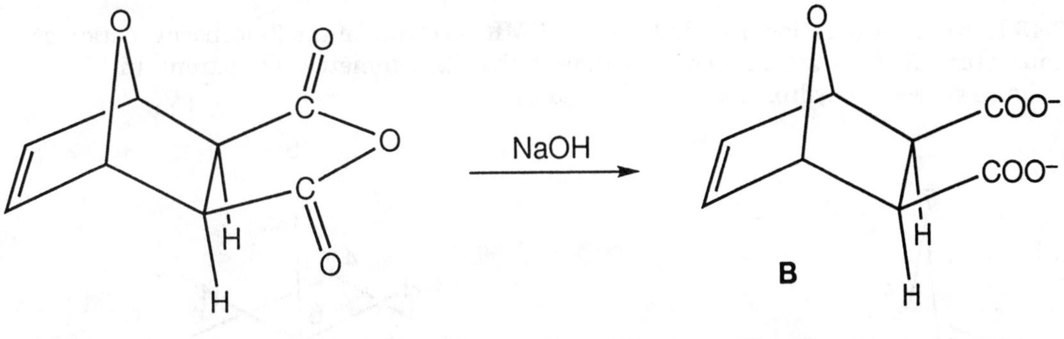

NaOH → **B**

Bromination of the double bond of **B** from the exo face still occurs. This time the product is bromonium ion **C**. However, now neither carboxylate is in a position to open the three-membered ring. Neither can reach.

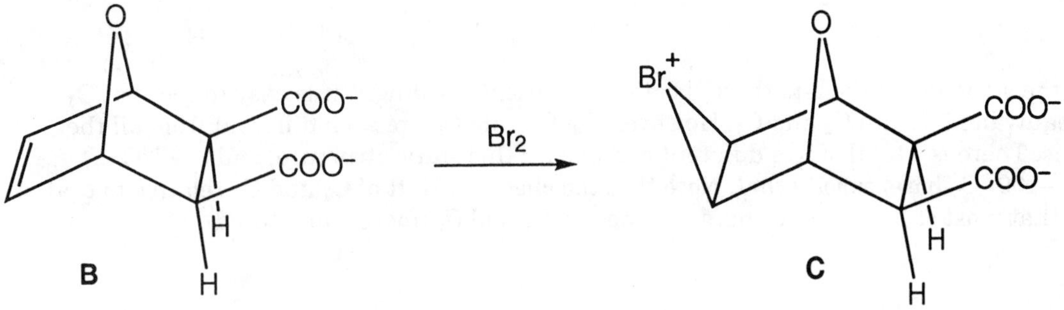

B Br_2 → **C**

What can happen instead? Note that one of the adjacent carbon–carbon σ bonds is positioned to open the bromonium ion from the backside (bold face line). The resulting carbocation **D** is resonance stabilized by the adjacent oxygen atom, as shown by the (+). In addition, one of the carboxylate groups is now poised to capture the carbocation. Acidification of the carboxylate affords the observed product.

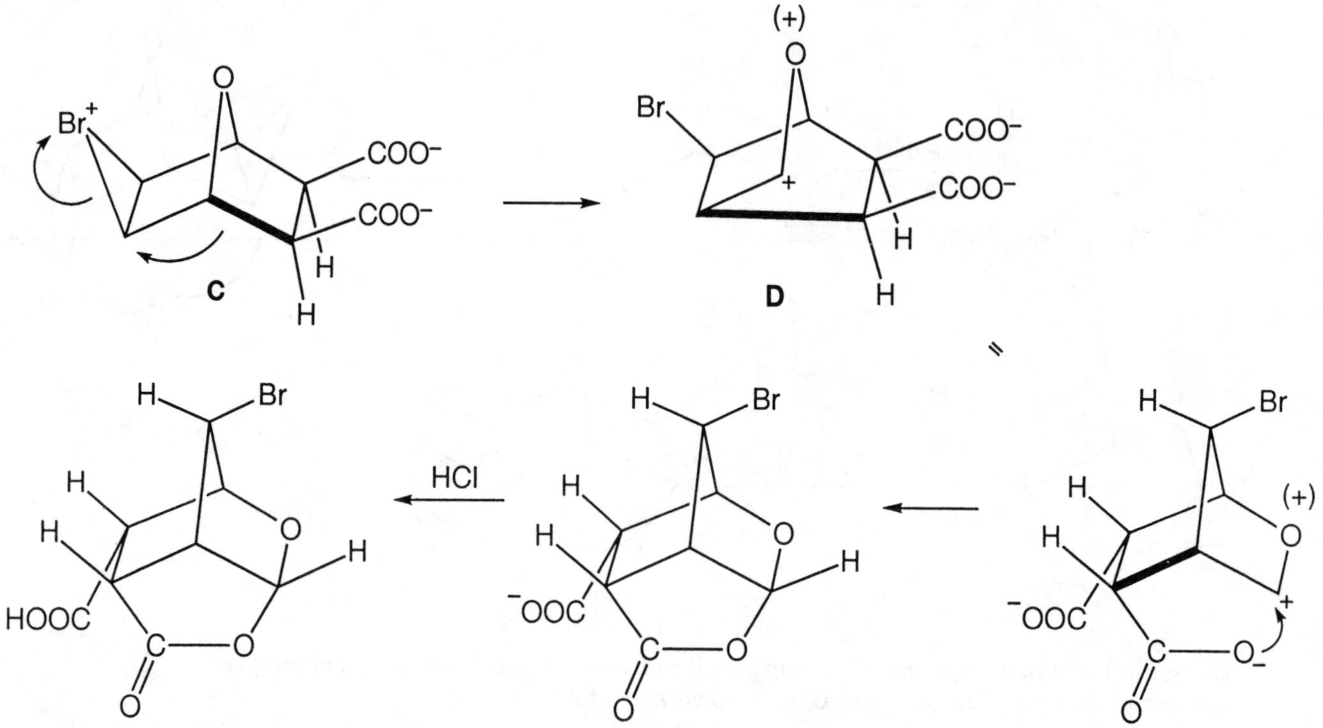

Problem 23.49 It is easiest to rationalize first the ^{13}C NMR spectrum of the 2-norbornyl cation at the lower temperature. Redraw the 2-norbornyl cation so that its symmetry is apparent. In addition, the figure shows the hydrogens at C_1, C_2, and C_6.

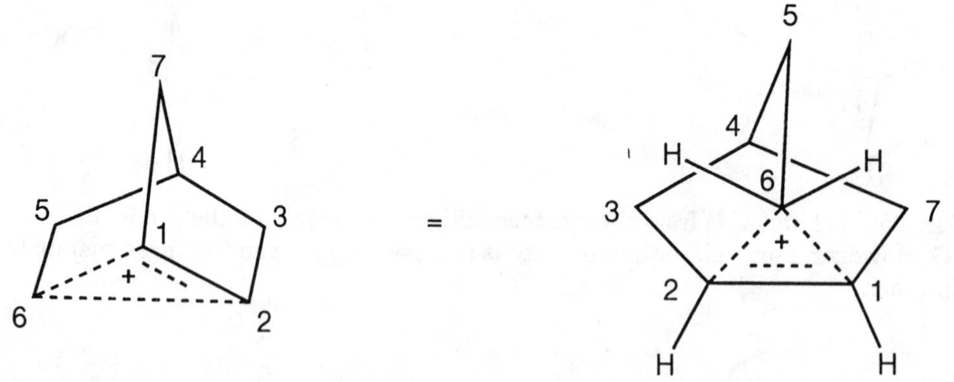

At −159 °C, the structure of the 2-norbornyl cation is static. Accordingly, it is easy to see that C_1 and C_2 are equivalent, as are C_3 and C_7. However, C_4, C_6, and C_5 are each different from all the other carbons. There is a total of five different carbons for this static structure, and the ^{13}C spectrum at −159 °C shows exactly that. Note that the chemical shift of C_1 and C_2 relative to that of C_6 shows that most of the positive charge resides on C_1 and C_2 (more substituted).

At –80 °C there is a simpler, three-line spectrum in which somehow C_1, C_2, and C_6 have become equivalent, as have C_3, C_5, and C_7. Only C_4 is unique. These equivalencies have been rationalized by postulating a series of fast hydride shifts between the 6-, 1-, and 2-positions. These shifts do not occur at the lower temperature—there is not enough energy. The barrier to these hydride shifts has been determined to be about 6 kcal/mol.

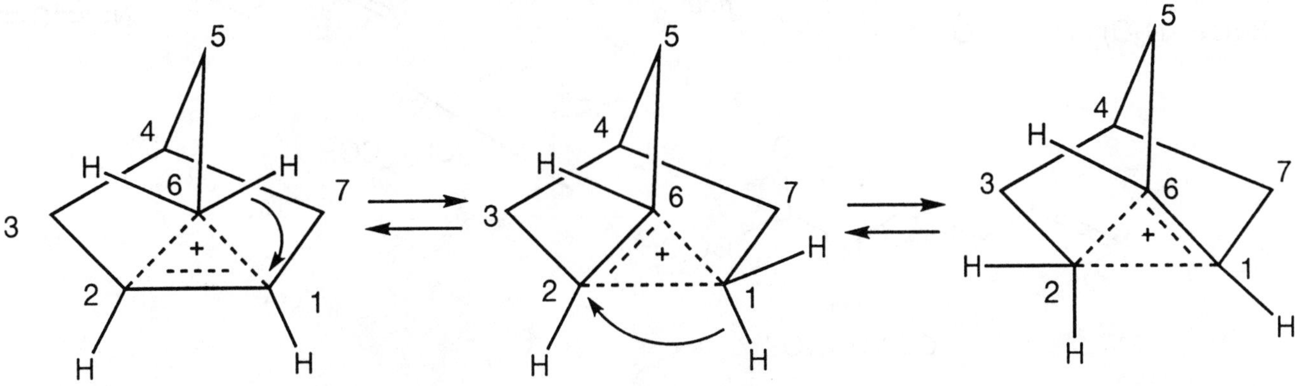

Problem 23.50 In the first step the dienoic acid is deprotonated by the amine base. Reaction with ethyl chloroformate through an addition–elimination process gives **A**.

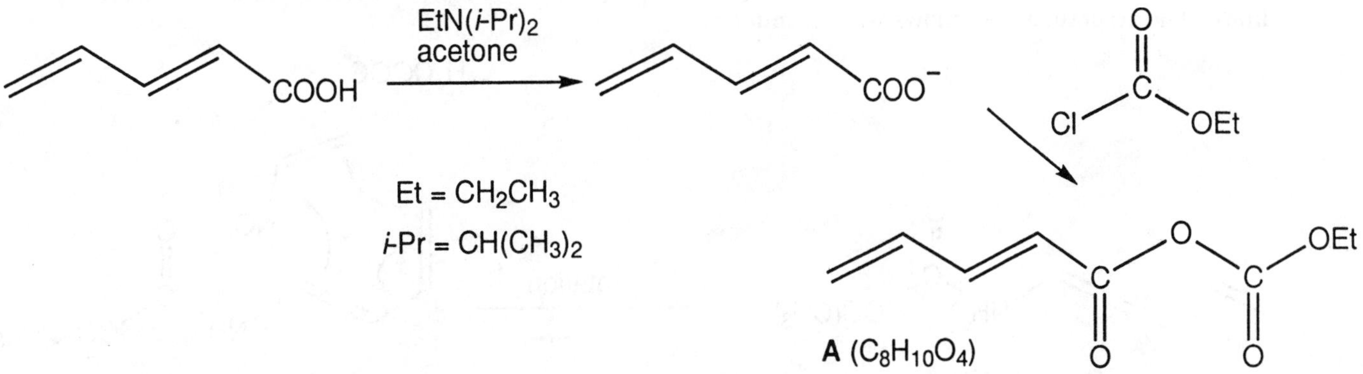

EtN(*i*-Pr)$_2$
acetone

Et = CH_2CH_3
i-Pr = $CH(CH_3)_2$

A ($C_8H_{10}O_4$)

In Section 23.5b, we saw that acyl azides are available from the reaction of acid chlorides and azide ion. Here a similar reaction occurs, with **A** playing the part of an acid chloride. Once again, the mechanism is addition–elimination.

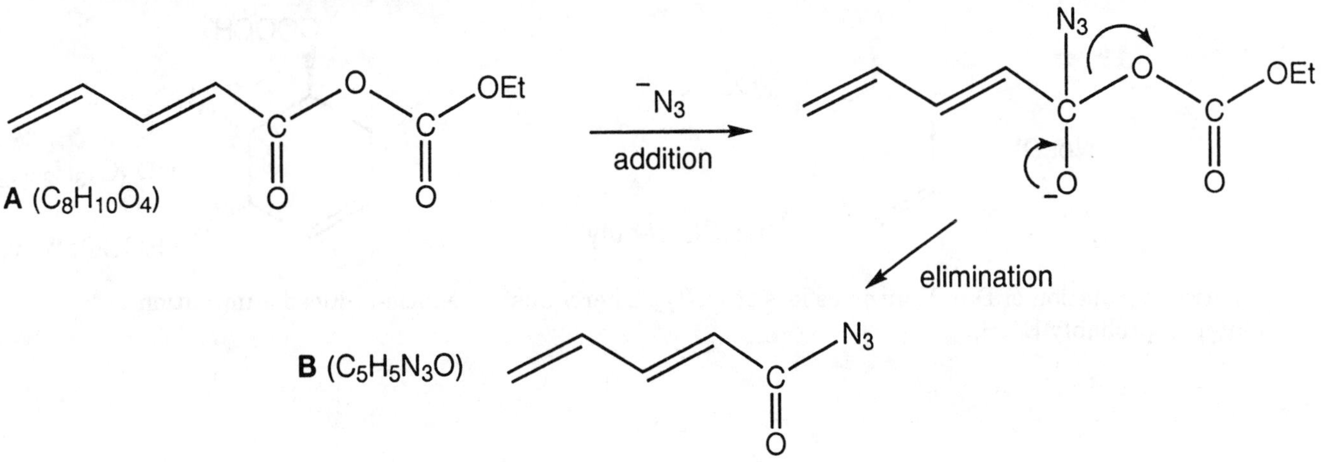

A ($C_8H_{10}O_4$)

$^-N_3$
addition

elimination

B ($C_5H_5N_3O$)

A Curtius rearrangement of the acyl azide B leads to an isocyanate that is captured by the solvent *tert*-butyl alcohol to produce carbamate C.

B ($C_5H_5N_3O$) —toluene, heat→ N_2 + isocyanate; —$(CH_3)_3COH$→ C ($C_9H_{15}NO_2$)

Carbamate C then undergoes a Diels–Alder reaction with methyl β-nitroacrylate to give cyclo-adduct D. You should be able to rationalize the stereochemistry of D through principles first encountered in Chapter 12. However, the regiochemistry (D not D') is not obvious from what we know. The structure of G shows what it must be.

C ($C_9H_{15}NO_2$) ⇌ rotation; H_3COOC, NO_2, NH, $OC(CH_3)_3$; Diels-Alder

Not D': NO_2, $COOCH_3$, NHCOO*t*-butyl

D ($C_{13}H_{20}N_2O_6$): $COOCH_3$, NO_2, NHCOO*t*-butyl

The transformation of D to E involves loss of HNO_2. There must be a base-induced elimination reaction, probably E1cB.

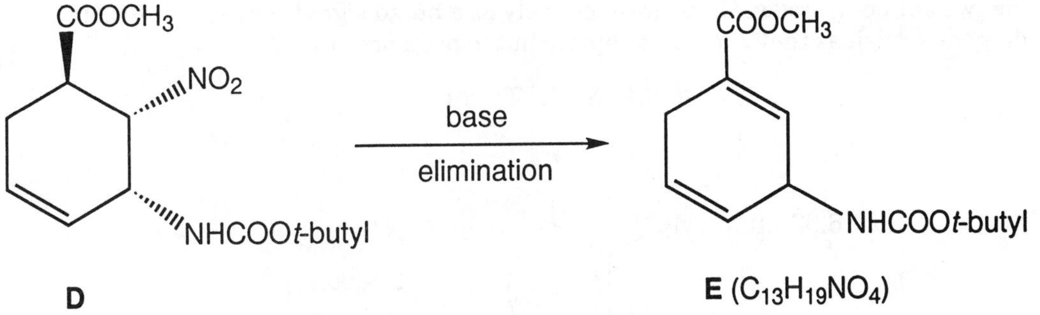

COOCH$_3$ NO$_2$ NHCOO*t*-butyl **D** —base, elimination→ COOCH$_3$ NHCOO*t*-butyl **E** ($C_{13}H_{19}NO_4$)

Two hydrolysis steps follow. The ester group of compound **E** is hydrolyzed in base to the carboxylic acid salt that is protonated in the second step to acid **F**. A final hydrolysis in HCl/H_2O leads to the product **G**.

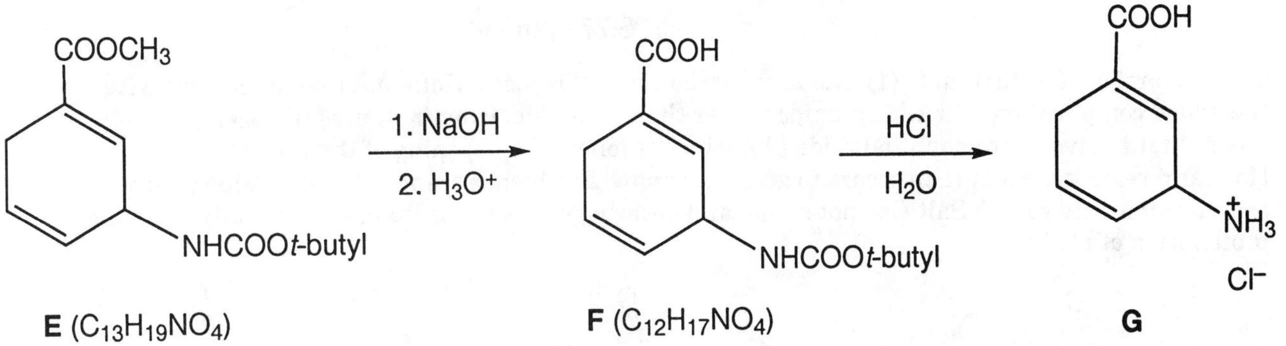

E ($C_{13}H_{19}NO_4$) —1. NaOH, 2. H_3O^+→ **F** ($C_{12}H_{17}NO_4$) —HCl, H_2O→ **G** (COOH, $\overset{+}{N}H_3$ Cl^-)

Problem 23.51 From the mass spectrum of **2**, we see a parent ion of m/z = 137. The ^{13}C NMR spectrum of **2** indicates a minimum of seven different carbons, and the ^{1}H NMR spectrum suggests a minimum of seven hydrogens. The compound C_7H_7 has a mass of 91 g/mol. So, we need to account for the missing mass of 46 g/mol. The ^{1}H NMR spectrum of **2** shows four aromatic hydrogens between δ 6.5 and 7.7 ppm. The coupling pattern suggests the possibility of an unsymmetrically 1,2-disubstituted benzene in which H$_3$ and H$_6$ appear as doublets and H$_4$ and H$_5$ appear as triplets. Remember that no meta or para couplings were observed.

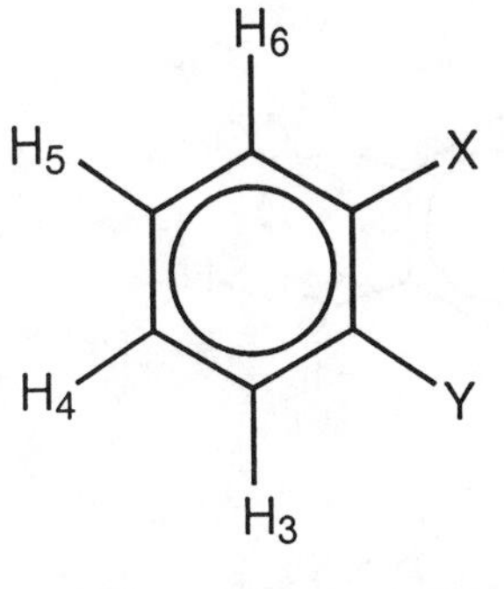

H$_6$, H$_5$, X, H$_4$, Y, H$_3$

Now, what are the substituents X and Y? The IR spectrum of **2** helps in this regard. First, note the broad band at 3200–2400 cm^{-1} and the intense band at 1665 cm^{-1}. These two bands are consistent with the O—H and C=O stretches of a carboxylic acid. The singlet at δ 169.5 ppm in the ^{13}C NMR spectrum of **2** is also consistent with the carbonyl carbon of a carboxylic acid. If we subtract the mass of two oxygen atoms (2 × 16 = 32) from the missing mass of 46, we still need to account for a mass of 14, which could be the result of a nitrogen atom. Once again, the IR spectrum of **2** helps to confirm this conjecture. The two bands at 3490 and 3380 cm^{-1} are consonant with the N—H stretching doublet expected for a primary amine. Thus, the spectral data suggest that compound **2** is an aminobenzoic acid—an anthranilic acid. The ^{1}H NMR chemical shifts of the aromatic hydrogens are summarized below. Because of rapid exchange reactions the carboxylic acid hydrogen and the two amine hydrogens appear collectively as a broad signal centered at δ 8.60 ppm. These hydrogens vanish as they exchange rapidly in the presence of D_2O.

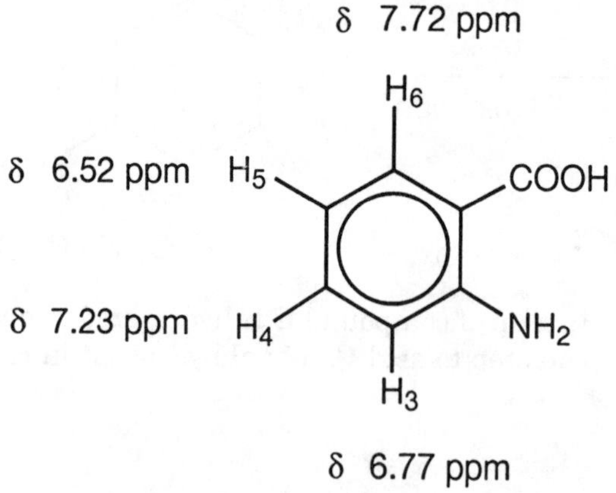

δ 7.72 ppm
H_6
δ 6.52 ppm H_5 COOH
δ 7.23 ppm H_4 NH_2
H_3
δ 6.77 ppm

The conversion of phthalimide (**1**) into anthranilic acid (**2**) is just a Hofmann rearrangement with the added complication that **1** is an imide rather than an amide. Bromination of nitrogen probably occurs first to give *N*-bromophthalimide (**A**), which is followed by opening of the imide ring. Hofmann rearrangement then occurs to give isocyanate **B**, which is trapped by hydroxide to give the carbamic acid salt **C**. Salt **C** is not stable and decarboxylates to the dianion **D**. Finally, a double protonation of **D** gives **2**.

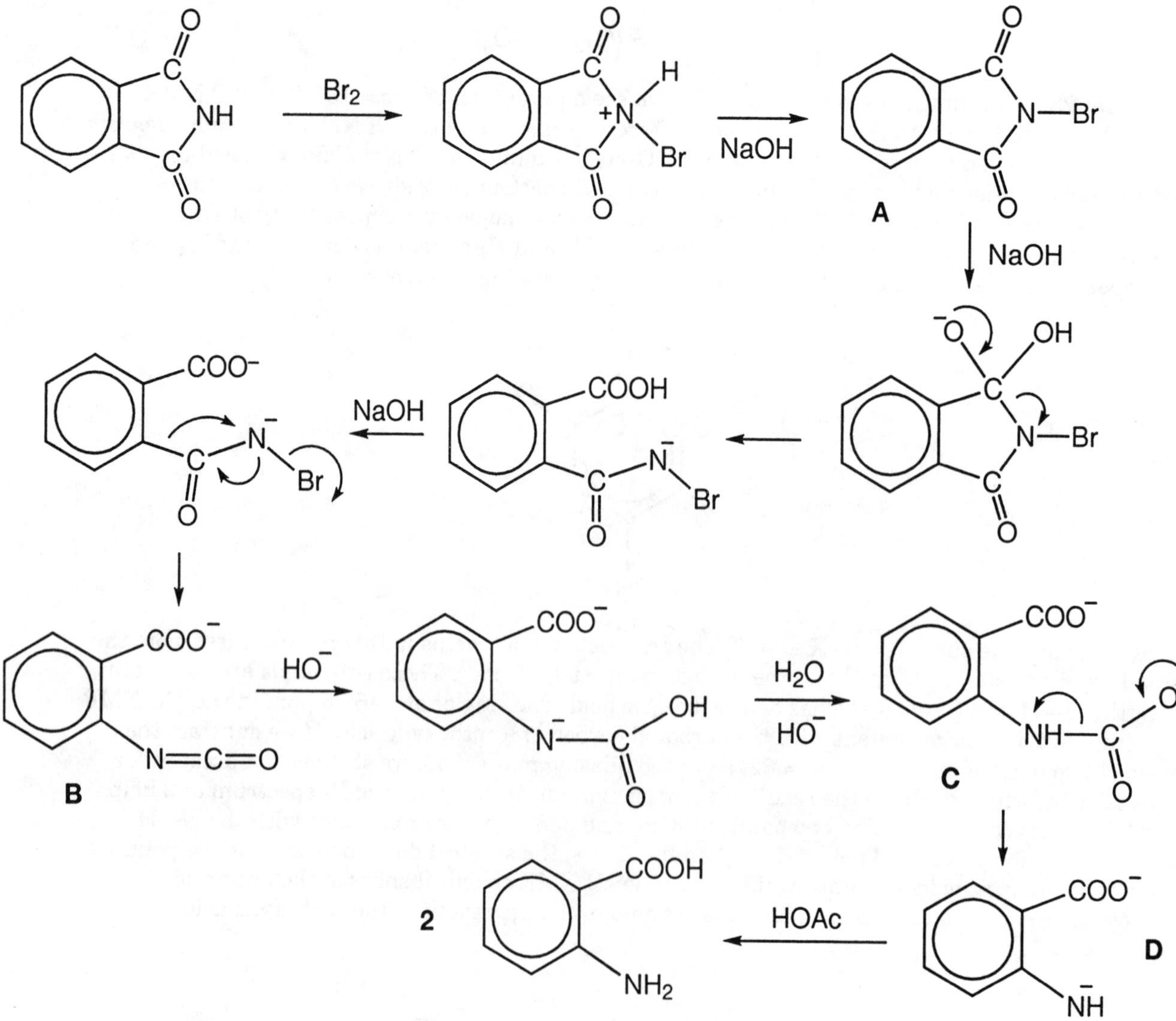

Br_2 NaOH A NaOH NaOH B HO^- H_2O HO^- C D HOAc 2

Chapter 24 Outline

24.1 Nomenclature and Structure
24.2 Reactions of Sugars
24.3 The Ring Size in Glucose
24.4 The Fischer Determination of the Structure of D-Glucose (and the 15 Other Aldohexoses)
24.5 Something More: Di- and Polysaccharides

24

Polyfunctional Natural Products: Carbohydrates

Even though there are few new reactions in this chapter, sugar chemistry can certainly be vexing. The profuse functionality makes for many possible choices in most reactions, and the stereochemical complexity is certainly serious. There is even a new stereochemical convention, the Fischer projection, to master. Still, if you keep your wits about you, go slowly and always keep in mind that what you are doing is applying old knowledge in a new setting, even sugar problems become easy. Always pay close attention to stereochemistry, however. Here, the Additional Problems section (Section 24.8) picks up with some building block problems before going on to more detailed, challenging exercises.

Problem 24.2 Remember: All vertical bonds are heading away from you and all horizontal bonds are coming toward you. The process of turning these Fischer projections into three-dimensional drawings begins with a transformation that puts in the wedges. Then, translate this into a "sawhorse form" and finally, draw this sawhorse in the staggered, energy minimum form.

CHO, H—OH, H—OH, CH_2OH = CHO, H—OH, H—OH, CH_2OH = HO, H, OHC, C-C, OH, H, CH_2OH = *R* *R* HO, H, OHC, C—C, CH_2OH, H, OH

Fischer projection

Fischer projection with more stereochemical detail

Eclipsed, energy maximum form

Staggered, energy minimum form

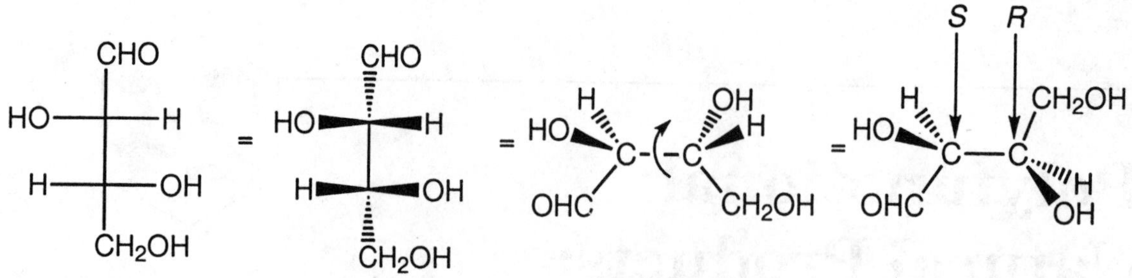

Fischer projection

Fischer projection with more stereochemical detail

Eclipsed, energy maximum form

Staggered, energy minimum form

Problem 24.3 In any Fischer projection, the CHO is drawn at the top and the CH_2OH group at the bottom. The OH groups are then filled in. First, draw the molecule in an energy maximum, eclipsed arrangement. Next, translate to a "wedge" Fischer projection and last, convert this into the Fischer form itself. This process is shown in detail for isomer a.

For isomer "a"

Energy maximum form

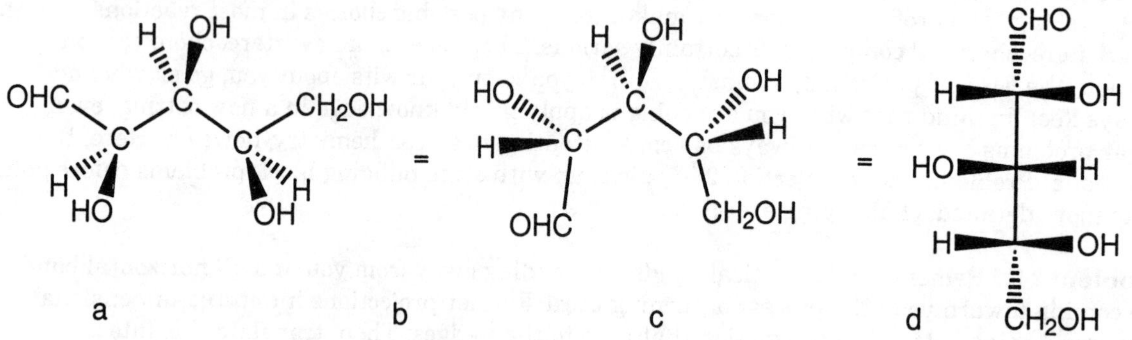

a

b

c

d

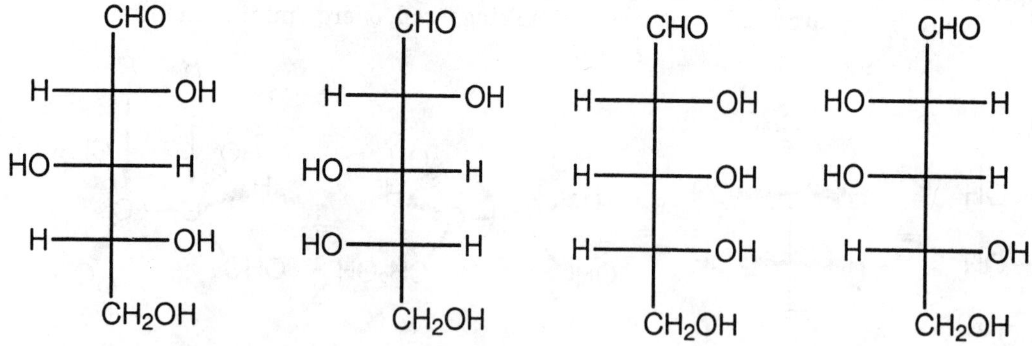

Problem 24.4 This problem is easy as long as you take a systematic approach. The question asks for L sugars so the "bottom" OH must be on the left. Start by putting all three OH groups on one side and then work out the three possible ways to have two OH groups on one side and one on the other (without making the molecule a D sugar).

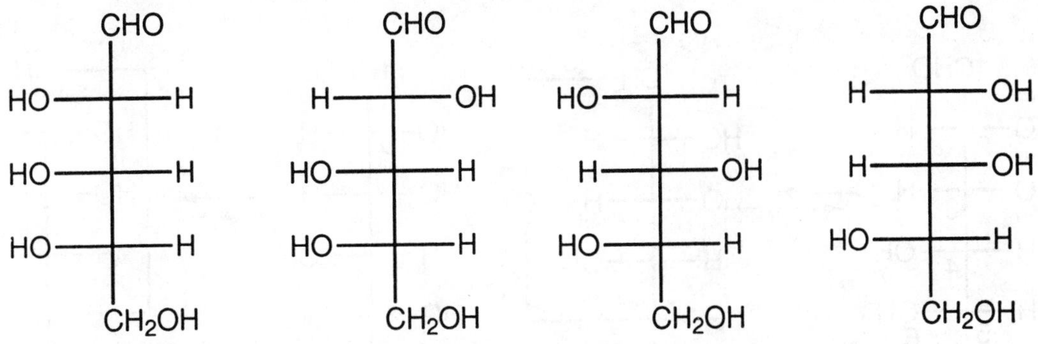

L-Ribose L-Arabinose L-Xylose L-Lyxose

Problem 24.5 There are five stereogenic carbons in an aldoheptose, so there will be $2^5 = 32$ stereoisomers, 16 in the D series and 16 mirror-image L isomers.

Problem 24.6 Sodium borohydride reduces carbonyl compounds through addition of hydride to give the alkoxide. Subsequent protonation yields the alcohol. Glucitol is a hexa-alcohol.

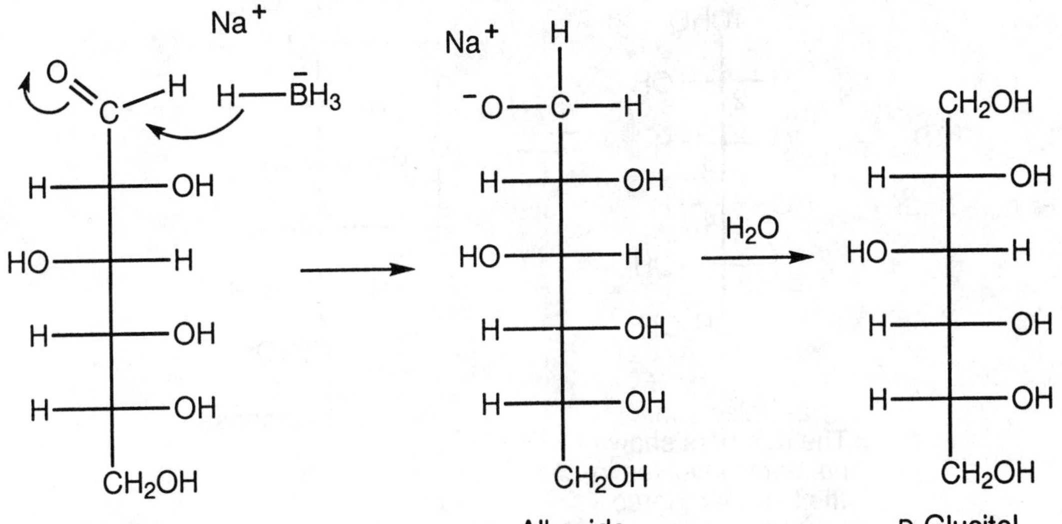

D-Glucose Alkoxide D-Glucitol

Problem 24.7 The two pyranoses have six-membered rings. Thus the connection must be made through the OH at C_5. The furanose contains a five-membered ring and thus it is the OH group at C_4 that is involved in hemiacetal formation. Watch out for the "around-the-corner" bonds in these projections. The "corners" are not really there.

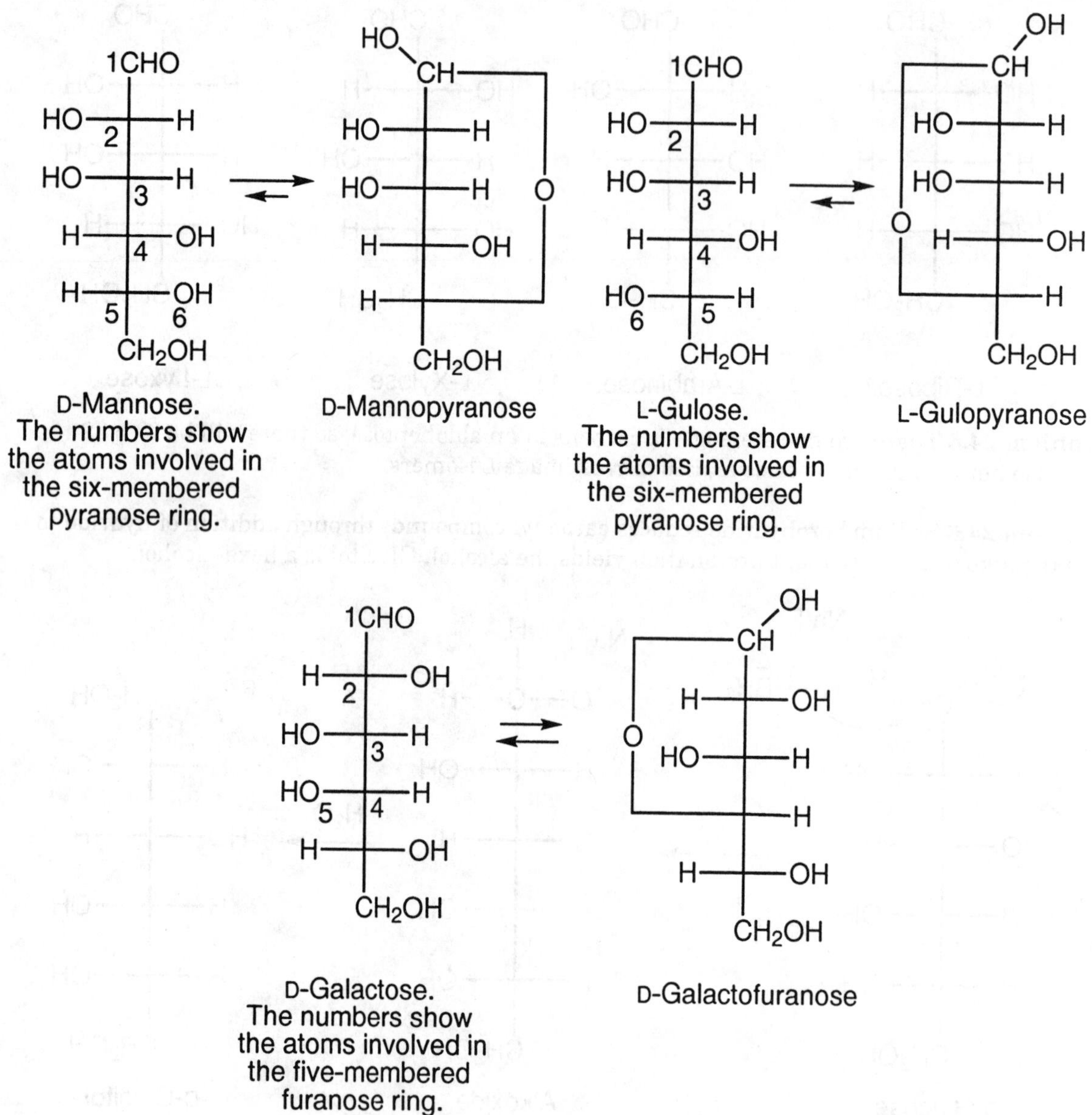

Problem 24.8 First draw Fischer projections of the two anomers. Now we can see the stereochemical relationships clearly.

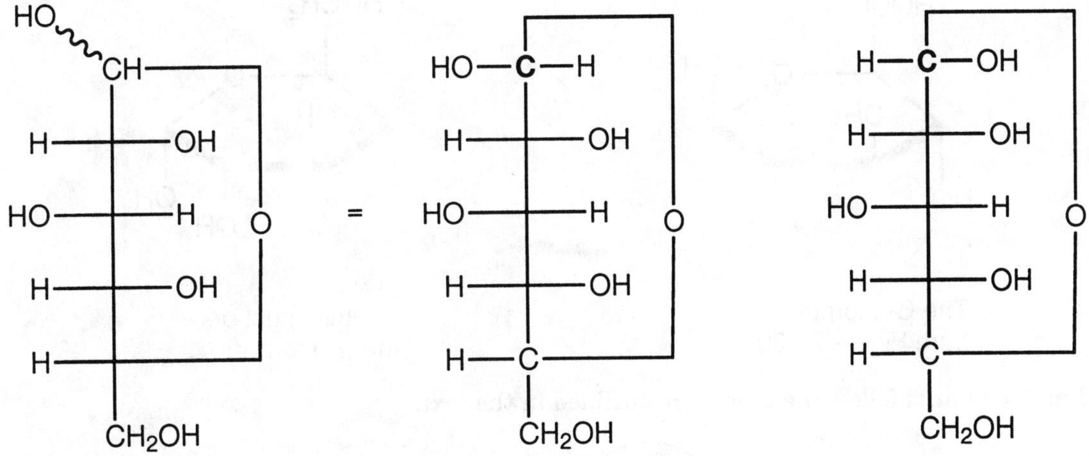

β–Anomer. The OH at **C** is on the opposite side from the O involved in the hemiacetal link (C).

α–Anomer. The OH at **C** is on the same side as the O involved in the hemiacetal link (C).

Next, add the wedges showing the groups at C_1 as either coming toward you or retreating from you. Now the configuration of the stereogenic carbon can easily be determined. The figure shows the priorities. If this figure is obscure, please go back and review Chapter 5 (p. 160).

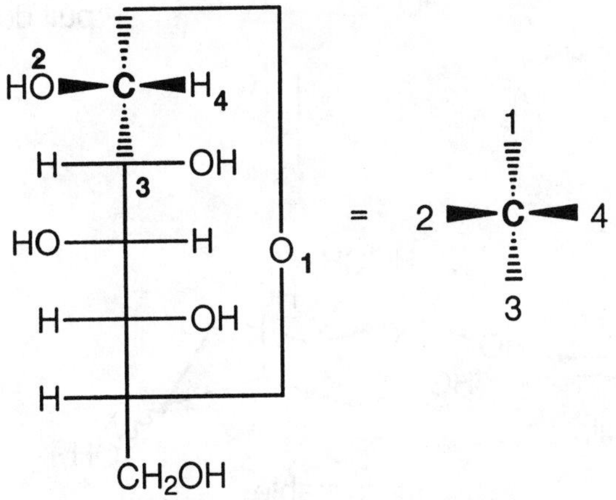

β-Anomer is (*R*)

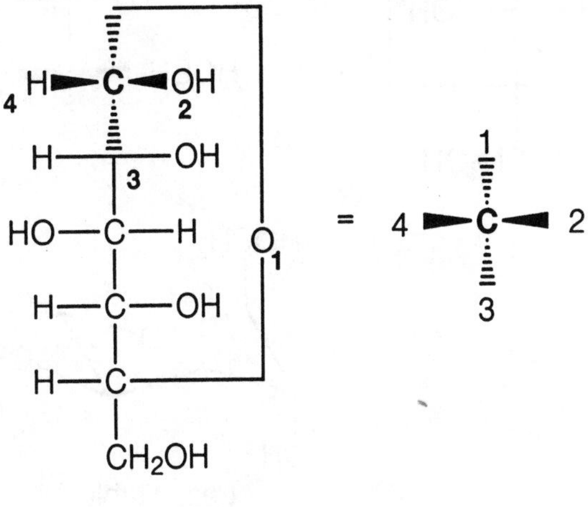

α-Anomer is (*S*)

Problem 24.9 The α–anomer will have the anomeric OH on the opposite side from the isomer in Figure 24.20.

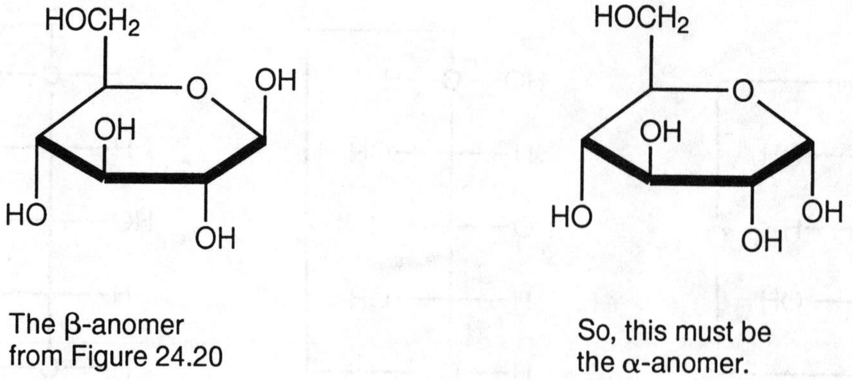

The β-anomer from Figure 24.20

So, this must be the α-anomer.

Problem 24.11 Just follow the procedure outlined in the text.

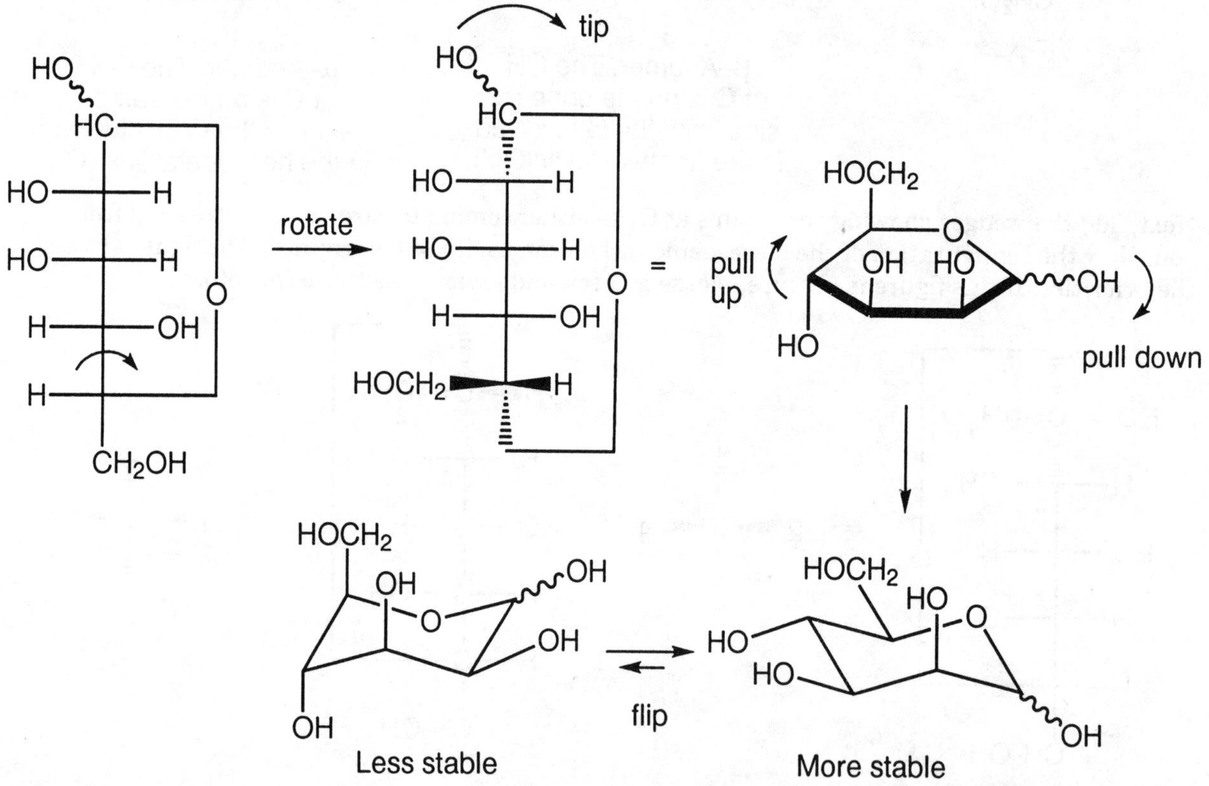

Less stable

More stable

Problem 24.13 The furanose form contains a five-membered ring and thus it is the OH group attached to C_5 that is involved in the ring (remember that the carbonyl group is at C_2 in fructose). In the six-membered pyranose form, it is the OH group at C_6 that participates in ring formation.

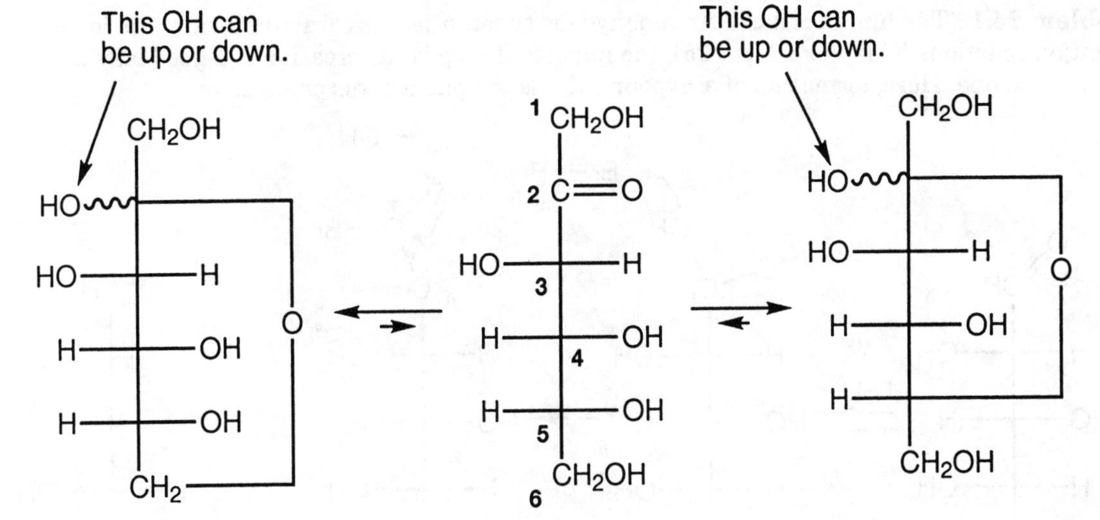

This OH can be up or down.

This OH can be up or down.

α- and β-D-Fructopyranose (~67%)

D-Fructose (open form)

α- and β-D-Fructofuranose (~31%)

Problem 24.14 The enolate is planar, as all enolates must be in order to insure maximum overlap of the 2*p* orbitals on the three atoms. Protonation can take place from either the top or bottom (a and b) to produce the two sugars, glucose and mannose. These two sugars differ only in the stereochemistry at C_2.

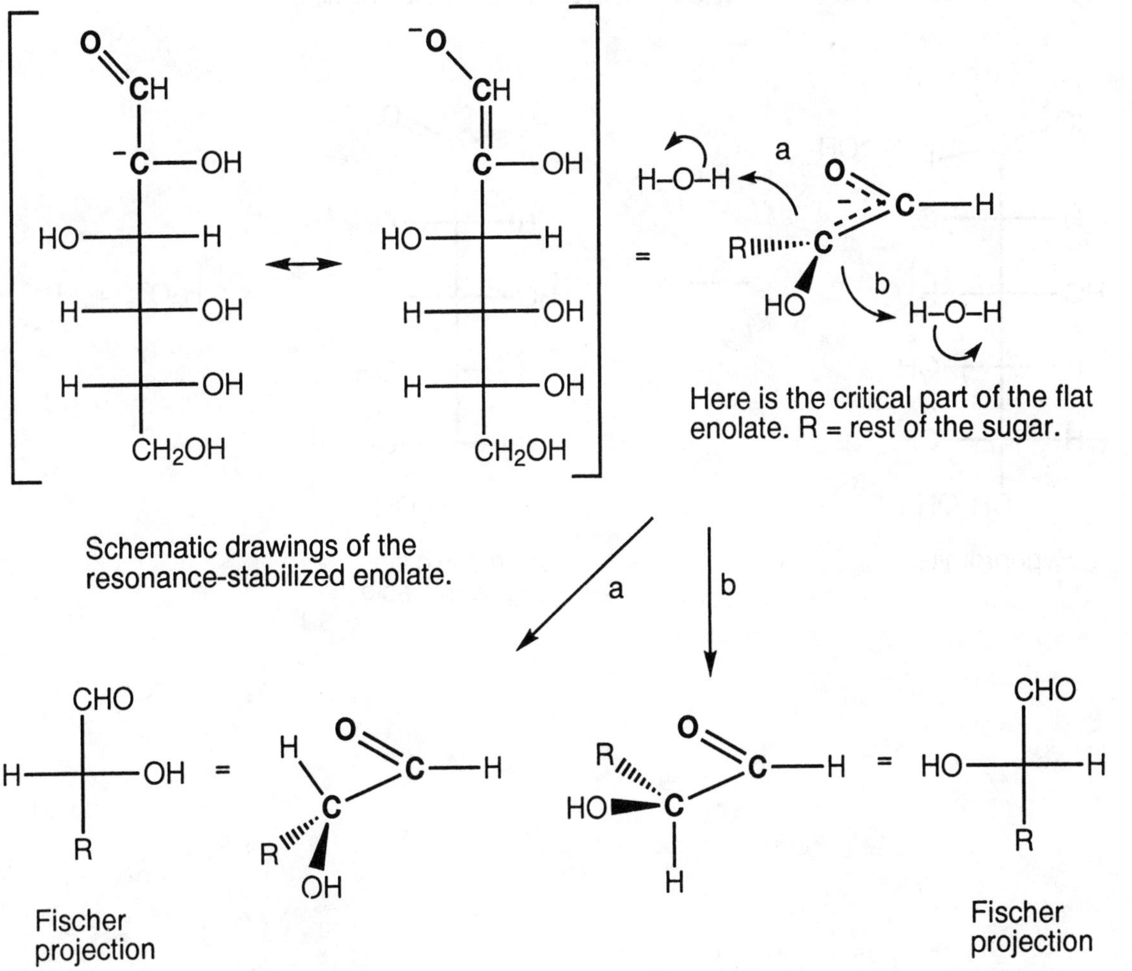

Here is the critical part of the flat enolate. R = rest of the sugar.

Schematic drawings of the resonance-stabilized enolate.

Fischer projection

Fischer projection

Problem 24.16 The hint is critical. An aldehyde is substantially hydrated in water. As in most oxidation reactions (Chapter 16, p. 795), the important step is the creation of a good leaving group out of a bad one. Here, formation of a hypobromite accomplishes this change.

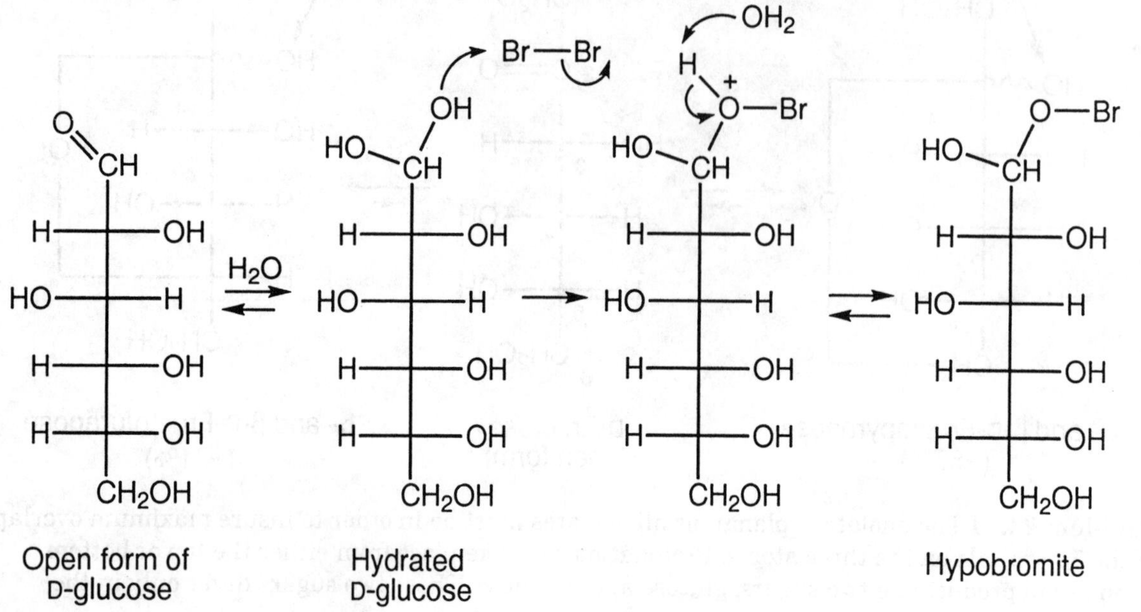

Open form of D-glucose

H_2O

Hydrated D-glucose

Hypobromite

Now, bromide functions as a leaving group and the oxidation is finished.

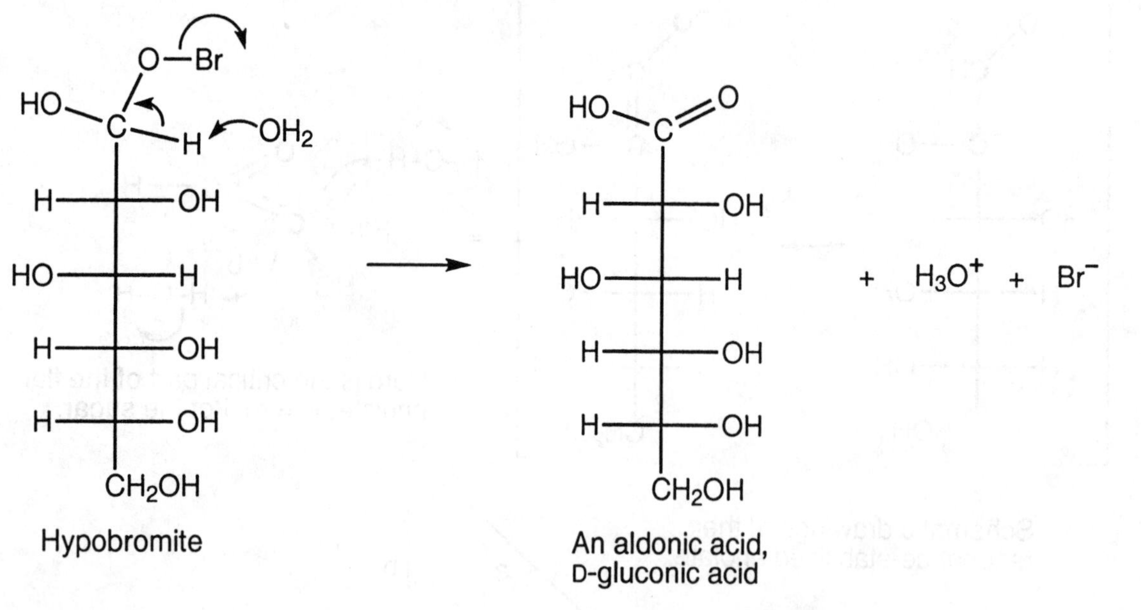

Hypobromite

$+ \; H_3O^+ \; + \; Br^-$

An aldonic acid, D-gluconic acid

Problem 24.17 Remember that aldohexoses show little or no evidence of an aldehyde group in their IR and NMR spectra. The reason is that there is little of the open aldehyde present; most of the molecule is tied up as the stable hemiacetal. A similar phenomenon occurs with aldonic acids. There is an intramolecular reaction between one of the OH groups and the acid to form a cyclic ester, a lactone. Here is an example.

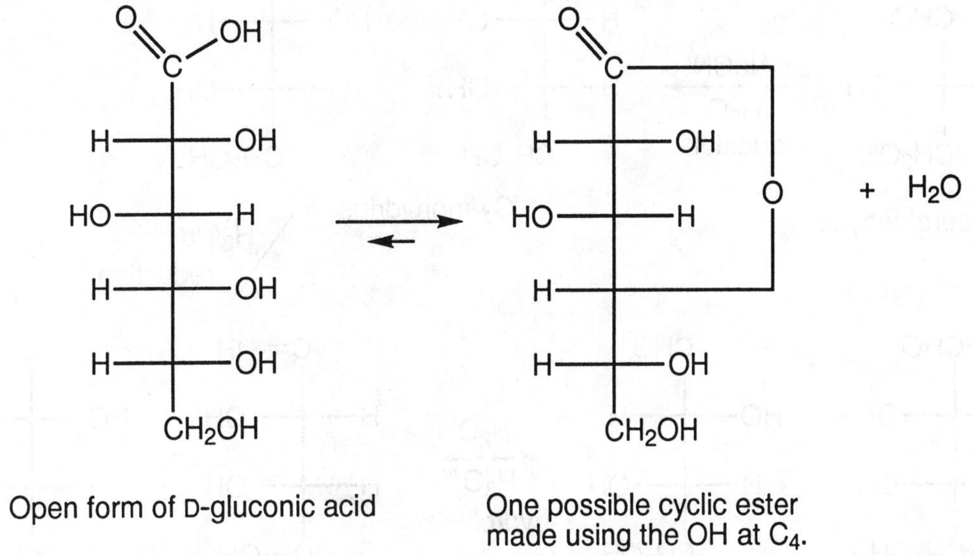

Open form of D-gluconic acid

One possible cyclic ester made using the OH at C_4.

Problem 24.19 The mechanism follows the general mechanism for acid-catalyzed imine formation (Chapter 16, p. 780). Addition of phenyl hydrazine to the protonated carbonyl group is followed by proton transfers and loss of water.

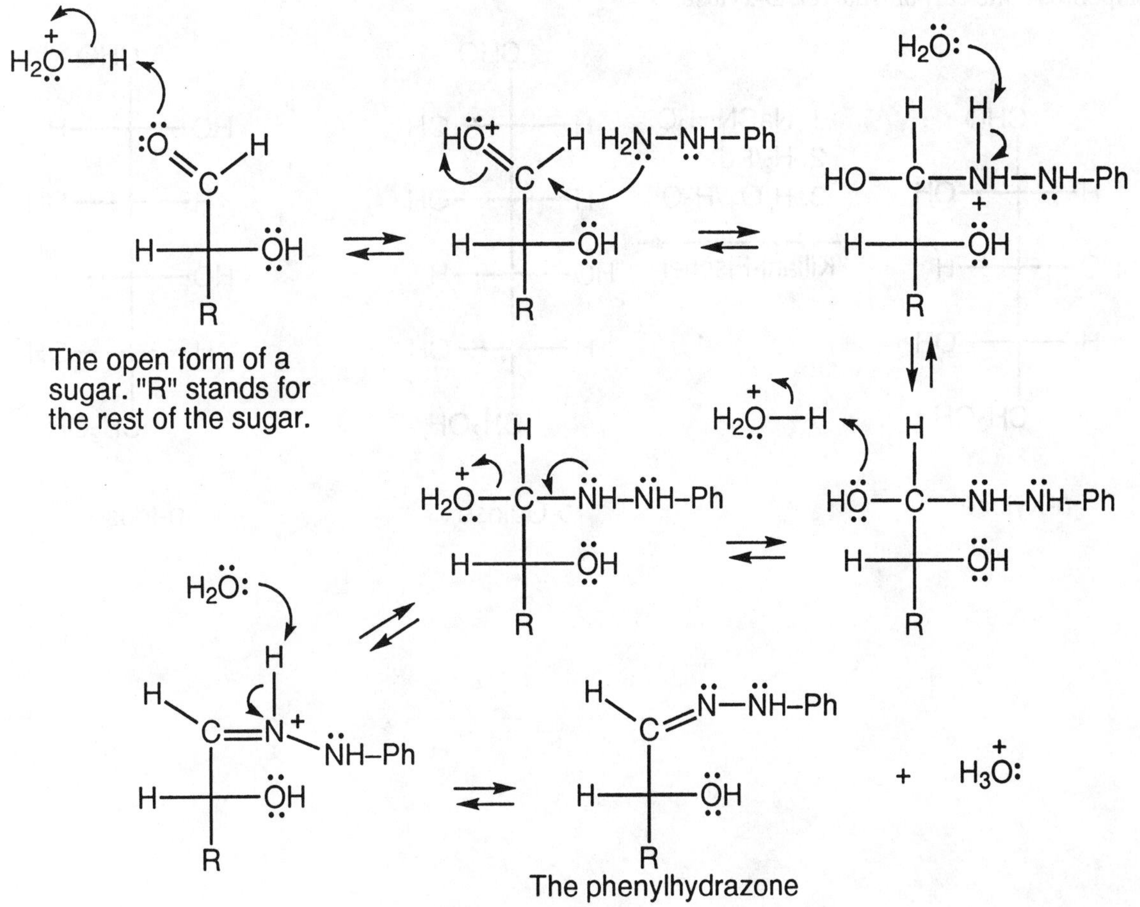

The open form of a sugar. "R" stands for the rest of the sugar.

The phenylhydrazone

Problem 24.20 The Kiliani–Fischer synthesis generates two new sugars, stereoisomeric at the new C_2, each one carbon longer than the starting sugar. In the first step, a pair of cyanohydrins is formed. These are reduced to the imines, and the imines are hydrolyzed to give the pair of aldehydes.

D-Glyceraldehyde —NaCN, H_2O, addition→ Cyanohydrins —H_2/Pd, reduction→ Imines —H_2O, H_3O^+, hydrolysis→ Aldehydes

Problem 24.21 The two sugars, D-gulose and D-idose, are exactly the same except for the stereochemistry at C_2. They can be made through the Kiliani–Fischer synthesis from the aldopentose one carbon shorter, D-xylose.

D-Xylose —1. NaCN/H_2O; 2. H_2/Pd; 3. H_3O^+/H_2O; Kiliani-Fischer→ D-Gulose + D-Idose

Problem 24.23 This problem seems really hard, but it is not so tough if you take it stepwise. The trick is to follow the procedure (pp. 1244-1246) for transforming Fischer projections into three-dimensional chair forms. The only real problem is to deal with the attachment point, and that turns out to take care of itself in most cases. Start with the right-hand sugar, D-galactose. Let "Glu" stand for the attached glucose. The Haworth form can be generated as shown.

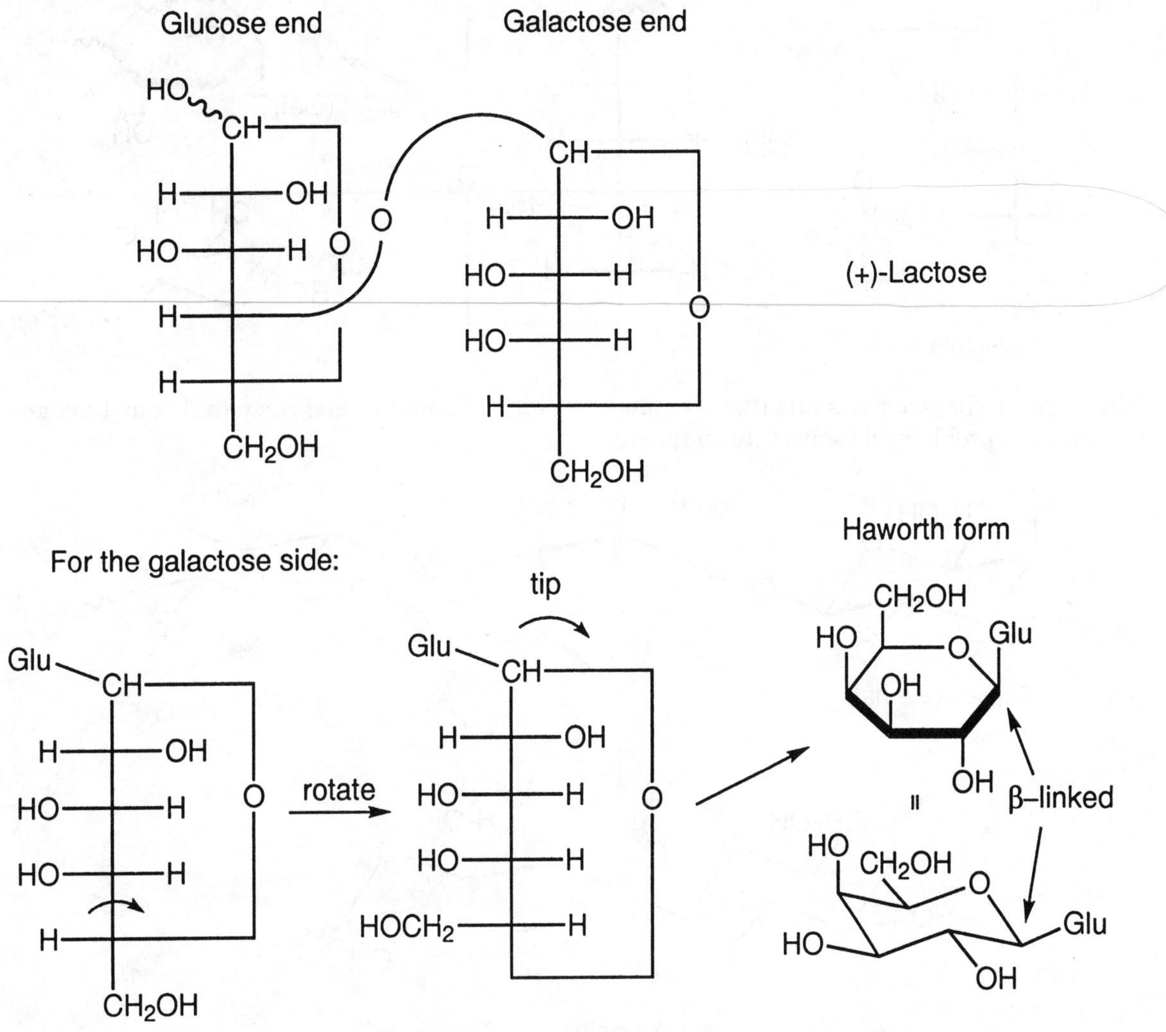

The important points are to be sure to get most groups equatorial and to get the correct (β = equatorial) link to the other sugar. Now do the same with the glucose half of (+)-lactose. Here Gal stands for galactose.

For the glucose side:

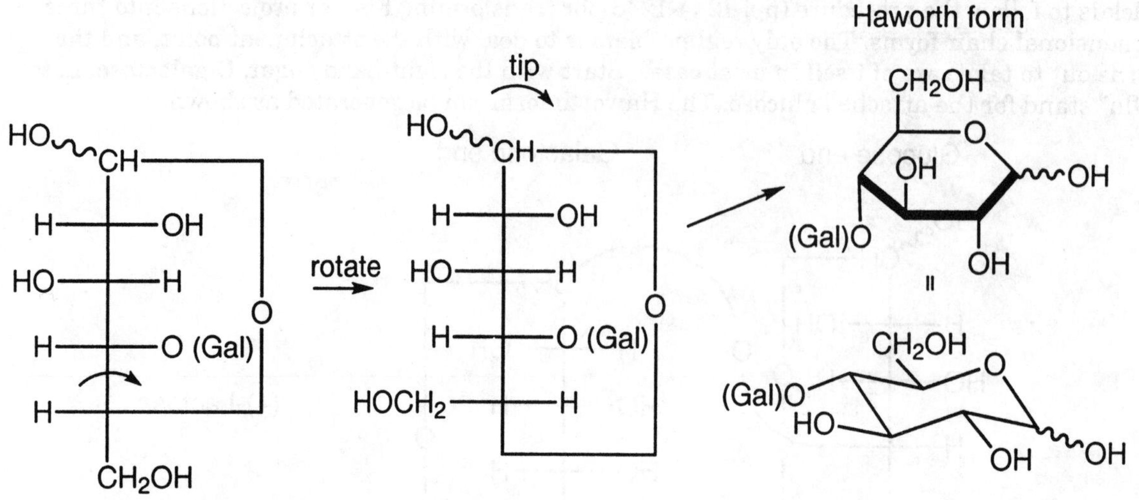

Now combine the two parts and this seemingly vexing problem is done! Keep to this method and all disaccharide problems dissolve into simplicity.

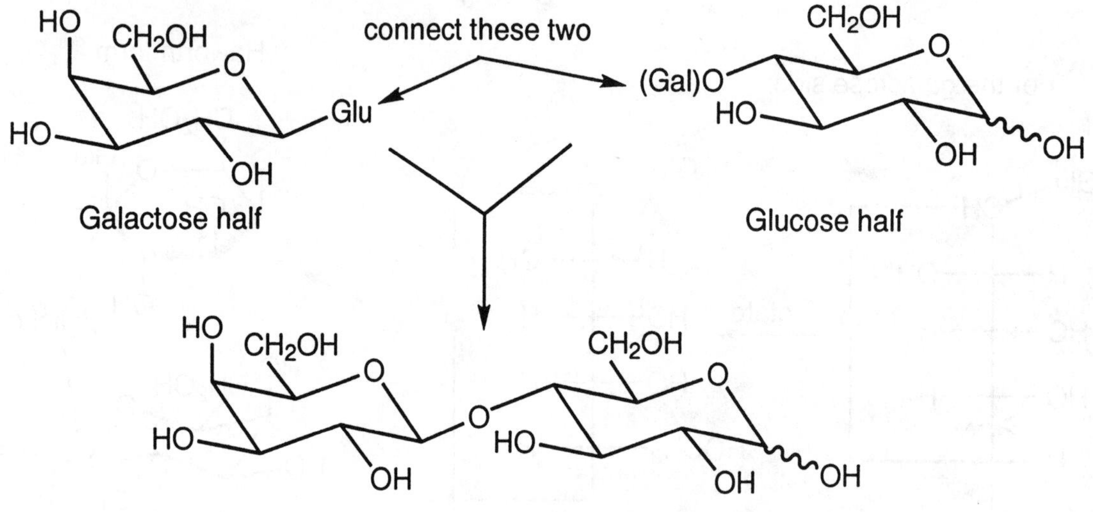

(+)-Lactose

Problem 24.24 Two glucose molecules are attached in this disaccharide. Methylation with dimethyl sulfate, followed by acid hydrolysis, pinpoints the position at which the two sugars were attached. This position appears as a bare OH group, in this case at C_4.

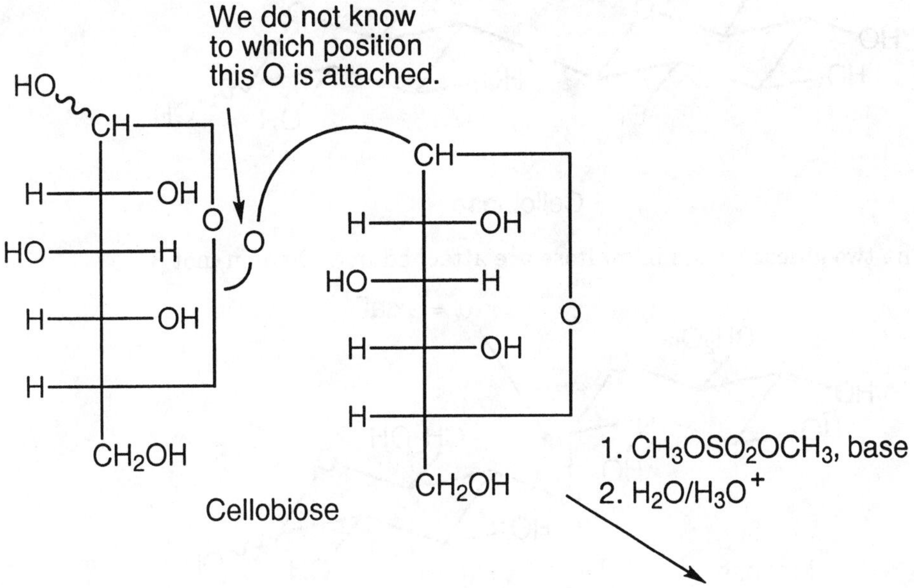

We do not know to which position this O is attached.

Cellobiose

1. $CH_3OSO_2OCH_3$, base
2. H_2O/H_3O^+

This shows that the attachment between the two glucoses was at C_4 of one glucose and C_1 of the other.

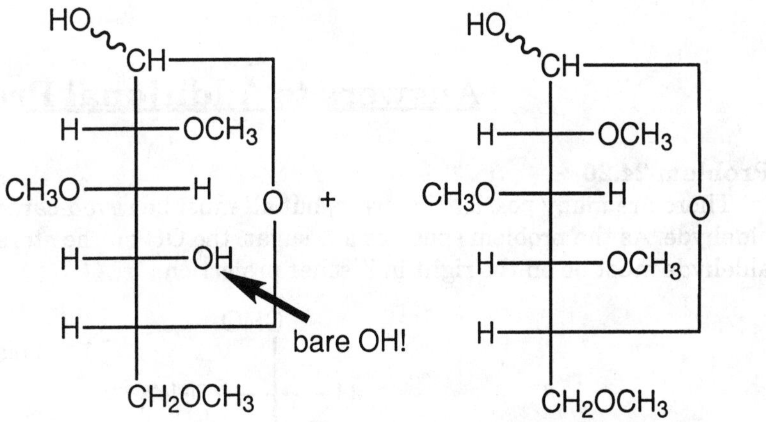

bare OH!

2,3,6-Trimethyl-D-glucose + 2,3,4,6-Tetramethyl-D-glucose

Octamethylcellobiose must be

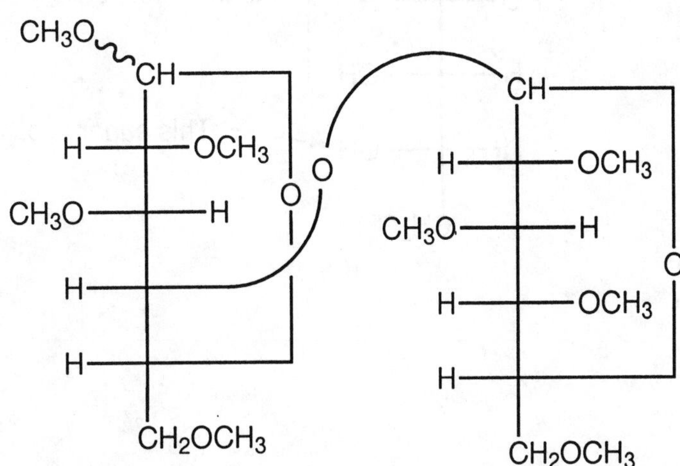

As cellobiose is cleaved by lactase, the two sugars are attached in β-fashion. Here is a three-dimensional drawing.

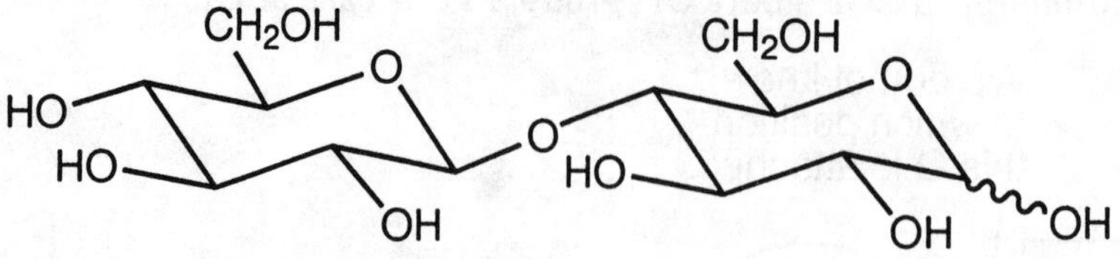

Cellobiose

Problem 24.25 The two glucose units in maltose are attached in α–fashion, not β.

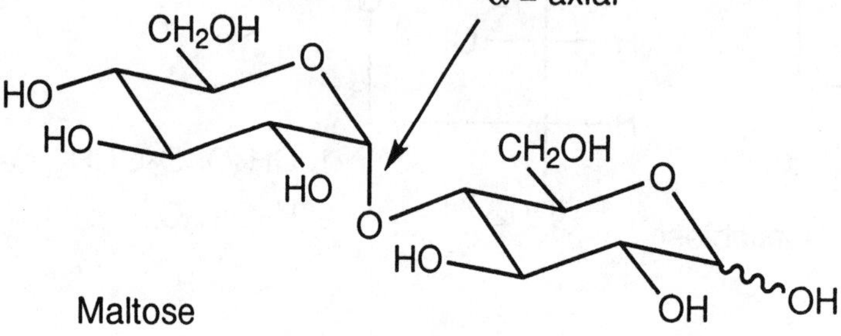

Maltose

Answers to Additional Problems

Problem 24.26
(a) There are many possible answers, but all must be seven-carbon C_7, sugars containing an aldehyde. As the problem specifies a D sugar, the OH on the stereogenic carbon furthest from the aldehyde must be on the right in Fischer projection.

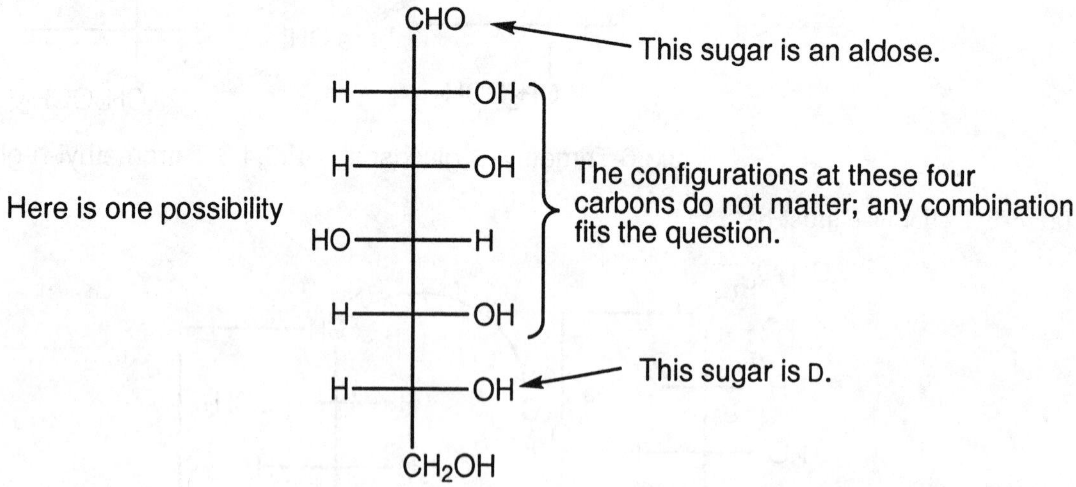

(b) This answer must show a five-carbon sugar containing a ketone. The OH group adjacent to the lower primary alcohol must be on the left in Fischer projection.

Here is one possibility:

CH_2OH
H—C—OH
C=O
HO—C—H
CH_2OH

This sugar is a ketose.

This sugar is L.

(c) There are only two D-aldotetroses: D-threose and D-erythrose. Only D-erythrose will give a meso diacid (*meso*-tartaric acid) on oxidation with nitric acid.

CHO
H—C—OH
H—C—OH
CH_2OH

D-Erythrose

$\xrightarrow{HNO_3}$

COOH
H—C—OH
H—C—OH
COOH

meso-Tartaric acid

(d) The problem tells us that the sugar galactose is in its pyranose form, so we must draw it as a six-membered ring. The sugar is present as a methyl glycoside, so the anomeric OH (the one newly created in the six-membered ring) is methylated. The α-anomer is specified, and that means that the OCH_3 must be on the same side in Fischer projection as the oxygen in the six-membered ring.

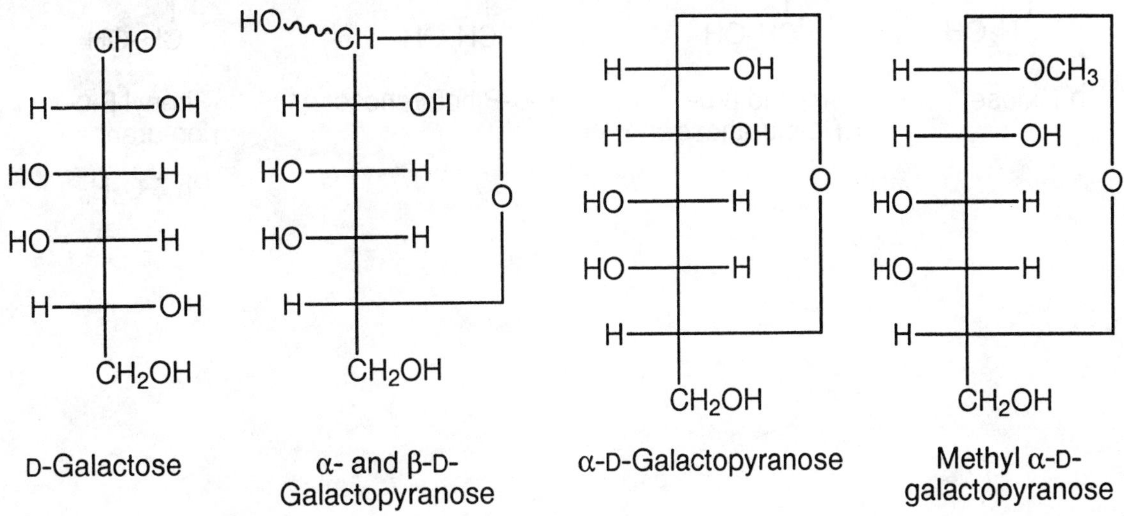

D-Galactose

α- and β-D-Galactopyranose

α-D-Galactopyranose

Methyl α-D-galactopyranose

(e) D-Allose is the aldohexose with all secondary OH groups on the right in Fischer projection. It is present as the osazone, and that means that both the aldehydic carbon and C_2 have been converted into phenylhydrazones.

CHO, H–C–OH, H–C–OH, H–C–OH, H–C–OH, CH_2OH

D-Allose

HC=N–NHPh, C=N–NHPh, H–C–OH, H–C–OH, H–C–OH, CH_2OH

Osazone of D-allose

(f) In this part, the aldopentose ribose is present as a five-membered ring. The anomeric OH is on the left in Fischer projection, on the side opposite to the ring oxygen (the β-position), and is phenylated.

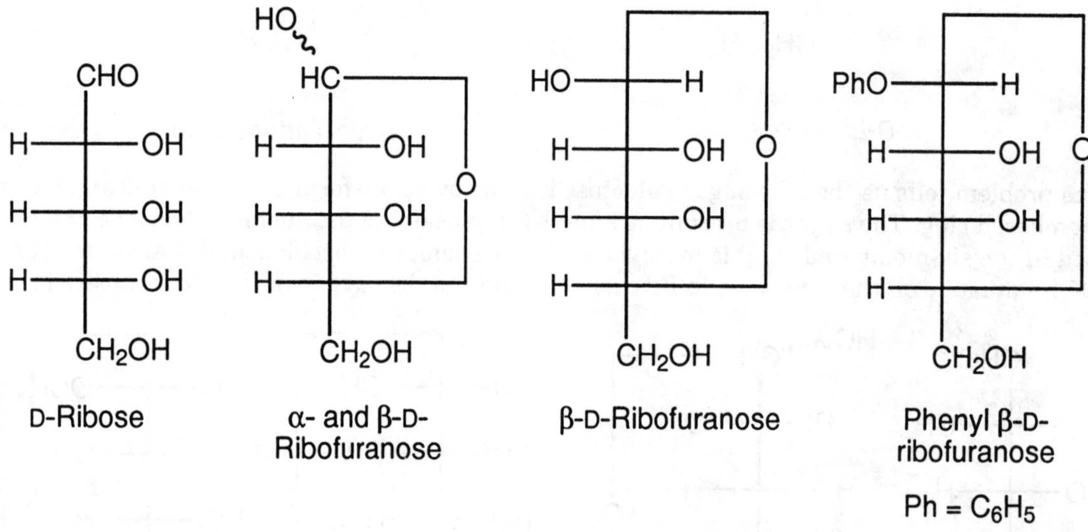

D-Ribose

α- and β-D-Ribofuranose

β-D-Ribofuranose

Phenyl β-D-ribofuranose

Ph = C_6H_5

Problem 24.27
First of all, there will be four pyranose forms, as both the α- and β-forms can exist as mixtures of two chairs. Follow the procedure in the chapter (or any other method that you devise) to draw the molecules. First, draw the Fischer projections.

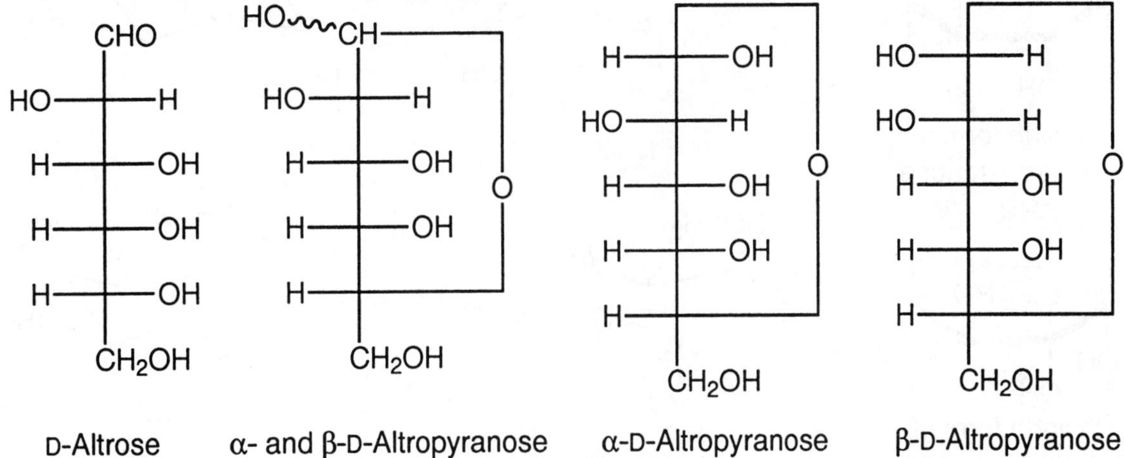

Now convert these into Haworth forms through rotation and tipping.

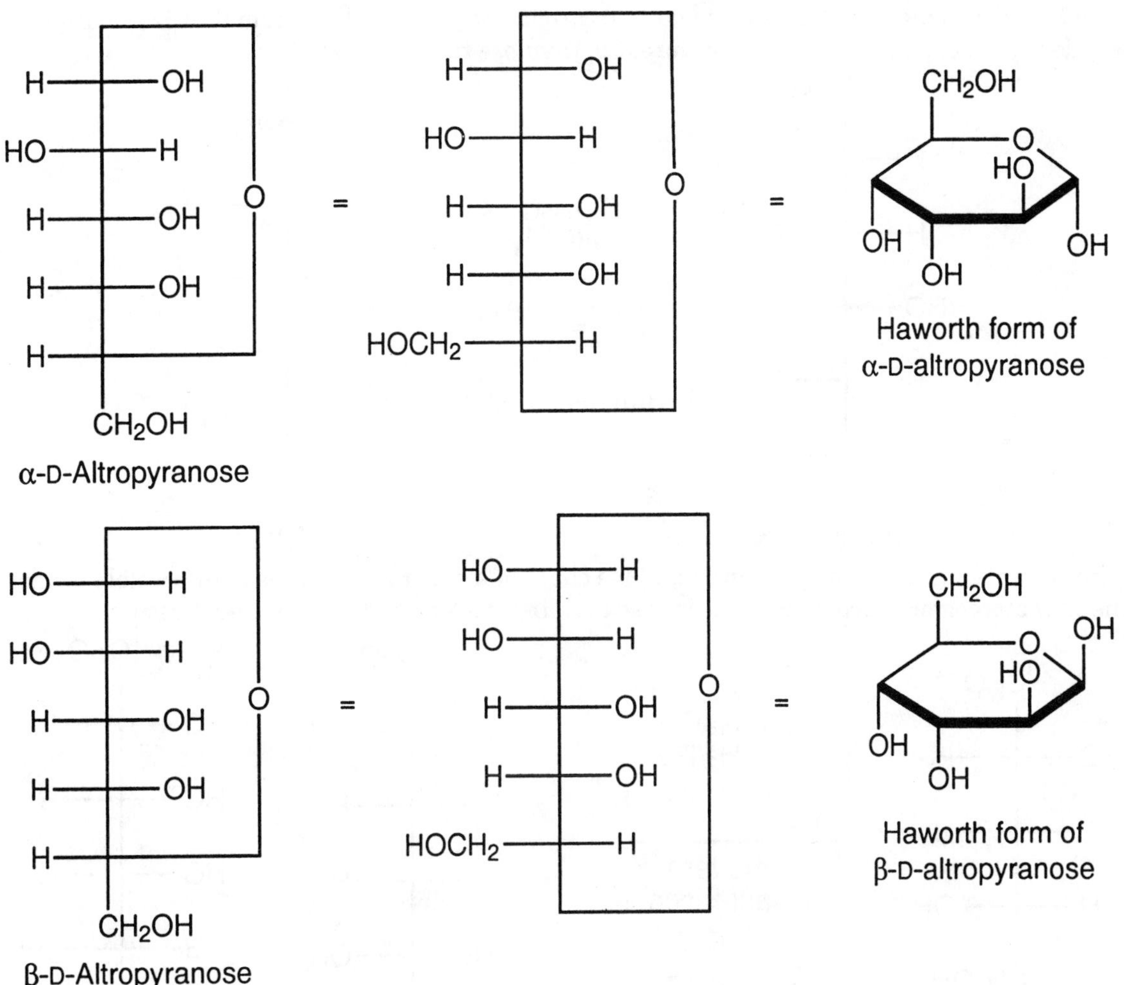

Now let the Haworth forms relax to pairs of equilibrating chair forms.

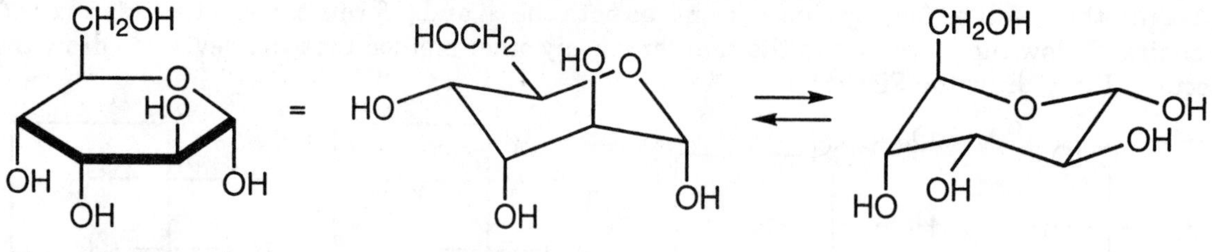

Haworth form of α-D-altropyranose

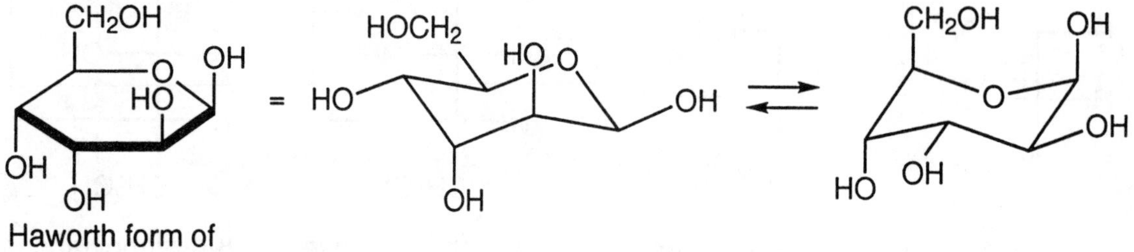

Haworth form of β-D-altropyranose

Problem 24.28

(a) The Ruff degradation results in a sugar one carbon shorter than the original. It is the original aldehyde that is clipped off in the procedure. So, D-gulose results in D-xylose.

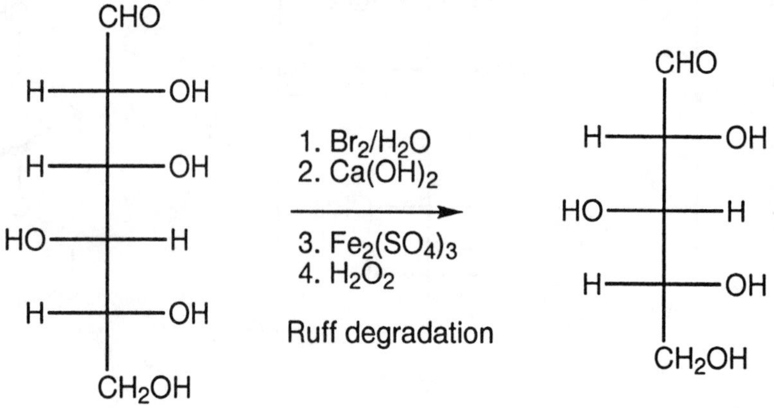

1. Br_2/H_2O
2. $Ca(OH)_2$
3. $Fe_2(SO_4)_3$
4. H_2O_2

Ruff degradation

D-Gulose D-Xylose

(b) The Kiliani–Fischer synthesis lengthens the chain of a sugar and creates both possible isomers at the new stereogenic carbon. So, in this case both D-galactose and D-talose must result.

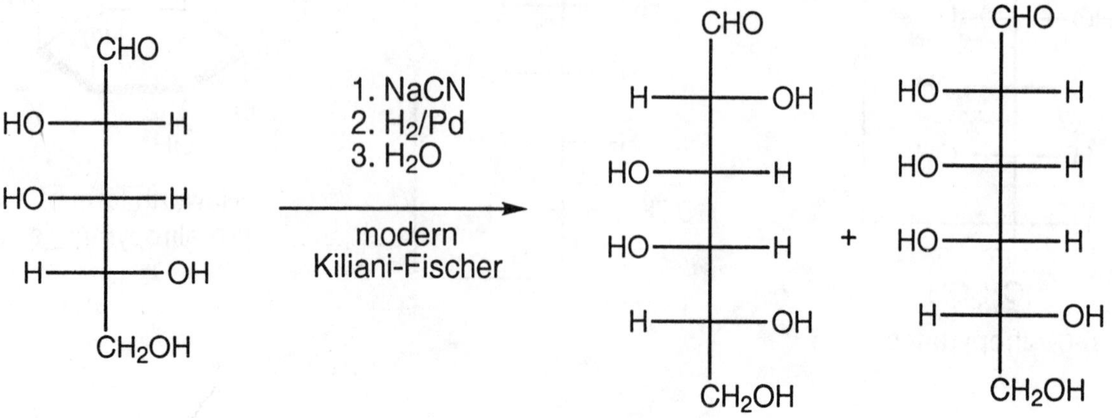

1. NaCN
2. H_2/Pd
3. H_2O

modern Kiliani-Fischer

D-Lyxose D-Galactose D-Talose

Problem 24.29 Nitric acid oxidizes both the aldehyde and the primary alcohol to carboxylic acids. The ends of the sugar are no longer differentiated in an aldaric acid.

(a) There is one other sugar, **A** (written with the aldehyde at the bottom and the primary alcohol at the top), that will give the same aldaric acid as D-talose. Now flip **A** over into proper Fischer form with the aldehyde group at the top, and we can see that this sugar is D-altrose.

D-Talose		Aldaric acid		A (flip)	D-Altrose
CHO		COOH		CH_2OH	CHO
HO—H		HO—H		HO—H	HO—H
HO—H	$\xrightarrow{HNO_3}$	HO—H	$\xleftarrow{HNO_3}$	HO—H	H—OH
HO—H		HO—H		HO—H	H—OH
H—OH		H—OH		H—OH	H—OH
CH_2OH		COOH		CHO	CH_2OH

(b) Here we use the same technique as in (a). In this case, it is L-xylose, the enantiomer of the original sugar, that is the answer.

D-Xylose				(flip)		L-Xylose
CHO		COOH		CH_2OH		CHO
H—OH		H—OH		H—OH		HO—H
HO—H	$\xrightarrow{HNO_3}$	HO—H	$\xleftarrow{HNO_3}$	HO—H	=	H—OH
H—OH		H—OH		H—OH		HO—H
CH_2OH		COOH		CHO		CH_2OH

(c) Oxidation of D-idose leads to **B**. If we flip idose, we get a "new" sugar that also would give **B**. But, this is not really a new sugar, it is D-idose all over again. There is no other sugar that can give **B**, only D-idose.

D-Idose		Aldaric acid **B**		(flip)		D-Idose
CHO		COOH		CH_2OH		CHO
HO—H		HO—H		HO—H		HO—H
H—OH	$\xrightarrow{HNO_3}$	H—OH	$\xleftarrow{HNO_3}$	H—OH	=	H—OH
HO—H		HO—H		HO—H		HO—H
H—OH		H—OH		H—OH		H—OH
CH_2OH		COOH		CHO		CH_2OH

Problem 24.30

Osazone formation destroys stereochemistry at C_2 through phenyhydrazone formation, as well as transforming the aldehyde at C_1 into a phenylhydrazone. Hence, the other sugar must be the same as L-talose in every respect, except at C_2. It is the C_2 epimer of L-talose, L-galactose.

L-Talose: CHO; H–C–OH; H–C–OH; H–C–OH; HO–C–H; CH_2OH → ($PhNHNH_2$) → Osazone: HC=NNHPh; C=NNHPh; H–C–OH; H–C–OH; HO–C–H; CH_2OH ← ($PhNHNH_2$) ← L-Galactose: CHO; HO–C–H; H–C–OH; H–C–OH; HO–C–H; CH_2OH

L-Talose Osazone L-Galactose

Problem 24.31

(a) Reduction of the aldehyde group to an alcohol. (b) Oxidation of only the aldehyde to the aldonic acid.

COOH; HO–C–H; HO–C–H; H–C–OH; CH_2OH ← (Br_2, H_2O; b) ← CHO; HO–C–H; HO–C–H; H–C–OH; CH_2OH → ($NaBH_4$, H_2O; a) → CH_2OH; HO–C–H; HO–C–H; H–C–OH; CH_2OH

(c) Oxidation of both the aldehyde and primary alcohol to give the aldaric acid.

(d) Methylation of all free OH groups. The OH involved in the furanose ring cannot be methylated.

COOH; HO–C–H; HO–C–H; H–C–OH; COOH ← (HNO_3; c) ← CHO; HO–C–H; HO–C–H; H–C–OH; CH_2OH → (CH_3I, Ag_2O; d) → CH_3O~C(ring O); CH_3O–C–H; CH_3O–C–H; H–C–(ring O); CH_2OCH_3

Problem 24.32

(a) Acetylation will occur at every free OH, leaving only the oxygen involved in the ring untouched.

(b) Under these conditions only the hemiacetal OH will be converted into OCH_3.

CH_3O ring: H–C–OH, H–C–OH, H–C–(O ring), CH_2OH ← (CH_3OH, H_3O^+; b) — CHO, H–C–OH, H–C–OH, H–C–OH, CH_2OH — (Ac_2O, H_3O^+; a) → AcO ring: H–C–OAc, H–C–OAc, H–C–(O ring), CH_2OAc

(c) Osazone formation; C_1 and C_2 will be turned into phenylhydrazones.

(d) This reaction is the Kiliani–Fischer synthesis. Two new sugars, one carbon longer, and epimeric at the new C_2, will be formed.

HC=NNHPh, C=NNHPh, H–C–OH, H–C–OH, CH_2OH ← (c; $PhNHNH_2$) — CHO, H–C–OH, H–C–OH, H–C–OH, CH_2OH — (1. NaCN 2. H_2/Pd 3. H_2O; d) → CHO, H–C–OH, H–C–OH, H–C–OH, H–C–OH, CH_2OH + CHO, HO–C–H, H–C–OH, H–C–OH, H–C–OH, CH_2OH

(e and f) Both the Wohl and Ruff degradations lead to the sugar one carbon shorter. In this case, the product is D-erythrose.

CHO, H–C–OH, H–C–OH, H–C–OH, CH_2OH → CHO, H–C–OH, H–C–OH, CH_2OH

e: 1. NH_2OH, CH_3OH, CH_3ONa 2. Ac_2O, 100 °C 3. CH_3OH, CH_3ONa 4. H_2O, H_3O^+

or

f: 1. Br_2/H_2O 2. $Ca(OH)_2$ 3. $Fe_2(SO_4)_3$ 4. H_2O_2

Problem 24.33 Only in glucose can the substituents on the six-membered β-pyranose ring be all equatorial.

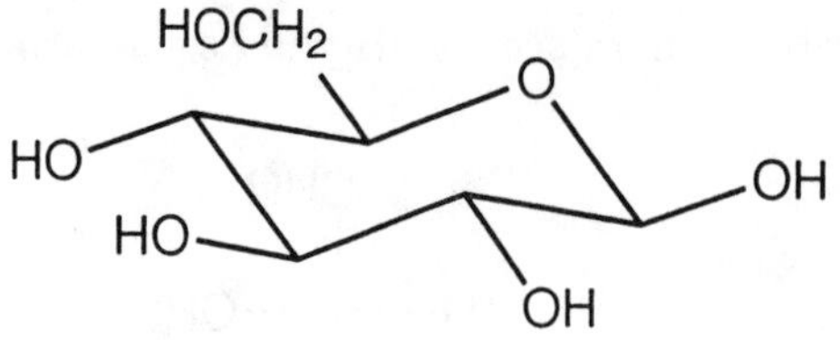

β-D-Glucopyranose

Problem 24.34 It's easy. Just take the ^{13}C NMR spectra. The diester from D-galactose will show four signals, whereas that from D-talose will show eight. There are other, similar answers.

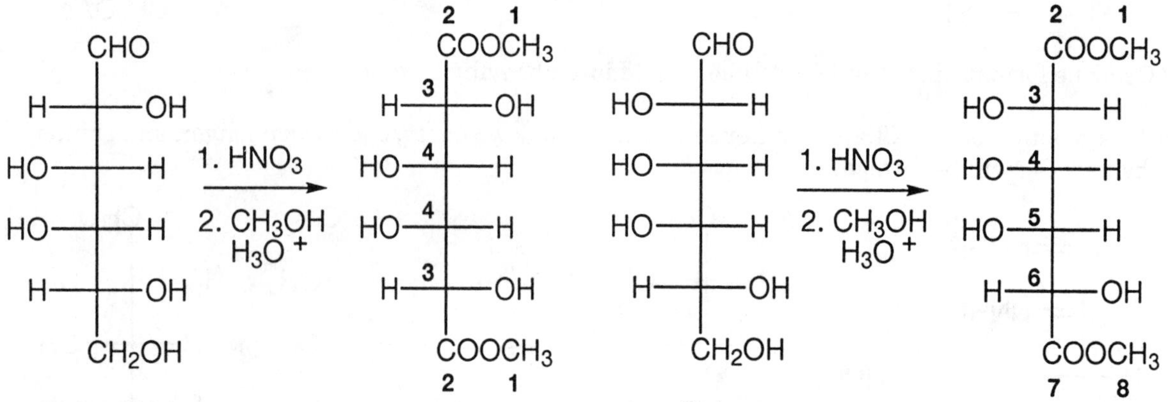

D-Galactose

D-Talose

Problem 24.35 Table 24.2 shows that at equilibrium D-allose exists 16% in the α-pyranose form and 76% in the β-pyranose form. Thus D-allose is 76/92 = 82.6% in the β-pyranose form and 16/92 = 17.4% in the α-pyranose form. The equilibrium constant is 82.6/17.4 = 4.75.
If $\Delta G = -2.3\ RT \log K$, and recalling that at 25 °C 2.3 RT is just about 1.36, then:
$\Delta G = -1.36 \log 4.75 = -0.92$ kcal/mol.

Problem 24.36
let X = fraction α, Y = fraction β.
X(112) + Y(18.7) = 52.7
and X + Y = 1, Y = 1 – X

So, 112X + 18.7 – 18.7X = 52.7,
93.3X = 34
X = 0.364, or 36.4%; Y = 0.636, or 63.6%. Table 24.2 shows that this answer is right on the nose.

Problem 24.37 First of all, it is clear that maltose is composed of two molecules of D-glucose. The problem is only to unravel how the two monosaccharides are attached, and then to make a good drawing.

Bromine in water oxidizes only the free aldehyde group to the acid. The aldehyde of one glucose, bound up in the linkage to the other monosaccharide, will not be affected. So, using Fischer projections for the two D-glucopyranoses, we have a very preliminary structure for maltose, and MBA.

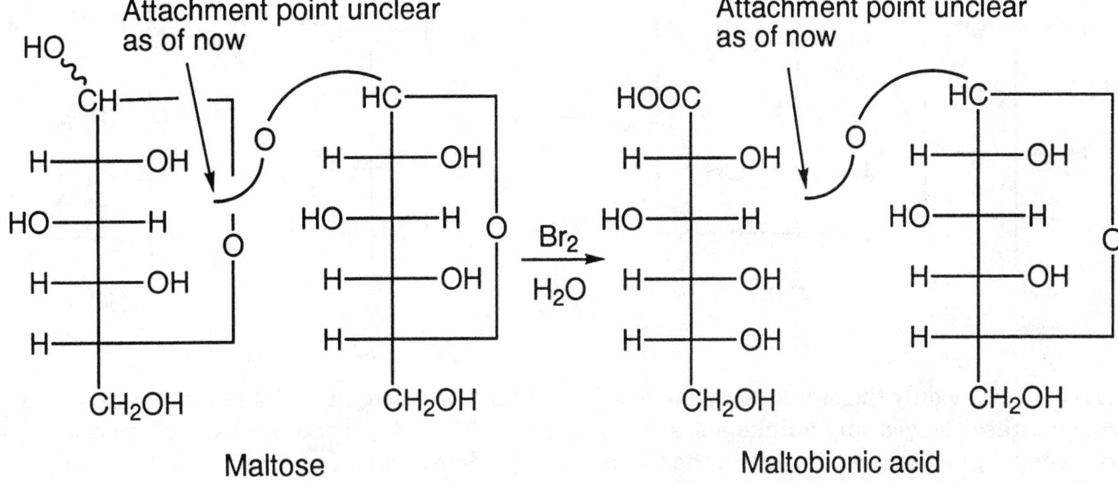

The strategy used is to uncover the attachment point by methylating all free hydroxyl groups in MBA. Hydrolysis will reveal the attachment point as an unmethylated hydroxyl. So,

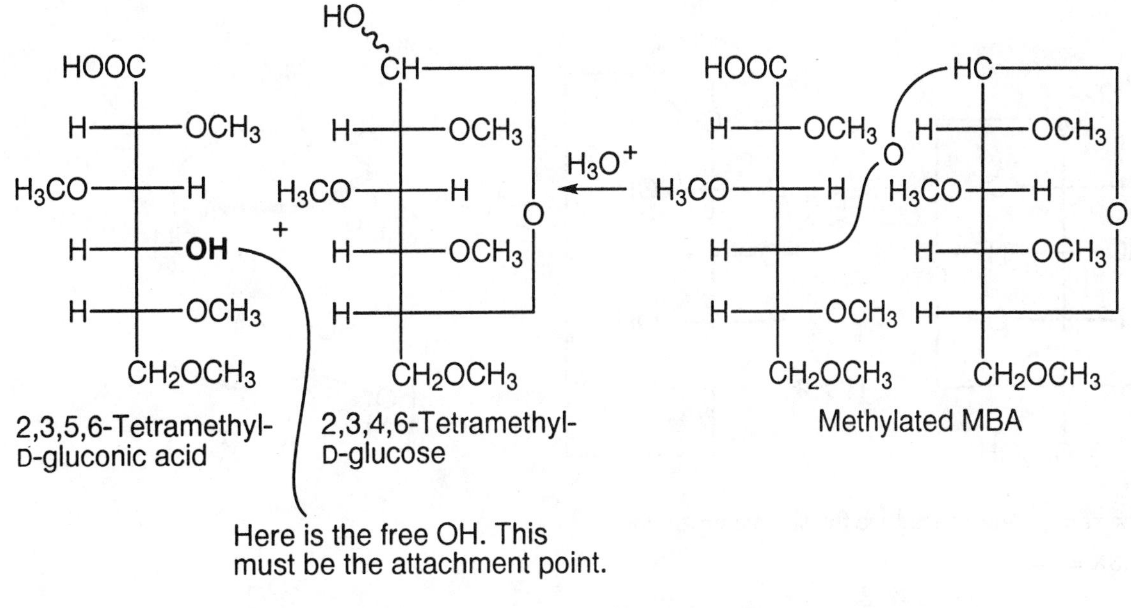

H_3O^+

2,3,5,6-Tetramethyl-D-gluconic acid + 2,3,4,6-Tetramethyl-D-glucose ← Methylated MBA

Here is the free OH. This must be the attachment point.

Now we know the structure of methylated MBA, and, by implication, MBA itself and maltose.

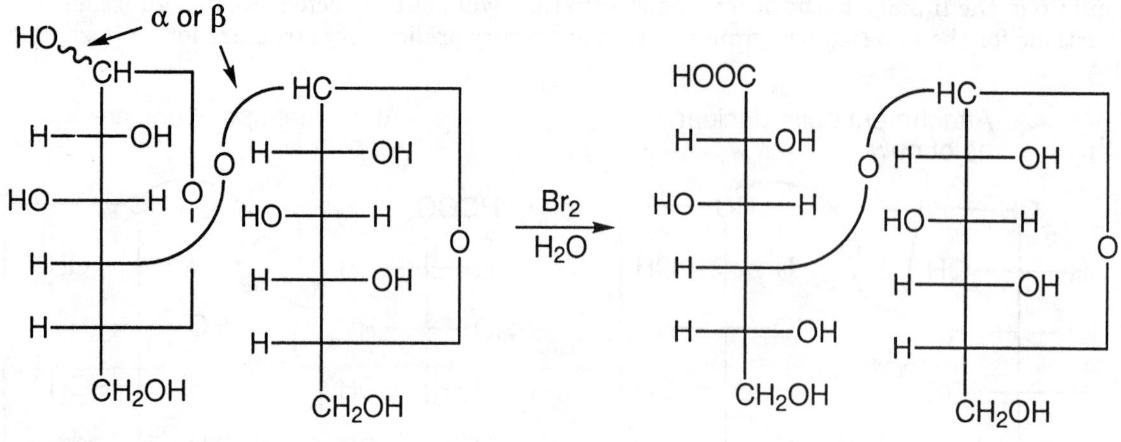

Maltose

MBA

What remains are only the stereochemical details and the translation into three dimensions. The enzyme maltase cleaves only α linkages, so the attachment from C_4 of the left-hand glucose to C_1 of the right-hand glucose is α. Of course, the OH at C_1 of the left-hand glucose can be either α or β.

To make the drawing, start with the right-hand glucose, and let Glu stand for the other glucose for the moment.

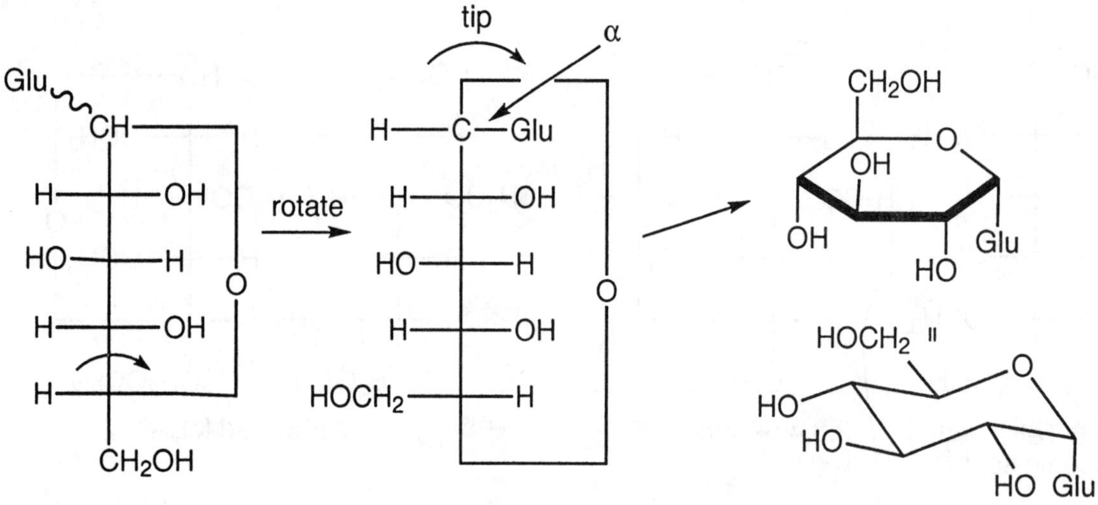

Now repeat the procedure for the other glucose.

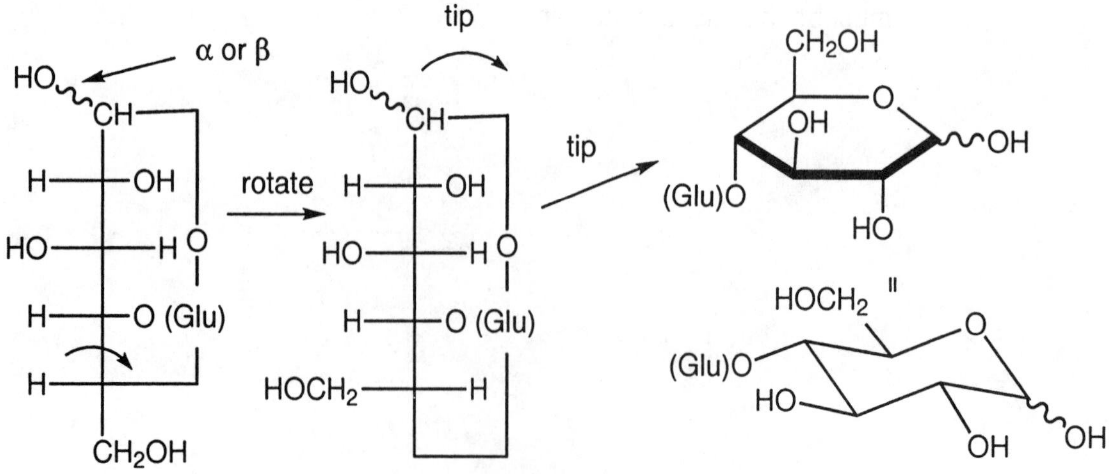

Now it only remains to combine the two three-dimensional forms for the monosaccharides, being careful to attach them in α-fashion.

Maltose

Problem 24.38 The question specifies a nonreducing disaccharide. Thus, there can be no amount of free aldehyde present. The two sugars must be attached at C_1 as parts of a full acetal. Otherwise, one sugar would always have some amount of free aldehyde present, and the disaccharide would be a reducing sugar.

First, let's draw D-altrose and D-allose in Fischer projection as their pyranose forms.

D-Altrose α- and β-D-Altropyranose D-Allose α- and β-D-Allopyranose

These must be connected by an acetal linkage of the two C_1 positions. The squiggly bonds still mean that we are not specifying the stereochemistry of the linkage (α or β).

Here is a schematic view.

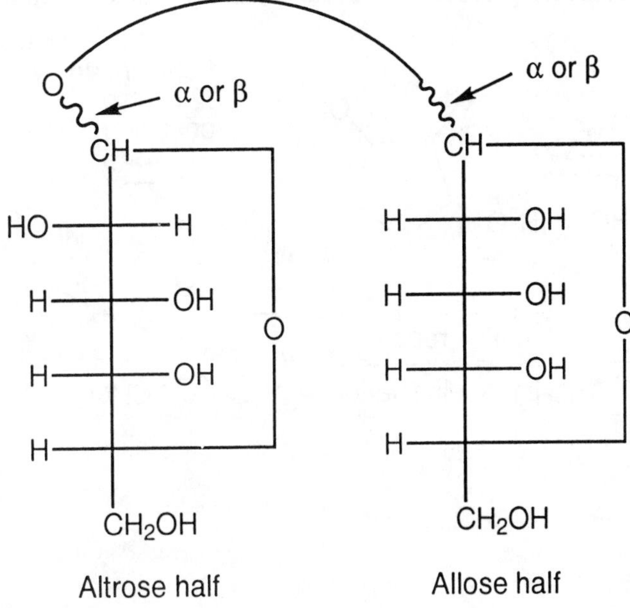

As each connection involves two α or β bonds (axial or equatorial in each ring) there are four possibilities: αα, αβ, βα, ββ.

Problem 24.39 Let's think a bit about strategy and what must happen in this reaction. An aldehyde group is still present in the product, so, at least for a start, it seems best to leave the aldehyde alone in the starting material. On the other hand, the primary alcohol, CH_2OH, has vanished. We need to close a ring and to remove some OH groups. Presumably, that removal will take the form of losses of water in elimination reactions generating the double bonds in the product. There are many ways to do this reaction. One good way is to set up ring formation by converting the primary OH into another aldehyde. This reaction is done through an elimination reaction to generate an enol.

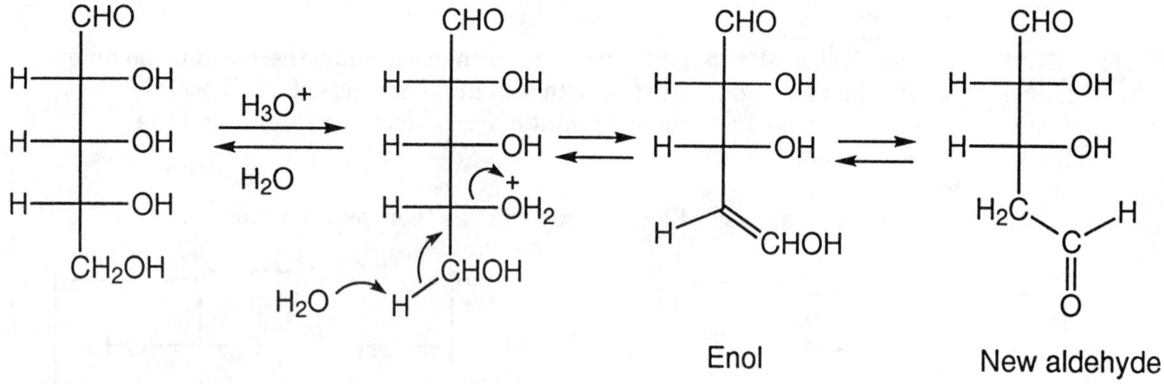

Enol

New aldehyde

Now we need to close a five-membered ring. The ring size dictates which oxygen atom to use in what is nothing more than an acid-catalyzed hemiacetal formation.

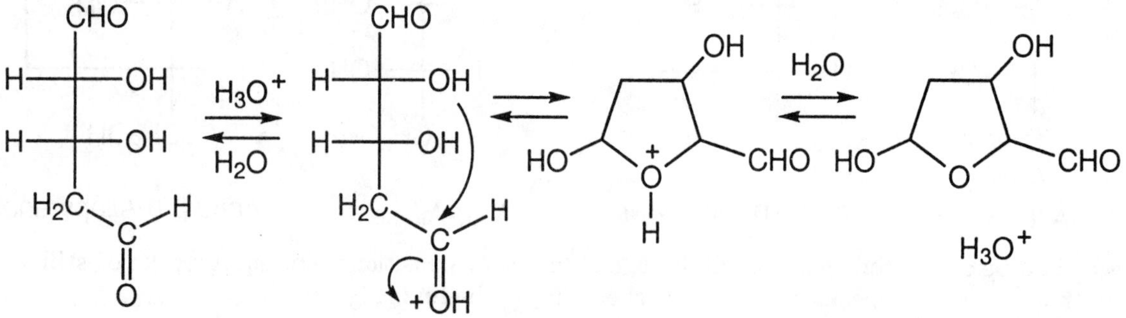

Now we need only protonate an OH and eliminate water twice to get furfural. In the figure, only one E1 reaction is shown. An E2 process is also reasonable. A second elimination gives the product.

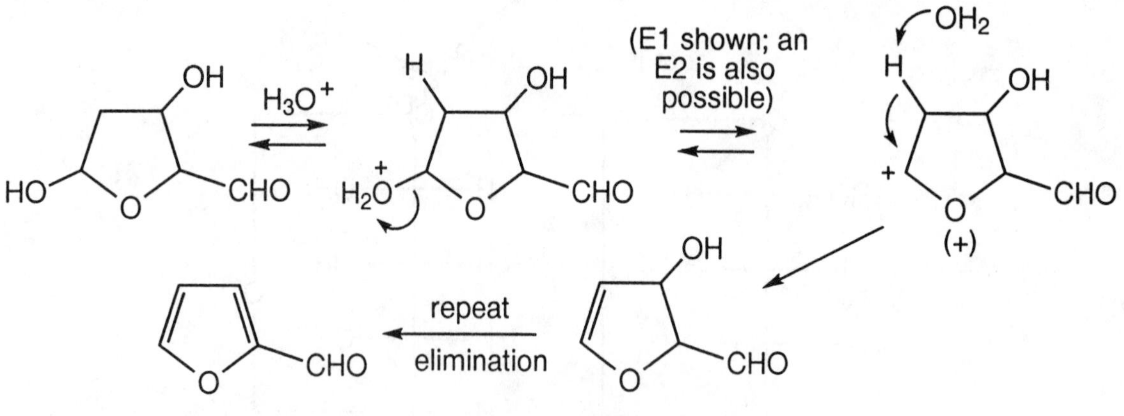

(E1 shown; an E2 is also possible)

repeat elimination

Problem 24.40 In base, not only will the enolate shown in Figures 24.24 and 24.25 form, but alkoxides will be generated as well. In one of these, a hydride shift (intramolecular Cannizzaro reaction) leads to fructose. Reversal of the hydride shift can occur in two ways to regenerate D-glucose or make the isomerized molecule, D-mannose.

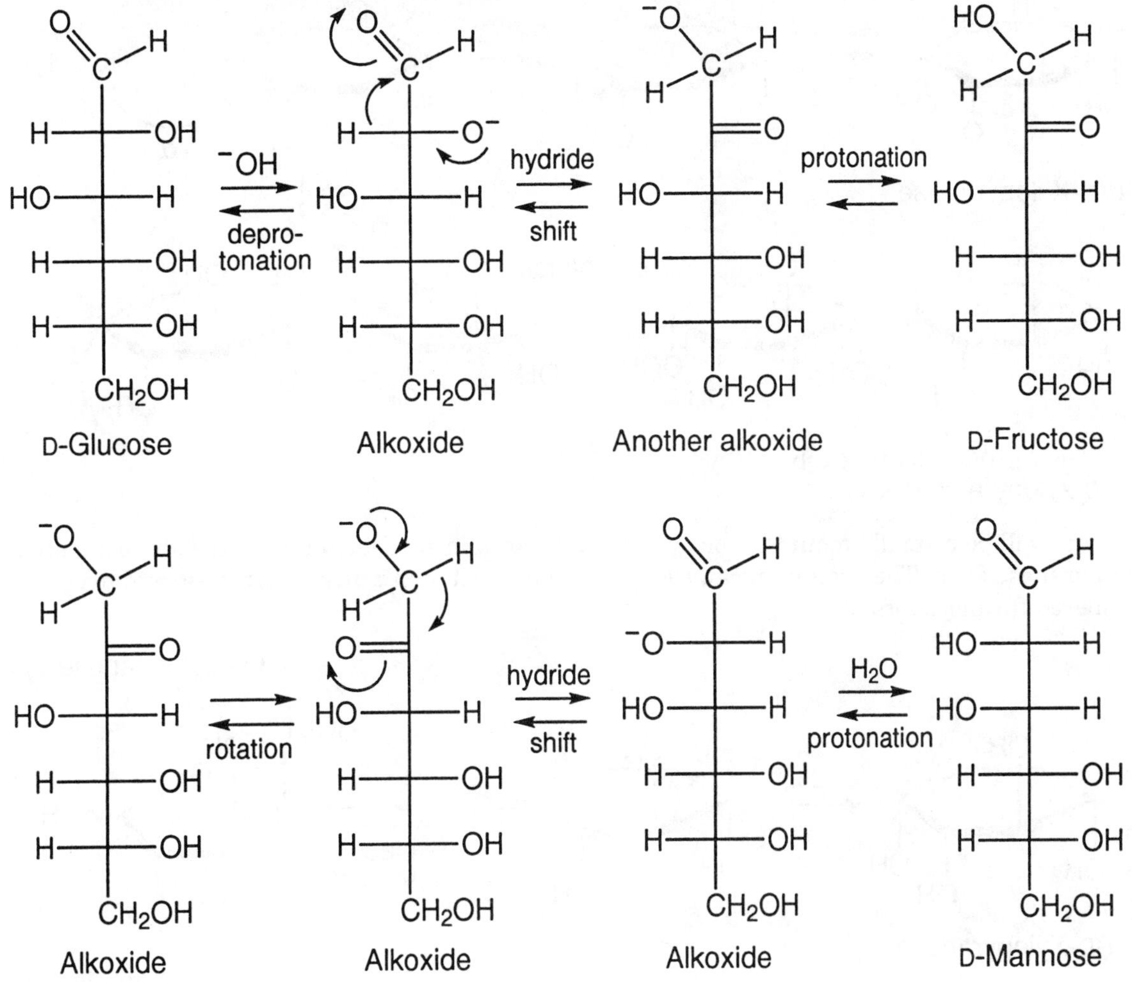

D-Glucose ⇌ ($^-$OH, deprotonation) Alkoxide ⇌ (hydride shift) Another alkoxide ⇌ (protonation) D-Fructose

Alkoxide ⇌ (rotation) Alkoxide ⇌ (hydride shift) Alkoxide ⇌ (H_2O, protonation) D-Mannose

Problem 24.41

(a) First of all, a xylopyranose must be a six-membered ring, and there is only one possible α-pyranose for D-xylose. Here it is in Haworth form.

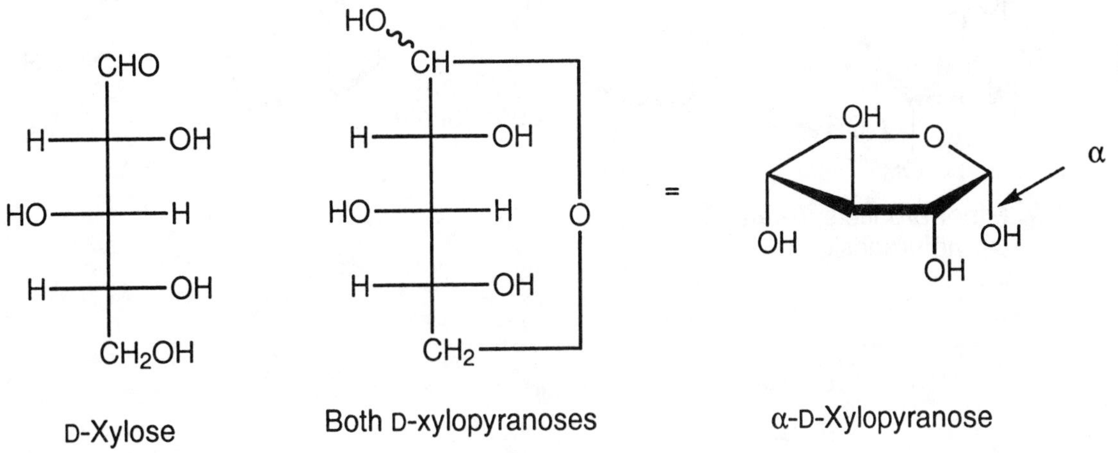

D-Xylose

Both D-xylopyranoses

α-D-Xylopyranose

The mechanism for methyl glycoside formation goes through a carbocation to which methyl alcohol can add in two ways to give the major products, α- and β-methyl D-xylopyranoside.

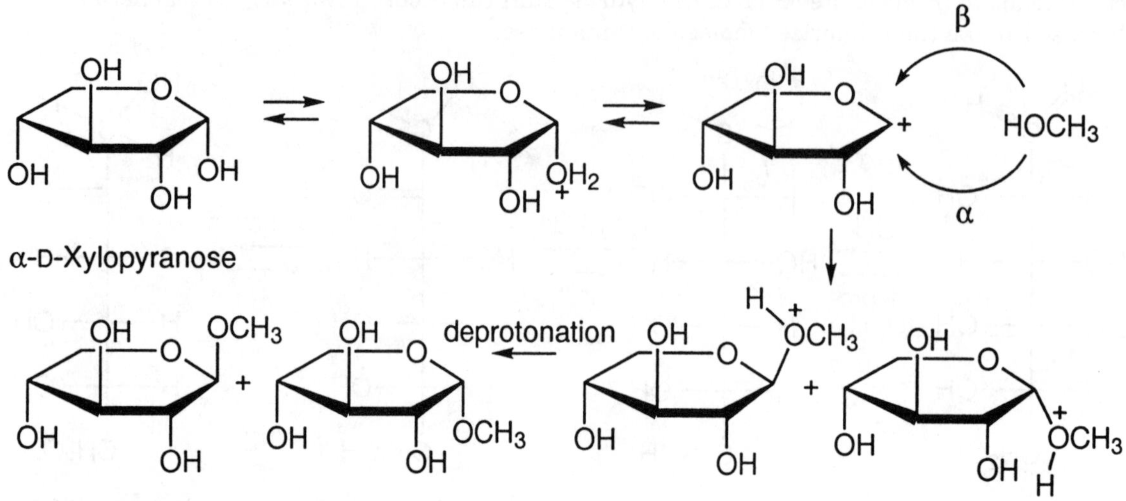

α-D-Xylopyranose

deprotonation

Major products, α- and β-methyl D-xylopyranosides

(b) There will be a small amount of the open pentose form in equilibrium with the six-membered ring pyranose form. The open form can close to an equilibrium mixture of pyranose and five-membered furanose forms.

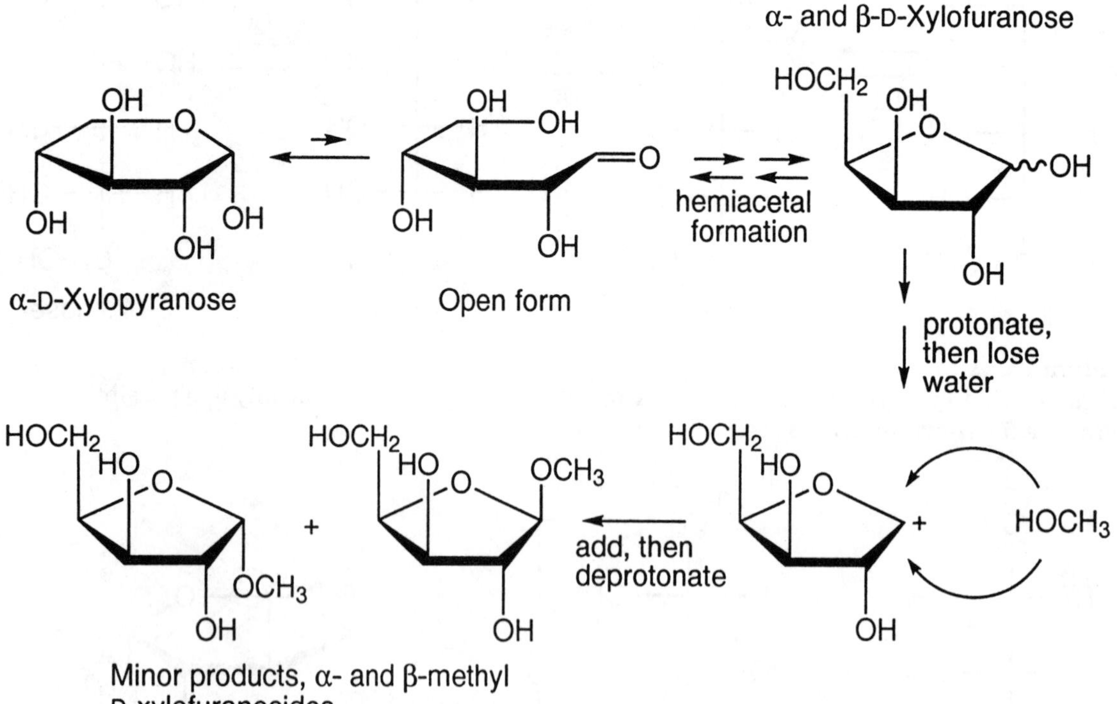

α- and β-D-Xylofuranose

hemiacetal formation

α-D-Xylopyranose

Open form

protonate, then lose water

add, then deprotonate

Minor products, α- and β-methyl D-xylofuranosides

Problem 24.42 Carbocations are not likely intermediates under these basic conditions, so please don't protonate and form a carbocation as in Problem 24.41. The pyranose form of glucose is in equilibrium with the open form, and this can react with ammonia to give an imine.

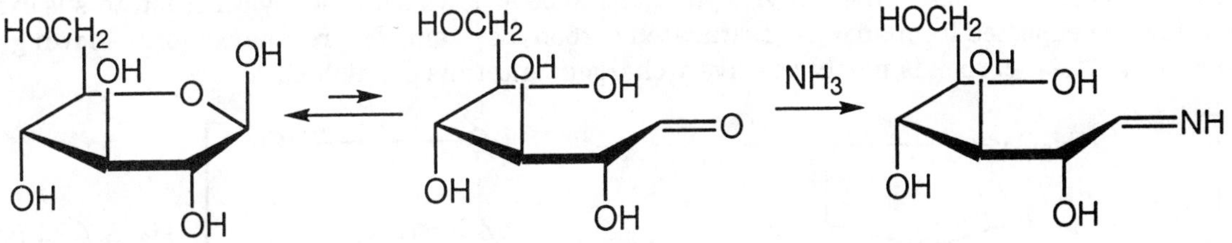

NH_3

β-D-Glucopyranose Open form Imine

Now reclosure can occur to give the amino sugar. The β-form will be favored.

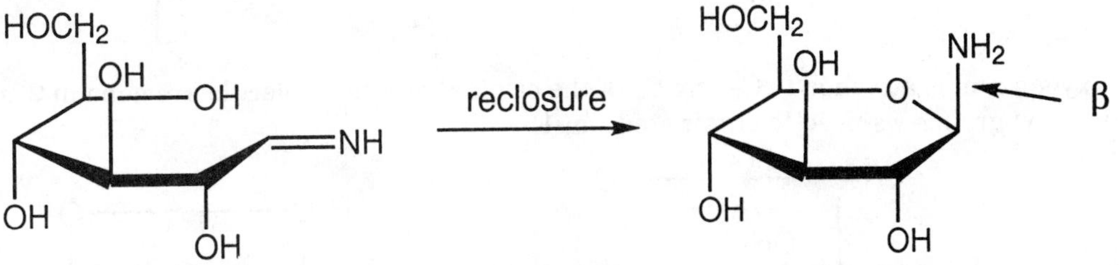

reclosure β

Problem 24.43 This problem is very similar to Problem 24.42. The pyranose equilibrates with a small amount of the open form, which reacts with ethyl mercaptan to give the thioacetal.

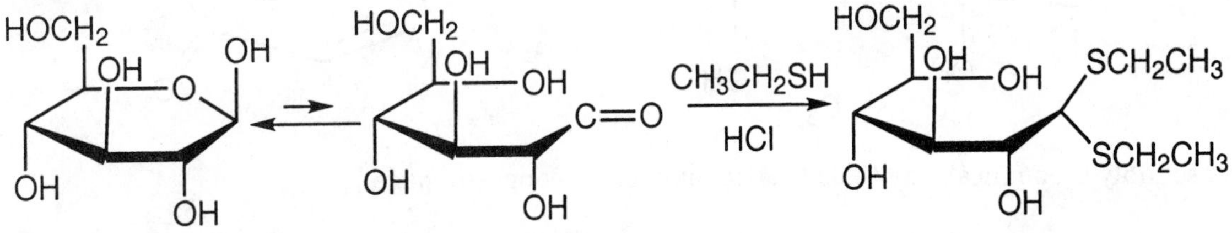

CH_3CH_2SH / HCl

β-D-Glucopyranose Open form

Problem 24.44 First, draw D-glucose in its β-pyranose form, then flip the six-membered ring. Of course this flipped form is less stable than the unflipped form, but some of it will be present at equilibrium. Now protonation, followed by an S_N2 displacement reaction, can take place to make **1**.

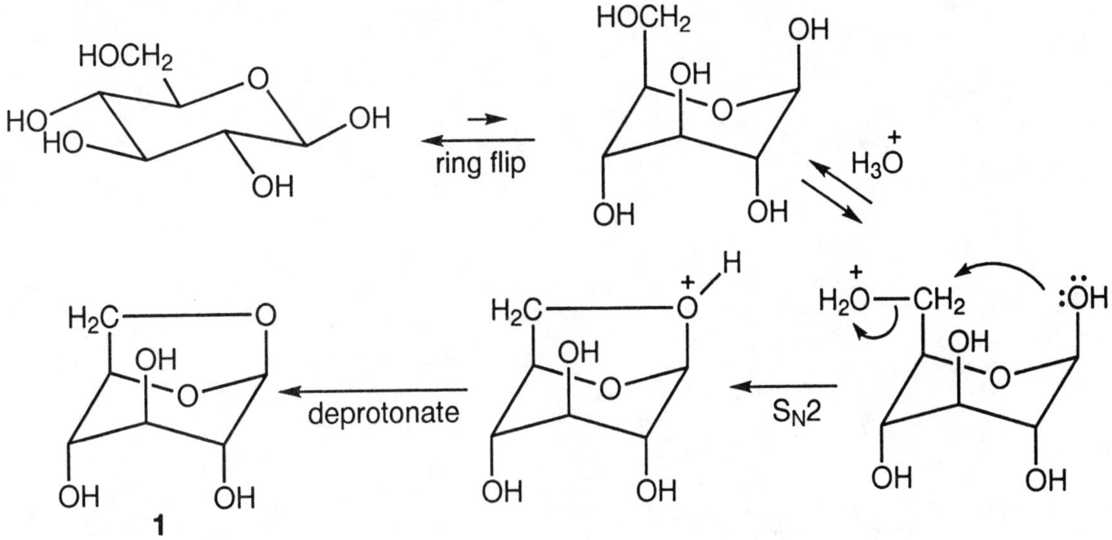

ring flip H_3O^+ S_N2 deprotonate

1

Problem 24.45

(a) Just about everything is incompatible. The UV and IR spectra indicate a conjugated carbonyl group, and this is absent in **3**. The ^{1}H NMR spectrum is more subtle, but just as inconsistent. In **3**, there is no reason for one of the "vinyl" hydrogens to be as low as δ 7.34. By contrast, this is exactly what is to be expected from an α,β-unsaturated carbonyl system. The resonance form shown gives the reason: The β-carbon is partly positively charged, and thus deshielded.

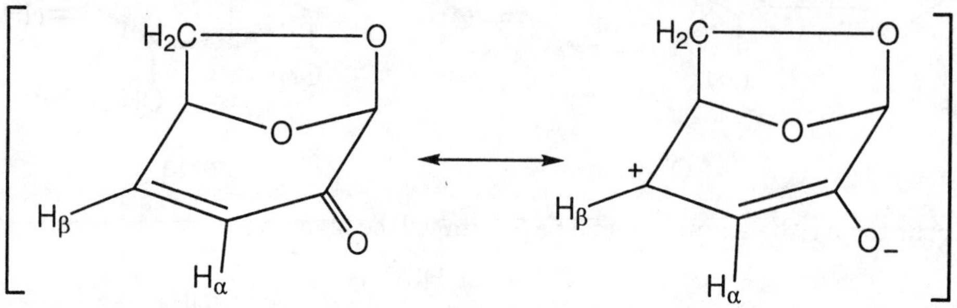

H_β δ 7.34 ppm

H_α δ 6.09 ppm

(b) The oxygen atoms attached to C_3 and C_4 of the original glucose molecule are gone in **2**, so it is those hydroxyl groups we need to eliminate from **1**.

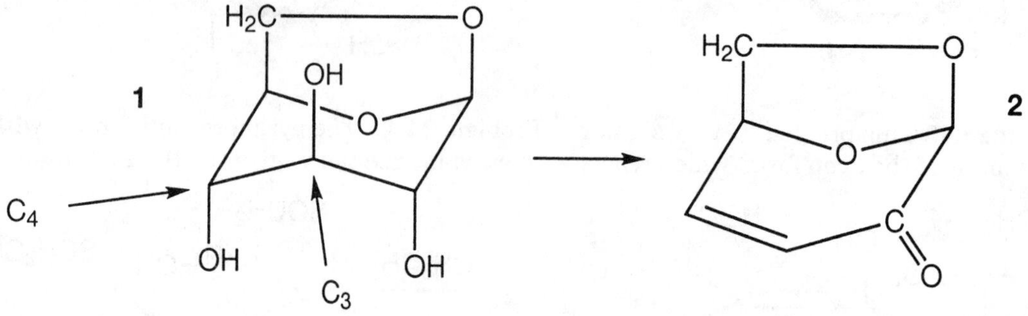

A reasonably economical way to do this appears on the opposite page.

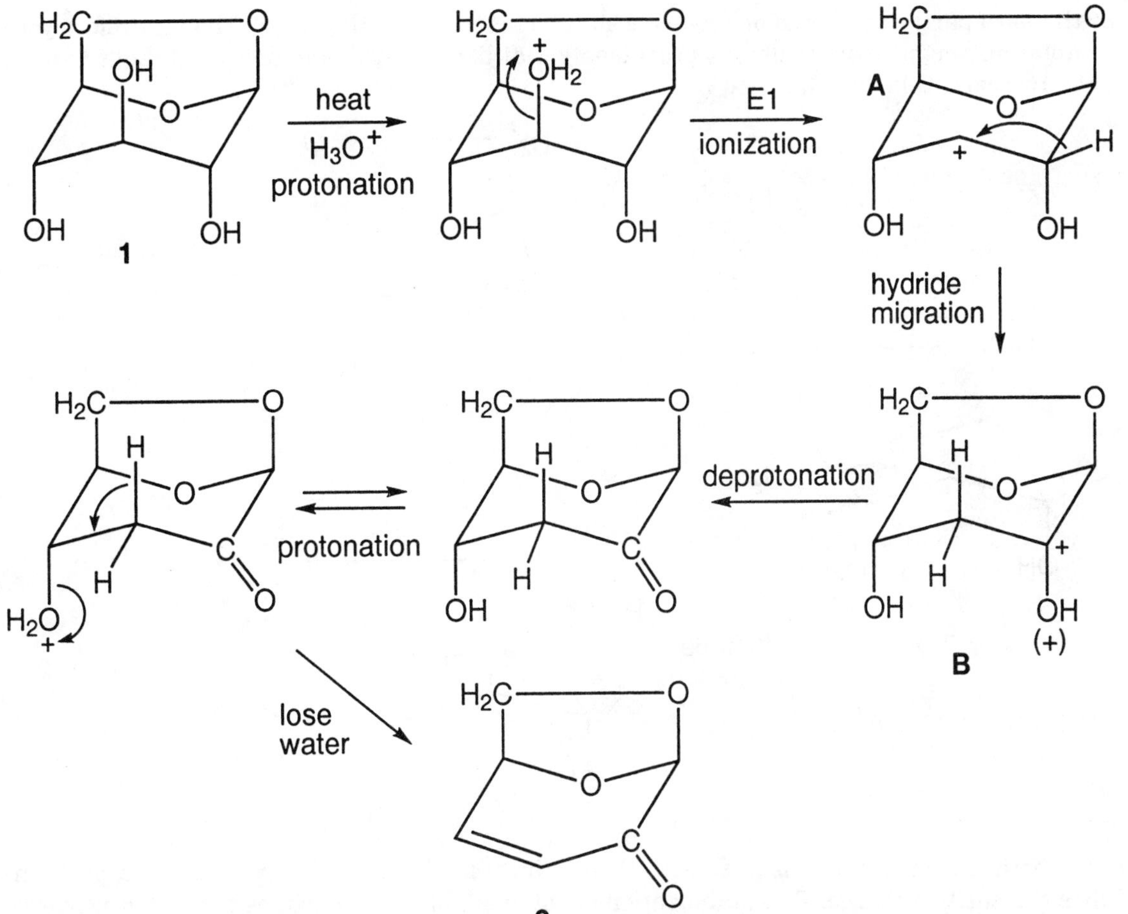

heat
H_3O^+
protonation
E1
ionization
A
hydride
migration
deprotonation
B
protonation
lose
water
1
2

Notice that the hydride migration at the heart of this mechanism produces a resonance-stabilized carbocation (**B**) from an ordinary secondary carbocation, **A**. Deprotonation of **B** generates the necessary carbonyl group. Elimination of water leads to **2**.

(c) First of all, how might **4** be formed? In secondary carbocation, **A**, there are two possible hydride shifts. The one shown above leads to the observed product, **2**, but the other one would give **4**, through the intermediate, resonance-stabilized cation **C**.

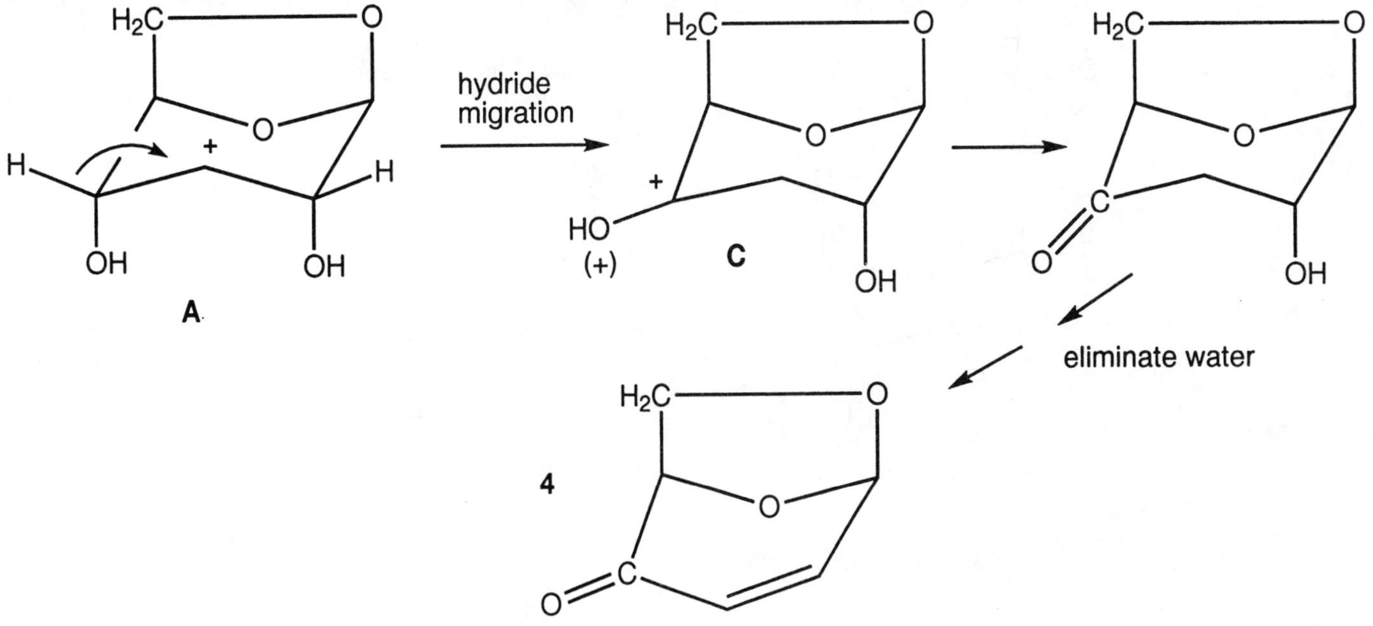

hydride
migration
A
C
eliminate water
4

Now for the hard part; why should one hydride shift be favored over the other? The hint refers you to the chapter on neighboring groups, so presumably that has something to do with it. Let's first look at the two cations involved, **C** and **B**.

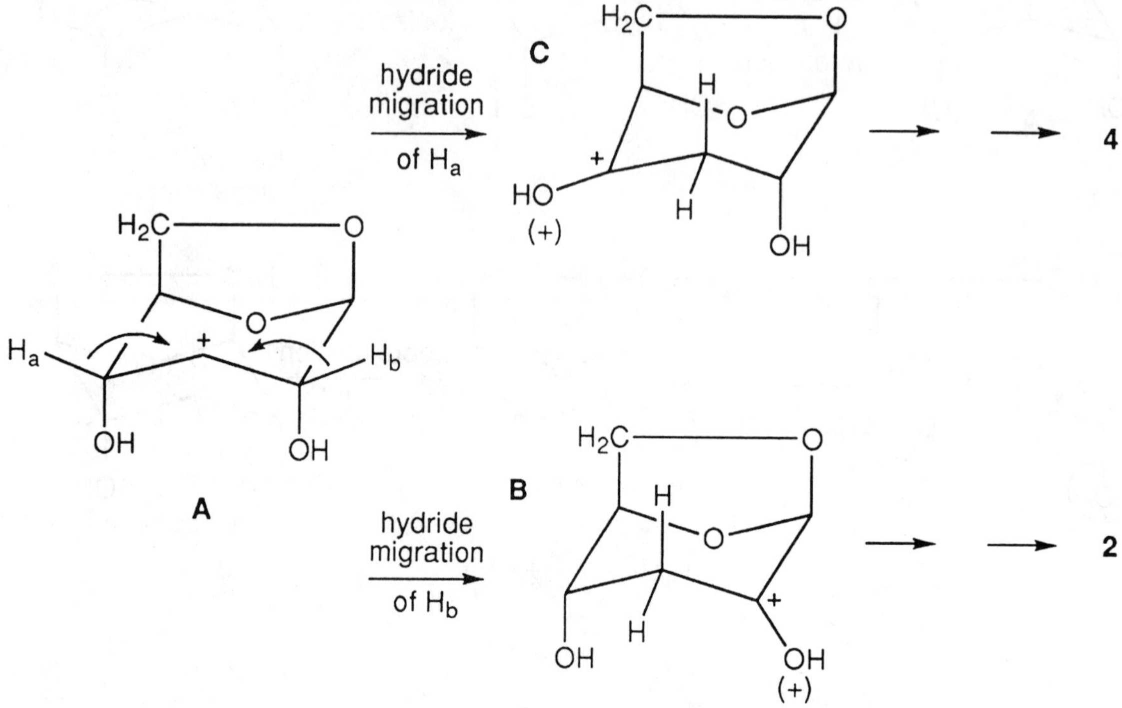

H_2C O C hydride migration of H_a H O + HO (+) H OH → → **4**

H_2C O O + H_a H_b OH OH **A**

H_2C O B H O hydride migration of H_b + H OH OH (+) → → **2**

Notice that both ring oxygen atoms in **B**, with their pairs of nonbonding electrons, are in a position to stabilize the positive charge. This stabilization is not available in **C**. Perhaps two different views of **C** and **B** show this better.

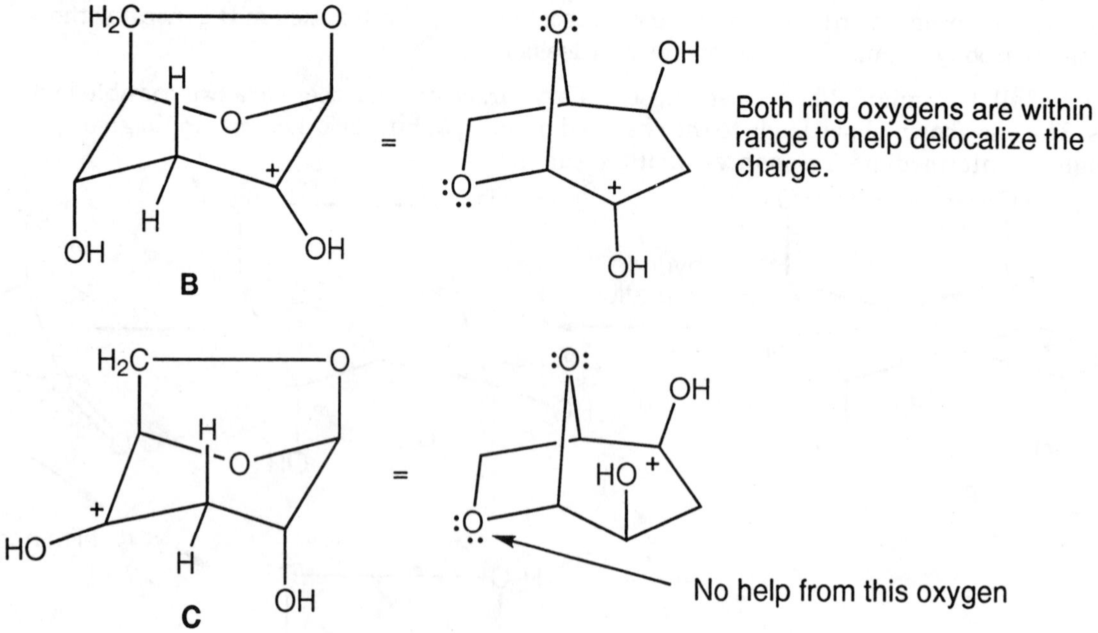

H_2C O H O + H OH OH **B** = :O: OH :O + OH

Both ring oxygens are within range to help delocalize the charge.

H_2C O H O + HO H OH **C** = :O: OH HO + :O

No help from this oxygen

Problem 24.46 The formula shows you that parts of two molecules of acetone have become incorporated into the new molecule (six carbons have been added to the sugar). What is the acid-catalyzed reaction between a diol and a carbonyl compound? Acetal formation (see Chapter 16). In this case, a double acetal is formed. As we know that the primary alcohol group is not involved (the question tells us this), there are only two possible 1,2-diols from which to make the acetals.

H_3C, $C{=}O$, H_3C; HO, CH_2OH, O, HO, OH, OH; O, C, H_3C, CH_3 → H_3O^+ → H_3C, O, C, H_3C, O, CH_2OH, O, O, O, C, H_3C, CH_3

$C_{12}H_{20}O_6$

A double acetal

Problem 24.47 All OH groups but the primary alcohol are protected as parts of acetals, which leaves the primary alcohol free to react, and this is where all the action is. First, a tosylate (**A**) is formed (remember that OTs is an excellent leaving group, Chapter 7). Displacement by iodide gives **B**, and reduction leads to a methyl group, as in **C**. Finally, treatment with a catalytic amount of acid removes the acetals and frees the four OH groups. The result is **D**.

$C_{12}H_{20}O_6$ (with CH_2OH) —TsCl, pyridine→ **A**, $C_{12}H_{19}O_6Ts$ (with CH_2OTs)

A —NaI→ **B**, $C_{12}H_{19}IO_5$ (with CH_2I)

B —RaNi→ **C**, $C_{12}H_{20}O_5$ (with CH_3)

C —1% H_2SO_4→ **D**, $C_6H_{12}O_5$ (HO, CH_3, O, HO, OH, OH)

Problem 24.48 Here is a detailed picture of the chair form of the six-membered acetal formed from reaction with benzaldehyde. Notice that the phenyl group can lie in the equatorial position thus avoiding destabilizing 1,3-diaxial interactions. The axial position is occupied by the small hydrogen atom.

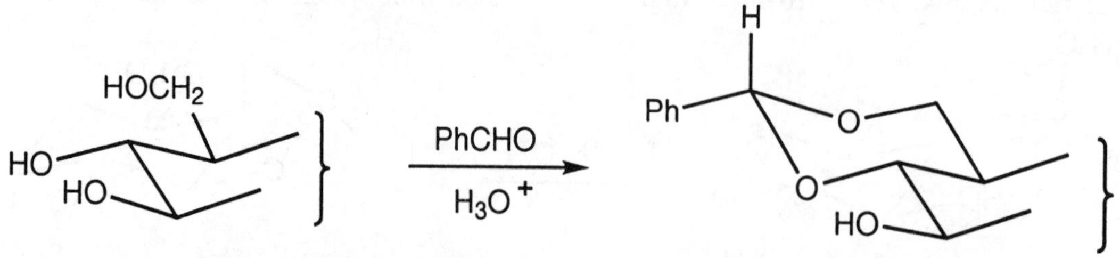

HOCH$_2$ HO HO → PhCHO / H_3O^+ → H Ph O O HO

By contrast, the six-membered acetal formed from a ketone such as acetone must have one group in the axial position. Presumably, this is why the six-membered acetals are disfavored in such cases in favor of the five-membered rings.

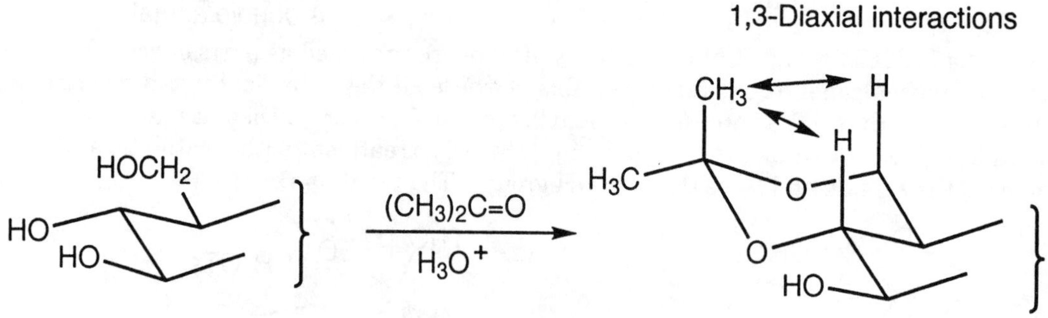

1,3-Diaxial interactions

HOCH$_2$ HO HO → $(CH_3)_2C{=}O$ / H_3O^+ → CH_3 H H H_3C O O HO

Chapter 25 Outline

25.1 Three-Membered Heterocycles: Oxiranes (Epoxides), Thiiranes (Episulfides), and Aziridines
25.2 Four-Membered Heterocycles: Oxetanes, Thietanes, and Azetidines
25.3 Five- and Six-Membered Heterocycles
25.4 Nonaromatic Heterocycles Containing More Than One Heteroatom
25.5 Nitrogen-Containing Aromatic Heterocyclic Compounds
25.6 Other Aromatic Heterocycles: Furan and Thiophene
25.7 Aromatic Heterocycles Containing More Than One Heteroatom
25.8 Two-Ring Heterocycles
25.9 Something More: Penicillin and Related Antibiotics

25

Introduction to the Chemistry of Heterocyclic Molecules

Problem 25.1 The more substituted an alkene, the better a nucleophile it is. So, it is the tetrasubstituted double bond that is the more reactive site in this diene.

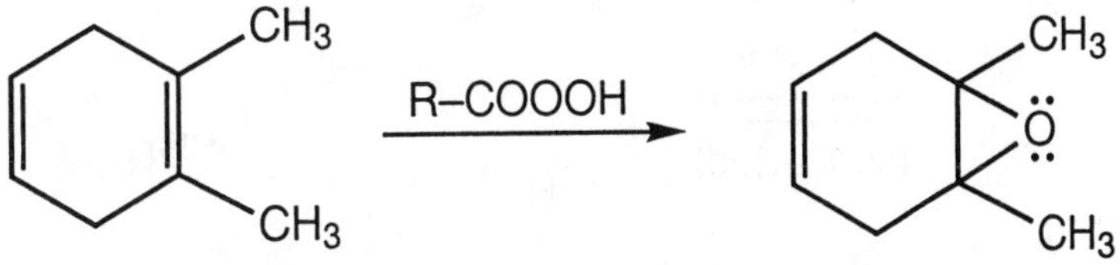
CH3
CH3
R–COOOH
CH3
CH3

But wait! There is another effect we haven't considered. Is not the tetrasubstituted double bond more hindered sterically? Yes it is, and the steric difficulties should slow the reaction. So, there are two factors working in opposite directions. It is the electronic effect that dominates in this case. A good answer to this question mentions both effects.

Problem 25.2 Additions of halogens in nucleophilic solvents lead to halohydrins (water solvent) and halo ethers (alcohol solvent). Reaction of cyclohexene with aqueous chlorine will do the job in this example.

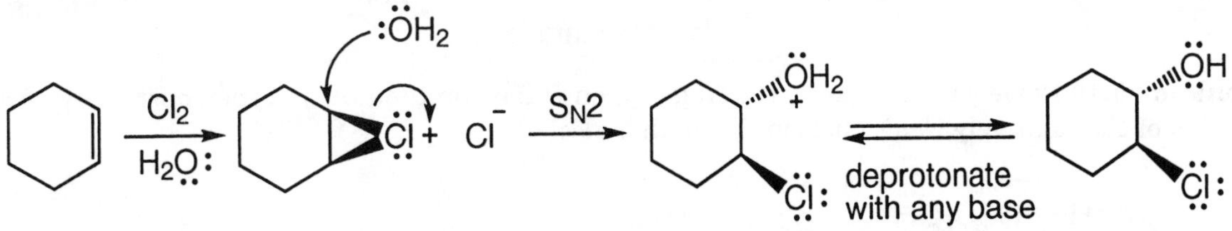
Cl2
H2O:
:OH2
Cl + Cl
SN2
OH2
Cl
deprotonate
with any base
OH
Cl

Problem 25.3 The electronegative chlorine atom is electron-withdrawing and will inductively stabilize the buildup of negative charge on the adjacent carbon.

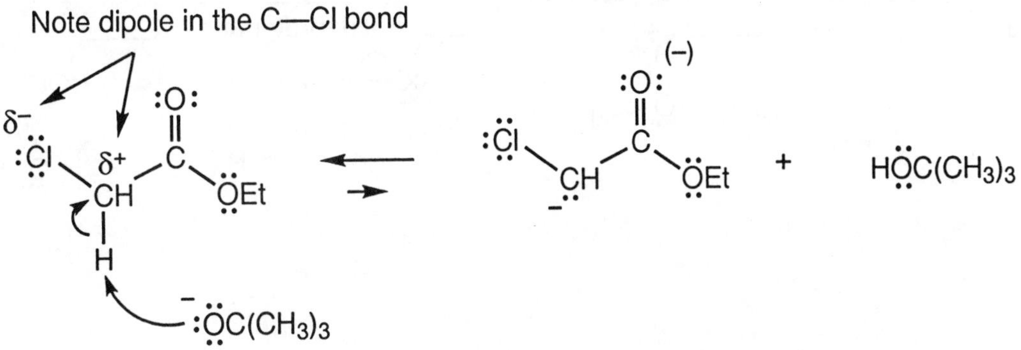
Note dipole in the C—Cl bond
δ−
δ+
:Cl
CH
C
:O:
OEt
H
:OC(CH3)3
(−)
:O:
:Cl
CH
C
OEt
+
HOC(CH3)3

Problem 25.7 It is the same old story. Hydroxide is a poor leaving group. Conversion of the OH into sulfate (OSO_2OH) makes the leaving group better, and the reaction possible.

H, :$NHCH_2CH_3$, H_3C, C, C, CH_3, HO:, H, vs. H, :$NHCH_2CH_3$, H_3C, C, C, CH_3, $HOSO_2O$:, H

Displacement is difficult because OH is a poor leaving group

S_N2 displacement

$HOSO_2O^-$ + Good leaving group; $\overset{+}{N}HCH_2CH_3$; H, H_3C, C, C, CH_3, H; deprotonation; NCH_2CH_3; H, H_3C, C, C, CH_3, H

Problem 25.8 Use an alkene with stereochemistry. If the addition is concerted, the stereochemistry present in the alkene will be preserved in the triazoline. If a stepwise mechanism is followed, a mixture of stereoisomeric products will be found.

cis; H_3C, CH_3; R, N, N≡N:, +; one step (concerted); H_3C, CH_3, H, H, R–N, N, N: Still cis

close

cis; H_3C, CH_3; R, N, N≡N:, +; form one bond; H_3C, CH_3, +, H, H, R, N, N, N:; Intermediate; rotate and close the other bond; H, CH_3, H_3C, H, R–N, N, N: trans

Problem 25.10 Review here. In acid, the ring oxygen will be protonated and then opened by addition of the relatively weak nucleophile methyl alcohol.

H–OSO_2OH; O: protonation; H, +, O:; H_3C–O–H; S_N2; OH; :O–H, +, H_3C; H_3C–O–H; deprotonation; OH; :OCH_3; $H_2\overset{+}{O}CH_3$

In base, the strong nucleophile methoxide opens the strained ring in S_N2 fashion. Protonation gives the product. Notice that the catalyst is regenerated at the end of this reaction.

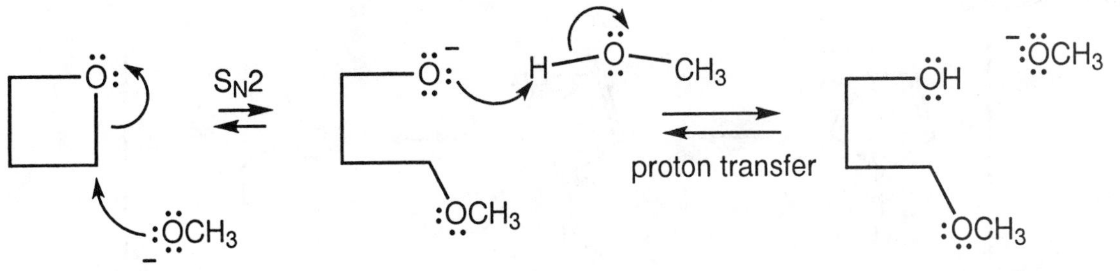

S_N2

proton transfer

Problem 25.11 Protonation of the alkene will take place so as to generate the more stable, resonance-stabilized carbocation.

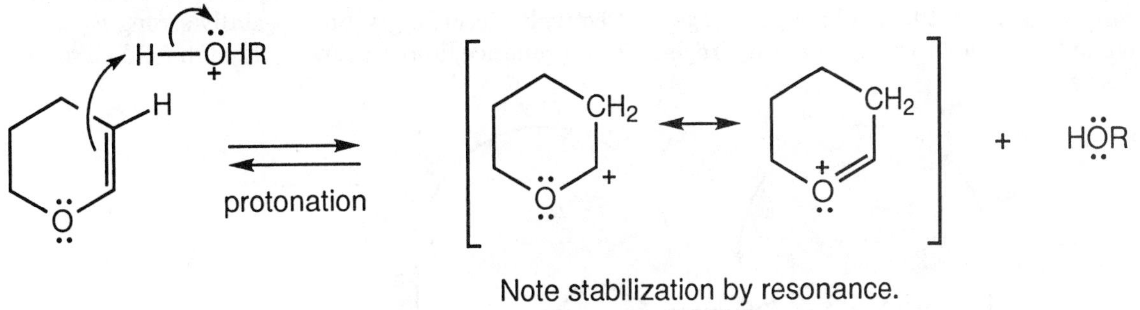

protonation

Note stabilization by resonance.

Addition of the alcohol now takes place. Deprotonation gives the THP derivative.

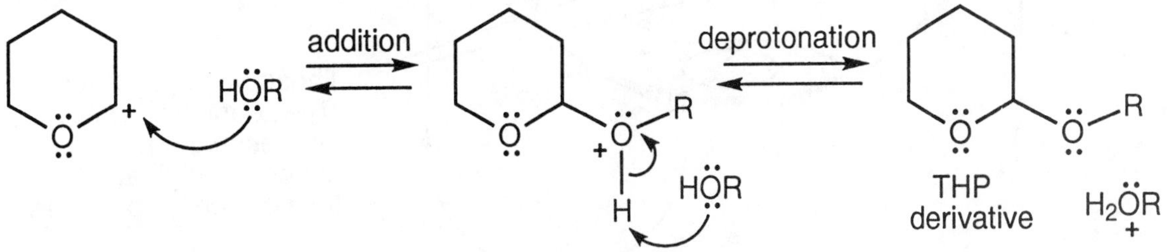

addition

deprotonation

THP derivative

To write the mechanism for the regeneration of the alcohol, simply read the steps you have already written backward. Protonate the THP derivative with H_3O^+ and lose alcohol.

Problem 25.12 Like chlorine (Problem 25.3, p. 1286), sulfur is electronegative and substantially electron withdrawing. Removal of electron density from carbon makes it easier to put a negative charge there.

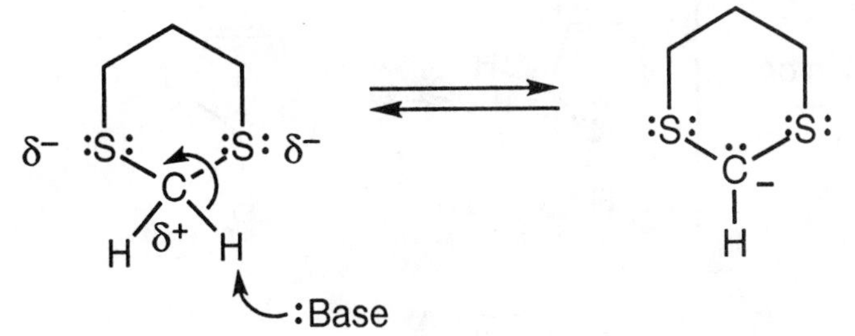

:Base

Problem 25.14 Not at all. As Figure 25.35 shows, both kinds of ring carbon in pyrrole are partially negative. Those additional electrons will serve to shield nearby hydrogens and shift them to higher applied field (B_0).

Problem 25.15 This kind of extreme high field signal is diagnostic for the ring currents present in aromatic compounds. In a cyclophane of this kind, one or more hydrogens will lie directly, or nearly directly over the center of the ring. At this point, the induced magnetic field (B_i) will be especially strong, and will act to shield the hydrogens effectively. Accordingly, an especially strong B_0 will have to be applied to bring these hydrogens into resonance, and an especially high field signal is observed.

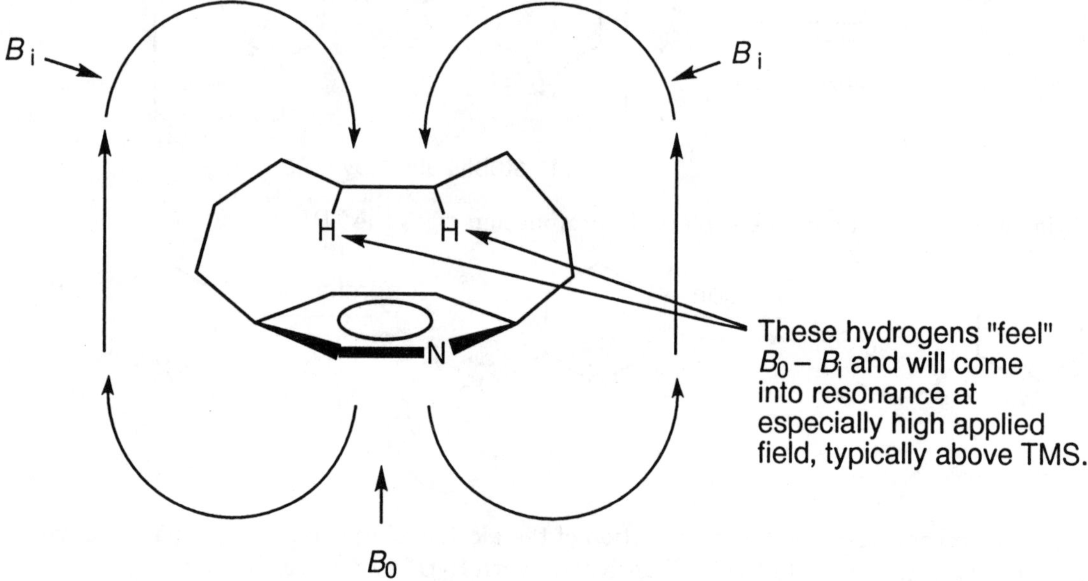
B_i
B_i
H
H
N
These hydrogens "feel" $B_0 - B_i$ and will come into resonance at especially high applied field, typically above TMS.
B_0

Problem 25.16 Protonation on carbon generates a delocalized, resonance-stabilized cation, whereas protonation on nitrogen does not, and destroys the aromatic sextet of electrons.

Problem 25.18 In each case, the active ingredient is the resonance-stabilized α-anion. The nitrogen helps to bear the negative charge along with three carbons.

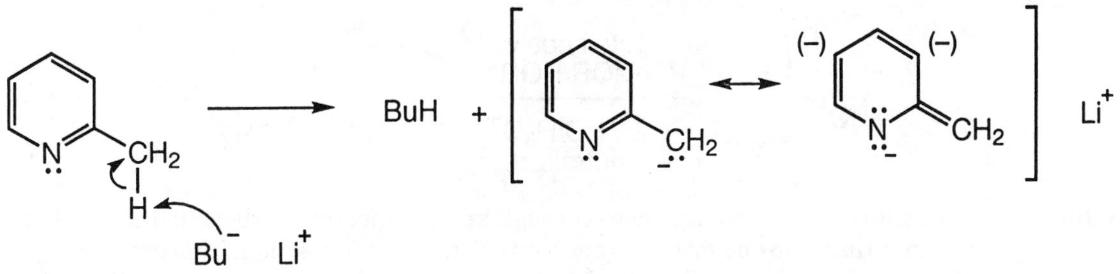

BuH + [↔] Li^+

Bu^- Li^+

The problem shows this anion carrying out two typical additions to carbonyl groups. In (a), it adds to an aldehyde to produce an alcohol (Chapter 16, p. 791). In (b), carbon dioxide plays the part of the Lewis acid, and the result is a carboxylic acid (Chapter 19, p. 991).

(a)

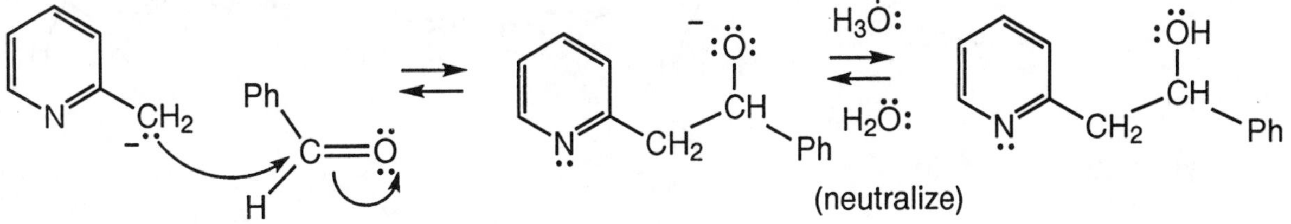

H_3O^+ / H_2O (neutralize)

(b)

H_3O^+ / H_2O (neutralize)

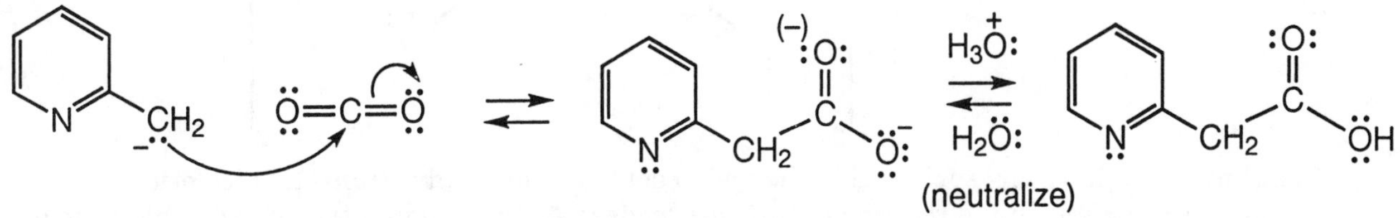

Problem 25.20 The easiest way to do this problem, by far, is to work backward. The first question to ask is, What is the intermediate immediately preceding the target, 2,4,6-trimethylpyridine? The answer is **A**, the molecule that hydrolyzes and decarboxylates to 2,4,6-trimethylpyridine.

A $\xrightarrow[\text{heat}]{KOH/H_2O}$ 2,4,6-trimethylpyridine + CO_2

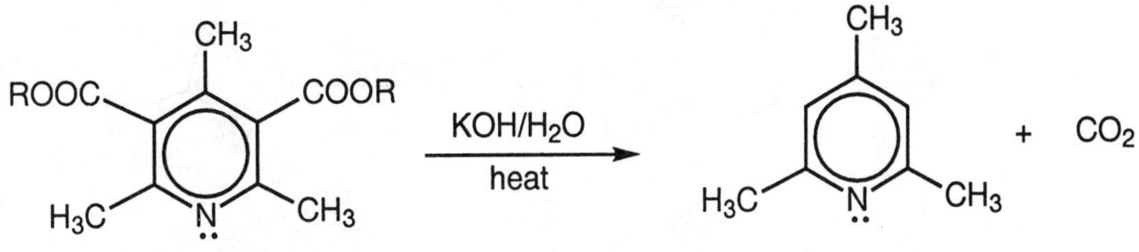

The target now becomes the more complicated pyridine, **A**. The general outline of the Hantzsch synthesis (Fig. 25.56) allows us to supply the components.

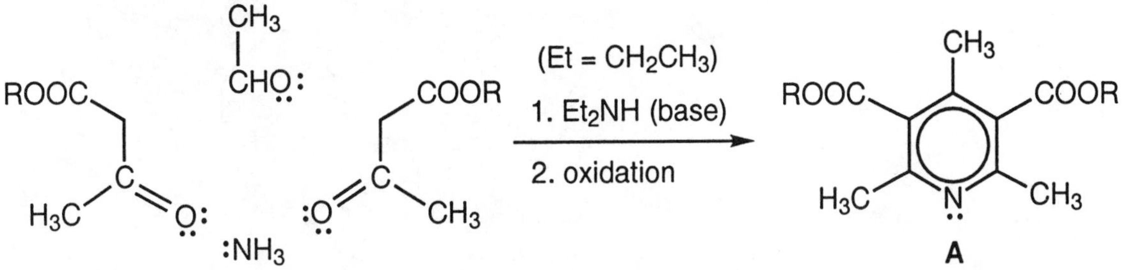

($Et = CH_2CH_3$)
1. Et_2NH (base)
2. oxidation

A

We still need to make the β-ketoester, and for that we have to recall Chapter 20 and the Claisen condensation.

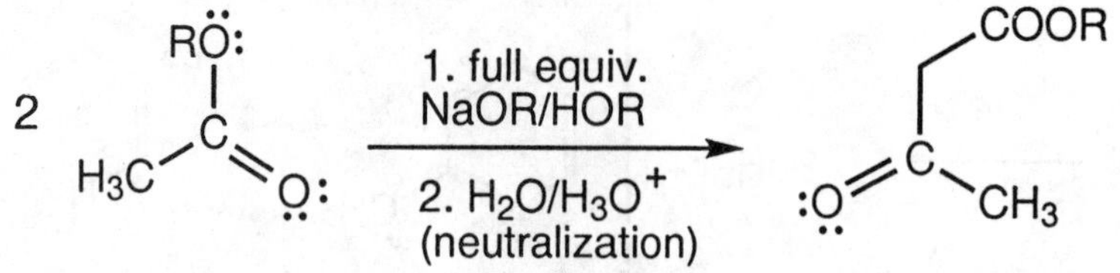

Problem 25.22 Substitution at position 2 goes through an intermediate carbocation that is better stabilized by resonance than the one formed by substitution at the 3-position. This greater resonance stabilization of the intermediate, and of the transition state leading to it, will favor the former pathway.

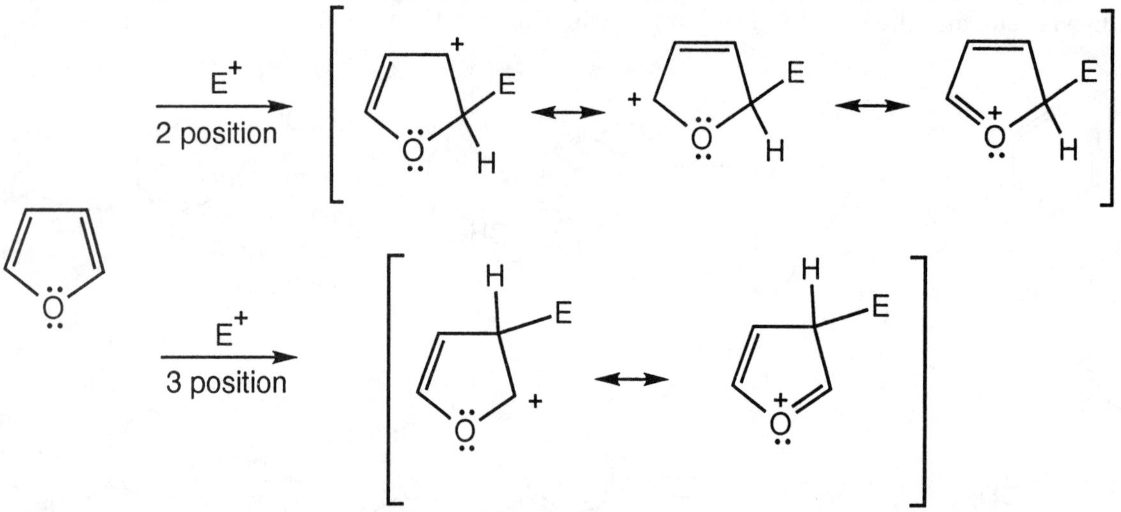

Problem 25.23 Reagents such as chlorine add to double bonds to give *trans*-1,2-dichlorides (Chapter 10, p. 393). Such a reaction in this case leads to **A**. Now, elimination of hydrochloric acid would give the product of apparent aromatic substitution.

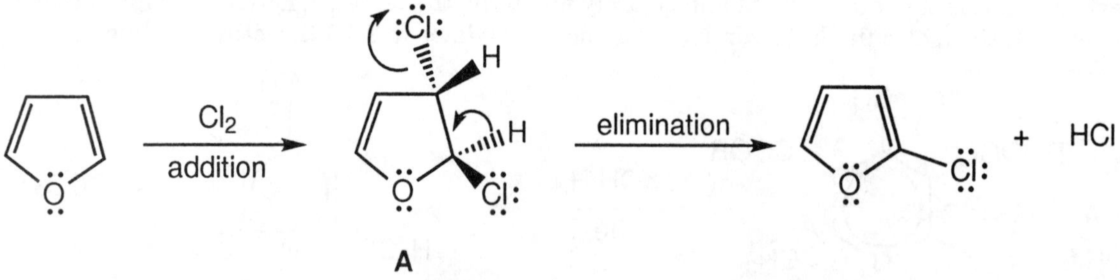

Problem 25.24 The E2 reaction shown in the answer to Problem 25.23 is a syn elimination, and not likely to be easy. However, ionization of one chloride in **A** leads to the same intermediate as is formed in the aromatic substitution reaction. This E1 reaction should be quite facile.

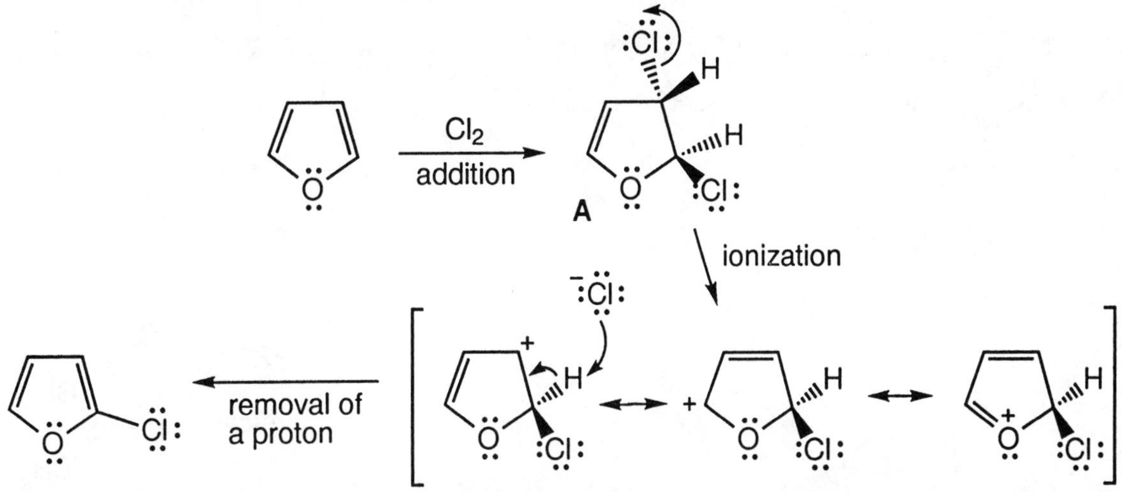

Cl_2 addition

A

ionization

removal of a proton

Problem 25.25 The nitro group is essential because it helps to stabilize the anionic intermediate formed by proton removal from the alkyl group. Without this nitro group the benzylic position is not acidic enough to allow proton removal.

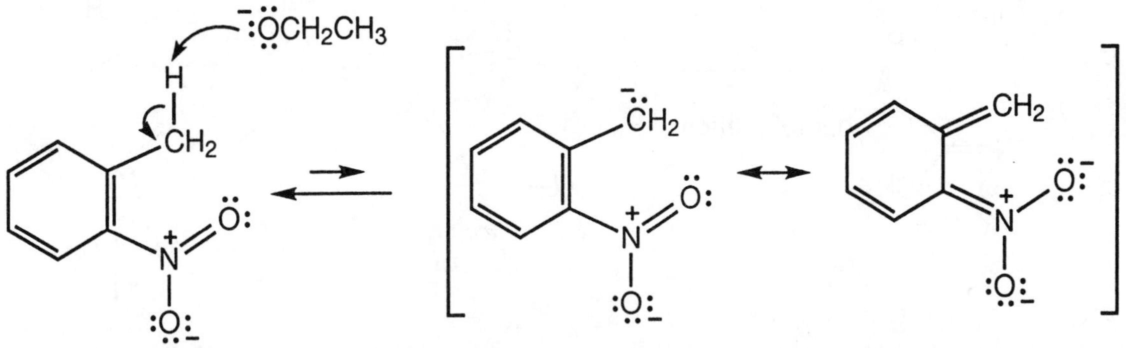

Problem 25.26 It is just a series of acid-catalyzed eliminations of water, with one crucial conversion of an enol into an aldehyde.

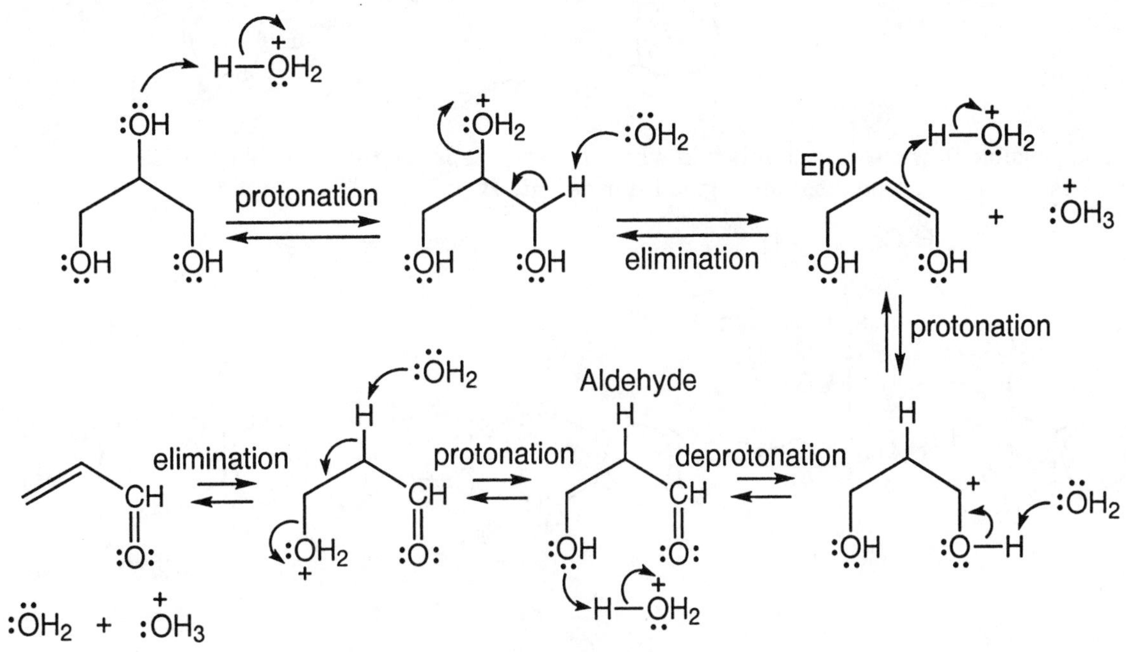

protonation

elimination

Enol

protonation

Aldehyde

deprotonation

protonation

elimination

Problem 25.27(a) This synthesis starts with formation of the imine **A**.

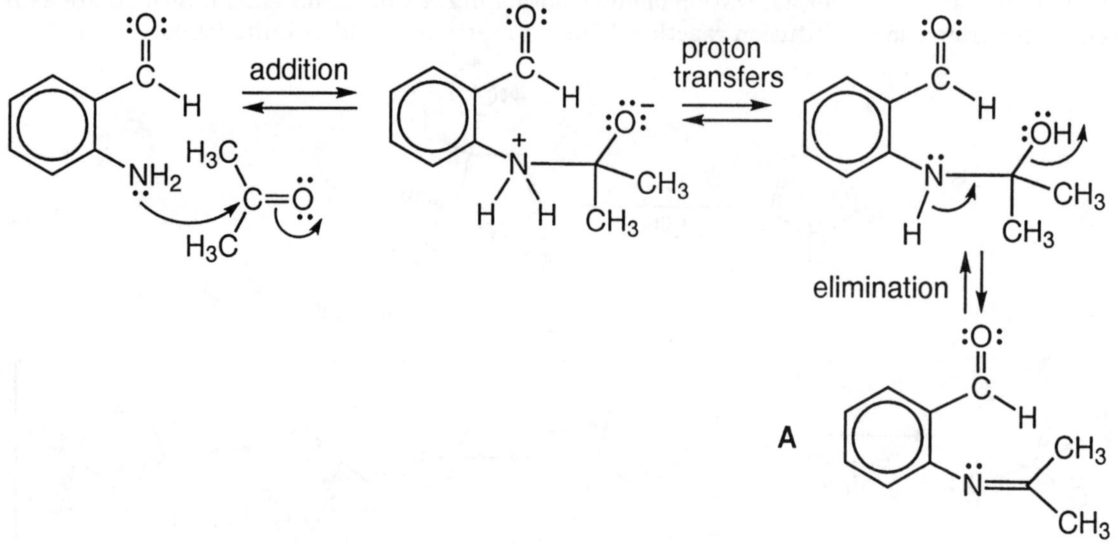

addition

proton transfers

elimination

A

Now, a base-catalyzed condensation closes the ring, and elimination of water leads to product.

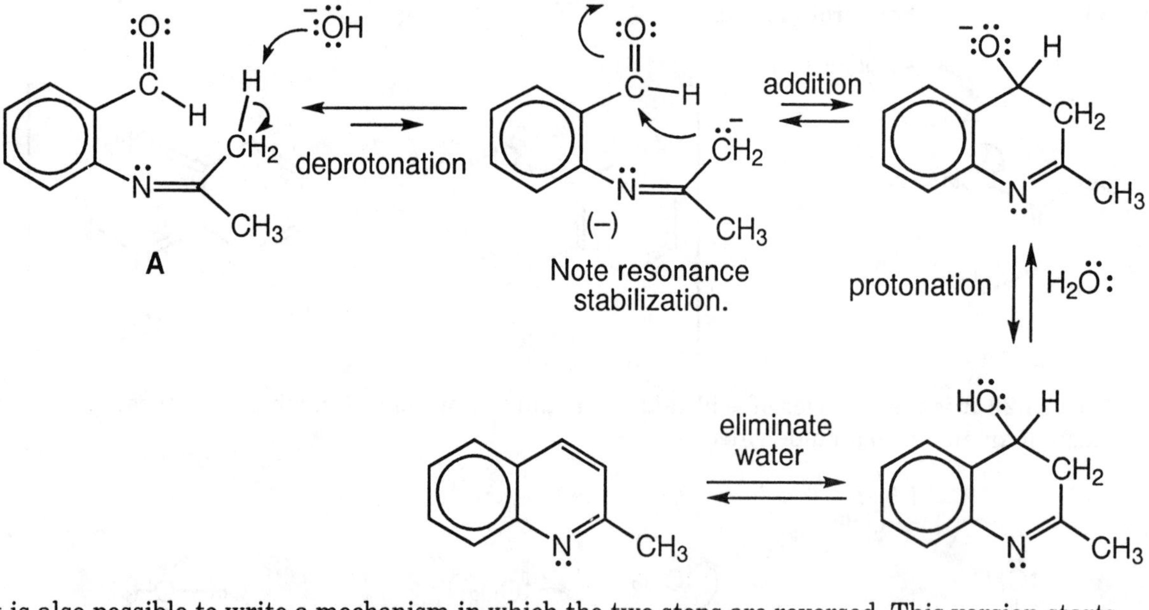

deprotonation

A

(−)

Note resonance stabilization.

addition

protonation $H_2\ddot{O}$:

eliminate water

It is also possible to write a mechanism in which the two steps are reversed. This version starts with a base-catalyzed condensation to give intermediate **B**.

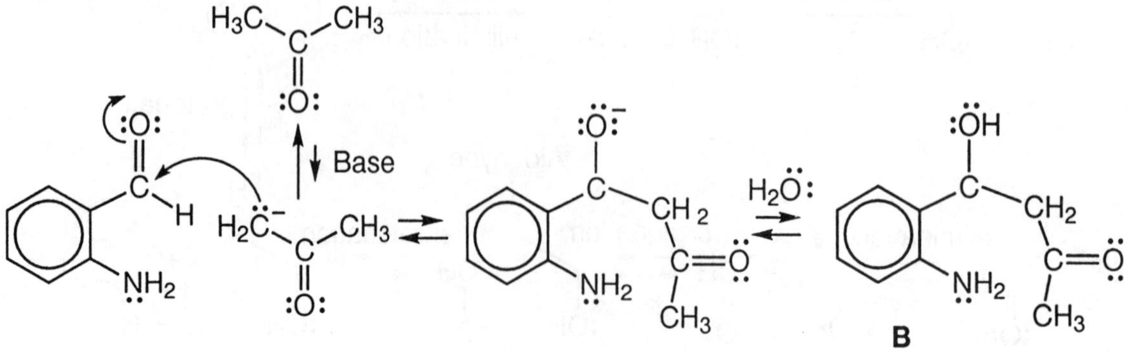

Base

$H_2\ddot{O}$:

B

Intermediate **B** now eliminates water and closes the ring in a standard addition of an amine to a carbonyl group.

:OH NH2 CH2 C=O CH3 B deprotonation :OH NH2 CH C=O CH3 elimination NH2 C=O CH3 addition protonation elimination N CH3 OH N H CH3 HO: deprotonation OH N+ H H CH3 HO:

(b) This acid-catalyzed cyclization is related, at least conceptually, to the Robinson annulation reaction (Chapter 18, p. 923). The first step is to align the partners in the reaction so that you can see where the new bonds must be made. This procedure is quite easy if you use the methyl group of methyl vinyl ketone as an anchor. Surely this methyl becomes the methyl group of the product.

Make a bond here. CH3 :O=C NH2 Make bonds here. CH3 N

There are Lewis acids aplenty here (HCl, $ZnCl_2$, or $FeCl_3$) and we will abbreviate them as A for acid. In the first step, methyl vinyl ketone is activated by reaction with a Lewis acid, and the amine adds in Michael fashion ("always do the Michael first", Chapter 18, p. 924).

A O+ C CH3 NH2 Michael addition A :O CH3 N+ H H deproton-ation A :O CH3 N H

Now protonation of the double bond leads to an intermediate that can add to the aromatic ring in a straightforward intramolecular aromatic substitution reaction.

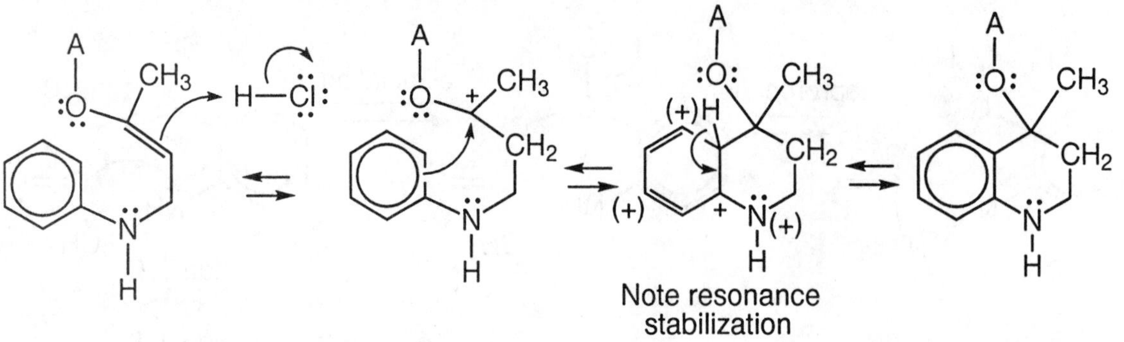

Note resonance stabilization

An elimination reaction leads to a dihydroquinoline, and now the nitrobenzene and/or iron takes over, oxidizing this intermediate to the aromatic compound.

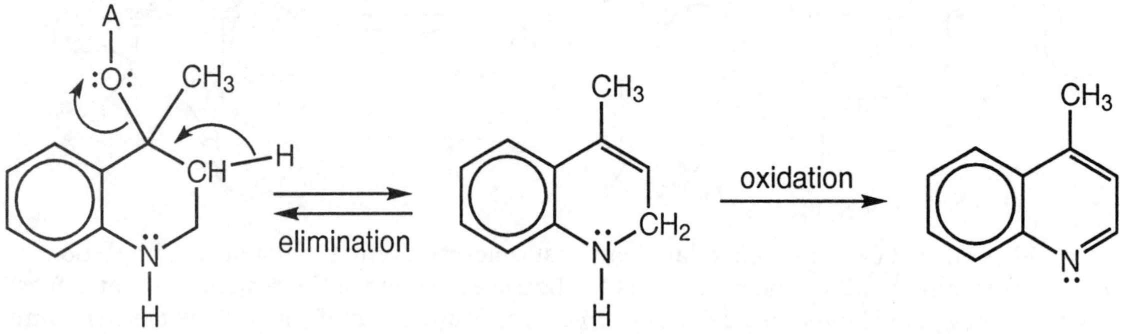

elimination

oxidation

Answers to Additional Problems, Chapter 25.

Problem 25.28 There are, of course, many possible answers to each part of this problem. The point is to construct molecules with $4n + 2$ π electrons. In the answer to (a), boron contributes no π electrons, but does maintain orbital connectivity, the cycle of conjugation, because of its empty $2p_z$ orbital. In (b), one of the oxygen lone pairs is in the π system, but the other is in the perpendicular σ system. The same is true of (c), and the nitrogen lone pair is also in the σ system. In (d), only the lone pair on the NH is in the π system. In (e), each nitrogen contributes an electron pair to the π system, and each boron maintains conjugation through its empty $2p$ orbital.

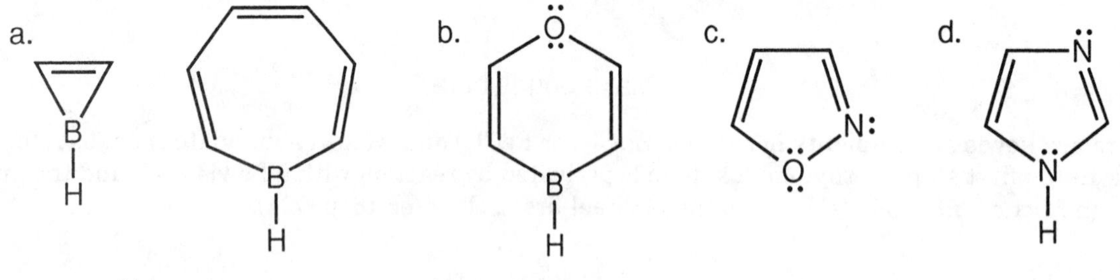

a.

b.

c.

d.

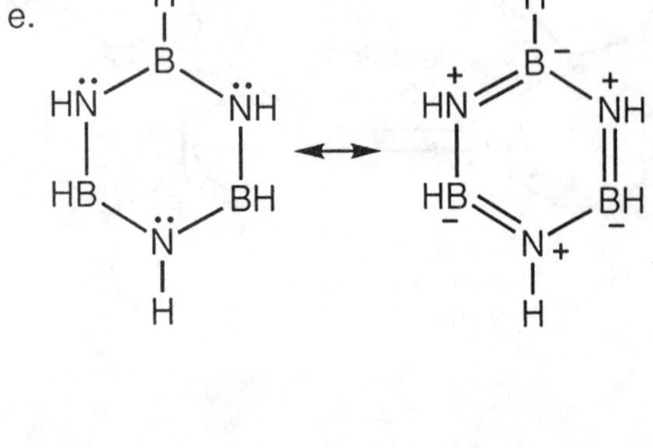

e.

Problem 25.29 Well, it's at least very hard to do what this problem asks. The difficulty is that either there are too many π electrons as in (a) or that conjugation isn't maintained as in (b). The question doesn't say "only" oxygen and nitrogen, so a possible answer is to introduce a boron to complete the conjugation, as in (c). Now there are six π electrons in a fully conjugated system. The two electrons on nitrogen and one electron pair on oxygen are in the σ system.

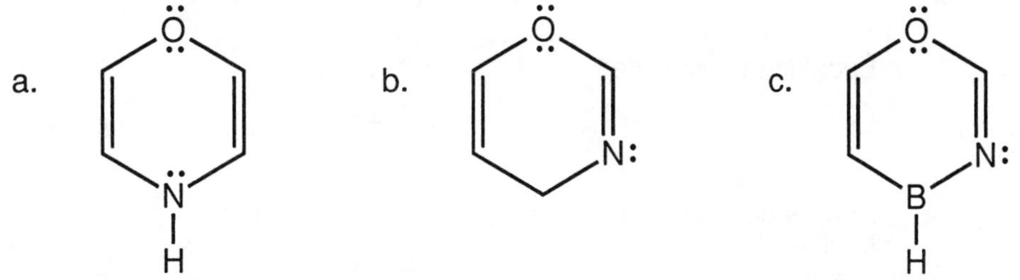

a. Eight π electrons

b. Not fully conjugated

c.

Problem 25.30 Both azirine isomers would surely be highly strained, as the three-membered ring bond angles will be far from the ideal sp^3 or sp^2 angles. In addition, isomer **1** will be destabilized by the presence of two electrons in antibonding orbitals (see Problem 25.9). By contrast, isomer **2** is not even a conjugated system, as the sp^3 carbon does not maintain the required orbital connectivity. Moreover, the nitrogen lone-pair electrons are in an approximately sp^2 orbital perpendicular to the π system. Thus, the two π electrons of **2** are nicely accommodated in the π bonding molecular orbital.

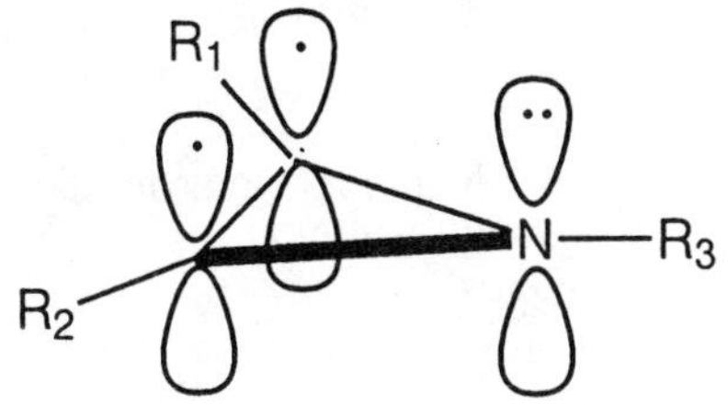

1

4 π electrons

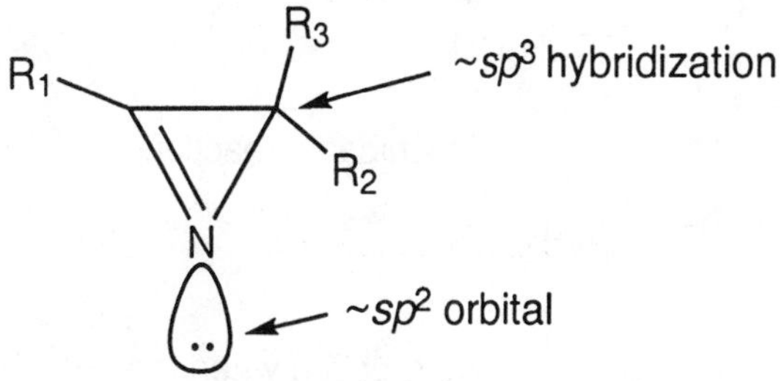

2

~sp^3 hybridization

~sp^2 orbital

Problem 25.31

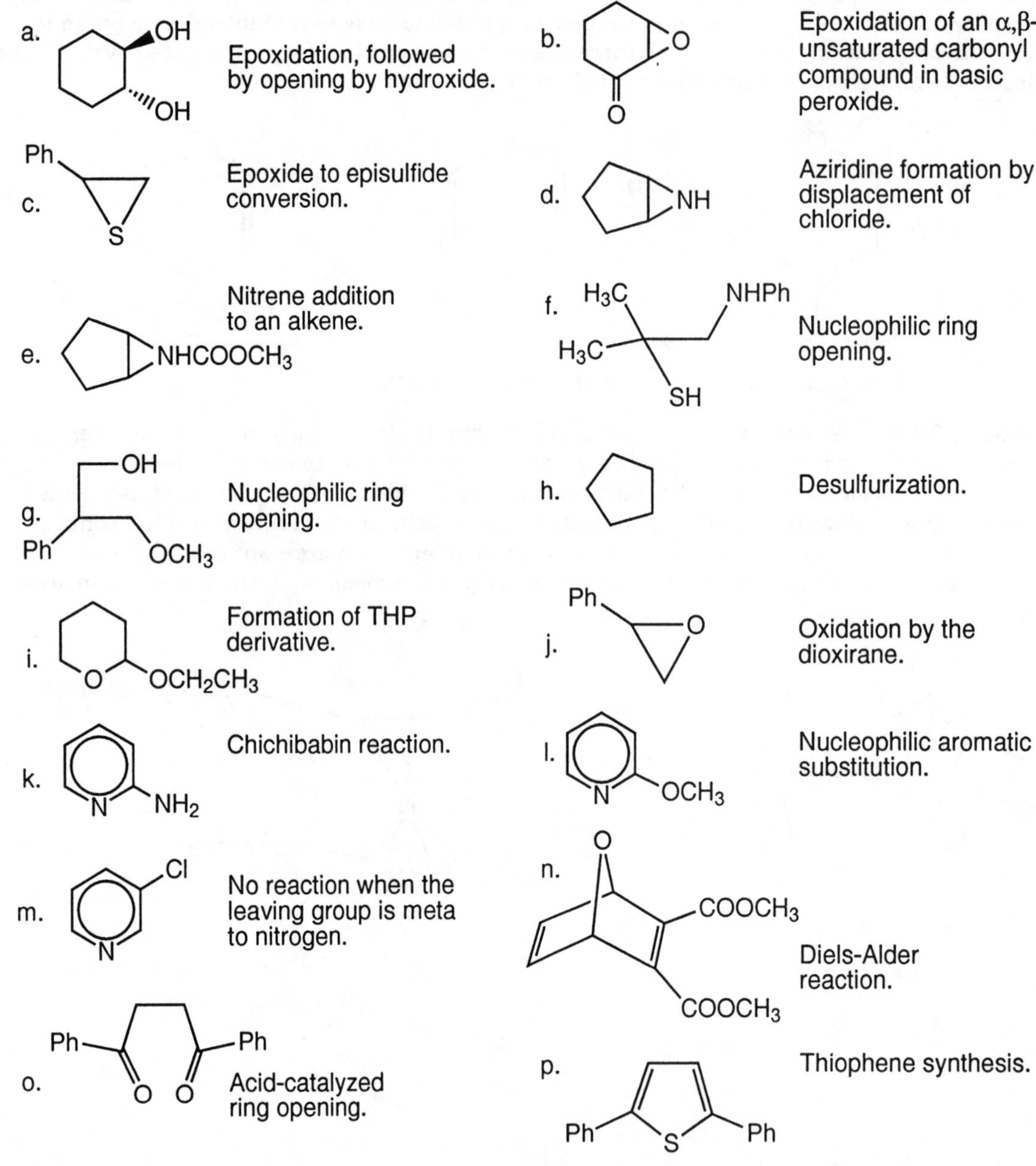

Problem 25.32 Once again, there are several possibilities. Here are some suggestions.

(a) Simple epoxidation with trifluoroperacetic acid would do the trick.
(b) The epoxide can be converted into the episulfide through reaction with thiocyanate. Remember the ring-opening, ring-closing, ring-opening, ring-closing sequence of the mechanism (Figs. 25.7 and 25.8).

CF_3COOOH → O → KSCN / H_2O → S

(c) There is a variety of possibilities. Treatment with the pseudohalogen IN_3, followed by reduction with $LiAlH_4$, is one method.

1. IN_3
2. $LiAlH_4$

N
H

(d) Photolysis of diazomethane in 2,3-dimethyl-2-butene will do the job through a carbene (divalent carbon, $:CH_2$) addition reaction.

CH_2N_2
$h\nu$

CH_2

Problem 25.33 The dipole moment of THF reflects the relative electronegativity of oxygen. Of course, the oxygen in furan is also more electronegative than carbon, but this effect is attenuated by resonance forms such as **A** in which the dipole is in the other direction. The net result is a reduced dipole.

δ^+ δ^- 1.7 D

A

δ^+ δ^- 0.7 D

Problem 25.34 Protonation of the NH leads to a simple ammonium ion, in which aromaticity is lost, whereas protonation of the doubly bonded N leads to a more stable, resonance-stabilized species. The transition state for protonation of this nitrogen will be partially delocalized and thus stabilized relative to that for protonation of the NH.

$H_3\overset{+}{O}$ faster

1

$H_3\overset{+}{O}$ slower

Problem 25.35 Ring opening by the nucleophilic thiocyanate ion is followed by decarboxylation to give **A**.

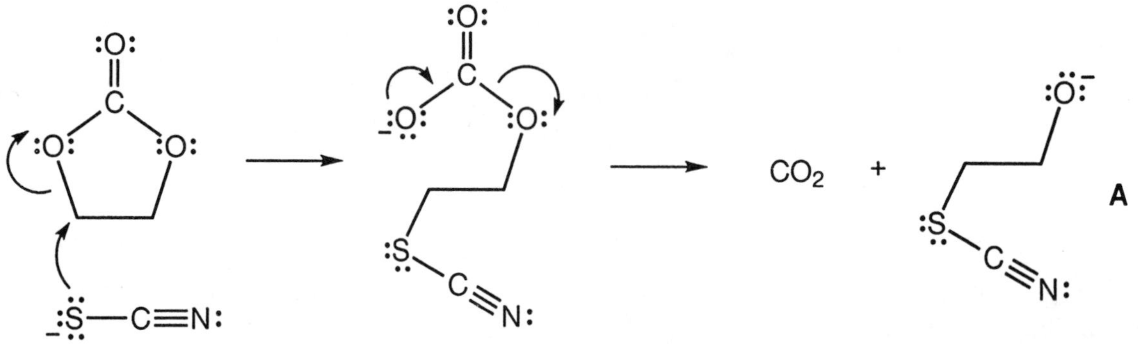

CO_2 + **A**

Oxyanion **A** is exactly the same intermediate as appears in Figures 25.7 and 25.8. Addition to the CN group is followed by opening to **B** and displacement of cyanate to give the episulfide.

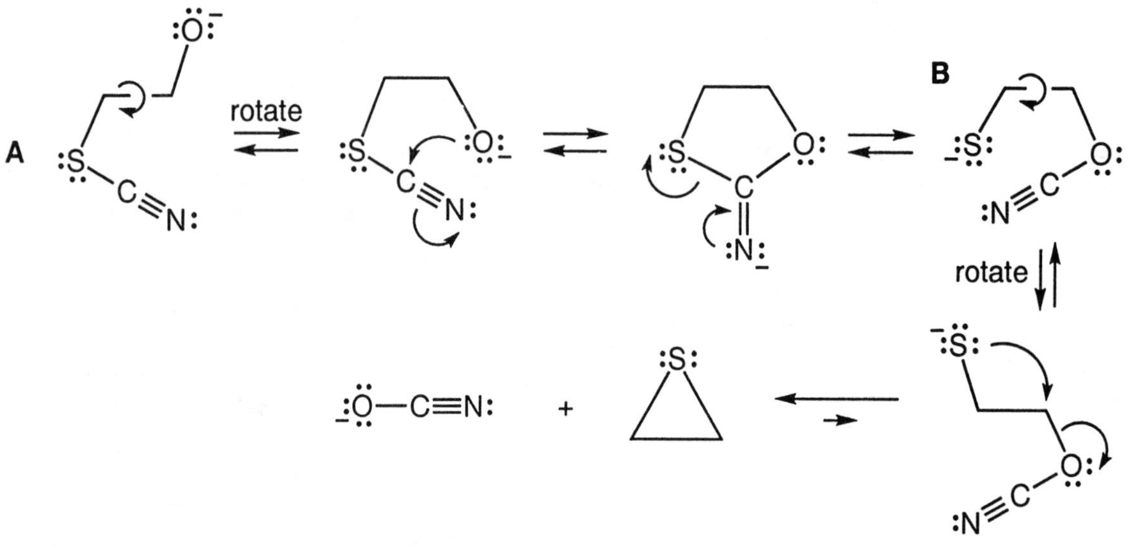

rotate

rotate

Problem 25.36

(a) This problem may be quite hard because it is so indirect. Perhaps the crucial insight is that a desulfurization must be involved (the sulfur atom is gone in the product, after all). What compound can be desulfurized to give the product? The answer to this question requires that you remember that desulfurization of thiophenes with Raney nickel also reduces the double bonds (Fig. 25.64).

So, the retrosynthetic analysis recognizes that the final product comes from Raney nickel desulfurization of 2-acetylthiophene, a molecule that can be made from thiophene itself.

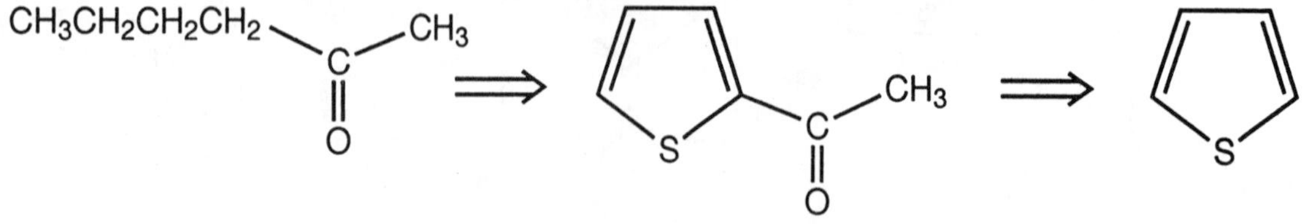

$CH_3CH_2CH_2CH_2COCH_3$ ⟹ 2-acetylthiophene ⟹ thiophene

The synthesis in a forward direction, with reagents, follows:

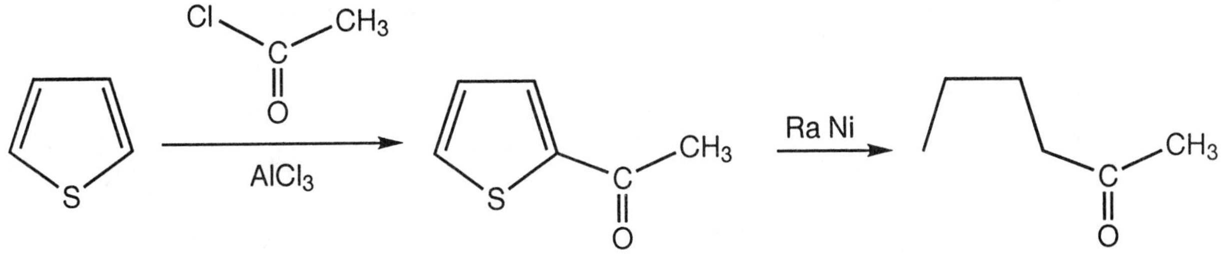

Cl—C(=O)—CH_3, $AlCl_3$; Ra Ni

(b) This part is easier. Still, it requires you to imagine that quinoline, like pyridine, can undergo the Chichibabin reaction. Quinoline can be easily made from aniline through a Skraup synthesis.

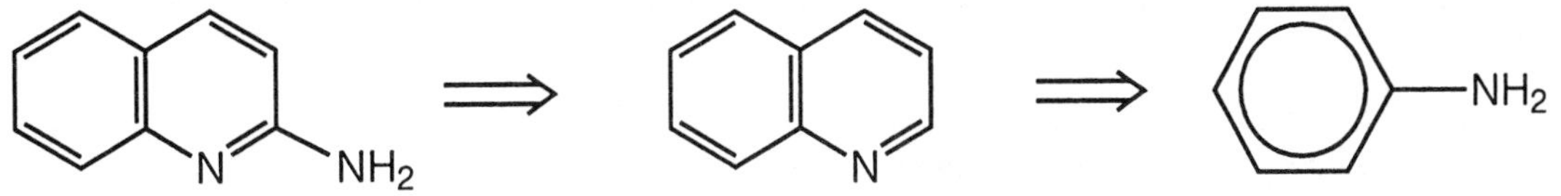

NH_2 ⟹ ⟹ NH_2

Here is the synthesis in a forward direction.

OH, HOH_2C, CH_2OH; NH_2; H_2SO_4, heat; $Fe_2(SO_4)_3$, nitrobenzene; 1. KNH_2/NH_3 2. H_2O; NH_2

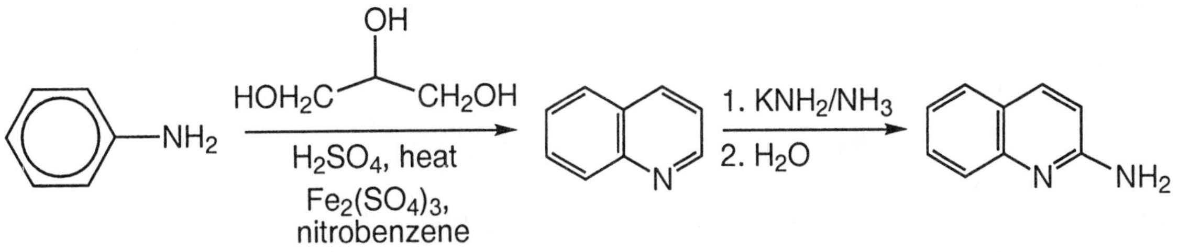

Problem 25.37 At least compound **A** should be easy. A straightforward Chichibabin reaction, right from Figure 25.47, makes 2-aminopyridine. Now you have to recall what nitrous acid does with aromatic amines. It produces the diazonium salts, in this case **B**. Treatment with CuBr leads to 2-bromopyridine in a Sandmeyer reaction.

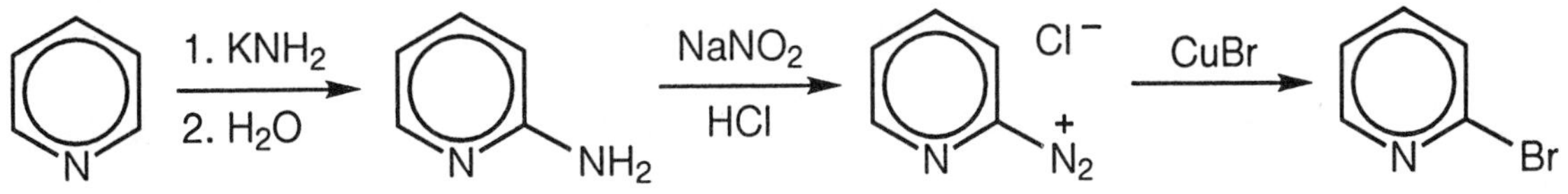

1. KNH_2 2. H_2O; NH_2 **A**; $NaNO_2$, HCl; Cl^-, $\overset{+}{N_2}$ **B**; CuBr; Br **C**

Problem 25.38 In a way, this problem is mindless; there is no real intellectual challenge. Still, it emphasizes the need to keep all the synthetic possibilities of the various "name" procedures of heterocyclic chemistry in mind.

(a) The Fischer indole synthesis involves the acid-induced cyclization of a phenylhydrazone. To get the products, one must only position the substituents correctly on the hydrazine and the carbonyl fragments.

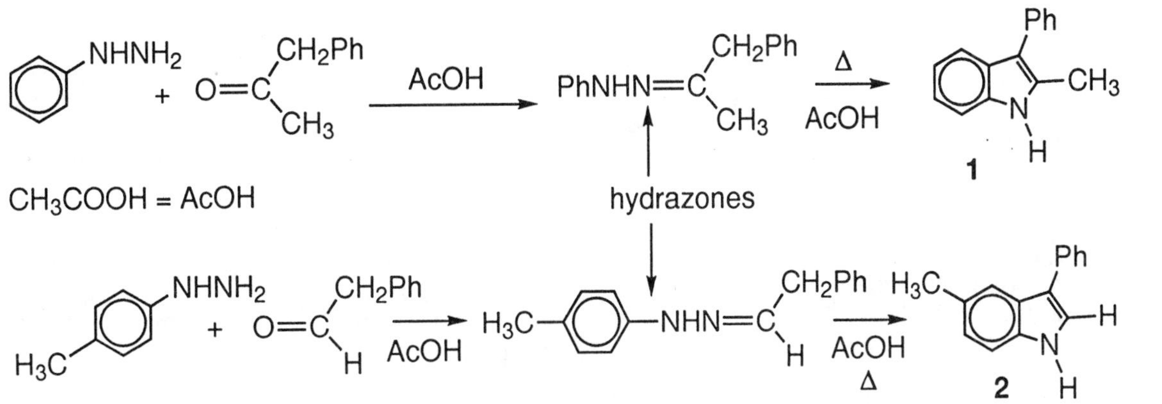

$NHNH_2$ + O=C(CH_2Ph)(CH_3) —AcOH→ PhNHN=C(CH_2Ph)(CH_3) —Δ, AcOH→ **1** (Ph, CH_3, N–H)

CH_3COOH = AcOH

hydrazones

H_3C–C$_6$H$_4$–$NHNH_2$ + O=C(CH_2Ph)(H) —AcOH→ H_3C–C$_6$H$_4$–NHN=C(CH_2Ph)(H) —AcOH, Δ→ **2** (H_3C, Ph, H, N–H)

(b) In the Reissert synthesis, an ester is first made and then decarboxylated. A condensation reaction first leads to **A**. Reduction then converts the nitro group into an amine, and a further condensation, followed by decarboxylation, gives **3**.

CH_3 COOR COOR NO_2 $\xrightarrow[ROH]{RO^-}$ CH_3 C COOR NO_2 O **A** $\xrightarrow{H_2/Pt}$ CH_3 COOR N H $\xrightarrow{H_2O,\ H_3O^+}$ CH_3 N H **3**

(c) The Skraup synthesis uses a reaction of a glycerol with an aniline to give a quinoline. The role of the glycerol is to produce an acrolein that can undergo a Michael reaction with the aniline. The initial product is **A**, and oxidation, usually by nitrobenzene or iron, leads to the quinoline.

OH OH OH Ph $\xrightarrow{H_2SO_4}$ O CH Ph

H_3C NH_2 $\xrightarrow{H_2SO_4}$ H_3C Ph N H **A** $\xrightarrow[Fe_2(SO_4)_3]{PhNO_2}$ H_3C Ph N **4**

(d) The Hantzsch synthesis uses ammonia as the source of nitrogen, and β-keto esters and an aldehyde as the blocks from which the pyridine ring is constructed. The initial product is **A**, and decarboxylation leads to **5**.

ROOC C Ph O Ph CHO COOR O C Ph NH_3 $\xrightarrow{R_2NH}$ ROOC Ph COOR Ph N Ph **A** $\xrightarrow[heat]{KOH}$ Ph Ph N Ph **5**

Problem 25.39

$\xrightarrow{I\text{-}NCO}$ **A** I NCO $\xrightarrow[HOCH_3]{NaOCH_3}$ **B** I O C N H OCH_3

A comes from anti addition of the pseudohalogen to the alkene.

Methanolysis of the isocyanate gives the carbamate ester.

$\xrightarrow{KOH/H_2O}$ I O C N H O^- $\xrightarrow{-CO_2}$ I $\underline{N}H$ $\longrightarrow$ **C** NH $\xrightarrow{NH_2CH_3}$ **D** NH_2 $NHCH_3$

Opening of **C** by the nucleophilic methylamine leads to **D**. Note trans stereochemistry.

Hydroxide hydrolyses the carbamate ester to an unstable carbamic acid anion that decarboxylates to give an intermediate amide that undergoes an intramolecular S_N2 displacement of iodide to give aziridine **C**.

Problem 25.40 This reaction is a variation of the Darzens condensation. Here it is used to make an aziridine. There is only one enolate possible, that formed from ethyl chloroacetate. If you count the carbons in the starting material and product, you can see that the product is a combination of the two starting molecules. So, the only reaction possible seems to be the addition of the enolate to the carbon–nitrogen double bond to give intermediate **A**.

$ClCH_2COOCH_2CH_3$ ⇌ ($(CH_3)_3COK$, enolate formation) $Cl\ddot{C}^{-}HCOOCH_2CH_3$ + PhC=NPh ⇌ (addition) **A**

The newly formed amide ion in **A** is now in perfect position to displace chloride and give the product aziridine. The trans product is more stable than the corresponding cis compound.

A → (S_N2) trans-aziridine (N-Ph, Ph, $COOCH_2CH_3$)

Problem 25.41 This reaction involves a Diels–Alder cycloaddition with the nitrile acting as dienophile. In this case, two initial cycloadducts are possible because both the nitrile and the diene are unsymmetrical. Loss of carbon monoxide from the primary adducts gives the two observed pyridines.

Diels-Alder reactions

Reverse Diels-Alder reactions

+ CO

+ CO

Problem 25.42 A quick check of the molecular formula of compound **3** reveals that it is a 1:1 adduct of **1** and **2**. We are most likely dealing with a Diels–Alder reaction here. However, 1,3-diphenylisobenzofuran possesses two different diene systems. Which one will react with the dienophile **2**? Reaction of **2** and the furan diene would lead to **A**, whereas reaction with the cyclohexadiene would give **B**. In the formation of **A** a benzene ring is generated, but in **B** it is not. We might well expect the path to **A** to be preferred. In addition, **A** has the proper carbon framework to give **4**, but **B** does not.

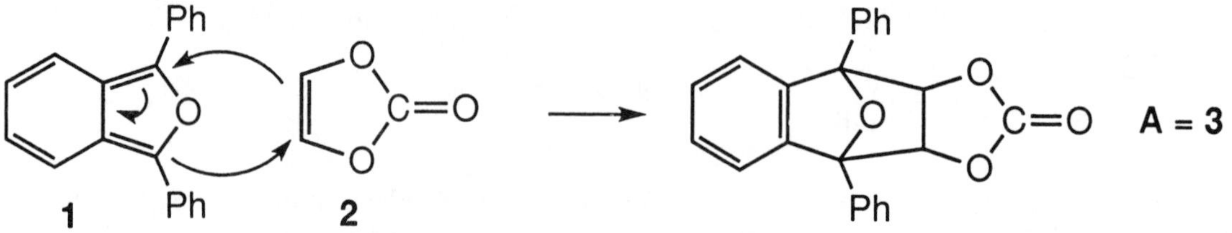

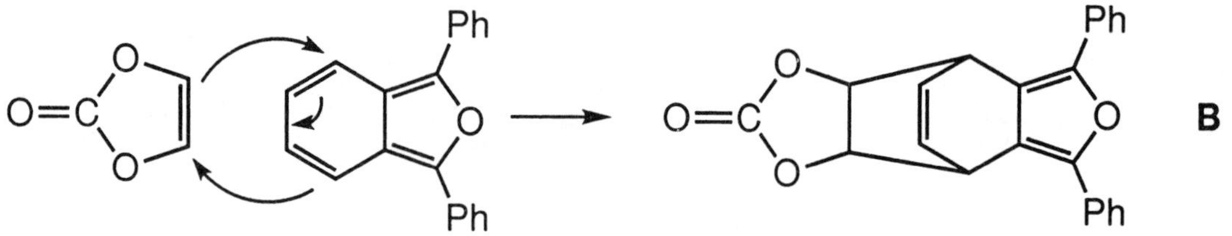

In fact, compound **A** is formed, mainly in its endo form, although small amounts of the exo compound are also isolated.

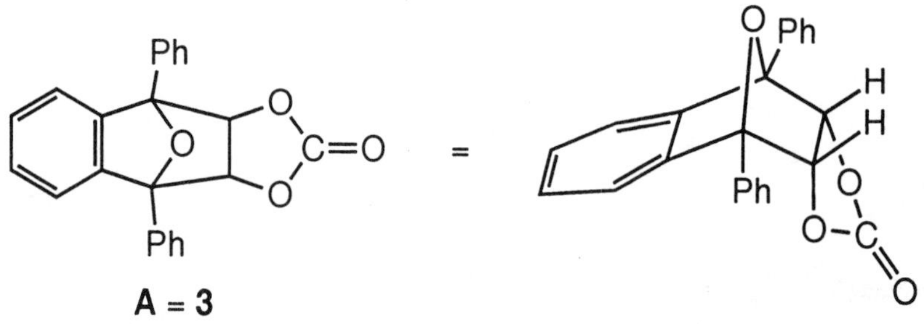

The transformation of cycloadduct **3** to naphthalene **4** involves aromatization and hydrolysis of the carbonate. These reactions can occur in either order, but let's take the aromatization first. Protonation of the bridge oxygen, followed by ring opening, gives carbocation **C**. Deprotonation gives one of the required carbon–carbon double bonds. Now, protonation of the alcohol, followed by dehydration, completes formation of the aromatic compound **D**.

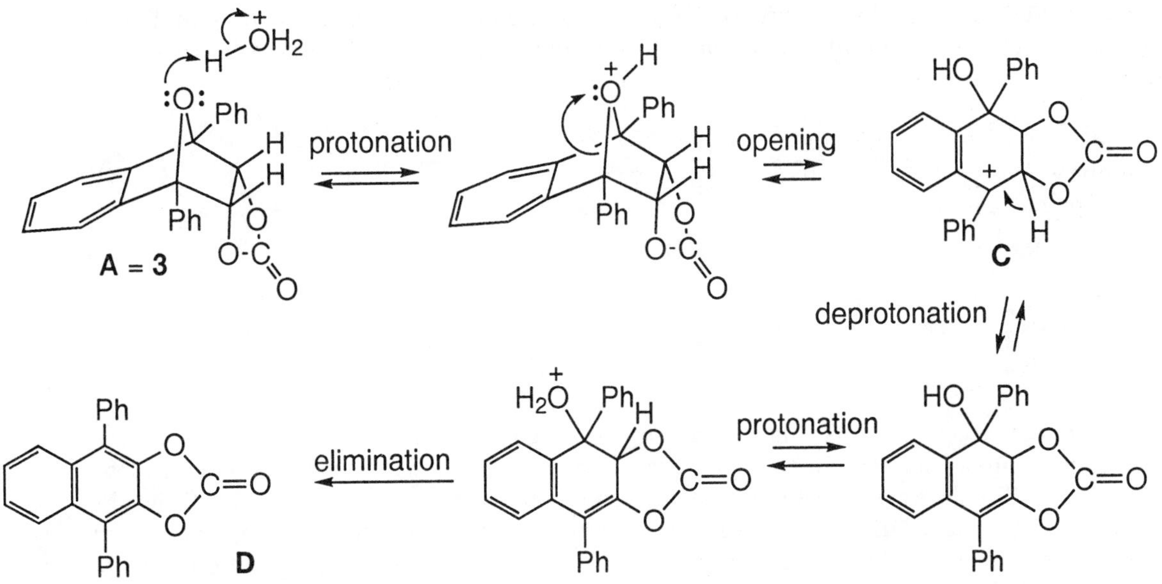

Now hydrolysis of the carbonate, a dazzling sequence of protonations, deprotonations, and additions of water, generates the final product, **4**.

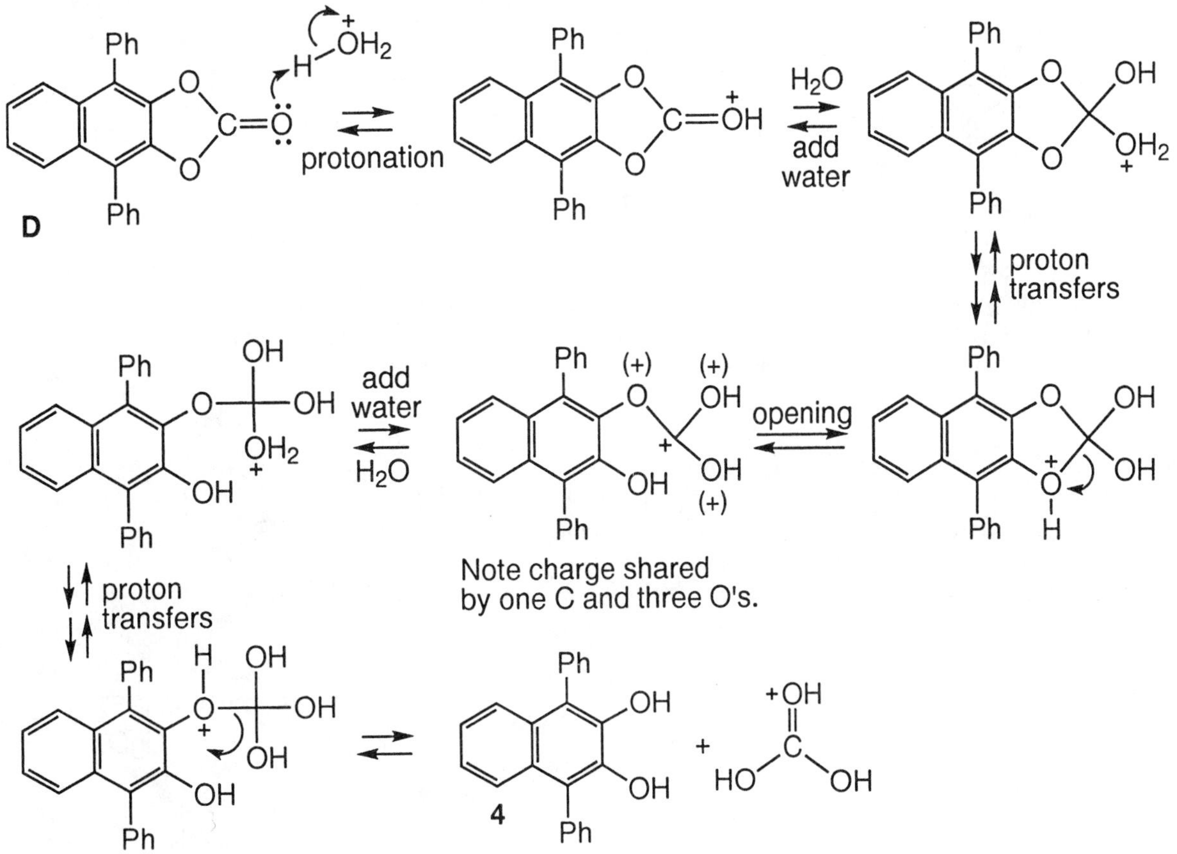

Problem 25.43 The reaction begins with addition of the nucleophilic oxygen of quinoline *N*-oxide to benzoyl chloride. The elimination phase of this addition–elimination process expels chloride and leads to **A**.

addition → elimination → **A**

What has been accomplished? It is an old story. An impossibly bad leaving group has been transformed into a good one. In addition, a good Lewis acid lies waiting for cyanide, a good nucleophile. Why does cyanide add to the 4-position and not to the 2-position? That's hard to answer; perhaps steric inhibition by the benzoyl group is to blame. In any event, the structure of the product tells you that it is the 4-position that is the more reactive one. Addition of cyanide and elimination of the good leaving group completes the reaction.

A —addition→ —elimination→

Problem 25.44 This problem involves formation of a thioacetal and three of its subsequent reactions; alkylation, hydrolysis to a carbonyl compound, and reduction.

CH_3CH_2CHO —$HS(CH_2)_3SH$, H_3O^+→ **A**, $C_6H_{12}S_2$ —alkylation: 1. BuLi, 2. $PhCH_2Br$→ **B**, $C_{13}H_{18}S_2$

B —hydrolysis: H_2O, HBF_4, HgO→ **C**, $C_{10}H_{12}O$ ($CH_3CH_2COCH_2Ph$)

B —Ra Ni, reduction→ **D**, $C_{10}H_{14}$ $CH_3CH_2CH_2CH_2Ph$

Problem 25.45 Notice that the desired aziridine **2** has a molecular formula of $C_{12}H_{17}N$. Accordingly, isomers **3** and **4** have incorporated the additional elements of SO_2. If we examine the chlorosulfite intermediate for this reaction, it is apparent that a five-membered heterocycle **A** with the correct molecular formula is available upon nucleophilic attack of nitrogen on the sulfinyl group to give **B.** (Alternatively, amino alcohol **1** could first react with thionyl chloride via the more nucleophilic amine nitrogen, followed by alcohol cyclization to give **C.**)

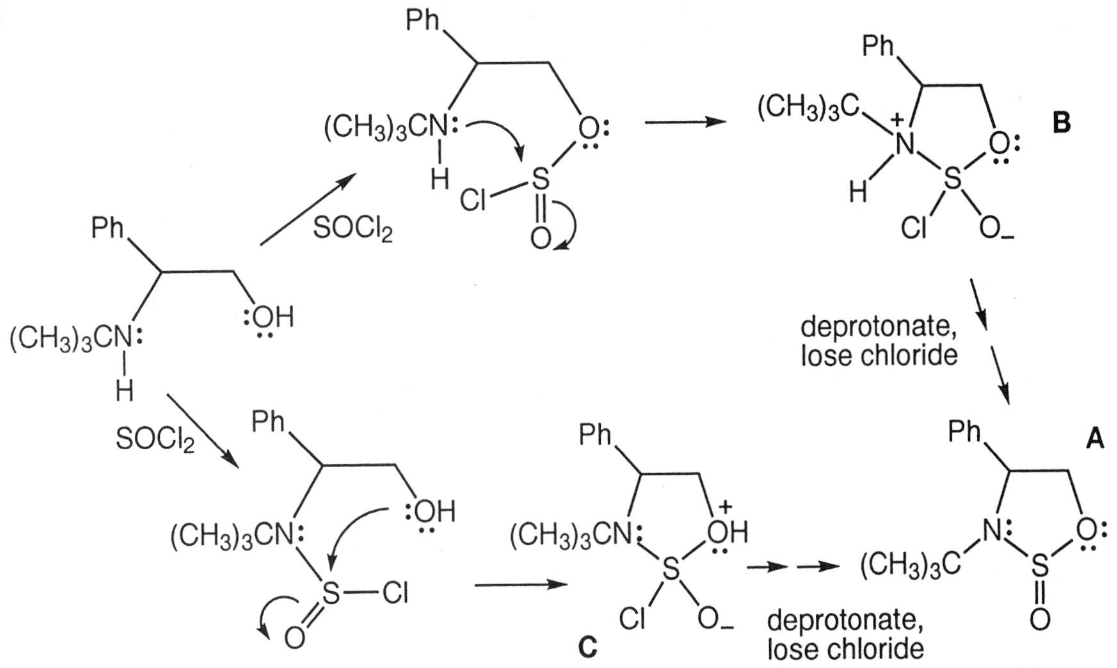

SOCl$_2$ · B · deprotonate, lose chloride · A · SOCl$_2$ · C · deprotonate, lose chloride

So far, so good. But where is the second compound $C_{12}H_{17}NO_2S$? If you examine heterocycle **A** closely, you will notice that there are two stereogenic atoms. The stereogenic carbon is probably obvious, but the second atom, a stereogenic sulfur, is not so clear. However, **A** is actually a pair of diastereomers, **3** and **4.**

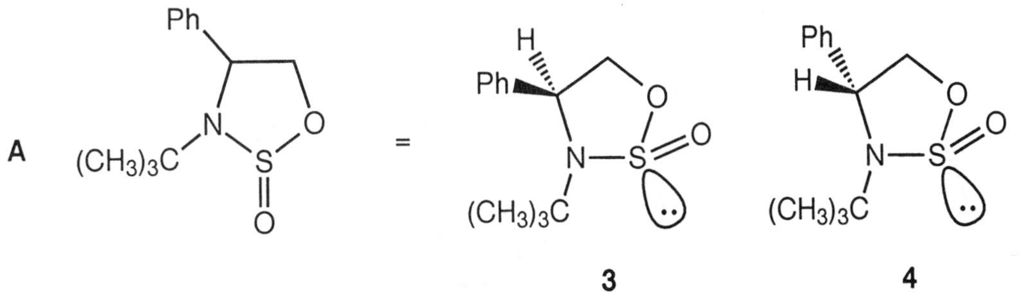

A = **3** **4**

Problem 25.46 The species :CHCl is a carbene, and carbenes make their living in large part by adding to double bonds. In this case, the most susceptible double bond is the one in the five-membered ring. Addition to this double bond does not disrupt the aromatic ring. This reaction gives intermediate **A**. Elimination of HCl leads to the product, and the formation of a second aromatic ring.

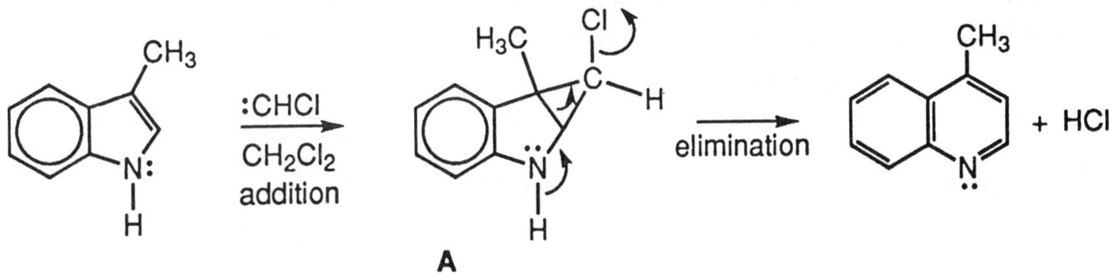

:CHCl, CH_2Cl_2 addition → **A** → elimination → + HCl

Problem 25.47 You know that carbonyl compounds react with the nucleophilic amines to form enamines (Chapter 16, p. 780). This problem is just a twofold example of that reaction. It starts with addition of the amine to one of the carbonyl groups. Dehydration of the carbinolamine gives the enamine.

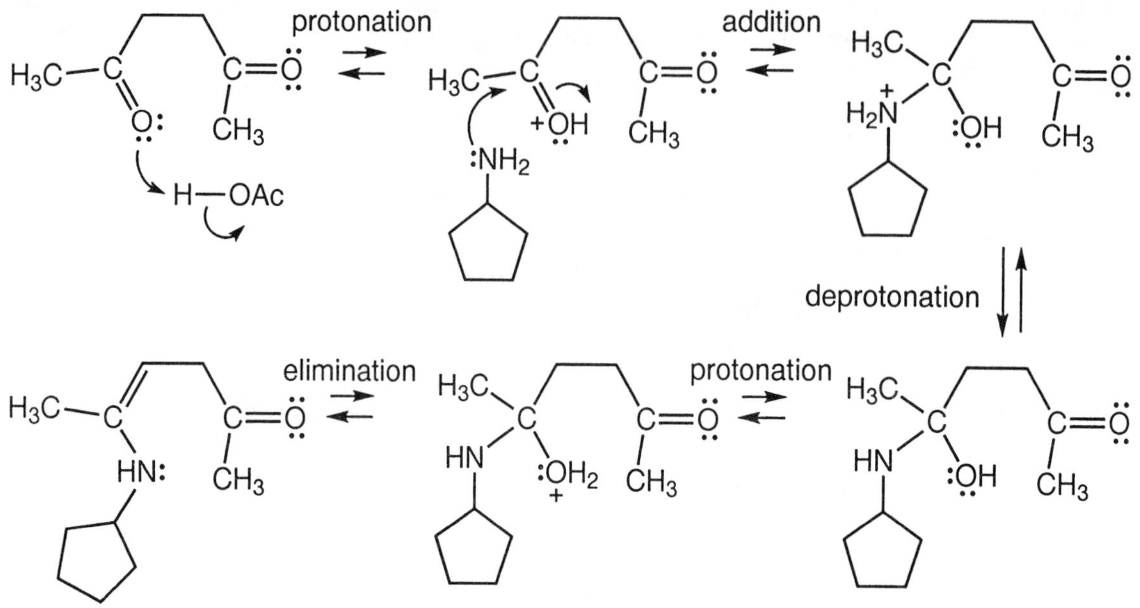

protonation addition deprotonation protonation elimination

Now the steps are repeated, as the intramolecular amine (the enamine) adds to the carbonyl group.

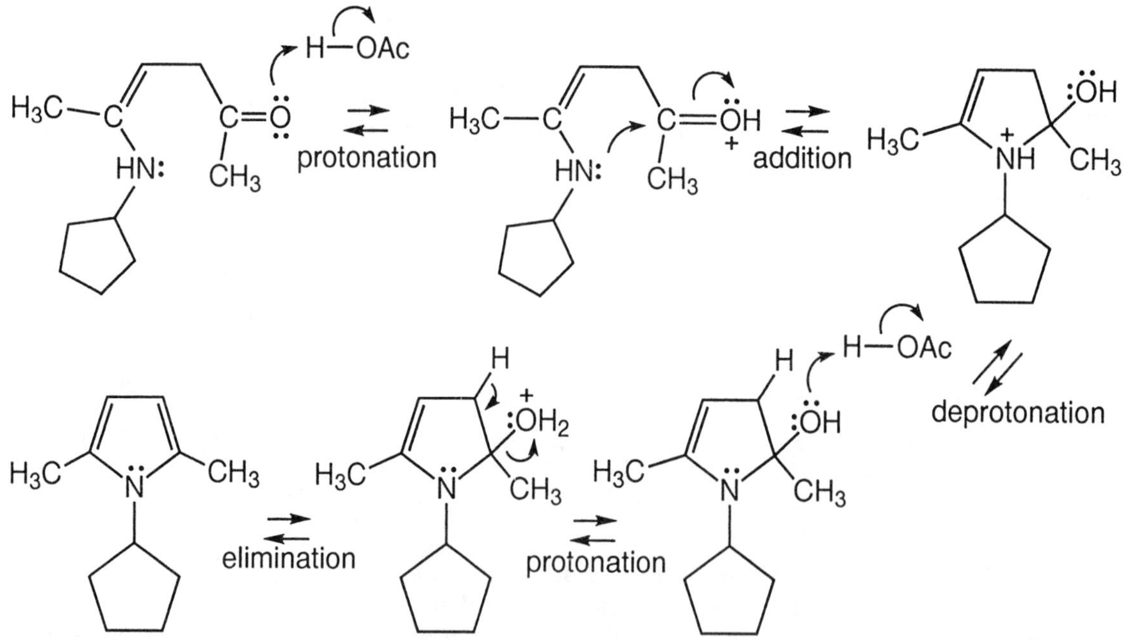

protonation addition deprotonation protonation elimination

Problem 25.48

(a) Deprotonation of oxazolines **1** and **3** is facilitated by the formation of a resonance-stabilized anion in which nitrogen shares the negative charge with carbon.

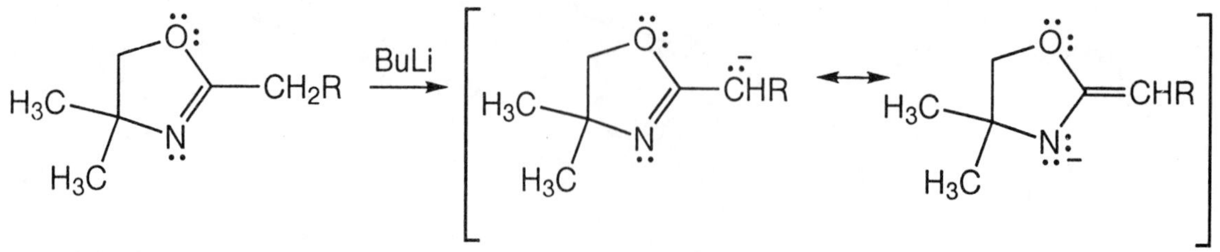

BuLi

(b) This reaction is yet another sequence of protonations, additions of water, and deprotonations.

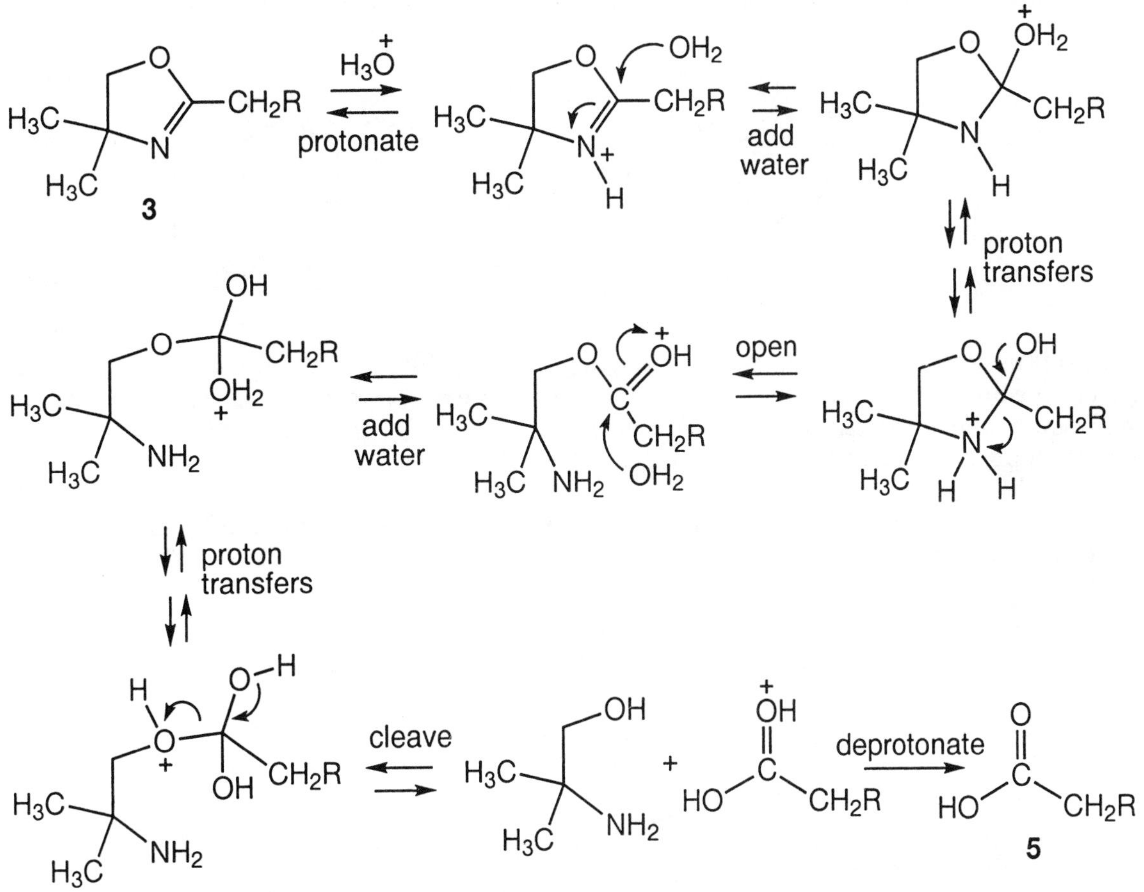

(c) As oxazoline **1** is derived from acetic acid, the sequence represents an indirect method for α-alkylation or α,α-dialkylation of acetic acid. We can accomplish the same type of transformation using the malonic ester synthesis (see Section 20.12).

Problem 25.49 As we saw for pyridine (see Fig. 25.48), the first step in the Chichibabin amination of isoquinoline (**1**) is the formation of **3**. It is this molecule that is observed at −10 °C by ^{1}H NMR spectroscopy.

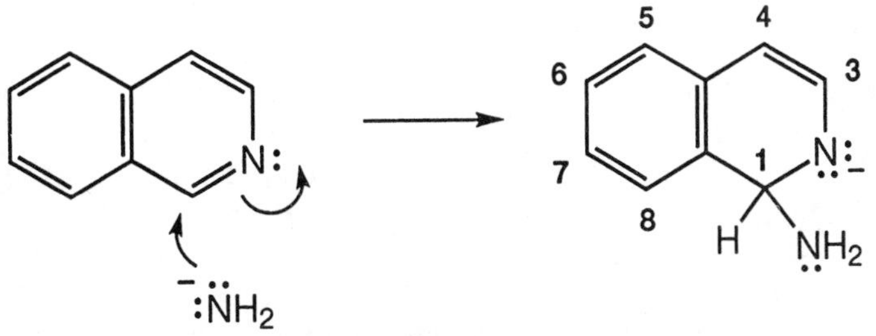

The hydrogen attached to the tetrahedral carbon (H_1) is easily recognized as the triplet at δ 5.34 ppm. It collapses to a singlet in the presence of excess amide ion. Apparently, when the amide ion concentration is low, coupling between H_1 and the hydrogens of the amino group is observed as a triplet. Excess amide ion catalyzes proton transfer between the amino group of **3** and the solvent ammonia, leading to effective decoupling of H_1 and the amino hydrogens (see Chapter 15).

The doublet at δ 4.87 ppm could be the result of either H_3 or H_4 of **3**. Proton H_4 seems the better choice as this hydrogen should be shielded by the adjacent partial negative charge.

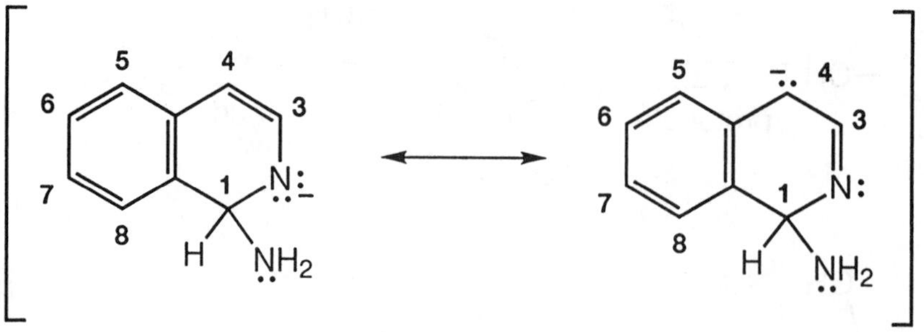

In fact, this assignment was confirmed by studying a 4-substituted isoquinoline. Thus, H_3 must be buried in the broad multiplet associated with the rest of the carbocyclic hydrogens. Notice that when isoquinoline forms anion **3**, the chemical shifts of H_1 and H_4 are moved upfield by about 4 and about 3 ppm, respectively.

Problem 25.50 The first step in this condensation reaction is an acid-catalyzed Michael addition of the enol of methyl ketone **2** to the protonated α,β-unsaturated ketone **1**. This reaction ultimately leads to 1,5-diketone **A**.

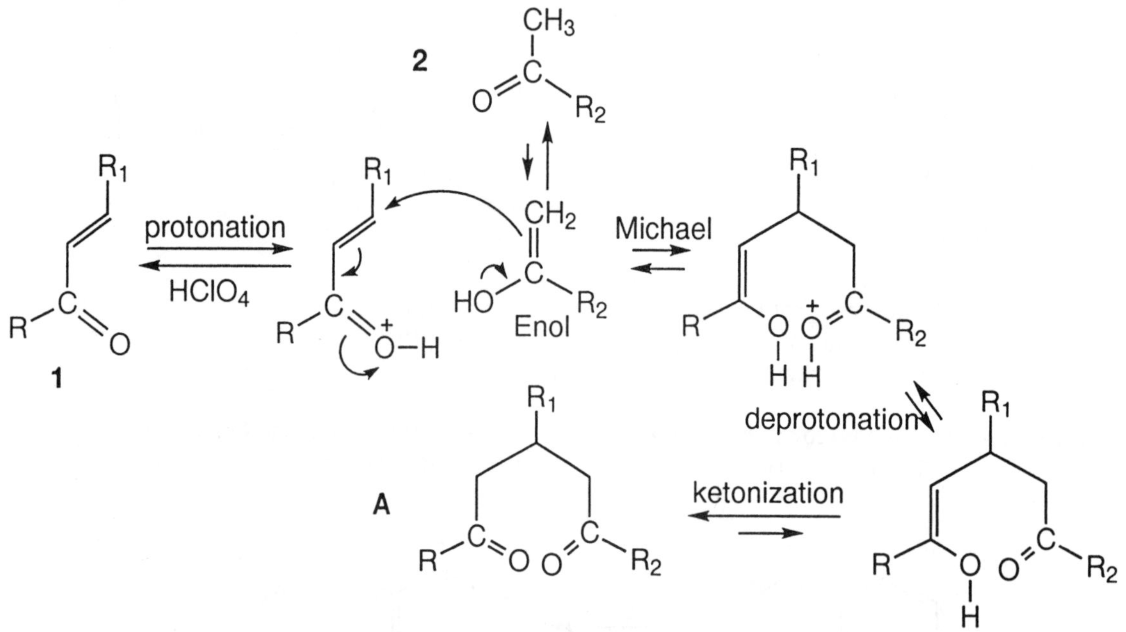

We now need to convert **A** into pyrilium salt **3**. Two things need to be accomplished in this transformation. One is obvious, but the other is more subtle. The obvious step is a cyclization with dehydration—we clearly must close a ring somehow if we are to make **3**. The other step is an oxidation. Unfortunately, we need to accomplish the subtle step before we can do the easy one. The oxidation can best be accomplished by a hydride transfer from **A** to the conjugate acid of **1**. This transfer gives carbocation **B**, the key to the cyclization step, and ketone **5**.

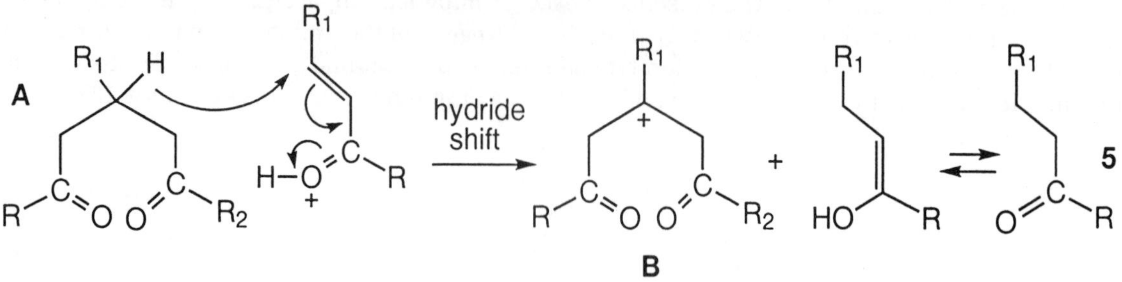

The enol of carbocation **B** leads to **3** through a cyclodehydration.

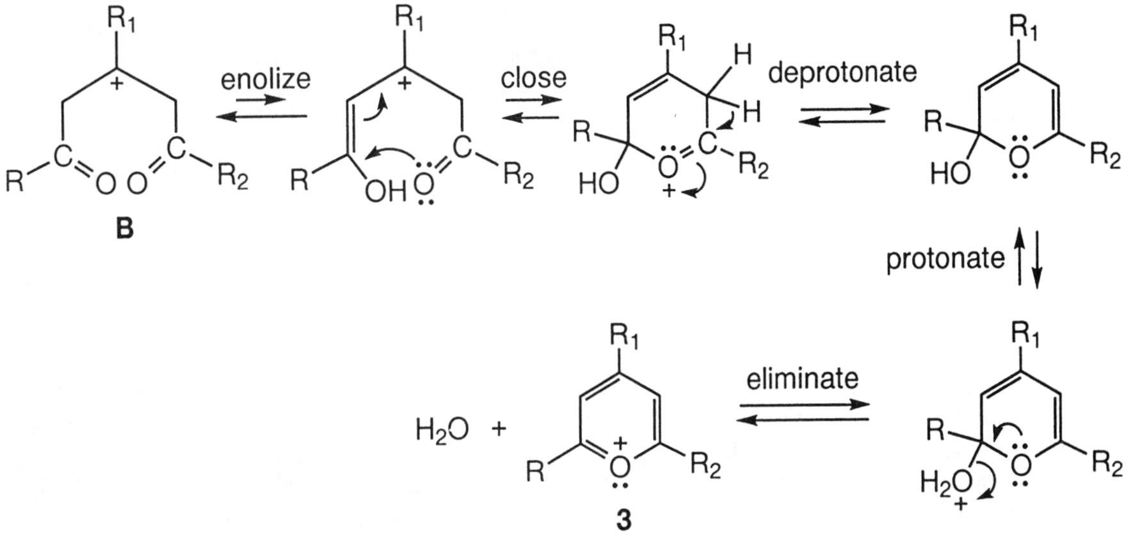

Now the pyrilium salt **3** is easily converted into pyridine **4** as shown below, in a classic "open–close" process. Ammonia first adds to C_2 and then the ring-opening, ring-closing sequence takes place. The final steps are a dehydration and deprotonation.

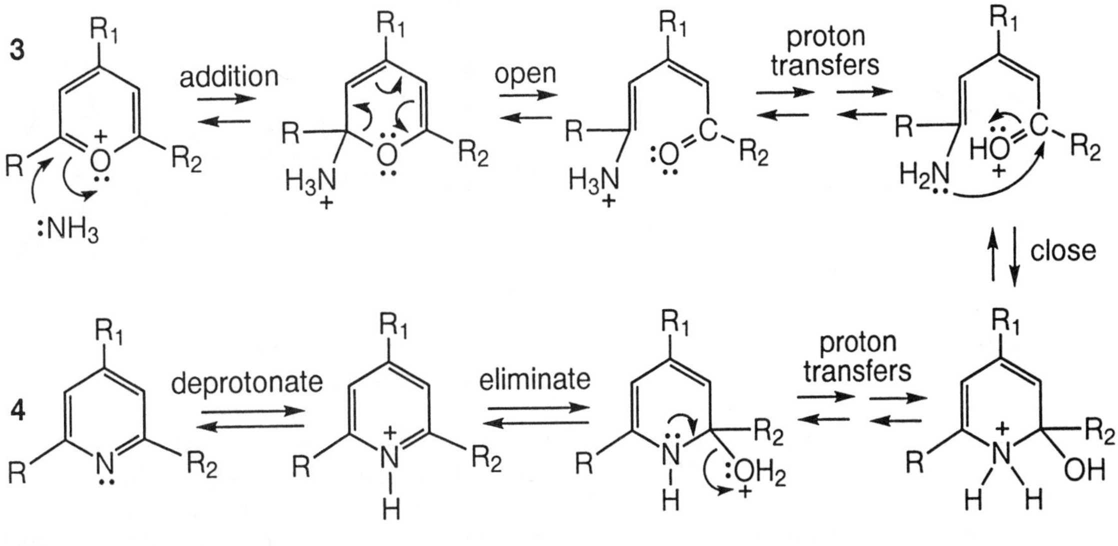

Problem 25.51

(a) This reaction was originally thought to involve a 1,3-dipolar cycloaddition of the *N*-oxide and the isocyanate C=N to give initially the 1,2-dihydropyridine **A**. Aromatization, accompanied by decarboxylation, would give the observed 2-anilinopyridine (**3**). The base (B^- in the drawing) in this step is presumably the oxygen of the pyridine *N*-oxide (**1**).

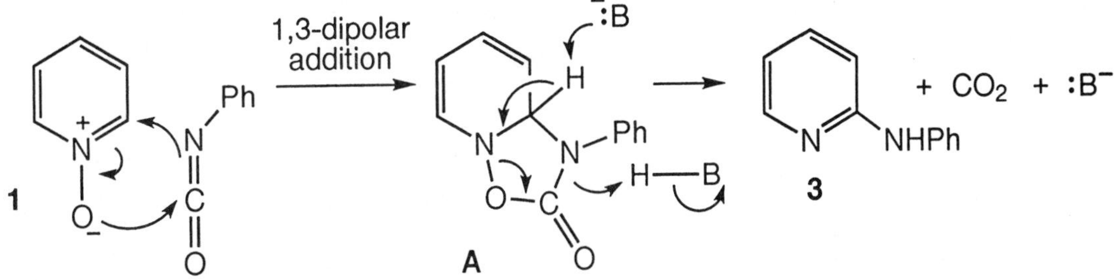

(b) Based on the mechanism of (a), it is not at all unreasonable to assume that the 1:1 adduct **5** is **B**, in direct analogy to **A** in the answer to (a). In fact, this was the structure assigned in the early literature on this reaction. Reduction of **B** by sodium borohydride might give $C_{14}H_{16}N_2O_2$ if a carbon–carbon double bond could be reduced. One possibility is shown below.

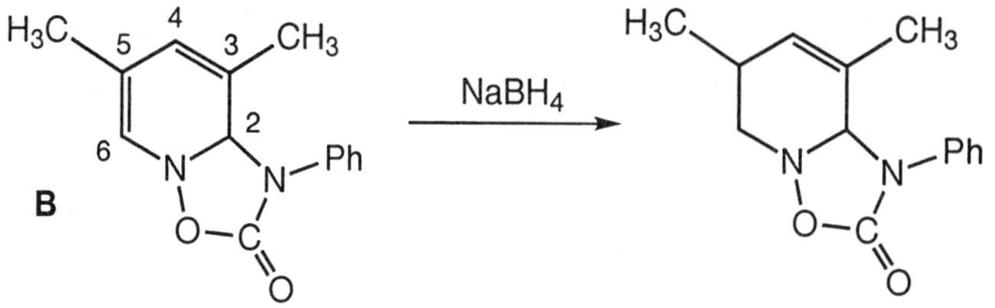

Could this be **7**?

However, a close examination of the ^1H NMR data revealed some problems with the proposed structure, **B**. For example, note that one of the methyl signals is a doublet (J = 1.8 Hz), whereas the other is a singlet. The small coupling of only 1.8 Hz suggests allylic coupling with either H_4 or H_6. Why should only one methyl group of **B** show allylic coupling? *Both* methyls should show allylic coupling. In addition, the sodium borohydride reduction is troublesome. Sodium borohydride will not reduce carbon–carbon double bonds, and the reaction shown in the figure above should not happen.

How do we rationalize these discrepancies? It has been proposed that the initially formed 1,2-dihydropyridine **B** undergoes a 1,5-sigmatropic shift of oxygen to give the 2,3-dihydropyridine **5**. Reduction of **5** with sodium borohydride would then give **7**.

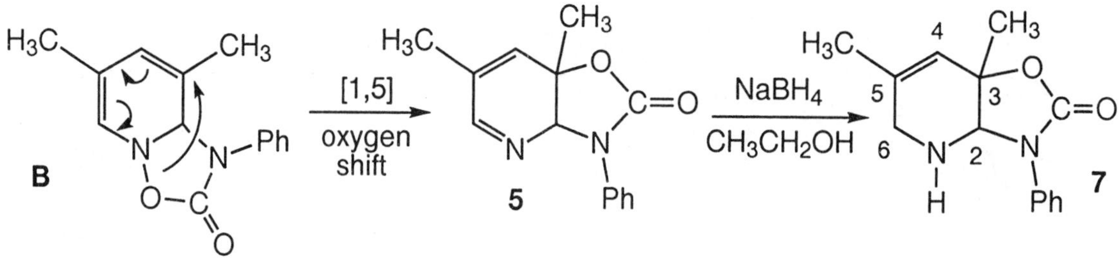

Note that the C_3 methyl group of **5** should appear as a singlet in the ^1H NMR spectrum, and reduction of the carbon–nitrogen double bond, not a carbon–carbon double bond, gives **7**.

(c) The mechanism in (a) should be revised to include rearrangement of the 1,2-dihydropyridines **B** to the 2,3-dihydropyridines **C** (if R = CH_3, **C** = **5**). Calculations have suggested that the 2,3-dihydropyridines **C** are as much as 35 kcal/mol more stable than the corresponding 1,2-dihydropyridines **B** largely because of the greater bond strength of the carbon–oxygen bond versus the nitrogen–oxygen bond. When R = H or CH_3, base-induced decarboxylation [path (a), the base can be the pyridine *N*-oxide or hydroxide] yields the 2-anilinopyridines **3** and **6**. When R = Br, dehydrobromination occurs [path (b)] to give the oxazolopyridine **9**.

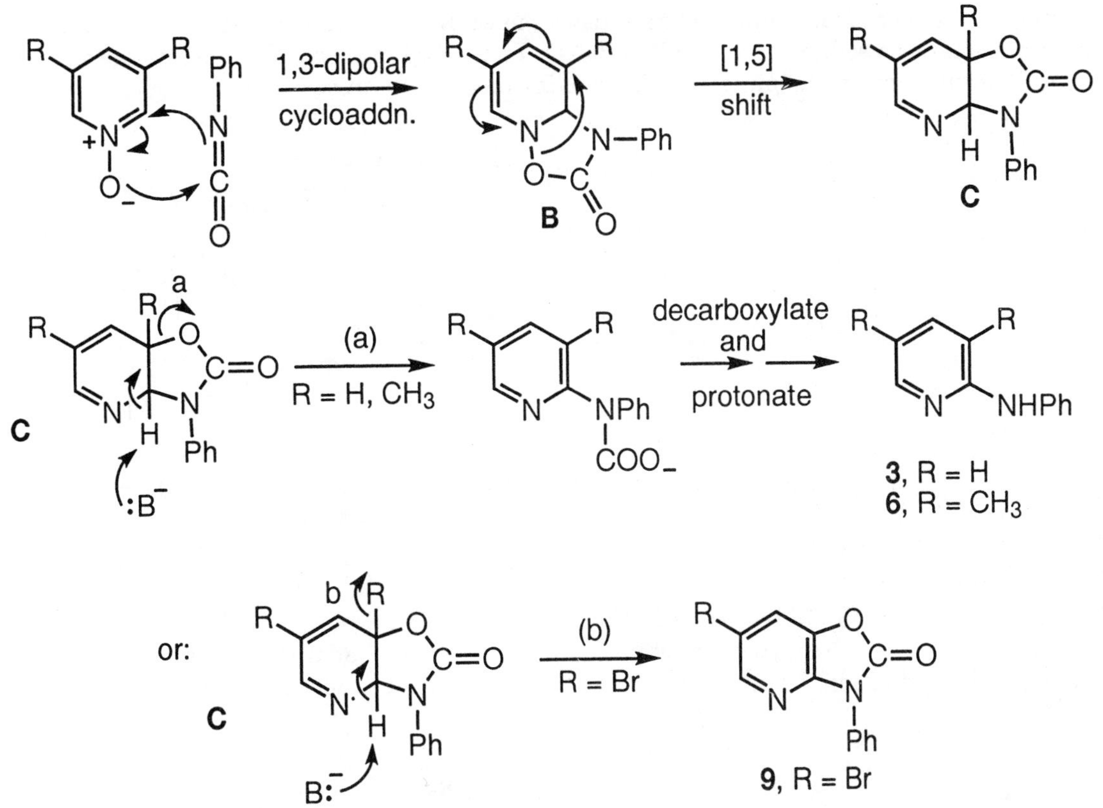

1,3-dipolar cycloaddn.
[1,5] shift
B
C

C
(a) R = H, CH3
decarboxylate and protonate
3, R = H
6, R = CH3

or:
C
(b) R = Br
9, R = Br

Problem 25.52 As benzoylacetone (**1**) has two carbonyl groups, we first need to decide which carbonyl will react with hydroxylamine. The higher energy carbonyl group is the one with the methyl substituent. (The other carbonyl group is lower in energy because it is conjugated with the adjacent benzene ring.) So, let's start with formation of oxime **A** by reaction with the higher energy carbonyl.

proton transfers
deprotonation
A

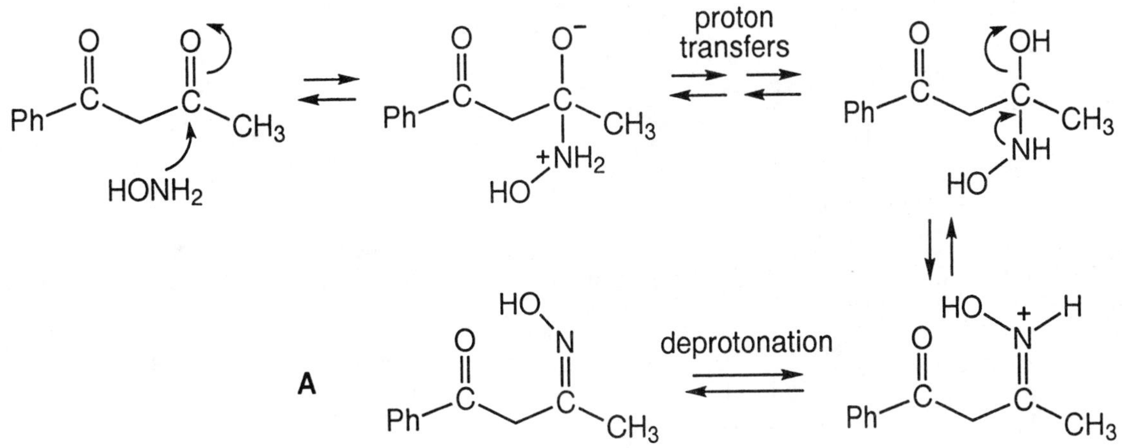

Now what can happen? Notice that oxime **A** has a molecular formula of $C_{10}H_{11}NO_2$, whereas compound **X** has a molecular formula of $C_{10}H_9NO$. Thus, H_2O must be lost in a "cyclodehydration" reaction. The oxime oxygen is perfectly poised to initiate such a process.

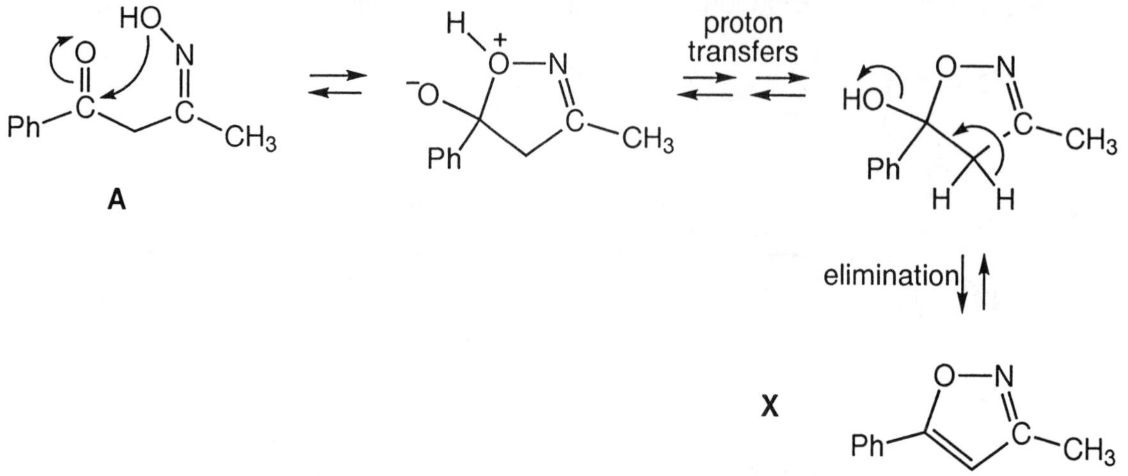

Thus, compound **X** is reasonably assigned 3-methyl-5-phenylisoxazole. However, given the available information, the isomeric 5-methyl-3-phenylisoxazole (**B**) is not a bad answer either. This molecule would be produced if you had started by making the oxime of the other carbonyl group of **1**.

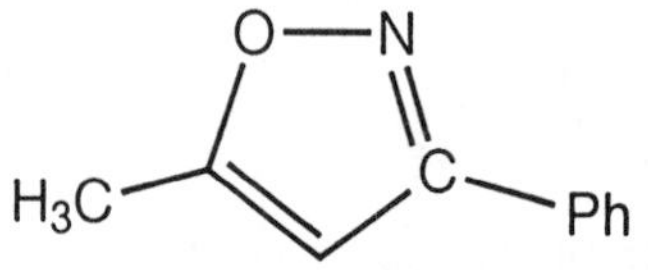

Problem 25.53 It is also possible to rationalize the observed regiochemistry of this reaction in a slightly different manner. Recall that β-diketones, such as benzoylacetone, exist largely in their enol forms in solution. In the case of benzoylacetone (**1**), two equilibrating enol forms are possible, **C** and **D**.

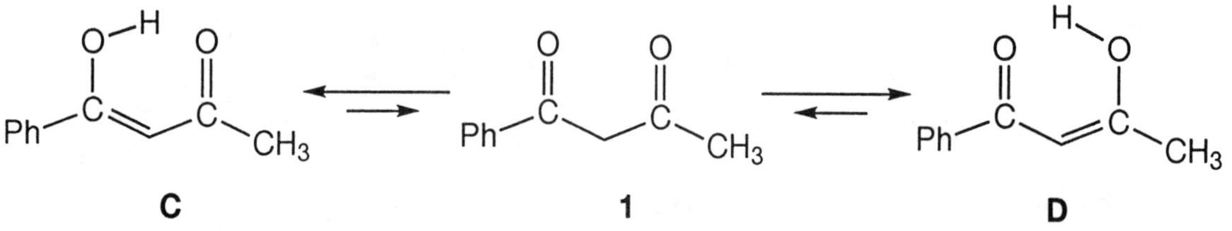

Michael addition of hydroxylamine to the more sterically accessible carbon–carbon double bond of **D**, followed by proton transfers and dehydration, leads to oxime **A**. Now the mechanism proceeds as before.

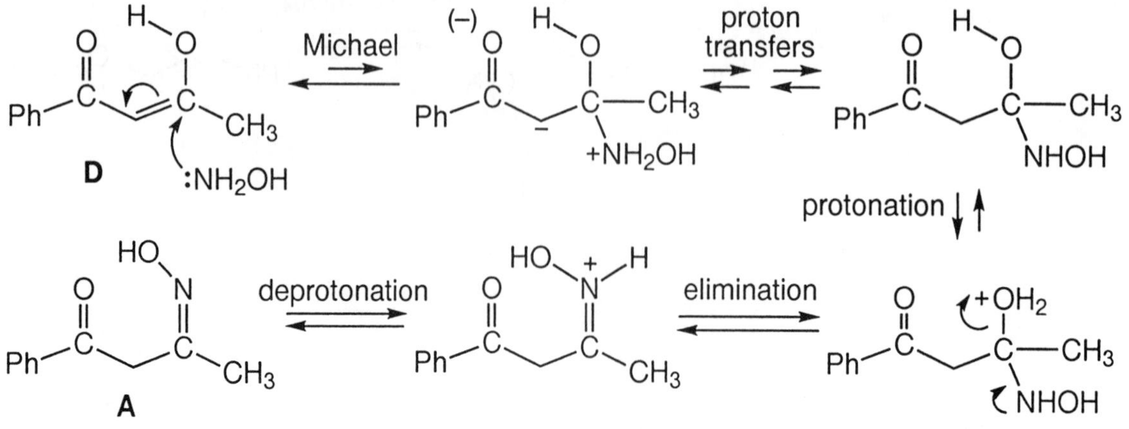

Chapter 26 Outline

26.1 Amino Acids
26.2 Reactions of Amino Acids
26.3 Peptide Chemistry
26.4 Something More: Nucleosides, Nucleotides, and Nucleic Acids

26

Introduction to Amino Acids and Polyamino Acids (Peptides and Proteins)

Polyamino acids are complex; there can be no doubt of that. Structural matters become more severe as questions of secondary and tertiary structure arise. Nonetheless, the microchemistry of the amino acids and their oligomers and polymers can still be understood through normal reactions, usually Lewis acids and Lewis bases reacting in familiar ways. "Round up the usual suspects!" says Claude Raines at the end of the movie *Casablanca,* and he's exactly right. That's what you should do in the problems in this chapter at the end of this book. Don't be put off by the polyfunctionality, forge ahead with the usual reactions, and they will work out. This section starts, as usual, with relatively simple structural and mechanistic questions, and then becomes more complicated.

Problem 26.3 The $^{+}NH_3$ group is strongly electron-withdrawing inductively, and will stabilize a neighboring carboxylate anion, which increases the acidity of the acid and lowers the pK_a. The small "dipole arrow" shows the polarization of the electrons in the carbon–nitrogen σ bond.

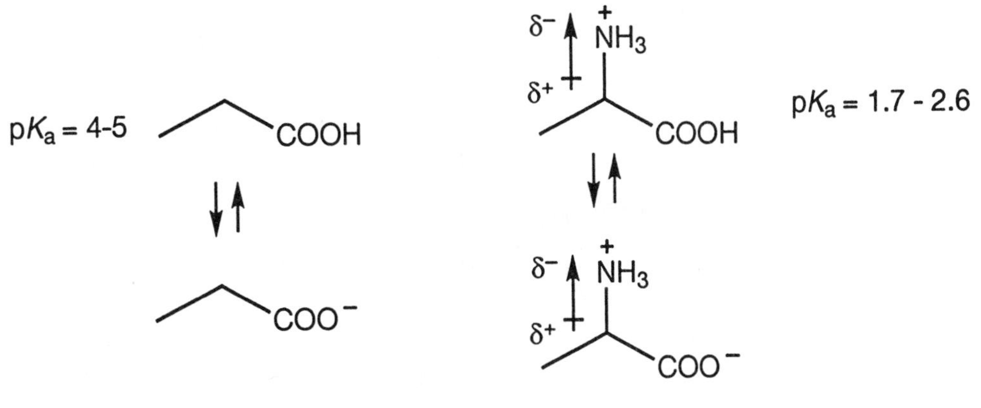

pK_a = 4-5 COOH ⇅ COO^-

δ− δ+ $\overset{+}{N}H_3$ COOH ⇅ δ− δ+ $\overset{+}{N}H_3$ COO^- pK_a = 1.7 - 2.6

Problem 26.4 The isoelectric point pH is the average of the two pK_a values.

$$\underset{\mathbf{A}}{H_3\overset{+}{N}CH(CH_3)COOH} \underset{}{\overset{K_{a1}}{\rightleftharpoons}} \underset{\mathbf{Z}}{H_3\overset{+}{N}CH(CH_3)COO^-} \qquad K_{a1} = \frac{[\mathbf{Z}][H_3O^+]}{[\mathbf{A}]}$$

$$\underset{\mathbf{Z}}{H_3\overset{+}{N}CH(CH_3)COO^-} \overset{K_{a2}}{\rightleftharpoons} \underset{\mathbf{B}}{H_2NCH(CH_3)COO^-} \qquad K_{a2} = \frac{[\mathbf{B}][H_3O^+]}{[\mathbf{Z}]}$$

so, $[K_{a1}]\,[K_{a2}] = \dfrac{[H_3O^+]^2\,[\mathbf{Z}]\,[\mathbf{B}]}{[\mathbf{A}]\,[\mathbf{Z}]}$ or, at the isoelectric point where by definition, $[\mathbf{A}] = [\mathbf{B}]$

$$[K_{a1}]\,[K_{a2}] = [H_3O^+]^2$$

and, taking the negative log of each side: $pK_{a1} + pK_{a2} = -\log\,[H_3O^+]^2 = 2\text{ pH}$

The pH at the isoelectric point is the average of the two pK_a values:

$$\frac{pK_{a1} + pK_{a2}}{2} = (2.3 + 9.7)/2 = 6.0$$

Problem 26.5 Protonation at the doubly bonded nitrogen leads to a resonance-stabilized cation. Protonation on either of the other nitrogens does not.

Arginine: $H_2\ddot{N}–C(=\ddot{N}H)–\ddot{N}H–(CH_2)_3–CH(^+NH_3)COO^-$ $\xrightarrow[H_3O^+]{H_2O}$

$$\left[\; H_2\ddot{N}–C(=\overset{+}{N}H_2)–\ddot{N}H–(CH_2)_3–CH(^+NH_3)COO^- \leftrightarrow H_2\ddot{N}–\overset{+}{C}(\ddot{N}H_2)–\ddot{N}H–(CH_2)_3–CH(^+NH_3)COO^- \leftrightarrow H_2\ddot{N}–C(\ddot{N}H_2)=\overset{+}{N}H–(CH_2)_3–CH(^+NH_3)COO^- \leftrightarrow H_2\overset{+}{N}=C(\ddot{N}H_2)–\ddot{N}H–(CH_2)_3–CH(^+NH_3)COO^- \;\right]$$

Problem 26.6 Remember that the pK_a value for amines refer to the donation of a proton by the conjugate acid, the ammonium ion (p. 1086). So the reaction referred to in the question is:

:B⁻ Aromatic! COO^- $+NH_3$ → Aromatic! COO^- $+NH_3$ + BH

This seems an odd question, as both histidine itself and its conjugate acid have an aromatic sextet of electrons. It is not really easy to identify special factors that make protonated histidine especially acidic. Could the authors of this question, whoever they may be, have written "acidic" on p. 1350 when they meant "basic"? We'll never tell.

Problem 26.7 All acids with an acidic α-hydrogen can be brominated in the α-position with PBr_3 or the equivalent, a mixture of phosphorus and bromine (the Hell–Volhard–Zelinsky reaction, Chapter 19, p. 987), so the last step is the same in each example. The first acid can be made by oxidation of the corresponding alcohols.

(a) This part of the problem is simple, as we are allowed to start with the four-carbon alcohol butanol.

Butanol —(1. basic $KMnO_4$; 2. H_2O/H_3O^+)→ Butyric acid —(P/Br_2, 100 °C)→ α-Bromobutyric acid

(b) This part is a bit harder because it is necessary to build up to a five-carbon acid: straightforward Grignard chemistry does the trick (Section 19.7, p. 991).

—(PBr_3)→ —(Mg, ether)→ MgBr —(1. CO_2; 2. H_2O/H_3O^+)→

—(P/Br_2, 100 °C)→

Problem 26.8 A real oldie. The transition state for the S_N2 reaction will benefit from delocalization, and the adjacent carbonyl group provides that delocalization. If the transition state (the top of the activation barrier) is stabilized, the reaction will go faster.

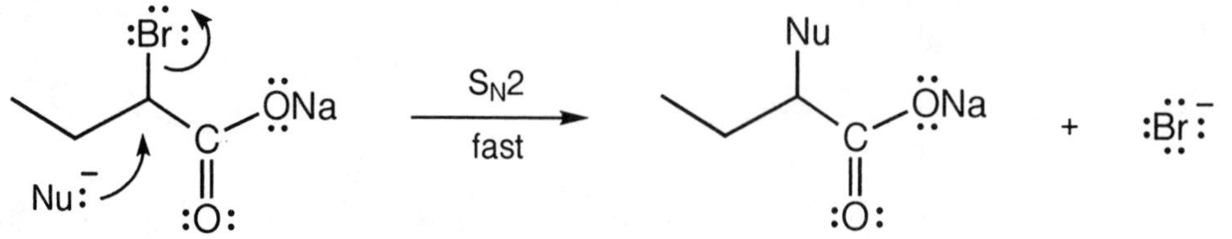

:Br: ONa Nu: :O: S_N2 fast Nu ONa :O: + :Br:

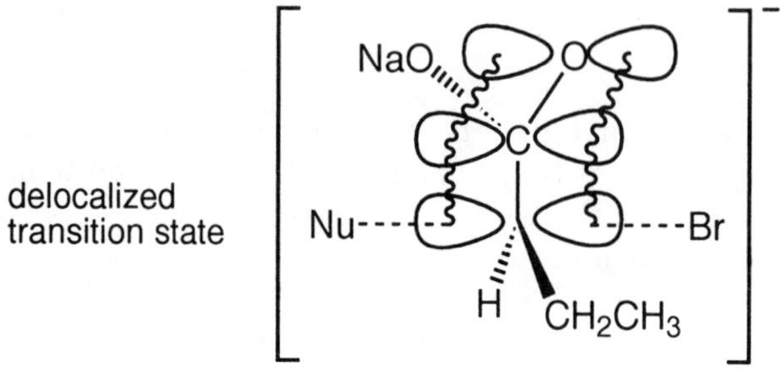

delocalized transition state NaO Nu Br H CH_2CH_3

Problem 26.9 The first step is, as advertised in the figure, a simple S_N2 displacement of bromide by the phthalimide anion.

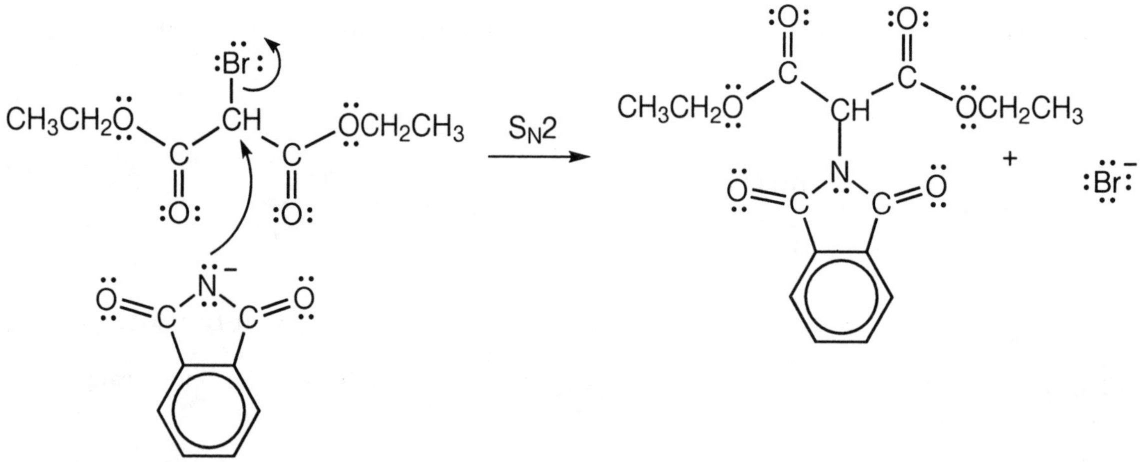

CH_3CH_2O :Br: CH OCH_2CH_3 :O: :O: S_N2 :O: :O: CH_3CH_2O CH OCH_2CH_3 N + :Br:

Now, an alkylation reaction takes place as the doubly α proton is removed in base to give an anion that can be used to displace chloride from the alkylating agent.

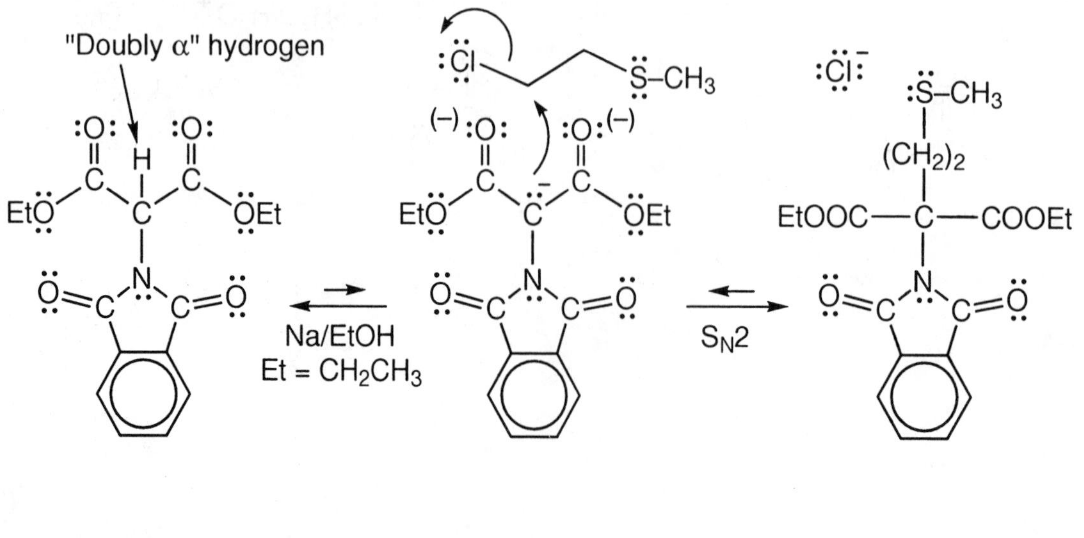

"Doubly α" hydrogen

:O: H :O: EtO C OEt N Na/EtOH Et = CH_2CH_3 :Cl S–CH_3 (–) :O: :O:(–) EtO C OEt N S_N2 :Cl: :S–CH_3 $(CH_2)_2$ EtOOC—C—COOEt N

In base, a set of four base-induced hydrolyses takes place. Hydroxide hydrolyzes the esters to carboxylate anions, and at the same time removes phthalic acid as the dianion, phthalate. In the figure, only one of these four essentially identical steps is shown in detail.

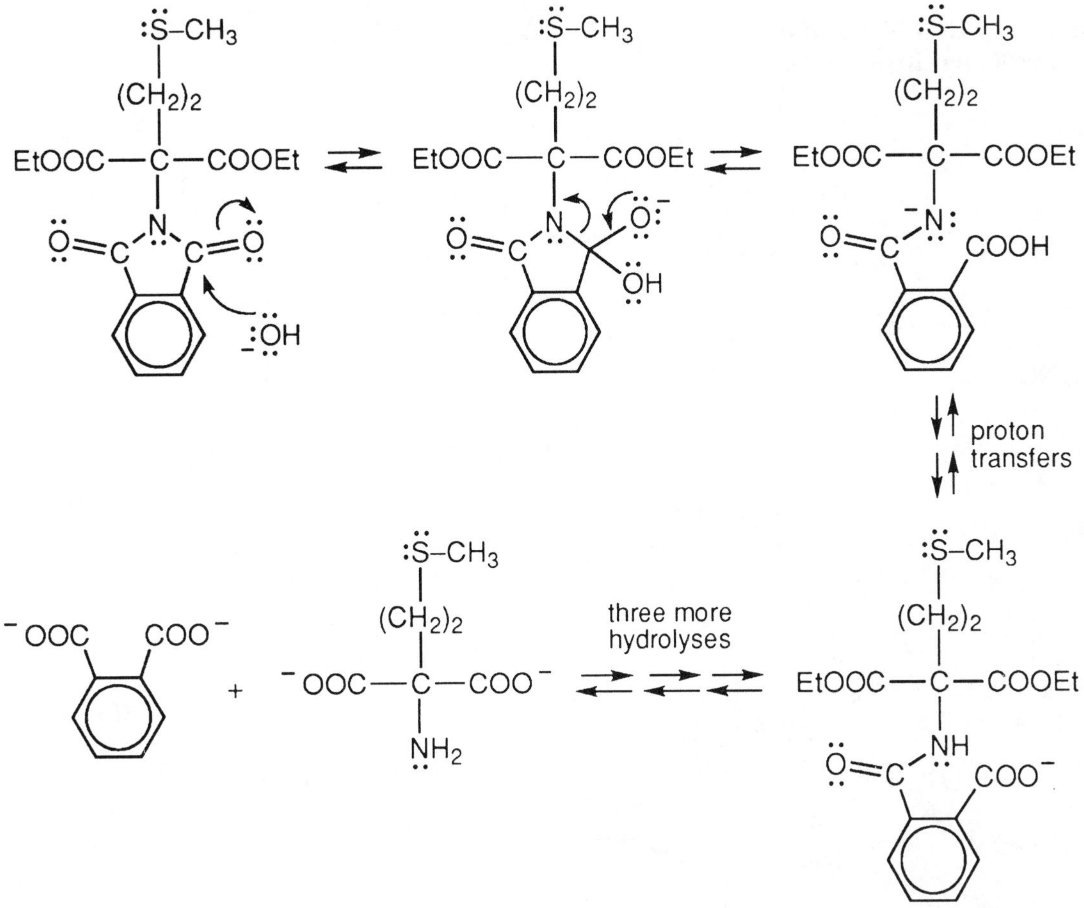

proton transfers

three more hydrolyses

The acid neutralizes the anions, forming phthalic acid and a substituted malonic acid.

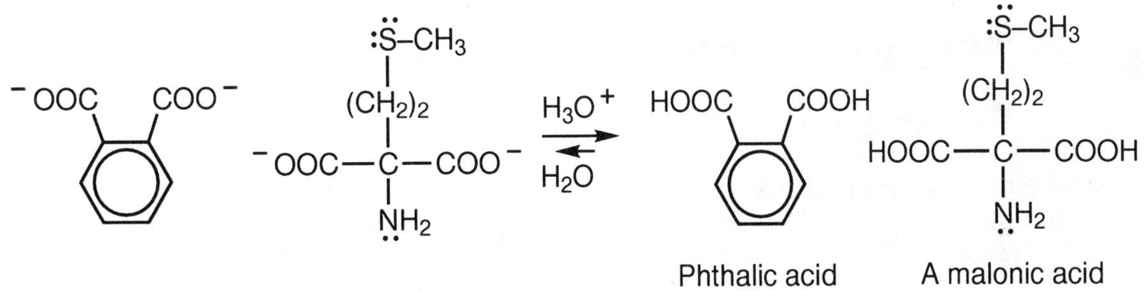

H_3O^+

H_2O

Phthalic acid

A malonic acid

Gentle heating decarboxylates the malonic acid (Chapter 20, p. 1052) to give the product.

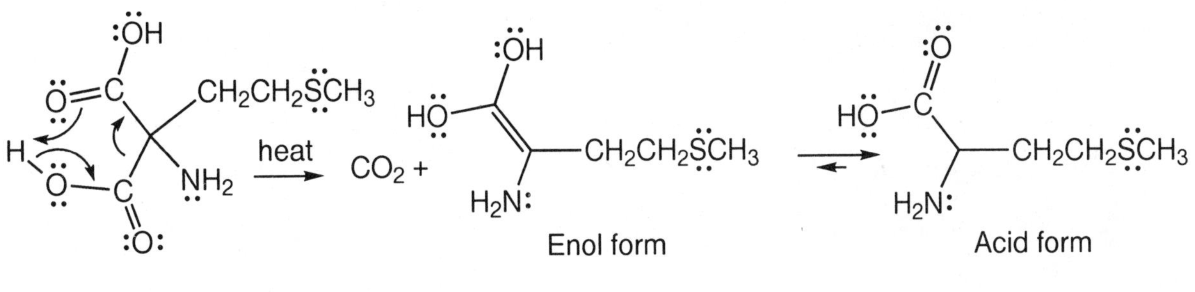

heat

CO_2 +

Enol form

Acid form

Problem 26.11 It is the ability to rotate the plane of plane-polarized light that differentiates the two. One enantiomer will rotate the plane of plane-polarized light to the right, the other by the same amount to the left (Chapter 5, p. 158).

Problem 26.12 Both reactions are examples of straightforward addition–elimination processes (Chapter 20, pp. 1016, 1021).

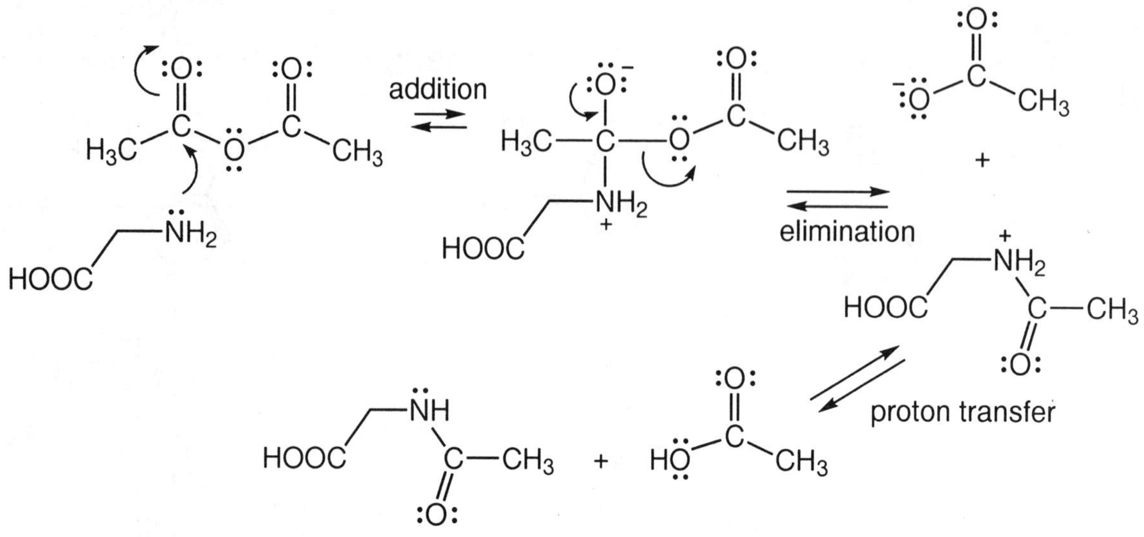

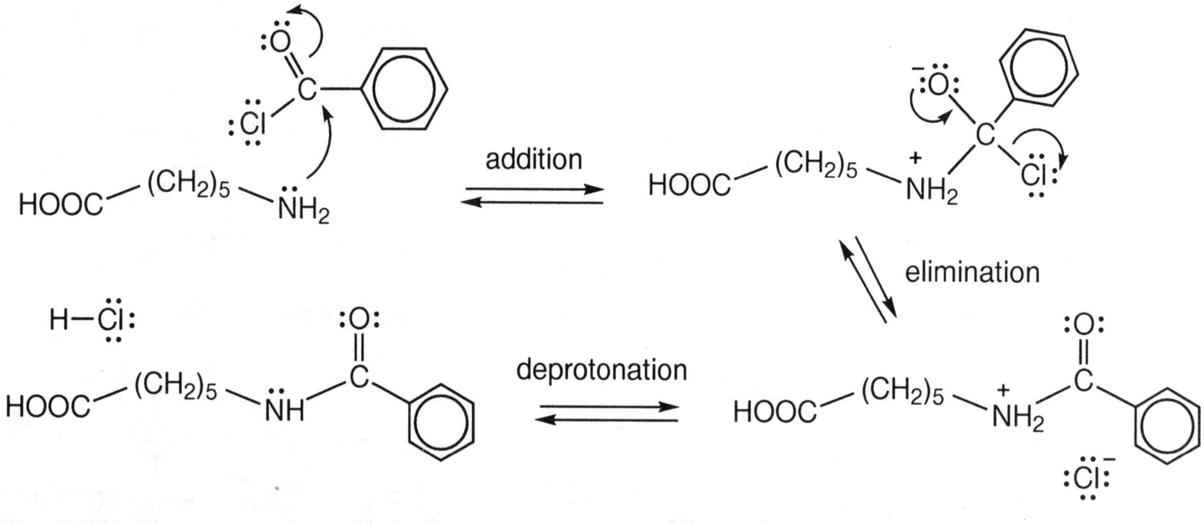

addition

elimination

proton transfer

addition

elimination

deprotonation

Problem 26.13 It is true that the amino acids are mostly in their zwitterionic forms, but there is always enough of the free acid and free amine groups present so that typical acid and amine chemistry is possible. The amino acids are not *exclusively* in their zwitterionic forms.

Problem 26.14 Vicinal (1,2) diones are destabilized by the repulsions of the adjacent dipoles in which like charges are opposed. In the trione, there are two 1,2-dipolar repulsions. Hydration at the central carbonyl removes both 1,2-interactions, whereas hydration at the side carbonyl leaves one 1,2-repulsion intact.

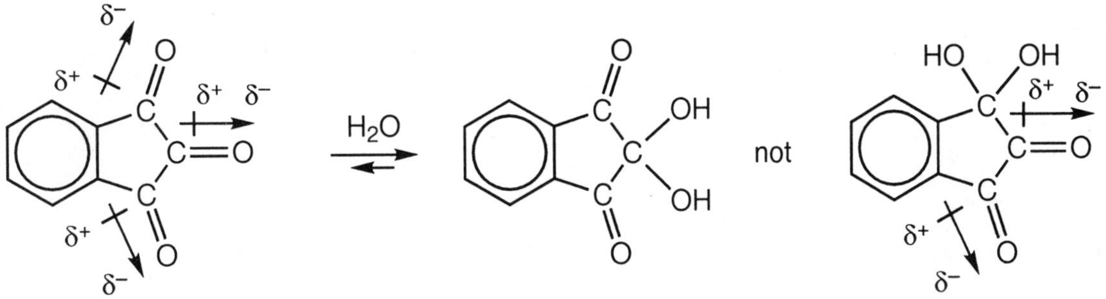
δ−
δ+
δ+ δ−
H₂O
not
HO
OH

Problem 26.15

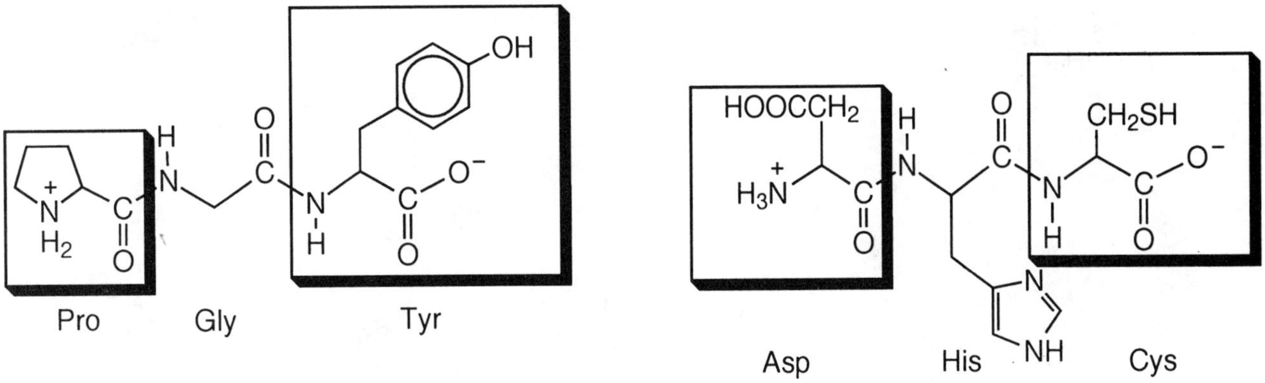
Pro
Gly
Tyr
Asp
His
Cys
HOOCCH₂
CH₂SH

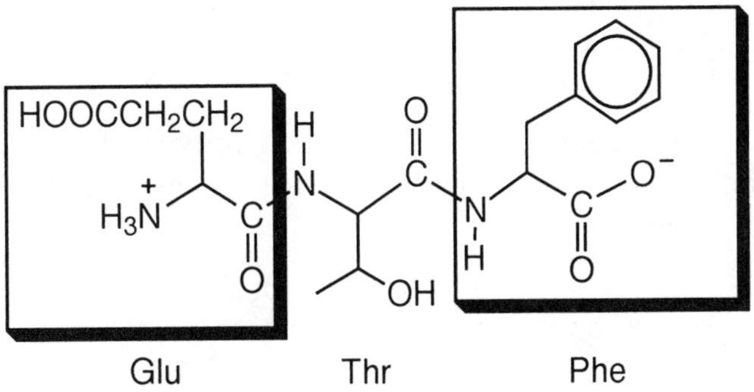
HOOCCH₂CH₂
Glu
Thr
Phe

Problem 26.16 (a) Phe·Lys·Trp (b) Asp·Met·Ile.

Problem 26.18 It's nucleophilic aromatic substitution. Let's abbreviate the tripeptide of Figure 26.39 as H_2N—R. Addition to the benzene ring at the same position as the F (ipso attack) leads to an ion in which both nitro groups help stabilize the negative charge. Without the nitro groups there is no such stabilization and the reaction fails. Loss of fluoride completes the reaction as the aromatic benzene ring is regenerated.

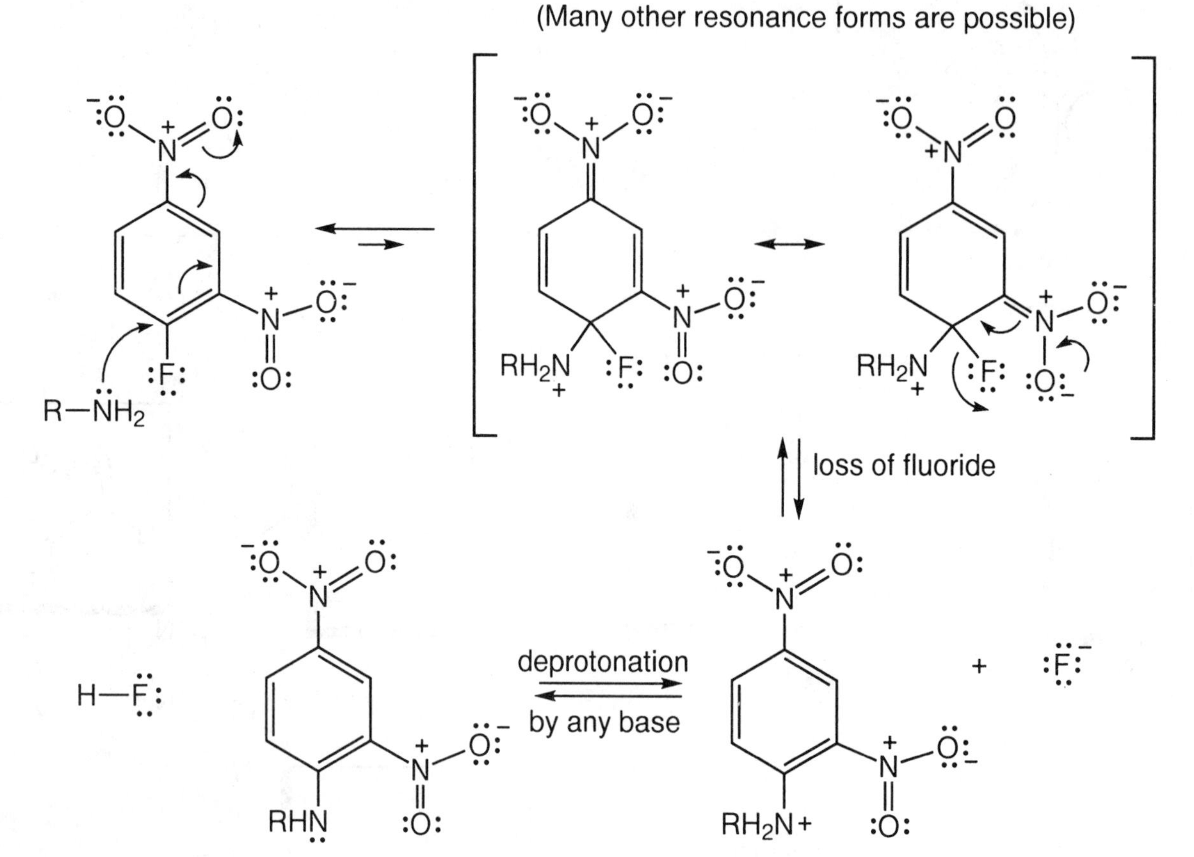

(Many other resonance forms are possible)

R—NH₂ :F: RH₂N⁺ :F: :O: loss of fluoride H—F: RHN deprotonation by any base RH₂N+ + :F:⁻

Problem 26.19 It's a Curtius rearrangement (Chapter 23, p. 1222). As the nitrogen departs, the "R" group migrates and the isocyanate is formed. No nucleophiles are present, and so the isocyanate survives.

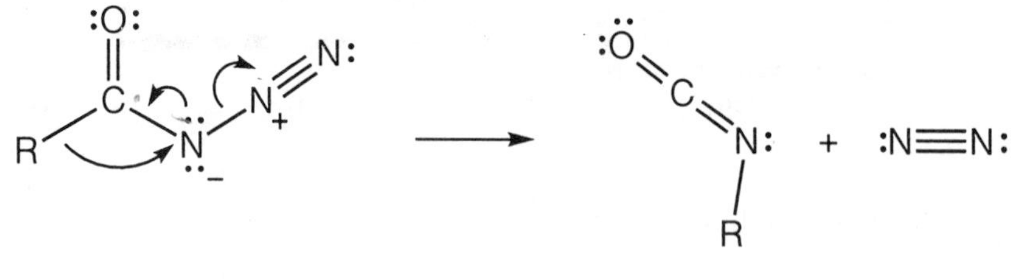

R :O: C N N N: → :O=C=N: R + :N≡N:

Problem 26.20 The critical, if obvious, realization is that this must be an "open–close" problem. After all, the sulfur is in the ring in the starting material and out of the ring in the product. We have to find a way to open the ring by breaking a carbon–sulfur bond. Here's a possibility.

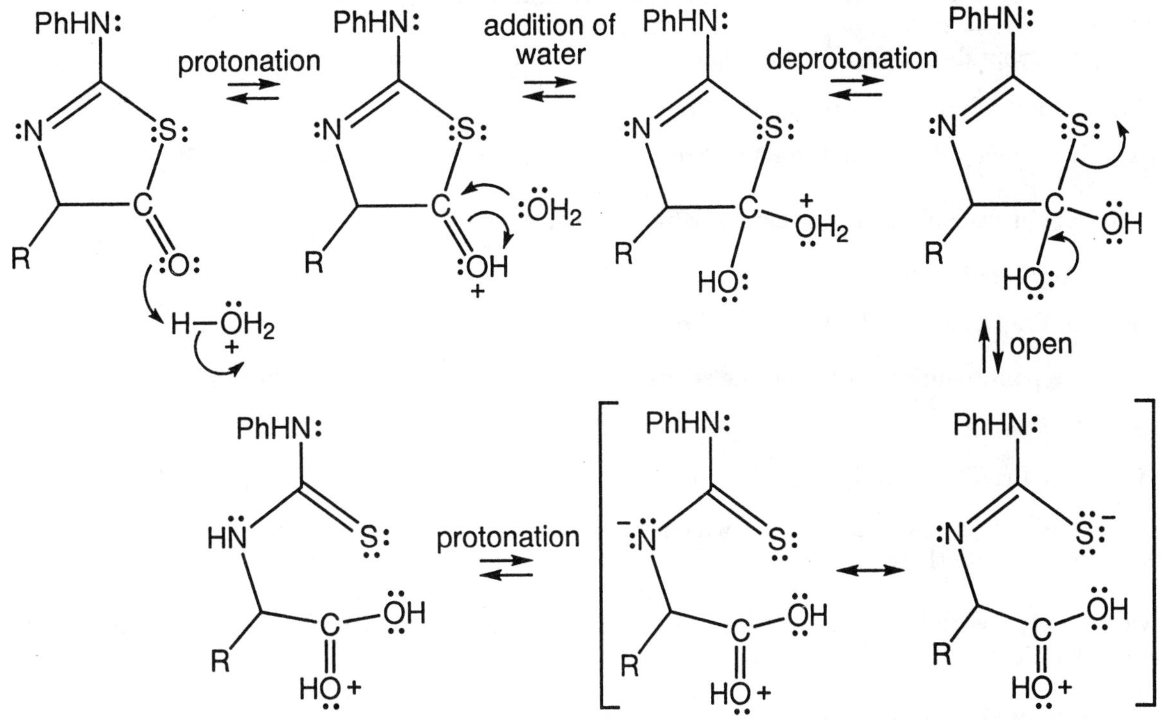

PhHN: protonation PhHN: addition of water PhHN: deprotonation PhHN:

:N :S: :N :S: :ÖH₂ :N :S: :N :S:

R C :O: R C :OH + R C $\overset{+}{O}H_2$ HO: R C ÖH HO:

H—ÖH₂ +

open

PhHN: protonation PhHN: PhHN:

HN S: :N S: :N S:

R C ÖH HO+ R C ÖH HO+ R C ÖH HO+

The ring is reclosed through amide formation, and a series of protonation–deprotonation steps finishes the reaction.

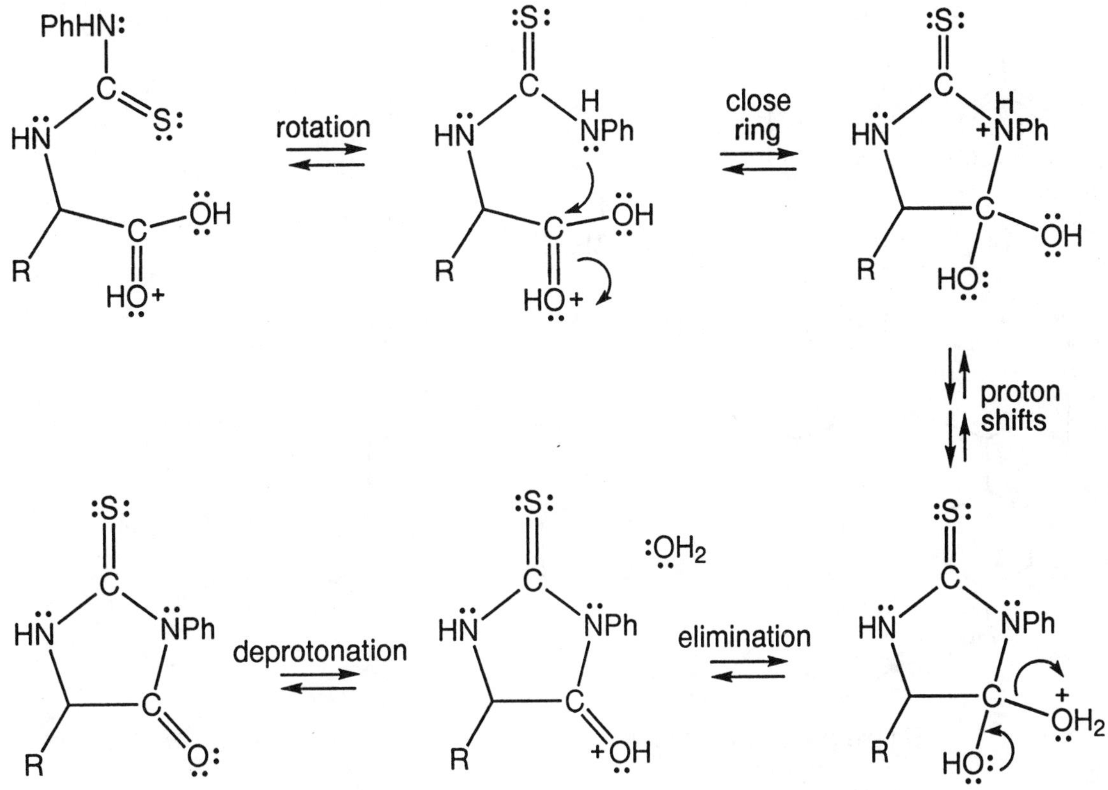

PhHN: :S: :S:

C S: rotation C H close ring C H

HN HN NPh HN +NPh

R C ÖH HO+ R C ÖH HO+ R C ÖH HO:

proton shifts

:S: :S: :ÖH₂ :S:

C C C

HN NPh deprotonation HN NPh elimination HN NPh

R C O: R C +OH R C $\overset{+}{O}H_2$ HO:

Problem 26.21 The acid hydrolysis gives the constituent amino acids of bradykinin. The key observation is that the only amino acid present in bradykinin after which chymotrypsin cleaves is Phe.

There are several ways in which the pieces could be fit together.

(a) the pentapeptide could start the sequence. If so, there are two possibilities.

1. (Arg·Pro·Pro·Gly·Phe) (Arg) (Ser·Pro·Phe)
2. (Arg·Pro·Pro·Gly·Phe) (Ser·Pro·Phe) (Arg)

The pentapeptide could be in the center, which gives two more possibilities.

3. (Arg) (Arg·Pro·Pro·Gly·Phe) (Ser·Pro·Phe)
4. (Ser·Pro·Phe) (Arg·Pro·Pro·Gly·Phe) (Arg)

Finally, the pentapeptide could end the sequence. Again there are two possibilities.

5. (Ser·Pro·Phe) (Arg) (Arg·Pro·Pro·Gly·Phe)
6. (Arg) (Ser·Pro·Phe) (Arg·Pro·Pro·Gly·Phe)

But the only amino acid present after which chymotrypsin cleaves is Phe. So, only possibilities 2 and 4 remain. Only these two would give the three pieces found.

However, only possibility 2 has the same amino acid at the amino and carboxy terminus, and so it must be the structure of bradykinin.

Problem 26.22 If the polymer is based on benzene rings, the chloromethyl groups are benzylic chlorides. A cyclohexane ring-based polymer would have simple primary chlorides:

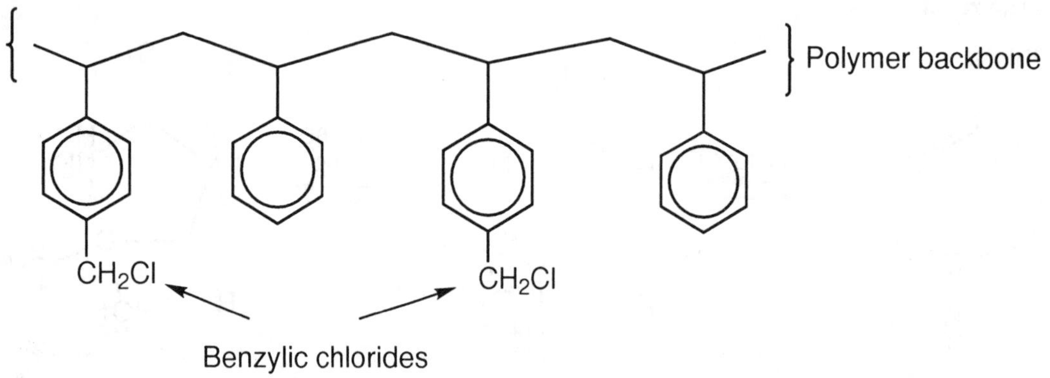

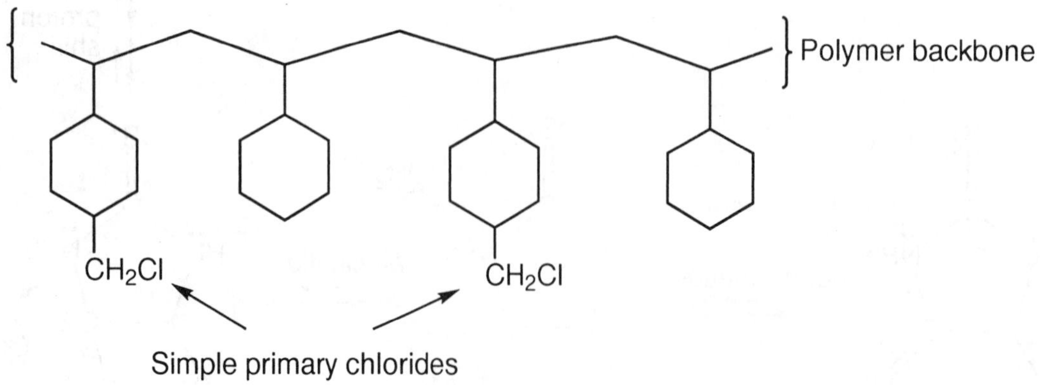

Attachment of the amino acid is through S_N2 displacement of chloride. These S_N2 reactions are much faster at the benzylic position than at a unconjugated primary position.

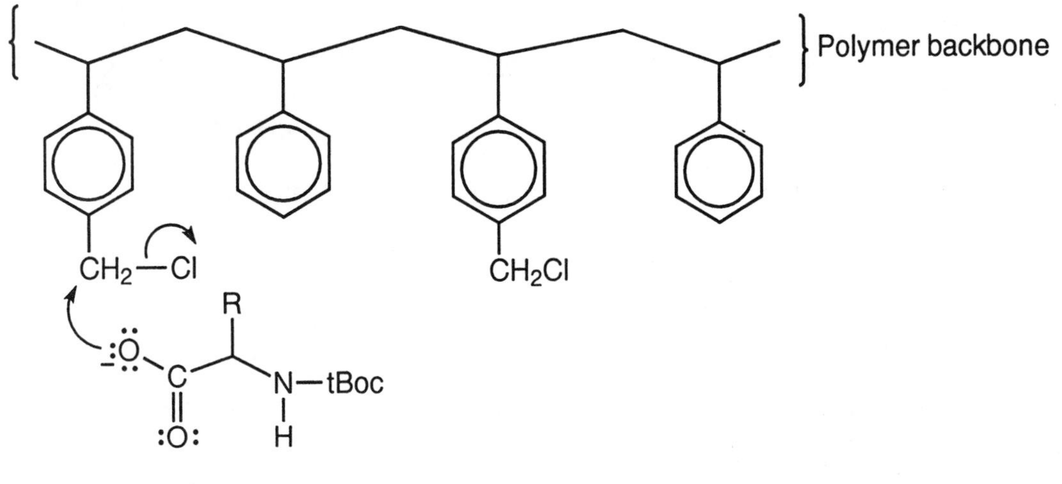

Polymer backbone

CH_2—Cl

CH_2Cl

R

N—tBoc

H

Answers to Additional Problems, Chapter 26.

Problem 26.24

(a) Protonation on oxygen leads to a resonance-stabilized cation, whereas protonation on nitrogen does not.

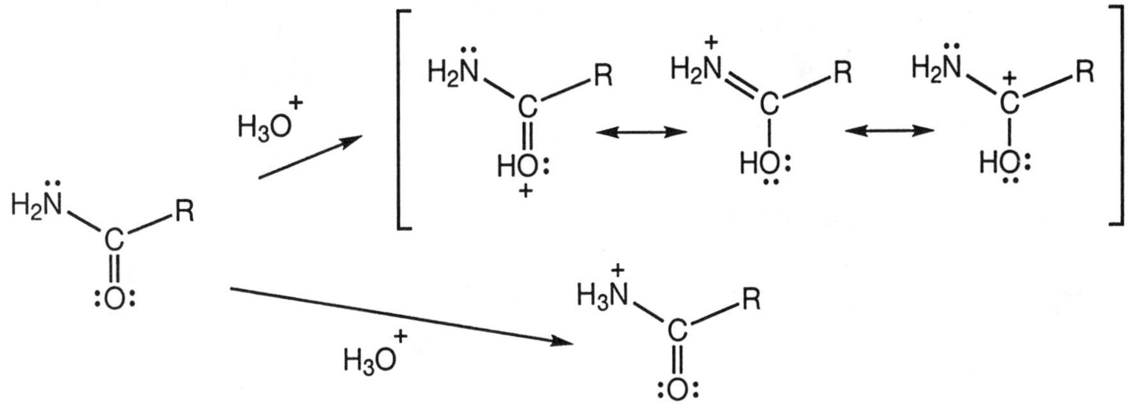

H_3O^+

H_3O^+

(b) The resonance stabilization in amides makes them more stable, relative to their conjugate acids, than are amines. This stability makes them less reactive—weaker bases.

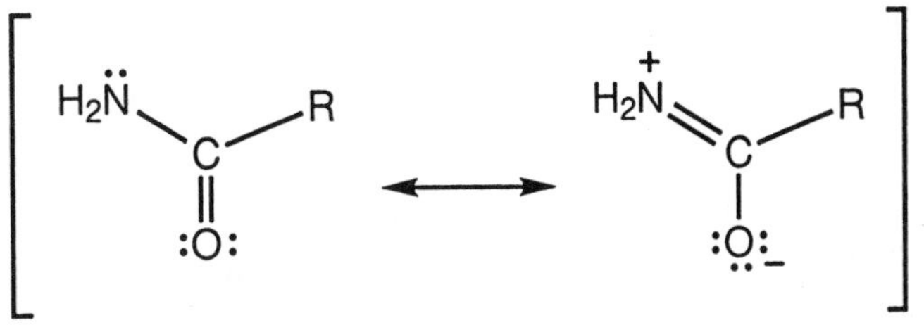

Problem 26.25

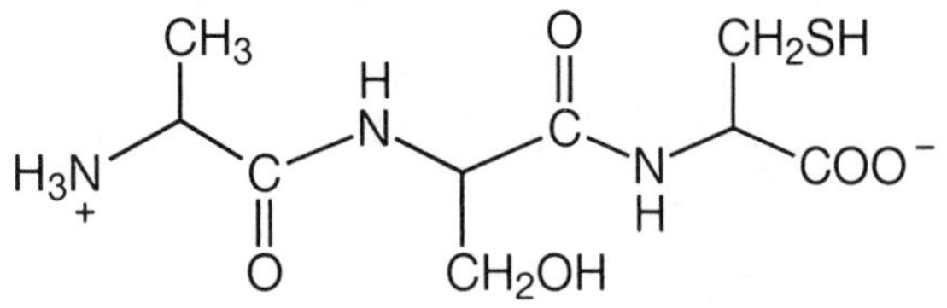

Alanylserinylcysteine

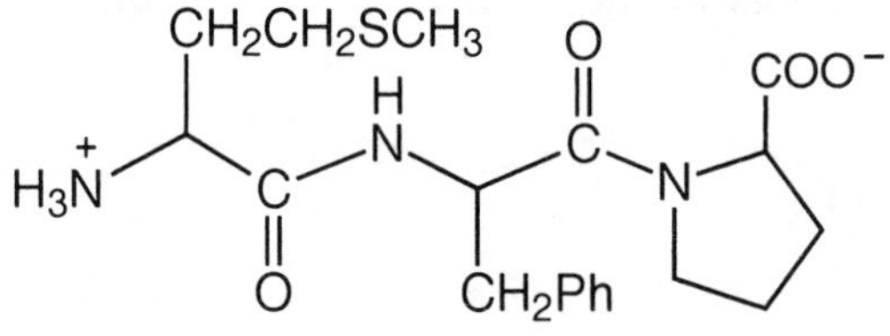

Methionylphenylalanylproline

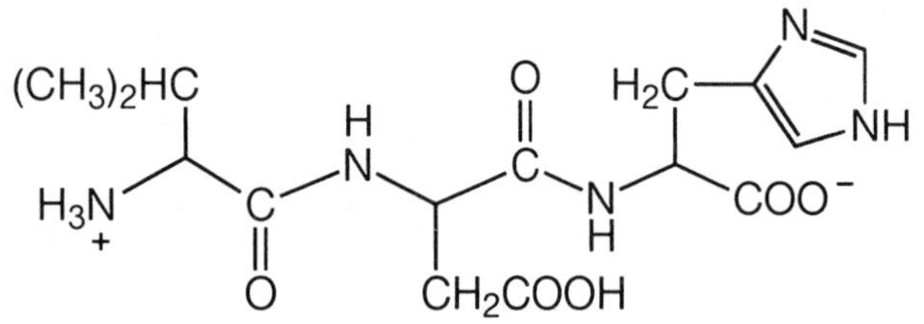

Valinylaspartylhistidine

Problem 26.26 (a) Thr·Leu, Threonylleucine and (b) Lys·Tyr, Lysinyltyrosine

Problem 26.27 The numbers in the following drawing show the priorities. Remember, if the 1 ⟶ 2 ⟶ 3 arrow runs clockwise, the compound is (*R*); if it runs counterclockwise, it is (*S*).

(*S*)-Serine

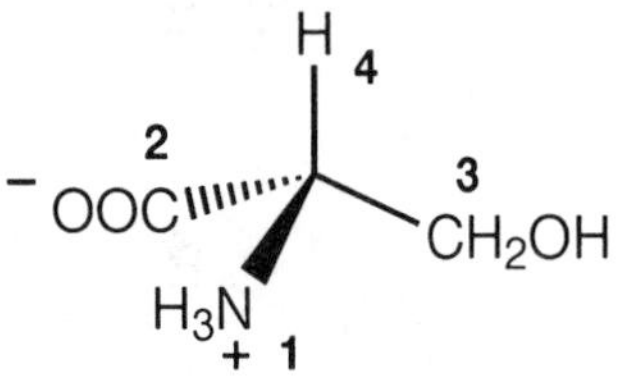

(*S*)-Proline

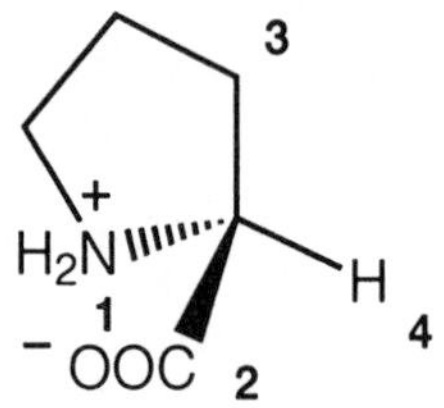

Problem 26.28

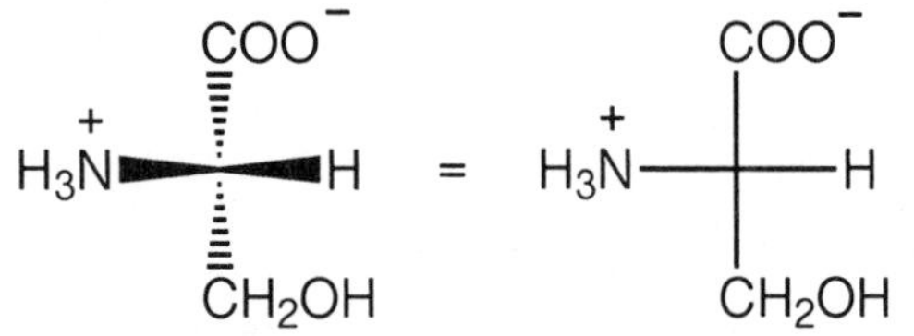

(*S*)-Serine = L-Serine

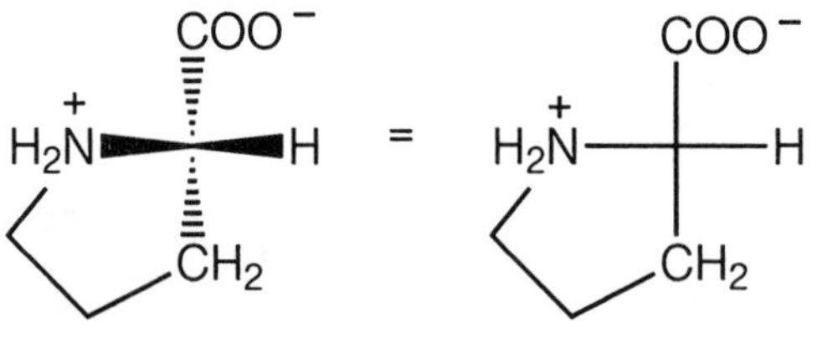

(*S*)-Proline = L-Proline

Problem 26.29 The isoelectric points are His = 7.6, Arg = 10.8, and Phe = 5.9. Above the isoelectric point the amino acids will be net negatively charged. Below the isoelectric point, they will be net positively charged. The structures will be.

At pH = 3

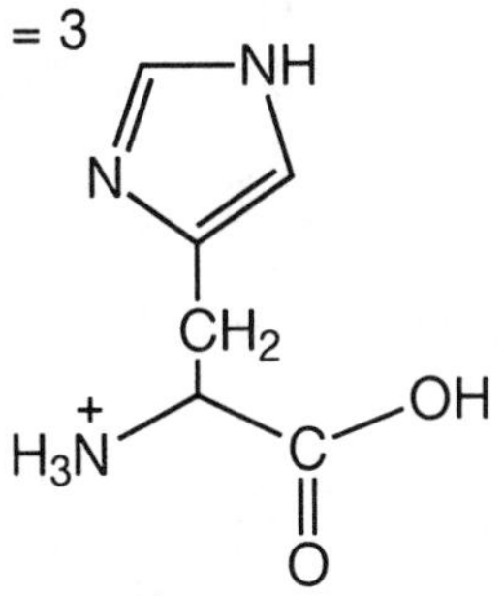

His (isoelectric point 7.6) 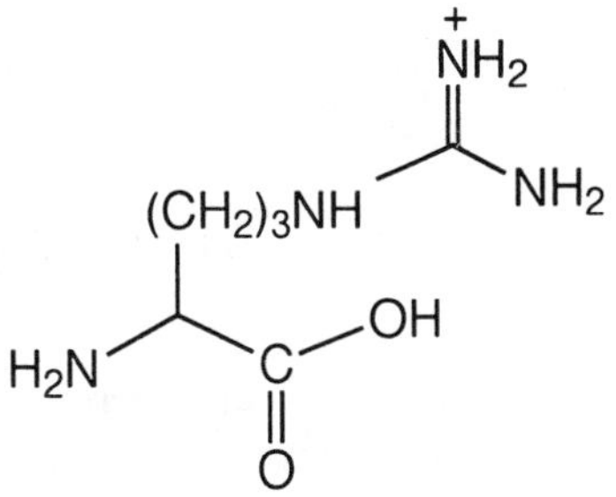 Arg (isoelectric point 10.8) 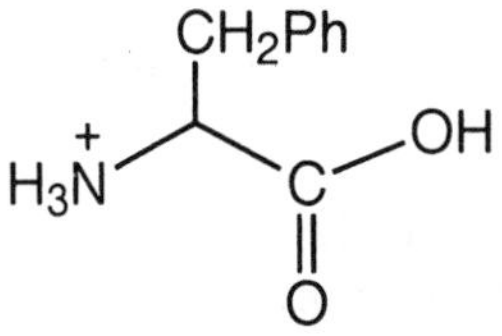 Phe (isoelectric point 5.9)

At pH = 12

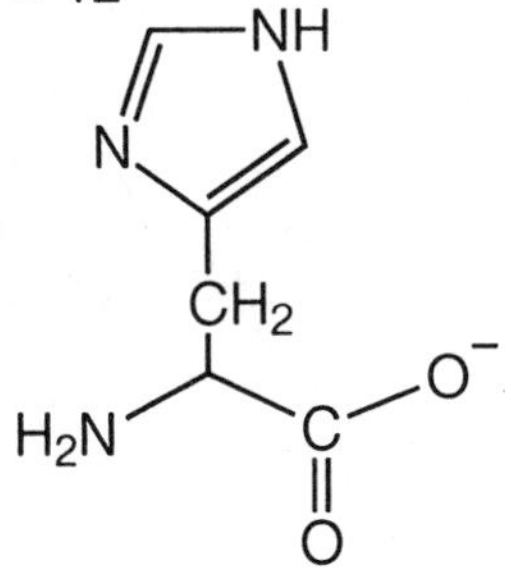

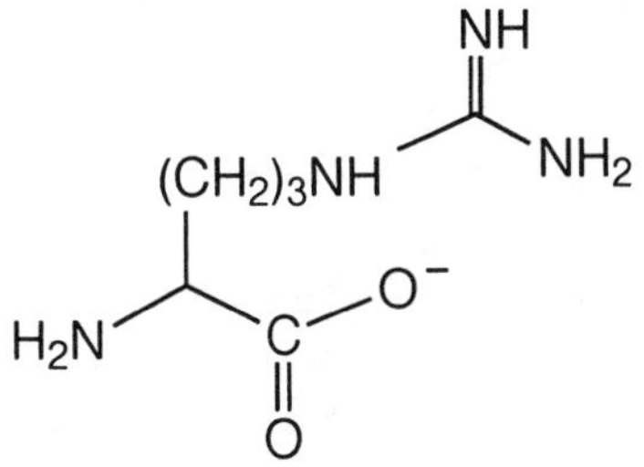

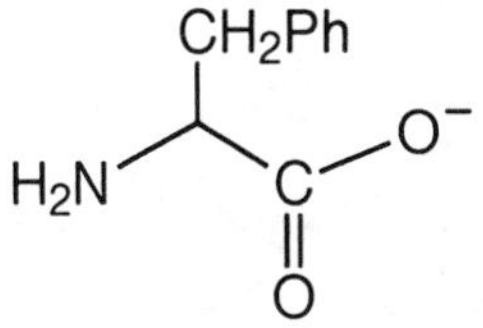

Problem 26.30 At pH = 7, the structures will be.

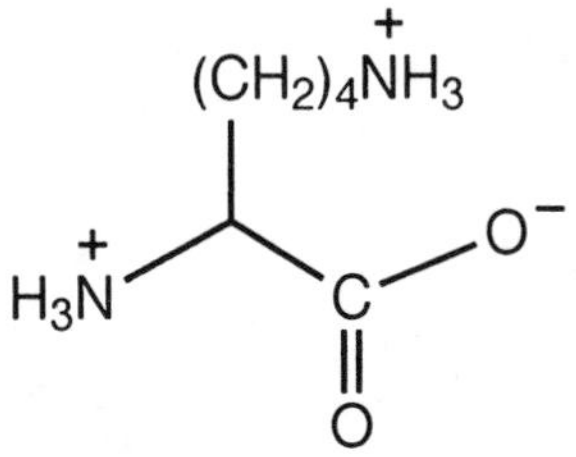

Lys (isoelectric point 9.7) 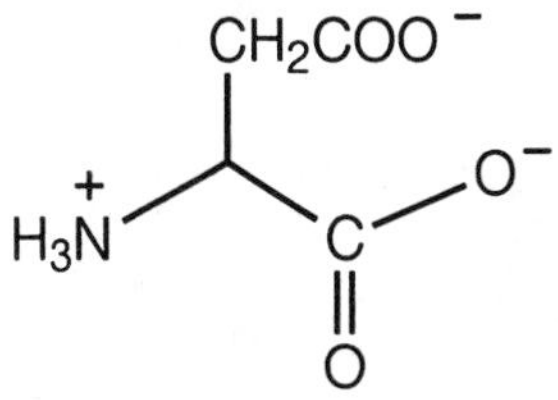 Asp (isoelectric point 2.9)

So Lys, net positively charged, will migrate to the negatively charged electrode, the cathode; and Asp, net negatively charged, will migrate to the positive electrode, the anode.

Problem 26.31 The L amino acid with the (*R*) configuration is cysteine. For the other amino acids in Table 26.1, the second priority group is always the carboxylic acid, as illustrated below for L-valine.

L-Valine = = (*S*)-Valine

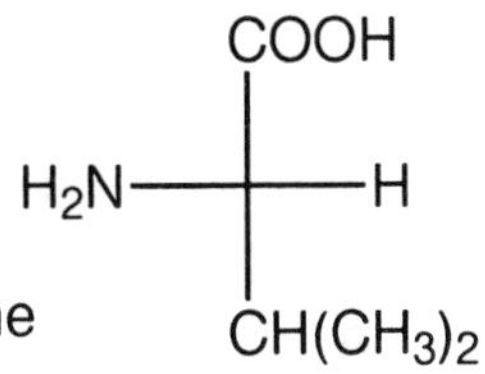

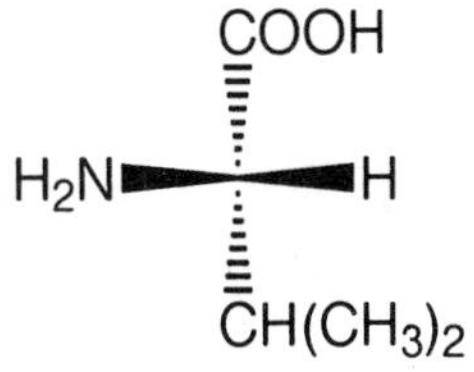

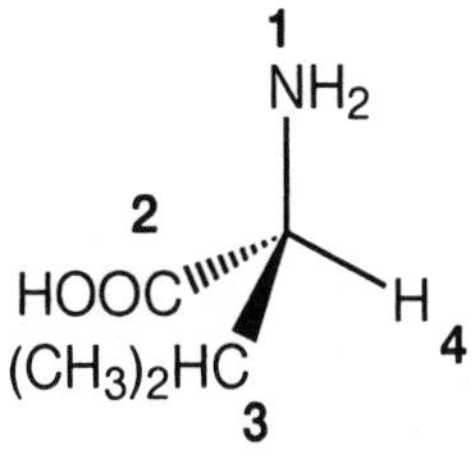

However, for cysteine the second priority group is the CH_2SH as sulfur has a higher atomic number, and hence a higher priority, than oxygen. Thus, L-cysteine has the (*R*) configuration.

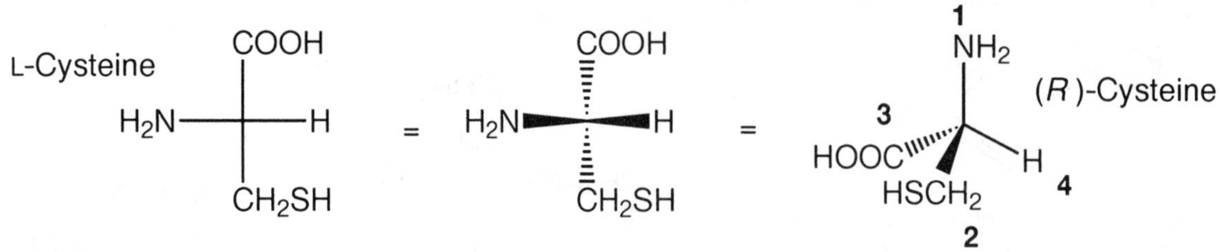

L-Cysteine

(*R*)-Cysteine

Problem 26.32 By analogy to the reaction of primary amino acids with ninhydrin (see Fig. 26.24), the free amine form of the secondary amino acid proline, a small amount of which is present at equilibrium, reacts with indan-1,2,3-trione to yield, after dehydration, **A**. Intermediate **A** then undergoes a decarboxylation to produce compound **1**. Unlike the related imine of Figure 26.24, compound **1** appears to be stable under normal reaction conditions.

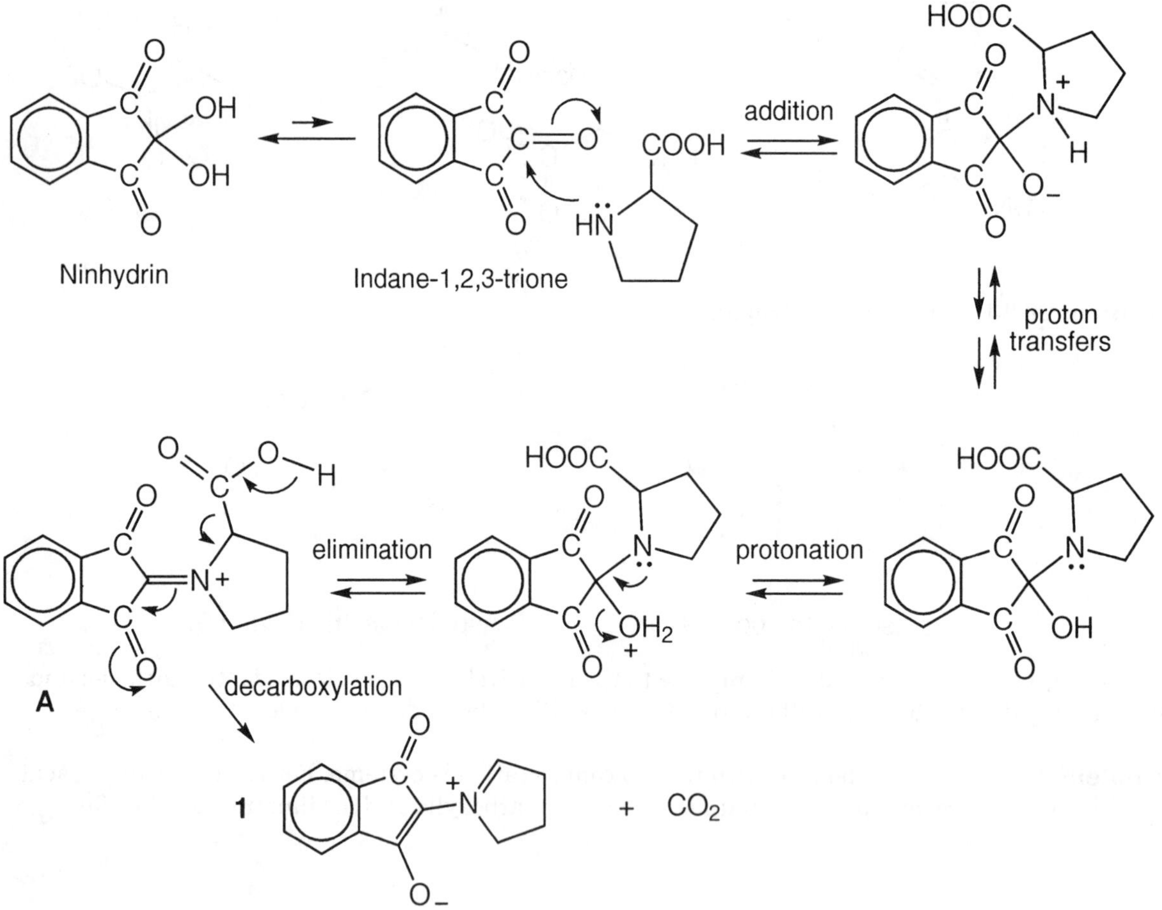

Ninhydrin

Indane-1,2,3-trione

addition

proton transfers

elimination

protonation

decarboxylation

A

1

$+ \; CO_2$

Problem 26.33 This problem involves the use of 2,4-dinitrofluorobenzene to label the amino end terminus of the peptide through a process of nucleophilic aromatic substitution.

(a) Of the common amino acids, only Phe brings with it five aromatic hydrogens. Together with the three remaining ring hydrogens of the 2,4-dinitrobenzene group, this makes eight. The amino terminus of this peptide must be phenylalanine, Phe.

(b) The region δ 3.5 ppm is appropriate for a hydrogen adjacent to an ether or alcohol oxygen atom. Only threonine would show an eight-line signal at this position. The amino end of this peptide must be threonine, Thr.

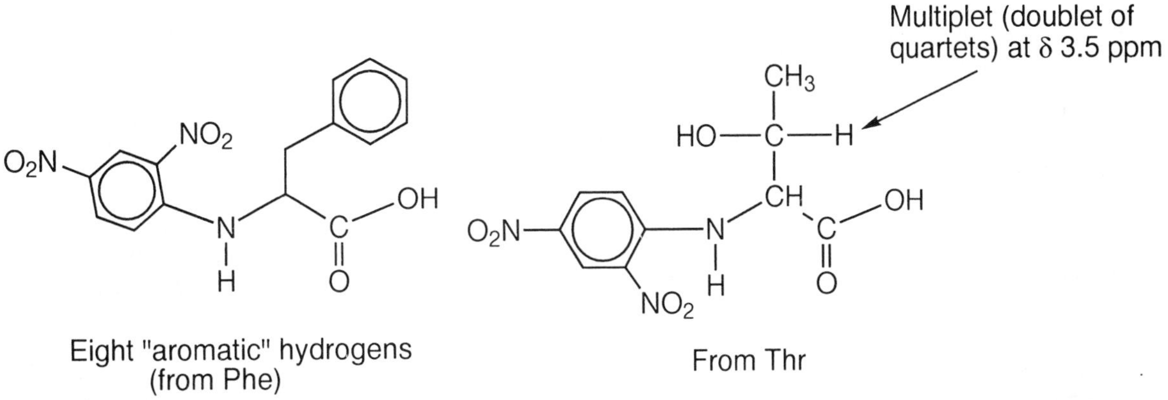

Multiplet (doublet of quartets) at δ 3.5 ppm

Eight "aromatic" hydrogens (from Phe)

From Thr

Problem 26.34 The Edman reaction determines the amino terminus. The only phenylthiohydantoin that would show *only* a methyl doublet at relatively high field is **A**, and the amino terminus of this tripeptide must be Ala.

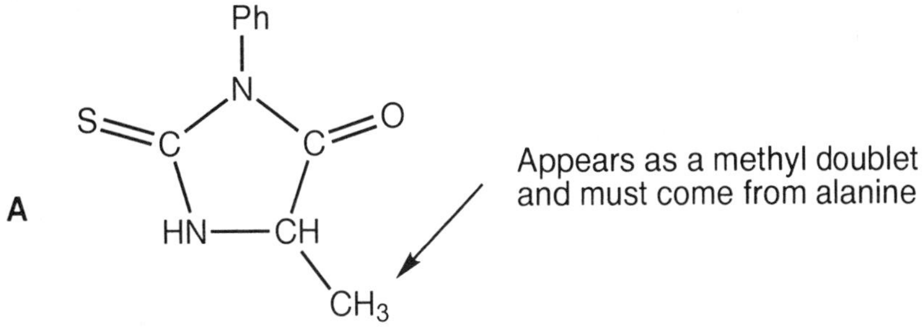

A

Appears as a methyl doublet and must come from alanine

The other two amino acids are Met and Cys. If the middle amino acid were Met, BrCN would cleave the tripeptide after the Met. As there is no reaction, the structure must be Ala·Cys·Met.

Problem 26.35 First of all, the amino terminus must be one of the Val residues, as the structure of the phenylthiohydantoin produced in the Edman procedure is **A**.

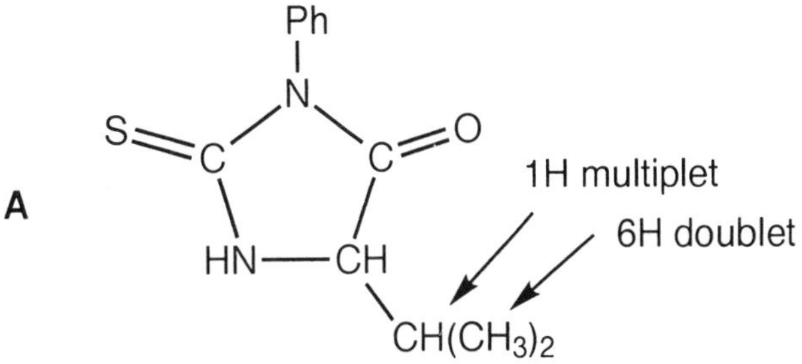

A

1H multiplet

6H doublet

Second, trypsin cleaves after Arg or Lys. Given that the polypeptide must start with Val, this means that there are only four possibilities. The arrows show the trypsin cleavage points.

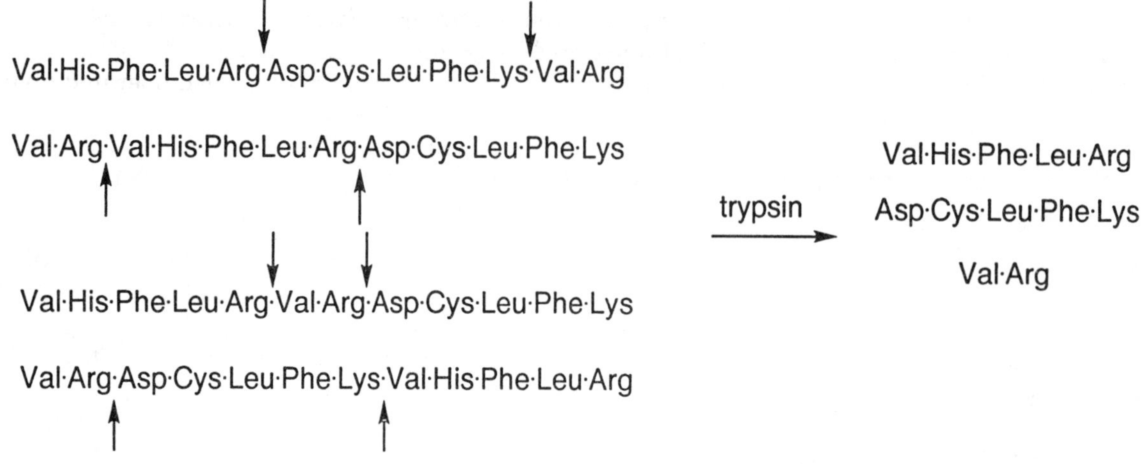

However, chymotrypsin, which cleaves after Phe, leads to a fragment, Lys·Val·Arg contained only in the first possibility. The arrows show the chymotrypsin cleavage points. The structure of **1** must be:

1
Val·His·Phe·Leu·Arg·Asp·Cys·Leu·Phe·Lys·Val·Arg —chymotrypsin→ Val·His·Phe; Leu·Arg·Asp·Cys·Leu·Phe; Lys·Val·Arg

Problem 26.36

(a) This problem is just another example of the standard addition–elimination mechanism. Note that the amino group, a better nucleophile than the carboxylate anion, adds to the dicarbonate **1** rather than to the carboxylate of the amino acid. It is also worth observing that half of the dicarbonate is wasted, as CO_2 and *tert*-butyl alcohol are lost in the process. The actual product of the reaction is the tBoc carboxylate. The free tBoc carboxylic acid is liberated by careful neutralization with acid.

NaOH
1 $(H_3C)_3CO-C(=O)-O-C(=O)-OC(CH_3)_3$
addition
elimination
CO_2 + $(H_3C)_3CO-C(=O)-N^+H_2-CHR-COO^-$
$(CH_3)_3CO^-$
deprotonation
$(CH_3)_3COH$ + $(H_3C)_3CO-C(=O)-NH-CHR-COO^-$ —neutralize→ $(H_3C)_3CO-C(=O)-NH-CHR-COOH$

(b) The removal of the tBoc group occurs by an uncommon hydrolysis mechanism, which operates for *tert*-butyl, and other *tert*-alkyl, esters. The first step is protonation of the "ester" carbonyl oxygen. The protonated ester then undergoes a straightforward E1 ionization to give a carbamic acid and a tertiary carbocation. The key to this mechanism is the formation of the relatively stable tertiary carbocation. The carbocation is deprotonated to form isobutene, and the carbamic acid decarboxylates to give the trifluoroacetate salt of the amine. The free amine is then liberated upon reaction with triethylamine.

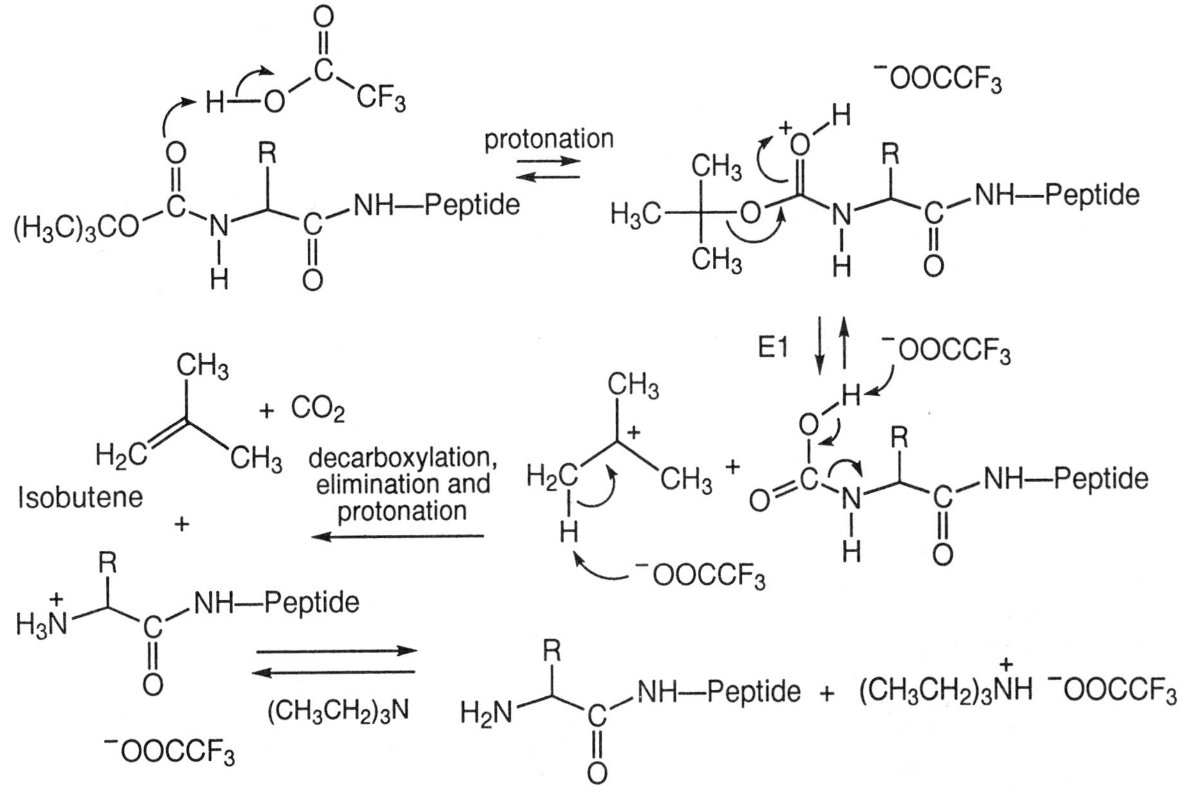

protonation

E1

Isobutene

decarboxylation, elimination and protonation

$(CH_3CH_2)_3N$

Problem 26.37 It is possible that the initially formed *O*-acylurea **1** could function as the acylating agent in some cases.

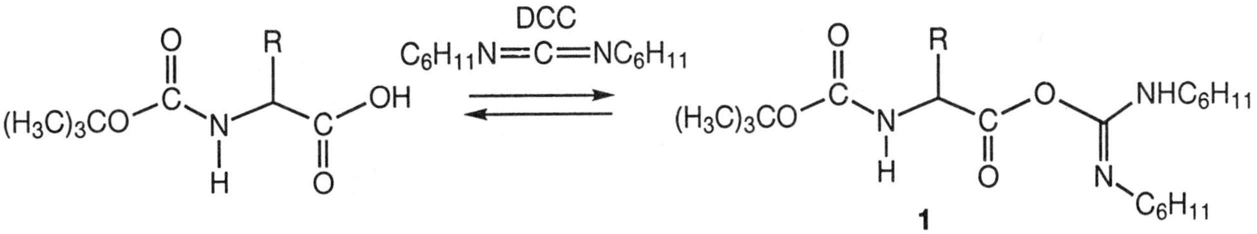

DCC

$C_6H_{11}N{=}C{=}NC_6H_{11}$

1

Addition of the new amino ester to the right-hand carbonyl group of **1** by the standard addition–elimination mechanism would ultimately produce the dipeptide and DCU as shown below.

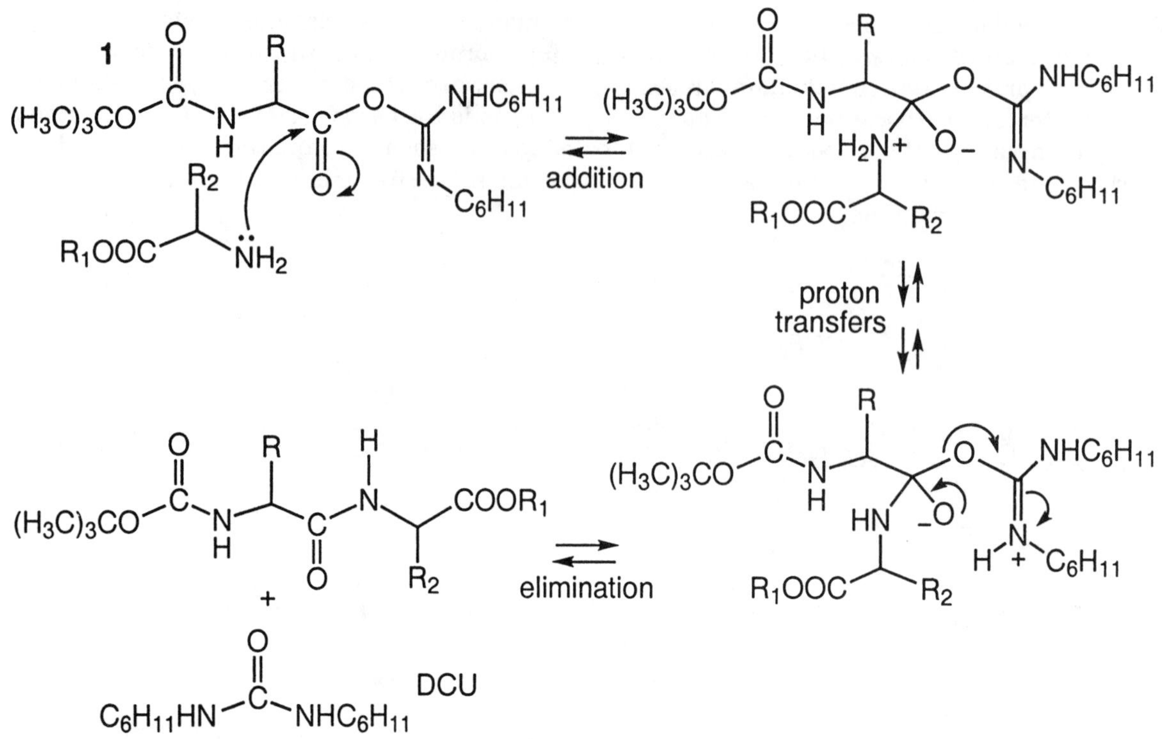

Problem 26.38 Secondary amines such as **1** form nitrosoamines (here, **2**) upon treatment with nitrous acid (p. 1097).

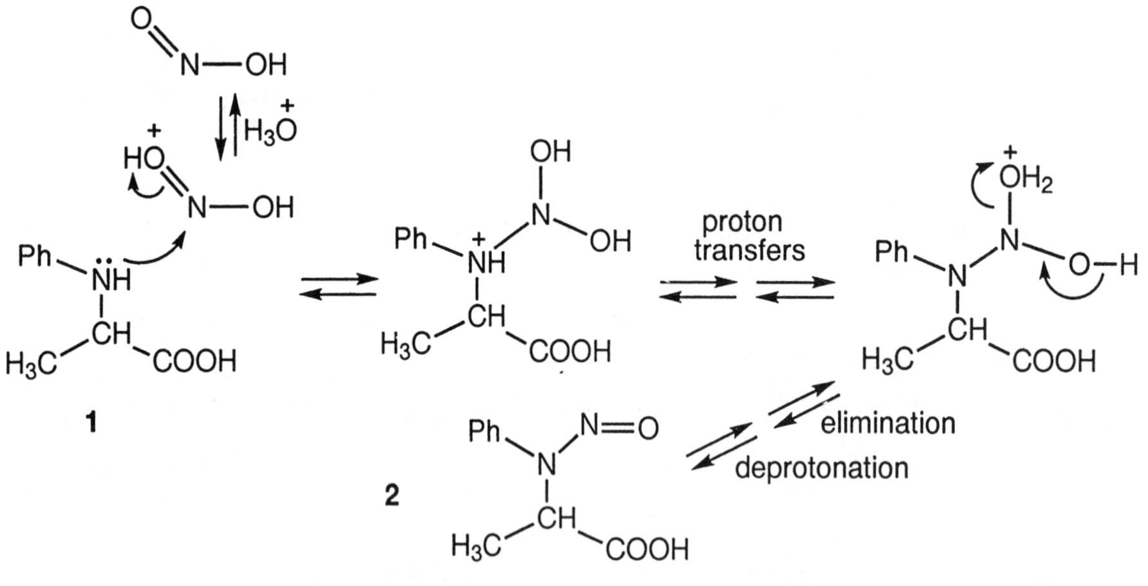

As we saw previously (Chapter 20), carboxylic acids react with acetic anhydride to give mixed anhydrides. In this case, the mixed anhydride **A** undergoes a cyclization reaction involving the nitroso oxygen atom to give, after deprotonation, the sydnone **3**.

addition

elimination

deprotonation

A

cyclization

elimination

deprotonation

3

+ HOAc

The formation of pyrazole **4** from the reaction of **3** and DMAD involves a 1,3-dipolar cycloaddition reaction, followed by a loss of CO_2 in a retrocycloaddition reaction. It is easier to see the 1,3-dipole present in **3** by looking at another Lewis structure. Although the addition can be written using any valid resonance form, we will use the 1,3-dipolar form. You might "push the arrows" for the other forms.

3

1,3-Dipole

Here's the rest of the mechanism

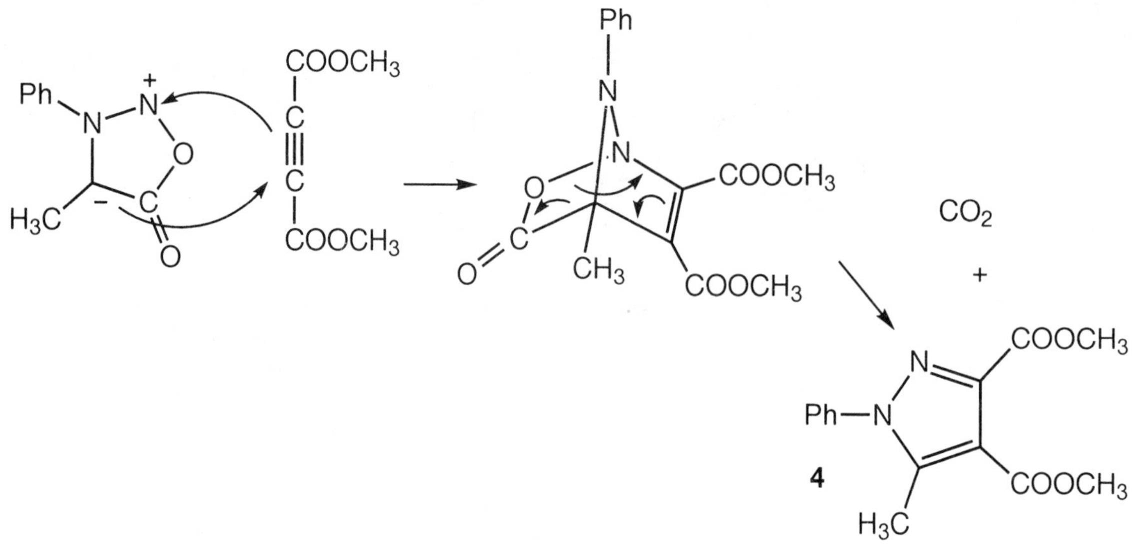

Problem 26.39 The mechanism for the formation of *N*-acetylglycine in the answer to Problem 26.12 is fine as far as it goes. In the presence of excess acetic anhydride, however, the carboxylic acid reacts further with acetic anhydride to form the mixed anhydride **A**. The amide oxygen of **A** is now poised to displace acetate (addition–elimination once again) to give, after deprotonation, the azlactone **1**. Thus, *N*-acetylglycine undergoes a "cyclodehydration" reaction to give azlactone **1**.

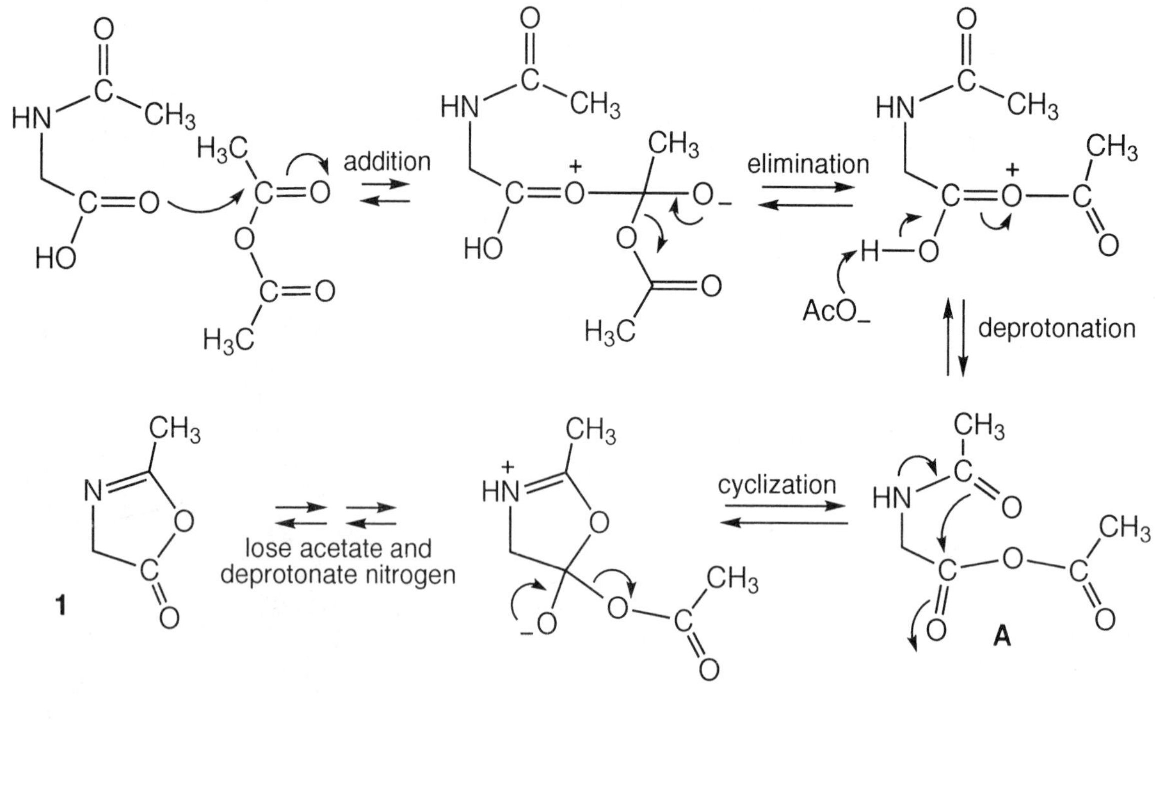

Hydrolysis of azlactone **1** to *N*-acetylglycine is essentially the reverse of this process.

CH3 N O C O H2O addition CH3 N O H2O+ O− proton transfers H N+ CH3 O HO O− elimination O C HN CH3 COOH

Problem 26.40 As we saw in Problem 26.39, reaction of glycine with acetic anhydride (Ac_2O) first gives *N*-acetylglycine. This product then reacts further with acetic anhydride to yield azlactone **1**.

NH2 COOH Ac2O O C HN CH3 COOH CH3 N O C O 1

Azlactone **1** can readily be deprotonated at C_4 by sodium acetate to give the resonance-stabilized anion **A**. Anion **A** then adds to the carbonyl group of benzaldehyde to give, after dehydration, the benzylidene azlactone **2**. This reaction sequence is known as the Erlenmeyer azlactone synthesis. Note the formation of the resonance-stabilized intermediate **B**.

1 CH3 N O H H O AcO− CH3 N O H O− CH3 N − O H O A CH3 N O H − O Ph C H O CH3 N O H Ph H O −O H—OAc

B

2 CH3 N O Ph H O CH3 N (−) O Ph − H OH O (−) AcO− CH3 N O H Ph H OH O

Problem 26.41 The first step in both of these reactions is the *N*-acetylation of **1** to form *N*-acetyl-*N*-methylalanine **2**. *N*-Acylamino acid (**2**) then reacts with acetic anhydride to form the mixed anhydride **A**. We won't write the mechanisms for these reactions, as they were written in the preceding problems, but this provides you with an opportunity to practice.

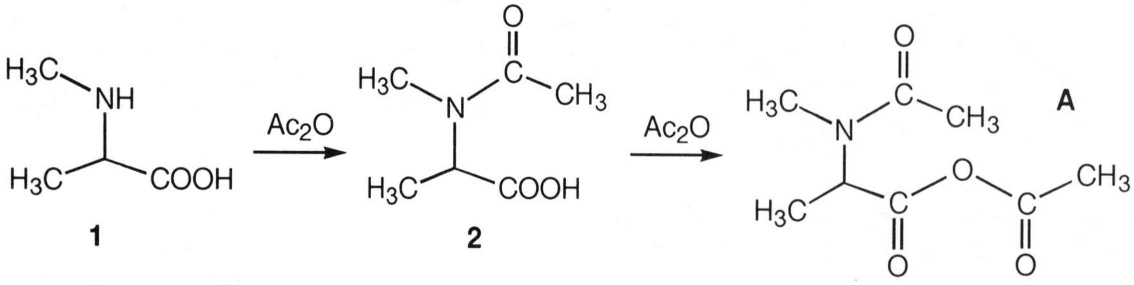

The amide oxygen of **A** is poised to displace acetate in an addition–elimination sequence to yield oxazolium salt **B**. Salt **B** cannot form an azlactone because there is no hydrogen on nitrogen to lose. However, loss of the hydrogen at C_4 affords the compound **C**, commonly known as a munchnone. Munchnones were first investigated by Professor Rolf Huisgen and his collaborators at the University of Munich, hence the name.

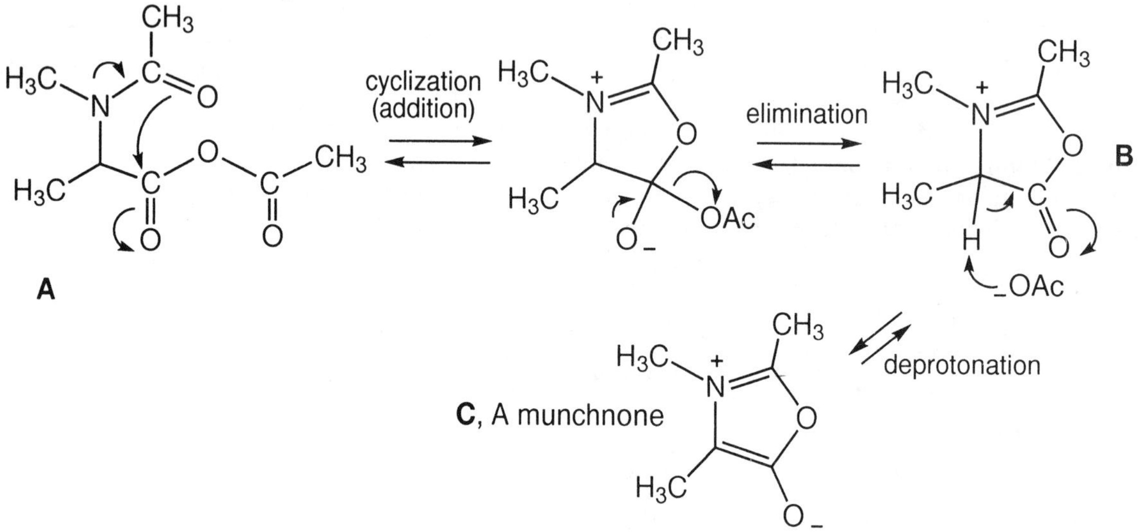

C, A munchnone

Munchnone **C** is the common intermediate that is responsible for both the racemization of **2** and the formation of pyrrole **3**. Note that the hydrogen at the stereogenic carbon C_4 has been lost. When **C** undergoes hydrolysis to form *N*-acetyl-*N*-methylalanine (**2**), this hydrogen can be readded to either face of the achiral munchnone **C**.

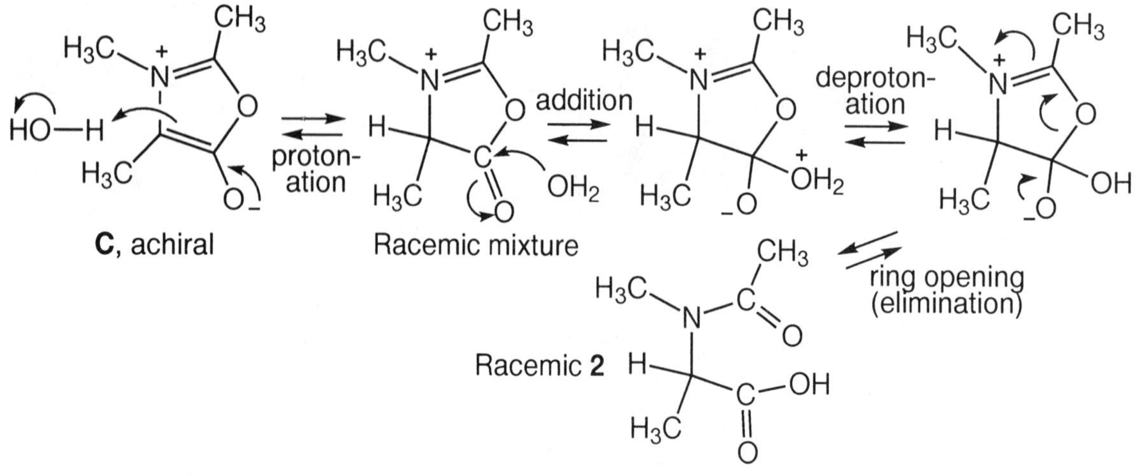

C, achiral

Racemic mixture

Racemic **2**

The ease of cyclodehydration and, hence, racemization (via munchnones or azlactones) upon carboxy activation with reagents such as DCC explains why simple *N*-acylamino acids are not employed in peptide synthesis. *N*-Alkoxycarbonylamino acids (that is, tBoc- or Cbz-protected amino acids) do not undergo cyclodehydration reactions as easily.

Munchnone **C** is also a 1,3-dipole, similar to the sydnone of Problem 26.38, and can be trapped in a 1,3-dipolar addition reaction by DMAD. The primary cycloadduct **D** easily loses CO_2 to give the observed pyrrole **3**. This reaction is aptly known as the Huisgen pyrrole synthesis, and constitutes a pyrrole synthesis of broad scope.

CH_3 H_3C N + O C O H_3C — $COOCH_3$ C≡C $COOCH_3$ DMAD → CH_3 N CH_3 O C O H_3C $COOCH_3$ $COOCH_3$ → CO_2 + H_3COOC $COOCH_3$ H_3C N CH_3 CH_3

C **D** **3**

Problem 26.42

(a) The facile base pairing of A with T (or with U in RNA) and G with C means that the mRNA sequence must be A U G C C C A A A U A G.

(b) Just look up the meanings in Table 26.2 (p. 1384). This message translates to Start, add Pro, add Lys, Stop.

Key Terms

Absolute configuration The arrangement in space of the atoms in an enantiomer.

Acetal The final product in the acid-catalyzed reaction of an aldehyde with an alcohol.

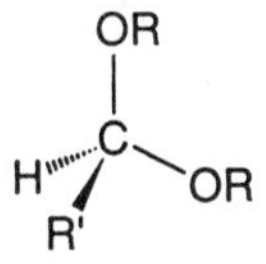

Acetoacetic ester The product of the Claisen condensation of an acetate, such as

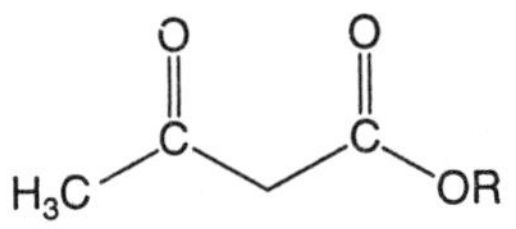

Acetone Dimethyl ketone, the simplest ketone.

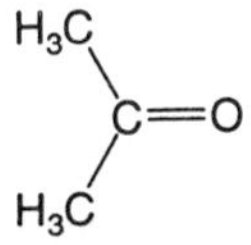

Acetylenes Hydrocarbons of the general formula C_nH_{2n-2}. These molecules, also called alkynes, contain carbon–carbon triple bonds. The parent compound, HC≡CH, is called acetylene, or ethyne.

Acetylide The anion formed by removal by base of a terminal hydrogen from an acetylene.

Achiral Not chiral.

Acid chloride A compound of the structure

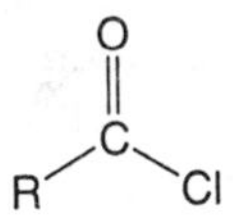

Acid halide A compound of the structure

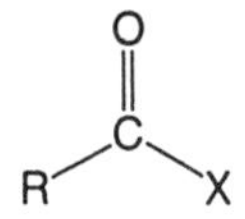

R = F, Cl, Br, or I

Activation energy ($\Delta G^{\ddagger}$) The difference in free energy between the starting material and the transition state in a reaction. It is this amount of energy that is required for a molecule of starting material to be transformed into product.

Active site The region in a protein in which a substrate molecule is bound and in which chemical reaction often takes place.

Acyl compound A compound of the structure

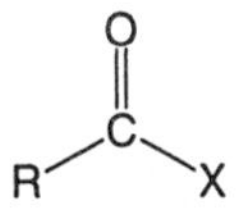

Acylating agent Any of a number of reagents capable of transferring an acyl group (RC=O) to a nucleophilic site, usually N or O. Examples are acetyl chloride and acetic anhydide.

Acylium ion $(RC{\equiv}O)^+$.

Addition–elimination reaction A carbonyl group can be attacked to give a tetrahedral intermediate. If the carbon atom of the carbonyl was originally attached to a good leaving group, it can now be lost with regeneration of the carbonyl.

Alcohol (R–OH) A molecule containing a simple hydroxyl group.

Aldaric acid A diacid derived from an aldohexose by oxidation with nitric acid. It has the structure $HOOC-(CHOH)_4-COOH$. In an aldaric acid, the old aldehyde and primary alcohol ends of the sugar have become identical.

Aldehydes Compounds containing a monosubstituted carbon–oxygen double bond.

$$\mathrm{R(H)C{=}O}$$

Aldohexose A hexose of the structure $OCH-(CHOH)_4-CH_2OH$

Aldol condensation The acid- or base-catalyzed conversion of a ketone or aldehyde into a β-hydroxy aldehyde or β-hydroxy ketone. In acid, the enol is an intermediate; in base, it is the enolate anion.

Aldonic acid A monoacid derived from an aldohexose through oxidation with bromine in water. Only the aldehyde group is oxidized to the acid state. It has the structure $HOOC-(CHOH)_4-CH_2OH$

Aldopentose A pentose of the structure $OCH-(CHOH)_3-CH_2OH$

Aldotetrose A tetrose of the structure $OCH-(CHOH)_2-CH_2OH$

Aldotriose A triose of the structure $OCH-CHOH-CH_2OH$

Alkaloids A nitrogen-containing compound, often polycyclic and generally of plant origin. The term is more loosely applied to other naturally occurring amines.

Alkanes The series of saturated hydrocarbons of the general formula C_nH_{2n+2}.

Alkenes Hydrocarbons of the general formula, C_nH_{2n}. These molecules, also called "olefins," contain carbon–carbon double bonds.

Alkoxide ion The conjugate base of an alcohol, RO^-.

Alkyl compounds Substituted alkanes. One or more hydrogens is replaced by another atom or group of atoms.

Alkyl halides Compounds of the formula $C_nH_{2n+1}X$, where X = F, Cl, Br, or I.

Alkynes Hydrocarbons of the general formula C_nH_{2n-2}. These molecules, also called acetylenes, contain carbon–carbon triple bonds.

Allene A 1,2-diene. A compound containing a carbon atom that is part of two double bonds.

Allyl The group $H_2C{=}CH-CH_2$.

Allylic halogenation Specific formation of a carbon–halogen bond at the position adjacent to a carbon–carbon double bond.

Amide A compound of the structure

$$\mathrm{RO-C(=O)-NR_2}$$

R can be H

Remember that this term also refers to the ions H_2N^-, RHN^-, R_2N^-, $RR'N^-$,

Amine A compound of the structure R_3N:, where R can be H or another group. Cyclic and aromatic amines are common.

Amine inversion The conversion of one pyramidal form of an amine into the other through a planar, sp^2 hybridized transition state.

Amine oxide A compound of the structure $R_3N^+-O^-$. Amine oxides are produced from the reaction of tertiary amines with hydrogen peroxide. See also ***N*-oxide**.

α-Amino acids 2-Amino acetic acids, the monomeric constituents of the polymeric peptides and proteins.

Amino terminus The amino acid at the free amine end of the peptide polymer.

Ammonia, H_3N: The simplest of all amines.

Ammonium ion Tetravalent nitrogen is R_4N^+, R = alkyl, aryl, or H.

Anchimeric assistance The increase in rate of a reaction that proceeds through intramolecular displacement over that expected of an intermolecular displacement.

Angle strain The increase in energy caused by the deviation of an angle from the ideal demanded by the hybridization.

Anhydride A compound formed by formal loss of water from two molecules of a carboxylic acid. The structure is

$$\mathrm{R-C(=O)-O-C(=O)-R}$$

Aniline Aminobenzene.

Anion A negatively charged atom or molecule.

Anisole Methoxybenzene, $PhOCH_3$.

Annulene A cyclic polyene that is at least formally fully conjugated.

Anomers For aldoses, these are sugars differing only in the stereochemistry at C(1). They are C(1) stereoisomers.

Antarafacial motion Migration of a group from one side of a π system to the other in a sigmatropic shift.

anti Elimination An elimination reaction in which the dihedral angle between the breaking bonds, usually C–H and C–L, is 180°.

Antibonding molecular orbital Occupation of an antibonding molecular orbital by an electron destabilizes a molecule.

Arene An aromatic compound containing a benzene ring or rings.

Arndt–Eistert reaction The use of the Wolff rearrangement to elongate the chain of a carboxylic acid by one carbon.

Aromatic character See **aromaticity**.

Aromaticity The special stability of planar, cyclic, fully conjugated molecules with $4n + 2$ π electrons. Such molecules will have molecular orbital systems with all bonding molecular orbitals completely filled and all antibonding and nonbonding molecular orbitals empty.

Arrow formalism A mapping device for chemical reactions. The electron pairs (lone pairs or bond pairs) are "pushed" using curved arrows that show the bonds that are forming and breaking in the reaction.

Atom A neutral atom consists of a nucleus, or core of protons and neutrons, orbited by a number of electrons equal to the number of protons.

Atomic orbital One of the energy levels allowed for an electron in an atom. These orbitals result from the solution of Schrödinger's equation describing the motion of an electron in the vicinity of a nucleus. Atomic orbitals have different shapes, which are determined by quantum numbers. The *s* orbitals are spherically symmetric, *p* orbitals roughly dumbbell shaped, and the *d* and *f* orbitals are even more complicated.

Aufbau principle When adding electrons to a system of orbitals, first fill the lowest energy orbital available before filling any higher energy orbitals. Electron-electron repulsion is minimized by filling systems of equi-energetic orbitals by singly occupying all orbitals with electrons of the same spin before doubly occupying any of them. See **Hund's rule**.

Axial hydrogens The set of six, straight up and down hydrogens in chair cyclohexane. Ring flip interconverts these hydrogens with the set of equatorial hydrogens.

Azetidine A saturated four-membered ring containing one nitrogen atom.

Azides Compounds of the structure

$$R-\ddot{N}^{-}-\overset{+}{N}\equiv N:$$

Aziridine A saturated three-membered ring containing one nitrogen atom; an azacyclopropane.

Azo compounds Compounds of the structure, R–N=N–R.

Bartlett–Condon–Schneider reaction An acid-catalyzed deuterium exchange reaction of alkanes. In the classic example, a trace of alkene catalyzes the exchange of all nine primary hydrogens of isobutane, but leaves the tertiary position unexchanged.

Base In biochemistry, one of the five heterocyclic molecules, adenine (A), thymine (T), uracil (U), cytosine (C), and guanine (G), attached to the 1′ position of the sugar (ribose or deoxyribose) in a nucleotide or nucleoside.

Base pair A hydrogen-bonded pair of bases, always A-T in DNA or A-U in RNA, and C-G in both DNA and RNA.

Base peak The largest peak in a mass spectrum, to which all other peaks are referred.

Benzaldehyde The common (and always used) name for the simplest aromatic aldehyde, "benzenecarboxaldehyde."

Benzene The archetypal aromatic compound; a planar, regular hexagon of sp^2 hybridized carbons. The six $2p$ orbitals overlap to form a six-electron cycle above and below the plane of the ring. The molecular orbital system has three fully occupied bonding molecular orbitals and three unoccupied antibonding orbitals.

Benzhydryl group The group Ph_2CH.

Benzofuran A compound incorporating the structure

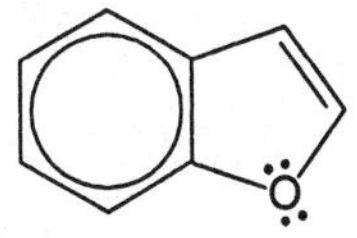

Benzoic acid Ph–COOH.

Benzoin condensation An aldehyde containing no α-hydrogens reacts with cyanide ion to form a cyanohydrin. The old aldehydic hydrogen is now acidic because of the resonance stabilization afforded by the cyano group. Its removal leads to a condensation reaction and the formation of an α-hydroxy ketone.

Benzothiophene A compound incorporating the structure

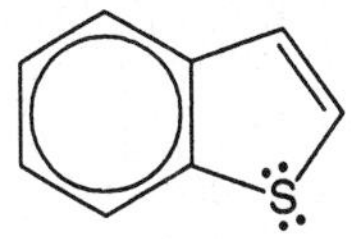

Benzyl group The group $Ph-CH_2$.

Benzyne 1,2-Dehydrobenzene, C_6H_4.

Benz[*a*]pyrene A powerfully carcinogenic polynuclear aromatic compound composed of five fused benzene rings.

Birch reduction The conversion of aromatic compounds into 1,4-cyclohexadienes through treatment with sodium in liquid ammonia–ethyl alcohol. Radical anions are the first formed intermediates.

tBOC A protecting group for the amino end of an amino acid that works by transforming the amine into a less basic carbamate.

Boltzmann distribution The range of energies of a set of molecules at a given temperature.

Bond dissociation energy (BDE) The amount of energy that must be applied to break a bond into two neutral species. See **homolytic bond cleavage.**

Bonding molecular orbital Occupation of a bonding orbital by an electron stabilizes a molecule.

Bredt's rule Bredt noticed that there were no examples of bicyclic molecules with double bonds at the bridgehead position.

Bridged rings In a bridged bicyclic molecule, two rings share more than two atoms.

Bridgehead position The bridgehead positions are shared by the rings in a bicyclic molecule. In a bicyclic molecule, the three bridges emanate from the bridgehead positions.

Bromonium ion A three-membered ring containing bromine that is formed by the reaction of an alkene with Br_2. The bromine atom in the ring is positively charged.

***N*-Bromosuccinimide (NBS)** An effective agent for allylic bromination.

Brønsted acid A donor of a proton.

Brønsted base An acceptor of a proton.

Bullvalene Bullvalene is a molecule with a fluxional structure. Every carbon of this $(CH)_{10}$ compound is bonded on time average to each of the other nine carbons.

Butyl group The group $CH_3CH_2CH_2CH_2$.

***sec*-Butyl group** The group $CH_3CH_2CH(CH_3)$.

***tert*-Butyl group** The group $(CH_3)_3C$.

Cahn–Ingold–Prelog priority system An arbitrary system for naming stereoisomers. It determines a priority system for ordering groups.

Cannizzaro reaction The redox reaction of an aldehyde containing no α-hydrogens with hydroxide ion. Addition of hydroxide to the aldehyde is followed by hydride transfer to another aldehyde. Protonation generates a molecule of the carboxylic acid and the alcohol related to the original aldehyde.

Carbamates Esters of carbamic acid. These molecules do not decarboxylate (lose CO_2) easily.

$$RO{-}C({=}O){-}NR_2$$

Carbamic acid A compound of the structure.

$$HO{-}C({=}O){-}NR_2$$

These acids easily decarboxylate to give amines.

Carbanion A compound containing a negatively charged carbon atom. A carbon-based anion.

Carbene A short-lived neutral intermediate containing a divalent carbon atom. See also **singlet** and **triplet carbene.**

Carbenium ion A name suggested for a molecule containing a trivalent, positively charged carbon. It is not widely used. See **carbocation.**

Carbinolamine The initial product in the reaction between a carbonyl-containing molecule, $R_2C{=}O$, and an amine. It is analogous to a hemiacetal.

$$R_2C(NHR)(OH)$$

Carbocation The compromise and currently widely used name for a molecule containing a trivalent, positively charged carbon atom.

Carbohydrate A molecule whose formula can be factored into $C_x(H_2O)_y$. A "sugar" or "saccharide."

Carbonic acid An unstable acid, H_2CO_3.

$$HO{-}C({=}O){-}OH$$

Carbonium ion The traditional name for a molecule containing a trivalent, positively charged carbon. It has fallen into disuse.

Carbonyl compound A compound containing a carbon–oxygen double bond.

Carbonyl group The carbon–oxygen double bond.

$$C{=}O$$

Carboxy terminus The amino acid at the free carboxylic acid end of the peptide polymer.

Carboxylate anion The resonance-stabilized anion formed on deprotonation of a carboxylic acid.

$$\left[R{-}C({=}O){-}O^- \longleftrightarrow R{-}C(O^-){=}O \right]$$

Carboxylic acid A compound of the structure

$$R{-}C({=}O){-}OH$$

Catalyst A catalyst functions to increase the rate of a chemical reaction. It is ultimately unchanged by the reaction and functions not by changing the energy of the starting material or product but by providing a lower energy pathway. Thus it operates to lower the energies of the transition states involved in the reaction.

Cation A positively charged atom or molecule.

Cationic polymerization A reaction in which an initially formed carbocation adds to an alkene that in turn adds to another alkene. Repeated additions can lead to polymer formation.

Cbz A protecting group for the amino end of an amino acid that works by transforming the amine into a less basic carbamate.

Cellulose A polymer of glucose in which C(4) of one glucose is linked in β fashion to C(1) of another.

Cephalosporin A powerful antibiotic related to penicillin incorporating a β-lactam ring fused to a six-membered ring containing sulfur.

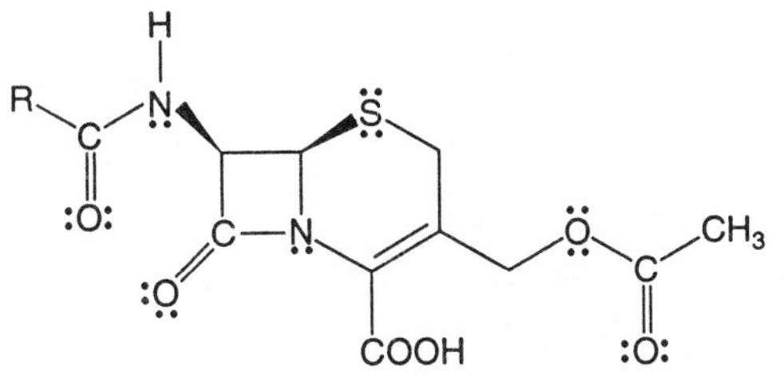

Chain reaction A cycling reaction in which the species necessary for the first step of the reaction is produced in the last step. This intermediate then recycles and starts the process over again.

Chemical shift (δ) The position, on the ppm scale, of a peak in an NMR spectrum. The chemical shift for ^1H and ^{13}C is given relative to a standard, TMS, and is determined by the chemical environment surrounding the nucleus.

Chichibabin reaction Nucleophilic addition of an amide ion to pyridine (or a related heteroaromatic compound) leading to an aminopyridine. A key step involves hydride transfer.

Chiral A chiral molecule is not superimposable on its mirror image.

Chirality The ability of a molecule to exist in two nonsuperimposable mirror-image forms; "handedness."

Chromatogram A plot of amount of component versus time for any chromatographic technique.

Chymotrypsin An enzyme that cleaves sequences of amino acids after any amino acid containing an aromatic side chain.

cis "On the same side." Applied to specify stereochemical (spatial) relationships in ring compounds and alkenes.

s-cis The less stable, coiled form of a 1,3-diene.

Claisen condensation A condensation reaction of esters in which an ester enolate adds to the carbonyl group of another ester. The result of this addition–elimination process is a β-keto ester.

Claisen–Schmidt condensation A crossed aldol condensation of an aldehyde without α-hydrogens with a ketone that does have at least one α-hydrogen.

β-Cleavage The fragmentation of a radical into a new radical and an alkene through breaking of the α–β bond.

Codon A three-base sequence in a polynucleotide that directs the addition of a particular amino acid to a growing chain of amino acids. Some codons also give directions: "start assembly" "stop assembly," and so on.

Coenzyme A molecule able to carry out a chemical reaction with another molecule only in cooperation with an enzyme. The enzyme's function is often to bring the substrate and the coenzyme together.

Concerted reaction A single-barrier process. In a concerted reaction, starting material is converted into product with no intermediate structures.

Conformation The three-dimensional structure of a molecule. Conformational isomers are interconverted by rotations about bonds.

Conformational analysis The study of the relative energies of conformational isomers.

Conformational enantiomers Enantiomers interconvertable through (generally easy) rotations around bonds within the molecule.

Conjugate acid Some molecule plus a proton. Conjugate acids and bases are related by the gain and loss of a proton.

Conjugate base Some molecule less a proton. Conjugate acids and bases are related by the gain and loss of a proton.

Conjugated double bonds Double bonds in a 1,3–relationship are conjugated.

Conrotation In a conrotatory process, the end *p* orbitals of a polyene HOMO rotate in the same sense (both clockwise or both counterclockwise).

Cope elimination The thermal formation of alkenes through the pyrolysis of *N*-oxides.

Cope rearrangement This [3,3] sigmatropic shift converts one 1,5-diene into another.

Coupling constant (*J*) The magnitude (in hertz) of *J*, the measure of the spin–spin interaction between two nuclei.

Covalent bond A bond formed by the sharing of electrons through the overlap of atomic or molecular orbitals.

Crossed aldol condensation An aldol condensation between two different carbonyl compounds. This reaction is not very useful unless strategies are employed to limit the number of possible products.

Crown ether A cyclic polyether often capable of forming complexes with metal ions. The ease of complexation depends on the size of the ring and the number of heteroatoms in the ring.

Cryptand A three-dimensional, bicyclic, counterpart of a crown ether. Various heteroatoms (O, N, S) act to complex metal ions that fit into the cavity.

Cumulene Any molecule containing at least three consecutive double bonds, $R_2C{=}C{=}C{=}CR_2$.

Curtius rearrangement The thermal or photochemical decomposition of an acyl azide to give an isocyanate.

Cyanides Compounds of the structure RCN. These compounds are also commonly called nitriles.

Cyanogen bromide (BrCN) A reagent able to cleave peptide chains after a methionine residue.

Cyanohydrin The product of addition of hydrogen cyanide to a carbonyl compound.

R R
C
NC OH

Cycloaddition reaction A reaction in which two π systems are converted into a ring. The Diels–Alder reaction and the 2 + 2 reaction of a pair of ethylenes to give a cyclobutane are examples.

Cycloalkenes Ring compounds containing a double bond within the ring.

Cycloheptatrienylium ion See **tropylium ion.**

Cyclopentadienyl anion A five-carbon aromatic anion containing 6 π electrons ($4n + 2$, $n = 1$).

Daughter ion In MS, an ion formed by the fragmentation of the first-formed parent ion.

Decarboxylation The loss of carbon dioxide, a common reaction of 1,1-diacids and β-keto acids.

Decoupling The removal of coupling between hydrogens or other nuclei (see **coupling constant**) through either chemical exchange or electronic means.

Degenerate reaction In a degenerate reaction, the starting material and product have the same structure.

Degree of unsaturation (Ω) In a hydrocarbon, this is the total number of π bonds and rings.

Dehydrobenzene See **benzyne.**

Delocalization energy The energy lowering conferred by the delocalization of electrons. In benzene, this is the amount by which benzene is more stable than the hypothetical 1,3,5-cyclohexatriene containing three localized double bonds. See **resonance energy.**

Denaturing The destruction of the higher order structures of a protein, sometimes reversible, sometimes not.

Deoxyribonucleic acid (DNA) A polymer of nucleotides made up of deoxyribose units connected by phosphoric acid links. Each sugar is attached to one of the bases, A, T, G, or C.

Detergent A long-chain alkyl sulfonic acid salt.

Dewar forms Resonance forms for benzene in which overlap between 2*p* orbitals on two para carbons is emphasized. These forms superficially resemble Dewar benzene (bicyclo[2.2.0]hexa-2,5-diene).

Dextrorotatory The rotation of the plane of plane-polarized light in the clockwise direction.

Dial A molecule containing two aldehyde groups, a dialdehyde.

Diastereomers Stereoisomers that are not mirror images.

Diastereotopic Diastereotopic hydrogens are different both chemically and spectroscopically under all circumstances.

Diazirine A three-membered ring containing both a double bond and two nitrogen atoms.

Diazo compounds Compounds of the structure $R_2C{=}N_2$.

Diazo ketone A compound of the structure

O
C
R CHN$_2$

Diazonium ion The group N_2^+ as in RN_2^+.

Diazotic acid An intermediate in the nitrosation reactions of amines, R–N=N–OH. The enol form of a nitroso compound.

Dicyclohexylcarbodiimide (DCC) A dehydrating agent effective in the coupling of amino acids through the formation of amide bonds.

N=C=N

Dieckmann condensation An intramolecular, or cyclic, Claisen condensation.

Diels–Alder reaction The concerted reaction of an alkene or alkyne with a 1,3-diene to form a six-membered ring.

Diene Any molecule containing two double bonds.

Dihedral angle The torsional, or twisting angle between two bonds. In an X–C–C–X system, the dihedral angle is the angle between the X–C–C and C–C–X planes.

Diimide (HN=NH) A nonmetallic hydrogenating agent.

Diol A molecule containing two OH groups. Also called a glycol.

Dione A compound containing two ketone groups, a diketone.

Dioxirane A three-membered ring containing two oxygen atoms. A versatile oxidizing agent.

Dipole moment A dipole moment in a molecule results when two opposite charges are separated.

1,3-Dipoles A class of reactive molecule, containing both positive and negative charges. These species undergo addition to π systems to give five-membered rings.

Diradical A species containing two unpaired electrons, usually on different atoms.

Disproportionation The reaction of a pair of radicals to give a saturated and unsaturated molecule by abstraction of a hydrogen by one radical from the position adjacent to the free electron of the other radical.

Disrotation In a disrotatory process, the end *p* orbitals of a polyene HOMO rotate in opposite senses (one clockwise and the other counterclockwise).

Disulfide bridges The attachment of amino acids through sulfur–sulfur bonds formed from the oxidation of cysteine CH_2SH side chains. Disulfide bridges can be formed within a single peptide or between two peptides.

Double bond Two atoms can be attached by a double bond composed of one σ bond and one π bond.

E1 Reaction The unimolecular elimination reaction. The ionization of the starting material is followed by the loss of a proton to base.

E1cB Reaction An elimination reaction in which the first step is loss of a proton to give an anion. The anion then internally displaces the leaving group in a second step.

E2 Reaction The bimolecular elimination reaction. The proton and leaving group are lost in a single, base-induced step.

Eclipsed ethane The conformation of ethane in which all carbon–hydrogen bonds are as close as possible. This conformation is not an energy minimum, but the top of the barrier separating two molecules of the stable, staggered conformation of ethane.

Edman degradation The phenyl isothiocyanate-induced cleavage of the amino acid at the amino terminus of a peptide. Successive applications of the Edman technique can determine the sequence of a peptide.

Electrocyclic reaction The interconversion of a polyene and a ring compound. The end *p* orbitals of the polyene rotate so as to form the new σ bond of the ring compound.

Electron A particle of tiny mass (1/1845 of a proton) and a single negative charge.

Electron affinity A measure of the tendency for an atom or molecule to accept an electron.

Electronegativity The tendency for an atom to attract electrons.

Electronic spectroscopy The measurement of the absorption of energy when electromagnetic radiation of the proper energy is provided. An electron is promoted from the HOMO to the LUMO.

Electrophile A lover of electrons, a Lewis acid.

Electrophilic aromatic substitution The classic substitution reaction of aromatic compounds with Lewis acids. A hydrogen attached to the benzene ring is replaced by the Lewis acid and the aromatic ring is retained in the overall reaction.

Electrophoresis A technique for separating amino acids or chains of amino acids that takes advantage of the different charge states of different amino acids (or their polymers) at a given pH.

Enamine The nitrogen analogue of an enol, a vinyl amine. These compounds are nucleophilic and useful in alkylation reactions.

Enantiomers Nonsuperimposable mirror images.

Enantiotopic Enantiotopic hydrogens are chemically and spectroscopically equivalent except in the presence of optically active (single enantiomer) reagents.

Endergonic In an endergonic reaction the products are less stable than the starting materials.

endo Aimed "inside" the cage in a bicyclic molecule. In a Diels–Alder reaction, the endo product generally has the substituents aimed toward the newly produced double bond.

Energy level See **atomic orbital** and **wave function**.

Enol A vinyl alcohol. These compounds usually equilibrate with the more stable keto forms.

Enolate The resonance-stabilized anion formed on treatment of an aldehyde or ketone containing an α-hydrogen with base.

Enthalpy change ($\Delta H°$) The difference in total bond energies between starting material and product in their standard states.

Entropy change ($\Delta S°$) The difference in disorder between the starting material and product in their standard states.

Episulfide A saturated three-membered ring containing a single sulfur atom. See **Thiirane**.

Episulfonium ion A three-membered ring containing a trivalent, positively charged sulfur atom.

Epoxide A saturated three-membered ring containing a single oxygen atom. See **Oxirane**.

Equatorial hydrogens The set of six hydrogens, also "up and down," but more or less in the plane of the ring in chair cyclohexane. These hydrogens are interconverted with the set of axial hydrogens through ring "flipping" of the chair.

Equilibrium constant (*K*) The equilibrium constant is related to the difference in energy between starting material and products ($\Delta G°$) in the following way: $K = e^{-\Delta G°/RT}$.

Essential amino acid Any of the 10 amino acids that cannot be synthesized by humans and must be ingested directly.

Ester A compound of the structure

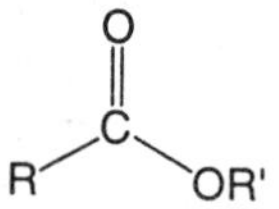

$R-C(=O)-OR'$

Ester hydrolysis The conversion of an ester into an acid through treatment with an acid catalyst in excess water. The reverse of Fischer esterification. This reaction also occurs in base, and is called "saponification."

Ethene The simplest alkene is $H_2C{=}CH_2$. It is usually known as ethylene.

Ether A compound of the general structure, ROR, or ROR′.

Ethyl compounds Substituted ethanes; $CH_3CH_2{-}X$ compounds.

Ethylene The simplest alkene is $H_2C{=}CH_2$. It is more properly known as ethene, a name that is rarely used.

Exergonic In an exergonic reaction, the products are more stable than the starting materials.

exo Aimed "outside" the cage in a bicyclic molecule. In a Diels–Alder reaction, the exo product generally has the substituents aimed away from the newly produced double bond.

Extinction coefficient The proportionality constant ε in Beer's law, $A = \log I_0/I = \varepsilon lc$.

Fatty acids Long-chain carboxylic acids generated by the hydrolysis of fats. Fatty acids are derived from acetic acid and always contain an even number of carbons.

First-order reaction A reaction for which the rate depends on the product of a rate constant and the concentration of a single reagent.

Fischer esterification The conversion of a carboxylic acid into an ester by treatment with an acid catalyst in excess alcohol. The reverse of ester hydrolysis.

Fischer indole synthesis A synthesis of indoles starting from phenylhydrazones.

Fischer projection A schematic stereochemical representation. In sugars, the aldehyde group is placed at the top and the primary alcohol at the bottom. Horizontal bonds are taken as coming toward the viewer and vertical bonds as retreating. If the OH of the H−C−OH adjacent to the CH_2OH group is on the right, the molecule is a D sugar; if it is on the left, it is an L sugar.

Fluxional structure In a molecule with a fluxional structure, a given atom does not have fixed nearest neighbors. Instead, the framework atoms move about over time, each being bonded on time average to all the others.

Force constant A property of a bond related to the bond strength: to the stiffness of the bond. Bonds with high force constants absorb at high frequency in the IR.

Formaldehyde The simplest possible aldehyde.

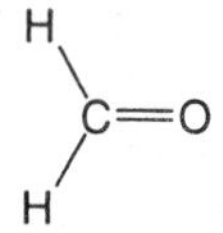

$H_2C{=}O$

Four-center transition state The chair-like transition state for the Cope rearrangement in which carbons 1 and 1′ and 3 and 3′ are within bonding distance (eclipsed) but carbons 2 and 2′ are as far apart as possible.

Fragmentation pattern The characteristic spectrum of ions formed by decomposition of a parent ion produced in a mass spectrometer when a molecule is bombarded by high-energy electrons.

Free radical A neutral molecule containing an odd, unpaired electron. Also simply called "radical."

Friedel–Crafts acylation The electrophilic substitution of aromatic molecules with acyl chlorides facilitated by strong Lewis acids, usually $AlCl_3$.

Friedel–Crafts alkylation The electrophilic substitution of aromatic molecules with alkyl chlorides catalyzed by strong Lewis acids, usually $AlCl_3$.

Frost circle A device used to find the relative energies of the molecular orbitals of planar, cyclic, fully conjugated molecules. A polygon corresponding to the ring size of the molecule is inscribed in a circle, vertex down. The intersections of the polygon with the circle give the relative positions of the molecular orbitals.

Fully conjugated In a fully conjugated molecule, every carbon has a *p* orbital that overlaps effectively with the *p* orbitals on the adjacent atoms.

Fulvene A 5-methylene-1,3-cyclopentadiene. These are often synthesized through a Knoevenagel condensation using the cyclopentadienide anion and an aldehyde or ketone.

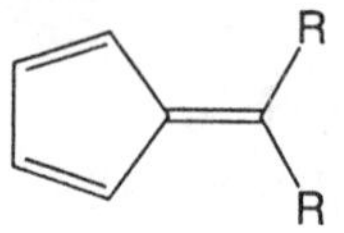

R
R

Functional group An atom or group of atoms that generally reacts the same way no matter what molecule it is in.

Furan An aromatic five-membered ring compound containing four CH units and one O atom.

Furanose A sugar containing a five-membered cyclic ether.

Furanoside A furanose in which the anomeric OH has been converted into an acetal.

Fused rings Two rings sharing only two carbons.

Gabriel amine synthesis A method of forming primary amines without overalkylation to give more substituted amines. Phthalimide is used as a masked amine to introduce the nitrogen atom.

Gabriel malonic ester synthesis A synthesis of amino acids that uses phthalimide as a source of the amine nitrogen. Overalkylation is avoided by decreasing the nucleophilicity of the nitrogen in this way.

Gas chromatogaphy (GC) A method of separation in which molecules are forced to equilibrate between the moving gas phase and a stationary phase packed in a column. The less easily a molecule is adsorbed in the stationary phase, the faster it moves through the column.

GC/IR The combination of gas chromatography and infrared spectroscopy in which the molecules separated by a gas chromatograph are led directly into an infrared spectrometer for analysis.

GC/MS The combination of gas chromatography and mass spectrometry in which the molecules separated by a gas chromatograph are led directly into a mass spectrometer for analysis.

Gel-filtration chromatography A chromatographic technique that relies on polymeric beads containing molecule-sized holes. Molecules that fit easily into the holes pass more slowly down the column than larger molecules, which fit less well into the holes.

Geminal 1,1-Disubstituted.

Gibbs free energy change ($\Delta G°$) The difference in free energy during a reaction. The parameter $\Delta G°$ is composed of an enthalpy ($\Delta H°$) term and an entropy ($\Delta S°$) term. $\Delta G° = \Delta H° - T\Delta S°$.

Glyceraldehyde The aldotriose, OCH–CHOH–CH_2OH

1,2-Glycol A vicinal dialcohol. A molecule bearing hydroxyl groups on adjacent carbons. 1,2-Glycols come from the treatment of alkenes with osmium tetroxide or potassium permanganate, followed by hydrolysis, or by the acid- or base-catalyzed opening of epoxides.

Glycol A dialcohol. See **diol.**

Glycoside A sugar in which the anomeric OH at C(1) has been converted into an OR group.

Grignard reagent (RMgX) A strongly basic organometallic reagent formed from a halide and magnesium in an ether solvent. An important and characteristic reaction is the addition to carbonyl groups.

Haloform A compound of the structure HCX_3, where X is F, Cl, Br, or I.

Haloform reaction The conversion of a methyl ketone into a molecule of a carboxylic acid and a molecule of haloform. The trihalo carbonyl compound is formed, base adds to the carbonyl group, and the trihalomethyl anion ($^-{:}CX_3$) is eliminated. The reaction works for X = Cl, Br, or I.

Hammond postulate The transition state for an endothermic reaction will resemble the product. It can be equivalently stated as, The transition state for an exothermic reaction will resemble the starting material.

Hantzsch synthesis A synthesis of pyridines that uses multiple condensation reactions of β-ketoesters, ammonia, and an aldehyde. A dihydropyridine is an intermediate.

Haworth form The representation of sugars in which they are shown as planar rings.

Heat of formation ($\Delta H_f°$) The heat evolved or required for the formation of a molecule from its constituent elements in their standard states. The more negative the heat of formation, the more stable the molecule.

Heisenberg uncertainty principle For an electron, the uncertainty in position times the uncertainty in momentum (or speed) is a constant. We cannot know the exact position and momentum (speed) of an electron at the same time.

α-Helix A right-handed coiled form adopted by many proteins in their secondary structures.

Hell–Volhard–Zelinsky reaction The conversion of a carboxylic acid into either the α-bromo acid or the α-bromo acid bromide through reaction with PBr_3/Br_2.

Hemiacetal The initial product when an alcohol adds to an aldehyde or ketone.

Hemoglobin The protein in human blood responsible for oxygen transport.

Heteroaromatic compound See **heterobenzene.**

Heterobenzene A benzene ring in which one (or more) ring carbons is replaced with another heavy (non-hydrogen) atom.

Heterocyclic compound A ring compound containing at least one heteroatom in the ring.

Heterogeneous catalysis A catalytic process in which the catalyst is insoluble.

Heterolytic bond cleavage The breaking of a bond to produce a pair of oppositely charged ions.

$$\text{X—Y} \longrightarrow \text{X}^+ + \,^-\!:\text{Y}$$

Hexose A six-carbon sugar.

High-performance liquid chromatography (HPLC) An especially effective version of column chromatography in which the stationary phase consists of many tiny spheres that provide an immense surface area for absorption.

Hofmann elimination The formation of the less substituted alkene in an elimination reaction.

Hofmann rearrangement The formation of amines through the treatment of amides with bromine and base. An intermediate isocyanate is hydrolyzed to a carbamic acid that decarboxylates.

HOMO The highest occupied molecular orbital.

Homogeneous catalysis A catalytic process in which the catalyst is soluble.

Homolytic bond cleavage The breaking of a bond to form two neutral species.

$$\text{X—Y} \longrightarrow \text{X}\cdot + \cdot\text{Y}$$

Homotopic Homotopic hydrogens are identical, both chemically and spectroscopically, under all circumstances.

Hückel's rule All planar, cyclic, fully conjugated molecules with $4n + 2$ π electrons will be aromatic (especially stable). The rule works because such molecules will have molecular orbital systems in which all bonding molecular orbitals are completely full and in which no antibonding or nonbonding molecular orbitals are occupied.

Hund's rule For a set of equi-energetic orbitals, the electronic configuration with the maximum number of parallel spins is the lowest in energy. That is,

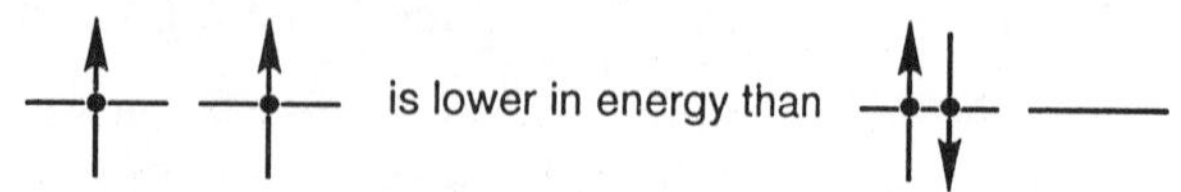

Hunsdiecker reaction The conversion of silver salts of carboxylic acids into alkyl halides, usually bromides.

Hybridization A mathematical model in which atomic orbital wave functions are combined to produce new, combination, or hybrid orbitals. The new orbitals are made up of fractions of the pure atomic orbital wave functions. Thus, an sp^3 hybrid is made of three parts *p* wave function and one part *s* wave function.

Hydrate The product of the reaction of a carbonyl compound with water.

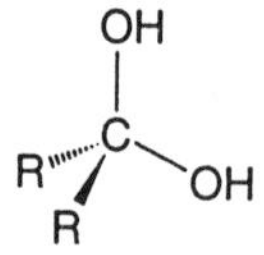

Hydration The addition of water to a molecule. This reaction is generally acid catalyzed.

Hydride A negatively charged hydrogen ion ($H{:}^-$) bearing a pair of electrons.

Hydride shift The migration of hydrogen with a pair of electrons ($H{:}^-$).

Hydroboration The addition of BH_3 (in equilibrium with the dimer, B_2H_6) to π systems to form alkylboranes. These alkylboranes can be further converted into alcohols. Addition to an unsymmetrical alkene proceeds so as to give the less substituted alcohol.

Hydrocarbon cracking The thermal treatment of high molecular weight hydrocarbons to give lower molecular weight fragments. Bonds are broken to give radicals that abstract hydrogen to give alkanes, undergo β-cleavage, and disproportionate to give alkenes and alkanes.

Hydrocarbons Molecules containing only carbon and hydrogen.

Hydrogen bonding A low-energy bond between a pair of electrons, usually on oxygen, and a hydrogen.

Hydrogenation Addition of hydrogen (H_2) to the π bond of an alkene to give an alkane. A soluble or insoluble metallic catalyst is necessary. Alkynes also undergo hydrogenation to give alkanes, or under special conditions, alkenes.

Hydroperoxide (ROOH) The prefix *per* means "extra oxygen."

Hydrophilic "Water-loving," hence a polar group soluble in water.

Hydrophilic group Literally a "water-loving" group, that is, a polar side chain or group.

Hydrophobic "Water-hating," hence a nonpolar group insoluble in water.

Hyperconjugation The stabilization of carbocations through the overlap of a filled σ orbital with the empty $2p$ orbital on carbon.

Imide A compound containing the structure O=C–NH–C=O, such as

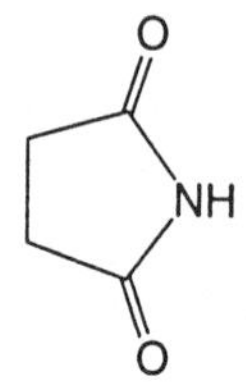

Imine The nitrogen analogue of a ketone or aldehyde.

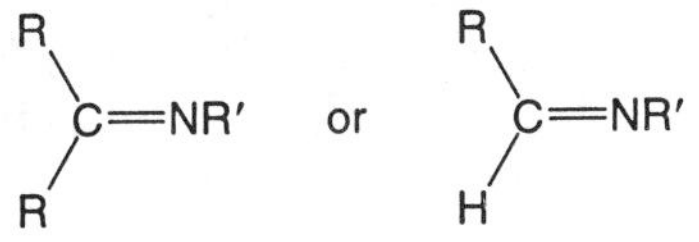

$R_2C{=}NR'$ or $RHC{=}NR'$

Immonium ion A compound of the structure

$RR'C{=}NR_2^+$

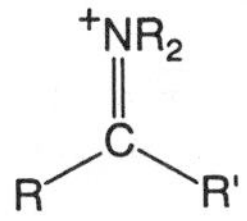

Indole alkaloids A family of alkaloids, all of which contain the indole ring system.

Indole Any compound containing the following ring system:

NH

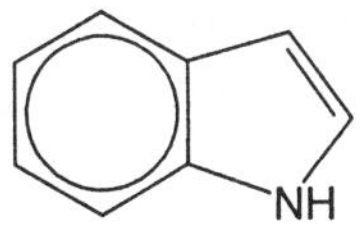

Inductive effects Electronic effects transmitted through the σ bonds. All bonds between different atoms are polar, and thus many molecules contain dipoles. These dipoles can affect reactions through induction.

Infrared spectroscopy (IR) In IR, absorption of IR energy causes bonds to vibrate. The characteristic vibrational frequencies can be used to determine the functional groups present in a molecule.

Inhibitors A species that can react with a radical, thus destroying it. Inhibitors interrupt chain reactions.

Initiation step The first step in a radical chain reaction in which a free radical is produced that can start the chain-carrying or propagation steps.

Integration The determination of the relative numbers of hydrogens corresponding to the signals in an NMR spectrum.

Inversion of stereochemistry The change in the handedness of a molecule during certain reactions. Generally, inversion of stereochemistry means that an (*R*) starting material would be transformed into an (*S*) product.

Ion A charged atom or molecule.

Ion exchange chromatography A separation technique that relies on differences in the affinity of molecules for a charged substrate.

Ionic bond The electrostatic attraction between a positively charged atom or group of atoms and a negatively charged atom or group of atoms.

Ionization potential A measure of the tendency of an atom or molecule to lose an electron.

Ipso attack Addition to an aromatic ring at a position already occupied by a non-hydrogen substituent.

Isobutyl group The group $(CH_3)_2CHCH_2$.

Isocyanate A compound of the structure

$R{-}N{=}C{=}O$

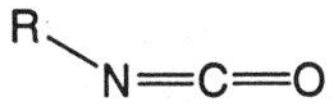

Isoelectric point The pH at which the amount of an amino acid present as the ammonium ion exactly equals the amount present as the carboxylate anion.

Isomers Molecules of the same formula but different structures.

Isoprene 2-Methyl-1,3-butadiene.

Isoprene rule Most terpenes are composed of isoprene units combined in "head-to-tail" fashion.

Isoprenoids Compounds whose carbon skeleton is composed of isoprene units. Terpenes.

Isopropyl group The group $(CH_3)_2CH$.

Isoquinoline Any compound containing the following ring system:

N:

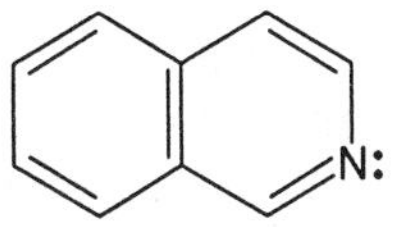

Isotope effect The effect on the rate of a reaction of the replacement of an atom with an isotope. The replacement of hydrogen with deuterium is especially common.

Karplus curve The dependence between the coupling constant of two hydrogens on adjacent atoms and the dihedral angle between the carbon–hydrogen bonds.

Kekulé forms Resonance forms for benzene in which overlap between adjacent carbons is emphasized. These forms superficially resemble 1,3,5-cyclohexatriene.

Ketene A compound of the structure

$RR'C{=}C{=}O$

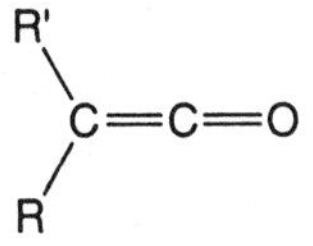

Keto sugar A sugar containing not the usual aldehyde group, but a ketone.

Ketone A compound containing a carbon–oxygen double bond in which the carbon is attached to two other carbons (not hydrogen).

Ketose A sugar containing not the usual aldehyde group, but a ketone.

Kiliani–Fischer synthesis A method of lengthening the chain of an aldose by one carbon atom. A pair of sugars, epimeric at the new C(1) is produced.

Kinetic control A reaction in which the product distribution is determined by the heights of the different transition states leading to products.

Kinetic enolate The most easily formed enolate. It may or may not be the same as the most stable possible enolate, the thermodynamic enolate.

Kinetic resolution A technique for separating a pair of enantiomers based on the selective transformation of one of them, often by an enzyme.

Kinetics The determination of the rates of reactions.

Knoevenagel condensation Any of a number of condensation reactions related to the crossed aldol condensation. A stabilized anion, often an enolate, is first formed and then adds to the carbonyl group of another molecule. Dehydration often leads to the formation of the final product.

Kolbe electrolysis The electrochemical conversion of carboxylate anions into hydrocarbons. The carboxyl radical produced loses CO_2 to give an alkyl radical that dimerizes.

Lactam A cyclic amide.

Lactone A cyclic ester.

α-Lactone A three-membered lactone (cyclic ester).

Le Chatelier's principle A system at equilibrium adjusts so as to relieve any stress upon it.

Least-motion mechanism The reaction mechanism involving the minimum translational and rotational motions for the atoms involved.

Leaving group The departing group in a substitution or elimination reaction.

Levorotatory The rotation of the plane of plane-polarized light in the counterclockwise direction.

Lewis acid Any species that reacts with a Lewis base.

Lewis base Any species with a reactive pair of electrons.

Lewis structure In a Lewis structure, every electron (except the 1*s* electrons for atoms other than hydrogen or helium) is shown as a dot. In slightly more abstract Lewis structures, electrons in bonds are shown as lines connecting atoms.

Lithium diisopropylamide (LDA) An effective base for alkylation reactions of carbonyl compounds. This compound is a strong base, but a poor nucleophile, and thus is effective at removing α-hydrogens but not at additions to carbonyl compounds.

Lobry de Bruijn–Alberda van Ekenstein reaction The base-catalyzed interconversion of aldo and keto sugars. The key intermediate is the double enol formed by protonation of an enolate.

Lone-pair electrons Electrons in an orbital that is not involved in binding atoms. See **nonbonding electrons**.

Long-range coupling Any coupling between nuclei separated by more than three bonds. It is not usually observed in ^1H NMR spectroscopy.

LUMO The lowest unoccupied molecular orbital

Lysergic acid diethylamide (LSD) An indole alkaloid of potent hallucinogenic properties.

Magid's Second Rule "Always try the Michael reaction first."

Magid's Third Rule "In times of desperation and/or despair, when all attempts at solution seem to have failed, try a hydride shift."

Malonic ester synthesis The alkylation of malonic esters to produce substituted 1,1-diesters, followed by hydrolysis and decarboxylation of the malonic acids to give substituted acetic acids.

Malonic ester
The 1,1-diester, $ROOC{-}CH_2{-}COOR$.

Markovnikov's rule In additions of Lewis acids to alkenes, the Lewis acid will add to the less substituted end of the alkene. The rule works because additions of Lewis acids produce the more substituted, more stable carbocation.

Mass spectrometry (MS) The analysis of the ions formed by bombardment of a molecule with high-energy electrons. High-resolution MS can give the molecular formula of an ion, and an analysis of the fragmentation pattern can give information about the structure.

McLafferty rearrangement A common process in MS through which a hydrogen in the γ-position of a carbonyl compound is transferred to oxygen to give an enol radical cation and an alkene.

Meerwein–Ponndorf–Verley–Oppenauer (MPVO) equilibration A method of oxidizing alcohols to carbonyl compounds, or, equivalently, of reducing carbonyls to alcohols, that uses an aluminum atom to clamp the partners in the reaction together. A hydride shift is involved in the crucial step.

Meisenheimer complex The intermediate in nucleophilic aromatic substitution formed by the addition of a nucleophile to an aromatic compound activated by electron-withdrawing groups, often NO_2.

Mercaptan (RSH) A thiol. The sulfur counterpart of an alcohol.

Mercaptide (RS^-) The sulfur counterpart of an alkoxide.

Meso compound An achiral compound containing stereogenic atoms.

Messenger RNA (mRNA) A polynucleotide produced from DNA whose base sequence codes for amino acid assembly.

Meta substitution 1,3-Substitution on a benzene ring.

Methane The simplest stable hydrocarbon, CH_4.
Methine group The group CH.
Methyl anion $^-:CH_3$.
Methyl cation $^+CH_3$.
Methyl compounds Substituted methanes; CH_3-X compounds.
Methyl radical $\cdot CH_3$.
Methylene group The group CH_2.
Micelle An aggregated group of fatty acid salts (or other molecules) in which a hydrophobic center consisting of the hydrocarbon chains is protected from an aqueous environment by a spherical skin of polar hydrophilic groups.
Michael reaction The addition of a nucleophile, often an enolate, to the β-position of a double bond of an α,β-unsaturated carbonyl group.
Microscopic reversibility The notion that the lowest energy path for a reaction in one direction will also be the lowest energy path in the other direction.
Molecular orbital An orbital not restricted to the region of space surrounding an atom, but extending over several atoms in a molecule. Molecular orbitals are formed through the overlap of atomic or molecular orbitals. Molecular orbitals can be bonding, nonbonding, or antibonding.
Molozonide The initial product of the reaction of ozone and an alkene. This molecule contains a five-membered ring with three oxygen atoms in a row. See **primary ozonide**.
Morphine An important medicinal alkaloid isolated from the opium poppy.
Mutarotation The interconversion of anomeric sugars in which an equilibrium mixture of α- and β-forms is reached.

NAD$^+$ Nicotinamide adenine dinucleotide, a biological oxidizing agent. Nature uses pyrophosphates as leaving groups.
Neighboring group effect A general term for the influence of an internal nucleophile (very broadly defined) on the rate of a reaction, or on the structures of the products produced.
Newman projection A convention used to draw what one would see if one could look down a bond. The groups attached to the front atom are drawn in as three lines. The back atom is represented as a circle to which its attached bonds are drawn. Newman projections are enormously useful in seeing spatial (stereochemical) relationships in molecules.
Ninhydrin The hydrate of indan-1,2,3-trione, a molecule that reacts with amino acids to give the purple dye used in quantitative analysis of amino acids.
Nitration The electrophilic substitution of aromatic compounds by nitro groups using a source of the nitronium ion, $^+NO_2$.
Nitrene A reactive intermediate containing a neutral, monovalent nitrogen atom. The nitrogen counterpart of a carbene.
Nitrile A compound of the structure RCN, also commonly called cyanides.
Nitroso compound A compound containing the N=O group.
Node The region of zero electron density separating regions of opposite sign in an orbital. At a node the sign of the wave function is zero.
Nonbonding electrons Electrons in an orbital that is not involved in binding atoms. See **lone pair**.
Nonreducing sugar A carbohydrate containing no amount of an oxidizable aldehyde group.
Nor A prefix meaning "no methyl groups."
Norbornyl system The bicyclo[2.2.1]heptyl system.
Nuclear magnetic resonance spectroscopy (NMR) Spectroscopy that detects the absorption of energy as the lower energy nuclear spin state in which the nuclear spin is aligned with an applied magnetic field flips to the higher energy spin state in which it is aligned against the field. See **chemical shift** and **coupling constant**.
Nucleic acid The polynucleotides DNA and RNA.
Nucleophile Generally, a Lewis base. A strong nucleophile has a high affinity for a carbon $2p$ orbital.
Nucleoside A sugar, either ribose or deoxyribose, bonded to a heterocyclic base at its 1′ position.
Nucleotide A phosphorylated nucleoside; one of the monomers of which DNA and RNA are composed.
Nucleus The positively charged core of an atom containing the protons and neutrons.

Octet rule The notion that special stability attends the filling of the $2s$ and $2p$ atomic orbitals to achieve the electronic configuration of neon, a noble gas.
Off-resonance decoupling A technique used in ^{13}C NMR in which coupling between ^{13}C and ^1H is restricted to the hydrogens directly attached to the carbon. The number of directly attached hydrogens can be determined from the multiplicity of the observed signal for the carbon.
Olefins Hydrocarbons of the general formula, C_nH_{2n}. These molecules, also called alkenes, contain carbon–carbon double bonds.
Optical activity The rotation by a molecule of the plane of plane-polarized light.
Orbital The volume of space in which an electron is likely to be found. Another word for "wave function." These are three-dimensional representations of the solutions to the Schrödinger equation.

π Orbitals Molecular orbitals made from the overlap of *p* orbitals. Such orbitals have a plane of symmetry.

Organocuprates An organometallic reagent (R_2CuLi) notable for its ability to react with primary or secondary halides (R'−X) to generate hydrocarbons, R−R', and to add to the β-position of α, β-unsaturated carbonyl compounds.

Organolithium reagent (R−Li) A strongly basic organometallic reagent formed from a halide and lithium. A characteristic reaction is addition to carbonyl groups.

Organometallic reagents Molecules that contain both carbon and a metal. Usually carbon is at least partially covalently bonded to the metal. Examples are Grignard reagents, organolithium reagents, and lithium organocuprates.

Ortho esters Compounds of the structure

$$R'-C(OR)_3$$

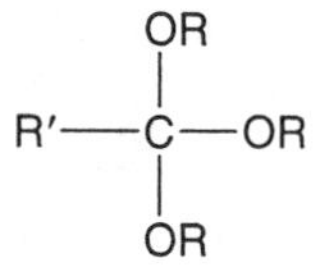

Ortho substitution 1,2-Substitution on a benzene ring.

Orthogonal orbitals Two noninteracting orbitals. The bonding interactions are exactly balanced by antibonding interactions

Osazone A 1,2-phenylhydrazone formed by treatment of a sugar with three equivalents of phenylhydrazine.

Oxacyclopropane A three-membered ring containing oxygen, also called an epoxide or oxirane.

Oxaphosphetane The four-membered intermediate in the Wittig reaction that contains two carbons, a phosphorus, and an oxygen.

Oxetane A saturated, four-membered ring containing a single oxygen atom.

***N*-Oxide** A compound of the structure $R_3N^+-O^-$. They are produced from the reaction of tertiary amines with hydrogen peroxide. See also **Amine oxide**.

Oxirane A three-membered ring containing oxygen, also called an epoxide or oxacyclopropane.

Oxonium ion A molecule containing a trivalent, positively charged, oxygen atom (R_3O^+).

Oxymercuration The mercury-catalyzed conversion of alkenes into alcohols. Addition is in the Markovnikov sense, and there are no rearrangements. A three-membered ring containing mercury is an intermediate in the reaction. Alkynes also undergo oxymercuration to give enols that are rapidly converted into carbonyl compounds under the reaction conditions.

Ozonide The product of rearrangement of the primary ozonide formed on reaction of ozone with an alkene.

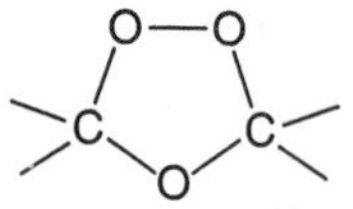

Ozonolysis The reaction of ozone with π systems. The intermediate products are ozonides that can be transformed into carbonyl-containing compounds of various kinds.

p + 1 peak An ion with a mass one greater than that of the parent ion. It appears because molecules contain atoms with isotopes of mass one greater than that of the most commonly observed isotope.

Paracyclophane An aromatic molecule bridged across the ring with one or more chains.

Para substitution 1,4-Substitution on a benzene ring.

Parent ion The ion directly formed in the mass spectrometer by ejection of an electron from the sample molecule. The parent ion commonly undergoes many fragmentation reactions before detection.

Pauli principle No two electrons in an atom or molecule may have the same values of the four quantum numbers.

Penicillin A powerful antibiotic. The key feature is the reactive β-lactam ring.

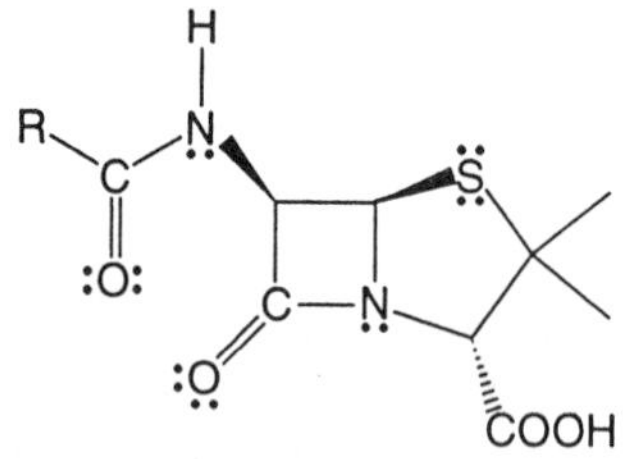

Pentose A five-carbon sugar.

Peptide A polyamino acid in which the constituent amino acids are linked through amide bonds. They are distinguished from proteins only by the length of the polymer.

Pericyclic reaction A reaction in which the maintenance of bonding overlap between the lobes of the orbitals involved in bond making and breaking is controlling. All single-barrier (concerted) reactions can be regarded as pericyclic reactions.

Peroxides Compounds of the structure R−O−O−R.

Phase-transfer catalysis A synthetic method that often takes advantage of the ability of alkylammonium ions to transport ions from aqueous to organic media.

Phenol

Phenonium ion The benzenonium ion produced through intramolecular displacement of a leaving group by the π system (not the σ system) of an aromatic ring.

Phenyl isothiocyanate The active ingredient, PhNCS, in the Edman degradation that converts the amino terminus of a peptide into a phenyl thiohydantoin.

Photohalogenation The formation of alkyl halides through the irradiation of molecular X_2 (X = Br or Cl) in an alkane.

Phthalimide The cyclic imide of phthalic acid.

Pinacol rearrangement An acid-catalyzed rearrangement reaction leading from 1,2-diols to carbonyl compounds. The rearrangement is driven by the conversion of a localized carbocation into a resonance-stabilized carbocation.

Piperidine A six-membered ring amine: azacyclohexane.

pK_a The negative logarithm of the acidity constant. A strong acid has a low pK_a, a weak acid has a high pK_a.

Pleated sheet One of the common forms of secondary peptide structure in which hydrogen bonding holds chains of amino acids roughly in parallel lines.

Polar covalent bond Any shared electron bond between two different atoms must be polar. Two nonidentical atoms must have different electronegativities and will attract the shared electrons to different extents, creating a dipole.

Polarimeter A device for measuring the amount of rotation of the plane of plane-polarized light.

Polynuclear aromatic compound An aromatic molecule composed of two or more fused aromatic rings.

ppm scale The chemical shift scale commonly used in NMR spectroscopy in which the positions of absorptions are quoted as parts per million (ppm) of applied magnetic field. The chemical shift is independent of the magnetic field on the ppm scale.

Primary amines An amine bearing only one R group and two hydrogens.

Primary carbon A carbon atom attached to one other carbon.

Primary ozonide The initial product of the reaction of ozone and an alkene. This molecule contains a five-membered ring with three oxygen atoms in a row. See **molozonide**.

Primary structure The sequence of amino acids in a protein.

Product-determining step The step in a multistep reaction that determines the structure or structures of the products.

Propagation steps The product-producing steps of a chain reaction. In the last propagation step, a molecule of product is formed along with a new radical that can carry the chain.

Propargyl group The H−C≡C−CH_2 group.

Propyl group The group $CH_3CH_2CH_2$.

Protecting group Sensitive functional groups in a molecule can be protected by a reaction that converts them into a less reactive group, called a protecting group. A protecting group must be removable to regenerate the original functionality.

Protein A polyamino acid in which the constituent amino acids are linked through amide bonds. They are distinguished from peptides only through the length of the polymer.

Protic solvent A solvent containing a hydrogen easily lost as a proton. Examples are water and most alcohols.

Protonium ion A three-membered ring in which a hydrogen atom bridges two carbons. In this ion two electrons bind three atoms, the two carbons and the hydrogen. It is very different from a bromonium ion in which the three atoms in the ring are connected by normal two-electron bonds.

Pseudo-first-order reaction A bimolecular reaction in which the concentration of one reagent (usually the solvent) does not change appreciably.

Pseudohalogen Any of a variety of X−Y compounds that add to alkenes as do the halogens. Examples include I−NCO and I−N_3.

Pyranose A sugar containing a six-membered cyclic ether.

Pyranoside A pyranose in which the anomeric OH at C(1) has been converted into an OR group.

Pyridine Any compound containing the following "azabenzene" ring system:

Pyridine *N*-oxide A pyridine in which the nitrogen has been oxidized to the oxide stage.

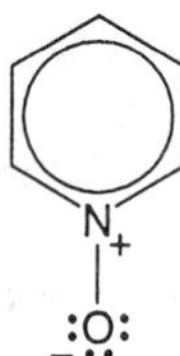

Pyridinium ion A quaternary ammonium ion formed from a pyridine.

Pyrolysis The process of inducing chemical change by supplying heat energy. See **Thermolysis**.

Pyrophosphate The group

O—P(=O)(OH)—O—P(=O)(OH)—OH

Pyrrole Any compound containing the following ring structure:

Pyrrolidine A saturated five-membered ring amine: azocyclopentane.

Quantum numbers These numbers evolve from the Schrödinger equation and characterize the various solutions to it. They may have only certain values, and these values determine the distance of an electron from the nucleus (n), the shape (l), and orientation (m_l), of the orbitals, and the electron spin (s).

Quaternary ammonium ions A compound of the structure $R_4N^{+}\,^{-}X$.

Quaternary carbon A carbon attached to four other carbons.

Quaternary structure A self-assembled aggregate of two or more protein units.

Quinine An important antimalarial alkaloid containing the quinoline ring structure.

Quinoline Any compound containing the following ring structure:

Racemic mixture (racemate) A mixture containing equal amounts of two enantiomeric forms of a chiral molecule.

Radical A neutral molecule containing an odd, unpaired electron. Also called a "free radical."

Radical anion A negatively charged molecule containing both a pair of electrons and an odd, unpaired electron.

Radical cation A species that is positively charged yet contains a single unpaired electron. These are commonly formed by ejection of an electron when a molecule is bombarded with high-energy electrons in a mass spectrometer.

Random coil Disordered portions of a chain of amino acids.

Raney nickel A good reducing agent composed of finely divided nickel on which hydrogen has been adsorbed.

Rate constant (k) A fundamental property of a reaction that depends on the temperature, pressure, and solvent, but not on the concentrations of the reactants.

Rate-determining step The step in a reaction with the highest activation energy.

Reaction mechanism Loosely speaking, How does the reaction occur? How do the reactants come together? Are there any intermediates? What do the transition states look like? More precisely, a determination in terms of structure and energy of the stable molecules, reaction intermediates, and transition states involved in the reaction, along with a consideration of how the energy changes as the reaction progresses.

Reactive intermediates Molecules of great instability, and hence fleeting existence under normal conditions. Most carbon-centered anions, cations, and radicals are examples.

Rearrangement The migration of an atom or group of atoms from one place to another in a molecule. Rearrangements are exceptionally common in reactions involving carbocations.

Reducing sugar A sugar containing some amount of an oxidizable free aldehyde group.

Reductive amination A synthetic method in which an imine or enamine is generated from a carbonyl compound and an amine only to be reduced in situ to give the substituted amine.

Regiochemistry The orientation of a reaction taking place on an unsymmetrical substrate.

Regioselectivity (regiospecificity) If a reaction may produce one or more isomers and one predominates, the reaction is called regioselective. If one product strongly predominates, the reaction is regiospecific.

Reissert synthesis A synthesis of indoles starting from a nitroalkylbenzene and an ester of oxalic acid.

Resolution The separation of a racemic mixture into its constituent enantiomeric molecules.

Resonance energy The energy lowering conferred by the delocalization of electrons. In benzene, this is the amount by which benzene is more stable than the hypothetical 1,3,5-cyclohexatriene containing three localized double bonds. See **delocalization energy**.

Resonance forms Many molecules cannot be represented adequately by a single Lewis form. Instead, two or more different electronic representations must often be combined to give a good description of the molecule. These different representations are called resonance forms.

Retention of stereochemistry The preservation of the handedness of a molecule in a reaction. Generally, retention of stereochemistry means that an (*R*) starting material would be transformed into an (*R*) product.

Retention time The time necessary for elution of a component of a mixture in chromatography.

Ribonucleic acid (RNA) A polymer of nucleotides made up of ribose units connected by phosphoric acid links. Each sugar is attached to one of the bases, A, U, G, or C.

Robinson annulation A classic method of construction for six-membered rings using a ketone and methyl vinyl ketone. It involves a two-step sequence of a Michael reaction between the enolate of the original ketone and methyl vinyl ketone, followed by an intramolecular aldol condensation.

Rosenmund reduction The reduction of acid chlorides to aldehydes using a poisoned catalyst.

Ruff degradation A method for shortening the carbon backbone of a sugar by one carbon. The aldehyde carbon at C(1) is lost and a new aldehyde created at the old C(2).

Saccharide A molecule whose formula can be factored into $C_x(H_2O)_y$. A "sugar" or "carbohydrate."

Sandmeyer reaction The reaction of an aromatic diazonium ion with cuprous salts to form substituted aromatic compounds.

Sanger degradation A method for determining the amino acid at the amino terminus of a chain; using 2,4-dinitrofluorobenzene. Unfortunately, the Sanger degradation hydrolyzes, and thus destroys, the entire peptide.

Saponification Base-induced hydrolysis of an ester, usually a fatty acid.

Saturated alkanes Alkanes of the molecular formula C_nH_{2n+2}.

Saytzeff elimination Formation of the more substituted alkene in an elimination reaction.

Schiff base The nitrogen analogue of a ketone or aldehyde.

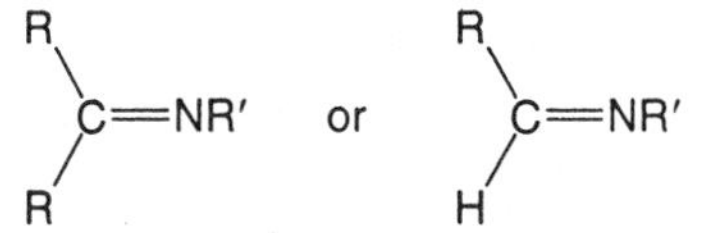

Schmidt reaction The formation of amines through the reaction of carboxylic acids with hydrazoic acid, HN_3. The key step is an intramolecular rearrangement to generate an isocyanate that is then converted into an unstable carbamic acid under the reaction conditions. Decarboxylation gives the amine.

Second-order reaction A reaction for which the rate depends on the product of a rate constant and the concentrations of two reagents.

Secondary amine An amine bearing one hydrogen and two R groups.

Secondary carbon A carbon attached to two other carbons.

Secondary structure Ordered regions of a protein chain. The two most common types of secondary structure are the α-helix and the β-pleated sheet.

Selenoxide A compound of the structure

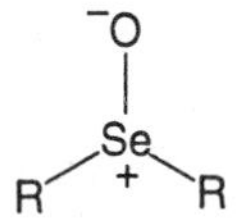

Sigma bond (σ) Any bond with cylindrical symmetry.

Sigmatropic shift The migration of an atom, under orbital symmetry control, along a π system.

Singlet carbene A singlet carbene contains only paired electrons. In a singlet carbene, the two nonbonding electrons have opposite spins and occupy the same orbital.

Six-center transition state The boat-like transition state for the Cope rearrangement in which carbons 1 and 1′, 2 and 2′, and 3 and 3′ are all eclipsed.

Skraup synthesis A synthesis of quinolines starting from aniline and glycerol (1,2,3-trihydroxypropane).

S_N1 Reaction Substitution, nucleophilic, unimolecular. An initial ionization is followed by attack of the nucleophilic solvent.

S_N2 Reaction Substitution, nucleophilic, bimolecular. In this reaction, the nucleophile dispaces the leaving group by attack from the rear. The stereochemistry of the starting material is inverted.

S_N2' Reaction Nucleophilic displacement of a leaving group through attack at an allylic position.

$$Nu:^- + \text{CH}_2{=}\text{CH–CH}_2\text{–L} \longrightarrow \text{Nu–CH}_2\text{–CH}{=}\text{CH}_2 + L:^-$$

Soap The salt of a fatty acid.

Sodium bicarbonate

$$Na^+ \; {}^-O\text{–C(=O)–OH}$$

Sodium carbonate

$$Na^+ \; {}^-O\text{–C(=O)–}O^- \; Na^+$$

Solvated electron The species formed when sodium is dissolved in ammonia. The product is a sodium ion and an electron surrounded by ammonia molecules. This electron can add to alkynes (and some other π systems) in a reduction step.

Solvolysis reaction In such reactions, the solvent acts as the nucleophile.

***sp* Hybrid** A hybrid orbital made by the combination of one *s* and one *p* atomic orbital.

sp^2 Hybrid A hybrid orbital made by the combination of one *s* orbital and two *p* orbitals.

sp^3 Hybrid A hybrid orbital made by the combination of one *s* orbital and three *p* orbitals.

Spectroscopy The study of the interactions between electromagnetic radiation and atoms and molecules.

Spiro rings Two rings that share a single carbon.

Staggered ethane The energy minimum conformation of ethane in which the carbon–hydrogen bonds (and the electrons in them) are as far apart as possible.

Starch A nonlinear polymer of glucose containing amylose, an α-linked linear polymer of glucose.

Stereochemistry The physical and chemical consequences of the arrangement in space of the atoms in molecules.

Stereogenic center An atom, usually carbon, of such nature and bearing groups of such nature that it can have two nonequivalent configurations.

Steric A generic word referring to the arrangement in space of atoms and groups of atoms.

Steroid A class of four-ring compounds, always containing three six-membered rings and one five-membered ring. The ring system can be substituted in many ways, but the rings are always connected in the following fashion:

Strecker synthesis A synthesis of amino acids using an aldehyde, cyanide, and ammonia, followed by hydrolysis.

Structural isomers Molecules of the same formula but with different connectivities among the constituent atoms.

Substitution reaction The replacement of a leaving group in R–L by a nucleophile, Nu:⁻. Typically,

$$Nu:^- + R\text{—}L \longrightarrow Nu\text{—}R + L:^-$$

Sugar A molecule whose formula can be factored into $C_x(H_2O)_y$. A "saccharide" or "carbohydrate."

Sulfide (RSR) The sulfur counterpart of an ether.

Sulfonation The electrophilic substitution of aromatic compounds by $SO_3/HOSO_2OH$ to give benzenesulfonic acids.

Sulfone ($R–SO_2–R$) The final product of oxidation of a sulfide with hydrogen peroxide.

Sulfonic acid A compound of the general structure RSO_2OH.

Sulfonium ion (R_3S^+) A positively charged sulfur ion.

Sulfoxide A compound of the structure

$$R\text{–}\overset{+}{S}(\text{–}O^-)\text{–}R$$

Superacid A number of highly polar, highly acidic, but weakly or nonnucleophilic solvent systems. Superacids are often useful in stabilizing carbocations, at least at low temperature.

Suprafacial motion Migration of a group in a sigmatropic shift in which bond breaking and bond making take place on the same side of the π system.

syn Elimination An elimination reaction in which the dihedral angle between the breaking bonds, usually C–H and C–L, is 0°.

Termination step The mutual annihilation of two radicals. Termination steps destroy chain-carrying radicals and end chains through bond formation.

Terpenes Compounds whose carbon skeleton is composed of isoprene units.

Tertiary amine An amine bearing no hydrogens and three R groups.

Tertiary carbon A carbon attached to three other carbons.

Tertiary structure The structure of a protein induced by its folding pattern.

Tetramethylsilane (TMS) The standard "zero point" on the ppm scale in NMR spectroscopy. The chemical shift of a nucleus is quoted in parts per million of applied magnetic field relative to the position of TMS.

Tetrose A four-carbon sugar.

Thermal eliminations A group of intramolecular, syn β-elimination reactions in which an ester or related group acts both as base and leaving group.

Thermodynamic control In a reaction under thermodynamic control, the product distribution depends on the energy differences between the products.

Thermodynamic enolate The most stable enolate. It may or may not be the same as the most rapidly formed enolate, the "kinetic enolate."

Thermodynamics The study of energetic relationships.

Thermolysis The process of inducing chemical change by supplying heat energy. See **Pyrolysis**.

Thietane A saturated four-membered ring containing a single sulfur atom.

Thiirane A saturated three-membered ring containing a single, unoxidized sulfur atom. See **Episulfide**.

Thioacetal The sulfur counterpart of an acetal.

Thioether (RSR) The sulfur counterpart of an ether. A sulfide.

Thiol The sulfur counterpart of an alcohol, RSH.

Thionyl chloride ($SOCl_2$) An effective reagent for converting alcohols into chlorides and carboxylic acids into acid chlorides.

Thiophene An aromatic five-membered ring compound containing four CH units and one S atom.

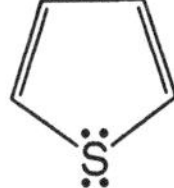

Thiourea A compound of the structure

:S:
RHN C NHR′

THP derivatives Protecting groups for alcohols of the structure

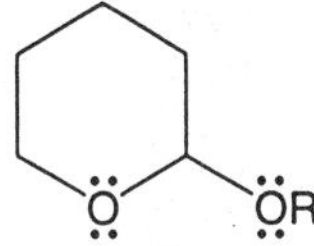

Three-center, two-electron bonding A bonding system in which only two electrons bind three atoms. This arrangement is common in electron-deficient molecules such as boranes and carbocations.

Torsional strain Destabilization caused by the proximity of bonds (usually eclipsing) and the electrons in them.

trans "On opposite sides." Used to specify stereochemical (spatial) relationships in ring compounds and alkenes.

s-trans The more stable, extended form of a 1,3-diene,

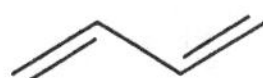

Transesterification The formation of one ester from another. The reaction can be catalyzed by either acid or base.

Transition state The high point in energy between starting material and product. The transition state is an energy maximum and not an isolable compound.

Triazoline Any compound containing the following ring system, usually formed by the 1,3-dipolar cycloaddition of an azide to an alkene:

R N N N

Triple bond Two atoms can be attached by a triple bond composed of one σ bond and two π bonds.

Triplet carbene In a triplet carbene, the two nonbonding electrons have the same spin and must occupy different orbitals.

Trityl cation The triphenylmethyl cation.

Tropane alkaloids A family of alkaloids, all of which contain the tropane ring system.

NCH_3

Tropylium ion The 1,3,5-cycloheptatrienylium ion. This ion has $4n + 2$ π electrons ($n = 1$) and is aromatic.

Trypsin An enzyme capable of cleaving a peptide after lysine or arginine.

Ultraviolet (UV) spectroscopy Electronic spectroscopy using light of wavelength 200–400 nm.

Unsaturated compound A compound not having the maximum number of hydrogens.

Unsaturated Containing less than the maximum number of hydrogens. Alkenes and alkynes (as well as ring compounds) are unsaturated molecules.

Urea

O
C
H_2N NH_2

Valence electrons The outermost, or most loosely held, electrons.

van der Waals forces Intermolecular forces in molecules caused by induced dipole–induced dipole interactions.

Vicinal 1,2-Disubstituted.

Vinyl group The $H_2C{=}CH$ group.

Visible (vis) spectroscopy Electronic spectroscopy using light of wavelength 400–800 nm.

Wagner–Meerwein rearrangement The 1,2-migration of an alkyl group in a carbocation.

Wave function (ψ) A solution to the Schrödinger equation, another word for "orbital."

Wavenumber The wavenumber ($\tilde{\nu}$) equals $1/\lambda$ or ν/c.

Williamson ether synthesis The S_N2 reaction of an alkoxide with R–L to give an ether.

Wittig reaction The reaction of an ylide with a carbonyl group to give, ultimately, an alkene.

Wohl degradation A method for shortening the carbon backbone of a sugar by one carbon. The aldehyde carbon at C(1) is lost and a new aldehyde created at the old C(2).

Wolff rearrangement The formation of a ketene through the thermal or photochemical decomposition of a diazo ketone.

Woodward–Hoffmann theory The notion that one must take account of the phase relationships between orbitals in order to understand reaction mechanisms. In a concerted reaction, bonding relationships must be maintained at all times.

Xanthate ester A compound of the structure

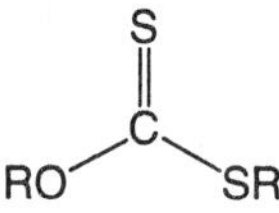

Ylide A compound containing opposite charges on adjacent atoms.

Zwitterion A dipolar species in which full plus and minus charges coexist within the same molecule.